Calculus

Calculus

STANLEY I. GROSSMAN

University of Montana

ACADEMIC PRESS New York San Francisco London
A Subsidiary of Harcourt Brace Jovanovich, Publishers

ACADEMIC PRESS, INC.
111 Fifth Avenue, New York, New York 10003

United Kingdom Edition published by
ACADEMIC PRESS, INC. (LONDON) LTD.
24/28 Oval Road, London NW1

Library of Congress Cataloging in Publication Data

Grossman, Stanley I
    Calculus.

    1.   Calculus.
QA303.G89   515   76-9157
ISBN 0–12–304350–6

To Arlene, Kerstin, Erik, and Aaron

PREFACE

The study of calculus has been of central importance to scientists throughout most of recorded history. Our present understanding of the subject owes much to Archimedes of Syracuse (287–212 B.C.) who developed what were, for his time, incredibly ingenious techniques for calculating the areas enclosed by a great variety of curves. Archimedes' work led directly to the modern concept of the integral. Much later, toward the end of the seventeenth century, Sir Isaac Newton and Gottfried Leibniz independently showed how to calculate instantaneous rates of change and slopes to curves. This development inspired an enormous variety of mathematical techniques and theorems that could be used to solve problems in a number of diverse, and often unrelated, fields.

Many of these techniques and theorems are considered to be part of "the calculus." No textbook could discuss all the myriad results in calculus discovered over the past several hundred years. Fortunately, a weeding out process has occurred and there is now fairly widespread agreement as to what topics should properly be included in a three semester (or four- to five-quarter) introduction to the subject. This book includes all the standard topics. In the process of writing it, I have worked to achieve certain goals that make the text unique.

Most mathematicians agree that the only way to learn calculus well is by solving problems, and this book is designed to encourage students to learn by this proven method. Thus, in the body of the text, there are approximately one thousand worked out examples and at the ends of sections there are almost six thousand problems to be worked by the student. In addition, there are review exercises at the end of each chapter. The problems are of varying difficulty and the most difficult are marked (*).

In the worked out examples, especially in the earlier chapters, I have tried to include all the algebraic steps needed to complete the solution. I remember, as a student, being infuriated by statements like "it now easily follows that . . ." when it was not at all easy for me. I feel that students have a right to see the "whole hand," so to speak, so that they always know how to get from "a" to "b." They will have many opportunities to test their skills on the problem sets.

As a second goal, I have tried to make the examples and problems interesting, varied, and meaningful. Many calculus books draw examples exclusively from the physical sciences even though calculus today is also used as a tool in the biological sciences, the social sciences, economics, and business. Thus, the examples and problems in this text, while including a great number from the physical sciences, also cover a wide range of other fields. Moreover, I have included "real-world" data wherever possible, to make the examples more meaningful. For example, students are asked to find the escape velocity from Mars, the effective interest rate of a large installment purchase, and the optimal branching angle between two blood vessels. Finally, as most of the world uses the metric system, and even the United States is reluctantly following suit, the majority of the applied examples and problems in the book make use of metric units.

One aim of any calculus book should be to make basic material available to the student as early as possible. This text does a number of things to accomplish this goal. First, integration, the tool initiated by Archimedes, is introduced early, in Chapter 4, so that students can use simple integration techniques in other courses by the end of the first semester. Second, intuition rather than rigor is stressed in the early parts of the book. For instance, in the introduction to the limit in Chapter 2, the student's intuition is appealed to when he is told what a limit is, and the concept is introduced by means of examples—together with detailed tables. I believe that the "ϵ–δ" or "neighborhood" approach to limits can be appreciated only after some feeling for the limit has been developed. However, I also feel strongly that, as mathematics depends on rigorous, unambiguous proofs, standard definitions must be included. To resolve this apparent contradiction I have put much of the one-variable theory in a separate chapter (Chapter 10). This chapter can be included at any stage, as the instructor sees fit, or can be omitted without loss of continuity if time is a problem.

Today many students who study calculus own or have access to a scientific, hand-held calculator. As well as being useful in solving computational problems, a calculator can be employed as a learning device. For example, I feel that a student will develop his feeling for limits more quickly if he can literally *see* the limit being approached. In Chapter 2 there are numerous tables which make use of a calculator to illustrate particular limits. At several places in the text a calculator has been used for illustrative purposes. In Chapter 5 (Section 5.5) there is a discussion of Newton's method for finding roots of equations. This method is very easy to employ with the aid of a calculator. In Section 6.5 Newton's method is used to estimate effective interest rates. Consistent with this increased attention to computation, there is a separate chapter on polynomial approximation (Chapter 13) including an optional section on the Lagrange interpolating polynomial.

Examples, problems, and sections employing a hand calculator are marked with the symbol $\square$. Note, however, that the availability of a calculator is *not* a prerequisite for use of this text. The vast majority of problems do not require a calculator and the illustrative tables can be appreciated without independent verification. Finally, a note for the student who does wish to use a calculator in his calculus work: find a calculator that has function keys for the three basic trigonometric functions (sin, cos, tan), common (log) and natural (ln) logarithm keys, and an exponential key (usually denoted y^x), which allows easy calculation of powers and roots; in addition, a memory unit (so that numbers can be stored for easy retrieval) is a very useful feature.

Students today enter colleges and universities with a very wide range of mathematical experiences. Some start calculus with as little as a year or two of high school algebra, while others have a thorough knowledge of algebra, trigonometry, and, perhaps, some college level mathematics. Obviously, then, an introductory calculus text has to be flexible. Three features in this text have been included to increase flexibility. First, more algebraic detail than is usual has been included. Second, Chapter 1 includes general discussions of several topics which are commonly taught in an intermediate algebra–college algebra course, including absolute value and inequalities, circles and lines, and an introduction to functions. Moreover, that chapter also includes some detailed discussion of graphing techniques—including a unique section on shifting the graphs of functions (an extremely useful procedure). There is also, in Section 6.1, a detailed introduction (without calculus) to exponential and logarithmic functions. Third, the trigonometric (or circular) functions are introduced so as to be accessible to the student who has never before seen them. The first four sections of Chapter 7 do not involve the calculus of trigonometric functions and can be covered at any time in the course. These four sections provide a not too brief summary of the trigonometric techniques a calculus student needs to know and includes, as elsewhere in the text, a variety of applications.

In an attempt to make the calculus of several variables more accessible to the student, the subject is introduced gradually. Vectors in the plane are discussed in Chapters 15 and 16, allowing the student to become comfortable with easily visualized vectors before generalizing to higher dimensions in Chapter 17. Functions of several variables are developed through the use of the gradient as a higher dimension generalization of the derivative of a function of one variable. Moreover, the gradient is presented via several geometric interpretations, one of which provides a natural introduction to Lagrange multipliers.

Analytic geometry is traditionally part of a standard calculus course. I have found, however, that many instructors (including myself) do not have the time to provide detailed discussions of the analytic geometric topics such as the conic sections and the translation and rotation of axes. As a compromise, I have included a brief discussion of the parabola, ellipse, and hyperbola in Chapter 1. This material is sufficient for the applications in the text. However, for those instructors who have the time and wish to include a traditional approach to analytic geometry, Appendix 3 includes in full detail a discussion of the conic sections, translation, and rotation.

Numbering in the book is fairly standard. Within each section examples, problems, theorems, and equations are numbered consecutively starting with 1. Reference to an example, problem, theorem, or equation outside the section in which it appears is referenced by chapter, section, and number. Thus, for example, Example 2 in Section 2.3 is called, simply, Example 2 in that section, but outside the section is referred to as Example 2.3.2. In addition, let me repeat that more difficult problems are marked (*) or occasionally (**) and problems where the use of a calculator is advisable are marked ($^\square$). Finally, sections which are more difficult are also marked (*).

Acknowledgments

I am very grateful to many people who helped me during the time this book was written. I was fortunate to obtain detailed, sometimes critical, and always helpful suggestions from the reviewers: Professor Neil Berger at the University of Illinois, Chicago; Professor Thomas Cairns at the University of Tulsa; Professor Albert J. Herr at Drexel University; Professor Stanley M. Lukawecki at Clemson University; Professor Nelson G. Rich at Monroe Community College; Professor Bobby L. Sanders at Texas Christian University; and Professor William R. Derrick at the University of Montana. I received much help in the preparation of problems and solution sets from Richard Lane and Michael Ragbourn, graduate students at the University of Montana. The entire manuscript was typed by Carolyn Chase who not only worked with great accuracy but also helped to clear up obscurities (and obvious errors) in the text. This book would not have appeared when it did without her expert help. I am also grateful to the Addison-Wesley Publishing Company, Inc., for permission to use material from my book (with William R. Derrick) *Elementary Differential Equations with Applications* in Chapter 20 and Appendix 2.

Finally, I owe a considerable debt to the editorial and production staffs of Academic Press who provided help in a great number of ways in the writing and production of this book.

STANLEY I. GROSSMAN

TO THE INSTRUCTOR

As an aid to the instructor, the following table indicates the interrelationships of the various parts of the text.

Chapter	Chapter dependence	Comments
1	—	Section 1.10 on geometric progressions is very useful for a variety of applications, the first of which appears in Section 2.6.
2	Chapter 1	Section 2.6 is optional. Derivatives are first introduced geometrically in Section 2.8, with the analytic definition postponed until Section 2.9. The derivative as a rate of change is introduced in Section 2.10.
3	Chapters 1 and 2	Curve plotting (Sections 3.5 and 3.9) is introduced early—before the traditional "applications" chapter.
4	Chapters 1 and 2 and Sections 3.1–3.4	Section 4.1 provides an historical introduction to the subject. The Σ notation is avoided and is not introduced until Chapter 14.
5	Chapters 1–4	Sections 5.4, 5.5, and 5.7 are optional. Section 5.5 makes heavy use of the calculator.
6	Section 6.1 is algebra and depends only on Sections 1.6 and 1.7. The rest of the chapter depends on Chapters 1–4.	Sections 6.4 and 6.5 are optional. But Section 6.4 should be covered in courses with science majors and Section 6.5 should be covered in courses with economics and business majors. Section 6.1 can be covered after Chapter 1 and the other sections can be covered after Chapter 4.
7	Sections 7.1–7.4 contain the bulk of high school trigonometry and require only Sections 1.6 and 1.7. Sections 7.5–7.8 depend on Chapters 1–4, and Sections 7.9 and 7.10 depend on Chapter 6.	Section 7.7 is optional but should be covered in courses with science majors. Sections 7.1–7.4 can be covered at any time in the course and Sections 7.5–7.8 can be covered after Chapter 4.
8	Chapters 1–4, 6, and 7	The material on partial fractions in Sections 8.6 and 8.7 is lengthy and complete. Section 8.7 should be omitted or covered only briefly if time is a problem.
9	Chapters 1–4, 6–8, and parts of Chapter 5	Each section presents a particular geometric or physical concept and is optional.
10	Chapters 1–4	This chapter provides much of the theory behind limits, differentiation, and integration. It can be covered any time after Chapter 4 or it can be incorporated with the material in Chapters 2–4. In especially applied courses with little time, the entire chapter could be omitted with no loss of continuity.

Chapter	Chapter dependence	Comments
11	Chapters 1–4, 6, 7, and parts of Chapter 8	Section 11.5 is optional.
12	Chapters 1–4, 6, 7, and parts of Chapter 8. Section 12.2 depends on Section 10.6	Section 12.2 can be omitted without loss of continuity.
13	Chapters 1–4, 6, 7, and parts of Chapters 8, 10, and 12	Section 13.1 introduces Taylor's theorem as an approximation technique before a discussion of infinite series. Section 13.2 can be omitted without loss of continuity. Section 13.3 makes use of the calculator. Section 13.4 discusses the Lagrange interpolating polynomial and should be covered only if sufficient time is available.
14	Chapters 1–4, 6, 7, and parts of Chapters 8, 10, 12, and 13	The Σ notation is introduced for the first time in Section 14.3. It can be covered at any time as it is independent of any section that precedes it. All sections in this chapter should be covered.
15	Chapters 1 and 2 only	An introduction to vectors in the plane—includes no calculus.
16	Chapters 1–4 and 6–8, Section 9.2, and parts of Chapter 11	An introduction to vector functions in the plane and parametric equations. Sections 16.6–16.8 are optional.
17	Chapters 1–4, 6–8, and 16, Section 9.2, and parts of Chapter 11	An extension of the material in Chapters 15 and 16 to three dimensions. Sections 17.6 and 17.8 are optional.
18	Chapters 1–8, 10, 12, and 15–17	An introduction to calculus of several variables. The gradient is introduced in Section 18.6 as the natural extension of the ordinary derivative, and this is done in such a way so as to allow a discussion of functions of n variables (although the discussion is limited to $\mathbb{R}^2$ and $\mathbb{R}^3$). Section 18.10 is optional but should be covered by students of physics and engineering. Section 18.14 on Lagrange multipliers is motivated by the geometry of the gradient.
19	Chapters 1–9, and parts of Chapters 11, 12, and 15–18	Introduction to multiple integration with an emphasis on applications. Sections 19.3, 19.5, and 19.7–19.9 are optional. Section 19.9 includes a discussion of Green's theorem in the plane from both an analytic and a geometric point of view. The curl and divergence are introduced there and the section can be used as an introduction to vector analysis.
20	Sections 20.1, 20.2, and 20.4–20.7 depend on Chapters 1–8. Section 20.3 also depends on Section 18.12 but can be covered independently of that section if necessary. Section 20.8 depends on Chapter 14.	This chapter can be used as a one-quarter introduction to ordinary differential equations—especially as the fifth quarter or fourth semester of a calculus sequence. With time limitations, Sections 20.1, 20.2, and 20.4–20.6 should be covered.

CONTENTS

*Asterisk denotes a more difficult and optional section.

□This symbol denotes a section in which a hand calculator is useful.

EIGHT   TECHNIQUES OF INTEGRATION

NINE   FURTHER APPLICATIONS OF THE DEFINITE INTEGRAL

TEN   THE MATHEMATICAL UNDERPINNINGS OF CALCULUS

ELEVEN   POLAR COORDINATES

ONE

PRELIMINARIES

1.1 Sets of Real Numbers

In this chapter we present some basic information that you will need for your study of calculus. We begin by discussing the *real number system*. Real numbers are familiar from high school mathematics and we assume that you are familiar with the basic techniques of addition, subtraction, multiplication, division, roots, and exponents.

The real numbers fall into several categories. The *positive integers* (sometimes called the *natural numbers*) are the numbers of counting: 1, 2, 3, 4, 5, ... The three dots indicate that the string of numbers goes on indefinitely. The *integers* consist of the positive integers, the negative integers, and the number 0. It is often convenient to represent the integers on a *number line* as indicated in Figure 1.

$$-4 \quad -3 \quad -2 \quad -1 \quad 0 \quad 1 \quad 2 \quad 3 \quad 4 \qquad \text{Figure 1}$$

A *rational number* is a real number which can be written as the quotient of two integers, where the integer in the denominator is not zero:

$$r = \frac{m}{n} \quad \text{where} \quad n \neq 0. \tag{1}$$

Every integer n is also a rational number since $n = n/1$.

EXAMPLE 1. The following are rational numbers:

(a) $\frac{1}{2}$ (b) $-\frac{3}{4}$ (c) -5 (d) $0 = \frac{0}{1}$ (e) $-\frac{127}{105}$ (f) $\frac{4521}{7132}$.

Any terminating decimal is a rational number as the following example suggests.

EXAMPLE 2. $r = 0.721$ is a rational number since $r = \frac{721}{1000}$.

All rational numbers can be represented in an infinite number of ways. For example,

$$\tfrac{1}{2} = \tfrac{2}{4} = \tfrac{4}{8} = \tfrac{3}{6} = \tfrac{125}{250} = \cdots.$$

Usually, however, we will write a rational number as m/n where m and n have no common factors.

Any real number which is not rational is called *irrational*. Examples of irrational numbers are $\pi = 3.14159265\ldots$ and $\sqrt{2} = 1.41421356\ldots$. (A proof that $\sqrt{2}$ is irrational is suggested in Problem 7.)

1

Every rational number can be written as a *repeating decimal* while no irrational number can be written in this way. For example, $\frac{1}{3} = 0.33333\ldots$, $\frac{3}{11} = 0.272727\ldots$, and $\frac{2}{7} = 0.285714285714\ldots$ are examples of rational numbers written as repeating decimals. In this chapter we will not prove that a rational number can be written in this way, but shall do so in Chapter 14 in our discussion of infinite series.

The following fact will be very useful to us in later chapters:

> Between any two real numbers there is a rational number and an irrational number.   (2)

Rational and irrational numbers can be represented on the number line used to depict integers (see Figure 2).

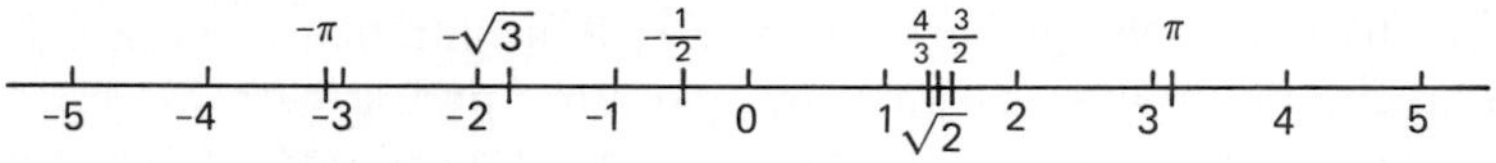

Figure 2

The number line can be used to give us a sense of order. We say that *the number a is greater than the number b if a is further to the right on the number line.* We then write this inequality as

$$a > b. \tag{3}$$

Similarly, if $b > a$ then a is less than b and we write the inequality as

$$a < b. \tag{4}$$

We use the notation

$$a \leq b \tag{5}$$

to indicate that a is less than or equal to b; that is, $a < b$ or $a = b$. Finally, we write

$$a \geq b \tag{6}$$

to indicate that a is greater than or equal to b.

EXAMPLE 3. The following inequalities are easily seen to be true by glancing at the number line in Figure 2.

(a) $2 < 3$ (b) $\frac{1}{3} < \frac{1}{2}$ (c) $\frac{4}{3} < \sqrt{2} < \frac{3}{2}$
(d) $-3 < 1$ (e) $-\sqrt{3} < -1 < 0$ (f) $-4 < -\pi < -3$.

The notation $a < c < b$ indicates that c lies between a and b on the number line. See Example 3(c), (e), and (f).

We will often be interested in *sets* of numbers. A set of numbers is any well-defined collection of numbers. For example, the integers and rational numbers are each sets of numbers. The set of all real numbers is denoted by $\mathbb{R}$. The collection of "large numbers" does not comprise a set since it is not well-defined. There is no universal agreement as to whether a given number is or is not in this collection. The numbers in a set are called *members* or *elements* of that set. If x is an element of the set A, we write $x \in A$ and read this as "x is an element of A" or "x belongs to A."

The elements of a set can be written in a bracket notation. For example, the set A of integers between $\frac{1}{2}$ and $\frac{11}{2}$ can be written as

$$A = \{x \in \mathbb{R}: x \text{ is an integer and } \tfrac{1}{2} < x < \tfrac{11}{2}\}.$$

This is read "A is the set of real numbers x such that x is an integer and x is between $\frac{1}{2}$ and $\frac{11}{2}$." Alternatively, we can write

$$A = \{1, 2, 3, 4, 5\}.$$

The *empty set,* denoted by $\emptyset$, is the set containing no elements.

We say that A is a *subset* of B if every element of A is an element of B and write

$$A \subseteq B. \tag{7}$$

Two sets are *equal* if they contain the same elements. The *union* of two sets of real numbers A and B is the set of numbers in A *or* B (or both); we write

$$\boxed{A \cup B = \{x: x \in A \text{ or } x \in B\}.} \tag{8}$$

The *intersection* of two sets A and B is the set of numbers in A *and* B; we write

$$\boxed{A \cap B = \{x: x \in A \text{ and } x \in B\}.} \tag{9}$$

The sets A and B are said to be *disjoint* if $A \cap B = \emptyset$. That is, A and B are disjoint if they have no elements in common. The *complement* of the set A is the set of real numbers not in A; we write

$$\boxed{\bar{A} = \{x: x \notin A\}.} \tag{10}$$

The symbol $\notin$ is read "is not an element of."

EXAMPLE 4.   Let $A = \{\text{positive integers}\}$, $B = \{\text{positive even integers}\}$ and $C = \{\text{negative integers and zero}\}$. Then

(a) $B \subset A$;
(b) $A \cup C = \{\text{integers}\}$;
(c) $\bar{A} = \{x: x \in \mathbb{R} \text{ and } x \text{ is not a positive integer}\}$;
(d) $A \cap C = \emptyset$.

We will often be interested in the set of numbers between two numbers a and b. We have the following important definitions:

(i)   The *open interval* (a, b) is the set of numbers between a and b, *not including* the numbers a and b. We have

$$\boxed{(a, b) = \{x: a < x < b\}.} \tag{11}$$

Note that $a \notin (a, b)$ and $b \notin (a, b)$. The numbers a and b are called *endpoints* of the interval. This is depicted in Figure 3.

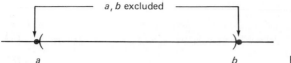

a b Figure 3

(ii)   The *closed interval* $[a, b]$ is the set of numbers between a and b, *including* the numbers a and b. We have

$$[a, b] = \{x: a \leq x \leq b\}. \tag{12}$$

Note that $a \in [a, b]$ and $b \in [a, b]$.

As before, the numbers a and b are called endpoints of the interval. This is depicted in Figure 4.

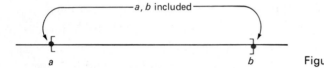

a b Figure 4

EXAMPLE 5. (a) $(-1, 5) = \{x: -1 < x < 5\}$ (b) $[0, 8] = \{x: 0 \leq x \leq 8\}$.

Sometimes we will need to include one endpoint but not the other. We can then define

(iii)   The half-open interval

$$[a, b) = \{x: a \leq x < b\}. \tag{13}$$

We include the endpoint $x = a$ but not the endpoint $x = b$. That is, $a \in [a, b)$ but $b \notin [a, b)$. Similarly,

(iv)   The half-open interval $(a, b]$ is given by

$$(a, b] = \{x: a < x \leq b\}. \tag{14}$$

We have $a \notin (a, b]$ but $b \in (a, b]$. Here b is included but a is not. See Figure 5.

(a)   (b)

Figure 5

Intervals may be infinite. We have

$$[a, \infty) = \{x: x \geq a\} \tag{15}$$
$$(a, \infty) = \{x: x > a\} \tag{16}$$
$$(-\infty, a] = \{x: x \leq a\} \tag{17}$$
$$(-\infty, a) = \{x: x < a\} \tag{18}$$
$$(-\infty, \infty) = \mathbb{R} \tag{19}$$

The symbols ∞ and $-\infty$, denoting infinity and minus infinity, respectively, are *not* real numbers and do not obey the usual laws of algebra, but can be used for notational convenience. We could also write, for example

$$[a, \infty) = \{x: a \leq x < \infty\} \qquad \text{and} \qquad (-\infty, a) = \{x: -\infty < x < a\}.$$

Finally, we discuss the notion of *boundedness*. The set A is *bounded above* if there is a number M, called an *upper bound* for A, such that $x \leq M$ for every $x \in A$. Likewise, the set A is *bounded below* if there is a number m, called a *lower bound* for A, such that $m \leq x$ for every $x \in A$. The set A is *bounded* if it is bounded above and below.

EXAMPLE 6. (a) The interval $[-6, 2]$ is bounded. An upper bound is 2 and a lower bound is -6.

(b) The infinite interval $(0, \infty)$ is bounded below, but not above. A lower bound is 0.

(c) The infinite interval $(-\infty, 2]$ is bounded above, but not below. An upper bound is 2.

(d) The set of positive integers is bounded below, but not above. A lower bound is 1.

(e) The set of rational numbers is neither bounded above nor bounded below.

Note that when a set is bounded above and/or below, the upper or lower bound is not unique. For example, $[-6, 2]$ is bounded above by 2.5, 7, 100, etc.

PROBLEMS 1.1

In Problems 1–6 insert either $<$ or $>$ in place of the comma so that the resulting inequality is true.

1. $-3, 2$
2. $4, -6$
3. $\pi/3, 1$
4. $\sqrt{2}, \frac{1}{4}$
5. $\sqrt{8}, 3.1$
6. $-10, -3\pi$.

*7. Prove that $\sqrt{2}$ is irrational. [*Hint:* Assume that $\sqrt{2}$ is rational. Then $\sqrt{2} = m/n$ where m and n are integers having no common factors. Show that this implies that $2n^2 = m^2$ and explain why this indicates that both m and n are even. Finally, explain why this proves that $\sqrt{2}$ must be irrational.]

8. Let $A = \{\text{positive integers}\}$, $B = \{\text{negative integers}\}$, and $C = \{\text{multiples of } 3\} = \{3n: n \text{ is an integer}\}$. Describe

   (a) $A \cup B$

(b) $A \cup B \cup C$ [*Hint:* first define what you mean by the union of three sets.]
(c) $B \cap C$ (d) $A \cap B$
(e) $\bar{C}$. [*Hint:* Treat C as a subset of $\mathbb{R}$.]

9.   Let $A = \{$multiples of 2$\}$, $B = \{$multiples of 3$\}$, and $C = \{$multiples of 5$\}$. Describe
     (a) $A \cup C$ (b) $B \cap \bar{C}$ (c) $A \cup B \cup C$.

10.   Let $A = (-\infty, 5)$ and $B = [-1, 10]$. Calculate $A \cup B$ and $A \cap B$.

11.   Show that $\overline{A \cap B} = \bar{A} \cup \bar{B}$. [*Hint:* $A = B$ if and only if $A \subseteq B$ and $B \subseteq A$.]

12.   Show that $\overline{A \cup B} = \bar{A} \cap \bar{B}$.

13.   Determine whether each of the following sets is bounded, bounded above, bounded below, or neither bounded above nor below.
     (a) $(1, 1000)$ (b) $[-10^{37}, 10^{37}]$ (c) $[4, \infty)$
     (d) $(-\infty, 6)$ (e) $\{$positive odd integers$\}$
     (f) $\{$negative rationals larger than $-20\}$.

14.   The set $A - B$ is defined by

$$A - B = \{x : x \in A \text{ and } x \notin B\}.$$

That is, $A - B$ is the set of numbers which are in A but not in B. If $A = [1, 3]$ and $B = [2, 4]$, calculate
     (a) $A - B$ (b) $B - A$.

15.   For the sets in Problem 8 calculate
     (a) $A - B$ (b) $B - A$ (c) $A - C$
     (d) $C - B$.

16.   For the sets of Problem 9 calculate
     (a) $A - C$ (b) $C - B$ (c) $B - C$.

17.   For the intervals of Problem 10 calculate
     (a) $A - B$ (b) $B - A$.

1.2 Absolute Value and Inequalities

We will often have need to deal with inequalities. The following arithmetic facts will prove to be very useful: Let a, b, and c be real numbers.

If $a < b$, then $a + c < b + c$.   **(1)**

If $a < b$ and $c > 0$, then $ac < bc$.   **(2)**

If $a < b$ and $c < 0$, then $ac > bc$.   **(3)**

Fact (3) can be restated as: *multiplying both sides of an inequality by a negative number reverses the direction of the inequality.*

EXAMPLE 1.   Since $2 < 5$, if we multiply by -1 we obtain $-2 > -5$. This is true since -2 is to the right of -5 on the number line.

EXAMPLE 2.   Solve the inequalities $-3 < (7 - 2x)/3 \le 4$.

SOLUTION. We need to find the set of numbers for which the inequalities are satisfied. This is called the *solution set* of the inequality. First, we multiply by 3: $-9 < 7 - 2x \le 12$. Then we subtract 7 (add -7): $-16 < -2x \le 5$. Then we multiply by $-\frac{1}{2}$ (which reverses the inequalities):

$$8 > x \ge -\tfrac{5}{2} \quad \text{or} \quad x \in [-\tfrac{5}{2}, 8).$$

Thus the solution set is the half-open interval $[-\frac{5}{2}, 8)$. Note that each of the steps in the computation served to simplify the term containing x.

EXAMPLE 3. Solve the inequality $x^2 - 4x - 12 > 0$.

SOLUTION. Factoring, we have $(x - 6)(x + 2) > 0$. In order that the product be positive, we must have either

$$(x - 6) > 0 \quad \text{and} \quad (x + 2) > 0$$

or

$$(x - 6) < 0 \quad \text{and} \quad (x + 2) < 0.$$

In the first case, we have $x > 6$. In the second, we must have $x < -2$. Therefore

$$x^2 - 4x - 12 > 0 \quad \text{if} \quad x > 6 \quad \text{or} \quad x < -2$$

and the solution set is given by $(-\infty, -2) \cup (6, \infty)$.

We now define the absolute value of a number. Look at Figure 1. The *absolute value* of a number a is defined as the distance from that number to zero and is written

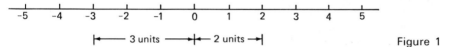

|←——— 3 units ———→|←— 2 units —→| Figure 1

$|a|$. Thus 2 is 2 units from zero so that $|2| = 2$. The number -3 is 3 units from zero so that $|-3| = 3$. Alternatively, we may define

$$|a| = a \quad \text{if} \quad a \geq 0 \tag{4}$$

and

$$|a| = -a \quad \text{if} \quad a < 0. \tag{5}$$

The absolute value of a number is a nonnegative number. Note that, for example, $|5| = 5$ and $|-5| = -(-5) = 5$ so that numbers which are negatives of one another have the same absolute value. Another way to calculate absolute value is to observe that

$$|x| = \sqrt{x^2} \tag{6}$$

where, of course, the positive square root is taken. Thus, for example, $|-3| = \sqrt{(-3)^2} = \sqrt{9} = 3$, $|3| = \sqrt{(3)^2} = \sqrt{9} = 3$, etc.

For all real numbers a and b, the following facts can be proven:

$$|-a| = |a| \tag{7}$$

$$|ab| = |a||b| \tag{8}$$

$$|a + b| \leq |a| + |b| \quad \text{(triangle inequality)} \tag{9}$$

Property (7) is obviously true. A proof of property (8) is suggested in Problem 12 and a proof of the triangle inequality is suggested in Problem 16.

EXAMPLE 4. $3 = |-2 + 5| \leq |-2| + |5| = 2 + 5 = 7$, which shows that the triangle inequality holds in this case.

In solving inequalities using absolute values, the following property will be very useful.

$$|x| < a \quad \text{is equivalent to} \quad -a < x < a; \qquad\qquad (10)$$

that is, $\{x : |x| < a\} = \{x : -a < x < a\}$ (see Figure 2), which indicates that for any x in the open interval $(-a, a)$ the distance from x to zero is less than a.

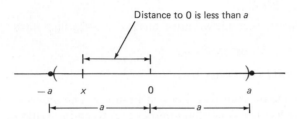

Figure 2

EXAMPLE 5. Solve the inequality $|x| \leq 3$.

SOLUTION. $|x| \leq 3$ implies that the distance from x to 0 is less than or equal to 3. Thus $-3 \leq x \leq 3$ and the solution set is the closed interval $[-3, 3]$. See Figure 3.

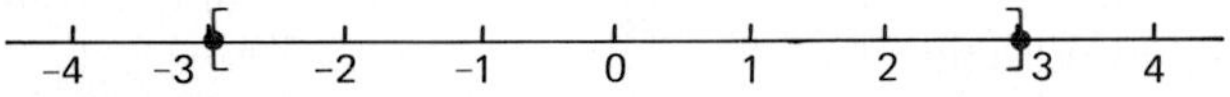

Figure 3

EXAMPLE 6. Solve the inequality $|x - 4| < 5$.

SOLUTION. The distance from $x - 4$ to 0 is less than 5 so that $-5 < x - 4 < 5$ and, adding 4 to each term, we see that $-1 < x < 9$. Thus the solution set is the open interval $(-1, 9)$. See Figure 4. Another way to think of this set is as the set of all x such that x is within 5 units of 4.

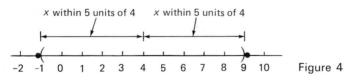

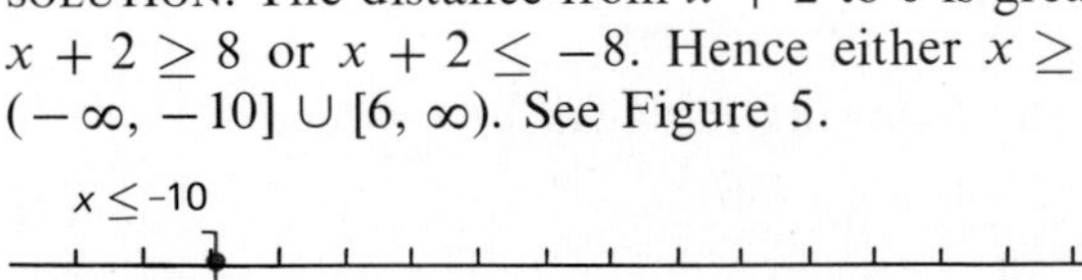

Figure 4

EXAMPLE 7. Solve the inequality $|x + 2| \geq 8$.

SOLUTION. The distance from $x + 2$ to 0 is greater than or equal to 8 so that either $x + 2 \geq 8$ or $x + 2 \leq -8$. Hence either $x \geq 6$ or $x \leq -10$. The solution set is $(-\infty, -10] \cup [6, \infty)$. See Figure 5.

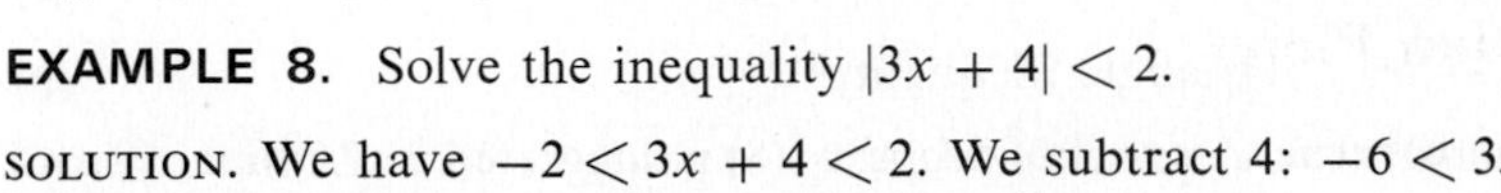

Figure 5

EXAMPLE 8. Solve the inequality $|3x + 4| < 2$.

SOLUTION. We have $-2 < 3x + 4 < 2$. We subtract 4: $-6 < 3x < -2$. Then we divide by 3: $-2 < x < -\frac{2}{3}$ or $x \in (-2, -\frac{2}{3})$.

EXAMPLE 9. Solve the inequality $|5 - 3x| \geq 1$.

SOLUTION. We have either $5 - 3x \geq 1$ or $5 - 3x \leq -1$. In the first case $-3x \geq -4$ which implies that $x \leq \frac{4}{3}$. In the second case $-3x \leq -6$ so that $x \geq 2$. The solution set is, therefore, $(-\infty, \frac{4}{3}] \cup [2, \infty)$.

PROBLEMS 1.2

In Problems 1–9, solve the given inequalities and display the solution sets on the number line.

1. $x - 2 < 5$ 2. $x + 3 > -6$ 3. $1 \leq 2x + 2 \leq 4$

4. $3x + 6 \geq -3$ 5. $3 < 2 - 5x < 13$ 6. $-4 < \dfrac{2x - 4}{3} \leq 7$

7. $2 \geq \dfrac{4 - 2x}{5} > -4$ 8. $\dfrac{1}{x} > 3$ 9. $\dfrac{4}{3x - 2} \leq -2.$

10. Solve the inequality $a \leq \dfrac{bx + c}{d} < e, \quad b, \, d > 0$.

11. Solve for x:
     (a) $x = |4 - 5|$ (b) $x = |-6 - (-2)|$ (c) $x = |2| - |-3|$.
12. Show that $|xy| = |x||y|$. [*Hint:* Deal with each of four cases separately: (1) $x \geq 0$, $y \geq 0$, (2) $x \geq 0$, $y < 0$, etc.]
13. Show that if $x \geq 0$ and $y \geq 0$, then $|x + y| = |x| + |y|$.
14. If $x > 0$ and $y < 0$, show that $|x + y| < |x| + |y|$.
15. If $x < 0$ and $y < 0$, show that $|x + y| = |x| + |y|$.
16. Using Problems 13–15, prove the triangle inequality (9).
*17. Show that $||x| - |y|| \leq |x - y|$. [*Hint:* Write $x = (x - y) + y$ and apply the triangle inequality.]

In Problems 18–37 solve the given inequalities and graph the solution sets.

18. $|x| \leq 4$ 19. $|x| > 2$ 20. $|x - 2| < 1$

21. $|x + 3| \leq 4$ 22. $|x + 6| > 3$ 23. $|2x + 4| < 3$

*24. $|6 - 4x| \geq |x - 2|$ 25. $\left|\dfrac{8 - 3x}{2}\right| \leq 3$ 26. $\left|\dfrac{3x + 17}{4}\right| > 9$

27. $x^2 - 4x + 4 \geq 0$ 28. $x^2 + 2x - 15 < 0$ 29. $x^2 + 8x + 15 > 0$

30. $3 - \sqrt{x} < 8$

31. $\dfrac{1}{x - 2} > \dfrac{2}{x + 3}$ [*Hint:* Keep track of whether you are multiplying by a positive or negative number.]

32. $\dfrac{2x + 3}{4x - 5} \leq -4$ 33. $|ax + b| < c, \quad a > 0$ 34. $|ax + b| \geq c, \quad a < 0, \quad c > 0$

*35. $x \leq |x|$ 36. $x > \dfrac{1}{x}$ *37. $|2x| > |5 - 2x|$.
                                                        [*Hint:* Use Eq. (6).]

1.3 The Cartesian Plane

To this point we have been concerned with single numbers. We shall now discuss properties of pairs of numbers. An *ordered pair* of numbers consists of two numbers;

one is called the *first element* and the other is called the *second element*. Ordered pairs are written in the form

$$(a, b) \tag{1}$$

where a and b are real numbers. Two ordered pairs are equal if their first elements are equal and their second elements are equal. Note that $(1, 0)$ and $(1, 1)$ are *different* ordered pairs since their second elements are different.

EXAMPLE 1. The following are ordered pairs of numbers:

(a) $(1, 3)$; (b) $(-2, 5)$; (c) $(-\sqrt{2}, 4)$; (d) $(0, 0)$; (e) $(1, 0)$; (f) $(0, 1)$.

DEFINITION 1. The set of all ordered pairs of real numbers is called the *Cartesian*† *plane* and is denoted $\mathbb{R}^2$. Hence

$$\boxed{\mathbb{R}^2 = \{(x, y) : x \in \mathbb{R} \text{ and } y \in \mathbb{R}.}$$

Any element (x, y) of the Cartesian plane is called a *point* in the plane.

Any point in $\mathbb{R}$ can be represented graphically on a number line. Analogously any point (x, y) can be represented graphically as a point in the Cartesian plane. We call the first number in the ordered pair the *x-coordinate* and the second number the *y-coordinate*. Following the technique invented by Descartes,‡ we draw a horizontal line called the *x-axis* and a vertical line perpendicular to it called the *y-axis*. The point at which the two lines meet is called the *origin* and is labeled $(0, 0)$ or 0. With this orientation for the point denoted (a, b), a is the x-coordinate and b is the y-coordinate. Along the x-axis, positive distances are measured to the right of the origin, and along the y-axis, positive distances are measured in the upward direction. This is illustrated in Figure 1. The arrows indicate the direction of increasing x and y. In Figure 1, a typical point (a, b) is drawn, where $a > 0$ and $b > 0$. We simply measure a units in the positive direction along the x-axis and b units in the positive direction along the y-axis to arrive at the point (a, b). In Figure 2 several different points are depicted. Note that $(1, 2) \neq (2, 1)$ since they represent different points in the plane.

> **Warning.** Do not become confused by the symbol (a, b) which can represent either a point in the plane or an open interval. The meaning will always be clear from the context.

> **Notation.** With the above construction, we will usually refer to the Cartesian plane as the *xy-plane*.

---

† This is named after the great French mathematician and philosopher René Descartes (1596–1650) who is considered to be the inventor of analytic geometry. He is known for the statement "Cogito ergo sum."—"I think, therefore I am." which played a central role in his philosophical writings.

‡ This first appeared in *La Geometrie*, published in 1637. One of the major successes of this work was that it connected figures of geometry with the equations of algebra. Our development of calculus will also follow this path.

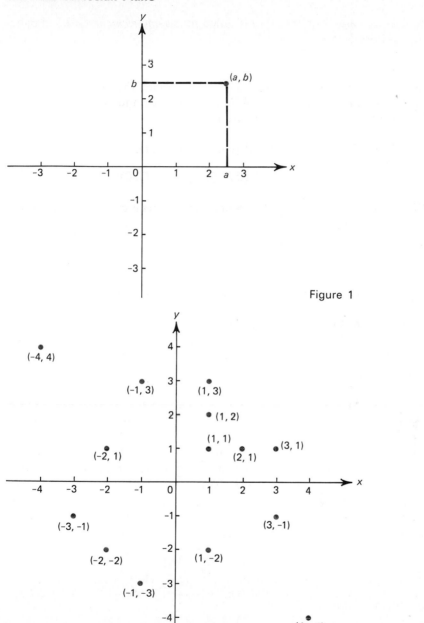

Figure 1

Figure 2

A glance at Figure 3 indicates that the x- and y-axes divide the xy-plane into four regions. These regions are called *quadrants* and are denoted as in the figure.

EXAMPLE 2. (a) $(1, 3)$ is in the first quadrant since $1 > 0$ and $3 > 0$.
(b) $(-4, -7)$ is in the third quadrant since $-4 < 0$ and $-7 < 0$.
(c) $(-2, 5)$ is in the second quadrant.
(d) $(7, -3)$ is in the fourth quadrant.

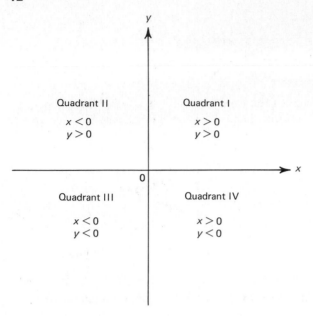

Figure 3

We will often need to calculate the distance between two points in the xy-plane. Consider the two points (x_1, y_1) and (x_2, y_2) as in Figure 4. The distance d between

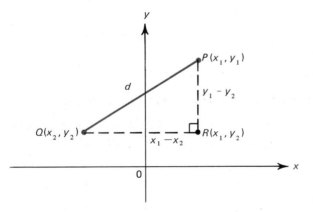

Figure 4

(x_1, y_1) and (x_2, y_2) is the length of the line segment PQ. Since PQR is a right triangle, we have, by the Pythagorean theorem,

$$\overline{PQ}^2 = \overline{PR}^2 + \overline{QR}^2.\dagger$$

But $\overline{PR} = |y_1 - y_2|$ and $\overline{QR} = |x_1 - x_2|$ so that $d = \overline{PQ} = \sqrt{\overline{PR}^2 + \overline{QR}^2}$ or

$$d = \sqrt{(x_1 - x_2)^2 + (y_1 - y_2)^2}. \tag{2}$$

EXAMPLE 3. Find the distance between the points $(2, 5)$ and $(-3, 7)$.

†The symbol $\overline{PQ}$ denotes the distance between the points P and Q.

SOLUTION. Let $(x_1, y_1) = (2, 5)$ and $(x_2, y_2) = (-3, 7)$ so that from (2),

$$d = \sqrt{(2 - (-3))^2 + (5 - 7)^2} = \sqrt{5^2 + (-2)^2} = \sqrt{29}.$$

EXAMPLE 4. Show that the points $P_1 = (0, -3)$, $P_2 = (1, -1)$ and $P_3 = (2, 1)$ are collinear.

SOLUTION. Three points are collinear if the lines joining them form a straight line. It is easy to check that 3 points are collinear. Look at Figure 5. If the points are collinear

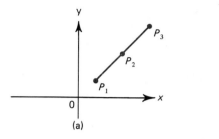
(a)

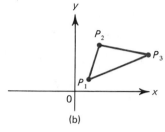
(b)   Figure 5

(Figure 5a), then $\overline{P_1 P_2} + \overline{P_2 P_3} = \overline{P_1 P_3}$. If they are not (Figure 5b), then $\overline{P_1 P_2} + \overline{P_2 P_3} > \overline{P_1 P_3}$ since a straight line is the shortest distance between two points. In our example,

$$\overline{P_1 P_2} = \sqrt{(0 - 1)^2 + (-3 - (-1))^2} = \sqrt{5},$$
$$\overline{P_2 P_3} = \sqrt{(1 - 2)^2 + (-1 - 1)^2} = \sqrt{5}$$

and

$$\overline{P_1 P_3} = \sqrt{(0 - 2)^2 + (-3 - 1)^2} = \sqrt{20} = 2\sqrt{5}$$

so that $\overline{P_1 P_3} = \overline{P_1 P_2} + \overline{P_2 P_3}$ and the points are collinear.

EXAMPLE 5.   Prove that the diagonals of a rectangle have equal lengths.

SOLUTION. We place the rectangle as in Figure 6. The lengths of the two diagonals are $\overline{AC}$ and $\overline{BD}$. But $\overline{AC} = \sqrt{x^2 + y^2}$ and $\overline{BD} = \sqrt{(-x)^2 + y^2}$ so that $\overline{AC} = \overline{BD}$.

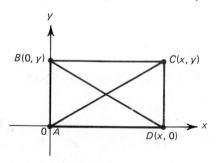
Figure 6

PROBLEMS 1.3

1. Determine the quadrant of each point and draw the point in the xy-plane.
   (a) $(3, -2)$ (b) $(-\sqrt{2}, -\sqrt{2})$ (c) $(-\frac{1}{3}, -7)$
   (d) $(\pi, 97)$.

In Problems 2–8 find the distance between the given points.

2. (1, 3), (4, 7)   3. (−7, 2), (4, 3)   4. (8, −1), (−2, 0)
5. $(\frac{1}{2}, \frac{1}{3})$, $(\frac{1}{3}, \frac{1}{2})$ 6. (−3, −7), (−1, −2)   7. (a, b), (b, a)
8. (a, b), (0, 0).

9. Are the points (−2, 5), (1, −1) and (3, −5) collinear? Illustrate your answer in the xy-plane.

10. Are the points (−3, 2), (0, 1) and (2, −1) collinear? Illustrate your answer in the xy-plane.

11. Let L denote the line segment joining the points $P_1 = (x_1, y_1)$ and $P_2 = (x_2, y_2)$. Show that the midpoint of L is the point

$$P_3 = \left(\frac{x_1 + x_2}{2}, \frac{y_1 + y_2}{2} \right).$$

[Hint: Show that $\overline{P_1 P_3} = \overline{P_3 P_2}$ and that $\overline{P_1 P_2} = \overline{P_1 P_3} + \overline{P_3 P_2}$.]

12. Find the midpoint of the line segment joining the points
(a) (2, 5) and (5, 12).   (b) (−3, 7) and (4, −2).

*13. Show that the area A of a triangle with vertices $P_1 = (x_1, y_1)$, $P_2 = (x_2, y_2)$ and $P_3 = (x_3, y_3)$ is given by

$$A = \tfrac{1}{2}|x_1 y_2 + x_2 y_3 + x_3 y_1 - x_1 y_3 - x_2 y_1 - x_3 y_2|.$$

[Hint: Look at Figure 7, calculate the area of the rectangle, and subtract from it the sum of the areas of triangles I, II, and III. Remember, the area of a triangle is equal to half the length of its base times its height.]

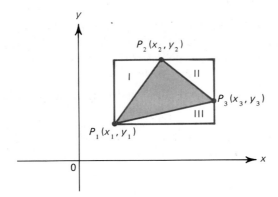

Figure 7

14. Using the result of Problem 13, calculate the area of the triangle with the given vertices.
(a) (2, 1), (0, 4), (3, −6)   (b) (4, 2), (−1, −5), (7, 3).

1.4 Lines

Lines will be very important to us in the study of calculus. Since a straight line (or, simply, a line) is the shortest distance between two points, any two points (x_1, y_1) and (x_2, y_2) determine a line. The *slope* of a line is a measure of the relative rate of change of the x- and y-coordinates of points on the line as we move along the line. We define

$$\boxed{\text{slope of } L = \frac{y_2 - y_1}{x_2 - x_1} = \frac{\Delta y}{\Delta x},} \tag{1}$$

where (x_1, y_1) and $x_2, y_2)$ are *any* two points on the line. Here Δy and Δx denote the changes in y and x, respectively. See Figure 1. An *equation* of a line is the equation in

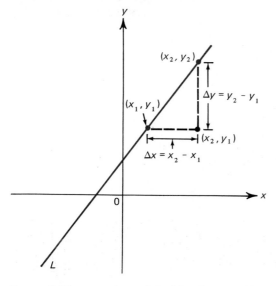

Figure 1

the variables x and y satisfied by the coordinates of every point on the line. The *graph* of an equation is the set of all points drawn in the xy-plane whose coordinates satisfy the equation. If we know two points on a line, then we can calculate the equation of the line and obtain its graph, as the next example shows.

EXAMPLE 1.   Calculate the equation of the line passing through the points $(-1, -2)$ and $(2, 5)$ and draw its graph.

SOLUTION. Look at Figure 2. Let (x, y) be another point on the line. We can calculate the slope of the line in several ways. Between the points $(-1, -2)$ and $(2, 5)$,

$$\frac{\Delta y}{\Delta x} = \frac{5 - (-2)}{2 - (-1)} = \frac{7}{3}.$$

Between $(2, 5)$ and (x, y),

$$\frac{\Delta y}{\Delta x} = \frac{y - 5}{x - 2}.$$

Since the slope of a straight line must be the same no matter which two points on the line are chosen, we have

$$\frac{y - 5}{x - 2} = \frac{7}{3}.$$

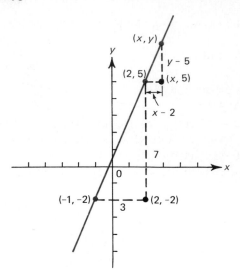

Figure 2

We then multiply both sides of this equation by $3(x - 2)$ to obtain

$$3(y - 5) = 7(x - 2) \qquad \text{or} \qquad 3y - 15 = 7x - 14$$

and

$$3y = 7x + 1$$

or

$$y = \tfrac{7}{3}x + \tfrac{1}{3}. \tag{2}$$

   From this example, we see that there are several ways to write the equation of a line. The value of the y-coordinate of the point at which the line intersects the y-axis is called the *y-intercept* of the line. This is easy to calculate since it is the value of y that we obtain when $x = 0$. Note that for any point on the y-axis the x-coordinate is zero. In (2), the y-intercept is $\tfrac{1}{3}$. In Equation (2), we have written the equation of the line in its *slope–intercept form* as the slope $\tfrac{7}{3}$ and the y-intercept $\tfrac{1}{3}$ are clearly displayed. More generally, the equation of any line, except the vertical lines of the form $x = a$ (see Figure 5b), can be written in the slope–intercept form

$$\boxed{y = mx + b} \tag{3}$$

where m is the slope and b is the y-intercept.
   In (2), we can also write the equation of the line as

$$3y - 7x = 1 \qquad \text{or} \qquad -7x + 3y - 1 = 0. \tag{4}$$

However, in the earlier parts of this text, it will usually be more convenient to write a line in its slope–intercept form (3).
   In general, to calculate the equation of the line passing through the points

(x_1, y_1) and (x_2, y_2) where $x_1 \neq x_2$

    (i)   calculate the slope $m = \dfrac{y_2 - y_1}{x_2 - x_1}$, and

    (ii) set $\dfrac{y - y_2}{x - x_2} = m$ and simplify.

To graph the line, it is only necessary to plot two points which lie on the line in the xy-plane, draw the line segment between them, and extend it in both directions.

    We say that two lines are *parallel* if they have the same slope. Thus, for example, the lines $y = 2x + 3$ and $y = 2x - 2$ are parallel since they both have the slope 2. This is illustrated in Figure 3.

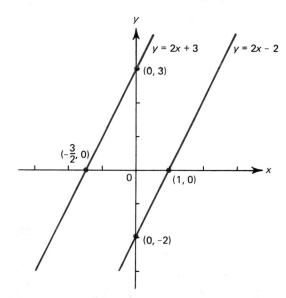

Figure 3

EXAMPLE 2. Find the equation of the line L passing through the point $(2, 3)$ and parallel to the line $y = -3x + 5$.

SOLUTION. L has slope -3 so that if (x, y) is any point on L,

$$\frac{\Delta y}{\Delta x} = \frac{y - 3}{x - 2} = -3$$

or, simplifying,

$$y - 3 = -3x + 6 \quad \text{and} \quad y = -3x + 9.$$

The simplest way to draw the graph of this line is to note that the y-intercept is 9 and when $y = 0$, $x = 3$ ($(3, 0)$ is called the *x-intercept*). Therefore the line passes through the points $(0, 9)$ and $(3, 0)$. This is illustrated in Figure 4. Example 2 shows us how to calculate the equation of a line given its slope and one point on it. That is, given a point (x_0, y_0) and a slope m, the equation of the line L is $(y - y_0)/(x - x_0) = m$.

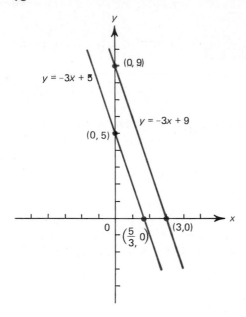

Figure 4

In Figure 2 we notice that the line $y = \frac{7}{3}x + \frac{1}{3}$ points upward as x increases while in Figure 4 the line $y = -3x + 9$ points downward as x increases. This should not be surprising. If $y = \frac{7}{3}x + \frac{1}{3}$, then as x increases, y increases (this is what is meant by a positive slope). If $y = -3x + 9$, then as x increases, y decreases (this is what is meant by a negative slope). In general,

---

(i)   if $m > 0$ the graph of the line will point upward in the direction of increasing x.

(ii)   if $m < 0$, the graph of the line will point downward in the direction of increasing x.

There are two cases that must be treated separately. In Figure 5a, we have drawn the line $y = a$ which is parallel to the x-axis. Here, as x changes, y doesn't change at all (since y is equal to the constant a). Therefore $\Delta y / \Delta x = 0/\Delta x = 0$. *Lines*

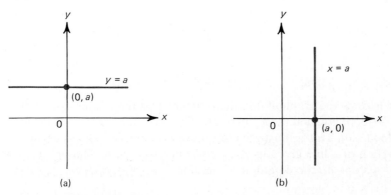

(a)   (b)   Figure 5

parallel to the x-axis have a slope of zero. In Figure 5b we have drawn the line $x = a$ which is parallel to the y-axis. Here, when y changes, x doesn't change at all. In this case we say that the line has an infinite slope. *Lines parallel to the y-axis have an infinite slope* and we write $m = \infty$.

We say that two lines L_1 and L_2 are *perpendicular* if they meet at right angles, and we write $L_1 \perp L_2$.

THEOREM 1. If $L_1 \perp L_2$, if m_1 and m_2 are their slopes, and if $m_2 \neq 0$ or ∞, then

$$m_1 = -\frac{1}{m_2}. \tag{5}$$

If $m_2 = 0$, then $m_1 = \infty$ and if $m_1 = 0$, then $m_2 = \infty$.

PROOF. Assume that $m_2 \neq 0$ and $m_2 \neq \infty$. Since parallel lines have the same slopes, we move L_1 parallel to itself and L_2 parallel to itself so that they intersect at the origin, obtaining the lines L_1' and L_2', respectively (see Figure 6). Labeling the various

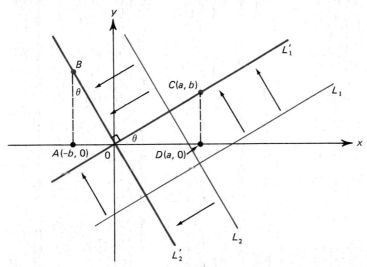

Figure 6

points as in the figure, we calculate the coordinates of the point B. By elementary geometry, we can see that the two angles denoted by θ are equal (this follows since the sum of the angles in a triangle is 180° and because the sum of the angles on one side of a line is 180°). Thus the right triangles $A0B$ and $C0D$ are congruent. In particular, $\overline{AB} = \overline{0D} = a$. Thus B must have the coordinates $(-b, a)$. Then L_2' contains the points $(-b, a)$ and $(0, 0)$ so that $m_2 = a/(-b) = -a/b$. L_1' contains the points $(0, 0)$ and (a, b) so that

$$m_1 = \frac{b}{a} \quad \text{and} \quad m_2 = -\frac{1}{b/a} = -\frac{1}{m_1}$$

which proves the first part of the theorem. The second part follows from the fact that if $m_2 = 0$, then L_2 is horizontal so that L_1 must be vertical, implying that $m_1 = \infty$. If $m_2 = \infty$, then $m_1 = 0$ by the same argument. Note that the proof we have given is independent of the points we chose on the line.

EXAMPLE 3. Find the equation of the line perpendicular to the line $2x + 3y = 4$ which passes through the point $(-1, 3)$.

SOLUTION. We rewrite the line $2x + 3y = 4$ as $y = -\frac{2}{3}x + \frac{4}{3}$, which has the slope $-\frac{2}{3}$. Thus the line we seek has the slope $-1/(-\frac{2}{3}) = \frac{3}{2}$. Then

$$\frac{y - 3}{x - (-1)} = \frac{y - 3}{x + 1} = \frac{3}{2} \qquad \text{or} \qquad 2(y - 3) = 3(x + 1)$$

and we simplify to find the equation

$$y = \tfrac{3}{2}x + \tfrac{9}{2}.$$

There is another type of problem we will encounter.

EXAMPLE 4. Find the point of intersection of the lines $2x + 3y = 7$ and $-x + y = 4$.

SOLUTION. The point of intersection, which we will label (a, b), must satisfy both equations. For the first equation we have

$$y = -\tfrac{2}{3}x + \tfrac{7}{3}$$

and for the second

$$y = x + 4.$$

Therefore, at (a, b),

$$b = -\tfrac{2}{3}a + \tfrac{7}{3} = a + 4 \qquad \text{or} \qquad \tfrac{5}{3}a = \tfrac{7}{3} - 4 = -\tfrac{5}{3} \quad \text{and} \quad a = -1.$$

Then

$$b = a + 4 = 3.$$

The lines intersect at $(-1, 3)$.

PROBLEMS 1.4

In Problems 1–12 find the slope–intercept form of the equation of the straight line when either two points on the line or a point and the slope of the line are given. Sketch the graph of the line in the xy-plane.

1. $(1, 2), (3, 6)$	**2.** $(-2, 3), (4, -1)$	**3.** $(3, 7), m = \frac{1}{2}$
4. $(4, -7), m = 0$	**5.** $(-3, -8), m = \infty$	**6.** $(3, -\frac{1}{2}), (\frac{1}{3}, 0)$
7. $(-2, -4), (3, 7)$	**8.** $(5, -1), (8, 2)$	**9.** $(7, -3), m = -\frac{4}{3}$
10. $(-5, 1), m = \frac{3}{7}$	**11.** $(a, b), (c, d)$	**12.** $(a, b), m = c.$

13. Find the equation of the line parallel to the line $2x + 5y = 6$ which passes through the point $(-1, 1)$.

14. Find the equation of the line parallel to the line $5x - 7y = 3$ which passes through the point $(2, 5)$.

15. Find the equation of the line perpendicular to the line $x + 3y = 7$ which passes through the point $(0, 1)$.

16. Find the equation of the line perpendicular to the line $2x - \frac{3}{2}y = 7$ which passes through the point $(-1, 4)$.

*17. Find the equation of the line perpendicular to the line $ax + by = c$ which passes through the point (α, β). Assume that $a \neq 0$ and $b \neq 0$.

In Problems 18–23 find the point of intersection (if there is one) of the two lines.

18. $x - y = 7;\ 2x + 3y = 1$	*19. $y - 2x = 4;\ 4x - 2y = 6$
*20. $4x - 6y = 7;\ 6x - 9y = 12$	*21. $4x - 6y = 10;\ 6x - 9y = 15$
22. $3x + y = 4;\ y - 5x = 2$	**23.** $3x + 4y = 5;\ 6x - 7y = 8.$

Let L be a line and let $L_{\perp}$ denote the line perpendicular to L which passes through a given point P. Then the *distance* from L to P is defined to be the distance between P and the point of intersection of L and $L_{\perp}$. In Problems 24–29, find the distance between the given line and point.

24. $x - y = 6;\ (0, 0)$	**25.** $2x + 3y = -1;\ (0, 0)$
26. $3x + y = 7;\ (1, 2)$	**27.** $5x - 6y = 3;\ (2, \frac{16}{5})$
28. $2y - 5x = -2;\ (5, -3)$	**29.** $6y + 3x = 3;\ (8, -1).$

30. Find the distance between the line $2x - y = 6$ and the point of intersection of the lines $2x - 3y = 1$ and $3x + 6y = 12$.

1.5 Circles

A *circle* is defined as the set of points equidistant from a given point. The given point is called the *center* of the circle and the common distance from the center is called the *radius*.

EXAMPLE 1. Find the equation of the circle centered at the origin with radius 1.

SOLUTION. If (x, y) is any point on the circle, then the distance from (x, y) to $(0, 0)$ is 1 so that $\sqrt{(x - 0)^2 + (y - 0)^2} = 1$ or

$$\boxed{x^2 + y^2 = 1.}$$ (1)

This is sketched in Figure 1. This circle is called the *unit circle* and will be of central importance in our study of the trigonometric functions in Chapter 7.

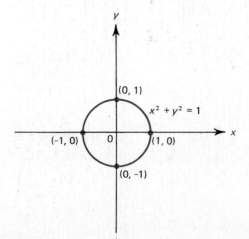

Figure 1

EXAMPLE 2. Find the equation of the circle centered at $(3, -5)$ with radius 4.

SOLUTION. If (x, y) is on the circle, then the distance from (x, y) to $(3, -5) = 4$ so that

$$\sqrt{(x-3)^2 + (y+5)^2} = 4 \qquad \text{or} \qquad (x-3)^2 + (y+5)^2 = 16.$$

This is sketched in Figure 2a.

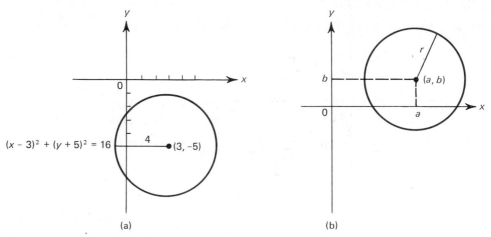

(a)   (b)

Figure 2

In general, the equation of the circle with center at (a, b) and radius r is given by

$$\boxed{(x-a)^2 + (y-b)^2 = r^2.} \tag{2}$$

This is sketched in Figure 2b.

EXAMPLE 3. Show that the equation $x^2 - 6x + y^2 + 2y - 17 = 0$ represents a circle and find its center and radius.

SOLUTION. We use the technique of completing the squares.† We have

$$x^2 - 6x + y^2 + 2y - 17 = (x^2 - 6x + 9) + (y^2 + 2y + 1) - 10 - 17$$
$$= (x-3)^2 + (y+1)^2 - 27 = 0.$$

This implies that

$$(x-3)^2 + (y+1)^2 = 27$$

which is the equation of a circle with center at $(3, -1)$ and radius $\sqrt{27}$.

EXAMPLE 4. Find the points of intersection of the circle $(x-1)^2 + (y+1)^2 = 4$ and the line $y = 2x - 1$.

SOLUTION. See Figure 3. To find the points of intersection, we substitute $y = 2x - 1$ into the equation of the circle to obtain

$$(x-1)^2 + (2x-1+1)^2 = 4$$

† Recall that this technique is often used to solve quadratic equations.

or, squaring,

$$x^2 - 2x + 1 + 4x^2 = 4 \qquad \text{and} \qquad 5x^2 - 2x - 3 = 0.$$

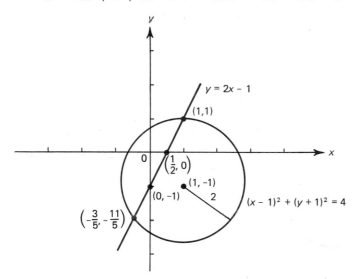

Figure 3

From the quadratic formula†

$$x = \frac{2 \pm \sqrt{4 - 4(5)(-3)}}{2(5)} = \frac{2 \pm \sqrt{64}}{10} = \frac{2 \pm 8}{10}.$$

Thus $x = 1$ and $x = -\frac{3}{5}$. When $x = 1$, $y = 2 \cdot 1 - 1 = 1$. When $x = -\frac{3}{5}$, $y = 2(-\frac{3}{5}) - 1 = -\frac{11}{5}$. Thus the points of intersection are $(1, 1)$ and $(-\frac{3}{5}, -\frac{11}{5})$.

EXAMPLE 5.   Show that the unit circle and the line $y = 3x - 4$ do not intersect.

SOLUTION.   If (x, y) is on the line and circle, then

$$x^2 + y^2 = x^2 + (3x - 4)^2 = 1$$

or

$$x^2 + 9x^2 - 24x + 16 = 10x^2 - 24x + 16 = 1$$

and we have

$$10x^2 - 24x + 15 = 0.$$

But the discriminant $(24)^2 - 4(10)(15) = 576 - 600 = -24 < 0$ so that the quadratic equation has no real solutions‡ meaning that there are no points of intersection. This is illustrated in Figure 4.

---

† The solutions to the quadratic equation $ax^2 + bx + c = 0$ are given by the quadratic formula

$$x = \frac{-b + \sqrt{b^2 - 4ac}}{2a} \qquad \text{and} \qquad x = \frac{-b - \sqrt{b^2 - 4ac}}{2a}.$$

The expression $b^2 - 4ac$ is called the *discriminant* of the equation.
‡ Since there is no real number which is the square root of a negative number.

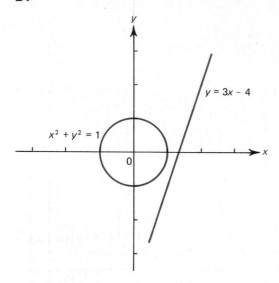

$y = 3x - 4$

$x^2 + y^2 = 1$

Figure 4

PROBLEMS 1.5

In Problems 1–9 find the equation of the circle with the given center and radius and sketch its graph.

1. $(1, 0)$; $r = 1$ **2.** $(0, 1)$; $r = 1$ **3.** $(1, 1)$; $r = \sqrt{2}$
4. $(1, -1)$; $r = 2$ **5.** $(-1, 4)$; $r = 5$ **6.** $(\frac{1}{2}, \frac{1}{3})$; $r = \frac{1}{2}$
7. $(\pi, 2\pi)$; $r = \sqrt{\pi}$ **8.** $(4, -5)$; $r = 7$ **9.** $(3, -2)$; $r = 4$.

In Problems 10–15 find the points of intersection, if any, of the given line and circle. Indicate what is happening on a graph.

10. $x^2 + y^2 = 1$; $y = x$
11. $(x - 2)^2 + (y - 3)^2 = 4$; $y = -x + 2$
12. $(x + 1)^2 + (y + 1)^2 = 1$; $y = 2x + 5$
13. $(x + 3)^2 + (y - 4)^2 = 25$; $y = -2x + 3$
14. $x^2 + (y - 3)^2 = 9$; $3x + 2y = 6$
15. $(x + 4)^2 + y^2 = 16$; $x = 1$.

16. Show that the equation $x^2 - 6x + y^2 + 4y - 12 = 0$ is the equation of a circle and find its center and radius.
*17. Show that the equation $x^2 + ax + y^2 + by + c = 0$ is the equation of a circle if and only if† $a^2 + b^2 - 4c > 0$.

In Example 3.4.3, we will prove that the tangent line T to a circle at a given point is perpendicular to the radial line R at that point (see Figure 5). In Problems 18–21 find the equation of the tangent line to the given circle at the given point.

18. $x^2 + y^2 = 1$; $\left(\dfrac{1}{\sqrt{2}}, \dfrac{1}{\sqrt{2}}\right)$ **19.** $(x - 1)^2 + (y + 1)^2 = 4$; $(1, 1)$
20. $(x + 2)^2 + (y - 3)^2 = 9$; $(0, 3 + \sqrt{5})$ **21.** $(x - 3)^2 + (y + 2)^2 = 5$; $(4, 0)$.

---

†The words "if and only if" mean that each of the two statements implies the other. For example, the statement "An integer is even if and only if it is a multiple of two." means that (i) if an integer is even, then it is a multiple of two and (ii) if an integer is a multiple of two, then it is even.

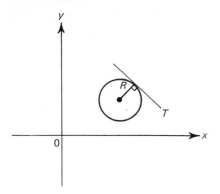

Figure 5

1.6 Functions

Let us return to the equation of a straight line. For example, if $y = 3x + 5$, then the line can be thought of as the set of all ordered pairs (or points) (x, y) such that $y = 3x + 5$. The important fact here is that for every real number x, there is a *unique* real number y such that $y = 3x + 5$ and the ordered pair (x, y) is on the line. We generalize this idea in the following definition.

DEFINITION 1. A *function f* is a set of ordered pairs (x, y) such that for every x there is a *unique y*. That is, if (x, y_1) and (x, y_2) are two pairs in f, then $y_1 = y_2$. The set of x values is called the *domain* of the function and the set of y values is called the *range* of the function. Instead of writing functions as ordered pairs, we will often use the notation†

$$y = f(x) \tag{1}$$

which is read "y is a function of x" or, more simply, "y equals f of x." This notation can be thought of in the following way: equation (1) gives us a *rule* which assigns to every x in the domain of f, a unique number y in the range of f.

To determine the domain of a function, we need to ask the question "for what values of x does the rule given in equation (1) make sense?" For example, if f is the function written as $f(x) = 1/x$, then since the expression $1/x$ is not defined for $x = 0$, the number 0 is not in the domain of f. However $1/x$ is defined for any $x \neq 0$, so that the domain of f is the set of all real numbers except zero. This can be written as $\mathbb{R} - \{0\}$.

To determine the range of a function, it is necessary to ask "what values do I obtain for y as x takes on all values in the domain of f?" For example, if f is the set of ordered pairs (x, x^2), then the domain of f is $\mathbb{R}$ since any real number can be squared. The range of f is the set of nonnegative real numbers, denoted $\mathbb{R}^+$, because the square of any real number is nonnegative.

In the examples that follow, we shall write our functions in the form (1) since that is the way functions are most commonly denoted.

---

†This notation was first used by the great Swiss mathematician Leonhard Euler (1707–1783) in his *Petersburg Commentaries,* published in 1734–1735.

Warning. In writing $y = f(x)$, it is necessary to distinguish between the symbol f, which stands for the function and is a *set* of ordered pairs, and the symbol $f(x)$, which is the *value* the function takes on for a given number x in the domain of f. Here $f(x)$ is simply the second element of the pair $(x, y) = (x, f(x))$.

EXAMPLE 1. The equation $y = f(x) = 3x + 5$ can be thought of as a function since for every real number x, there is a unique real number y which is equal to $3x + 5$. Here the domain of f is $\mathbb{R}$. To calculate the range of f, we note that if y is any real number, then there is a unique number x such that $y = 3x + 5$. To see this, simply solve for x: $x = (y - 5)/3$. Thus every real number y is in the range of f. In this example, the function f is called a *straight line function*. In general, linear functions have the form $y = f(x) = ax + b$, where a and b are real numbers.

EXAMPLE 2. Let $y = f(x) = x^2$. Then the set of ordered pairs $(x, y) = (x, x^2)$ constitutes a function whose domain is $\mathbb{R}$ since each real number has a unique square. We also see, as mentioned above, that the range of f equals $\{x: x \geq 0\} = \mathbb{R}^+$, the set of nonnegative real numbers, since the square of any real number is nonnegative. The graph of this function is obtained by plotting all points of the form $(x, y) = (x, x^2)$.†
First we note that as $f(x) = x^2$, $f(x) = f(-x)$ since $(-x)^2 = x^2$. Thus it is only necessary to calculate $f(x)$ for $x \geq 0$. For every $x > 0$, there is a value of $x < 0$ which gives the same value of y. In this situation we say that the function is *symmetric* about the y-axis. Some values for $f(x)$ are shown in Table 1. This function, which is graphed in Figure 1, is called a *parabola*.

TABLE 1

x	0	$\frac{1}{2}$	1	$\frac{3}{2}$	2	$\frac{5}{2}$	3	4	5
$f(x) = x^2$	0	$\frac{1}{4}$	1	$\frac{9}{4}$	4	$\frac{25}{4}$	9	16	25

Remark. We repeat that care must be taken to distinguish between the symbols f and $f(x)$. In this example, f is the *rule* that assigns to each real number the square of that number. The symbol $f(x)$ is the *number* x^2.

EXAMPLE 3. Consider the set of ordered pairs $(u, 1/(u^2 - 1))$. For a fixed number $u \neq \pm 1$, there is a unique number v which is equal to $1/(u^2 - 1)$. Thus this set of ordered pairs constitutes a function g whose domain is

$$\{u \in \mathbb{R}: u \neq \pm 1\} = \mathbb{R} - \{1, -1\}.$$

Here we have used the letters u, v, and g instead of x, y, and f to illustrate the fact that there is nothing special about any particular set of letters. Often, we will use the letter t to denote time and the letter s to denote distance. The definition of a function is independent of the letters used to denote the function and variables.
We now calculate the range of the function $v = g(u) = 1/(u^2 - 1)$. We note that as u becomes large, $1/(u^2 - 1)$ becomes very small. If $u > 1$ but is close to 1, then $1/(u^2 - 1)$ is very large in the positive direction. For example, $1/[(1.0001)^2 - 1] \approx 5000$. If $u < 1$ and u is close to 1, then $u^2 - 1 < 0$

† Of course, we can't plot *all* points (there are an infinite number of them). Rather, we plot some sample points and assume they can be connected to obtain the sketch of the graph.

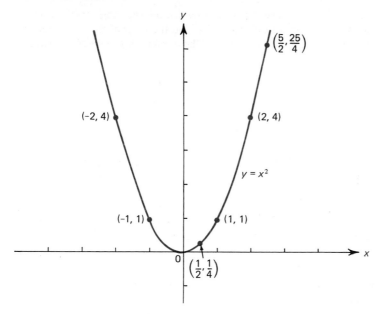

$\left(\frac{5}{2}, \frac{25}{4}\right)$

$(-2, 4)$ $(2, 4)$

$y = x^2$

$(-1, 1)$ $(1, 1)$

$\left(\frac{1}{2}, \frac{1}{4}\right)$

Figure 1

and $1/(u^2 - 1)$ is very large in the negative direction. For example, $1/[(.9999)^2 - 1] \approx -5000$. Therefore $1/(u^2 - 1)$ can take on any positive real value. It is never equal to 0. On the other hand, if $-1 < u < 1$, then $u^2 - 1 < 0$. But $u^2 - 1 \geq -1$ so that $1/(u^2 - 1) \leq -1$. Thus the range of $g = \mathbb{R} - (-1, 0]$. We will not sketch the graph of this function here, but will wait until our discussion of curve sketching in Chapter 3.

EXAMPLE 4. Consider the equation $y^2 = x$. The set of all ordered pairs (x, y) which satisfy this equation does *not* constitute a function since for every $x > 0$, there are *two* values of y such that $y^2 = x$; namely $y = \sqrt{x}$ and $y = -\sqrt{x}$ (for example, if $x = 4$, then we obtain the two pairs $(4, 2)$ and $(4, -2)$ both of which satisfy the equation). However, if we specify one of these values, say $y = \sqrt{x}$, then the set of ordered pairs $\{(x, \sqrt{x})\}$ is a function. Here the domain of f is $\mathbb{R}^+$ and the range of f is $\mathbb{R}^+$. We could obtain a second function g by choosing the negative square root. That is, $g(x) = -\sqrt{x}$ is a function with domain $\mathbb{R}^+$ and range $\mathbb{R}^-$ (the non-positive real numbers).

EXAMPLE 5. Consider the graph of the unit circle $x^2 + y^2 = 1$ given in Figure 2. It is clear that for every real number x in the open interval $(-1, 1)$ there are two values of y given by $y = \pm\sqrt{1 - x^2}$. Hence we do not have a function. We can obtain two separate functions by defining

$$y_1 = f_1(x) = \sqrt{1 - x^2} \quad \text{and} \quad y_2 = f_2(x) = -\sqrt{1 - x^2}.$$

Then the domain of $f_1 = $ the domain of $f_2 = [-1, 1]$, the range of $f_1 = [0, 1]$ and the range of $f_2 = [-1, 0]$.

Sometimes a function may be defined in a more complicated way.

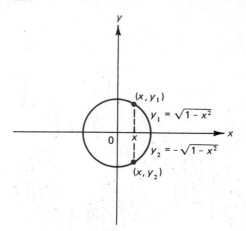

Figure 2

EXAMPLE 6. Let

$$f(x) = \begin{cases} 1, & x \geq 0 \\ 2, & x < 0 \end{cases}.$$

It is perfectly legitimate to define a function in "pieces," as we have done, so long as for each x in the domain of f, there is a unique y in the range. A graph of this function is given in Figure 3. We have domain of $f = \mathbb{R}$ and range of $f = \{1, 2\}$.

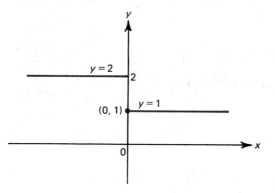

Figure 3

We can, if we wish, deliberately restrict the domain of a function.

EXAMPLE 7. Let $g(x) = 1 - x^2, x \geq 1$. This is a function with domain $\{x: x \geq 1\}$ and range $\mathbb{R}^- = (-\infty, 0]$ since, if $x \geq 1, 1 - x^2 \leq 0$. This function is *not* the same as the function $f(x) = 1 - x^2$ which has the domain $\mathbb{R}$ and the range $(-\infty, 1]$. The functions f and g cannot be the same because they do not contain the same ordered pairs. For example, the pair $(0, 1)$ is in f but not in g (see Figure 4).

PROBLEMS 1.6

In Problems 1–12 an equation involving x and y is given. Determine whether or not y is a function of x.

1. $2x + 3y = 6$ **2.** $\dfrac{x}{y} = 2$ **3.** $x^2 - 3y = 4$

4. $x - 3y^2 = 4$ **5.** $x^2 + y^2 = 4$ **6.** $x^2 - y^2 = 1$

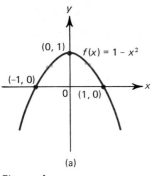

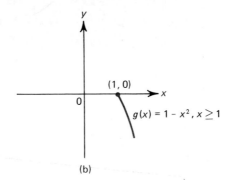

(a)   (b)

Figure 4

7. $\sqrt{x + y} = 1$ 8. $y^2 + xy + 1 = 0$ [*Hint:* Use the quadratic formula.]

9. $y^3 - x = 0$ 10. $y^4 - x = 0$ 11. $y = |x|$.

12. $y^2 = \dfrac{x}{x + 1}$.

13. Explain why the equation $y^n - x = 0$ allows us to write y as a function of x if n is an odd integer but does not if n is an even integer. [*Hint:* First solve Problems 9 and 10.]

In Problems 14–41 find the domain and range of the given function.

14. $y = f(x) = 2x - 3$ 15. $s = g(t) = 4t - 5$ 16. $y = f(x) = 3x^2 - 1$

17. $v = h(u) = \dfrac{1}{u^2}$ 18. $y = f(x) = x^3$ 19. $y = f(x) = \dfrac{1}{x + 1}$

20. $s = g(t) = t^2 + 4t + 4$ 21. $y = f(x) = \sqrt{x^3 - 1}$ 22. $v = h(u) = |u - 2|$

23. $y = f(x) = \dfrac{1}{|x|}$ 24. $y = f(x) = \dfrac{1}{|x + 2|}$ 25. $y = \begin{cases} x, & x \geq 0 \\ -x, & x < 0 \end{cases}$

26. $y = \begin{cases} x, & x \geq 1 \\ 1, & x < 1 \end{cases}$ 27. $y = \begin{cases} x^3, & x > 0 \\ x^2, & x \leq 0 \end{cases}$ 28. $y = \dfrac{x^2 + 3x}{x + 3}$

29. $y = \dfrac{x^5 + 5x^2}{x^2}$ 30. $y = \dfrac{1}{\sqrt{x - 1}} + 2$ 31. $y = \dfrac{1}{x^2 + 1} + 3$

*32. $y = \dfrac{1}{x^2 - 1} + 3$ 33. $y = \dfrac{1}{1 + x^2 + x^4 + x^6}$ *34. $y = \dfrac{x}{x + 1}$

*35. $y = \dfrac{x^2}{x^2 + 3}$ *36. $y = \dfrac{2x + 3}{3x + 4}$

*37. $y = \dfrac{x^n}{x^n + 1}$, n an integer ≥ 1 *38. $y = \sqrt{x^2 - 3x + 2}$

*39. $y = \sqrt[3]{x^2 - 3x + 2}$ 40. $y = 3x^2 - 5$; $x \geq 2$ 41. $y = \dfrac{1}{x}$; $x < 3$.

42. If $f(x) = \dfrac{1}{x + 1}$, find $f(0)$, $f(1)$, $f(-2)$, and $f(-5)$.

43. If $f(x) = 1 + \sqrt{x}$, find $f(0)$, $f(1)$, $f(16)$ and $f(25)$.

44. If $f(x) = \dfrac{x}{x - 2}$, find $f(0)$, $f(1)$, $f(-1)$, $f(3)$ and $f(10)$.

45. Let $f(x) = \sqrt{x + 1}$. Find $f(t^3)$, $f(\sqrt[3]{t})$ and $f\left(\dfrac{1}{t}\right)$.

*46.   Let $f(x) = \dfrac{1}{x-1}$. Find $f(t^2)$, $f(3t + 2)$, $f(\text{John})$ and $f(\text{Mary})$. [*Note:* This problem is not as silly as it seems. It simply illustrates the fact that to evaluate a function, it is only necessary to "plug in" the expression in parentheses—no matter how complicated (or ridiculous) that expression may seem.]

47.   Let $f(x) = x^2$. Find $f(x + \Delta x)$ and $\dfrac{f(x + \Delta x) - f(x)}{\Delta x}$, where Δx denotes an arbitrary real number. Simplify your answer.

48.   Let $f(x) = \sqrt[3]{x^2 - 1}$. Find $f(1)$, $f(2x)$, $f(3x^2 - 6)$, $f(g(t))$, and $f(x + \Delta x)$.

49.   For $f(x) = \dfrac{x}{x^5 + 3}$, calculate $f(-5x)$, $f\left(\dfrac{1}{x}\right)$, $f(x^{10} - 25)$, $f(g(t))$, and $f(x + \Delta x)$.

*50.   Let $f(x) = \sqrt{x}$. Show, assuming that $\Delta x \neq 0$, that

$$\frac{f(x + \Delta x) - f(x)}{\Delta x} = \frac{1}{\sqrt{x + \Delta x} + \sqrt{x}}.$$

[*Hint:* Multiply and divide by $\sqrt{x + \Delta x} + \sqrt{x}$.]

51.   Let $f(x) = \dfrac{|x|}{x}$. Show that $f(x) = \begin{cases} 1, & x > 0 \\ -1, & x < 0 \end{cases}$. Find the domain and range of f.

1.7 Operations with Functions

We begin this section by showing how functions can be added, subtracted, multiplied, and divided. To simplify notation, we will use the symbol $D(f)$ to denote the domain of the function f.

DEFINITION 1.   Let f and g be two functions. Then

     (i)   the sum $f + g$ is defined by

$$\boxed{(f + g)(x) = f(x) + g(x)} \tag{1}$$

     (ii)   the difference $f - g$ is defined by

$$\boxed{(f - g)(x) = f(x) - g(x)} \tag{2}$$

     (iii)   the product $f \cdot g$ is defined by

$$\boxed{(f \cdot g)(x) = f(x)g(x)} \tag{3}$$

     (iv)   the quotient f/g is defined by

$$\boxed{\left(\frac{f}{g}\right)(x) = \frac{f(x)}{g(x)}.} \tag{4}$$

Furthermore, $f + g$, $f - g$, and $f \cdot g$ are defined for all x for which both f and g are defined. That is

$$\boxed{D(f + g) = D(f - g) = D(f \cdot g) = D(f) \cap D(g).}$$

Finally, f/g is defined whenever both f and g are defined and $g(x) \neq 0$ (so that we do not divide by zero). That is

$$D(f/g) = D(f) \cap D(g) - \{x: g(x) = 0\}.$$

EXAMPLE 1. Let $f(x) = \sqrt{x+1}$ and $g(x) = \sqrt{4-x^2}$. Since $D(f) = [-1, \infty)$ and $D(g) = [-2, 2]$, we have $D(f+g) = D(f-g) = D(f \cdot g) = [-1, \infty) \cap [-2, 2] = [-1, 2]$, and $D(f/g) = [-1, 2] - \{x: \sqrt{4-x^2} = 0\} = [-1, 2] - \{-2, 2\} = [-1, 2)$. The functions are

(i) $(f+g)(x) = \sqrt{x+1} + \sqrt{4-x^2}$

(ii) $(f-g)(x) = \sqrt{x+1} - \sqrt{4-x^2}$

(iii) $(f \cdot g)(x) = \sqrt{x+1} \cdot \sqrt{4-x^2} = \sqrt{(x+1)(4-x^2)}$

(iv) $\left(\dfrac{f}{g}\right)(x) = \dfrac{\sqrt{x+1}}{\sqrt{4-x^2}} = \sqrt{\dfrac{x+1}{4-x^2}}.$

Definition 1 above can lead to some confusion if we are not careful.

EXAMPLE 2. Let $f(x) = \sqrt{x}$ and $g(x) = 2\sqrt{x}$. Then

$$D(f) = [0, \infty), \qquad D(g) = [0, \infty),$$

and

$$D(f \cdot g) = D(f) \cap D(g) = [0, \infty).$$

But $(f \cdot g)(x) = \sqrt{x} \cdot 2\sqrt{x} = 2x$ which is defined for every real number x—or is it? The apparent problem here lies in the definition of a function. The function $f \cdot g$, with domain $[0, \infty)$, is defined as the set of ordered pairs (x, y) with $x \geq 0$ and $y = 2x$. This is *not* the same function as the function $h(x) = 2x$ which is the set of ordered pairs (x, y) with $y = 2x$, *without* the restriction that $x \geq 0$. These two functions are not the same since they have different domains. We will, for most of the remainder of this book, not worry about such fine distinctions—but the reader should be aware that some care has to be taken when dealing with functions.

We will often need to deal with functions of functions.

DEFINITION 2. If f and g are functions, then the *composite function* $f \circ g$ is defined by

$$(f \circ g)(x) = f(g(x)) \tag{5}$$

and $D(f \circ g) = \{x: x \in D(g) \text{ and } g(x) \in D(f)\}$. That is, $(f \circ g)(x)$ is defined for every x such that $g(x)$ and $f(g(x))$ are defined

EXAMPLE 3. Let $f(x) = \sqrt{x}$ and $g(x) = x^2 + 1$. Then

$$(f \circ g)(x) = f(g(x)) = \sqrt{g(x)} = \sqrt{x^2 + 1}$$

and

$$(g \circ f)(x) = g(f(x)) = (f(x))^2 + 1 = (\sqrt{x})^2 + 1 = x + 1.$$

Now $D(f) = \mathbb{R}^+, D(g) = \mathbb{R}$, and we have

$$D(f \circ g) = \{x : g(x) = x^2 + 1 \in D(f)\}.$$

But since $x^2 + 1 > 0$, $x^2 + 1 \in D(f)$ for every real x so that $D(f \circ g) = \mathbb{R}$. On the other hand, $D(g \circ f) = \mathbb{R}^+$ since f is only defined for $x \geq 0$.

> **Warning.** Although $(f \cdot g)(x) = (g \cdot f)(x)$, it is *not* true, in general, that $(f \circ g)(x) = (g \circ f)(x)$.

EXAMPLE 4. Let $f(x) = 3x - 4$ and $g(x) = x^3$. Then

$$(f \circ g)(x) = f(g(x)) = f(x^3) = 3x^3 - 4$$

and

$$(g \circ f)(x) = g(f(x)) = g(3x - 4) = (3x - 4)^3.$$

Here $D(f \circ g) = D(g \circ f) = \mathbb{R}$.

EXAMPLE 5. Let $f(x) = 2x + 1$. Find a function $g(x)$ such that $(f \circ g)(x) = x^3$.

SOLUTION.   We   must   have $(f \circ g)(x) = f(g(x)) = 2g(x) + 1 = x^3$.   Then $2g(x) = x^3 - 1$ and $g(x) = \dfrac{x^3 - 1}{2}$.

PROBLEMS 1.7

In Problems 1–8 two functions f and g are given. Calculate $f + g$, $f - g$, $f \cdot g$ and f/g and determine their respective domains.

1. $f(x) = 2x - 5$, $g(x) = -4x$.

2. $f(x) = x^2$, $g(x) = x + 1$.

3. $f(x) = \sqrt{x + 2}$, $g(x) = \sqrt{2 - x}$.

4. $f(x) = x^3 + x$, $g(x) = \dfrac{1}{\sqrt{x + 1}}$.

5. $f(x) = 1 + x^5$, $g(x) = 1 - |x|$.

6. $f(x) = \sqrt{1 + x}$, $g(x) = \dfrac{1}{x^5}$.

7. $f(x) = \sqrt[5]{x + 2}$, $g(x) = \sqrt[4]{x - 3}$.

8. $f(x) = \dfrac{x}{x + 1}$, $g(x) = \dfrac{x - 1}{x}$.

In Problems 9–20 find $f \circ g$ and $g \circ f$ and determine the domain of each.

9. $f(x) = x + 1$, $g(x) = 2x$.

10. $f(x) = x^2$, $g(x) = 2x + 3$.

11. $f(x) = 3x + 5$, $g(x) = 5x + 2$.

12. $f(x) = \sqrt{x + 1}$, $g(x) = x^4$.

13. $f(x) = \dfrac{x}{x + 2}$, $g(x) = \dfrac{x - 1}{x}$.

14. $f(x) = |x|$, $g(x) = -x$.

15. $f(x) = \sqrt{1 - x}$, $g(x) = \sqrt{x - 1}$.

16. $f(x) = \dfrac{|x|}{x}$, $g(x) = x^2$.

17. $f(x) = \dfrac{1}{\sqrt{x + 1}}$, $g(x) = x^5$.

18. $f(x) = \sqrt[3]{x + 1}$, $g(x) = \sqrt[4]{x - 1}$.

19. $f(x) = \begin{cases} x, & x \geq 0 \\ 2x, & x < 0 \end{cases}$, $g(x) = \begin{cases} -3x, & x \geq 0 \\ 5x, & x < 0 \end{cases}$.

*20. $f(x) = \begin{cases} x^2, & x > 0 \\ x^3, & x \le 0 \end{cases}$, $g(x) = \begin{cases} \sqrt{x}, & x > 0 \\ \sqrt{-x}, & x \le 0 \end{cases}$.

21. Let $f(x) = 2x + 4$ and $g(x) = \frac{1}{2}x - 2$. Show that $(f \circ g)(x) = (g \circ f)(x) = x$. (When this occurs, we say that f and g are *inverse functions*. These are discussed in detail in Section 10.7.)

22. If $f(x) = -3x + 2$, find a function g, such that $(f \circ g)(x) = (g \circ f)(x) = x$.

23. If $f(x) = x^2$, find two functions g, such that $(f \circ g)(x) = x^2 - 10x + 25$.

*24. If $f(x) = ax + b$, find a function g, such that $(f \circ g)(x) = (g \circ f)(x) = x$. Assume that $a \ne 0$.

*25. If $f(x) = \sqrt[3]{x} - 1$, find a function g, such that $(f \circ g)(x) = (g \circ f)(x) = x$.

26. Let $h(x) = 1/\sqrt{x^2 + 1}$. Determine two functions f and g such that $f \circ g = h$.

1.8 Shifting the Graphs of Functions

While more advanced methods are needed to obtain the graphs of most functions (without plotting a large number of points), there are some techniques which make it a relatively simple matter to sketch certain functions. As an illustration of what we have in mind, consider the six functions

(a) $f(x) = x^2$ (b) $g(x) = x^2 + 1$ (c) $h(x) = x^2 - 1$
(d) $j(x) = (x - 1)^2$ (e) $k(x) = (x + 1)^2$ (f) $l(x) = -x^2$.

These are all graphed in Figure 1. In (a), we have used the graph of $y = x^2$ obtained in Figure 1.6.1. For (b), in order to graph $y = x^2 + 1$ we simply add 1 unit to every y value obtained in (a); that is, we shift the graph of $y = x^2$ *up* one unit. Analogously, for (c), we simply shift the graph of $y = x^2$ *down* one unit to obtain the graph of $y = x^2 - 1$. The analysis of the graph in (d) is a little trickier. Since, for example, $y = 0$ when $x = 0$ for the function $y = x^2$, then $y = 0$ when $x = 1$ for the function $y = (x - 1)^2$. Similarly, $y = 4$ when $x = -2$ if $y = x^2$, and $y = 4$ when $x = -1$ if $y = (x - 1)^2$. By continuing in this manner, we can see that y values in the graph of $y = x^2$ are the same as y values in the graph of $y = (x - 1)^2$ except that they are achieved one unit later (that is, $x^2 = [(x - 1) + 1]^2$). Thus we find that the graph of $y = (x - 1)^2$ is the graph of $y = x^2$ *shifted one unit to the right*. Similarly, in (e) we find that the graph of $y = (x + 1)^2$ is the graph of $y = x^2$ shifted *one unit to the left*. Finally, in (f), to obtain the graph of $y = -x^2$, we note that each y value is replaced by its negative so that the graph of $y = -x^2$ is the graph of $y = x^2$ *reflected through the x-axis* (that is, turned upside down).

In general, we have the following rules. Let $y = f(x)$. Then to obtain the graph of
  (i) $y = f(x) + c$, shift the graph of $y = f(x)$ *up* c units if $c \ge 0$ and *down* c units if $c < 0$.
  (ii) $y = f(x - c)$, shift the graph of $y = f(x)$ *to the right* c units if $c > 0$ and *to the left* c units if $c < 0$.
  (iii) $y = -f(x)$, reflect the graph of $y = f(x)$ *through the x-axis.*
  (iv) $y = f(-x)$, reflect the graph of $y = f(x)$ *through the y-axis.*

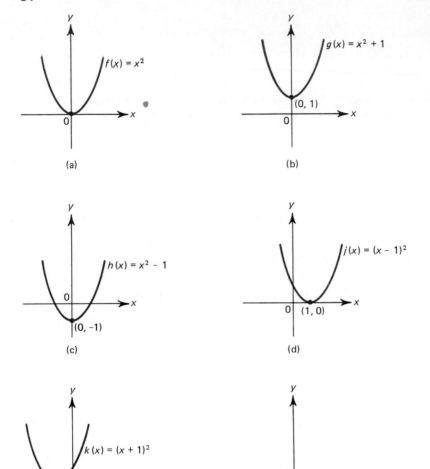

(a) (b) (c) (d) (e) (f) Figure 1

EXAMPLE 1. The graph of $y = \sqrt{x}$ is given in Figure 2a. Then, using the above rules, the graphs of $\sqrt{x} + 3$, $\sqrt{x} - 2$, $\sqrt{x} - 4$, $\sqrt{x} + 3 = \sqrt{x - (-3)}$, $-\sqrt{x}$ and $\sqrt{-x}$ are given in the other parts of Figure 2.

EXAMPLE 2. The graph of a certain function $f(x)$ is given in Figure 3a. To obtain the graph of $-f(3 - x)$ we

> (i)   reflect through the y-axis to obtain the graph of $f(-x)$ (Figure 3b).

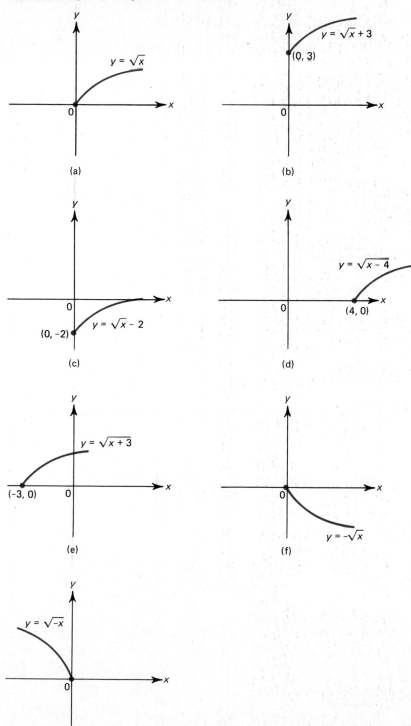

Figure 2

(ii) shift to the right 3 units to obtain the graph of $f(-(x-3)) = f(3-x)$ (Figure 3c).

(iii) reflect through the x-axis to obtain the graph of $y = -f(3-x)$ (Figure 3d).

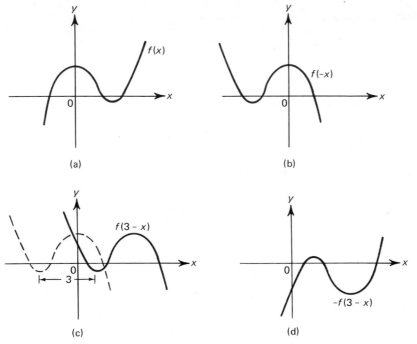

(a)   (b)

(c)   (d)

Figure 3

EXAMPLE 3. Graph the parabola $y = x^2 - 10x + 22$.

SOLUTION. Completing the square we see that

$$x^2 - 10x + 22 = x^2 - 10x + 25 - 3 = (x-5)^2 - 3.$$

Thus the graph of $f(x)$ is obtained by shifting the graph of $y = x^2$ to the right 5 units and then down 3 units as in Figure 4.

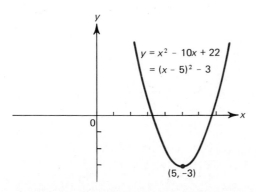

Figure 4

PROBLEMS 1.8

1. The graph of $f(x) = x^3$ is given in Figure 5. Sketch the graph of $f(x) =$
    (a) $(x - 2)^3$ (b) $-x^3$ (c) $(4 - x)^3 + 5$.
2. The graph of $f(x) = 1/x$ is given in Figure 6. Sketch the graph of $f(x) =$

    (a) $\dfrac{1}{x + 3}$ (b) $2 - \dfrac{1}{x}$.

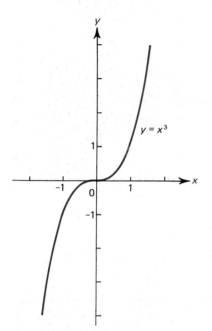

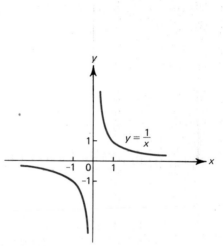

Figure 5   Figure 6

3. After completing the squares, sketch the graphs of the parabolas
    (a) $y = x^2 - 4x + 7$.   (b) $y = x^2 + 8x + 2$.
    (c) $y = x^2 + 3x + 4$.
    (d) $y = -x^2 + 2x - 3$ [*Hint:* Write $-x^2 + 2x - 3 = -(x^2 - 2x + 3)$]
    (e) $y = -x^2 - 5x + 8$.

In Problems 4–12 the graph of a function is sketched. Obtain the graph of (a) $f(x - 2)$, (b) $f(x + 3)$, (c) $-f(x)$, (d) $f(-x)$, (e) $f(2 - x) + 3$.

4.

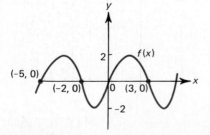

Figure 7

5.

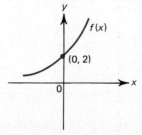

Figure 8

6.

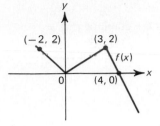

Figure 9

7.

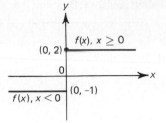

Figure 10

8.

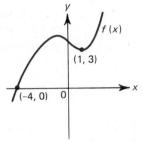

Figure 11

9.

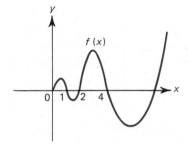

Figure 12

10.

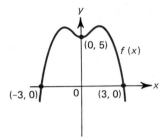

Figure 13

11.

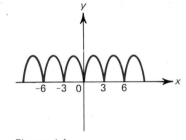

Figure 14

12.

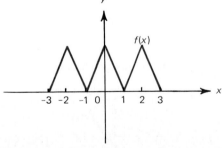

Figure 15

1.9 Second-Degree Equations

A *second-degree equation* in the variables x and y is an equation of the form

$$ax^2 + by^2 + cxy + dx + ey + f = 0. \tag{1}$$

Second degree equations can be written in four basic forms. They are:

(i) $x^2 + y^2 = r^2$		*Circle*
(ii) $y = ax^2, \quad a > 0$		*Parabola*
(iii) $\dfrac{x^2}{a^2} + \dfrac{y^2}{b^2} = 1$	$a, b > 0$	*Ellipse*
(iv) $\dfrac{x^2}{a^2} - \dfrac{y^2}{b^2} = 1$		*Hyperbola*

In Appendix 3 we will show how most second-degree equations can be written in one of these forms. In this section our goal will be to learn how to sketch the graphs of these equations so that we can refer to them in later chapters. More detailed descriptions will appear in Appendix 3.

We discussed the circle in Section 1.5.

(ii)   The parabola: $y = ax^2, \quad a > 0.$ **(2)**

In Example 1.6.2 we discussed the parabola $y = x^2$. The parabola $y = ax^2$ has the same general shape. It is symmetric about the y-axis since $f(-x) = a(-x)^2 = ax^2 = f(x)$. The point $(0, 0)$ is called the *vertex* of the parabola. We sketch a typical parabola in Figure 1.

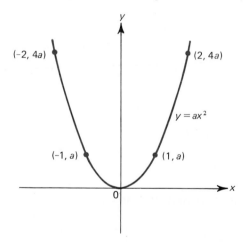

$(-2, 4a)$ $(2, 4a)$

$y = ax^2$

$(-1, a)$ $(1, a)$

Figure 1

(iii)   The ellipse: $\dfrac{x^2}{a^2} + \dfrac{y^2}{b^2} = 1.$ **(3)**

We first notice that if we replace x by $-x$ and y by $-y$ in (3) the equation remains unchanged. Thus the graph is symmetric about the x- and y-axes. To sketch it, therefore, we need only sketch some sample points in the first quadrant and then reflect through the x- and y-axes. When we do this, we obtain the oval-shaped curves in Figure 2. We are aided by the fact that it is easy to calculate the x- and y-intercepts. When $x = 0$, $y = \pm b$ and when $y = 0$, $x = \pm a$. If $a > b$, the line extending from $(-a, 0)$ to $(a, 0)$ is called the *major axis* of the ellipse and the line extending from

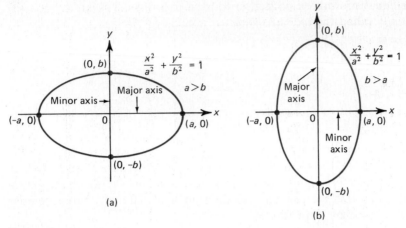

Figure 2

$(0, -b)$ to $(0, b)$ is called the *minor axis* of the ellipse. (See Figure 2a.) If $b > a$ then the situation is reversed as in Figure 2b.

> **Note.** If $a = b$, we obtain the circle $x^2 + y^2 = a^2$. Thus we can think of a circle as a special kind of ellipse.

(iv)   The hyperbola: $\dfrac{x^2}{a^2} - \dfrac{y^2}{b^2} = 1.$ (4)

Like the ellipse, this curve is symmetric about the x- and y-axes. Also, since

$$\frac{x^2}{a^2} = 1 + \frac{y^2}{b^2} \geq 1,$$

we find that $x^2 \geq a^2$ which implies, if $x \geq 0$, that $x \geq a$. When $y = 0$, $x = \pm a$. For $x \neq 0$, we can divide both sides of equation (4) by x^2 and multiply by b^2 to obtain

$$\frac{b^2}{a^2} - \frac{y^2}{x^2} = \frac{b^2}{x^2}.$$ (5)

If $|x|$ becomes large, then b^2/x^2 becomes small so that, for large x, we have

$$\frac{y^2}{x^2} \approx \frac{b^2}{a^2} \dagger$$

or, taking square roots,

$$y \approx \pm \frac{b}{a} x.$$ (6)

The two straight lines given in (6) are called *asymptotes* for the hyperbola. This means that as $|x|$ becomes large, the curve given by (4) gets closer and closer to these lines. We will discuss this idea in greater detail in our discussion of limits in Chapter 2.

†The symbol $\approx$ stands for the words "is approximately equal to."

Putting all this information together, we obtain the sketch given in Figure 3. Here, the x-axis is called the *axis of symmetry* of the hyperbola.

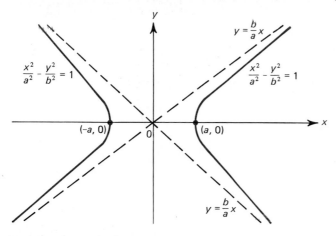

Figure 3

EXAMPLE 1.   Sketch the graph of $4x^2 + 9y^2 = 36$.

SOLUTION. Dividing by 36, we obtain

$$\frac{x^2}{9} + \frac{y^2}{4} = 1$$

which is the equation of an ellipse with major axis along the x-axis ($a = 3$) and minor axis along the y-axis ($b = 2$). It is sketched in Figure 4.

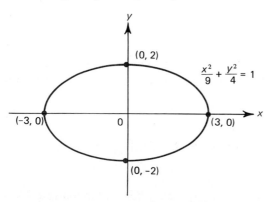

Figure 4

EXAMPLE 2.   Sketch the curve $x^2 - 4y^2 = 16$.

SOLUTION. Dividing by 16, we obtain

$$\frac{x^2}{16} - \frac{y^2}{4} = 1$$

which is the equation of a hyperbola with $a = 4$ and $b = 2$. The curve passes through the points $(-4, 0)$ and $(4, 0)$ and is asymptotic to the lines $y = \frac{1}{2}x$ and $y = -\frac{1}{2}x$. The curve is sketched in Figure 5.

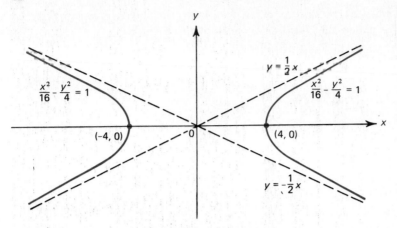

Figure 5

EXAMPLE 3.   Sketch the curve $x = 2y^2$.

SOLUTION. This is not an equation in one of our standard forms. However, we see that it is the equation of a parabola (with $a = 2$) with the roles of x and y reversed. Taking this into account (that is, interchanging the roles of the x- and y-axes), we obtain the sketch in Figure 6.

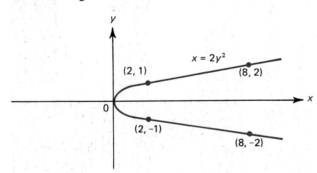

Figure 6

EXAMPLE 4.   Sketch the curve $y^2 - 4x^2 = 16$.

SOLUTION. Dividing by 16, we obtain

$$\frac{y^2}{16} - \frac{x^2}{4} = 1.$$

This is the hyperbola of Example 2 except that x and y have been interchanged. Thus we interchange the roles of the x- and y-axes to obtain the sketch in Figure 7.

EXAMPLE 5.   Discuss the curve $4x^2 - 9y^2 = 0$.

SOLUTION. Dividing by 36, we have

$$\frac{x^2}{9} - \frac{y^2}{4} = 0 \quad \text{or} \quad y^2 = \frac{4}{9}x^2$$

which implies that $y = \pm\frac{2}{3}x$. This is the equation of two straight lines. This kind of second degree equation is called a *degenerate* equation.

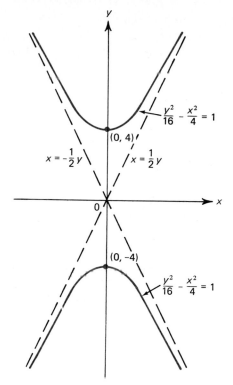

$$\frac{y^2}{16} - \frac{x^2}{4} = 1$$

(0, 4)

$x = -\frac{1}{2}y$ $x = \frac{1}{2}y$

(0, -4)

$$\frac{y^2}{16} - \frac{x^2}{4} = 1$$

Figure 7

PROBLEMS 1.9

In the following problems identify and sketch the graphs of the given equations.

1. $y = 3x^2$	**2.** $x = 3y^2$	**3.** $4x^2 + y^2 = 4$
4. $x^2 + 4y^2 = 4$	**5.** $4x^2 + 4y^2 = 4$	**6.** $4x^2 - y^2 = 4$
7. $x^2 - 4y^2 = 4$	**8.** $4y^2 - x^2 = 4$	**9.** $9y^2 + 4x^2 = 36$
10. $x^2 - 4y^2 = 1$	**11.** $4x^2 - y^2 = 1$	**12.** $4y^2 + x^2 = 1$
13. $y^2 - 4x^2 = 1$	**14.** $4x^2 - y^2 = 0$	**15.** $x^2 - 4y^2 = 0$
16. $x^2 + 4y^2 = 0$	**17.** $y = 9x^2$	**18.** $y = -9x^2$
19. $x = 9y^2$	**20.** $x = -9y^2$	**21.** $4x^2 + 7y^2 = 28$
22. $7x^2 + 4y^2 = 28$	**23.** $2x^2 - 5y^2 = 10$	**24.** $2y^2 - 5x^2 = 10$
25. $ay^2 - bx^2 = 0, \quad (a, b > 0)$		
26. $ay^2 + bx^2 = 0, \quad (a, b > 0)$		
27. $3x^2 - 3y^2 = 3$	**28.** $3y^2 - 3x^2 = 3$	**29.** $10x^2 + 4y^2 = 40$
30. $10x^2 + 2y^2 = 1.$		

1.10 Geometric Progressions

Consider the sum

$$S_7 = 1 + 2 + 4 + 8 + 16 + 32 + 64 + 128.$$

This can be written as

$$S_7 = 1 + 2 + 2^2 + 2^3 + 2^4 + 2^5 + 2^6 + 2^7.$$

In general, the sum of a *geometric progression* is a sum of the form

$$S_n = 1 + r + r^2 + r^3 + \cdots + r^{n-1} + r^n \tag{1}$$

where r is a real number and n is a fixed positive integer. We will frequently encounter such sums in our study of calculus. We now obtain a formula for the sum in (1).

THEOREM 1. If $r \neq 1$, the sum of a geometric progression (1) is given by

$$S_n = \frac{1 - r^{n+1}}{1 - r}. \tag{2}$$

PROOF. We write

$$S_n = 1 + r + r^2 + r^3 + \cdots + r^{n-1} + r^n \tag{3}$$

and then multiply both sides of (3) by r:

$$rS_n = r + r^2 + r^3 + r^4 + \cdots + r^n + r^{n+1}. \tag{4}$$

We now subtract (4) from (3) and note that all terms except the first and the last cancel:

$$S_n - rS_n = 1 - r^{n+1}$$

or

$$(1 - r)S_n = 1 - r^{n+1}. \tag{5}$$

Finally, we divide both sides of (5) by $1 - r$ (which is nonzero) to obtain equation (2).

Note. If $r = 1$, we obtain $S_n = \overbrace{1 + 1 + \cdots + 1}^{n+1 \text{ terms}} = n + 1$.

EXAMPLE 1. Calculate $S_7 = 1 + 2 + 4 + 8 + 16 + 32 + 64 + 128$ using formula (2).

SOLUTION. Here $r = 2$ and $n = 7$ so that

$$S_7 = \frac{1 - 2^8}{1 - 2} = 2^8 - 1 = 256 - 1 = 255.$$

EXAMPLE 2. Calculate

$$S_{10} = 1 + \frac{1}{2} + \frac{1}{4} + \cdots + \frac{1}{2^{10}}.$$

SOLUTION. Here $r = \frac{1}{2}$ and $n = 10$ so that

$$S_{10} = \frac{1 - \left(\frac{1}{2}\right)^{11}}{1 - \frac{1}{2}} = \frac{1 - \frac{1}{2048}}{\frac{1}{2}} = 2\left(\frac{2047}{2048}\right) = \frac{2047}{1024}.$$

EXAMPLE 3. Calculate

$$S_6 = 1 - \frac{2}{3} + \left(\frac{2}{3}\right)^2 - \left(\frac{2}{3}\right)^3 + \left(\frac{2}{3}\right)^4 - \left(\frac{2}{3}\right)^5 + \left(\frac{2}{3}\right)^6.$$

SOLUTION. Here $r = -\frac{2}{3}$ and $n = 6$ so that

$$S_6 = \frac{1 - \left(-\frac{2}{3}\right)^7}{1 - \left(-\frac{2}{3}\right)} = \frac{1 + \frac{128}{2187}}{\frac{5}{3}} = \frac{3}{5}\left(\frac{2315}{2187}\right) = \frac{463}{729}.$$

EXAMPLE 4. Calculate the sum $2 + \frac{2}{5} + \frac{2}{5^2} + \cdots + \frac{2}{5^{12}}$.

SOLUTION. The required sum is equal to

$$2\left(1 + \frac{1}{5} + \frac{1}{5^2} + \cdots + \frac{1}{5^{12}}\right) = 2\left(\frac{1 - \left(\frac{1}{5}\right)^{13}}{1 - \frac{1}{5}}\right) = \frac{5}{2}\left(1 - \frac{1}{5^{13}}\right).$$

EXAMPLE 5. Calculate the sum $1 + b^2 + b^4 + b^6 + \cdots + b^{20}$, for $b \neq \pm 1$.

SOLUTION. Note that the sum can be written $1 + b^2 + (b^2)^2 + (b^2)^3 + \cdots + (b^2)^{10}$. Here $r = b^2 \neq 1$ and $n = 10$ so that

$$S_{10} = \frac{1 - (b^2)^{11}}{1 - b^2} = \frac{b^{22} - 1}{b^2 - 1}.$$

PROBLEMS 1.10

In Problems 1–11 calculate the sum of the given geometric progression.

1. $1 + 3 + 9 + 27 + 81 + 243$

2. $1 + \frac{1}{4} + \frac{1}{16} + \cdots + \frac{1}{4^8}$

3. $1 - 5 + 25 - 125 + 625 - 3125$
4. $0.2 + 0.2^2 + 0.2^3 + \cdots + 0.2^9$
5. $0.3^2 - 0.3^3 + 0.3^4 - 0.3^5 + 0.3^6 - 0.3^7 + 0.3^8$
6. $1 + b^3 + b^6 + b^9 + b^{12} + b^{15} + b^{18} + b^{21}$

7. $1 - \frac{1}{b^2} + \frac{1}{b^4} - \frac{1}{b^6} + \frac{1}{b^8} - \frac{1}{b^{10}} + \frac{1}{b^{12}} - \frac{1}{b^{14}}$

8. $\pi - \pi^3 + \pi^5 - \pi^7 + \pi^9 - \pi^{11} + \pi^{13}$
9. $1 + \sqrt{2} + 2 + 2^{3/2} + 4 + 2^{5/2} + 8 + 2^{7/2} + 16$

10. $1 - \frac{1}{\sqrt{3}} + \frac{1}{3} - \frac{1}{3\sqrt{3}} + \frac{1}{9} - \frac{1}{9\sqrt{3}} + \frac{1}{27} - \frac{1}{27\sqrt{3}} + \frac{1}{81}$

11. $-16 + 64 - 256 + 1024 - 4096.$
12. A bacteria population initially contains 1000 organisms and proceeds to double every two hours. How many organisms will be alive after 12 hours if none of the bacteria die during the growth period?

Review Exercises for Chapter One

1. Let $A = (-\infty, 2]$, $B = (-4, \infty)$, and $C = [-1, 1]$. Calculate $A \cup B$, $A \cup C$, $B \cup C$, $A \cap B$, $A \cap C$, $B \cap C$, $\bar{A}$, $\bar{B}$, $\bar{C}$, $A - B$, $B - A$, $A - C$, $C - A$, $B - C$, and $C - B$.

In Exercises 2–9 find and graph the solution sets of the given inequalities.

2. $|x| < 2$

3. $|x| \geq 4$

4. $|x - 1| \leq 5$

5. $|x + 3| > 1$

6. $\left| \dfrac{2x - 3}{4} \right| \leq 4$

7. $x^2 + 6x + 9 \geq 0$

8. $x^2 + x - 2 < 0$

9. $\dfrac{4}{x + 3} > \dfrac{1}{x - 2}$.

10. Find the distance between the points
    (a) $(1, 5)$, $(-3, 2)$.   (b) $(-6, 1)$, $(-11, -4)$.
11. Find the midpoint of the line joining the points $(-3, 2)$ and $(5, -8)$.
12. Find the area of the triangle with vertices at $(1, 4)$, $(1, 8)$ and $(7, 8)$.

In Exercises 13–18 find the equation of a straight line when either two points on it or its slope and one point are given.

13. $(2, 5)$, $(-1, 3)$

14. $(-2, 4)$, $m = 3$

15. $(3, -1)$, $(1, -3)$

16. $(-1, 4)$, $m = 2$

17. $(1, 4)$, $(1, 7)$

18. $(3, -8)$, $(-8, -8)$.

19. Find the equation of the line parallel to the line $2x - 5y = 6$ which passes through the point $(4, -2)$.
20. Find the equation of the line perpendicular to the line $3x + 2y = 7$ which passes through the point $(-2, 3)$.
21. Find the distance between the line $-x + 3y = 4$ and the point $(2, 3)$.
22. Find the equation of the circle with radius 3 centered at the point $(-1, 2)$.
23. Find the equation of the line tangent to the circle $(x - 1)^2 + y^2 = 10$ at the point $(2, 3)$.

In Exercises 24–32 determine whether the given equation defines a function and, if so, find its domain and range.

24. $4x - 2y = 5$

25. $\dfrac{x^2 - y}{2} = 4$

26. $\dfrac{y}{x} = 1$

27. $(x - 1)^2 + (y - 3)^2 = 4$ 28. $y = \sqrt{x + 2}$

29. $3 = \dfrac{1 + x^2 + x^4}{2y}$

30. $y = \dfrac{x}{x^2 + 1}$

31. $y = \dfrac{x}{x^2 - 1}$

32. $y = \sqrt{x^2 - 6}$.

33. For $y = f(x) = \sqrt{x^2 - 4}$, calculate $f(2)$, $f(-\sqrt{5})$, $f(x + 4)$, $f(x^3 - 2)$, $f(-1/x)$ and $f(\text{Carlos})$.
34. If $y = f(x) = 1/x$, show that for $\Delta x \neq 0$,

$$\frac{f(x + \Delta x) - f(x)}{\Delta x} = -\frac{1}{x(x + \Delta x)}.$$

35. Let $f(x) = \sqrt{x + 1}$ and $g(x) = x^3$. Find $f + g$, $f - g$, $f \cdot g$, g/f, $f \circ g$, and $g \circ f$ and determine their respective domains.
36. Do the same for $f(x) = 1/x$ and $g(x) = x^2 - 4x + 3$.
37. For $f(x) = 4x - 6$, find a function $g(x)$ such that $(f \circ g)(x) = (g \circ f)(x) = x$.
38. The graph of the function $y = f(x)$ is given in Figure 1. Sketch the graph of $f(x - 3)$, $f(x) - 5$, $f(-x)$, $-f(x)$ and $4 - f(1 - x)$.

39. Do the same for the function graphed in Figure 2.

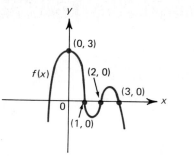

Figure 1   Figure 2

In Exercises 40–51 sketch the graph of the given equation.

40. $y = 4x^2$ **41.** $x = 4y^2$ **42.** $y = -4x^2$

43. $x = -4y^2$ **44.** $9x^2 + 16y^2 = 144$ **45.** $16x^2 + 9y^2 = 144$

46. $9x^2 - 16y^2 = 144$ **47.** $9y^2 - 16x^2 = 144$ **48.** $16x^2 - 9y^2 = 144.$

49. $16x^2 - 9y^2 = 1$ **50.** $16x^2 - 9y^2 = 0$ **51.** $16x^2 + 9y^2 = 0.$

52. Find the sum of the geometric progressions:

(a) $1 + \frac{1}{3} + \frac{1}{9} + \frac{1}{27} + \frac{1}{81} + \frac{1}{243}$ (b) $1 - \frac{1}{2} + \frac{1}{4} - \frac{1}{8} + \frac{1}{16} - \frac{1}{32} + \frac{1}{64}$

(c) $b^3 + b^6 + b^9 + b^{12} + b^{15} + b^{18} + b^{21}.$

TWO

LIMITS AND DERIVATIVES

2.1 Introduction to Limits

The notion of a limit, which we shall discuss extensively in this chapter and in Chapter 10, plays a central role in calculus and in much of modern mathematics. However, although mathematics dates back over 3000 years, limits were not really understood until the monumental work of the great French mathematician Augustin-Louis Cauchy† (1789–1857) in the nineteenth century.

Some of the early work on limits was motivated by unresolved questions that had been posed by Greek mathematicians. For example, the fifth century B.C. philosopher and mathematician Zeno (495?–435? B.C.) posed four problems which came to be known as *Zeno's paradoxes*. In the second of these, Zeno claimed that the legendary Greek hero Achilles could never overtake a tortoise. Suppose that the tortoise starts 100 yards ahead and that Achilles can run 10 times as fast as the tortoise. Then when Achilles has run 100 yards, the tortoise has run 10 yards, and when Achilles has covered this distance, the tortoise is still a yard ahead, and so on. Clearly the tortoise will stay ahead!

We may formulate this seeming paradox in another way which is equally contradictory of common sense. Suppose that a man is standing a certain distance, say 10 feet, from a door (see Figure 1). Using Zeno's reasoning, we may claim that it

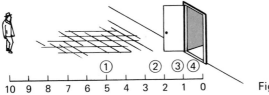

Figure 1

is mathematically impossible for the man to walk to the door. In order to reach the door, the man must walk half the distance (5 feet) to the door. He then reaches point ① on Figure 1. From point ①, 5 feet from the door, he must again walk half way ($2\frac{1}{2}$ feet) to the door, to point ②. Continuing in this manner, no matter how close he comes to the door, he must walk half way to the door and halfway from there and halfway from there, . . . , and so on. Thus, no matter how close the man gets to the door, he still has half of some remaining distance to cover. It seems that the man will never actually reach the door. Of course, this contradicts our common sense. But where is the flaw in Zeno's reasoning?

† *Resumé of Lessons on Infinitesimal Calculus,* Fourth Lesson, Paris, 1823.

It took more than 2000 years to provide a satisfactory answer to this question, and in order to do so it was necessary to use the notion of a limit. Intuitively, Zeno's man is indeed covering an infinite number of intervals in his walk towards the door, but each interval is over a shorter and shorter distance and, therefore, takes less and less time. Indeed, the time necessary to walk over each succeeding interval "approaches" the limit zero, thus allowing the man to reach the door. We shall discuss what we mean by something approaching a limit in the next four sections and shall prove, in Section 6, that Zeno's man really can reach the door in a finite time. If at this point the idea of a limit seems fuzzy, remember that it took Western civilization 2000 years to discover it!

2.2 The Calculation of Limits

In this section we show by example how some limits can be calculated.

EXAMPLE 1. We begin by looking at the function

$$y = f(x) = x^2 + 3 \tag{1}$$

This function is graphed in Figure 1 (see Section 1.8). What happens to $f(x)$ as x gets closer and closer to the value $x = 2$? To get an idea, look at Table 1, keeping in mind the fact that x can get close to 2 from the right of 2 and from the left of 2 along the

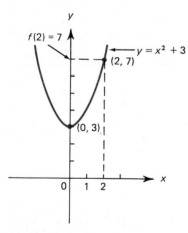

Figure 1

TABLE 1

x	$f(x) = x^2 + 3$	x	$f(x) = x^2 + 3$
3	12	1	4
2.5	9.25	1.5	5.25
2.3	8.29	1.7	5.89
2.1	7.41	1.9	6.61
2.05	7.2025	1.95	6.8025
2.01	7.0401	1.99	6.9601
2.001	7.004001	1.999	6.996001
2.0001	7.00040001	1.9999	6.99960001

x-axis. It appears from the table that as x gets close to $x = 2$, $f(x) = x^2 + 3$ gets close to 7. This is not surprising since if we now calculate $f(x)$ at $x = 2$, we obtain $f(2) = 2^2 + 3 = 4 + 3 = 7$. In mathematical symbols we write

$$\lim_{x \to 2} (x^2 + 3) = 7.$$

This is read "the limit as x approaches 2 (or tends to 2) of $x^2 + 3$ is equal to 7."

Note. In order to calculate this limit we did *not* have to evaluate $x^2 + 3$ at $x = 2$.

□**EXAMPLE 2.** Let $y = f(x) = x^3 + 2x - 9$. What happens to this function as x gets close to 1?

SOLUTION. Values of $f(x)$ for x near 1 are given in Table 2. From Table 2 it appears that as x gets close to 1, $f(x) = x^3 + 2x - 9$ gets close to -6. As in Example 1, this is not surprising since

$$f(1) = 1^3 + 2 \cdot 1 - 9 = 1 + 2 - 9 = -6.$$

TABLE 2

x	x^3	$2x$	$x^3 + 2x - 9$	x	x^3	$2x$	$x^3 + 2x - 9$
2.0	8	4	3.0	0	0	0	-9
1.5	3.375	3	-2.625	0.5	0.125	1.0	-7.875
1.1	1.331	2.2	-5.469	0.9	0.729	1.8	-6.471
1.05	1.157625	2.1	-5.742375	0.95	0.857375	1.9	-6.242625
1.01	1.030301	2.02	-5.949699	0.99	0.970299	1.98	-6.049701
1.001	1.003003001	2.002	-5.994996999	0.999	0.997002999	1.998	-6.004997001
1.0001	1.00030003	2.0002	-5.99949997	0.9999	0.99970003	1.9998	-6.00049997

In mathematical symbols we write our conclusion as

$$\lim_{x \to 1} (x^3 + 2x - 9) = -6.$$

This is read "the limit as x approaches 1 (or tends to 1) of $x^3 + 2x - 9$ is equal to -6."

□**EXAMPLE 3.** What happens to the function $y = f(x) = \sqrt{2x - 6}$ as x gets close to $x = 5$?

SOLUTION. Since when $x = 5$, $\sqrt{2x - 6} = \sqrt{2 \cdot 5 - 6} = \sqrt{10 - 6} = \sqrt{4} = 2$, we might guess that as x gets close to 5, $\sqrt{2x - 6}$ gets close to 2. That this is indeed true is suggested by the computations in Table 3.

EXAMPLE 4. Consider the function

$$f(x) = \frac{x(x + 1)}{x}. \tag{2}$$

Since we cannot divide by zero, this function is defined for every real number except

TABLE 3

x	$2x$	$2x - 6$	$\sqrt{2x - 6}$	x	$2x$	$2x - 6$	$\sqrt{2x - 6}$
6.0	12.0	6.0	2.449489743	4.0	8.0	2.0	1.414213562
5.5	11.0	5.0	2.236067977	4.5	9.0	3.0	1.732050808
5.1	10.2	4.2	2.049390153	4.9	9.8	3.8	1.949358869
5.01	10.02	4.02	2.004993766	4.99	9.98	3.98	1.994993734
5.001	10.002	4.002	2.000499938	4.999	9.998	3.998	1.999499937
5.0001	10.0002	4.0002	2.000049999	4.9999	9.9998	3.9998	1.999949999

for $x = 0$. Since $x/x = 1$, we see that $f(x) = x + 1$ for all $x \neq 0$. Let

$$g(x) = x + 1.$$

Then $f(x) = g(x)$ for all $x \neq 0$. However we emphasize that $f(x)$ and $g(x)$ are *not* the same function since $g(x)$ *is* defined at $x = 0$ while $f(x)$ is not. Nevertheless, for $x \neq 0$, $f(x) = g(x)$. These functions are graphed in Figure 2. What happens to $f(x)$ as x approaches 0? Again we illustrate this with a table of values (Table 4). As long as $x \neq 0$, we may use the fact that $f(x) = g(x) = x + 1$. It is clear that as x gets close to 0, $f(x)$ gets close to 1. In mathematical notation we write

$$\lim_{x \to 0} \frac{x(x + 1)}{x} = 1. \tag{3}$$

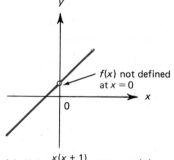

(a) $f(x) = \dfrac{x(x + 1)}{x}$: $D(f) = \mathbb{R} - \{0\}$

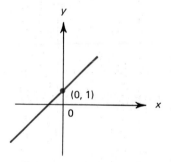

(b) $g(x) = x + 1$: $D(g) = \mathbb{R}$

Figure 2

TABLE 4

x	$f(x) = \dfrac{x(x + 1)}{x} = x + 1$	x	$f(x) = \dfrac{x(x + 1)}{x} = x + 1$
1	2	-1	0
0.5	1.5	-0.5	0.5
0.1	1.1	-0.1	0.9
0.05	1.05	-0.05	0.95
0.01	1.01	-0.01	0.99
0.001	1.001	-0.001	0.999

It is important to note that for $f(x) = x(x + 1)/x$, it is still not permissible to set $x = 0$ as this would imply division by zero. However we now know what happens to this function as x approaches zero. We can see why it is important that we are not required to evaluate $f(x)$ at $x = 0$ when we calculate the limit as x approaches zero.

☐**EXAMPLE 5.** Consider the function

$$f(x) = \frac{x^2 + 3x + 2}{13x + 8}.$$ (4)

To see what happens to this function as x approaches 4, look at Table 5. It seems that $f(x)$ approaches $0.5 = \frac{1}{2}$ as x approaches 4. This is reasonable since

$$f(4) = \frac{4^2 + 3 \cdot 4 + 2}{13 \cdot 4 + 8} = \frac{16 + 12 + 2}{52 + 8} = \frac{30}{60} = \frac{1}{2}.$$

TABLE 5

x	$f(x) = \dfrac{x^2 + 3x + 2}{13x + 8}$	x	$f(x) = \dfrac{x^2 + 3x + 2}{13x + 8}$
5	0.5753424658	3	0.4255319149
4.5	0.5375939850	3.5	0.4626168224
4.1	0.5075040783	3.9	0.4925042589
4.01	0.5007500416	3.99	0.4992500418
4.001	0.5000750004	3.999	0.4999250004
4.0001	0.5000075000	3.9999	0.4999925000

We write the limit we have discovered as

$$\lim_{x \to 4} \frac{x^2 + 3x + 2}{13x + 8} = \frac{1}{2}.$$

Before giving further examples, we shall give a more formal definition of a limit. The definition given below is meant to appeal to your intuition. It is *not* a precise mathematical definition. In this section and the ones that follow, we hope that you will begin to get comfortable with the notion of limits and will acquire some facility in calculating them. Later, in Section 10.1, we shall return to limits and give precise mathematical definitions which will enable us to prove the theorems cited in the next and subsequent sections.

DEFINITION 1. We say that the limit as x approaches x_0 of $f(x)$ is L, written

$$\lim_{x \to x_0} f(x) = L$$ (5)

if f gets arbitrarily close to L as x gets close to x_0 (from either direction) without ever being equal to x_0. Another way to write (5) is

$$f(x) \to L \quad \text{as} \quad x \to x_0.$$ (6)

Here the arrow stands for the word *approaches* or the words *tends to*.

Remark 1. We insist that f be defined on an open interval (see Section 1.1) containing the number x_0. This ensures that f is defined on both sides of x_0, even though it *may* not be defined at x_0 itself (see Figure 3). It is important that $f(x)$ get close to L when x gets

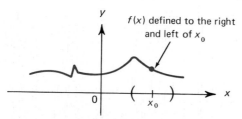

Figure 3

close to x_0 from either side. In Table 1, for example, $x^2 + 3$ gets close to 7 when x gets close to 2 from the right (the table on the left) and the left (the table on the right). Similarly, in Table 4, $f(x) = x(x + 1)/x$ gets close to 1 whether x approaches 0 from the left or the right.

Remark 2. It should be emphasized that while we do not actually need to know what $f(x_0)$ is (in fact, $f(x_0)$ need not even exist), it is nevertheless often very helpful to know $f(x_0)$ in the actual computation of $\lim_{x \to x_0} f(x)$, since it frequently happens that $\lim_{x \to x_0} f(x)$ indeed equals $f(x_0)$. However, we again emphasize that this is *not always* the case. In Example 4, we showed that $\lim_{x \to 0} f(x) = 1$ even though $f(0)$ did not exist.

□**EXAMPLE 6.**   Consider the function

$$f(x) = \sqrt{x}. \tag{7}$$

In Table 6 we calculate values of $f(x)$ as x approaches 0. It seems as if $\lim_{x \to 0} \sqrt{x} = 0$, but that is *not* the case since $f(x) = \sqrt{x}$ is *not even defined* for $x < 0$ (since we cannot

TABLE 6

x	$f(x) = \sqrt{x}$
1	1
0.5	0.7071
0.2	0.4472
0.1	0.3162
0.01	0.1000
0.001	0.0316
0.0001	0.01
0.00000001	0.0001

take the square root of a negative number). Therefore as $x \to 0$ from the left, $\sqrt{x}$ is not defined so that $\lim_{x \to 0} \sqrt{x}$ *does not exist*. Note also that there is *no* open interval containing 0 in which $f(x)$ is defined. The graph of $f(x) = \sqrt{x}$ is sketched in Figure 4. This example illustrates the reason for making Remark 1 above.

EXAMPLE 7.   Calculate

$$\lim_{h \to 0} \frac{(3 + h)^2 - 9}{h}. \tag{8}$$

Note. Here we use the letter h instead of the letter x. Nothing else is changed.

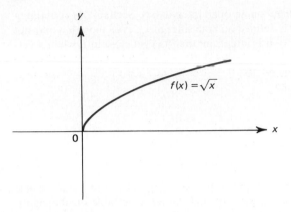

Figure 4

SOLUTION. The function

$$f(h) = \frac{(3 + h)^2 - 9}{h}$$

is not defined at $h = 0$ since, at $h = 0$,

$$\frac{(3 + 0)^2 - 9}{0} = \frac{9 - 9}{0} = \frac{0}{0},$$

which is an undefined expression. Nevertheless, we may calculate the limit in several steps:

$$\lim_{h \to 0} \frac{(3 + h)^2 - 9}{h} = \lim_{h \to 0} \frac{(9 + 6h + h^2) - 9}{h}$$

$$= \lim_{h \to 0} \frac{6h + h^2}{h} = \lim_{h \to 0} \frac{h(6 + h)}{h}$$

Now, for $h \neq 0$,

$$\frac{h(6 + h)}{h} = 6 + h$$

which is a function identical to $f(h)$ except at $h = 0$. Thus

$$\lim_{h \to 0} \frac{(3 + h)^2 - 9}{h} = \lim_{h \to 0} \frac{h(6 + h)}{h} = \lim_{h \to 0} (6 + h).$$

But, as h gets close to zero, $6 + h$ gets close to $6 + 0 = 6$. Hence

$$\lim_{h \to 0} \frac{(3 + h)^2 - 9}{h} = 6.$$

This is another example which illustrates the fact that $\lim_{x \to x_0} f(x)$ may exist even though $f(x_0)$ does not exist. Here $f(0)$ is not defined while $\lim_{h\to 0} f(h)$ exists and is equal to 6.

There is an interesting geometrical interpretation to the result we have just obtained. In Figure 5 we have drawn the graph of the function $y = x^2$. When $x = 3$,

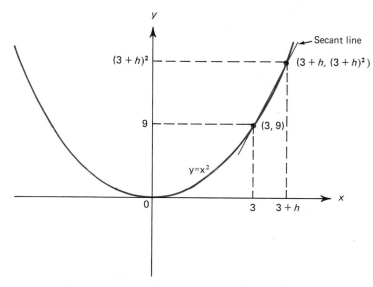

Figure 5

$y = 3^2 = 9$. If h represents some small number, then $x = 3 + h$ is close to the value $x = 3$. Moreover, $f(3 + h) = y = (3 + h)^2$ is the value of y at $x = 3 + h$. Consider the straight line joining the two points $(3, 9)$ and $(3 + h, (3 + h)^2)$. From the results of Section 1.4, we find that the slope of this line is given by

$$m = \frac{\Delta y}{\Delta x} = \frac{(3 + h)^2 - 9}{(3 + h) - 3} = \frac{(3 + h)^2 - 9}{h}.$$

We called this last function $f(h)$ in Example 7. The line joining the points $(3, 9)$ and $(3 + h, (3 + h)^2)$ is called a *secant line* to the parabola $y = x^2$. Now a glance at Figure 6 indicates that as h approaches zero, $3 + h$ approaches 3 and the secant line approaches the line which is tangent to the curve $y = x^2$ at the point $(3, 9)$. This line is called a *tangent line*. Since the secant lines are approaching the tangent line as $h \to 0$, it is reasonable to assert that the slopes of the secant lines approach the slope of the tangent line as $h \to 0$. But, since $f(h)$ represents the slope of the secant line joining the points $(3, 9)$ and $(3 + h, (3 + h)^2)$, we have

$$\begin{aligned}\text{slope of the tangent line to the} \\ \text{curve } y = x^2 \text{ at the point } (3, 9)\end{aligned} = \lim_{h\to 0} f(h) = \lim_{h\to 0} \frac{(3 + h)^2 - 9}{h} = 6.$$

Moreover, we can write the equation of the tangent line to the curve $y = x^2$ at the point $(3, 9)$:

$$\frac{y - 9}{x - 3} = 6 \qquad \text{or} \qquad y = 6x - 9.$$

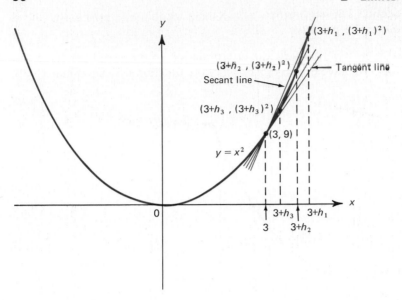

Figure 6

In Section 2.8 we shall discuss a more general way to calculate the slope of the tangent line to a given curve at a given point.

□**EXAMPLE 8.**   Calculate

$$\lim_{x \to 0} f(x) = \lim_{x \to 0} \frac{\sqrt{4 + x} - 2}{x}.$$ (9)

SOLUTION. Note that if we substitute $x = 0$ in (9), then

$$f(0) = \frac{\sqrt{4 + 0} - 2}{0} = \frac{2 - 2}{0} = \frac{0}{0}$$

which is an undefined expression. In Table 7 we tabulate values of $f(x)$ for x near 0. From the table, we conclude that

$$\lim_{x \to 0} \frac{\sqrt{4 + x} - 2}{x} = 0.25.$$

TABLE 7

x	$f(x) = \dfrac{\sqrt{4 + x} - 2}{x}$	x	$f(x) = \dfrac{\sqrt{4 + x} - 2}{x}$
1	0.2360679775	−1	0.2679491924
0.5	0.2426406871	−0.5	0.2583426132
0.1	0.2484567313	−0.1	0.2515823419
0.01	0.2498439448	−0.01	0.2501564457
0.001	0.2499843740	−0.001	0.2500156290
0.0001	0.2499984200	−0.0001	0.2500015900

We shall see in Section 2.8 that this value represents the slope of the line tangent to the curve $y = \sqrt{x}$ at the point $(4, 2)$. This is sketched in Figure 7.

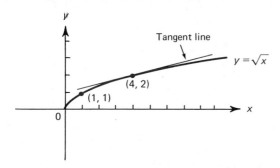

Figure 7

We close this section with two slightly different examples of limits.

EXAMPLE 9. Calculate $\lim_{x \to 0} |x|$.

SOLUTION. From Section 1.2, we have

$$|x| = \begin{cases} x, & x \geq 0 \\ -x, & x \leq 0 \end{cases}.$$

If $x > 0$, then $|x| = x$ which tends to zero as $x \to 0$ from the right of 0. If $x < 0$, then $|x| = -x$ which again tends to zero as $x \to 0$ from the left of 0. Then, since we get the same answer when we approach zero from the left and from the right, we have

$$\lim_{x \to 0} |x| = 0.$$

This is pictured in Figure 8.

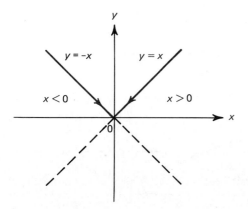

Figure 8

EXAMPLE 10. Calculate $\lim_{x \to 0} \dfrac{|x|}{x}$.

SOLUTION. If $x > 0$, then $|x| = x$ so that $|x|/x = x/x = 1$. On the other hand, if $x < 0$, then $|x| = -x$ so that $|x|/x = -x/x = -1$. Note that $|x|/x$ is not defined at

$x = 0$. The graph of $|x|/x$ is sketched in Figure 9. In sum, we have

$$\frac{|x|}{x} = \begin{cases} 1, & \text{if } x > 0 \\ -1, & \text{if } x < 0. \end{cases}$$

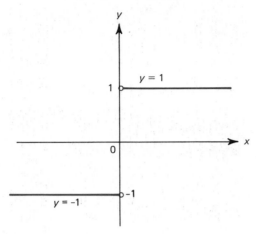

Figure 9

From Figure 9, we conclude that $f(x) = |x|/x$ has *no* limit as $x \to 0$; for if $x > 0$, then $f(x)$ remains constant at the value 1 while for $x < 0, f(x)$ remains constant at the value -1. Since the value of the limit has to be the same no matter from which direction we approach the value 0, we are left to conclude that there is no limit at 0. Of course, for any other value of x there is a limit. For example, $\lim_{x \to 2} |x|/x = 1$ since near $x = 2$, $|x| = x$ and $|x|/x = 1$. Similarly, $\lim_{x \to -2} |x|/x = -1$.

PROBLEMS 2.2

1. (a) Draw the graph of the function $f(x) = x + 7$.
    (b) Calculate $f(x)$ for $x = 3, 1, 2.5, 1.5, 2.1, 1.9, 2.01$ and 1.99.
    (c) Calculate $\lim_{x \to 2} (x + 7)$.
2. (a) Draw the graph of the function $f(x) = x^2 - 4$.
    (b) Calculate $f(x)$ for $x = 2, 0, 1.5, 0.5, 1.1, 0.9, 1.01$, and 0.99.
    (c) Calculate $\lim_{x \to 1} (x^2 - 4)$.
3. (a) Draw the graph of the function $f(x) = x^2 - 3x + 4$.
    (b) Calculate $f(x)$ for $x = -0.5, -1.5, -0.9, -1.1, -0.99$, and -1.01.
    (c) Calculate $\lim_{x \to -1} (x^2 - 3x + 4)$.

4. Let $f(x) = \dfrac{(x - 1)(x - 2)}{x - 1}$.

    (a) Explain why $f(x)$ is not defined for $x = 1$.
    (b) For $x \neq 1$, find another function which is equal to f for all $x \neq 1$ and *is* defined for $x = 1$.

    (c) Calculate $\lim_{x \to 1} \dfrac{(x - 1)(x - 2)}{x - 1}$.

5. Let $f(x) = \dfrac{x^3 - 8}{x - 2}$.

    (a) Explain why $f(x)$ is not defined at $x = 2$.

(b)   For $x \neq 2$, find another function which is equal to f for all $x \neq 2$ and is defined for $x = 2$. [*Hint:* Calculate $(x^3 - 8)/(x - 2)$ by long division.]

(c)   Calculate $\lim\limits_{x \to 2} \left(\dfrac{x^3 - 8}{x - 2} \right)$.

6.   Explain why $\lim\limits_{x \to -1} \sqrt{x + 1}$ does not exist.

7.   Does $\lim\limits_{x \to -1} \sqrt{x^2 + 1}$ exist? If so, calculate it.

In Problems 8–25 calculate each limit if it exists and explain why there is no limit if it does not exist.

8. $\lim\limits_{x \to 5} (x^2 - 6)$.

9. $\lim\limits_{x \to 0} (x^3 + 17x + 45)$.

10. $\lim\limits_{x \to 0} (-x^5 + 17x^3 + 2x)$.

11. $\lim\limits_{x \to 0} \dfrac{1}{x^5 + 6x + 2}$.

12. $\lim\limits_{x \to 2} (x^4 - 9)$.

13. $\lim\limits_{x \to -1} \dfrac{(x + 1)^2}{x + 1}$.

14. $\lim\limits_{x \to -4} \sqrt{x + 4}$.

15. $\lim\limits_{x \to 0} \dfrac{x^3}{x^2}$.

16. $\lim\limits_{x \to 5} \sqrt{x^2 - 25}$.

17. $\lim\limits_{x \to 4} \sqrt{25 - x^2}$.

18. $\lim\limits_{x \to 4} \sqrt{x^2 - 25}$.

19. $\lim\limits_{x \to 3} \dfrac{x^2 - 4x + 3}{x - 3}$. [*Hint:* Divide.]

20. $\lim\limits_{x \to 2} \sqrt[3]{x^3 - 8}$.

21. $\lim\limits_{x \to 2} \sqrt[4]{x^4 - 16}$.

22. $\lim\limits_{x \to -2} \dfrac{x^2 + 6x + 8}{x + 2}$.

23. $\lim\limits_{x \to 1} \dfrac{x^4 - x}{x^3 - 1}$.

24. $\lim\limits_{x \to 1} \dfrac{\sqrt{x} - 1}{x - 1}$. [*Hint:* $a^2 - b^2 = (a + b)(a - b)$.]

25. $\lim\limits_{x \to 2} \dfrac{1 - \sqrt{x/2}}{1 - (x/2)}$.

26.   Let $f(x) = \dfrac{x^3 - 6x + 2}{x^2 + x + 9}$.

(a)   Calculate $f(x)$ for $x = 3$, 1, 2.5, 1.5, 2.1, 1.9, 2.01, 1.99, 2.001, and 1.999.

(b)   Estimate $\lim\limits_{x \to 2} \dfrac{x^3 - 6x + 2}{x^2 + x + 9}$.

(c)   Calculate $f(2)$ and compare it with your estimate.

27.   Let $f(x) = \dfrac{\sqrt{x^3 + 13}}{x + 8}$.

(a)   Calculate $f(x)$ for $x = -1, -3, -1.5, -2.5, -1.9, -2.1, -1.99, -2.01, -1.999$ and -2.001.

(b)   Estimate $\lim\limits_{x \to -2} \dfrac{\sqrt{x^3 + 13}}{x + 8}$.

(c)   Calculate $f(-2)$ and compare it with your estimate.

28. (a)   Graph the curve $y = x^2 + 3$.

(b)   Draw (on your graph) the straight line joining the points $(1, 4)$ and $(2, 7)$.

(c)   Draw the straight line joining the points $(1, 4)$ and $(1.5, 5.25)$.

(d)   For any real number h, what is represented by the quotient

$$\frac{[(1 + h)^2 + 3] - 4}{h}?$$

(e)   Calculate $\lim\limits_{h \to 0} \dfrac{[(1 + h)^2 + 3] - 4}{h}$.

(f)   What is the slope of the line tangent to the curve $y = x^2 + 3$ at the point $(1, 4)$?

29. (a)   Graph the curve $y = 5 - x^2$.

(b)   Draw (on your graph) the straight line joining the points $(-3, -4)$ and $(-4, -11)$.

(c)   Draw the straight line joining the points $(-3, -4)$ and $(-3.5, -7.25)$.

(d)   For any real number h, what is represented by the quotient

$$\frac{[5 - (-3 - h)^2] + 4}{-h}?$$

(e)   Calculate $\lim\limits_{h \to 0} \dfrac{[5 - (-3 - h)^2] + 4}{-h}$.

(f)   What is the slope of the line tangent to the curve $y = 5 - x^2$ at the point $(-3, -4)$?

30. (a)   Graph the function $f(x) = |x - 3|$.   (b)   Calculate $\lim\limits_{x \to 3} |x - 3|$.

31. (a)   Graph the function $f(x) = |3x - 4|$.   (b)   Calculate $\lim\limits_{x \to 4/3} |3x - 4|$.

32. (a)   Graph the function $f(x) = \dfrac{|x + 3|}{x + 3}$.

(b)   Explain why $\lim\limits_{x \to -3} \dfrac{|x + 3|}{x + 3}$ does not exist.

(c)   Calculate $\lim\limits_{x \to 5} \dfrac{|x + 3|}{x + 3}$ and $\lim\limits_{x \to -5} \dfrac{|x + 3|}{x + 3}$.

33. (a)   Graph the function $f(x) = \dfrac{|2x - 4|}{2x - 4}$.

(b)   Explain why $\lim\limits_{x \to 2} \dfrac{|2x - 4|}{2x - 4}$ does not exist.

(c)   Calculate $\lim\limits_{x \to 3} \dfrac{|2x - 4|}{2x - 4}$ and $\lim\limits_{x \to 0} \dfrac{|2x - 4|}{2x - 4}$.

2.3 The Limit Theorems

In this section we state several theorems which will make our calculations of a number of limits a great deal easier. The proofs of these theorems will be deferred until Section 10.2.

We saw in the previous section that

$$\lim_{x \to 2} (x^2 + 3) = 2^2 + 3 = 7 \quad \text{(Example 2.2.1)}.$$

That is, the limit of $f(x) = x^2 + 3$ as x tends to 2 is equal to $f(x)$ evaluated at $x = 2$ (i.e., $f(2)$). Similarly, in Example 2, we saw that $\lim_{x \to 1} (x^3 + 2x - 9) = -6 = f(1)$. However, as we remarked earlier, this process of evaluation (that is, substituting the value x_0 into $f(x)$ to find a limit as $x \to x_0$) will not always work since $f(x)$ may not even be defined at x_0. Nevertheless, it is interesting that if f is a polynomial, then it is always possible to calculate the limit by evaluation.

THEOREM 1. Let $P(x) = c_0 + c_1x + c_2x^2 + c_3x^3 + \cdots + c_nx^n$ be a polynomial, where $c_0, c_1, c_2, c_3, \ldots, c_n$ are real numbers and n is a fixed positive integer. Then

$$\lim_{x \to x_0} P(x) = P(x_0) = c_0 + c_1x_0 + c_2x_0^2 + c_3x_0^3 + \cdots + c_nx_0^n. \qquad (1)$$

EXAMPLE 1. Calculate $\lim_{x \to 3} (x^3 - 2x + 6)$.

SOLUTION. $x^3 - 2x + 6$ is a polynomial. Hence

$$\lim_{x \to 3} (x^3 - 2x + 6) = 3^3 - 2 \cdot 3 + 6 = 27 - 6 + 6 = 27.$$

EXAMPLE 2. Calculate $\lim_{x \to -1} (x^5 - 4x^4 + 3x^3 + 8x^2 - 5x - 4)$.

SOLUTION. Since $x^5 - 4x^4 + 3x^3 + 8x^2 - 5x - 4$ is a polynomial, we have

$$\lim_{x \to -1} (x^5 - 4x^4 + 3x^3 + 8x^2 - 5x - 4)$$

$$= (-1)^5 - 4(-1)^4 + 3(-1)^3 + 8(-1)^2 - 5(-1) - 4$$
$$= -1 - 4 - 3 + 8 + 5 - 4 = 1.$$

EXAMPLE 3. Calculate $\lim_{x \to x_0} 4$ for any real number x_0.

SOLUTION. $f(x) = 4$ is a polynomial (of degree 0). Hence $\lim_{x \to x_0} 4 = P(x_0) = 4$. This result simply states that *the limit of a constant function is that constant.* See Figure 1.

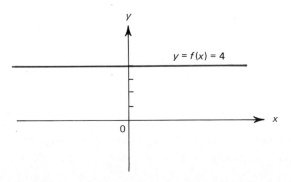

Figure 1

THEOREM 2. Let c be any real number and suppose that $\lim_{x \to x_0} f(x)$ exists. Then

$$\lim_{x \to x_0} cf(x) = c \lim_{x \to x_0} f(x). \tag{2}$$

Theorem 2 states that *the limit of a constant times a function is equal to the product of that constant and the limit of the function.*

EXAMPLE 4. Calculate $\lim_{x \to 3} 5(x^3 - 2x + 6)$.

SOLUTION. We can find this limit two ways. We can multiply to find that $5(x^3 - 2x + 6) = 5x^3 - 10x + 30$ and then use Theorem 1. However, in Example 1 we calculated

$$\lim_{x \to 3} (x^3 - 2x + 6) = 27.$$

Therefore, using Theorem 2 we have

$$\lim_{x \to 3} 5(x^3 - 2x + 6) = 5 \lim_{x \to 3} (x^3 - 2x + 6) = 5(27) = 135.$$

We can check this answer:

$$\lim_{x \to 3} 5(x^3 - 2x + 6) = \lim_{x \to 3} (5x^3 - 10x + 30)$$

$$= 5(3^3) - 10(3) + 30 = 5 \cdot 27 - 30 + 30 = 135.$$

EXAMPLE 5. Calculate $\lim_{x \to 0} -7 \dfrac{x(x + 1)}{x}$.

SOLUTION. From Example 2.2.4, we have

$$\lim_{x \to 0} \frac{x(x + 1)}{x} = 1.$$

Therefore

$$\lim_{x \to 0} -7 \frac{x(x + 1)}{x} = -7 \lim_{x \to 0} \frac{x(x + 1)}{x} = -7 \cdot 1 = -7.$$

THEOREM 3. If $\lim_{x \to x_0} f(x)$ and $\lim_{x \to x_0} g(x)$ both exist, then

$$\lim_{x \to x_0} [f(x) + g(x)] = \lim_{x \to x_0} f(x) + \lim_{x \to x_0} g(x). \tag{3}$$

Theorem 3 states that *the limit of the sum of two functions is equal to the sum of their limits.*

EXAMPLE 6. Calculate $\lim\limits_{x\to 0}\left[\dfrac{x(x+1)}{x}+4x^3+3\right]$.

SOLUTION. $\lim\limits_{x\to 0}\dfrac{x(x+1)}{x}=1$ and $\lim\limits_{x\to 0}[4x^3+3]=4\cdot 0^3+3=3$. Hence

$$\lim_{x\to 0}\left[\frac{x(x+1)}{x}+4x^3+3\right]=\lim_{x\to 0}\frac{x(x+1)}{x}+\lim_{x\to 0}(4x^3+3)=1+3=4.$$

EXAMPLE 7. Calculate $\lim_{x\to 8}(\sqrt{x}+\sqrt[3]{x})$.

SOLUTION. By the methods of the previous section, we can see that $\lim_{x\to 8}\sqrt{x}=\sqrt{8}$ and $\lim_{x\to 8}\sqrt[3]{x}=\sqrt[3]{8}=2$. Hence

$$\lim_{x\to 8}(\sqrt{x}+\sqrt[3]{x})=\lim_{x\to 8}\sqrt{x}+\lim_{x\to 8}\sqrt[3]{x}=\sqrt{8}+2.$$

THEOREM 4. If $\lim_{x\to x_0}f(x)$ and $\lim_{x\to x_0}g(x)$ both exist, then $\lim_{x\to x_0}[f(x)\cdot g(x)]$ exists and

$$\boxed{\lim_{x\to x_0}[f(x)\cdot g(x)]=\left[\lim_{x\to x_0}f(x)\right]\cdot\left[\lim_{x\to x_0}g(x)\right].}\qquad (4)$$

Theorem 4 says that the *limit of the product of two functions is the product of their limits*.

EXAMPLE 8. Calculate $\lim_{x\to 1}(x^2+3x+5)^2$.

SOLUTION. $\lim_{x\to 1}(x^2+3x+5)=1^2+3\cdot 1+5=9$. Hence

$$\lim_{x\to 1}(x^2+3x+5)^2=\lim_{x\to 1}(x^2+3x+5)(x^2+3x+5)$$

$$=\left[\lim_{x\to 1}(x^2+3x+5)\right]\cdot\left[\lim_{x\to 1}(x^2+3x+5)\right]$$

$$=9\cdot 9=81.$$

EXAMPLE 9. Calculate $\lim_{x\to 13}(\sqrt{x+3})(x-4)$.

SOLUTION. By the methods of the previous section we find that $\lim_{x\to 13}\sqrt{x+3}=\sqrt{13+3}=\sqrt{16}=4$ and $\lim_{x\to 13}(x-4)=13-4=9$. Therefore

$$\lim_{x\to 13}(\sqrt{x+3})(x-4)=\left[\lim_{x\to 13}\sqrt{x+3}\right]\left[\lim_{x\to 13}(x-4)\right]=4\cdot 9=36.$$

Note. In Section 4 (Theorem 2.4.2) we will show how limits like $\lim_{x\to 13}\sqrt{x+3}$ can easily be calculated.

EXAMPLE 10. Calculate $\lim_{x\to -1}(x^2-3)^{10}$.

SOLUTION. $\lim_{x \to -1}(x^2 - 3) = (-1)^2 - 3 = -2$. We now may apply Theorem 4 several times (nine times to be exact):

$$\lim_{x \to -1} (x^2 - 3)^{10} = \left[\lim_{x \to -1} (x^2 - 3)\right]\left[\lim_{x \to -1} (x^2 - 3)\right] \cdots \left[\lim_{x \to -1} (x^2 - 3)\right]$$

$$\underbrace{\qquad\qquad\qquad\qquad\qquad\qquad\qquad\qquad}_{10 \text{ terms}}$$

$$= \underbrace{(-2)(-2) \cdots (-2)}_{10 \text{ terms}} = (-2)^{10} = 1024.$$

The last example can be generalized.

COROLLARY TO THEOREM 4. If $\lim_{x \to x_0} f(x)$ exists, then

$$\boxed{\lim_{x \to x_0} [f(x)]^n = \left[\lim_{x \to x_0} f(x)\right]^n} \qquad\qquad (5)$$

where n is any positive integer.

The limit theorem for quotients is just what you would expect except that we have to be careful not to divide by zero.

THEOREM 5. If $\lim_{x \to x_0} f(x)$ and $\lim_{x \to x_0} g(x)$ both exist and $\lim_{x \to x_0} g(x) \ne 0$, then $\lim_{x \to x_0} f(x)/g(x)$ exists and

$$\boxed{\lim_{x \to x_0} \frac{f(x)}{g(x)} = \frac{\lim_{x \to x_0} f(x)}{\lim_{x \to x_0} g(x)}.} \qquad\qquad (6)$$

Theorem 5 says that *the limit of the quotient of two functions is the quotient of their limits, provided that the limit in the denominator function is not zero.*

EXAMPLE 11. Calculate $\lim_{x \to 3} \dfrac{\sqrt{x + 1}}{x^2 - 2}$.

SOLUTION

$$\lim_{x \to 3} \sqrt{x + 1} = \sqrt{3 + 1} = \sqrt{4} = 2 \quad \text{and} \quad \lim_{x \to 3} (x^2 - 2) = 3^2 - 2 = 7.$$

Therefore

$$\lim_{x \to 3} \frac{\sqrt{x + 1}}{x^2 - 2} = \frac{\lim_{x \to 3} \sqrt{x + 1}}{\lim_{x \to 3} (x^2 - 2)} = \frac{2}{7}.$$

EXAMPLE 12. Calculate $\lim_{x \to -2} \dfrac{x^3 - 3x + 6}{-x^2 + 15}$.

SOLUTION

$$\lim_{x \to -2} (x^3 - 3x + 6) = (-2)^3 - 3(-2) + 6 = -8 + 6 + 6 = 4$$

and

$$\lim_{x \to -2} (-x^2 + 15) = -(-2)^2 + 15 = -4 + 15 = 11.$$

Thus

$$\lim_{x \to -2} \frac{x^3 - 3x + 6}{-x^2 + 15} = \frac{\lim_{x \to -2} (x^3 - 3x + 6)}{\lim_{x \to -2} (-x^2 + 15)} = \frac{4}{11}.$$

DEFINITION 1. A *rational function* $r(x)$ is the quotient of two polynomials; that is

$$r(x) = \frac{p(x)}{q(x)} \tag{7}$$

where $p(x)$ and $q(x)$ are both polynomials.

For example, the function

$$r(x) = \frac{x^3 - 3x + 6}{-x^2 + 15}$$

given in Example 12 is a rational function. As another example, the function

$$x^2 + \frac{x^3}{1 + x}$$

is a rational function since

$$x^2 + \frac{x^3}{1 + x} = \frac{x^2(1 + x)}{1 + x} + \frac{x^3}{1 + x} = \frac{x^2 + x^3 + x^3}{1 + x} = \frac{x^2 + 2x^3}{1 + x}$$

which is the quotient of two polynomials.

COROLLARY TO THEOREMS 1 AND 5. Let $r(x) = p(x)/q(x)$ be a rational function with $q(x_0) \neq 0$. Then

$$\lim_{x \to x_0} r(x) = \lim_{x \to x_0} \frac{p(x)}{q(x)} = \frac{p(x_0)}{q(x_0)} = r(x_0). \tag{8}$$

EXAMPLE 13. Calculate $\lim\limits_{x \to 4} \dfrac{x^3 - x^2 - 3}{x^2 - 3x + 5}$.

SOLUTION. Here $q(x) = x^2 - 3x + 5$ and $q(4) = 16 - 12 + 5 = 9 \neq 0$. Therefore

$$\lim_{x \to 4} \frac{x^3 - x^2 - 3}{x^2 - 3x + 5} = \frac{4^3 - 4^2 - 3}{4^2 - 3 \cdot 4 + 5} = \frac{64 - 16 - 3}{16 - 12 + 5} = \frac{45}{9} = 5.$$

EXAMPLE 14. Calculate $\lim\limits_{x \to 1} \dfrac{x^2 - 1}{x - 1}$.

SOLUTION. We cannot use the corollary since $q(x) = x - 1$ and $q(1) = 0$. However, for $x \neq 1$,

$$\frac{x^2 - 1}{x - 1} = \frac{(x - 1)(x + 1)}{x - 1} = x + 1$$

so that

$$\lim_{x \to 1} \frac{x^2 - 1}{x - 1} = \lim_{x \to 1} (x + 1) = 2.$$

PROBLEMS 2.3

In the following problems use the limit theorems to calculate the given limits.

1. $\lim\limits_{x \to 3} (x^2 - 2x - 1)$

2. $\lim\limits_{x \to -2} (-x^3 - x^2 - x - 1)$

3. $\lim\limits_{x \to 1} (x^{50} - 1)$

4. $\lim\limits_{x \to -1} (x^{49} + 1)$

5. $\lim\limits_{x \to 5} 3\sqrt{x - 1}$

6. $\lim\limits_{x \to 3} 5\sqrt{x^2 + 7}$

7. $\lim\limits_{x \to -2} -4\sqrt{x + 3}$

8. $\lim\limits_{x \to 1} 8(x^{100} + 2)$

9. $\lim\limits_{x \to 5} (\sqrt{x - 1} + \sqrt{x^2 - 9})$

10. $\lim\limits_{x \to -2} (1 + x + x^2 + x^3 + \sqrt{x^2 - 3})$

11. $\lim\limits_{x \to 0} (3\sqrt{x + 1} - 5\sqrt{x^2 - 3x + 4})$

12. $\lim\limits_{x \to 27} (\sqrt{x + 9} + \sqrt[3]{x})$

13. $\lim\limits_{x \to 0} (\sqrt{x + 1})(\sqrt{x^2 - 3x + 4})$

14. $\lim\limits_{x \to 5} (\sqrt{x - 1})(\sqrt{x^2 - 9})$

15. $\lim\limits_{x \to 2} (x^3 - 4x^2 + 5x - 3)(x^2 + 17x - 4)$

16. $\lim\limits_{x \to 2} (x^2 - 1)^5$

17. $\lim\limits_{x \to -1} (x^9 + 2)^{53}$

18. $\lim\limits_{x \to 4} (x^2 - x - 10)^7$

19. $\lim\limits_{x \to 0} \dfrac{\sqrt{x + 1}}{\sqrt{x^2 - 3x + 4}}$

20. $\lim\limits_{x \to -2} \dfrac{\sqrt{x^2 - 3}}{1 + x + x^2 + x^3}$

21. $\lim\limits_{x \to 5} \sqrt{\dfrac{x - 1}{x^2 - 9}}$

22. $\lim\limits_{x \to 3} \dfrac{1}{x^3 - 8}$

23. $\lim\limits_{x \to 0} \dfrac{3}{x^5 + 3x^2 + 3}$

24. $\lim\limits_{x \to -4} \dfrac{x^3 - x^2 - x + 1}{x^2 + 3}$

25. $\lim\limits_{x \to 1} \dfrac{x^8 + x^6 + x^4 + x^2 + 1}{x^7 + x^5 + x^3 + x}$

26. $\lim_{x \to 0} \dfrac{x^{28} - 17x^{14} + x^2 - 3}{x^{51} + x^{31} - 23x^2 + 2}$

27. $\lim_{x \to 0} \dfrac{2x^2 + 5x + 1}{3x^5 - 9x + 2}$

28. $\lim_{x \to 0} \dfrac{x^{81} - x^{41} + 3}{23x^4 - 8x^7 + 5}$

29. $\lim_{x \to 2} \dfrac{x^2 - x - 12}{x^2 - 5x + 4}$

2.4 Continuity

The concept of continuity is one of the central notions in mathematics. Intuitively, a function is continuous at a point if it is defined at that point and if its graph moves "smoothly" through that point. Figure 1 shows the graphs of 5 functions, three of which are continuous at x_0 and two of which are not.

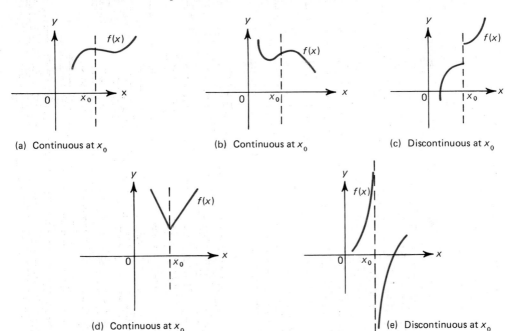

(a) Continuous at x_0

(b) Continuous at x_0

(c) Discontinuous at x_0

(d) Continuous at x_0

(e) Discontinuous at x_0

Figure 1

There are several equivalent definitions of continuity. The one we give here depends explicitly on the theory of limits developed earlier in this chapter.

DEFINITION 1. Let $f(x)$ be defined for every x in an open interval containing the number x_0. Then f is *continuous* at x_0 if all of the following three conditions hold:

    (i) $f(x_0)$ exists (that is, x_0 is in the domain of f).
    (ii) $\lim_{x \to x_0} f(x)$ exists.
    (iii) $\lim_{x \to x_0} f(x) = f(x_0)$.   **(1)**

If one of these three conditions fails to hold, then f is said to be *discontinuous* at x_0.

    Remark. Condition (iii) tells us that if a function f is continuous at x_0, then we can calculate $\lim_{x \to x_0} f(x)$ by evaluation. This is only one of the reasons continuous functions are so important. In the next few chapters we will see that a large majority of the functions we encounter in applications are indeed continuous.

EXAMPLE 1. Let $f(x) = x^2$. Then, for any real number x_0,

$$\lim_{x \to x_0} f(x) = \lim_{x \to x_0} x^2 = x_0{}^2 = f(x_0)$$

so that f is continuous at every real number (see Figure 2).

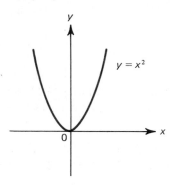

Figure 2

EXAMPLE 2. Let $p(x) = c_0 + c_1 x + c_2 x^2 + c_3 x^3 + \cdots + c_n x^n$ be a polynomial. By Theorem 2.3.1,

$$\lim_{x \to x_0} p(x) = p(x_0) \tag{2}$$

for every real number x_0. Therefore every polynomial is continuous at every real number. (Note that this also shows that any constant function is continuous.)

EXAMPLE 3. Let $r(x) = p(x)/q(x)$ be a rational function ($p(x)$ and $q(x)$ are polynomials). Then from Theorem 2.3.5, we have, if $q(x_0) \neq 0$, $\lim_{x \to x_0} r(x) = p(x_0)/q(x_0) = r(x_0)$ so that any rational function is continuous at all points x_0 at which the denominator, $q(x_0)$, is nonzero.

EXAMPLE 4. Let

$$f(x) = \frac{x^5 + 3x^3 - 4x^2 + 5x - 2}{x^2 - 5x + 6}.$$

Here f is a rational function and therefore is continuous at any x for which the denominator, $x^2 - 5x + 6$, is not zero. Since $x^2 - 5x + 6 = (x - 3)(x - 2) = 0$ only when $x = 2$ or 3, f is continuous at all real numbers except at these two points.

EXAMPLE 5. Let $f(x) = (x^2 - 1)/(x - 1)$. Although

$$\lim_{x \to 1} \frac{x^2 - 1}{x - 1} = 2,$$

f is *not* continuous at 1 since $f(x)$ is not defined there so that condition (i) is violated.

We can construct a continuous function g by defining

$$g(x) = \begin{cases} \dfrac{x^2 - 1}{x - 1}, & \text{if } x \neq 1 \\ 2, & \text{if } x = 1 \end{cases}.$$

This function is continuous at every real value of x and $g(x) = f(x)$ for all $x \neq 1$. Notice that any other choice for $g(1)$ would make g discontinuous at 1. For example, if we define

$$g(x) = \begin{cases} \dfrac{x^2 - 1}{x - 1}, & \text{if } x \neq 1 \\ 1, & \text{if } x = 1 \end{cases}$$

then condition (iii) is violated since $\lim_{x \to 1} g(x) = 2$ which is not equal to $g(1)$. [*Note:* We will often encounter functions which are defined by two or more "pieces" (as above).]

DEFINITION 2. A function f is continuous over an open interval† (a, b) if f is continuous at every point in that interval (a may be $-\infty$ and/or b may be $+\infty$).

EXAMPLE 6. Let $f(x) = |x|$. Since for $x > 0, f(x) = x$ and for $x < 0, f(x) = -x, f$ is continuous at every $x \neq 0$. If $x = 0$, then $\lim_{x \to 0} f(x) = \lim_{x \to 0} |x| = 0 = f(0)$ so that f is continuous at 0 also (see Figure 3). Thus f is continuous in every open interval (a, b).

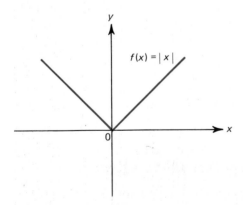

$f(x) = |x|$

Figure 3

EXAMPLE 7. Consider the function

$$f(x) = [x] \tag{3}$$

where $[x]$ is defined‡ as the greatest integer smaller than or equal to x. Thus $[3] = 3$, $[\frac{1}{2}] = 0$, $[2.16] = 2$, $[-5.6] = -6$ and so on. A graph of this function is given in Figure 4. Let us try to calculate $\lim_{x \to 2} [x]$. If $1 < x < 2$, then $[x] = 1$. But if

---

† We will define continuity over a closed interval $[a, b]$ in Section 2.7 (see Definition 2.7.3).
‡ This function is called the *greatest integer function*.

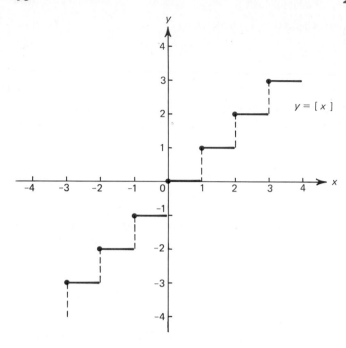

Figure 4

$2 < x < 3, [x] = 2$. Hence if we approach $x = 2$ from the right, the functional values approach 2, but if we approach from the left the functional values approach 1. Therefore $\lim_{x \to 2} [x]$ does not exist and so $f(x) = [x]$ is *not* continuous at $x = 2$. The same argument shows that f is discontinuous at any integer. On the other hand, if x_0 is not an integer, then $[x]$ is continuous at x_0 (why?). Thus f is continuous in any interval (a, b) which does not contain an integer.

The kind of discontinuity exhibited in the above example is called a *jump discontinuity* because the function "jumps" from one *finite* value to another as is depicted in Figure 4. The behavior of the function $f(x) = [x]$ gives rise to another definition.

DEFINITION 3. The function $f(x)$ is called *piecewise continuous* on the interval (a, b) if it is continuous there except for, at most, a finite number of jump discontinuities.

The function $f(x) = [x]$ is piecewise continuous over any finite interval (a, b). The function $f(x) = 1/x$ is *not* piecewise continuous in $(-1, 1)$ since the function becomes *infinitely* large near $x = 0$. The graph of $f(x) = 1/x$ is sketched in Figure 5. We will discuss functions which become infinitely large in great detail in the next section.

EXAMPLE 8. Let

$$f(x) = \begin{cases} x, & x \neq 2 \\ 1, & x = 2 \end{cases}.$$

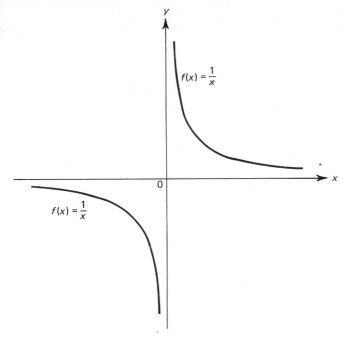

$f(x) = \dfrac{1}{x}$

$f(x) = \dfrac{1}{x}$

Figure 5

The graph of this function is sketched in Figure 6. From the figure we see that f has a jump discontinuity at $x = 2$, since $f(x)$ is close to 2 for x close to (but not equal to) 2 but "jumps" down to the value 1 for $x = 2$. Therefore f is piecewise continuous over any interval. However, if we redefine f at the single point 2 by

$$f(x) = \begin{cases} x, & \text{if } x \neq 2 \\ 2, & \text{if } x = 2, \end{cases}$$

then this new function f is continuous in the interval $(-\infty, \infty)$.

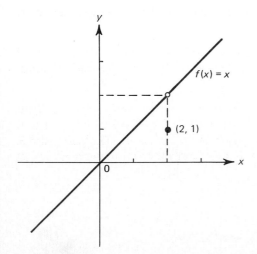

$f(x) = x$

$(2, 1)$

Figure 6

EXAMPLE 9. Let

$$f(x) = \frac{1}{(x-1)(x-2)}.$$

Then f is not defined at $x = 1$ and $x = 2$. Since f is a rational function, it is continuous at all other points. Hence f is continuous in the three disjoint intervals $(-\infty, 1)$, $(1, 2)$, and $(2, \infty)$. It is *not* piecewise continuous in any open interval containing 1 or 2. (Explain why.)

Before leaving this section, we state the following theorem about the sum, product, and quotient of continuous functions which is an immediate consequence of the definition of continuity and the limit theorems of the last section.

THEOREM 1. Let the functions f and g be continuous at x_0. Then the following functions are continuous at x_0:

   (i) *cf,*   (ii) $f + g$,   (iii) $f \cdot g$.

If, in addition, $g(x_0) \neq 0$, then

   (iv) f/g is continuous at x_0.

For example, to prove part (iv), we have, if $g(x_0) \neq 0$,

$$\lim_{x \to x_0} \frac{f(x)}{g(x)} = \frac{\lim_{x \to x_0} f(x)}{\lim_{x \to x_0} g(x)} = \frac{f(x_0)}{g(x_0)} = \left(\frac{f}{g}\right)(x_0)$$

so that f/g is continuous at x_0.

The following theorem, whose proof will be given in Section 10.4, is very useful for the calculation of limits.

THEOREM 2.   If f is continuous at a and if $\lim_{x \to x_0} g(x) = a$, then $\lim_{x \to x_0} f(g(x)) = f(a)$; in other words,

$$\boxed{\lim_{x \to x_0}(f \circ g)(x) = \lim_{x \to x_0} f(g(x)) = f(\lim_{x \to x_0} g(x))} \qquad (4)$$

EXAMPLE 10.   Calculate $\lim_{x \to 3} \left\{ \frac{x^2 - 9}{x - 3} \right\}^3$.

SOLUTION. Let $f(x) = x^3$ and

$$g(x) = \frac{x^2 - 9}{x - 3}.$$

Then $\lim_{x \to 3} g(x) = 6$ so that $\lim_{x \to 3} f(g(x)) = 6^3 = 216$.

PROBLEMS 2.4

In Problems 1–8 find the open interval or intervals in which the given function is continuous.

1. $f(x) = x^{17} - 3x^{15} + 2$

2. $f(x) = x^{1/3}$

3. $f(x) = x^{1/4}$

4. $f(x) = \dfrac{1}{x + 2}$

5. $f(x) = \dfrac{-17x}{x^2 - 1}$

6. $f(x) = \dfrac{1}{(x - 10)^{15}}$

7. $f(x) = \dfrac{2x}{x^3 - 8}$

8. $f(x) = \dfrac{|x + 2|}{x + 2}$.

*9. Graph the function $f(x) = x[x]$. Find the open intervals in which f is continuous.

10. Show that the function

$$f(x) = \begin{cases} \dfrac{x^3 - 1}{x - 1}, & x \neq 1 \\ 3, & x = 1 \end{cases}$$

is continuous in $(-\infty, \infty)$.

11. For what value of α is the function

$$f(x) = \begin{cases} \dfrac{x^4 - 1}{x - 1}, & x \neq 1 \\ \alpha, & x = 1 \end{cases}$$

continuous at $x = 1$?

12. Let

$$f(x) = \begin{cases} x, & x \neq \text{an integer} \\ x^2, & x = \text{an integer} \end{cases}.$$

Graph the function for $-3 \leq x \leq 3$. For what integer values of x is f continuous?

13. Let

$$f(x) = \begin{cases} 0, & x < 0 \\ x^2, & 0 \leq x < 2. \\ 4, & x \geq 2 \end{cases}$$

Graph the function. Show that f is continuous in $(-\infty, \infty)$.

14. Let

$$f(x) = \begin{cases} x, & x < 0 \\ x^2, & 0 \leq x < 1. \\ x^3, & x \geq 1 \end{cases}$$

Graph the function. Show that f is continuous in $(-\infty, \infty)$.

15. Show that the function

$$f(x) = \begin{cases} x, & x < 0 \\ 2x, & 0 \leq x \leq 1 \\ 3x^3, & x > 1 \end{cases}$$

is piecewise continuous over any interval.

16. Is the function

$$f(x) = \frac{x + 1}{x - 2}$$

piecewise continuous over the interval $[-3, 3]$? Over the interval $[-1, 1]$?

17. Explain why every continuous function is piecewise continuous.

18. Use Theorem 2 to calculate $\lim_{x \to 4} \sqrt{x^2 - x + 13}$.

19. Use Theorem 2 to calculate $\lim_{x \to -1} (3x^3 + 8x^2 - 9x + 2)^{3/4}$.

**20. Let

$$f(x) = \begin{cases} 1, & x \text{ rational} \\ 0, & x \text{ irrational} \end{cases}.$$

Explain why f is not continuous at any point in any interval $[a, b]$.

2.5 Infinite Limits and Limits at Infinity

Consider the problem of calculating

$$\lim_{x \to 0} \frac{1}{x^2}. \tag{1}$$

Values of $1/x^2$ for x "near" 0 are given in Table 1. We see that as x gets closer and closer to zero, $f(x) = 1/x^2$ gets larger and larger. In fact, $1/x^2$ grows *without bound* as x approaches zero from either side. The graph of the function $f(x) = 1/x^2$ is given in Figure 1. In this situation, we say that $f(x)$ *tends to infinity* as x *approaches zero* and we write

$$\lim_{x \to 0} \frac{1}{x^2} = \infty. \tag{2}$$

TABLE 1

x	x^2	$\dfrac{1}{x^2}$	x	x^2	$\dfrac{1}{x^2}$
1	1	1	-1	1	1
0.5	0.25	4	-0.5	0.25	4
0.1	0.01	100	-0.1	0.01	100
0.01	0.0001	10,000	-0.01	0.0001	10,000
0.001	0.000001	1,000,000	-0.001	0.000001	1,000,000
0.0001	0.00000001	100,000,000	-0.0001	0.00000001	100,000,000

In general, we have the following definition:

DEFINITION 1. If $f(x)$ grows without bound in the positive direction as x gets close to the number x_0 from either side, then we say that $f(x)$ *tends to infinity* as x approaches x_0 and we write

$$\lim_{x \to x_0} f(x) = \infty.$$

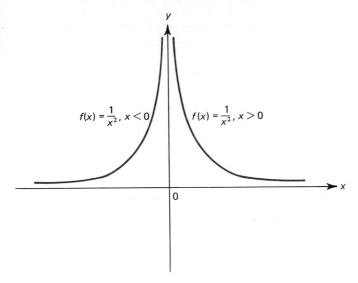

$f(x) = \frac{1}{x^2}, \; x < 0$ $f(x) = \frac{1}{x^2}, \; x > 0$

Figure 1

EXAMPLE 1.   Calculate $\lim_{x \to -1} 1/(x+1)^2$.

SOLUTION. The graph of the function $1/(x+1)^2$ is given in Figure 2. As x gets closer and closer to -1 from either side, $1/(x+1)^2$ grows without bound. Hence

$$\lim_{x \to -1} \frac{1}{(x+1)^2} = \infty. \tag{3}$$

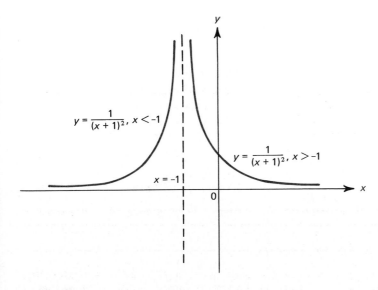

$y = \dfrac{1}{(x+1)^2}, \; x < -1$

$y = \dfrac{1}{(x+1)^2}, \; x > -1$

$x = -1$

Figure 2

☐ **EXAMPLE 2.**   Calculate $\lim_{x \to 0} 1/(x^2 + x^3)$

SOLUTION. With the aid of a calculator we find values for $1/(x^2 + x^3)$ for x near zero (see Table 2). We can see that as x gets close to zero, $1/(x^2 + x^3)$ grows without

TABLE 2

x	x^2	x^3	$x^2 + x^3$	$\dfrac{1}{x^2 + x^3}$
0.5	0.25	0.125	0.375	2.66667
0.1	0.01	0.001	0.011	90.90909
0.01	0.0001	0.000001	0.000101	9,900.99
0.001	0.000001	0.000000001	0.000001001	999,000.999
−0.5	0.25	−0.125	0.125	8
−0.1	0.01	−0.001	0.009	111.1111
−0.01	0.0001	−0.000001	0.000099	10,101.01
−0.001	0.000001	−0.000000001	0.000000999	1,001,001

bound. Notice that for x negative, x^3 is also negative. This poses no problem since for $|x| < 1$, $|x^3| < |x^2|$ so that $x^2 + x^3$ is a positive number. Thus, we have

$$\lim_{x \to 0} \frac{1}{x^2 + x^3} = \infty. \tag{4}$$

To calculate this limit without making a table, we first divide the numerator and denominator by x^2. This can be done since, in the calculation of the limit, $x \neq 0$ so that $x^2 \neq 0$. We then have

$$\lim_{x \to 0} \frac{1}{x^2 + x^3} = \lim_{x \to 0} \frac{1/x^2}{(x^2 + x^3)/x^2} = \lim_{x \to 0} \frac{1/x^2}{1 + x}.$$

Now $\lim_{x \to 0} (1/x^2) = \infty$ and $\lim_{x \to 0} (1 + x) = 1$. Thus the numerator grows without bound while the denominator approaches 1, implying that $1/(x^2 + x^3)$ does tend to infinity. This example illustrates how a difficult calculation can be greatly simplified by a few algebraic manipulations.

We now consider the slightly different problem of calculating $\lim_{x \to 0} (-1/x^2)$. The values of $-1/x^2$ near $x = 0$ are given in Table 3. The graph of $f(x) = -1/x^2$ is given in Figure 3. As x gets close to zero (from either side), $f(x) = -1/x^2$ grows without bound in the *negative* direction. In this case we say that $-1/x^2$ *tends to minus infinity as x tends to zero* and write

$$\lim_{x \to 0} \frac{-1}{x^2} = -\infty. \tag{5}$$

TABLE 3

x	x^2	$\dfrac{-1}{x^2}$	x	x^2	$\dfrac{-1}{x^2}$
1	1	−1	−1	1	−1
0.5	0.25	−4	−0.5	0.25	−4
0.1	0.01	−100	−0.1	0.01	−100
0.01	0.0001	−10,000	−0.01	0.0001	−10,000
0.001	0.000001	−1,000,000	−0.001	0.000001	−1,000,000
0.0001	0.00000001	−100,000,000	−0.0001	0.00000001	−100,000,000

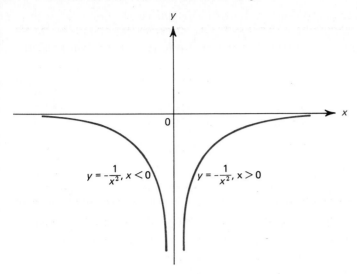

Figure 3

DEFINITION 2. If $f(x)$ grows without bound in the negative direction as x gets close to the number x_0 from either side, then we say that $f(x)$ *tends to minus infinity* as x approaches x_0 and we write

$$\lim_{x \to x_0} f(x) = -\infty.$$

EXAMPLE 3. Following the reasoning of Example 1, we find that

$$\lim_{x \to -1} \frac{-1}{(x + 1)^2} = -\infty.$$

☐**EXAMPLE 4.** Following the reasoning of Example 2, we find that

$$\lim_{x \to 0} \frac{-1}{x^2 + x^3} = -\infty. \tag{6}$$

EXAMPLE 5. Calculate $\lim\limits_{x \to 0} 1/x$.

SOLUTION. To see what is happening in this problem, we tabulate in Table 4 a few values of $1/x$ for x near zero. If $x > 0$, then as x gets close to zero, $1/x$ grows without

TABLE 4

x	$\dfrac{1}{x}$	x	$\dfrac{1}{x}$
1	1	-1	-1
.5	2	$-.5$	-2
.1	10	$-.1$	-10
.01	100	$-.01$	-100
.001	1,000	$-.001$	$-1,000$
.0001	10,000	$-.0001$	$-10,000$

bound in the positive direction. But if $x < 0$, then as x gets close to zero, $1/x$ grows without bound in the negative direction (see Figure 4). Since the behavior of $1/x$ depends on the way in which x approaches zero (i.e., from the right or from the left), we must conclude that $1/x$ *has no limit* as $x \to 0$, or equivalently that $\lim_{x \to 0} 1/x$ *does not exist.*

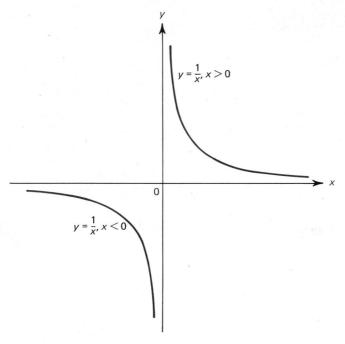

$y = \frac{1}{x}, x > 0$

$y = \frac{1}{x}, x < 0$

Figure 4

□ **EXAMPLE 6.**   Calculate $\lim_{x \to 0} -1/(x + x^2)$.

SOLUTION. Before drawing any conclusions, we tabulate, using a calculator, some values of $-1/(x + x^2)$ for x near zero (Table 5). We see that, as in the previous example, the behavior of $f(x) = -1/(x + x^2)$ as x approaches zero depends on the

TABLE 5

x	x^2	$x + x^2$	$\dfrac{-1}{x + x^2}$
0.5	0.25	0.75	−1.33333
0.1	0.01	0.11	−9.09091
0.01	0.0001	0.0101	−99.00990
0.001	0.000001	0.001001	−999.00100
0.0001	0.00000001	0.00010001	−9,999.00010
−0.5	0.25	−0.25	4.00000
−0.1	0.01	−0.09	11.11111
−0.01	0.0001	−0.0099	101.01010
−0.001	0.000001	−0.000999	1,001.00100
−0.0001	0.00000001	−0.00009999	10,001.00010

way in which x approaches zero. Here, if x is positive, then $-1/(x + x^2)$ grows without bound in the negative direction while for x negative, $f(x)$ grows without bound in the positive direction. This can also be seen by dividing both numerator and denominator by x:

$$\lim_{x \to 0} \frac{-1}{x + x^2} = \lim_{x \to 0} \frac{-1/x}{(x + x^2)/x} = \lim_{x \to 0} \frac{-1/x}{1 + x}.$$

Since the denominator tends to 1, we see that the function $-1/(x + x^2)$ behaves like the function $-1/x$ near $x = 0$; that is, $\lim_{x \to 0} -1/(x + x^2)$ does not exist.

To this point, we have considered limits as $x \to x_0$ where x_0 is a finite number. But in many important applications it is necessary to determine what happens to $f(x)$ as x becomes very large. For example, we may ask what happens to the function $f(x) = 1/x$ as x becomes large. To illustrate, we again make use of a table (Table 6). It is evident that as x gets large, $1/x$ approaches zero. We then write

$$\lim_{x \to \infty} \frac{1}{x} = 0. \tag{7}$$

Similarly, if x grows large in the negative direction, then $1/x$ approaches zero (see Table 7); that is

$$\lim_{x \to -\infty} \frac{1}{x} = 0. \tag{8}$$

Look again at Figure 4 which illustrates the behavior of $1/x$ as $x \to \pm\infty$.

TABLE 6

x	$f(x) = \dfrac{1}{x}$
1	1
10	0.1
100	0.01
1,000	0.001
10,000	0.0001

TABLE 7

x	$\dfrac{1}{x}$
-1	-1
-10	-0.1
-100	-0.01
$-1,000$	-0.001
$-10,000$	-0.0001

These examples suggest the following definition:

DEFINITION 3. (i) The *limit as x approaches infinity of $f(x)$ is L,* written

$$\lim_{x \to \infty} f(x) = L, \tag{9}$$

if $f(x)$ is defined for all large values of x and if $f(x)$ gets arbitrarily close to L as x increases without bound.

   (ii) *The limit as x approaches minus infinity of $f(x)$ is L,* written

$$\lim_{x \to -\infty} f(x) = L, \tag{10}$$

if $f(x)$ is defined for all values of x which are large in the negative direction and $f(x)$ gets arbitrarily close to L as x increases without bound in the negative direction.

We emphasize that the definitions in this section, like the ones in Section 2.2, are *not* precise mathematical statements but are only intended to appeal to your intuition. Precise mathematical definitions will appear in Section 10.1.

The limit theorems of Section 2.3 apply in the same way when $x \to \infty$ or $x \to -\infty$ as they do when $x \to x_0$, where x_0 is a finite number.

EXAMPLE 7. Calculate $\lim_{x \to \infty} 1/x^2$?

SOLUTION. We can calculate this limit in one of three ways. First we could construct a table of values as in Table 8. It seems from this table that

$$\lim_{x \to \infty} \frac{1}{x^2} = 0.$$

TABLE 8

x	x^2	$\dfrac{1}{x^2}$
1	1	1
10	100	0.01
100	10,000	0.0001
1,000	1,000,000	0.000001

Second, we could use the corollary to Theorem 2.3.4. Since $\lim_{x \to \infty} 1/x = 0$, we have

$$\lim_{x \to \infty} \frac{1}{x^2} = \lim_{x \to \infty} \left(\frac{1}{x}\right)^2 = 0^2 = 0. \tag{11}$$

Third, we note that for $x > 1$, $x^2 > x$, so that $0 < 1/x^2 < 1/x$. Then, since $1/x \to 0$ as $x \to \infty$, $1/x^2$, which is between 0 and $1/x$, must also approach 0 as $x \to \infty$. This is illustrated in Figure 5.

The third method used above is an example of the "squeezing" theorem, whose proof will be given in Section 10.2.

THEOREM 1 (Squeezing Theorem). Suppose that $f(x) \le g(x) \le h(x)$ for x near x_0 and $\lim_{x \to x_0} f(x) = \lim_{x \to x_0} h(x) = L$, where the number x_0 may be a real number or $+\infty$ or $-\infty$ (if $x_0 = \infty$, then the phrase "for x near x_0" means "for x very large"). Then

$$\lim_{x \to x_0} g(x) = L. \tag{12}$$

This theorem states that if $g(x)$ is "squeezed" between $f(x)$ and $h(x)$ near x_0 and if $f(x)$ and $h(x)$ have the same limit L, then $g(x)$ must also have the limit L. In Example 7, $f(x) = 0$, $g(x) = 1/x^2$ and $h(x) = 1/x$. Then, for $x > 1$, $f(x) \le g(x) \le h(x)$. Here $\lim_{x \to \infty} f(x) = \lim_{x \to \infty} 0 = 0$ and $\lim_{x \to \infty} h(x) = \lim_{x \to \infty} 1/x = 0$ so that $L = 0$ and $\lim_{x \to \infty} 1/x^2 = 0$ (look again at Figure 5).

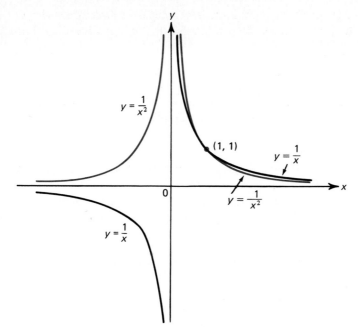

Figure 5

EXAMPLE 8. Calculate $\lim_{x\to\infty} 1/(x^3 + 8)$.

SOLUTION. For $x > 0$, $x^3 + 8 > x$ and, therefore, $0 < 1/(x^3 + 8) < 1/x$. Then, by the squeezing theorem, since $\lim_{x\to\infty} 1/x = 0$ and $\lim 0 = 0$, we have

$$\lim_{x\to\infty} \frac{1}{x^3 + 8} = 0. \tag{13}$$

EXAMPLE 9. Calculate $\lim_{x\to\infty} 1/\sqrt{x}$.

SOLUTION. $\sqrt{x}$ grows without bound as $x \to \infty$—only it does so at a slower rate than x. For example, when $x = 100$, $\sqrt{x} = 10$; when $x = 1{,}000{,}000$, $\sqrt{x} = 1{,}000$. Nevertheless, since $\sqrt{x}$ does grow without bound, we have

$$\lim_{x\to\infty} \frac{1}{\sqrt{x}} = 0.$$

EXAMPLE 10. Calculate $\lim_{x\to\infty} x^2$.

SOLUTION. Here as x grows, x^2, which is bigger than x for $x > 1$, grows even faster (see Figure 6). Thus

$$\lim_{x\to\infty} x^2 = \infty.$$

☐**EXAMPLE 11.** Calculate

$$\lim_{x\to\infty} \frac{x^2 - 2x + 3}{x^2 + 4x + 4}. \tag{14}$$

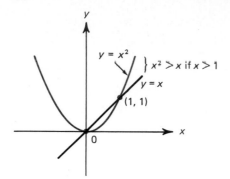

Figure 6

SOLUTION. There is an easy way to find this limit but first we will use a calculator to see if we can find a pattern. From Table 9, we might be led to the conclusion that the

TABLE 9

x	x^2	$-2x$	$x^2 - 2x + 3$	$4x$	$x^2 + 4x + 4$	$\dfrac{x^2 - 2x + 3}{x^2 + 4x + 4}$
1	1	-2	2	4	9	0.22222
10	100	-20	83	40	144	0.57639
100	10,000	-200	9,803	400	10,404	0.94223
1,000	1,000,000	$-2,000$	998,003	4,000	1,004,004	0.99402
10,000	100,000,000	$-20,000$	99,980,003	40,000	100,040,004	0.99940

limit is 1. To calculate this more easily we could try to use the limit theorem for quotients (Theorem 2.3.5). But

$$\lim_{x \to \infty} x^2 - 2x + 3 = \infty, \qquad \lim_{x \to \infty} x^2 + 4x + 4 = \infty$$

and ∞/∞ is an undefinied quantity. Fortunately, we may simplify this problem by dividing the numerator and denominator of (14) by x^2. We have

$$\lim_{x \to \infty} \frac{x^2 - 2x + 3}{x^2 + 4x + 4} = \lim_{x \to \infty} \frac{(x^2 - 2x + 3)/x^2}{(x^2 + 4x + 4)/x^2}$$

$$= \lim_{x \to \infty} \frac{\dfrac{x^2}{x^2} - \dfrac{2x}{x^2} + \dfrac{3}{x^2}}{\dfrac{x^2}{x^2} + \dfrac{4x}{x^2} + \dfrac{4}{x^2}} = \lim_{x \to \infty} \frac{1 - \dfrac{2}{x} + \dfrac{3}{x^2}}{1 + \dfrac{4}{x} + \dfrac{4}{x^2}}.$$

But $2/x$, $3/x^2$, $4/x$ and $4/x^2$ all approach 0 as $x \to \infty$. Therefore

$$\lim_{x \to \infty} \frac{1 - \dfrac{2}{x} + \dfrac{3}{x^2}}{1 + \dfrac{4}{x} + \dfrac{4}{x^2}} = \frac{1 - 0 + 0}{1 + 0 + 0} = 1.$$

Note. We made use of several of our limit theorems here. Can you name which ones?

EXAMPLE 12. Calculate

$$\lim_{x \to -\infty} \frac{4x^5 - 3x^2 + 17x + 2}{2x^5 + x^2 + 5} \tag{15}$$

SOLUTION. We divide numerator and denominator of (15) by x^5, the highest power of x that appears in the rational function, to obtain

$$\lim_{x \to -\infty} \frac{\dfrac{4x^5}{x^5} - \dfrac{3x^2}{x^5} + \dfrac{17x}{x^5} + \dfrac{2}{x^5}}{\dfrac{2x^5}{x^5} + \dfrac{x^2}{x^5} + \dfrac{5}{x^5}} = \frac{\lim\limits_{x \to -\infty} \left(4 - \dfrac{3}{x^3} + \dfrac{17}{x^4} + \dfrac{2}{x^5}\right)}{\lim\limits_{x \to -\infty} \left(2 + \dfrac{1}{x^3} + \dfrac{5}{x^5}\right)}$$

$$= \frac{4 - 0 + 0 + 0}{2 + 0 + 0} = \frac{4}{2} = 2.$$

In general, the limits as $x \to \pm\infty$ of rational expressions (like (14) or (15)) can be found by first dividing numerator and denominator by the highest power of x that appears in the expression, and then calculating the limit at $x \to \infty$ (or $-\infty$) of both the numerator and denominator.

EXAMPLE 13. Calculate $\lim\limits_{x \to \infty} \dfrac{x^3 + 3x + 6}{x^5 + 2x^2 + 9}$.

SOLUTION. Divide by x^5:

$$\lim_{x \to \infty} \frac{x^3 + 3x + 6}{x^5 + 2x^2 + 9} = \frac{\lim\limits_{x \to \infty} \left(\dfrac{1}{x^2} + \dfrac{3}{x^4} + \dfrac{6}{x^5}\right)}{\lim\limits_{x \to \infty} \left(1 + \dfrac{2}{x^3} + \dfrac{9}{x^5}\right)} = \frac{0 + 0 + 0}{1 + 0 + 0} = 0.$$

EXAMPLE 14. Calculate $\lim\limits_{x \to \infty} \dfrac{x^3 + 2x + 3}{5x^2 + 1}$.

SOLUTION. Divide by x^3.

$$\lim_{x \to \infty} \frac{x^3 + 2x + 3}{5x^2 + 1} = \lim_{x \to \infty} \frac{1 + \dfrac{2}{x^2} + \dfrac{3}{x^3}}{\dfrac{5}{x} + \dfrac{1}{x^3}}.$$

Here the numerator approaches 1 but the denominator approaches 0 (and stays positive). Thus

$$\lim_{x \to \infty} \frac{x^3 + 2x + 3}{5x^2 + 1} = \infty.$$

In general, if $r(x) = p(x)/q(x)$ and the degree of the polynomial $p(x)$ is greater than the degree of $q(x)$, then $\lim_{x\to\infty} p(x)/q(x) = +\infty$ or $-\infty$.

If the degree of $q(x)$ is greater than the degree of $p(x)$, then $\lim_{x\to\infty} p(x)/q(x) = 0$.

Finally, if the degree of $p(x)$ is equal to the degree of $q(x)$, then $\lim_{x\to\infty} p(x)/q(x) = c_m/d_n$ where c_m is the coefficient of the highest power of x in $p(x)$ and d_n is the coefficient of the highest power of x in $q(x)$. (See Problem 33.)

PROBLEMS 2.5

In Problems 1–29 find each limit (if it exists).

1. $\lim\limits_{x\to 0} \dfrac{1}{x^4}$.

2. $\lim\limits_{x\to 0} \dfrac{1}{x^5}$.

3. $\lim\limits_{x\to 0} \dfrac{1}{x^6}$.

4. $\lim\limits_{x\to 5} \dfrac{1}{(x-5)^2}$.

5. $\lim\limits_{x\to 5} \dfrac{1}{(x-5)^3}$.

6. $\lim\limits_{x\to -3} \dfrac{x-4}{(x+3)^2}$.

7. $\lim\limits_{x\to \pi} \dfrac{\pi}{(x-\pi)^6}$.

8. $\lim\limits_{x\to 0} \dfrac{1}{x^4 + x^8 + x^{12}}$.

9. $\lim\limits_{x\to 0} \dfrac{x+x^2}{x^2 + x^3}$.

10. $\lim\limits_{x\to\infty} \dfrac{1}{x+x^3}$.

11. $\lim\limits_{x\to -\infty} \dfrac{x}{1+x}$.

12. $\lim\limits_{x\to\infty} \dfrac{1}{1-\sqrt{x}}$.

13. $\lim\limits_{x\to\infty} \dfrac{2x}{3x^3+4}$.

14. $\lim\limits_{x\to -\infty} \dfrac{2x+3}{3x+2}$.

15. $\lim\limits_{x\to\infty} \dfrac{5x-x^2}{3x+x^2}$.

16. $\lim\limits_{x\to\infty} \dfrac{1+\sqrt{x}}{1-\sqrt{x}}$.

17. $\lim\limits_{x\to\infty} \dfrac{2x^2+3x+5}{3x^2-x+2}$.

18. $\lim\limits_{x\to\infty} \dfrac{4x^4+1}{1+5x^4}$.

19. $\lim\limits_{x\to\infty} \dfrac{x^5-3x+4}{7x^6+8x^4+2}$.

20. $\lim\limits_{x\to\infty} \dfrac{x^8-2x^5+3}{5x^4+3x+1}$.

21. $\lim\limits_{x\to\infty} \dfrac{x^8-1}{x^9+1}$.

22. $\lim\limits_{x\to -\infty} \dfrac{x^{16}+5x^8+2}{23x^{15}-8x^{13}+x^{11}}$.

23. $\lim\limits_{x\to\infty} \dfrac{\sqrt{x}+2}{x+3}$.

24. $\lim\limits_{x\to\infty} \dfrac{\sqrt[3]{x}-1}{\sqrt{x}+3}$.

25. $\lim\limits_{x\to\infty} \dfrac{3x^{5/3}+2\sqrt{x}-3}{7x^{5/3}-3x+6}$.

26. $\lim\limits_{x\to\infty} \dfrac{7x^{1/7}-1}{4x^{1/7}-x^{1/9}}$.

*27. $\lim\limits_{x\to\infty} \dfrac{\sqrt{x}}{\sqrt[3]{x}}$.

*28. $\lim\limits_{x\to\infty} \dfrac{\sqrt[3]{x}}{\sqrt{x}}$.

*29. $\lim\limits_{x\to\infty} \dfrac{\sqrt{x}+1}{\sqrt{x}+3}$.

30. We have seen that $\lim_{x\to 0} 1/x^2 = \infty$. How small in absolute value must x be chosen in order that $1/x^2 > 1{,}000{,}000$? $10{,}000{,}000$? $100{,}000{,}000$?

31. We have seen that $\lim_{x\to\infty} 1/\sqrt{x} = 0$. How large must x be in order that $1/\sqrt{x} < 0.01$? 0.001? 0.0001?

□32. Show that

$$\lim_{x\to -\infty} \frac{2x^5 - 3x^3 + 17x + 2}{x^5 + 4x^2 + 16} = 2.$$

For $x < -2$, how small must x be (i.e., large in the negative direction) in order that this quotient be greater than 1.999?

*33. Let $r(x) = p(x)/q(x)$ where

$$p(x) = c_0 + c_1 x + \cdots + c_m x^m \qquad \text{and} \qquad q(x) = d_0 + d_1 x + \cdots + d_n x^n.$$

(a) If $m < n$, show that $\lim_{x \to \infty} p(x)/q(x) = 0$. [*Hint:* Divide top and bottom by x^n.]
(b) If $m > n$, show that $\lim_{x \to \infty} p(x)/q(x) = +\infty$ or $-\infty$.
(c) If $m = n$, show that $\lim_{x \to \infty} p(x)/q(x) = c_m/d_n$.

*34. Explain why we had to stipulate that $x < -2$ in Problem 32.

*2.6 An Explanation of Zeno's Paradox

In Section 1.8 we discussed the geometric progression

$$S_n = 1 + r + r^2 + r^3 + \cdots + r^{n-1} + r^n \tag{1}$$

which had the sum

$$S_n = \frac{1 - r^{n+1}}{1 - r}, \qquad \text{if } r \neq 1. \tag{2}$$

We will now examine what happens to the value of S_n as $n \to \infty$. It is not difficult to see that if $r \geq 1$, then $S_n \to \infty$ as $n \to \infty$, since we are adding together more and more terms, each of which is ≥ 1. If $r = -1$, then $S_0 = 1$, $S_1 = 1 - 1 = 0$, $S_2 = 1 - 1 + 1 = 1$, $S_3 = 0$, and so on so that S_n does not approach any fixed finite limit even though $|S_n| \leq 1$ for every n. If $r < -1$, then the terms r^n grow without bound as n becomes large so again there will be no finite limit. What happens if $|r| < 1$ (that is, if $-1 < r < 1$)? To determine this, we calculate $\lim_{n \to \infty} r^{n+1}$.

☐**EXAMPLE 1.** If $r = \frac{1}{2}$, then $r^{n+1} = (\frac{1}{2})^{n+1} = 1/2^{n+1}$. Table 1 gives values of $1/2^{n+1}$ for various values of n. It is clear that as n increases, $r^{n+1} = 1/2^{n+1}$ approaches zero.

TABLE 1

n	$n + 1$	2^{n+1}	$\dfrac{1}{2^{n+1}}$
1	2	4	0.25
5	6	64	0.015625
10	11	2,048	0.0004882812
20	21	2,097,152	0.0000004768
100	101	2.535×10^{30}	3.945×10^{-31}

☐**EXAMPLE 2.** If $r = -\frac{1}{3}$, then $r^{n+1} = (-\frac{1}{3})^{n+1} = 1/(-3)^{n+1}$. Table 2 gives values of $1/(-3)^{n+1}$ for several values of n. Again, it is clear that as n increases, $r^{n+1} = 1/(-3)^{n+1}$ approaches zero, even though this time r^{n+1} is alternatively positive and negative.

TABLE 2

n	$n + 1$	$(-3)^{n+1}$	$\dfrac{1}{(-3)^{n+1}}$
1	2	9	0.1111111111
5	6	729	0.0013717421
10	11	$-177,147$	-0.0000056450
25	26	2.542×10^{12}	3.934×10^{-13}
100	101	-1.546×10^{48}	-6.468×10^{-49}

With these two examples before us, it is not difficult to believe the following:

$$\lim_{n \to \infty} r^{n+1} = 0, \quad \text{if} \quad -1 < r < 1. \tag{3}$$

Now, returning to the sum of the geometric progression (1) with $-1 < r < 1$, we may take the limit in (2) to obtain

$$S = \lim_{n \to \infty} S_n = \lim_{n \to \infty} \frac{1 - r^{n+1}}{1 - r} = \frac{1 - 0}{1 - r} = \frac{1}{1 - r}, \quad -1 < r < 1. \tag{4}$$

Note. Here again we used several of our limit theorems. Which ones?

The sum

$$S = 1 + r + r^2 + r^3 + \cdots + r^n + \cdots \tag{5}$$

is called a *geometric series* and we have shown that for $|r| < 1$, the sum of the geometric series is $1/(1 - r)$; that is,

$$\lim_{n \to \infty} S_n = 1 + r + r^2 + \cdots + r^n + \cdots = \frac{1}{1 - r} \quad \text{for} \quad |r| < 1. \tag{6}$$

This is a special case of an *infinite series* which we shall discuss in detail in Chapter 14.

EXAMPLE 3. Calculate the sum of the geometric series

$$S = 1 + \tfrac{1}{2} + \tfrac{1}{4} + \tfrac{1}{8} + \tfrac{1}{16} + \cdots.$$

SOLUTION. Here $r = \tfrac{1}{2}$ so that $S = 1/(1 - \tfrac{1}{2}) = 2$.

EXAMPLE 4. Calculate the sum of the geometric series

$$S = 1 - \tfrac{1}{3} + \tfrac{1}{9} - \tfrac{1}{27} + \cdots.$$

SOLUTION. Here $r = -\tfrac{1}{3}$ so that $S = \dfrac{1}{1 - (-\tfrac{1}{3})} = \dfrac{1}{\tfrac{4}{3}} = \tfrac{3}{4}$.

Many people are surprised to learn that an infinite number of terms can add up to a finite number. It takes a bit of experience to be comfortable with this fact. However it is precisely the notion that an infinite sum can have a finite limit which enables us to explain Zeno's paradox, as we shall now do. We shall prove that the man will really reach the door in a finite time.

Suppose the man in Figure 2.1.1 starts walking toward the door at the fixed velocity of 5 feet per second. Let us calculate the time it takes him to walk to the door, using Zeno's argument. Since (velocity) $\times$ (time) = distance, we have $t = $ distance/velocity, where t stands for time. Thus, it takes the man $t = (5 \text{ ft})/(5 \text{ ft/sec}) = 1$ sec to walk to the point 5 feet from the door (recall that he starts 10 feet from the door). To walk to the next point $2\frac{1}{2}$ ft from the door, it takes $(2\frac{1}{2} \text{ ft})/(5 \text{ ft/sec}) = \frac{1}{2}$ sec. The next point takes $(1\frac{1}{4} \text{ ft})/(5 \text{ ft/sec}) = \frac{1}{4}$ sec to reach. It is clear that to reach succeeding points each half the distance to the door from the preceding point, the time taken will be $\frac{1}{8}$ sec, $\frac{1}{16}$ sec. . . . , $(\frac{1}{2})^n$ sec, ... Thus the total time he takes to walk to the door is

$$t = 1 + \tfrac{1}{2} + \tfrac{1}{4} + \tfrac{1}{8} + \tfrac{1}{16} + \cdots = 2 \quad \text{seconds}$$

since this is nothing but the sum of a geometric series with $r = \frac{1}{2}$. Hence the man will reach the door in 2 seconds, as is certainly not surprising. Therefore, we see that with the concept of a limit at infinity, Zeno's paradox is really no paradox at all.

In Problem 15 we ask you to "explain" the seeming paradox in the original version of Zeno's paradox: the race between Achilles and the hare.

PROBLEMS 2.6

In Problems 1–10 calculate the sums of the given geometric series.

1. $1 + \dfrac{1}{4} + \dfrac{1}{4^2} + \dfrac{1}{4^3} + \cdots$

2. $1 - \dfrac{1}{2} + \dfrac{1}{4} - \dfrac{1}{8} + \dfrac{1}{16} - \cdots$

3. $1 + \dfrac{1}{10} + \dfrac{1}{100} + \dfrac{1}{1000} + \cdots$

4. $1 - \dfrac{1}{10} + \dfrac{1}{100} - \dfrac{1}{1000} + \cdots$

5. $1 + \dfrac{1}{\pi} + \dfrac{1}{\pi^2} + \dfrac{1}{\pi^3} + \cdots$

6. $1 + 0.7 + 0.7^2 + 0.7^3 + \cdots$

7. $1 - 0.62 + 0.62^2 - 0.62^3 + 0.62^4 - \cdots.$

8. $\dfrac{1}{4} + \dfrac{1}{16} + \dfrac{1}{64} + \cdots \left[\textit{Hint: factor out the term } \dfrac{1}{4}. \right]$

9. $\dfrac{3}{5} - \dfrac{3}{25} + \dfrac{3}{125} - \cdots$

10. $\dfrac{1}{9} + \dfrac{1}{27} + \dfrac{1}{81} + \cdots$

11. How large must n be in order that $(\frac{1}{2})^n < 0.01$?

☐ 12. How large must n be in order that $(0.8)^n < 0.01$?

☐ 13. How large must n be in order that $(.99)^n < 0.01$?

14. If $x > 1$, show that

$$1 + \frac{1}{x} + \frac{1}{x^2} + \frac{1}{x^3} + \cdots = \frac{x}{x - 1}.$$

*15. Suppose that in the original version of Zeno's paradox the tortoise is moving at a rate of 1 km/hr while Achilles is running at a rate of 201 km/hr. Give the tortoise a 40 km head start.

(a) Show, using the arguments of this section, that Achilles will really overtake the tortoise.

(b) How long will it take?

*2.7 One-Sided Limits

In Example 2.2.6 we tried to calculate $\lim_{x\to 0} \sqrt{x}$ but concluded that this limit did not exist since $\sqrt{x}$ was not defined for $x < 0$. On the other hand, if we *require* x to approach zero from the right only (that is, x stays >0), then a limit does exist. This leads to the following definition.

DEFINITION 1. (i) Suppose that $f(x)$ is defined near x_0 for $x > x_0$ and that as x gets close to x_0 (with $x > x_0$), $f(x)$ gets arbitrarily close to L. Then we say that L is the *right-hand limit* of $f(x)$ as x approaches x_0 and we write

$$\lim_{x\to x_0^+} f(x) = L. \tag{1}$$

(ii) Suppose that $f(x)$ is defined near x_0 for $x < x_0$ and that as x gets close to x_0 (with $x < x_0$), $f(x)$ gets arbitrarily close to L. Then we say that L is the *left-hand limit* of $f(x)$ as x approaches x_0 and we write

$$\lim_{x\to x_0^-} f(x) = L. \tag{2}$$

As in Section 2.2, we stress that these definitions are not mathematically precise and are only intended to appeal to your intuition. We will return to one-sided limits in Section 10.3 and will give more precise definitions there.

EXAMPLE 1. In the example discussed above, we have $\lim_{x\to 0^+} \sqrt{x} = 0$ and $\lim_{x\to 0^-} \sqrt{x}$ does not exist.

EXAMPLE 2. In Example 2.5.5, we showed that $\lim_{x\to 0} 1/x$ does not exist. However, from Table 2.5.4, we easily see that

$$\lim_{x\to 0^+} \frac{1}{x} = \infty \quad \text{and} \quad \lim_{x\to 0^-} \frac{1}{x} = -\infty.$$

EXAMPLE 3. In Example 2.4.7, we considered the function $f(x) = [x]$ whose graph was given in Figure 2.4.4. Let n be any integer. Then, from that graph, we find that

$$\lim_{x\to n^+} [x] = n \quad \text{and} \quad \lim_{x\to n^-} [x] = n - 1.$$

For example, if $n = 5$ we have $\lim_{x\to 5^+} [x] = 5$ and $\lim_{x\to 5^-} [x] = 4$.

EXAMPLE 4. In Example 2.2.10 we discussed the function

$$f(x) = \frac{|x|}{x} = \begin{cases} 1, & x > 0 \\ -1, & x < 0 \end{cases}.$$

Then

$$\lim_{x\to 0^+} \frac{|x|}{x} = 1 \quad \text{and} \quad \lim_{x\to 0^-} \frac{|x|}{x} = -1.$$

In Example 3, $\lim_{x \to n} [x]$ does not exist since we get different values when $x \to n$ from the left and from the right. For the same reason $\lim_{x \to 0} |x|/x$ does not exist. In general, we have the following theorem whose proof is given in Section 10.3.

THEOREM 1. $\lim_{x \to x_0} f(x) = L$ exists if and only if $\lim_{x \to x_0^+} f(x) = L$ and $\lim_{x \to x_0^-} f(x) = L$. That is, the limit exists if and only if the right and left hand limits exist and are equal.

EXAMPLE 5. Let

$$f(x) = \begin{cases} x^2, & x < 1 \\ x^3, & x \geq 1 \end{cases}.$$

This function is graphed in Figure 1. Then, since for $x > 1$, $f(x) = x^3$, we have

$$\lim_{x \to 1^+} f(x) = \lim_{x \to 1^+} x^3 = 1.$$

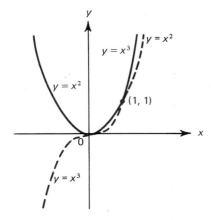

Figure 1

Similarly,

$$\lim_{x \to 1^-} f(x) = \lim_{x \to 1^-} x^2 = 1.$$

Since these two limits are equal, we have

$$\lim_{x \to 1} f(x) = 1.$$

EXAMPLE 6. Let

$$f(x) = \begin{cases} x + 1, & x > 0 \\ x - 1, & x < 0 \end{cases}.$$

This function is graphed in Figure 2. We immediately see that

$$\lim_{x \to 0^+} f(x) = \lim_{x \to 0^+} x + 1 = 1$$

and

$$\lim_{x \to 0^-} f(x) = \lim_{x \to 0^-} x - 1 = -1.$$

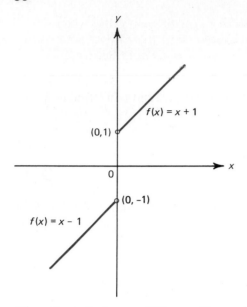

Figure 2

Since these limits are different, $\lim_{x \to 0} f(x)$ does not exist.

Using the notion of right- and left-hand limits, we can give a more precise definition of a jump discontinuity at x_0.

DEFINITION 2. The function $f(x)$ has a *jump discontinuity* at x_0 if

(i) $\lim_{x \to x_0^+} f(x)$ and $\lim_{x \to x_0^-} f(x)$ both exist, are finite, and are unequal, or

(ii) $f(x_0)$ exists and is finite, both $\lim_{x \to x_0^+} f(x)$ and $\lim_{x \to x_0^-} f(x)$ exist, are finite, and are equal, but $\lim_{x \to x_0^+} f(x) \neq f(x_0)$.

Remark. In Example 2.4.8 the given function has a jump discontinuity of the type (ii).

To conclude this section, we use the notion of one-sided limits to define continuity over the closed interval $[a, b]$.

DEFINITION 3. The function f is continuous in the closed interval $[a, b]$ if

(i) f is continuous in (a, b).

(ii) $f(a)$ and $f(b)$ both exist and are finite.

(iii) $\lim_{x \to a^+} f(x) = f(a)$ and $\lim_{x \to b^-} f(x) = f(b)$.

EXAMPLE 7. The function $f(x) = \sqrt{x}$ is continuous in $[0, 1]$ since $\lim_{x \to 0^+} \sqrt{x} = 0 = f(0)$ and $\lim_{x \to 1^-} \sqrt{x} = 1 = f(1)$.

EXAMPLE 8. The function $1/x$ is not continuous in $[0, 1]$ since $1/x$ is not defined at 0. However, it is continuous in $(0, 1)$.

PROBLEMS 2.7

In Problems 1–24 find the indicated limits, if they exist.

1. $\lim_{x \to 2^+} \sqrt{x - 2}$

2. $\lim_{x \to 2} \sqrt{x - 2}$

3. $\lim_{x \to 0^+} \sqrt[3]{x}$

4. $\lim\limits_{x\to 0^-} \sqrt[3]{x}$

5. $\lim\limits_{x\to 0} \dfrac{1}{x^3}$

6. $\lim\limits_{x\to 0^+} \dfrac{1}{x^3}$

7. $\lim\limits_{x\to 0^-} \dfrac{1}{x^3}$

8. $\lim\limits_{x\to -2^+} \sqrt{x+2}$

9. $\lim\limits_{x\to -2^-} \sqrt{x+2}$

10. $\lim\limits_{x\to -1^+} \sqrt{-1-x}$

11. $\lim\limits_{x\to -1^-} \sqrt{-1-x}$

12. $\lim\limits_{x\to 1^+} \dfrac{|x-1|}{x-1}$

13. $\lim\limits_{x\to 1^-} \dfrac{|x-1|}{x-1}$

14. $\lim\limits_{x\to 1^+} \sqrt{(x-1)(x-2)}$

15. $\lim\limits_{x\to 1^-} \sqrt{(x-1)(x-2)}$

16. $\lim\limits_{x\to 1} \sqrt{(x-1)(x-2)}$

17. $\lim\limits_{x\to -2^+} \dfrac{3x\,|x+2|}{x+2}$

18. $\lim\limits_{x\to -2^-} \dfrac{3x\,|x+2|}{x+2}$

19. $\lim\limits_{x\to (3/2)^+} [x]$

20. $\lim\limits_{x\to 2^-} [x]$

21. $\lim\limits_{x\to -7^+} [x]$

22. $\lim\limits_{x\to 1^+} \dfrac{1}{x^2-1}$

23. $\lim\limits_{x\to 1^-} \dfrac{1}{x^2-1}$

24. $\lim\limits_{x\to 1} \dfrac{1}{x^2-1}$.

25. Let

$$f(x) = \begin{cases} x-2, & x>2 \\ 0, & x\le 2 \end{cases}.$$

Find $\lim_{x\to 2^+} f(x)$ and $\lim_{x\to 2^-} f(x)$.

26. Let

$$f(x) = \begin{cases} x^4, & x<1 \\ x^5, & x\ge 1 \end{cases}.$$

Show that $\lim_{x\to 1} f(x) = 1$ by showing that the right- and left-hand limits exist and are equal.

27. Let

$$f(x) = \begin{cases} |x|, & \text{for } x\le 2 \\ [x], & \text{for } x>2 \end{cases}.$$

Show that $\lim_{x\to 2} f(x)$ exists.

28. Let

$$f(x) = \begin{cases} 0, & x<0 \\ x^2, & 0\le x<2 \\ 4, & x\ge 2 \end{cases}.$$

Calculate $\lim_{x\to 0^-} f(x)$, $\lim_{x\to 0^+} f(x)$, $\lim_{x\to 2^-} f(x)$, and $\lim_{x\to 2^+} f(x)$.

29. Show that the function in Problem 27 is continuous in $[0, 2]$ and $(0, 3)$ but not in $[0, 3]$.

30. Show that the function in Problem 28 is continuous in $[0, 4]$.

2.8 Tangent Lines and Derivatives

Dating from before the time of the great Greek scientist Archimedes† (287–212 B.C.), mathematicians were concerned with the problem of finding the unique tangent line

†We shall refer to the work of Archimedes a great deal in Chapter 4.

(if one exists) to a given curve at a given point on the curve.† Some typical tangent lines are drawn in Figure 1.

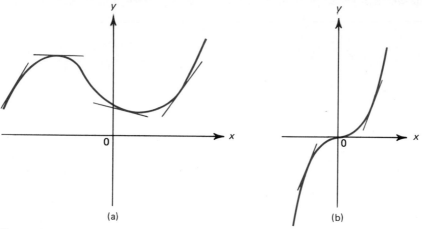

(a)   (b)

Figure 1

The Greeks knew how to find the line tangent to a circle at a given point, using the fact that for a circle, the tangent line is perpendicular to the radius at the given point (see Figure 2). The Greeks also discovered how to construct tangent lines to other particular curves, and Archimedes himself devoted a major part of one of his books (*On Spirals*) to the tangent problem for a special curve that is called the *spiral of Archimedes*.‡

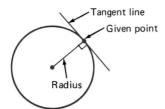

Figure 2

However, as more and more curves were studied, it became increasingly difficult to treat the large number of special cases and a general method was sought for solving all such problems. Unfortunately, these early attempts met with failure and it wasn't until the independent discoveries of Isaac Newton (1642–1727) and Gottfried Leibniz (1646–1716) that the problem was resolved.

Before discussing this discovery which is the basis of modern calculus, we will describe some of the geometric properties of tangent lines. We recall from our discussion of slope in Section 1.4 that if the straight line $y = mx + b$ has a positive slope ($m > 0$), then y is an *increasing* function of x, and as a point moves along the

---

† We will not formally define a tangent line here, but will rely on our intuition. A formal definition of a tangent line will be given in Section 2.9.

‡ We will discuss the spiral of Archimedes in Chapter 11 (see Example 11.2.9).

line from left to right, it also moves *upward*. If the line has a negative slope ($m < 0$), then y is a *decreasing* function of x, and as a point moves along the line from left to right, it also moves *downward*. This is illustrated in Figure 3.

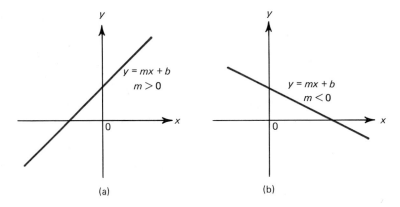

(a)   (b)   Figure 3

Now consider the graph of the function f drawn in Figure 4. For each point on this graph, there is a unique tangent line.† Each of these tangent lines has a slope. This allows us to define a new function.

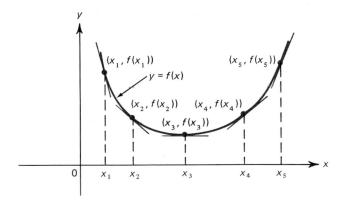

Figure 4

DEFINITION 1 (First Definition of the Derivative). Let f be a function whose graph has a unique tangent line at the point $(x, f(x))$. Then the *derivative* of f, written f', is the function that assigns to the value x the slope of the line tangent to the graph of f at the point $(x, f(x))$. That is,

$$f'(x) = \text{slope of line tangent to the graph of } f \text{ at the point } (x, f(x)). \qquad \textbf{(1)}$$

The remainder of this chapter and most of the next one will be concerned with the calculation of derivatives. However we can get an idea of what the function f' looks like by examining the graph of the function f.

†There are functions whose graphs do *not* have unique tangent lines at a given point. See Example 3.

EXAMPLE 1. Graph the derivative of the function whose graph is sketched in Figure 4.

SOLUTION. At the point $(x_1, f(x_1))$, the tangent line has a negative slope so that $f'(x_1) < 0$. At the point $(x_2, f(x_2))$ the slope of the tangent line is still negative, but it is less negative (i.e., closer to 0) than at x_1 since the curve is not decreasing as sharply. Thus $f'(x_1) < f'(x_2) < 0$. At $(x_3, f(x_3))$ the tangent line is horizontal which means that $f'(x_3) = 0$. At $(x_4, f(x_4))$ and $(x_5, f(x_5))$ the tangent lines have positive slopes and since the curve is climbing more steeply at $(x_5, f(x_5))$, we have $0 < f'(x_4) < f'(x_5)$. Thus, as we move from left to right, the derivative function f' increases from negative values to zero to positive values. This gives us enough information to roughly draw its graph (Figure 5).

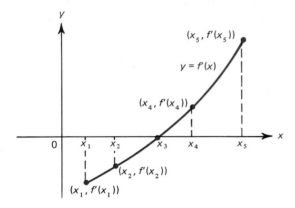

Figure 5

EXAMPLE 2. Draw the graph of the derivative of the function whose graph is sketched in Figure 6.

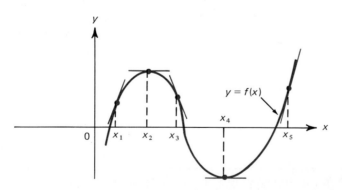

Figure 6

SOLUTION. To the left of x_1 the tangents have positive slopes. When $x = x_2$ the slope of the tangent line is zero and between x_2 and x_4, the slopes are negative. At x_4 the slope is again zero while to the right of x_4, the slopes are positive. Hence, the derivative $f'(x)$ starts positive then decreases to negative values and then increases to again take positive values. This is depicted in Figure 7.

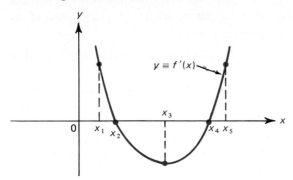

Figure 7

EXAMPLE 3. Consider the function f whose graph is given in Figure 8. At the point x_0, the derivative $f'(x_0)$ *is not defined* since the curve does not have a unique tangent

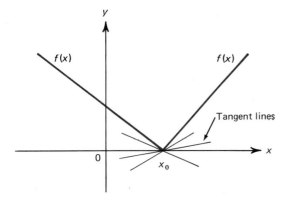

Figure 8

line at that point (it has many of them). According to Definition 1, f' is not *defined* at x_0. At any other point, the tangent line coincides with part of the curve. If $x < x_0$, then $f'(x)$ is a negative constant c_1 (the slope of the line making up the left-hand part of the function $f(x)$). For $x > x_0$, $f'(x)$ is a positive constant c_2. See Figure 9.

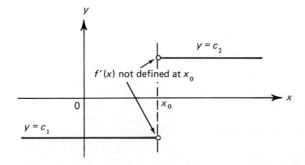

Figure 9

Remark. When we formally define a tangent line in the next section, our definition will imply that f has a tangent line at a point only if there is a unique such line at the point.

Thus, since the graph of the function in Figure 8 has more than one such line when $x = 0$, we say that the function has *no* tangent at 0.

EXAMPLE 4. Consider the function f whose graph is given in Figure 10a. At $(0, 0)$ the tangent line is the x-axis, which has the slope zero. For all other values of x the slope of the tangent line is positive. The graph of the function f' is sketched in Figure 10b. Note that here we have an example of a tangent line which *crosses* the curve at its point of tangency.

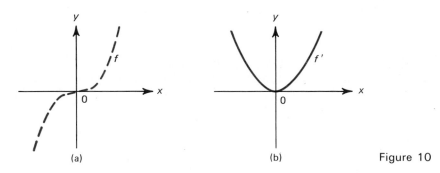

(a)   (b)   Figure 10

PROBLEMS 2.8

1. Determine whether the slopes of the straight lines in Figure 11 are positive, negative, or zero.

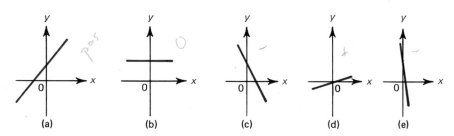

(a)   (b)   (c)   (d)   (e)

Figure 11

2. Sketch the graph of the derivative of the function whose graph is the semicircle in Figure 12.

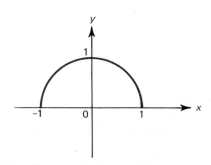

Figure 12

In Problems 3–10 sketch the graph of the derivative function f'.

3.

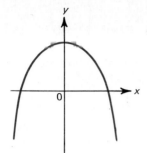

Figure 13

4.

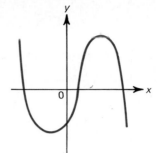

Figure 14

5.

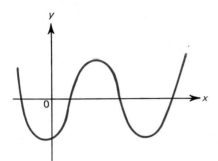

Figure 15

6.

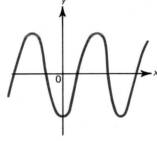

Figure 16

7.

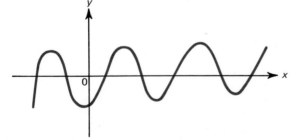

Figure 17

8.

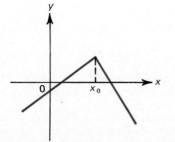

Figure 18

*9.

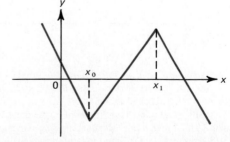

Figure 19

*10.

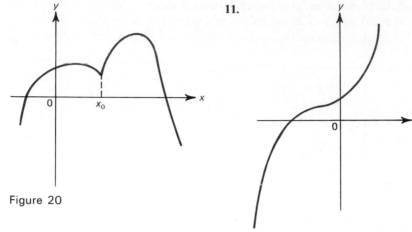

Figure 20

11.

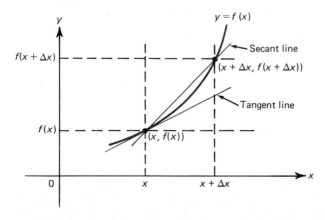

Figure 21

2.9 Tangent Lines and Derivatives (Continued)

Having defined the derivative of a function f intuitively, we now begin the task of calculating it. The method we give below is essentially the method of Newton[†] and Leibniz[‡] which resolved the tangent problem posed so long ago by the Greek mathematicians.[§]

    Let us consider the function $y = f(x)$, a part of whose graph is given in Figure 1. In order to calculate the derivative function, we must calculate the slope of the line tangent to the curve at each point of the curve at which there is a unique tangent line. Let $(x, f(x))$ be such a point. From now on we will assume that f is defined "near" x.

Figure 1

[†] *Mathematical Principles of Natural Philosophy* (*Principia*), published in 1687.
[‡] *A New Method for Maxima and Minima, and Also for Tangents, Which is not Obstructed by Irrational Quantities,* published in 1684.
[§] Actually, this method was first used by the great French mathematician Pierre de Fermat (1601–1655) in his *Method of Finding Maxima and Minima* published in 1629. However Fermat did not explain his procedure satisfactorily and it was left to Newton and Leibniz to explain the method and apply it to the calculation of tangent lines.

If Δx is a small number (positive or negative) then $x + \Delta x$ will be close to x. In moving from x to $(x + \Delta x)$, the values of f will move from $f(x)$ to $f(x + \Delta x)$. Now look at the straight line in Figure 1, called a *secant line,* joining the points $(x, f(x))$ and $(x + \Delta x, f(x + \Delta x))$. What is its slope? If we define $\Delta y = f(x + \Delta x) - f(x)$ and if we use m_s to denote the slope of such a secant line, we have, from Section 1.4,

$$m_s = \frac{\text{change in } y}{\text{change in } x} = \frac{f(x + \Delta x) - f(x)}{(x + \Delta x) - x} = \frac{f(x + \Delta x) - f(x)}{\Delta x} = \frac{\Delta y}{\Delta x}. \tag{1}$$

What does this have to do with the slope of the tangent line? The answer is suggested in Figure 2. From this illustration, we see that as Δx gets smaller, the secant line gets

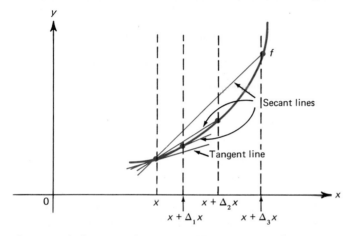

Figure 2

closer and closer to the tangent line. Put another way, as Δx approaches zero, the slope of the secant line approaches the slope of the tangent line. But the slope of the tangent line at the point $(x, f(x))$ is the derivative, $f'(x)$. We therefore have

$$f'(x) = \lim_{\Delta x \to 0} m_s = \lim_{\Delta x \to 0} \frac{\Delta y}{\Delta x} = \lim_{\Delta x \to 0} \frac{f(x + \Delta x) - f(x)}{\Delta x}. \tag{2}$$

DEFINITION 1 (Second Definition of the Derivative).   The derivative f' of the function f is the function defined by

$$\boxed{f'(x) = \lim_{\Delta x \to 0} \frac{\Delta y}{\Delta x} = \lim_{\Delta x \to 0} \frac{f(x + \Delta x) - f(x)}{\Delta x}.} \tag{3}$$

The derivative f' is defined for every value of x such that $f(x)$ is defined and the limit in (3) exists. The process of taking the derivative of a function is called *differentiation,* and a function which possesses a derivative at x is said to be *differentiable* at x.

We emphasize that *the derivative of a function is another function.* The value of the derivative at a given number x is the limit obtained in (3). Note that the above definition does *not* say anything about tangent lines. Simply put, f is differentiable at x if the limit in (3) exists. However, we can now formally define what we mean by a tangent line.

DEFINTION 2. If the limit in (3) exists, we say that the function *f* has a *tangent line* at the point $(x, f(x))$. This is the line passing through the point $(x, f(x))$ with slope $f'(x)$.

> **Remark.** According to this definition, a graph *cannot* have more than one tangent line at the point $(x, f(x))$ since this would imply that $f'(x)$ took on more than one value, thereby contradicting the fact that if the limit in (3) exists, then it must be unique. This explains why the graph of the function in Example 2.8.3 has *no* tangent at 0.

EXAMPLE 1. Let $y = f(x) = 3x + 5$. Calculate $f'(x)$.

SOLUTION. To solve this problem and the ones that follow, we simply use formula (3). For

$$f(x) = 3x + 5, \qquad f(x + \Delta x) = 3(x + \Delta x) + 5.$$

Then

$$f'(x) = \lim_{\Delta x \to 0} \frac{f(x + \Delta x) - f(x)}{\Delta x} = \lim_{\Delta x \to 0} \frac{[3(x + \Delta x) + 5] - [3x + 5]}{\Delta x}$$

$$= \lim_{\Delta x \to 0} \frac{3x + 3\Delta x + 5 - 3x - 5}{\Delta x} = \lim_{\Delta x \to 0} \frac{3\Delta x}{\Delta x} = \lim_{\Delta x \to 0} 3 = 3.$$

This answer is not surprising. It simply says that the slope of the line $y = 3x + 5$ is equal to the constant function 3.

Before giving further examples, if $y = f(x)$ we introduce the additional symbols dy/dx or df/dx to denote the derivative.

$$f'(x) = \frac{df}{dx} = \frac{dy}{dx} = \lim_{\Delta x \to 0} \frac{\Delta y}{\Delta x} = \lim_{\Delta x \to 0} \frac{f(x + \Delta x) - f(x)}{\Delta x}. \tag{4}$$

The symbol dy/dx is read *"the derivative of y with respect to x."* We emphasize that dy/dx is *not* a fraction. At this point the symbols dy and dx have no meaning of their own. (We will define these symbols in Section 3.6.) There are other symbols for the derivative. We will often use the symbol y' or $y'(x)$ in place of f' or $f'(x)$. Thus, if $y = f(x)$, we may denote the derivative in four different ways:†

$$f'(x) = y'(x) = \frac{df}{dx} = \frac{dy}{dx}. \tag{5}$$

EXAMPLE 2. Calculate the derivative of the function $y = x^2$. What is the equation of the line tangent to this curve at the point $(3, 9)$?

---

† Newton (in England) and Leibniz (in Germany) independently discovered (in the 1670s) the equation for the slope of the tangent line given in this section. Newton used the symbol $\dot{y}$ (read "y dot") and Leibniz used the symbol dy/dx to indicate the derivative. There was a raging controversy, never resolved, over who had made this momentous discovery first. This controversy was so intense that, to some extent, British mathematicians were alienated from mathematicians in the rest of Europe until well into the eighteenth century.

SOLUTION. For $y = f(x) = x^2$, $f(x + \Delta x) = (x + \Delta x)^2$. Then

$$\frac{dy}{dx} = \lim_{\Delta x \to 0} \frac{f(x + \Delta x) - f(x)}{\Delta x} = \lim_{\Delta x \to 0} \frac{(x + \Delta x)^2 - x^2}{\Delta x}$$

$$= \lim_{\Delta x \to 0} \frac{x^2 + 2x\,\Delta x + \Delta x^2 - x^2}{\Delta x} = \lim_{\Delta x \to 0} \frac{2x\,\Delta x + \Delta x^2}{\Delta x}$$

$$= \lim_{\Delta x \to 0} \frac{\Delta x(2x + \Delta x)}{\Delta x} = \lim_{\Delta x \to 0} (2x + \Delta x) = 2x.$$

At every point of the form $(x, f(x)) = (x, x^2)$, the slope of the line tangent to the curve is $2x$. For $x = 3$, $2x = 6$. Therefore the slope of the tangent line at the point $(3, 9)$ is 6; that is, $f'(3) = 6$. We can now find the equation of the tangent line since it passes through the point $(3, 9)$ and has the slope 6. We have

$$\frac{y - 9}{x - 3} = 6 \qquad \text{or} \qquad y = 6x - 9.$$

Knowing the derivative of $y = x^2$ helps us to graph the curve. First, we note that x^2 is always positive. If $x < 0$, then $y' = dy/dx = 2x < 0$ so the tangent lines have negative slopes. For $x = 0$, $dy/dx = 2 \cdot 0 = 0$ so that the tangent line is horizontal. For $x > 0$, $dy/dx = 2x > 0$ and the tangent lines have positive slopes. Using this information, we obtain the graph of Figure 3.

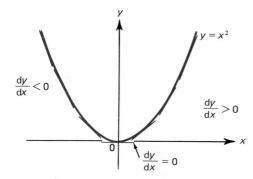

Figure 3

It is interesting to see how the slopes of the secant lines approach the slope of the tangent line in this problem. For $x_0 = 3$, we have, from (3),

$$f'(3) = \lim_{\Delta x \to 0} \frac{(3 + \Delta x)^2 - 9}{\Delta x}. \tag{6}$$

The slope of a secant line is $[(3 + \Delta x)^2 - 9]/3$. Table 1 illustrates how quickly the slopes of secant lines approach the value 6, which is the slope of the tangent line, as $x \to 0$.

TABLE 1

Δx	$3 + \Delta x$	$(3 + \Delta x)^2$	$(3 + \Delta x)^2 - 9$	$\dfrac{(3 + \Delta x)^2 - 9}{\Delta x} = $ slope of secant line
0.5	3.5	12.25	3.25	6.5
−0.5	2.5	6.25	−2.75	5.5
0.1	3.1	9.61	0.61	6.1
−0.1	2.9	8.41	−0.59	5.9
0.01	3.01	9.0601	0.0601	6.01
−0.01	2.99	8.9401	−0.0599	5.99
0.0001	3.0001	9.00060001	0.00060001	6.0001
−0.0001	2.9999	8.99940001	−0.00059999	5.9999

□**EXAMPLE 3.**   Find the equation of the line tangent to the curve $y = 2x^3$ at the point (2, 16).

SOLUTION. Before calculating the derivative dy/dx explicitly, we will try to estimate the slope of the tangent line $f'(2)$. Since $f(x) = 2x^3$, $f(x + \Delta x) = 2(x + \Delta x)^3$. Then, from (3) we have

$$f'(2) = \lim_{\Delta x \to 0} \frac{f(2 + \Delta x) - f(2)}{\Delta x} = \lim_{\Delta x \to 0} \frac{2(2 + \Delta x)^3 - 2 \cdot 2^3}{\Delta x}$$

$$= \lim_{\Delta x \to 0} \frac{2(2 + \Delta x)^3 - 16}{\Delta x}.$$

We tabulate values for $[2(2 + \Delta x)^3 - 16]/\Delta x$ in Table 2. It appears from the table

TABLE 2

Δx	$2 + \Delta x$	$(2 + \Delta x)^3$	$2(2 + \Delta x)^3$	$2(2 + \Delta x)^3 - 16$	$\dfrac{2(2 + \Delta x)^3 - 16}{\Delta x}$
0.5	2.5	15.625	31.25	15.25	30.5
−0.5	1.5	3.375	6.75	−9.25	18.5
0.1	2.1	9.261	18.522	2.522	25.22
−0.1	1.9	6.859	13.718	−2.282	22.82
0.05	2.05	8.615125	17.23025	1.23025	24.605
−0.05	1.95	7.414875	14.82975	−1.17025	23.405
0.01	2.01	8.120601	16.241202	0.241202	24.1202
−0.01	1.99	7.880599	15.761198	−0.238802	23.8802
0.001	2.001	8.012006001	16.024012	0.024012	24.012
−0.001	1.999	7.988005999	15.976012	−0.023988	23.988
0.0001	2.0001	8.00120006	16.00240012	0.00240012	24.0012
−0.0001	1.9999	7.99880006	15.99760012	−0.00239988	23.9988

that $f'(2) = 24$. Then the sought after tangent line is the line passing through the point (2, 16) with slope 24.

$$\frac{y - 16}{x - 2} = 24 \quad \text{or} \quad y - 16 = 24x - 48 \quad \text{or} \quad y = 24x - 32.$$

We now calculate $dy/dx = f'(x)$ and verify that $f'(2) = 24$. To do this, we need the following fact which follows from the binomial theorem (see Appendix 4): for any real numbers a and b,

$$(a + b)^3 = a^3 + 3a^2b + 3ab^2 + b^3. \tag{7}$$

Now, for $f(x) = 2x^3$,

$$f'(x) = \lim_{\Delta x \to 0} \frac{2(x + \Delta x)^3 - 2x^3}{\Delta x} = \lim_{\Delta x \to 0} \frac{2(x^3 + 3x^2\,\Delta x + 3x\,\Delta x^2 + \Delta x^3) - 2x^3}{\Delta x}$$

$$= \lim_{\Delta x \to 0} \frac{6x^2\,\Delta x + 6x\,\Delta x^2 + 2\,\Delta x^3}{\Delta x} = \lim_{\Delta x \to 0} \frac{\Delta x(6x^2 + 6x\,\Delta x + 2\,\Delta x^2)}{\Delta x}$$

$$= \lim_{\Delta x \to 0} (6x^2 + 6x\,\Delta x + 2\,\Delta x^2) = 6x^2. \tag{8}$$

Hence $dy/dx = 6x^2$ and, when $x = 2$, $dy/dx = 6 \cdot 2^2 = 6 \cdot 4 = 24$, as expected.

We can again use our knowledge of the derivative to obtain the graph of the curve. First we note that if $x < 0$, then $2x^3 < 0$ and if $x > 0$, then $2x^3 > 0$. However, $df/dx = d(2x^3)/dx = 6x^2$ which is always positive. This means that all tangent lines to the curve have positive slopes except at the point $(0, 0)$. Thus the curve is increasing. We put these facts together to arrive at the graph of $y = 2x^3$ in Figure 4.

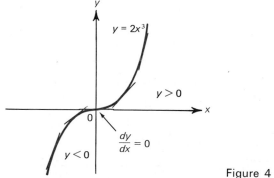

Figure 4

☐ **EXAMPLE 4.** Find the derivative of $y = \sqrt{x}$ and calculate the slope of the tangent line at the point $(9, 3)$.

SOLUTION. We first estimate the slope of the tangent line at the point $(9, 3)$. For $f(x) = \sqrt{x}$, $f(x + \Delta x) = \sqrt{x + \Delta x}$. Then

$$f'(9) = \lim_{\Delta x \to 0} \frac{f(9 + \Delta x) - f(9)}{\Delta x} = \lim_{\Delta x \to 0} \frac{\sqrt{9 + \Delta x} - \sqrt{9}}{\Delta x}$$

$$= \lim_{\Delta x \to 0} \frac{\sqrt{9 + \Delta x} - 3}{\Delta x}. \tag{9}$$

Values of the quotient in the limit are tabulated in Table 3. It seems as if the slopes of

TABLE 3

Δx	$9 + \Delta x$	$\sqrt{9 + \Delta x}$	$\sqrt{9 + \Delta x} - 3$	$\dfrac{\sqrt{9 + \Delta x} - 3}{\Delta x}$
0.5	9.5	3.082207001	0.0822070014	0.1644140030
−0.5	8.5	2.915475947	−0.0845240525	0.1690481052
0.1	9.1	3.016620626	0.0166206258	0.1662062580
−0.1	8.9	2.983286778	−0.0167132219	0.1671322197
0.01	9.01	3.001666204	0.0016662039	0.1666203960
−0.01	8.99	2.99833287	−0.0016671298	0.1667129890
0.001	9.001	3.000166662	0.000166662	0.1666620350
−0.001	8.999	2.999833329	−0.0001666713	0.1666713000

the secant lines are approaching the limiting value $0.16666 \cdots = \frac{1}{6}$. Using our estimate $\frac{1}{6}$ as the slope of the tangent line passing through the point $(9, 3)$, we obtain

$$\frac{y - 3}{x - 9} = \frac{1}{6} \quad \text{or} \quad 6y - 18 = x - 9.$$

Hence

$$6y = x + 9 \quad \text{or} \quad y = \frac{x}{6} + \frac{3}{2}.$$

We now calculate dy/dx formally. To do this, we need the fact that for any real numbers a and b,

$$(a - b)(a + b) = a^2 - b^2. \tag{10}$$

We have

$$f'(x) = \lim_{\Delta x \to 0} \frac{f(x + \Delta x) - f(x)}{\Delta x} = \lim_{\Delta x \to 0} \frac{\sqrt{x + \Delta x} - \sqrt{x}}{\Delta x}.$$

We now multiply numerator and denominator of this quotient by $\sqrt{x + \Delta x} + \sqrt{x}$. Notice that setting $a = \sqrt{x + \Delta x}$ and $b = \sqrt{x}$ in (10) yields

$$(\sqrt{x + \Delta x} - \sqrt{x})(\sqrt{x + \Delta x} + \sqrt{x}) = (\sqrt{x + \Delta x})^2 - (\sqrt{x})^2$$
$$= x + \Delta x - x = \Delta x.$$

Thus

$$f'(x) = \lim_{\Delta x \to 0} \frac{\sqrt{x + \Delta x} - \sqrt{x}}{\Delta x} = \lim_{\Delta x \to 0} \frac{(\sqrt{x + \Delta x} - \sqrt{x})(\sqrt{x + \Delta x} + \sqrt{x})}{\Delta x(\sqrt{x + \Delta x} + \sqrt{x})}$$

$$= \lim_{\Delta x \to 0} \frac{\Delta x}{\Delta x(\sqrt{x + \Delta x} + \sqrt{x})} = \lim_{\Delta x \to 0} \frac{1}{\sqrt{x + \Delta x} + \sqrt{x}} = \frac{1}{2\sqrt{x}}. \tag{11}$$

If $x = 9$, then

$$\frac{1}{2\sqrt{x}} = \frac{1}{2\sqrt{9}} = \frac{1}{2 \cdot 3} = \frac{1}{6},$$

as predicted. Note too that although the function $f(x) = \sqrt{x}$ is defined for $x \geq 0$, its derivative $1/2\sqrt{x}$ is only defined for $x > 0$ since for $x = 0$, the limit taken in the last step of (11) does not exist (explain why).

We can again make use of the derivative to graph the curve. First note that $\sqrt{x} > 0$ if $x > 0$ and $\sqrt{x}$ is not defined if $x < 0$. Look at the derivative $f'(x) = 1/2\sqrt{x}$. Since $\sqrt{x} > 0$, $1/2\sqrt{x} > 0$ so that all tangents to the curve have a positive slope and the function is increasing. But we also have

$$\lim_{x \to \infty} f'(x) = \lim_{x \to \infty} \frac{1}{2\sqrt{x}} = 0.$$

This tells us that as x increases, the derivative approaches zero, or, equivalently, the tangent lines approach the horizontal (a slope of zero). Thus we conclude that for large values of x, the curve $y = \sqrt{x}$ is nearly flat. On the other hand, since $\lim_{x \to 0^+} (1/2\sqrt{x}) = \infty$, the tangent line to the curve near $x = 0$ is nearly vertical.† We combine all this information in the graph in Figure 5.

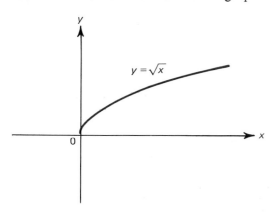

$y = \sqrt{x}$

Figure 5

EXAMPLE 5. Consider the function $y = |x|$. Since

$$|x| = \begin{cases} x, & x \geq 0 \\ -x, & x \leq 0 \end{cases},$$

we obtain the graph in Figure 6. To see if the graph f has a tangent line at the point $(0, 0)$, we calculate

$$f'(0) = \lim_{\Delta x \to 0} \frac{f(0 + \Delta x) - f(0)}{\Delta x} = \lim_{\Delta x \to 0} \frac{|0 + \Delta x| - |0|}{\Delta x} = \lim_{\Delta x \to 0} \frac{|\Delta x|}{\Delta x}.$$

But as we saw in Examples 2.2.10 and 2.7.4, this limit does not exist so that, as we have already seen, $|x|$ does not have a tangent line at $(0, 0)$. On the other hand, if $x \neq 0$ then the derivative does exist (see Problem 20).

Before leaving this section we point out that there is a close connection between continuity and differentiability.

---

† From Section 1.4: if a line has an infinite slope, then the line is parallel to the y-axis.

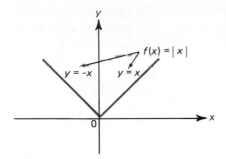

Figure 6

THEOREM 1. Let f be differentiable at x_0. Then f is continuous there.

This result appeals to our intuition. For if a function is differentiable at x_0 then its graph has a tangent line at the point $(x_0, f(x_0))$ and the curve would seem to move "smoothly" through that point.

PROOF. We need to show that $\lim_{x \to x_0} f(x) = f(x_0)$. This is the same as showing that

$$\lim_{x \to x_0} [f(x) - f(x_0)] = 0. \quad \text{(Why?)}$$

Define $\Delta x = x - x_0$. Then $f(x) = f(x_0 + \Delta x)$ and when $x \to x_0$ we have $\Delta x \to 0$. Then

$$\lim_{x \to x_0} [f(x) - f(x_0)] = \lim_{\Delta x \to 0} [f(x_0 + \Delta x) - f(x_0)]$$

$$= \lim_{\Delta x \to 0} \frac{[f(x_0 + \Delta x) - f(x_0)] \Delta x}{\Delta x}$$

$$= \lim_{\Delta x \to 0} \frac{f(x_0 + \Delta x) - f(x_0)}{\Delta x} \lim_{\Delta x \to 0} \Delta x$$

$$= f'(x_0) \cdot 0 = 0.$$

In the last step we used the fact that $f'(x_0)$ exists by assumption.

Note that the converse of this theorem is not true; that is, a function which is continuous at a point is *not* necessarily differentiable at that point. For example, in Example 2.4.6 we showed that $f(x) = |x|$ is continuous at 0. However, as Example 5 above shows, it is not differentiable there.

PROBLEMS 2.9

1. Consider the function $f(x) = 3x^2$.
   (a) For $x = 2$, calculate $f(x + \Delta x) = f(2 + \Delta x)$ for $\Delta x = 0.5$, $\Delta x = 0.1$, $\Delta x = 0.01$, $\Delta x = 0.001$, $\Delta x = -0.01$ and $\Delta x = -0.001$.
   (b) Calculate $[f(2 + \Delta x) - f(2)]/\Delta x$ for the values of Δx in part (a) and "guess" the value of $f'(2)$.
   (c) From the definition, calculate $f'(x)$, use this to compute $f'(2)$, and compare this with the answer you obtained in part (b).
   (d) What is the equation of the line tangent to the curve at the point $(2, 12)$?
   (e) Using the derivative, graph the curve.

▢**2.**   Carry out the steps in Problem 1 for the function $f(x) = 1/x$ at the point $(1, 1)$.
▢**3.**   Carry out the steps in Problem 1 for the function $f(x) = 5\sqrt{x}$ at the point $(1, 5)$.

In Problems 4–19 find the derivative of the given function and the equation of the tangent line to the curve at the given point.

4. $f(x) = 15x - 2$, $(1, 13)$ **5.** $f(x) = -4x + 6$, $(3, -6)$
6. $f(x) = (x + 1)^2$, $(1, 4)$ **7.** $f(x) = 2(x - 3)^2$, $(4, 2)$
8. $f(x) = x^2 - 1$, $(-2, 3)$ **9.** $f(x) = -x^2 + 3x + 5$, $(0, 5)$
10. $f(x) = -(x + 2)^3$, $(-1, -1)$ **11.** $f(x) = x^3 + 6$, $(-1, 5)$
12. $y = x^3 + x^2 + x + 1$, $(1, 4)$ **13.** $y = -\sqrt{x + 3}$, $(6, -3)$
14. $y = \sqrt{x - 2}$, $(6, 2)$ **15.** $y = \sqrt{2x}$, $(8, 4)$
16. $y = x^{3/2}$, $(1, 1)$ [*Hint:* $(a^{3/2} - b^{3/2})(a^{3/2} + b^{3/2}) = a^3 - b^3$.]

17. $y = \dfrac{x + x^2}{\sqrt{x}}$, $(1, 2)$.   **18.** $y = \dfrac{1}{\sqrt{x}}$, $(4, \tfrac{1}{2})$.

19. $y = x^4$, $(2, 16)$. [*Hint:* Use the binomial theorem.]

20. Let $y = |x|$. Calculate dy/dx for $x \neq 0$.
21. Show from the definition that if $y = mx + b$, then $dy/dx = m$.
22. Show that for any constants a, b, and c

$$\frac{d}{dx}(ax^2 + bx + c) = 2ax + b.$$

23. Show that for any constant α

$$\frac{d}{dx}\alpha x^3 = 3\alpha x^2.$$

24. Calculate the derivative of $\alpha x^3 + ax^2 + bx + c$ for any constants a, b, c, and α.
25. Let $f(x) = x$ and $g(x) = x^2$. For what values of x are the tangents to these curves parallel?
26. Let $f(x) = x^2$ and $g(x) = x^3$. For what values of x are the tangents to these curves parallel?
27. Let $y = f(x) = ax^2$ and $g(x) = 6x - 5$. For what value of a is the tangent to the graph of f parallel to the tangent to the graph of g when $x = 2$?
28. Let $f(x) = 1/\sqrt{x}$. For what value of x is the tangent to this curve parallel to the line $x + 8y = 10$?

2.10 The Derivative as a Rate of Change

In this section we look at the derivative from a different point of view. To begin the discussion, we again consider the straight line given by the equation

$$y = mx + b, \tag{1}$$

where m is the slope and b is the y-intercept (see Figure 1). Let us examine the slope more closely. It is defined as the change in y divided by the change in x:

$$m = \frac{\Delta y}{\Delta x}. \tag{2}$$

Implicit in this definition is the understanding that no matter which two points are chosen on the line, we obtain the same value for this ratio. Thus the slope of a straight

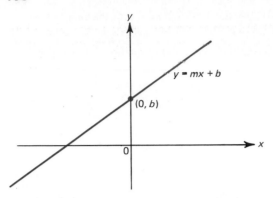

Figure 1

line could instead be referred to as *the rate of change of y with respect to x*. It tells us how many units y changes for every one unit that x changes. In fact, instead of following Euclid in defining a straight line as the shortest distance between two points, we could define a straight line as the curve whose rate of change is constant.

EXAMPLE 1. Let $y = 3x - 5$. Then in moving from the point $(1, -2)$ to the point $(2, 1)$ along the line, we see that as x has changed (increased) one unit, y has increased three units, corresponding to the slope $m = 3$.

We would like to be able to calculate rates of change for functions which are not straight lines. As a simple example of why such a thing is useful, suppose that an object is dropped from rest from a given height. The distance s the object has dropped after t seconds (ignoring air resistance) is given by the formula

$$s = \tfrac{1}{2}gt^2 \tag{3}$$

where $g = 9.8 \text{ m/sec}^2 = 32 \text{ ft/sec}^2$ is the acceleration due to gravity. We now ask: What is the velocity of the object after 2 seconds? To answer this question, it is first necessary to note that velocity can be regarded as a rate of change. Whether measured in meters per second, feet per second, or miles per hour, velocity is the ratio of change in distance (meters, feet, miles) with respect to the change in time (seconds, hours).

Table 1 gives (in column 3) the distance the object has fallen (in meters) after t seconds, tabulated every 0.2 seconds. Column 4 shows the distance the object has fallen in the previously elapsed 0.2 seconds. Column 5 indicates the average velocity over each 0.2-second interval of time. We have

$$\text{average velocity} = \frac{\text{distance fallen}}{\text{elapsed time}} = \frac{\text{change in distance}}{\text{change in time}} = \frac{\Delta s}{\Delta t}, \tag{4}$$

where in this problem $\Delta t = 0.2$ sec. From the table, we see that the average velocity of the object is constantly increasing (which is certainly not surprising). What is the exact velocity at $t = 2$? That is, instead of an average velocity we want to know the velocity after precisely 2 seconds. Table 1 will help answer this question. When $t = 2$, the object is falling faster than at any previous time. Let $v(t)$ denote the velocity at time t. Since 18.62 m/sec is the *average* velocity for the time $t = 1.8$ sec to $t = 2$ sec,

TABLE 1

t	t^2	$s = \frac{1}{2}gt^2 = 4.9t^2$ (m)	Δs = change in s since last measurement (m)	average velocity over last 0.2 sec, $\Delta s/\Delta t$ (m/sec)
0.	0.	0.	—	—
0.2	0.04	0.196	0.196	0.98
0.4	0.16	0.784	0.588	2.94
0.6	0.36	1.764	0.98	4.9
0.8	0.64	3.136	1.372	6.86
1.0	1.0	4.9	1.764	8.82
1.2	1.44	7.056	2.156	10.78
1.4	1.96	9.604	2.548	12.74
1.6	2.56	12.544	2.94	14.7
1.8	3.24	15.876	3.332	16.66
2.0	4.0	19.6	3.724	18.62
2.2	4.84	23.716	4.116	20.58
2.4	5.76	28.224	4.508	22.54
2.6	6.76	33.124	4.9	24.5
2.8	7.84	38.416	5.292	26.46
3.0	9.0	44.1	5.684	28.42

we have

$$\text{(velocity at } t = 2) = v(2) > 18.62 \quad \text{m/sec.} \tag{5}$$

Similarly, at $t = 2$, the object is falling more slowly than when $t > 2$. Therefore

$$v(2) < 20.58 \quad \text{m/sec} \tag{6}$$

which is the average velocity between the time $t = 2$ sec and $t = 2.2$ sec. Combining (4) and (5) we obtain the estimate

$$18.62 < v(2) < 20.58 \quad \text{m/sec.} \tag{7}$$

To narrow this down, we choose a smaller value for Δt. If $\Delta t = 0.05$ sec, we calculate the average velocity, $\Delta s/\Delta t$, for t between 1.95 and 2.0 sec and for t between 2.0 and 2.05 sec. Then $v(2)$ will lie between these two values (see Table 2). We then have the new estimate

$$19.355 < v(2) < 19.845 \quad \text{m/sec.} \tag{8}$$

TABLE 2

t	t^2	$s = 4.9t^2$ (m)	Δs (m)	$\dfrac{\Delta s}{\Delta t} = \dfrac{\Delta s}{0.05}$ (m/sec)
1.95	3.8025	18.63225	—	—
2.0	4.0	19.6	.96775	19.355
2.05	4.2025	20.59225	.99225	19.845

We continue this process by choosing an even smaller value for Δt, say $\Delta t = 0.001$. Then we calculate the values in Table 3. Therefore

$$19.5951 < v(2) < 19.6049 \quad \text{m/sec.} \tag{9}$$

TABLE 3

t	t^2	$s = 4.9t^2$ (m)	Δs (m)	$\dfrac{\Delta s}{\Delta t} = \dfrac{\Delta s}{0.001}$ (m/sec)
1.999	3.996001	19.5804049	—	—
2.0	4.0	19.6	0.0195951	19.5951
2.001	4.004001	19.6196049	0.0196049	19.6049

As Δt becomes smaller and smaller, we see that the average velocity $\Delta s/\Delta t$ gets closer and closer to the actual velocity at the value $t = 2$. This enables us to make the following definition.

DEFINITION 1. Let $s(t)$ denote the distance traveled by a moving object in t seconds. Then the *instantaneous velocity* after t seconds, denoted ds/dt or $s'(t)$, is given by

$$s'(t) = \frac{ds}{dt} = \lim_{\Delta t \to 0} \frac{\Delta s}{\Delta t} = \lim_{\Delta t \to 0} \frac{s(t + \Delta t) - s(t)}{\Delta t}. \tag{10}$$

In our example, $s(t) = 4.9t^2$ and $s(t + \Delta t) = 4.9(t + \Delta t)^2$. Thus

$$\frac{ds}{dt} = \lim_{\Delta t \to 0} \frac{4.9(t + \Delta t)^2 - 4.9t^2}{\Delta t} = 4.9 \lim_{\Delta t \to 0} \frac{(t + \Delta t)^2 - t^2}{\Delta t}$$

$$= 4.9 \lim_{\Delta t \to 0} \frac{t^2 + 2t\,\Delta t + \Delta t^2 - t^2}{\Delta t} = 4.9 \lim_{\Delta t \to 0} \frac{\Delta t(2t + \Delta t)}{\Delta t}$$

$$= 4.9 \lim_{\Delta t \to 0} (2t + \Delta t) = (4.9)(2t) = 9.8t.$$

After 2 seconds have elapsed, the velocity is $9.8(2) = 19.6$ m/sec which agrees with our estimates (7), (8) and (9).

As you have probably noticed, the velocity ds/dt given by (10) is the derivative of s with respect to t. Although our previous definitions of derivative involved the variables x and y, there is no change in this concept when we insert t in place of x and s in place of y. Thus we can think of a derivative as a velocity, or, more generally, as a rate of change. After Newton discovered the derivative, he used the word *fluxion* instead of velocity in his discussion of a moving object. (In more technical terminology, a moving particle is a particle "in flux.")

EXAMPLE 2. Let $P(t)$ denote the population of a colony of bacteria after t hours. If $P(t) = 100 + t^4$, how fast is the population growing after 3 hours?

SOLUTION. This problem is again a "rate of change" problem even though it does not make sense to speak about "velocity." We are asking for the "instantaneous rate of growth" of the population when $t = 3$ hours. This is given by

$$P'(t) = \frac{dP}{dt} = \lim_{\Delta t \to 0} \frac{P(t + \Delta t) - P(t)}{\Delta t}$$

$$= \lim_{\Delta t \to 0} \frac{(100 + (t + \Delta t)^4) - (100 + t^4)}{\Delta t} = \lim_{\Delta t \to 0} \frac{(t + \Delta t)^4 - t^4}{\Delta t}. \tag{11}$$

From the binomial theorem (see Appendix 4),

$$(t + \Delta t)^4 = t^4 + 4t^3\,\Delta t + 6t^2\,\Delta t^2 + 4t\,\Delta t^3 + \Delta t^4. \tag{12}$$

Inserting (12) into (11), we have (after cancelling the t^4 terms)

$$P'(t) = \frac{dP}{dt} = \lim_{\Delta t \to 0} \frac{4t^3\,\Delta t + 6t^2\,\Delta t^2 + 4t\,\Delta t^3 + \Delta t^4}{\Delta t}.$$

Dividing numerator and denominator by Δt we have

$$\frac{dP}{dt} = \lim_{\Delta t \to 0} (4t^3 + 6t^2\,\Delta t + 4t\,\Delta t^2 + \Delta t^3) = 4t^3.$$

When $t = 3$ hours, $dP/dt = 4 \cdot 3^3 = 4 \cdot 27 = 108$, and the population is growing at a rate of 108 individuals/hour. (We know that it is growing because $dP/dt > 0$. It would be declining if dP/dt were negative.)

EXAMPLE 3. The volume of a growing spherical cell is proportional to the cube of the radius of the cell. If r denotes radius and V denotes volume, then

$$V = \tfrac{4}{3}\pi r^3. \tag{13}$$

What is the rate of growth of the volume when the radius is $10\ \mu\text{m}$ ($1\ \mu\text{m} = 1$ micrometer $= 1/1,000,000$ meter)?

SOLUTION. We must calculate the instantaneous rate of change $V'(r) = dV/dr$. We have

$$V'(r) = \lim_{\Delta r \to 0} \frac{V(r + \Delta r) - V(r)}{\Delta r} = \lim_{\Delta r \to 0} \frac{\tfrac{4}{3}\pi(r + \Delta r)^3 - \tfrac{4}{3}\pi r^3}{\Delta r}.$$

$$= \frac{4}{3}\pi \lim_{\Delta r \to 0} \frac{(r^3 + 3r^2\,\Delta r + 3r\,\Delta r^2 + \Delta r^3) - r^3}{\Delta r}$$

$$= \frac{4}{3}\pi \lim_{\Delta r \to 0} (3r^2 + 3r\,\Delta r + \Delta r^2) = \left(\frac{4}{3}\pi\right)(3r^2) = 4\pi r^2.$$

Hence $dV/dr = 4\pi r^2$. When $r = 10\ \mu\text{m}$, $dV/dr = 4\pi \cdot 100 = 400\pi$ so that the volume is increasing at a rate of $400\pi \approx 1257$ cubic micrometers (μm^3) for every $1\ \mu\text{m}$ increase in r.

EXAMPLE 4. The mass of a 4-meter long nonuniform metal beam varies with the distance along the beam measured from the left end (see Figure 2). The mass μ is given by the formula

$$\mu(x) = x^{3/2} \quad \text{kg}. \tag{14}$$

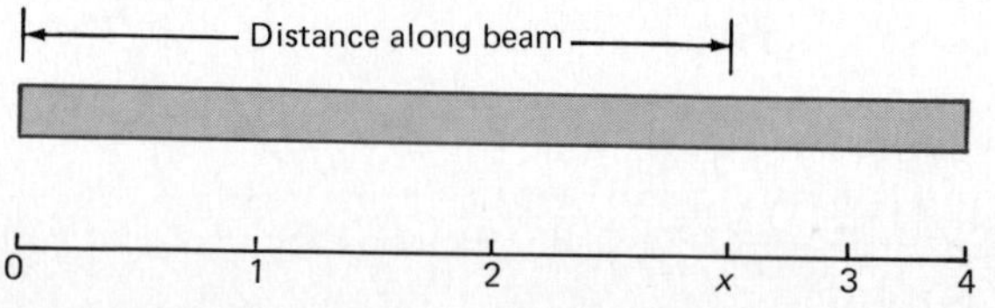

Figure 2

This tells us how much the beam weighs from the left end to the point x units from the left end. For example, at the end $x = 4$, $\mu = 4^{3/2} = 8$ kg so that the entire beam has a mass of 8 kg. On the other hand, when $x = 2$, $\mu = 2^{3/2} = 2\sqrt{2} \approx 2.83$ kg so that the left-hand half of the beam carries only approximately 35% of the mass of the beam ($2\sqrt{2}/8 \approx 0.35$). What is the density ρ of the beam for $x = 1$, $x = 2$, $x = 3$, and $x = 4$ meters?

SOLUTION. Density = mass per unit of length = kilograms per meter = kg/m. The density is changing as we move from left to right along the beam. Since the right side carries more of the mass, the density *increases* as we move from left to right. Between $x = 0$ and $x = 2$, the "average" density is $(2\sqrt{2} \text{ kg})/(2 \text{ m}) = \sqrt{2}$ kg/m while along the entire length of the beam (i.e., between $x = 0$ and $x = 4$) the average density is $(4 \text{ kg})/(2 \text{ m}) = 2$ kg/m. The problem asks for the "instantaneous" density at $x = 1, 2, 3$, and 4, or equivalently, for the instantaneous rate of change of mass per unit of length. We therefore have

$$\rho(x) = \frac{d\mu}{dx} = \lim_{\Delta x \to 0} \frac{\mu(x + \Delta x) - \mu(x)}{\Delta x} = \lim_{\Delta x \to 0} \frac{(x + \Delta x)^{3/2} - x^{3/2}}{\Delta x}. \tag{15}$$

Since, for any numbers a and b, $(a - b)(a + b) = a^2 - b^2$, we have,

$$[(x + \Delta x)^{3/2} - x^{3/2}][(x + \Delta x)^{3/2} + x^{3/2}] = [(x + \Delta x)^{3/2}]^2 - (x^{3/2})^2$$
$$= (x + \Delta x)^3 - x^3 = x^3 + 3x^2\,\Delta x + 3x\,\Delta x^2 + \Delta x^3 - x^3$$
$$= 3x^2\,\Delta x + 3x\,\Delta x^2 + \Delta x^3.$$

Thus we may multiply numerator and denominator of the last expression in (15) by $(x + \Delta x)^{3/2} + x^{3/2}$ to obtain

$$\frac{d\mu}{dx} = \lim_{\Delta x \to 0} \frac{[(x + \Delta x)^{3/2} - x^{3/2}][(x + \Delta x)^{3/2} + x^{3/2}]}{\Delta x[(x + \Delta x)^{3/2} + x^{3/2}]}$$

$$= \lim_{\Delta x \to 0} \frac{3x^2\,\Delta x + 3x\,\Delta x^2 + \Delta x^3}{\Delta x[(x + \Delta x)^{3/2} + x^{3/2}]}$$

$$= \lim_{\Delta x \to 0} \frac{3x^2 + 3x\,\Delta x + \Delta x^2}{(x + \Delta x)^{3/2} + x^{3/2}}$$

$$= \frac{3x^2}{x^{3/2} + x^{3/2}} = \frac{3x^2}{2x^{3/2}} = \frac{3}{2}x^{1/2} = \frac{3}{2}\sqrt{x}.$$

Thus the density ρ is given by the formula

$$\rho(x) = \tfrac{3}{2}\sqrt{x} \quad \text{kg/m}$$

and

$$\rho(1) = \frac{3}{2} = 1.5, \quad \rho(2) = \frac{3}{2}\sqrt{2} = \frac{3}{\sqrt{2}} \approx 2.12, \quad \rho(3) = \frac{3\sqrt{3}}{2} \approx 2.6,$$

and

$$\rho(4) = \frac{3\sqrt{4}}{2} = 3 \quad \text{(all in kg/m).}$$

We can clearly see how the density increases as we move from left to right.

EXAMPLE 5 (Marginal Cost).   A manufacturer buys large quantities of a certain machine replacement part. He finds that his cost depends on the number of cases bought at the same time, and the cost per unit decreases as the number of cases bought increases. He determines that a reasonable model for this is given by the formula

$$C(x) = 100 + 5x - 0.01x^2 \tag{16}$$

where x is the number of cases bought (up to 300 cases) and $C(x)$, measured in dollars, is the *total* cost of purchasing x cases. The 100 in (16) is a *fixed* cost which does not depend on the number of cases bought. What is the marginal cost for various levels of purchase?

SOLUTION. *Marginal cost* is the cost per additional unit at a given level of purchase. For example, if the manufacturer buys 25 cases, the marginal cost is the cost for *one more* case, i.e., the 26th case. This cost is not a constant. We can see this by calculating that one case costs $100 + 5 \cdot 1 - 0.01 = \$104.99$ and two cases cost $100 + 5 \cdot 2 - (0.01)4 = \$109.96$. Thus the additional cost of buying the second case is $C(2) - C(1) = \$4.97$. On the other hand, for 100 cases, it costs

$$100 + 5 \cdot 100 - (0.01)(100)^2 = 600 - 100 = \$500$$

and it costs

$$100 + 5 \cdot 101 - (0.01)(101)^2 = 100 + 505 - (.01)(10,201)$$
$$= 605 - 102.01 = \$502.99$$

to buy 101 cases. The additional cost is now $C(101) - C(100) = \$2.99$. The hundred and first case is cheaper than the second. If we look closely, we see that marginal cost is cost per additional units which is measured in dollars/unit. Alternatively, marginal cost is the *rate of change* of the cost with respect to the number of units purchased. Thus

$$\text{marginal cost} = \frac{dC}{dx} = \lim_{\Delta x \to 0} \frac{C(x + \Delta x) - C(x)}{\Delta x}$$

$$= \lim_{\Delta x \to 0} \frac{[100 + 5(x + \Delta x) - 0.01(x + \Delta x)^2] - [100 + 5x - 0.01x^2]}{\Delta x}$$

$$= \lim_{\Delta x \to 0} \frac{5\,\Delta x - 0.01(2x\,\Delta x + \Delta x^2)}{\Delta x} = 5 - 0.02x.$$

Thus at $x = 10$ cases, the marginal cost is $5 - 0.2 = \$4.80$ while at $x = 100$ cases the marginal cost is $5 - 2 = \$3$. This confirms the manufacturer's statement that "the more he buys, the cheaper it gets."

PROBLEMS 2.10

1. If a ball is thrown straight up into the air with an initial velocity of 75 ft/sec, its height (measured in feet) after t seconds is given by

    $$h = 75t - 16t^2.$$

    (a) Tabulate the heights of the ball from t between 0 and 2 seconds in increments of 0.1 second and construct a table similar to Table 1.
    (b) What is the average velocity between 1.4 and 1.5 seconds?
    (c) What is the average velocity between 1.5 and 1.6 seconds?
    (d) Find an estimate for the instantaneous velocity $v(1.5)$ after exactly 1.5 seconds.
    (e) Using $\Delta t = 0.01$, find a better estimate for $v(1.5)$.
    (f) Find $v(1.5)$ exactly and compare this with your estimates.

2. The distance an accelerating race car travels is given by $s = 0.6t^3$ where t is measured in minutes and s is measured in kilometers.

    (a) Tabulate the distance traveled by the car in increments of 0.1 minutes from $t = 0$ to $t = 2.5$ minutes and construct a table similar to Table 1.
    (b) What is the average velocity of the car between 1.9 and 2.0 minutes?
    (c) What is the average velocity of the car between 2.0 and 2.1 minutes?
    (d) Estimate the instantaneous velocity $v(2)$ at $t = 2.0$ minutes.
    (e) Using $t = 0.001$, find a better estimate for $v(2)$.
    (f) Calculate $v(2)$ exactly and compare this with your estimates.

In Problems 3–8 distance is given as a function of time. Find the instantaneous velocity at the indicated time.

3. $s = 1 + t + t^2, \quad t = 4$
4. $s = t^3 - t^2 + 3, \quad t = 5$
5. $s = 1 + \sqrt{2t}, \quad t = 8$
6. $s = (1 + t)^2, \quad t = 2.5$ [*Hint:* Expand $(1 + t)^2$.]
7. $s = 100t - 5t^2, \quad t = 6$
8. $s = t^4 - t^3 + t^2 - t + 5, \quad t = 3.$

9. Fuel in a rocket burns for $3\frac{1}{2}$ minutes. In the first t seconds the rocket reaches a height of $70t^2$ feet above the earth (for any t from 0 to 210 seconds). What is the velocity of the rocket (in ft/sec) after 3 seconds? after 10 seconds?

10. For the bacteria population of Example 2, how fast is the population growing after 5 hours? after 1 day? Is this model realistic for large values of t?

11. A colony of bacteria is dying out. It starts with an initial population of 10,000 organisms and after t days the population $P(t)$ is $10{,}000 - 2.5t^2$.

    (a) How fast is the population declining after 1 week? [*Hint:* The minus sign in your answer indicates a decline in the population.]
    (b) After how many days is the colony extinct?

12. In Example 3, how fast is the volume increasing when $r = 20\ \mu m$? 50 μm?

13. The surface area of a spherical cell is proportional to the square of the radius of the cell: $S = 4\pi r^2$. What is the rate of growth of the surface area when $r = 10\ \mu m$? 20 μm?

14. (a) Combining Example 3 and Problem 13, find an expression for the volume of the cell as a function of the surface area.
    *(b) How fast is the volume increasing when the surface area is 100 μm^2?

15. The density of the spherical head of a comet, 2 kilometers in diameter, varies according to distance r from its center. If $M(r)$ denotes the mass of a spherical "core" of radius r (where r is measured in meters) of the comet, then $M(r) = \sqrt{3r}$ kg (see Figure 3).
    (a) What is the density of the comet in kg/m^3 at its surface?

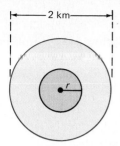

2 km

r

Figure 3

(b) What is the density $\frac{1}{2}$ km below the surface?

(c) What happens to the density as we approach the center?

16. The manufacturer in Example 5 finds that his cost function for another machine part is given by

$$C(x) = 100x + 55.$$

What can you say about his marginal cost?

17. Assume that the cost function of Example 5 is given by

$$C(x) = 200 + 6x - 0.01x^2 + 0.01x^3.$$

(a) Find the marginal cost.

(b) Is the manufacturer better off buying in large quantities?

*18. In Example 5 the difference between the cost of buying 101 units and buying 100 units is given by

$$C(101) - C(100) = \$2.99.$$

However, at $x = 100$, $dC/dx = \$3$. Explain this apparent discrepancy of one cent.

Review Exercises for Chapter Two

1. Tabulate values of $f(x) = x^2 - 3x + 6$ for $x = 3, 1, 2.5, 1.5, 2.1, 1.9, 2.01$ and 1.99. What does your table tell you about $\lim_{x \to 2} (x^2 - 3x + 6)$?

2. Tabulate values of $f(x) = x^2 + 10x + 8$ for $x = -4, -2, -3.5, -2.5, -3.1, -2.9, -3.01$, and -2.99. What does your table tell you about $\lim_{x \to -3} (x^2 + 10x + 8)$?

3. Calculate

(a) $\lim_{x \to 1} (x^3 - 3x + 2)$

(b) $\lim_{x \to 5} (-x^3 + 17)$

(c) $\lim_{x \to 3} \dfrac{x^4 - 2x + 1}{x^3 + 3x - 5}$

(d) $\lim_{x \to -1} \dfrac{x^3 + x^2 + x + 1}{x^4 + x^3 + x^2 + x + 1}$.

4. Calculate

(a) $\lim_{x \to 0} |x + 2|$

(b) $\lim_{x \to 1} |x - 3|$

(c) $\lim_{x \to -3} |x + 4|$

(d) $\lim_{x \to 1} \dfrac{|x|}{x}$.

5. Calculate

(a) $\lim_{x \to 3} \dfrac{(x - 3)(x - 4)}{x - 3}$

(b) $\lim_{x \to 5} \dfrac{x^2 - 6x + 5}{x - 5}$.

6. Do the following limits exist? If not, explain why. If so, calculate them.

(a) $\lim_{x \to 1} \sqrt{x - 1}$

(b) $\lim_{x \to 2} \sqrt{x - 1}$

(c) $\lim_{x \to -3} \dfrac{|x + 2|}{x + 2}$

(d) $\lim_{x \to -2} \dfrac{|x + 2|}{x + 2}$

(e) $\lim_{x \to 1} \sqrt[3]{x - 1}$

(f) $\lim_{x \to 1} \sqrt[4]{x - 1}.$

7. Calculate

(a) $\lim_{x \to 1} 23 \sqrt{x - 17}$

(b) $\lim_{x \to -1} (1 - x + x^2 - x^3 + x^4)$

(c) $\lim_{x \to 4} \dfrac{x^2 + 9}{x^2 - 9}$

(d) $\lim_{x \to -1} 5x^{250}$

(e) $\lim_{x \to -1} 6x^{251}$

(f) $\lim_{x \to 3} (x^2 + x - 8)^5$

(g) $\lim_{x \to 0} \dfrac{x^8 - 7x^5 + x^3 - x^2 + 3}{x^{23} - 2x + 9}$

(h) $\lim_{x \to 0} \dfrac{ax^2 + bx + c}{dx^2 + ex + f}$, $(a, b, c, d, e, f$ are all nonzero real numbers$)$.

8. Find open intervals in which each of the following functions is continuous.

(a) $f(x) = 2\sqrt{x}$

(b) $f(x) = 3\sqrt[3]{x}$

(c) $f(x) = \dfrac{1}{x - 6}$

(d) $f(x) = \dfrac{x}{x^2 - 4}$

(e) $f(x) = |x + 2|$

(f) $f(x) = \dfrac{|x + 3|}{x + 3}.$

9. Explain why $\lim_{x \to 0} (1/x^3)$ does not exist but that $\lim_{x \to 0} (1/x^4)$ does exist.

10. Calculate

(a) $\lim_{x \to 2} \dfrac{x - 3}{(x - 2)^2}$

(b) $\lim_{x \to -1} \dfrac{x + 10}{(x + 1)^{10}}.$

11. Calculate

(a) $\lim_{x \to \infty} \dfrac{1}{x^3}$

(b) $\lim_{x \to \infty} \dfrac{1}{\sqrt{x + 2}}$

(c) $\lim_{x \to \infty} \dfrac{\sqrt{x}}{x^2 + 3}$

(d) $\lim_{x \to \infty} \dfrac{x^3 + 6x^2 + 4x + 2}{3x^3 - 9x^2 + 11}$

(e) $\lim_{x \to \infty} \dfrac{3x^5 - 6x^2 + 3}{x^7 - 2}$

(f) $\lim_{x \to \infty} \dfrac{x^7 - 9}{30x^5 + x^4 + 161}.$

12. How large must x be in order that $1/\sqrt[3]{x} < 0.01$?

13. Draw a rough sketch of the derivative function for each of the curves in Figure 1.

14. Find the equation of the tangent line to the given curve at the given point.

(a) $y = x^2 - 3$, $(2, 1)$

(b) $y = 7x^3 - 8$, $(1, -1)$

(c) $y = \sqrt{x + 1}$, $(3, 2)$

(d) $y = \dfrac{1}{x + 1}$, $(0, 1).$

15. Explain why the curve $y = |x + 1|$ does not have a derivative at the point $x = -1$.

16. If the distance a particle travels is given by $s(t) = t^3 + t^2 + 6$ kilometers after t hours, how fast is it traveling (in km/hr) after 2 hours?

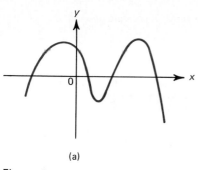
(a)

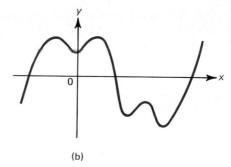
(b)

Figure 1

17. The mass of a nonuniform 8-meter metal bar is given by $\mu = \sqrt{x + 1}$ kg, where x is measured in meters from the left in Figure 2. What is the density of the bar at a point 3 m from the left end?

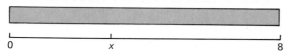

0   x 8   Figure 2

18. The volume of a sphere is $\frac{4}{3}\pi r^3$. How is the volume changing when the radius r is equal to 2 ft?

19. A slaughterhouse purchases cattle from a ranch at a cost of $C(x) = 200 + 8x - 0.02x^2$ where x is the number of head of cattle bought at one time up to a maximum of 150. What is the house's marginal cost as a function of x? Does it pay to buy in large quantities?

THREE

MORE ABOUT DERIVATIVES

In the previous chapter we introduced the concepts of the limit and the derivative but found that the calculation of derivatives could be extremely tedious. An even more annoying problem was the seeming necessity to come up with a special trick (like multiplying and dividing by some quantity) each time we took the limit in the process of computing a derivative.

In the present chapter we continue the discussion of properties of functions. We will be principally concerned with simplifying the process of differentiation so that it will no longer be necessary to deal with complicated limits. We shall also show (expanding on the material in Chapter 2) how derivatives can be used to help us obtain a reasonably accurate graph of a given function.

In many of the sections of this chapter we will be deriving formulas for calculating derivatives. Formulas are usually not very exciting and we ask you to bear with us here. By the time you have completed the chapter, you will find that differentiation is not nearly as complicated as it now seems. You should be assured that the work involved in memorizing the appropriate formulas will pay dividends in the chapters to come.

3.1 Some Differentiation Formulas

In this section we begin the task of simplifying the calculation of derivatives. We begin with a theorem that states the obvious fact that a constant function does not change.

THEOREM 1. Let $f(x) = c$, a constant. Then $f'(x) = 0$.

> **Note.** This result is evident from looking at the graph of a constant function—a horizontal line with a slope of zero (see Figure 1).

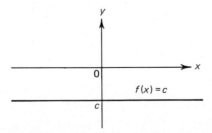

Figure 1

PROOF

$$f'(x) = \lim_{\Delta x \to 0} \frac{f(x + \Delta x) - f(x)}{\Delta x} = \lim_{\Delta x \to 0} \frac{c - c}{\Delta x}$$

$$= \lim_{\Delta x \to 0} \frac{0}{\Delta x} = \lim_{\Delta x \to 0} 0 = 0.$$

The converse of this theorem is also true. Namely that if $f' = 0$, then f is a constant function. The proof of this theorem is more difficult and will be deferred until Section 10.6.

THEOREM 2. (i) Let c be a constant. Then if f is differentiable, cf is also differentiable and

$$\boxed{\frac{d}{dx} cf = c \frac{df}{dx}.}$$

(1)

(ii) Let f and g be differentiable. Then $f + g$ is also differentiable and

$$\boxed{(f + g)' = \frac{d}{dx}(f + g) = \frac{df}{dx} + \frac{dg}{dx} = f' + g'.}$$

(2)

That is, *the derivative of the sum of two differentiable functions is the sum of their derivatives.*

PROOF

(i) $\dfrac{d}{dx} cf = \lim\limits_{\Delta x \to 0} \dfrac{(cf)(x + \Delta x) - (cf)(x)}{\Delta x}$

$= c \lim\limits_{\Delta x \to 0} \dfrac{f(x + \Delta x) - f(x)}{\Delta x} = c \dfrac{df}{dx}.$

$\left[\text{We have used the (limit) Theorem 2.3.2 here } \left(\lim\limits_{x \to x_0} cf = c \lim\limits_{x \to x_0} f \right). \right]$

(ii) $\dfrac{d}{dx}(f + g) = \lim\limits_{\Delta x \to 0} \dfrac{(f + g)(x + \Delta x) - (f + g)(x)}{\Delta x}$

$= \lim\limits_{\Delta x \to 0} \dfrac{f(x + \Delta x) + g(x + \Delta x) - f(x) - g(x)}{\Delta x}$

$= \lim\limits_{\Delta x \to 0} \left[\left(\dfrac{f(x + \Delta x) - f(x)}{\Delta x} \right) + \left(\dfrac{g(x + \Delta x) - g(x)}{\Delta x} \right) \right]$

$= \lim\limits_{\Delta x \to 0} \dfrac{f(x + \Delta x) - f(x)}{\Delta x} + \lim\limits_{\Delta x \to 0} \dfrac{g(x + \Delta x) - g(x)}{\Delta x}$

$= \dfrac{df}{dx} + \dfrac{dg}{dx}.$

(We have made use of the fact that the limit of the sum is the sum of the limits—Theorem 2.3.3.)

EXAMPLE 1. Calculate the derivative of $y = x^3$.

SOLUTION. From Example 2.9.3 we know that

$$\frac{d}{dx} \frac{4}{3}\pi x^3 = 4\pi x^2$$

Then, using part (i) of the theorem,

$$\frac{d}{dx} \frac{4}{3}\pi x^3 = \frac{4}{3}\pi \frac{d}{dx} x^3 = 4\pi x^2$$

so that

$$\frac{d}{dx} x^3 = \frac{4\pi x^2}{(4/3)\pi} = 3x^2$$

EXAMPLE 2. Let $f(x) = 4x^3 + 3\sqrt{x}$. Find $f'(x)$.

SOLUTION. We saw in Section 2.9 that

$$\frac{d}{dx} \sqrt{x} = \frac{1}{2\sqrt{x}} \quad \text{(Example 2.9.4)}.$$

Thus

$$\frac{d}{dx}(4x^3 + 3\sqrt{x}) = 4\frac{d}{dx}x^3 + 3\frac{d}{dx}\sqrt{x} = 4 \cdot 3x^2 + \frac{3}{2\sqrt{x}} = 12x^2 + \frac{3}{2\sqrt{x}}.$$

We have shown that the derivative of the sum of two functions is the sum of the derivatives of the two functions. It is not too difficult to show that this fact applies to more than two functions. The theorem which states this is given below. The proof is left as an exercise (see Problem 19).

THEOREM 3. Let $f_1, f_2, \ldots, f_n$ be n differentiable functions. Then

$$\boxed{\frac{d}{dx}(f_1 + f_2 + \cdots + f_n) = \frac{df_1}{dx} + \frac{df_2}{dx} + \cdots + \frac{df_n}{dx}.} \tag{3}$$

EXAMPLE 3. Let $f(x) = 1 + x + x^2 + x^3$. Calculate df/dx.

SOLUTION

$$\frac{df}{dx} = \frac{d}{dx}1 + \frac{d}{dx}(x) + \frac{d}{dx}(x^2) + \frac{d}{dx}(x^3) = 0 + 1 + 2x + 3x^2$$

$$= 1 + 2x + 3x^2.$$

Here we used the facts that $d1/dx = 0$ from Theorem 1, $dx/dx = 1$ since 1 is the slope of the line $y = x$, $dx^2/dx = 2x$ from Example 2.9.2, and $dx^3/dx = 3x^2$ from Example 1 above.

EXAMPLE 4.   Let $s(t) = 17 - 4t^2 + 8t^3 - 9\sqrt{t}$. Calculate ds/dt.

SOLUTION. We find the derivative of each term separately and then add. We have

(i) $\dfrac{d}{dt} 17 = 0$ since 17 is a constant.

(ii) $\dfrac{d}{dt}(-4t^2) = -4\dfrac{d}{dt} t^2 = -4 \cdot 2t = -8t$.

(iii) $\dfrac{d}{dt} 8t^3 = 8\dfrac{d}{dt} t^3 = 8 \cdot 3t^2 = 24t^2$.

(iv) $\dfrac{d}{dt}(-9\sqrt{t}) = -9\dfrac{d}{dt} \sqrt{t} = -9\left(\dfrac{1}{2\sqrt{t}}\right) = \dfrac{-9}{2\sqrt{t}}$.

Therefore

$$\frac{ds}{dt} = \frac{d}{dt}(17 - 4t^2 + 8t^3 - 9\sqrt{t}) = \frac{d}{dt} 17 - 4\frac{d}{dt} t^2 + 8\frac{d}{dt} t^3 - 9\frac{d}{dt}\sqrt{t}$$

$$= -8t + 24t^2 - \frac{9}{2\sqrt{t}}.$$

We now derive a formula for calculating the derivative of $y = f(x) = x^n$ where n is a positive integer. First, let us look for a pattern. We have already calculated the following derivatives:

(i) $\dfrac{d}{dx} x = 1$ (since $y = x = 1 \cdot x + 0$ is the equation of a straight line with slope 1)

(ii) $\dfrac{d}{dx} x^2 = 2x$

(iii) $\dfrac{d}{dx} x^3 = 3x^2$

(iv) $\dfrac{d}{dx} x^4 = 4x^3$ (see Example 2.10.2).

Do you see a pattern? The answer is given in Theorem 4.

THEOREM 4.   If n is a positive integer, then

$$\boxed{\frac{d}{dx} x^n = nx^{n-1}.}$$ (4)

PROOF. We first note that for any real numbers a and b (such that $a \neq b$)

$$\frac{a^2 - b^2}{a - b} = \frac{(a - b)(a + b)}{a - b} = a + b.$$

$$\frac{a^3 - b^3}{a - b} = \frac{(a - b)(a^2 + ab + b^2)}{a - b} = a^2 + ab + b^2$$

$$\frac{a^4 - b^4}{a - b} = \frac{(a - b)(a^3 + a^2b + ab^2 + b^3)}{a - b} = a^3 + a^2b + ab^2 + b^3$$

$$\vdots$$

and, for any positive integer n,

$$\frac{a^n - b^n}{a - b} = \frac{(a - b)(a^{n-1} + a^{n-2}b + a^{n-3}b^2 + \cdots + ab^{n-2} + b^{n-1})}{a - b}$$

$$= a^{n-1} + a^{n-2}b + a^{n-3}b^2 + \cdots + ab^{n-2} + b^{n-1}. \tag{5}$$

For each n, the equality can be verified by multiplying $(a - b)$ by $a^{n-1} + a^{n-2}b + a^{n-3}b^2 + \cdots + ab^{n-2} + b^{n-1}$ and noting that all terms except a^n and $-b^n$ cancel. Now

$$\frac{d}{dx}x^n = \lim_{\Delta x \to 0} \frac{(x + \Delta x)^n - x^n}{\Delta x} = \lim_{\Delta x \to 0} \frac{(x + \Delta x)^n - x^n}{(x + \Delta x) - x}. \tag{6}$$

Letting $a = x + \Delta x$ and $b = x$ in (5), we have

$$\lim_{\Delta x \to 0} \frac{(x + \Delta x)^n - x^n}{(x + \Delta x) - x} = \lim_{\Delta x \to 0} [(x + \Delta x)^{n-1} + (x + \Delta x)^{n-2}x + (x + \Delta x)^{n-3}x^2$$

$$+ \cdots + (x + \Delta x)x^{n-2} + x^{n-1}]$$

$$= x^{n-1} + x^{n-2}x + x^{n-3}x^2 + \cdots + x \cdot x^{n-2} + x^{n-1}$$

$$= \underbrace{x^{n-1} + x^{n-1} + \cdots + x^{n-1}}_{n \text{ times}} = nx^{n-1}$$

EXAMPLE 5. Let $f(x) = x^{17}$. Then $f'(x) = 17x^{16}$.

EXAMPLE 6. Let $f(x) = 3x^{10} + 8x^7 - 14x^4 + 9x^3 - x + 3$. Calculate df/dx.

SOLUTION. Using theorems 3 and 4, we obtain

$$\frac{df}{dx} = 3\frac{d}{dx}x^{10} + 8\frac{d}{dx}x^7 - 14\frac{d}{dx}x^4 + 9\frac{d}{dx}x^3 - \frac{d}{dx}x + \frac{d}{dx}3$$

$$= 3 \cdot 10x^9 + 8 \cdot 7x^6 - 14 \cdot 4x^3 + 9 \cdot 3x^2 - 1 + 0$$

$$= 30x^9 + 56x^6 - 56x^3 + 27x^2 - 1.$$

EXAMPLE 7. Find the equation of the line tangent to the curve of Example 6 at the point $(1, 8)$.

SOLUTION. Since $f'(x) = 30x^9 + 56x^6 - 56x^3 + 27x^2 - 1$, we have

$$f'(1) = 30 \cdot 1^9 + 56 \cdot 1^6 - 56 \cdot 1^3 + 27 \cdot 1^2 - 1$$
$$= 30 + 56 - 56 + 27 - 1 = 56.$$

This is the slope of the tangent line so that the equation of the line is

$$\frac{y - 8}{x - 1} = 56 \quad \text{or} \quad y = 56x - 48.$$

Consider the curve depicted in Figure 2. The tangent line T is drawn. The line N which is perpendicular to T at the point $(x, f(x))$ is called a *normal line*.

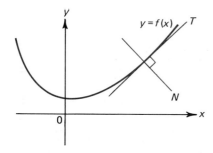

Figure 2

In general a *normal line to a curve at a point is a line which is perpendicular to the tangent line at that point.*

If the slope of the tangent line is m and if $m \neq 0$, then from Section 1.4, the slope of the normal line is $-1/m$. Normal lines are very important in applications as we shall see in some of the later chapters of this book (especially Chapters 16, 17, 18, and 19).

EXAMPLE 8. Calculate the normal line to the curve $y = x^2$ at the point $(2, 4)$.

SOLUTION. We have $dy/dx = 2x$ and when $x = 2$, $dy/dx = 2 \cdot 2 = 4$. This is the slope of the tangent line at the point $(2, 4)$. Hence the slope of the normal line is $-\frac{1}{4}$ and the normal line has the equation

$$\frac{y - 4}{x - 2} = -\frac{1}{4} \qquad \text{or} \qquad x + 4y = 18.$$

PROBLEMS 3.1

In Problems 1–10 calculate the derivative of the given function.

1. $f(x) = x^5$
2. $f(x) = 27$
3. $f(x) = 3x^2 + 19x + 2$
4. $g(t) = t^{10} - t^3$
5. $g(t) = t^5 + \sqrt{t}$
6. $g(t) = 1 - t + t^4 - t^7$
7. $h(z) = z^{100} + 100z^{10} + 10$
8. $h(z) = 27z^6 + 3z^5 + 4z$
9. $v(r) = 3r^8 - 8r^6 - 7r^4 + 2r^2 + 3$
10. $v(r) = -3r^{12} + 12r^3$.

In Problems 11–17 find the line which is tangent to the given curve at the given point.

11. $y = x^4$, $(1, 1)$
12. $y = 3x^5 - 3x^3 + 1$, $(-1, 1)$
13. $y = 2x^7 - x^6 - x^3$, $(1, 0)$
14. $y = 5x^6 - x^4 + 2x^3$, $(1, 6)$
15. $y = 1 + x + x^2 + x^3 + x^4 + x^5$, $(0, 1)$
16. $y = 1 - x + 2x^2 - 3x^3 + 4x^4$, $(1, 3)$
17. $y = x^6 - 6\sqrt{x}$, $(1, -5)$.

18. Show that if $y = x^n$ where n is a positive integer, then the tangent line to the curve at the point $(0, 0)$ is the x-axis.

19. (a) Prove Theorem 3 when $n = 3$ and $n = 4$.
    *(b) Use mathematical induction to prove Theorem 3 for general n (see Appendix 1).

In Problems 20–24 find the normal line to the given curve at the given point.

20. $y = x^3$, $(1, 1)$ 21. $y = x^7 - 6x^5$, $(1, -5)$
22. $y = 2x^4 - 3x^3 + 2x + 1$, $(2, 13)$ 23. $y = 4x^{10} + 3x^2 + 6$, $(0, 6)$
24. $y = x^6 - 6\sqrt{x}$, $(1, -5)$.

25. Find the points on the graph of $y = 2x^3 + 3x^2 - 6x + 1$ where the curve has a horizontal tangent.

26. Let $f(x) = x^3 + 9x^2 + 2x + 2$. Find the values of x for which $f(x) = f'(x)$.

*27. Find the equation of the two lines tangent to the curve $y = x^2 + 3$ which pass through the point $(1, 0)$. [*Hint:* Any point on the curve has the coordinates $(a, a^2 + 3)$. If L is a tangent line at the point $(a, a^2 + 3)$, then the slope of L is $2a$.]

28. Suppose that when an airplane takes off (starting from rest), the distance (in feet) it travels during the first few seconds is given by the formula

$$s = 1 + 4t + 6t^2.$$

How fast (in ft/sec) is the plane traveling after 10 sec? after 20 sec?

29. A petri dish contains two colonies of bacteria. The population of the first colony is given by $P_1(t) = 1000 + 50t - 20\sqrt{t}$ and the population of the second is given by $P_2(t) = 2000 + 30t^2 - 80t$, where t is measured in hours.
     (a) Find a function that represents the *total* population of the 2 species.
     (b) What is the instantaneous growth rate of the total population?
     (c) How fast is the total population growing after 4 hr? after 16 hr?

*30. For the model of Problem 29, how fast is the first population growing when the second population is growing at a rate of 160 individuals per hour?

31. Show that the rate of change of the area of a circle with respect to its radius is equal to its circumference.

32. *Stefan's law* for the amount of radiant energy emitted from the surface of a body is given by $R = \sigma T^4$ where R is the rate of emission per unit area, T is the temperature measured in degrees Kelvin and σ is a universal constant called the *Stefan–Boltzmann constant.*† At a temperature of $400°$ K, what is the instantaneous rate of change of R with respect to the temperature?

*33. A growing grapefruit with a diameter of $2k$ inches has a skin which is $k/12$ inches thick (the skin is included in the diameter of the grapefruit). What is the rate of growth of the volume of the skin (per unit growth in the radius) when the radius of the grapefruit is 3 inches?

3.2 The Product and Quotient Rules

In this section we develop some additional rules to simplify the calculation of derivatives. To see why additional rules are needed, consider the problem of calculating the derivatives of

$$f(x) = \sqrt{x}(x^4 + 3) \qquad \text{or} \qquad g(x) = \frac{x^4 + 3}{\sqrt{x}}.$$

To carry out the calculations from the definition would be very tedious. However, we will shortly see that these calculations are rather simple.

Let f and g be two differentiable functions of x. What is the derivative of the

†$\sigma = 5.67 \times 10^{-8}$ watts/m²(°K)⁴.

product fg? It is easy to be led astray here. Theorem 2.3.4 states that

$$\lim_{x \to x_0} f(x)g(x) = \lim_{x \to x_0} f(x) \lim_{x \to x_0} g(x).$$

Is it true that the derivative of the product is equal to the product of the derivatives? That is, is it true that

$$\frac{d}{dx}fg = \frac{df}{dx} \cdot \frac{dg}{dx}?$$

Originally Leibniz, the codiscoverer of the derivative, thought so. However, an easy example shows that this is false. Let $f(x) = x$ and $g(x) = x^2$. Then

$$(fg)(x) = f(x)g(x) = x^3 \quad \text{and} \quad \frac{d}{dx}fg = 3x^2.$$

But

$$\frac{df}{dx} = 1 \quad \text{and} \quad \frac{dg}{dx} = 2x$$

so that

$$\frac{df}{dx} \cdot \frac{dg}{dx} = 1 \cdot 2x = 2x$$

which is not equal to

$$\frac{d}{dx}fg = 3x^2.$$

The correct formula, discovered after many false steps by both Leibniz and Newton, is given below.

THEOREM 1 (Product Rule). Let f and g be differentiable. Then fg is differentiable and

$$(fg)' = \frac{d}{dx}fg = f(x)\frac{dg}{dx} + g(x)\frac{df}{dx} = fg' + gf'. \tag{1}$$

Verbally, the product rule says that *the derivative of the product of two functions is equal to the first times the derivative of the second plus the second times the derivative of the first.*

PROOF

$$\frac{d}{dx}fg = \lim_{\Delta x \to 0} \frac{f(x + \Delta x)g(x + \Delta x) - f(x)g(x)}{\Delta x}.$$

This doesn't look very much like the derivative of anything. To continue, we will use the trick of adding and subtracting the term $f(x)g(x + \Delta x)$ in the numerator. As we shall see, this makes everything come out nicely. We have

(here are the additional terms)

$$\frac{d}{dx}fg = \lim_{\Delta x \to 0} \frac{f(x + \Delta x)g(x + \Delta x) - f(x)g(x + \Delta x) \overbrace{+ f(x)g(x + \Delta x) - f(x)g(x)}}{\Delta x}$$

$$= \lim_{\Delta x \to 0} \frac{g(x + \Delta x)(f(x + \Delta x) - f(x))}{\Delta x} + \lim_{\Delta x \to 0} \frac{f(x)(g(x + \Delta x) - g(x))}{\Delta x}$$

$$= \lim_{\Delta x \to 0} g(x + \Delta x) \lim_{\Delta x \to 0} \frac{f(x + \Delta x) - f(x)}{\Delta x} + \lim_{\Delta x \to 0} f(x) \lim_{\Delta x \to 0} \frac{g(x + \Delta x) - g(x)}{\Delta x}$$

$$= g(x)\frac{df}{dx} + f(x)\frac{dg}{dx}.$$

Here we have used the fact that $\lim_{\Delta x \to 0} g(x + \Delta x) = g(x)$† since g is continuous (every differentiable function is continuous by Theorem 2.9.1).

EXAMPLE 1. Let $h(x) = \sqrt{x}\,(x^4 + 3)$. Calculate dh/dx.

SOLUTION. $h(x) = f(x)g(x)$ where $f(x) = \sqrt{x}$ and $g(x) = x^4 + 3$. Then

$$\frac{dh}{dx} = f\frac{dg}{dx} + g\frac{df}{dx} = \sqrt{x}(4x^3) + (x^4 + 3)\frac{1}{2\sqrt{x}}.$$

This is the correct answer but we will use some algebra to simplify the result.

$$\frac{dh}{dx} = \frac{2\sqrt{x}\sqrt{x}(4x^3) + x^4 + 3}{2\sqrt{x}} = \frac{2x(4x^3) + x^4 + 3}{2\sqrt{x}} = \frac{9x^4 + 3}{2\sqrt{x}}.$$

EXAMPLE 2. Let $h(t) = (2t + 3)(t^9 - 8t + 1)$. Calculate $h'(t)$.

SOLUTION. This could be computed "directly" by first carrying out the multiplication. However the product rule is easier to use. Here $h(t) = f(t) \cdot g(t)$ where $f(t) = 2t + 3$ and $g(t) = t^9 - 8t + 1$. Then

$$h' = fg' + gf' \text{ or } h'(t) = (2t + 3)(9t^8 - 8) + (t^9 - 8t + 1)2.$$

We now multiply through and combine terms to simplify this expression (although the answer already given is correct).

$$h'(t) = 18t^9 - 16t + 27t^8 - 24 + 2t^9 - 16t + 2$$
$$= 20t^9 + 27t^8 - 32t - 22.$$

Having now discussed the product of two functions, we turn to their quotient.

---

† This looks a bit different from our definition of continuity given in Section 2.4. There we said that g was continuous at x_0 if $\lim_{x \to x_0} g(x) = g(x_0)$. But since $\Delta x \to 0$ is equivalent to $x + \Delta x \to x$, the statement $\lim_{\Delta x \to 0} g(x + \Delta x) = g(x)$ is equivalent to $\lim_{x + \Delta x \to x} g(x + \Delta x) = g(x)$. This last statement can be thought of as an alternative definition of continuity.

THEOREM 2 (Quotient Rule). Let f and g be differentiable at x. Then, if $g(x) \neq 0$, we have

$$\frac{d}{dx}\frac{f}{g} = \frac{g(x)(df/dx) - f(x)(dg/dx)}{g^2(x)} = \frac{gf' - fg'}{g^2} \tag{2}$$

The quotient rule states that *the derivative of the quotient is equal to the denominator times the derivative of the numerator minus the numerator times the derivative of the denominator all over the denominator squared.*

PROOF

$$\frac{d}{dx}\frac{f}{g} = \lim_{\Delta x \to 0} \frac{\dfrac{f(x + \Delta x)}{g(x + \Delta x)} - \dfrac{f(x)}{g(x)}}{\Delta x} = \lim_{\Delta x \to 0} \frac{f(x + \Delta x)g(x) - f(x)g(x + \Delta x)}{\Delta x\, g(x + \Delta x)g(x)}$$

$$= \lim_{\Delta x \to 0} \frac{f(x + \Delta x)g(x) - \overbrace{f(x)g(x) + f(x)g(x)}^{\text{these terms sum to zero}} - f(x)g(x + \Delta x)}{\Delta x\, g(x + \Delta x)g(x)}$$

(the term $f(x)g(x)$ was added and subtracted)

$$= \lim_{\Delta x \to 0} \left\{ \frac{g(x)\left[\dfrac{f(x + \Delta x) - f(x)}{\Delta x}\right] - f(x)\left[\dfrac{g(x + \Delta x) - g(x)}{\Delta x}\right]}{g(x + \Delta x)g(x)} \right\}$$

$$= \frac{\left\{ g(x) \displaystyle\lim_{\Delta x \to 0}\left[\dfrac{f(x + \Delta x) - f(x)}{\Delta x}\right] - f(x) \displaystyle\lim_{\Delta x \to 0}\left[\dfrac{g(x + \Delta x) - g(x)}{\Delta x}\right]\right\}}{g(x) \displaystyle\lim_{\Delta x \to 0} g(x + \Delta x)}$$

$$= \frac{g(x)f'(x) - f(x)g'(x)}{[g(x)]^2}.$$

EXAMPLE 3. Let $h(x) = (x^4 + 3)/\sqrt{x}$. Calculate dh/dx.

SOLUTION. $h(x) = f(x)/g(x)$ where $f(x) = x^4 + 3$ and $g(x) = \sqrt{x}$. Thus

$$\frac{dh}{dx} = \frac{g(df/dx) - f(dg/dx)}{g^2} = \frac{\sqrt{x}\,(4x^3) - (x^4 + 3)(1/2\sqrt{x})}{(\sqrt{x})^2}$$

$$= \frac{2\sqrt{x}\,\sqrt{x}(4x^3) - x^4 - 3}{2\sqrt{x}} \cdot \frac{1}{x} = \frac{8x^4 - x^4 - 3}{2x^{3/2}} = \frac{7x^4 - 3}{2x^{3/2}}.$$

EXAMPLE 4. Let $h(x) = \dfrac{x^3 + x + 1}{x^2 - 5}$. Calculate $\dfrac{dh}{dx}$.

SOLUTION

$$\frac{dh}{dx} = \frac{(x^2 - 5)\frac{d}{dx}(x^3 + x + 1) - (x^3 + x + 1)\frac{d}{dx}(x^2 - 5)}{(x^2 - 5)^2}$$

$$= \frac{(x^2 - 5)(3x^2 + 1) - (x^3 + x + 1)(2x)}{(x^2 - 5)^2} = \frac{x^4 - 16x^2 - 2x - 5}{x^4 - 10x^2 + 25}.$$

In the last section we proved that

$$\frac{d}{dx}x^n = nx^{n-1}$$

if n is a positive integer. With the aid of the quotient rule we can extend this result to the case in which n is a negative integer. [Note that when $n = 0$, $x^0 = 1$ which we already know has a derivative of 0.]

THEOREM 3. Let $y = x^{-n}$ where n is a positive integer (so that $-n$ is a negative integer). Then

$$\boxed{\frac{dy}{dx} = \frac{d}{dx}(x^{-n}) = -nx^{-n-1}.}$$

(3)

PROOF. $x^{-n} = 1/x^n$. Then we use the quotient rule (2) to obtain

$$\frac{dy}{dx} = \frac{d}{dx}\frac{1}{x^n} = \frac{x^n\frac{d}{dx}(1) - (1)\frac{d}{dx}x^n}{(x^n)^2} = \frac{x^n \cdot 0 - nx^{n-1}}{x^{2n}}$$

$$= \frac{-nx^{n-1}}{x^{2n}} = -nx^{(n-1)-2n} = -nx^{-n-1}.$$

Note. We now know that $d(x^n)/dx = nx^{n-1}$ holds for any integer.

EXAMPLE 5. Let $y = x^{-7}$. Calculate dy/dx.

SOLUTION. Using (3) we obtain

$$\frac{dy}{dx} = -7x^{-8}.$$

EXAMPLE 6. Let $y = 1/x$. Calculate dy/dx. x^{-1}

SOLUTION. $1/x = x^{-1}$, so that

$$\frac{d}{dx}\frac{1}{x} = \frac{d}{dx}x^{-1} = -1 \cdot x^{-2} = \frac{-1}{x^2}.$$

EXAMPLE 7. Let $y = 7/x^2$. Calculate dy/dx.

SOLUTION. $1/x^2 = x^{-2}$, so that

$$\frac{d}{dx}\frac{7}{x^2} = 7\frac{d}{dx}x^{-2} = 7(-2)x^{-3} = -\frac{14}{x^3}.$$

EXAMPLE 8. Newton's law of universal gravitation states that *the force between any two particles having masses m_1 and m_2 separated by a distance d is an attraction acting along the line joining the particles and has a magnitude*

$$F = G\frac{m_1 m_2}{r^2} \tag{4}$$

where G is a universal constant having the same value for all pairs of particles† (see Figure 1).

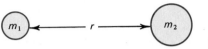

Figure 1

Two asteroids are approaching each other. The first has a mass of 1000 kg and the second a mass of 3000 kg.

(a)   What is the gravitational force between the two asteroids when they are 10 km apart?

(b)   How is this force changing at that distance?

SOLUTION. (a) $r = 10$ km $= 10{,}000$ m. Then

$$F = G\frac{m_1 m_2}{r^2} = G\frac{(1000)(3000)}{10{,}000^2} = 0.03G \quad \text{nt.}$$

(b)   The rate of change of F when $r = 10{,}000$ m is $F'(10{,}000)$. But

$$\frac{dF}{dr} = \frac{d}{dr}\,Gm_1 m_2 r^{-2} = Gm_1 m_2 \frac{d}{dr}r^{-2} = Gm_1 m_2(-2r^{-3}) = \frac{-2Gm_1 m_2}{r^3}$$

$$= \frac{(-2)(1000)(3000)G}{(10{,}000)^3} = -0.000006G = -6 \times 10^{-6}G \quad \text{nt/m.}$$

when $r = 10{,}000$. Here the minus sign indicates that as r increases, the force F decreases and vice versa. But since the asteroids are approaching one another, r is decreasing and therefore F is increasing at a rate of $6 \times 10^{-6}G$ nt for every 1 m decrease in r.

PROBLEMS 3.2

In Problems 1–24 find the derivative of the given function.

1. $f(x) = 2x(x^2 + 1)$

2. $g(t) = t^3(1 + \sqrt{t}\,)$

3. $s(t) = \dfrac{t^3}{1 + \sqrt{t}}$

4. $f(z) = \dfrac{1 + \sqrt{z}}{z^3}$

5. $f(x) = (1 + x + x^5)(2 - x + x^6)$

6. $f(x) = \dfrac{1 + x + x^5}{2 - x + x^6}$

†In metric units $G = 6.673 \times 10^{-11}$ nt-m²/kg² (nt = Newton) where 1 Newton is the force that will accelerate a 1 kg mass at the rate of 1 m/sec² (1 nt = 1 kg/sec²).

7. $f(x) = \dfrac{2 - x + x^6}{1 + x + x^5}$

8. $g(t) = (1 + \sqrt{t})(1 - \sqrt{t})$

9. $g(t) = \dfrac{1 + \sqrt{t}}{1 - \sqrt{t}}$

10. $g(t) = \dfrac{1 - \sqrt{t}}{1 + \sqrt{t}}$

11. $p(v) = (v^3 - \sqrt{v})(v^2 + 2\sqrt{v})$

12. $p(v) = \dfrac{v^3 - \sqrt{v}}{v^2 + 2\sqrt{v}}$

13. $p(v) = \dfrac{v^2 + 2\sqrt{v}}{v^3 - \sqrt{v}}$

14. $p(v) = v^{3/2} = v \cdot \sqrt{v}$

15. $p(v) = v^{5/2} = v \cdot v^{3/2}$ (see Problem 14)

16. $p(v) = v^{7/2} = v \cdot v^{5/2}$

17. $m(r) = \dfrac{1}{\sqrt{r}}$

18. $f(x) = \dfrac{1}{x^5 + 3x}$

19. $g(t) = \dfrac{1}{\sqrt{t}}\left(\dfrac{1}{t^4 + 2}\right)$

20. $g(t) = \dfrac{\sqrt{t} - \dfrac{2}{\sqrt{t}}}{3t^3 + 4}$

21. $f(x) = \dfrac{1}{x^6}$

22. $g(t) = t^{-100}$

23. $f(x) = \dfrac{5}{7x^3}$

24. $p(v) = \dfrac{v}{v^5}$.

In Problems 25–28 find the equation of the tangent line to the curve passing through the given point.

25. $f(x) = 4x(x^5 + 1)$, $(1, 8)$

26. $g(t) = \dfrac{t^2}{1 + \sqrt{t}}$, $\left(4, \dfrac{16}{3}\right)$

27. $h(u) = \dfrac{1 + \sqrt{u}}{u^2}$, $(1, 2)$

28. $p(v) = (1 + \sqrt{v})(1 - \sqrt{v})$, $(1, 0)$.

In Problems 29–32 find the equation of the normal line passing through the given point.

29. $f(x) = \dfrac{1}{\sqrt{x}}$, $(1, 1)$

30. $g(t) = \dfrac{1}{t^7}$, $(1, 1)$

31. $f(x) = \dfrac{1 - \sqrt{x}}{1 + \sqrt{x}}$, $(1, 0)$

32. $g(t) = \dfrac{t^3 + 2}{t^2 + 2}$, $(0, 1)$.

33. Let $v(x) = f(x)g(x)h(x)$.

(a) Using the product rule, show that

$$\frac{dv}{dx} = f\frac{d(gh)}{dx} + gh\frac{df}{dx}.$$

(b) Use part (a) to show that

$$\frac{dv}{dx} = fg\frac{dh}{dx} + fh\frac{dg}{dx} + gh\frac{df}{dx}.$$

In Problems 34–36 use the result of Problem 33 to calculate the derivative of the given function.

34. $v(x) = x(1 + \sqrt{x})(1 - \sqrt{x})$

35. $v(x) = (x^2 + 1)(x^3 + 2)(x^4 + 3)$

36. $v(x) = x^{-2}(2 - 3\sqrt{x})(1 + x^3)$.

37. Find the derivative of $fg/(f + g)$ where f and g are differentiable functions.

*38. The mass of the earth is 5.983×10^{24} kg. A meteorite with a mass of 10,000 kg is moving towards a collision with the earth.

(a) When the meteorite is 100 km from the earth, what is the force (in Newtons) of gravitational attraction between the meteorite and the earth?

(b) At that distance, how fast is this force increasing (in nt/m) as the meteorite continues on its collision course?

39. According to Poiseuille's† law, the resistance R of a blood vessel of length l and radius r is given by $R = \alpha l/r^4$ where α is a constant of proportionality determined by the viscosity of blood. Assuming that the length of the vessel is kept constant while the radius increases, how fast is the resistance decreasing when $r = 0.2$ mm?

*40. Find

$$\frac{d}{dv} v^{n/2}$$

where n is a positive integer. [*Hint:* See Problems 14, 15, and 16.]

3.3 The Derivative of Composite Functions—The Chain Rule

In this section we derive a result which greatly increases the number of functions whose derivatives can be easily calculated. The idea behind the result is illustrated below

Suppose that $y = f(u)$ is a function of u and $u = g(x)$ is a function of x.‡ Then $du/dx = g'(x)$ is the rate of change of u with respect to x while $dy/du = f'(u)$ is the rate of change of y with respect to u. We now may ask: what is the rate of change of y with respect to x? That is, what is dy/dx?

As an illustration of this idea, suppose that a particle is moving in the xy plane such that its x-coordinate is given by $x = 3t$, where t stands for time. For example, if t is measured in seconds and x is measured in feet, then, in the x-direction, we suppose that the particle is moving with a velocity of 3 ft/sec. That is, $dx/dt = 3$. In addition, suppose that we know that for every one unit change in the x-direction, the particle moves 4 units in the y-direction; that is, $dy/dx = 4$ (see Figure 1). Now we ask, what is the velocity of the particle, in feet per second, in the y-direction; that is, what is dy/dt? It is clear that for every one unit change in t, x will change 3 ft and so y will change $4 \cdot 3 = 12$ ft. That is, $dy/dt = 12$. We may write this as

$$\frac{dy}{dt} = \frac{dy}{dx}\frac{dx}{dt} = 4 \cdot 3 = 12 \quad \text{ft/sec.} \tag{1}$$

This result states simply that if x is changing 3 times as fast as t and if y is changing 4 times as fast as x, then y is changing $4 \cdot 3 = 12$ times as fast as t.

We now generalize the idea behind this example. Let $y(x) = f(g(x))$. That is, y is the composite function $f \circ g$. It will be very useful to be able to express y' in terms of f' and g'. As will be seen shortly, it can be shown that

$$(f \circ g)'(x) = y'(x) = f'(g(x)) \cdot g'(x). \tag{2}$$

†Jean Louis Poiseuille (1799–1869) was a French physiologist.
‡So that $y(x) = (f \circ g)(x)$ where $f \circ g$ denotes the composition of f and g. See Section 1.7.

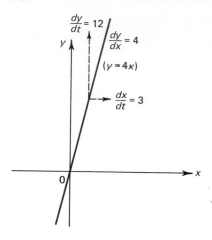

Figure 1

In most of the functions one encounters in applications, $g'(x)$ and $f'(g(x))$ exist for almost all real numbers x.

If we write $u = g(x)$, then $f'(g(x)) = f'(u) = dy/du$, $g'(x) = du/dx$ and we can write Equation (2) in the form

$$\frac{dy}{dx} = \frac{dy}{du}\frac{du}{dx}. \tag{3}$$

The result given by Equation (2) or (3) is called the *chain rule*. Before proving the chain rule, we will illustrate its use with some examples.

EXAMPLE 1.   Let $u = 3x - 6$ and let $y = 7u + 10$. Then $du/dx = 3$ and $dy/du = 7$, so that

$$\frac{dy}{dx} = \frac{dy}{du}\frac{du}{dx} = 7 \cdot 3 = 21.$$

To calculate dy/dx directly, we see that $y = 7u + 10 = 7(3x - 6) + 10 = 21x - 32$. Therefore

$$\frac{dy}{dx} = 21 = \frac{dy}{du}\frac{du}{dx}.$$

EXAMPLE 2.   Let $u = g(x) = \sqrt{x}$ and let $y = f(u) = 2u^2$. Then $y = 2u^2 = 2(\sqrt{x})^2 = 2x$ so that $dy/dx = 2$ by direct calculation. Alternatively, using the chain rule (3), we calculate

$$\frac{du}{dx} = \frac{1}{2\sqrt{x}} \qquad \text{and} \qquad \frac{dy}{du} = 4u$$

so that

$$\frac{dy}{dx} = \frac{dy}{du}\frac{du}{dx} = 4u \cdot \frac{1}{2\sqrt{x}} = 4\sqrt{x} \cdot \frac{1}{2\sqrt{x}} = 2.$$

In these two examples we were able to calculate dy/dx directly. It is usually very difficult to do this and this is why the chain rule is so useful. For example, let us calculate dy/dx where $y = \sqrt{x + x^2}$. If we define $u = x + x^2$, then $y = \sqrt{x + x^2} = \sqrt{u}$. Hence, by the chain rule (3)

$$\frac{dy}{dx} = \frac{dy}{du}\frac{du}{dx} = \frac{1}{2\sqrt{u}}(1 + 2x) = \frac{1}{2\sqrt{x + x^2}}(1 + 2x) = \frac{1 + 2x}{2\sqrt{x + x^2}}.$$

In this case dy/dx could not be calculated directly and there is no other way to calculate this derivative except by using the original definition (Definition 2.9.1), which in this example is very tedious.

Now let $u = g(x)$ and $y = f(g(x)) = f(u)$. We assume that for every point x_0 such that $g(x_0)$ is defined, $f(g(x_0))$ is also defined so that it makes sense to talk about the function $f(g(x))$ at x_0.

THEOREM 1 (Chain Rule). Let g and f be differentiable functions such that the above assumptions hold. Then with $u = g(x)$, the composite function $y = (f \circ g)(x) = f(g(x)) = f(u)$ is a differentiable function of x and

$$\frac{dy}{dx} = \frac{d}{dx}(f \circ g)(x) = \frac{d}{dx}f(g(x)) = f'(g(x))g'(x) = \frac{df}{du}\frac{du}{dx}. \qquad (4)$$

Remark. To simplify the proof we assume that for Δx small, $g(x + \Delta x) - g(x) \neq 0$ (so that we can multiply and divide by it). The case where this assumption is not valid can be treated separately but makes the proof much more complicated. We shall not worry about this point here, but shall give a complete proof of the chain rule in Section 10.5.

PROOF

$$\frac{d}{dx}f(g(x)) = \lim_{\Delta x \to 0} \frac{f(g(x + \Delta x)) - f(g(x))}{\Delta x}$$

$$= \lim_{\Delta x \to 0} \left[\frac{f(g(x + \Delta x)) - f(g(x))}{\Delta x} \cdot \frac{g(x + \Delta x) - g(x)}{g(x + \Delta x) - g(x)}\right]$$

$$= \lim_{\Delta x \to 0} \frac{f(g(x + \Delta x)) - f(g(x))}{g(x + \Delta x) - g(x)} \cdot \lim_{\Delta x \to 0} \frac{g(x + \Delta x) - g(x)}{\Delta x}. \qquad (5)$$

The second limit in (5) is $g'(x) = du/dx$. To evaluate the first limit, we note that as $\Delta x \to 0$, $g(x + \Delta x) \to g(x)$ since the differentiable function $g(x)$ is continuous (Theorem 2.9.1). Then, defining the difference $\Delta g = g(x + \Delta x) - g(x)$, we may write

$$g(x + \Delta x) = g(x) + \Delta g. \qquad (6)$$

Also, $\Delta g \to 0$ as $\Delta x \to 0$ since

$$\lim_{\Delta x \to 0} \Delta g = \lim_{\Delta x \to 0} [g(x + \Delta x) - g(x)] = \left[\lim_{\Delta x \to 0} g(x + \Delta x)\right] - g(x)$$

$$= g(x) - g(x) = 0.$$

Thus, using (6), the first limit of (5) is

$$\lim_{\Delta g \to 0} \frac{f(g(x) + \Delta g) - f(g(x))}{g(x) + \Delta g - g(x)} = \lim_{\Delta g \to 0} \frac{f(g(x) + \Delta g) - f(g(x))}{\Delta g}$$

$$= f'(g(x)) = f'(u) = \frac{df}{du}$$

and the theorem is proved.

The trick in using the chain rule to calculate the derivative dy/dx is to find a function $u(x)$ which has the property that $y(x) = f(u(x))$, where both df/du and du/dx can be calculated without too much difficulty.

EXAMPLE 3. Let $y = f(x) = 1/(x^3 + 1)^5$. Find dy/dx.

SOLUTION. We define $u = g(x) = x^3 + 1$. (There really is no other possibility.) Then $1/(x^3 + 1)^5 = 1/u^5$ and we know how to differentiate $1/u^5$ with respect to u. We have

$$\frac{dy}{du} = \frac{d}{du}\left(\frac{1}{u^5}\right) = \frac{-5}{u^6} \quad \text{and} \quad g'(x) = \frac{d}{dx}(x^3 + 1) = 3x^2.$$

We now use the chain rule to obtain

$$\frac{dy}{dx} = \frac{dy}{du}\frac{du}{dx} = \left(-\frac{5}{u^6}\right)(3x^2) = -\frac{15x^2}{(x^3 + 1)^6}.$$

Another equivalent way to solve this problem is to write $y = f(g(x)) = 1/[g(x)]^5 = [g(x)]^{-5}$, where $g(x) = x^3 + 1$. Then $f'(g(x)) = -5/[g(x)]^6$ and

$$\frac{dy}{dx} = f'(g(x))g'(x) = -\frac{5}{[g(x)]^6}g'(x) = \frac{-15x^2}{(x^3 + 1)^6}.$$

Doing the problem this way avoids the need to introduce the variable u.

EXAMPLE 4. Let $y = f(x) = (\sqrt{x} + 3)^{15}$. Calculate $dy/dx = f'(x)$.

SOLUTION. We choose $u = g(x) = \sqrt{x} + 3$ (again, there is really no other choice) so that $du/dx = 1/2\sqrt{x}$, $y = f(u) = u^{15}$, $dy/du = 15u^{14}$ and

$$f'(x) = \frac{dy}{dx} = \frac{dy}{du}\frac{du}{dx} = \frac{15u^{14}}{2\sqrt{x}} = \frac{15(\sqrt{x} + 3)^{14}}{2\sqrt{x}}.$$

Alternatively, we may write

$$y = f(g(x)) = [g(x)]^{15} \quad \text{where} \quad g(x) = \sqrt{x} + 3.$$

Then

$$\frac{dy}{dx} = f'(g(x))g'(x) = 15[g(x)]^{14} \cdot \frac{1}{2\sqrt{x}} = \frac{15(\sqrt{x} + 3)^{14}}{2\sqrt{x}}.$$

EXAMPLE 5. Let $y = f(x) = \left[\frac{(1 + x)}{(1 + x^2)}\right]^3$. Find $f'(x)$.

SOLUTION. We choose $u = (1 + x)/(1 + x^2)$. Then $y = u^3$ and

$$y' = 3u^2$$

$$\frac{du}{dx} = \frac{(1 + x^2) - (1 + x)(2x)}{(1 + x^2)^2} = \frac{1 - 2x - x^2}{(1 + x^2)^2}$$

so that

$$\frac{dy}{dx} = \frac{dy}{du}\frac{du}{dx} = 3u^2\left[\frac{1 - 2x - x^2}{(1 + x^2)^2}\right] = 3\left[\frac{1 + x}{1 + x^2}\right]^2\left[\frac{1 - 2x - x^2}{(1 + x^2)^2}\right]$$

$$= \frac{3(1 + x)^2(1 - 2x - x^2)}{(1 + x^2)^4}.$$

Note that in each of the last three examples the derivative f' was written as a function of x only. It simply would not make sense to find that the function f' contained functions of u. Remember that $f'(x)$ is the rate of change of f with respect to x and the introduction of the function $u = g(x)$ merely serves as a useful but artificial device for calculating f'. In fact, you will find that after some practice you will be able to calculate the derivatives of the type of functions we encounter in this section without needing to introduce additional functions.

The use of the chain rule may involve other variables and more complicated expressions requiring several of the rules of differentiation. However, it will usually be clear how to proceed.

EXAMPLE 6.   Let $s(t) = [(t^3 + 1)/(t^2 - 5)]^4$. Find ds/dt.

SOLUTION. $s(t)$ is written as a function of t raised to the fourth power. Since we can easily differentiate u^4, we set u equal to this function. That is

$$u = \frac{t^3 + 1}{t^2 - 5}.$$

$$4\left(\qquad\right)^3$$

Then, by the quotient rule,

$$\frac{du}{dt} = \frac{(t^2 - 5)3t^2 - (t^3 + 1)2t}{(t^2 - 5)^2} = \frac{3t^4 - 15t^2 - 2t^4 - 2t}{(t^2 - 5)^2}$$

$$= \frac{t^4 - 15t^2 - 2t}{(t^2 - 5)^2}.$$

Now $s = u^4$ and $ds/du = 4u^3$ so that

$$\frac{ds}{dt} = \frac{ds}{du}\frac{du}{dt} = 4u^3\frac{(t^4 - 15t^2 - 2t)}{(t^2 - 5)^2} = 4\left[\frac{t^3 + 1}{t^2 - 5}\right]^3\left[\frac{t^4 - 15t^2 - 2t}{(t^2 - 5)^2}\right].$$

This expression could be further simplified but that is algebra, not calculus, and we will leave well enough alone here.

There is a formula which can be derived from the chain rule and which is very useful for calculations. This formula is a generalization of the last example.

THEOREM 2 (Power Rule). Let $u(x)$ be a differentiable function of x. Then if n is an integer

$$\frac{d}{dx} u^n = nu^{n-1} \frac{du}{dx}. \tag{7}$$

Alternatively, we may write

$$\frac{d}{dx} [g(x)]^n = n[g(x)]^{n-1} g'(x). \tag{8}$$

PROOF. If $f(u) = u^n$, then, from the chain rule,

$$\frac{d}{dx} u^n = \frac{df}{du} \frac{du}{dx} = nu^{n-1} \frac{du}{dx}.$$

EXAMPLE 7. Find the derivative of $(1 + \sqrt{x})^{100}$.

SOLUTION. Setting $u = 1 + \sqrt{x}$, we have

$$\frac{d}{dx} u^{100} = 100u^{100-1} \frac{du}{dx} = 100(1 + \sqrt{x})^{99} \cdot \frac{1}{2\sqrt{x}} = \frac{50(1 + \sqrt{x})^{99}}{\sqrt{x}}$$

Alternatively, if $g(x) = 1 + \sqrt{x}$, then

$$\frac{dy}{dx} = \frac{d}{dx} [g(x)]^{100} = 100[g(x)]^{99} g'(x) = 100(1 + \sqrt{x})^{99} \cdot \frac{1}{2} \sqrt{x}$$

$$= \frac{50(1 + \sqrt{x})^{99}}{\sqrt{x}}.$$

PROBLEMS 3.3

In Problems 1–25 find the derivative of the given function.

1. $f(x) = (x + 1)^3$

2. $f(x) = (x^2 - 1)^2$

3. $f(x) = (\sqrt{x} + 2)^4$

4. $f(x) = (x^2 - x^3)^4$

5. $f(x) = (1 + x^6)^6$

6. $f(x) = (1 - x^2 + x^5)^3$

7. $f(x) = (x^2 - 4x + 1)^5$

8. $f(x) = \dfrac{1}{(\sqrt{x} - 3)^4}$

9. $s(t) = \left(\dfrac{t + 1}{t - 1}\right)^3$

10. $s(t) = (\sqrt{t} - t)^7$

11. $g(u) = (u^5 + u^4 + u^3 + u^2 + u + 1)^2$

12. $g(u) = \dfrac{5}{u^3 + u + 1}$

13. $h(y) = (y^2 + 3)^{-4}$

14. $h(y) = (y^3 - \sqrt{y} + 1)^{-17}$

15. $f(x) = (x^2 + 2)^5 (x^4 + 3)^3$

16. $f(x) = (x^4 + 1)^{1/2} (x^3 + 3)^4$

17. $s(t) = \dfrac{\sqrt{t^2 + 1}}{(t + 2)^4}$

18. $s(t) = \left(\dfrac{t^4 + 1}{t^4 - 1}\right)^{1/2}$

19. $g(u) = \dfrac{(u^2 + 1)^3(u^2 - 1)^2}{\sqrt{u - 2}}$

20. $g(u) = \dfrac{(x^2 + 1)^2(x^3 + 2)^3}{(x^4 + 3)^{1/2}}$

21. $f(x) = \sqrt{x + \sqrt{1 + \sqrt{x}}}$. [*Hint:* Use the chain rule twice.]

22. $f(x) = \sqrt{x^2 + \sqrt{1 + x^2}}$

23. $h(y) = (y^{-2} + y^{-3} + y^{-7})^{-5}$

24. $p(s) = \left\{\left(1 - \dfrac{1}{s}\right)^{-1} + 4\right\}^{-1}$

25. $w(q) = (q^{-7} - q^{-3})^{-1/2}$.

26. A missile travels along the path $y = 6(x - 3)^3 + 3x$. When $x = 1$ the missile flies off this path tangentially (that is, along the tangent line). Where is the missile when $x = 4$? [*Hint:* Find the equation of the tangent line when $x = 1$.]

27. The mass μ of a 4 m metal rod is given by $\mu = 4(1 + \sqrt{x})^3$ kg where x is the distance along the rod measured from the left. What is the density of the rod (in kg/m) when $x \doteq 1$ m?

*28. Let f and g be differentiable functions with $f(g(x)) = x$. Show that

$$f'(g(x)) = \frac{1}{g'(x)}. \tag{9}$$

(This formula is called the *differentiation rule for inverse functions*. This topic is discussed in detail in Section 10.7.)

3.4 The Derivative of a Power Function

In this section we derive a general formula for finding the derivative of $y = x^r$, where r is a rational number (remember that a rational number is a number of the form $r = m/n$ where m and n are integers and $n \neq 0$). Such a function is called a *power function*.

If r is a positive or negative integer, then we have shown that

$$\frac{d}{dx} x^r = r x^{r-1}. \tag{1}$$

We shall show that formula (1) holds for *all* rational numbers r and shall suggest an extension of that result to all real numbers. This will be done in several steps.

THEOREM 1. Let n be a positive integer. Then

$$\frac{d}{dx} x^{1/n} = \frac{1}{n} x^{(1/n)-1}. \tag{2}$$

PROOF. Recall Equation (3.1.5)

$$\frac{a^n - b^n}{a - b} = a^{n-1} + a^{n-2}b + a^{n-3}b^2 + \cdots + ab^{n-2} + b^{n-1}. \tag{3}$$

Then

$$\frac{a - b}{a^n - b^n} = \frac{1}{a^{n-1} + a^{n-2}b + a^{n-3}b^2 + \cdots + ab^{n-2} + b^{n-1}}. \tag{4}$$

Let $a = (x + \Delta x)^{1/n}$ and $b = x^{1/n}$. Then

$$a^n - b^n = [(x + \Delta x)^{1/n}]^n - [x^{1/n}]^n = x + \Delta x - x = \Delta x, \tag{5}$$

and

$$\frac{a - b}{a^n - b^n} = \frac{(x + \Delta x)^{1/n} - x^{1/n}}{[(x + \Delta x)^{1/n}]^n - (x^{1/n})^n} = \frac{(x + \Delta x)^{1/n} - x^{1/n}}{\Delta x}$$

$$= \frac{1}{[(x + \Delta x)^{1/n}]^{n-1} + [(x + \Delta x)^{1/n}]^{n-2}x^{1/n} + \cdots \atop + (x + \Delta x)^{1/n}(x^{1/n})^{n-2} + (x^{1/n})^{n-1}} \tag{6}$$

Returning to our original problem, we have

$$\frac{d}{dx}x^{1/n} = \lim_{\Delta x \to 0} \frac{(x + \Delta x)^{1/n} - x^{1/n}}{\Delta x}.$$

Then using Equation (6), we obtain

$$\lim_{\Delta x \to 0} \frac{(x + \Delta x)^{1/n} - x^{1/n}}{\Delta x}$$

$$= \frac{1}{x^{(n-1)/n} + x^{(n-2)/n}x^{1/n} + x^{(n-3)/n}x^{2/n} + \cdots + x^{1/n}x^{(n-2)/n} + x^{(n-1)/n}}. \tag{7}$$

The denominator contains n terms, each of which is equal to $x^{(n-1)/n}$ (since $x^{(n-2)/n}x^{1/n} = x^{(n-2)/n+(1/n)} = x^{(n-1)/n}$, and so on). Therefore (7) is equal to

$$\frac{1}{nx^{(n-1)/n}} = \frac{1}{nx^{1-(1/n)}} = \frac{1}{n} \cdot \frac{1}{x^{1-(1/n)}} = \frac{1}{n}x^{(1/n)-1}$$

and the theorem is proved.†

EXAMPLE 1.   Let $y = x^{1/5}$. Calculate dy/dx.

SOLUTION

$$\frac{dy}{dx} = \frac{1}{5}x^{(1/5)-1} = \frac{1}{5}x^{-4/5}.$$

EXAMPLE 2.   Let $y = f(x) = (x^3 + 3)^{1/4}$. Find dy/dx.

SOLUTION. We define $u = x^3 + 3$. Then $y = u^{1/4}$ so that by the chain rule,

$$\frac{dy}{dx} = \frac{dy}{du}\frac{du}{dx} = \frac{1}{4}u^{-3/4}(3x^2) = \frac{1}{4}(x^2 + 3)^{-3/4}(3x^2).$$

EXAMPLE 3.   Prove that the tangent line to a circle is perpendicular to the radius.

SOLUTION. We assume that the circle is centered at the origin (see Figure 1). We may

---

†A much easier proof of this theorem can be given after we have discussed implicit differentiation (see Problem 3.7.33).

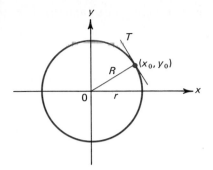

Figure 1

assume this since moving a circle so that its center is shifted to the origin does not affect the angle between the tangent line and the radius. From Section 1.5 we know that the equation of the circle is

$$x^2 + y^2 = r^2, \tag{8}$$

where r is the radius. Then, solving for y in terms of x, we have

$$y = \pm\sqrt{r^2 - x^2}$$

where the plus sign is taken for the parts of the circle in the first two quadrants (the upper half plane) and the minus sign is taken in quadrants 3 and 4 (the lower half plane). We will assume that the point at which the tangent is taken is in the first quadrant so that

$$y = \sqrt{r^2 - x^2}. \tag{9}$$

The proof for the other quadrants is similar. Now let (x_0, y_0) be a point on the circle (in the first quadrant). Then the radius r passes through the points $(0, 0)$ and (x_0, y_0) and therefore has the slope

$$m_r = \frac{y_0 - 0}{x_0 - 0} = \frac{y_0}{x_0}. \tag{10}$$

To find the slope of the tangent line at the point (x_0, y_0) we must calculate dy/dx. Setting $u = r^2 - x^2$ in (9), we have $du/dx = -2x$ and $y = u^{1/2}$. Then

$$\frac{dy}{dx} = \frac{dy}{du}\frac{du}{dx} = \frac{1}{2}u^{-1/2}\frac{du}{dx} = \frac{1}{2\sqrt{u}}(-2x) = \frac{-x}{\sqrt{u}} = \frac{-x}{\sqrt{r^2 - x^2}} = \frac{-x}{y}$$

(since $\sqrt{r^2 - x^2} = y$). Therefore at the point (x_0, y_0), the slope of the tangent line T is given by

$$m_T = -\frac{x_0}{y_0}. \tag{11}$$

Combining (10) and (11) we see that

$$m_T = -\frac{1}{m_R} \tag{12}$$

so that the slopes of the radius and tangent line are negative reciprocals and the lines are therefore perpendicular.

We now extend Theorem 1 to rational numbers.

THEOREM 2. Let $y = x^r$ where $r = m/n$ is a rational number (m and n are integers and $n \neq 0$). Then

$$\boxed{\frac{dy}{dx} = \frac{d}{dx} x^r = rx^{r-1}.}$$

(13)

PROOF

$$\frac{dy}{dx} = \frac{d}{dx} x^r = \frac{d}{dx} x^{m/n} = \frac{d}{dx} (x^{1/n})^m.$$

Let $u = x^{1/n}$. Then $y = u^m$ and

$$\frac{dy}{dx} = \frac{dy}{du}\frac{du}{dx} = mu^{m-1} \cdot \frac{1}{n} x^{(1/n)-1} = m(x^{1/n})^{m-1} \cdot \frac{1}{n} x^{(1-n)/n}$$

$$= \frac{m}{n} (x^{1/n})^{m-1}(x^{1/n})^{1-n} = \frac{m}{n} (x^{1/n})^{m-1+1-n} = \frac{m}{n} x^{(m-n)/n}$$

$$= \frac{m}{n} x^{(m/n)-1} = rx^{r-1}$$

and the theorem is proved.

EXAMPLE 4. Let $y = x^{2/3}$. Calculate dy/dx.

SOLUTION

$$\frac{dy}{dx} = \frac{2}{3} x^{(2/3)-1} = \frac{2}{3} x^{-1/3} = \frac{2}{3x^{1/3}}.$$

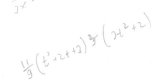

EXAMPLE 5. Let $s = (t^3 + 2t + 3)^{11/9}$. Find ds/dt.

SOLUTION. By now, the reader should recognize that the chain rule is called for in this problem. Let $u = t^3 + 2t + 3$. Then $s = u^{11/9}$ so that

$$\frac{ds}{dt} = \frac{ds}{du}\frac{du}{dt} = \frac{11}{9} u^{2/9}(3t^2 + 2) = \frac{11}{9} (t^3 + 2t + 3)^{2/9}(3t^2 + 2).$$

EXAMPLE 6. Let $g(u) = [\sqrt{u + 3}/(u^2 + 4)]^{4/7}$. Find dg/du.

SOLUTION. Again we use the chain rule but in order to avoid confusion we must pick another letter (other than u) for the function being raised to the $\frac{4}{7}$th power. Let $v(u) = \sqrt{u + 3}/(u^2 + 4)$. Then $g = v^{4/7}$, $dg/dv = \frac{4}{7}v^{-3/7}$ and

$$\frac{dv}{du} = \frac{(u^2 + 4)\dfrac{1}{2\sqrt{u + 3}} - (\sqrt{u + 3})(2u)}{(u^2 + 4)^2} = \frac{4 - 3u^2 - 12u}{2\sqrt{u + 3}(u^2 + 4)^2}.$$

Thus

$$\frac{dg}{du} = \frac{dg}{dv}\frac{dv}{du} = \frac{4}{7}\left(\frac{\sqrt{u+3}}{u^2+4}\right)^{-3/7}\left[\frac{4-3u^2-12u}{2\sqrt{u+3}(u^2+4)^2}\right].$$

We shall rarely, in practice, encounter derivatives as messy as the one in this last example. However it is instructive to see how the chain rule allows us to calculate the derivatives of some very complicated functions.

It is true that the formula

$$\frac{d}{dx}x^\alpha = \alpha x^{\alpha-1} \tag{14}$$

holds when α is any real number. A proof of this fact is beyond the scope of this book. However we may obtain a rough outline of the proof by noting that if α is rational, (14) has already been proven, and if α is irrational, then it can be approximated as closely as possible by a rational number (for example, the irrational number π can be approximated by 3.14, 3.142, 3.1416, 3.14159, etc.).

EXAMPLE 7. Let $y = x^\pi$. Calculate dy/dx.

SOLUTION. Using the formula (14) we have $dy/dx = \pi x^{\pi-1}$.

EXAMPLE 8. Let $y = [2x(1 + \sqrt{x})]^{\sqrt{2}}$. Find dy/dx.

SOLUTION. Let $u = 2x(1 + \sqrt{x}) = 2x + 2x^{3/2}$. Then $y = u^{\sqrt{2}}$, $dy/du = \sqrt{2}u^{\sqrt{2}-1}$ and $du/dx = 2 + 3\sqrt{x}$ so that

$$\frac{dy}{dx} = \frac{dy}{du}\frac{du}{dx} = \sqrt{2}u^{\sqrt{2}-1}(2 + 3\sqrt{x}) = \sqrt{2}[2x(1 + \sqrt{x})]^{\sqrt{2}-1}(2 + 3\sqrt{x}).$$

At this point we see that it is possible to differentiate a wide variety of functions. To aid the reader we give, in Table 1, a summary of the differentiation rules we have so far discussed. In the notation of the table, c stands for an arbitrary constant, and $u(x)$ and $v(x)$ denote differentiable functions.

TABLE 1

Function $y = f(x)$	Its derivative $\dfrac{dy}{dx}$
I. c	$\dfrac{dc}{dx} = 0$
II. $cu(x)$	$\dfrac{d}{dx}cu(x) = c\dfrac{du}{dx}$
III. $u(x) + v(x)$	$\dfrac{d}{dx}(u + v) = \dfrac{du}{dx} + \dfrac{dv}{dx}$
IV. x^r, r a real number	$\dfrac{d}{dx}x^r = rx^{r-1}$
V. $u(x)\cdot v(x)$	$\dfrac{d}{dx}uv = u(x)\dfrac{dv}{dx} + v(x)\dfrac{du}{dx}$

TABLE 1   (*continued*)

Function $y = f(x)$	Its derivative $\dfrac{dy}{dx}$
VI. $u(x)/v(x)$, $v(x) \neq 0$	$\dfrac{d}{dx}(u/v) = \dfrac{v\dfrac{du}{dx} - u\dfrac{dv}{dx}}{v^2}$
VII. $u^r(x)$	$\dfrac{d}{dx}u^r(x) = ru^{r-1}\dfrac{du}{dx}$ (from the chain rule)
VIII. $f(g(x))$	$\dfrac{d}{dx}f(g(x)) = f'(g(x))g'(x)$

PROBLEMS 3.4

In the Problems 1–20 find the derivative of the given function.

1. $f(x) = x^{2/5} + 2x^{1/3}$ 2. $f(x) = 5x^{-2/9}$

3. $f(x) = (x^2 + 1)^{5/3}$ 4. $f(x) = (x^2 - 1)^{-2/3}$

5. $f(x) = (x^3 + 3)^{1/10}$ 6. $f(x) = (x^5 + 2x + 1)^{3/2}$

7. $s(t) = (t^{10} - 2)^{4/7}$ 8. $s(t) = (t^3 + t^2 + 1)^{5/8}$

9. $g(u) = \sqrt[3]{u^3 + 3u + 1}$ 10. $g(u) = \left(\dfrac{u + 2}{u - 1}\right)^{7/2}$

11. $h(z) = \left(\dfrac{z^2 - 1}{z^2 + 1}\right)^{-7/17}$ 12. $h(z) = (z^5 - 1)^{3/4}(z^3 + 2)^{8/9}$

13. $v(r) = (r^2 + 1)^{5/6}(r - 1)^{1/2}$ 14. $v(r) = (r + 1)^{1/3}(r - 1)^{1/4}(r + 2)^{1/5}$

15. $f(x) = 3x^{\sqrt{2}}$ 16. $f(x) = \sqrt{3}\, x^{\sqrt{3}}$

17. $s(t) = (t^2 + 1)^{\sqrt{3}}$ 18. $s(t) = (t^4 - \pi)^{\pi^2}$

19. $g(u) = (u^{\sqrt{2}} - u^{\sqrt{3}})^4$ 20. $g(u) = \dfrac{u^{\sqrt{5}} - 3}{u^{\sqrt{5}} + 1}$.

In Problems 21–24 find the equations of the tangent and normal lines to each of the following curves at the indicated point.

21. $y = x^{1/2} + x^{3/2}$, $(1, 2)$ 22. $y = \dfrac{(x + 2)^3}{(x - 1)^{1/3}}$, $(0, -8)$

23. $y = (x - 1)^{\sqrt{2}}$, $(1, 0)$ 24. $y = (x + 11x^2 + 3x^3 + x^4)^{5/4}$, $(1, 32)$.

*25.   Recall from Section 1.9 that the equation of an ellipse centered at the origin is

$$\frac{x^2}{a^2} + \frac{y^2}{b^2} = 1$$

where a and b are positive real numbers (see Figure 2 and Section 1.9).
   (a)   Write y as two different functions of x (one for quadrants 1 and 2 and one for quadrants 3 and 4).
   (b)   For a point (x_0, y_0) on the ellipse in the first quadrant, find the equation of the radial line R and the tangent line T.
   (c)   Show that R and T are always perpendicular if and only if $a = b$ (in which case the ellipse is really a circle).

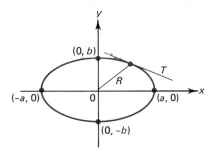

Figure 2

26. In the study of soil mechanics, it is found that the total runoff of rainfall from an area to an inlet reaches its maximum after t minutes, where t is given by

$$t = C\left(\frac{L}{mi^2}\right)^{1/3}.$$

Here C is a constant depending on the nature of the soil, L is the distance from the most remote area, m is the slope of the land, and i is the rain intensity (in in./hr).

(a) What is the instantaneous rate of change of total runoff time as a function of rain intensity if all the other variables are kept constant?

▢(b) Calculate this rate of change when $L = 1500$ ft, $m = 0.1$, $i = 1.1$ in./hr, and $C = 2$.

27. In the kinetic theory of gases, the *root-mean-square* speed of gas molecules is related to the average or typical molecular speed of the molecules comprising the gas and is given by

$$v_{rms} = \sqrt{\frac{3P}{\rho}},$$

where P is the pressure (measured in atmospheres) of the gas and ρ is its density (measured in kg/m³).

▢(a) Calculate v_{rms} of hydrogen molecules at 0 °C and 1-atm of pressure. At that temperature, the density of hydrogen is 8.99×10^{-2} kg/m².

(b) With P fixed, calculate the instantaneous rate of change of v_{rms} with respect to ρ. (It is known that P is inversely proportional to the temperature T with P fixed so that ρ will increase as the gas cools and ρ will decrease as the gas is heated.)

▢(c) Calculate the instantaneous rate of change of v_{rms} as a function of ρ (with P fixed) for the values of P and ρ given in part (a).

28. In astronomy, the *luminosity* of a star is the star's total energy output. Loosely speaking, a star's luminosity is a measure of how bright the star would appear at the surface of the star. The *mass-luminosity relation* gives the approximate luminosity of a star as a function of its mass. It has been found experimentally† that, approximately,

$$\frac{L}{L_0} = \left(\frac{M}{M_0}\right)^r,$$

where L and M are the luminosity and mass of the star and L_0 and M_0 denote the luminosity and mass of the sun. The exponent r depends on the mass of the star as shown in Table 2.

† These data are based on stellar models computed by D. Ezer and A. Cameron in their paper "Early and main sequence evolution of stars in the range 0.5 to 100 solar masses," *Canadian Journal of Physics,* **45,** 3429–3460 (1967).

TABLE 2

Mass range, M/M_0	r
1.0–1.4	4.75
1.4–1.7	4.28
1.7–2.5	4.15
2.5–5	3.95
5–10	3.38
10–20	2.80
20–50	2.30
50–100	1.90

(a) In this model, how is the luminosity changing as a function of mass when the mass increases from 2 solar masses?

(b) How is it changing at a mass of 8 solar masses?

(c) At a mass of 30 solar masses?

*(d) Writing L as a function of M, for what values of M does dL/dM not exist? How would you suggest altering the model so as to avoid discontinuities in this derivative?

3.5 Increasing and Decreasing Functions and Elementary Curve Sketching

In Section 2.8 we saw how we could draw a rough sketch of the derivative of a function by looking at the graph of that function and observing where the function was increasing (the tangent line has a positive slope) and decreasing (the tangent line has a negative slope). In Section 2.9 (Examples 2.9.2, 2.9.3, and 2.9.4) we "reversed" the process and showed how information about the derivative of some simple functions could help us to graph those functions. Now that we know how to calculate quite a few derivatives, we will begin the important task of learning how to draw the graphs of a much wider class of functions. In this section we will make use of a theorem about increasing and decreasing functions. This will enable us to get a fairly good picture about the appearance of the graph of the function in question. Later, in Section 3.9, we will use further facts about derivatives to get even more information about these graphs.

Consider the function whose graph is depicted in Figure 1. In the interval (x_0, x_1), the function increases. In (x_1, x_2) it decreases. From x_2 to x_3, the function again increases, and so on. Moreover, when $x = x_1, x_2, x_3$, and x_4, the tangent line is horizontal so that the derivative is zero at those points. To be more precise, we have the following definitions.

DEFINITION 1. The function f defined on an interval $[a, b]$ is said to be *increasing* on that interval if whenever $a \le x_1 < x_2 \le b$, we have

$$f(x_1) < f(x_2). \tag{1}$$

EXAMPLE 1. In Figure 1, the function is increasing on the intervals $[x_0, x_1], [x_2, x_3]$ and $[x_4, x_5]$.

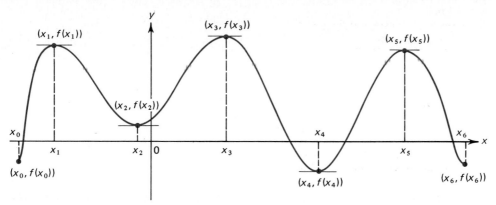

Figure 1

DEFINITION 2. The function f defined on an interval $[a, b]$ is said to be *decreasing* on that interval if whenever $a \le x_1 < x_2 \le b$, we have

$$f(x_1) > f(x_2). \tag{2}$$

EXAMPLE 2. In Figure 1 the function is decreasing on the intervals $[x_1, x_2], [x_3, x_4]$ and $[x_5, x_6]$.

DEFINITION 3. Let f be defined and continuous on the interval $[a, b]$ and let x_0 be in (a, b). Then the number x_0 is called a *critical point* of f if $f'(x_0) = 0$ or if $f'(x_0)$ does not exist.

EXAMPLE 3. In Figure 1 the critical points are x_1, x_2, x_3, x_4, and x_5 because f' evaluated at these points is zero.

EXAMPLE 4. Let $f(x) = |x|$. Then 0 is a critical point since $f'(0)$ does not exist. (See Example 2.9.5.)

Remark. The critical point x_0 is sometimes called a *critical number* or a *critical value*.

In Section 2.8 we inferred that a function was increasing when its derivative was positive and decreasing when its derivative was negative. We now state a theorem that asserts this. We will then indicate why the theorem is true. A simpler and more complete proof will be given in Section 10.6.

THEOREM 1. Let $f(x)$ be differentiable for $a < x < b$ and continuous for $a \le x \le b$.
Then

(i) if $f'(x) > 0$ for every x in (a, b) then f is increasing on $[a, b]$.

(ii) if $f'(x) < 0$ for every x in (a, b), then f is decreasing on $[a, b]$.

INDICATION OF PROOF. We will indicate a proof for part (i). Part (ii) is very similar and is left to the reader. We refer to Figure 2. Let x_0 be in the interval (a, b); that is

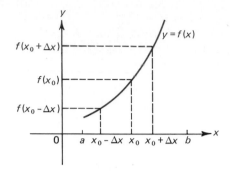

Figure 2

$a < x_0 < b$. We now show that for Δx small and positive, the function $f(x)$ increases when we move from $x_0 - \Delta x$ to x_0 to $x_0 + \Delta x$. That is, we will show that

$$f(x_0 - \Delta x) < f(x_0) < f(x_0 + \Delta x). \tag{3}$$

This will prove that $f(x)$ increases as x increases over small intervals, suggesting that $f(x)$ is increasing over the entire interval. By definition we have

$$f'(x_0) = \lim_{\Delta x \to 0} \frac{f(x_0 + \Delta x) - f(x_0)}{\Delta x}.$$

The definition of the limit tells us that as Δx gets close to zero, the quotient $[f(x_0 + \Delta x) - f(x_0)]/\Delta x$ gets close to $f'(x_0)$ which is greater than 0. Thus, if Δx is small enough, $[f(x_0 + \Delta x) - f(x_0)]/\Delta x$ will also be greater than zero. That is

$$\frac{f(x_0 + \Delta x) - f(x_0)}{\Delta x} > 0. \tag{4}$$

Multiplying both sides of (4) by Δx (which is positive) yields

$$f(x_0 + \Delta x) - f(x_0) > 0 \cdot \Delta x = 0, \quad \text{or} \quad f(x_0) < f(x_0 + \Delta x). \tag{5}$$

Now, since Δx can approach zero from the left or the right in the definition of the derivative, we may write

$$f'(x_0) = \lim_{-\Delta x \to 0} \frac{f(x_0 - \Delta x) - f(x_0)}{-\Delta x}$$

so that for $-\Delta x$ sufficiently close to zero,

$$\frac{f(x_0 - \Delta x) - f(x_0)}{-\Delta x} > 0.$$

Multiplying both sides of this last inequality by $-\Delta x$ (which is negative), we obtain

$$f(x_0 - \Delta x) - f(x_0) < 0 \quad \text{or} \quad f(x_0 - \Delta x) < f(x_0). \tag{6}$$

Finally, (5) and (6) together imply (3).

EXAMPLE 5. For what values of x is the function $f(x) = x^2 - 2x + 4$ increasing and decreasing? Sketch the curve.

SOLUTION. $f'(x) = 2x - 2 = 2(x - 1)$. If $x > 1$, $x - 1 > 0$ and $2(x - 1) > 0$ and if $x < 1$, $x - 1 < 0$ and $2(x - 1) < 0$. Thus for $x < 1$, f is decreasing and for $x > 1$, f is increasing. When $x = 1$, $f'(x) = 0$ so that 1 is a critical point. This is summarized in Table 1. We can use this information to obtain a rough sketch of the function. When

TABLE 1

	$f'(x)$	$f(x)$ is
$-\infty < x < 1$	$-$	decreasing
$x = 1$	0	(critical point)
$1 < x < \infty$	$+$	increasing

$x = 1$, $f(x) = 3$. Table 1 tells us that $f(x)$ decreases until $x = 1$ and increases thereafter. Also, the y-intercept is found by setting $x = 0$ and is at the point $(0, 4)$. We put this information together in Figure 3. We notice that $f(x)$ takes its *minimum value*

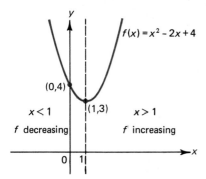

Figure 3

when $x = 1$. In Section 3.9 we will see how maximum and minimum values of a function can be found before drawing a sketch of the curve and shall analyze critical points more closely.

EXAMPLE 6. Let $y = x^3 + 3x^2 - 9x - 10$. For what values of x is this function increasing and decreasing? Sketch the curve.

SOLUTION. $dy/dx = 3x^2 + 6x - 9 = 3(x^2 + 2x - 3) = 3(x + 3)(x - 1)$. Now look at Table 2. The critical points are the numbers -3 and 1. At $x = -3$, $y = 17$ and at

TABLE 2

	$x + 3$	$x - 1$	$\dfrac{dy}{dx} = 3(x + 3)(x - 1)$	f is
$-\infty < x < -3$	$-$	$-$	$+$	increasing
$x = -3$	0	2	0	(critical point)
$-3 < x < 1$	$+$	$-$	$-$	decreasing
$x = 1$	4	0	0	(critical point)
$1 < x < \infty$	$+$	$+$	$+$	increasing

$x = 1, y = -15$. Setting $x = 0$, we obtain the y-intercept $y = -10$. Hence, the curve passes through the point $(0, -10)$. Using this information we obtain the graph given in Figure 4. We notice from the curve that the point $(-3, 17)$ is a maximum point in

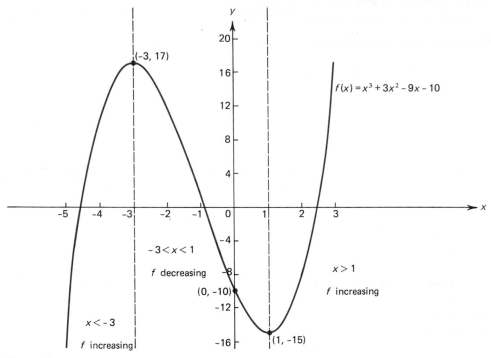

Figure 4

the sense that "near" $x = -3$, y takes its largest value at $x = -3$. However, there is no *global* (or *absolute*) maximum value for the function since as x increases beyond the value 1, y increases without bound. For example, if $x = 10$, $y = 10^3 + 3 \cdot 10^2 - 9 \cdot 10 - 10 = 1200$ which is much bigger than 17. In this setting, the point $(-3, 17)$ is called a *local maximum* (or *relative maximum*) in the sense that the function achieves its maximum value there for points *near* $(-3, 17)$. Similarly, we call the point $(1, -15)$ a *local minimum* (or *relative minimum*).

In general, we have the following definition.

DEFINITION 4.   The function f has a

    (i)   *local maximum* at x_0 if there is an open interval (c, d) containing x_0 such that $f(x_0) \geq f(x)$ for every x in (c, d).

    (ii)   *local minimum* at x_0 if there is an open interval (c, d) containing x_0 such that $f(x_0) \leq f(x)$ for every x in (c, d).

    (iii) *global maximum* at x_0 if $f(x_0) \geq f(x)$ for every x in the domain of f.

    (iv) *global minimum* at x_0 if $f(x_0) \leq f(x)$ for every x in the domain of f.

Note 1. (i)   In Example 6, f has a local, but not a global maximum at $x = -3$. It also has a local, but not global, minimum at $x = 1$. This is evident from Figure 4.

(ii) In Example 5, f has a local *and* global minimum at $x = 1$. There is no local or global maximum.

Note 2. We should point out that the interval (c, d) in Definition 4 could be very small. In Figure 5, $f(x)$ has a relative minimum at x_0 and a relative maximum at x_1

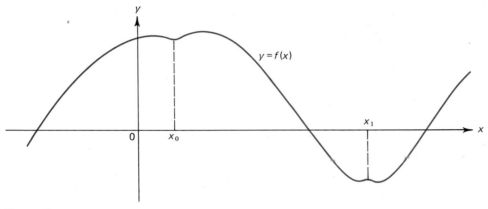

Figure 5

even though these are minimum and maximum values of f over very small intervals. This indicates why calculus methods are superior to other methods for curve sketching. Small "wiggles" such as the two in Figure 5 could easily be missed if we were to try to sketch the curve, say, by plotting points.

As we saw in Examples 5 and 6, whenever we had a local maximum or minimum, we also had a critical point. This is not a coincidence.

THEOREM 2. Let f have a local maximum or minimum at x_0. Then x_0 is a critical point of f.

PROOF. Let f have a local maximum or local minimum at x_0. If $f'(x_0)$ does not exist, x_0 is a critical point. Otherwise we may assume that f has a local maximum at x_0 and that $f'(x_0)$ exists. Then it is necessary to show that $f'(x_0) = 0$. The proof for a local minimum is the same. We will prove this theorem by assuming that $f'(x_0) \neq 0$ to obtain a contradiction of the hypothesis that there is a local maximum at x_0. If $f'(x_0) \neq 0$, then either $f'(x_0) > 0$ or $f'(x_0) < 0$ (see Figure 6). If $f'(x_0) > 0$, then if

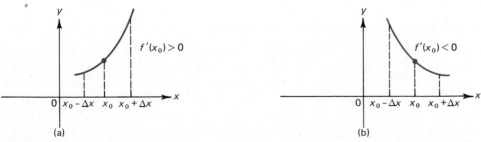

(a)

(b)

Figure 6

Δx is sufficiently small we see from the "proof" of Theorem 1 that

$$f(x_0 - \Delta x) < f(x_0) < f(x_0 + \Delta x)$$

and f cannot have a local maximum at x_0 because $f(x_0 + \Delta x) > f(x_0)$. Similarly, if $f'(x_0) < 0$, then for x_0 sufficiently small, we would have

$$f(x_0 - \Delta x) > f(x_0) > f(x_0 + \Delta x)$$

which again contradicts our hypothesis. This proves the theorem since the only possibility left is $f'(x_0) = 0$.

Remark. The converse of this theorem is not true as the next example shows.

EXAMPLE 7. Let $y = f(x) = x^3$. Then $f'(x) = 3x^2$ which is always positive except at the critical point $x = 0$. If $x < 0$, then $f(x) < 0$ and if $x > 0$, then $f(x) > 0$ so that the function increases from negative values to zero to positive values at $x = 0$ (see Figure 7) and has neither a local maximum nor a local minimum there. Thus, as this example shows, a critical point may be neither a local maximum nor a local minimum.

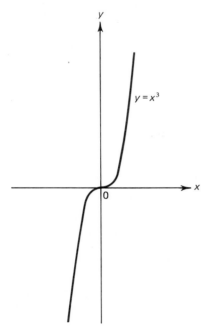

Figure 7

EXAMPLE 8. Sketch the curve $y = x^4 - 8x^2$.

SOLUTION. $dy/dx = 4x^3 - 16x = 4x(x^2 - 4) = 4x(x + 2)(x - 2)$. The critical points are $x = 0$ ($y = 0$), $x = -2$ ($y = -16$), and $x = 2$ ($y = -16$). The behavior of the function is given in Table 3. When $x = 0$, $y = 0$ so the curve passes through the origin. Moreover, we can find the x-intercepts—the points on the curve which meet the x-axis—by setting $y = 0$ to obtain $0 = x^4 - 8x^2 = x^2(x^2 - 8)$. This is equal to zero when $x = 0$ or when $x = \pm\sqrt{8}$. Using all this information, we obtain the curve drawn in Figure 8.

TABLE 3

x	$x + 2$	$x - 2$	$f'(x) =$ $4x(x + 2)(x - 2)$	f Is
$-\infty < x < -2$	$-$	$-$	$-$	$-$	decreasing
$x = -2$	$-$	0	$-$	0	(critical point)
$-2 < x < 0$	$-$	$+$	$-$	$+$	increasing
$x = 0$	0	$+$	$-$	0	(critical point)
$0 < x < 2$	$+$	$+$	$-$	$-$	decreasing
$x = 2$	$+$	$+$	0	0	(critical point)
$2 < x < \infty$	$+$	$+$	$+$	$+$	increasing

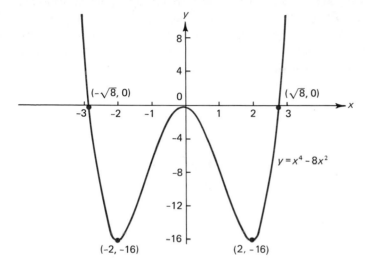

Figure 8

EXAMPLE 9. Let $f(x) = x + (1/x)$. When is this function increasing and decreasing? Sketch the curve.

SOLUTION.

$$f'(x) = \frac{d}{dx}x + \frac{d}{dx}x^{-1} = 1 - x^{-2} = 1 - \frac{1}{x^2} = \frac{x^2}{x^2} - \frac{1}{x^2} = \frac{x^2 - 1}{x^2}.$$

First we notice that the denominator is always positive (for $x \neq 0$). The derivative is not defined for $x = 0$. Thus $f'(x) < 0$ if $x^2 < 1$ and $f'(x) > 0$ if $x^2 > 1$ (see Table 4).

TABLE 4

	x^2	$x^2 - 1$	$\dfrac{x^2 - 1}{x^2}$	f is
$-\infty < x < -1$	> 1	$+$	$+$	increasing
$x = -1$	1	0	0	(critical point)
$-1 < x < 0$	< 1	$-$	$-$	decreasing
$0 < x < 1$	< 1	$-$	$-$	decreasing
$x = 1$	1	0	0	(critical point)
$1 < x < \infty$	> 1	$+$	$+$	increasing

To draw the graph, we need more information. We have

$$(i) \quad \lim_{x \to -\infty} \left(x + \frac{1}{x} \right) = \lim_{x \to -\infty} x + \lim_{x \to -\infty} \frac{1}{x} = -\infty,$$

since the second limit is zero.

$$(ii) \quad \lim_{x \to \infty} \left(x + \frac{1}{x} \right) = \lim_{x \to \infty} x + \lim_{x \to \infty} \frac{1}{x} = \infty.$$

This tells us what happens as x gets large in the positive or negative direction. But what happens near zero? In Example 2.7.2, we saw that $\lim_{x \to 0^+} (1/x) = \infty$ and $\lim_{x \to 0^-} (1/x) = -\infty$. Also, $\lim_{x \to 0} [(x^2 - 1)/x^2] = -\infty$ so the tangent line to the curve becomes vertical as $x \to 0$, since its slope approaches ∞. Finally, as $x \to \infty$, $1/x \to 0$ so that $x + (1/x)$ gets very close to the line $y = x$. This line is called an *asymptote* for the function $y = x + (1/x)$. We now have enough information to sketch the curve (see Figure 9). Here the critical points are the points $(-1, -2)$ and $(1, 2)$. The first is a local maximum and the second is a local minimum.

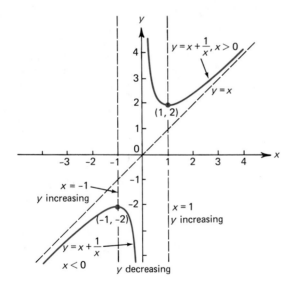

Figure 9

In general, an *asymptote* to a curve is a line that the curve approaches as x approaches some fixed number, $+\infty$, or $-\infty$. In the last example the line $y = x$ is an asymptote since $f(x) \to x$ as $x \to \infty$. Also, the y-axis (the line $x = 0$) is an asymptote since as $x \to 0$ (from the right or left), the slope of the function approaches $\pm\infty$ and so approaches the vertical line $x = 0$. We will discuss asymptotes in greater detail in Section 3.9.

PROBLEMS 3.5

In Problems 1–12: (a) find the intervals over which the given function is increasing or decreasing; (b) find all critical points; (c) find the y-intercept (and the x-intercepts if convenient); (d) sketch the curve; (e) locate all relative maxima and minima on the sketch.

1. $y = x^2 + x - 30$ 2. $y = x^3 - 12x + 10$ 3. $y = x^3 - 3x^2 - 45x + 25$
4. $y = 4x^3 - 3x + 2$ 5. $y = x^4$ 6. $y = x^4 - 18x^2$
7. $y = x^4 - 4x^3 + 4x^2 + 1$ 8. $y = x^5$ 9. $y = x^3 + 3$
10. $y = x^4 + 1$ 11. $y = (x + 1)^{1/3}$ 12. $y = (x - 2)^{2/3}$.

13. Let $f(x) = \sqrt{x + 1}$.
    (a) Show that f is increasing wherever it is defined.
    (b) Show that there are no points on the curve for which $f'(x) = 0$.
    (c) Find the x- and y-intercepts.
    (d) Sketch the curve. What happens as x gets close to -1?
14. Answer the questions of Problem 13 for the function $y = \sqrt[4]{x + 1}$.
15. Answer the questions of Problem 13 for the decreasing function $y = 1/\sqrt{x + 1}$.
16. Sketch the curve $y = x^2 - (1/x^2)$. Be careful near $x = 0$. What function does this curve resemble for large values of x?
17. Sketch the curve $y = (x - 1)/(x - 2)$. Be careful near $x = 2$.
18. Consider the function $y = x/(1 + x)$.
    (a) Show that this function is always increasing except at $x = -1$ where it is not defined.
    (b) What happens as $x \to \pm\infty$?   (e) Show that $y > 1$ if $x < -1$.
    (c) What happens as $x \to -1$?   (f) Sketch the graph.
    (d) Show that $y < 1$ if $x > 0$.
19. Suppose that $f'(x_0) = 0$ and that there is a number δ such that $f'(x) > 0$ for $x \in (x_0 - \delta, x_0)$ and $f'(x_0) < 0$ for $x \in (x_0, x_0 + \delta)$. Show that f has a local maximum at x_0. If $f'(x) < 0$ for $x \in (x_0 - \delta, x_0)$ and $f'(x) > 0$ for $x \in (x_0, x_0 + \delta)$, then show that f has a local minimum at x_0. The rule given here is often called the *first-derivative test for local maxima and minima*.

3.6 Approximation and Differentials

In Definition 2.9.1 we defined the derivative of a function $y = f(x)$ as

$$\frac{dy}{dx} = \lim_{\Delta x \to 0} \frac{\Delta y}{\Delta x} = f'(x). \tag{1}$$

From the meaning of the limit in (1), we see that for Δx "small", $\Delta y/\Delta x$ is close to $f'(x)$. From Figure 1 we see that this means that the slope of the secant line joining

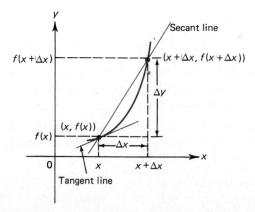

Figure 1

the points $(x, f(x))$ and $(x + \Delta x, f(x + \Delta x))$ and the slope of the tangent line at the point $(x, f(x))$ are approximately the same.

We can use this information to find the approximate value of a function at certain points. If Δx is small, then

$$\frac{\Delta y}{\Delta x} \approx f'(x) \tag{2}$$

or

$$\boxed{\Delta y \approx f'(x)\, \Delta x} \tag{3}$$

where the symbol $\approx$ stands for "is approximately equal to." Equation (3) tells us that if x changes by the *small* amount Δx, then y will change by (approximately) the amount $f'(x)\, \Delta x \approx \Delta y$. The amounts Δx and Δy in this context are called *increments*. How good the approximation is depends, in general, on how small we choose Δx. The approximation will be exact if f is a constant or linear function (since in these cases $dy/dx = \Delta y/\Delta x$).

EXAMPLE 1.   We illustrate in Table 1 the "closeness" of the estimate for the function $y = x^2$ near the value 2. Since $f'(x) = 2x$, we have $f'(2) = 4$. In Table 1, the

TABLE 1

Δx	$2 + \Delta x$	$(2 + \Delta x)^2$	Actual value, $\Delta y = (2 + \Delta x)^2 - 4$	Approximation, $f'(2)\,\Delta x = 4\,\Delta x$	Error
1	3	9	5	4	1.0
0.5	2.5	6.25	2.25	2	0.25
0.1	2.1	4.41	0.41	0.4	0.01
0.05	2.05	4.2025	0.2025	0.2	0.0025
0.01	2.01	4.0401	0.0401	0.04	0.0001
0.001	2.001	4.004001	0.004001	0.004	0.000001

actual value of

$$\Delta y = f(x + \Delta x) - f(x) = (2 + \Delta x)^2 - 2^2$$

is given in column 4 while the approximate value $f'(2)\, \Delta x$ is given in column 5. The error between the actual value and the approximation is given in column 6. We see that the smaller we take Δx, the better our approximation becomes.

□**EXAMPLE 2.**   We do the same thing in Table 2 for the function $y = f(x) = \sqrt{x}$ near the value 4. Here $f'(x) = 1/2\sqrt{x}$ so that $f'(4) = \frac{1}{4}$. Again we see how the approximation gets better and better as Δx decreases.

We now illustrate how this approximation may be useful with a number of examples, always keeping in mind that *the approximation is only valid when Δx is small*. How small Δx must be depends on the individual function and choice of x.

EXAMPLE 3.   Find an approximate value for $(2.01)^3$.

TABLE 2

Δx	$4 + \Delta x$	$\sqrt{4 + \Delta x}$	Actual value, $\Delta y = \sqrt{4 + \Delta x} - 2$	Approximation, $f'(4)\,\Delta x = \dfrac{\Delta x}{4}$	Error
1	5	2.236068	0.236068	0.25	0.013932
0.5	4.5	2.121320	0.121320	0.125	0.00368
0.1	4.1	2.024846	0.024846	0.025	0.000154
0.05	4.05	2.012461	0.012461	0.0125	0.000039
0.01	4.01	2.002498	0.002498	0.0025	0.000002

SOLUTION. Let $y = f(x) = x^3$. Then we are asked to find $f(2.01)$. Since $f(2) = 8$ is known, *we may look upon the number 2.01 as a deviation from 2*. That is, $2.01 = 2 + .01 = x + \Delta x$. From Equation (3) we see that if x changes 0.01 units, then the change in y is given by

$$f(2.01) - f(2) = \Delta y \approx f'(x)\,\Delta x = f'(2)(0.01).$$

But $f'(x) = 3x^2$ so that $f'(2) = 3 \cdot 2^2 = 12$. Hence

$$\Delta y \approx 12(.01) = 0.12.$$

We then have $\Delta y = f(2.01) - f(2)$ so that

$$(2.01)^3 = f(2.01) = f(2) + \Delta y \approx 8 + 0.12 = 8.12.$$

The exact value is 8.120601 so that 8.12 is a good approximation.

EXAMPLE 4. Calculate an approximate value for $\sqrt{16.2}$.

SOLUTION. We choose $f(x) = \sqrt{x}$, $x = 16$ and $\Delta x = 0.2$. (We choose $x = 16$ since $\sqrt{16} = 4$ is easily calculated). Since $f'(x) = 1/2\sqrt{x}$, we have $f'(16) = \frac{1}{8}$ so that

$$\Delta y \approx \tfrac{1}{8}\Delta x = 0.2/8 = 0.025.$$

Thus $\sqrt{16.2} = f(16.2) \approx f(16) + \Delta y = 4.025$. This value of $\sqrt{16.2}$, correct to 5 decimal places is 4.02492 so that this is again a good approximation.

EXAMPLE 5. Calculate an approximate value for $\sqrt[3]{7.95}$.

SOLUTION. We choose $f(x) = x^{1/3}$, $x = 8$ and $\Delta x = -.05$ (since 7.95 is less than 8). Then $f'(x) = 1/3x^{2/3}$ and $f'(8) = \frac{1}{12}$ so that

$$\Delta y \approx \tfrac{1}{12}(-0.05) = -0.004167.$$

Thus

$$\sqrt[3]{7.95} = f(7.95) \approx f(8) - 0.004167 = 2 - 0.004167 \approx 1.995833.$$

The correct value is 1.9958246 . . .

EXAMPLE 6. In this example we illustrate that Δx has to be small in order to obtain a good approximation for Δy. Suppose we try to calculate $(2.4)^2$ by the method of this section. Then choosing $f(x) = x^2$, $x = 2$ and $\Delta x = 0.4$, we obtain, since $f'(x) = 2x = 4$ at $x = 2$,

$$\Delta y \approx 4(0.4) = 1.6$$

so that $f(2.4) \approx 5.6$. But $(2.4)^2 = 5.76$ so that our answer is not even correct to one decimal place. In this example Δx is $\frac{1}{5}$ of x. In general, a good rule of thumb is: *the smaller the ratio $\Delta x/x$, the better the approximate value of Δy.*

EXAMPLE 7. A water tank is built in the shape of a hemisphere (see Figure 2) and is filled with water. The radius of the tank is measured to be 3 meters with a possible

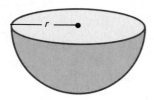

Figure 2

error in the measurement of as much as 0.02 m $= 2$ cm (high or low). Given that the density of water is approximately 1 gm/cm³ $= 1000$ kg/m³ at $4\,°C$

(a)   Calculate the approximate mass of the water in the tank.
(b)   What is the approximate error in your calculation due to the error in measuring the radius of the tank?

SOLUTION. (a)   The volume of a sphere is given by the formula $V = \frac{4}{3}\pi r^3$. The volume of the hemisphere is therefore given by the formula $V(r) = \frac{2}{3}\pi r^3$. If $r = 3$ m, then $V = \frac{2}{3}\pi(3)^3 = \frac{2}{3}\pi(27) = 18\pi$ m³. Since

$$\mu = \text{mass} = \text{volume} \times \text{density},$$

the mass of the water in the tank is approximately

$$\mu = 18\pi \text{ m}^3 \times 1000 \text{ kg/m}^3 = 18{,}000\pi \text{ kg} \approx 56{,}549 \text{ kg}.$$

(b)   If the measurement of the radius is off by an amount Δr where Δr is small, then the calculation of the volume will be off by approximately

$$\Delta V \approx V'(r)\,\Delta r.$$

Here $V'(r) = dV/dr = 2\pi r^2$, $r = 3$ m and Δr is at most ± 0.02 m. Thus

$$\Delta V \approx \pm(2\pi 3^2)(0.02) = \pm(18\pi)(0.02) = \pm 0.36\pi \text{ m}^3.$$

This is our approximate error in the calculation of V. Then the error in the calculation of the mass, $\Delta\mu$, is approximately

$$\Delta\mu = \Delta V \times \text{density} \approx 0.36\pi \text{ m}^3 \times 1000 \text{ kg/m}^3 = 360\pi \text{ kg } (\approx 1131 \text{ kg}).$$

A more interesting problem here is to find the relative error in our calculation.

We define

> relative error in the measurement of a quantity $=$ actual error in measurement / measured value of the quantity.

For example, if the water weighs 1000 kg, then an error of 360π kg ≈ 1131 kg would be a relative error of $1131/1000 = 1.131 = 131.1$ percent. This would be a significant error indeed! However, if the water weighs $18,000\pi$ kg, then the relative error is $(360\pi$ kg$)/(18,000\pi$ kg$) = 0.02 = 2$ percent, a much more tolerable error. In most applications it is usually important to reduce the relative error, while the actual value of the error is far less crucial.

In using the chain rule, we obtained the formula

$$\frac{dy}{dx} = \frac{dy}{du}\frac{du}{dx}. \tag{4}$$

It seems as if in the right-hand side of (4), we have "cancelled off" the terms du to obtain the left-hand side. But the expressions dx, dy and du have been given no meaning by themselves and are only part of a larger expression like dy/dx which stands for "the derivative of y with respect to x." However, it is often convenient to treat the expressions dx, dy and the like as separate entities. We now show how this can be done.

As a basis for further discussion, we return again to the formula (3) which states that if Δx is small and if f is differentiable at x, then

$$\Delta y \approx f'(x)\,\Delta x.$$

That is, given a change in x (small) we can find an approximate value for the change in y. We now generalize this notion.

DEFINITION 1 (The Differential). Let $y = f(x)$ where f is a differentiable function and let Δx be any nonzero real number. Then

(i)   the differential dx is a function given by $dx = \Delta x$.
(ii) the differential dy of f is a function given by

> $$dy = f'(x)\,dx. \tag{5}$$

Note that dx does not have to be a small number in accordance with this definition. In fact, dx can take on any value between $-\infty$ and ∞.

The first thing to notice about these definitions is that (since dx is assumed to be nonzero)

$$\frac{dy}{dx} = \frac{f'(x)\,dx}{dx} = f'(x) \tag{6}$$

which is certainly not surprising since dy was chosen so that (6) would be satisfied. We should also note that the definitions here are artificial in the sense that they have been created so that we can manipulate the symbols dx and dy. It must be emphasized that formulas like the chain rule (4) are true not because differentials can automatically be cancelled, but because they were proven true before we even had such things as differentials around.

EXAMPLE 8. Let $y = x^2$. Calculate the differential dy.

SOLUTION. Since $f'(x) = 2x$, we have $dy = 2x\, dx$.

EXAMPLE 9. Let $y = \sqrt{x + 1}$. Find dy.

SOLUTION. Here

$$f'(x) = \frac{1}{2\sqrt{x + 1}} = \frac{dy}{dx}$$

so that

$$dy = \frac{dx}{2\sqrt{x + 1}}.$$

The differentiation formulas given in Table 3.4.1 can be extended to differential formulas. This is done in Table 3.

TABLE 3

	$y = f(x)$	dy
I.	c	$dc = 0$
II.	$cu(x)$	$d(cu) = c\, du$
III.	$u(x) + v(x)$	$d(u + v) = du + dv$
IV.	x^r, r real	$d(x^r) = rx^{r-1}\, dx$
V.	$u(x) \cdot v(x)$	$d(uv) = u\, dv + v\, du$
VI.	$u(x)/v(x)$, $v(x) \neq 0$	$d(u/v) = \dfrac{v\, du - u\, dv}{v^2}$
VII.	$u^r(x)$	$d(u^r) = ru^{r-1}\, du$
VIII.	$f(g(x))$	$d(f(g(x))) = f'(g(x))g'(x)\, dx$

We close this section by illustrating the relationship between the increment Δy and the differential dy. Consider the function $y = f(x)$, let the number x_0 be fixed and suppose that $f'(x_0)$ exists. For any value of Δx, we have $dx = \Delta x$. The equation

$$dy = f'(x_0)\, dx \tag{7}$$

is the equation of a *straight line* with slope $f'(x_0)$ which coincides with the tangent line to the curve $f(x)$. (This is a straight line in the coordinates (dx, dy) instead of (x, y). The origin in these coordinates is the point $(x_0, f(x_0))$ since $dx = dy = 0$ there.) Now look at Figure 3. We see that $\Delta y = f(x_0 + \Delta x) - f(x_0)$ changes along the curve while dy changes along the tangent line. In this illustration, dy is only close to Δy for small values of $dx = \Delta x$. Tables 1 and 2 given earlier in this section illustrate this phenomenon. In those tables the fourth column is Δy and the fifth column is dy.

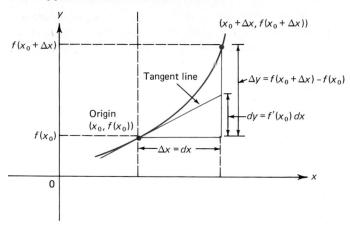

Figure 3

PROBLEMS 3.6

1. Let $y = x^3$. For $x = 3$, tabulate values of Δy and dy for $\Delta x = 1, 0.5, 0.1, 0.05, 0.01$, and 0.001. Show how the approximating error decreases as Δx decreases.

2. Let $y = x^2$. For $x = 1$, tabulate values of Δy and dy for $\Delta x = -1, -0.5, -0.1, -0.05$, -0.01 and -0.001. Show how the approximating error decreases as Δx gets closer to zero.

3. Let $y = x^{3/2}$. For $x = 4$, tabulate values of Δy and dy for $\Delta x = 1, -1, 0.5, -0.5, 0.1$, $-0.1, 0.01$, and -0.01. Show how the approximating error decreases as Δx approaches zero.

In Problems 4–17 calculate an approximate value for the given expression using differentials.

4. $\sqrt{25.03}$

5. $(0.99)^4$

6. $1 + (4.02)^2$

7. $\dfrac{1}{5.1}$

8. $\dfrac{1}{\sqrt{3.98}}$

9. $\dfrac{1}{35^{1/5}}$

10. $(2.02)^3 - 4(2.02)^2$

11. $(1.03)^{20}$

12. $(0.97)^{5/8}$

13. $(15.95)^{3/4}$

14. $(1.01)^2 + (1.01)^4 + (1.01)^6$

15. $(0.98)^7 - (0.98)^3$

16. $\sqrt[4]{80}$

17. $(5 + \sqrt{3.9})^3$.

In Problems 18–32 find the differential of the given function.

18. $y = 3x + 6$

19. $y = 2$

20. $y = x^4$

21. $y = x^{1/3}$

22. $y = (1 + x^2)^4$

23. $y = (1 + x^3)^{1/4}$

24. $y = \dfrac{x + 1}{x - 1}$

25. $y = \dfrac{1}{x}$

26. $y = \sqrt{1 + \sqrt{x}}$

27. $y = \dfrac{1 - \sqrt{x}}{1 + \sqrt{x}}$

28. $y = \dfrac{x + x^2}{1 - x^3}$

29. $y = \sqrt{\dfrac{1 + x}{1 - x}}$

30. $y = (1 + x^3)^{17/2}$

31. $y = \dfrac{1}{x^2 + 2}$

32. $y = (x + 3)^{4/3}(x^2 + 2)^{3/4}$

33. A tank filled with water is in the shape of a right circular cone (Figure 4). The volume of

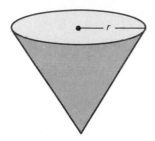

Figure 4

the cone is given by $V = \frac{1}{3}\pi r^2 h$. The radius of the cone is measured to be 2 m with a maximum error of 0.01 m. The height is exactly 3 m.
(a)   What is the approximate mass of the water in the tank?
(b)   By how much could this calculation be off?
(c)   How large a relative error does this represent?

*34.   According to Poiseuille's law, the resistance R of a blood vessel of length l and radius r is given by $R = \alpha l/r^4$, where α is a constant of proportionality. If l remains fixed, how does a small change in r affect the resistance R?

35.   One side of a square field was measured to be 75 ft. What is the relative error in the calculation of the area of the field if the true length of the side is 74.8 ft?

36.   What is the largest allowable relative error allowed in the measurement of the radius of a right circular cone if the error in the calculation of its volume is to be less than 0.8%?

37.   Stefan's law for the emission of radiant energy from the surface of a body is given by $R = \sigma T^4$, where R is the rate of emission per unit area, T is the temperature measured in degrees Kelvin and σ is a universal constant. A relative error of at most 0.02 in T will result in what maximum relative error in R?

38.   A grapefruit with a diameter of 6 inches has a skin $\frac{1}{5}$ of an inch thick (the skin is included in the diameter).
(a)   What is the approximate volume of the skin?
(b)   What percentage of the total volume is skin?

3.7 Implicit Differentiation

In most of the problems we have encountered the variable y was given *explicitly* as a function of the variable x. For example, for each of the functions

$$y = 3x + 6, \quad y = x^2, \quad y = \sqrt{x + 3}, \quad y = 1 + 2x + 4x^3, \quad y = (1 + 8x)^{3/2}$$

the variable y appears alone on the left-hand side. Thus we may say "you give me an x and I'll tell you the value of $y = f(x)$." One exception to this was given in Example 3.4.3 where the variables x and y were given *implicitly* in the equation of the circle of radius r centered at the origin:

$$x^2 + y^2 = r^2. \tag{1}$$

Here x and y are not given separately. In general, we say that x and y are given *implicitly* if neither one is expressed as an explicit function of the other.

Note. This is *not* to say that one variable can *not* be solved explicitly in terms of the other.

EXAMPLE 1. The following are examples in which the variables x and y are given implicitly.

(a) $x^3 + y^3 = 6xy^4$

(b) $(2x^{3/2} + y^{5/3})^{17} - 6y^5 = 2$

(c) $2xy(x + y)^{4/3} = 6x^{17/9}$

(d) $\dfrac{x + y}{\sqrt{x^2 - y^2}} = 16y^5$

(e) $xy = 1$ (Here it is easy to solve for one variable in terms of the other.)

　　For the example of the circle, it was possible to solve the equation (1) explicitly for y in order to calculate dy/dx. However, it is impossible to do the same thing for the functions (a)–(d) given above in Example 1 (try it!). Nevertheless, the derivative dy/dx may exist. Can we calculate it?

　　To illustrate the answer to this question, let us again return to equation (1)

$$x^2 + y^2 = r^2.$$

In Example 3.4.3 we found that, for $y > 0$

$$\frac{dy}{dx} = -\frac{x}{\sqrt{r^2 - x^2}} = -\frac{x}{y}.$$

We now calculate this another way. Assuming that y can be written as a function of x, we can write $y^2 = (f(x))^2$ for some function $f(x)$ (which we assume to be unknown†). Then, by the chain rule,

$$\frac{d}{dx}(y^2) = \frac{d}{dx}(f(x))^2 = 2f(x) \cdot f'(x) = 2y\frac{dy}{dx}. \tag{2}$$

Now taking the derivatives of both sides of (1) with respect to x and using (2), we obtain

$$2x + 2y\frac{dy}{dx} = \frac{d}{dx}r^2 = 0.\ddagger \tag{3}$$

We can now solve for dy/dx in (3):

$$\frac{dy}{dx} = -\frac{x}{y}. \tag{4}$$

If we do not know y as a function of x, then this is as far as we can go. However, since in this case we may choose $y = \sqrt{r^2 - x^2}$, we may write equation (4) as

$$\frac{dy}{dx} = \frac{-x}{\sqrt{r^2 - x^2}}.$$

The method we have used in the above calculation is called *implicit differentiation* and is the only way to calculate derivatives when it is impossible to solve for one variable in terms of the other. We illustrate this method with a number of examples.

---

†In this case we know that $f(x) = \sqrt{r^2 - x^2}$ or $-\sqrt{r^2 - x^2}$, but ordinarily (like the functions in Example 1) $f(x)$ will *really* be unknown.

‡We used the fact that both sides on an equation can always be differentiated provided that the derivatives of the functions involved exist.

Note. We should keep in mind that what makes this technique work is that we are *assuming* that y is a differentiable function of x. Thus we may calculate

$$\frac{d}{dx}(y^2) = 2y \frac{dy}{dx}$$

as in the last example.

EXAMPLE 2. Suppose that

$$x^2 + x^3 = y^4 + y^4. \qquad (5)$$

Find dy/dx.

SOLUTION. By the chain rule,

$$\frac{d}{dx}y^4 = 4y^3 \frac{d}{dx}y = 4y^3 \frac{dy}{dx}.$$

Thus we may differentiate both sides of (5) with respect to x to obtain

$$\frac{d}{dx}x^2 + \frac{d}{dx}x^3 = \frac{d}{dx}y + \frac{d}{dx}y^4$$

or

$$2x + 3x^2 = \frac{dy}{dx} + 4y^3 \frac{dy}{dx} = (1 + 4y^3)\frac{dy}{dx}.$$

Then

$$\frac{dy}{dx} = \frac{2x + 3x^2}{1 + 4y^3}.$$

At the point $(-1, 0)$, for example,

$$\frac{dy}{dx} = \frac{2(-1) + 3 \cdot 1}{1 + 0} = 1$$

and the equation of the tangent line at that point is

$$y = x + 1.$$

In this example, there was no special reason to calculate dy/dx. We may have just as well been asked to calculate dx/dy.† We again use the chain rule to find that

$$\frac{d}{dy}x^2 = 2x \frac{dx}{dy} \qquad \text{and} \qquad \frac{d}{dy}x^3 = 3x^2 \frac{dx}{dy}.$$

Then differentiating both sides of (5) with respect to y yields

$$\frac{d}{dy}x^2 + \frac{d}{dy}x^3 = \frac{d}{dy}y + \frac{d}{dy}y^4$$

†Assuming, of course, that x is a differentiable function of y.

or

$$2x \frac{dx}{dy} + 3x^2 \frac{dx}{dy} = 1 + 4y^3 \quad \text{and} \quad \frac{dx}{dy} = \frac{1 + 4y^3}{2x + 3x^2}.$$

Note that $dx/dy = 1/(dy/dx)$. Although we will not prove this here, it is true that under certain hypotheses $dy/dx = 1/(dx/dy)$ (see Problem 3.3.28 and Section 10.7).

EXAMPLE 3. Find the equation of the tangent line to the curve

$$x^4 + y^4 = 17 \tag{6}$$

at the point $(2, 1)$.

SOLUTION. Since

$$\frac{d}{dx} y^4 = 4y^3 \frac{dy}{dx},$$

we may differentiate both sides of (6) with respect to x to obtain

$$\frac{d}{dx}(x^4) + \frac{d}{dx}(y^4) = \frac{d}{dx}(17) \quad \text{or} \quad 4x^3 + 4y^3 \frac{dy}{dx} = 0$$

(the derivative of a constant is zero). Then

$$4y^3 \frac{dy}{dx} = -4x^3 \quad \text{and} \quad \frac{dy}{dx} = -\frac{x^3}{y^3}.$$

At the point $(2, 1)$, $dy/dx = -8$ so that the equation of the tangent line is

$$\frac{y - 1}{x - 2} = -8 \quad \text{or} \quad 8x + y = 17.$$

EXAMPLE 4. Find the equation of the line tangent to the curve

$$(1 + x^2y)^3 + x\sqrt{y} = 9 \tag{7}$$

at the point $(1, 1)$.

SOLUTION. We first calculate dy/dx. We do this in several steps. First

$$\frac{d}{dx}(1 + x^2y)^3 = 3(1 + x^2y)^2 \frac{d}{dx}(1 + x^2y).$$

Now,

$$\frac{d}{dx} 1 = 0$$

and using the product rule

$$\frac{d}{dx} x^2y = x^2 \frac{d}{dx} y + y \left(\frac{d}{dx} x^2\right) = x^2 \frac{dy}{dx} + 2xy.$$

Thus

$$\frac{d}{dx}(1 + x^2y)^3 = 3(1 + x^2y)^2 \left(2xy + x^2 \frac{dy}{dx}\right).$$

Second, we again use the product rule to obtain

$$\frac{d}{dx} x \sqrt{y} = x \frac{d}{dx} \sqrt{y} + \sqrt{y} \left(\frac{d}{dx} x\right) = x \cdot \frac{1}{2\sqrt{y}} \frac{dy}{dx} + \sqrt{y}.$$

Thus, taking the derivative of both sides of (7), we obtain

$$3(1 + x^2y)^2(2xy) + 3(1 + x^2y)^2x^2 \frac{dy}{dx} + \frac{x}{2\sqrt{y}} \frac{dy}{dx} + \sqrt{y} = 0.$$

We then keep all terms with dy/dx on the left and move everything else to the right

$$\frac{dy}{dx} \left(3(1 + x^2y)^2x^2 + \frac{x}{2\sqrt{y}}\right) = -3(1 + x^2y)^2(2xy) - \sqrt{y},$$

and

$$\frac{dy}{dx} = \frac{-\sqrt{y} - 3(1 + x^2y)^2(2xy)}{3(1 + x^2y)^2x^2 + (x/2\sqrt{y})}.$$

At the point $(1, 1)$, $dy/dx = -25/(12\frac{1}{2}) = -2$. Therefore, the equation of the tangent line is

$$\frac{y - 1}{x - 1} = -2 \qquad \text{or} \qquad 2x + y = 3.$$

We return briefly to the circle $x^2 + y^2 = r^2$. For $y > 0$ we calculated $dy/dx = -x/y$ which is zero when $x = 0$ (and $y = r$). Thus for $x = 0$, the tangent line has slope zero and is horizontal (see Figure 1). Now let us consider x as a function

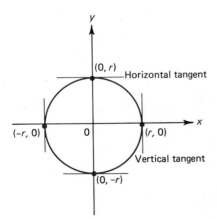

Figure 1

of y (for $x > 0$) and differentiate implicitly with respect to y to obtain

$$2x\frac{dx}{dy} + 2y\frac{dy}{dy} = \frac{d}{dy}r^2 = 0 \quad \text{and} \quad \frac{dx}{dy} = -\frac{y}{x}$$

which is zero at the point $(r, 0)$. If $dx/dy = 0$, then x is a *constant* function of y. Therefore the tangent line to the curve $x^2 + y^2 = r^2$ at point $(0, r)$ is the line $x = r$. This line is vertical as depicted in Figure 1.

We generalize this example with the following rule:

(i)   If $dy/dx = 0$ at the point (x_0, y_0), then the curve $y = f(x)$ has a *horizontal tangent* at that point, given by the straight line $y = y_0$.

(ii)   If $dx/dy = 0$ at the point (x_0, y_0), then the curve $y = f(x)$ has a *vertical tangent* at that point given by the straight line $x = x_0$.

Note. In both of these cases $dy/dx \neq 1/(dx/dy)$ since we cannot divide by zero.

PROBLEMS 3.7

In Problems 1–25 find dy/dx and dx/dy by implicit differentiation.

1. $x^3 + y^3 = 3$ 2. $x^3 + y^3 = xy$ 3. $\sqrt{x} + \sqrt{y} = 2$

4. $xy + x^2y^2 = x^5$ 5. $\dfrac{1}{x} + \dfrac{1}{y} = 1$ 6. $\dfrac{1}{x^2} - \dfrac{1}{y^2} = x + y$

7. $(x + y)^{1/2} = (x^2 + y)^{1/3}$ 8. $xy + x^2y^2 + x^3y^3 = 2$ 9. $\dfrac{1}{\sqrt{x^2 + y^2}} = 4$

10. $\sqrt{x^3} + xy + y^3 = 6$ 11. $(3xy + 1)^5 = x^2$ 12. $\dfrac{x + y}{x - y} = 2$

13. $(x + y)(x - y) = 7$ 14. $\sqrt{xy^2 + yx^2} = 0$ 15. $\dfrac{x^2 + y^2}{x^2 - y^2} = 4$

16. $x^{3/4} + y^{3/4} = 2$ 17. $\dfrac{2xy + 1}{3xy - 1} = 2$ 18. $x^{-1/2} + y^{-1/2} = 4$

19. $xy + x^2y^2 = 2$ 20. $\dfrac{x}{y} + \dfrac{x^2}{y^2} = 3$ 21. $x^{-7/8} + y^{-7/8} = \dfrac{7}{8}$

22. $x^2 - \sqrt{xy} + y^2 = 6$ 23. $(4x^2y^2)^{1/5} = 1$

24. $(\sqrt{x} + \sqrt{y})(\sqrt[3]{x} - \sqrt[3]{y}) = -3$ 25. $x^2y^3 + x^3y^2 = xy$.

26. Find the equation of the line tangent to the curve $x^5 + y^5 = 2$ at the point $(1, 1)$.

For Problems 27–32 find the points where the given curve has a vertical tangent. Also find the points where it has a horizontal tangent.

27. $\sqrt{x} + \sqrt{y} = 1$ 28. $\dfrac{1}{x} + \dfrac{1}{y} = 1$ 29. $xy = 1$

30. $(x - 3)^2 + (y - 4)^2 = 9$

31. $\dfrac{x^2}{a^2} - \dfrac{y^2}{b^2} = 1$ 32. $\dfrac{x^2}{a^2} + \dfrac{y^2}{b^2} = 1$.

*33.   Use implicit differentiation to prove that if n is a positive integer, then

$$\frac{d}{dx} x^{1/n} = \frac{1}{n} x^{(1/n)-1}.$$

(You may use the fact that

$$\frac{d}{dx} x^n = nx^{n-1}.) \ [\textit{Hint:} \ x^{1/n} = y \ \text{means that} \ y^n = x.]$$

3.8 Higher Order Derivatives

Let $s(t)$ represent the distance a falling object has dropped after t units of time have elapsed. Then, as we have seen, the derivative, ds/dt, evaluated at a time t, may be interpreted as the instantaneous velocity of the object at that time. The velocity is the rate of change of position with respect to time. By definition, acceleration is the rate of change of velocity with respect to time. Thus acceleration can be thought of as the derivative of the derivative or, more simply, as the *second derivative* of the function representing position. If, for example, $s = \frac{1}{2}gt^2$ represents the position of a falling object (as in Section 2.10), then

$$\frac{ds}{dt} = gt \quad \text{and} \quad \frac{d}{dt}\left(\frac{ds}{dt}\right) = g,$$

which is the acceleration due to gravity.

We now generalize these ideas. Let $y = f(x)$ be a differentiable function. Then the derivative $y' = dy/dx = f'$ is also a function of x. This new function of x, f' may or may not be a differentiable function. If it is, we call the derivative of f' the *second derivative* of f (i.e., the derivative of the derivative) and denote it by

$$f''. \tag{1}$$

There are other commonly used notations as well. We will write

$$f'' = (f')' = \left(\frac{dy}{dx}\right)' = \frac{d}{dx}\left(\frac{dy}{dx}\right) = \frac{d^2y}{dx^2}. \tag{2}$$

The notations

$$f'' = \frac{d^2y}{dx^2} = y'' \tag{3}$$

will be used interchangeably in this book to denote the second derivative.

Similarly, if f'' exists, it might or might not be differentiable. If it is, then the derivative of f'' is called the *third derivative* of f and is denoted

$$f'''. \tag{4}$$

Alternate notations are

$$f''' = \frac{d^3y}{dx^3} = y'''. \tag{5}$$

We can continue this definition indefinitely as long as each successive derivative is differentiable. After the third derivative, we avoid a clumsy succession of primes by using small Roman numerals to denote higher derivatives:

$$f^{(iv)}, f^{(v)}, f^{(vi)}, \ldots \tag{6}$$

Alternate notations are, for the successive derivatives,

$$f^{(iv)} = \frac{d^4 y}{dx^4} = y^{(iv)} \tag{7}$$

$$f^{(v)} = \frac{d^5 y}{dx^5} = y^{(v)} \tag{8}$$

$$\vdots$$

$$f^{(n)} = \frac{d^n y}{dx^n} = y^{(n)}. \tag{9}$$

We emphasize that *each derivative of $y = f(x)$ is a new function of x* (if it exists).

EXAMPLE 1.   Let $y = f(x) = x^3$. Then

$$\frac{dy}{dx} = f'(x) = 3x^2$$

The second derivative is simply the derivative of the first derivative so that

$$\frac{d^2 y}{dx^2} = f''(x) = \frac{d}{dx} 3x^2 = 6x.$$

Similarly, the third derivative is the derivative of the second derivative and we have

$$\frac{d^3 y}{dx^3} = f'''(x) = \frac{d}{dx} 6x = 6.$$

Finally,

$$\frac{d^4 y}{dx^4} = f^{(iv)}(x) = \frac{d}{dx} 6 = 0.$$

Note that for $k \geq 4$, $f^{(k)}(x) = 0$ since the derivative of the zero function is zero.

EXAMPLE 2.   Let $y = f(x) = 1/x$. Then

$$\frac{dy}{dx} = f'(x) = -\frac{1}{x^2}, \quad \frac{d^2 y}{dx^2} = f''(x) = \frac{2}{x^3}, \quad \frac{d^3 y}{dx^3} = f'''(x) = -\frac{6}{x^4},$$

and so on. In general $d^n y / dx^n = f^{(n)}(x) = (-1)^n \, n! / x^{n+1}$ (see Problem 18). The symbol $n!$, which is read "n factorial," is defined (for the nonnegative integers) by

$$n! = 1 \cdot 2 \cdot 3 \cdot 4 \cdots \cdot n, \quad \text{for} \quad n = 1, 2, 3, \ldots \tag{10}$$

and

$$0! = 1$$

For example, $3! = 1 \cdot 2 \cdot 3 = 6$ and $7! = 1 \cdot 2 \cdot 3 \cdot 4 \cdot 5 \cdot 6 \cdot 7 = 5040$.

Although we have now defined derivatives of all orders, almost all of our major applications will involve first and/or second derivatives. Second derivatives are important for several reasons. In the next section, for example, we shall show that the sign of the second derivative of a function tells us something about the shape of the graph of that function. Perhaps an even more interesting application is the following: from elementary Newtonian physics we may represent a force acting on a particle as the mass of the particle times its acceleration: $F = ma$. Now as we have seen, acceleration is merely the second derivative of the position function. Thus Newton's law states that $F = m\, d^2s/dt^2$. We can then use Newton's laws of motion to derive equations whose solutions tell us how the particle moves. Such applications will be given in Chapters 16 and 17.

PROBLEMS 3.8

In Problems 1–15 find d^2y/dx^2 and d^3y/dx^3.

1. $y = 3$

2. $y = 17x + 1$

3. $y = 4x^2$

4. $y = 9x^3$

5. $y = \sqrt{x}$

6. $y = \dfrac{1}{\sqrt{x}}$

7. $y = (x + 1)^{2/3}$

8. $y = (x^2 + 1)^{1/2}$

9. $y = \sqrt{1 - x^2}$

10. $y = \dfrac{1 + x}{1 - x}$

11. $y = x^r$
     (r a real number)

12. $y = \dfrac{1}{x^2}$

13. $y = ax^2 + bx + c$

14. $y = ax^3 + bx^2 + cx + d$

15. $y = \dfrac{1}{(x + 1)^5}.$

*16.   Use mathematical induction (Appendix 1) to show that the nth derivative,

$$\frac{d^n}{dx^n} x^n = n! = n(n - 1)(n - 2) \cdots 3 \cdot 2 \cdot 1.$$

17.   Let $p(x) = a_n x^n + a_{n-1} x^{n-1} + \cdots + a_1 x + a_0$. Using Problem 16, show that

$$\frac{d^n p}{dx^n} = n! a_n \quad \text{and} \quad \frac{d^{n+1} p}{dx^{n+1}} = 0.$$

*18.   Let $y = 1/x$. Show, using mathematical induction, that

$$\frac{d^n y}{dx^n} = \frac{(-1)^n n!}{x^{n+1}}.$$

19.   A rocket is shot upward in the earth's gravitational field so that its velocity at any time t is given by $v = 50t$. What is its acceleration?

20.   A particle moves along a line so that its position along the line at time t is given by

$$s = 2t^3 - 4t^2 + 2t + 3.$$

The initial position is the position at $t = 0$.
(a)   What is its initial position?
(b)   What is its initial velocity?
(c)   What is its initial acceleration?
(d)   Show that the particle is initially decelerating.
(e)   For what value of t does the particle stop decelerating and begin accelerating?

*21. A 1000 kg asteroid moves so that its distance from a certain planet is given (in meters) by $s = 64 + 8t^2 - 3t^3$. With what force (measured in Newtons) will it strike the planet? [*Hint:* Force in Newtons (nt) $= ma =$ kg $\times$ m/sec^2. The asteroid will hit the planet when $s = 0$.]

*22. Find a function f, continuous on $\mathbb{R}$, such that

$$D(f'') \subsetneq D(f') \subsetneq D(f).$$

Note that $\subsetneq$ means "contained in but not equal to."

3.9 The Shape of a Curve—Concavity

In this section we continue the discussion of curve sketching begun in Section 3.5. Consider the graph of $f(x) = x^2 - 2x + 4$ given in Figure 1a (see Figure 3.5.3). For

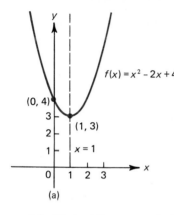

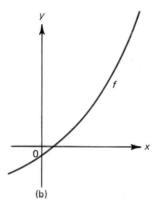

(a)   (b)   Figure 1

$x < 1, f'(x) < 0$; at $x = 1, f'(x) = 0$; and for $x > 1, f'(x) > 0$. Thus the derivative function f' *increases* continuously from negative to positive values. When this occurs, the function is said to be *concave up*. Functions which are concave up have the shape of the curve in Figure 1a or 1b. We may think of the graph of a function which is concave up as a graph which "holds water."

Now consider the function $y = -x^2$ whose graph is given in Figure 2. For $x < 0, f'(x) > 0$, at $x = 0, f'(x) = 0$, and for $x > 0, f'(x) < 0$. Thus the derivative

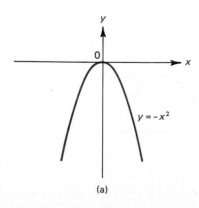

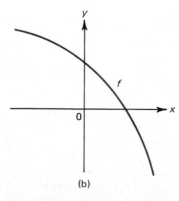

(a)   (b)   Figure 2 decreases continuously from positive to negative values and, in this case, the function is called *concave down*. Functions which are concave down have the shape of the curve in Figure 2a or 2b. We may think of the graph of a function which is concave down as a graph which "spills water."

The more general definition of this concept is given below.

DEFINITION 1.   Let f be differentiable in the interval (a, b). Then

> (i) f is concave up in (a, b) if over that interval f' is an increasing function.
> (ii) f is concave down in (a, b) if over that interval f' is a decreasing function.

Remark. In this definition it is essential that f' exist in (a, b). For example, the function $f(x) = |x|$ "holds water" and has a derivative which goes from negative values (-1) to positive values $(+1)$; yet it has "no concavity" in any interval containing 0.

Many functions will alternate between concave up and concave down.

DEFINITION 2.   The number x_0 is a *point of inflection* of f if $f'(x_0)$ exists and if f changes its direction of concavity at x_0. That is, if there is an interval (c, d) containing x_0 such that

> (i) f is concave down in (c, x_0) and concave up in (x_0, d) or
> (ii) f is concave up in (c, x_0) and concave down in (x_0, d).

EXAMPLE 1.   Consider the function graphed in Figure 3. The function is concave

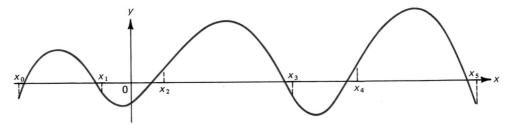

Figure 3

down in the intervals (x_0, x_1), (x_2, x_3) and (x_4, x_5) and concave up in the intervals (x_1, x_2) and (x_3, x_4). The points x_1, x_2, x_3 and x_4 are points of inflection.

Another way to think of concavity is suggested by Figure 4. Here we have drawn

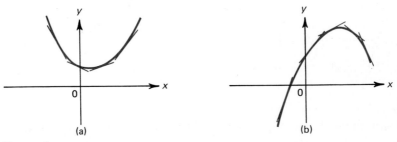

Figure 4

in some tangent lines. In Figure 4a f is concave up and all points on the curve lie *above* the tangent lines. In 4b, f is concave down and all points on the curve lie *below* the tangent lines. This information is very useful in curve plotting. Suppose that f is concave down in (a, b). Then the type of behavior exhibited in Figure 5 is impossible.

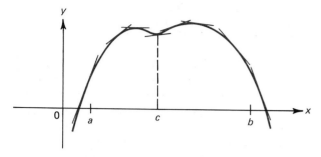

Figure 5

At the point c the curve lies above the tangent line but this is impossible since f is concave down. In general, if a curve is concave up or down in a certain interval, then it cannot "wiggle around" in that interval (i.e., it cannot behave like the curve in Figure 5).

Before looking at other examples, let us make some observations. If f is concave up then f' is increasing. This means that the derivative of f', f'', is positive (that is, $f'' > 0$). Similarly, if f is concave down, then f' is decreasing and f'' is negative. On the other hand, if the second derivative $f'' > 0$, then f' is increasing and f is concave up. If $f'' < 0$, then f' is decreasing and f is concave down. Thus if f'' exists, *we can determine the direction of concavity by simply looking at the sign of the second derivative:*

> $f'' > 0$ means concave up;
> $f'' < 0$ means concave down.

Now suppose that f has a point of inflection at x_0. That means that at x_0, f' goes from increasing to decreasing (Figure 6a) or from decreasing to increasing (Figure 6b). In the first case, f' has a local maximum at x_0 and in the second case,

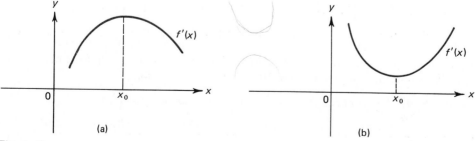

Figure 6

f' has a local minimum at x_0. In either case, by Theorem 3.5.2 $(f')' = f'' = 0$ at x_0. This implies the following result.

THEOREM 1. At a point of inflection x_0, if $f''(x_0)$ exists, then $f''(x_0) = 0$.

> **Note:** The converse of this theorem is not true in general. That is, if $f''(x_0) = 0$, then x_0 is not necessarily a point of inflection (see Example 7).

Finally, suppose that $f'(x_0) = 0$. If $f''(x_0) > 0$, then the curve is concave up at x_0. A glance at Figure 7a suggests that f has a local minimum at x_0. Similarly, if

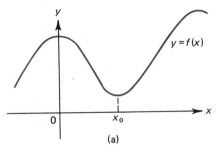

(a)

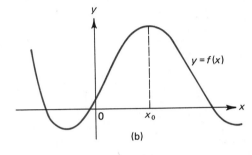

(b)

Figure 7

$f'(x_0) = 0$ and $f''(x_0) < 0$, then f has a local maximum at x_0. It is not difficult to prove these statements (see Problems 49 and 50). In sum, we have

THEOREM 2. Let f be differentiable on an open interval containing x_0. Then

> (i) if $f'(x_0) = 0$ and $f''(x_0) > 0$, then f has a local minimum at x_0.
> (ii) if $f'(x_0) = 0$ and $f''(x_0) < 0$, then f has a local maximum at x_0.

> **Remark.** If $f'(x_0) = 0$ and $f''(x_0) = 0$, then f may have a local maximum, a local minimum, or neither at x_0. For example, $y = x^3$ has neither at $x = 0$.

We now give some examples of the technique of curve sketching making use of information derived from the second derivative.

EXAMPLE 2. If $f(x) = x^2 - 2x + 4$, then $f'(x) = 2x - 2$ and $f'' = 2$ which is greater than 0. Thus the curve is concave up for $-\infty < x < \infty$. This justifies the accuracy of the graph in Figure 1(a). The function f has a local (and global) minimum at the point (1, 3).

EXAMPLE 3. If $f(x) = -x^2$, then $f''(x) = -2$ which is less than 0. This curve is concave down for $-\infty < x < \infty$. This justifies Figure 2(a). The function f has a local (and global) maximum at the point (0, 0).

EXAMPLE 4. Let $f(x) = 2x^3 - 3x^2 - 12x + 5$. Sketch the curve.

SOLUTION. $f'(x) = 6x^2 - 6x - 12 = 6(x^2 - x - 2) = 6(x - 2)(x + 1)$, and

$$f''(x) = 12x - 6 = 6(2x - 1).$$

The curve is

concave down if $x < \tfrac{1}{2}$ and   concave up if $x > \tfrac{1}{2}$.

The point $x = \frac{1}{2}(y = -\frac{3}{2})$ is a point of inflection. The critical points are $x = 2$ ($y = -15$) and $x = -1$ ($y = 12$). When $x = 2, f'' = 6(4 - 1) = 18 > 0$ so that f has a local minimum at $(2, -15)$. When $x = -1, f'' = -18 < 0$ so that f has a local maximum at $(-1, 12)$. From an examination of f', we see that f increases up to $x = -1$, decreases for $-1 < x < 2$, and increases thereafter. When $x = 0, y = 5$ so the y-intercept is the point $(0, 5)$. Putting this all together we obtain the curve in Figure 8. From this graph we can see that the graph crosses the x-axis at three places

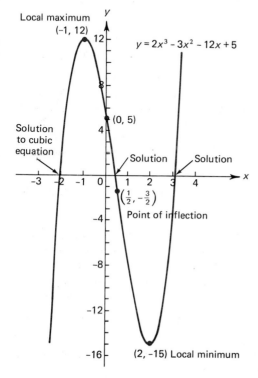

Figure 8

(the three x-intercepts). This tells us that the cubic equation

$$2x^3 - 3x^2 - 12x + 5 = 0$$

has three real solutions. Moreover, we see that two of these solutions are positive and one is negative. This information, which comes without additional work, can often be very useful.

EXAMPLE 5. Graph the curve $y = x^4 - 4x^3 + 4x^2 + 1$.

SOLUTION. We have

$$\frac{dy}{dx} = 4x^3 - 12x^2 + 8x = 4x(x^2 - 3x + 2) = 4x(x - 1)(x - 2)$$

and

$$\frac{d^2y}{dx^2} = 12x^2 - 24x + 8 = 4(3x^2 - 6x + 2).$$

The second derivative is zero when $3x^2 - 6x + 2 = 0$ or

$$x = \frac{6 \pm \sqrt{36 - 24}}{6} = \frac{6 \pm \sqrt{12}}{6} = \frac{3 \pm \sqrt{3}}{3} = 1 \pm \frac{\sqrt{3}}{3} = 1 \pm \frac{1}{\sqrt{3}}.$$

If $x < 1 - (1/\sqrt{3})$ then $d^2y/dx^2 > 0$. If $x > 1 + (1/\sqrt{3})$ then $d^2y/dx^2 > 0$ also. If $1 - (1/\sqrt{3}) < x < 1 + (1/\sqrt{3})$, then $d^2y/dx^2 < 0$. To see this simply plug in some sample points (try the points $x = 0$, $x = 2$, and $x = 1$). Hence the curve is

concave up for $x < 1 - \dfrac{1}{\sqrt{3}}$,

concave down for $1 - \dfrac{1}{\sqrt{3}} < x < 1 + \dfrac{1}{\sqrt{3}}$, and

concave up for $x > 1 + \dfrac{1}{\sqrt{3}}$.

The points of inflection are the points $x = 1 - (1/\sqrt{3})(y = \frac{13}{9})$ and $x = 1 + (1/\sqrt{3})(y = \frac{13}{9})$. The critical points are $x = 0(y = 1)$, $x = 1(y = 2)$, and $x = 2(y = 1)$. When $x = 0$, $d^2y/dx^2 = 8$ so there is a local minimum at $(0, 1)$. Similarly, there is a local maximum at $(1, 2)$ and a local minimum at $(2, 1)$. Putting all this together, we obtain the graph in Figure 9. Since the graph *never* crosses the

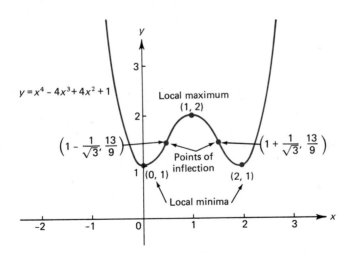

Figure 9

x-axis, we see immediately that the quartic equation

$$x^4 - 4x^3 + 4x^2 + 1 = 0$$

has *no* real solutions.

EXAMPLE 6.   Let $f(x) = x/(1 + x)$. Graph the function.

SOLUTION. Before doing any calculations, we first note that we will have problems at $x = -1$ since the function is not defined there. We can expect that the function will "blow up" as x approaches -1 from the right or left. Now

$$f'(x) = \frac{d}{dx} \frac{x}{1 + x} = \frac{(1 + x)1 - x}{(1 + x)^2} = \frac{1}{(1 + x)^2}.$$

Therefore f is always increasing (except at $x = -1$ where it is not even defined). Also f has *no* critical points (-1 is *not* a critical point because f is not defined at -1). Now

$$f''(x) = \frac{d}{dx} (1 + x)^{-2} = \frac{-2}{(1 + x)^3}.$$

If $x < -1$, then $1 + x < 0$,

$$(1 + x)^3 < 0 \quad \text{and} \quad \frac{-2}{(1 + x)^3} > 0.$$

If $x > -1$, then $1 + x > 0$,

$$(1 + x)^3 > 0 \quad \text{and} \quad \frac{-2}{(1 + x)^3} < 0.$$

Hence f is

concave up if $x < -1$ and   concave down if $x > -1$.

Consider

$$\lim_{x \to \infty} \frac{x}{1 + x} = \lim_{x \to \infty} \frac{1}{(1/x) + 1} = 1$$

and

$$\lim_{x \to -\infty} \frac{x}{1 + x} = \lim_{x \to -\infty} \frac{1}{(1/x) + 1} = 1.$$

Therefore as $x \to \pm\infty$, $f(x) \to 1$. The line $x = 1$ is called a *horizontal asymptote* for the function $x/(1 + x)$. We also observe that if

$x < -1$,   then $f(x) > 1$ (a negative over a negative with the numerator more negative than the denominator)

$-1 < x < 0$,   then $f(x) < 0$

$x = 0$,   then $f(x) = 0$

$x > 0$,   then $0 < f(x) < 1$ (denominator larger than the numerator).

Also $\lim_{x \to -1^+} [x/(1 + x)] = -\infty$ (since the numerator is negative and the denominator is positive) and $\lim_{x \to -1^-} [x/(1 + x)] = \infty$. Finally

$$\lim_{x \to \pm\infty} f'(x) = \lim_{x \to \pm\infty} \frac{1}{(1 + x)^2} = 0$$

so that the tangent lines become flat (horizontal) as $x \to \infty$ and

$$\lim_{x \to -1} f'(x) = \lim_{x \to -1} \frac{1}{(1 + x)^2} = \infty$$

so that the tangent lines become vertical (that is what an infinite slope means) as $x \to -1$ (from either side); the line $x = -1$ is called a *vertical asymptote*. We put this all together in Figure 10.

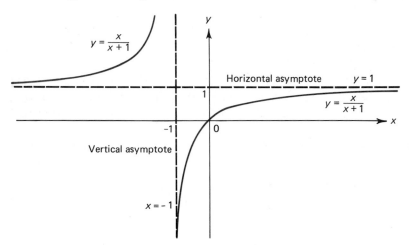

Figure 10

EXAMPLE 7.   Graph the curve $y = x^4$.

SOLUTION. First we note that $x^4 \geq 0$ for every x. Now $dy/dx = 4x^3$ which is greater than 0 for $x > 0$ and less than 0 for $x < 0$. The only critical point is $x = 0$ ($y = 0$). Since $x^4 > 0$ if $x \neq 0$, $x = 0$ is a minimum. However $d^2y/dx^2 = 12x^2$ and this is zero at $x = 0$ so the second derivative does not tell us that the function has a minimum at $x = 0$ (fortunately, that was obvious). Since $d^2y/dx^2 > 0$ if $x \neq 0$, the function is concave up. The point $x = 0$ is *not* a point of inflection even though d^2y/dx^2 is zero there (since the direction of concavity doesn't change). The curve is graphed in Figure 11.

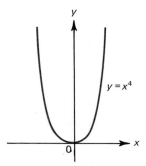

Figure 11

We now review the kinds of information that are available to help us sketch curves. The following steps should be carried out in order to obtain an accurate picture. *To sketch a curve* $y = f(x)$:

> (i)   Calculate the derivative $dy/dx = f'$ and determine where the curve is increasing and decreasing. Find all points at which $f' = 0$.
>
> (ii)   Calculate the second derivative $f'' = d^2y/dx^2$ and determine where the curve is concave up and concave down.
>
> (iii)   If x_0 is a point such that $f'(x_0) = 0$ and $f''(x_0) < 0$, then f has a relative maximum at x_0. If $f'(x_0) = 0$ and $f''(x_0) > 0$, then f has a relative minimum at x_0.
>
> (iv)   Find all points of inflection.
>
> (v)   Determine the y-intercept and (if possible) the x-intercept(s).
>
> (vi)   Determine $\lim_{x \to \infty} f(x)$ and $\lim_{x \to -\infty} f(x)$. If either of these is finite, then we obtain horizontal asymptotes (as in Example 6).
>
> (vii)   If f is not defined at x_0, determine $\lim_{x \to x_0^+} f(x)$ and $\lim_{x \to x_0^-} f(x)$.
>
> (viii)   Look for vertical asymptotes. These are the lines of the form $x = x_0$, where $\lim_{x \to x_0^+} f'(x) = \infty$ (or $-\infty$) or $\lim_{x \to x_0^-} f'(x) = \infty$ (or $-\infty$) (see Example 6).

Note that if $x = x_0$ is a vertical asymptote and if f is continuous in an interval containing x_0, then x_0 is a critical point of f since $f'(x_0)$ does not exist.

PROBLEMS 3.9

In Problems 1–32 follow the steps outlined in this section to graph the given function.

1. $y = x^2 + x - 30$

2. $y = x^3 - 12x + 10$

3. $y = x^3 - 3x^2 - 45x + 25$

4. $y = 4x^3 - 3x + 2$

5. $y = x^3 + 3$

6. $y = x^4 + 1$

7. $y = 2x^3 - 9x^2 + 12x - 3$

8. $y = \dfrac{x^3}{3} + \dfrac{x^2}{2} - 2x - \dfrac{2}{3}$

9. $y = 2x(x + 4)^3$

10. $y = x^3 - x^2$

11. $y = x^4 - x^3$

12. $y = 3x^5 - 5x^3$

13. $y = \sqrt{1 + x}$

14. $y = \sqrt[3]{1 + x}$

15. $y = x\sqrt{1 + x}$

16. $y = x\sqrt[3]{1 + x}$

17. $y = (x - 2)^2(x + 2)^2$

18. $y = \dfrac{1}{x - 3}$

19. $y = 1 + \dfrac{1}{x^2}$

20. $y = \dfrac{x^2 - 1}{x^2}$

21. $y = \dfrac{1}{(x - 1)(x - 2)}$

22. $y = \dfrac{x - 1}{x - 2}$

23. $y = \dfrac{1}{x^2 - 1}$

24. $y = \dfrac{x}{x^2 - 1}$

25. $y = x - \sqrt[3]{x}$

26. $y = \dfrac{x + 1}{\sqrt{x - 1}}$

27. $y = \sqrt[3]{x^2 - 1}$

28. $y = \sqrt[3]{x^2 + 1}$

29. $y = \dfrac{x^2 + x + 7}{\sqrt{2x + 1}}$

30. $y = \dfrac{4x - 4}{3x + 6}$

31. $y = x^{2/3}(x - 3)$

32. $y = (x - 1)^{1/5}$.

33.   Show that if $f(x) = x^5$, then f has a critical point at $x = 0$ but neither a local maximum nor local minimum there. Is $x = 0$ a point of inflection?

34.   Show that $f(x) = x^2 - 6x + 13$ is always positive. [*Hint:* Graph the curve.]

35. Show that the cubic equation $x^3 + x^2 + 5x - 15 = 0$ has exactly one real root. Is it positive or negative?

36. How many real roots do each of the following cubic equations have?
    (a) $x^3 + 1 = 0$ (b) $x^3 + x^2 + x + 1 = 0$
    (c) $4x^3 - 3x^2 + 2x - 1 = 0$.

37. Show that the equation $x^{11} + x^3 + x + 3 = 0$ has exactly one real root. [*Hint:* Consider its graph.]

38. Show that the quartic equation $x^4 + x^3 - 5x^2 - 6 = 0$ has exactly two real roots. Find open intervals that contain them.

39. Show that the quartic equation $x^4 + 10x^2 + 21 = 0$ has no real roots.

40. Explain why the cubic equation $ax^3 + bx^2 + cx + d = 0$ (a, b, c, d real numbers) always has at least one real root. [*Hint:* Consider its graph.]

41. Explain why the quintic equation $ax^5 + bx^4 + cx^3 + dx^2 + ex + f = 0$ with all coefficients real, has at least one real root.

A function is *piecewise linear* if it is continuous and its graph is made up by joining together a number of straight line segments. For example $y = |x|$ is piecewise linear (see Figure 2.2.8). In general, a piecewise linear function looks something like what is shown in Figure 12.

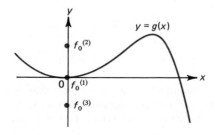

Figure 12

42. Show that the function $f(x) = x + |x|$ is piecewise linear and graph it.
43. Graph $f(x) = |x + 3| + x + 3$.
44. Graph $f(x) = |x + 3| + |x - 3| + 2$. How many "pieces" does it take?
45. Graph $f(x) = |x| + |x - 1| + |x - 2| + |x - 3|$. How many pieces does it take?
46. Graph $f(x) = |2x - 4| + |3x - 3| + 2x - 7$.
*47. Graph $f(x) = |x^2 - 3|$.
*48. Graph $f(x) = |x^2 - 3| + |x^2 - 1|$.
49. Prove that if $f'(x_0) = 0$ and $f''(x_0) < 0$, then f has a local maximum in some interval $[c, d]$. [*Hint:* Show that near x_0, $f'(x_0) > 0$ for $x < x_0$ and $f'(x_0) < 0$ for $x > x_0$. Explain why this implies that f has a local maximum at x_0.]
50. Explain why if $f'(x_0) = 0$ and $f''(x_0) > 0$, then f has a local minimum at x_0.
*51. A function $f(x)$ satisfies the *differential equation*

$$\frac{df}{dx} = g(x), \qquad f(0) = f_0.$$

Figure 13

The graph of $y = g(x)$ is sketched in Figure 13 and three values for f_0 are shown. Draw a rough sketch of the function $f(x)$ for each of the three possible values for f_0. In your sketch indicate where the function is increasing and decreasing, and indicate maxima, minima, concavity, points of inflection, and asymptotes.

51. Show that if $f''(x_0) = 0$ and $f'''(x_0) \neq 0$, then x_0 is a point of inflection.

Review Exercises for Chapter Three

In Exercises 1–20 calculate dy/dx.

1. $y = 3x + 4$
2. $y = 3x^2 - 6x + 2$

3. $x + y + xy = 3$
4. $y = \dfrac{x + 1}{x - 2}$

5. $y = (3x^2 + 2)\sqrt{x + 1}$
6. $y = x^5 - \sqrt{x}$

7. $x^{3/4} + y^{3/4} = 4$
8. $y = (x^2 - 1)^{1/2}(x^4 - 2)^{1/3}$

9. $y = \sqrt[3]{\dfrac{x + 2}{x - 3}}$
10. $xy^2 + yx^2 = xy$

11. $y = \dfrac{x^2 - 2x + 3}{x^3 + 5x - 6}$
12. $y = \dfrac{3}{x - \sqrt{x^3 - 2}}$

13. $y = (x^3 - 3)^{5/6}(x^3 + 4)^{6/7}$
14. $y = \dfrac{1 + x + x^5}{3 - x^2 + x^7}$

15. $x + \sqrt{x + y} = y^{1/3}$
16. $y = \dfrac{1 - \sqrt[3]{x}}{1 + \sqrt[3]{x}}$

17. $y = (1 + x^3)^{\sqrt{2}}$
18. $(1 + x^2 + y^2)^{3/2} = 6$

19. $y = \dfrac{(x^3 - 1)^{2/3}}{(x^5 + 3)^{3/4}}$
20. $y = \sqrt{1 - \sqrt{1 - \sqrt{x}}}.$

In Exercises 21–24 find the equations of the tangent and normal lines to the given curve at the given point.

21. $y = x^3 - x + 4$; (1, 4)
22. $y = \dfrac{\sqrt{x^2 + 9}}{x^2 - 6}$; $\left(4, \dfrac{1}{2}\right)$

23. $y = (x^2 - 4)^2(\sqrt{x} + 3)^{1/2}$; (1, 18)
24. $xy^2 - yx^2 = 0$; (1, 1).

In Exercises 25–30 calculate the second and third derivatives of the given function.

25. $y = x^7 - 7x^6 + x^3 + 3$
26. $y = \sqrt{1 + x}$
27. $y = \dfrac{1}{1 + x}$

28. $y = \dfrac{x + 1}{x - 1}$
29. $y = \dfrac{x^2 - 3}{x^2 + 5}$
30. $y = \sqrt{1 + \sqrt{x}}.$

In Exercises 31–36, find the differential dy.

31. $y = 17x^3 - 6$
32. $y = \sqrt{x^2 - 3}$
33. $y = \dfrac{x + 4}{x - 7}$

34. $y = \dfrac{-3}{1 + x}$
35. $y = (\sqrt{x^2 - 3})\sqrt[3]{x^4 + 5}$
36. $y = \sqrt{1 + \sqrt{x}}.$

37. Using differentials, find an approximate value for (a) $(1.99)^5$.   (b) $1/\sqrt[3]{8.02}$.

38. The radius of a sphere is 1 m with a possible error in measurement of 0.01 m = 1 cm. What is the maximum error you would expect in the calculation of the volume?

In Exercises 39–50 graph the given curves by following the eight steps outlined in Section 3.9.

39. $y = x^2 - 3x - 4$ 40. $y = x^3 - 3x^2 - 9x + 25$

41. $y = x^5 + 2$ 42. $y = \sqrt[3]{x}$

43. $y = \sqrt[4]{x}$ 44. $y = \dfrac{1}{x + 2}$

45. $y = x(x - 1)(x - 2)(x - 3)$ 46. $y = \dfrac{x + 1}{x + 2}$

47. $y = \dfrac{1}{x^2 - 1}$ 48. $y = -x(x - 1)^3$

49. $y = |x - 4|$ 50. $y = |x^2 - 4|.$

FOUR

THE INTEGRAL

4.1 The Area Problem

Modern calculus has its origins in two mathematical problems of antiquity. The first of these, the problem of finding the line tangent to a given curve, was, as we have noted, not solved until the seventeenth century. Its solution (by Newton and Leibniz) gave rise to what is known as *differential calculus*. The second of these problems was to find the area enclosed by a given curve. The solution of this problem led to what is now termed *integral calculus*.

It is not known how long scientists have been concerned with finding the area bounded by a curve. In 1858 Henry Rhind, an Egyptologist from Scotland, discovered fragments of a papyrus manuscript written (approximately) in 1650 B.C. which came to be known as the *Rhind papyrus*.† The Rhind papyrus contains 85 problems and was written by the Egyptian scribe Ahmes, who wrote that he copied the problems from an *earlier* manuscript. In Problem 50, Ahmes assumed that the area of a circular field with a diameter of nine units is the same as the area of a square with a side of eight units. If we compare this with the correct formula for the area of a circle, we find that

$$A = \pi r^2 = \pi \left(\frac{9}{2}\right)^2 = \text{(according to Ahmes) } 8^2$$

or

$$\pi \approx \frac{64}{(4.5)^2} = \frac{64}{20.25} \approx 3.16$$

Thus we see that *before 1650* B.C. the Egyptians could calculate the area of a circle from the formula

$$A = 3.16r^2.$$

Since $\pi \approx 3.1416$, we see that this is remarkably close, considering that the Egyptian formula dates back nearly 4000 years!

In ancient Greece there was much interest in obtaining methods for calculating the areas bounded by curves other than circles and rectangles. The problem was solved for a wide variety of curves by Archimedes of Syracuse (287–212 B.C.) who is

† For an interesting discussion of the Rhind papyrus and other similar finds, see C. B. Boyer, *A History of Mathematics,* Wiley, New York, 1968.

considered by many to be the greatest mathematician who ever lived. Archimedes used what he called the *method of exhaustion* to calculate the shaded area A bounded by a parabola (see Figure 1). The shaded area A bounded by the parabola $y = x^2$ the

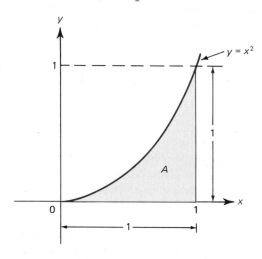

Figure 1

x-axis, and the line $x = 1$, was approximated by rectangles under the curve and over the curve (see Figure 2). Since the equation of the parabola is $y = x^2$, the height of each rectangle (which is the y-value on the curve) is the square of the x-value. If we let s denote the sum of the areas of the rectangles in Figure 2a and S the sum of the

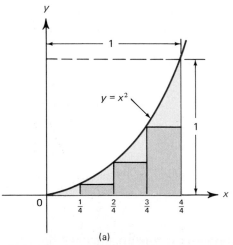

(a)   (b)

Figure 2

areas of the rectangles in Figure 2b, then

$$s < A < S. \tag{1}$$

Since Archimedes knew how to calculate the area of a rectangle (by multiplying its base by its height), he was able to calculate s and S exactly, thereby obtaining an estimate for the area A. We have

$$s = \frac{1}{4}\left(\frac{1}{4}\right)^2 + \frac{1}{4}\left(\frac{2}{4}\right)^2 + \frac{1}{4}\left(\frac{3}{4}\right)^2 = \frac{1}{4}\left\{\left(\frac{1}{4}\right)^2 + \left(\frac{2}{4}\right)^2 + \left(\frac{3}{4}\right)^2\right\}$$

$$= \frac{1}{4 \cdot 4^2}(1^2 + 2^2 + 3^2) = \frac{1^2 + 2^2 + 3^2}{4^3} = \frac{14}{64} = \frac{7}{32}$$

and

$$S = \frac{1}{4}\left(\frac{1}{4}\right)^2 + \frac{1}{4}\left(\frac{2}{4}\right)^2 + \frac{1}{4}\left(\frac{3}{4}\right)^2 + \frac{1}{4}\left(\frac{4}{4}\right)^2 = \frac{1^2 + 2^2 + 3^2 + 4^2}{4^3}$$

$$= \frac{30}{64} = \frac{15}{32}.$$

Thus, from (1), with $s = 7/32$ and $S = 15/32$ we obtain

$$0.22 \approx \frac{7}{32} < A < \frac{15}{32} \approx 0.47. \tag{2}$$

It was clear to Archimedes that this estimate could be improved by increasing the number of rectangles so that the error in the estimate becomes smaller. By doubling the number of rectangles, we obtain the approximation depicted in Figure 3. Here

$$s = \frac{1}{8}\left(\frac{1}{8}\right)^2 + \frac{1}{8}\left(\frac{2}{8}\right)^2 + \cdots + \frac{1}{8}\left(\frac{7}{8}\right)^2 = \frac{1^2 + 2^2 + 3^2 + 4^2 + 5^2 + 6^2 + 7^2}{8^3}$$

$$= \frac{140}{512} = \frac{35}{128}$$

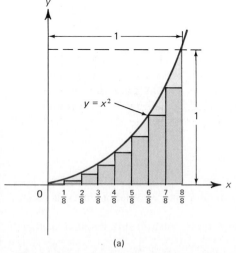

(a)

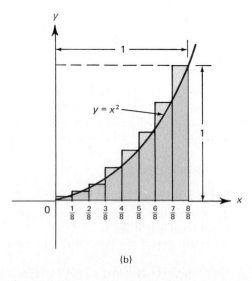

(b)

Figure 3

and

$$S = \frac{1}{8}\left(\frac{1}{8}\right)^2 + \frac{1}{8}\left(\frac{2}{8}\right)^2 + \cdots + \frac{1}{8}\left(\frac{8}{8}\right)^2$$

$$= \frac{1^2 + 2^2 + 3^2 + 4^2 + 5^2 + 6^2 + 7^2 + 8^2}{8^3} = \frac{204}{512} = \frac{51}{128}.$$

Hence, using $s = 35/128$ and $S = 51/128$ in (1), we have

$$0.27 \approx \frac{35}{128} < A < \frac{51}{128} \approx 0.40. \tag{3}$$

We see that the approximations have "squeezed down" somewhat. We now double the number of rectangles again to obtain

$$s = \frac{1}{16}\left(\frac{1}{16}\right)^2 + \frac{1}{16}\left(\frac{2}{16}\right)^2 + \cdots + \frac{1}{16}\left(\frac{15}{16}\right)^2$$

$$= \frac{1^2 + 2^2 + \cdots + 14^2 + 15^2}{16^3} = \frac{1240}{4096} = \frac{155}{512}$$

and

$$S = \frac{1}{16}\left(\frac{1}{16}\right)^2 + \frac{1}{16}\left(\frac{2}{16}\right)^2 + \cdots + \frac{1}{16}\left(\frac{15}{16}\right)^2 + \frac{1}{16}\left(\frac{16}{16}\right)^2$$

$$= \frac{1^2 + 2^2 + \cdots + 15^2 + 16^2}{16^3} = \frac{1496}{4096} = \frac{187}{512}.$$

Then

$$0.30 \approx \frac{155}{512} < A < \frac{187}{512} \approx 0.37$$

and the calculation of area is squeezed down further.

Using his method of exhaustion, Archimedes continued the process and was able to show that $s \to \frac{1}{3}$ and $S \to \frac{1}{3}$ as the number of rectangles increased, without bound, from which he concluded that $A = \frac{1}{3}$.

Over the years since the monumental work of Archimedes, the method of exhaustion has been generalized into a system called *integral calculus,* and the method used to calculate areas is called *integration.* The symbol that we use to indicate an integration is a large German s, $\int$, first used by Leibniz in the seventeenth century.

PROBLEMS 4.1

Problems 1–3 refer to the function $f(x) = x^2$.

1. Calculate s and S where the interval $[0, 1]$ is divided into 32 smaller intervals (each having length $\frac{1}{32}$).

2. Show that if the interval $[0, 1]$ is divided into n pieces each having length $1/n$, then

$$s = \frac{1}{n^3}(1^2 + 2^2 + \cdots + (n-1)^2) \qquad \text{and} \qquad S = \frac{1}{n^3}(1^2 + 2^2 + \cdots + n^2).$$

3. In Appendix 1 we show that for any positive integer n

$$1^2 + 2^2 + 3^2 + \cdots + n^2 = \frac{n(n+1)(2n+1)}{6}.$$

Use this to show that $\lim_{n\to\infty} s = \lim_{n\to\infty} S = \frac{1}{3}$ and conclude, as did Archimedes, that $A = \frac{1}{3}$.

*4. Using the method of exhaustion, find the area bounded by the curve $y = x^3$, the x-axis, and the line $x = 1$. [*Hint:* You will need the fact that $1^3 + 2^3 + 3^3 + \cdots + (n-1)^3 + n^3 = n^2(n+1)^2/4$.]

4.2 Approximations to Area

Let $y = f(x)$ be a function which is positive and continuous on the interval $[a, b]$ (see Figure 1). We will now show how the area bounded by this curve, the x-axis, and the

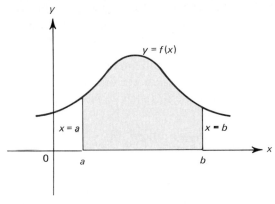

Figure 1

lines $x = a$ and $x = b$ can be approximated. The method we will give here is very similar to Archimedes' method of exhaustion discussed in Section 1.

As before, we begin by dividing the interval $[a, b]$ into a number of smaller subintervals. Such a division is called a *partition*. We label the *partition points* $x_0, x_1, x_2, \ldots, x_n$ where $x_0 = a$ and $x_n = b$ and it is assumed that

$$x_0 < x_1 < x_2 < \cdots < x_n$$

(see Figure 2). If we define $\Delta x_i = x_i - x_{i-1}$, then $\Delta x_1 = x_1 - x_0$,

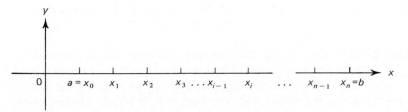

Figure 2

$\Delta x_2 = x_2 - x_1, \ldots, \Delta x_n = x_n - x_{n-1}$ and $\Delta x_1 + \Delta x_2 + \cdots + \Delta x_n = b - a$. In other words, the sum of the lengths of the subintervals equals the length of the entire interval.

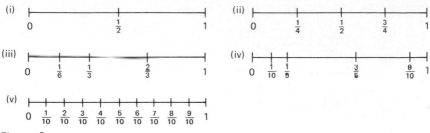

Figure 3

EXAMPLE 1. Let $[a, b] = [0, 1]$. Then (i)–(v) in Figure 3 are all partitions of $[0, 1]$. In partitions (i), (ii), and (v) the partition points are equally spaced and each subinterval has the same length. In partitions (iii) and (iv) the partition points are not equally spaced and the subintervals have varying lengths. In (iii) for example, $\Delta x_1 = \frac{1}{6}$, $\Delta x_2 = \frac{1}{6}$, $\Delta x_3 = \frac{1}{3}$, $\Delta x_4 = \frac{1}{3}$, and $\Delta x_1 + \Delta x_2 + \Delta x_3 + \Delta x_4 = 1$. In (iv) $\Delta x_1 = \frac{1}{10}$, $\Delta x_2 = \frac{1}{10}$, $\Delta x_3 = \frac{4}{10}$, $\Delta x_4 = \frac{3}{10}$, $\Delta x_5 = \frac{1}{10}$, and we have $\Delta x_1 + \Delta x_2 + \Delta x_3 + \Delta x_4 + \Delta x_5 = 1$.

We will be interested in seeing what happens to a certain quantity when each of the subintervals in a partition becomes small and the number of subintervals increases. Let P denote a given partition. Then we define the *norm of P*, written $|P|$, as the length of the longest subinterval in P.

EXAMPLE 2. If we denote the five partitions in Figure 3 by $P_{(i)}$, $P_{(ii)}$, $P_{(iii)}$, $P_{(iv)}$ and $P_{(v)}$, respectively, then

$$|P_{(i)}| = \tfrac{1}{2}, \quad |P_{(ii)}| = \tfrac{1}{4}, \quad |P_{(iii)}| = \tfrac{1}{3}, \quad |P_{(iv)}| = \tfrac{4}{10}; \quad \text{and} \quad |P_{(v)}| = \tfrac{1}{10}.$$

Partitions in which the partition points are equally spaced are particularly easy to deal with. Suppose that $[a, b]$ is partitioned into the subintervals $[x_0, x_1]$, $[x_1, x_2], \ldots, [x_{n-2}, x_{n-1}], [x_{n-1}, x_n]$. If the partition points are equally spaced, then since there are n subintervals of equal length, the length of each subinterval is $(b - a)/n$ because the length of the entire interval is $b - a$. In this case,

$$|P| = \frac{b - a}{n}. \tag{1}$$

EXAMPLE 3. In Figure 3, $b - a = 1$. Then, for the equally spaced partitions,

$$|P_{(i)}| = \frac{1 - 0}{2} = \frac{1}{2}, \quad |P_{(ii)}| = \frac{1 - 0}{4} = \frac{1}{4}, \quad \text{and} \quad |P_{(v)}| = \frac{1 - 0}{10} = \frac{1}{10}.$$

Suppose that the interval $[a, b]$ is partitioned into n subintervals (not necessarily of equal length). We denote the area bounded by the curve $y = f(x)$, the x-axis, and the lines $x = a$ and $x = b$ by A_a^b. We will approximate A_a^b by drawing rectangles whose total area is "close" to the actual area (see Figure 4). We first choose a point *arbitrarily* in each interval $[x_{i-1}, x_i]$ and label it x_i^*. This gives us n points x_1^*, $x_2^*, \ldots, x_n^*$. Next, we locate the points $(x_i^*, f(x_i^*))$ on the curve for $i = 1, 2, \ldots, n$. The numbers $f(x_i^*)$ give us the heights of our n rectangles. The base of each rectangle has length $x_i - x_{i-1} = \Delta x_i$. This is all illustrated in Figure 4. The area, A_i, of the ith

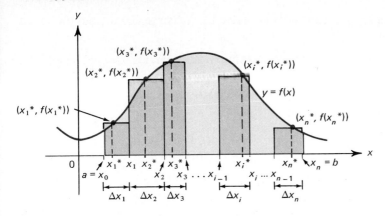

Figure 4

rectangle is the height of the rectangle times its length:

$$A_i = f(x_i^*)\,\Delta x_i. \tag{2}$$

Let us take a closer look at one of these rectangles. In Figure 5 the shaded area

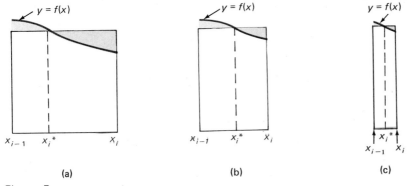

(a)   (b)   (c)

Figure 5

depicts the differences between the region whose area we wish to calculate and the rectangle. We see that as the rectangle becomes "thinner and thinner," the area of the rectangle seems to get closer and closer to the area under the curve. But the length of the base of each rectangle is Δx_i and so, if Δx_i is reasonably small, we have the approximation

$$A_a^b \approx A_1 + A_2 + \cdots + A_i + \cdots + A_n \tag{3}$$

or, using (2)

$$A_a^b \approx f(x_1^*)\,\Delta x_1 + f(x_2^*)\,\Delta x_2 + \cdots + f(x_i^*)\,\Delta x_i + \cdots + f(x_n^*)\,\Delta x_n. \tag{4}$$

In the last section, we saw an example of this process at work. There the subintervals had equal length and we chose x_i^* to be either the right or the left endpoint of the subinterval. Before continuing, we will further illustrate our approximation technique.

☐**EXAMPLE 4.** Approximate the area bounded by the curve $y = x^2$ and the x-axis from $x = 0$ to $x = 1$.

SOLUTION. We will partition the interval $[0, 1]$ into n subintervals of equal length and will choose $x_i^* = x_{i-1}$ the left-hand endpoint of each subinterval. Since $(b - a)/n = 1/n$, the length of each subinterval is $1/n$ and the partition points are (see Figure 6)

$$0 = \frac{0}{n} < \frac{1}{n} < \frac{2}{n} < \frac{3}{n} < \cdots < \frac{n-1}{n} < \frac{n}{n} = 1.$$

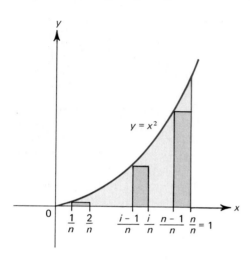

Figure 6

Then for $i = 1, 2, \ldots, n$, $f(x_i^*) = f(x_{i-1}) = f[(i-1)/n] = [(i-1)/n]^2$, $\Delta x_i = 1/n$, and formula (4) becomes

$$A_0^1 \approx \left(\frac{0}{n}\right)^2 \cdot \frac{1}{n} + \left(\frac{1}{n}\right)^2 \cdot \frac{1}{n} + \left(\frac{2}{n}\right)^2 \cdot \frac{1}{n} + \cdots + \left(\frac{n-1}{n}\right)^2 \cdot \frac{1}{n}$$

$$= \frac{1}{n^3}(1^2 + 2^2 + 3^2 + \cdots + (n-1)^2). \tag{5}$$

In Appendix 1, we prove using mathematical induction that the sum of the first n squares is given by the formula

$$1^2 + 2^2 + 3^2 + \cdots + n^2 = \frac{n(n+1)(2n+1)}{6}. \tag{6}$$

If we stop at $(n-1)^2$, we obtain

$$1^2 + 2^2 + 3^2 + \cdots + (n-1)^2 = \frac{(n-1)(n-1+1)(2(n-1)+1)}{6}$$

$$= \frac{(n-1)(n)(2n-1)}{6}. \tag{7}$$

You should "check" this formula by verifying it for a few values of n. Substituting (7) in (5), we have

$$A_0^1 \approx \frac{1}{n^3}\left\{\frac{(n-1)(n)(2n-1)}{6}\right\} = \frac{(n-1)(2n-1)}{6n^2}. \tag{8}$$

In Table 1 we tabulate values of this quantity for several different values of n. It seems

TABLE 1

n	$n-1$	$2n-1$	$(n-1)(2n-1)$	$6n^2$	$\dfrac{(n-1)(2n-1)}{6n^2}$
1	0	1	0	6	0
2	1	3	3	24	0.125
5	4	9	36	150	0.24
10	9	19	171	600	0.285
25	24	49	1,176	3,750	0.3136
50	49	99	4,851	15,000	0.3234
100	99	199	19,701	60,000	0.32835
1,000	999	1,999	1,997,001	6,000,000	0.3328335
10,000	9999	19,999	199,970,001	600,000,000	0.333283335

as if the approximation is tending to the limit $0.3333\ldots = \frac{1}{3}$ as the number of subintervals grows. Actually, this is easy to prove since

$$\lim_{n\to\infty}\frac{(n-1)(2n-1)}{6n^2} = \frac{1}{6}\lim_{n\to\infty}\left[\left(\frac{n-1}{n}\right)\left(\frac{2n-1}{n}\right)\right]$$

$$= \frac{1}{6}\lim_{n\to\infty}\left[\left(1-\frac{1}{n}\right)\left(2-\frac{1}{n}\right)\right]$$

$$= \frac{1}{6}\lim_{n\to\infty}\left(1-\frac{1}{n}\right)\cdot\lim_{n\to\infty}\left(2-\frac{1}{n}\right) = \frac{1}{6}\cdot 1\cdot 2 = \frac{2}{6} = \frac{1}{3}.$$

Note that as $n\to\infty$, $|P|\to 0$.

EXAMPLE 5. Approximate the area bounded by the curve $y=3x$ and the x-axis from $x=0$ to $x=2$.

SOLUTION. The area sought is depicted in Figure 7. Since the area of a triangle is $\frac{1}{2}bh$ where b denotes its base and h its height, we can immediately calculate that $A_0^2 = \frac{1}{2}(2)(6) = 6$. We will show how our approximations lead to the same answer. To do this, we partition $[0,2]$ into n equal subintervals, each having length $(b-a)/n = 2/n$. The partition points are then

$$0 = \frac{0}{n} < \frac{2}{n} < \frac{4}{n} < \cdots < \frac{2(i-1)}{n} < \frac{2i}{n} < \cdots < \frac{2(n-1)}{n} < \frac{2n}{n} = 2.$$

If we choose $x_i^* = 2i/n$, the right-hand endpoint of each subinterval, we obtain the situation depicted in Figure 8. Here $f(x_i^*) = f(2i/n) = 3\cdot 2i/n = 6i/n$. Then (4)

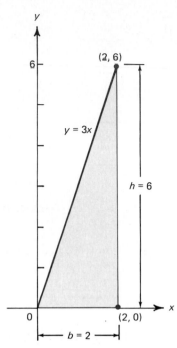

Figure 7   Figure 8

becomes

$$A_0^2 \approx f(x_1{}^*)\,\Delta x_1 + f(x_2{}^*)\,\Delta x_2 + \cdots + f(x_n{}^*)\,\Delta x_n$$

$$= \frac{6 \cdot 2}{n}\left(\frac{1}{n}\right) + \frac{6 \cdot 4}{n}\left(\frac{1}{n}\right) + \cdots + \frac{6(2(n-1))}{n}\left(\frac{1}{n}\right) + \frac{6(2n)}{n}\left(\frac{1}{n}\right)$$

$$= \frac{12}{n^2}[1 + 2 + 3 + \cdots + (n-1) + n]. \tag{9}$$

Now, the formula for the sum of the first n integers (proven in Appendix 1) is

$$1 + 2 + 3 + \cdots + n - 1 + n = \frac{n(n+1)}{2}. \tag{10}$$

This formula should also be checked by verifying it for some sample values of n. Using (10) in (9), we obtain

$$A_0^1 \approx \frac{12}{n^2}\left[\frac{n(n+1)}{2}\right] = \frac{6(n+1)}{n} = 6\left(1 + \frac{1}{n}\right). \tag{11}$$

Values of this approximation are tabulated in Table 2. It is apparent that these approximations are tending to the limit 6. Indeed, we have

$$\lim_{n\to\infty} 6\left(1 + \frac{1}{n}\right) = 6 \lim_{n\to\infty}\left(1 + \frac{1}{n}\right) = 6 \cdot 1 = 6.$$

TABLE 2

n	$n + 1$	$6(n + 1)$	$\dfrac{6(n + 1)}{n}$
1	2	12	12
5	6	36	7.2
10	11	66	6.6
100	101	606	6.06
1,000	1,001	6,006	6.006
10,000	10,001	60,006	6.0006

Note that as $n \to \infty$, $|P| \to 0$.

EXAMPLE 6. Approximate the area bounded by the curve $y = x^3$ and the x-axis from $x = 0$ to $x = 3$.

SOLUTION. For simplicity, we again divide the interval into n equal subintervals, each having length $(b - a)/n = 3/n$. The partition points are

$$0 = \frac{0}{n} < \frac{3}{n} < \frac{6}{n} < \frac{9}{n} < \cdots < \frac{3(n - 1)}{n} < \frac{3n}{n} = 3.$$

For convenience, we choose $x_i{}^* = x_i$ (the right-hand endpoint of the subinterval), so that

$$f(x_i{}^*) = f(x_i) = x_i{}^3 = \left(\frac{3i}{n}\right)^3 = \frac{27i^3}{n^3} \qquad \text{for} \quad i = 1, 2, \ldots, n.$$

(see Figure 9). Then

$$A_0^3 \approx \left(\frac{3}{n}\right)^3\left(\frac{3}{n}\right) + \left(\frac{6}{n}\right)^3\left(\frac{3}{n}\right) + \left(\frac{9}{n}\right)^3\left(\frac{3}{n}\right) + \cdots + \left(\frac{3n}{n}\right)^3\left(\frac{3}{n}\right)$$

$$= 27\left(\frac{1^3}{n^3}\right)\left(\frac{3}{n}\right) + 27\left(\frac{2^3}{n^3}\right)\left(\frac{3}{n}\right) + \cdots + 27\left(\frac{(n - 1)^3}{n^3}\right)\left(\frac{3}{n}\right) + \frac{27n^3}{n^3}\left(\frac{3}{n}\right)$$

$$= \frac{81}{n^4}(1^3 + 2^3 + 3^3 + \cdots + (n - 1)^3 + n^3). \tag{12}$$

Yes, there is a formula for the sum of the first n cubes. In Appendix 1 (see Problem A.1.1) we prove that

$$1^3 + 2^3 + 3^3 + \cdots + (n - 1)^3 + n^3 = \left[\frac{n(n + 1)}{2}\right]^2. \tag{13}$$

Again you should satisfy yourself that this formula is correct. Substituting (13) in (12), we obtain

$$A_0^3 \approx \frac{81}{n^4} \cdot \frac{n^2(n + 1)^2}{4} = \frac{81}{4} \cdot \frac{(n + 1)^2}{n^2}. \tag{14}$$

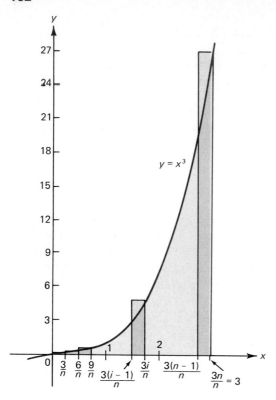

Figure 9

Some of these approximations are tabulated in Table 3. It seems that as $n \to \infty$, $A_0^3 \to 20.25 = 81/4$. Indeed,

$$\lim_{n\to\infty} \frac{81}{4} \cdot \frac{(n+1)^2}{n^2} = \frac{81}{4} \lim_{n\to\infty} \frac{n^2 + 2n + 1}{n^2} = \frac{81}{4} \lim_{n\to\infty} \left(1 + \frac{2}{n} + \frac{1}{n^2} \right) = \frac{81}{4}.$$

TABLE 3

n	n^2	$n+1$	$(n+1)^2$	$\dfrac{(n+1)^2}{n^2}$	$\dfrac{81}{4} \cdot \dfrac{(n+1)^2}{n^2}$
1	1	2	4	4.0	81.0
5	25	6	36	1.44	29.16
10	100	11	121	1.21	24.5025
100	10,000	101	10,201	1.0201	20.657025
1,000	1,000,000	1,001	1,002,001	1.002001	20.29052025
10,000	100,000,000	10,001	100,020,001	1.00020001	20.2540502

Again note that as $n \to \infty$, $|P| \to 0$.

We will put these calculations on a more formal basis in the next section and shall show how the techniques we have introduced here can also be used to solve problems that do not involve areas.

PROBLEMS 4.2

In Problems 1–4 estimate the area bounded by the straight line, the x-axis, and the lines $x = a$ and $x = b$, where a and b are the endpoints of the given interval, by dividing the interval into 2, 4 and 8 subintervals of equal length. Then calculate the actual area using the fact that the area of a triangle is $\frac{1}{2}bh$ and the area of a rectangle is $b \cdot h$.

1. $y = 7x$; [0, 4].
2. $y = 4x$; [1, 2]. [*Hint:* To calculate this area exactly, divide the trapezoid you obtain into a triangle and a rectangle.]

3. $y = 3x + 2$; [0, 3].
4. $y = 7x - 3$; [1, 4].

In Problems 5–18 (a) Estimate the area bounded by the given curve, the x-axis and the lines $x = a$ and $x = b$ by dividing the interval into a number (given in the braces) of subintervals of equal length and choosing a convenient point in each subinterval. (b) Repeat the process by doubling the number of subintervals. (c) Obtain a formula for the area enclosed by n estimating rectangles where n is any positive integer. *(d) Find the exact area by taking the limit as $n \to \infty$ of the formula obtained in part (c).

5. $y = \frac{1}{2}x^2$; [0, 2], {4}
6. $y = 2x^2$; [−1, 1], {3}
7. $y = x^3$; [0, 5], {4}
8. $y = 17x^2$; [0, 8], {5}
9. $y = 3x^3$; [−1, 0], {8}
10. $y = (1 - x)^2$; [0, 1], {6}
11. $y = 1 - x^2$; [0, 1], {4}
12. $y = 1 - x^3$; [0, 1], {5}
13. $y = x^2$; [1, 2], {8}
14. $y = x^3$; [1, 2], {4}
15. $y = x + x^2$; [0, 1], {6}
16. $y = x^3 + 3x$; [0, 1], {5}
17. $y = 1 + 2x + 3x^2$; [0, 1], {4}
18. $y = 1 + x + x^2 + x^3$; [0, 1], {4}

19. Consider the problem of calculating the area bounded by the curve $y = x^2$, the x-axis and the lines $x = a$ and $x = b$.
    (a) Divide the interval $[a, b]$ into n subintervals of equal length. That is, $\Delta x_i = (b - a)/n$. What are the endpoints of each subinterval?
    (b) Letting $x_i^* = $ the right-hand endpoint of each subinterval, calculate $f(x_i^*) = (x_i^*)^2$.
    (c) Use (4) to estimate A_a^b for 4, 8, 16, and 32 subintervals.
    *(d) Obtain a formula for the total area under the curve.
*20. Following the steps of Problem 19, obtain a formula for the area bounded by the curve $y = x^3$, the x-axis, and the lines $x = a$ and $x = b$ with a and b positive.

4.3 The Definite Integral

In this section we put the discussion of Section 2 on a more formal basis. The assumptions needed in this section are:

(i) f is defined on the interval $[a, b]$. (Here we do *not* require that f be positive on $[a, b]$.)

(ii) P is a partition of $[a, b]$:

$$a = x_0 < x_1 < x_2 < \cdots < x_{n-1} < x_n = b,$$

with the partition points x_i not necessarily equally spaced.

(iii) $x_i{}^*$ is any point in the interval $[x_{i-1}, x_i]$ and $\Delta x_i = x_i - x_{i-1} =$ the length of the subinterval.

(iv)   The norm of P, $|P|$, is defined as the length of the longest subinterval in the partition.

Then we have the following definition which is central to the study of calculus.

DEFINITION 1 (The Definite Integral).   Let $a < b$ and suppose that

$$\lim_{|P| \to 0} [f(x_1{}^*) \Delta x_1 + f(x_2{}^*) \Delta x_2 + \cdots + f(x_n{}^*) \Delta x_n]$$

exists and is independent of the way in which the interval $[a, b]$ is partitioned and the way in which the numbers $x_i{}^*$ are chosen. Then the *integral of $f(x)$ on the interval* $[a, b]$, written $\int_a^b f(x)\, dx$ or $\int_a^b f$, is defined by

$$\boxed{\int_a^b f(x)\, dx = \lim_{|P| \to 0} [f(x_1{}^*) \Delta x_1 + f(x_2{}^*) \Delta x_2 + \cdots + f(x_n{}^*) \Delta x_n].} \tag{1}$$

The process of calculating an integral is called *integration* and the numbers a and b are called the *lower and upper limits of integration,* respectively. If the limit in (1) exists, then f is said to be *integrable* on the interval $[a, b]$.

> **Note.** The integral can be thought of as an infinite sum since we are adding up more and more little rectangles, each of which is getting thinner and thinner.

Before citing examples, some comments about integration are in order.

> **Remark 1.** As we indicated in Section 1, the symbol $\int$ was introduced by Leibniz and is really an elongated German S, standing for the German word *Summe* (which means "sum").

> **Remark 2.** The integral defined by (1) is called a *definite* integral because if the limit in (1) exists, the integral is a *number*. The integral takes on a definite value. This contrasts sharply with the indefinite integral that we will define in the next section. As we shall see, an indefinite integral is a *function* and is *not* unique.

> **Remark 3.** The clause "and is independent of the way in which the interval $[a, b]$ is partitioned and the way the numbers $x_i{}^*$ are chosen" cannot be omitted. It is possible to find a function $f(x)$ which takes values between 0 and 1, say, on the interval $[0, 1]$ and for which the limit in (1) will be equal to 0 for one choice of the $x_i{}^*$'s and will be equal to 1 for another choice of the $x_i{}^*$'s (see Problem 46). In that case we cannot say that f is integrable since we have no way to choose between these two or more possible limits.

> **Remark 4.** The variable x in (1) is called a *dummy variable* since it could be replaced by *any other* variable without changing the value of the integral. This is true because the definite integral is a number. We could, instead, subdivide $[a, b]$ by

$$a = t_0 < t_1 < t_2 < \cdots < t_{n-1} < t_n = b.$$

Then we could choose t_i^* in $[t_{i-1}, t_i]$ and (1) would become

$$\int_a^b f(t)\, dt = \lim_{|P| \to 0} [f(t_1^*)\, \Delta t_1 + f(t_2^*)\, \Delta t_2 + \cdots + f(t_n^*)\, \Delta t_n]. \tag{2}$$

This is, of course, the same definition as in (1). We can also substitute any other variable for t. Thus

$$\int_a^b f(x)\, dx = \int_a^b f(t)\, dt = \int_a^b f(z)\, dz = \int_a^b f(\text{Dummy})\, d\,\text{Dummy}. \tag{3}$$

This is why we will often simply write $\int_a^b f$, since the symbol for the dummy variable is not important.

Remark 5. Suppose that f is integrable on $[a, b]$. (See Theorem 1.) Since the limit in (1) is independent of the way in which the interval $[a, b]$ is partitioned, we may assume from now on that the subintervals are equally spaced. If the subintervals in P are equally spaced, then $\Delta x_i = \Delta x = (b - a)/n$ for every i and $|P| = \Delta x$ so that $|P| \to 0$ is the same as $\Delta x \to 0$ or $n \to \infty$. Then (1) becomes

$$\int_a^b f(x)\, dx = \lim_{\Delta x \to 0} [f(x_1^*) + f(x_2^*) + \cdots + f(x_n^*)]\, \Delta x$$

or

$$\int_a^b f(x)\, dx = \lim_{n \to \infty} [f(x_1^*) + f(x_2^*) + \cdots + f(x_n^*)]\, \frac{b - a}{n}. \tag{4}$$

Remark 6. The definition in (1) says nothing about area under a curve. Of course, as we saw in Section 2, the limit *may* represent the area under the curve, but it may not. If f is positive, then $\int_a^b f$ does represent area. However, as we shall see in this section and later in the book, the definite integral has many applications having nothing at all to do with the problem of finding areas.

Remark 7. If a function is integrable according to Definition 1, then it must be bounded. For if f is not bounded and $a = x_0 < x_1 < \cdots < x_n = b$ is any partition, there is a subinterval $[x_{i-1}, x_i]$ in which f is not bounded. Consequently the sum (1) for this partition can be made arbitrarily large in absolute value by an appropriate choice of x_i^* in $[x_{i-1}, x_i]$.

Following these remarks, a basic question remains: What functions are integrable? After all, the limit in (1) is usually difficult to obtain (as you could imagine after working through the problems of the previous section) and so it would be nice to know *before* doing any calculations that a given integral does exist. Recall from Section 2.4 that a function is *piecewise continuous* if it is continuous everywhere in the interval $[a, b]$ except at a finite number of points at which it has jump discontinuities (a continuous function is piecewise continuous). The following theorem, whose proof is beyond the scope of this book,† tells us that a great number of the functions we have already discussed are integrable.

---

†For a proof, consult any book in advanced calculus such as R. C. Buck, *Advanced Calculus,* McGraw-Hill, New York, 1965, Chapter 3.

THEOREM 1 (Existence of Definite Integrals). Let f be piecewise continuous on the interval $[a, b]$. Then f is integrable on $[a, b]$. That is, $\int_a^b f(x)\, dx$ exists.

EXAMPLE 1. The following integrals all exist since the functions being integrated are all piecewise continuous on the interval of integration.

(a) $\displaystyle\int_1^7 x^7\, dx$ 　　　　　(b) $\displaystyle\int_1^3 \sqrt{x}\, dx$ 　　　　　(c) $\displaystyle\int_{-10}^{10} |x|\, dx$

(d) $\displaystyle\int_0^3 \frac{1}{\sqrt{(x+1)(x+2)}}\, dx$ 　　　　　(e) $\displaystyle\int_3^5 \frac{1}{x}\, dx$

(f) $\displaystyle\int_2^4 t^{3/5}\, dt$ 　　　　　(g) $\displaystyle\int_0^{1,000,000} s^{100}\, ds$ 　　　　　(h) $\displaystyle\int_{-3}^0 \frac{1}{z-1}\, dz$

(i) $\displaystyle\int_{-5}^{-1} \frac{1}{cabin}\, d\,cabin$ 　　　　　(j) $\displaystyle\int_{-5}^{10} [y]\, dy$ 　(see Example 2.4.7).

(Note that all the functions being integrated in this example are continuous except the one in (j).

EXAMPLE 2. $\int_{-1}^8 (1/x)\, dx$ does not exist. The function $1/x$ "blows up" as $x \to 0^+$ or $x \to 0^-$. However *there are functions which are not piecewise continuous on $[a, b]$ for which $\int_a^b f(x)\, dx$ does exist*. In Example 12.4.15 we shall show that $\int_{-1}^8 (1/\sqrt[3]{x})\, dx$ does exist (and is equal to $\frac{9}{2}$) even though $1/\sqrt[3]{x}$ is not piecewise continuous in the interval. This means that in our definition of the definite integral it is *not* necessary to assume that f is defined at every point in $[a, b]$.†

We now calculate a number of definite integrals.

EXAMPLE 3. Calculate $\displaystyle\int_0^1 x^2\, dx$.

SOLUTION. We have done this already in Example 4.2.4. In the subinterval $[(i-1)/n, i/n]$, $\Delta x = 1/n$ and we chose $x_i^* = (i-1)/n$ (the left-hand endpoint of the subinterval). Then we calculated

$$\int_0^1 x^2\, dx = \lim_{n\to\infty} \frac{1}{n^3}(1^2 + 2^2 + 3^2 + \cdots + (n-1)^2)$$

$$= \lim_{n\to\infty} \frac{(n-1)(2n-1)}{6n^2} = \frac{1}{3}.$$

EXAMPLE 4. Calculate $\displaystyle\int_0^3 x^3\, dx$.

SOLUTION. This was done in Example 4.2.6. There we chose the point $x_i^* = 3i/n$ the

---

† However, before calculating $\int_{-1}^8 (1/\sqrt{x})\, dx$, it will be necessary to define what we mean by the integral of a function over an interval containing points at which the function becomes infinite.

right-hand endpoint of the subinterval $[3(i - 1)/n, 3i/n]$, $\Delta x = 3/n$, and we found that $\Delta x = 3/n$, and we found that

$$\int_0^3 x^3 \, dx = \lim_{\Delta x \to 0} \left(\frac{3 \cdot 1}{n}\right)^3 \left(\frac{3}{n}\right) + \left(\frac{3 \cdot 2}{n}\right)^3 \left(\frac{3}{n}\right) + \cdots + \left(\frac{3 \cdot n}{n}\right)^3 \left(\frac{3}{n}\right)$$

$$= \lim_{n \to \infty} 81 \frac{\cdot (1^3 + 2^3 + \cdots + n^3)}{n^4} = \lim_{n \to \infty} \frac{81(n + 1)^2}{4n^2} = \frac{81}{4}.$$

EXAMPLE 5. Let $f(x) = 5$. Calculate $\int_1^3 f(x) \, dx$.

SOLUTION. We divide the interval $[1, 3]$ into n equal subintervals each having length $\Delta x = 2/n$ (see Figure 1). A typical subinterval is $[1 + (2(i - 1)/n), 1 + (2i/n)]$. We

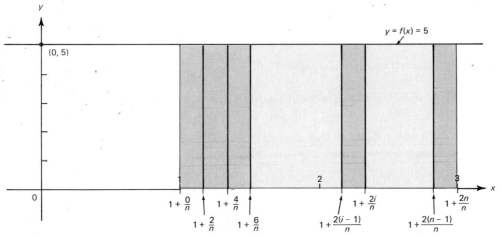

Figure 1

may choose x_i^* to be any point in the interval. We will have $f(x_i^*) = 5$ since f is a constant function. Then

$$\int_1^3 f(x) \, dx = \int_1^3 5 \, dx = \lim_{n \to \infty} \left\{ \left[f(x_1^*) + f(x_2^*) + \cdots + f(x_n^*) \right] \frac{2}{n} \right\}$$

$$= \lim_{n \to \infty} \left[\underbrace{(5 + 5 + \cdots + 5)}_{n \text{ times}} \left(\frac{2}{n}\right) \right] = \lim_{n \to \infty} \left[5n \left(\frac{2}{n}\right) \right] = \lim_{n \to \infty} [5(2)] = 10.$$

Thus $\int_1^3 5 \, dx = 10$. Of course, we did not need calculus to solve this problem since we needed only to calculate the area of the rectangle of base 2 and height 5. However the example is useful to motivate the following theorem.

THEOREM 2. Let c be a constant. Then

$$\int_a^b c \, dx = c(b - a).\tag{5}$$

In particular, for $c = 1$ we have

$$\int_a^b 1 \, dx = \int_a^b dx = b - a.\tag{6}$$

This theorem makes sense geometrically if c is a positive constant. Look at Figure 2.

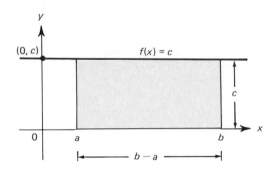

Figure 2

Then $c(b - a)$ is equal to the area of the rectangle under the curve $y = c$ from $x = a$ to $x = b$.

To this point we have required that a be less than b in the definition of the integral. The case $a = b$ and $a > b$ is defined below.

DEFINITION 2. For any real number a,

$$\int_a^a f(x) \, dx = 0.\tag{7}$$

Note that if $\int_a^a f(x) \, dx$ represents area, then this definition states that the area of a region with a width of zero is zero. (The interval $[a, a]$ has a width of zero.)

DEFINITION 3. If $a < b$,

$$\int_b^a f(x) \, dx = -\int_a^b f(x) \, dx.\tag{8}$$

This definition says that reversing the order of integration changes the sign of the integral, thus enabling us to apply Definition 1 when $a > b$.

EXAMPLE 6. Calculate $\displaystyle\int_1^0 x^2 \, dx$.

SOLUTION. $\int_1^0 x^2 \, dx = -\int_0^1 x^2 \, dx = -\frac{1}{3}$.

As we stated in Remark 6, the integral does not always represent the area under a curve. The following example illustrates this fact.

EXAMPLE 7.   Calculate $\int_{-4}^{7} (-2)\, dx$.

SOLUTION. $b - a = 7 - (-4) = 11$ so that

$$\int_{-4}^{7} (-2)\, dx = -2[7 - (-4)] = (-2)11 = -22 \qquad \text{(from Theorem 2)}.$$

Here the integral is negative. To see why look at Figure 3. The area enclosed in the

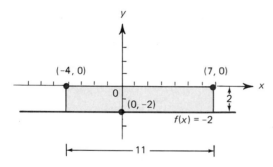

Figure 3

figure is equal to 22 square units. However, we see that *the integral treats areas below the x-axis as negative.* Therefore we cannot calculate an area under the curve $y = f(x)$ by simple integration unless $f(x) \geq 0$ on $[a, b]$. To handle the more general case, we define area in terms of the definite integral as follows.

DEFINITION 4.   The *area* bounded by the function $y = f(x)$, the x-axis, and the lines $x = a$ and $x = b$ (for $a < b$) is denoted by A_a^b and given by the formula

$$A_a^b = \int_a^b |f(x)|\, dx. \qquad\qquad (9)$$

Note. If $a < b$ and if $f(x)$ is positive for $a \leq x \leq b$, then $|f(x)| = f(x)$, and, in this case, $A_a^b = \int_a^b f(x)\, dx$.

EXAMPLE 8.   Calculate the area bounded by the curve $y = x^3$, the x-axis, and the lines $x = -1$ and $x = 1$ (see Figure 4).

SOLUTION. In $[-1, 0]$, x^3 is negative so that $|x^3| = -x^3$. In $[0, 1]$, x^3 is positive. Thus

$$A_{-1}^1 = \int_{-1}^1 f(x)\, dx, \qquad \text{where } f(x) = \begin{cases} x^3, & x \geq 0 \\ -x^3, & x < 0 \end{cases}.$$

Geometrically, it appears that the area of region I in Figure 4 is equal to the area of region II. To verify this, we have

$$A_{-1}^0 = \int_{-1}^0 -x^3\, dx = -\lim_{|P| \to 0} [(x_1^*)^3\, \Delta x_1 + (x_2^*)^3\, \Delta x_2 + \cdots + (x_3^*)^3\, \Delta x_n].$$

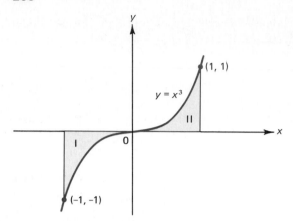

Figure 4

We now choose the partition points

$$-1 = \frac{-n}{n} < -\frac{(n-1)}{n} < \cdots < \frac{-i}{n} < \cdots < \frac{-1}{n} < \frac{0}{n} = 0 \text{ and } x_i^* = \frac{-i}{n}$$

so that

$$\Delta x_i = \Delta x = \frac{1}{n}, \quad (x_i^*)^3 = \frac{-i^3}{n^3}, \quad \text{and} \quad (x_i^*)^3 \Delta x_i = \frac{-i^3}{n^4}.$$

Then we have

$$\int_{-1}^{0} -x^3 \, dx = -\lim_{n\to\infty}\left[\frac{1}{n^4}(-1^3 - 2^3 - \cdots - n^3)\right]$$

$$= \lim_{n\to\infty}\left[\frac{1}{n^4}(1^3 + 2^3 + \cdots + n^3)\right]$$

$$= \lim_{n\to\infty}\frac{1}{n^4}\cdot\frac{n^2(n-1)^2}{4} = \frac{1}{4}\lim_{n\to\infty}\frac{(n-1)^2}{n^2} = \frac{1}{4}.$$

We have again made use of formula (4.2.13). We can in an analogous manner calculate $A_0^1 = \frac{1}{4}$ so that

$$A_{-1}^1 = \tfrac{1}{4} + \tfrac{1}{4} = \tfrac{1}{2}.$$

Note that this is *not* the same as $\int_{-1}^{1} x^3 \, dx$ which is equal to zero! (The "negative" area below the x-axis "cancels off" the positive area above it.) The student is urged to verify this by partitioning the interval $[-1, 1]$ in such a way that areas of rectangles below the x-axis "cancel" with areas of rectangles above it.

We now state three theorems which can be very useful for calculating integrals. The proofs of the theorems are not difficult. However, since these will be especially easy to prove after we have discussed the fundamental theorem of calculus in Section 5, we shall delay the proofs until that section (or see Problems 26, 27 and 28).

THEOREM 3. If f is integrable on $[a, b]$ and if $a < c < b$, then f is integrable on $[a, c]$ and on $[c, b]$ and

$$\int_a^b f = \int_a^c f + \int_c^b f. \qquad (10)$$

Theorem 3 is obvious if $\int_a^b f$ represents the area under a curve. In Figure 5, the area under the curve is given by $\int_a^b f$ which is equal to $A_1 + A_2 = \int_a^c f + \int_c^b f$.

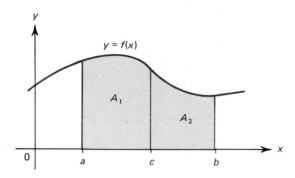

Figure 5

EXAMPLE 9. Calculate $\displaystyle\int_1^3 x^3\, dx$.

SOLUTION. We have already calculated (in Examples 4 and 8)

$$\int_0^1 x^3\, dx = \frac{1}{4} \qquad \text{and} \qquad \int_0^3 x^3\, dx = \frac{81}{4}.$$

Then, using (10), $\int_0^3 x^3\, dx = \int_0^1 x^3\, dx + \int_1^3 x^3\, dx$ or

$$\int_1^3 x^3\, dx = \int_0^3 x^3\, dx - \int_0^1 x^3\, dx = \frac{81}{4} - \frac{1}{4} = \frac{80}{4} = 20.$$

THEOREM 4 (Multiplication by a Constant). If f is integrable on $[a, b]$ and if k is any constant, then kf is integrable on $[a, b]$ and

$$\int_a^b kf = k \int_a^b f. \qquad (11)$$

This theorem is analogous to Theorem 3.1.2(i) for derivatives.

EXAMPLE 10. Calculate $\displaystyle\int_1^3 17x^3\, dx$.

SOLUTION. In Example 9 we saw that $\int_1^3 x^3\, dx = 20$. Then

$$\int_1^3 17x^3\, dx = 17 \int_1^3 x^3\, dx = 17(20) = 340.$$

THEOREM 5 (The Sum of Two Functions). If the functions f and g are both integrable on $[a, b]$, then $f + g$ is integrable on $[a, b]$ and

$$\int_a^b (f + g) = \int_a^b f + \int_a^b g. \tag{12}$$

That is, *the integral of the sum is the sum of the integrals.* (See Theorem 3.1.2(ii) for a comparable theorem about derivatives.)

> *Remark.* Theorem 5 can be easily extended to the integral of a finite sum of integrable functions.

EXAMPLE 11. Calculate $\displaystyle\int_0^1 (x^2 + x^3)\, dx$.

SOLUTION. $\int_0^1 (x^2 + x^3)\, dx = \int_0^1 x^2\, dx + \int_0^1 x^3\, dx = \frac{1}{3} + \frac{1}{4} = \frac{7}{12}$.

EXAMPLE 12. Calculate $\displaystyle\int_0^1 (x^2 - x^3)\, dx$.

SOLUTION. $\int_0^1 (x^2 - x^3)\, dx = \int_0^1 x^2\, dx + \int_0^1 (-x^3)\, dx$. But, from Theorem 4, $\int_0^1 (-x^3)\, dx = -\int_0^1 x^3\, dx = -\frac{1}{4}$. Thus

$$\int_0^1 (x^2 - x^3)\, dx = \frac{1}{3} - \frac{1}{4} = \frac{1}{12}.$$

The next two theorems give us a way to compare integrals without tedious calculations.

THEOREM 6. If f is integrable on $[a, b]$ and $f \geq 0$ there, then

$$\int_a^b f(x)\, dx \geq 0.$$

PROOF. This follows since every term in the limit (1) defining the integral is non-negative.

THEOREM 7 (Comparison Theorem). Let f and g be integrable on $[a, b]$ and suppose that for every x in $[a, b]$,

$$f(x) \leq g(x). \tag{13}$$

Then

$$\int_a^b f(x)\, dx \leq \int_a^b g(x)\, dx. \tag{14}$$

PROOF. Define $h(x) = g(x) - f(x)$. Then by (13), $h(x) \geq 0$ on $[a, b]$. By Theorem 6, $\int_a^b h(x)\, dx \geq 0$. But

$$0 \leq \int_a^b h(x)\, dx = \int_a^b [g(x) - f(x)]\, dx = \int_a^b g(x)\, dx + \int_a^b [-f(x)]\, dx$$

$$= \int_a^b g(x)\, dx - \int_a^b f(x)\, dx$$

(by Theorems 4 and 5). Therefore $\int_a^b f(x)\,dx \leq \int_a^b g(x)\,dx$.

EXAMPLE 13. On the interval [0, 1], $x^3 \leq x^2$. This tells us (from Theorem 7) that

$$\int_0^1 x^3\,dx \leq \int_0^1 x^2\,dx.$$

This is indeed the case since $\int_0^1 x^3\,dx = \frac{1}{4}$ and $\int_0^1 x^2\,dx = \frac{1}{3}$.

EXAMPLE 14. Since $x^5 \leq x^7$ on [2, 5], we have

$$\int_2^5 x^5\,dx \leq \int_2^5 x^7\,dx.$$

The following theorem is sometimes extremely useful for estimating integrals when calculations are difficult.

THEOREM 8 (Upper and Lower Bound Theorem). Suppose that on $[a, b]$

$$m \leq f(x) \leq M. \tag{15}$$

for all x in $[a, b]$. That is, m and M are lower and upper bounds, respectively, for the function f on the interval $[a, b]$ (see Section 1.1). Then if f is integrable on $[a, b]$,

$$\boxed{m(b - a) \leq \int_a^b f(x)\,dx \leq M(b - a).} \tag{16}$$

PROOF. If we define g by $g(x) = M$, then $f(x) \leq g(x)$ and by the comparison theorem

$$\int_a^b f(x)\,dx \leq \int_a^b M\,dx = M(b - a)$$

(from Theorem 2). Similarly

$$m(b - a) = \int_a^b m\,dx \leq \int_a^b f(x)\,dx$$

and the theorem is proved.

> **Remark.** We will cite a theorem in Section 10.4 which states that if f is continuous on $[a, b]$, then f takes on a maximum and minimum value in that interval so that, in this case, the numbers m and M do exist. The function $1/x$, for example, does not have such bounds in the interval $[-1, 1]$.

EXAMPLE 15. Estimate $\displaystyle\int_9^{10} x^2\,dx$.

SOLUTION. $f(x)$ is an increasing function of x on [9, 10] so that

$$f(9) = 9^2 \leq f(x) \leq 10^2 = f(10)$$

on [9, 10]. Then $m = 9^2 = 81$, $M = 10^2 = 100$. $b - a = 1$, and

$$81 \leq \int_9^{10} x^2\,dx \leq 100.$$

We close this section by showing how definite integrals can arise in an application having nothing at all to do with area.

Suppose that a particle is moving with velocity $v = v(t)$. If $v(t) = c$, a constant, then we can calculate the distance traveled between the initial time $t = t_0$ and the final time $t = t_f$ by

$$\text{distance} = \text{velocity} \times \text{elapsed time} = v \cdot (t_f - t_0). \tag{17}$$

However, if the velocity is changing (that is, if the particle is accelerating or decelerating) then formula (17) simply will not work. Still, we can calculate the distance traveled by a method identical to the one we used to calculate areas. The basic idea is simple: even though $v(t)$ is changing, over a very small period of time it will be almost constant. We begin by dividing the time interval $[t_0, t_f]$ into n equal subintervals of length $(t_f - t_0)/n$ (see Figure 6). Let t_i^* denote a time in the interval $[t_{i-1}, t_i]$. If n is

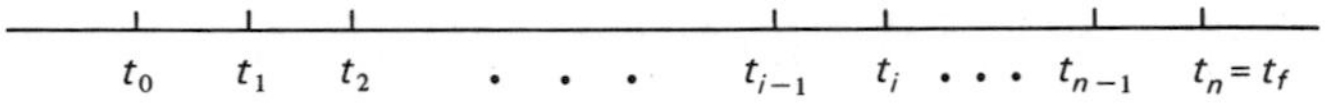

Figure 6

large enough, (that is, if the lengths of the intervals are small) then the velocity v_i of the particle in the time period $[t_{i-1}, t_i]$ is almost constant: $v_i \approx v(t_i^*)$ in the interval $[t_{i-1}, t_i]$. Let $\Delta t = t_i - t_{i-1}$. If s_i denotes the distance traveled by the particle between the time $t = t_{i-1}$ and the time $t = t_i$, then from (17)

$$s_i = v_i(t_i - t_{i-1}) \approx v(t_i^*)\,\Delta t. \tag{18}$$

The total distance traveled by the particle over the entire time interval $[t_0, t_f]$ is then given (approximately) by

$$s = s_1 + s_2 + \cdots + s_n \approx v(t_1^*)\,\Delta t + v(t_2^*)\,\Delta t + \cdots + v(t_n^*)\,\Delta t. \tag{19}$$

This should look familiar. Formula (19) is only an approximation because $v(t)$ is not really constant over $[t_{i-1}, t_i]$. However, as the length of the interval becomes smaller and smaller, the approximation gets better and better. We therefore can conclude that

$$s = \lim_{\Delta t \to 0} [v(t_1^*)\,\Delta t + v(t_2^*)\,\Delta t + \cdots + v(t_n^*)]\,\Delta t = \int_{t_0}^{t_f} v(t)\,dt. \tag{20}$$

Thus we obtain the distance traveled by "adding up" a great number of small distances just as we obtained the area under a curve by "adding up" a great number of areas of "thin" rectangles.

EXAMPLE 16. A race car starting at rest accelerates so that its velocity after t seconds is given by $v(t) = 4t^3$ ft/sec (for the first 4 seconds).

    (a)   What is the car's velocity after 3 seconds?
    (b)   How far does the car travel in the first 3 seconds?

SOLUTION. (a) $v(3) = 4 \cdot 3^3 = 4(27) = 108$ ft/sec (≈ 73.6 mi/hr).
    (b)   Here $t_0 = 0$, $t_f = 3$, and $v(t) = 4t^3$. Then, from Equation (20) we have

$$s = \int_{t_0}^{t_f} v(t)\,dt = \int_0^3 4t^3\,dt = 4\int_0^3 t^3\,dt = 4\left(\frac{81}{4}\right) = 81 \text{ feet.}$$

We have used the fact that $\int_0^3 t^3 \, dt = \int_0^3 x^3 \, dx = 81/4$ (Example 4).

There is an interesting interpretation of Theorem 3 in this use of the definite integral:

If $v(t)$ represents the velocity of a moving object, then

$$\int_a^b v(t) \, dt = \text{distance traveled from time } a \text{ to time } b,$$

$$\int_a^c v(t) \, dt = \text{distance traveled from time } a \text{ to time } c,$$

and

$$\int_c^b v(t) \, dt = \text{distance traveled from time } c \text{ to time } b.$$

It is easy to see why it should be true that, if $a < c < b$,

$$\int_a^b v(t) \, dt = \int_a^c v(t) \, dt + \int_c^b v(t) \, dt.$$

PROBLEMS 4.3

In Problems 1–24 calculate the given definite integral.

1. $\displaystyle\int_0^4 7x \, dx$

2. $\displaystyle\int_1^2 4x \, dx$

3. $\displaystyle\int_2^5 (3x + 2) \, dx$

4. $\displaystyle\int_1^4 (7t - 3) \, dt$

5. $\displaystyle\int_0^2 \tfrac{1}{2}t^2 \, dt$

6. $\displaystyle\int_{-1}^0 t^2 \, dt$

7. $\displaystyle\int_{-1}^1 2s^2 \, ds$

8. $\displaystyle\int_0^5 s^3 \, ds$

9. $\displaystyle\int_0^8 17s^2 \, ds$

10. $\displaystyle\int_{-1}^1 3z^3 \, dz$

11. $\displaystyle\int_1^0 z^3 \, dz$

12. $\displaystyle\int_3^0 (2z + 4) \, dz$

13. $\displaystyle\int_0^1 (1 - y^2) \, dy$

14. $\displaystyle\int_0^1 (1 - y^3) \, dy$

15. $\displaystyle\int_5^2 y^3 \, dy$

16. $\displaystyle\int_0^1 (1 + x + x^2 + x^3) \, dx$

17. $\displaystyle\int_{-2}^1 (x^3 - x^2) \, dx$

18. $\displaystyle\int_1^{-1} (x^3 + 3x) \, dx$

*19. $\displaystyle\int_{-1}^1 |x| \, dx$ [*Hint:* Draw a picture.]

*20. $\displaystyle\int_{-3}^2 |x + 1| \, dx$

*21. $\displaystyle\int_0^5 [x] \, dx$ [*Hint:* This is the greatest integer function. To calculate the area add the areas of the rectangles.]

*22. $\displaystyle\int_{-4}^4 [x] \, dx$

*23. $\int_a^b x^2 \, dx$ [*Hint:* See Problem 4.2.19.]

*24. $\int_a^b x^3 \, dx$. [*Hint:* See Problem 4.2.20.]

*25. Prove Theorem 2. [*Hint:* Follow the steps in Example 5.]

*26. Prove Theorem 3 by dividing the interval $[a, b]$ into $2n$ subintervals where the nth partition point is c.

*27. Prove Theorem 4. [*Hint:* Use the fact that $\lim_{x \to x_0} kf(x) = k \lim_{x \to x_0} f(x)$.]

*28. Prove Theorem 5. [*Hint:* Use the fact that $\lim_{x \to x_0} (f + g) = \lim_{x \to x_0} f + \lim_{x \to x_0} g$.]

29. Explain why $\int_{23}^{47} s^{17} \, ds < \int_{23}^{47} s^{55/3} \, ds$.

30. Explain why $\int_0^1 \sqrt{x} \, dx < \int_0^1 \sqrt[3]{x} \, dx$.

31. Which is greater: $\int_1^2 x \, dx$ or $\int_1^2 \sqrt{x} \, dx$?

32. Which is greater: $\int_{-1}^0 x^{1/3} \, dx$ or $\int_{-1}^0 x^{1/5} \, dx$?

*33. Show that $\int_0^1 \sqrt{1 + x^3} \, dx$ lies between 1 and $\frac{5}{4}$. [*Hint:* Write the integral as the sum of two integrals.]

In Problems 34–41 find upper and lower bounds for the given integrals. Do *not* try to calculate them.

34. $\int_{-1}^1 x^{10} \, dx$

35. $\int_1^4 4\sqrt{x} \, dx$

36. $\int_1^8 7x^{1/3} \, dx$

37. $\int_1^9 \frac{1}{\sqrt{x}} \, dx$

38. $\int_2^3 (x^2 + x^3) \, dx$

39. $\int_1^{100} \frac{1}{x} \, dx$

40. $\int_{-2}^2 3x^4 \, dx$

41. $\int_0^1 \frac{1}{1 + x^2} \, dx$.

42. A ball is dropped from a height of 400 ft. Its velocity after t seconds is given by the formula $v = 32t$ ft/sec.
  (a) How fast is the ball dropping after 4 sec?
  (b) How far has the ball dropped after 4 sec?
  *(c) After how many seconds will the ball hit the ground?

43. Gravity on a certain planet X is only half that of earth. Then the force of acceleration due to gravity is 16 ft/sec^2 = 4.91 m/sec^2 instead of the usual 32 ft/sec^2 = 9.81 m/sec^2. Answer the questions in Problem 42 assuming that the ball is dropped on planet X. [*Hint:* Find a new formula for v as a function of t.]

44. A bullet is shot straight into the air with an initial velocity of 500 m/sec. Its velocity after t sec have elapsed is $v(t) = (500 - 9.81t)$ m/sec.
  (a) After how many seconds will the bullet begin to fall?
  (b) How high will the bullet go?

45.   Show that if a, b, and c are any real numbers such that f is integrable on $[a, b]$, $[a, c]$ and $[c, b]$, then

$$\int_a^b f = \int_a^c f + \int_c^b f,$$

where it is *not* required that $a < c < b$. [*Hint:* There are 6 cases to consider: $a < c < b$, $a < b < c, b < a < c$, etc. Use Theorem 3 and Definition 3 to test each case separately.]

*46.   Let

$$f(x) = \begin{cases} 1, & \text{if } x \text{ is rational} \\ 0, & \text{if } x \text{ is irrational} \end{cases}$$

(This function was discussed in Problem 2.4.20 and you should look at that problem again now.)

(a)   Explain why, for any partition of the interval $[0, 1]$, it is possible to choose points x_i^* in $[x_{i-1}, x_i]$ such that $f(x_i^*) = 1$.

(b)   Explain why it is possible to choose points y_i^* in $[x_{i-1}, x_i]$ such that $f(y_i^*) = 0$.

(c)   Show that $\lim_{|P| \to 0} [f(x_1^*) \Delta x_1 + f(x_2^*) \Delta x_2 + \cdots + f(x_n^*) \Delta x_n]$ depends on the choice of the points x_i^*.

(d)   Explain why $\int_0^1 f(x)\,dx$ does not exist.

(e)   Is f piecewise continuous?

*47.   If $f(x)$ is continuous on $[a, b]$, show that

$$\left| \int_a^b f(x)\,dx \right| \le \int_a^b |f(x)|\,dx.$$

[*Hint:* Show from the definition that $\int_a^b f(x)\,dx \le \int_a^b |f(x)|\,dx$ and $-\int_a^b f(x)\,dx \le \int_a^b |f(x)|\,dx$.]

4.4 The Indefinite Integral

In this section we introduce another kind of integral which seemingly is very different from the definite integral discussed in the last three sections. We will wait until the next section to see the fundamental relationship between the two.

In Chapters 2 and 3 we discussed the problem of finding the derivative of a given function. We now consider the problem in reverse.

DEFINITION   1 (The Indefinite Integral).   Let f be defined on $[a, b]$. Then if there exists a function $y = F(x)$ such that F is continuous on $[a, b]$, differentiable on (a, b), and the derivative of F is f for every x in (a, b), that is, if

$$F'(x) = \frac{dF}{dx} = f(x), \quad x \text{ in } (a, b), \tag{1}$$

then F is called an *indefinite integral* or *antiderivative* of f on the interval $[a, b]$ and we write

$$F(x) = \int f(x)\,dx = \int f. \tag{2}$$

This is read "$F(x)$ is the integral of $f(x)$ with respect to x," or, simply, "F is an indefinite integral of f." If such a function F exists, then f is said to be *integrable* and the process of calculating an integral is called *integration*. The variable x is called the *variable of integration* and the function f is called the *integrand*.

> **Remark 1.** There is a great difference between the definite and indefinite integrals. The definite integral is a number while the indefinite integral is a function. Nevertheless we use the words "integral," "integrable," and "integration" in both cases. There should not be any confusion between the two uses of the word since the two integrals stand for such different things.

> **Remark 2.** The terms *antiderivative* or *primitive* are sometimes used in place of the term indefinite integral. All three terms stand for the same thing.

EXAMPLE 1. Find $\int 3x^2\, dx$.

$$\frac{3x^3}{3} = x^3 + C$$

SOLUTION. Since $d(x^3)/dx = 3x^2$ we have

$$\int 3x^2\, dx = x^3$$

But, the derivative of any constant is zero, so that $x^3 + C$ is also an indefinite integral of $3x^2$ for any constant C. To see this, we have

$$\frac{d}{dx}(x^3 + C) = \frac{d}{dx}(x^3) + \frac{d}{dx}(C) = 3x^2 + 0 = 3x^2.$$

This leads to the following.

> **Remark 3.** The reason that we refer to an *indefinite* integral is that there are always an infinite number of them. If F is an indefinite integral of f, then so is $F + C$ for every constant C since

$$\frac{d}{dx}(F + C) = \frac{dF}{dx} + \frac{dC}{dx} = f + 0 = f.$$

Of course, you may ask, how do we know we have found all the indefinite integrals of f? For example, are there any other functions F such that $F'(x) = 3x^2$, other than functions of the form $F(x) = x^3 + C$? The answer is no. The proof of the following theorem will be given in Section 10.8.

THEOREM 1. If F and G are two differentiable functions which have the same derivative, then they differ by a constant. That is

> if $F'(x) = G'(x)$, then $F(x) - G(x) = C$ for some constant C.

This theorem allows us to define the *most general indefinite integral* of f as $F(x) + C$ where F is some indefinite integral of f and C is an arbitrary constant. For the remainder of this section we will speak about "the integral," leaving out the word "indefinite" and will look for the most general indefinite integral in our calculations. We can tell the difference between a definite and indefinite integral because definite integrals always have lower and upper limits ($\int_a^b f(x)\,dx$) while indefinite integrals do not.

We now show how some integrals can be calculated. In Section 3.4 we showed that

$$\frac{d}{dx} x^r = rx^{r-1}, \tag{3}$$

where r is a nonzero real number. We therefore have

THEOREM 2

> If $r \neq -1$, then
>
> $$\int x^r\,dx = \frac{x^{r+1}}{r+1} + C. \tag{4}$$

PROOF

$$\frac{d}{dx}\left[\frac{x^{r+1}}{r+1} + C\right] = \frac{(r+1)x^r}{(r+1)} + 0 = x^r.$$

Note. The result does *not* hold for $r = -1$ because the denominator is $-1 + 1 = 0$ and we cannot divide by zero.

EXAMPLE 2. Calculate $\int x^9\,dx$.

$\int x^9\,dx$

SOLUTION. $r = 9$ so that

$$\int x^9\,dx = \frac{x^{9+1}}{9+1} + C = \frac{x^{10}}{10} + C.$$

$\dfrac{x^{10}}{10} + C$

EXAMPLE 3. Calculate $\int x^{1/3}\,dx$.

SOLUTION. $r = \frac{1}{3}$ so that

$\dfrac{x^{4/3}}{4/3} = 3 x^{4/3} \cdot \dfrac{1}{4} = \dfrac{1}{4} x^{4/3}$

$$\int x^{1/3}\,dx = \frac{x^{(1/3)+1}}{(\frac{1}{3})+1} + C = \frac{x^{4/3}}{\frac{4}{3}} + C = \frac{3}{4} x^{4/3} + C.$$

EXAMPLE 4. Calculate $\int (1/\sqrt{t})\,dt$.

SOLUTION. There is no difference between using x and t as the variable of integration. Then, since $1/\sqrt{t} = t^{-1/2}$,

$$\int \frac{1}{\sqrt{t}}\,dt = \int t^{-1/2}\,dt = \frac{t^{(-1/2)+1}}{(-\frac{1}{2})+1} + C = \frac{t^{1/2}}{\frac{1}{2}} + C = 2\sqrt{t} + C.$$

In the last three examples it was easy to check the answer by differentiating. This should always be done since it is always easier to differentiate than to integrate. Thus differentiation provides a method for verifying the result of calculating an integral.

As we have already mentioned, Theorem 2 does not hold if $r = -1$. The case

$t^{-\frac{1}{2}}$ $\dfrac{t^{\frac{1}{2}}}{\frac{1}{2}} = 2t^{\frac{1}{2}} = 2\sqrt{t}$

$r = -1$ is in some sense more interesting than the cases we have so far discussed. We will analyze this case in great detail in Chapter 6.

The following theorem follows easily from the analogous results for derivatives.

THEOREM 3. If f and g are integrable and if k is any constant, then kf and $f + g$ are integrable and we have

$$(i) \quad \int kf(x)\,dx = k \int f(x)\,dx. \tag{5}$$

and

$$(ii) \quad \int [f(x) + g(x)]\,dx = \int f(x)\,dx + \int g(x)\,dx. \tag{6}$$

PROOF. Let $F = \int f$ and $G = \int g$. Then

$$\frac{d}{dx}\,kF = k\,\frac{d}{dx}\,F = kf,$$

which proves (i). Similarly,

$$\frac{d}{dx}\,(F + G) = \frac{d}{dx}\,F + \frac{d}{dx}\,G = f + g,$$

which proves (ii).

Remark. This theorem can be extended to more than two functions. For example, it is easy to show that $\int (f + g + h) = \int f + \int g + \int h$.

EXAMPLE 5. Calculate $\int [(3/x^2) + 6x^2]\,dx.$

SOLUTION

$$\int \left(\frac{3}{x^2} + 6x^2\right) dx = \int \frac{3}{x^2}\,dx + \int 6x^2\,dx = 3\int x^{-2}\,dx + 6 \int x^2\,dx$$

$$= \frac{3x^{-2+1}}{-2+1} + \frac{6x^{2+1}}{2+1} + C = -3x^{-1} + 2x^3 + C$$

$$= \frac{-3}{x} + 2x^3 + C.$$

CHECK

$$\frac{d}{dx}\left(-\frac{3}{x} + 2x^3 + C\right) = -3(-x^{-2}) + 2\cdot 3x^2 = \frac{3}{x^2} + 6x^2.$$

EXAMPLE 6. Calculate $\int (2x^{7/5} - 3x^{-11/9} + 17x^{17})\,dx.$

SOLUTION

$$\int (2x^{7/5} - 3x^{-11/9} + 17x^{17})\,dx$$

$$= \int 2x^{7/5}\,dx + \int -3x^{-11/9}\,dx + \int 17x^{17}\,dx$$

$$= 2 \int x^{7/5} \, dx - 3 \int x^{-11/9} \, dx + 17 \int x^{17} \, dx$$

$$= 2 \, \frac{x^{(7/5)+1}}{(7/5) + 1} - \frac{3x^{(-11/9)+1}}{(-11/9) + 1} + \frac{17x^{17+1}}{17 + 1} + C$$

$$= \frac{2x^{12/5}}{12/5} - \frac{3x^{-2/9}}{-2/9} + \frac{17x^{18}}{18} + C$$

$$= \frac{5}{6} x^{12/5} + \frac{27}{2} x^{-2/9} + \frac{17}{18} x^{18} + C.$$

CHECK

$$\frac{d}{dx} \left(\frac{5}{6} x^{12/5} + \frac{27}{2} x^{-2/9} + \frac{17}{18} x^{18} + C \right)$$

$$= \frac{5}{6} \cdot \frac{12}{5} x^{(12/5)-1} + \frac{27}{2} \cdot \frac{-2}{9} x^{(-2/9)-1} + \frac{17}{18} \cdot x^{18-1} + 0$$

$$= 2x^{7/5} - 3x^{-11/9} + 17x^{17}.$$

Let us look more closely at the general integral of a function. Let $f(x) = 2x$. Then $\int f = x^2 + C$. For every value of C, we get a different integral. But these integrals are very similar geometrically. For example, the curve $y = x^2 + 1$ is obtained by "shifting" the curve $y = x^2$ up one unit. The curve $y = x^2 + C$ is obtained by shifting the curve $y = x^2$ up or down C units (up if $C > 0$ and down if $C < 0$; see Section 1.8). Some of these curves are plotted in Figure 1. These curves never intersect. To prove this, suppose that (x_0, y_0) is a point on both curves $y = x^2 + A$ and $y = x^2 + B$. Then $y_0 = x_0^2 + A = x_0^2 + B$ which implies that $A = B$. That is, if the two curves have a point in common, then the two curves are the same. Thus, if we specify one point through which the integral passes, then we know *the* integral.

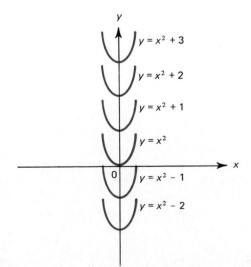

$y = x^2 + 3$

$y = x^2 + 2$

$y = x^2 + 1$

$y = x^2$

$y = x^2 - 1$

$y = x^2 - 2$

Figure 1

EXAMPLE 7. Find $\int 2x\, dx$ which passes through the point $(2, 7)$.

SOLUTION. $y = \int 2x\, dx = x^2 + C$. But when $x = 2$, $y = 7$, so that $7 = 2^2 + C = 4 + C$ or $C = 3$. The solution is the function $x^2 + 3$.

EXAMPLE 8. The slope of the tangent to a given curve is given by $f(x) = 6x^3 - 4x^2 + (1/x^2)$. The curve passes through the point $(1, 4)$. Find the curve.

SOLUTION. The slope of the curve $y = F(x)$ for any x is given by $dy/dx = F'(x) = 6x^3 - 4x^2 + (1/x^2)$. Then

$$F(x) = \int \left(6x^3 - 4x^2 + \frac{1}{x^2}\right) dx = \frac{6x^4}{4} - \frac{4x^3}{3} + \frac{x^{-1}}{-1} + C$$

$$= \frac{3}{2}x^4 - \frac{4}{3}x^3 - \frac{1}{x} + C.$$

When $x = 1$, $y = 4$, so that

$$4 = \frac{3}{2}(1)^4 - \frac{4}{3}(1)^3 - \frac{1}{1} + C = \frac{3}{2} - \frac{4}{3} - 1 + C$$

$$= \frac{9 - 8 - 6}{6} + C = -\frac{5}{6} + C.$$

Then

$$C = 4 + \frac{5}{6} = \frac{29}{6} \qquad \text{and} \qquad F(x) = \frac{3}{2}x^4 - \frac{4}{3}x^3 - \frac{1}{x} + \frac{29}{6}.$$

We may now ask: What functions are integrable? The answer is similar to the one we gave in Section 4.3 for the definite integral. The proof of the theorem below can be found in any standard advanced calculus text.†

THEOREM 4. If f is piecewise continuous on $[a, b]$, then f is integrable on that interval in the sense that there exists a function F such that $F'(x) = f(x)$ at every point x in $[a, b]$ at which f is continuous. (Here a and b may be $-\infty$ and ∞, respectively, in which case we refer to the interval $(-\infty, \infty)$.)

> *Remark.* This theorem does *not* say that functions which are not piecewise continuous are not integrable.

However, a serious problem remains. Consider the function $y = (1 + x^6)^{1/3}$. This function is certainly continuous (and even differentiable) everywhere. But it is impossible to find a function $F(x)$, written in terms of functions we recognize, which has the property that $F'(x) = (1 + x^6)^{1/3}$. (Try $(1 + x^6)^{4/3}/(\frac{4}{3})$ and see what happens.) This does not mean that the integral does not exist. It means that there is no way to express this integral in terms of simple functions. In fact, it is perhaps surprising that most continuous functions have integrals which cannot be represented

---

†Such as the one by R. C. Buck cited in Section 4.3.

in terms of recognizable functions. We will devote most of Chapter 8 to a discussion of how to find integrals of a wide variety of functions. For the rest, we will explore (also in Chapter 8) numerical methods for estimating definite integrals as closely as we wish.

In Section 2.10 we gave several examples of the applicability of the derivative. We close this section by looking at some of these examples from the "reversed" point of view, that of the applicability of the integral.

EXAMPLE 9. A ball is dropped from rest from a certain height. Its velocity after t seconds due to the earth's gravitational field is given by

$$v = 32t,$$

where v is measured in ft/sec. How far has the ball fallen after t seconds?

SOLUTION. If $s = s(t)$ represents distance, then

$$\frac{ds}{dt} = v \quad \text{or} \quad s(t) = \int v(t)\, dt = \int 32t\, dt = 16t^2 + C.$$

But $0 = s(0) = 16 \cdot 0^2 + C$ which implies that $C = 0$. Thus $s(t) = 16t^2$. For example, after 3 seconds, the ball has fallen $16 \cdot 3^2 = 16(9) = 144$ feet.

We can remember the rule:

> The velocity function is the derivative of the distance function and the distance function is an indefinite integral of the velocity function.

Similarly, since acceleration equals a equals dv/dt, we have:

> The acceleration function is the derivative of the velocity function and the velocity function is an indefinite integral of the acceleration function.

EXAMPLE 10. The acceleration under the pull of the earth's gravity is 9.81 m/sec². Find a formula for velocity and distance traveled by a particle under the influence of the force of gravitational attraction.

SOLUTION. $v(t) = \int a(t)\, dt = \int 9.81\, dt = 9.81t + C$. Let $v(0)$, the initial velocity of the particle being considered, be denoted by v_0. Then $v(0) = v_0 = (9.81)(0) + C$ or $C = v_0$. Hence

$$v(t) = 9.81t + v_0 \tag{7}$$

and

$$s(t) = \int v(t)\, dt = \int (9.81t + v_0)\, dt = \frac{9.81}{2}t^2 + v_0 t + C.$$

If we let s_0 denote $s(0)$, the initial position of the particle, we obtain

$$s(0) = s_0 = \frac{9.81}{2}(0)^2 + v_0(0) + C \quad \text{or} \quad C = s_0$$

so that

$$s(t) = \frac{9.81}{2} t^2 + v_0 t + s_0.$$ (8)

This is called the *equation of motion* of the particle.

EXAMPLE 11. A population is growing with an instantaneous growth rate of $100 + 12\sqrt{t} - 5t$ organisms per hour after t hours. If the initial population is 10,000, what is the population after t hours?

SOLUTION. Since the rate of growth is the derivative of the population size $P(t)$ (see Example 2.10.2), $P(t)$ is the indefinite integral of population growth so that

$$P(t) = \int (100 + 12t^{1/2} - 5t)\, dt = 100t + 12\frac{t^{3/2}}{\frac{3}{2}} - \frac{5t^2}{2} + C$$

$$= 100t + 8t^{3/2} - \frac{5}{2}t^2 + C.$$

But $P(0) = $ the initial population $= 10,000$ so that

$$P(0) = 100(0) + 8(0)^{3/2} - \frac{5}{2}(0)^2 + C = 10,000 \quad \text{or} \quad C = 10,000$$

and

$$P(t) = 100t + 8t^{3/2} - \frac{5}{2}t^2 + 10,000$$

organisms after t hours.

EXAMPLE 12. The density ρ of a 4-meter nonuniform metal beam, measured in kg/m, is given by

$$\rho(x) = 2\sqrt{x}$$ (9)

where x is the distance along the beam measured from the left end (see Figure 2).

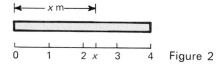

Figure 2

    (a)   What is the mass of the beam up to the point x units from the left?
    (b)   What is the total mass of the beam?

SOLUTION. (a) Since density is the rate of change of the mass μ (see Example 2.10.4), we have

$$\rho = \frac{d\mu}{dx} \quad \text{or} \quad \mu(x) = \int \rho(x)\, dx.$$ (10)

Then

$$\mu(x) = \int 2\sqrt{x}\, dx = 2 \int x^{1/2}\, dx = 2\frac{x^{3/2}}{3/2} + C = \frac{4}{3}x^{3/2} + C.$$

But $\mu(0) = 0$ since there is no mass zero units from the left side of the beam. Hence $C = 0$ and

$$\mu(x) = \tfrac{4}{3}x^{3/2} \quad \text{kg.}$$

(b)   When $x = 4$ m, $\mu(x) = \frac{4}{3}(4)^{3/2} = \frac{4}{3}(8) = \frac{32}{3}$ kg. This is the total mass of the beam.

EXAMPLE 13.   The marginal revenue that a manufacturer receives for his goods is given by

$$MR = 100 - 0.03x. \tag{11}$$

That is, MR is the revenue he receives for each additional unit sold when he has sold x units. How much money does he receive if he sells x units?

SOLUTION. Marginal revenue is similar to marginal cost. (See Example 2.10.5.) Equation (11) tells us that the manufacturer receives less and less per unit, the more units he sells. Marginal revenue is the rate of change of money received per unit change in the number of units sold. That is, if R denotes total revenue,

$$\frac{dR}{dx} = MR = 100 - 0.03x. \tag{12}$$

Then

$$R(x) = \int MR = \int (100 - 0.03x)\, dx = 100x - 0.03\frac{x^2}{2} + C.$$

But the manufacturer certainly receives nothing if he sells nothing. Therefore $R(0) = 0 = C$ and

$$R(x) = 100x - 0.03\frac{x^2}{2}.$$

PROBLEMS 4.4

In Problems 1–15 find the most general indefinite integral.

1. $\int dx$

2. $\int x\, dx$

3. $\int a\, dx,$ where a is a constant

4. $\int (ax + b)\, dx,$ a, b constants

5. $\int (1 + x + x^2 + x^3 + x^4)\, dx$

6. $\int \frac{1}{x^5}\, dx$

7. $\int \frac{4}{x^3}\, dx$

8. $\int \frac{7}{\sqrt{x}}\, dx$

9. $\int \left(\sqrt{x} + \frac{1}{\sqrt{x}}\right) dx$

10. $\int \left(3\sqrt[3]{x} + \frac{3}{\sqrt[3]{x}}\right) dx$

11. $\int (x^{1/2} + x^{3/2} + x^{3/4})\, dx$

12. $\int \left(\dfrac{4}{3x^4} - \dfrac{5}{7x^5} + \dfrac{6}{11x^6} \right) dx$

13. $\int \dfrac{x^3 + x^2 - x}{x^{3/2}}\, dx$

14. $\int \dfrac{4x^{2/3} - x^{5/8} + x^{17/5}}{5x^3}\, dx$

[*Hint:* Divide through.]

15. $\int \left(-\dfrac{17}{x^{13/17}} + \dfrac{3}{x^{4/9}} \right) dx.$

In Problems 16–20 the derivative of a function and one point on its graph are given. Find the function.

16. $\dfrac{dy}{dx} = x^3 + x^2 - 3;\quad (1, 5)$

17. $\dfrac{dy}{dx} = 2x(x + 1);\quad (2, 0)$

18. $\dfrac{dy}{dx} = \sqrt[3]{x} + x - \dfrac{1}{3\sqrt[3]{x}};\quad (-1, -8)$

19. $y' = 13x^{15/18} - 3;\quad (1, 14)$

20. $y' = 1 + x + x^2 + x^3 + x^4 + x^5;\quad (0, 83).$

21. A particle moves with the constant acceleration of 5.8 m/sec². It starts with an initial velocity of 0.2 m/sec and initial position of 25 m. Find the equation of motion of the particle.

22. A bullet is shot from ground level straight up into the air at an initial velocity of 2000 ft/sec. Write down a function which tells us the height of the bullet after t seconds, up until the time the bullet hits the earth (assume that the only force acting on the bullet is the force of gravitational attraction which imparts the constant acceleration $g = 32$ ft/sec²).

23. The acceleration of a moving particle starting at a position 100 m along the x-axis and moving with an initial velocity of $v_0 = 25$ m/min is given by

$$a(t) = 13\sqrt{t}\quad \text{m/min}^2.$$

Find the equation of motion of the particle.

24. If the instantaneous rate of change of a population is $50t^2 - 100t^{3/2}$ measured in individuals/year and the initial population is 25,000,
   (a) what is the population after t years?
   (b) what is the population after 25 years?

25. If the density of a 10-m beam is given by

$$\rho(x) = 3x + 2x^2 - x^{3/2}\quad \text{kg/m},$$

what is the total mass (in kg) of the beam?

26. If the density of a 64-ft beam is given by $\rho(x) = 3x^{5/6}$, measured in lb/ft, what is the total weight (in lb) of the beam?

27. If the marginal cost function for a certain manufacturer is $MC = 25x - 0.02x^2 + 20$, and if it costs $2000 to produce 10 units of a certain product, how much does it cost to produce 100 units? 500 units?

28. If the marginal revenue function for a certain product is given by $MR = 3x - 2\sqrt{x} + 10$, how much money will the manufacturer receive if he sells 50 units? 100 units?

29. In Example 13, at what point does it become unprofitable to the manufacturer to produce additional units of the product he sells? (This is often called the point of *diminishing returns.*)

*30. Consider the function $F(x) = \int (1/x)\, dx$. We have not yet learned how to integrate $1/x = x^{-1}$, but we can still obtain some information about $F(x)$. Assuming that $F(1) = 0$, use information about the derivatives of F to graph the curve for $x > 0$. [*Hint:* Show that F is always increasing, is concave down, and has a vertical tangent as $x \rightarrow 0^+$.]

4.5 The Fundamental Theorem of Calculus

In Chapter 2 we saw how the tangent line problem, formulated long ago by Greek mathematicians and not solved until the seventeenth century, gave rise to the modern theory of derivatives. In Sections 4.1–4.3 we saw how the area problem whose origins are lost in antiquity gave rise to the idea behind the definite integral. Finally, in the last section we introduced the indefinite integral which was seemingly unrelated to the definite integral and was defined as the "inverse" operation to differentiation.

In this section we show how these three operations on functions (the derivative and the two integrals) are intimately related. The remarkable theorem linking these together is called the *fundamental theorem of calculus*.

We now make a start by showing the relationship between the definite and indefinite integrals. We assume that the function f is continuous on $[a, b]$ (piecewise continuity would do just as well here). Then, by Theorem 4.3.1, $\int_a^b f(t)\, dt$ exists. But we have even more than that. If x is any number in $[a, b]$, then f is certainly continuous on the smaller interval $[a, x]$ and so, again by Theorem 4.3.1, $\int_a^x f(t)\, dt$ exists. For every value of x in $[a, b]$, this integral is a real number. Now we define a new function G by

$$G(x) = \int_a^x f(t)\, dt. \tag{1}$$

We emphasize that $G(x)$ is the *function* which assigns to every number x in $[a, b]$ the value of the definite integral of f over the interval $[a, x]$. The theorem below tells us that $G(x)$ is really an *indefinite* integral for the function f over the interval $[a, b]$. Its proof is difficult and will be delayed until Section 10.8. However, we will indicate by a graph why the theorem is plausible.

THEOREM 1. If f is continuous on $[a, b]$ then the function $G(x) = \int_a^x f(t)\, dt$ is continuous on $[a, b]$, differentiable on (a, b), and for every x in (a, b)

$$G'(x) = f(x). \tag{2}$$

That is, G is an indefinite integral of f on the interval $[a, b]$.

GRAPHICAL INDICATION OF PROOF. Consider Figure 1, paying particular attention to the various areas under the curve $f(x)$. By definition of the derivative,

$$G'(x) = \lim_{\Delta x \to 0} \frac{G(x + \Delta x) - G(x)}{\Delta x}.$$

Since $G(x) = \int_a^x f(t)\, dt =$ area under the curve $f(x)$ between a and x, and assuming

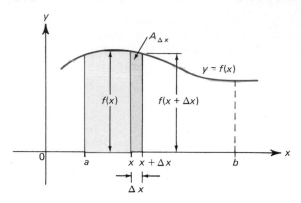

Figure 1

that $f > 0$ on (a, b), we have

$G(x + \Delta x) =$ area between a and $x + \Delta x$

$G(x) =$ area between a and x

$G(x + \Delta x) - G(x) =$ area between x and $x + \Delta x$.

This last area is denoted $A_{\Delta x}$ in Figure 1. Now if Δx is small, then x and $x + \Delta x$ are close and since f is continuous, $f(x + \Delta x)$ is close to $f(x)$. That is, the shaded area $A_{\Delta x}$ is almost rectangular. We therefore see that

$$G(x + \Delta x) - G(x) = A_{\Delta x} \approx f(x) \, \Delta x. \tag{3}$$

Then, for Δx small,

$$\frac{G(x + \Delta x) - G(x)}{\Delta x} \approx \frac{f(x) \, \Delta x}{\Delta x} = f(x). \tag{4}$$

Since as $\Delta x \to 0$, this approximation gets better and better, we may assert that

$$G'(x) = \lim_{\Delta x \to 0} \frac{G(x + \Delta x) - G(x)}{\Delta x} = f(x),$$

which indicates that the derivative of G is f, as we wanted to show.

We now know that we can obtain an indefinite integral of f by considering the definite integral. We are now ready to state and prove the fundamental theorem of calculus which allows us to find the definite integral of f by calculating an indefinite integral of f.

THEOREM 2 (Fundamental Theorem of Calculus). Let f be continuous on $[a, b]$. If F is any indefinite integral of f on $[a, b]$, then

$$\boxed{\int_a^b f(t) \, dt = F(b) - F(a).} \tag{5}$$

Remark. The theorem simply asserts that we may calculate a definite integral by evaluating some indefinite integral at the endpoints of the interval of integration and then subtracting.

PROOF. If F is an indefinite integral of f, then $F'(x) = f(x)$ for all x in (a, b). Also, if $G(x) = \int_a^x f(t)\, dt$, then $G'(x) = f(x)$ for all x in (a, b). From Theorem 4.4.1, we know that since F and G have the same derivative, they differ by a constant. Thus F and G are continuous on $[a, b]$, differentiable on (a, b) and

$$G(x) = F(x) + C, \qquad \text{for every } x \text{ in } [a, b].$$

In particular, $G(a) = F(a) + C$. But $G(a) = \int_a^a f(t)\, dt = 0$ by Definition 4.3.2. Thus

$$0 = F(a) + C \text{ or } C = -F(a).$$

Hence

$$G(x) = F(x) - F(a), \qquad \text{for } x \text{ in } [a, b]. \tag{6}$$

Since formula (6) is true for every x in $[a, b]$, it is true when $x = b$. That is,

$$G(b) = F(b) - F(a). \tag{7}$$

But $G(b) = \int_a^b f(t)\, dt$ and so the theorem is proved.

EXAMPLE 1.   Calculate $\displaystyle\int_0^1 x^2\, dx$.

SOLUTION. We have seen that $x^3/3$ is an indefinite integral for x^2. Thus

$$\int_0^1 x^2\, dx = \left(\frac{x^3}{3} \text{ evaluated at } x = 1\right) - \left(\frac{x^3}{3} \text{ evaluated at } x = 0\right) = \frac{1}{3} - 0 = \frac{1}{3}.$$

There is a simple notation we will use to avoid writing out the words "evaluated at" each time.

Notation. $F(x)\big|_a^b = F(b) - F(a)$.
In the above example we could have written

$$\int_0^1 x^2\, dx = \frac{x^3}{3}\bigg|_0^1 = \frac{1}{3} - 0 = \frac{1}{3}.$$

Remark. It doesn't make any difference which indefinite integral we choose to evaluate the definite integral. For example, if C is any constant, then

$$\int_0^1 x^2\, dx = \frac{x^3}{3} + C\bigg|_0^1 = \left(\frac{1}{3} + C\right) - (0 + C) = \frac{1}{3} + C - C = \frac{1}{3}.$$

The constants will always "disappear" in this manner. Thus we will use the "easiest" indefinite integral in our evaluation of $\int_a^b f$, which will almost always be the one in which $C = 0$.

EXAMPLE 2.   Calculate $\displaystyle\int_0^3 x^3\, dx$.

SOLUTION. $\displaystyle\int x^3\, dx = \frac{x^4}{4} + C$. Then

$$\int_0^3 x^3\, dx = \frac{x^4}{4}\bigg|_0^3 = \frac{3^4}{4} - 0 = \frac{81}{4}.$$

EXAMPLE 3. Calculate $\int_1^2 (3x^4 - x^5)\,dx$.

SOLUTION. $\int (3x^4 - x^5)\,dx = \dfrac{3x^5}{5} - \dfrac{x^6}{6} + C$. Thus

$$\int_1^2 (3x^4 - x^5)\,dx = \frac{3x^5}{5} - \frac{x^6}{6}\Big|_1^2 = \left(\frac{3(2)^5}{5} - \frac{2^6}{6}\right) - \left(\frac{3(1)^5}{5} - \frac{1^6}{6}\right)$$

$$= \left(\frac{96}{5} - \frac{64}{6}\right) - \left(\frac{3}{5} - \frac{1}{6}\right) = \left(\frac{576}{30} - \frac{320}{30}\right) - \left(\frac{18}{30} - \frac{5}{30}\right)$$

$$= \frac{243}{30} = \frac{81}{10}.$$

Warning. It is easy to lose track of minus signs when performing these calculations. Often the minus signs will cancel. Be careful!

EXAMPLE 4. Calculate $\int_{-2}^2 x^2\,dx$.

SOLUTION

$$\int_{-2}^2 x^2\,dx = \frac{x^3}{3}\Big|_{-2}^2 = \frac{2^3}{3} - \frac{(-2)^3}{3} = \frac{8}{3} - \left(-\frac{8}{3}\right) = \frac{16}{3}.$$

EXAMPLE 5. Calculate

$$\int_1^4 \left(\frac{3}{\sqrt{s}} - 5\sqrt{s}\right) ds.$$

SOLUTION

$$\int \left(\frac{3}{\sqrt{s}} - 5\sqrt{s}\right) ds = \int (3s^{-1/2} - 5s^{1/2})\,ds = 6\sqrt{s} - \frac{10}{3}s^{3/2} + C.$$

Therefore

$$\int_1^4 \left(\frac{3}{\sqrt{s}} - 5\sqrt{s}\right) ds = \left(6\sqrt{s} - \frac{10}{3}s^{3/2}\right)\Big|_1^4$$

$$= \left(6\sqrt{4} - \frac{10}{3}(4)^{3/2}\right) - \left(6\sqrt{1} - \frac{10}{3}(1)^{3/2}\right)$$

$$= 6(2) - \frac{10}{3}(8) - 6 + \frac{10}{3} = 6 - \frac{70}{3} = -\frac{52}{3}.$$

EXAMPLE 6. Calculate $\int_a^b x^r\,dx$ where $r \neq -1$ is a real number.

SOLUTION

$$\int x^r\,dx = \frac{x^{r+1}}{r+1} + C$$

so that

$$\int_a^b x^r \, dx = \frac{x^{r+1}}{r+1} \bigg|_a^b = \frac{1}{r+1}\{b^{r+1} - a^{r+1}\}. \tag{8}$$

In Section 4.4 we stated three theorems whose proofs were promised here. They were

(i) $\displaystyle\int_a^b f = \int_a^c f + \int_c^b f;$ (ii) $\displaystyle\int_a^b kf = k\int_a^b f;$ (iii) $\displaystyle\int_a^b (f+g) = \int_a^b f + \int_a^b g.$

Let F and G be indefinite integrals of f and g, respectively. Then kF is an indefinite integral of kf and $F + G$ is an indefinite integral of $f + g$ (check this by differentiating). We now have

(i) $\displaystyle\int_a^c f + \int_c^b f = [F(c) - F(a)] + [F(b) - F(c)]$

$$= F(b) - F(a) = \int_a^b f.$$

(ii) $\displaystyle\int_a^b kf = kF(b) - kF(a) = k[F(b) - F(a)] = k\int_a^b f.$

(iii) $\displaystyle\int_a^b (f+g) = (F + G) \bigg|_a^b = (F + G)(b) - (F + G)(a)$

$$= [F(b) + G(b)] - [F(a) + G(a)]$$

$$= [F(b) - F(a)] + [G(b) - G(a)] = \int_a^b f + \int_a^b g.$$

EXAMPLE 7.   Calculate $\displaystyle\int_{-5}^3 |x + 1| \, dx.$

SOLUTION. We know that

$$|x + 1| = \begin{cases} x + 1, & \text{if } x \geq -1 \\ -(x + 1), & \text{if } x \leq -1 \end{cases}.$$

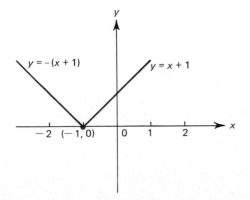

Figure 2

(See Figure 2.) Then

$$\int_{-5}^{3} |x + 1| \, dx = \int_{-5}^{-1} -(x + 1) \, dx + \int_{-1}^{3} (x + 1) \, dx \quad \text{(from Theorem 4.3.3)}$$

$$= \left(-\frac{x^2}{2} - x\right)\Big|_{-5}^{-1} + \left(\frac{x^2}{2} + x\right)\Big|_{-1}^{3}$$

$$= \left(-\frac{1}{2} - (-1) + \frac{25}{2} - 5\right) + \left(\frac{9}{2} + 3 - \frac{1}{2} - (-1)\right) = 16.$$

EXAMPLE 8. Calculate the area bounded by the curve $y = x^3 - 6x^2 + 11x - 6$ and the x-axis.

SOLUTION. We have $y = x^3 - 6x^2 + 11x - 6 = (x - 1)(x - 2)(x - 3)$. The curve is

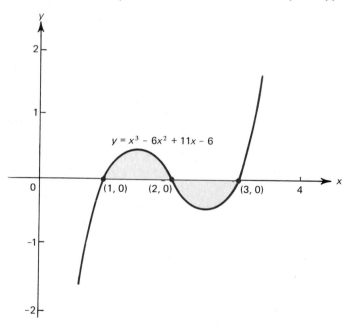

Figure 3

graphed in Figure 3. The desired area is the shaded part of the graph. We know that

$$A = \int_{1}^{3} |x^3 - 6x^2 + 11x - 6| \, dx$$

$$= \int_{1}^{2} (x^3 - 6x^2 + 11x - 6) \, dx + \int_{2}^{3} -(x^3 - 6x^2 + 11x - 6) \, dx$$

$$= \left(\frac{x^4}{4} - 2x^3 + \frac{11x^2}{2} - 6x\right)\Big|_{1}^{2} - \left(\frac{x^4}{4} - 2x^3 + \frac{11x^2}{2} - 6x\right)\Big|_{2}^{3}$$

$$= \left(4 - 16 + 22 - 12 - \frac{1}{4} + 2 - \frac{11}{2} + 6\right)$$

$$- \left(\frac{81}{4} - 54 + \frac{99}{2} - 18 - 4 + 16 - 22 + 12\right) = \frac{1}{2}.$$

$\int \cos \, s\hbar\phi$
$\int \cos x \int s \, u \, y$

Note that

$$\int_1^3 (x^3 - 6x^2 + 11x - 6)\, dx = \left(\frac{x^4}{4} - 2x^3 + \frac{11x^2}{2} - 6x \right)\Big|_1^3 = 0.$$

This illustrates why in the process of calculating area, care must be taken so that "positive" areas and "negative" areas don't cancel each other out.

PROBLEMS 4.5

In Problems 1–23 calculate the given integral.

1. $\int_{-1}^2 x^4\, dx$

2. $\int_2^5 3s^3\, ds$

3. $\int_1^9 \frac{\sqrt{t}}{2}\, dt$

4. $\int_1^3 (x^2 + 3x + 5)\, dx$

5. $\int_a^b (c_1 y^2 + c_2 y + c_3)\, dy$

6. $\int_2^4 \left(\frac{1}{z^3} - \frac{1}{z^2} \right) dz$

7. $\int_1^8 \left(\frac{1}{\sqrt[3]{x}} + 7\sqrt[3]{x} \right) dx$

8. $\int_0^1 (1 + s^8 + s^{16} + s^{32})\, ds$

9. $\int_{-1}^1 (p^9 + p^{17})\, dp$

10. $\int_{-a}^a x^{2n+1}\, dx$, where n is a positive integer and a is a real number

11. $\int_9^{16} \frac{v+1}{\sqrt{v}}\, dv$ [*Hint:* Divide.]

12. $\int_2^3 (y-1)(y+2)\, dy$ [*Hint:* Multiply.]

13. $\int_1^4 \frac{z^2 + 2z + 5}{z^{3/2}}\, dz$

14. $\int_{-2}^2 (v^2 - 4)(v^5 + 6)\, dv$

15. $\int_0^1 (t^{3/2} - t^{2/3})(t^{4/3} - t^{3/4})\, dt$

16. $\int_0^1 (\sqrt{x} - \sqrt[3]{x})^2\, dx$

17. $\int_1^0 s^{100}\, ds$

18. $\int_{-1}^{-2} \frac{1}{s^{100}}\, ds$

19. $\int_2^4 \frac{x^2 + 7x + 6}{x+1}\, dx$ [*Hint:* Divide.]

20. $\int_0^1 \frac{(x+1)(x-2)^3}{x^2 - x - 2}\, dx$

21. $\int_{-3}^3 |x - 1|\, dx$

22. $\int_{-5}^{15} |2x + 6|\, dx$

*23. $\int_{-2}^5 f(x)\, dx$, where $f(x) = \begin{cases} 1 + x^2, & x \ge 2 \\ 13 - x^3, & x \le 2 \end{cases}$

In Problems 24–35 find the area bounded by the given curve and the given lines. [*Hint:* Sketch the curves.]

24. $y = x^2 - 4$; x-axis
25. $y = 4 - x^2$; x-axis
26. $y = x^2 + 2x - 3$; $x = 1$; $x = 3$; x-axis
27. $y = x^2 - 6x + 5$; x-axis
28. $y = x^2 - 15x - 34$; x-axis
29. $y = x^4$; $x = -2$; $x = 4$; x-axis

30. $y = x^3 + 2x^2 - x - 2$; x-axis
31. $y = (x^2 - 1)(x^2 - 4)$; x-axis
32. $y = x^3 + 2x^2 - 13x + 10$; $x = 0$; $x = 3$; x-axis
33. $y = (x - a)(x - b)$, $a < b$; x-axis
34. $y = (x - a)(x - b)(x - c)$; x-axis, $a < b < c$
35. $y = 3x^7$; $x = -2$, $x = 2$; x-axis.

36. An *even function* is a function f which has the property that $f(-x) = f(x)$ for every real number x. (For example, 1, x^2, x^4, $|x|$, and $1/(1 + x^2)$ are all even functions.) Show that if f is even, then

$$\int_{-a}^{a} f = 2 \int_{0}^{a} f,$$

for every real number a. Can you explain this fact geometrically?

37. An *odd function* f has the property that $f(-x) = -f(x)$ for every real number x. (For example, x, x^3, $1/x^5 + x^7$ are all odd functions.) Show that if f is odd, then

$$\int_{-a}^{a} f = 0$$

for any real number a. Can you explain this fact geometrically?

38. Calculate

(a) $\displaystyle\int_{-50}^{50} (x + x^3 + x^{17})\, dx.$

(b) $\displaystyle\int_{-100}^{100} (1 + x^{1/3} + x^{5/3} + x^{11/3})\, dx.$

In Problems 39–42 use Theorem 1 to calculate the derivative $F'(x)$. Then evaluate $F'(x_0)$ for the given x_0.

39. $F(x) = \displaystyle\int_{3}^{x} \dfrac{dt}{1 + t^3}$; $x_0 = 2$

40. $F(x) = \displaystyle\int_{-1}^{x} \dfrac{s^2}{s^2 + 5}\, ds$; $x_0 = 3$

41. $F(x) = \displaystyle\int_{0}^{x} \sqrt{\dfrac{u - 1}{u + 1}}\, du$; $x_0 = 1$

42. $F(x) = \displaystyle\int_{2}^{x} \left(\dfrac{1 + 2t - 3t^2}{t^3 + t^{7/9}}\right) dt$; $x_0 = 1.$

43. The velocity of a moving particle after t seconds is given by $v(t) = t^{3/2} + 16t + 1$ where v is measured in m/sec. How far does the particle travel between $t = 4$ and $t = 9$ sec?

44. A ball is thrown down from a tower with an initial velocity of 25 ft/sec. The tower is 500 ft high.
(a) How fast is the ball traveling after 3 sec?
(b) How far does the ball travel in the first 3 sec?
*(c) How long does it take for the ball to hit the ground? [*Hint:* Use $a = 32$ ft/sec².]

*45. The density of a 3-m metal bar is given by $\rho(x) = 1/\sqrt{x + 1}$ kg/m, where x is measured in units from the left.
(a) What is the total mass of the bar? [*Hint:* It is not difficult to find a function whose derivative is $1/\sqrt{x + 1}$.]
(b) For what value of x is the mass of the bar from 0 to x units equal to half the total mass of the bar?

46. The marginal revenue a manufacturer receives is given by $MR = 2 - 0.02x + 0.003x^2$ dollars per additional unit sold. How much additional money does the manufacturer receive if he increases sales from 50 to 100 units?

47. The marginal cost to produce a widget is given by $MC = 2 - 0.003x + 0.00005x^2$ dollars. How much does it cost to increase production from 100 to 200 widgets?

48. The population of a species of insects is increasing at a rate of $3000/\sqrt{t}$ individuals per week. How many individuals are added to the population between the ninth and the twenty-fifth week?

49. A charged particle enters a linear accelerator. Its velocity increases with a constant acceleration from an initial velocity of 500 m/sec to a velocity of 10,500 m/sec in 1/100 sec.
    (a) What is the acceleration?
    (b) How far does the particle travel in 1/100 sec?

If $f(x)$ takes on the n values $x_1, x_2, \ldots, x_n$, then the *average* of these values is $(x_1 + x_2 + \cdots + x_n/n)$. That is, we add up the values the function takes and divide the sum by n, the number of points. If f is continuous over the interval $[a, b]$ then the *average value of f over* $[a, b]$ is defined by

$$\text{average value} = \frac{1}{b-a} \int_a^b f(x) \, dx.$$

In Problems 50–61 find the average value of the given function over the given interval.

50. $f(x) = 3x + 5$; $[1, 2]$

51. $f(x) = C$, C a constant; $[a, b]$

52. $f(x) = x^2$; $[0, 1]$

53. $f(x) = x^3$; $[0, 1]$

54. $f(x) = x^r$; $[0, 1]$ $(r \neq -1)$

55. $f(x) = x^2 - 2x + 5$; $[-2, 2]$

56. $f(x) = \sqrt{x} + 5x^3$; $[0, 4]$

57. $f(x) = 1 + x + x^2 + x^3 + x^4$; $[0, 1]$

58. $f(x) = \dfrac{1}{x^2}$; $[1, 2]$

59. $f(x) = x^{2/3} + \dfrac{1}{\sqrt[3]{x}}$; $[1, 8]$

60. $f(x) = x^{17/11}$; $[0, 1]$

61. $f(x) = x^{10} - \dfrac{1}{x^{10}}$; $[1, 2]$.

62. The density of a 4-m metal bar is given by $\rho(x) = 1 + x - \sqrt{x}$ kg/m. What is the average density over the length of the entire bar?

63. If the marginal cost of a given product is $50 - 0.05x$ dollars, what is the average cost per unit of production if 200 units are produced?

64. One day the temperature of the air t hours after noon was found to be $60 + 4t - t^2/3$ degrees (Fahrenheit). What was the average temperature between noon and 5 P.M.?

*65. A ball was dropped from rest from a height of 400 feet. What was its average velocity on the way to the ground?

4.6 Integration by Substitution

As we saw in the last section, the problem of finding a definite or indefinite integral is solved if we can find one indefinite integral for the function f. However, as we have already mentioned, this can often be very difficult. So far, the only functions we know how to integrate are functions of the form x^r where $r \neq -1$.

There are many other functions for which indefinite integrals can be found, but learning to recognize them takes a lot of experience. We shall devote an entire chapter (Chapter 8) to the problem of recognizing the types of functions for which integrals can be expressed in terms of certain elementary functions. In this section we enlarge the class of functions whose integrals can be immediately determined.

We start with the formula

$$\int u^r \, du = \frac{u^{r+1}}{r+1} + C, \qquad \text{if } r \neq -1. \tag{1}$$

Formula (1) can be proved by differentiating:

$$\frac{d}{du} \frac{u^{r+1}}{r+1} = u^r.$$

Consider the integral

$$\int (1 + x^2)^3 2x \, dx. \tag{2}$$

If we define the *function* $u = 1 + x^2$, then $du/dx = 2x$ and, taking differentials, we have $du = 2x \, dx$. Then $(1 + x^2)^3 = u^3$ and equation (2) becomes

$$\int (1 + x^2)^3 2x \, dx = \int u^3 \, du = \frac{u^4}{4} + C = \frac{(1 + x^2)^4}{4} + C.$$

In general, to calculate $\int f(x) \, dx$, perform the following steps:

(i)   Make a substitution $u = g(x)$ so that the integral can be expressed in the form $u^r \, du$ (if possible).
(ii)   Calculate $du = g'(x) \, dx$.
(iii)   Write $\int f(x) \, dx$ as $\int u^r \, du$ for some number $r \neq -1$.
(iv)   Integrate using formula (1).
(v)   Substitute $g(x)$ for u to obtain the answer in terms of x.

EXAMPLE 1. Calculate $\int \sqrt{3 + x} \, dx$.

SOLUTION. Let $u = g(x) = 3 + x$, then $du = dx$ and

$$\int \sqrt{3 + x} \, dx = \int \sqrt{u} \, du = \int u^{1/2} \, du = \tfrac{2}{3} u^{3/2} + C = \tfrac{2}{3}(3 + x)^{3/2} + C.$$

EXAMPLE 2. Calculate $\int_0^1 (x^3 - 1)^{11/5} 3x^2 \, dx$.

SOLUTION. Let $u = g(x) = x^3 - 1$. Then $du = 3x^2 \, dx$ and

$$\int (x^3 - 1)^{11/5} 3x^2 \, dx = \int u^{11/5} \, du = \tfrac{5}{16} u^{16/5} + C = \tfrac{5}{16}(x^3 - 1)^{16/5} + C.$$

Then

$$\int_0^1 (x^3 - 1)^{11/5} 3x^2 \, dx = \tfrac{5}{16}(x^3 - 1)^{16/5} \Big|_0^1 = \tfrac{5}{16}(0 - (-1)^{16/5}) = -\tfrac{5}{16}.$$

In the two examples above, the integrand was already in the form $\int u^r \, du$ for the appropriate value of r. Sometimes this is not the case, but we can salvage the problem by multiplying and dividing by an appropriate constant.

EXAMPLE 3. Calculate $\int x \sqrt[3]{1 + x^2}\, dx$.

SOLUTION. Let $u = 1 + x^2$. Then $du = 2x\, dx$. All we have is $x\, dx$. We therefore multiply and divide by 2 to obtain

$$\int x \sqrt[3]{1 + x^2}\, dx = \tfrac{1}{2} \int \sqrt[3]{1 + x^2}\, 2x\, dx = \tfrac{1}{2} \int u^{1/3}\, du$$

$$= \tfrac{1}{2} \cdot \tfrac{3}{4} u^{4/3} + C = \tfrac{3}{8}(1 + x^2)^{4/3} + C.$$

We were allowed to multiply and divide by the constant 2 because of the theorem which states that $\int kf = k\int f$ for any constant k.

Warning. You cannot multiply and divide in the same way by functions which are not constants. If we try, we get incorrect answers such as

$$\int \sqrt{1 + x^2}\, dx = \frac{1}{2x} \int \sqrt{1 + x^2}\, 2x\, dx \qquad \text{(we multiplied and divided by } 2x)$$

$$= \frac{1}{2x} \int u^{1/2}\, du \qquad\qquad \text{(where } u = 1 + x^2)$$

$$= \frac{1}{2x} \cdot \frac{2}{3} u^{3/2} + C = \frac{1}{3x}(1 + x^2)^{3/2} + C.$$

But the derivative of $(1 + x^2)^{3/2}/3x$ is not even close to $\sqrt{1 + x^2}$. (Check this!) It is possible to calculate $\int \sqrt{1 + x^2}\, dx$ but we will have to wait until Chapter 8 to see how to do it.

EXAMPLE 4. Calculate $\displaystyle\int_1^2 \frac{s^2}{(1 + s^3)^{3/4}}\, ds$.

SOLUTION. Let $u = 1 + s^3$. Then $du = 3s^2\, ds$ and we multiply and divide by 3 to obtain

$$\int \frac{s^2}{(1 + s^3)^{3/4}}\, ds = \frac{1}{3} \int \frac{3s^2\, ds}{(1 + s^3)^{3/4}} = \frac{1}{3} \int u^{-3/4}\, du = \frac{4}{3} u^{1/4} + C.$$

Therefore

$$\int_1^2 \frac{s^2}{(1 + s^3)^{3/4}}\, ds = \frac{4}{3}(1 + s^3)^{1/4} \Big|_1^2 = \frac{4}{3}(9^{1/4} - 2^{1/4}).$$

EXAMPLE 5. Calculate $\displaystyle\int \frac{(1 + (1/t))^5}{t^2}\, dt$.

SOLUTION. Let $u = 1 + (1/t)$. Then $du = -(1/t^2)\, dt$. We then multiply and divide by -1 to obtain

$$\int \frac{[1 + (1/t)]^5}{t^2}\, dt = -\int \left(1 + \frac{1}{t}\right)^5 \left(\frac{-1}{t^2}\right) dt = -\int u^5\, du$$

$$= -\frac{u^6}{6} + C = -\frac{[1 + (1/t)]^6}{6} + C.$$

EXAMPLE 6. Calculate the area bounded by the curve $y = (1 + x^3)/\sqrt[4]{4x + x^4}$, the x-axis, and the lines $x = 1$ and $x = 3$.

SOLUTION. It is not necessary to draw this curve to see that $(1 + x^3)/\sqrt[4]{4x + x^4} > 0$ for $1 \le x \le 3$. Thus the area is represented by the definite integral

$$\int_1^3 \frac{1 + x^3}{\sqrt[4]{4x + x^4}} \, dx.$$

Let $u = 4x + x^4$. Then $du = (4 + 4x^3) \, dx$ and we multiply and divide by 4 to obtain

$$\int \frac{1 + x^3}{\sqrt[4]{4x + x^4}} \, dx = \frac{1}{4} \int \frac{4(1 + x^3)}{\sqrt[4]{4x + x^4}} \, dx = \frac{1}{4} \int u^{-1/4} \, du$$

$$= \frac{1}{3} u^{3/4} + C = \frac{1}{3}(4x + x^4)^{3/4} + C.$$

Thus

$$\int_1^3 \frac{1 + x^3}{\sqrt[4]{x + x^4}} \, dx = \frac{1}{3}(4x + x^4)^{3/4} \bigg|_1^3 = \frac{1}{3}(12 + 81)^{3/4} - \frac{1}{3}(5)^{3/4}$$

$$= \frac{1}{3}(93^{3/4} - 5^{3/4}).$$

PROBLEMS 4.6

In Problems 1–23 carry out the indicated integration by making an appropriate substitution $u = g(x)$.

1. $\int_0^7 \sqrt{9 + x} \, dx$

2. $\int \sqrt{10 + 3x} \, dx$

3. $\int_0^1 \sqrt{10 - 9x} \, dx$

4. $\int_0^2 x^2 \sqrt{1 + x^3} \, dx$

5. $\int x^3 \sqrt[5]{1 + 3x^4} \, dx$

6. $\int_0^2 \frac{t}{(1 + 2t^2)^{5/2}} \, dt$

7. $\int (s^4 + 1) \sqrt{s^5 + 5s} \, ds$

8. $\int_{-2}^2 \sqrt{1 + |s|} \, ds$

9. $\int_1^2 \frac{t + 3t^2}{\sqrt{t^2 + 2t^3}} \, dt$

10. $\int_0^1 (3w - 2)^{99} \, dw$

11. $\int \frac{dx}{\sqrt{x}(1 + \sqrt{x})^5}$

12. $\int_1^2 \frac{w + 1}{\sqrt{w^2 + 2w - 1}} \, dw$

13. $\int \frac{[1 + (1/v^2)]^{5/3}}{v^3} \, dv$

14. $\int_0^3 \left(\frac{x}{3} - 1\right)^{77} \, dx$

15. $\int (ax + b) \sqrt{ax^2 + 2bx + c} \, dx$

16. $\int (ax^2 + bx + c) \sqrt{2ax^3 + 3bx^2 + 6cx + d} \, dx$

17. $\int \frac{ax + b}{(ax^2 + 2bx + c)^{3/7}} \, dx$

18. $\int_0^\alpha t \sqrt{t^2 + \alpha^2} \, dt, \ \alpha > 0$

19. $\int_0^\alpha t^n \sqrt{\alpha^2 + t^{n+1}} \, dt; \ \alpha > 0, n \ge 0$

20. $\int \frac{s^{n-1}}{\sqrt{a + bs^n}} \, ds$

21. $\int_{-\alpha}^\alpha p^2 \sqrt{\alpha^3 - p^3} \, dp, \ \alpha > 0$

22. $\int_{-\alpha}^\alpha p^5 \sqrt{\alpha^6 - p^6} \, dp, \ \alpha > 0$

23. $\int_{-\alpha}^\alpha p^{7/3}(\alpha^{10/3} - p^{10/3})^{19/7} \, dp.$

In Problems 24–27 find the areas bounded by the given curves and lines.

24. $y = \sqrt{x + 2}$; x-axis; y-axis; $x = 7$

25. $y = x \sqrt{x^2 + 7}$; x-axis; y-axis; $x = 3$

26. $y = x^2 \sqrt{\dfrac{x^3}{3} + 1}$; x-axis; y-axis; $x = 3$

27. $y = \dfrac{x + 1}{x^3} = \dfrac{1 + (1/x)}{x^2}$; x-axis; $x = \dfrac{1}{3}$; $x = \dfrac{1}{2}$.

28. A particle moves with velocity $v(t) = 1/(3\sqrt{2 + t})$ m/min after t minutes. How far does the particle travel between the times $t = 2$ and $t = 7$?

29. The density of a 19-m metal beam is given by $\rho(x) = 1/(3\sqrt[3]{8 + x})$, where x is measured in meters from the left end of the beam and ρ is measured in kg/m.
   (a)   What is the mass of the beam?
   *(b)   For what value x does the interval $[0, x]$ contain exactly half the mass of the beam?

30. If the marginal cost incurred in the manufacture of flidgets is given by $MC = 4/\sqrt{x + 4}$, what is the cost incurred by raising production from 60 to 77 units?

31. What is the average velocity of the particle in Problem 28? [*Hint:* See the comment preceding Problem 4.5.50.]

32. What is the average density of the beam in Problem 29?

33. What is the average cost per flidget to manufacture 96 flidgets where the marginal cost is given in Problem 30?

Review Exercises for Chapter Four

In Exercises 1–10 calculate the given definite or indefinite integral.

1. $\displaystyle\int x^5 \, dx$

2. $\displaystyle\int_0^2 (t^3 + 3t + 5) \, dt$

3. $\displaystyle\int_1^8 \dfrac{ds}{\sqrt[3]{s}}$

4. $\displaystyle\int \dfrac{du}{(u + 3)^3}$

5. $\displaystyle\int_0^1 x\sqrt{x^2 + 1} \, dx$

6. $\displaystyle\int \dfrac{t^2 \, dt}{(t^3 + 3)^{3/4}}$

7. $\displaystyle\int_{-1}^1 (x^2 - 3)(x^5 + 2) \, dx$

8. $\displaystyle\int_{-3}^4 |s + 2| \, ds$

9. $\displaystyle\int_1^{\sqrt{2}} \dfrac{[1 - (1/t^2)]^4}{t^3} \, dt$

10. $\displaystyle\int (3ax^2 + 2bx + c)(ax^3 + bx^2 + cx + d)^{-1/9} \, dx$.

In Exercises 11–15 find the areas bounded by the given curves and lines.

11. $y = 3x - 7$; x-axis; $x = -2$; $x = 5$
                                  12. $y = \sqrt{x + 1}$; x-axis; y-axis; $x = 15$

13. $y = -x^2 - x + 2$; x-axis
                                    14. $y = x^3 - 7x^2 + 7x + 15$; x-axis

15. $y = 1/(x + 2)^2$; x-axis; y-axis; $x = 3$.

In Exercises 16–18 calculate the derivative of the given function and evaluate it at the given point.

16. $F(x) = \displaystyle\int_0^x \dfrac{t^3}{\sqrt{t^2 + 17}} \, dt$; $x_0 = 8$ (Recall that $x_0 = x(0)$.)

17. $F(t) = \displaystyle\int_3^t (s^5 - 17s + 8)^{11/3} \, ds$; $t_0 = 1$ **18.** $F(s) = \displaystyle\int_{-1}^s (1 + u^2)^{2001} \, du$; $s_0 = 0$.

19. A particle is moving with the velocity $v(t) = t + 1/\sqrt{1 + t}$ m/sec.
   (a)   How far does the particle move in the first 15 sec?
   (b)   What is the average velocity of the particle?

20. The density of a tree is given by $\rho(h) = 50/\sqrt{h + 1}$ (measured in kg/m) where h denotes the distance in meters above the ground. The height of the tree is 24 m.
   (a)   What is the total mass of the tree?
   (b)   For what value of h does the first h meters of the tree contain half the total mass of the tree?

FIVE

APPLICATIONS

In the first four chapters of this book we discussed the basic tools of differentiation and integration. Before continuing with the development of the theory, we pause in this chapter to look at some of the ways the derivatives and integrals we have calculated can be used to solve some interesting problems. The problems we can solve here will be somewhat limited as we still have not discussed some extremely important functions. Some of these will be treated in Chapters 6 and 7. When this is done we shall be able to add to the list of applications included here, and shall continue to do so in many of the chapters to follow.

5.1 Related Rates of Change

We have seen, beginning in Section 2.10, that the derivative can be interpreted as a rate of change. There are many problems involving two or more variables in which it is necessary to calculate the rate of change with respect to time of one or more of these variables. After giving an example, we will suggest a procedure for handling problems of this type.

EXAMPLE 1. A rope is attached to a pulley mounted on a 15-ft tower. The end of the rope is attached to a heavily loaded cart (see Figure 1). A worker can pull in rope at a rate of 2 ft/sec. How fast is the cart approaching the tower when it is 8 ft from the tower?

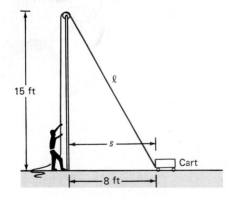

15 ft

ℓ

s

Cart

8 ft

Figure 1

SOLUTION. We let s denote the horizontal distance of the cart from the tower and l the length of rope from the top of the tower to the cart, as in Figure 1. Since the speed of the cart is ds/dt, the question asks us to determine ds/dt when $s = 8$ ft. We are told

230

that $dl/dt = 2$. To calculate ds/dt, we must first find a relationship between s and l. From the Pythagorean theorem we immediately obtain

$$15^2 + s^2 = l^2. \tag{1}$$

We now differentiate (1) implicitly with respect to t to find that

$$\frac{d}{dt}(15^2) + \frac{d}{dt}(s^2) = \frac{d}{dt}(l^2)$$

or

$$0 + 2s\frac{ds}{dt} = 2l\frac{dl}{dt}. \tag{2}$$

When $s = 8$, $l^2 = 15^2 + 8^2 = 225 + 64 = 289$, and $l = 17$. Then inserting $s = 8$, $l = 17$ and $dl/ds = 2$ into equation (2) gives us

$$\frac{ds}{dt} = \frac{17}{8}(2) = \frac{17}{4} = 4\frac{1}{4} \quad \text{ft/sec.}$$

The solution given above suggests that the following steps be taken to solve a problem involving the rates of change of related variables:

(i)   If feasible, draw a picture of what is going on.
(ii)   Determine the important variables in the problem and find an equation relating them.
(iii)   Differentiate the equation obtained in (ii) with respect to t.
(iv)   Solve for the derivative sought.
(v)   Evaluate that derivative by substituting given and calculated values of the variables in the problem.

EXAMPLE 2.   An oil storage tank is built in the form of an inverted right circular cone with a height of 6 m and a base radius of 2 m (see Figure 2). Oil is being pumped into the tank at a rate of 2 liters/min $= 0.002$ m³/min (since 1m³ $= 1000$ liters). How fast is the level of the oil rising when the tank is filled to a height of 3 m?

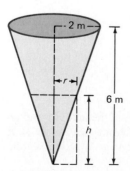

Figure 2

SOLUTION. In mathematical terms, we are asked to calculate dh/dt when $h = 3$ m, where h denotes the height of the oil at a given time and r the radius of the cone filled with oil (see Figure 2). The volume of a right circular cone is $V = \frac{1}{3}\pi r^2 h$. From the data given in the problem we see from Figure 2 (using similar triangles) that $h/r = 6/2$ or $r = h/3$. Then

$$V = \frac{1}{3}\pi \left(\frac{h}{3}\right)^2 h = \frac{1}{27}\pi h^3 = \text{volume of oil at height } h.$$

Differentiating with respect to t and using the fact (given to us) that $dV/dt = 0.002$, we obtain

$$\frac{dV}{dt} = \frac{1}{9}\pi h^2 \frac{dh}{dt} = 0.002 \quad \text{or} \quad \frac{dh}{dt} = \frac{9(0.002)}{\pi h^2} = \frac{0.018}{\pi h^2} \text{ m/min.}$$

Then, for $h = 3$,

$$\frac{dh}{dt} = \frac{0.018}{\pi \cdot 9} = \frac{0.002}{\pi} \approx 6.37 \times 10^{-4} \text{ m/min.}$$

EXAMPLE 3. In chemistry, an *adiabatic* process is one in which there is no gain or loss of heat. During an adiabatic process, the pressure P and volume V of certain gases such as hydrogen or oxygen in a container are related by the formula

$$PV^{1.4} = \text{constant.} \tag{3}$$

At a certain time the volume of hydrogen in a closed container is 4 m³ and the pressure is 0.75 kg/m². Suppose that the volume is increasing at a rate of $\frac{1}{2}$ m³/sec. How fast is the pressure decreasing?

SOLUTION. We use the product rule to differentiate (3) with respect to t.

$$P\frac{d}{dt}(V^{1.4}) + \frac{dP}{dt} \cdot V^{1.4} = 0, \qquad P(1.4)V^{0.4}\frac{dV}{dt} + \frac{dP}{dt} \cdot V^{1.4} = 0$$

or

$$\frac{dP}{dt} = -\frac{1.4PV^{0.4}}{V^{1.4}}\frac{dV}{dt} = -1.4\frac{P}{V} \cdot \frac{dV}{dt}.$$

Using the data given in the problem, we find that

$$\frac{dP}{dt} = \frac{-(1.4)(0.75)}{4 \text{ m}^3} \frac{\text{kg/m}^2}{} \cdot (0.5) \text{ m}^3/\text{sec} = -0.13125 \text{ kg/m}^2/\text{sec.}$$

PROBLEMS 5.1

1. Let $xy = 6$. If $dx/dt = 5$, find dy/dt when $x = 3$.
2. Let $x/y = 2$. If $dx/dt = 4$, find dy/dt when $x = 2$.
3. A 10-ft ladder is leaning against the side of a house. As the foot of the ladder is pulled away from the house, the top of the ladder slides down along the side (see Figure 3). If the foot of the ladder is pulled away at a rate of 2 ft/sec, how fast is the ladder sliding down when the foot is 8 ft from the house?
4. A cylindrical water tank 6-m high with a radius of 2 m is being filled at a rate of 10 liters/min. How fast is the water rising when the water level is at a height of 0.5 m? [*Hint:* 1 liter = 1000 cm³ or 1 m³ = 1000 liters.]

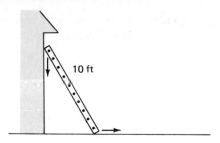

10 ft

Figure 3

5. A baseball player can run at a top speed of 35 ft/sec. A catcher can throw a ball at a speed of 140 ft/sec. The player attempts to steal third base. The catcher (who is 90 ft from third base) throws the ball toward the third baseman when the player is 30 ft from third base. What is the rate of change of the distance between the ball and the runner at the instant the ball is thrown? (See Figure 4.)

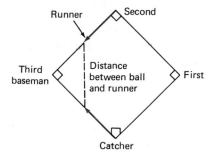

Figure 4

6. Sand is being dropped in a conical pile at a rate of 15 m³/min. The height of the pile is always equal to its diameter. How fast is the height increasing when the pile is 7-m high?

7. When helium expands adiabatically, its pressure is related to its volume by the formula $PV^{1.67} = $ constant. At a certain time, the volume of the helium in a balloon is 18 m³ and the pressure is 0.3 kg/m². If the pressure is increasing at a rate of 0.01 kg/m²/sec, how fast is the volume changing? Is the volume increasing or decreasing?

8. A man standing on a pier 15 ft above the water is pulling in his boat by means of a rope attached to the boat's bow. He can pull in the rope at a rate of 5 ft/min. How fast is the boat approaching the foot of the pier when the boat is 20 ft away?

9. An airplane at a height of 1000 m is flying horizontally at a velocity of 500 km/hr and passes directly over a civil defense observer. How fast is the plane receding from the observer when it is 1500-m away from the observer?

10. At 2 P.M. on a certain day, ship A is 100-km due north of ship B. At that moment, ship A begins to sail due east at a rate of 15 km/hr while ship B sails due north at a rate of 20 km/hr. How fast is the distance between the two ships changing at 5 P.M.? Is it increasing or decreasing?

11. A storage tank is 20-ft long and its ends are isosceles triangles having bases and altitudes of 3 ft. Water is poured into the tank at a rate of 4 ft³/min. How fast is the water level rising when the water in the tank is 6-in. deep?

12. Two roads intersect at right angles. A car traveling 80 km/hr reaches the intersection half an hour before a bus that is traveling on the other road at 60 km/hr. How fast is the distance between the car and the bus increasing 1 hr after the bus reaches the intersection (see Figure 5)?

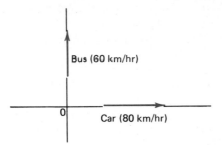

Bus (60 km/hr)

0

Car (80 km/hr)

Figure 5

13. A rock is thrown into a pool of water. A circular wave leaves the point of impact and travels so that its radius increases at a rate of 25 cm/sec. How fast is the circumference of the wave increasing when the radius of the wave is 1 m?

14. The body of a snowman is in the shape of a sphere and is melting at a rate of 2 ft³/hr. How fast is the radius changing when the body is 3-ft in diameter (assuming that the body stays spherical)?

15. In Problem 14, how fast is the surface area of the body changing when $d = 3$ ft?

*16. Water is leaking out of the bottom of a hemispherical tank with a radius of 6 m at a rate of 3 m³/hr. If the tank was full at noon, how fast is the height of the water level in the tank changing at 3 P.M.? [*Hint:* The volume of a segment of a sphere of radius r is $\pi h^2 [r - (h/3)]$, where h is the height of the segment (see Figure 6).]

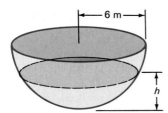

6 m

h

Figure 6

17. A light is affixed to the top of a 12-ft lamppost. If a 6-ft man walks away from the lamppost at a rate of 5 ft/sec, how fast is the length of his shadow increasing when he is 5 ft away?

18. Bacteria grow in circular colonies. The radius of one colony is increasing at a rate of 4 mm/day. On Wednesday, the radius of the colony is 1 mm. How fast is the area of the colony changing one week later?

19. A pill is in the shape of a right circular cylinder with a hemisphere on each end. The height (excluding its hemispherical ends) of the cylinder is half its radius. What is the rate of change of the volume of the pill with respect to the radius of the cylinder?

20. A spherical mothball is dissolving at a rate of 8π cc/hr (1 cc = 1 cubic centimeter). How fast is the radius of the mothball decreasing when the radius is 3 cm?

*21. A particle moves along the parabola $y = x^2$ such that $dx/dt = 3$. Find the rate of change of the distance between the particle and the point $(2, 5)$ when the particle is at $(-1, 1)$; when it is at $(3, 9)$?

22. In soil mechanics the vertical stress σ_z in a soil at a point Z-feet deep and located r feet (horizontally) from a concentrated surface load of P pounds is given by

$$\sigma_z = \frac{3P}{2\pi Z^2} \left[1 + \left(\frac{r}{Z} \right)^2 \right]^{2.5}.$$

(a)   Calculate the vertical stress in a soil at a point 2-feet deep and located 4-feet horizontally from a concentrated surface load of 10,000 lb.

(b)   If the load is moved horizontally toward the point at a rate of 1 ft/hr, how is the vertical stress changing when the load is 3 feet from the point?

5.2 The Theory of Maxima and Minima

In this section we consider the problem of finding the maximum (largest) and minimum (smallest) values of a function $y = f(x)$ over a finite interval $[a, b]$. In the next section we show how this theory can be used in applications.

In Section 3.5 we defined what we meant by a function having a local maximum or minimum at a point x_0. Roughly, f has a local maximum at x_0 if among all the values of $f(x)$ in an open interval containing x_0, f takes its largest value at x_0. A similar statement can be made for a local minimum at x_0. We showed (in Theorem 3.5.2) that if f has a local maximum or minimum at x_0, then x_0 is a critical point of f. That is, either $f'(x_0) = 0$ or $f(x_0)$ exists but $f'(x_0)$ does not exist. We further demonstrated (Theorem 3.9.2) that

if $f'(x_0) = 0$ and $f''(x_0) < 0$, then f has a local maximum at x_0

and

if $f'(x_0) = 0$ and $f''(x_0) > 0$, then f has a local minimum at x_0.

But there could be more than one local maximum or minimum in an interval. Moreover, there are other ways a maximum or minimum could be reached. Look at Figure 1. The function depicted there has local maxima at x_0, x_2, and x_4 and local

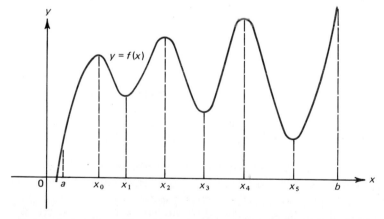

Figure 1

minima at x_1, x_3, and x_5. However, in the interval $[a, b]$ the maximum and minimum values of f are taken at none of these critical points. The maximum value of f is taken at $x = b$ and the minimum is taken at $x = a$. Thus it is necessary to check both endpoints as well as all the critical points to find the maximum and minimum.

There is another problem that could arise. Consider the function

$$f(x) = \begin{cases} x, & x \neq 1, \\ 5, & x = 1. \end{cases}$$

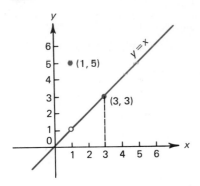

Figure 2

This is graphed in Figure 2. We see that in the interval $[0, 3]$ f takes on its maximum value at $x = 1$, which is not an endpoint of the interval and at which $f'(1)$ is not zero. In fact, $f'(1)$ does not even exist since f is not continuous at $1.$† Thus we have the following rule:

  If $f(x)$ is defined on the finite interval $[a, b]$, then in order to find the maximum and minimum values of f over that interval

---

  (i) find all local maxima and minima in (a, b) and let M_1 and m_1 denote the largest and smallest values of f over all these local maxima and minima, respectively;
  (ii) evaluate $f(a)$ and $f(b)$;
  (iii) evaluate f at all points in $[a, b]$ where the derivative of f does not exist.

Then

  (a) $f_{max}[a, b]$ = the largest of the values calculated in (i), (ii), and (iii), and
  (b) $f_{min}[a, b]$ = the smallest of the values calculated in (i), (ii), and (iii).

Remark 1. $f_{max}[a, b]$ is read as the largest value of f over the interval $[a, b]$, and analogously for $f_{min}[a, b]$.

Remark 2. In an infinite interval, the function may have neither a maximum nor a minimum. For example $f(x) = x^2$ has a minimum (zero) but no maximum on $[0, \infty)$. The function $f(x) = x^3$ has neither a maximum nor a minimum in $(-\infty, \infty)$.

EXAMPLE 1.   Find the maximum and minimum values of $f(x) = 2x^3 - 3x^2 - 12x + 5$ in the interval $[0, 4]$.

SOLUTION. $f'(x) = 6x^2 - 6x - 12 = 6(x^2 - x - 2) = 6(x - 2)(x + 1)$. The only place in the interval $[0, 4]$ where $f'(x) = 0$ is at $x = 2$. At $x = 2$, $f(x) = -15$. We then find that $f''(x) = 12x - 6 = 18$ when $x = 2$ so that f has a local minimum at $(2, -15)$ and this is the only critical point in $[0, 4]$. We also calculate $f(0) = 5$ and $f(4) = 37$. There are no points in $[0, 4]$ at which f' is not defined. Therefore $f_{max}[0, 4] = 37$ and $f_{min}[0, 4] = -15$. This is sketched in Figure 3.

---

† Recall, from Theorem 2.9.1, that if f is differentiable at x_0, then f is continuous there. Thus if f is not continuous at x_0, it cannot be differentiable at x_0.

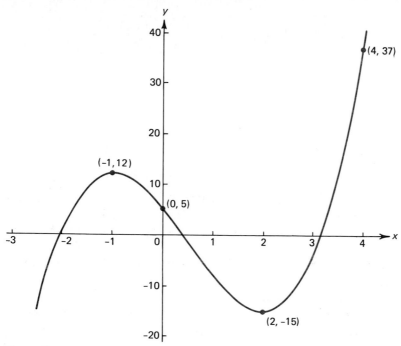

Figure 3

EXAMPLE 2.   Find the maximum and minimum values of $f(x) = x/(1 + x)$ in the interval $[1, 5]$.

SOLUTION, $f'(x) = 1/(1 + x)^2$ which is never zero so that there are no critical points.† Since f' is defined everywhere in $[1, 5]$, the maximum and minimum values are taken at the endpoints of the interval. We have $f(1) = \frac{1}{2}$ and $f(5) = \frac{5}{6}$, so that $f_{max}[1, 5] = \frac{5}{6}$ and $f_{min}[1, 5] = \frac{1}{2}$. This is sketched in Figure 4.

EXAMPLE 3.   Find the maximum and minimum values of $f(x) = -2x\sqrt{x + 3}$ in the interval $[-\frac{5}{2}, 1]$.

SOLUTION

$$f'(x) = -\frac{2x}{2\sqrt{x + 3}} - 2\sqrt{x + 3} = -\frac{3x + 6}{\sqrt{x + 3}}.$$

We see that $f'(x) = 0$ at $x = -2$. For $x = -2, f(x) = 4$. Also, -3 is a critical point since $f(-3)$ exists ($f(-3) = 0$) but $f'(-3)$ does not exist. However this critical point does not concern us since -3 is not in the interval $[-\frac{5}{2}, 1]$. We then calculate

$$f''(x) = \frac{-3\sqrt{x + 3} + \dfrac{3x + 6}{2\sqrt{x + 3}}}{x + 3}$$

†$f'(-1)$ does not exist but -1 is not a critical point because $f(-1)$ does not exist either.

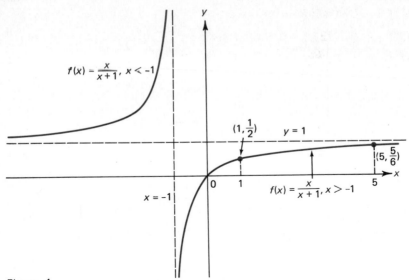

Figure 4

which is equal to -3 when $x = -2$. Therefore f has a local maximum at $x = -2$. At the endpoints,

$$f\left(-\frac{5}{2}\right) = 5\sqrt{\frac{1}{2}} = \frac{5}{\sqrt{2}} \qquad \text{and} \qquad f(1) = -4.$$

Since $\dfrac{5}{\sqrt{2}} \approx 3.5 < 4$, $f_{\max}\left[-\dfrac{5}{2}, 1\right] = 4$ and $f_{\min}\left[-\dfrac{5}{2}, 1\right] = -4$.

This is sketched in Figure 5.

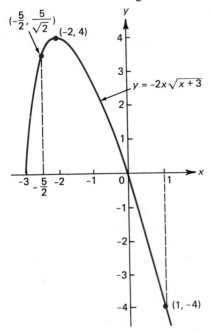

Figure 5

EXAMPLE 4.   Find the maximum and minimum values of $f(x) = \sqrt[3]{x}$ in $[-1, 1]$.

SOLUTION. $f'(x) = \frac{1}{3}x^{-2/3}$ which is not defined at $x = 0$. There are no other critical points. We therefore check the endpoints and the point $x = 0$. We have $f(1) = 1$, $f(-1) = -1$ and $f(0) = 0$. Therefore $f_{\max}[-1, 1] = 1$ and $f_{\min}[-1, 1] = -1$. This is sketched in Figure 6.

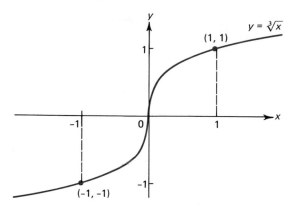

Figure 6

PROBLEMS 5.2

In each of the following problems find the maximum and minimum values for the given function over the indicated interval.

1. $f(x) = x^2 + x - 30$; $[0, 2]$
2. $f(x) = x^2 + x - 30$; $[-2, 0]$
3. $f(x) = x^3 - 12x + 10$; $[-10, 10]$
4. $f(x) = 4x^3 - 3x + 2$; $[-5, 5]$
5. $f(x) = x^5$; $(-\infty, 1]$
6. $f(x) = x^7$; $[-1, \infty)$
7. $f(x) = x^3 - 3x^2 - 45x + 25$; $[-5, 5]$
8. $f(x) = \frac{x^3}{3} + \frac{x^2}{2} - 2x - \frac{2}{3}$; $[-3, 3]$
9. $f(x) = x^{20}$; $[0, 1]$
10. $f(x) = x^4 - 18x^2$; $[-2, 2]$
11. $f(x) = (x + 1)^{1/3}$; $[-2, 7]$
12. $f(x) = (x - 2)^{2/3}$; $[-14, 3]$
13. $f(x) = 2x(x + 4)^3$; $[-1, 1]$
14. $f(x) = x\sqrt[3]{1 + x}$; $[-2, 26]$
15. $f(x) = \dfrac{1}{(x - 1)(x - 2)}$; $[3, 5]$
16. $f(x) = \dfrac{1}{(x - 1)(x - 2)}$; $[-3, 0]$
17. $f(x) = \dfrac{x - 1}{x - 2}$; $[-3, 1]$
18. $f(x) = \dfrac{1}{x^2 - 1}$; $[2, 5]$
19. $f(x) = x^{2/3}(x - 3)$; $[-1, 1]$
20. $f(x) = (x - 1)^{1/5}$; $[-31, 33]$
21. $f(x) = 1 + x + \dfrac{1}{2x^2}$; $\left[\dfrac{1}{2}, \dfrac{3}{2}\right]$
22. $f(x) = \sqrt{x} + \dfrac{1}{\sqrt{x}}$; $\left[\dfrac{1}{2}, \dfrac{3}{2}\right]$
23. $f(x) = (1 + \sqrt{x} + \sqrt[3]{x})^9$; $[0, \infty)$
24. $f(x) = \dfrac{1}{x - 2}$; $[3, \infty)$
25. $f(x) = \begin{cases} x^2, & x \neq 3 \\ 20, & x = 3 \end{cases}$; $[0, 4]$
*26. $f(x) = \begin{cases} x^2, & x \leq 1 \\ x^3, & x \geq 1 \end{cases}$; $[0, 5]$
*27. $f(x) = \begin{cases} x - 3, & x < 1 \\ 2x + 4, & 1 \leq x \leq 3 \\ 3x - 7, & x > 3 \end{cases}$; $[-5, 5]$
28. $f(x) = [x]$; $[-5, 5]$
29. $f(x) = |x^3 - 12x|$; $[-3, 3]$
30. $f(x) = |x^4 - 16x|$; $[-2, 2]$.

5.3 Maxima and Minima: Applications

In this section and the next one we show how the theory of maxima and minima can be applied to a great variety of problems. Before citing general rules for dealing with these problems we begin with a simple example that is historically one of the first applications of the theory to a practical problem.

EXAMPLE 1. Suppose that a farmer has 1000 yards of fence which he wishes to use to fence off a rectangular plot along the bank of a river (see Figure 1). What are the dimensions of the maximum area he can enclose?

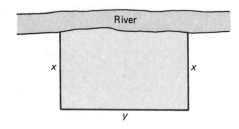

Figure 1

SOLUTION. Since one side of the rectangular plot is taken up by the river, the farmer can use the fence for the other three sides. From the figure, the area of the plot is

$$A = xy \tag{1}$$

where y is the length of the side parallel to the river. Without other information, it would be impossible to solve this problem since the area, A, is a function of the *two* variables x and y. However, we are also told, since the farmer will obviously use all the available fence, that

$$2x + y = 1000 \tag{2}$$

or

$$y = 1000 - 2x. \tag{3}$$

Then

$$A = x(1000 - 2x) = 1000x - 2x^2. \tag{4}$$

To find the maximum value for A, we set dA/dx equal to zero and use the results of the preceding section. We have

$$\frac{dA}{dx} = 1000 - 4x = 0$$

when $x = 250$. Also,

$$\frac{d^2A}{dx^2} = -4 < 0$$

so that A is a maximum when $x = 250$. This is depicted in Figure 2. When $x = 250$ yards, then $y = 500$ yards and $A = 125{,}000$ square yards which is the maximum area that the farmer can enclose. We note that, as is evident from Figure 2, A is positive

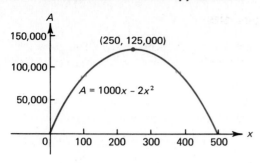

Figure 2

when $0 \le x \le 500$ and at the endpoints $x = 0$ and $x = 500$ the area is zero, so the local maximum we have obtained is a maximum over the entire interval $[0, 500]$.

This example suggests that the following steps be taken to solve a maximum or minimum problem:

> (i)   Draw a picture if it makes sense to do so.
> (ii)   Write all the information in the problem in mathematical terms, giving a letter name to each variable.
> (iii)   Determine the variable that is to be maximized or minimized and write this as a function of the other variables in the problem.
> (iv)   Using information given in the problem, eliminate all variables except one, so that the function to be maximized or minimized is written in terms of *one* of the variables of the problem.
> (v)   Determine the interval over which this one variable can be defined.
> (vi)   Follow the steps of the previous section to maximize or minimize the function over this interval.

EXAMPLE 2.   Suppose that the farmer in Example 1 wishes to build his rectangular plot away from the river (so that he must use his fence for all four sides of the rectangle). How large an area can he enclose in this case?

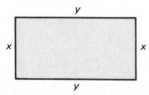

Figure 3

SOLUTION. The situation is now as in Figure 3. The area is again given by $A = xy$, but now

$$2x + 2y = 1000 \tag{5}$$

or

$$y = 500 - x. \tag{6}$$

Hence the problem is to maximize

$$A = x(500 - x) = 500x - x^2, \qquad \text{for } x \text{ in } [0, 500]. \tag{7}$$

Now

$$\frac{dA}{dx} = 500 - 2x = 0$$

when $x = 250$. Also, $d^2A/dx^2 = -2$ so that a local maximum is achieved when $x = y = 250$ yards and the answer is $A = 62,500$ square yards. (Note that the plot is, in this case, a square.) At 0 and 500, $A = 0$ so that $x = 250$ is indeed the maximum.

Remark 1. The reasoning in Example 2 can be used to prove: *For a given perimeter, the rectangle containing the greatest area is a square.*

Remark 2. If we do not require that the plot be rectangular, then we can enclose an even greater area. Although the proof of this fact is beyond the scope of this book, it can be shown that *for a given perimeter, the geometric shape with the largest area is a circle.* For example, if the 1000 yards of fence in Example 2 are formed in the shape of a circle, then the circle has a circumference of $2\pi r = 1000$ so that the radius of the circle $r = 1000/2\pi$ yards. Then $A = \pi r^2 = \pi(1,000,000/4\pi^2) = 1,000,000/4\pi \approx 79,577$ square yards.

EXAMPLE 3.   Find the point on the straight line $x + 2y = 5$ which is closest to the origin.

SOLUTION. The distance D from a point (x, y) to the origin $(0, 0)$ is given by (see Figure 4)

$$D = (x^2 + y^2)^{1/2}. \tag{8}$$

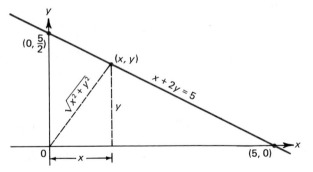

(0, $\frac{5}{2}$)

(x, y)

$\sqrt{x^2 + y^2}$

$x + 2y = 5$

y

0   $\leftarrow x \rightarrow$ (5, 0)

Figure 4

But $x = 5 - 2y$ so that

$$D = [(5 - 2y)^2 + y^2]^{1/2} = [25 - 20y + 4y^2 + y^2]^{1/2} = [25 - 20y + 5y^2]^{1/2}. \tag{9}$$

Then

$$\frac{dD}{dy} = \frac{1}{2}(25 - 20y + 5y^2)^{-1/2}(10y - 20). \tag{10}$$

We may assume that $25 - 20y + 5y^2 > 0$ since otherwise the distance from (x, y) to $(0, 0)$ would be zero, implying that $(0, 0)$ is on the line (which is false). Therefore, setting $dD/dy = 0$ we obtain

$$\frac{dD}{dy} = \frac{10y - 20}{2(25 - 20y + 5y^2)^{1/2}} = 0$$

which can only occur if $10y = 20$ or $y = 2$. But we find that

$$\frac{d^2D}{dy^2} = \frac{2(25 - 20y + 5y^2)^{1/2}10}{4(25 - 20y + 5y^2)}$$

$$- \frac{(10y - 20)(25 - 20y + 5y^2)^{-1/2}(10y - 20)}{4(25 - 20y + 5y^2)} = \sqrt{5} > 0$$

(check!) when $y = 2$. Thus a minimum occurs at $y = 2$. Then $x = 5 - 2 \cdot 2 = 1$ and at the point $(1, 2)$, $D = \sqrt{5}$.

> **Note.** The interval under consideration here is $(-\infty, \infty)$. Explain why it is not necessary to check the endpoints.

EXAMPLE 4. A cylindrical barrel is to be constructed to hold 32π m^3 (cubic meters) of liquid. The cost per square meter of constructing the side of the barrel is half the cost per square meter of constructing the top and bottom. What are the dimensions of the barrel which costs the least to construct?

SOLUTION. Consider Figure 5. Let h be the height of the barrel and let r be the radius of the top and bottom. Then the volume of the barrel is given by $V = \pi r^2 h = 32\pi$ m^3.

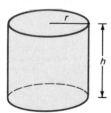

Figure 5

Assuming that the cost of constructing the material on the side is k cents per m^2, then the cost of the top and bottom is $2k$ cents per m^2. The area of the top or the bottom is πr^2 m^2 while that of the side is $2\pi rh$ m^2 (since the circumference of the side is $2\pi r$). Thus the total cost is given by

$$C = 2\pi r^2(2k) + 2\pi rhk = k(4\pi r^2 + 2\pi rh). \tag{11}$$

To write C as a function of one variable only (which we must do in order to solve the problem), we use $32\pi = \pi r^2 h$ to obtain $h = 32\pi/\pi r^2 = 32/r^2$ so that

$$C = k\left(4\pi r^2 + 2\pi r \cdot \frac{32}{r^2}\right) = k\left(4\pi r^2 + \frac{64\pi}{r}\right). \tag{12}$$

Then, since k is a constant,

$$\frac{dC}{dr} = k\left(8\pi r - \frac{64\pi}{r^2}\right).$$

Setting this equal to zero, we obtain

$$8\pi r = \frac{64\pi}{r^2}, \quad 8r^3 = 64, \quad r^3 = 8, \quad \text{and} \quad r = 2 \text{ m}.$$

In addition, $d^2C/dr^2 = k[8\pi + (128\pi/r^3)]$ which is > 0 when $r = 2$. Hence there is a local minimum when $r = 2$ m. When $r = 2$, $h = 32/4 = 8$ m. Note that this local minimum is a true minimum since the only endpoint of the interval occurs at $r = 0$ which makes no practical sense since, in that case, the barrel can hold nothing.

EXAMPLE 5. A cardboard box with a square base and an open top is to be constructed from a square piece of cardboard 10 cm on a side by cutting out four squares at the corners and folding up the sides. What should be the dimensions of the box in order to make the volume enclosed as large as possible?

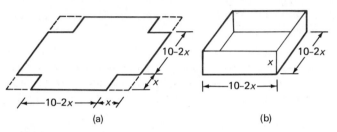

(a)   (b)   Figure 6

SOLUTION. We refer to Figures 6a and 6b. If we cut out squares with sides of x cm, the base of the box will have each side equal to $10 - 2x$ cm and the height of the box will be x cm. Then the volume of the box is given by

$$V = x(10 - 2x)^2, \quad \text{for } x \text{ in } [0, 5].$$

Now,

$$\frac{dV}{dx} = (10 - 2x)^2 + 2x(10 - 2x)(-2) = (10 - 2x)^2 - 4x(10 - 2x)$$

which is equal to zero when

$$4x(10 - 2x) = (10 - 2x)^2.$$

If $(10 - 2x) \neq 0$, then $4x = 10 - 2x$ or $x = \frac{5}{3}$. If $10 - 2x = 0$, then $x = 5$. The second derivative of V is

$$\frac{d^2V}{dx^2} = -8(10 - 2x) + 8x$$

so that

$$\frac{d^2V}{dx^2} = 40 \quad \text{when} \quad x = 5, \quad \text{and} \quad \frac{d^2V}{dx^2} = -40 \quad \text{when} \quad x = \frac{5}{3}.$$

Thus the maximum occurs when $x = 5/3$ cm, and the maximum volume is $V = 2000/27$ cm³. (The minimum volume is zero, which is obtained by setting x equal to 5 cm; that is, leave a base of zero).

EXAMPLE 6. In chemistry, a *catalyst* is defined as a substance that alters the rate of a chemical reaction without itself undergoing a change; the phenomenon is called *catalysis*. If, in a chemical reaction, the product of the reaction serves as a catalyst for the reaction, then the process is called *autocatalysis*. Suppose that in the autocatalytic process we start with an amount A of a given substance. Let x be the amount of the product (i.e., the result of the process). It is reasonable to assume that the rate of reaction depends both on the amount of the product x, and the amount of remaining substance $(A - x)$. If this rate is given by

$$R = \alpha x(A - x) \tag{13}$$

where α is a known positive constant, for what concentration x is the rate of reaction greatest?

SOLUTION. $R = \alpha(Ax - x^2)$ so that

$$\frac{dR}{dx} = \alpha(A - 2x) = 0 \quad \text{when} \quad x = \frac{A}{2}.$$

In addition, $d^2R/dx^2 = -2\alpha < 0$ so that a local maximum is reached when $x = A/2$. The endpoints are $x = 0$ and $x = A$, both of which give a reaction rate of zero. Thus we may conclude that the rate of reaction is greatest when exactly half the original substance has been catalyzed.

EXAMPLE 7. A man is on a lake in a canoe one kilometer from the closest point P of a straight shore line (see Figure 7). He wishes to get to a point Q, 10-km along the

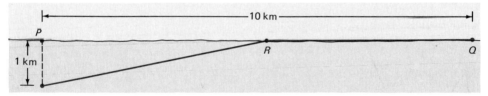

Figure 7

shore from P. In order to accomplish this, he paddles to a point R between P and Q and then walks the remaining distance to Q. He can paddle 3 km/hour and walk 5 km/hour. How should he pick the point R so as to get to Q as quickly as possible?

SOLUTION. To simplify matters we draw Figure 8, where the starting point is placed at

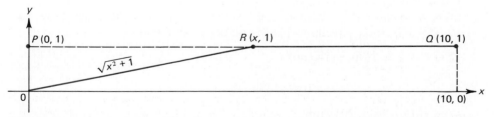

Figure 8

the origin. Then, as drawn, P has the coordinates $(0, 1)$, R the coordinates $(x, 1)$ and Q the coordinates $(10, 1)$. The problem is to find the value of x (the point where he reaches the shore) which minimizes the total travel time. The distance from the origin to R is given by

$$d_1 = \sqrt{x^2 + 1}.$$

Since (velocity) $\times$ (time) = distance, and the velocity v_1 in this case is 3 km/hour, we have

$$t_1 = \text{time from 0 to } R = \frac{d_1}{v_1} = \frac{1}{3}\sqrt{x^2 + 1}. \tag{14}$$

The distance, d_2, from R to Q is $10 - x$ and the walking velocity is $v_2 = 5$ km/hour so that

$$t_2 = \text{time from } R \text{ to } Q = \frac{1}{5}(10 - x).$$

Thus, the total time in getting from 0 to Q is given by

$$T = \frac{\sqrt{x^2 + 1}}{3} + \frac{10 - x}{5} \tag{15}$$

To find the minimum, we calculate dT/dx and set it equal to zero:

$$\frac{dT}{dx} = \frac{x}{3\sqrt{x^2 + 1}} - \frac{1}{5} = 0 \quad \text{or} \quad \frac{x}{3\sqrt{x^2 + 1}} = \frac{1}{5} \quad \text{or} \quad 5x = 3\sqrt{x^2 + 1}.$$

We now square both sides to obtain

$$25x^2 = 9x^2 + 9 \text{ or } x = \pm\tfrac{3}{4}.$$

We use the value $x = \tfrac{3}{4}$ since the value $x = -\tfrac{3}{4}$ has no meaning in this problem. But

$$\frac{d^2T}{dx^2} = \frac{3\sqrt{x^2 + 1} - (3x^2/\sqrt{x^2 + 1})}{9(x^2 + 1)} = \frac{3(x^2 + 1) - 3x^2}{9(x^2 + 1)^{3/2}} = \frac{1}{3(x^2 + 1)^{3/2}}$$

which is always positive. Therefore there is a local minimum at $x = \tfrac{3}{4}$. When $x = \tfrac{3}{4}$, we obtain from (15)

$$T = \frac{\sqrt{(9/16) + 1}}{3} + \frac{10 - (3/4)}{5} = \frac{5}{12} + 2 - \frac{3}{20} = \frac{136}{60} \text{ hr} = 136 \text{ min.}$$

We still have to check the endpoints $x = 0$ (the man rows straight to the shore) and $x = 10$ (the man rows directly to his destination). At $x = 0$,

$$T = \tfrac{1}{3} + 2 = 2\tfrac{1}{3} \text{ hr} = 140 \text{ min.}$$

At $x = 10$,

$$T = \frac{\sqrt{101}}{3} \text{ hr} \approx 201 \text{ min.}$$

Thus the minimum time needed is 136 min.

EXAMPLE 8. The growth rate per individual in a population is the difference between the average birth rate and the average death rate. We assume that the birth rate is a positive constant β, independent of time t and the population size $P(t)$. Suppose that the average death rate is proportional to the size of the population and is given by $\delta P(t)$, where δ is a positive constant. The increasing mortality (with increasing population) may be due to the effects of crowding or to increased competition for the available food resources. The rate of growth of the population is given by dP/dt and the average growth is $(1/P)dP/dt$. Therefore, the equation governing population growth is

$$\frac{1}{P}\frac{dP}{dt} = \beta - \delta P \qquad \text{or} \qquad \frac{dP}{dt} = P(\beta - \delta P). \tag{16}$$

This type of equation is called a *differential equation*.† Differential equations will be discussed in detail in Chapter 20. The question here is: At what population level will the growth be a maximum?

SOLUTION. To avoid cumbersome notation, we define $U = dP/dt$. Then

$$U = P(\beta - \delta P) = \beta P - \delta P^2 \qquad \text{and} \qquad \frac{dU}{dP} = \beta - 2\,\delta P = 0$$

when $P = \beta/2\delta$. Also, $d^2U/dP^2 = -2\delta < 0$ so that there is a local maximum at $P = \beta/2\delta$. P can (theoretically) take on any value between 0 and ∞. At $P = 0$ the growth rate is 0. For $P > \beta/\delta$, the growth rate will be negative. Therefore the maximum growth is indeed reached when $P = \beta/2\delta$. This fact confirms our intuitive feeling that population growth must depend on the relative sizes of the birth and death rates.

In the examples we have given, the maximum or minimum did not occur at an endpoint. The next example shows that this can happen.

□**EXAMPLE 9.** A wire, 20-cm long, is cut into two pieces. One piece is bent in the shape of an equilateral triangle and the other is bent in the shape of a circle. How should the wire be cut so as to maximize the total area enclosed by the shapes? How should it be cut to minimize the total area?

SOLUTION. Let x denote the length of wire used for the equilateral triangle. Then the length of each side is $x/3$ and the circumference of the circle is $20 - x$ (see Figure 9).

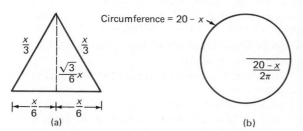

(a) (b) Figure 9

---

† This specific equation is called the *logistic equation*.

Using the Pythagorean theorem, we find that the height of the triangle is $\sqrt{3}\,x/6$ so that

$$\text{area of the triangle} = \frac{1}{2}(\text{base}) \times (\text{height}) = \frac{1}{2}\left(\frac{x}{3}\right)\left(\sqrt{3}\frac{x}{6}\right) = \sqrt{3}\,\frac{x^2}{36}.$$

Moreover, since the circumference of the circle is $2\pi r = 20 - x$, we find that the radius of the circle is $(20 - x)/2\pi$ and its area is

$$\pi r^2 = \pi\left(\frac{20 - x}{2\pi}\right)^2 = \frac{(20 - x)^2}{4\pi}.$$

Thus the total area (as a function of x) is

$$A(x) = \frac{\sqrt{3}}{36}x^2 + \frac{(20 - x)^2}{4\pi}.$$

Differentiating, we obtain

$$\frac{dA}{dx} = \frac{\sqrt{3}}{18}x - \frac{(20 - x)}{2\pi} = \left(\frac{\sqrt{3}}{18} + \frac{1}{2\pi}\right)x - \frac{10}{\pi}.$$

This is equal to 0 when

$$x = \frac{10/\pi}{(\sqrt{3}/18) + (1/2\pi)} \approx 12.46 \text{ cm}.$$

Also, $d^2A/dx^2 = (\sqrt{3}/18) + (1/2\pi) > 0$ so that there is a local minimum at $x \approx 12.46$. For that value of x, $A \approx 11.99$ cm². Checking the endpoints of the interval $[0, 20]$, we find that $A(0) = 400/4\pi \approx 31.83$ cm², and $A(20) = (\sqrt{3}/36)(400) \approx 19.25$ cm². Thus the maximum area is enclosed when we do not cut the wire at all and use it to form a circle. If the wire must be cut, then we would cut as small a piece as possible for the triangle. The local minimum at $x = 12.46$ is indeed the minimum over the entire interval.

PROBLEMS 5.3

1. A rectangle has a perimeter of 300 m. What length and width will maximize its area?
2. A farmer wishes to set aside one acre of his land for corn and wheat. To keep out the cows, the field is enclosed by a fence (costing 50¢ per running foot). In addition, a fence running down the middle of the field is needed with a cost per foot of $1. Given that 1 acre = 43,560 square feet, what dimensions should the field have so as to minimize his total cost? The field is rectangular.
3. An isosceles triangle has a perimeter of 24 cm. What are the lengths of the sides of such a triangle which maximizes the area of the triangle?
4. Show that among all isosceles triangles with a given perimeter, the area enclosed is greatest when the triangle is equilateral.
5. Find the point on the straight line $2x + 3y = 6$ which is closest to the origin.
6. Find the point on the straight line $3x - 4y = 12$ which is closest to the point $(1, 2)$.
7. Find the point on the circle $x^2 + y^2 = 1$ which is closest to the point $(2, 5)$.
8. Find the points on the hyperbola $x^2 - y^2 = 1$ which are closest to the origin.
9. A wire, 35 cm long, is cut into two pieces. One piece is bent in the shape of a square and

the other is bent in the shape of a circle. How should the wire be cut so as to minimize and maximize the total area enclosed by the shapes?

10. Answer question 9 if the two pieces are to be formed in the shape of a circle and an equilateral triangle.

11. What is the area of the largest rectangle that can be inscribed in a circle of radius 10?

12. Find the positive number which exceeds its square by the largest amount.

13. Find the two positive numbers whose sum is 20 having the maximum product.

14. A cylindrical tin can is to hold 50 cm^3 of tomato juice. How should the can be constructed in order to minimize the amount of material needed in its construction?

15. A Transylvanian submarine is traveling due east and heading straight for a point P. A battleship from Luxembourg is traveling due south and heading for the same point P. Both ships are traveling at a velocity of 30 km/hr. Initially, their distances from P are 210 km for the submarine and 150 km for the battleship. The range of the submarine's torpedoes is 3 km. How close will the two vessels come? Does the submarine have a chance to torpedo the battleship?

16. Find the dimensions of the right circular cylinder of largest volume that can be inscribed in a sphere of radius 10; of radius r.

17. A Norman window is constructed from a rectangular sheet of glass surmounted by a semicircular sheet of glass. (Stained glass windows are often of this type). The light that enters through a window is proportional to the area of the window. What are the dimensions of the Norman window having a perimeter of 30 feet which admits the most light?

18. Answer the question of Problem 17 if the glass in the rectangular part of the window is colored while that of the circular part is clear, given that colored glass admits only half as much light as clear glass.

19. Two saplings, 6 and 8 feet high, are planted 10 feet apart. To prevent bending, poles the height of the trees are pounded in, then attached to each tree, and a rope tied to the top of each pole is then fixed to the ground (between the two trees) after being pulled taut. How close to the tallest tree will the rope be fixed if the total length of rope is to be minimized?

20. The strength of a rectangular beam is proportional to the product of the breadth and the square of its depth. Find the dimensions of the strongest such beam that can be cut from a cylindrical log of radius $\frac{1}{2}$ m.

21. The stiffness of a rectangular beam is proportional to the product of its breadth and the cube of its depth. Find the dimensions of the stiffest beam that can be cut from a cylindrical log of radius $\frac{1}{2}$ m.

22. A clear rectangle of glass is inserted in a colored semicircular glass window (see Figure 10). If the radius of the window is 3 feet and if the clear glass passes twice as much light as the colored glass, find the dimensions of the rectangular insert which passes the maximum light (through the entire window).

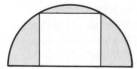

Figure 10

23. The rate of population growth of pigeons in a large city is given by

$$R = 400P^2 - \frac{1}{5}P^3,$$

where P is the population of pigeons. For what population level is this rate maximized?

24.   The position of a moving object is given by

$$s(t) = 4t - 6t^2 + 6, \qquad t \geq 0.$$

For what value of t is the velocity of the object a maximum?

25.   In Problem 24, for what value of t is the acceleration a maximum?

26.   A ball is thrown in the air with an initial vertical velocity of 64 feet/second. How high will the ball go (ignoring air resistance)? After how many seconds will the ball reach its maximum height?

27.   A chemical substance is produced at the rate of r moles per minute in a chemical reaction. At the same time, it is consumed at a rate of c moles per minute per mole of s. Let $s(t)$ be the number of moles of the chemical present at time t.
      (a)   Find an equation for the net rate of change of $s(t)$.
      (b)   When is this net rate of change a maximum?

28.   A farmer wishes to divide 20 acres of land along a river into 6 smaller plots by using one fence parallel to the river and 7 fences perpendicular to it. Show that the total amount of fencing is minimized if the sum of the lengths of the 7 cross fences equals the length of the one fence parallel to the river.

29.   In water, the product of the concentration of hydrogen ions and hydroxl ions is approximately equal to 10^{-14} mole. Find the ratio of hydrogen ions to hydroxl ions which minimizes the *sum* of the concentrations.

30.   An oil pipeline is built with two different kinds of tubing. From a point P on one side of a river, the line must cross the river and then proceed to a point Q along the bank on the other side (see Figure 11). The tubing used in crossing the river costs 50 percent

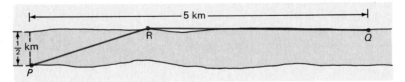

Figure 11

more than the tubing which can be used on dry land. The river is $\frac{1}{2}$ km wide and the point Q is 5 km down river from P. To what point R across the river must the pipe be directed so as to minimize the total cost?

*31.   Find a differentiable function f which maximizes the integral $\int_0^1 f(x)\, dx$ and satisfies the conditions $f(0) = 0$ and $f'(x) \leq x$ for every real number x. [*Hint:* Draw a sketch.]

5.4 Some Applications in Economics

We have seen in Chapters 2–4 that the derivative can be used to represent the marginal cost and marginal revenue in producing or selling a given product. The idea of margin can also be applied to other important notions in economics such as demand, consumption, profit and savings. The following is a summary of some of the important terms we will need in this section. Other terms will be introduced later.

      (i)   The *total cost function* gives the cost C of producing x units of a given product. A typical cost function is

$$C = ax^2 + bx + c \tag{1}$$

where the number c represents the *fixed cost* that will have to be spent even if nothing is produced (for rent, depreciation, utilities, etc.). Fixed cost is often referred to as *overhead*.

(ii)   The *total revenue function* gives the amount R received for selling x units of the product. A typical revenue function is

$$R = ax + bx^2. \tag{2}$$

Revenue can often be calculated by multiplying the price times the number sold.

(iii)   The *profit function* gives the profit P received when x units of the product are sold. In simple models we will have

$$P = R - C. \tag{3}$$

(iv)   The *demand function* expresses the relationship between the unit price that a product can sell for and the number of units that can be sold at that price. Typically, the more units sold, the lower the price so that if p represents price, dp/dx will be negative. A typical demand function is

$$p = a - bx. \tag{4}$$

We now give some examples of how these four functions can be used.

EXAMPLE 1.   A toy manufacturer finds that the cost in dollars of producing x copies of a certain doll is given by

$$C = 250 + 3x + 0.01x^2.$$

The dolls can be sold for $14 each. How many should he produce to maximize his profit, assuming that he can sell all he produces?

SOLUTION. Since the price does not vary, the revenue is $14x$ dollars. Thus

$$P = R - C = 14x - (250 + 3x + 0.01x^2)$$
$$= 11x - 0.01x^2 - 250.$$

Then the marginal profit $dP/dx = 11 - 0.02x = 0$ when $x = 11/0.02 = 550$ units. Since $d^2P/dx^2 = -0.02$, there is a local maximum at $x = 550$. When $x = 0$, $P = -250$. Also, as x gets very large the profit becomes negative (since the term $0.01x^2$ becomes larger than the x term). Therefore the maximum profit is $P = 11 \cdot 550 - (0.01)(550)^2 - 250 = 6050 - 3025 - 250 = \$2775$, and the answer to the problem is 550 dolls.

EXAMPLE 2.   A manufacturer of men's shirts figures that her exclusive "Parisian" model will cost $500 for overhead plus $9 for each shirt made. The price she can get for the shirts depends on how exclusive they are. From experience, her accountant has estimated the following demand function:

$$p = 30 - 0.2\sqrt{x}$$

where x is the number of shirts sold. How many shirts should the manufacturer produce in order to maximize her profit? (Assume that all the shirts produced will be sold.)

SOLUTION. From the information given, we have

$$C = 500 + 9x,$$
$$R = (\text{price}) \times (\text{number sold}) = (30 - 0.2\sqrt{x})x = 30x - 0.2x^{3/2}$$

and

$$P = R - C = 21x - 0.2x^{3/2} - 500. \tag{5}$$

Then the marginal profit $dP/dx = 21 - 0.3\sqrt{x} = 0$ when $\sqrt{x} = 21/0.3 = 70$ or when $x = 70^2 = 4900$ shirts. Since $d^2P/dx^2 = -0.3/2\sqrt{x}$ which is < 0, there is a local maximum at $x = 4900$. This is easily seen to be a true maximum as well and is therefore the answer to the problem. At a production level of 4900,

$$P = 21 \cdot 4900 - (0.2)(4900)^{3/2} - 500$$
$$= 102{,}900 - (0.2)(343{,}000) - 500 = \$33{,}800.$$

Note that if 4900 shirts are sold, then they will be sold at a price of $p = 30 - 0.2(70) = 30 - 14 = \$16$ each.

EXAMPLE 3. A certain manufacturer has a steady demand for 50,000 refrigerators each year. The machines are not made continuously, but rather in equally sized batches. Production costs are $10,000 to set up the machinery plus $100 for each refrigerator made. In addition, there is a storage (inventory) charge of $2.50 per year per refrigerator. If the demand is steady throughout the year, how should the manufacturer schedule his production runs so as to minimize his total costs?

SOLUTION. We have

$$C = \text{total cost} = \text{production cost} + \text{inventory cost} = PC + IC.$$

If batches are in lots of x refrigerators at a time, then there must be $50{,}000/x$ runs each year. The production cost will then be

$$PC = \left(\frac{50{,}000}{x}\right) \cdot 10{,}000 + (50{,}000)(100) = \frac{500{,}000{,}000}{x} + 5{,}000{,}000.$$

Now we calculate inventory costs. It is reasonable to assume that production is scheduled so that one new batch is complete just as the previous batch has run out. Therefore inventory starts at x units and decreases steadily (because of the steady demand) to 0 units. The average number of units in storage at any one time will then be $x/2$ units. Therefore, the inventory costs are

$$IC = 2.50\frac{x}{2} = 1.25x.$$

Then we find that

$$C = \frac{500{,}000{,}000}{x} + 5{,}000{,}000 + 1.25x$$

and the marginal cost $dC/dx = -(500,000,000/x^2) + 1.25$. This is zero when

$$1.25x^2 = 500,000,000 \quad \text{or when} \quad x^2 = \frac{500,000,000}{1.25} = 400,000,000$$

and $x = \pm 20,000$.

Of course, the value $x = -20,000$ is meaningless. We can then verify that a minimum is indeed reached when $x = 20,000$. If $x = 20,000$, then $50,000/x = 2\frac{1}{2}$ which means that costs will be minimized when there are $2\frac{1}{2}$ runs a year, which works out (practically) to 5 runs every two years. The minimum annual cost is \$5,050,000 (check this).

Often a person in business will be faced with the choice between increasing or not increasing prices. Generally the demand function will show that an increase in price will cause a drop in sales. But how much of a drop? If there would be a very small drop, then profit would increase. On the other hand, if a large drop in sales would result, then profit would fall. The *average price elasticity of demand* is defined as the *relative change* in quantity demanded divided by the *relative change* in price:

$$\boxed{\text{average price elasticity of demand} = E_{AV} = -\frac{\Delta x/x}{\Delta p/p}.} \tag{6}$$

The relative change in x is the change Δx divided by x and likewise for the relative change in p. Note that $\Delta x/x$ can also be thought of as the percent change in x. The minus sign in (6) is put there so that E_{AV} will turn out to be positive. E_{AV} will be positive since if $\Delta p > 0$, Δx will usually be < 0 (why?).

In general, if $E_{AV} < 1$ then the percent decrease in demand is less than the percent increase in price so that an increase in price will lead to an increase in profits. If $E_{AV} > 1$, then the opposite is true. If $E_{AV} = 1$, then the price increase will not make any difference. The loss in demand will just offset the revenue gain by the increase in price.

EXAMPLE 4.   In Example 2 suppose that the demand function is given by

$$p(x) = 30 - 0.2\sqrt{x}. \tag{7}$$

Calculate the elasticity of demand if the price per shirt is increased from \$20 to \$22.

SOLUTION. At a price of \$20, we calculate from (7) that $20 = 30 - 0.2\sqrt{x}$, $\sqrt{x} = 10/0.2 = 50$, and $x = 2500$. At $p = 22$, $x = 1600$. Therefore we have $p = 20$, $\Delta p = 2$, $x = 2500$, $\Delta x = -900$ and

$$E_{AV} = -\frac{(-900)/2500}{2/20} = \frac{90}{25} = 3.6.$$

Since this number is > 1, it means that the percent loss in demand (the numerator of E_{AV}) is greater than the percent increase in price. Therefore the increase of \$2 would result in a net *decrease* in profits. To verify this, we use the profit formula (5).

$$P(2500) = 21 \cdot 2500 - 0.2(2500)^{3/2} - 500 = \$27,000$$

and

$$P(1600) = 21 \cdot 1600 - 0.2(1600)^{3/2} - 500 = \$20,300.$$

There is a decrease in profits of $6,700 due to the price increase.

We return to the question of whether the person in business should increase or decrease prices. Put another way, will a very small increase in price lead to a greater or smaller profit? To answer this, let $p = p(x)$ represent the demand function. If $p(x)$ is continuous (as it is assumed to be), a small change Δp in price will be caused by a small change Δx in demand. Then we define the *price elasticity of demand* when the demand is x and the price is $p(x)$ as the limit of E_{AV} as $\Delta x \to 0$. That is,

$$E(x) = \lim_{\Delta x \to 0} E_{AV} = \lim_{\Delta x \to 0} - \frac{\Delta x / x}{\Delta p / p} = \lim_{\Delta x \to 0} \frac{-p(x)}{x \, \Delta p / \Delta x} = \frac{-p}{x p'(x)}. \qquad (8)$$

EXAMPLE 5.   The demand function for a certain electric toaster is $p(x) = 35 - 0.05 \sqrt{x}$. Will a rise in price increase or decrease profit if 10,000 toasters are in demand?

SOLUTION. We calculate E. If $E > 1$, then, as before, an increase in price will cause a decrease in revenue (and profit) while if $E < 1$, then an increase in price will cause an increase in profit. Here

$$p'(x) = -\frac{0.05}{2\sqrt{x}} = -\frac{0.05}{2\sqrt{10,000}} = -\frac{0.05}{200} = -0.00025.$$

When $x = 10,000$, $p = 35 - 0.05\sqrt{10,000} = 30$ and

$$E = -\frac{p}{xp'(x)} = \frac{-30}{(10,000)(-0.00025)} = \frac{30}{2.5} = 12.$$

Since $E > 1$, there will be a *decrease* in profit if the price is raised and, therefore, prices should *not* be increased.

EXAMPLE 6.   Answer the question in Example 5 if 250,000 toasters are being produced.

SOLUTION. When $x = 250,000$, $p = 35 - 0.05\sqrt{250,000} = 35 - (0.05)(500) = 10$, $p'(250,000) = -0.05/2(500) = -0.00005$ and

$$E = \frac{-10}{(250,000)(-0.00005)} = \frac{-10}{-12.5} = 0.8 < 1$$

so that there would be an increase in profit if prices were increased.

The notion of elasticity of demand has an interesting interpretation in global economics. Suppose that certain economists in the United States are concerned about an imbalance in the balance of payments. That is, there are more dollars going out than are coming in. To offset this problem, they suggest a devaluation of the dollar. This would make dollars cheaper abroad, thereby making American goods cheaper to foreign consumers. Therefore there would be an increase in the foreign purchase of

U.S. goods, thereby leading to an increase in the number of export dollars flowing back to the United States. Or will something else happen? The economists must be careful. For if the elasticity of demand for American exports is less than one, more American products would indeed be purchased abroad but there would be a net *decrease* in the value of the dollars paid for these goods. Things are never simple in the area of international trade.

The examples given so far in this section illustrate the use of the derivative in economics. There are also some very interesting applications of the integral. We saw some of these in Chapter 4. Others will be given in Sections 5.6 and 6.5.

PROBLEMS 5.4

1. The cost of producing x color television sets is given by $C = 5000 + 250x - 0.01x^2$. The revenue received is $R = 400x - 0.02x^2$. Assuming that all sets produced will be sold, how many should be produced so as to maximize the profit?

2. What is the demand function in Problem 1?

3. In Problem 1, at a production level of 10,000 sets, will an increase in price generate an increase or decrease in profits?

4. Bottles of whiskey cost a distiller in Scotland $2 a bottle to produce. In addition he has fixed costs of $500. The demand function worldwide for the whiskey is given by $p = 12 - 0.001x$. How many bottles should he produce to maximize his profit?

5. The distiller of Problem 4 now sells exclusively to the United States. He must pay $0.50 per bottle duty. The demand function does not change. How does he maximize his profit in this case?

*6. The distiller of Problem 4 shifts his sales to France where an import duty of 20 percent of the sales price is charged. How does he now maximize his profit?

7. In Problems 4, 5, and 6, determine whether a price increase will result in an increase or decrease in profits at a production level of 10,000 bottles.

8. Show for any problem of the type we have considered that whenever profit is maximized, the marginal cost and the marginal revenue are equal.

9. A manufacturer of kitchen sinks finds that if he produces x sinks per week he has fixed costs of $1000, labor and materials costs of $5/sink, and advertising costs of $10,000/x$. How many sinks should he manufacture weekly to minimize costs?

10. The demand function for the sinks of Problem 9 is $p(x) = 25 - 0.01x$. How many sinks should be manufactured weekly to maximize profits?

11. In Problem 10, at a level of production of 2,000 sinks, can the manufacturer afford to raise the prices?

12. If the demand function for a certain manufactured good is $p(x) = 75 - 0.1\sqrt{x} - 0.002x^2$, at what level of production will it not make any difference whether the price is increased or decreased?

13. A Detroit manufacturer has a steady annual demand for 50,000 pickup trucks. The trucks are made in batches. The costs of production include a $20,000 setup cost per batch and a cost of $2,500 per truck. Inventory charges are $50 per truck per year. How is production to be scheduled so as to minimize total costs?

14. Suppose that the manufacturer in Problem 13 produces trucks continuously (so that there is only one setup cost) but the demand is not constant and is given by $x = 5000 + (300,000/\sqrt{p})$, where p is the price in dollars and x is the demand. It is assumed that if fewer than 10,000 trucks are sold, the manufacturer will go bankrupt. What should be the price charged for each truck so as to maximize the total profits? [*Hint:* Find the demand function by writing p as a function of x.]

15. In Problem 14 determine the elasticity function.
16. In Problem 14, at a level of production of 25,000 trucks, will it be profitable to increase prices?
17. A woman has $10,000 to invest in two companies. The return from investing x dollars in Company 1 is $4\sqrt{x}$ dollars and the return from investing x dollars in Company 2 is $2\sqrt{x}$ dollars. How should she invest her money so as to maximize her return?
18. The Grosshop Company does public opinion polls. They have observed that the cost of conducting a national survey of n people is

$$C(n) = 25,000 + 0.02(n - 1500)^2.$$

Of course, the more people surveyed, the better are the results (up to a point). They have estimated that the value (in dollars, since better accuracy ensures greater profits) is given by

$$V(n) = 500,000 - 0.01(n - 8,000)^2.$$

If "profit" is defined by $P(n) = V(n) - C(n)$, what is the optimal number of people to be polled (to the nearest person)?

*19. Show that if the demand law is given as $x = x(p)$ (that is demand as a function of price rather than vice versa), then

$$E(p) = -\frac{px'(p)}{x(p)}.$$

(This expresses the elasticity in terms of price rather than in terms of demand.)
20. Show that if $x(p) = 5/p^4$ then $E(p) = 4$.
21. Show in general that if $x(p) = a/p^\alpha$ with $\alpha > 0$, then $E(p) = \alpha$.
*22. Show that if $p = p(x)$ is the demand function, then at a level of production which maximizes total revenue, the elasticity is 1.
*23. Let the cost function be $C = ax^2 + bx + c$ and the demand function $p = \alpha - \beta x$, where a, b, c, α, and β are positive.
  (a) At what level of production is profit maximized?
  (b) What is the elasticity?
  (c) At what level of production does it make no difference whether the price is increased or decreased?

☐ 5.5 Newton's Method for Solving Equations

In this section we look at a very different kind of application of the derivative. Consider the equation

$$f(x) = 0, \tag{1}$$

where f is assumed to be differentiable in some interval $[a, b]$. It is often important to calculate the *roots* of equation (1); that is, the values of x which satisfy the equation. For example, if $f(x)$ is a polynomial of degree 5, say, then the roots of $f(x)$ could be as in Figure 1. (We have already used graphs to determine the number of roots of certain polynomials; see Section 3.9.)

In the seventeenth century Newton discovered a method for estimating a solution or root by defining a sequence of numbers which become successively closer and closer to the root sought. His method is best illustrated graphically. Let $y = f(x)$

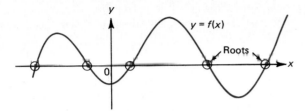

Figure 1

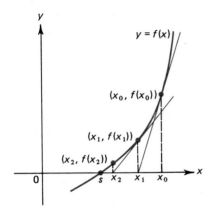

Figure 2

as in Figure 2. A number x_0 is chosen arbitrarily. We then locate the point $(x_0, f(x_0))$
on the graph and draw the tangent line to the curve at that point. Next, we follow the
tangent line down until it hits the x-axis. The point of intersection of the tangent line
and the x-axis is called x_1. We then repeat the process to arrive at the next point x_2.
On the graph, we have labeled the solution to $f(x) = 0$ as s. That is, $f(s) = 0$. For our
graph at least, it seems as if the points $x_0, x_1, x_2, \ldots$ are approaching the point $x = s$.
In fact, this happens for quite a few functions and the rate of approach to the solution
is quite rapid.

Having briefly looked at a graphical representation of Newton's method, let us
next develop a formula for giving us x_1 from x_0, x_2 from x_1, etc. The slope of the
tangent line at the point $(x_0, f(x_0))$ is $f'(x_0)$. Two points on this line are $(x_1, 0)$ and
$(x_0, f(x_0))$. Therefore

$$\frac{0 - f(x_0)}{x_1 - x_0} = f'(x_0). \qquad (2)$$

Solving (2) for x_1 gives us

$$x_1 = x_0 - \frac{f(x_0)}{f'(x_0)}. \qquad (3)$$

Similarly,

$$x_2 = x_1 - \frac{f(x_1)}{f'(x_1)}. \qquad (4)$$

In general, we obtain

$$x_{n+1} = x_n - \frac{f(x_n)}{f'(x_n)}. \tag{5}$$

This last step tells us how to obtain the $(n + 1)$st point if the nth point is given, as long as $f'(x_n) \neq 0$ so that (5) is defined. Thus, if we start with a given value x_0, we can obtain $x_1, x_2, x_3, x_4, \ldots$. The formula (5) is called *Newton's formula*. The set of numbers $x_0, x_1, x_2, x_3, \ldots$ is called a *sequence*.† If the numbers in the sequence get closer and closer to a certain number s as n gets larger and larger, then we say that the sequence *converges* to s and we write

$$\lim_{n \to \infty} x_n = s. \tag{6}$$

Before stating the theorem which tells us when the sequence defined by (5) converges to a solution of $f(x) = 0$, we give a simple example.

EXAMPLE 1. Let $r > 1$. Formulate a rule for calculating the square root of r.

SOLUTION. We must find an x such that $x = \sqrt{r}$, or $x^2 = r$, or $x^2 - r = 0$. Let $f(x) = x^2 - r$. Then if $f(s) = 0$, s will be a square root of r. ($f(s) = 0$ means that $s^2 - r = 0$ or $s^2 = r$.) By Newton's formula, since $f'(x) = 2x$, we obtain the sequence $x_0, x_1, x_2, \ldots$ where x_0 is arbitrary and

$$x_{n+1} = x_n - \frac{f(x_n)}{f'(x_n)} = x_n - \frac{(x_n^2 - r)}{2x_n} = \frac{2x_n^2 - x_n^2 + r}{2x_n} = \frac{1}{2}\left(x_n + \frac{r}{x_n}\right). \tag{7}$$

☐**EXAMPLE 2.** Calculate $\sqrt{2}$ by Newton's method.

SOLUTION. In formula (7), $r = 2$ so that

$$x_{n+1} = \frac{1}{2}\left(x_n + \frac{2}{x_n}\right). \tag{8}$$

Using a calculator, we obtain the sequence in Table 1 starting with $x_0 = 1$. We can see here the remarkable accuracy of Newton's method. An answer correct to 9 decimal places was obtained after only 4 steps! We were limited in accuracy only by the fact that our calculator could display only 10 digits.

TABLE 1

n	x_n	$\dfrac{2}{x_n}$	$x_n + \dfrac{2}{x_n}$	$x_{n+1} = \dfrac{1}{2}\left(x_n + \dfrac{2}{x_n}\right)$
0	1.0	2.0	3.0	1.5
1	1.5	1.333333333	2.833333333	1.416666667
2	1.416666667	1.411764706	2.828431373	1.414215686
3	1.414215686	1.414211438	2.828427125	1.414213562
4	1.414213562	1.414213562	2.828427125	1.414213562

† We shall discuss sequences extensively in Chapter 14.

We now state a theorem which gives conditions that guarantee that Newton's method will work. The important fact is that we must usually choose a "starting value" x_0 reasonably close to the solution s. The proof of this theorem cannot be given here. For a proof, the reader should consult the footnoted text.†

THEOREM 1.   Suppose that on some interval $[a, b]$, $f(x)$ is defined, f'' exists and is continuous, and

(i) $f(a)$ and $f(b)$ have different signs.
(ii) $f'(x) \neq 0$ for every x in $[a, b]$.
(iii) $f''(x)$ does not change sign in $[a, b]$.
(iv)   If $f'(a) \leq f'(b)$, then $\left| \dfrac{f(a)}{f'(a)} \right| \leq b - a$,

or

$$\text{if } f'(b) \leq f'(a), \text{ then } \left| \frac{f(b)}{f'(b)} \right| \leq b - a.$$

Then the sequence $x_0, x_1, x_2, x_3, \ldots$ defined by (5) converges to the *unique* solution s of $f(x) = 0$ for every initial choice x_0 in $[a, b]$.

EXAMPLE 3.   In Example 2, $f(x) = x^2 - 2$. Let us verify that the conditions of Theorem 1 hold. If $[a, b] = [1, 10]$, for example, then $f(1) = -1$ and $f(10) = 98$ so that $f(a)$ and $f(b)$ have different signs. Also, $f'(x) = 2x$ which is nonzero in the interval. Clearly (iii) holds since $f''(x) = 2$, a constant. Finally, $f'(1) < f'(10)$ and

$$\left| \frac{f(1)}{f'(1)} \right| = \left| \frac{-1}{2} \right| = \frac{1}{2} < b - a = 9.$$

Therefore all conditions are satisfied.

EXAMPLE 4.   Formulate a rule for calculating the kth root of a given number r.

SOLUTION. We must find an x such that $x = r^{1/k}$ or $x^k = r$ or $f(x) = x^k - r = 0$. Then $f'(x) = kx^{k-1}$ and

$$x_{n+1} = x_n - \frac{x_n^k - r}{kx_n^{k-1}} = x_n - \frac{1}{k} \frac{x_n^k}{x_n^{k-1}} + \frac{r}{kx_n^{k-1}} = \left(1 - \frac{1}{k}\right)x_n + \frac{r}{kx_n^{k-1}}. \tag{9}$$

EXAMPLE 5.   Calculate $\sqrt[3]{17}$.

SOLUTION. By (9) with $k = 3$ and $r = 17$, we have $x_{n+1} = \frac{2}{3}x_n + (17/3x_n^2)$. Since $\sqrt[3]{17}$ is between 2 and 3, we choose $x_0 = 2$ (3 would do just as well). If $[a, b] = [1, 10]$, then since

$$f(x) = x^3 - 17, \quad f(1) = -16, \quad f(10) = 983$$

and (i) is satisfied. Also $f'(x) = 3x^2$ and $f''(x) = 6x$ so that (ii) and (iii) are satisfied on $[1, 10]$. Finally, $f'(1) = 3$, $f'(10) = 300$ and

$$\left| \frac{f(1)}{f'(1)} \right| = \left| -\frac{16}{3} \right| = \frac{16}{3} < b - a = 9.$$

†Peter Henrici, *Elements of Numerical Analysis*, Wiley, New York, 1969, p. 79.

Thus we know that our sequence will converge to the unique cube root of 17 in $[1, 10]$. Values of x_n are tabulated in Table 2. The last number is correct to 9 decimal places. Again, the rapid convergence of Newton's method is illustrated.

TABLE 2

n	x_n	$\frac{2}{3}x_n$	x_n^2	$\frac{17}{3x_n^2}$	$x_{n+1} = \frac{2}{3}x_n + \frac{17}{3x_n^2}$
0	2.0	1.333333333	4.0	1.416666667	2.75
1	2.75	1.833333333	7.5625	0.7493112948	2.582644628
2	2.582644628	1.721763085	6.670053275	0.8495684267	2.571331512
3	2.571331512	1.714221008	6.611745745	0.8570605836	2.571281592
4	2.571281592	1.714187728	6.611489025	0.8570938627	2.571281591

□**EXAMPLE 6.** Find the real roots of $p(x) = x^3 + x^2 + 7x - 3$.

SOLUTION. We differentiate to find that $p'(x) = 3x^2 + 2x + 7$. This polynomial has no real roots (explain why) so there are no local maxima or minima. Also, $p''(x) = 6x + 2 = 0$ when $x = -1/3$ so that $(-1/3, -142/27)$ is a point of inflection. The graph of $p(x)$ is given in Figure 3. From the graph, we see that there is exactly one

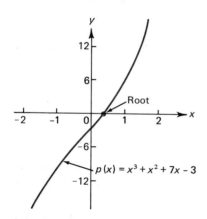

Figure 3

real root. Also, since $p(0) = -3$ and $p(1) = 6$, the root must lie between 0 and 1. In the interval $[0, 1]$, the conditions of Theorem 1 can be easily verified. Now,

$$x_{n+1} = x_n - \frac{p(x_n)}{p'(x_n)} = x_n - \frac{x_n^3 + x_n^2 + 7x_n - 3}{3x_n^2 + 2x_n + 7}$$

$$= \frac{x_n(3x_n^2 + 2x_n + 7) - (x_n^3 + x_n^2 + 7x_n - 3)}{3x_n^2 + 2x_n + 7} = \frac{2x_n^3 + x_n^2 + 3}{3x_n^2 + 2x_n + 7}.$$

If we choose $x_0 = 0$, we obtain the results in Table 3. The root is $s = 0.3970992165$, correct to 10 decimal places.

TABLE 3

n	x_n	$2x_n^3 + x_n^2 + 3$	$3x_n^2 + 2x_n + 7$	$\dfrac{2x_n^3 + x_n^2 + 3}{3x_n^2 + 2x_n + 7}$
0	0	3.0	7.0	0.4285714286
1	0.4285714286	3.341107872	8.408163265	0.3973647712
2	0.3973647712	3.283385572	8.268425827	0.3970992352
3	0.3970992352	3.282923214	8.267261878	0.3970992164
4	0.3970992164	3.282923182	8.267261796	0.3970992165
5	0.3970992165	3.282923182	8.267261796	0.3970992165

Let us make some final remarks about Newton's method. First, if x_n is a solution to $f(x) = 0$, then

$$x_{n+1} = x_n - \frac{f(x_n)}{f'(x_n)} = x_n - 0 = x_n .$$

This explains why, in the last table, the value of x_n did not change after the desired accuracy was reached. Second, it is not necessary to check the conditions of Theorem 1 every time Newton's method is used. However failure to do so could result in one or both of the following problems:

(a) The sequence generated by (5) could grow without bound, never even getting close to a solution.

(b)   The sequence could converge to a solution, but there could be *other* solutions in the interval $[a, b]$ which would be missed. This last problem can be avoided if a graph is drawn which gives an idea as to where (approximately) the roots are located.

An interesting example of the need to find the root of a polynomial will be given in Example 6.5.15.

PROBLEMS 5.5

In the problems below calculate all answers to 4 decimal places of accuracy if you do not have a hand calculator. If you have one, calculate your answers to as many decimal places of accuracy as are displayed on the machine.

1.   Find an interval over which Theorem 1 applies to the calculation of $\sqrt{90}$. Perform the calculation.
2.   Calculate $\sqrt[4]{25}$ using Newton's method.
3.   Calculate $\sqrt[5]{10}$ using Newton's method.
4.   Calculate $\sqrt[6]{100}$ using Newton's method.
5.   Use Newton's method to calculate the roots of $x^2 - 7x + 5 = 0$. It will be necessary to do two separate calculations using two distinct intervals. Compare this with the answers obtained by the Quadratic formula.
6.   Show that there is *no* interval $[a, b]$ for which Theorem 1 applies when $f(x) = x^2 + 5x + 7$. Explain why this must be the case. Try Newton's method with $x_0 = 0$. What happens?
7.   Find all solutions of the equation $x^3 - 6x^2 - 15x + 4 = 0$. [*Hint:* Draw a sketch and

estimate the roots to the nearest integer. Then use this estimate as your initial choice of x_0 for each of the three roots.]

8. Find all solutions of the equation $x^3 + 14x^2 + 60x + 105 = 0$.
9. Find all solutions of $x^3 - 8x^2 + 2x - 15 = 0$,
*10. Find all solutions of $x^3 + 3x^2 - 24x - 40 = 0$.
11. Using Newton's method, find a formula for finding reciprocals without dividing (the reciprocal of x is $1/x$).
12. Using the formula found in Problem 11, calculate $\frac{1}{7}$ and $\frac{1}{81}$.

5.6 The Area between Two Curves

We developed the definite integral in Chapter 4 by calculating the area under a curve. We now show that by using similar methods, the area between two curves can be calculated.

Consider the two functions f and g as graphed in Figure 1. We shall calculate

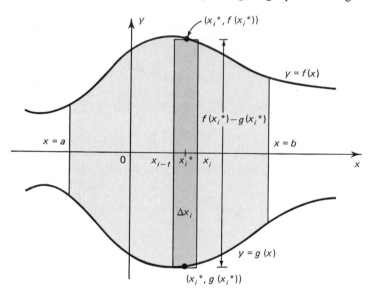

Figure 1

the area between the curves $y = f(x)$ and $y = g(x)$ and the lines $x = a$ and $x = b$ by adding up the areas of a large number of rectangles. Let P be a partition of the interval $[a, b]$. Then a typical rectangle (see Figure 1) has the area

$$[f(x_i^*) - g(x_i^*)] \Delta x_i$$

where x_i^* is a point in the interval $[x_{i-1}, x_i]$ and $\Delta x_i = x_i - x_{i-1}$. Then, proceeding exactly as we have proceeded before, we find that

$$\text{area} = \lim_{|P| \to 0} \{[f(x_1^*) - g(x_1^*)] \Delta x_1 + [f(x_2^*) - g(x_2^*)] \Delta x_2 + \cdots$$
$$+ [f(x_n^*) - g(x_n^*)] \Delta x_n\} = \int_a^b [f(x) - g(x)]\, dx. \qquad (1)$$

This formula is valid as long as $f(x) \geq g(x)$ in $[a, b]$. We will deal with more general cases later in this section.

Remark. There is no difficulty if one (or both) of the functions takes on negative values in $[a, b]$. If you are uncomfortable with these negative values, simply *shift upward*. If we define $\bar{f}(x) = f(x) + C$ and $\bar{g}(x) = g(x) + C$, where C is a positive constant, then

$$\text{area between } \bar{f} \text{ and } \bar{g} = \int_a^b (\bar{f} - \bar{g}) = \int_a^b [(f + C) - (g + C)] = \int_a^b (f - g)$$

$$= \text{area between } f \text{ and } g.$$

This fact is illustrated in Figure 2. The shaded areas retain the same area.

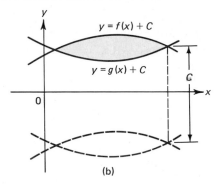

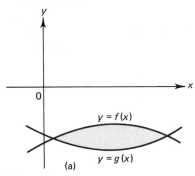

(a) $y = g(x)$ (b)

Figure 2

EXAMPLE 1. Find the area bounded by $y = x^2$ and $y = 4x$, for x between 0 and 1.

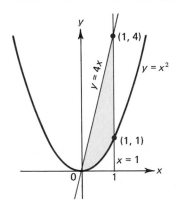

Figure 3

SOLUTION. In any problem of this type, it is helpful to draw a graph. In Figure 3 the required area is shaded. We have

$$A = \int_0^1 (4x - x^2)dx = 2x^2 - \frac{x^3}{3} \Big|_0^1 = 2 - \frac{1}{3} = \frac{5}{3}.$$

EXAMPLE 2. Find the area bounded by the curves $y = x^3$ and $y = \sqrt{x}$.

SOLUTION. This problem only makes sense if the curves intersect at at least two points. (Otherwise we would have to be given other bounding lines.) See Figure 4. We need

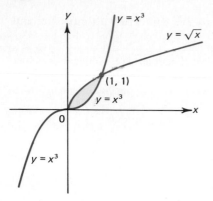

Figure 4

to find the points of intersection of these two curves. To find them, we set the two functions equal. If $x^3 = \sqrt{x}$, then $x^6 = x$ or $x^6 - x = x(x^5 - 1) = 0$. This occurs when $x = 0$ and $x = 1$. Then

$$A = \int_0^1 (\sqrt{x} - x^3)\, dx = \frac{2x^{3/2}}{3} - \frac{x^4}{4}\Big|_0^1 = \frac{2}{3} - \frac{1}{4} = \frac{5}{12}.$$

Note that in these last two examples we have had no trouble deciding which function came first in the expression $f(x) - g(x)$. We always put the larger function first. Thus, in $[0, 1]$, $\sqrt{x} \geq x^3$ so that $\sqrt{x}$ comes first.

EXAMPLE 3. Find the area bounded by the two curves $y = x^2 + 3x + 5$ and $y = -x^2 + 5x + 9$.

SOLUTION. We first sketch the two curves (see Figure 5). To find the points of

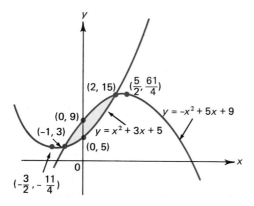

Figure 5

intersection, we set $x^2 + 3x + 5 = -x^2 + 5x + 9$. This leads to the equation $2x^2 - 2x - 4 = 0$ which has roots $x = -1$ and $x = 2$. Thus

$$A = \int_{-1}^2 [(-x^2 + 5x + 9) - (x^2 + 3x + 5)]\, dx$$

$$= \int_{-1}^2 (-2x^2 + 2x + 4)\, dx = -\frac{2x^3}{3} + x^2 + 4x \Big|_{-1}^2 = 9.$$

EXAMPLE 4.   Find the area bounded by the two curves $y = x^2 + 3x + 5$ and $y = -x^2 + 5x + 9$ and the line $x = 4$.

SOLUTION. This is more complicated than Example 3. See Figure 6. There are now 2

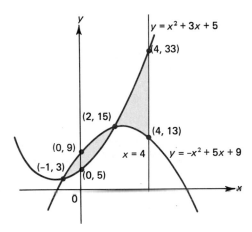

Figure 6

areas to be added together. In the first (calculated in Example 3),

$$-x^2 + 5x + 9 \geq x^2 + 3x + 5$$

In the second

$$x^2 + 3x + 5 \geq -x^2 + 5x + 9.$$

We therefore break the calculation into two parts:

$$A = \int_{-1}^{2} [(-x^2 + 5x + 9) - (x^2 + 3x + 5)]\, dx$$

$$+ \int_{2}^{4} [(x^2 + 3x + 5) - (-x^2 + 5x + 9)]\, dx$$

$$= 9 + \int_{2}^{4} (2x^2 - 2x - 4)\, dx = 9 + \left(\frac{2x^3}{3} - x^2 - 4x \right)\Big|_{2}^{4} = 9 + \frac{52}{3} = \frac{79}{3}.$$

We note that $(-x^2 + 5x + 9) - (x^2 + 3x + 5)$ in $[-1, 2]$ and $(x^2 + 3x + 5) - (-x^2 + 5x + 9)$ in $[2, 4]$ can be written as $|(x^2 + 3x + 5) - (-x^2 + 5x + 9)|$ in $[-1, 4]$. (Why?) We can generalize this to obtain the general rule.

The area between the curves $y = f(x)$ and $y = g(x)$ between $x = a$ and $x = b$ $(a < b)$ is given by

$$A = \int_{a}^{b} |f(x) - g(x)|\, dx.$$ (2)

This rule forces the integrand to be positive so that we cannot run into the problem of adding negative areas (however it does not make the calculation any easier).

EXAMPLE 5.   Calculate the area bounded by the curves $y = 2x^5 + 3$ and $y = 32x + 3$.

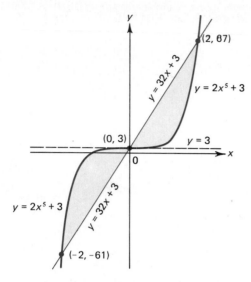

Figure 7

SOLUTION. Look at Figure 7. The two curves intersect when

$$2x^5 + 3 = 32x + 3 \qquad \text{or} \qquad x^5 - 16x = x(x^4 - 16) = 0;$$

that is, when $x = -2$, 0 and 2. In $[-2, 0]$, $2x^5 + 3 \geq 32x + 3$ and in $[0, 2]$, $32x + 3 \geq 2x^5 + 3$. Thus

$$A = \int_{-2}^{0} (2x^5 - 32x)\, dx + \int_{0}^{2} (32x - 2x^5)\, dx$$

$$= \left(\frac{x^6}{3} - 16x^2 \right) \Big|_{-2}^{0} + \left(16x^2 - \frac{x^6}{3} \right) \Big|_{0}^{2} = \frac{128}{3} + \frac{128}{3} = \frac{256}{3}.$$

EXAMPLE 6.   Find the area bounded by the curves $y^2 = 4 + x$ and $x + 2y = 4$.

SOLUTION. This problem is more complicated than it might at first appear. Look at Figure 8. We can easily calculate that the two curves intersect at $(0, 2)$ and $(12, -4)$

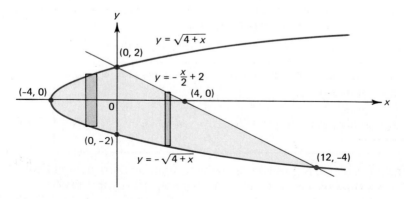

Figure 8

by setting $x = y^2 - 4$ equal to $x = 4 - 2y$ to find that $y = 2$ and $y = -4$. Calculating the area is more complicated. We see that x ranges from -4 to 12. But we can't simply integrate from -4 to 12 with respect to x because the integrand changes. In $[-4, 0]$, a typical rectangle has a height of $\sqrt{4 + x_i^*} - (-\sqrt{4 + x_i^*}) = 2\sqrt{4 + x_i^*}$. But between 0 and 12, a typical rectangle has the height

$$\left(-\frac{x_i^*}{2} + 2\right) - (-\sqrt{4 + x_i^*}) = \sqrt{4 + x_i^*} - \frac{x_i^*}{2} + 2.$$

Thus

$$A = \int_{-4}^0 2\sqrt{4 + x}\, dx + \int_0^{12} \left(\sqrt{4 + x} - \frac{x}{2} + 2\right) dx$$

$$= \frac{4}{3}(4 + x)^{3/2} \bigg|_{-4}^0 + \left[\frac{2}{3}(4 + x)^{3/2} - \frac{x^2}{4} + 2x\right]\bigg|_0^{12} = \frac{32}{3} + \frac{76}{3} = 36.$$

There is a much easier way to solve this problem. We are in the habit of writing every functional relationship between x and y in terms of y as a function of x. However, in this problem it is clearly more convenient to treat x as a function of y. It is, for example, an easy matter to obtain the graph of $x = y^2 - 4$. This is just like the graph of $y = x^2 - 4$ with the x and y axes interchanged. We again draw the graph with the only change being that everything is labeled as a function of y. This is

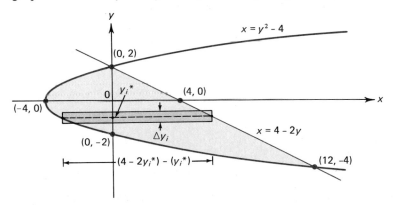

$x = y^2 - 4$

$(0, 2)$

$(4, 0)$

$(-4, 0)$

$(0, -2)$

$x = 4 - 2y$

$(4 - 2y_i^*) - (y_i^*)$

$(12, -4)$

Figure 9

done in Figure 9. We now construct our rectangles horizontally instead of vertically by partitioning along the y-axis. The height of a typical rectangle is Δy_i and the width is $(4 - 2y_i^*) - (y_i^{*2} - 4)$. We may therefore write

$$A = \int_{-4}^2 [(4 - 2y) - (y^2 - 4)]\, dy$$

$$= \int_{-4}^2 (8 - 2y - y^2)\, dy = \left(8y - y^2 - \frac{y^3}{3}\right)\bigg|_{-4}^2 = 36.$$

In general, if it is easier to treat y as the independent variable, then for $a < b$

$$A = \int_{y=a}^{y=b} |f(y) - g(y)|\, dy. \tag{3}$$

Whether to integrate with respect to x or y is a matter of judgment. In most cases a glance at a sketch of the area being considered will indicate which is preferable.

We close this section by showing how the calculation of the area between two curves can be applied to a problem in economics.

Let $p = p(x)$ be the demand function for a certain product (see Section 5.4 for an explanation of terminology). A typical demand function is like the function graphed in Figure 10. There is a point x_{max} at which the demand is exhausted so that

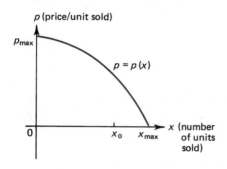

Figure 10

no more of the product can even be given away. Likewise, there is a price p_{max} at which no items can be sold.

In a typical market situation, prices do not fluctuate instantly with demand but rather are held constant over a certain period of time (after which, in today's marketplace, they will usually rise—even if demand falls). Let this fixed price be denoted by $\bar{p}$ (see Figure 11). Choose a small subinterval $[x_{i-1}, x_i]$ as shown. If the

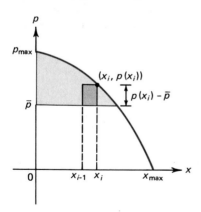

Figure 11

demand is x_i, then the correct price should be $p(x_i)$ (according to the demand function). Since the price is only $\bar{p}$, the consumer is saving the amount $p(x_i) - \bar{p}$ for each unit purchased. If Δx_i units are purchased, the amount saved is $[p(x_i) - \bar{p}]\Delta x_i$. We now "add up" these rectangles to define *consumers' surplus* as the area bounded by $p(x)$, $x = 0$ and $y = \bar{p}$. This is the "triangular" region in Figure 11.

EXAMPLE 7. Let the demand function for a product be given by $p = 30 - 0.2\sqrt{x}$. If the price per unit is fixed at $\$10$, calculate the consumers' surplus.

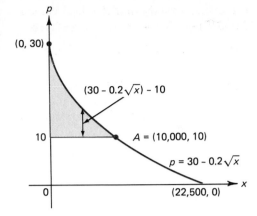

Figure 12

SOLUTION. Look at Figure 12. The consumers' surplus is the shaded area of the graph. To calculate this area, we need to find the value of x which corresponds to $p = 10$ (the point A). We set $30 - 0.2\sqrt{x} = 10$ and find that $x = 10{,}000$. Therefore

$$CS = \int_{0}^{10{,}000} [(30 - 0.2\sqrt{x}) - 10]\, dx = \left(20x - \frac{2}{15}x^{3/2}\right)\Big|_{0}^{10{,}000} = \$66{,}666.67.$$

Suppose now that we are given a *supply function* $s = s(x)$. This function gives the relationship between the expected price of a product and the number of units the manufacturer will produce. It is reasonable to assume that as the expected price increases, the number of units the manufacturer will produce also increases so that $s(x)$ is an increasing function. One supply function is sketched in Figure 13. The amount $s_{\min}$ is the minimum price that must be paid before the manufacturer will produce anything. It is related to the manufacturer's fixed cost (or overhead). If items

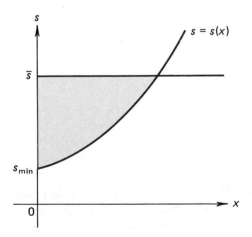

Figure 13

are sold for $\$\bar{s}$ per unit and if $\bar{s} > s_{\min}$, then the manufacturer will earn more money initially than if he had sold at his expected price per unit. This gain (at least on paper) is called *producer's surplus*. It is represented, graphically, by the shaded area of Figure 13.

EXAMPLE 8.   Let $s(x) = 250 + 3x = 0.01x^2$. If items are sold for \$425 each, calculate the producer's surplus.

SOLUTION. The producer's surplus is the shaded region of Figure 14. When $\bar{s} = \$425$, $425 - 250 + 3x + 0.01x^2$

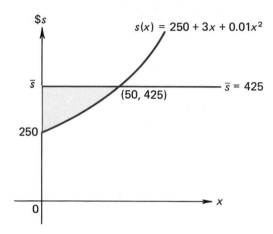

Figure 14

which has the positive solution $x = 50$. Then the calculation of the shaded area in the figure gives us

$$PS = \int_0^{50} [425 - (250 + 3x + 0.01x^2)] \, dx = \int_0^{50} (175 - 3x - 0.01x^2) \, dx$$

$$= \left(175x - \frac{3x^2}{2} - \frac{0.01x^3}{3}\right)\Big|_0^{50} = \$4583.33$$

If we put these two ideas together, we may define *pure competition* as the situation that obtains when the price is set so that supply equals demand. Typically, we have a picture like the one in Figure 15. We can see, graphically, that in a pure market competitive system there is a level of production which guarantees "profit" to the consumer as well as the producer.

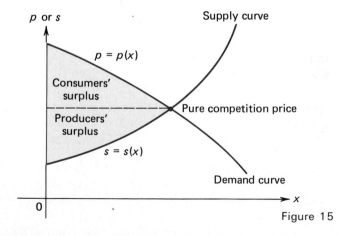

Figure 15

PROBLEMS 5.6

In Problems 1–20 calculate the area bounded by the given curves and lines. If possible, find the areas by integration with respect to both x and y.

1. $y = x^2, y = x$
2. $y = x^2, y = x^3$
3. $y = x^2, y = x^3, x = 3$
4. $y = 2x^2 + 3x + 5,$ $y = x^2 + 3x + 6$
5. $y = 3x^2 + 6x + 8, y = 2x^2 + 9x + 18$
6. $y = x^2 - 7x + 3,$ $y = -x^2 - 4x + 5$
*7. $y = 2x, y = \sqrt{4x - 24}, y = 0, y = 10$
8. $y = x^3, y = x^3 + x^2 + 6x + 5$
9. $y = x^4 + x - 81, y = x$
10. $xy^2 = 1, x = 5, y = 5$
11. $\sqrt{x} + \sqrt{y} = 4, x = 0, y = 0$
12. $x + y^2 = 8, x + y = 2$
13. $x = y^2, x^2 = 6 - 5y$
*14. $xy^2 = 1, y = 3 - 2\sqrt{x}$
15. $y = x^{1/3}, y = x$
16. $y = \dfrac{2}{x^2} - 1, y = x^2, x\text{-axis}$
17. $x = y^3 - y^2, x = 5y^2$
18. $y = x^3 + x + 1, y = x^2 + x + 1$
19. $y = x + 6, y = 2x - 2, y = 3x + 4$
20. $y = ax + b,$ $y = ax^2, y\text{-axis}; a, b > 0$

21. Find the area of the triangle with vertices at $(1, 6), (2, 4), (-3, 7)$. [*Hint:* Find the equations of the straight lines forming the sides and draw a sketch.]
22. Find the area of the triangle with vertices at $(2, 0), (3, 2), (6, 7)$.
23. Find the area of the triangle with sides given by $x + y = 3, 2x - y = 4, 6x + 3y = 3$.
24. Find the area of the triangle with sides given by $y = x - 3, y = 2x + 4, y = -3x + 8$.
25. If the demand function is given by $p = 10 - 0.01x$, find the consumers' surplus if the product is sold for $5.
*26. If the demand function is $p(x) = 5 + (180/\sqrt{1 + x})$ and the product is sold for $10, find the consumers' surplus.
27. If the supply function for a certain product is $s(x) = 400 + 2x + 0.02x^2$ and the product sells for $462.50, find the producer's surplus.
28. Given the demand function $p(x) = 175 - x$ and the supply function $s(x) = 50 + x + 0.01x^2$ for a certain product
    (a) calculate the pure competition price.
    (b) calculate the consumers' and producers' surplus at that price.
29. Answer the questions of Problem 28 for the demand function $p = 30 - \sqrt{x}$ and the supply function $s = 10 + 3\sqrt{x}$.

*5.7 Work, Power, and Energy

Consider a particle acted on by a force. In the simplest case the force is constant and the particle moves in a straight line in the direction of the force (see Figure 1). In this

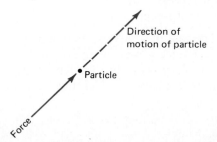

Figure 1

situation we define the *work W done by the force on the particle* as the product of the magnitude of the force F and the distance s through which the particle travels:

$$W = Fs.$$ (1)

One unit of work is the work done by a unit force in moving a body a unit distance in the direction of the force. In the metric system the unit of work is 1 newton†-meter, called 1 *joule*. In the British system, the unit of work is the foot-pound.‡

EXAMPLE 1.   A block of mass 10.0 kg is raised 5 m off the ground. How much work is done to accomplish this?

SOLUTION. Force = (mass) × (acceleration). The acceleration here is opposing acceleration due to gravity and is therefore equal to 9.81 m/sec². We have $F = ma = (10 \text{ kg})(9.81 \text{ m/sec}^2) = 98.1 \text{ nt}$.   Therefore, $W = Fs = (98.1) \times 5 \text{ m} = 490.5$ joules.

EXAMPLE 2.   How much work is done in lifting a 25-lb weight 5 ft off the ground?

SOLUTION. W (in ft-lb) $= Fs = 25 \text{ lb} \times 5 \text{ ft} = 125$ ft-lb.

Since we will be dealing with mass and force in many problems in this text, we take a moment here to explain units of weight, mass, and force in both the English and metric systems.

In the English system we start with the unit of force which is the *pound*. (This is also the standard unit of weight.) Since $F = mg$ by Newton's second law of motion, we obtain $m = F/g$. In the English system the unit of mass is the *slug* and a 1-lb weight (force) has a mass of $(1/32.2) \text{ lb}/(\text{ft/sec}^2) = 1$ slug. Or we can equivalently define a slug as the mass of an object whose acceleration is 1 ft/sec² when it is subjected to a force of 1 lb. This explains why, in Example 2, we did not multiply the weight of 25 lb by the acceleration to obtain the force. The weight (in pounds) was *already* given as a force.

In the metric system, we start with the standard unit of mass which is the *kilogram*. This is also the standard unit of weight. In this case, in order to obtain the force, it is necessary to multiply the mass by the acceleration:

$$F \text{ (in newtons)} = [\text{mass or weight (in kilograms)}] \times g(= 9.81 \text{ m/sec}^2).$$

In sum, we have

The pound is a weight and a force, but not a mass.
The kilogram is a weight and a mass, but not a force.

We now return to the calculation of work. The case in which the motion of the particle is not in a straight line in the direction of the force is more complicated than the case we discussed in Examples 1 and 2. We will discuss this in Chapters 15, 16, and 17. However, we can use the integral to handle the case in which the force is variable.

---

† 1 newton is the force that will accelerate a 1 kg mass at the rate of 1 m/sec². 1 nt = 0.2248 lb.
‡ 1 joule = 0.7376 ft-lb or 1 ft-lb = 1.356 joules.

We wish to calculate how much work is done in moving from the point $x = a$ to the point $x = b$. As usual, we partition the interval into n subintervals, a typical interval being $[x_{i-1}, x_i]$. Now, if the force is given by $F(x)$ (which is assumed to be continuous), then over a very small interval, the force will be almost constant. Therefore, if x_i^* is a point in the interval $[x_{i-1}, x_i]$, then the work done in moving from x_{i-1} to x_i is, approximately,

$$\Delta W \approx F(x_i^*)(x_i - x_{i-1}) = F(x_i^*)\,\Delta x_i. \tag{2}$$

Adding the work done over these small intervals and taking the limit as $|P| \to 0$ leads to the formula

$$W = \lim_{|P|\to 0} [F(x_1^*)\,\Delta x_1 + F(x_2^*)\,\Delta x_2 + \cdots + F(x_n^*)\,\Delta x_n] = \int_a^b F(x)\, dx. \tag{3}$$

EXAMPLE 3. A spring is stretched 10 cm by a force of 2 nt. How much work is needed to stretch it 50 cm $= \frac{1}{2}$ m (see Figure 2) assuming that the spring satisfies Hooke's law?

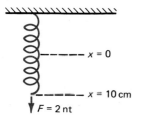

$x = 0$

$x = 10$ cm

$F = 2$ nt

Figure 2

SOLUTION. *Hooke's law*† states that the force needed to stretch a spring is proportional to the amount of the spring displaced: that is,

$$F(x) = kx \tag{4}$$

where k is a constant of proportionality, called the *spring constant,* and x is the amount displaced. From the information given, we have $2 \text{ nt} = k(0.1 \text{ m})$ or $k = 2/0.1 = 20$. Therefore the force is given by $F(x) = 20x$. Then

$$W = \int_0^{1/2} F(x)\, dx = \int_0^{1/2} 20x\, dx = 10x^2 \Big|_0^{1/2} = \frac{10}{4} = \frac{5}{2} \text{ joules.}$$

We should point out that Hooke's law is only valid if the spring is not stretched too far. After a certain point the spring will become stretched out of shape and Hooke's law will fail to hold.

EXAMPLE 4. A cylindrical tank 12-feet in diameter and 20-feet high is filled with water. How much work will be done in pumping out the water?

SOLUTION. Look at Figure 3. We define the variable x as the distance to the top of the tank. We then partition the interval [0, 20] in the usual way. Consider the shaded amount of water. The work needed to get it to the top of the tank is its weight (in

† Named for Robert Hooke (1635–1703) who was an English philosopher, physicist, chemist, and inventor.

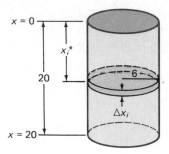

Figure 3

pounds) times the distance to the top. The density of water is 62.4 lb/ft³. The water layer of width Δx_i has a volume given by

$$\text{volume} = (\text{area of cross section}) \times (\Delta x_i) = \pi r^2 \Delta x_i = \pi \left(\frac{12}{2}\right)^2 \Delta x_i = \pi 36 \Delta x_i \quad \text{ft}^3.$$

Therefore the weight of the water in the shaded region is equal to

$$\left(62.4 \frac{\text{lb}}{\text{ft}^3}\right)(36\pi \Delta x_i) \text{ ft}^3 = 2246.4\pi \Delta x_i \text{ lb}.$$

If x_i^* is in the interval $[x_{i-1}, x_i]$, then x_i^* is the approximate distance from the water in the layer to the top of the tank. Hence the work needed to get the water in the ith layer to the top is given by $W_i = 2246.4\pi x_i^* \Delta x_i$ and the total work is

$$W = \int_0^{20} 2246.4\pi x \, dx = 1123.2\pi x^2 \Big|_0^{20} = (1123.2)(400)\pi = 449{,}280\pi \quad \text{ft-lb}.$$

EXAMPLE 5. A storage tank in the shape of an inverted right circular cone has a radius of 4 meters and a height of 8 meters. It is filled to a height of 6 meters with olive oil (density = 920 kg/m³). In order to bottle the olive oil it is first necessary to pump it to the top of the tank. How much work is done in accomplishing this?

SOLUTION. This problem involves a lot of things going on at once. First, we draw a sketch (see Figure 4). It helps if we put in some identifying coordinates. We place the

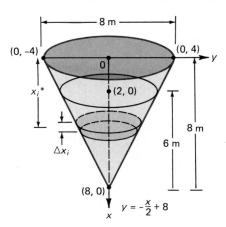

Figure 4

origin 8-m up and in the center of the cone. Then if the x-axis is the vertical axis and the y-axis the horizontal axis, we can label various points on the cone as in the figure. Now the "x-axis" in the cone includes the interval $[0, 8]$ and the olive oil fills the interval $[2, 8]$. We partition that last interval in the usual way and look at a layer of oil in the ith subinterval $[x_{i-1}, x_i]$. The work needed to get this layer of oil to the top is the force needed to overcome gravity times the distance the layer must travel. If Δx_i is small, then this distance is approximately equal to x_i^* for any x_i^* in $[x_{i-1}, x_i]$. What is the force that must be overcome to lift the layer of oil? We have $F = ma$ where $a = 9.81$ m/sec^2. The mass m of the oil in the layer is the volume of the oil times its density $(= 920$ kg/m$^3)$. The volume of the layer is the cross-sectional area times the thickness Δx_i. Thus the only thing left to calculate is the cross-sectional area. This is πy_i^2 where y_i is the radius of the layer. We can think of this radius as varying along the straight line from $(8, 0)$ to $(0, 4)$ (see Figure 4 again). The equation of this straight line is

$$\frac{y - 4}{x - 0} = \frac{4}{-8} \quad \text{or} \quad y = -\frac{x}{2} + 4,$$

Then $y_i \approx -(x_i^*/2) + 4$. Now we put everything together.

$$\text{cross-sectional area of layer} = \pi\left(4 - \frac{x_i^*}{2}\right)^2 \quad \text{m}^2;$$

$$\text{volume of layer} = \pi\left(4 - \frac{x_i^*}{2}\right)^2 \Delta x_i \quad \text{m}^3;$$

$$\text{mass of layer} = 920\pi\left(4 - \frac{x_i^*}{2}\right)^2 \Delta x_i \quad \text{kg};$$

$$\begin{aligned}\text{force to overcome force} \\ \text{of gravity to raise layer}\end{aligned} = (9.81)(920)\pi\left(4 - \frac{x_i^*}{2}\right)^2 \Delta x_i \quad \text{nt}$$

$$= 9025.2\pi\left(4 - \frac{x_i^*}{2}\right)^2 \Delta x_i \quad \text{nt};$$

$$\text{work to raise layer } x_i^* \text{ m} = 9025.2\pi\left(4 - \frac{x_i^*}{2}\right)^2 x_i^* \Delta x_i \quad \text{joules.}$$

Therefore, the total work is given by

$$W = \int_2^8 9025.2\pi\left(4 - \frac{x}{2}\right)^2 x \, dx = 9025.2\pi \int_2^8 \left(16x - 4x^2 + \frac{x^3}{4}\right) dx$$

$$= (9025.2\pi)\left(8x^2 - \frac{4x^3}{3} + \frac{x^4}{16}\Big|_2^8\right) = (9025.2\pi)(63) = 568,587.6\pi \quad \text{joules.}$$

You should go through the steps in this long calculation again. The key to solving the problem is to look at a typical layer and determine, using the definition of work, what it takes to move that layer to the top.

So far we have not considered the time spent in doing work. The work it takes to pump oil out of a tank is the same whether it is done in one minute or in one month.

Sometimes, however, the rate at which work is done is more important than the actual amount of work done. The rate at which work is done is called *power*. If the work is done at a constant rate, then the power produced in t units of time is $P = W/t$. The *instantaneous power* is the rate of change of work with respect to time:

$$P = \frac{dW}{dt}. \tag{5}$$

In the metric system, the unit of power is the *watt*.†

$$1 \text{ watt} = 1 \text{ joule/sec.}$$

In the British system, the unit of power is 1 ft-lb/sec. Since this is very small, another unit called the *horsepower* is often used. We have

$$1 \text{ h.p.} = 550 \text{ ft-lb/sec} = 746 \text{ watts} \approx \tfrac{3}{4} \text{ kilowatt.}$$

Sometimes work is expressed in terms of power × time. (This explains the use of the term kilowatt-hour (kwh) which is the work done by something (such as 20 50-watt light bulbs) working 1 hr at a constant rate of 1 kw.

EXAMPLE 6. A man works at a constant rate and performs 129,600 joules of work in 4 hr. Then his power output

$$P = \frac{\text{work done}}{\text{time in seconds}} = \frac{129,600 \text{ joules}}{4(3600) \text{ sec}} = \frac{129,600}{14,400} = 9 \text{ watts.}$$

EXAMPLE 7. The instantaneous power of a machine is $50 - 3\sqrt{t+4}$ watts. How much work does the machine do in 1 minute?

SOLUTION. Since $P = dW/dt$, we have

$$W = \int_0^{60} P(t)\, dt = \int_0^{60} (50 - 3\sqrt{t+4})\, dt$$

$$= (50t - 2(t+4)^{3/2}) \Big|_0^{60} = 1992 \text{ joules.}$$

(Here the 60 is for the 60 sec in 1 min.)

Let $x(t)$ represent the position of a moving particle at time t and assume that the particle starts at the origin. Then $x'(t) = v(t)$ (velocity) and $x''(t) = a(t)$ (acceleration). If a is constant, then the velocity at time t is given by $v = \int a\, dt = at + c$. But $v(0) = v_0$, the initial velocity, so that $c = v_0$ and $v = at + v_0$ or

$$a = \frac{v - v_0}{t}. \tag{6}$$

Since velocity changes at a constant rate ($a = $ constant), the average velocity of the particle is simply the average of the initial velocity v_0 and the final velocity v. That is

$$v_{av} = \frac{v + v_0}{2}.$$

†Named after James Watt (1736–1819), the inventor of the steam engine.

Therefore the distance the particle travels is $x = v_{av} \cdot t$ or

$$x = \frac{v + v_0}{2} t. \tag{7}$$

Then the work done on the particle in moving from 0 to x is (using (6) and (7) and Newton's second law $F = ma$)

$$W = Fx = max = m \left(\frac{v - v_0}{t} \right) \left(\frac{v + v_0}{2} \right) t = \frac{1}{2} mv^2 - \frac{1}{2} mv_0^2. \tag{8}$$

The product $\frac{1}{2} mv^2$ is called the *kinetic energy* of the particle. It is denoted by the symbol K. We have

$$\boxed{K = \tfrac{1}{2} mv^2.} \tag{9}$$

Equation (8) states that

> *The work done by the force acting on a particle is equal to the change in kinetic energy of the particle.*

This result is known as the *work–energy law*.

We have only shown why the work–energy law is true when a is a constant. If a is not constant, we can still illustrate it as follows: From (3), the work done in moving from x_0 to x_1 is

$$W = \int_{x_0}^{x_1} F\, dx.$$

Now

$$a = \frac{dv}{dt} = \frac{dv}{dx} \frac{dx}{dt} \text{ (by the chain rule)} = \frac{dv}{dx} v = v \frac{dv}{dx}. \tag{10}$$

Then

$$W = \int_{x_0}^{x_1} F\, dx = \int_{x_0}^{x_1} ma\, dx = \int_{x_0}^{x_1} mv \frac{dv}{dx} dx.$$

We know how to integrate $mv\, dv$ ($\int mv\, dv = mv^2/2$ since m is constant). Using a technique we will develop in Section 8.4, we change the variable of integration. When $x = x_0$, $v = v(x_0) = v_0$ and when $x = x_1$, $v = v(x_1) = v_1$. Furthermore,

$$\frac{dv}{dx} dx = dv.$$

Therefore, we have

$$W = \int_{x_0}^{x_1} F\, dx = \int_{x_0}^{x_1} mv \frac{dv}{dx} dx = \int_{v_0}^{v_1} mv\, dv = \frac{1}{2} mv^2 \Big|_{v_0}^{v_1} = \frac{1}{2} mv_1^2 - \frac{1}{2} mv_0^2$$
$$= \text{change in kinetic energy.}$$

EXAMPLE 8. A block weighing 5 kg slides on a horizontal table with a speed of 0.3 m/sec. It is stopped by a spring with a spring constant $k = \frac{1}{4}$ placed in its path. How much is the spring compressed?

SOLUTION. We assume (for simplicity) that until the block stops, its velocity is constant. Then its kinetic energy is the constant $\frac{1}{2}mv^2 = \frac{1}{2}(5)(0.3)^2 = 0.45/2$ joules. The work done in compressing the spring x units is

$$W = \int_0^x kx\,dx = \frac{kx^2}{2} = \frac{x^2}{8},$$

since $k = \frac{1}{4}$. Therefore, equating work to the change in kinetic energy (from $0.45/2$ to 0), we have

$$\frac{x^2}{8} = \frac{0.45}{2} \quad \text{or} \quad x^2 = 1.8 \quad \text{and} \quad x \approx 1.34 \text{ m}.$$

EXAMPLE 9. Neglecting air resistance, find the velocity v_0 at which a missile at the surface of the earth would have to be fired in order to escape the earth's gravitational field.

SOLUTION. The only force the missile must overcome is the force of gravity. Let δ denote the distance from the center of the earth to its surface ($\delta \approx 6378$ km ≈ 3963 miles). The force of gravity is, by Newton's law of gravitational attraction, inversely proportional to the square of the distance u between the missile and the center of the earth. That is $F = \alpha/u^2$, where α is a constant of proportionality. To find α, we observe that when $u = \delta$ (so that the missile is at the surface of the earth), $F = mg$ (where m denotes the mass of the missile). Then $F(\delta) = mg = \alpha/\delta^2$ or $\alpha = mg\delta^2$ and

$$F = mg\frac{\delta^2}{u^2}.$$

The work needed to lift the missile from a distance of δ to a height of x is

$$W = \int_\delta^x F(u)\,du = \int_\delta^x mg\frac{\delta^2}{u^2}\,du = -\frac{mg\delta^2}{u}\Big|_\delta^x = mg\delta^2\left(\frac{1}{\delta} - \frac{1}{x}\right). \tag{11}$$

To escape the earth's gravitational pull, the initial velocity of the missile must be sufficient to thrust the missile to an "infinite" height (once a certain height is reached, the missile keeps "rising" indefinitely). We therefore let $x \to \infty$ in (11) to find that the work needed to allow the missile to escape is given by

$$W = mg\delta^2\frac{1}{\delta} = mg\delta. \tag{12}$$

Now, the missile starts at rest and then moves with a velocity of v_0. The change in kinetic energy is, therefore,

$$\Delta K = \frac{1}{2}mv_0^2. \tag{13}$$

Using the work–energy law, we equate (12) and (13) to obtain

$$v_0 = \sqrt{2g\delta}. \tag{14}$$

This value of v_0 is called the *escape velocity*. Note that the escape velocity is independent of the mass of the missile. In metric units,

$$v_0 = \sqrt{(2)\frac{9.81 \text{ m}}{\text{sec}^2} \times 6378 \times 10^3 \text{ m}} \approx 11{,}186 \quad \text{m/sec} \approx 11.2 \quad \text{km/sec.}$$

In English units,

$$v_0 = \sqrt{(2)32.2 \text{ ft/sec}^2 \times 3963 \times 5280 \text{ ft}} \approx 36{,}709 \quad \text{ft/sec} \approx 6.95 \quad \text{mi/sec.}$$

PROBLEMS 5.7

1. How much work is done in vertically lifting a 5-kg rock 2 meters?
2. How much work is done in lifting an 8-lb weight 6 ft?
3. A force of 5 lb stretches a spring 6 in. How much work is done in stretching the spring 2 ft?
4. A force of 3 nt stretches a spring 50 cm. How much work is done in stretching the spring 3 m?
5. A force of 10 nt stretches a spring 25 cm. How much work is done in stretching the spring $1\frac{1}{2}$ m?
6. How much work is done to stretch the spring in Problem 5 from 1 to 2 m?
7. A force of 25 lb stretches the spring in Problem 3 how many feet?
8. A force of 8 nt stretches the spring in Problem 4 how many meters?
9. A cylindrical tank 10-m high and 10-m in diameter is filled with water. How much work does it take to pump the water out? [*Hint:* The density of water at 0 °C is 1000 kg/m³.]
10. A tank in the shape of an inverted cone has a radius of 3 m and a height of 9 m and is filled with water. How much work is needed to pump the water out?
11. How much work is needed in Problem 9 if only half the water is to be pumped out?
*12. How much work is needed in Problem 10 if only half the water is to be pumped out?
13. The cylindrical tank of Problem 9 is now placed 10-m above the ground. How much work is needed to fill the tank from a hose which runs from ground level to the top of the tank?
14. The conical tank of Problem 10 is placed 8-m above the ground. How much work is needed to fill the tank from a hose which runs from ground level to a hole at the vertex (bottom) of the tank?

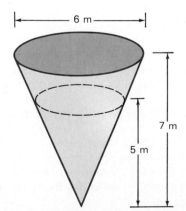

6 m

7 m

5 m

Figure 5

15. Answer Problem 9 if the tank is filled with olive oil instead of water.
16. Answer Problem 10 if the tank is filled with olive oil instead of water.
17. A conical tank (inverted) has a height of 7 m and a diameter of 6 m (see Figure 5). The tank is filled with ethyl alcohol to a level of 5 m. How much work is needed to pump the alcohol out of the tank? [*Hint:* Density of ethyl alcohol $= 810 \text{ kg/m}^3$.]
18. A 10 ft metal chain weighs $1\frac{1}{2}$ lb/ft. The chain is lifted vertically from one end until the other end is 5-ft off the ground. How much work is done?
19. A 40-ft chain which weighs $1\frac{1}{2}$ lb/ft is hanging from the top of a building. How much work is needed to pull 10 ft of it to the top? [*Hint:* Integrate from 30 to 40 since it is the *last* 10 ft which are being pulled to the top.]
20. A 25-m rope weighing $\frac{3}{4}$ kg/m is hanging from the top of a cliff. How much work is needed to pull 7 m of the rope to the top of the cliff?
21. A swimming pool is 20-ft on each side and 6-ft deep. If it is filled with water, how much work is done by the water as it empties out of a hole in the bottom of the pool?
22. How much work is needed in Problem 21 to fill the pool up to a level of 2 ft?
23. How long will it take a 2-h.p. motor to do the work required in Problem 22?
24. A 100-m long chain is attached to a drum. The chain weighs 2 kg/m. If the chain initially does not touch the floor, how much work is done in winding the chain up?
*25. The moon is approximately 380,000 km from the earth. By Newton's law of gravitational attraction, the attraction between two masses is Gm_1m_2/r^2, where G is a universal constant (see Example 3.2.8) and r is the distance between them. If M denotes the mass of the earth, then the mass of the moon is $0.01228M$. In revolving about the earth, a force is created, called *centrifugal force,* which counteracts the force of gravity and keeps the moon in orbit around the earth.

  A popular science fiction series on television portrays the moon, blown out of orbit by nuclear explosions, wandering aimlessly through space. Using the information above, calculate how much work would be required to accomplish this feat. (You can express your answer in terms of G and M, assuming the loss of mass to be negligible, and the final position of the moon to be essentially an infinite distance from the Earth.)
26. A 40-watt light bulb burns for $2\frac{1}{2}$ hours. How much work is performed during that time?
27. The instantaneous power of a machine is $P = 30 + 4t - t^2/2$ watts. How much work is performed in 15 seconds?
28. The instantaneous power of a machine is $P = t + (30/\sqrt{1 + t})$ watts. How much work is performed in the second minute of its operation?
*29. An arrow weighing 0.0322 lb is shot from a 35-lb draw bow at full draw $d = 16$ in. Assume that the force of the bow (in pounds) is proportional to the draw (in inches). At what velocity does the arrow leave the bow?

$$\left[\textit{Hint: } \text{mass in slugs} = \frac{\text{weight in lbs}}{\text{acceleration of gravity}}.\right]$$

30. A block weighing 12 kg slides on a horizontal table with a speed of 4 m/sec. It is stopped by a spring with a spring constant of 2 placed in its path. How much is the spring compressed?
31. The diameter of Mars is 6860 km. The acceleration due to gravity at its surface is 3.92 m/sec². What is the escape velocity from Mars?
32. An object is dropped from a distance x_0 meters above the surface of the earth. Assuming that the force of gravity is constant for small distances above the earth, what is the kinetic energy of the object when it hits the earth? What is its final velocity?
33. From what height would you have to drop an automobile in order that it gain the same kinetic energy it has from going 50 mi/hr?

Review Exercises for Chapter Five

1. A rope is attached to a pulley mounted on top of a 5-m tower. One end of the rope is attached to a heavy mass. If the rope is pulled in at a rate of $1\frac{1}{2}$ m/sec, how fast is the mass approaching the tower when it is 3 m from the base of the tower?

2. A storage tank is in the shape of an inverted right circular cone with radius 6 ft and height 14 ft. If water is pumped into the tank at a rate of 20 gal/min, how fast is the water level rising when it has reached a height of 5 ft? [*Hint:* 1 gal ≈ 0.1337 ft³.]

3. Find the maximum and minimum values for the following functions over the indicated intervals:
   (a) $y = 2x^3 + 9x^2 - 24x + 3$; $[-2, 5]$
   (b) $y = (x - 2)^{1/3}$; $[0, 4]$
   (c) $y = (x + 1)/(x^2 - 4)$; $[-1, 1]$.

4. What is the maximum area that can be enclosed with 800 m of wire fencing?

5. Find the point on the line $2x - 3y = 6$ which is nearest the origin.

6. A cylindrical barrel is to be constructed to hold 128π ft³ of liquid. The cost per square foot of constructing the side of the barrel is three times that of constructing the top and bottom. What are the dimensions of the barrel which cost the least to construct?

7. The population of a certain species is given by $P(t) = 1000 + 800t + 96t^2 + 12t^3 - t^4$, where t is measured in weeks. After how many weeks is the instantaneous rate of population increase a maximum?

8. A producer of dog food finds that the cost in dollars of producing x cans of dog food is $C = 200 + 0.2x + 0.001x^2$. If the cans sell for $0.35 each, how many cans should be produced to maximize the profit?

9. If the cost function for a certain product is $C(x) = 100 + (1000/x)$ and the demand function is $p(x) = 50 - 0.02x$, at what level of production is profit maximized?

10. Calculate the price elasticity of demand for the product of Exercise 9. At what level of production will it make no difference whether prices are increased or decreased?

11. Calculate $\sqrt[6]{135}$ to 4 decimal places using Newton's method.

12. Use Newton's method to find all roots of $x^3 - 2x^2 + 5x - 8 = 0$.

In Exercises 13–16 calculate the areas bounded by the given curves and lines.

13. $y = 3x^3$; $y = x$
14. $x^{1/3} + y^{1/3} = 1$; $x = 0$; $y = 0$
15. $y^2 + x = 0$; $y = 2x + 1$
16. $y = x^3$; $y = x^4$; $x = 1$.

17. A force of 8 nt stretches a spring 30 cm. How much work is done in stretching the spring 60 cm?

18. A tank in the shape of an inverted cone has a radius of 6 ft and a height of 12 ft. It is filled with water. How much work is needed to pump it out?

19. The instantaneous power of a certain machine is given by $P = (t/10) + (t^2/100) + 45\sqrt{t}$ watts, where t is measured in seconds. How much work is done in the first 25 seconds of operation?

20. An object is dropped from a height of 500 m. What is the kinetic energy of the object when it hits the earth? What is its final velocity?

SIX

EXPONENTIALS AND LOGARITHMS

In this chapter we discuss two of the most important functions of mathematics: the exponential and logarithmic functions. Perhaps you are already familiar with these functions through some previous course. Perhaps you will be seeing these functions for the first time. In any event, the first section of this chapter will be devoted to an introductory discussion (or review) of the two functions. In Sections 6.2 and 6.3 we will differentiate and integrate them, and in the remainder of the chapter we will illustrate, through applications, why these functions are so important.

6.1 The Exponential and Logarithmic Functions

Let a be a positive number. Then an *exponential function* is a function of the form

$$f(x) = a^x, \tag{1}$$

where x can be any real number. Before discussing properties of this function, we must explain what we mean by a^x. (We shall assume that $a \neq 1$, for if $a = 1$, then $a^x = 1^x = 1$ for every x.) There are several cases to consider:

CASE 1. $x = n$, a positive integer. Then

$$a^x = a^n = \underbrace{a \cdot a \cdot a \cdots a}_{n \text{ times}}. \tag{2}$$

This is of course the familiar rule for raising a number to the nth power.

CASE 2. $x = -n$, a negative integer. Then

$$a^x = a^{-n} = \frac{1}{a^n}. \tag{3}$$

CASE 3. $x = 1/n$, where n is a positive integer. Then

$$a^x = a^{1/n} = \text{the } n\text{th root of } a, \tag{4}$$

i.e., $a^{1/n} = b$ is equivalent to $a = b^n$. For example, $64^{1/6} = 2$ since $2^6 = 64$.

In the last two cases we see why it is necessary to require that $a > 0$. If $a = 0$, then $a^{-n} = 1/a^n = 1/0$ is not defined. If $a < 0$, then $a^{1/2}, a^{1/4}, a^{1/6}$, etc. are not defined (for example, look at $(-2)^{1/2}$).

CASE 4. $x = -1/n$, n a positive integer. Then

$$a^x = a^{-1/n} = \frac{1}{a^{1/n}}. \tag{5}$$

CASE 5. x is a rational number. Then $x = m/n$ where m and n are integers and

$$a^x = a^{m/n} = (a^{1/n})^m. \tag{6}$$

This takes care of many cases of interest. But how do we interpret expressions like $a^{\sqrt{2}}$ and a^{π}? To do so we make use of a fundamental fact of mathematical analysis that any irrational number can be approximated as closely as desired by a rational number.

CASE 6. x is an irrational number. Then

$$a^x = \lim_{r \to x} a^r, \qquad r \text{ a rational number.} \tag{7}$$

☐ **EXAMPLE 1.**   Find an approximate value for 3^{π}.

SOLUTION. To 6 decimal places, $\pi = 3.141592$. Using a calculator, we arrive at Table 1. We can continue this calculation to the number of digits displayed on our calculating machine. To 6 decimal places, $3^{\pi} = 31.544281$.

TABLE 1

r	3^r
3.0	27.0
3.1	30.1353257
3.14	31.48913565
3.141	31.52374901
3.1415	31.54106996
3.14159	31.54418874
3.141592	31.54425805

The basic properties of exponential functions that we shall need are contained in Theorem 1.

THEOREM 1.   Let $a > 0$ and let x and y be any real numbers. Then

(i) $a^x > 0$.

(ii) $a^{-x} = \dfrac{1}{a^x}$.

(iii) $a^{x+y} = a^x \cdot a^y$.

(iv) $a^{x-y} = a^x \cdot a^{-y} = \dfrac{a^x}{a^y}$.

(v) $a^0 = a^{x-x} = \dfrac{a^x}{a^x} = 1$.

(vi) $(a^x)^y = a^{xy}$.

(vii)   If $a > 1$, then a^x is an increasing function in any interval.
(viii)   If $0 < a < 1$, then a^x is a decreasing function in any interval.

We shall not prove this theorem here. The first 6 properties are familiar if x and y are integers. If x and y are rational numbers, then it is not difficult to show that properties (i)–(vi) hold using the fact that they hold for integers. Then since (i)–(vi) are true for rational numbers, we can prove them for irrational numbers by taking limits. Properties (vii) and (viii) will follow easily once we have calculated the derivative of a^x. However, it is easy to see that these facts are true by examining the cases $a > 1$ and $a < 1$ separately and observing what happens as x grows. (Try $a = 2$ and $a = \frac{1}{2}$.) The graphs of $y = 2^x$ and $y = (\frac{1}{2})^x$ are given in Figure 1.

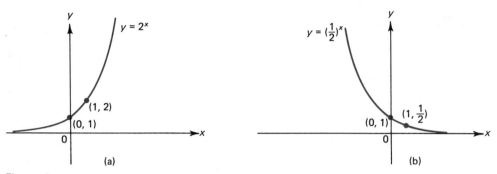

Figure 1

Before continuing, we should note that there is a substantial difference between the function a^x and the function x^a. The first is an exponential function. It is the *exponent* which varies. The second is a *power function* (see Section 3.4). The exponent in this case remains fixed.

If $y = a^x$, we may ask the following question: which value of x corresponds to a given y?

EXAMPLE 2. Solve for x:

(a) $2^x = 8$ (b) $3^x = \frac{1}{9}$ (c) $(\frac{1}{2})^x = 8$.

SOLUTION. (a) It is clear that $2^x = 8$ if and only if $x = 3$. Similarly, in (b) $3^x = \frac{1}{9}$ if and only if $x = -2$ and, in (c), $(\frac{1}{2})^x = 8$ if and only if $x = -3$. $[(\frac{1}{2})^3 = \frac{1}{8}]$.

We now make use of the important fact that an exponential function always increases or always decreases (from Theorem 1). Let $y = a^x$ where $a > 1$ (see Figure 2). For any positive number y, there is a unique x such that $y = a^x$. The same fact holds if $0 < a < 1$. A function which has this property is called *one-to-one*, written 1-1. An increasing or decreasing function is always 1-1. Since there is only one value of x for every $y > 0$, we can make the following important definition.

DEFINITION 1 (The Logarithm to the Base a). If $x = a^y$, then the *logarithm* to the base a of x is y. This is written

$$y = \log_a x. \tag{8}$$

The relationship between the exponential and logarithmic functions is illustrated in Figure 3. We will discuss the graphs of logarithmic functions later but the graph of $\log_a x$ can be immediately obtained by turning the graph of $y = a^x$ on its

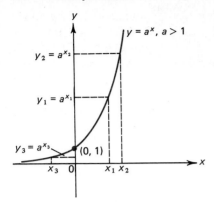

Figure 2

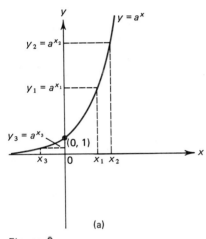 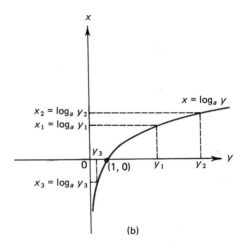

Figure 3

side and flipping it over so as to interchange the positions of the x- and y-axes (see Figure 3). Figure 3b shows a typical graph of the logarithmic function for $a > 1$. The two functions $\log_a x$ and a^x are called *inverse* functions. Each is the inverse of the other. We will discuss inverse functions in detail in Section 10.7.

The logarithmic function $f(x) = \log_a x$ often causes a lot of confusion when it is first encountered. We emphasize that

> (a) $y = \log_a x$ means that $x = a^y$

and

> (b) $y = a^x$ means that $x = \log_a y$.

Think of $y = \log_a x$ as an answer to the question: To what power must a be raised to obtain the number x? This immediately implies that

> (c) $a^{\log_a x} = x$ for every positive real number x

y = aˣ
and *log a y*

x =

(d) $\log_a a^x = x$ for every real number x.

Remember that $\log_a x$ is only defined for $x > 0$ since it is impossible to raise a positive number to a power and obtain a number ≤ 0.

EXAMPLE 3. In Example 2, we found that

(a) $\log_2 8 = 3$,   (b) $\log_3 \tfrac{1}{9} = -2$,   and (c) $\log_{1/2} 8 = -3$.

We will draw graphs of several logarithmic functions in Section 6.2 using properties of their derivatives. Graphs of $y = \log_2 x$ and $y = \log_{1/2} x$ are given in Figure 4.

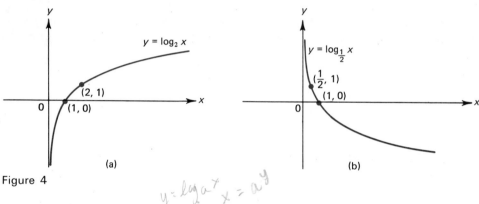

(a)

(b)

Figure 4

y = log aˣ   *x = aʸ*

Properties of the function $\log_a x$ are summarized in the theorem below.

THEOREM 2. If $a > 0$ $(a \neq 1)$ and x and y are positive, then

(i) $\log_a xy = \log_a x + \log_a y$.   (ii) $\log_a(x/y) = \log_a x - \log_a y$.
(iii) $\log_a 1 = 0$.   (iv) $\log_a(1/x) = -\log_a x$.
(v) $\log_a a = 1$.   (vi) $\log_a x^y = y \log_a x$.
        (vii) $\log_a b = 1/(\log_b a)$, $b > 0$, $b \neq 1$.
        (viii) If $a > 1$, $\log_a x$ is an increasing function of x for $x > 0$.
        (ix) If $0 < a < 1$, $\log_a x$ is a decreasing function of x for $x > 0$.
        (x) $f(x) = \log_a x$ is continuous for every $x > 0$.

PROOF. (i) Let $u = \log_a x$ and $v = \log_a y$. Then $a^u = x$ and $a^v = y$ so that $xy = a^u \cdot a^v = a^{u+v}$. By definition of the logarithm, we have $u + v = \log_a xy$ and this proves (i).

(ii) With u and v as above, $x/y = a^u/a^v = a^{u-v}$ so that $u - v = \log_a(x/y)$, which proves (ii).

(iii) Since $a^0 = 1$, $0 = \log_a 1$.

(iv) Follows from (ii) and (iii).

(v) Since $a^1 = a$, $1 = \log_a a$.

(vi) Let $u = \log_a x$. Then $x = a^u$ and $x^y = (a^u)^y = a^{uy}$. Thus $uy = \log_a x^y$ which proves (vi).

(vii)   Let $u = \log_a b$ for $b > 0$. Then $a^u = b$. Now, taking logarithms with respect to b, we have $\log_b a^u = \log_b b = 1$. But $\log_b a^u = u \log_b a = (\log_a b)(\log_b a) = 1$ which proves (vii).

Properties (viii) and (ix) will follow easily once we have calculated the derivative of $\log_a x$.

(x)   A proof of the continuity of $\log_a x$ is suggested in Problem 48.

While we have discussed logarithms to any positive base, there are only two bases which are commonly used. Logarithms to base 10 are called *common logarithms* and are very useful for computations since our number system is based on powers of 10. For simplicity, we will write $\log_{10} x$ as $\log x$, the base 10 being understood. We will not discuss here the way in which calculations can be simplified using base 10. A discussion of this subject can be found in any advanced algebra or college algebra text. Values of $\log x$ are given in Table A.3.

For the remainder of this book we will be concerned with another base for logarithms, called *natural logarithms*. These are logarithms defined by using the base number e.

DEFINITION 2.   The number e is defined by†

$$e = \lim_{n \to \infty} \left(1 + \frac{1}{n}\right)^n. \tag{9}$$

Before discussing this number, we indicate (but do not prove) that the limit in (9) exists. This limit is different from the ones we encountered in Chapter 2. We can see that the exponent is becoming infinite but that $1 + (1/n)$ is approaching 1. Since 1 to any power is 1, your first guess might be that this limit is equal to 1. Table 2 illustrates what happens as n grows. The last number in the table is correct to 9

TABLE 2

n	$\dfrac{1}{n}$	$1 + \dfrac{1}{n}$	$\left(1 + \dfrac{1}{n}\right)^n$
1	1	2	2
2	0.5	1.5	2.25
5	0.2	1.2	2.48832
10	0.1	1.1	2.59374246
100	0.01	1.01	2.704813829
1000	0.001	1.001	2.716923932
10,000	0.0001	1.0001	2.718145926
100,000	0.00001	1.00001	2.718268237
1,000,000	0.000001	1.000001	2.718280469
1,000,000,000	10^{-9}	$1 + 10^{-9}$	2.718281828

decimal places. For convenience, we will use the value 2.72 or 2.718 as an approximation to e. The number e is the second *transcendental* number we have encountered. The first was π. A transcendental number is an irrational number which cannot be

†The number e was discovered by and is named after the great Swiss mathematician and physicist Leonhard Euler (1707–1783).

written as the root of a polynomial with integer coefficients. (The irrational number $\sqrt{2}$ is not transcendental because it is a root of the polynomial $x^2 - 2 = 0$.)

There are other definitions for e. For example

$$e = \lim_{n \to -\infty} \left(1 + \frac{1}{n}\right)^n. \tag{10}$$

(See Problem 37.) If $n \to \pm\infty$, then $u = 1/n \to 0$. This leads to another definition for e.

$$e = \lim_{u \to 0} (1 + u)^{1/u}. \tag{11}$$

Two other definitions for e will be given in Sections 6.2 and 14.8.

We will see in the next section why the number e is so useful and important. As we have already stated, logarithms to the base e are called *natural logarithms* and have a special notation:

$$\ln x = \log_e x. \tag{12}$$

The symbol $\ln x$ should be read "the natural logarithm of x."

Tables of the exponential functions e^x and e^{-x} are given in Table A.1 and a table of natural logarithms is given in Table A.2 (both at the end of the book).

PROBLEMS 6.1

In Problems 1–32 solve for the unknown variable. (Do not use any tables.)

1. $y = \log_2 4$ **2.** $y = \log_4 16$ **3.** $y = \log_{1/3} 27$

4. $y = \frac{1}{3}\log_7(\frac{1}{7})$ **5.** $y = \pi \log_\pi \left(\frac{1}{\pi^4}\right)$ **6.** $y = \log_{81} 3$

7. $y = \ln e^5$ **8.** $y = \ln \frac{1}{e^{3.7}}$ **9.** $y = \log_{1/4} 2$

10. $y = \log_a a \cdot \log_b b^2 \cdot \log_c c^3$ **11.** $y = \log_6 36 \log_{25} \frac{1}{5}$

12. $64 = x \log_{1/4} 64$ **13.** $2^{x^2} = 64$ **14.** $y = 1.3^{\log_{1.3} 48}$

15. $y = e^{\ln \sqrt{2}}$ **16.** $y = e^{\ln 14.6}$ **17.** $y = e^{\ln e^\pi}$

18. $\log_2 x^4 = 4$ **19.** $\log_x 64 = 3$ **20.** $\log_x 125 = -3$

21. $\log_x 32 = -5$ **22.** $y = \log 0.01$ **23.** $\log x = 10^{-23}$

24. $\log x = 3 \log 2 + 2 \log 3$ **25.** $\log x = 4 \log \frac{1}{2} - 3 \log \frac{1}{3}$

26. $\log x = a \log b + b \log a$ **27.** $\log x = \log 1 + \log 2 + \log 3 + \log 4$

28. $\log x = \log 2 - \log 3 + \log 5 - \log 7$ **29.** $\log x^3 = 2 \log 5 - 3 \log 2$

30. $\log \sqrt{x} = 4 \log 2 - 3 \log \frac{1}{2}$ **31.** $\ln x^2 - \ln x = \ln 18 - \ln 6$

32. $\log_2 x^{3/2} - \log_2 \sqrt{x} = \log_2 9 - \log_2 3$.

In Problems 33–36 simplify the given expressions by using Theorem 2.

33. $\log x^3 \sqrt{x}$ **34.** $\log[\sqrt[3]{x^2 + 3}\,(x^5 - 9)^{1/8}]$

35. $\ln \dfrac{x^5}{(1 + x)^{18}}$ **36.** $\ln \sqrt[5]{\dfrac{(x + 1)x^2}{(x - 12)}}$.

□ **37.**   Using a calculator, calculate $[1 - (1/n)]^{-n}$ for $n = 2, 5, 10, 1000, 10{,}000$ and $1{,}000{,}000$ to illustrate the fact that $\lim_{n \to -\infty}[1 + (1/n)]^n = e$.

****38.**   Show that if $a > 1$, then $\lim_{x \to 0^+} \log_a x = -\infty$ and $\lim_{x \to \infty} \log_a x = \infty$.

****39.**   If $0 < a < 1$, show that $\lim_{x \to 0^+} \log_a x = \infty$ and $\lim_{x \to \infty} \log_a x = -\infty$.

40.   The acidity of a substance is measured by the concentration of the hydrogen ion $[H^+]$ in the substance. This is usually measured in terms of moles/liter. A standard way to describe this is to define the pH of a substance by

$$pH = -\log[H^+].$$

Distilled water has an approximate H^+ concentration of 10^{-7} moles/liter so that its $pH = -\log 10^{-7} = 7$. A substance with a pH under 7 is termed an *acid* while one with a pH above 7 is called a *base*. Determine the pH of the following substances with the indicated hydrogen ion concentrations.

(a) $[H^+] = 4.2 \times 10^{-6}$ (b) $[H^+] = 8 \times 10^{-6}$ (c) $[H^+] = 0.6 \times 10^{-7}$.

41.   Using the table of common logarithms (Table A.3) determine which is greater: 999^{1000} or 1000^{999}.

42.   A general psychophysical relation was established in 1834 by the physiologist Weber[†] and given a more precise phrasing later by Fechner.[‡] By the Weber–Fechner law, $S = c \log(R + d)$, where S is the intensity of a sensation, R is the strength of the stimulus producing it, and c and d are constants. The Greek astronomer Ptolemy catalogued stars according to their visual brightness in six categories or *magnitudes*. A star of first magnitude was about $2\frac{1}{2}$ times as bright as a star of the second magnitude which in turn was about 2.5 times as bright as a star of third category, and so on. Let b_n and b_m denote the apparent brightness of two stars having magnitudes n and m, respectively. Then modern astronomers have established the Weber–Fechner law relating the relative brightness to the difference in magnitudes as

$$(m - n) = 2.5 \log\left(\frac{b_n}{b_m}\right).$$

(a)   Using this formula, calculate the ratio of brightness for two stars of the second and fifth magnitudes, respectively.

(b)   If star A is five times as bright to the naked eye as star B, what is the difference in their magnitudes?

(c)   How much brighter is Sirius (magnitude 1.4) than a star of magnitude 21.5?

(d)   The Nova Aquilae in a 2–3 day period in June, 1918, increased in brightness about 45,000 times. How many magnitudes did it rise?

**(e)   The bright star Castor appears to the naked eye as a single star but can be seen with the aid of a telescope to be really two stars whose magnitudes have been calculated to be 1.97 and 2.95. What is the magnitude of the two combined? [*Hint:* Brightnesses, but not magnitudes, can be added.]

43.   The subjective impression of loudness can be described by a Weber–Fechner law. Let I denote the intensity of a sound. The least intense sound that can be heard is $I_0 = 10^{-12}$ watt/m^2 at a frequency of 1000 cycles/second (this is called the *threshold of audibility*). If L denotes the loudness of a sound, measured in decibels,[§] then $L = 10 \log(I/I_0)$.

(a)   If one sound has twice the intensity of another, what is the ratio of the perceived loudness of the two sounds?

†Ernest Weber (1796–1878) was a German physiologist.
‡Gustav Fechner (1801–1887) was a German physicist.
§ 1 decibel (dB) = $\frac{1}{10}$ Bel, named after Alexander Graham Bell (1847–1922), inventor of the telephone.

(b)   If one sound appears to be twice as loud as another, what is the ratio of their intensities?

(c)   Ordinary conversation sounds six times as loud as a low whisper. What is the actual ratio of intensity of their sounds?

44.   Natural logarithms can be calculated on a hand calculator even if the calculator does not have an $\boxed{\ln}$ key. If $\frac{1}{2} \le x \le \frac{3}{2}$ and if $A \equiv (x - 1)/(x + 1)$, then a good approximation to $\ln x$ is given by

$$\ln x \approx \left[\left(\frac{3A^2}{5} + 1\right) \cdot \frac{A^2}{3} + 1\right] 2A, \qquad \frac{1}{2} \le x \le \frac{3}{2}.$$

(a)   Use this formula to calculate $\ln 0.8$ and $\ln 1.2$.

(b)   Using facts about logarithms, use the formula to calculate (approximately) $\ln 2 = \ln\left(\frac{3}{2} \cdot \frac{4}{3}\right)$.

(c)   Using (b), calculate $\ln 3$ and $\ln 8$.

45.   The exponential e^x can be estimated for x in $[-\frac{1}{2}, \frac{1}{2}]$ by the formula

$$e^x \approx \left(\left\{\left[\left(\frac{x}{5} + 1\right)\frac{x}{4} + 1\right]\frac{x}{3} + 1\right\}\frac{x}{2} + 1\right)x + 1.$$

(a)   Calculate an approximate value for $e^{0.13}$.

(b)   Calculate an approximate value for $e^{-0.37}$.

(c)   Calculate an approximate value for $e^{4.13}$.

(d)   Calculate an approximate value for $e^{-2.63}$. [*Hint:* Use part (b).]

(e)   Calculate approximate values for $e^{4.82}$ and $e^{-1.44}$.

46.   Show using logarithms that $a^b = b^a$ where $a = [1 + (1/n)]^n$ and $b = [1 + (1/n)]^{n+1}$. What does this prove in the case $n = 1$?

47.   The quantity $n! = n(n - 1)(n - 2) \cdots 3 \cdot 2 \cdot 1$ grows very rapidly as n increases. According to *Stirling's formula*, when n is large

$$n! \approx \sqrt{2\pi n}\left(\frac{n}{e}\right)^n.$$

Use Stirling's formula to estimate 100! and 200!. [*Hint:* Use common logarithms.]

*48.   Prove that if $a > 0$, then the function $y = \log_a x$ is continuous at every $x > 0$. [*Hint:* Show that $\lim_{\Delta x \to 0} [\log_a(x + \Delta x)] = \log_a x$ by showing that

(1) $\log_a(x + \Delta x) - \log_a x = \log_a \dfrac{x + \Delta x}{x}$;

(2)   if $y_{\Delta x} = \log_a \dfrac{x + \Delta x}{x} = \log_a\left(1 + \dfrac{\Delta x}{x}\right)$, then $a^{y_{\Delta x}} = 1 + \dfrac{\Delta x}{x}$;

(3) $\lim_{\Delta x \to 0} y_{\Delta x} = 0$.]

6.2 The Derivatives and Integrals of $\log_a x$ and a^x

In this section we calculate the derivatives of $\log_a x$ and a^x.

THEOREM 1

$$\boxed{\frac{d}{dx}\log_a x = \frac{1}{x}\log_a e.}$$ (1)

PROOF. If $y = \log_a x$, then

$$\frac{dy}{dx} = \lim_{\Delta x \to 0} \frac{\log_a(x + \Delta x) - \log_a x}{\Delta x} = \lim_{\Delta x \to 0} \left[\frac{1}{\Delta x} \log_a \left(\frac{x + \Delta x}{x} \right) \right]$$

(from Theorem 6.1.2(ii)). We now multiply and divide by x:

$$\lim_{\Delta x \to 0} \left[\frac{1}{\Delta x} \log_a \left(\frac{x + \Delta x}{x} \right) \right] = \lim_{\Delta x \to 0} \left[\frac{x}{x \, \Delta x} \log_a \left(1 + \frac{\Delta x}{x} \right) \right]$$

$$= \frac{1}{x} \lim_{\Delta x \to 0} \log_a \left(1 + \frac{\Delta x}{x} \right)^{x/\Delta x}$$

(from Theorem 6.1.2(vi)). We give a new name to the variable $x/\Delta x$. If $u = \Delta x/x$, then $x/\Delta x = 1/u$ and, for each fixed x, $u \to 0$ as $\Delta x \to 0$ so that

$$\lim_{\Delta x \to 0} \left[\frac{1}{x} \log_a \left(1 + \frac{\Delta x}{x} \right)^{x/\Delta x} \right] = \lim_{u \to 0} \left[\frac{1}{x} \log_a (1 + u)^{1/u} \right]$$

$$= \frac{1}{x} \lim_{u \to 0} [\log_a (1 + u)^{1/u}].$$

It follows from the continuity of $\log_a x$, Theorem 2.4.2, and equation (6.1.11) that

$$\lim_{u \to 0} \log_a (1 + u)^{1/u} = \log_a \lim_{u \to 0} (1 + u)^{1/u} = \log_a e.$$

This completes the proof.

Note that since $\log_a e = 1/\log_e a = 1/\ln a$, (1) can be written

$$\boxed{\frac{d}{dx} \log_a x = \frac{1}{x \ln a}.}$$
(2)

If $a = e$, then (1) becomes

$$\boxed{\frac{d}{dx} \log_e x = \frac{d}{dx} \ln x = \frac{1}{x} \ln e = \frac{1}{x}.}$$
(3)

The differentiation formula (2) immediately proves parts (viii) and (ix) of Theorem 6.1.2. For if $a > 1$, then $\ln a > 0$ so that $1/(x \ln a) > 0$ (remember that x must always be > 0). Thus $\log_a x$ is an increasing function. On the other hand, if $0 < a < 1$, then $\ln a < 0$ so that $1/(x \ln a) < 0$ and $\log_a x$ is a decreasing function. Also,

$$\frac{d^2}{dx^2} \log_a x = - \frac{1}{x^2 \ln a}$$

so that if $a > 1$, $\log_a x$ is concave down while if $a < 1$, $\log_a x$ is concave up. Finally, $\lim_{x \to \infty} 1/(x \ln a) = 0$ so that $\log_a x$ has a horizontal asymptote at ∞ (and a vertical asymptote at 0). We draw some typical graphs in Figure 1.

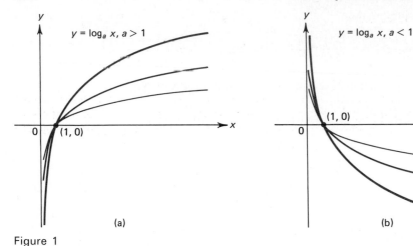

Figure 1

Recall that in Section 4.4 (Theorem 4.4.2) we showed that $\int x^r \, dr = [x^{r+1}/(r+1)] + C$ if $r \neq -1$. We can now handle the case $r = -1$ since we can easily see from (3) that

$$\int \frac{1}{x} \, dx = \ln x + C.$$

(4)

We now turn to the differentiation of the exponential function.

THEOREM 2. If $a > 0$,

$$\frac{d}{dx} a^x = a^x \ln a.$$

(5)

PROOF. If $y = a^x$, then $x = \log_a y$. We now differentiate implicitly with respect to x to obtain

$$1 = \frac{d}{dx} \log_a y.$$

But, using the chain rule

$$\frac{d}{dx} \log_a y = \frac{1}{y \ln a} \frac{dy}{dx}.$$

Hence $dy/dx = y \ln a = a^x \ln a$ and the theorem is proved.

Using (5) we immediately obtain the integration formula

$$\int a^x \, dx = \frac{1}{\ln a} a^x + C.$$

(6)

When $a = e$, (5) becomes

$$\frac{d}{dx} e^x = e^x. \tag{7}$$

This is a remarkable fact. We have shown that *the function e^x is its own derivative!* This is the major reason why the number e is so important. In fact, except for the zero function ($f = 0$), e^x is the only continuous function which has this property. We will exploit the property a great deal in Section 4.

From (7), we immediately obtain

$$\int e^x \, dx = e^x + C. \tag{8}$$

We can now draw the graphs of $y = a^x$. If $a > 1$, then

$$\frac{d}{dx} a^x = a^x \ln a > 0 \qquad \text{and if} \qquad 0 < a < 1, \quad \frac{d}{dx} a^x < 0.$$

Also,

$$\frac{d^2}{dx^2} a^x = \frac{d}{dx} a^x \ln a = (a^x \ln a) \ln a = a^x \ln^2 a > 0.$$

Therefore $\log_a x$ is always concave up. If $0 < a < 1$, then $\lim_{x \to \infty} a^x = 0$ and $\lim_{x \to -\infty} a^x = \infty$. If $a > 1$, then $\lim_{x \to \infty} a^x = \infty$ and $\lim_{x \to -\infty} a^x = 0$. Since a^x is always positive, we obtain the typical graphs drawn in Figure 2.

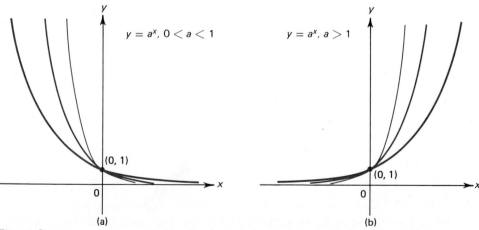

$y = a^x, \, 0 < a < 1$

$y = a^x, \, a > 1$

(0, 1)

(0, 1)

(a)

(b)

Figure 2

We summarize in Table 1 the differentiation and integration formulas obtained in this section.

Let us now look at the graph of $y = 1/t$ drawn in Figure 3. The shaded area under the curve can be written

$$A = \int_1^x \frac{1}{t} \, dt = \ln t \Big|_1^x = \ln x - \ln 1 = \ln x.$$

TABLE 1   Summary of Differentiation and
          Integration Formulas

I. $\dfrac{d}{dx} \log_a x = \dfrac{1}{x \ln a}$.

II. $\dfrac{d}{dx} \ln x = \dfrac{1}{x}$.

[a]III. $\displaystyle\int \log_a x \, dx = x \log_a x - \dfrac{x}{\ln a} + C$.

[a]IV. $\displaystyle\int \ln x \, dx = x \ln x - x + C$.

V. $\displaystyle\int \dfrac{1}{x} \, dx = \ln x + C$.

VI. $\dfrac{d}{dx} a^x = a^x \ln a$.

VII. $\dfrac{d}{dx} e^x = e^x$.

VIII. $\displaystyle\int a^x \, dx = \dfrac{a^x}{\ln a} + C$.

IX. $\displaystyle\int e^x \, dx = e^x + C$.

[a] Formulas III and IV were not proven in this
section but were inserted for completeness.
They can be easily verified by differentiation. When we discuss integration by parts
(in Chapter 8), we will show how these integrals are calculated.

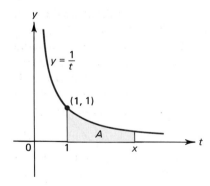

Figure 3

Thus we may think of the function $\ln x$ as representing the area bounded by the curve $y = 1/t$ and the lines $t = 1, t = x$, and the t-axis. In some calculus books the natural logarithm function is *defined* in this way. Then we may define the number e as the number with the property that the area under the curve $y = 1/t$ between $t = 1$ and $t = e$ is 1. (This corresponds to the fact that $\ln e = 1$.)

   We could define the exponential function by $y = \exp x$ if $x = \ln y$. It can then be shown that all the properties of Theorems 6.1.1 and 6.1.2 continue to hold and that $\exp x = e^x$ using the new definition of e. This alternate definition of $\ln x$ avoids the need to discuss a^x or $\log_a x$ for $a \neq e$ and also avoids the definition of e as a limit. However, no matter which definition is used, we arrive at the same functions which can be integrated or differentiated as in Table 1. In the next section we will show how some very complicated functions involving logarithms and exponentials can be differentiated, integrated, and graphed.

There is an interesting application of the natural logarithm function in thermodynamics. *Thermodynamics* is the study of the relationship between heat and other forms of energy. Heat is energy that flows from one body to another because of a temperature difference between them. In the metric system the unit of heat energy is the *calorie*. One calorie is the quantity of heat needed to raise the temperature of 1 gram of water from 14.5 to 15.5 °C. In terms of mechanical energy, it has been calculated that

1 cal = 4.186 joules.

Work, like heat, involves a transfer of energy. We can distinguish between work and heat energy by defining work as energy that passes from one system to another such that a difference of temperature is not involved. To illustrate this difference, rub your hands together vigorously. The heat you feel is simply a conversion from mechanical energy to heat energy.

Consider a gas in a cylindrical container with a movable piston (see Figure 4).

Piston   Wall

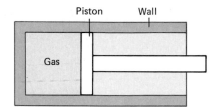

Gas

Figure 4

Heat can flow into or out of the cylinder through the wall and work can be done on the gas by having the piston compress it, or work can be done by the gas by having the gas expand against the piston. Let P and V denote the pressure and volume of the gas. Suppose the gas expands slightly so that the piston moves a distance Δs. Since pressure is given in terms of force per unit area, the force against the piston is $F = P \times A$, where A is the area of the face of the piston (A is also the cross-sectional area of the interior of the cylinder). Then the work done is $F \times \Delta s$ (see Section 5.7) or

$$\Delta W = PA\,\Delta s = P\,dV. \tag{9}$$

If the gas expands from an initial volume V_0 to a final value V_f, then (9) yields

$$W = \int_{V_0}^{V_f} P\,dV. \tag{10}$$

The *ideal gas law* relates pressure, volume, and temperature of an ideal gas (that is, a gas that follows the gas law exactly for all conditions of pressure and temperature. No gas actually does this but some come close). It is given by

$$PV = nRT, \tag{11}$$

where T is the temperature, n is the number of moles of the gas, and R is a constant.†

†$R = 8.314$ joules/(mole)(K°) $= 1.987$ cal/(mole)(K°).

Returning to our gas–piston system, suppose that the gas expands with no change in the temperature (this is called an *isothermal expansion*). The *first law of thermodynamics* states that the change in energy ΔE of any system is given by

$$\Delta E = Q + W, \tag{12}$$

where Q is the heat absorbed by the system and W is the work done on the system (by convention W is written as negative). In our system, no energy is being added (since the temperature does not change) so that $\Delta E = 0$ and, therefore

$$Q = -W.$$

To calculate W, we use equations (10) and (11) and the fact that the work is negative.

$$W = -\int_{V_0}^{V_f} P\, dV = -\int_{V_0}^{V_f} \frac{nRT}{V}\, dV = -nRT \int_{V_0}^{V_f} \frac{dV}{V} \qquad \text{(since T is constant)}$$

$$= nRT \ln V \Big|_{V_0}^{V_f} = -nRT(\ln V_f - \ln V_0) = -nRT \ln \frac{V_f}{V_0}. \tag{13}$$

EXAMPLE 1. How much heat is absorbed by one mole of an ideal gas which doubles in volume at a constant temperature of $0\ ^\circ\text{C}$ ($= 273\ ^\circ\text{K}$)?

SOLUTION. Here $n = 1$ and $T = 273\ ^\circ\text{K}$. If V_0 denotes the initial volume, then $V_f = 2V_0$ and

$$W = -273R \ln \frac{2V_0}{V_0}$$

$$= -273(1.987)(\ln 2) \approx -376 \quad \text{cal}$$

$$= (-273)(8.314)(\ln 2) \approx -1573 \quad \text{joules.}$$

Then the heat energy Q absorbed by the system is $Q = -W = 376\ \text{cal} = 1573$ joules.

There are many other applications of the logarithmic function in thermo-dynamics, as you will find by consulting any standard chemistry textbook.

PROBLEMS 6.2

In Problems 1–8 calculate the given definite integral.

1. $\displaystyle\int_1^4 2^x\, dx$
2. $\displaystyle\int_{-1}^3 \left(\frac{1}{2}\right)^x dx$
3. $\displaystyle\int_1^{e^3} \frac{1}{x}\, dx$

4. $\displaystyle\int_1^e \ln x\, dx$
5. $\displaystyle\int_{\log_3(5\ln 3)}^{\log_3(7\ln 3)} 3^x\, dx$
6. $\displaystyle\int_1^{\ln 4} e^x\, dx$

7. $\displaystyle\int_{\ln 8}^{\ln 23} e^x\, dx$
8. $\displaystyle\int_{\log_{1/2}(-8\ln\frac{1}{4})}^{\log_{1/2}(-3\ln\frac{1}{4})} \left(\frac{1}{4}\right)^x dx.$

In Problems 9–24 draw a graph of the given function.

9. $y = \ln|x|$
10. $y = \ln(x + 1)$
11. $y = 3\ln(x - 4)$

12. $y = e^{-x}$
13. $y = \ln(-x)$
14. $y = -2e^{(2+x)}$

15. $y = e^{3x}$
16. $y = \ln x^4$
17. $y = 3^{2x}$

18. $y = 3 \log_3(x + 1)$ **19.** $y = \left(\dfrac{1}{2}\right)^{-x}$ **20.** $y = \log_{1/2} \dfrac{x}{2}$

21. $y = 10^{x+3}$ **22.** $y = \log x^{17/4}$ **23.** $y = 1 - \ln \sqrt{x}$

24. $y = 7(\log_7 x) - 7.$

*25.   For $x > 0$, express $\displaystyle\int_{-1}^{-x} (1/t)\, dt$ in terms of the natural logarithm. [*Hint:* Draw a picture.]

*26.   Do the same for $\displaystyle\int_{2}^{x} [1/(t - 1)]\, dt$ for $x > 1$. [Same hint.]

*27.   Calculate

$$\lim_{n \to \infty} \left(\frac{1}{n+1} + \frac{1}{n+2} + \frac{1}{n+3} + \cdots + \frac{1}{2n} \right).$$

[*Hint:* Partition the interval $[1, 2]$ into n subintervals. Then this limit represents the integral of some function over the interval $[1, 2]$.]

28.   Calculate the heat absorbed by 2 moles of an ideal gas expanding isothermally at a temperature of 373 °K when the volume is tripled.

29.   Suppose that 500 cal are added to the system depicted in Figure 4 (containing 1 mole) by an external heating device and that the volume of the gas doubles without any change in its temperature of 300 °K. How much work is done on the piston?

*30.   If 500 cal are externally added to the system of Figure 4 (containing 1 mole of gas) and if the temperature of the gas remains at 273 °K, what must be the ratio of V_f to V_0 in order that no heat energy be absorbed by the system?

6.3 The Differentiation and Integration of Exponential and Logarithmic Functions

Once we know how to differentiate the basic functions $\log_a x$ and a^x we can use the chain rule and implicit differentiation to greatly expand the number of new functions we can differentiate and integrate. We will illustrate this in this section with a number of examples. Since for the rest of the book we will only be concerned with the base e, we will use that base in all our examples.

EXAMPLE 1.   Differentiate $y = \ln(x^3 + 3x + 1)$.

SOLUTION. Using the chain rule, we set $u = x^3 + 3x + 1$. Then

$$y = \ln u \quad \text{and} \quad \frac{dy}{dx} = \frac{dy}{du}\frac{du}{dx} = \frac{1}{u}(3x^2 + 3) = \frac{3x^2 + 3}{x^3 + 3x + 1}.$$

(Note: $\ln(x^3 + 3x + 1)$ is only defined when $x^3 + 3x + 1 > 0$.)

EXAMPLE 2.   Differentiate $y = \ln(1 + e^x)$.

SOLUTION. Letting $u = 1 + e^x$, we have

$$\frac{dy}{dx} = \frac{dy}{du}\frac{du}{dx} = \frac{1}{u}e^x = \frac{e^x}{1 + e^x}.$$

Generalizing the last two examples, we obtain the rule

$$\frac{d}{dx}\ln u = \frac{1}{u}\frac{du}{dx}. \tag{1}$$

We must always remember that $\ln u$ is only defined if $u > 0$. Now consider

$$\frac{d}{dx}\ln|u|, \qquad \text{for } u \neq 0.$$

If $u \neq 0$, then $|u| > 0$ and $\ln|u|$ is defined. There are two cases to consider:

CASE 1. $u > 0$. Then

$$|u| = u \qquad \text{and} \qquad \frac{d}{dx}\ln|u| = \frac{d}{dx}\ln u = \frac{1}{u}\frac{du}{dx}.$$

CASE 2. $u < 0$. Then

$$|u| = -u \qquad \text{and} \qquad \frac{d}{dx}\ln|u| = \frac{d}{dx}\ln(-u)$$

$$= \frac{1}{-u}\frac{d(-u)}{dx} = -\frac{1}{u}\left(-\frac{du}{dx}\right) = \frac{1}{u}\frac{du}{dx}.$$

Thus, in either case we have

$$\boxed{\frac{d}{dx}\ln|u| = \frac{1}{u}\frac{du}{dx}, \qquad u \neq 0.} \tag{2}$$

This fact is useful for now we only have to worry about zero values of u. Negative values of u pose no problem.

Using (2), we immediately obtain

$$\boxed{\int \frac{du}{u} = \ln|u| + C.} \tag{3}$$

Otherwise put, *if the numerator is the derivative of the denominator then the integral of the fraction is the natural logarithm of the absolute value of the denominator.*

EXAMPLE 3. Calculate $\int \dfrac{2x}{x^2+1}\,dx$.

SOLUTION. Since $2x$ is the derivative of $x^2 + 1$,

$$\int \frac{2x}{x^2+1}\,dx = \ln|x^2+1| + C = \ln(x^2+1) + C$$

(since $x^2 + 1$ is always positive).

EXAMPLE 4. Calculate $\int \dfrac{1}{x \ln x}\,dx$.

SOLUTION. We write the integral as $\int [(1/x)/(\ln x)]\, dx$. Then the numerator is the derivative of the denominator and

$$\int \frac{1/x}{\ln x}\, dx = \ln |\ln x| + C.$$

EXAMPLE 5. Calculate

$$\int \frac{x^3 + 1}{x^4 + 4x}\, dx.$$

SOLUTION. Here the numerator is not the derivative of the denominator. The derivative of the denominator is $4x^3 + 4 = 4(x^3 + 1)$. Thus we need to multiply and divide by 4 to obtain

$$\int \frac{x^3 + 1}{x^4 + 4x}\, dx = \frac{1}{4} \int \frac{4x^3 + 4}{x^4 + 4x}\, dx = \frac{1}{4} \ln |x^4 + 4x| + C.$$

EXAMPLE 6. Differentiate $y = e^{2+3x}$.

SOLUTION. If we let $u = 2 + 3x$, then $y = e^u$ and, using the chain rule, we obtain

$$\frac{dy}{dx} = \frac{dy}{du}\frac{du}{dx} = e^u \cdot 3 = 3e^{2+3x}.$$

EXAMPLE 7. Differentiate $y = e^{\sqrt{(x+1)/(x-1)}}$.

SOLUTION. Setting $u = \sqrt{(x + 1)/(x - 1)}$, we obtain

$$\frac{dy}{dx} = \frac{dy}{du}\frac{du}{dx} = e^u \frac{d}{dx}\sqrt{\frac{x + 1}{x - 1}} = e^{\sqrt{(x+1)/(x-1)}} \cdot \frac{1}{2}\left(\frac{x + 1}{x - 1}\right)^{-1/2} \frac{d}{dx}\left(\frac{x + 1}{x - 1}\right)$$

$$= \frac{e^{\sqrt{(x+1)/(x-1)}}}{2\sqrt{\dfrac{x + 1}{x - 1}}} \cdot \frac{(x - 1) - (x + 1)}{(x - 1)^2} = \frac{-e^{\sqrt{(x+1)/(x-1)}}}{\left(\sqrt{\dfrac{x + 1}{x - 1}}\right)(x - 1)^2}.$$

We now generalize Examples 6 and 7 to see that

$$\boxed{\frac{d}{dx}e^u = e^u \frac{du}{dx}.} \tag{4}$$

If we integrate (4) we get the rule

$$\boxed{\int e^u\, du = e^u + C.} \tag{5}$$

That is, *the integral of e raised to a power times the derivative of the power is simply e to that power.*

EXAMPLE 8. Calculate $\int e^{x^3} \cdot 3x^2\, dx$.

SOLUTION. If $u = x^3$, then $du = 3x^2\, dx$ so that

$$\int e^{x^3} \cdot 3x^2\, dx = \int e^u\, du = e^u + C = e^{x^3} + C.$$

EXAMPLE 9. Calculate $\int e^{-2x}\,dx$.

SOLUTION. If $u = -2x$, then $du = -2\,dx$ so we multiply and divide by -2 to obtain

$$\int e^{-2x}\,dx = -\tfrac{1}{2}\int e^{-2x}(-2)\,dx = -\tfrac{1}{2}\int e^u\,du = -\tfrac{1}{2}e^u + C = -\tfrac{1}{2}e^{-2x} + C.$$

EXAMPLE 10. Calculate

$$\int \frac{e^{\sqrt{x}}}{\sqrt{x}}\,dx.$$

SOLUTION. If $u = \sqrt{x}$, then $du = (1/2\sqrt{x})\,dx$ and we multiply and divide by $\tfrac{1}{2}$ to obtain

$$\int \frac{e^{\sqrt{x}}}{\sqrt{x}}\,dx = 2\int \frac{e^{\sqrt{x}}}{2\sqrt{x}}\,dx = 2\int e^u\,du = 2e^u + C = 2e^{\sqrt{x}} + C.$$

We can make use of our ability to integrate and differentiate certain logarithmic and exponential functions to do all the things we have done in earlier chapters: namely, sketch curves, find the area between curves, find maxima and minima and so on.

EXAMPLE 11. Sketch the curve $y = (\ln^2 x)/x$. [*Note:* $\ln^2 x = (\ln x)^2$.]

SOLUTION. We differentiate to obtain

$$\frac{dy}{dx} = \frac{x \cdot \dfrac{d}{dx}(\ln^2 x) - \ln^2 x}{x^2} = \frac{x \cdot 2(\ln x)\dfrac{1}{x} - \ln^2 x}{x^2} = \frac{2\ln x - \ln^2 x}{x^2}.$$

The function and its derivative are only defined for $x > 0$. Now $dy/dx = 0$ when $2\ln x - \ln^2 x = \ln x\,(2 - \ln x) = 0$. This occurs when $\ln x = 0$ and when $\ln x = 2$. When $\ln x = 0$, $x = 1$. When $\ln x = 2$, $x = e^2 \approx 7.4$. If $0 < x < 1$, then $dy/dx < 0$ and y is decreasing. For $1 < x < e^2$, $dy/dx > 0$ and y is increasing. For $x > e^2$, y is again decreasing. Now

$$\frac{d^2y}{dx^2} = \frac{x^2 \dfrac{d}{dx}(2\ln x - \ln^2 x) - (2\ln x - \ln^2 x)(2x)}{x^4}$$

$$= \frac{x^2\left(\dfrac{2}{x} - \dfrac{2\ln x}{x}\right) - 4x\ln x + 2x\ln^2 x}{x^4}$$

$$= \frac{2x - 6x\ln x + 2x\ln^2 x}{x^4} = \frac{2 - 6\ln x + 2\ln^2 x}{x^3}.$$

If $x = 1$, $\ln x = 0$ and $d^2y/dx^2 = 2$. If $x = e^2$, $\ln x = 2$ and $d^2y/dx^2 = -2/e^6$. Therefore there is a local minimum at $(1, 0)$ and a local maximum at $(e^2, 4/e^2) \approx (7.4, .54)$. To find the points of inflection, it is necessary to solve the equation $2 - 6\ln x + 2\ln^2 x = 0$. This is a quadratic equation in $\ln x$. We have

$$\ln^2 x - 3\ln x + 1 = 0 \qquad \text{or} \qquad \ln x = \frac{3 \pm \sqrt{9 - 4}}{2} = \frac{3 \pm \sqrt{5}}{2}.$$

If $\ln x = (3 + \sqrt{5})/2 \approx 2.62$, then $x \approx e^{2.62} \approx 13.74$. If $\ln x = (3 - \sqrt{5})/2 \approx 0.38$, then $x \approx e^{0.38} \approx 1.46$. These give the two points of inflection. If $x < 1.46$, then the curve is concave up. For $1.46 < x < 13.74$ the curve is concave down and for $x > 13.74$ the curve is again concave up. Finally, since $\ln x$ seems to increase much more slowly than x itself, it is reasonable to believe that

$$\lim_{x \to 0^+} \frac{\ln^2 x}{x} = \infty \quad \text{and} \quad \lim_{x \to \infty} \frac{\ln^2 x}{x} = 0.$$

We will prove these facts in Chapter 12. Putting all this together we obtain the curve in Figure 1.

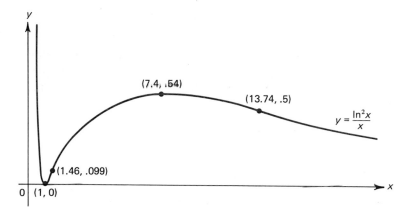

(7.4, .64)

(13.74, .5)

$y = \dfrac{\ln^2 x}{x}$

(1.46, .099)

0 (1, 0)

Figure 1

EXAMPLE 12. Find the area between the curve $y = e^{x+1}$, the y-axis, the curve $y = x^2$, and the line $x = 1$.

SOLUTION. After graphing, we can depict the area to be as in Figure 2. Then

$$A = \int_0^1 (e^{x+1} - x^2)\, dx = \left(e^{x+1} - \frac{x^3}{3}\right)\Big|_0^1 = e^2 - \frac{1}{3} - e \approx 4.34.$$

Before leaving this section we show how the use of logarithms can sometimes simplify the process of differentiation. We need only recall the rules given in Theorem 6.1.2.

EXAMPLE 13. Differentiate $y = \sqrt[4]{\dfrac{x^3 + 1}{x^{7/9}}}$.

SOLUTION. We first take the natural logarithm of both sides:

$$\ln y = \ln\left(\frac{x^3 + 1}{x^{7/9}}\right)^{1/4} = \frac{1}{4}\ln\left(\frac{x^3 + 1}{x^{7/9}}\right) = \frac{1}{4}\ln(x^3 + 1) - \frac{7}{36}\ln x.$$

Now we differentiate implicitly:

$$\frac{1}{y}\frac{dy}{dx} = \frac{3x^2}{4(x^3 + 1)} - \frac{7}{36x}$$

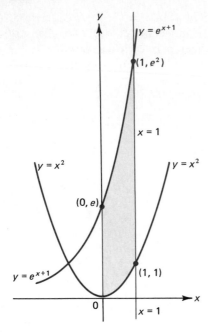

Figure 2

or

$$\frac{dy}{dx} = y\left[\frac{3x^2}{4(x^3 + 1)} - \frac{7}{36x}\right] = \left[\sqrt[4]{\frac{x^3 + 1}{x^{7/9}}}\right]\left[\frac{3x^2}{4(x^3 + 1)} - \frac{7}{36x}\right].$$

Here it was not necessary to use the quotient rule at all. The process which we used to simplify the differentiation is called *logarithmic differentiation*.

EXAMPLE 14. Differentiate $y = x^x$.

SOLUTION. If $y = x^x$, then $\ln y = x \ln x$ and

$$\frac{1}{y}\frac{dy}{dx} = x \cdot \frac{1}{x} + \ln x = 1 + \ln x$$

so that

$$\frac{dy}{dx} = y(1 + \ln x) = x^x(1 + \ln x).$$

PROBLEMS 6.3

In Problems 1–34 perform the indicated integration or differentiation.

1. $\dfrac{d}{dx}(1 - \ln 3x)$

2. $\dfrac{d}{dx}(1 + \ln x + \ln^2 x + \ln^3 x)$

3. $\displaystyle\int \frac{x^3}{x^4 + 1}\,dx$

4. $\displaystyle\int \frac{e^{2x}}{1 + e^{2x}}\,dx$

5. $\displaystyle\int \frac{e^{\sqrt[3]{x}}}{x^{2/3}}\,dx$

6. $\dfrac{d}{dx}(e^{\sqrt[3]{x}} + 4)$

7. $\dfrac{d}{dx}e^{\ln(x^5 + 6)}$ [*Hint:* This is easier than it looks.]

8. $\dfrac{d}{dx} e^{\ln(1+e^{\ln x})}$

9. $\dfrac{d}{dx} x^{\sqrt{x}}$

10. $\dfrac{d}{dx} x^{2(1-x^3)}$

11. $\dfrac{d}{dx} \dfrac{x^2}{3 \ln x}$

12. $\displaystyle\int (1 + x + x^2) e^{2x+x^2+(2x^3/3)} \, dx$

13. $\dfrac{d}{dx} e^{\sqrt{x-1}/x}$

14. $\displaystyle\int \dfrac{e^{1/x}}{x^2} \, dx$

15. $\displaystyle\int \dfrac{\ln(1/x)}{x} \, dx$

16. $\displaystyle\int -\dfrac{\ln x}{x} \, dx$

17. $\displaystyle\int_0^{\ln 2} e^{3x} \, dx$

18. $\displaystyle\int_0^3 \dfrac{x^2}{x^3 + 1} \, dx$

19. $\displaystyle\int_{\ln 3}^{\ln 5} \dfrac{e^x}{e^x + 4} \, dx$

20. $\displaystyle\int_1^2 \dfrac{e^{\sqrt{x}}}{8\sqrt{x}} \, dx$

21. $\dfrac{d}{dx} \ln|x^3 + 3|$

22. $\dfrac{d}{dx} \ln|e^x - 8|$

23. $\dfrac{d}{dx} \ln|1 - \ln x|$

24. $\dfrac{d}{dx} \ln\left|\dfrac{x-2}{x+3}\right|$

25. $\dfrac{d}{dx} \ln \ln x$

26. $\dfrac{d}{dx} \ln \ln \ln x$

27. $\displaystyle\int_1^3 \dfrac{1}{\ln e^x} \, dx$

28. $\displaystyle\int 2^x e^x \, dx$

29. $\displaystyle\int \dfrac{1}{x(\ln x)^5} \, dx$

30. $\displaystyle\int \dfrac{(\ln x)^{5/9}}{x} \, dx$

31. $\displaystyle\int \dfrac{1}{x \ln x \, [\ln(\ln x)]} \, dx$

32. $\dfrac{d}{dx} 2 e^{2e^{2x}}$

33. $\dfrac{d}{dx} e^{x+e^x}$

*34. $\displaystyle\int \dfrac{1}{e^x + e^{-x} + 2} \, dx.$ [*Hint:* Multiply numerator and denominator by e^x and then simplify.]

In Problems 35–44 sketch the given curve, indicating maxima, minima and points of inflection.

35. $y = e^{x^2}$

36. $y = e^{2x+3}$

37. $y = \ln(1 + e^x)$

38. $y = \ln(1 + x^2)$

39. $y = e^{e^x}$

*40. $y = \dfrac{\ln x}{x}$

41. $y = x \ln x$

42. $y = \ln(\ln(x + 2))$

*43. $y = e^x \ln x$

44. $y = x/(\ln^2 x).$

In Problems 45–50 find the areas bounded by the given curves and lines.

45. $y = e^{-x}; \ x = 0, \ x = 1, \ y = 2.$

46. $y = \ln x; \ x = 1; \ x = 2; \ y = 0$

47. $y = \ln x; \ y = e^x; \ x = 0; \ y = 0; \ x = 3.$

48. $y = (\ln x)/x; \ x = 2; \ x = 4; \ y = 0$

49. $y = e^x; \ y = xe^{x^2}; \ x = 0.$

*50. $y = \ln(x + 1); \ y = 3; \ x = 0.$

In Problems 51–54 use logarithmic differentiation to evaluate the derivative.

51. $y = \sqrt[5]{\dfrac{x^3 - 3}{1 + x^2}}$

52. $y = \dfrac{x^5(x^4 - 3)}{x^8 \sqrt{x^5 + 1}}$

53. $y = (\sqrt{x})(\sqrt[3]{x + 2})(\sqrt[5]{x} - 1)$

54. $y = \left[\dfrac{x^3(x + 3)(x^{11/9} + 2)^{1/2}}{(x^5 - 3)\sqrt{x^7 + 2}}\right]^{-9/14}$

55. Show that $\dfrac{d}{dx} a^u = a^u \ln a \dfrac{du}{dx}.$

56. Show that $\dfrac{d}{dx} \log_a u = \dfrac{1}{u} \dfrac{du}{dx} \log_a e.$

In Problems 57–70 use the results of problems 55 and 56 to evaluate the given derivatives and integrals.

57. $\dfrac{d}{dx} 3^{x^2+5}$

58. $\dfrac{d}{dx} \log_2(x^3 + 2x - 3)$

59. $\displaystyle\int \dfrac{\log_5 2x}{x} \, dx$

60. $\displaystyle\int_2^3 \frac{\sqrt{\log_2 x}}{x}\,dx$　　　　**61.** $\displaystyle\int \frac{3\sqrt{x}}{\sqrt{x}}\,dx$　　　　**62.** $\displaystyle\frac{d}{dx}\log_3(1+2^{x^2})$

63. $\displaystyle\int \frac{\log_{1/2} x^3}{x}\,dx$　　　　**64.** $\displaystyle\int_2^4 \left(\frac{1}{5}\right)^{x/2}\,dx$　　　　**65.** $\displaystyle\int \frac{3^{\ln x}}{x}\,dx$

66. $\displaystyle\frac{d}{dx}\log_3\{\log_{10}[\log_5(x+1)]\}$　　　　　　　　**67.** $\displaystyle\int \frac{1}{x\log_{14}^3 x}\,dx$

68. $\displaystyle\int 13^{\log_{13} e^{2x}}\,dx$　　　　**69.** $\displaystyle\frac{d}{dx}\log_8 \frac{x^3}{x^4+3}$　　　　**70.** $\displaystyle\int e^x 7^{-3e^x}\,dx.$

71. The revenue a manufacturer receives when x units of a given product are sold is given by

$$R = 0.50x(e^{-0.001x}).$$

    (a) What is the marginal revenue when $x = 100$ units?
    (b) At what level of sales is revenue maximized?

72. What is the acceleration after 10 minutes of a particle whose equation of motion is $s(t) = 30 + 3t + 0.01t^2 + \ln t + e^{-2t^2}$ (t is measured in minutes)?

73. The density of a 5-m bar is given by $\rho(x) = xe^{-x^2/3}$ kg/m where x is measured in meters from the left.

    (a) At what point is the bar most dense?
    (b) What is the total mass of the bar?

6.4 Simple Differential Equations: Exponential Growth and Decay

In this section we begin to illustrate the great importance of the exponential function e^x in applications. Before citing examples, we will discuss a very basic type of mathematical model.

    Let $y = f(x)$ represent some physical quantity such as the volume of a substance, the population of a certain species, the mass of a decaying radioactive substance, the number of dollars invested in bonds, and so on. Then the growth of $f(x)$ is given by its derivative dy/dx. Thus if $f(x)$ is growing at a constant rate, $dy/dx = k$ and $y = kx + C$; that is, $y = f(x)$ is a straight line function.

    It is sometimes more interesting and more appropriate to consider the *relative rate of growth* defined by

$$\boxed{\begin{array}{c}\text{relative rate}\\ \text{of growth}\end{array} = \frac{\text{actual rate of growth}}{\text{size of } f(x)} = \frac{f'(x)}{f(x)} = \frac{dy/dx}{y}.} \qquad (1)$$

The relative rate of growth indicates the percentage increase or decrease in f. For example, an increase of 100 individuals for a species with a population size of 500 would probably have a significant impact, being an increase of 20 percent. On the other hand, if the population were 1,000,000, then the addition of 100 would hardly be noticed, being an increase of only 0.01 percent.

    In many applications we are told that the relative rate of growth of the given physical quantity is constant. That is,

$$\frac{dy/dx}{y} = \alpha \tag{2}$$

or

$$\frac{dy}{dx} = \alpha y, \tag{3}$$

where α is the constant percentage increase or decrease in the quantity. If $\alpha > 0$ the quantity is increasing while if $\alpha < 0$ it is decreasing. Equation (3) is called a *differential equation* because it is an equation involving a derivative. Differential equations arise in a great variety of settings in the physical, biological, engineering and social sciences. It is a vast subject and many of you will take courses in differential equations after completing your basic calculus sequence. We will encounter differential equations periodically throughout this book and we will discuss a certain class of these equations in Chapter 20.

We return to equation (3). In order to solve it,† we

  (i)   write it in the form $(1/y) \, dy/dx = \alpha$,
  (ii) multiply both sides by dx‡: $(1/y) \, dy = \alpha \, dx$, and
  (iii) integrate both sides of the equation $\int (1/y) \, dy = \int \alpha \, dx$

From the rules of Section 3 we then obtain

$$\ln |y| = \alpha x + C_1. \tag{4}$$

We now use the fact that if $a = b$ then $e^a = e^b$. Then, taking the exponentials of both sides

$$e^{\ln |y|} = e^{\alpha x + C_1} = e^{\alpha x} e^{C_1}.$$

Since $e^{\ln a} = a$ for any value of a, $e^{\ln |y|} = |y|$, and assigning the letter k to the constant e^{C_1} we obtain

$$|y| = k e^{\alpha x} \qquad \text{or} \qquad y = \pm k e^{\alpha x}.$$

The number k can take on any positive real value (since e^{C_1} is positive), so that $\pm k$ can take on any nonzero real number. But $y = 0$ is obviously a solution to (3) also since $0' = \alpha \cdot 0 = 0$. Thus any solution to (3) can be written as

$$\boxed{y = c e^{\alpha x},} \tag{5}$$

---

† By a solution to this differential equation we mean a function f which is differentiable and for which $f'(x) = \alpha f(x)$.

‡ We must keep in mind the fact that the differentials dx and dy are functions (see Section 3.6) and as such can not be arbitrarily "multiplied." However, if we follow these steps without worrying about this technicality, we will obtain the solution to the differential equation.

where c can be any real number. To check this, we simply differentiate:

$$\frac{dy}{dx} = \frac{d}{dx} c e^{\alpha x} = c \frac{d}{dx} e^{\alpha x} = c \alpha e^{\alpha x} = \alpha (c e^{\alpha x}) = \alpha y$$

so that $y = c e^{\alpha x}$ does satisfy (5). We therefore have proven

THEOREM 1. If α is any real number, then there are an infinite number of solutions to the differential equation $y' = \alpha y$ which take the form $y = c e^{\alpha x}$ for any real number α.

If $\alpha > 0$, we say that the quantity described by $f(x)$ is *growing exponentially*. If $\alpha < 0$, it is *decaying exponentially* (see Figure 1). Of course, if $\alpha = 0$, then there is no growth and $f(x)$ remains constant.

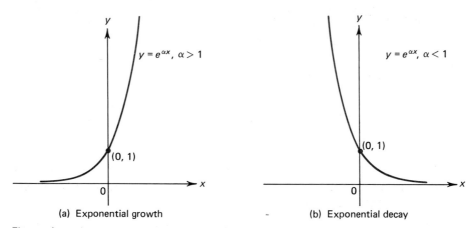

(a) Exponential growth   (b) Exponential decay

Figure 1

For a physical problem it would not make sense to have an infinite number of solutions. We can usually get around this difficulty by specifying the value of y for one particular value of x; say $y(x_0) = y_0$. This is called an *initial condition* and it will give us a unique solution to the problem. We will see this illustrated in the examples that follow.

EXAMPLE 1. (a)   Find all solutions to $dy/dx = 3y$.
   (b)   Find the solution that satisfies the initial condition $y(0) = 2$.

SOLUTION. (a)   Since $\alpha = 3$, all solutions are of the form

$$y = c e^{3x}.$$

   (b) $2 = y(0) = c e^{3 \cdot 0} = c \cdot 1 = c$ so that $c = 2$ and the unique solution is

$$y = 2 e^{3x}.$$

The situation is illustrated in Figure 2. We can see that while there are indeed an infinite number of solutions, there is only one which passes through the point $(0, 2)$.

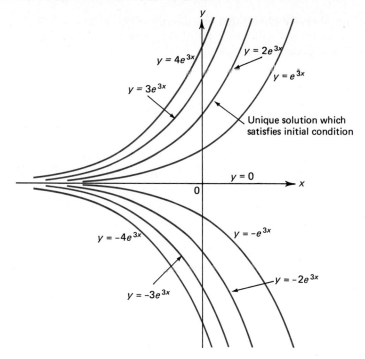

Figure 2

EXAMPLE 2. A bacterial population is growing at a rate equal to 10% of its population each day. Its initial size is 10,000 organisms. How many bacteria are present after 10 days? After 30 days?

SOLUTION. Since the percentage growth of the population is $10\% = 0.1$, we have

$$\frac{dP/dt}{P} = 0.1 \qquad \text{or} \qquad \frac{dP}{dt} = 0.1P.$$

Here $\alpha = 0.1$ and all solutions have the form

$$P(t) = ce^{0.1t}$$

where t is measured in days. Since $P(0) = 10{,}000$, we have

$$ce^{(0.1)(0)} = c = 10{,}000 \quad \text{and} \quad P(t) = 10{,}000e^{0.1t}$$

After 10 days, $P(10) = 10{,}000e^{(0.1)(10)} = 10{,}000e \approx 27{,}183$ and after 30 days $P(30) = 10{,}000e^{0.1(30)} = 10{,}000e^3 = 200{,}855$ bacteria.

EXAMPLE 3. *Newton's law of cooling* states that the rate of change of the temperature difference between an object and its surrounding medium is proportional to the temperature difference. If $D(t)$ denotes this temperature difference at time t and if α denotes the constant of proportionality, then we obtain

$$dD/dt = -\alpha D. \tag{6}$$

The minus sign indicates that this difference decreases. (If the object is cooler than the surrounding medium—usually air—it will warm up; if it is hotter, it will cool.) The solution to this differential equation is

$$D(t) = ce^{-\alpha t}.$$

If we denote the initial ($t = 0$) temperature difference by D_0, then

$$D(t) = D_0 e^{-\alpha t} \tag{7}$$

is the formula for the temperature difference for any $t > 0$. Notice that for t large, $e^{-\alpha t}$ is very small so that, as we have all observed, temperature differences tend to die out rather quickly.

We now may ask: in terms of the constant α, how long does it take for the temperature difference to decrease to half its original value?

SOLUTION. The original value is D_0. We are therefore looking for a value of t for which $D(t) = \frac{1}{2}D_0$. That is, $\frac{1}{2}D_0 = D_0 e^{-\alpha t}$ or $e^{-\alpha t} = \frac{1}{2}$. Taking natural logarithms, we obtain

$$-\alpha t = \ln\frac{1}{2} = -\ln 2 = -0.6931 \quad \text{(from Table A.2)} \quad \text{and} \quad t = \frac{0.6931}{\alpha}.$$

Notice that this value of t does *not* depend on the initial temperature difference D_0.

EXAMPLE 4. With the air temperature equal to 30 °C, an object with an initial temperature of 10 °C warmed to 14 °C in one hour.

   (a)   What was its temperature after 2 hr?
   (b)   After how many hours was its temperature 25 °C?

SOLUTION. Let $T(t)$ denote the temperature of the object. Then $D(t) = 30 - T(t) = D_0 e^{-\alpha t}$ (from (7)). But $D_0 = 30 - T(0) = 30 - 10 = 20$ so that

$$D(t) = 20e^{-\alpha t}.$$

We are given that $T(1) = 14$ so that $D(1) = 30 - T(1) = 16$ and

$$16 = D(1) = 20e^{-\alpha \cdot 1} = 20e^{-\alpha} \quad \text{or} \quad e^{-\alpha} = 0.8$$

and, taking natural logarithms,

$$-\alpha = \ln 0.8 = -0.223.$$

Thus,

$$D(t) = 30 - T(t) = 20e^{-0.223t} \quad \text{and} \quad T(t) = 30 - 20e^{-0.223t}.$$

We can now answer the two questions:

   (a) $T(2) = 30 - 20e^{-(0.223)\cdot 2} = 30 - 20e^{-0.446} \approx 17.2$ °C.
   (b)   We need to find t such that $T(t) = 25$. That is,

$$25 = 30 - 20e^{-(.223)t} \quad \text{or} \quad e^{-0.223t} = \frac{1}{4}, \quad -0.223t = -\ln 4 = -1.3863$$

and

$$t = \frac{1.3863}{0.223} \approx 6.2 \text{ hr} = 6 \text{ hr } 12 \text{ min.}$$

EXAMPLE 5. *Carbon dating* is a technique used by archaeologists, geologists and others who want to estimate the ages of certain artifacts and fossils they uncover. The technique is based on certain properties of the carbon atom. In its natural state, the nucleus of the carbon atom C^{12} has 6 protons and 6 neutrons. An *isotope* of carbon C^{12} is C^{14} which has 2 additional neutrons in its nucleus. C^{14} is *radioactive*. That is, it emits neutrons until it reaches the stable state C^{12}. We make the assumption that the ratio of C^{14} to C^{12} in the atmosphere is constant. This assumption has been shown experimentally to be valid for although C^{14} is being constantly lost through *radioactive decay* (as this process is often termed) new C^{14} is constantly being produced by the cosmic bombardment of nitrogen in the upper atmosphere. Living plants and animals do not distinguish between C^{12} and C^{14} so, at the time of death, the ratio of C^{12} to C^{14} in an organism is the same as the ratio in the atmosphere. However, this ratio changes after death since C^{14} is converted to C^{12} but no further C^{14} is taken in.

   It has been observed that C^{14} decays at a rate proportional to its mass and that its *half-life* is approximately 5580 years.† That is, if a substance starts with 1 gram of C^{14}, then 5580 years later it would have $\frac{1}{2}$ gram of C^{14}, the other $\frac{1}{2}$ g having been converted to C^{12}.

   We may now pose a question typically asked by an archaeologist: A fossil is unearthed and it is determined that the amount of C^{14} present is 40 percent of what it would be for a similarly sized living organism. What is the approximate age of the fossil?

SOLUTION. Let $M(t)$ denote the mass of C^{14} present in the fossil. Then since C^{14} decays at a rate proportional to its mass, we have

$$\frac{dM}{dt} = -\alpha M,$$

where α is the constant of proportionality. Then $M(t) = ce^{-\alpha t}$, where $c = M_0$, the initial amount of C^{14} present. When $t = 0$, $M(0) = M_0$; when $t = 5580$ years, $M(5580) = \frac{1}{2}M_0$, since half the original amount of C^{14} has been converted to C^{12}. We can use this fact to solve for α since we have

$$\tfrac{1}{2}M_0 = M_0 e^{-\alpha \cdot 5580} \qquad \text{or} \qquad e^{-5580\alpha} = \tfrac{1}{2}$$

and, taking natural logarithms,

$$-5580\alpha = -\ln 2 = -0.6931$$

which yields

$$\alpha = \frac{0.6931}{5580} \approx 0.0001242.$$

---

†This number was first determined by the chemist W. S. Libby who based his calculations on the wood from sequoia trees whose ages were determined by rings marking years of growth.

Thus

$$M(t) = M_0 e^{-0.0001242t}.$$

Now, we are told that after t years (from the death of the organism fossilized to the present) $M(t) = 0.4M_0$ and we are asked to determine t. Then

$$0.4M_0 = M_0 e^{-0.0001242t} \qquad \text{or} \qquad 0.4 = e^{-0.0001242t}$$

and again taking natural logarithms we obtain

$$-0.0001242t = \ln 0.4 \approx -0.9163$$

so that

$$t \approx \frac{-0.9163}{-0.0001242} \approx 7378 \quad \text{years.}$$

The carbon dating method has been used successfully on numerous occasions. It was this technique which established that the Dead Sea Scrolls were prepared and buried about 2000 years ago.

PROBLEMS 6.4†

1. The growth rate of a bacteria population is proportional to its size. Initially the population is 10,000, while after 10 days its size is 25,000. What is the population size after 20 days? After 30 days?
2. In Problem 1, suppose instead that the population after 10 days is 6000. What is the population after 20 days? After 30 days?
3. The population of a certain city grows 6% a year. If the population in 1970 was 250,000, what would be the population in 1980? In 2000?
4. When the air temperature is 70 °F, an object cools from 170 °F to 140 °F in $\frac{1}{2}$ hr.
    (a) What will be the temperature after 1 hr?
    (b) When will the temperature be 90 °F? [*Hint:* Use Newton's law of cooling.]
5. A hot coal (temperature 150 °C) is immersed in ice water (temperature -10 °C). After 30 sec the temperature of the coal is 60 °C. Assuming that the ice water is kept at -10 °C,
    (a) What is the temperature of the coal after 2 min?
    (b) When will the temperature of the coal be 0 °C?
**6. The President and Vice-President sit down for coffee. They are both served a cup of hot black coffee (at the same temperature). The President takes a container of cream and immediately adds it to his coffee, stirs it and waits. The Vice-President waits ten minutes and then adds the same amount of cream (which has been kept cool) to his coffee and stirs it in. Then they both drink. Assuming that the temperature of the cream is lower than that of the air, who drinks the hotter coffee? [*Hint:* Use Newton's law of cooling. It is necessary to treat each case separately and to keep track of the volumes of coffee, cream, and the coffee–cream mixture.]‡

† To complete most of these problems it will be necessary either to consult Tables A.1 and A.2 or to use a hand calculator with ln and e^x function keys.

‡ This is a famous old problem that keeps on popping up (with an ever-changing pair of characters) in books on games and puzzles in mathematics. The problem is hard and has stymied many a mathematician. Do not get frustrated if you cannot solve it. The trick is to write everything down and to keep track of all the variables. The fact that the air is warmer than the cream is critical.

7. A fossilized leaf contains 70% of a "normal" amount of C^{14}. How old is the fossil?

8. 40% of a radioactive substance disappears in 100 years.
   (a) What is its half-life?
   (b) After how many years will 90% be gone?

9. Salt decomposes in water into sodium $[Na^+]$ and chloride $[Cl^-]$ ions at a rate proportional to its mass. If there were initially 25 kg of salt, and 15 kg after 10 hr,
   (a) how much salt would be left after 1 day?
   (b) after how many hours would there be less than $\frac{1}{2}$ kg of salt left?

10. X-rays are absorbed into a uniform, partially opaque body as a function not of time but of penetration distance. The rate of change of the intensity $I(x)$ of the X-ray is proportional to the intensity. Here x measures the distance of penetration. The more the X-ray penetrates, the lower the intensity. The constant of proportionality is the density D of the medium being penetrated.
    (a) Formulate a differential equation describing this phenomenon.
    (b) Solve for $I(x)$ in terms of x, D and the initial (surface) intensity $I(0)$.

11. Radioactive beryllium is sometimes used to date fossils found in deep-sea sediment. The decay of beryllium satisfies the equation

$$\frac{dA}{dt} = -\alpha A \qquad \text{where} \qquad \alpha = 1.5 \times 10^{-7}.$$

What is the half-life of beryllium?

12. In a certain medical treatment a tracer dye is injected into the pancreas to measure its function rate. A normally active pancreas will secrete 4% of the dye each minute. 0.3 g of the dye is injected and 30 min later 0.1 g remains. How much dye would remain if the pancreas were functioning normally?

13. Atmospheric pressure is a function of altitude above sea level and is given by $dP/da = \beta P$, where β is a constant. The pressure is measured in millibars. At sea level $(a = 0)$, $P(0)$ is 1013.25 millibars (mb) which means that the atmosphere at sea level will support a column of mercury 1013.25 mm high at a standard temperature of 15 °C. At an altitude of $a = 1500$ m, the pressure is 845.6 mb.
    (a) What is the pressure at $a = 4000$ m?
    (b) What is the pressure at 10 km?
    (c) In California, the highest and lowest points are Mount Whitney (4418 m) and Death Valley (86 m below sea level). What is the difference in their atmospheric pressures?
    (d) What is the atmospheric pressure at Mount Everest (elevation 8848 m)?
    (e) At what elevation is the atmospheric pressure equal to 1 mb?

14. A bacteria population is known to grow exponentially. The following data were collected.

Number of days	Number of bacteria
5	936
10	2190
20	11,986

    (a) What was the initial population?
    (b) If the present growth rate were to continue, what would be the population after 60 days?

15. A bacteria population is declining exponentially. The following data were collected.

Number of hours	Number of bacteria
12	5969
24	3563
48	1269

(a) What was the initial population?
(b) How many bacteria are left after 1 week?
(c) When will there be no bacteria left? (i.e., $P(t) < 1$)

16. The models of population growth that we have so far discussed are not always realistic over long periods of time since an exponentially growing population would overrun the earth if it continued to grow unchecked (although exponential growth is often a reasonable approximation to reality over short periods of time). To model the growth situation more accurately, we assume that the average birth rate β is a positive constant (independent of the population $P(t)$) and the average death rate is proportional to the size of the population and is given by $\delta P(t)$, where δ is another positive constant (see Example 5.3.8). Then the average population growth is the difference between the birth and death rates and is given by

$$\frac{dP/dt}{P} = \beta - \delta P \qquad \text{or} \qquad \frac{dP}{dt} = P(\beta - \delta P).$$

This is called the *logistic equation* and the growth exhibited by the population is called *logistic growth*.

(a) Show by differentiation that if $P(0)$ is the initial population, then a solution to the logistic equation is

$$P(t) = \frac{\beta}{\delta + [(\beta/P(0)) - \delta]e^{-\beta t}}.$$

(b) Show that when $P = \beta/\delta$, $dP/dt = 0$. What does this mean?
(c) Show that $\lim_{t \to \infty} P(t) = \beta/\delta$.
(d) Graph the population curve for $\delta = 0.001$, $\beta = 0.4$, and $P(0) = 300$.
(e) Graph the population curve for $\delta = 0.001$, $\beta = 0.4$, and $P(0) = 500$.

17. Consider the simple electric circuit in Figure 3.

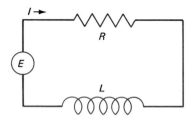

Figure 3

(i) An electromotic force (emf) E (volts), usually a battery or generator, drives an electric charge Q (coulombs) and produces a current I (amps). A current is a rate of flow of charge: $I = dQ/dt$.

(ii) A resistor of resistance R (ohms) opposes the current, converting some of the electrical energy to heat. It produces a drop in the voltage given by *Ohm's Law*

$$E_R = RI.$$

(iii) An inductor of inductance L (henrys) opposes any change in the current by causing a drop in the voltage of

$$E_L = L\frac{dI}{dt}.$$

A basic principle of electric circuits is given by *Kirchoff's voltage law* which states that *the algebraic sum of all voltage drops around a closed circuit is zero*. In the circuit above, the resistor and the inductor cause voltage drops of E_R and E_L, respectively. The emf *provides* a voltage of E (which is a voltage *drop* of $-E$). Thus $E_R + E_L - E = 0$ or

$$E = E_R + E_L = RI + L\frac{dI}{dt}.$$

(a) Verify that the solution to the above differential equation is

$$I(t) = \frac{E}{R} + \left[I(0) - \frac{E}{R}\right]e^{-Rt/L},$$

where $I(0)$ is the initial current.

(b) Show that $\lim_{t\to\infty} I(t) = E/R$. Here E/R is called the *steady state* part of the current and $[I(0) - (E/R)]e^{-Rt/L}$ is called the *transient part* of the current.

18. In Problem 17 let $E = 50$ volts, $R = 8$ ohms, and $L = 2$ henrys. Suppose initially that the current is zero.

(a) Calculate the current after 0.05 sec.
(b) Calculate the current after 0.2 sec.
(c) What is the steady state current?
(d) How long does it take for the current to reach 99.9% of the steady state current?

*6.5 Applications in Economics

In this section we show in a variety of ways how the exponential function can arise in economics. We begin with the problem of calculating compound interest.

Suppose A_0 dollars are invested in an enterprise (which may be a bank, bonds, a common stock, etc.) with an annual interest rate of i. *Simple interest* is the amount earned on the $\$A_0$ over a period of time. If the $\$A_0$ are invested for t years, then the simple interest I is given by

$$\boxed{I = A_0 ti.} \tag{1}$$

EXAMPLE 1. If $\$1000$ is invested for 5 years with an interest rate of 6%, then $i = 0.06$ and the simple interest earned is

$$I = (\$1000)(5)(0.06) = \$300.$$

Compound interest is interest paid on the interest previously earned as well as on the original investment. Suppose that interest is paid annually. Then if $\$A_0$ are invested, the interest after one year is $\$iA_0$ and the original investment is now worth $A_0 + iA_0 = A_0(1 + i)$ dollars. After two years, the compound interest paid would be $i[A_0(1 + i)]$ dollars and the investment would then be worth

$$A_0(1 + i) + iA_0(1 + i) = A_0(1 + i)(1 + i) = A_0(1 + i)^2$$

dollars. Continuing in this fashion, we see that after t years the investment would be worth

$$A(t) = A_0(1 + i)^t \quad \text{dollars.} \tag{2}$$

We have used the notation $A(t)$ to denote the value of the investment after t years.

EXAMPLE 2. If the interest in Example 1 is compounded annually, then after 5 years the investment is worth

$$A(5) = 1000(1 + 0.06)^5 = 1000(1.33823) = \$1338.23.$$

The actual interest paid is $338.23.

In practice, interest is compounded more frequently than annually. If it is paid m times a year, then in each interest period the rate of interest is i/m and in t years there are tm pay periods. Then, according to formula (2) we obtain:

The value of an investment of $\$A_0$ compounded m times a year with an annual interest rate of i after t years is

$$A(t) = A_0 \left(1 + \frac{i}{m}\right)^{mt}. \tag{3}$$

□**EXAMPLE 3.** If the interest in Example 1 is compounded quarterly (4 times a year) then after 5 years the investment is worth

$$A(5) = 1000 \left(1 + \frac{0.06}{4}\right)^{(4)(5)} = 1000(1.015)^{20} = 1000(1.34686) = \$1346.86.$$

The interest paid is now $346.86.

It is clear from the above examples that the interest actually paid increases as the number of interest periods increases. We are naturally led to ask: What is the value of an investment if interest is compounded continuously; that is, when the number of interest payments approaches infinity? To answer this we calculate, from (3),

$$A(t) = \lim_{m \to \infty} A_0 \left(1 + \frac{i}{m}\right)^{mt} = A_0 \lim_{m \to \infty} \left[\left(1 + \frac{i}{m}\right)^m\right]^t = A_0 \left[\lim_{m \to \infty} \left(1 + \frac{i}{m}\right)^m\right]^t$$

(from the corollary to Theorem 2.3.4.). We will show in Chapter 12 that

$$\lim_{m \to \infty} [1 + (i/m)]^m = e^i.$$

Then

$$A(t) = A_0 e^{it} \tag{4}$$

is the value of an original investment after t years with an interest rate of i compounded continuously.

There is another way to arrive at formula (4). In the time period t to $t + \Delta t$, the interest earned is $A(t + \Delta t) - A(t)$. If Δt is small, then the interest paid on $A(t)$ dollars would be (from formula (1)) approximately equal to $A(t) \Delta t\, i$. We say "approximately" because $A(t) \Delta t\, i$ represents simple interest between t and $t + \Delta t$. However the difference is small if Δt is small. Thus

$$A(t + \Delta t) - A(t) \approx A(t) \Delta t\, i$$

or, dividing by Δt,

$$\frac{A(t + \Delta t) - A(t)}{\Delta t} \approx iA(t).$$

Then, taking the limit as $\Delta t \to 0$ we obtain

$$\frac{dA}{dt} = iA(t)$$

and, from the last section,

$$A(t) = ce^{it}.$$

But $A(0) = A_0$ which tells us that $c = A_0$ and we have again obtained formula (4).

EXAMPLE 4. If the interest in Example 1 is compounded continuously, then

$$A(5) = 1000e^{(0.06)(5)} = 1000e^{0.3} = \$1349.86$$

and the interest paid is $349.86. Compare this to the result in Examples 2 and 3.

▢**EXAMPLE 5.** $5000 is invested in a bond yielding $8\frac{1}{2}\%$ annually. What will the bond be worth in 10 years if interest is compounded continuously?

SOLUTION. $A(t) = A_0 e^{it} = 5000e^{(0.085)(10)} = 5000e^{0.85} = \$11,698.23$.

EXAMPLE 6. How long does it take for an investment to double if the annual interest rate is 6% compounded continuously?

SOLUTION. We need to determine a value of t such that $A(t) = 2A_0$. That is,

$$A_0 e^{0.06t} = 2A_0 \qquad \text{or} \qquad e^{0.06t} = 2.$$

Taking natural logarithms, $0.06t = \ln 2 = 0.6931$ and $t = 0.6931/0.06 = 11.55$ years.

EXAMPLE 7. What must be the interest rate in order that an investment double in 7 years when interest is compounded continuously?

SOLUTION. We need to determine i such that $A_0 e^{7i} = 2A_0$. Then $e^{7i} = 2$, $7i = \ln 2$ and $i = 0.6931/7 = 0.099 = 9.9\%$.

EXAMPLE 8. If money is invested at 8% compounded continuously, what is the effective rate of interest?

SOLUTION. The *effective interest rate* is the rate of simple interest received over a one year period. Starting with A_0 dollars, there will be $A_0 e^{0.08} \approx A_0(1.0833)$ after one year. Thus the effective interest rate is about $8.33 = 8\frac{1}{3}\%$.

An extremely important concept in economics is that of the *present value* of money. Since $1000 invested today will be worth more in the future, say two years hence, it follows that $1000 which will be collected in 2 years is worth *less* today. We can use the compound interest formulas (3) and (4) to calculate present value.

If we invest A_0 dollars today, then after t years our investment is worth $A(t) = A_0[1 + (i/m)]^{mt}$ if compounded periodically and $A(t) = A_0 e^{it}$ if compounded continuously. $A(t)$ dollars after t years has the present value of A_0. Thus present value may be calculated by solving for A_0.

If compounded m times a year,

$$\text{present value} = A_0 = \frac{A(t)}{[1 + (i/m)]^{mt}} = A(t)\left(1 + \frac{i}{m}\right)^{-mt}. \tag{5}$$

If compounded continuously,

$$\text{present value} = A_0 = \frac{A(t)}{e^{it}} = A(t)e^{-it}. \tag{6}$$

EXAMPLE 9. What is the present value of $1000 after 5 years at 6% compounded semiannually?

SOLUTION. We have $t = 5$, $A(t) = \$1000$, $i = 0.06$, and $m = 2$ so that

$$A_0 = 1000(1.03)^{-10} = \$744.09.$$

Put another way, we would have to invest $744.09 now at 6% interest to have $1000 in 5 years.

☐**EXAMPLE 10.** A manufacturer receives a note promising payment of $25,000 in 3 years. He needs capital now so he sells the note to a bank who reimburses him based on an annual interest rate of 7% compounded continuously. How much money does the manufacturer receive? (*Note:* The act of purchasing something for its present value is sometimes called *discounting* and the interest rate i is called the *discount rate.*)

SOLUTION. He will receive the present value of the $25,000 based on an interest rate of 7%.

$$A_0 = A(3)e^{-(0.07)3} = 25{,}000e^{-0.21} = \$20{,}264.61.$$

EXAMPLE 11. If the manufacturer in Example 10 receives $18,000 from the bank, how much interest is he effectively paying for the loan?

SOLUTION. Here $A_0 = 18{,}000$ so that

$$18{,}000 = 25{,}000e^{-3i} \quad \text{or} \quad e^{-3i} = \frac{18}{25} = 0.72.$$

Then

$$-3i = \ln 0.72 = -0.3285 \quad \text{and} \quad i = 0.1095 = 10.95 \quad \text{percent.}$$

EXAMPLE 12. A stamp dealer calculates that the value of one of his collections will increase according to the formula $V(t) = 1000e^{\sqrt{t/3}}$, where t is measured in years. If he were to sell today he could receive $1000 and the longer he holds the collection the more he could receive. However, he calculates that money he invests earns 10% compounded continuously. When should he sell the collection to maximize his return?

SOLUTION. The present value of the money $V(t)$ he receives t years in the future is

$$V(t)e^{-0.1t} = (1000e^{\sqrt{t/3}})e^{-0.1t} = 1000e^{\sqrt{t/3}-0.1t}.$$

This is the quantity to be maximized. Then

$$\frac{dV}{dt} = (1000e^{\sqrt{t/3}-0.1t}) \left(\frac{1}{6\sqrt{t/3}} - 0.1\right)$$

which is equal to zero when

$$\frac{1}{6\sqrt{t/3}} = 0.1, \quad \frac{1}{\sqrt{t/3}} = 0.6, \quad \frac{3}{t} = 0.36, \quad \text{and} \quad t = \frac{3}{0.36} = 8\frac{1}{3} \text{ years.}$$

In all the above examples, we have calculated values of investments assuming that capital was invested at one fixed time. But in many important problems money is invested periodically (for example, monthly deposits in a bank, annual life insurance premiums, installment loan payments, etc.). This type of payment is called an *annuity*. If we deposit a fixed amount B dollars each year, then after one year we deposit B dollars; after two years the B dollars have now become $B(1 + i)$ (compounded annually) and we deposit another B dollars to obtain $A(2) = B + B(1 + i)$. After 3 years we have $A(3) = B + B(1 + i) + B(1 + i)^2$ and, after t years,

$$A(t) = B + B(1 + i) + B(1 + i)^2 + \cdots + B(1 + i)^{t-1}$$
$$= B[1 + (1 + i) + (1 + i)^2 + \cdots + (1 + i)^{t-1}].$$

The expression in brackets is the sum of a geometric progression (see Section 1.10). Since $1 + a + a^2 + \cdots + a^{n-1} = (1 - a^n)/(1 - a)$, we have

$$A(t) = B\left[\frac{1 - (1 + i)^t}{1 - (1 + i)}\right] = \frac{B[(1 + i)^t - 1]}{i}. \tag{7}$$

If the money is compounded m times during every interval of deposit, then B dollars are worth $B[1 + (i/m)]^m$ dollars after one such interval (from our compound interest formula). Then, after t intervals of deposit, we have

$$A(t) = B + B\left(1 + \frac{i}{m}\right)^m + B\left(1 + \frac{i}{m}\right)^{2m} + \cdots + B\left(1 + \frac{i}{m}\right)^{(t-1)m}$$
$$= B\left[1 + \left(1 + \frac{i}{m}\right)^m + \left(\left(1 + \frac{i}{m}\right)^m\right)^2 + \cdots + \left(\left(1 + \frac{i}{m}\right)^m\right)^{t-1}\right]$$

or

$$A(t) = B\left[\frac{(1 + (i/m))^{mt} - 1}{(1 + (i/m))^m - 1}\right]. \tag{8}$$

☐**EXAMPLE 13.** If a man deposits $500 every 6 months and this is compounded quarterly at 6%, how much does he have after 10 years?

SOLUTION. Here $B = 500$ and $i = 0.03$ since the interval of deposit is $\frac{1}{2}$ year. Then $m = 2$ (2 payments every $\frac{1}{2}$ year), $t = 20$ (there are 20 semiannual deposits) and

$$A(20) = 500 \left[\frac{(1 + (0.03/2))^{2(20)} - 1}{(1 + (0.03/2))^2 - 1} \right]$$

$$= 500 \left[\frac{1.015^{40} - 1}{1.015^2 - 1} \right] \approx 500 \left[\frac{1.8140 - 1}{1.0302 - 1} \right] \approx \$13{,}465.82.$$

Suppose money is received, instead of being paid out at periodic intervals. (For example, you may own an apartment building where monthly rents are received.) What is the present value of such payments? We suppose that an interest rate of i is paid over a fixed time period and assume that $\$B_1$ is expected after the first time period, $\$B_2$ after the second, and so on. Then the present values are $B_1 = B_1/(1 + i)$, $B_2 = B_2/(1 + i)^2$, and so on so that the total present value over t time periods is

$$A_0 = \frac{B_1}{1 + i} + \frac{B_2}{(1 + i)^2} + \cdots + \frac{B_t}{(1 + i)^t}. \tag{9}$$

If all payments are equal, then $B_1 = B_2 = \cdots = B_t = B$ and

$$A_0 = B \left[\frac{1}{(1 + i)} + \frac{1}{(1 + i)^2} + \cdots + \frac{1}{(1 + i)^t} \right]$$

$$= \frac{B}{1 + i} \left[1 + \frac{1}{(1 + i)} + \frac{1}{(1 + i)^2} + \cdots + \frac{1}{(1 + i)^{t-1}} \right]$$

or

$$A_0 = \frac{B}{1 + i} \left[\frac{1 - (1/(1 + i))^t}{1 - (1/(1 + i))} \right] = \frac{B}{i} \left[1 - \left(\frac{1}{1 + i} \right)^t \right]. \tag{10}$$

☐**EXAMPLE 14.** What is the present value of an annuity that would pay $3000 a year for 20 years assuming an interest rate of 6% compounded annually?

SOLUTION. $A(20) = \dfrac{3000}{0.06} \left(1 - \dfrac{1}{1.06^{20}} \right) = \$34{,}409.76.$

☐**EXAMPLE 15.** A new car costs $4000. If you purchase it in 36 monthly installments of $150/month, what is the effective annual interest you are paying?

SOLUTION. The dealer is receiving an annuity of $150/month in exchange for a present value of $4000. Thus in formula (10)

$$A_0 = 4000 = \frac{150}{i} \left[1 - \frac{1}{(1 + i)^{36}} \right].$$

Rewriting, we obtain

$$4000i = 150 - \frac{150}{(1 + i)^{36}}, \qquad 4000i(1 + i)^{36} - 150(1 + i)^{36} + 150 = 0$$

or, dividing by 50,

$$80i(1 + i)^{36} - 3(1 + i)^{36} + 3 = 0.$$

There is, of course, no formula for finding the roots of this 37th degree polynomial. However we can estimate the root i by Newton's method (Section 5.5). (Note that $i = 0$ is a root. We seek the first positive root.) Defining

$$P(i) = 80i(1 + i)^{36} - 3(1 + i)^{36} + 3,$$

we have

$$P'(i) = 80(1 + i)^{36} + (80 \cdot 36)i(1 + i)^{35} - 108(1 + i)^{35}$$

and Newton's method yields the sequence

$$i_{n+1} = i_n - \frac{P(i_n)}{P'(i_n)}$$

$$= i_n - \left[\frac{80i_n(1 + i_n)^{36} - 3(1 + i_n)^{36} + 3}{80(1 + i_n)^{36} + 2880i_n(1 + i_n)^{35} - 108(1 + i_n)^{35}} \right]$$

or, dividing numerator and denominator by $(1 + i_n)^{35}$ to simplify, we obtain

$$i_{n+1} = i_n - \left[\frac{80i_n(1 + i_n) - 3(1 + i_n) + (3/(1 + i_n)^{35})}{80(1 + i_n) + 2880i_n - 108} \right]$$

$$= i_n - \left[\frac{80i_n^2 + 77i_n - 3 + (3/(1 + i_n)^{35})}{2960i_n - 28} \right].$$

Starting with $i_0 = 0.02$ (representing a monthly rate of 2%) we obtain the values in Table 1. The required interest rate is $i \approx 0.0172 = 1.72\%$ a month (correct to 4

TABLE 1

i_n	$\dfrac{3}{(1 + i_n)^{35}}$	A, $80i_n^2 + 77i_n - 3 + \dfrac{3}{(1 + i_n)^{35}}$	B, $2960i_n - 28$	$\dfrac{A}{B}$	$i_{n+1} = i_n - \dfrac{A}{B}$
0.02	1.50008284	0.07208284	31.2	0.0023103474	0.0176896526
0.0176896526	1.62399129	0.0111284445	24.3613717	0.0004568069	0.0172328456
0.0172328456	1.649712021	0.0003988095	23.00922298	0.0000173325	0.017215513
0.017215513	1.650696151	0.0000005635	22.9579185	0.0000000245	0.0172154884

decimal places). The annual interest is $1.72 \times 12 = 20.64\%$. You would be better off borrowing the money from a bank!

To conclude this section we consider the case where income is received continuously, rather than at fixed intervals. Let $a(t)$ denote a *stream of income*. Then the present value of the money received at time t is $a(t)e^{-it}$. Between the time t and $t + \Delta t$ the present value of the money received is approximately $a(t)e^{-it} \Delta t$ and, using a now familiar argument, the present value of the money received up to the time

$t = T$ years is given by

$$\text{present value} = \int_0^T a(t)e^{-it}\, dt. \tag{11}$$

☐**EXAMPLE 16.** An investment broker has a choice between investing $100,000 in a growing business or putting it into bonds yielding an 8% annual return, compounded continuously for 20 years. The annual income from the business is projected to be $4000(1 + e^{t/10})$ dollars after t years. Where should he put his money?

SOLUTION. Here $a(t) = 4000(1 + e^{0.1t})$. To the banker, the money can earn 8%. Thus the present value of the investment in the business is

$$\int_0^{20} 4000(1 + e^{0.1t})e^{-0.08t}\, dt = 4000 \int_0^{20} (e^{-0.08t} + e^{0.02t})\, dt$$

$$= 4000 \left\{ \frac{e^{-0.08t}}{-0.08} \bigg|_0^{20} + \frac{e^{0.02t}}{0.02} \bigg|_0^{20} \right\} = 4000 \left\{ \frac{1 - e^{-1.6}}{0.08} + \frac{e^{0.4} - 1}{0.02} \right\}$$

$$\approx 4000(9.9763 + 24.5912) = (4000)(34.5675) = \$138,270.$$

Thus, if he invests in the business, he can increase the present value of his money by about 38%. Of course, this is under the assumption that there is no greater risk in the business investment.

The notion of discounting and present value has recently been applied to the management of our dwindling supply of natural resources. For example, what is the value to us now of oil that will be worth more per barrel in the future? Harder to evaluate is the present value of an animal nearing extinction. In order to manage wisely, it is necessary to "discount the future" to have a real understanding of what we have now. The more we discount into the future, the more apparent it becomes that what we have is less than we thought we had. There is a very strong interdependence between economic and biological forces in problems of resource management. We will not discuss this idea any further but the interested reader is urged to look at one or more of the references listed below.†

PROBLEMS 6.5

1.   A sum of $5000 is invested for 8 years at a return of 7% per year. What simple interest is paid over that period of time?
2.   If the money in Problem 1 is compounded annually, what is the investment worth after 8 years?
3.   If the money in Problem 1 is compounded monthly, what is the investment worth after 8 years?
4.   If the money in Problem 1 is compounded continuously, what is the investment worth after 8 years?

†(1) S. V. Ciriacy-Wantrup, *Resource Conservation: Economics and Policies,* Univ. of California Press, Berkeley, 1963; (2) C. W. Clark, "Profit Maximization and the Extinction of Animal Species," *Journal of Political Economy* **81,** 950–961 (1973a); (3) C. W. Clark, "Mathematical Bioeconomics" in *Mathematical Problems in Biology–Victoria Conference,* Springer-Verlag, New York, 1974, pp. 29–45.

5. Calculate the percentage increase in return on investment if $\$A_0$ are invested for 10 years at 6% compounded annually and quarterly.

6. As a gimmick to lure depositors, some banks offer 5% interest compounded continuously in comparison to their competitors who offer $5\frac{1}{8}$% compounded annually. Which bank would you choose?

7. Suppose a competitor in Problem 6 now compounds $5\frac{1}{8}$% semiannually. Which bank would you choose?

8. $10,000 is invested in bonds yielding 9% compounded continuously. What will the bonds be worth in 8 years?

9. A certain government bond sells for $750 and can be redeemed for $1000 in 8 years. Assuming continuous compounding, what is the rate of interest paid?

10. How long would it take an investment to increase by half if it is invested at 4% compounded continuously?

11. What must be the interest rate in order that an investment triple in 15 years if interest is continuously compounded?

12. If money is invested at 10% compounded continuously, what is the effective interest rate?

13. What is the most a banker should pay for a $10,000 note due in 5 years if he can invest a like amount of money at 9% compounded annually?

14. What is the present value of $5,000 in 3 years at 5%, compounded quarterly?

15. What is the present value of $8,000 in 10 years, compounded continuously at 4%?

16. A sales firm receives a cash payment of $8,000 for a $15,000 note due in 5 years. Assuming continuous compounding, what interest are they paying?

*17. A real estate dealer owns recreation land which she calculates will appreciate in value according to the formula

$$V(t) = 100,000(1 + e^{\sqrt{0.2t}}),$$

where t is measured in years. At any time she sells, she can reinvest the money and be guaranteed a 12% return, compounded continuously. When should she sell?

18. If a $2,000 bond is bought every year and the money earns 7% compounded annually, how much is available after 20 years?

*19. A man wishes to prepare for his child's education. He calculates that he will need $25,000 in 15 years. He wishes to make monthly deposits into an account that pays 5% compounded quarterly. How large must his payments be if he is to meet his goal?

20. A decision must be made for a capital equipment purchase. Using an interest rate of 7%, what is the maximum reasonable purchase price for a machine whose earnings over the next 5 years are projected to be $1000, $1500, $1800, $1300, and $700? Assume a value of zero after 5 years.

21. A novelist writes a novel loaded with adventure and sex which centers around the story of a 5-year old beagle named Bennette who can turn lead into gold. A publisher is eager to publish this book as it is expected to be a best seller. The author demands a cash advance based on first year royalties which are anticipated to be $250,000. Because of the time needed to revise the novel and publish the book, the first royalty payments are not due for three years. The publisher must pay 11% compounded continuously on any money he pays out now in anticipation of future earnings. If the author's demand is to be met, what is a reasonable cash advance?

*22. A broker invests $10,000 in second mortgages and receives $100/month for 20 years. At the end of that time his investment is worthless. What is his effective rate of return on capital?

23. An investment counselor has the opportunity to purchase an apartment building which

yields, after expenses, $6,000 a month. The purchase price is $750,000. He calculates that he will hold the property for 20 years, after which it will begin to deteriorate and thereafter be essentially worthless. Assuming that he could earn 6% compounded monthly in another venture,

   (a) Calculate the present value of the investment in the building.
   (b) Should he make the purchase?
   [*Hint:* Here $i = \frac{1}{2}\%$ and t is the number of months.]

24. Answer the questions in Problem 23 assuming that the counselor can realize a return of 12% compounded monthly.

*25. In Problems 23 and 24, for what annual interest rate (compounded monthly) will it make no difference whether or not the counselor invests in the building?

*26. A new refrigerator costs $400. At 24 monthly installments of $20, what is the effective annual interest being paid?

27. In order to achieve an effective annual interest of 6% in Problem 26, what monthly payment should be paid?

28. A $30,000 mortgage is obtained for a house with monthly payments of $230 spread over a 30 year period. What is the interest rate?

29. What would be the monthly payments on a $25,000, 25-year mortgage at $7\frac{1}{2}\%$?

30. What is the present value of an annuity if the stream of income is $a(t) = 10,000(1 + e^{0.13t})$ after 10 years, assuming continuous compounding at 8%?

Review Exercises for Chapter Six

In Exercises 1–6 solve for the given variable.

1. $y = \log_3 9$ 2. $y = \log_{1/3} 9$ 3. $4 = \log_x \frac{1}{16}$

4. $y = e^{\ln 17.2}$ 5. $\log x = 10^{-9}$ 6. $\log_x 32 = -5$.

7. Simplify $\ln \left[\dfrac{\sqrt{x^3 - 1}(x^5 + 1)^{4/3}}{\sqrt{(x - 3)(x - 2)}} \right]^{-5/6}$.

8. If $y = 3 \ln x$, what happens to y if x doubles?

9. Sketch the curve $y = \ln|x + 1|$.   10. Sketch the curve $y = e^{x+2}$.

In Exercises 11–25 calculate the given derivative or integral.

11. $\dfrac{d}{dx} \ln(1 + x^2)$ 12. $\displaystyle\int \dfrac{3x}{1 + x^2}\, dx$ 13. $\dfrac{d}{dx} e^{x^2 + 2x + 1}$

14. $\dfrac{d}{dx} e^{\sqrt{x^3 - 3}}$ 15. $\displaystyle\int \dfrac{e^{\sqrt{x}}}{5\sqrt{x}}\, dx$ 16. $\displaystyle\int \dfrac{1}{3x \ln x}\, dx$

17. $\dfrac{d}{dx} \ln|e^x - 5|$ 18. $\dfrac{d}{dx} \ln[x + \ln(x + 3)]$ 19. $\displaystyle\int \dfrac{e^{-1/x^2}}{x^3}\, dx$

20. $\dfrac{d}{dx} x^{x+1}$ 21. $\dfrac{d}{dx} \left[\dfrac{(\sqrt[3]{x + 3})(x - 5)}{\sqrt{(x + 1)(x + 2)}} \right]$ 22. $\displaystyle\int \dfrac{4^{\ln x}}{x}\, dx$

23. $\displaystyle\int \dfrac{1}{x} \sqrt{\log_5 x}\, dx$ 24. $\dfrac{d}{dx} \log_3 2^{x+5}$ 25. $\dfrac{d}{dx} e^{\ln(x^3 + 1)}$.

In Exercises 26–29 sketch the given curve, indicating local maxima and minima and points of inflection.

26. $y = xe^{x^2 + 1}$ 27. $y = \dfrac{x}{\ln x}$

28. $y = xe^{-x}$ **29.** $y = \ln |\ln x|$.

30. The relative annual rate of growth of a population is 15%. If the initial population is 10,000, what is the population after 5 years? After 10 years?

31. In Exercise 30, how long will it take for the population to double?

32. When a cake is taken out of the oven, its temperature is 125 °C. Room temperature is 23 °C. If the temperature of the cake is 80 °C after 10 minutes
(a)   what will be its temperature after 20 minutes?
(b)   how long will the cake take to cool to 25 °C?

33. A fossil contains 35% of the normal amount of C^{14}. What is its approximate age?

34. What is the half-life of an exponentially decaying substance that loses 20% of its mass in one week?

35. How long will it take the substance in Exercise 34 to lose 75% of its mass? 95% of its mass?

36. A sum of 10,000 is invested at an interest rate of 6% compounded quarterly. What is the investment worth in 8 years?

37. What is the investment in Problem 36 worth if interest is compounded continuously?

38. What is the present value of a $5000 note payable in 4 years based on 7% compounded semiannually?

39. What is the present value of the note in Exercise 39 compounded continuously?

40. What is the present value of an annuity yielding $4000(1 + e^{0.15t})$ dollars over the next ten years assuming continuous compounding at 9%?

SEVEN

THE TRIGONOMETRIC AND HYPERBOLIC FUNCTIONS

In this chapter we complete the discussion of the most commonly used functions of calculus: the trigonometric or circular functions and the hyperbolic functions. We begin the chapter with a discussion of some basic facts from trigonometry. For those of you who have had an earlier course in trigonometry, the first four sections should be read as a review and can be covered quickly. In Sections 7.5 and 7.6 we differentiate and integrate the trigonometric functions and in Section 7.7 we introduce the inverse trigonometric functions. We show, in Section 7.8, how the trigonometric functions arise in some very basic applications. Finally, in Sections 7.9 and 7.10, we discuss the hyperbolic functions.

7.1 Angles and Radian Measure

We begin by drawing the unit circle—the circle with radius 1 centered at the origin. *Angles* are measured starting at the positive x-axis. An angle is positive if it is measured in the counterclockwise direction, and it is negative, if it is measured in the clockwise direction. We measure angles in degrees using the fact that the circle contains $360°$. Then we can describe any angle by comparison with the circle. Some

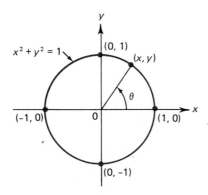

Figure 1

angles are depicted in Figure 2. In Figure 2e, we obtained the angle $-90°$ by moving in the negative (clockwise) direction. In part (f) we obtained an angle of $720°$ by moving around the circle twice in the counter-clockwise direction.

324

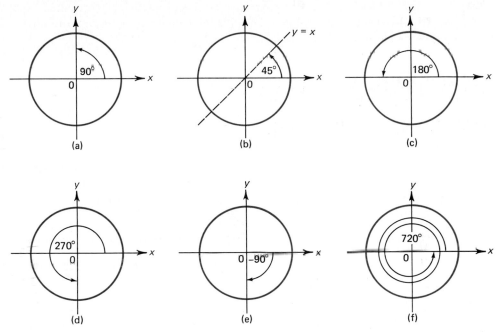

Figure 2

Figure 2 illustrates the great advantage of using circles rather than triangles to measure angles. Any angle in a triangle must be between $0°$ and $180°$. In a circle, there is no such restriction.

There is another way to measure angles which, in many instances, is more useful than measurement in degrees. Let R denote the radial line which makes an angle of θ with the positive x-axis. See Figure 3. Let (x, y) denote the point at which this radial line intersects the unit circle. Then the *radian measure* of the angle is the length of the arc of the unit circle from the point $(1, 0)$ to the point (x, y).

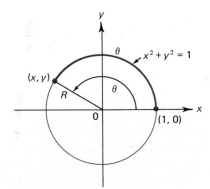

Figure 3

Since the circumference of a circle is $2\pi r$ where r is the radius, the circumference of the unit circle is 2π. Thus

$$360° = 2\pi \quad \text{radians.} \tag{1}$$

Then since $180° = \frac{1}{2}(360°)$, $180° = \frac{1}{2}(2\pi) = \pi$ radians. In general θ (in degrees) is to $360°$ as θ (in radians) is to 2π radians. Thus $(\theta/360)$ (degrees) $= (\theta/2\pi)$ (radians) or

$$\theta \text{ (degrees)} = \frac{180}{\pi}\theta \text{ (radians)} \tag{2}$$

and

$$\theta \text{ (radians)} = \frac{\pi}{180}\theta \text{ (degrees)}. \tag{3}$$

We can calculate that 1 radian $= 180/\pi \approx 57.3°$ and $1° = \pi/180 \approx 0.0175$ radians. Representative values of θ in degrees and radians are given in Table 1.

TABLE 1

θ (degrees)	0	90	180	270	360	45	30	60	-90	135	120	720
θ (radians)	0	$\frac{\pi}{2}$	π	$\frac{3\pi}{2}$	2π	$\frac{\pi}{4}$	$\frac{\pi}{6}$	$\frac{\pi}{3}$	$\frac{-\pi}{2}$	$\frac{3\pi}{4}$	$\frac{2\pi}{3}$	4π

The radian measure of an angle does not refer to "degrees," but instead refers to distance measured along an arc of the unit circle. This is an advantage when discussing trigonometric functions which arise in applications having nothing at all to do with angles (see Section 7.8).

Let C_r denote the circle of radius r centered at the origin (see Figure 4). If $0P$

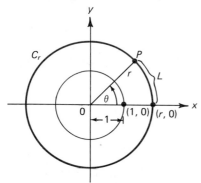

Figure 4

denotes a radial line as pictured in the figure, then $0P$ cuts an arc from C_r of length L. Let θ be the positive angle between $0P$ and the positive x-axis. If $\theta = 360°$, then $L = 2\pi r$. If $\theta = 180°$, then $L = \pi r$. In fact, it is evident from the figure that

$$\frac{\theta}{360°} = \frac{L}{2\pi r} \tag{4}$$

or

$$\theta° = 360° \frac{L}{2\pi r}. \tag{5}$$

If we measure θ in radians, then (4) becomes

$$\frac{\theta}{2\pi} = \frac{L}{2\pi r} \qquad (6)$$

or

$$\theta = \frac{2\pi L}{2\pi r} = \frac{L}{r}. \qquad (7)$$

Finally, rewriting (7), we obtain

$$\boxed{L = r\theta.} \qquad \theta = \frac{L}{r} \qquad (8)$$

That is, if θ is measured in radians, then the angle θ "cuts" from the circle of radius r centered at the origin an arc of length $r\theta$. Note that if $r = 1$, then (8) reduces to $L = \theta$ which is the definition of the radian measure of an angle.

EXAMPLE 1. What is the length of an arc cut from the circle of radius 4 centered at the origin by an angle of

(a) $45°$ (b) $60°$ (c) $270°$.

SOLUTION. From (8), we find that $L = 4\theta$ where θ is the radian measure of the angle. We therefore have

(a) $L = 4 \cdot \dfrac{\pi}{4} = \pi$ (b) $L = 4 \cdot \dfrac{\pi}{3} = \dfrac{4\pi}{3}$ (c) $L = 4 \cdot \dfrac{3\pi}{2} = 6\pi.$

PROBLEMS 7.1

In Problems 1–6 convert from degrees to radians.

1. $\theta = 150°$ 2. $\theta = -45°$ 3. $\theta = 300°$
4. $\theta = 72°$ 5. $\theta = 144°$ 6. $\theta = 1080°$.

In Problems 7–12 convert from radians to degrees.

7. $\dfrac{\pi}{12}$ 8. $\dfrac{7\pi}{12}$ 9. $\dfrac{\pi}{8}$

10. 3π 11. $-\dfrac{\pi}{3}$ 12. $\dfrac{5\pi}{4}$.

13. Let C denote the circle of radius 2 centered at the origin. If a radial line cuts an arc of length π (starting at the point $(2, 0)$), what is the angle (in degrees) between this line and the positive x-axis?

14. If the radial line in Problem 13 makes an angle of $75°$ with the positive x-axis, what is the length of the arc it cuts from the circle?

7.2 The Trigonometric Functions and Basic Identities

We again begin with the unit circle (see Figure 1). An angle θ uniquely determines a point (x, y) where the radial line intersects the circle. We then define

$$\text{cosine } \theta = x \quad \text{and} \quad \text{sine } \theta = y. \qquad (1)$$

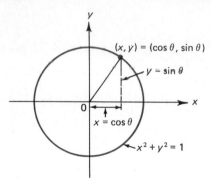

Figure 1

These are the two basic trigonometric (or circular) functions, usually written cos θ and sin θ. Since the equation of the circle is $x^2 + y^2 = 1$, we see that sin θ and cos θ satisfy the equation

$$\sin^2 \theta + \cos^2 \theta = 1.$$ (2)

We emphasize that cos θ is the x-coordinate of the point (x, y) and sin θ is the y-coordinate. As θ varies, cos θ and sin θ oscillate between $+1$ and -1. For example, if $\theta = 0$, then the radial line intersects the circle at the point $(1, 0)$ and we have cos $0 = 1$ and sin $0 = 0$. If $\theta = 90° = \pi/2$, then the radial line intersects the circle at the point $(0, 1)$ and we have cos $90° = \cos \pi/2 = 0$ and sin $90° = \sin \pi/2 = 1$. If $\theta = 45° = \pi/4$, then the radial line is the line $y = x$. Since $x^2 + y^2 = 1$ and $y = x$ at the point of intersection, we see that $x^2 + x^2 = 2x^2 = 1$ or $x = y = 1/\sqrt{2} = \sqrt{2}/2$. Thus cos $45° = \cos \pi/4 = \sqrt{2}/2$ and sin $45° = \sin \pi/4 = \sqrt{2}/2$. It will be shown in Section 7.4 that cos $30° = \cos \pi/6 = \sqrt{3}/2$, sin $30° = \sin \pi/6 = \frac{1}{2}$, cos $60° = \cos \pi/3 = \frac{1}{2}$, and sin $60° = \sin \pi/3 = \sqrt{3}/2$. The most commonly used values of cos θ and sin θ are given in Table 1.

TABLE 1

θ	0	$\dfrac{\pi}{6}$	$\dfrac{\pi}{4}$	$\dfrac{\pi}{3}$	$\dfrac{\pi}{2}$	π	$\dfrac{3\pi}{2}$	2π
cos θ	1	$\dfrac{\sqrt{3}}{2}$	$\dfrac{\sqrt{2}}{2}$	$\dfrac{1}{2}$	0	-1	0	1
sin θ	0	$\dfrac{1}{2}$	$\dfrac{\sqrt{2}}{2}$	$\dfrac{\sqrt{3}}{2}$	1	0	-1	0

Some basic facts about the functions sin θ and cos θ can be derived by simply looking at the graph of the unit circle. First, we note that if we add 360° to the angle θ in Figure 1, then we end up with the same point (x, y) on the circle. This means that

$$\cos(\theta + 360°) = \cos(\theta + 2\pi) = \cos \theta$$ (3)

and

$$\sin(\theta + 360°) = \sin(\theta + 2\pi) = \sin\theta.$$ (4)

In general, if α is the *smallest* positive number such that $f(x + \alpha) = f(x)$, we say that f is *periodic* of *period* α. Thus, from (3) and (4), we see that the functions $\cos\theta$ and $\sin\theta$ are periodic of period 2π.

A glance at Figure 2 tells us the sign of the two basic functions. With all the

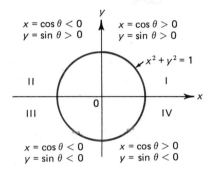

Figure 2

above information we can draw a sketch of $y = \cos\theta$ and $y = \sin\theta$. This is done in Figures 3 and 4. Now look at Figure 5a. We see that in comparing θ and $-\theta$, the

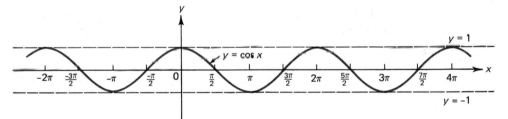

Figure 3

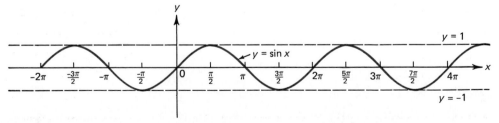

Figure 4

x-coordinates are the same while the y-coordinates have opposite signs. This suggests that

$$\cos(-\theta) = \cos\theta$$ (5)

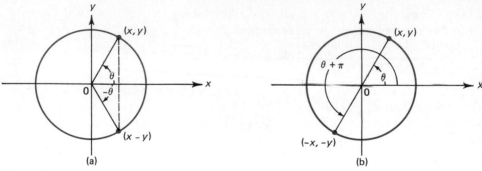

Figure 5

and

$$\sin(-\theta) = -\sin\theta. \tag{6}$$

To obtain another identity, we add $180° = \pi$ to θ (Figure 5b). Then the x and y coordinates of $\theta + \pi$ have signs opposite to those of the x and y coordinates of θ. Thus

$$\cos(\theta + 180°) = \cos(\theta + \pi) = -\cos\theta \tag{7}$$

and

$$\sin(\theta + 180°) = \sin(\theta + \pi) = -\sin\theta. \tag{8}$$

Several other identities can be obtained by simply glancing at a graph of the unit circle.

We now obtain another identity which is very useful in computations.

THEOREM 1. $\cos(\theta + \varphi) = \cos\theta\cos\varphi - \sin\theta\sin\varphi.$

PROOF. We first prove the theorem in the case θ and φ are between 0 and $\pi/2$. We will leave a more general case as a problem (see Problem 16). From Figure 6, we see that the arc P_1P_3 has the same length as the arc P_2P_4 (they are both equal to the

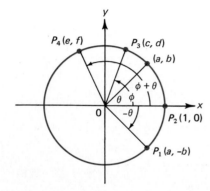

Figure 6

radian measure of the angle $\theta + \varphi$). Then the distance from P_1 to P_3 is the same as the distance from P_2 to P_4. Using the distance formula (see equation (1.3.2)) we obtain

$$\overline{P_1 P_3}\dagger = \sqrt{(c - a)^2 + (d + b)^2} = \sqrt{(e - 1)^2 + f^2} = \overline{P_0 P_4}. \tag{9}$$

But

$$\begin{array}{ccc} \cos\theta = a, & \cos\varphi = c, & \cos(\theta + \varphi) = e, \\ \sin\theta = b, & \sin\varphi = d, & \sin(\theta + \varphi) = f. \end{array} \tag{10}$$

Then we square both sides of (9):

$$c^2 - 2ac + a^2 + d^2 + 2bd + b^2 = e^2 - 2e + 1 + f^2.$$

Since $a^2 + b^2 = c^2 + d^2 = e^2 + f^2 = 1$ (why?), we have

$$-2ac + 2bd + 2 = -2e + 2$$

or

$$e = ac - bd. \tag{11}$$

Substituting (10) into (11) proves the theorem.

There are many other identities which can be proven using Theorem 1. We indicate in Table 2 some of the identities we will find useful in other parts of this text. The proofs of these identities are suggested in the problems. We use x and y instead of θ and φ in this table. These identities will be very useful to us when we discuss techniques of integration in Chapter 8.

TABLE 2   Basic Identities Involving $\cos x$ and $\sin x$

- (i) $\sin^2 x + \cos^2 x = 1.$
- (ii) $\cos(-x) = \cos x.$
- (iii) $\sin(-x) = -\sin x.$
- (iv) $\cos(x + \pi) = -\cos x.$
- (v) $\sin(x + \pi) = -\sin x.$
- (vi) $\cos(\pi - x) = -\cos x.$
- (vii) $\sin(\pi - x) = \sin x.$

(viii) $\cos\left(\dfrac{\pi}{2} + x\right) = -\sin x.$

(ix) $\sin\left(\dfrac{\pi}{2} + x\right) = \cos x.$

- (x) $\cos\left(\dfrac{\pi}{2} - x\right) = \sin x.$

- (xi) $\sin\left(\dfrac{\pi}{2} - x\right) = \cos x.$

- (xii) $\cos(x + y) = \cos x \cos y - \sin x \sin y.$
- (xiii) $\sin(x + y) = \sin x \cos y + \cos x \sin y.$
- (xiv) $\cos(x - y) = \cos x \cos y + \sin x \sin y.$
- (xv) $\sin(x - y) = \sin x \cos y - \cos x \sin y.$
- (xvi) $\cos 2x = \cos^2 x - \sin^2 x = 2\cos^2 x - 1 = 1 - 2\sin^2 x.$

†The symbol $\overline{P_1 P_3}$ denotes the distance between the points P_1 and P_3.

TABLE 2   (*continued*)

___

(xvii) $\sin 2x = 2 \sin x \cos x.$

(xviii) $\cos \dfrac{x}{2} = \pm \sqrt{\dfrac{1 + \cos x}{2}}.$

(xix) $\sin \dfrac{x}{2} = \pm \sqrt{\dfrac{1 - \cos x}{2}}.$

These are the *half-angle* formulas.

(xx) $\cos^2 x = \dfrac{1 + \cos 2x}{2}.$

(xxi) $\sin^2 x = \dfrac{1 - \cos 2x}{2}.$

(xxii) $\cos x - \cos y = 2 \sin \dfrac{x + y}{2} \sin \dfrac{y - x}{2}.$

(xxiii) $\sin x - \sin y = 2 \sin \dfrac{x - y}{2} \cos \dfrac{x + y}{2}.$

___

PROBLEMS 7.2

In Problems 1–15, use the basic identities to calculate $\sin \theta$ and $\cos \theta$.

1. $\theta = 6\pi$

2. $\theta = -30°$

3. $\theta = \dfrac{7\pi}{6}$

4. $\theta = \dfrac{5\pi}{6}$

5. $\theta = 75°$

6. $\theta = 15°$

7. $\theta = \dfrac{13\pi}{12}$

8. $\theta = -150°$

9. $\theta = -\dfrac{\pi}{12}$

10. $\theta = \dfrac{\pi}{8}$

11. $\theta = \dfrac{\pi}{16}$ [*Hint:* Use the result of Problem 10.]

12. $\theta = \dfrac{3\pi}{8}$

13. $\theta = 67\dfrac{1}{2}°$

14. $\theta = \dfrac{7\pi}{24}$

15. $\theta = -7\dfrac{1}{2}°.$

16. Prove Theorem 1 in the case $0° < \theta < 90°$ and $90° < \varphi < 180°$.

17. Prove that $\cos[(\pi/2) + x] = -\sin x.$

18. Show that $\sin[(\pi/2) + x] = \cos x.$ [*Hint:* Use Problem 17 to show that $\sin[x + (\pi/2)] = -\cos(x + \pi)$ and then use identity (iv).]

19. Use Problems 17 and 18 to show that $\sin(x + y) = \sin x \cos y - \cos x \sin y.$ [*Hint:* Start with $\sin(x + y) = -\cos[(\pi/2) + x + y]$ and then apply Theorem 1.]

20. Prove identities (xiv) and (xv). [*Hint:* Use Theorem 1, Problem 19 and identities (ii) and (iii).]

21. Prove identities (vi), (vii), (x), and (xi). [*Hint:* Use Problem 20.]

22. Prove identities (xvi) and (xvii). [*Hint:* Use Theorem 1 and Problem 19.]

23. Prove that $\cos(x/2) = \pm\sqrt{(1 + \cos x)/2}.$ [*Hint:* Use identity (xvi) to show that $\cos x = 2 \cos^2(x/2) - 1.$]

24. Prove identity (xix). [*Hint:* Use identity (i) and Problem 23.]

25. Prove identities (xx) and (xxi). [*Hint:* Use identities (xviii) and (xix).]

26. Prove that $\cos x - \cos y = 2 \sin[(x + y)/2] \sin[(y - x)/2].$ [*Hint:* Expand the right side using identities (xiii), (xv), and (xix).]

27. Prove identity (xxiii).

28. Graph the function $y = 3 \sin x$. The greatest value the function takes is called the *amplitude* of the function. Show that in this case the amplitude is equal to 3.
29. Graph the function $y = -2 \cos x$. What is the amplitude?
30. Show that the function $y = \sin 2x$ is periodic of period π. Graph the function.
31. Show that the function $y = 4 \cos(x/2)$ is periodic of period 4π. What is its amplitude? Graph the curve.
32. Graph the curve $y = 3 \sin(x/3)$.
33. Graph the curve $y = \sin(x - 1)$. [*Hint:* See Section 1.7.] Show that its period is 2π.
34. Graph the curve $y = 2 \sin[(x/2) + 3]$. What is its period? What is its amplitude?
35. Graph the curve $y = 3 \cos(3x - \frac{1}{2})$. What is its period? What is its amplitude?

7.3 Other Trigonometric Functions

Besides the two functions we have already discussed, there are four other trigonometric functions defined in terms of $\sin x$ and $\cos x$:

$$
\begin{array}{lll}
\text{(i)} & \text{tangent } x = \tan x = \dfrac{\sin x}{\cos x}, & \text{for } \cos x \neq 0. \\[2.5ex]
\text{(ii)} & \text{cotangent } x = \cot x = \dfrac{\cos x}{\sin x} = \dfrac{1}{\tan x}, & \text{for } \sin x \neq 0. \\[2.5ex]
\text{(iii)} & \text{secant } x = \sec x = \dfrac{1}{\cos x}, & \text{for } \cos x \neq 0. \\[2.5ex]
\text{(iv)} & \text{cosecant } x = \csc x = \dfrac{1}{\sin x}, & \text{for } \sin x \neq 0.
\end{array}
$$

Each of these four functions grows without bound as x approaches certain values. When $x \to 0^+$, $\cot x$ and $\csc x$ approach $+\infty$ and when $x \to 0^-$, $\cot x$ and $\csc x$ approach $-\infty$. This is true because $\sin x > 0$ for x near 0 and positive and $\sin x < 0$ for x near zero and negative (see Figure 7.2.4). Also, $\cos x$ is near 1 for x near 0. Similarly, we obtain

$$
\lim_{x \to \pi/2^-} \tan x = +\infty, \qquad \lim_{x \to \pi/2^+} \tan x = -\infty,
$$

$$
\lim_{x \to \pi/2^-} \sec x = +\infty, \qquad \lim_{x \to \pi/2^+} \sec x = -\infty.
$$

These facts hold because $\cos x$ is positive for $x < \pi/2$ and x near $\pi/2$ and $\cos x$ is negative for $x > \pi/2$ and x near $\pi/2$. (See Figure 7.2.3.)

We also observe that

$$
\tan 0 = \frac{\sin 0}{\cos 0} = \frac{0}{1} = 0 \quad \text{and} \quad \cot \frac{\pi}{2} = \frac{\cos(\pi/2)}{\sin(\pi/2)} = \frac{0}{1} = 0.
$$

We note that since $-1 \leq \sin x \leq 1$ and $-1 \leq \cos x \leq 1$, we see that $|\sec x| \geq 1$ and $|\csc x| \geq 1$. That is, $\sec x$ and $\csc x$ can never take values in the open interval $(-1, 1)$. Finally,

$$
\tan(x + \pi) = \frac{\sin(x + \pi)}{\cos(x + \pi)} = \frac{-\sin x}{-\cos x} = \frac{\sin x}{\cos x} = \tan x
$$

so that $\tan x$ is periodic of period π. Similarly $\cot x$ is periodic of period π. Also

$$\sec(x + 2\pi) = \frac{1}{\cos(x + 2\pi)} = \frac{1}{\cos x} = \sec x$$

so that $\sec x$ (and $\csc x$) are periodic of period 2π.

Values of $\tan x$, $\cot x$, $\sec x$ and $\csc x$ are given in Table 1. Putting this all

TABLE 1

x	0	$\dfrac{\pi}{6}$	$\dfrac{\pi}{4}$	$\dfrac{\pi}{3}$	$\dfrac{\pi}{2}$	π	$\dfrac{3\pi}{2}$	2π
$\tan x$	0	$\dfrac{1}{\sqrt{3}}$	1	$\sqrt{3}$	undefined	0	undefined	0
$\cot x$	undefined	$\sqrt{3}$	1	$\dfrac{1}{\sqrt{3}}$	0	undefined	0	undefined
$\sec x$	1	$\dfrac{2}{\sqrt{3}}$	$\sqrt{2}$	2	undefined	-1	undefined	1
$\csc x$	undefined	2	$\sqrt{2}$	$\dfrac{2}{\sqrt{3}}$	1	undefined	-1	undefined

together and using our knowledge of the functions $\sin x$ and $\cos x$, we obtain the graphs given in Figure 1.

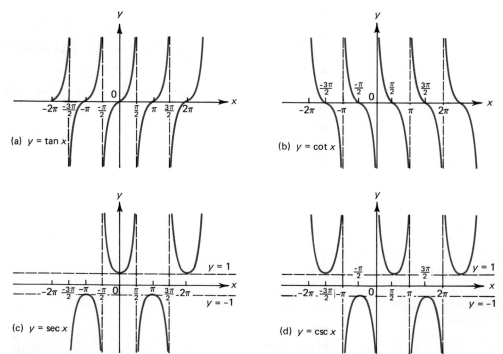

(a) $y = \tan x$

(b) $y = \cot x$

(c) $y = \sec x$

(d) $y = \csc x$

Figure 1

There are many identities involving the four functions introduced in this section. However, for our purposes, there are only two that will prove especially useful:

$$1 + \tan^2 x = \sec^2 x, \quad \text{★} \tag{1}$$

$$1 + \cot^2 x = \csc^2 x. \quad \text{⋈} \tag{2}$$

Both can be obtained by starting with the identity

$$\sin^2 x + \cos^2 x = 1$$

and then dividing both sides by $\cos^2 x$ to obtain (1) and by $\sin^2 x$ to obtain (2).

PROBLEMS 7.3

In Problems 1–15 calculate $\tan x$, $\cot x$, $\sec x$ and $\csc x$.

1. $x = 6\pi$

2. $x = -30°$

3. $x = \dfrac{7\pi}{6}$

4. $x = \dfrac{5\pi}{6}$

5. $x = 75°$

6. $x = 15°$

7. $x = \dfrac{13\pi}{12}$

8. $x = -150°$

9. $x = -\dfrac{\pi}{12}$

10. $x = \dfrac{\pi}{8}$

11. $x = \dfrac{\pi}{16}$

12. $x = \dfrac{3\pi}{8}$

13. $x = 67\dfrac{1}{2}°$

14. $x = \dfrac{7\pi}{24}$

15. $x = -7\dfrac{1}{2}°.$

In Problems 16–21, find the period of the given function and sketch the graph of the function.

16. $y = \tan 2x$

17. $y = 3 \sec \dfrac{x}{3}$

18. $y = 4 \cot(4x + 1)$

19. $y = 2 \csc 6x$

20. $y = -2 \tan\left(\dfrac{x}{5} + 1\right)$

21. $y = 8 \sec(5x + 5).$

22. Using the corresponding formulas for sine and cosine, show that

$$\tan(x - y) = \frac{\tan x - \tan y}{1 + \tan x \tan y}.$$

23. Show that

$$\tan(x + y) = \frac{\tan x + \tan y}{1 - \tan x \tan y}.$$

7.4 Triangles

In many elementary courses in trigonometry, the six trigonometric functions are introduced in terms of the ratios of sides of a right triangle. Consider the angle θ in the right triangle in Figure 1. The side opposite θ is labeled "*op*," the side adjacent to

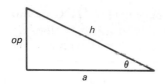

Figure 1

θ (which is not the hypotenuse) is labeled "a" and the hypotenuse (the side opposite the right angle) is labeled "h." Then we define

$$\sin \theta = \frac{\text{opposite}}{\text{hypotenuse}} = \frac{op}{h}, \qquad \cos \theta = \frac{\text{adjacent}}{\text{hypotenuse}} = \frac{a}{h},$$

$$\tan \theta = \frac{\text{opposite}}{\text{adjacent}} = \frac{op}{a}, \qquad \cot \theta = \frac{\text{adjacent}}{\text{opposite}} = \frac{a}{op},$$

$$\sec \theta = \frac{\text{hypotenuse}}{\text{adjacent}} = \frac{h}{a}, \qquad \csc \theta = \frac{\text{hypotenuse}}{\text{opposite}} = \frac{h}{op}.$$

Of course, these definitions are limited to angles between $0°$ and $90°$ (since the sum of the angles of a triangle is $180°$ and the right angle is $90°$). We now show that these "triangular" definitions give the same values as the "circular" definitions given earlier. It is only necessary to show this for the functions $\sin \theta$ and $\cos \theta$ as the other four functions are defined in terms of them. To verify this for $\sin \theta$ and $\cos \theta$, we place the triangle as in Figure 2. We then draw the circle with radius h that is centered at

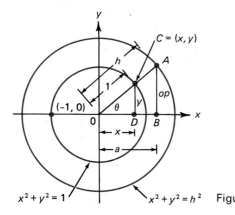

Figure 2

the origin and draw the unit circle. The triangles $0AB$ and $0CD$ are similar (since they have the same angles). Therefore, the ratios of corresponding sides are equal. This fact tells us that

$$\frac{op}{h} = \frac{y}{1} \qquad \text{and} \qquad \frac{a}{h} = \frac{x}{1}.$$

But op/h is the "triangle" definition of $\sin \theta$ while $y/1 = y$ is the circular definition of $\sin \theta$. Thus the two definitions lead to the same function. In a similar fashion we see that the two definitions of $\cos \theta$ lead to the same function.

Let L be any straight line. Its slope is the tangent of the angle θ that the line makes with the positive x-axis. To see this, we look at the line parallel to the given line that passes through the origin, and draw the unit circle around that new line (see Figure 3). Then the slope of the new line (which is parallel to L) which contains the points $(0, 0)$ and $(\cos\theta, \sin\theta)$ is given by

$$m = \frac{\Delta y}{\Delta x} = \frac{\sin\theta - 0}{\cos\theta - 0} = \tan\theta.$$

Triangles are often useful for computations of value of trigonometric functions. For example, we can use a triangle to prove that $\sin 30° = \frac{1}{2}$. Look at the equilateral triangle in Figure 4. Let the sides of the triangle have lengths of 1 unit and let BD be the angle bisector of angle B which is also the perpendicular bisector of side BD (this can be proven since the triangles ABD and DBC are congruent). The length of side BD is, from the Pythagorean theorem, equal to $\sqrt{3}/2$. Then

$$\sin 30° = \frac{op}{h} = \frac{\frac{1}{2}}{1} = \frac{1}{2}, \qquad \cos 30° = \frac{a}{h} = \frac{\sqrt{3}/2}{1} = \frac{\sqrt{3}}{2}, \qquad \text{and so on.}$$

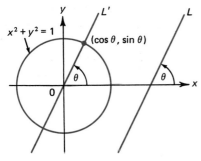

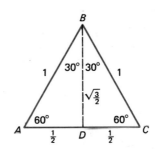

Figure 3   Figure 4

Another use of triangles is given in the example below.

EXAMPLE 1. If $\sin\theta = \frac{3}{5}$, calculate $\cos\theta$, $\tan\theta$, $\cot\theta$, $\sec\theta$ and $\csc\theta$.

SOLUTION. We draw a triangle (see Figure 5). Since $\sin\theta = op/h = \frac{3}{5}$, we set $op = 3$

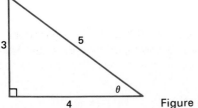

Figure 5

and $h = 5$. Then $a = \sqrt{5^2 - 3^2} = 4$ and, from the triangle, $\cos\theta = \frac{4}{5}$, $\tan\theta = \frac{3}{4}$, $\cot\theta = \frac{4}{3}$, $\sec\theta = \frac{5}{4}$ and $\csc\theta = \frac{5}{3}$. There is another possible answer. The function $\sin\theta$ is positive in the second quadrant. In that quadrant $\cos\theta$ is negative, $\tan\theta$ is negative, $\cot\theta$ is negative, $\sec\theta$ is negative and $\csc\theta$ is positive. Thus another possible set of answers is $\cos \theta = -\frac{4}{5}$, $\tan \theta = -\frac{3}{4}$, $\cot \theta = -\frac{4}{3}$, $\sec \theta = -\frac{5}{4}$ and $\csc \theta = \frac{5}{3}$. In order to get a single answer, we must indicate the quadrant to which θ belongs.

The procedure outlined in the above example will be useful in Chapter 8. There are many other things that can be discussed using triangles and which are covered in any trigonometry course. Two interesting rules, the *law of cosines* and the *law of sines*, are stated in Problems 14 and 16.

EXAMPLE 2.   According to *Fermat's principle†: when light travels through various media, it takes the path which requires the least time.* To illustrate this principle, we suppose that light is traveling from a point beneath the surface of water to a point in the air, above the surface. In each medium (water and air), the light ray will, according to Fermat's principle, travel in a straight line. To illustrate this phenomenon, we assume that the light emanates from a point P one unit of distance below the water surface to a point Q one unit above the surface and two units away in horizontal distance (see Figure 6). At what point S on the surface of the water between points R and U (in the figure) will the light ray pass in order that it reach Q in minimum time?

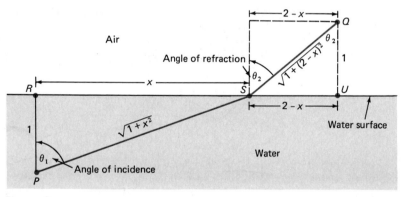

Figure 6

SOLUTION. If the distance from R to S is denoted by x, then the distance from S to U is $2 - x$ so that

$$\overline{PS} = \sqrt{1 + x^2} \qquad \text{and} \qquad \overline{SQ} = \sqrt{1 + (2 - x)^2}. \tag{1}$$

Since distance = velocity × time, time = distance/velocity and if v_1 and v_2 represent the velocities of the light ray in water and air, respectively, then the total time for the ray to go from P to Q is given by

$$T = \frac{\sqrt{1 + x^2}}{v_1} + \frac{\sqrt{1 + (2 - x)^2}}{v_2}. \tag{2}$$

† Pierre Fermat (1601–1665) was a celebrated French mathematician who helped develop a wide variety of topics of modern mathematics.

Then

$$\frac{dT}{dt} = \frac{x}{v_1\sqrt{1+x^2}} - \frac{(2-x)}{v_2\sqrt{1+(2-x)^2}} \tag{3}$$

and

$$\frac{d^2T}{dt^2} = \frac{1}{v_1(1+x^2)^{3/2}} + \frac{1}{v_2(1+(2-x)^2)^{3/2}}.$$

Since $d^2T/dt^2 > 0$, the critical points of (3) will all be minima. Setting $dT/dt = 0$ in (3), we obtain

$$\frac{x}{v_1\sqrt{1+x^2}} = \frac{2-x}{v_2\sqrt{1+(2-x)^2}}$$

or

$$\frac{v_1}{v_2} = \frac{x/\sqrt{1+x^2}}{(2-x)/\sqrt{1+(2-x)^2}}. \tag{4}$$

But, again from Figure 6, if θ_1 and θ_2 denote the angles at which the ray leaves the source and approaches the object, respectively, then

$$\frac{x}{\sqrt{1+x^2}} = \sin\theta_1 \quad \text{and} \quad \frac{2-x}{\sqrt{1+(2-x)^2}} = \sin\theta_2. \tag{5}$$

Then, from (4), the minimum time is achieved when

$$\frac{v_1}{v_2} = \frac{\sin\theta_1}{\sin\theta_2}. \tag{6}$$

The angles θ_1 and θ_2 are called the angles of *incidence* and *refraction*, respectively. The equation (6) is known as *Snell's law* and it states that, according to Fermat's principle, the ratio of the sines of the angles of incidence and refraction is equal to the ratio of the velocities of light in the two media. Note that if $v_1 = v_2$, then $\theta_1 = \theta_2$ so that the light ray will move in a straight line.

PROBLEMS 7.4

In Problems 1–10 the value of one of the six circular functions is given. Find the values of the other five functions in the indicated quadrant.

1. $\cos\theta = \frac{5}{11}$; first quadrant
2. $\tan\theta = 3$; first quadrant
3. $\sec\theta = 2$; fourth quadrant
4. $\cot\theta = -1$; second quadrant
5. $\csc\theta = 5$; second quadrant
6. $\sin\theta = -\frac{2}{3}$; third quadrant
7. $\sin\theta = -\frac{2}{3}$; fourth quadrant
8. $\sec\theta = -7$; second quadrant
9. $\tan\theta = 10$; third quadrant
10. $\cot\theta = 3$; first quadrant.

11. If the angle of incidence of light emanating from one medium to another is 45°, and if the light travels twice as fast in the first medium as in the second, calculate the angle of refraction.

12.   If the angle of incidence in the above example is 45° and the angle of refraction is 60°, how much faster does light travel in the second medium?

13.   The speed of light in air is, approximately,

$$c = 3 \times 10^{10} \quad \text{cm/sec.}$$

Light enters a certain medium at an angle of 60° and "reflects" at an angle of 30°. What is the speed of light in the second medium?

*14.   Let A, B, and C be the vertices of a triangle and let a, b, and c denote the corresponding sides (see Figure 7a). Then the law of cosines states that

$$c^2 = a^2 + b^2 - 2ab \cos C.$$

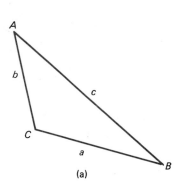

(a)

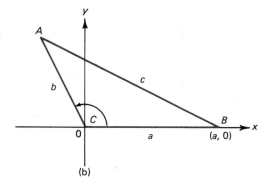

(b)

Figure 7

Prove this result. [*Hint:* Place the triangle as in Figure 7b. Use the circular definition of the functions $\sin x$ and $\cos x$ to show that the coordinates of the vertex A are $(b \cos C, b \sin C)$. Then use the distance formula to find $c = \overline{AB}$.]

15.   The sides of a triangle are 6, 6, and 2. Use the law of cosines to calculate the angles of the triangle.

16.   Let ABC be the vertices of a triangle. The *law of sines* states that

$$\frac{a}{\sin A} = \frac{b}{\sin B} = \frac{c}{\sin C}.$$

Prove this result. [*Hint:* Drop a perpendicular line from A to the side a (see Figure 8a or 8b). Show that, in either case,

$$\frac{h}{c} = \sin B \qquad \text{and} \qquad \frac{h}{b} = \sin C$$

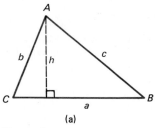

(a)

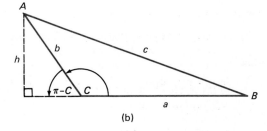

(b)

Figure 8

(in 8b, we need the fact that $\sin(\pi - C) = \sin C$). Use these facts to complete the proof.]

17.   Two angles of a triangle are $23°$ and $85°$. The side between them has a length of 5. Use the law of sines to find the lengths of the other two sides.

7.5 Differentiation of Trigonometric Functions

In this section we show how the derivatives of the six circular functions can be calculated. All our calculations are based on the theorem below.

THEOREM 1

$$\lim_{\theta \to 0} \frac{\sin \theta}{\theta} = 1. \tag{1}$$

Remark. The limit in (1) cannot be calculated by evaluation since $(\sin \theta)/\theta$ is not defined at 0.

PROOF. We prove the theorem in the case $0 < \theta < \pi/2$. The case $-\pi/2 < \theta < 0$ is similar and is left as an exercise (see Problem 59). Consider Figure 1. From the figure, we see that

$$\text{area of } \Delta 0AC < \text{area of sector } 0AC < \text{area of } \Delta 0BC. \tag{2}$$

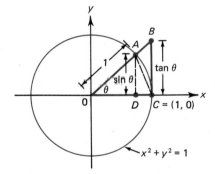

Figure 1

To calculate the area of a sector, we use the fact that if $\theta = 2\pi$ radians, then the sector is the area of the entire unit circle which is π. If an angle of 2π radians corresponds to a sector area of π, then θ radians corresponds to a sector area of $\theta/2$. Therefore, the area of the sector $0AC = \theta/2$. We also have, $\sin \theta = \overline{AD}$ since A is a point on the unit circle. Similarly, $\tan \theta = \overline{BC}/\overline{OC} = \overline{BC}$ since $\overline{OC} = 1$. Thus,

$$\text{area of } \Delta 0AC = \tfrac{1}{2}(\text{base} \times \text{height}) = \tfrac{1}{2}\overline{OC} \times \overline{AD} = \tfrac{1}{2}\sin \theta$$

and

$$\text{area of } \Delta 0BC = \tfrac{1}{2}\overline{OC} \times \overline{BC} = \tfrac{1}{2}\tan \theta.$$

Thus, we obtain from (2)

$$\tfrac{1}{2}\sin \theta < \tfrac{1}{2}\theta < \tfrac{1}{2}\tan \theta. \tag{3}$$

For $\theta \neq 0$, we multiply the inequalities in (3) by 2 and divide by $\sin \theta$ to obtain

$$1 < \frac{\theta}{\sin \theta} < \frac{1}{\cos \theta} \tag{4}$$

or

$$\cos \theta < \frac{\sin \theta}{\theta} < 1 \tag{5}$$

(since $0 < a < b$ implies that $1/a > 1/b$). Now, a glance at Figure 7.2.3 tells us that $\lim_{\theta \to 0} \cos \theta = 1$. Since $0 < \theta < \pi/2$, we let $\theta \to 0^+$ in (5) and use the squeezing theorem (Theorem 2.5.1) to conclude that $\lim_{\theta \to 0^+}[(\sin \theta)/\theta] = 1$. Similarly, $\lim_{\theta \to 0^-}[(\sin \theta)/\theta] = 1$ and the proof is complete. Actually the proof is intuitively reasonable since, in Figure 1, as $\theta \to 0^+$ the length of the line AD and the length of the arc AC (denoted $\widehat{AC}$) get closer and closer together. Then the theorem follows since $\overline{AD} = \sin \theta$ and $\widehat{AC} = \theta$.

To illustrate this important result in another way, we calculate, in Table 1, some values of $(\sin \theta)/\theta$ for θ near 0. It seems clear, from the table, that

$$\lim_{\theta \to 0} \frac{\sin \theta}{\theta} = 1.$$

TABLE 1

θ (radians)	$\sin \theta$	$\dfrac{\sin \theta}{\theta}$
1	0.8414709848	0.8414709848
0.5	0.4794255386	0.9588510772
0.1	0.0998334166	0.9983341665
0.01	0.0099998333	0.9999833334
0.001	0.0009999998	0.999999833
0.0001	0.0000999999	0.9999999995

EXAMPLE 1. Calculate $\lim\limits_{\theta \to 0} \dfrac{\sin 2\theta}{\theta}$.

SOLUTION

$$\lim_{\theta \to 0} \frac{\sin 2\theta}{\theta} = \lim_{\theta \to 0} \frac{2 \sin 2\theta}{2\theta} = 2 \lim_{\theta \to 0} \frac{\sin 2\theta}{2\theta}.$$

But as $\theta \to 0$, $2\theta \to 0$ so that

$$2 \lim_{\theta \to 0} \frac{\sin 2\theta}{2\theta} = 2 \lim_{2\theta \to 0} \frac{\sin 2\theta}{2\theta} = 2 \cdot 1 = 2.$$

EXAMPLE 2. Calculate $\lim\limits_{\theta \to 0} \dfrac{\cos \theta - 1}{\theta}$.

SOLUTION

$$\lim_{\theta \to 0} \frac{\cos \theta - 1}{\theta} = \lim_{\theta \to 0} \frac{(\cos \theta - 1)(\cos \theta + 1)}{\theta(\cos \theta + 1)} = \lim_{\theta \to 0} \frac{\cos^2 \theta - 1}{\theta(\cos \theta + 1)}$$

$$= \lim_{\theta \to 0} \frac{-\sin^2 \theta}{\theta(\cos \theta + 1)} \qquad \text{(since } \sin^2 \theta + \cos^2 \theta = 1\text{)}$$

$$= -\lim_{\theta \to 0} \frac{\sin \theta}{\theta} \lim_{\theta \to 0} \frac{\sin \theta}{(\cos \theta + 1)} = -1 \cdot \frac{\lim_{\theta \to 0} \sin \theta}{\lim_{\theta \to 0}(\cos \theta + 1)} = 1 \cdot \frac{0}{2} = 0.$$

THEOREM 2. $\dfrac{d}{dx} \sin x = \cos x.$ (6)

PROOF

$$\frac{d}{dx} \sin x = \lim_{\Delta x \to 0} \frac{\sin(x + \Delta x) - \sin x}{\Delta x}.$$

We now will use a basic trigonometric identity to simplify this expression. From identity (xiii) in Table 7.2.2, $\sin(x + \Delta x) = \sin x \cos \Delta x + \cos x \sin \Delta x$ so that

$$\frac{d}{dx} \sin x = \lim_{\Delta x \to 0} \frac{\sin x \cos \Delta x + \cos x \sin \Delta x - \sin x}{\Delta x}$$

$$= \lim_{\Delta x \to 0} \sin x \left(\frac{\cos \Delta x - 1}{\Delta x} \right) + \lim_{\Delta x \to 0} \cos x \cdot \frac{\sin \Delta x}{\Delta x}$$

$$= \sin x \lim_{\Delta x \to 0} \frac{\cos \Delta x - 1}{\Delta x} + \cos x \lim_{\Delta x \to 0} \frac{\sin \Delta x}{\Delta x}$$

$$= \sin x \cdot 0 + \cos x \cdot 1 = \cos x.$$

We can use the derivative of $\sin x$ to calculate the derivatives of the other five trigonometric functions. First we show how the chain rule applies in this setting.

THEOREM 3.   If $u(x)$ is a differentiable function of x, then

$$\boxed{\frac{d}{dx} \sin u = \cos u \frac{du}{dx}.}$$ (7)

PROOF. Let $f(u) = \sin u$. Then from the chain rule

$$\frac{d}{dx} f(u) = \frac{df}{du} \frac{du}{dx} = \cos u \frac{du}{dx}.$$

EXAMPLE 3.   Calculate $\dfrac{d}{dx} \sin x^3.$

SOLUTION. If $u = x^3$, then $du/dx = 3x^2$ and

$$\frac{d}{dx} \sin x^3 = (\cos x^3) 3x^2.$$

THEOREM 4

$$\frac{d}{dx}\cos x = -\sin x. \tag{8}$$

PROOF. From identity (xi) in Table 7.2.2, $\cos x = \sin[(\pi/2) - x]$ so that

$$\frac{d}{dx}\cos x = \frac{d}{dx}\sin\left(\frac{\pi}{2} - x\right).$$

If $u = (\pi/2) - x$, then $du/dx = -1$ and

$$\frac{d}{dx}\sin\left(\frac{\pi}{2} - x\right) = -\cos\left(\frac{\pi}{2} - x\right) = -\sin x.$$

The last step follows from identity (x) in Table 7.2.2.

The following result can be proven by using the chain rule as in Theorem 3.

THEOREM 5. If u is a differentiable function of x, then

$$\boxed{\frac{d}{dx}\cos u = -\sin u\,\frac{du}{dx}.} \tag{9}$$

EXAMPLE 4. Calculate $\dfrac{d}{dx}\cos e^{x^2}$.

SOLUTION. If $u = e^{x^2}$, then $du/dx = 2xe^{x^2}$ and

$$\frac{d}{dx}\cos e^{x^2} = -(\sin e^{x^2})(2xe^{x^2}).$$

We can now calculate the derivatives of the other four trigonometric functions.

THEOREM 6

(i) $\dfrac{d}{dx}\tan x = \sec^2 x.$ (10)

(ii) $\dfrac{d}{dx}\cot x = -\csc^2 x.$ (11)

(iii) $\dfrac{d}{dx}\sec x = \sec x \tan x.$ (12)

(iv) $\dfrac{d}{dx}\csc x = -\csc x \cot x.$ (13)

PROOF. We will prove (i) and (iii) and leave (ii) and (iv) as exercises (see Problems 60 and 61).

(i) $\tan x = \sin x/\cos x$, so that using the quotient rule,

$$\frac{d}{dx}\tan x = \frac{d}{dx}\frac{\sin x}{\cos x} = \frac{\cos x\,\dfrac{d}{dx}\sin x - \sin x\,\dfrac{d}{dx}\cos x}{\cos^2 x}$$

$$= \frac{\cos x \cos x - \sin x (-\sin x)}{\cos^2 x} = \frac{\cos^2 x + \sin^2 x}{\cos^2 x}$$

$$= \frac{1}{\cos^2 x} = \sec^2 x.$$

(iii) $\dfrac{d}{dx} \sec x = \dfrac{d}{dx} \dfrac{1}{\cos x} = \dfrac{d}{dx}(\cos x)^{-1}$

$$= -(\cos x)^{-2} \frac{d}{dx} \cos x = -\frac{1}{\cos^2 x} \cdot -\sin x \qquad \text{(by the chain rule)}$$

$$= \frac{\sin x}{\cos^2 x} = \frac{1}{\cos x} \cdot \frac{\sin x}{\cos x} = \sec x \tan x.$$

For convenience, we summarize in Table 2 the rules for differentiating trigonometric functions.

TABLE 2

(i)	$\dfrac{d}{dx} \sin x = \cos x$	$\dfrac{d}{dx} \sin u = \cos u \dfrac{du}{dx}$
(ii)	$\dfrac{d}{dx} \cos x = -\sin x$	$\dfrac{d}{dx} \cos u = -\sin u \dfrac{du}{dx}$
(iii)	$\dfrac{d}{dx} \tan x = \sec^2 x$	$\dfrac{d}{dx} \tan u = \sec^2 u \dfrac{du}{dx}$
(iv)	$\dfrac{d}{dx} \cot x = -\csc^2 x$	$\dfrac{d}{dx} \cot u = -\csc^2 u \dfrac{du}{dx}$
(v)	$\dfrac{d}{dx} \sec x = \sec x \tan x$	$\dfrac{d}{dx} \sec u = \sec u \tan u \dfrac{du}{dx}$
(vi)	$\dfrac{d}{dx} \csc x = -\csc x \cot x$	$\dfrac{d}{dx} \csc u = -\csc u \cot u \dfrac{du}{dx}$

EXAMPLE 5.   Calculate $\dfrac{d}{dx} \tan^2 \sqrt{x}.$

SOLUTION. We use the chain rule twice. We know that

$$\frac{d}{dx} u^2 = 2u \frac{du}{dx}.$$

Then

$$\frac{d}{dx} \tan^2 \sqrt{x} = 2 \tan \sqrt{x} \frac{d}{dx} \tan \sqrt{x} = 2 \tan \sqrt{x} \cdot \sec^2 \sqrt{x} \frac{d}{dx} \sqrt{x}$$

$$= \frac{2 \tan \sqrt{x} \sec^2 \sqrt{x}}{2\sqrt{x}} = \frac{\tan \sqrt{x} \sec^2 \sqrt{x}}{\sqrt{x}}.$$

EXAMPLE 6.   Calculate $\dfrac{d}{dx} \sec(\ln x).$

SOLUTION

$$\frac{d}{dx} \sec(\ln x) = [\sec(\ln x)][\tan(\ln x)]\left[\frac{d}{dx} \ln x\right] = \frac{[\sec(\ln x)][\tan(\ln x)]}{x}.$$

EXAMPLE 7. Find $\dfrac{d}{dx}\ln(\sec x + \tan x)$.

SOLUTION

$$\frac{d}{dx}\ln(\sec x + \tan x) = \frac{1}{\sec x + \tan x} \cdot \frac{d}{dx}(\sec x + \tan x)$$

$$= \frac{\sec x \tan x + \sec^2 x}{\sec x + \tan x} = \frac{\sec x(\sec x + \tan x)}{\sec x + \tan x} = \sec x.$$

EXAMPLE 8. Let $y = \sin(x + y)$. Find dy/dx.

SOLUTION. By implicit differentiation we find that

$$\frac{dy}{dx} = \cos(x + y)\left(1 + \frac{dy}{dx}\right) = \cos(x + y) + \frac{dy}{dx}\cos(x + y)$$

and

$$\frac{dy}{dx} = \frac{\cos(x + y)}{1 - \cos(x + y)}.$$

Note that dy/dx is only defined if $\cos(x + y) \neq 1$; that is, when

$$x + y \neq \pm\frac{\pi}{2}, \pm\frac{3\pi}{2}, \ldots \qquad x + y \neq 0, \pm 2\pi, \pm 4\pi, \pm 6\pi, \ldots$$

EXAMPLE 9. Estimate $\sin 31°$ using differentials.

SOLUTION. We use the fact that we know $\sin 30°$. If $f(x) = \sin x$, $x = 30° = \pi/6$, and $dx = 1° \approx 0.0175$ radians; then $\sin x = \tfrac{1}{2}$ and

$$dy = f'(x)\,dx \approx \left(\cos\frac{\pi}{6}\right)(0.0175) = \left(\frac{\sqrt{3}}{2}\right)(0.0175) \approx (0.866)(0.0175) = 0.015$$

so that $\sin 31° \approx .5 + .015 = .515$. The actual value, correct to 6 decimal places, is .515038. (What would happen if we used degrees instead of radians? Explain why this doesn't work.)

EXAMPLE 10. Graph the curve $y = \cos^2 x$.

SOLUTION. Since $\cos^2 x$ is periodic of period π, we need only draw the graph between $x = 0$ and $x = \pi$ and then continue it periodically. We have

$$\frac{dy}{dx} = 2\cos x(-\sin x) = 0$$

when either $\sin x$ or $\cos x$ is zero. This occurs at $x = 0, \pi/2$, and π. Also, from identity (xvii) in Table 7.2.2, $dy/dx = -2\sin 2x$ and $d^2y/dx^2 = -4\cos 2x$. When $x = 0$, $f'' = -4$, when $x = \pi/2$, $f'' = 4$, and when $x = \pi$, $f'' = -4$. Therefore, there are local maxima at $x = 0$ and $x = \pi$ and a local minimum at $x = 2$. Also, $\cos 2x = 0$ at $x = \pi/4$ and $3\pi/4$ and these are the points of inflection. For $0 \le x \le \pi/4$, $f''(x) < 0$ and the curve is concave down. It is concave up between $\pi/4$ and $3\pi/4$ and again concave down for $3\pi/4 < x \le \pi$. Putting this all together, we obtain the curve in Figure 2.

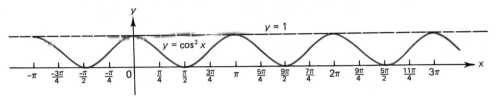

Figure 2

EXAMPLE 11. Graph the curve $y = e^{-x} \sin x$.

SOLUTION. First notice that since $-1 \le \sin x \le 1$ and since $e^{-x} \to 0$ as $x \to \infty$, $\lim_{x \to \infty} e^{-x} \sin x = 0$. (Be careful, however. The limit $\lim_{x \to \infty} \sin x$ *does not exist* since $\sin x$ oscillates between -1 and 1 but does not approach any single number.) Also, since $e^{-x} > 0$, $-1 \le \sin x \le 1$ implies that $-e^{-x} \le e^{-x} \sin x \le e^{-x}$ so that the graph of the function $e^{-x} \sin x$ stays between the graphs of the functions $-e^{-x}$ and e^{-x}. We then calculate

$$\frac{dy}{dx} = e^{-x} \cos x - e^{-x} \sin x.$$

This is zero when $e^{-x} \cos x = e^{-x} \sin x$ or when $\sin x = \cos x$ and $\sin x / \cos x = \tan x = 1$. Since $\tan x = 1$ when $x = \pi/4$ and since $\tan x$ is periodic of period π, we see that the critical points are the points $\pi/4, -3\pi/4, 5\pi/4, -7\pi/4, 9\pi/4, \ldots$. Now $d^2y/dx^2 = -e^{-x} \sin x - e^{-x} \cos x - e^{-x} \cos x + e^{-x} \sin x = -2e^{-x} \cos x$ and there are points of inflection at $x = \pi/2, 3\pi/2, 5\pi/2, \ldots$. When $x = \pi/4, d^2y/dx^2 < 0$ and when $x = 5\pi/4$, $\cos x < 0$ (since $5\pi/4$ is in the third quadrant) so that $d^2y/dx^2 > 0$. Hence when $x = \pi/4$ we have a local maximum and at $x = 5\pi/4$ there is a local minimum. We now draw the curve using the fact that $e^{-x} \sin x$ always remains between $-e^{-x}$ and e^{-x}. See Figure 3.

The periodic motion exemplified by the function $y = \sin x$ is often called *harmonic motion* and the motion depicted in Figure 3 is called *damped harmonic motion*. We will discuss harmonic motion in detail in Section 7.8.

EXAMPLE 12. A ladder 8-m long leans against a wall 4-m long. If the lower end of the ladder is pulled away from the wall at a rate of 2 m/sec, how fast is the angle between the top of the ladder and the wall changing when the angle is $60° = \pi/3$ radians?

SOLUTION. We refer to Figure 4. We are asked to calculate $d\theta/dt$ when $\theta = \pi/3$. From the figure, we see that $\sin \theta = x/8$, where x is the distance from the foot of the ladder to the wall. We are told that $dx/dt = 2$. Then, using the chain rule,

$$\cos \theta \, \frac{d\theta}{dt} = \frac{d}{dt} \sin \theta = \frac{d}{dt} \frac{x}{8} = \frac{1}{8} \frac{dx}{dt} = \frac{2}{8} = \frac{1}{4}.$$

Thus $d\theta/dt = 1/(4 \cos \theta) = (\sec \theta)/4$ and, when $\theta = \pi/3$, $\sec \theta = 2$ so that $d\theta/dt = \frac{2}{4} = \frac{1}{2}$ radian/sec.

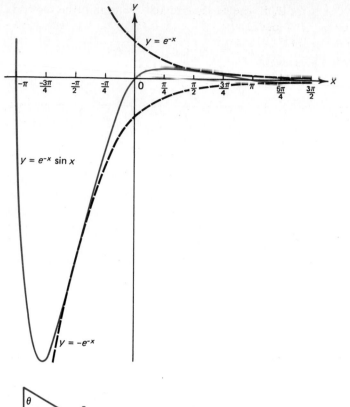

Figure 3

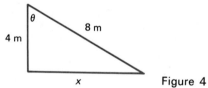

Figure 4

EXAMPLE 13. Blood is transported in the body by means of arteries, veins, arterioles, and capillaries. This procedure is ideally carried out in such a way so as to minimize the energy required to transport the blood from the heart to the organs and back again. Consider the two blood vessels depicted in Figure 5. There are many ways to try to compute the minimum energy required to pump blood from the point

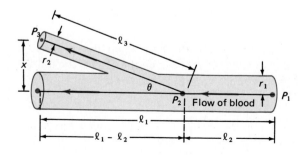

Figure 5

P_1 in the main vessel to the point P_3 in the smaller vessel.† We will try to find the minimum total resistance of the blood along the path $P_1P_2P_3$. According to Poiseuille's law (see Problem 3.2.39) the resistance is given by

$$R = \frac{\alpha l}{r^4}, \tag{14}$$

where l is the length of the vessel, r is its radius, and α is a constant of proportionality. The problem, then, is to find the "optimal" branching angle θ at which R is a minimum. According to Figure 5, $\sin\theta = x/l_3$ so that $l_3 = x/\sin\theta = x\csc\theta$. Also, $x/(l_1 - l_2) = \tan\theta$ so that $l_1 - l_2 = x/(\tan\theta) = x\cot\theta$ and $l_2 = l_1 - x\cot\theta$. We now calculate the total resistance along the path $P_1P_2P_3$. From P_1 to P_2,

$$R_{1,2} = \frac{\alpha l_2}{r_1^{\,4}}.$$

From P_2 to P_3,

$$R_{2,3} = \frac{\alpha l_3}{r_2^{\,4}}.$$

Then $R = R_{1,2} + R_{2,3} = \alpha[(l_2/r_1^{\,4}) + (l_3/r_2^{\,4})]$. But $l_3 = x\csc\theta$ and $l_2 = l_1 - x\cot\theta$ so that

$$R = \alpha\left(\frac{l_1 - x\cot\theta}{r_1^{\,4}} + \frac{x\csc\theta}{r_2^{\,4}}\right). \tag{15}$$

We now simplify matters by using the fact that x is fixed since x depends on P_1, P_3, and l_1, which are given in the problem, but not on P_2, l_2, l_3, or θ. (P_2, l_2, and l_3 all vary as θ varies but P_1, P_3, and l_1 do not.) Since R is a function of θ, we can find the minimum value of R by calculating $dR/d\theta$ and setting it to zero. But

$$\frac{dR}{d\theta} = \alpha\left(\frac{x\csc^2\theta}{r_1^{\,4}} - \frac{x\csc\theta\cot\theta}{r_2^{\,4}}\right)$$

$$= \alpha\left(\frac{x}{\sin^2\theta\, r_1^{\,4}} - \frac{x\cos\theta}{\sin^2\theta\, r_2^{\,4}}\right) = \frac{\alpha x}{\sin^2\theta}\left(\frac{1}{r_1^{\,4}} - \frac{\cos\theta}{r_2^{\,4}}\right).$$

Now, $\alpha x/(\sin^2\theta) \neq 0$ so that $dR/d\theta = 0$ when $1/r_1^{\,4} = (\cos\theta)/r_2^{\,4}$ or when

$$\cos\theta = \left(\frac{r_2}{r_1}\right)^4.$$

That this is indeed a minimum can be verified by calculating $d^2R/d\theta^2$. Thus we can calculate the optimum branching angle by merely considering the ratio of the radii of the blood vessels. In the cited reference by Rosen, it is shown that branching angles of blood vessels do, in many cases, obey the optimization rule we have just derived.

---

†For a more complete discussion of this topic, consult the book by R. Rosen, *Optimality Principles in Biology*, Butterworth, London, 1967.

PROBLEMS 7.5

In Problems 1–10 calculate the indicated limits.

1. $\displaystyle\lim_{x\to0}\frac{\sin\frac12 x}{x}$

2. $\displaystyle\lim_{x\to0}\frac{\sin^2 x}{x^2}$

3. $\displaystyle\lim_{x\to0}\frac{\sin^2 4x}{x^2}$ $\left[\textit{Hint: First find }\lim_{x\to0}\frac{\sin 4x}{x}.\right]$

4. $\displaystyle\lim_{x\to0}\frac{3x}{\sin 2x}$

5. $\displaystyle\lim_{x\to0}\frac{\sin^7 2x}{4x^7}$ $\left[\textit{Hint: First find }\lim_{x\to0}\frac{\sin 2x}{x}.\right]$

6. $\displaystyle\lim_{x\to0}\frac{\sin^2 x}{x}$

7. $\displaystyle\lim_{x\to0}\frac{\sin^2 3x}{4x}$

8. $\displaystyle\lim_{x\to0}\frac{\sin^3 4x}{3x^3}$

9. $\displaystyle\lim_{x\to0}\frac{\sin^3 2x}{x^2}$

10. $\displaystyle\lim_{x\to0}\frac{x^3}{3\sin^2 2x}.$

In Problems 11–50 calculate dy/dx.

11. $y = \sin 3x^2$

12. $y = \tan\dfrac{1}{x}$

13. $y = \sec^3 x^3$

14. $y = \sin x \cos x$

15. $y = \sqrt{1 + \sin x^2}$

16. $y = \sqrt{\sin x + \cos x}$

17. $y = \ln|\sin x|$

18. $y = \ln|\cos x|$

19. $y = -\ln|\csc x + \cot x|$

20. $y = x\tan^2 x$

21. $y = \cot\sqrt{x+1}$

22. $y = e^x \sin x$

23. $y = e^x \cos x$

24. $y = e^{-x}\sin\dfrac{\pi x}{2}$

25. $y = \sin(\cos x)$

26. $y = \cos(\sin x)$

27. $y = \cos e^x$

28. $y = x\cos x - \sin x$

29. $y = e^{\sin x}$

30. $y = \sin e^x$

31. $y = (\sec x + \tan x)^{10}$

32. $y = \sec^3\dfrac{\pi x}{2}$

*33. $y = (\sin x)^x$

34. $y = \sin x - \sqrt{\cos x}$

35. $y = \sqrt{x} + \cot\sqrt{x}$

36. $y = \dfrac{\tan x}{1 + \sec x}$

37. $y = x\sec x$

38. $y = (\csc x + 2\tan x)^3$

39. $y = (\cos 2x)^{5/7}$

40. $y = \sqrt{\tan^2 x + 1}$

41. $\sin x = \cos y$

42. $\sec x = \tan y$

43. $\sin(x+y) = \cos(x-y)$

44. $(\sin x + \sin y)^2 = 1$

45. $\sin x \cos y = \tan x \sec y$

46. $\sin y = \ln(x+y)$

47. $y = \cot(x+y)$

48. $\dfrac{\sin(x^2 + y^2)}{\cos(x^2 - y^2)} = 3$

49. $\csc(x+y) = \cot(2x+y)$

50. $y\sin x = x\tan y$

In Problems 51–58 sketch the given curve including all maxima, minima, and points of inflection.

51. $y = \sin x + \cos x$

52. $y = 2\sin x + \sin 2x$

53. $y = e^x \cos x$ **54.** $y = e^{-2x} \sin 3x$

55. $y = x + \cos x$ **56.** $y = x - \sin x$

57. $y = \tan^2 x$ **58.** $y = \sec^2 x.$

59. Prove Theorem 1 in the case $-\pi/2 < \theta < 0$. [*Hint:* Draw a sketch similar to the sketch in Figure 1 but with θ negative.]

60. Prove that $d \cot x/dx = -\csc^2 x$. [*Hint:* Write $\cot x = \cos x/\sin x$.]

61. Prove that $d \csc x/dx = -\csc x \cot x$.

62. Show graphically that there are an infinite number of solutions to the equation $x = -\tan x$.

63. Show that the solutions of $x = -\tan x$ are critical points of the curve $y = x \sin x$. Use this information to sketch $y = x \sin x$.

*64. Let k be any positive integer. Show by mathematical induction that

$$\frac{d^n \sin x}{dx^n} = \begin{cases} \sin x & \text{if } n = 4k, \\ \cos x & \text{if } n = 4k + 1, \\ -\sin x & \text{if } n = 4k + 2, \\ -\cos x & \text{if } n = 4k + 3. \end{cases}$$

65. Derive a formula analogous to the one in Problem 64 for the nth derivative of $\cos x$.

*66. (a) Find a formula for the angle at which the tangent line to $y = \sin x$ intersects the positive x-axis for $0 < x < \pi$. What is this angle when $x = \pi/6$?

(b) Find a formula for the angle at which the tangent line to $y = \cos x$ intersects the positive x-axis for $-\pi/2 < x < \pi/2$. What is this angle when $x = \pi/4$?

(c) Find the angle at which the tangents to the curves $y = \sin x$ and $y = \cos x$ intersect for $0 < x < \pi/2$.

In Problems 67–70 use differentials to approximate the value of the given function at the given point. In each problem, convert degrees to radians if appropriate.

67. $y = \sin 59°$ **68.** $y = \tan \dfrac{11\pi}{40}$ $\left[Hint: \dfrac{11\pi}{40} = \dfrac{\pi}{4} + \dfrac{\pi}{40}. \right]$

69. $y = \sec 151°$ **70.** $y = \csc 224°.$

71. A circular wheel centered at the origin with a radius of 13 cm is rotating in the counter-clockwise direction at a constant rate of 10 revolutions per minute. How fast are x and y changing when $x = 12$ and $y = 5$? [*Hint:* Express everything in terms of θ, the angle between the radius of the circle and the positive x-axis.]

72. A prison guard tower has a light which is 300 m from a straight wall. Its light revolves at a rate of 5 revolutions per minute. How fast is the light beam moving along the wall at a point 300 m along the wall (measured from the point on the wall which is closest (300 m) to the tower)?

73. In Problem 72, how fast is the light moving at a point $300\sqrt{3}$ m along the wall?

74. Show that if two blood vessels have equal radii, then blood resistance is minimized when the branching angle between them is zero (see Example 13).

75. Using Table A.4, find the branching angle which minimizes resistance if one blood vessel has twice the diameter of a second.

□76. The amount of fluid flowing across a weir (or dam) with a V-shaped notch has been determined to be

$$Q = 2.505 \left(\tan \frac{\theta}{2} \right) h^{2.47},$$

where Q is the volume of flow measured in m³/sec, θ is the angle of the notch, and h is the height in meters of the liquid, measured from the bottom edge of the notch. Assuming that h is kept constant at 2 m while the angle of the weir is increasing, how fast is Q increasing when $\theta = 40°$?

77. Can you explain why the derivatives of $\sec^2 x$ and $\tan^2 x$ are the same? What about $\cot^2 x$ and $\csc^2 x$?

7.6 Integration of Trigonometric Functions

We can reverse the process of the last section to calculate the integrals of certain trigonometric functions. The formulas in Table 1 follow from Table 7.5.2.

TABLE 1

(i) $\int \cos u \, du = \sin u + C.$

(ii) $\int \sin u \, du = -\cos u + C.$

(iii) $\int \sec^2 u \, du = \tan u + C.$

(iv) $\int \csc^2 u \, du = -\cot u + C.$

(v) $\int \sec u \tan u \, du = \sec u + C.$

(vi) $\int \csc u \cot u \, du = -\csc u + C.$

  In this section we will integrate functions which can be written in one of the above forms. In Chapter 8 we will see how other trigonometric functions can be integrated.

EXAMPLE 1. Calculate $\int 2 \cos 2x \, dx.$

SOLUTION. If $u = 2x$, then $du = 2 \, dx$ so that

$$\int 2 \cos 2x \, dx = \int \cos u \, du = \sin u + C = \sin 2x + C.$$

EXAMPLE 2. Calculate $\int x^2 \sec x^3 \tan x^3 \, dx.$

SOLUTION. If $u = x^3$, then $du = 3x^2 \, dx$ so we multiply and divide by 3 to obtain

$$\tfrac{1}{3} \int \sec x^3 \tan x^3 \cdot 3x^2 \, dx = \tfrac{1}{3} \int \sec u \tan u \, du = \tfrac{1}{3} \sec u + C = \tfrac{1}{3} \sec x^3 + C.$$

EXAMPLE 3. Calculate $\int \tan^5 x \sec^2 x \, dx.$

SOLUTION. This problem is a little different. If we set $u = \tan x$, then $du = \sec^2 x \, dx$ so that

$$\int \tan^5 x \sec^2 x \, dx = \int u^5 \, du = \frac{u^6}{6} = \frac{\tan^6 x}{6} + C.$$

EXAMPLE 4. Calculate $\displaystyle\int_{\pi/4}^{\pi/2} \sqrt{\cot x} \, \csc^2 x \, dx.$

SOLUTION. If $u = \cot x$, then $du = -\csc^2 x\,dx$ so we multiply and divide by -1 to obtain

$$-\int_{\pi/4}^{\pi/2} \sqrt{\cot x}(-\csc^2 x)\,dx = -\int u^{1/2}\,du = -\tfrac{2}{3}u^{3/2}$$

$$= -\tfrac{2}{3}\cot x^{3/2}\Big|_{\pi/4}^{\pi/2} = -\tfrac{2}{3}(0^{3/2} - 1^{3/2}) = \tfrac{2}{3}.$$

EXAMPLE 5. Calculate $\displaystyle\int \frac{\cot x}{\csc^4 x}\,dx$.

SOLUTION. If $u = \csc x$, then $du = -\csc x \cot x\,dx$. Then

$$\int \frac{\cot x}{\csc^4 x}\,dx = -\int \frac{-\csc x \cot x}{\csc^5 x}\,dx = -\int u^{-5}\,du = \frac{u^{-4}}{4} + C = \frac{1}{4\csc^4 x} + C.$$

EXAMPLE 6. Calculate $\displaystyle\int \tan x\,dx$.

SOLUTION

$$\int \tan x\,dx = \int \frac{\sin x}{\cos x}\,dx.$$

$$\int \tan x = \int \frac{\sin x}{\cos x}\,dx \qquad u = \cos x$$
$$\qquad du = -\sin x\,dx$$
$$-\int \frac{du}{u} = -\ln|u| + c$$

If $u = \cos x$, then $du = -\sin x\,dx$ so that

$$\int \frac{\sin x\,dx}{\cos x} = -\int \frac{-\sin x\,dx}{\cos x} = -\int \frac{du}{u} = -\ln|u| + C = -\ln|\cos x| + C.$$

Since $-\ln x = \ln 1/x$, we also have

$$\int \frac{\sin x}{\cos x}\,dx = \ln|\sec x| + C.$$

EXAMPLE 7. Calculate $\displaystyle\int_0^{\pi/2} \sin^2 x\,dx$.

SOLUTION. By identity (xxi) in Table 7.2.2,

$$\int_0^{\pi/2} \sin^2 x\,dx = \int_0^{\pi/2} \frac{1 - \cos 2x}{2}\,dx = \int_0^{\pi/2} \frac{1\,dx}{2} - \int_0^{\pi/2} \frac{\cos 2x}{2}\,dx$$

$$= \frac{x}{2}\Big|_0^{\pi/2} - \frac{\sin 2x}{4}\Big|_0^{\pi/2} = \frac{\pi}{4}.$$

EXAMPLE 8. Calculate the area of the region in the first quadrant bounded by $y = \sin x$, $y = \cos x$, and the y-axis.

SOLUTION. The area requested is drawn in Figure 1. The curves intersect when $\sin x = \cos x$, $\tan x = 1$, and $x = \pi/4$. Then

$$A = \int_0^{\pi/4} (\cos x - \sin x)\,dx = (\sin x + \cos x)\Big|_0^{\pi/4} = \sqrt{2} - 1.$$

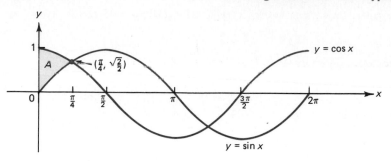

Figure 1

PROBLEMS 7.6

In Problems 1–36 calculate the integrals.

1. $\displaystyle\int \sin 3x\, dx$

2. $\displaystyle\int \cos \frac{x}{5}\, dx$

3. $\displaystyle\int_0^{\pi/12} 3 \cos 2x\, dx$

4. $\displaystyle\int_{-\pi/8}^{\pi/8} \frac{\sin 4x}{2}\, dx$

5. $\displaystyle\int \sin 2x \cos 2x\, dx$

6. $\displaystyle\int \cos^3 x \sin x\, dx$

7. $\displaystyle\int_0^{\pi/6} \sqrt{\sin x}\, \cos x\, dx$

8. $\displaystyle\int \sin^{3/2} \frac{x}{2} \cos \frac{x}{2}\, dx$

9. $\displaystyle\int \sec 4x \tan 4x\, dx$

10. $\displaystyle\int_0^{3\pi/4} \tan \frac{x}{3}\, dx$

11. $\displaystyle\int_0^{\pi/9} (\sin^2 x + \cos^2 x)\, dx$

12. $\displaystyle\int_0^{\pi/2} \sec^2(2 + x) \tan(2 + x)\, dx$

13. $\displaystyle\int \sqrt{\cos x}\, \sin x\, dx$

14. $\displaystyle\int_0^{\pi/10} \sin 5x \cos 5x\, dx$

15. $\displaystyle\int x \csc^2 x^2\, dx$

16. $\displaystyle\int \frac{\csc \sqrt{x}\, \cot \sqrt{x}}{\sqrt{x}}\, dx$

17. $\displaystyle\int \cos^2 x\, dx$

18. $\displaystyle\int 3 \sin^2 \frac{x}{2}\, dx$

19. $\displaystyle\int (\sec x)^{6/5} \tan x\, dx$

20. $\displaystyle\int_0^{\pi/4} (\sec^2 x + \tan^2 x)\, dx$ [*Hint:* Use the identity $\tan^2 x = \sec^2 x - 1$.]

21. $\displaystyle\int x \cot x^2 \csc x^2\, dx$

22. $\displaystyle\int_0^1 \sec x \cos x\, dx$

23. $\displaystyle\int_0^{\pi/4} \sec^5 x \tan x\, dx$

24. $\displaystyle\int \csc^{3/2} x \cot x\, dx$

25. $\displaystyle\int \sec^2 x \cot x\, dx \left[Hint:\ \cot x = \frac{1}{\tan x}. \right]$

26. $\displaystyle\int_{\pi/4}^{\pi/2} \frac{1 + \cot^2 x}{\csc^2 x}\, dx$

*27. $\displaystyle\int \frac{1}{\sqrt{\csc x \tan x}}\, dx$

28. $\displaystyle\int \sin^2 5x\, dx$

*29. $\displaystyle\int \frac{\cos^2 \sqrt{x}}{\sqrt{x}}\, dx$

30. $\displaystyle\int \cot \frac{x}{8}\, dx$

31. $\displaystyle\int \frac{\cos x}{1 + \sin x}\, dx$

32. $\displaystyle\int x \tan x^2\, dx$

33. $\displaystyle\int \cot(2 + x)\, dx$

34. $\displaystyle\int e^{\cos x} \sin x\, dx$

35. $\displaystyle\int e^{-2 \tan x} \sec^2 x\, dx$

36. $\displaystyle\int \sin(\cos x) \sin x\, dx.$

*37. Calculate $\int \sec x\, dx$. [*Hint:* Multiply the numerator and denominator by $\sec x + \tan x$.]

*38. Calculate $\int \csc x\, dx$.

39. To calculate $\int \sin x \cos x\, dx$, we first set $u = \sin x$. Then

$$\int \sin x \cos x\, dx = \int u\, du = \frac{u^2}{2} + C = \frac{\sin^2 x}{2} + C.$$

Next, we set $u = \cos x$ so that $du = -\sin x\, dx$ and

$$\int \sin x \cos x\, dx = -\int \cos x(-\sin x)\, dx = -\int u\, du = -\frac{u^2}{2} + C = -\frac{\cos^2 x}{2} + C.$$

Can you explain this apparent discrepancy?

40. What is wrong with the following calculation:

$$\int_0^\pi \sec^2 x\, dx = \tan x \Big|_0^\pi = \tan \pi - \tan 0 = 0?$$

41. Calculate the area bounded by the curve $y = \sin x$, the x-axis, the line $x = 0$ and the line $x = \pi$.

42. Calculate the area bounded by the curve $y = \cot x$, the x-axis, the line $x = \pi/3$ and the line $x = \pi/2$.

43. Calculate the area bounded by one arch of the curve $y = \sin^2 x$ and the x-axis.

44. Calculate the area bounded by the curve $y = \sec x$, the x- and y-axes, and the line $x = \pi/4$.

45. Calculate the area bounded by the x-axis and one arch of the curve $y = \cos^2(x/3)$.

46. Calculate the area between $y = \sin x$ and $y = \cos x$, and to the left of the line $x = \pi$.

47. Calculate the area bounded by the curve $y = \tan x$, the y-axis, and the line $y = 1$.

48. Calculate the area bounded by the curve $y = \csc^2 x$, the lines $x = \pi/4$ and $x = \pi/2$, and the x-axis.

49. Calculate the area bounded by the curves $y = \sin^2 x, y = \cos^2 x$, and the y-axis, $x \geq 0$.

50. Find one of the areas bounded by the curve $y = \csc^2 x$ and the line $y = 2$.

51. A manufacturer of machine parts finds that his marginal cost (in dollars) per unit fluctuates and is given by $\text{MC} = 7.5 + 2 \cos(\pi x/500)$, where x is the number of units manufactured. What is the total cost of producing 250 units if his fixed cost (overhead) is $50?

52. The density of a nonuniform 5-m metal bar is given by

$$\rho = 0.8 + \tan\left(1 + \frac{x}{8}\right)$$

(measured in kg/m). What is the total mass of the bar?

7.7 The Inverse Trigonometric Functions

Let $y = \sin x$. Can we solve for x as a function of y? At this point the answer is no. Suppose that $\sin x = \frac{1}{2}$. Then x could be $\pi/6$, or it could be $5\pi/6$, or it could be $13\pi/6$, etc. In fact, if y is any number in the interval $[-1, 1]$, then there are an *infinite* number of values of x for which $\sin x = y$. This is illustrated in Figure 1. At the

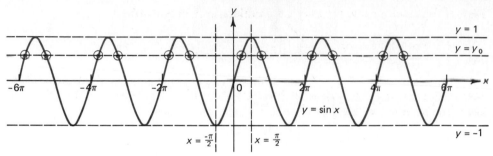

Figure 1

circled points, $\sin x = y_0$. We can alleviate this problem if we restrict x to lie in a certain interval, say $[-\pi/2, \pi/2]$. Then as in Figure 1, for each value of y in $[-1, 1]$, there is a unique value x in $[-\pi/2, \pi/2]$ such that $\sin x = y$.

DEFINITION 1. The *inverse sine function* is the function which assigns to each number x in $[-1, 1]$ the unique number y in $[-\pi/2, \pi/2]$ such that $x = \sin y$. We write this as

$$y = \sin^{-1} x. \tag{1}$$

Note. The -1 appearing in (1) does *not* mean $1/(\sin x)$ which is equal to $(\sin x)^{-1} = \csc x$.

Another commonly used notation for the inverse sine function is

$$y = \arcsin x. \tag{2}$$

EXAMPLE 1. Calculate $\sin^{-1} x$ when

(a) $x = 0$ (b) $x = 1$ (c) $x = \frac{1}{2}$ (d) $x = -1$.

SOLUTION. (a) $\sin^{-1} 0 = 0$ since $\sin 0 = 0$.
  (b) $\sin^{-1} 1 = \pi/2$ since $\sin \pi/2 = 1$.
  (c) $\sin^{-1} \frac{1}{2} = \pi/6$ since $\sin \pi/6 = \frac{1}{2}$.
  (d) $\sin^{-1}(-1) = -\pi/2$ since $\sin(-\pi/2) = -1$.

We emphasize that the function $y = \sin^{-1} x = \arcsin x$ is only defined if we restrict y to lie in $[-\pi/2, \pi/2]$. We will discuss this idea further in our discussion of inverse functions in Section 10.7. Note, for the moment, that if x is in $[-1, 1]$ and if y is in $[-\pi/2, \pi/2]$, then $\sin(\sin^{-1} x) = x$ and $\sin^{-1}(\sin y) = y$. (This follows from Definition 1.) However it is not true in general. For example, let $y = 2\pi$. Then $x = \sin y = 0$ and $y = \sin^{-1} x = \sin^{-1}(\sin y) = \sin^{-1}(0) = 0$ which is not equal to the original value of y.

We now differentiate $y = \sin^{-1} x$ for y in $[-\pi/2, \pi/2]$.

THEOREM 1. In the open interval $(-1, 1)$, $y = \sin^{-1} x = \arcsin x$ is differentiable and

$$\boxed{\frac{d}{dx} \sin^{-1} x = \frac{d}{dx} \arcsin x = \frac{1}{\sqrt{1-x^2}}, \quad -1 < x < 1.} \tag{3}$$

PROOF. Since $x = \sin y$, we may differentiate implicitly to obtain

$$1 = \cos y \frac{dy}{dx} \qquad \text{or} \qquad \frac{dy}{dx} = \frac{1}{\cos y}.$$

This is defined as long as $\cos y \neq 0$; that is, as long as $y \neq -\pi/2$ or $\pi/2$. Since $\cos^2 y + \sin^2 y = 1$, we see that $\cos y = \pm\sqrt{1 - \sin^2 y}$. But since $-\pi/2 < y < \pi/2$, $\cos y > 0$ so that, in this case,

$$\cos y = \sqrt{1 - \sin^2 y} = \sqrt{1 - x^2}.$$

Thus

$$\frac{d \sin^{-1} x}{dx} = \frac{1}{\sqrt{1 - x^2}}.$$

This is defined as long as $-1 < x < 1$.

For the remainder of this chapter we will use the notation $y = \sin^{-1} x$ because it expresses more clearly the fact that the function is the inverse of the sine function.

To graph $y = \sin^{-1} x$ for x in $(-1, 1)$ we first observe that $dy/dx = 1/\sqrt{1 - x^2} > 0$ so that the function is increasing. There are no critical points. In addition,

$$\frac{d^2 y}{dx^2} = \frac{d}{dx}(1 - x^2)^{-1/2} = \frac{x}{(1 - x^2)^{3/2}}$$

which is negative for $x < 0$ and positive for $x > 0$. The origin is a point of inflection. The graph is given in Figure 2b and, as a frame of reference, is placed next to the graph of $y = \sin x$.

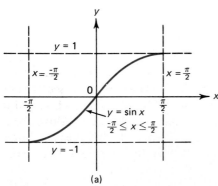

(a)

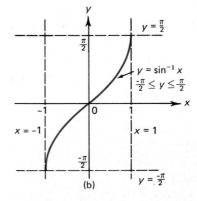
(b)

Figure 2

Next, consider the graph of $\cos x$ in Figure 7.2.3. If $y = \cos x$ and x is restricted to lie in the interval $[0, \pi]$, then for every y in $[-1, 1]$, there is a unique x such that $\cos x = y$.

DEFINITION 2. The *inverse cosine* function is the function which assigns to each number x in $[-1, 1]$ the unique number y in $[0, \pi]$ such that $x = \cos y$. We write this as $y = \cos^{-1} x$.

Note. $\cos^{-1} x$ is also written as arccos x.

EXAMPLE 2. Calculate $y = \cos^{-1} x$ for

(a) $x = 0$ (b) $x = 1$ (c) $x = \frac{1}{2}$ (d) $x = -1$.

SOLUTION. (a) $\cos^{-1} 0 = \pi/2$ since $\cos \pi/2 = 0$.
   (b) $\cos^{-1} 1 = 0$ since $\cos 0 = 1$.
   (c) $\cos^{-1} \frac{1}{2} = \pi/3$ since $\cos \pi/3 = \frac{1}{2}$.
   (d) $\cos^{-1}(-1) = \pi$ since $\cos \pi = -1$.

THEOREM 2

$$\frac{d}{dx} \cos^{-1} x = -\frac{1}{\sqrt{1 - x^2}}, \quad -1 < x < 1. \tag{4}$$

PROOF. If $y = \cos^{-1} x$, then $x = \cos y$ and $1 = (-\sin y)\, dy/dx$ or

$$\frac{dy}{dx} = -\frac{1}{\sin y} = -\frac{1}{\sqrt{1 - x^2}}.$$

$(\sin y = +\sqrt{1 - \cos^2 y}$ because $0 \le y \le \pi$.) The graphs of $y = \cos x$ and $y = \cos^{-1} x$ are given in Figure 3.

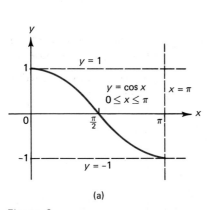

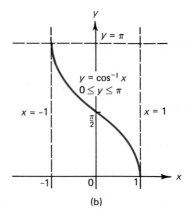

(a)   (b)

Figure 3

We next consider the graph of $y = \tan x$ given in Figure 7.3.1a. The function $\tan x$ can take on values in the interval $(-\infty, \infty)$. To get a unique x for a given y, we restrict y to the interval $(-\pi/2, \pi/2)$.

DEFINITION 3. The *inverse tangent* function is the function which assigns to each real number x the unique number y in $(-\pi/2, \pi/2)$ such that $x = \tan y$. We write this as $y = \tan^{-1} x = \arctan x$.

THEOREM 3

$$\frac{d}{dx}\tan^{-1}x = \frac{1}{1 + x^2}, \qquad -\infty < x < \infty.$$ (5)

PROOF. If $y = \tan^{-1}x$, then $x = \tan y$ and $1 = (\sec^2 y)\,dy/dx$ or

$$\frac{dy}{dx} = \frac{1}{\sec^2 y} = \frac{1}{1 + \tan^2 y} = \frac{1}{1 + x^2}.$$

The graphs of $\tan x$ and $\tan^{-1}x$ are given in Figure 4.

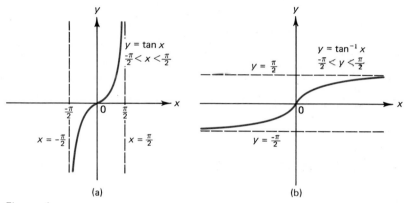

(a) (b)

Figure 4

There are, of course, three other inverse trigonometric functions. These arise less frequently in applications and so we will delay their introduction until the problems.

We summarize in Table 1 the formulas which can be obtained from the results of this section. The last three rows will be derived in the problem set.

TABLE 1

$\dfrac{d}{dx}\sin^{-1}u = \dfrac{1}{\sqrt{1 - u^2}}\dfrac{du}{dx}.$	$\displaystyle\int\frac{du}{\sqrt{1 - u^2}} = \sin^{-1}u + C.$
$\dfrac{d}{dx}\cos^{-1}u = -\dfrac{1}{\sqrt{1 - u^2}}\dfrac{du}{dx}.$	$\displaystyle -\int\frac{du}{\sqrt{1 - u^2}} = \cos^{-1}u + C.$
$\dfrac{d}{dx}\tan^{-1}u = \dfrac{1}{1 + u^2}\dfrac{du}{dx}.$	$\displaystyle\int\frac{du}{1 + u^2} = \tan^{-1}u + C.$
$\dfrac{d}{dx}\cot^{-1}u = -\dfrac{1}{1 + u^2}\dfrac{du}{dx}.$	$\displaystyle -\int\frac{du}{1 + u^2} = \cot^{-1}u + C.$
$\dfrac{d}{dx}\sec^{-1}u = \dfrac{1}{|u|\sqrt{u^2 - 1}}\dfrac{du}{dx}.$	$\displaystyle\int\frac{du}{u\sqrt{u^2 - 1}} = \sec^{-1}|u| + C.$
$\dfrac{d}{dx}\csc^{-1}u = -\dfrac{1}{|u|\sqrt{u^2 - 1}}\dfrac{du}{dx}.$	$\displaystyle -\int\frac{du}{u\sqrt{u^2 - 1}} = \csc^{-1}|u| + C.$

EXAMPLE 3. Calculate $y = \dfrac{d}{dx} \sin^{-1} x^2$.

SOLUTION. If $u = x^2$, then $du/dx = 2x$ and, using the chain rule,

$$\frac{dy}{dx} = \frac{dy}{du}\frac{du}{dx} = \frac{1}{\sqrt{1 - u^2}} \cdot 2x = \frac{2x}{\sqrt{1 - x^4}}.$$

EXAMPLE 4. Calculate $\dfrac{d}{dx} \cos^{-1}(\ln x)$.

SOLUTION. First note that this function is only defined for $1/e < x < e$ (so that $-1 < \ln x < 1$). Then for $u = \ln x$,

$$\frac{du}{dx} = \frac{1}{x} \quad \text{and} \quad \frac{d}{dx} \cos^{-1} \ln x = -\frac{1}{\sqrt{1 - \ln^2 x}} \cdot \frac{1}{x}.$$

EXAMPLE 5. Calculate $\displaystyle\int \frac{dx}{a^2 + x^2}$.

SOLUTION. We want to write the integral in the form

$$\int \frac{du}{1 + u^2}$$

since we can integrate this. To do so, we factor out the a^2 term in the denominator to obtain

$$\int \frac{dx}{a^2 + x^2} = \frac{1}{a^2} \int \frac{dx}{1 + (x^2/a^2)}.$$

Let $u = x/a$, then $du = (1/a) \, dx$ so that

$$\frac{1}{a^2} \int \frac{dx}{1 + (x^2/a^2)} = \frac{1}{a} \int \frac{(1/a) \, dx}{1 + (x/a)^2} = \frac{1}{a} \int \frac{du}{1 + u^2} = \frac{1}{a} \tan^{-1} u + C$$

$$= \frac{1}{a} \tan^{-1} \frac{x}{a} + C.$$

EXAMPLE 6. Calculate $\displaystyle\int \frac{x^2}{\sqrt{1 - x^6}} \, dx$.

SOLUTION. If $u = x^3$, then $du = 3x^2 \, dx$ so we multiply and divide by 3 to obtain

$$\int \frac{x^2}{\sqrt{1 - x^6}} \, dx = \frac{1}{3} \int \frac{3x^2}{\sqrt{1 - (x^3)^2}} \, dx = \frac{1}{3} \int \frac{du}{\sqrt{1 - u^2}} = \frac{1}{3} \sin^{-1} u + C$$

$$= \frac{1}{3} \sin^{-1} x^3 + C.$$

EXAMPLE 7. Calculate $\displaystyle\int \frac{e^{-x} \, dx}{\sqrt{1 - e^{-2x}}}$.

SOLUTION. Let $u = e^{-x}$. Then $u^2 = e^{-2x}$, $du = -e^{-x}\,dx$ and

$$\int \frac{e^{-x}\,dx}{\sqrt{1 - e^{-2x}}} = -\int \frac{-e^{-x}\,dx}{\sqrt{1 - e^{-2x}}} = -\int \frac{du}{\sqrt{1 - u^2}} = \cos^{-1} u + C$$

$$= \cos^{-1} e^{-x} + C.$$

In some applications it is necessary to evaluate expressions like $\sin(\tan^{-1} 3)$. To do this, we draw a triangle with an angle whose tangent is 3. See Figure 5. If $\tan \theta = 3$, then with the sides as in the figure, we use the Pythagorean theorem to find that the hypotenuse must be $\sqrt{10}$. Then $\sin \theta = 3/\sqrt{10}$, $\cos \theta = 1/\sqrt{10}$, and so on.

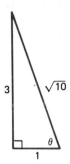

Figure 5

EXAMPLE 8.   Calculate $\tan(\cos^{-1}\frac{3}{7})$.

SOLUTION. By using the triangle in Figure 6 we see that $\tan(\cos^{-1}\frac{3}{7}) = \tan \theta = \sqrt{40}/3$.

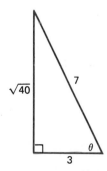

Figure 6

EXAMPLE 9.   Calculate $\cos(\sin^{-1} x)$.

SOLUTION. Look at Figure 7. It is clear that $\cos(\sin^{-1} x) = \cos \theta = \sqrt{1 - x^2}$.

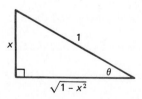

Figure 7

PROBLEMS 7.7

In Problems 1–18 calculate the indicated values.

1. $\sin^{-1}\dfrac{\sqrt{3}}{2}$ 2. $\cos^{-1}\left(-\dfrac{\sqrt{3}}{2}\right)$ 3. $\sin^{-1}\left(-\dfrac{1}{2}\right)$

4. $\tan^{-1}(-1)$ 5. $\tan^{-1}\dfrac{1}{\sqrt{3}}$ 6. $\tan^{-1}\left(-\dfrac{1}{\sqrt{3}}\right)$

7. $\sin^{-1}\left(-\dfrac{\sqrt{2}}{2}\right)$ 8. $\tan^{-1}\cos(-5\pi)$ 9. $\tan^{-1}(\sin 30\pi)$

10. $\sin(\cos^{-1}\tfrac{3}{5})$ 11. $\cos(\sin^{-1}(-\tfrac{3}{5}))$ 12. $\tan(\sin^{-1}\tfrac{3}{5})$

13. $\sin(\tan^{-1}\tfrac{3}{5})$ 14. $\sin(\tan^{-1}(-5))$ *15. $\tan(\sin^{-1}5)$
                                                                                                                            [*Hint:* Watch out.]

16. $\sin(\tan^{-1}x)$ 17. $\sin(\cos^{-1}x)$ 18. $\tan(\sin^{-1}x)$.

19. Show that if $-1 \le x \le 1$, then $\sin(\cos^{-1}x) = \cos(\sin^{-1}x)$.
*20. Show that $\lim_{x\to\infty}\tan^{-1}x = \pi/2$ and $\lim_{x\to-\infty}\tan^{-1}x = -\pi/2$.

In Problems 21–44 calculate the given derivative or integral.

21. $\dfrac{d}{dx}\sin^{-1}3x$ 22. $\dfrac{d}{dx}\cos^{-1}\dfrac{x^3+1}{x^5}$ 23. $\dfrac{d}{dx}\tan^{-1}(x-4)^2$

24. $\displaystyle\int\dfrac{dx}{\sqrt{1-9x^2}}$ 25. $\displaystyle\int\dfrac{x}{x^4+1}\,dx$ 26. $\displaystyle\int\dfrac{x^3}{x^8+4}\,dx$

27. $\dfrac{d}{dx}\tan^{-1}\ln x$ 28. $\dfrac{d}{dx}\ln(\tan^{-1}x)$

29. $\dfrac{d}{dx}\sqrt{\sin^{-1}x + \cos^{-1}x}$ 30. $\displaystyle\int_0^{\sqrt{3}/2}\dfrac{dx}{\sqrt{1-x^2}}$

31. $\displaystyle\int\dfrac{e^{-x}}{4+e^{-2x}}\,dx$ 32. $\displaystyle\int\dfrac{\sin x}{\sqrt{4-\cos^2 x}}\,dx$ 33. $\dfrac{d}{dx}\ln\sin^{-1}(e^{-x})$

34. $\dfrac{d}{dx}\tan^{-1}\dfrac{1}{x}$ 35. $\dfrac{d}{dx}x^2\cos^{-1}(1-x)$

36. $\displaystyle\int\dfrac{dx}{x^2+2x+2}$ [*Hint:* Use the fact that $(x+1)^2 = x^2 + 2x + 1$.]

37. $\displaystyle\int\dfrac{dx}{x^2-2x+2}$ 38. $\displaystyle\int_0^2\dfrac{dx}{1+x^2}$ 39. $\displaystyle\int_0^1\dfrac{dx}{\sqrt{1-x^2}}$

40. $\displaystyle\int\dfrac{\sin^{-1}x}{\sqrt{1-x^2}}\,dx$ 41. $\displaystyle\int\dfrac{\cos^{-1}x}{\sqrt{1-x^2}}\,dx$ 42. $\displaystyle\int_0^1\dfrac{\sqrt{\tan^{-1}x}}{1+x^2}\,dx$

43. $\displaystyle\int\dfrac{\sec x\tan x}{1+\sec^2 x}\,dx$ 44. $\displaystyle\int\dfrac{\sec^2 x}{\sqrt{1-\tan^2 x}}\,dx$.

45. Find the area bounded by the curve $y = 1/(1+x^2)$, the x- and y-axes, and the line $x = 1$.
46. Find the area bounded by the curve $y = 1/\sqrt{1-x^2}$, the x-axis the y-axis, and the line $x = \tfrac{1}{2}$.
47. Find the area bounded by the curve $y = -1/\sqrt{1-x^2}$ and the line $y = -2$. [*Hint:* Sketch the curve.]

48.   Show that

$$\int \frac{dx}{\sqrt{a^2 - x^2}} = \sin^{-1}\frac{x}{a} + C.$$

*49.   Show that $\sin^{-1} x + \cos^{-1} x = \pi/2$. [*Hint:* First show by differentiation that $\sin^{-1} x + \cos^{-1} x = C$. Then find C by evaluating at one value of x.]

50.   Show that

$$\frac{d}{dx}(x \sin^{-1} x + \sqrt{1 - x^2}) = \sin^{-1} x.$$

This gives us the integral of $\sin^{-1} x$.

51.   Show that

$$\frac{d}{dx}(x \cos^{-1} x - \sqrt{1 - x^2}) = \cos^{-1} x.$$

52.   Show that

$$\frac{d}{dx}(x \tan^{-1} x - \tfrac{1}{2}\ln(1 + x^2)) = \tan^{-1} x.$$

53.   Consider the graph of $y = \cot x$ in Figure 7.3.1b. Show that if $y = \cot x$, then for every y in $(-\infty, \infty)$ there is a unique x in $(0, \pi)$ such that $y = \cot x$. Then, define the function $y = \cot^{-1} x$ by $y = \cot^{-1} x$ if $x = \cot y$, $0 < y < \pi$.

54.   Calculate
  (a) $\cot^{-1} 0$ (b) $\cot^{-1}(-1)$
  (c) $\cot^{-1}(-\sqrt{3})$ (d) $\cot^{-1}(1/\sqrt{3})$.

55.   Show that $\dfrac{d}{dx}\cot^{-1} x = -\dfrac{1}{1 + x^2}$.

56.   Graph the function $y = \cot^{-1} x$, $0 < y < \pi$.

57.   Consider the graph of $y = \sec x$ in Figure 7.3.1c. Show that if $y = \sec x$, then for every y in $(-\infty, -1]$ or $[1, \infty)$, there is a unique x in $(\pi/2, \pi]$ or $[0, \pi/2)$ such that $y = \sec x$. Explain why the value $x = \pi/2$ must be excluded. Then define the function $y = \sec^{-1} x$ by $y = \sec^{-1} x$ if $x = \sec y$, $0 \le y < \pi/2$ or $\pi/2 < y \le \pi$.†

58.   Show that $\sec^{-1} x = \cos^{-1}(1/x)$, for $|x| \ge 1$.

59.   Calculate
  (a) $\sec^{-1} 2$ (b) $\sec^{-1}(-2)$
  (c) $\sec^{-1} \sqrt{2}$ (d) $\sec^{-1}(2/\sqrt{3})$
  (e) $\sec^{-1} 1$ (f) $\sec^{-1}(-1)$

60.   (a)   Show that if $x \ge 1$, then $\dfrac{d}{dx}\sec^{-1} x = \dfrac{1}{x\sqrt{x^2 - 1}}$.

   (b)   Show that if $x \le -1$, then $\dfrac{d}{dx}\sec^{-1} x = -\dfrac{1}{x\sqrt{x^2 - 1}}$.

---

† This problem suggests one way to define $\sec^{-1} x$. An alternative way is to restrict the range of $\sec^{-1} x$ to the intervals $[-\pi, -\pi/2)$ and $[0, \pi/2)$. Then for every y in $(-\infty, -1]$ or $[1, \infty)$ there is a unique x in $[-\pi, \pi/2)$ or $[0, \pi/2]$ such that $y = \sec x$.

(c)　Conclude that $\dfrac{d}{dx}\sec^{-1}x=\dfrac{1}{|x|\,\sqrt{x^2-1}}$.†

(d)　What is the domain of the function $y'=\dfrac{d}{dx}\sec^{-1}x$?

61.　Graph the function $y=\sec^{-1}x$, $0\le y<\dfrac{\pi}{2}$ and $\dfrac{\pi}{2}<y\le\pi$.

62.　Consider the graph of the function $y=\csc x$ in Figure 7.3.1d. Show that if $y=\csc x$, then for every y in $(-\infty,-1]$ or $[1,\infty)$, there is a unique x in $[-\pi/2,0)$ or $(0,\pi/2]$ such that $y=\csc x$. Explain why the value $x=0$ must be excluded. Then define the function $y=\csc^{-1}x$ by $y=\csc^{-1}x$ if $x=\csc y$, $-\pi/2\le y<0$ or $0<y\le\pi/2$.

63.　Show that $\csc^{-1}x=\sin^{-1}(1/x)$, for $|x|\ge1$.

64.　Calculate

(a)　$\csc^{-1}2$

(b)　$\csc^{-1}(-2)$

(c)　$\csc^{-1}\sqrt{2}$

(d)　$\csc^{-1}\left(\dfrac{2}{\sqrt{3}}\right)$

(e)　$\csc^{-1}1$

(f)　$\csc^{-1}(-1)$.

65.　(a)　Show that if $x\ge1$, then $\dfrac{d}{dx}\csc^{-1}x=-\dfrac{1}{x\sqrt{x^2-1}}$.

(b)　Show that if $x\le-1$, then $\dfrac{d}{dx}\csc^{-1}x=\dfrac{1}{x\sqrt{x^2-1}}$.

(c)　Conclude that $\dfrac{d}{dx}\csc^{-1}x=-\dfrac{1}{|x|\sqrt{x^2-1}}$.

66.　Calculate

(a)　$\sin(\csc^{-1}4)$.

(b)　$\sec(\cot^{-1}8)$.

(c)　$\cos[\sec^{-1}(-10)]$.

(d)　$\cot[\sec^{-1}(-3)]$.

(e)　$\tan(\csc^{-1}4)$.

(f)　$\sin(\cot^{-1}\tfrac{1}{5})$.

67.　Verify the integral formulas in Table 1.

68.　Show by differentiation that $\tan^{-1}x+\cot^{-1}x=\pi/2$.

In Problems 69–84 calculate the given derivative or integral.

69.　$\dfrac{d}{dx}\sec^{-1}(4x+2)$

70.　$\dfrac{d}{dx}\cot^{-1}(x^2+5)$

71.　$\dfrac{d}{dx}\csc^{-1}\sqrt{x}$

72.　$\dfrac{d}{dx}x^2\sec^{-1}x^2$

73.　$\displaystyle\int_{-6}^{-3\sqrt{2}}\dfrac{dx}{x\sqrt{x^2-9}}$

74.　$\displaystyle\int_{2/\sqrt{3}}^{2}\dfrac{dx}{x\sqrt{x^2-1}}$

75.　$\dfrac{d}{dx}\cot^{-1}(e^x)$

76.　$\dfrac{d}{dx}\csc^{-1}\ln x$

*77.　$\displaystyle\int\dfrac{dx}{x\sqrt{x^2-a^2}}$

78.　$\displaystyle\int\dfrac{dx}{1+(3-x)^2}$

79.　$\displaystyle\int_{0}^{\ln2}\dfrac{e^x\,dx}{e^x\sqrt{e^{2x}-1}}$

*80.　$\displaystyle\int\dfrac{dx}{x\sqrt{x-1}}$　[Hint: Let $u=\sqrt{x}$.]

81.　$\dfrac{d}{dx}\sec^{-1}(\cot x)$

82.　$\dfrac{d}{dx}\cot^{-1}(\sec x)$

*83.　$\dfrac{d}{dx}\sin^{-1}(\cot^{-1}\ln x)$

84.　$\displaystyle\int_{0}^{\pi/4}\dfrac{\sin x}{1+\cos^2 x}\,dx$.

†With the alternative definition of $\sec^{-1}x$ given in the last footnote, it is not difficult to show that

$$\dfrac{d}{dx}\sec^{-1}x=\dfrac{1}{x\sqrt{x^2-1}}.$$

Some mathematicians prefer this alternative definition because it is then not necessary to keep track of the absolute value bars in the derivative of $\sec^{-1}x$.

85.   Show that

$$\int \frac{dx}{x\sqrt{x^2 - a^2}} = \frac{1}{a}\sec^{-1}\left|\frac{x}{a}\right| + C = \frac{1}{a}\cos^{-1}\left|\frac{a}{x}\right| + C.$$

*86.   Show that $\sec^{-1}|x| + \csc^{-1}|x|$ is constant. [*Hint:* Differentiate.]

*87.   Prove that $\cot^{-1}[(x + 1)/(x - 1)] + \cot^{-1}x$ is a constant if $x > 1$ and is a different constant for $x < 1$. Then find the constants. [*Hint:* Differentiate and see what happens.]

*7.8 Periodic Motion

Periodic motion is a very common occurrence in physical or biological† settings. A simple example is given by the back and forth motion of a pendulum. Another is given by the oscillating size of the population of an animal species. Yet another is given by the rise and fall of average daily temperature in a fixed location over a time span of several years.

It often happens that the equations describing periodic, or *harmonic*,‡ motion (as it is often called) involve the sine and cosine functions. To see how such an equation could arise, we consider a model for the horizontal motion of a spring.

Consider a mass m attached to a spring resting on a horizontal frictionless surface. If the mass is pulled out beyond its equilibrium (resting) position, then the spring will generally exert a restoring force F which is proportional to distance pulled and is in the opposite direction. This can be expressed by

$$F = -kx \tag{1}$$

where k is a positive number (called the *spring constant;* see Section 5.7). The minus sign indicates that the force acts in such a way so as to bring the mass back to its equilibrium position. This is illustrated in Figure 1. The restoring force of the spring

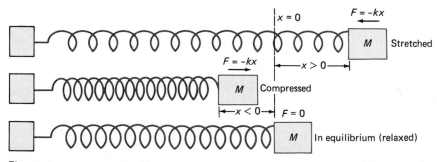

Figure 1

is the only force acting on the mass if we ignore air resistance since we have assumed that there is no friction. By Newton's second law of motion, the sum of the forces

---

† There are a great variety of periodic or rhythmic phenomena in biology. For example, diurnal (daily) or *circadian* rhythms have been studied for more than 100 years. The interested reader should consult the collection of papers in *Circadian Clocks,* Jürgen Aschoff, editor, North-Holland Publ. Co., Amsterdam, 1965.

‡ The term "harmonic" is applied to expressions containing certain combinations of the sine and cosine functions.

acting on the mass is given by $F = ma$ where a is the acceleration of the particle. But $a = d^2x/dt^2$. This leads to the equation $m\,d^2x/dt^2 = -kx$ or

$$m\frac{d^2x}{dt^2} + kx = 0. \tag{2}$$

Since k and m are positive numbers, we can write $k/m = \omega^2$ where ω is a positive number. Then if we divide (2) through by m, we obtain

$$\frac{d^2x}{dt^2} + \omega^2 x(t) = 0. \tag{3}$$

Equation (3) is called the *equation of the harmonic oscillator*. It is another kind of differential equation (we discussed an easier differential equation in Section 6.4). The equation of the harmonic oscillator arises in a great number of physical and biological applications. We will consider this type of equation in more detail in Chapter 20. It can be shown† that any solution to equation (3) can be written in the form

$$x(t) = A\cos(\omega t + \delta) \tag{4}$$

where A and δ are constants. We can check that a function of the form (4) is a solution of (3) by differentiating:

$$x'(t) = -A\omega \sin(\omega t + \delta) \quad \text{and} \quad x''(t) = -A\omega^2 \cos(\omega t + \delta) = -\omega^2 x(t)$$

so that

$$x'' + \omega^2 x = -\omega^2 x + \omega^2 x = 0.$$

Whatever the values A and δ, equation (4) tells us that the mass vibrates back and forth with a *period T* of $2\pi/\omega$ since

$$A\cos\left[\omega\left(t + \frac{2\pi}{\omega}\right) + \delta\right] = A\cos(\omega t + 2\pi + \delta) = A\cos(\omega t + \delta).$$

If t is measured in seconds, then the mass oscillates once every $2\pi/\omega$ seconds and therefore oscillates $\omega/2\pi$ times in 1 second. In this case we say that the *frequency f* is $\omega/2\pi$ cycles/sec.‡ In terms of m and k, since $\omega^2 = k/m$, the period measured in seconds is

$$T = \frac{2\pi}{\omega} = \frac{2\pi}{\sqrt{k/m}} = 2\pi\sqrt{\frac{m}{k}} \tag{5}$$

and the frequency

$$f = \frac{1}{T} = \frac{\omega}{2\pi} = \frac{1}{2\pi}\sqrt{\frac{k}{m}}. \tag{6}$$

---

†For a derivation of equation (4) see W. R. Derrick and S. I. Grossman, *Elementary Differential Equations with Applications,* Addison Wesley, Reading, Massachusetts, 1976, p. 101.

‡Cycles/sec are also called *hertz* (abbreviated Hz), named after the German physicist Heinrich Rudolph Hertz (1857–1894).

The constant A has a simple physical meaning. It is the *amplitude* of the oscillation (since $-1 \le \cos x \le 1$, $-A \le A \cos(\omega t + \delta) \le A$). The amplitude is the maximum value of the displacement from equilibrium. The number δ which can always be taken to be between $-\pi/2$ and $\pi/2$ is called the *phase constant*. From Section 1.8 we see that this constant has the simple effect of shifting the cosine curve δ units to the left if δ is positive and δ units to the right if δ is negative (see Figure 2).

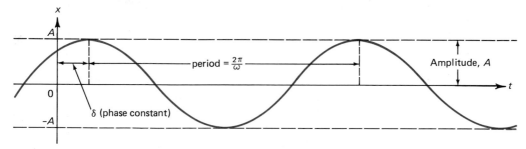

Figure 2

If $\delta = 0$ we obtain the cosine function itself. If $\delta = -\pi/2$, then we obtain the sine function since $\cos(x - \pi/2) = \cos(\pi/2 - x) = \sin x$ (from identity (x) in Table 7.2.2). To obtain the constants A and δ, it is necessary to have further information, as the next example illustrates.

EXAMPLE 1. A 2 kg mass is attached to a spring with spring constant 8 and is pulled out a distance of 25 cm and then released. Find the equation of motion of the mass on the spring. Where is the mass after $\frac{1}{2}$ sec? After 5 sec? What are the period and frequency of oscillation?

SOLUTION. Here $m = 2$ and $k = 8$ so that $\omega = \sqrt{k/m} = 2$. The equation of motion is $x(t) = A \cos(2t + \delta)$. When $t = 0$, $x(0) = 0.25$ m. At the instant the mass is released, it is not moving so that its velocity is zero. But $v(t) = x'(t) = -2A \sin(2t + \delta)$. Thus we obtain

$$x(0) = A \cos \delta = 0.25 \quad \text{and} \quad x'(0) = -2A \sin \delta = 0.$$

Since $-2A \ne 0$, we must have $\sin \delta = 0$ or $\delta = \sin^{-1} 0 = 0$. Then $x(0) = A \cos 0 = A = 0.25 = \frac{1}{4}$. Hence the equation of motion is

$$x(t) = \frac{\cos 2t}{4}.$$

If $t = \frac{1}{2}$, then $x(t) = (\cos 1)/4 \approx 0.135$ m. If $t = 5$, then $x(t) = (\cos 10)/4 \approx -0.21$ m. This means that after 5 sec the spring is compressed 21 cm. Finally, $T = 2\pi/\omega = \pi$ and $f = (1/\pi)$ cycles/sec $= 1/\pi$ Hz.

EXAMPLE 2. What is the maximum speed of the mass in Example 1?

SOLUTION. In general, if $x(t) = A \cos(\omega t + \delta)$, then $v = dx/dt = -A\omega \sin(\omega t + \delta)$. Since $-1 \le \sin(\omega t + \delta) \le 1$, we see that $v_{max} = A\omega$. In this problem $A = \frac{1}{4}$ and $\omega = 2$ so that $v_{max} = \frac{1}{2}$ m/sec.

EXAMPLE 3. In a predator–prey relationship the population growth of the predator species is proportional to the population of the prey species, while the rate of
decline of the prey species is proportional to the population of the predator species.
Find the populations of each species as a function of time.

SOLUTION. Let $x(t)$ represent the population of the predator species and $y(t)$ the
population of the prey species. Then

$$\frac{dx}{dt} = k_1 y \tag{7}$$

and

$$\frac{dy}{dt} = -k_2 x \tag{8}$$

where k_1 and k_2 are positive constants.

To solve these equations, we differentiate (7) with respect to t:

$$\frac{d^2x}{dt^2} = k_1 \frac{dy}{dt} = k_1(-k_2 x) \quad \text{(from (8))} \qquad \text{or} \qquad \frac{d^2x}{dt^2} + k_1 k_2 x = 0.$$

If we set $k_1 k_2 = \omega^2$, then we obtain the equation of the harmonic oscillator. This tells
us that

$$x(t) = A \cos(\omega t + \delta). \tag{9}$$

Then since $y(t) = (1/k_1)\, dx/dt$ (from (7)),

$$y(t) = -\frac{A\omega}{k_1} \sin(\omega t + \delta). \tag{10}$$

EXAMPLE 4. If, in Example 3, $k_1 = k_2 = 1$ and initially the population of the
predator species is 500 while that of the prey species is 2000, find the populations of
both species as a function of time if t is measured in weeks.

SOLUTION. Here $\omega = \sqrt{k_1 k_2} = 1$. Then from equations (9) and (10)

$$x(0) = A \cos \delta = 500 \qquad \text{and} \qquad y(0) = -A \sin \delta = 2000$$

(since $k_1 = 1$). Then $A = 500/\cos \delta$ so that $y(0) = (-500/\cos \delta)\sin \delta = 2000$ or
$\tan \delta = -4$, and, since

$$\tan(-x) = \frac{\sin(-x)}{\cos(-x)} = -\frac{\sin x}{\cos x} = -\tan x,$$

we obtain from Table A4 that $\delta = \tan^{-1}(-4) \approx -76° \approx -1.326$ radians. Then

$$A = \frac{500}{\cos \delta} = \frac{500}{\cos(-1.326)} \approx 2063.$$

Thus the population equations are

$$x(t) = 2063 \cos(t - 1.326)$$

and

$$y(t) = -\frac{2063}{1.326}\sin(t - 1.326) = -1556\sin(t - 1.326).$$

Note that the population of the prey species is zero when $t = 1.326$ weeks which means that the prey becomes extinct after that period of time. In Chapter 20 we will consider models of population growth where both the predator and prey species can continue to exist for all values of t and we will also consider other models which lead to harmonic motion.

PROBLEMS 7.8

1. Solve the equation of the harmonic oscillator when $\omega = 1$, $x(0) = 0$ and $x'(0) = 1$.
2. Find the equation of motion of a 3-kg mass attached to a spring which slides on a horizontal frictionless surface given that the amplitude of oscillation is 6, the spring constant is 2, and $x(1) = 3$.
3. A 5-kg mass attached to a spring oscillates with an amplitude of 8 m, a phase constant of $\frac{1}{4}$ and has the value $x = 4\sqrt{2}$ when $t = 2$ sec. What are the period and frequency of the oscillations? What is the spring constant.
4. A mass attached to a spring oscillates and it is known that, initially, $x(0) = 0$, and $v(0) = 10$ m/sec. If the period of oscillation is 5 sec, find the equation of motion of the mass.
*5. Find the equation of motion of a mass attached to a spring on a horizontal frictionless surface with initial position x_0, initial velocity v_0, and period T. [*Hint:* Write everything in terms of x_0, v_0, and T and assume that the mass will not move if $x_0 = 0$; that is, assume that equilibrium is at 0.]
6. If $x(t) = A\cos(\omega t + \delta)$, find
   (a) the time at which the velocity is a maximum.
   (b) the time at which the velocity is zero.
   (c) the time at which the acceleration is a maximum.
   (d) the time at which the acceleration is zero.
7. Show that the solution to the equation of the harmonic oscillator can also be written

   $$x(t) = c_1\cos\omega t + c_2\sin\omega t$$

   for appropriate constants c_1 and c_2.
8. If $k_1 = 8$ and $k_2 = 2$ in Example 3, find the population equations of the predator and prey species if the initial populations of the two species are 1000 and 10,000, respectively. When is the prey species extinct?
9. Answer the question of Problem 8 if $k_1 = 2$ and $k_2 = 8$ (with the same initial populations).

7.9 The Hyperbolic Functions

In many physical applications (especially those involving differential equations) functions arise that are combinations of e^x and e^{-x}. In fact, this happens so often that certain of the combinations that occur most frequently have been given special names.

DEFINITION 1. The *hyperbolic cosine* function is defined by

$$\cosh x = \frac{e^x + e^{-x}}{2}. \tag{1}$$

The domain of $\cosh x$ is $(-\infty, \infty)$.

DEFINITION 2. The *hyperbolic sine* function is defined by

$$\sinh x = \frac{e^x - e^{-x}}{2} \tag{2}$$

The domain of $\sinh x$ is $(-\infty, \infty)$.

Since $e^{-x} > 0$ for all x, we see that for every x,

$$\cosh x > \sinh x. \tag{3}$$

THEOREM 1.

$$\frac{d}{dx} \cosh x = \sinh x \tag{4}$$

and

$$\frac{d}{dx} \sinh x = \cosh x. \tag{5}$$

PROOF

$$\frac{d}{dx} \cosh x = \frac{d}{dx} \left(\frac{e^x + e^{-x}}{2} \right) = \frac{e^x - e^{-x}}{2} = \sinh x.$$

The proof for the derivative of sinh is just as easy.

It is not very difficult to draw the graphs of these functions. We immediately see that $\sinh x < 0$ if $x < 0$ and $\sinh x > 0$ if $x > 0$. If $x = 0$, $\sinh x = 0$ and $\cosh x = 1$. Therefore $\cosh x$ is decreasing if $x < 0$ and $\cosh x$ is increasing if $x > 0$, and the point $x = 0$ is a critical point for $\cosh x$. In addition,

$$\frac{d^2}{dx^2}(\cosh x) = \cosh x$$

so that $\cosh x$ is concave up and the point $(0, 1)$ is a minimum. This is all depicted in Figure 1.

Since

$$\frac{d}{dx} \sinh x = \cosh x > 0,$$

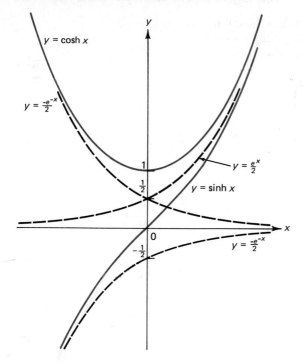

$y = \cosh x$

$y = \frac{-e^{-x}}{2}$

$y = \frac{e^x}{2}$

$y = \sinh x$

$y = \frac{-e^{-x}}{2}$

Figure 1

sinh x is always increasing (and has no critical points). We have

$$\frac{d^2}{dx^2} \sinh x = \sinh x$$

so that sinh x is concave down for $x < 0$ and concave up for $x > 0$. At $(0, 0)$ there is a point of inflection. This figure is also drawn in Figure 1.

Consider the expression $\cosh^2 x - \sinh^2 x$. From Definitions 1 and 2, it is easy to show that

$$\boxed{\cosh^2 x - \sinh^2 x = 1.} \tag{6}$$

(See Problem 1.) This is one of the reasons sinh x and cosh x are called *hyperbolic* functions. The equation $x^2 - y^2 = 1$ is the equation of a hyperbola (see Section 1.9). The curve $x^2 - y^2 = 1$ is graphed in Figure 2.

From Theorem 1, we obtain the differentiation and integration formulas shown in Table 1.

TABLE 1

$\dfrac{d}{dx} \cosh u = \sinh u \dfrac{du}{dx}.$	$\displaystyle\int \sinh u \, du = \cosh u + C.$
$\dfrac{d}{dx} \sinh u = \cosh u \dfrac{du}{dx}.$	$\displaystyle\int \cosh u \, du = \sinh u + C.$

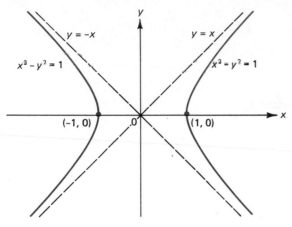

Figure 2

EXAMPLE 1. Calculate $\dfrac{d}{dx}\cosh\sqrt{x^3-1}$.

SOLUTION. If $u=\sqrt{x^3-1}$, then

$$\frac{du}{dx}=\frac{3x^2}{2\sqrt{x^3-1}}$$

and

$$\frac{d}{dx}\cosh\sqrt{x^3-1}=\sinh u\,\frac{du}{dx}=(\sinh\sqrt{x^3-1})\frac{3x^2}{2\sqrt{x^3-1}}.$$

EXAMPLE 2. Calculate $\displaystyle\int x\sinh 3x^2\,dx$.

SOLUTION. If $u=3x^2$, then $du=6x\,dx$ so that

$$\int x\sinh 3x^2\,dx=\tfrac{1}{6}\int\sinh 3x^2(6x)\,dx=\tfrac{1}{6}\int\sinh u\,du$$
$$=\tfrac{1}{6}\cosh u+C=\tfrac{1}{6}\cosh 3x^2+C.$$

EXAMPLE 3. A *catenary* is the curve which is formed by a uniform flexible cable hanging from two points under the influence of its own weight. If the lowest point of the catenary is at the point $(0, a)$, then its equation (which we shall not derive here) is given by

$$y=a\cosh\frac{x}{a},\qquad a>0.$$

This is graphed in Figure 3.

EXAMPLE 4. *Symbiosis* is defined as the relationship of two or more different organisms in a close association that may be of benefit to each. If two species are living in a symbiotic relationship, then we may suppose that the rate of growth of the first population is proportional to the size of the second, and vice versa. If $x(t)$ and $y(t)$ denote the populations of the two species, respectively, then we have

$$\frac{dx}{dt}=k_1 y(t) \tag{7}$$

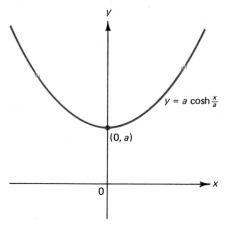

$y = a \cosh \frac{x}{a}$

$(0, a)$

Figure 3

and

$$\frac{dy}{dt} = k_2 x(t) \tag{8}$$

where k_1 and k_2 are positive constants (compare this with the model in Example 7.8.3). We then differentiate (7) and use (8) to obtain

$$\frac{d^2 x}{dt^2} = k_1 \frac{dy}{dt} = k_1 k_2 x.$$

If we define $\omega^2 = k_1 k_1 > 0$, then we have

$$\frac{d^2 x}{dt^2} - \omega^2 x(t) = 0. \tag{9}$$

The only difference between equation (9) and the equation of harmonic motion (7.8.3) is the minus sign before the $\omega^2 x$ term. It is easy to see that any function of the form

$$x(t) = c_1 \sinh \omega t + c_2 \cosh \omega t, \tag{10}$$

where c_1 and c_2 are constants, is a solution to (9) (see Problem 2). The fact that these are the only solutions to (9) will be discussed in Chapter 20. Also, from (7) we see that $y(t) = (1/k_1)\, dx/dt$ so that

$$y(t) = \frac{1}{k_1}(\omega c_1 \cosh \omega t + \omega c_2 \sinh \omega t)$$

or

$$y(t) = \frac{\omega}{k_1}(c_1 \cosh \omega t + c_2 \sinh \omega t). \tag{11}$$

This is the best we can do unless further information is provided.

EXAMPLE 5. Find the populations as a function of t of the species in Example 4 if $k_1 = 4$, $k_2 = 1$, and, initially, $x(0) = 500$ and $y(0) = 1000$, and t is measured in months. What are the populations after $\frac{1}{2}$ month?

SOLUTION. We have $\omega^2 = k_1 k_2 = 4$ so that $\omega = 2$. Then

$$x(t) = c_1 \sinh 2t + c_2 \cosh 2t \qquad \text{and} \qquad y(t) = \tfrac{1}{2}(c_1 \cosh 2t + c_2 \sinh 2t).$$

When $t = 0$, $\sinh 0 = 0$ and $\cosh 0 = 1$ so that

$$x(0) = c_2 = 500, \qquad y(0) = \tfrac{1}{2}c_1 = 1000, \qquad \text{and} \qquad c_1 = 2000.$$

Thus

$$x(t) = 2000 \sinh 2t + 500 \cosh 2t \qquad \text{and} \qquad y(t) = 1000 \cosh 2t + 250 \sinh 2t.$$

We then calculate $\sinh 1 \approx 1.175$ and $\cosh 1 \approx 1.543$ so that $x(\tfrac{1}{2}) \approx 3122$ and $y(\tfrac{1}{2}) \approx 1837$.

After looking at the several examples in this section and in Section 7.8, we see that the trigonometric and hyperbolic functions describe different types of behavior: Trigonometric functions describe oscillatory behavior while hyperbolic functions

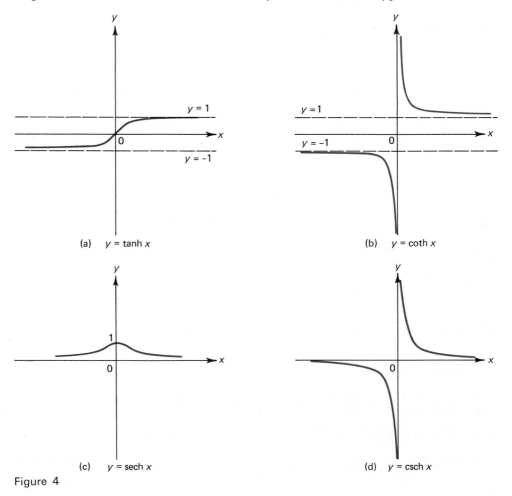

(a) $y = \tanh x$

(b) $y = \coth x$

(c) $y = \operatorname{sech} x$

(d) $y = \operatorname{csch} x$

Figure 4

describe situations in which there is exponential growth or decay (or an exponential approach to a steady state).

As you might expect, there are four other hyperbolic functions. They do not arise nearly as often in applications as the two we have already discussed.

DEFINITION 3

(i) $\tanh x = \dfrac{\sinh x}{\cosh x} = \dfrac{e^x - e^{-x}}{e^x + e^{-x}}.$

The domain of $\tanh x$ is $(-\infty, \infty)$.

(ii) $\coth x = \dfrac{1}{\tanh x} = \dfrac{\cosh x}{\sinh x} = \dfrac{e^x + e^{-x}}{e^x - e^{-x}}.$

The domain of $\coth x$ is all x except $x = 0$.

(iii) $\operatorname{sech} x = \dfrac{1}{\cosh x} = \dfrac{2}{e^x + e^{-x}}.$

The domain of $\operatorname{sech} x$ is $(-\infty, \infty)$.

(iv) $\operatorname{csch} x = \dfrac{1}{\sinh x} = \dfrac{2}{e^x - e^{-x}}.$

The domain of $\operatorname{csch} x$ is all x except $x = 0$.

The graphs of these four functions are given in Figure 4.
Integrals and derivatives of these four hyperbolic functions are given in Table 2. The proofs of these facts are easy, and are left as problems.

TABLE 2

$\dfrac{d}{dx}\tanh u = \operatorname{sech}^2 u \dfrac{du}{dx}.$	$\displaystyle\int \operatorname{sech}^2 u \, du = \tanh u + C.$
$\dfrac{d}{dx}\coth u = -\operatorname{csch}^2 u \dfrac{du}{dx}.$	$\displaystyle\int \operatorname{csch}^2 u \, du = -\coth u + C.$
$\dfrac{d}{dx}\operatorname{sech} u = -\operatorname{sech} u \tanh u \dfrac{du}{dx}.$	$\displaystyle\int \operatorname{sech} u \tanh u = -\operatorname{sech} u + C.$
$\dfrac{d}{dx}\operatorname{csch} u = -\operatorname{csch} u \coth u \dfrac{du}{dx}.$	$\displaystyle\int \operatorname{csch} u \coth u = -\operatorname{csch} u + C.$

PROBLEMS 7.9

1.   Show, from the definition, that $\cosh^2 x - \sinh^2 x = 1$.
2.   Verify that (10) is a solution to equation (9).
3.   Show that $\dfrac{d}{dx}\tanh x = \operatorname{sech}^2 x.$ $\left[\textit{Hint: Use } \tanh x = \dfrac{\sinh x}{\cosh x}.\right]$
4.   Show that $\dfrac{d}{dx}\coth x = -\operatorname{csch}^2 x.$

5. Show that $\dfrac{d}{dx}\,\text{sech}\,x = -\text{sech}\,x\,\tanh x$.

6. Show that $\dfrac{d}{dx}\,\text{csch}\,x = -\text{csch}\,x\,\coth x$.

In Problems 7–30 calculate the given derivatives and integrals.

7. $\dfrac{d}{dx}\,\sinh(4x + 2)$

8. $\dfrac{d}{dx}\,\cosh\dfrac{1}{x}$

9. $\dfrac{d}{dx}\,\text{csch}\,\ln x$

10. $\displaystyle\int \sinh 2x\, dx$

11. $\displaystyle\int \dfrac{\text{sech}^2 \sqrt{x}}{\sqrt{x}}\, dx$

12. $\displaystyle\int \dfrac{\text{csch}\,\ln x\,\coth \ln x}{x}\, dx$

13. $\dfrac{d}{dx}\,\sin(\sinh x)$

14. $\dfrac{d}{dx}\,\sinh(\sin x)$

15. $\dfrac{d}{dx}\,e^{\tanh^2 x}$

16. $\displaystyle\int e^{\sinh x}\,\cosh x\, dx$

*17. $\displaystyle\int \cosh^3 x\, dx$ [*Hint:* use Problem 1.]

18. $\displaystyle\int \tanh x\, dx$

19. $\displaystyle\int \coth x\, dx$

20. $\displaystyle\int \dfrac{\text{csch}^2 x\, dx}{\sqrt[3]{\coth x}}$

21. $\displaystyle\int \dfrac{\sinh (1/x)}{x^2}\, dx$

22. $\dfrac{d}{dx}\,\tanh(\text{csch}\,x)$

*23. $\dfrac{d}{dx}\,\sin^{-1}(\cosh x)$ [*Hint:* Be careful.]

24. $\dfrac{d}{dx}\,\tanh(\tan^{-1} x)$

25. $\displaystyle\int \dfrac{\cosh 2x}{1 + \sinh 2x}\, dx$

*26. $\displaystyle\int \dfrac{(\cos^{-1}(\tanh x))\text{sech}^2 x}{\sqrt{1 - \tanh^2 x}}\, dx$

27. $\displaystyle\int \text{sech}^{11/3} x\,\tanh x\, dx$

*28. $\dfrac{d}{dx}\,\cosh x^x$

*29. $\dfrac{d}{dx}\,x^{\cosh x}$

*30. $\dfrac{d}{dx}\,\sinh x^{\cosh x}$.

31. Find the area bounded by the catenary $y = a\cosh\dfrac{x}{a}$, the x-axis, and the lines $x = -a$ and $x = a$.

32. Show that $e^x = \cosh x + \sinh x$.

33. Show that $e^{-x} = \cosh x - \sinh x$.

34. Use the result of Problem 32 to prove that
$$(\sinh x + \cosh x)^n = \sinh nx + \cosh nx.$$

35. Use the result of Problem 33 to prove that
$$(\cosh x - \sinh x)^n = \cosh nx - \sinh nx.$$

36. Show that
$$\sinh(x + y) = \sinh x\,\cosh y + \cosh x\,\sinh y.$$
[*Hint:* $\sinh(x + y) = \dfrac{e^{x+y} - e^{-(x+y)}}{2} = \dfrac{e^x e^y - e^{-x}e^{-y}}{2}$. Now write $\sinh x\,\cosh y + \cosh x\,\sinh y$ in terms of exponentials.]

37.   Show that

$$\cosh(x + y) = \cosh x \cosh y + \sinh x \sinh y.$$

38.   Show that $\sinh(-x) = -\sinh x$.
39,   Show that $\cosh(-x) = \cosh x$.
40.   Using Problems 36–39, derive formulas for $\sinh(x - y)$ and $\cosh(x - y)$.
41.   Show that $\sinh 2x = 2 \sinh x \cosh x$.
42.   Show that $\cosh 2x = \cosh^2 x + \sinh^2 x = 2 \sinh^2 x + 1$.
43.   Using equation (6), show that

    (a) $\operatorname{sech}^2 x + \tanh^2 x = 1.$ (b) $\coth^2 x - \operatorname{csch}^2 x = 1.$

44.   Using equation (6), find $\sinh x$ if $\cosh x = 8$.
45.   Find $\cosh x$ if $\sinh x = -\frac{2}{5}$.
46.   Using Problem 43, calculate $\operatorname{sech} x$ if $\tanh x = \frac{1}{2}$.
47.   Calculate $\tanh x$ if $\operatorname{sech} x = \frac{3}{8}$.
48.   Calculate $\coth x$ if $\operatorname{csch} x = 3$.
49.   Calculate $\operatorname{csch} x$ if $\coth x = \frac{3}{2}$.
*50.   If $\sinh x = \frac{2}{3}$, calculate the other five hyperbolic functions.
*51.   If $\operatorname{csch} x = 5$, calculate the other five hyperbolic functions.

*7.10 The Inverse Hyperbolic Functions

From Figure 7.9.1 we see that for every real number y there is a unique x such that $y = \sinh x$.

DEFINITION 1.   The *inverse hyperbolic sine* function is defined by $y = \sinh^{-1} x$ if and only if $x = \sinh y$. The domain of $y = \sinh^{-1} x$ is $(-\infty, \infty)$.

> **Note.** Unlike the definitions of the inverse trigonometric functions, there is no need to restrict y to a fixed interval since the function $\sinh x$ is 1-1.

Also from Figure 7.9.1, we see that for each $y \geq 1$, there are two values of x for which $y = \cosh x$: one positive, one negative. Thus we have

DEFINITION 2.   The *inverse hyperbolic cosine* function is defined by $y = \cosh^{-1} x$ if and only if $\cosh y = x$, $y \geq 0$. The domain of $\cosh^{-1} x$ is the interval $[1, \infty)$.

> **Remark.** Here we need to restrict y to be nonnegative so that for each x, there will be a *unique* y such that $x = \cosh y$ (otherwise there would be two values for $\cosh^{-1} x$ and $\cosh^{-1} x$ would then *not* be a function.)

THEOREM 1

(i) $\dfrac{d}{dx} \sinh^{-1} x = \dfrac{1}{\sqrt{x^2 + 1}}.$ (1)

(ii) $\dfrac{d}{dx} \cosh^{-1} x = \dfrac{1}{\sqrt{x^2 - 1}}, \quad x > 1.$ (2)

PROOF. (i) If $y = \sinh^{-1} x$, then $x = \sinh y$ so that

$$1 = \cosh y \frac{dy}{dx} \quad \text{and} \quad \frac{dy}{dx} = \frac{1}{\cosh y}.$$

But $\cosh^2 y - \sinh^2 y = 1$ so that $\cosh y = \sqrt{1 + \sinh^2 y} = \sqrt{1 + x^2}$ which proves (i).

(ii)   If $y = \cosh^{-1} x$, then $x = \cosh y$ and

$$1 = \sinh y \frac{dy}{dx} \quad \text{or} \quad \frac{dy}{dx} = \frac{1}{\sinh y}.$$

But $\sinh y = \pm\sqrt{\cosh^2 y - 1}$. Since $y \geq 0$, $\sinh y \geq 0$ so that

$$\sinh y = \sqrt{\cosh^2 y - 1} = \sqrt{x^2 - 1}$$

which proves (ii).

The graphs of $y = \sinh^{-1} x$ and $y = \cosh^{-1} x$ are given in Figure 1.

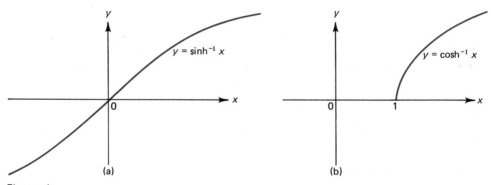

(a)   (b)

Figure 1

From the graph of $y = \tanh x$ in Figure 7.9.3a, we see that for every y in $(-1, 1)$ there is an x such that $y = \tanh x$.

DEFINITION 3.   The *inverse hyperbolic tangent* function is defined by $y = \tanh^{-1} x$ if and only if $x = \tanh y$. The domain of $\tanh^{-1} x$ is $(-1, 1)$.

THEOREM 2

$$\frac{d}{dx} \tanh^{-1} x = \frac{1}{1 - x^2}, \quad -1 < x < 1.$$

PROOF. If $y = \tanh^{-1} x$, then $x = \tanh y$ and

$$1 = \operatorname{sech}^2 y \frac{dy}{dx} \quad \text{or} \quad \frac{dy}{dx} = \frac{1}{\operatorname{sech}^2 y} = \frac{1}{1 - \tanh^2 y} = \frac{1}{1 - x^2}$$

(see Problem 7.9.43(a)).

A sketch of $y = \tanh^{-1} x$ appears in Figure 2.

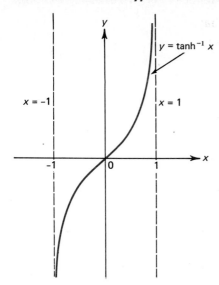

$$y = \tanh^{-1} x$$

$$x = -1 \qquad x = 1$$

Figure 2

We summarize in Table 1 the rules we have discovered in this section.

TABLE 1

$$\frac{d}{dx} \sinh^{-1} u = \frac{1}{\sqrt{u^2 + 1}} \frac{du}{dx}.$$

$$\frac{d}{dx} \cosh^{-1} u = \frac{1}{\sqrt{u^2 - 1}} \frac{du}{dx}, \; u > 1.$$

$$\frac{d}{dx} \tanh^{-1} u = \frac{1}{1 - u^2} \frac{du}{dx}, \; |u| < 1.$$

EXAMPLE 1. Calculate $\dfrac{d}{dx} \cosh^{-1}(x^2 + e^x)$.

SOLUTION. If $u = x^2 + e^x$, then $du/dx = 2x + e^x$ and

$$\frac{d}{dx} \cosh^{-1}(x^2 + e^x) = \frac{2x + e^x}{\sqrt{(x^2 + e^x)^2 - 1}}.$$

EXAMPLE 2. Calculate $\dfrac{d}{dx} \tanh^{-1} \tan x$.

SOLUTION. If $u = \tan x$, $du/dx = \sec^2 x$ so that

$$\frac{d}{dx} \tanh^{-1} \tan x = \frac{\sec^2 x}{1 - \tan^2 x}.$$

There are three other inverse hyperbolic functions but these are rarely used so we will not discuss them here. In fact, the inverse hyperbolic functions occur rarely in applications and are usually used as devices to help us to integrate, as we shall see in Chapter 8.

PROBLEMS 7.10

In Problems 1–12 calculate the derivative of the given function.

1. $y = \sinh^{-1}(3x + 2)$

2. $y = \cosh^{-1} \sqrt{x}$

3. $y = \tanh^{-1} \ln x$

4. $y = \sinh^{-1} \sec x$

5. $y = \sec \sinh^{-1} x$

6. $y = \sinh^{-1}(\cosh x)$

7. $y = \cosh(\sinh^{-1} x)$

8. $y = \tanh^{-1} \dfrac{1}{x^2}$

9. $y = \sqrt{\sinh^{-1} x}$

10. $y = [\sinh^{-1}(x + 1)] \cosh^{-1}(x + 2)$

11. $y = \dfrac{\sinh^{-1} x}{\cosh^{-1} x}$

12. $y = \cos(\sinh^{-1} x)$.

13. Show that

$$\frac{d}{dx} \sinh^{-1} \frac{x}{a} = \frac{1}{\sqrt{x^2 + a^2}} \quad \text{for every } x.$$

14. Show that

$$\frac{d}{dx} \cosh^{-1} \frac{x}{a} = \frac{1}{\sqrt{x^2 - a^2}} \qquad \text{for} \quad \cosh^{-1} \frac{x}{a} > 0 \quad \text{and} \quad \frac{x}{a} > 1.$$

15. Show that

$$\frac{d}{dx} \tanh^{-1} \frac{x}{a} = \frac{a}{a^2 - x^2} \qquad \text{for} \quad x^2 < a^2.$$

*16. Show that $\sinh^{-1} x = \ln(x + \sqrt{1 + x^2})$. [*Hint:* Differentiate both functions and show that they have the same derivative. This means that they differ by a constant. Then show that this constant is zero by evaluating both functions at $x = 0$.]

*17. Show that $\cosh^{-1} x = \ln(x + \sqrt{x^2 - 1})$.

*18. Show that $\tanh^{-1} x = \dfrac{1}{2} \ln\left(\dfrac{1 + x}{1 - x}\right)$.

19. Use Problems 16, 17 and 18 to evaluate
  (a) $\sinh^{-1} 4$ (b) $\cosh^{-1} 3$ (c) $\tanh^{-1} \frac{1}{2}$ (d) $\sinh^{-1}(-1)$.

20. Use the results of Problems 13 and 16 to show that

$$\int \frac{du}{\sqrt{u^2 + a^2}} = \ln(u + \sqrt{a^2 + u^2}) + C.$$

21. Use the results of Problems 14 and 17 to show that for $u > a$,

$$\int \frac{du}{\sqrt{u^2 - a^2}} = \ln(u + \sqrt{u^2 - a^2}) + C.$$

22. Use the results of Problems 15 and 18 to show that for $u^2 < a^2$,

$$\int \frac{a \, du}{a^2 - u^2} = \tfrac{1}{2} \ln\left(\frac{a + u}{a - u}\right) + C.$$

In Problems 23–34 calculate the integrals.

23. $\displaystyle \int \frac{x}{\sqrt{x^4 - 4}} \, dx$

24. $\displaystyle \int \frac{x}{\sqrt{x^4 + 4}} \, dx$

25. $\displaystyle \int \frac{x}{4 - x^4} \, dx$

26. $\int \dfrac{e^x}{\sqrt{1+e^{2x}}}\,dx$

27. $\int \dfrac{e^x}{\sqrt{e^{2x}-9}}\,dx$

28. $\int \dfrac{e^x}{16-e^{2x}}\,dx$

29. $\int \dfrac{\cos x}{\sqrt{1+\sin^2 x}}\,dx$

30. $\int \dfrac{\sin x}{\sqrt{1+\cos^2 x}}\,dx$

31. $\int \dfrac{\sec x\,\tan x}{\sqrt{4+\sec^2 x}}\,dx$

32. $\int \dfrac{dx}{x\sqrt{\ln^2 x-25}}$

33. $\int \dfrac{dx}{\sqrt{x^2+x}}$ [*Hint:*

34. $\int \dfrac{7}{4-3x^2}\,dx.$

Complete the square.]

Review Exercises for Chapter Seven

1. Convert to radians
 (a) 45°   (b) 150°   (c) −60°   (d) −75°.
2. Convert to degrees
 (a) $-\dfrac{\pi}{5}$ (b) $\dfrac{11\pi}{12}$ (c) $-\dfrac{7\pi}{4}$ (d) $\dfrac{13\pi}{8}$.

3. Calculate

 (a) $\sin\left(-\dfrac{\pi}{3}\right)$ (b) $\cos\left(\dfrac{\pi}{6}+\dfrac{\pi}{4}\right)$ (c) $\sin\dfrac{\pi}{12}$

 (d) $\tan\left(-\dfrac{\pi}{6}\right)$ (e) $\cot\dfrac{2\pi}{3}$ (f) $\csc\dfrac{4\pi}{3}$

 (g) $\sec\left(-\dfrac{\pi}{6}\right)$ (h) $\tan 37\pi$ (i) $\cos(-81\pi).$

4. Graph the curves
 (a) $y=4\sin(2\pi x+4)$.   (b) $y=-3\cos[(x/3)+2]$.
 Find the period and amplitude of each function.
5. Graph the curve $y=4\tan(5x+1)$. What is its period?
6. Calculate $\sin\theta$ if $\cos\theta=-\tfrac{4}{9}$ and θ is in the second quadrant.
7. Calculate $\tan\theta$ if $\csc\theta=\tfrac{8}{5}$ and θ is in the first quadrant.
8. Calculate

 (a) $\lim\limits_{x\to 0}\dfrac{\sin\frac{1}{3}x}{x}$.   (b) $\lim\limits_{x\to 0}-\dfrac{\sin 5x}{3x}$.

In Exercises 9–32 calculate the derivative or integral.

9. $\dfrac{d}{dx}\sin(8x^2+2)$

10. $\dfrac{d}{dx}\csc\sqrt{x}$

11. $\dfrac{d}{dx}\sin^{-1}(e^x+1)$

12. $\dfrac{d}{dx}\tan^{-1}\sqrt{\cos x}$

13. $\dfrac{d}{dx}(\sin x+\cos^{-1}x)^{1/3}$

14. $\dfrac{d}{dx}\ln|\tan x|$

15. $\int_0^{5\pi/6}\tan\dfrac{x}{5}\,dx$

16. $\int \sin 3x\cos 3x\,dx$

17. $\int \dfrac{\sec\sqrt{x}\,\tan\sqrt{x}}{\sqrt{x}}\,dx$

18. $\int \dfrac{dx}{\sqrt{1-x^2}}$

19. $\int \dfrac{3x\,dx}{x^4+1}$

20. $\int \dfrac{x^2}{\sqrt{1-x^6}}\,dx$

21. $\dfrac{d}{dx}\tan^{-1}\dfrac{1}{x^2}$

22. $\dfrac{d}{dx}\cot\left(\dfrac{1+x}{1-x}\right)$

23. $\dfrac{d}{dx}xe^x\sin^{-1}x$

24. $\int \dfrac{\sec^2(1/x)}{x^2}\,dx$

25. $\int e^x(\csc^2 e^x)\,dx$

26. $\int \dfrac{e^x}{\sqrt{1-e^{2x}}}\,dx$

27. $\dfrac{d}{dx}\tan(\cos x)$ **28.** $\dfrac{d}{dx}\ln|\sec x + \tan x|$ **29.** $\dfrac{d}{dx}\dfrac{\sin x}{\sin^{-1} x}$

30. $\displaystyle\int \dfrac{e^{5x}}{1 + e^{10x}}\,dx$ **31.** $\displaystyle\int \sin^2(3x + 2)$ **32.** $\displaystyle\int \cot\dfrac{x}{5}\,dx.$

33. Calculate
      (a) $\sin(\cos^{-1}\tfrac{7}{10})$.   (b) $\tan[\sec^{-1}(-4)]$.   (c) $\csc(\cot^{-1}\tfrac{2}{3})$.

34. A 2 kg mass is attached to a spring with spring constant $k = 2$. The spring slides on a horizontal frictionless surface. The mass is initially displaced 50 cm from the equilibrium position of the spring and is then released. Find the equation of motion of the mass for all future time.

In Problems 35–52 calculate the derivative or integral.

35. $\dfrac{d}{dx}(\sinh(x^2 + 3x))$ **36.** $\dfrac{d}{dx}\operatorname{csch}\sqrt[3]{x}$ **37.** $\dfrac{d}{dx}e^{\operatorname{sech}^3 x}$

38. $\displaystyle\int \coth x\,dx$ **39.** $\displaystyle\int x^3 \sinh x^4\,dx$ **40.** $\displaystyle\int \dfrac{\operatorname{sech}^2 \ln x}{x}\,dx$

41. $\dfrac{d}{dx}\sin(\cosh x)$ **42.** $\dfrac{d}{dx}\cos^{-1}(\operatorname{sech}^9 x)$ **43.** $\dfrac{d}{dx}\coth(\cot x)$

44. $\displaystyle\int \dfrac{\operatorname{csch}^2 x}{40 + 3\coth x}\,dx$ **45.** $\displaystyle\int \operatorname{csch}^{4/3} x \coth x\,dx$ **46.** $\displaystyle\int \dfrac{\cosh(1/x^5)}{x^6}\,dx$

47. $\dfrac{d}{dx}\sinh^{-1}(\cos x)$ **48.** $\dfrac{d}{dx}\cosh^{-1}\dfrac{1}{x}$ **49.** $\dfrac{d}{dx}\tanh^{-1}\sqrt{1 + x}$

50. $\displaystyle\int \dfrac{x}{9 + x^4}\,dx$ **51.** $\displaystyle\int \dfrac{\sec^2 x}{\sqrt{\tan^2 x - 4}}\,dx$ **52.** $\displaystyle\int \dfrac{e^{4x}}{\sqrt{16 + e^{8x}}}\,dx.$

53. Find $\sinh x$ if $\cosh x = 2$.
54. Find $\operatorname{sech} x$ if $\tanh x = \tfrac{3}{7}$.

EIGHT

TECHNIQUES OF INTEGRATION

In Chapters 3, 6, and 7 we developed methods for differentiating almost any conceivable function we could write down. However, as we have seen, the process of integration is much more difficult. There are many harmless looking functions which are continuous (and so have indefinite integrals) but for which it is impossible to express indefinite integrals in terms of the functions we have so far discussed. The functions e^{x^2} and $\sin(1/\sqrt{x})$ fall into this category. We simply do not have enough functions and must do the best we can with algebraic functions (i.e., functions like $[\sqrt{x^2 - 1}/(x^4 - 2\sqrt{x})]^{3/5}$), exponential and logarithmic functions, trigonometric functions and their inverses, and the hyperbolic functions (which are really exponential functions). A sad fact of life is that most continuous functions do not have integrals expressible in terms of the functions listed above.

  The aim of this chapter is to learn to recognize which functions can be integrated in the sense of actually finding a recognizable indefinite integral. There are many techniques which have evolved over a period of decades and we shall present the more general ones. More than any other chapter in the book, this one consists primarily of a collection of tricks. Each trick was discovered in response to the need to calculate a particular integral. There is no short, easy way to learn these techniques and it will be necessary to do great numbers of problems.†

  After having completed the first eight sections of this chapter, it will be apparent that there are many integrals for which none of our techniques will yield an answer. For such cases we can still evaluate the definite integral, since the definite integral of any continuous function can be calculated to any number of decimal places of accuracy by a variety of numerical techniques. Some of these techniques will be discussed in Section 9.

8.1 Review of the Basic Formulas of Integration

We give below the integration formulas that were developed in Chapters 4, 6, and 7.

$$\int u^n \, du = \frac{u^{n+1}}{n + 1} + C \qquad (n \neq -1) \tag{1}$$

$$\int \frac{1}{u} \, du = \ln |u| + C \tag{2}$$

$$\int a^u \, du = \frac{1}{\ln a} a^u + C \tag{3}$$

---

†Actually, there are theorems which give us some classes of functions that can be integrated in terms of the functions listed above. For an interesting discussion of this topic see the paper by D. G. Mead, "Integration," American Mathematical Monthly, 68 (1961), 152–156.

$$\int e^u \, du = e^u + C \qquad \int e^{-u}\, du = -e^{-u} + C \tag{4}$$

$$\int \cos u \, du = \sin u + C \tag{5}$$

$$\int \sin u \, du = -\cos u + C \tag{6}$$

$$\int \sec^2 u \, du = \tan u + C \tag{7}$$

$$\int \csc^2 u \, du = -\cot u + C \tag{8}$$

$$\int \sec u \tan u \, du = \sec u + C \tag{9}$$

$$\int \csc u \cot u \, du = -\csc u + C \tag{10}$$

$$\int \tan u \, du = -\ln |\cos u| + C \qquad \text{(see Example 7.6.6)} \tag{11}$$

$$\int \cot u \, du = \ln |\sin u| + C \tag{12}$$

$$\int \sec u \, du = \ln |\sec u + \tan u| + C \qquad \text{(see Example 7.5.7)} \tag{13}$$

$$\int \csc u \, du = -\ln |\csc u + \cot u| + C$$

$$\text{(see Problems 7.6.37 and 7.6.38)} \tag{14}$$

$$\int \frac{du}{\sqrt{1 - u^2}} = \sin^{-1} u + C \tag{15}$$

$$\int \frac{-du}{\sqrt{1 - u^2}} = \cos^{-1} u + C \tag{16}$$

$$\int \frac{du}{1 + u^2} = \tan^{-1} u + C \tag{17}$$

$$\int \frac{-du}{1 + u^2} = \cot^{-1} u + C \tag{18}$$

$$\int \frac{du}{u\sqrt{u^2 - 1}} = \sec^{-1} |u| + C \tag{19}$$

$$\int \frac{-du}{u\sqrt{u^2 - 1}} = -\csc^{-1} |u| + C \tag{20}$$

$$\int \sinh u \, du = \cosh u + C \tag{21}$$

$$\int \cosh u \, du = \sinh u + C \tag{22}$$

$$\int \operatorname{sech}^2 u \, du = \tanh u + C \tag{23}$$

$$\int \operatorname{csch}^2 u \, du = -\coth u + C \tag{24}$$

$$\int \operatorname{sech} u \tanh u = -\operatorname{sech} u + C \tag{25}$$

$$\int \operatorname{csch} u \coth u = -\operatorname{csch} u + C \tag{26}$$

$$\int \frac{du}{\sqrt{u^2 + 1}} = \sinh^{-1} u + C = \ln |u + \sqrt{1 + u^2}| + C \tag{27}$$

$$\int \frac{du}{\sqrt{u^2 - 1}} = \cosh^{-1} u + C = \ln |u + \sqrt{u^2 - 1}| + C, \qquad u > 1 \qquad (28)$$

$$\int \frac{du}{1 - u^2} = \tanh^{-1} u + C = \frac{1}{2} \ln \left| \frac{1 + u}{1 - u} \right| + C, \qquad u^2 < 1 \qquad (29)$$

A much more complete table of integrals appears in Table A.6.

8.2 Integration by Parts

Of all the methods to be discussed in this chapter, the most powerful is called *integration by parts* which is derived from the product rule of differentiation (written in terms of differentials)

$$d(uv) = u\, dv + v\, du. \qquad (1)$$

Integrating both sides of (1), we obtain

$$uv = \int u\, dv + \int v\, du$$

or, rearranging terms,

$$\boxed{\int u\, dv = uv - \int v\, du.} \qquad (2)$$

As we shall see, the trick in using integration by parts is to rewrite an expression which is difficult to integrate in terms of an expression for which an integral is readily obtainable.

EXAMPLE 1. Calculate $\int xe^x\, dx.$

SOLUTION. We cannot integrate xe^x directly because the x term gets in the way. However, if we set $u = x$ and $dv = e^x\, dx$, then $du = dx$, $v = \int e^x\, dx = e^x$ and

$$\int xe^x\, dx = \int u\, dv = uv - \int v\, du = xe^x - \int e^x\, dx = xe^x - e^x + C.$$

This can be checked by differentiation. Note how the x term "disappeared." Example 1 is typical of the type of problem that can be solved by integration by parts.

EXAMPLE 2. Calculate $\int_0^{\pi/2} x \cos x\, dx.$

SOLUTION. Again we could integrate directly were it not for the x term. We set $u = x$ (so that $du = dx$ causing the x term to "vanish") and $dv = \cos x\, dx$. Then $v = \sin x$ and

$$\int_0^{\pi/2} x \cos x\, dx = \int_0^{\pi/2} u\, dv = uv \Big|_0^{\pi/2} - \int_0^{\pi/2} v\, du$$

$$= x \sin x \Big|_0^{\pi/2} - \int_0^{\pi/2} \sin x\, dx = \frac{\pi}{2} + \cos x \Big|_0^{\pi/2} = \frac{\pi}{2} - 1.$$

EXAMPLE 3. Calculate $\int \ln x \, dx$.

SOLUTION. There are two terms here, $\ln x$ and dx. If we are to integrate by parts, the only choices we have for u and dv are $u = \ln x$ and $dv = dx$. Then $du = (1/x) \, dx$, $v = x$ and

$$\int \ln x \, dx = \int u \, dv = uv - \int v \, du = x \ln x - \int x \cdot \frac{1}{x} \, dx = x \ln x - x + C.$$

Note. In many of the integrations involving $\ln x$, we may take $u = \ln x$ so that $du = (1/x) \, dx$ and the ln term "vanishes."

EXAMPLE 4. Calculate $\int_0^1 x^3 e^{x^2} \, dx$.

SOLUTION. Here there are several choices for u and dv. The possibilities are shown in Table 1. From the table we see that while there are several choices for u and dv, only

TABLE 1

u	dv	du	v	$uv - \int v \, du$	Comments
x^3	$e^{x^2} dx$	$3x^2 \, dx$	$\int dv = \int e^{x^2} dx$ (we're stuck)	—	Try something else
e^{x^2}	$x^3 \, dx$	$2xe^{x^2} dx$	$\dfrac{x^4}{4}$	$\dfrac{x^4 e^{x^2}}{4} - \dfrac{1}{2}\int x^5 e^{x^2} \, dx$	We're worse off than when we started
xe^{x^2}	$x^2 \, dx$	$e^{x^2}(1 + 2x^2) \, dx$	$\dfrac{x^3}{3}$	$\dfrac{x^4 e^{x^2}}{3} - \dfrac{1}{3}\int x^3 e^{x^2}(1 + 2x^2) \, dx$	Ditto
x^2	$xe^{x^2} dx$	$2x \, dx$	$\dfrac{1}{2} e^{x^2}$	$\dfrac{x^2 e^{x^2}}{2} - \int xe^{x^2} \, dx$	We can integrate this
$x^2 e^{x^2}$	$x \, dx$	$e^{x^2}(2x^3 + 2x) \, dx$	$\dfrac{x^2}{2}$	$\dfrac{x^4}{2} e^{x^2} - \int e^{x^2}(x^5 + x^3) \, dx$	What a mess!
x	$x^2 e^{x^2} dx$	dx	$\int x^2 e^{x^2} dx$	—	Try something else
$x^3 e^{x^2}$	dx	$e^{x^2}(2x^4 + 3x^2) \, dx$	x	$x^4 e^{x^2} - \int e^{x^2}(2x^5 + 3x^3) \, dx$	This is the worst of all

one works. The others fail either because (a) dv cannot be readily integrated to find v, or (b) $\int v \, du$ is not any easier to integrate than the integral we started with. Hence, to complete the problem we set $u = x^2$, $dv = xe^{x^2} dx$ to obtain

$$\int_0^1 x^3 e^{x^2} \, dx = \frac{x^2 e^{x^2}}{2} \Big|_0^1 - \int_0^1 xe^{x^2} \, dx = \frac{e}{2} - \frac{1}{2} e^{x^2} \Big|_0^1 = \frac{e}{2} - \frac{e}{2} + \frac{1}{2} = \frac{1}{2}.$$

The above example points out two things to look for in choosing u and dv:

(1) It must be possible to evaluate $\int dv$.

(2) $\int v \, du$ should be easier to evaluate than $\int u \, dv$.

EXAMPLE 5. Calculate $\int x^2 e^{-x} \, dx$.

SOLUTION. We need to get rid of the x^2 term. Let $u = x^2$ and $dv = e^{-x}\,dx$. Then $du = 2x\,dx$ and $v = -e^{-x}$ so that $\int u\,dv = uv - \int v\,du$

$$\int x^2 e^{-x}\,dx = -x^2 e^{-x} + 2 \int xe^{-x}\,dx.$$

(The plus sign comes from $-\int -e^{-x}\,dx$.) We see that $\int xe^{-x}\,dx$ is simpler than $\int x^2 e^{-x}\,dx$ but it is still necessary to integrate by parts once more. Setting $u = x$ and $dv = e^{-x}\,dx$, we have $du = dx$ and $v = -e^{-x}$ so that

$$2 \int xe^{-x}\,dx = -2xe^{-x} + 2 \int e^{-x}\,dx = -2xe^{-x} - 2e^{-x} + C.$$

Therefore

$$\int x^2 e^{-x}\,dx = -x^2 e^{-x} - 2xe^{-x} - 2e^{-x} + C.$$

EXAMPLE 6. Evaluate $\int e^x \sin x\,dx$.

SOLUTION. We set $I = \int e^x \sin x\,dx$. (The reason for this will become apparent as we work the problem.) Let $u = e^x$ and $dv = \sin x\,dx$. Then $du = e^x\,dx, v = -\cos x$ and

$$I = -e^x \cos x + \int e^x \cos x\,dx.$$

It seems as if we have not accomplished anything, but this is not the case. We now set $u = e^x$ and $dv = \cos x\,dx$ so that $du = e^x\,dx$, $v = \sin x$ and

$$I = -e^x \cos x + e^x \sin x - \int e^x \sin x\,dx = e^x(\sin x - \cos x) - I$$

or

$$2I = e^x(\sin x - \cos x)$$

and

$$I = \int e^x \sin x\,dx = \frac{e^x(\sin x - \cos x)}{2} + C.$$

Note that in this example we were able to find the integral of $e^x \sin x$ without calculating any integrals directly.

EXAMPLE 7. Calculate $\int \sin 2x \cos 3x\,dx$.

SOLUTION. Since if we differentiate or integrate $\sin 2x$ and $\cos 3x$ twice we end up with constant multiples of the same functions, we suspect that the trick of Example 6 will work here. Set $I = \int \sin 2x \cos 3x\,dx$ and set $u = \sin 2x$ and $dv = \cos 3x\,dx$. Then $du = 2 \cos 2x\,dx, v = \frac{1}{3} \sin 3x$ and

$$I = \frac{1}{3} \sin 2x \sin 3x - \frac{2}{3} \int \cos 2x \sin 3x\,dx.$$

We now set $u = \cos 2x$ and $dv = -\frac{2}{3} \sin 3x\,dx$ so that $du = -2 \sin 2x\,dx$, $v = \frac{2}{9} \cos 3x$ and

$$I = \frac{1}{3} \sin 2x \sin 3x + \frac{2}{9} \cos 2x \cos 3x + \frac{4}{9} \int \sin 2x \cos 3x\,dx$$
$$= \frac{1}{3} \sin 2x \sin 3x + \frac{2}{9} \cos 2x \cos 3x + \frac{4}{9} I$$

or

$$\tfrac{5}{9}I = \tfrac{1}{3}\sin 2x \sin 3x + \tfrac{2}{9}\cos 2x \cos 3x$$

and

$$I = \int \sin 2x \cos 3x \, dx = \tfrac{9}{5}(\tfrac{1}{3}\sin 2x \sin 3x + \tfrac{2}{9}\cos 2x \cos 3x)$$
$$= \tfrac{1}{5}(3 \sin 2x \sin 3x + 2 \cos 2x \cos 3x) + C.$$

This should be checked by differentiation.

EXAMPLE 8. Calculate $\int \sin^{-1} x \, dx$.

SOLUTION. There is no other choice but to set $u = \sin^{-1} x$ and $dv = dx$. Then $du = dx/\sqrt{1 - x^2}$, $v = x$ and

$$\int \sin^{-1} x \, dx = x \sin^{-1} x - \int \frac{x}{\sqrt{1 - x^2}} \, dx = x \sin^{-1} x + \sqrt{1 - x^2} + C.$$

PROBLEMS 8.2

In Problems 1–27 evaluate the given integral.

1. $\int x e^{3x} \, dx$

2. $\int_0^1 x e^{-7x} \, dx$

3. $\int_0^4 x^2 e^{x/4} \, dx$

4. $\int_1^2 x \ln x \, dx$

5. $\int x^7 \ln x \, dx$

6. $\int_0^{\pi/2} x \sin x \, dx$

7. $\int x^2 \cos 2x \, dx$

8. $\int_0^5 x \sqrt{3x + 1} \, dx$

9. $\int x \sqrt{1 - \dfrac{x}{2}} \, dx$

10. $\int \cos^{-1} x \, dx$

11. $\int_0^{1/2} \tan^{-1} 2x \, dx$

12. $\int x \sinh x \, dx$

13. $\int_0^1 x^2 \cosh 2x \, dx$

14. $\int x 5^x \, dx$

15. $\int_0^1 \sin^{-1} x \, dx$

*16. $\int x \tan^{-1} x \, dx$

17. $\int \cos(\ln x) \, dx$

18. $\int e^x \cos x \, dx$

19. $\int \sin x \sin 4x \, dx$

20. $\int_0^\pi \cos \dfrac{x}{2} \cos \dfrac{x}{3} \, dx$

21. $\int_0^1 x^5 e^{x^3} \, dx$

22. $\int_1^2 \ln^2 x \, dx$

23. $\int e^{3x} \sin 2x \, dx$

24. $\int e^{-x} \cos \dfrac{x}{2} \, dx$

*25. $\int x^2 \tan^{-1} x \, dx$

*26. $\int \sinh 2x \cosh 3x \, dx$

*27. $\int \sinh 2x \sinh 5x \, dx$.

*28. Show that if $a \neq 0$,

$$\int x^3 \sin ax \, dx = \left(\frac{3x^2}{a^2} - \frac{6}{a^4}\right) \sin ax + \left(\frac{6x}{a^3} - \frac{x^3}{a}\right) \cos ax + C$$

*29. Using mathematical induction (Appendix 1) and integration by parts, show that if n is a positive integer and $a \neq 0$,

$$\int x^n e^{ax} \, dx = \frac{e^{ax}}{a}\left(x^n - \frac{nx^{n-1}}{a} + \frac{n(n-1)x^{n-2}}{a^2} - \cdots + \frac{(-1)^n n!}{a^n}\right) + C$$

where $n! = n(n - 1)(n - 2) \cdots 3 \cdot 2 \cdot 1$.

*30. Show that if $a \neq 0$,

$$\int x^2 \sinh ax \, dx = \left(\frac{x^2}{a} + \frac{2}{a^3}\right) \cosh ax - \frac{2x}{a^2} \sinh ax + C$$

31. The density of a 5-m bar is given by $\rho = xe^{-x}$ kg/m where x is measured in meters from the left. What is the total mass of the bar?

32. The marginal cost of a certain product fluctuates according to the rule

   $MC = 20 + e^{-x} \sin \pi x$ cents,

   where x is the number of items produced. What is the total cost of producing 100 units?

33. (a) Graph the curve $y = xe^{-x}$. [*Hint:* Use the fact (which we shall prove in Chapter 12) that $\lim_{x \to \infty} xe^{-x} = 0$.]
   (b) Calculate the area bounded by the curve $y = xe^{-x}$, the x-axis and the lines $x = -1$, $x = 1$.

34. Calculate the area bounded by the curve $y = \sin x \sin 4x$, the x-axis, the y-axis and the line $x = \pi/2$.

8.3 Integrals of Certain Trigonometric Functions

In this section we show how to integrate some classes of trigonometric functions. They are

   (i) $\int \sin^m x \cos^n x \, dx$, either m or n an odd positive integer

   (ii) $\int \sin^m x \cos^n x \, dx$, both m and n even positive integers

   (iii) $\int \sin mx \cos nx \, dx$

   (iv) $\int \sin mx \sin nx \, dx$

   (v) $\int \cos mx \cos nx \, dx$

   (vi) $\int \tan^n x \, dx$

   (vii) $\int \sec^n x \, dx$

   (viii) $\int \tan^m x \sec^n x \, dx$, n even

   (ix) $\int \tan^m x \sec^n x \, dx$, n and m odd

   (x) $\int \tan^m x \sec^n x \, dx$, n odd, m even.

We shall see that each case can be treated fairly easily by making use of some trigonometric identity.

CASE (i) $\int \sin^m x \cos^n x \, dx$, where either m or n is an odd integer.†

† Here if m is an odd integer then n need not be an integer, and vice versa.

EXAMPLE 1. Calculate $\int \sin^3 x \cos^{4/3} x \, dx$.

SOLUTION. Write $\sin^3 x = (\sin^2 x) \sin x = (1 - \cos^2 x) \sin x$. Then

$$\int \sin^3 x \cos^{4/3} x \, dx = \int (1 - \cos^2 x) \sin x \cos^{4/3} x \, dx$$
$$= \int \cos^{4/3} x \sin x \, dx - \int \cos^{10/3} x \sin x \, dx$$
$$= -\tfrac{3}{7} \cos^{7/3} x + \tfrac{3}{13} \cos^{13/3} x + C.$$

In Case (i), by breaking off $\sin^2 x$ terms (if m is odd) or $\cos^2 x$ terms (if n is odd), we can always obtain integrals of the form $\int \cos^p x \sin x \, dx$ or $\int \sin^q x \cos x \, dx$ and these can be integrated directly since

$$\int \cos^p x \sin x \, dx = -\frac{\cos^{p+1} x}{p+1} + C$$

and $\quad\displaystyle\int \sin^q x \cos x \, dx = \frac{\sin^{q+1} x}{q+1} + C.$

EXAMPLE 2. Calculate $\int \sqrt{\sin x} \cos^5 x \, dx.$

SOLUTION

$$\int \sqrt{\sin x} \cos^5 x \, dx = \int \sqrt{\sin x} \cos^3 x (1 - \sin^2 x) \, dx$$
$$= \int (\sin^{1/2} x - \sin^{5/2} x) \cos^3 x \, dx$$

$$= \int (\sin^{1/2} x - \sin^{5/2} x)(1 - \sin^2 x) \cos x \, dx$$

$$= \int \sin^{1/2} x \cos x \, dx - 2 \int \sin^{5/2} x \cos x \, dx$$

$$+ \int \sin^{9/2} x \cos x \, dx$$

$$= \tfrac{2}{3} \sin^{3/2} x - \tfrac{4}{7} \sin^{7/2} x + \tfrac{2}{11} \sin^{11/2} x + C.$$

CASE (ii). $\int \sin^m x \cos^n x \, dx$, both m and n are even. We use the identities

$$\cos^2 x = \frac{1 + \cos 2x}{2}, \qquad \sin^2 x = \frac{1 - \cos 2x}{2}. \tag{1}$$

(See (xx) and (xxi) in Table 7.2.2.)

EXAMPLE 3. Calculate $\displaystyle\int_0^{\pi/2} \sin^2 x \, dx$.

SOLUTION

$$\int_0^{\pi/2} \sin^2 x \, dx = \frac{1}{2} \int_0^{\pi/2} (1 - \cos 2x) \, dx = \left(\frac{x}{2} - \frac{\sin 2x}{4} \right) \Big|_0^{\pi/2} = \frac{\pi}{4}.$$

EXAMPLE 4. Calculate $\displaystyle\int_0^{\pi/4} \sin^2 x \cos^2 x \, dx$.

SOLUTION

$$\int_0^{\pi/4} \sin^2 x \cos^2 x \, dx = \frac{1}{4} \int_0^{\pi/4} (1 - \cos 2x)(1 + \cos 2x) \, dx$$

$$= \frac{1}{4} \int_0^{\pi/4} (1 - \cos^2 2x) \, dx$$

$$= \frac{1}{4} \int_0^{\pi/4} dx - \frac{1}{4} \int_0^{\pi/4} \cos^2 2x \, dx$$

$$= \frac{\pi}{16} - \frac{1}{8} \int_0^{\pi/4} (1 + \cos 4x) \, dx$$

$$= \frac{\pi}{16} - \frac{\pi}{32} - \frac{1}{8} \int_0^{\pi/4} \cos 4x \, dx$$

$$= \frac{\pi}{32} - \frac{\sin 4x}{32} \Big|_0^{\pi/4} = \frac{\pi}{32}.$$

Here we used the second identity of (1) twice.

EXAMPLE 5. Calculate $\displaystyle\int_0^{\pi/4} \sin^6 x \, dx$.

SOLUTION

$$\int_0^{\pi/4} \sin^6 x \, dx = \int_0^{\pi/4} (\sin^2 x)^3 \, dx = \int_0^{\pi/4} \left(\frac{1 - \cos 2x}{2} \right)^3 dx$$

$$= \frac{1}{8} \int_0^{\pi/4} (1 - 3 \cos 2x + 3 \cos^2 2x - \cos^3 2x) \, dx$$

$$= \frac{1}{8} \int_0^{\pi/4} dx - \frac{3}{8} \int_0^{\pi/4} \cos 2x \, dx + \frac{3}{16} \int_0^{\pi/4} (1 + \cos 4x) \, dx$$

$$- \frac{1}{8} \int_0^{\pi/4} \cos 2x (1 - \sin^2 2x) \, dx$$

$$= \frac{\pi}{32} - \frac{3}{16} \sin 2x \Big|_0^{\pi/4} + \frac{3}{16} \cdot \frac{\pi}{4} + \frac{3}{64} \sin 4x \Big|_0^{\pi/4}$$

$$- \frac{\sin 2x}{16} \Big|_0^{\pi/4} + \frac{\sin^3 2x}{48} \Big|_0^{\pi/4}$$

$$= \frac{\pi}{32} - \frac{3}{16} + \frac{3\pi}{64} + 0 - \frac{1}{16} + \frac{1}{48} = \frac{5\pi}{64} - \frac{11}{48}.$$

Here we used the technique of Case (i) to integrate $\cos^3 2x$.

CASES (iii), (iv) and (v). We use the three identities

$$\sin mx \cos nx = \tfrac{1}{2}\{\sin[(m + n)x] + \sin[(m - n)x]\} \qquad (2)$$

$$\sin mx \sin nx = \tfrac{1}{2}\{\cos[(m - n)x] - \cos[(m + n)x]\} \qquad (3)$$

$$\cos mx \cos nx = \tfrac{1}{2}\{\cos[(m - n)x] + \cos[(m + n)x]\} \qquad (4)$$

These identities are easily obtained from the addition formulas (see Problems 28, 29, 30).

EXAMPLE 6. Calculate $\int \sin mx \cos nx \, dx$.

SOLUTION. If $m = n \neq 0$, then $\int \sin mx \cos mx \, dx = (\sin^2 mx)/2m + C$. If $m \neq \pm n$, then

$$\int \sin mx \cos nx \, dx = \frac{1}{2} \int \{\sin[(m+n)x] + \sin[(m-n)x]\} \, dx$$

$$= -\frac{\cos[(m+n)x]}{2(m+n)} - \frac{\cos[(m-n)x]}{2(m-n)} + C.$$

Other integrals in Cases (iii), (iv) and (v) are handled just as easily.

CASE (vi). $\int \tan^n x \, dx$. We use the identity $1 + \tan^2 x = \sec^2 x$.

EXAMPLE 7. Calculate $\int \tan^2 x \, dx$.

SOLUTION. $\int \tan^2 x \, dx = \int (\sec^2 x - 1) \, dx = \tan x - x + C$.

EXAMPLE 8. Calculate $\int \tan^5 x \, dx$.

SOLUTION

$$\int \tan^5 x \, dx = \int \tan^3 x \tan^2 x \, dx = \int \tan^3 x (\sec^2 x - 1) \, dx$$

$$= \int \tan^3 x \sec^2 x \, dx - \int \tan^3 x \, dx$$

$$= \frac{\tan^4 x}{4} - \int \tan x (\sec^2 x - 1) \, dx$$

$$= \frac{\tan^4 x}{4} - \frac{\tan^2 x}{2} + \int \tan x \, dx$$

$$= \frac{\tan^4 x}{4} - \frac{\tan^2 x}{2} - \ln|\cos x| + C.$$

CASE (vii). $\int \sec^n x \, dx$. We use the same identity as in Case (vi).

EXAMPLE 9. Calculate $\int_0^{\pi/4} \sec^4 x \, dx$.

SOLUTION

$$\int_0^{\pi/4} \sec^4 x \, dx = \int_0^{\pi/4} \sec^2 x (1 + \tan^2 x) \, dx$$

$$= \int_0^{\pi/4} \sec^2 x \, dx + \int_0^{\pi/4} \tan^2 x \sec^2 x \, dx$$

$$= \tan x \Big|_0^{\pi/4} + \frac{\tan^3 x}{3} \Big|_0^{\pi/4} = \frac{4}{3}.$$

EXAMPLE 10. Calculate $\int \sec^3 x \, dx$.

SOLUTION. This problem is a bit trickier. We have

$$I = \int \sec^3 x \, dx = \int \sec x \sec^2 x \, dx = \int \sec x \, (1 + \tan^2 x) \, dx$$

$$= \int \sec x \, dx + \int \sec x \tan^2 x \, dx$$

$$= \ln |\sec x + \tan x| + \int \sec x \tan^2 x \, dx.$$

We cannot evaluate this last integral directly. We therefore set $u = \tan x$ and $dv = \sec x \tan x \, dx$. Then $du = \sec^2 x \, dx$, $v = \sec x$ so that

$$I = \ln |\sec x + \tan x| + \sec x \tan x - \int \sec^3 x \, dx$$

$$= \ln |\sec x + \tan x| + \sec x \tan x - I$$

and

$$I = \tfrac{1}{2}(\ln |\sec x + \tan x| + \sec x \tan x) + C.$$

CASE (viii). $\int \tan^m x \sec^n x \, dx$, n even. Use $1 + \tan^2 x = \sec^2 x$ and break off one $\sec^2 x$ term.

EXAMPLE 11. Calculate $\displaystyle\int_0^{\pi/3} \tan^7 x \sec^4 x \, dx.$

SOLUTION

$$\int_0^{\pi/3} \tan^7 x \sec^2 x \sec^2 x \, dx = \int_0^{\pi/3} (\tan^7 x)(1 + \tan^2 x) \sec^2 x \, dx$$

$$= \int_0^{\pi/3} \tan^7 x \sec^2 x \, dx + \int_0^{\pi/3} \tan^9 x \sec^2 x \, dx$$

$$= \frac{\tan^8 x}{8} + \frac{\tan^{10} x}{10} \Big|_0^{\pi/3}$$

$$= \frac{\sqrt{3}^8}{8} + \frac{\sqrt{3}^{10}}{10} = \frac{3^4}{8} + \frac{3^5}{10}$$

$$= \frac{81}{8} + \frac{243}{10} = \frac{1377}{40}.$$

CASE (ix). $\int \tan^m x \sec^n x \, dx$, m, n odd. In this case, break off a term of the form $\sec x \tan x$.

EXAMPLE 12. $\int \tan^3 x \sec^5 x \, dx = \int \tan^2 x \sec^4 x \cdot \sec x \tan x \, dx$. Since $\sec x \tan x$ is the derivative of $\sec x$, we write everything in terms of $\sec x$:

$$\int \tan^2 x \sec^4 x \cdot \sec x \tan x \, dx = \int (\sec^2 - 1) \sec^4 x \cdot \sec x \tan x \, dx$$

$$= \int (\sec^6 x - \sec^4 x) \sec x \tan x \, dx$$

$$= \frac{\sec^7 x}{7} - \frac{\sec^5 x}{5} + C.$$

CASE (x). $\int \tan^m x \sec^n x \, dx$, m even, n odd. In this case we can write everything in terms of $\sec x$ and proceed as in Case (vii).

EXAMPLE 13. Calculate $\int \tan^2 x \sec^3 x \, dx$.

SOLUTION

$$\int \tan^2 x \sec^3 x \, dx = \int (\sec^2 x - 1) \sec^3 x \, dx = \int \sec^5 x \, dx - \int \sec^3 x \, dx.$$

We have already calculated $\int \sec^3 x \, dx$. Now

$$I = \int \sec^5 x \, dx = \int (1 + \tan^2 x)^2 \sec x \, dx$$

$$= \int \sec x \, dx + 2 \int \sec x \tan^2 x \, dx + \int \tan^4 x \sec x \, dx.$$

In order to calculate $\int \sec x \tan^2 x \, dx$, let $u = \tan x$ and $dv = \sec x \tan x \, dx$ so that $du = \sec^2 x \, dx$, $v = \sec x$ and

$$2 \int \sec x \tan^2 x \, dx = 2 \sec x \tan x - 2 \int \sec^3 x \, dx.$$

To calculate $\int \tan^4 x \sec x \, dx$, let $u = \tan^3 x$ and $dv = \sec x \tan x$ so that $du = 3 \tan^2 x \sec^2 x \, dx$, $v = \sec x$ and

$$\int \tan^4 x \sec x \, dx = \sec x \tan^3 x - 3 \int \sec^3 x \tan^2 x \, dx$$

$$= \sec x \tan^3 x - 3 \int \sec^3 x (\sec^2 x - 1) \, dx$$

$$= \sec x \tan^3 x + 3 \int \sec^3 x \, dx - 3 \int \sec^5 x \, dx.$$

Thus

$$I = \int \sec x \, dx + 2 \sec x \tan x - 2 \int \sec^3 x \, dx$$

$$+ \sec x \tan^3 x + 3 \int \sec^3 x \, dx - 3I$$

or

$$I = \tfrac{1}{4} \left(\ln |\sec x + \tan x| + 2 \sec x \tan x + \sec x \tan^3 x + \int \sec^3 x \, dx \right).$$

Finally,

$$\int \tan^2 x \sec^3 x \, dx = I - \int \sec^3 x \, dx$$

$$= \tfrac{1}{4} \left(\ln |\sec x + \tan x| + 2 \sec x \tan x + \sec x \tan^3 x - 3 \int \sec^3 x \, dx \right)$$

$$= \tfrac{1}{4}(\ln |\sec x + \tan x| + 2 \sec x \tan x + \sec x \tan^3 x - \tfrac{3}{2} \ln |\sec x + \tan x|$$
$$- \tfrac{3}{2} \sec x \tan x)$$

$$= -\frac{\ln |\sec x + \tan x|}{8} + \frac{\sec x \tan x}{8} + \frac{\sec x \tan^3 x}{4} + C.$$

The last example illustrates that the computation of trigonometric integrals can be very complicated. Nevertheless, any integral of one of the forms (i)–(x) can be integrated. Sometimes it requires a bit of patience.

PROBLEMS 8.3

In Problems 1–27 calculate the integral.

1. $\int_0^\pi \sin^3 x \cos^2 x \, dx$

   2. $\int \cos^3 x \sqrt{\sin x} \, dx$

   3. $\int_0^{\pi/4} \cos^2 x \, dx$

4. $\int \sin^4 x \, dx$

   5. $\int \cos^4 2x \, dx$

   6. $\int_0^{\pi/2} \sin^2 3x \cos^4 3x \, dx$

7. $\int \sin 2x \cos 3x \, dx$

   8. $\int_0^{\pi/2} \sin \frac{x}{2} \cos \frac{x}{3} \, dx$

   9. $\int \sin \sqrt{2} x \cos \frac{x}{\sqrt{2}} \, dx$

10. $\int_0^\pi \sin 2x \sin \frac{x}{2} \, dx$

    11. $\int \cos 10x \cos 100x \, dx$

    12. $\int_0^{\pi/8} \tan^2 2x \, dx$

13. $\int \sec^3 5x \, dx$

    14. $\int_0^{\pi/6} \sec^3 2x \tan 2x \, dx$

    15. $\int_0^\pi \sin^6 x \cos^4 x \, dx$

16. $\int \cos^{10} x \sin^7 x \, dx$

    17. $\int_0^{\pi/4} \sec^3 x \tan^3 x \, dx$

    18. $\int \sec^6 \frac{x}{2} \, dx$

19. $\int \sin^5 x \cos^2 x \, dx$

    20. $\int \cos^5 x \sin^2 x \, dx$

    21. $\int_0^{\pi/2} \tan^5 \frac{x}{2} \, dx$

22. $\int_0^{\pi/3} \sec^6 x \, dx$

    23. $\int \tan^3 x \sec^4 x \, dx$

    24. $\int \frac{\sin^5 x}{\cos^2 x} \, dx$

25. $\int_0^{\pi/4} \sin^2 x \tan x \, dx$

    26. $\int \sec x \sin^3 x \, dx$

    27. $\int \cos^3 x \sqrt{\csc x} \, dx.$

28. Show that $\sin mx \cos nx = \frac{1}{2}\{\sin[(m+n)x] + \sin[(m-n)x]\}$. [*Hint:* Add the identities $\sin(a+b) = \sin a \cos b + \cos a \sin b$ and $\sin(a-b) = \sin a \cos b - \cos a \sin b$.]

29. Prove identity (3).

    30. Prove identity (4).

31. Show that $\int \sin mx \sin nx \, dx = \frac{1}{2} \left\{ \frac{\sin[(m-n)x]}{m-n} - \frac{\sin[(m+n)x]}{m+n} \right\} + C.$

32. Show that $\int \cos mx \cos nx \, dx = \frac{1}{2} \left\{ \frac{\sin[(m-n)x]}{m-n} + \frac{\sin[(m+n)x]}{m+n} \right\} + C.$

33. Show that $\int_0^{2\pi} \sin mx \cos nx \, dx = 0.$

34. Show that $\int_0^{2\pi} \cos mx \cos nx \, dx = 0$ if $m \neq n$.

35. Show that $\int_0^{2\pi} \sin mx \sin nx \, dx = 0$ if $m \neq n$.

Using the identity $1 + \cot^2 x = \csc^2 x$, we can evaluate integrals of the form $\int \cot^n x \, dx$, $\int \csc^n x \, dx$ and $\int \cot^m x \csc^n x \, dx$. We need only use the methods outlined for integrals of type (vi)–(x). In Problems 36–44 evaluate the integral.

36. $\int \cot^2 x \, dx$

    37. $\int \cot^5 x \, dx$

    38. $\int_{\pi/4}^{\pi/2} \csc^4 x \, dx$

39. $\int \csc^3 x \, dx$

    40. $\int_{\pi/6}^{\pi/2} \cot^7 x \csc^4 x \, dx$

    41. $\int \cot^3 x \csc^5 x \, dx$

42. $\int \cot^2 x \csc^6 x \, dx$

    43. $\int \csc^4 2x \, dx$

    44. $\int \cot^2 x \csc x \, dx.$

45.   Show that $\displaystyle\int_0^{\pi/2} \sin^2 x \, dx = \frac{\pi}{4} = \frac{1}{2} \cdot \frac{\pi}{2}$.

46.   Show that $\displaystyle\int_0^{\pi/2} \sin^4 x \, dx = \frac{3}{4} \cdot \frac{1}{2} \cdot \frac{\pi}{2}$.

47.   Show that $\displaystyle\int_0^{\pi/2} \sin^6 x \, dx = \frac{5}{6} \cdot \frac{3}{4} \cdot \frac{1}{2} \cdot \frac{\pi}{2}$.

48.   By breaking off a $\sin^2 x$ term, show that

$$\int_0^{\pi/2} \sin^{2n} x \, dx = \frac{2n-1}{2n} \int_0^{\pi/2} \sin^{2n-2} x \, dx.$$

*49.   Using Problems 45–48, show that

$$\int_0^{\pi/2} \sin^{2n} x \, dx = \frac{2n-1}{2n} \cdot \frac{2n-3}{2n-2} \cdot \frac{2n-5}{2n-4} \cdot \ldots \cdot \frac{5}{6} \cdot \frac{3}{4} \cdot \frac{1}{2} \cdot \frac{\pi}{2}.$$

This is called a *Wallis product.*

*50.   Show that

$$\int_0^{\pi/2} \sin^{2n+1} x \, dx = \frac{2n}{2n+1} \cdot \frac{2n-2}{2n-1} \cdot \frac{2n-4}{2n-3} \cdot \ldots \cdot \frac{4}{5} \cdot \frac{2}{3}.$$

This is another Wallis product.

51.   Show that $\displaystyle\left| \int_0^{\pi/2} \sin^{2n} x \, dx \right| \leq \pi/2$.

*52.   Show using the result of Problem 51 and Problem 54 below that

$$\lim_{n \to \infty} \frac{\displaystyle\int_0^{\pi/2} \sin^{2n} x \, dx}{\displaystyle\int_0^{\pi/2} \sin^{2n+1} x \, dx} = 1.$$

*53.   Divide the result of Problem 49 by the result of Problem 50 and use Problem 52 to derive the *Wallis formula*

$$\frac{\pi}{2} = \lim_{n \to \infty} \frac{2}{1} \cdot \frac{2}{3} \cdot \frac{4}{3} \cdot \frac{4}{5} \cdot \frac{6}{5} \cdot \frac{6}{7} \cdot \ldots \cdot \frac{2n}{2n-1} \cdot \frac{2n}{2n+1}.$$

54.   Using the techniques of this section and integration by parts, derive the *reduction formula*

$$\int \sin^n ax \, dx = -\frac{\sin^{n-1} ax \cos ax}{an} + \frac{n-1}{n} \int \sin^{n-2} ax \, dx.$$

55.   Derive the reduction formula

$$\int \cos^n ax \, dx = \frac{\sin ax \cos^{n-1} ax}{an} + \frac{n-1}{n} \int \cos^{n-2} ax \, dx.$$

56.   Derive the formula

$$\int \tan^n ax \, dx = \frac{\tan^{n-1} ax}{(n-1)a} - \int \tan^{n-2} ax \, dx.$$

8.4 The Idea behind Integration by Substitution

In Section 4.6 we saw how a substitution could be made to change $\int f(x)\, dx$ into an integral of the form $\int u^r\, du$. There are many different kinds of substitutions that will allow us to integrate more easily. Some of these will be illustrated in Sections 5 and 8. In this section we shall describe the steps involved in making a substitution in an integral.

Consider the definite integral

$$\int u^r\, du = \frac{u^{r+1}}{r+1} + C$$

$$\int_a^b f(x)\, dx. \tag{1}$$

Suppose we wish to make the substitution $x = g(u)$ for some function g. Then there are three things that must be done to change the integral in (1) from an integral with respect to the variable x to an integral with respect to the variable u:

> (i) $f(x)$ is replaced by $f(g(u))$.
> (ii) dx is replaced by $g'(u)\, du$ (since $dx = g'(u)\, du$).
> (iii) If c and d are numbers such that $a = g(c)$ and $b = g(d)$, then a is replaced by c and b is replaced by d.

EXAMPLE 1. Consider $\int_0^1 \sqrt{1 - x^2}\, dx$. Let $x = \sin u$ (the reason for making this substitution will be clear shortly). Then

$$g(u) = \sin u \quad \text{and} \quad f(g(u)) = \sqrt{1 - \sin^2 u} = \sqrt{\cos^2 u} = \cos u.$$

Also $dx = \cos u\, du$. When $x = 0$, $\sin u = 0$, so that $u = \sin^{-1} 0 = 0$. When $x = 1$, $u = \sin^{-1} 1 = \pi/2$. Thus $\int_0^1 \sqrt{1 - x^2}\, dx = \int_0^{\pi/2} \cos^2 u\, du$, which we can easily calculate.

EXAMPLE 2. Consider

$$\int_1^{64} \frac{x + 1}{\sqrt{x}(1 + \sqrt[3]{x})}\, dx.$$

If we make the substitution $x = u^6$, then $\sqrt{x} = u^3$, $\sqrt[3]{x} = u^2$ and $dx = 6u^5\, du$. When $x = 1$, $u = x^{1/6} = 1$ and when $x = 64$, $u = 2$. Thus

$$\int_1^{64} \frac{x + 1}{\sqrt{x}(1 + \sqrt[3]{x})}\, dx = \int_1^2 \frac{(u^6 + 1)6u^5}{u^3(1 + u^2)}\, du = 6 \int_1^2 \frac{u^8 + u^2}{1 + u^2}\, du.$$

We will see how to complete this integration in Section 8.7.

It is important to remember that in substituting a new variable in an integral, it is necessary to change everything—including the limits of integration. In Sections 5 and 8 we will show how substitutions can be chosen to make integration, in some cases, a good deal easier.

PROBLEMS 8.4

In each of the following problems transform the given integral to an integral with respect to a new variable by making the indicated substitution. Do not try to integrate.

1. $\displaystyle\int_0^3 e^{x^4}\, dx; \quad x = u^{1/4}$

2. $\displaystyle\int_1^e x^2 \ln x\, dx, \quad x - e^u$

3. $\displaystyle\int_0^1 \sqrt{1 + x^5}\, dx; \quad 1 + x^5 = z$

4. $\displaystyle\int_0^1 \sqrt{1 + x^2}\, dx; \quad x = \tan\theta$

5. $\displaystyle\int_0^1 \frac{dx}{3 + \sqrt{x}}; \quad x = v^2$

6. $\displaystyle\int_1^4 \frac{x^{1/3}}{2 + x^{2/3}}\, dx; \quad x = y^3$

7. $\displaystyle\int_0^1 (1 - x^2)^{3/2}\, dx; \quad x = \sin\theta$

8. $\displaystyle\int_{-2}^1 x^2 \sqrt{x + 3}\, dx; \quad \sqrt{x + 3} = s$

9. $\displaystyle\int_0^4 x^3 \sqrt{16 - x^2}\, dx; \quad \sqrt{16 - x^2} = y$

10. $\displaystyle\int_1^2 \frac{dx}{x^3(9 - x^3)^{1/3}}; \quad v = \left(\frac{9 - x^3}{x^3}\right)^{1/3}$

11. $\displaystyle\int_0^3 x^3 \sqrt{9 - x^2}\, dx; \quad x = 3\cos\theta$

12. $\displaystyle\int_1^{\sqrt{2}} (x^2 - 1)^{5/2}\, dx; \quad x = \sec\theta$

*13. $\displaystyle\int_0^1 \sqrt{1 + x^2}\, dx; \quad x = \sinh t$

14. $\displaystyle\int_0^1 \frac{x^7}{1 + x^2}\, dx; \quad x = \tan\theta.$

8.5 Integrals Involving $\sqrt{a^2 - x^2}$, $\sqrt{a^2 + x^2}$, and $\sqrt{x^2 - a^2}$—Trigonometric Substitutions

Integrals containing terms of the form $\sqrt{a^2 - u^2}$, $\sqrt{a^2 + u^2}$ and $\sqrt{u^2 - a^2}$ can often be calculated by making an appropriate trigonometric substitution.

If the term is of the form

(i) $\sqrt{a^2 - x^2}$, set $x = a\sin\theta$. Then

$\sqrt{a^2 - x^2} = \sqrt{a^2 - a^2\sin^2\theta} = \sqrt{a^2(1 - \sin^2\theta)} = \sqrt{a^2\cos^2\theta} = |a\cos\theta|$ †

(ii) $\sqrt{a^2 + x^2}$, set $x = a\tan\theta$. Then

$\sqrt{a^2 + x^2} = \sqrt{a^2 + a^2\tan^2\theta} = \sqrt{a^2(1 + \tan^2\theta)} = \sqrt{a^2\sec^2\theta} = |a\sec\theta|$

(iii) $\sqrt{x^2 - a^2}$, set $x = a\sec\theta$. Then

$\sqrt{x^2 - a^2} = \sqrt{a^2\sec^2\theta - a^2} = \sqrt{a^2(\sec^2\theta - 1)} = \sqrt{a^2\tan^2\theta} = |a\tan\theta|$

EXAMPLE 1. Calculate $\displaystyle\int_0^1 \sqrt{1 - x^2}\, dx.$

SOLUTION. Let $x = \sin\theta$. Then $dx = \cos\theta\, d\theta$ and

$$\sqrt{1 - x^2} = \sqrt{1 - \sin^2\theta} = \sqrt{\cos^2\theta} = \cos\theta.$$

---

† Since $\sqrt{a^2 - x^2}$ denotes the positive square root of $a^2 - x^2$, we choose $a\cos\theta$ rather than $-a\cos\theta$ as the square root of $a^2\cos^2\theta$. We will do this kind of thing several times in this section without discussing it further.

When $x = \sin \theta = 0, \theta = \sin^{-1} 0 = 0$. When $x = \sin \theta = 1, \theta = \sin^{-1} 1 = \pi/2$. Then

$$\int_0^1 \sqrt{1 - x^2}\, dx = \int_0^{\pi/2} \cos^2 \theta\, d\theta = \frac{1}{2} \int_0^{\pi/2} (1 + \cos 2\theta)\, d\theta$$

$$= \frac{1}{2} \left(\theta + \frac{\sin 2\theta}{2} \right) \Big|_0^{\pi/2} = \frac{\pi}{4}.$$

EXAMPLE 2. Calculate $\int \sqrt{1 - x^2}\, dx$.

SOLUTION. This problem is a little trickier than the one before. Setting $x = \sin \theta$, we obtain

$$\int \sqrt{1 - x^2}\, dx = \int \cos^2 \theta\, d\theta = \frac{\theta}{2} + \frac{\sin 2\theta}{4} + C.$$

But here we need an answer which is a function of x, not θ. If $x = \sin \theta$, $\theta = \sin^{-1} x$. Then

$$\sin 2\theta = 2 \sin \theta \cos \theta = 2 \sin(\sin^{-1} x) \cos(\sin^{-1} x) = 2x \cos(\sin^{-1} x).$$

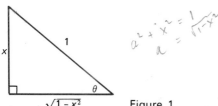

Figure 1

Now, from the triangle in Figure 1 (see Section 7.4), $\cos(\sin^{-1} x) = \sqrt{1 - x^2}$. Thus

$$\int \sqrt{1 - x^2}\, dx = \frac{\theta}{2} + \frac{\sin 2\theta}{4} = \frac{\sin^{-1} x}{2} + \frac{x\sqrt{1 - x^2}}{2} + C.$$

This can be checked by differentiation. Note that

$$\frac{\sin^{-1} x}{2} + \frac{x\sqrt{1 - x^2}}{2} \Big|_0^1 = \frac{\sin^{-1} 1}{2} = \frac{\pi}{4}.$$

    Examples 1 and 2 illustrate the fact that when making substitutions, it is easier to evaluate definite integrals, since in those cases it is not necessary to write the answer in terms of the original variable. In calculating indefinite integrals of this sort, it is often helpful to draw a triangle to yield a function of x for the answer.

EXAMPLE 3. Calculate $\int_0^4 x^3 \sqrt{16 - x^2}\, dx$.

SOLUTION. Let $x = 4 \sin \theta$. Then $\sqrt{16 - x^2} = \sqrt{16 - 16 \sin^2 \theta} = \sqrt{16 \cos^2 \theta} = 4 \cos \theta$, $dx = 4 \cos \theta$ and since $\theta = \sin^{-1}(x/4)$, the limits of integration become $\sin^{-1} 0 = 0$ and $\sin^{-1}(4/4) = \pi/2$. Then

$$\int_0^4 x^3 \sqrt{16 - x^2}\, dx = 4^5 \int_0^{\pi/2} \sin^3 \theta \cos^2 \theta\, d\theta$$

$$= 1024 \int_0^{\pi/2} (1 - \cos^2 \theta) \cos^2 \theta \sin \theta \, d\theta$$

$$= 1024 \int_0^{\pi/2} (\cos^2 \theta - \cos^4 \theta) \sin \theta \, d\theta$$

$$= 1024 \left(\frac{\cos^5 \theta}{5} - \frac{\cos^3 \theta}{3} \right) \Big|_0^{\pi/2}$$

$$= 1024 \left(\frac{1}{3} - \frac{1}{5} \right) = 1024 \cdot \frac{2}{15} = \frac{2048}{15}.$$

EXAMPLE 4. Calculate $\int \sqrt{x^2 - a^2} \, dx$.

SOLUTION. Let $x = a \sec \theta$. Then $dx = a \sec \theta \tan \theta \, d\theta$ and

$$\sqrt{x^2 - a^2} = \sqrt{a^2 \sec^2 \theta - a^2} = \sqrt{a^2 \tan^2 \theta} = a \tan \theta$$

so that

$$\int \sqrt{x^2 - a^2} \, dx = a^2 \int \tan^2 \theta \sec \theta \, d\theta$$

$$= a^2 \int (\sec^2 \theta - 1) \sec \theta \, d\theta = a^2 \int (\sec^3 \theta - \sec \theta) \, d\theta$$

$$= \frac{a^2}{2} (\sec \theta \tan \theta + \ln |\sec \theta + \tan \theta|)$$

$$- a^2 \ln |\sec \theta + \tan \theta| \qquad \text{(from Example 8.3.10)}$$

$$= \frac{a^2}{2} (\sec \theta \tan \theta - \ln |\sec \theta + \tan \theta|).$$

Now since $x = a \sec \theta$, $\sec \theta = x/a$ and we see from the triangle in Figure 2 that

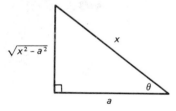

Figure 2

$\tan \theta = (\sqrt{x^2 - a^2})/a$. This also follows from the identity $a^2 \sec^2 \theta = a^2 + a^2 \tan^2 \theta$.
Thus

$$\int \sqrt{x^2 - a^2} \, dx = \frac{a^2}{2} \left(\frac{x}{a} \cdot \frac{\sqrt{x^2 - a^2}}{a} - \ln \left| \frac{x + \sqrt{x^2 - a^2}}{a} \right| \right)$$

or

$$\int \sqrt{x^2 - a^2} \, dx = \frac{x \sqrt{x^2 - a^2}}{2} - \frac{a^2}{2} \ln |x + \sqrt{x^2 - a^2}| + C.$$

Note that $\dfrac{a^2}{2} \ln \left| \dfrac{x + \sqrt{x^2 - a^2}}{a} \right| = \dfrac{a^2}{2} \ln |x + \sqrt{x^2 - a^2}| - \dfrac{a^2}{2} \ln |a|$

and we included the constant term $(a^2/2) \ln |a|$ in the general constant C.

EXAMPLE 5. Calculate $\displaystyle\int_0^1 \frac{x^3}{(3+x^2)^{5/2}} \, dx$.

SOLUTION. If $x^2 + 3 = x^2 + a^2$, we must have $a = \sqrt{3}$. We set $x = \sqrt{3} \tan \theta$. Then $dx = \sqrt{3} \sec^2 \theta \, d\theta$ and

$$(3 + x^2)^{5/2} = (3 + 3\tan^2 \theta)^{5/2} = (3\sec^2 \theta)^{5/2} = 9\sqrt{3} \sec^5 \theta.$$

Also, since $\theta = \tan^{-1}(x/\sqrt{3})$, the limits of integration are $\tan^{-1} 0 = 0$ and $\tan^{-1}(1/\sqrt{3}) = \pi/6$. Thus

$$\int_0^1 \frac{x^3}{(3+x^2)^{5/2}} \, dx = \int_0^{\pi/6} \frac{(3\sqrt{3} \tan^3 \theta)(\sqrt{3} \sec^2 \theta)}{9\sqrt{3} \sec^5 \theta} \, d\theta$$

$$= \frac{1}{\sqrt{3}} \int_0^{\pi/6} \frac{\tan^3 \theta}{\sec^3 \theta} \, d\theta$$

$$= \frac{1}{\sqrt{3}} \int_0^{\pi/6} \frac{\sin^3 \theta}{\cos^3 \theta} \cos^3 \theta \, d\theta$$

$$= \frac{1}{\sqrt{3}} \int_0^{\pi/6} \sin^3 \theta \, d\theta$$

$$= \frac{1}{\sqrt{3}} \int_0^{\pi/6} (1 - \cos^2 \theta) \sin \theta \, d\theta$$

$$= \frac{1}{\sqrt{3}} \left(\frac{\cos^3 \theta}{3} - \cos \theta \right) \Big|_0^{\pi/6}$$

$$= \frac{1}{\sqrt{3}} \left[\frac{(\sqrt{3}/2)^3}{3} - \frac{\sqrt{3}}{2} - \frac{1}{3} + 1 \right] = \frac{2}{3\sqrt{3}} - \frac{3}{8}.$$

EXAMPLE 6. Evaluate $\displaystyle\int \frac{x^3}{(3+x^2)^{5/2}} \, dx$.

SOLUTION. Proceding as in Example 5, we see that the integral is equal to

$$\frac{1}{\sqrt{3}} \left(\frac{\cos^3 \theta}{3} - \cos \theta \right)$$

where $\theta = \tan^{-1}(x/\sqrt{3})$. Then, using the triangle in Figure 3, we find that $\cos \theta = \sqrt{3}/(x^2 + 3)$ so that

$$\int \frac{x^3}{(3+x^2)^{5/2}} \, dx = \frac{1}{\sqrt{3}} \left\{ \frac{[3/(x^2 + 3)]^{3/2}}{3} - \sqrt{\frac{3}{3+x^2}} \right\} + C$$

$$= \frac{1}{\sqrt{3+x^2}} \left[\frac{1}{3+x^2} - 1 \right] + C = -\frac{2+x^2}{(3+x^2)^{3/2}} + C.$$

In problems this complicated, it is especially necessary to check the answer by differentiating.

EXAMPLE 7. Show by integration that $\sinh^{-1} x = \ln(x + \sqrt{1 + x^2})$.

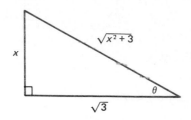

Figure 3

SOLUTION. Consider the integral

$$\int \frac{dx}{\sqrt{x^2 + 1}} = \sinh^{-1} x + C.$$

Let $x = \tan \theta$. Then $\sqrt{x^2 + 1} = \sec \theta$ and $dx = \sec^2 \theta \, d\theta$ so that

$$\int \frac{dx}{\sqrt{x^2 + 1}} = \int \frac{\sec^2 \theta \, d\theta}{\sec \theta} = \int \sec \theta \, d\theta$$

$$= \ln |\sec \theta + \tan \theta| = \ln (x + \sqrt{1 + x^2}) + C'.$$

Thus $\sinh^{-1} x$ and $\ln(x + \sqrt{1 + x^2})$ differ by a constant. However, since both expressions are zero when $x = 0$, it follows that the constant is zero and the result is proven.

EXAMPLE 8. Calculate $\displaystyle\int_0^{1/3} \sqrt{4 - 9x^2} \, dx$.

SOLUTION. Let $u = 3x$. Then $du = 3 \, dx$ so that $dx = \frac{1}{3} du$, $u = 0$ when $x = 0$, $u = 1$ when $x = \frac{1}{3}$, and

$$\int_0^{1/3} \sqrt{4 - 9x^2} \, dx = \frac{1}{3} \int_0^1 \sqrt{4 - u^2} \, du.$$

Now set $u = 2 \sin \theta$. Then

$$\frac{1}{3} \int_0^1 \sqrt{4 - u^2} \, du = \frac{1}{3} \int_0^{\pi/6} 4 \cos^2 \theta \, d\theta = \frac{2}{3} \int_0^{\pi/6} (1 + \cos 2\theta) \, d\theta$$

$$= \frac{2}{3} \left(\theta + \frac{\sin 2\theta}{2} \right) \Big|_0^{\pi/6} = \frac{2}{3} \left(\frac{\pi}{6} + \frac{\sqrt{3}}{4} \right) = \frac{\pi}{9} + \frac{\sqrt{3}}{6}.$$

PROBLEMS 8.5

In Problems 1–54 evaluate the given integral by making an appropriate trigonometric substitution.

1. $\displaystyle\int_0^1 \frac{x}{\sqrt{4 - x^2}} \, dx$

2. $\displaystyle\int \frac{dx}{\sqrt{4 - x^2}}$

3. $\displaystyle\int \frac{x^2}{\sqrt{4 - x^2}} \, dx$

4. $\displaystyle\int_0^1 (4 - x^2)^{3/2} \, dx$

5. $\displaystyle\int_0^1 x^3 \sqrt{4 - x^2} \, dx$

6. $\displaystyle\int \frac{x}{\sqrt{x^2 - 4}} \, dx$

7. $\displaystyle\int \frac{dx}{\sqrt{x^2 - 4}}$

8. $\displaystyle\int \frac{x^2}{\sqrt{x^2 - 4}} \, dx$

9. $\displaystyle\int_{4/\sqrt{3}}^4 \sqrt{x^2 - 4} \, dx$

*10. $\int x^3 \sqrt{x^2 - 4}\, dx$ **11.** $\int_0^2 \frac{x}{\sqrt{4 + x^2}}\, dx$ **12.** $\int \frac{dx}{\sqrt{4 + x^2}}$

13. $\int \frac{dx}{4 + x^2}$ **14.** $\int_0^2 \frac{x^2}{\sqrt{4 + x^2}}\, dx$ **15.** $\int (4 + x^2)^{3/2}\, dx$

16. $\int_0^2 x^3 \sqrt{4 + x^2}\, dx$ **17.** $\int_0^{\sqrt{3}} \frac{x^5}{1 + x^2}\, dx$ **18.** $\int_{\sqrt{2}}^2 (x^2 - 2)^{1/2}\, dx$

19. $\int_0^{\sqrt{2}} \frac{dx}{2 + x^2}$ **20.** $\int_0^{1/\sqrt{2}} \frac{x^3}{2 + x^2}\, dx$ **21.** $\int_2^4 \sqrt{4x^2 - 9}\, dx$

*22. $\int_1^3 \frac{dx}{x^4 \sqrt{x^2 + 3}}$ **23.** $\int_4^8 \frac{x^2}{(x^2 - 4)^{3/2}}\, dx$ **24.** $\int_1^4 \frac{dx}{x^4 \sqrt{x^2 + 16}}$

25. $\int_4^5 \frac{dx}{(x^2 - 9)^{3/2}}$ **26.** $\int \sqrt{a^2 - x^2}\, dx$ **27.** $\int x^2 \sqrt{a^2 - x^2}\, dx$

28. $\int \sqrt{a^2 + x^2}\, dx$ **29.** $\int \sqrt{1 + a^2 x^2}\, dx$ **30.** $\int \sqrt{1 - a^2 x^2}\, dx$

31. $\int \frac{\sqrt{5x^2 - 9}}{x}\, dx$ **32.** $\int \frac{\sqrt{5x^2 + 9}}{x}\, dx$ **33.** $\int \frac{\sqrt{9 - 5x^2}}{x}\, dx$

34. $\int (a^2 - x^2)^{3/2}\, dx$ **35.** $\int \frac{dx}{x \sqrt{x^2 - a^2}}$ **36.** $\int \frac{dx}{x^3 \sqrt{x^2 - a^2}}$

37. $\int x^3 (x^2 - a^2)^{3/2}\, dx$ **38.** $\int \frac{(x^2 - a^2)^{3/2}}{x^3}\, dx$ **39.** $\int \frac{x^3}{\sqrt{a^2 - x^2}}\, dx$

*40. $\int \frac{dx}{x \sqrt{a^2 - x^2}}$ **41.** $\int \frac{dx}{x^3 \sqrt{a^2 - x^2}}$ **42.** $\int \frac{dx}{(a^2 - x^2)^{3/2}}$

43. $\int \frac{dx}{(x^2 - a^2)^{3/2}}$ **44.** $\int x^2 (a^2 - x^2)^{3/2}\, dx$ **45.** $\int \frac{dx}{x \sqrt{x^2 + a^2}}$

46. $\int \frac{\sqrt{x^2 + a^2}}{x}\, dx$ **47.** $\int \frac{x^2}{(x^2 + a^2)^{3/2}}\, dx$ **48.** $\int \frac{dx}{x^3 (x^2 + a^2)^{3/2}}$

49. $\int \frac{x^2}{x^2 + a^2}\, dx$ ***50.** $\frac{dx}{x^3 (x^2 + a^2)}$ **51.** $\int \frac{dx}{x^2 (x^2 + a^2)^2}$

52. $\int \frac{dx}{\sqrt{a^2 - b^2 x^2}}$ **53.** $\int \frac{dx}{\sqrt{a^2 + b^2 x^2}}$ **54.** $\int \frac{dx}{\sqrt{b^2 x^2 - a^2}}.$

55. Prove by integration that $\cosh^{-1} x = \ln |x + \sqrt{x^2 - 1}|$.

56. Prove that $\tanh^{-1} x = \frac{1}{2} \ln \left| \frac{1 + x}{1 - x} \right|$.

57. The density of a 5-m bar is given by $\rho(x) = \sqrt{50 - 2x^2}$. Find the total mass of the bar.

*58. The marginal cost of a certain product is given by $MC = 20 + (30/\sqrt{5 + 4x^2})$ cents per unit. What is the total cost of producing 100 units?

59. (a) Graph the curve $y = \sqrt{1 - x^2}$ for $0 \le x \le 1$.
   (b) Find the area bounded by the curve and the x- and y-axes.

60. (a) Graph the curve $y = \sqrt{9 + x^2}$ for $x \ge 0$.
   (b) Find the area bounded by the curve, the x- and y-axes, and the line $x = 4$.

8.6 Integration by Partial Fractions—Linear Factors

We recall from Chapter 3 that a rational function $r(x)$ is the quotient of two polynomials. That is,

$$r(x) = \frac{p(x)}{q(x)} \tag{1}$$

where p and q are polynomials. From Example 2.4.3, we know that $r(x)$ is continuous at any point x at which $q(x) \neq 0$. Thus $\int_a^b r(x)\,dx$ exists over any interval $[a, b]$ in which $q(x) \neq 0$. It turns out that it is possible to find integrals for all rational functions. We will not prove this fact in general but will, instead, illustrate it with a number of examples.

EXAMPLE 1. Calculate $\int \dfrac{dx}{x^2 + x}$.

SOLUTION

$$\frac{1}{x^2 + x} = \frac{1}{x(x + 1)}.$$

We try to find constants A and B such that

$$\frac{1}{x(x + 1)} = \frac{A}{x} + \frac{B}{x + 1}. \tag{2}$$

The writing of $1/[x(x + 1)]$ as the sum of two simpler fractions is called a *decomposition into partial fractions*. Cross multiplying in the right-hand side of (2), we find that

$$\frac{A}{x} + \frac{B}{x + 1} = \frac{A(x + 1) + Bx}{x(x + 1)} = \frac{(A + B)x + A}{x(x + 1)}.$$

Thus we have

$$\frac{(A + B)x + A}{x(x + 1)} = \frac{1}{x(x + 1)} = \frac{0 \cdot x + 1}{x(x + 1)}.$$

This last equation holds only when

$$A + B = 0 \qquad \text{and} \qquad A = 1.$$

Thus

$$A = 1, \qquad B = -1, \qquad \frac{1}{x(x + 1)} = \frac{\overset{A}{1}}{x} - \frac{\overset{B}{1}}{x + 1},$$

and

$$\int \frac{1}{x(x + 1)}\,dx = \int \left(\frac{1}{x} - \frac{1}{x + 1} \right) dx = \ln |x| - \ln |x + 1| + C$$

$$= \ln \left| \frac{x}{x + 1} \right| + C.$$

EXAMPLE 2. Calculate $\int \dfrac{x^5 + 4x^4 - 14x^3 - 31x^2 + 57x - 72}{x^2 + 2x - 15}\, dx.$

SOLUTION. The first thing we notice in this problem is that the degree of the polynomial in the numerator is 5 while the degree of the polynomial in the denominator is 2. Thus if we divide the numerator by the denominator, we will obtain a polynomial of degree 3 plus a remainder term of the form

$$\frac{ax + b}{x^2 + 2x - 15}.$$

We carry out the division:

$$
\begin{array}{r}
x^3 + 2x^2 - 3x + 5 \\
x^2 + 2x - 15\,\overline{\big)\,x^5 + 4x^4 - 14x^3 - 31x^2 + 57x - 72} \\
\underline{x^5 + 2x^4 - 15x^3} \\
2x^4 + x^3 - 31x^2 \\
\underline{2x^4 + 4x^3 - 30x^2} \\
-3x^3 - x^2 + 57x \\
\underline{-3x^3 - 6x^2 + 45x} \\
5x^2 + 12x - 72 \\
\underline{5x^2 + 10x - 75} \\
2x + 3
\end{array}
$$

Therefore

$$\int \frac{x^5 + 4x^2 - 14x^3 - 31x^2 + 57x - 72}{x^2 + 2x - 15}\, dx$$

$$= \int \left(x^3 + 2x^2 - 3x + 5 + \frac{2x + 3}{x^2 + 2x - 15} \right) dx$$

$$= \frac{x^4}{4} + \frac{2x^3}{3} - \frac{3x^2}{2} + 5x + \int \frac{2x + 3}{(x + 5)(x - 3)}\, dx.$$

We now write (in partial fractions)

$$\frac{2x + 3}{(x + 5)(x - 3)} = \frac{A}{x + 5} + \frac{B}{x - 3} = \frac{A(x - 3) + B(x + 5)}{(x + 5)(x - 3)}$$

$$= \frac{(A + B)x + (-3A + 5B)}{(x + 5)(x - 3)} = 2x + 3.$$

For this equation to hold, we must have

$$A + B = 2$$
$$-3A + 5B = 3.$$

To solve this system of equations, we multiply the first equation by 3 and add it to the second:

$$
\begin{array}{r}
3A + 3B = 6 \\
\underline{-3A + 5B = 3} \\
8B = 9 \quad \text{or} \quad B = \tfrac{9}{8}.
\end{array}
$$

Then $A = \frac{7}{8}$ (from the first equation) and we obtain

$$\int \frac{2x+3}{(x+5)(x-3)}\,dx = \int \left[\frac{7}{8(x+5)} + \frac{9}{8(x-3)}\right]dx$$

$$= \frac{7}{8}\ln|x+5| + \frac{9}{8}\ln|x-3|$$

$$= \frac{1}{8}[7\ln|x+5| + 9\ln|x-3|]$$

$$= \frac{1}{8}\ln|(x+5)^7(x-3)^9| + C$$

so that

$$\int \frac{x^5 + 4x^4 - 14x^3 - 31x^2 + 57x - 72}{x^2 + 2x - 15}\,dx$$

$$= \frac{x^4}{4} + \frac{2x^3}{3} - \frac{3x^2}{2} + 5x + \frac{1}{8}\ln|(x+5)^7(x-3)^9| + C.$$

As the two examples above illustrate, the integration of rational functions by partial fractions involves a number of algebraic manipulations but very little actual integration. We now enumerate the steps that should be taken to integrate a rational function $r(x) = p(x)/q(x)$.

   (i)   If degree $p(x) \geq$ degree $q(x)$, divide to obtain

$$\frac{p(x)}{q(x)} = s(x) + \frac{t(x)}{q(x)},$$

       where degree $t(x) <$ degree $q(x)$, and $s(x)$ is a polynomial.
   (ii)   Integrate $s(x)$.
   (iii)   It is always possible to factor $q(x)$ into linear and quadratic factors (the proof of this algebraic fact would take us too far afield here). Then there are four cases to consider. Let

$$q(x) = a_0 x^n + a_1 x^{n-1} + a_2 x^{n-2} + \cdots + a_{n-1}x + a_n.$$

We can assume that $a_0 = 1$, for if not, we simply factor out a_0 and reintroduce it after we have carried out the integration.

We will devote the remainder of this section and all of Section 7 to a discussion of methods for carrying out step (iii).

CASE 1. Distinct Linear Factors

$$q(x) = (x - x_1)(x - x_2) \cdots (x - x_n)$$

where no two of the x_i's are equal.

CASE 2. Repeated Linear Factors. Same as in Case 1 except that some of the roots x_i are the same.

CASE 3. Distinct Quadratic Factors. Here $q(x)$ can be factored into linear terms (like $x - x_i$) and quadratic terms of the form $x^2 + ax + b$, where no two quadratic terms are the same. Quadratic terms occur only when a term of the form $x^2 + ax + b$ cannot be factored as $(x - x_1)(x - x_2)$, where x_1 and x_2 are real numbers. This occurs when $a^2 - 4b < 0$.

CASE 4. Repeated Quadratic Factors. Same as Case 3 except that some of the quadratic terms are repeated.

We will deal with Cases 1 and 2 in this section and will defer Cases 3 and 4 until Section 7.

CASE 1. $q(x) = (x - x_1)(x - x_2) \cdots (x - x_n)$. Then we can write

$$\frac{t(x)}{q(x)} = \frac{A_1}{x - x_1} + \frac{A_2}{x - x_2} + \cdots + \frac{A_n}{x - x_n} \tag{3}$$

We can find the constants $A_1, A_2, \ldots, A_n$ as in Examples 1 and 2 but there is a much easier way to do it.

THEOREM 1. If $q(x) = (x - x_1)(x - x_2) \cdots (x - x_n)$ where none of the factors are repeated, then (3) holds with

$$A_k = \left. \frac{t(x)(x - x_k)}{q(x)} \right|_{x = x_k}, \qquad k = 1, 2, \ldots, n.$$

That is, the kth coefficient A_k is obtained by multiplying $t(x)/q(x)$ by $(x - x_k)$, canceling the term $(x - x_k)$ in the numerator and denominator, and evaluating the resulting expression at $x = x_k$.

PROOF. We have

$$\frac{t(x)}{q(x)} = \frac{t(x)}{(x - x_1)(x - x_2) \cdots (x - x_n)}$$

$$= \frac{A_1}{x - x_1} + \frac{A_2}{x - x_2} + \cdots + \frac{A_k}{x - x_k} + \cdots + \frac{A_n}{x - x_n}. \tag{4}$$

We multiply both sides of (4) by $(x - x_k)$. Then the right-hand side becomes

$$A_1 \left\{ \frac{x - x_k}{x - x_1} \right\} + A_2 \left\{ \frac{x - x_k}{x - x_2} \right\} + \cdots + A_{k-1} \left\{ \frac{x - x_k}{x - x_{k-1}} \right\} + A_k \left\{ \frac{x - x_k}{x - x_k} \right\}$$

$$+ A_{k+1} \left\{ \frac{x - x_k}{x - x_{k+1}} \right\} + \cdots + A_n \left\{ \frac{x - x_k}{x - x_n} \right\}.$$

$$= A_1 \left\{ \frac{x - x_k}{x - x_1} \right\} + A_2 \left\{ \frac{x - x_k}{x - x_2} \right\} + \cdots + A_{k-1} \left\{ \frac{x - x_k}{x - x_{k-1}} \right\} + A_k$$

$$+ A_{k+1} \left\{ \frac{x - x_k}{x - x_{k+1}} \right\} + \cdots + A_n \left\{ \frac{x - x_k}{x - x_n} \right\}. \tag{5}$$

When $x = x_k$, the expression (5) reduces to A_k and the theorem is proved.

EXAMPLE 3. In Example 2, we have

$$\frac{2x + 3}{(x + 5)(x - 3)} = \frac{A_1}{x + 5} + \frac{A_2}{x - 3}.$$

Then

$$A_1 = \frac{(2x + 3)(x + 5)}{(x + 5)(x - 3)}\bigg|_{x=-5} = \frac{2x + 3}{x - 3}\bigg|_{x=-5} = \frac{-7}{-8} = \frac{7}{8}$$

and

$$A_2 = \frac{2x + 3}{x + 5}\bigg|_{x=3} = \frac{9}{8}.$$

Notice how much easier this is.

EXAMPLE 4. Calculate $\int \frac{2x^3 - 3x^2 - 5x + 11}{x^3 - 2x^2 - x + 2} dx$.

SOLUTION. Here degree p = degree q so we first divide:

$$
\begin{array}{r}
2 \\
x^3 - 2x^2 - x + 2\overline{\smash{\big)}\,2x^3 - 3x^2 - 5x + 11} \\
\underline{2x^3 - 4x^2 - 2x + 4} \\
x^2 - 3x + 7
\end{array}
$$

Thus

$$\frac{p}{q} = s + \frac{t}{q} = 2 + \frac{x^2 - 3x + 7}{x^3 - 2x^2 - x + 2}.$$

Now $x^3 - 2x^2 - x + 2 = (x + 1)(x - 1)(x - 2)$ with roots $x_1 = -1$, $x_2 = 1$ and $x_3 = 2$. Thus

$$A_1 = \frac{x^2 - 3x + 7}{(x - 1)(x - 2)}\bigg|_{x=-1} = \frac{11}{6}, \qquad A_2 = \frac{x^2 - 3x + 7}{(x + 1)(x - 2)}\bigg|_{x=1} = -\frac{5}{2},$$

and

$$A_3 = \frac{x^2 - 3x + 7}{(x + 1)(x - 1)}\bigg|_{x=2} = \frac{5}{3}$$

so that

$$\frac{x^2 - 3x + 7}{x^3 - 2x^2 - x + 2} = \frac{11}{6(x + 1)} - \frac{5}{2(x - 1)} + \frac{5}{3(x - 2)}$$

and

$$\int \frac{2x^3 - 3x^2 - 5x + 11}{x^3 - 2x^2 - x + 2} dx = \int \left[2 + \frac{11}{6(x + 1)} - \frac{5}{2(x - 1)} + \frac{5}{3(x - 2)} \right] dx$$

$$= 2x + \frac{11}{6} \ln|x + 1| - \frac{5}{2} \ln|x - 1|$$

$$+ \frac{5}{3} \ln|x - 2| + C.$$

CASE 2. Repeated Linear Factors

EXAMPLE 5. Calculate $\int \dfrac{dx}{x(x-1)^2}$.

SOLUTION. We write

$$\frac{1}{x(x-1)^2} = \frac{A_1}{x} + \frac{A_2}{x-1} + \frac{A_3}{(x-1)^2}$$

$$= \frac{A_1(x-1)^2 + A_2 x(x-1) + A_3 x}{x(x-1)^2}$$

$$= \frac{A_1(x^2 - 2x + 1) + A_2(x^2 - x) + A_3 x}{x(x-1)^2}$$

$$= \frac{(A_1 + A_2)x^2 + (-2A_1 - A_2 + A_3)x + A_1}{x(x-1)^2}.$$

In order that this equality hold, we must have

$$(A_1 + A_2)x^2 + (-2A_1 - A_2 + A_3)x + A_1 = 1.$$

Thus

$$\begin{aligned} A_1 + A_2 &= 0 \\ -2A_1 - A_2 + A_3 &= 0 \\ A_1 &= 1. \end{aligned}$$

We easily find the solution $A_1 = 1$, $A_2 = -1$, $A_3 = 1$, so that

$$\int \frac{dx}{x(x-1)^2} = \int \left[\frac{1}{x} - \frac{1}{x-1} + \frac{1}{(x-1)^2} \right] dx$$

$$= \ln|x| - \ln|x-1| - \frac{1}{x-1} + C$$

$$= \ln\left| \frac{x}{x-1} \right| - \frac{1}{x-1} + C.$$

As the above example illustrates, the only difference between Case 1 and Case 2 is that there is more algebraic work involved. In general, we have the following rule for integrating $t(x)/q(x)$ where degree $t(x) <$ degree $q(x)$: Let $x_1, x_2, \ldots, x_k$ be the k distinct roots of $q(x)$. Then if $q(x) = (x - x_1)^{m_1}(x - x_2)^{m_2} \cdots (x - x_k)^{m_k}$, we can write

$$\begin{aligned}
\frac{t(x)}{q(x)} &= \frac{A_{1,1}}{x - x_1} + \frac{A_{1,2}}{(x - x_1)^2} + \cdots + \frac{A_{1,m_1}}{(x - x_1)^{m_1}} \\
&\quad + \frac{A_{2,1}}{(x - x_2)} + \frac{A_{2,2}}{(x - x_2)^2} + \cdots + \frac{A_{2,m_2}}{(x - x_2)^{m_2}} + \cdots \\
&\quad + \frac{A_{k,1}}{x - x_k} + \frac{A_{k,2}}{(x - x_k)^2} + \cdots + \frac{A_{k,m_k}}{(x - x_k)^{m_k}}.
\end{aligned} \tag{6}$$

In equation (6) the numbers $m_1, m_2, \ldots, m_k$ are called the *multiplicities* of the roots and we have $m_1 + m_2 + \cdots + m_k = n$, the degree of $q(x)$, so that there are exactly n terms in the partial fraction expansion of $t(x)/q(x)$. As in Case 1, there are two ways to calculate the coefficients $A_{i,j}$. The first method was illustrated in Example 5. The second one will be discussed in Problems 25–28.

EXAMPLE 6. Calculate $\displaystyle\int \frac{x^3 + 2}{x^3(x - 2)^2} \, dx$.

SOLUTION. $q(x)$ has two roots $x_1 = 0$ (with multiplicity $m_1 = 3$) and $x_2 = 2$ (with multiplicity $m_2 = 2$). Then

$$\frac{x^3 + 2}{x^3(x - 2)^2}$$

$$= \frac{A_{1,1}}{x} + \frac{A_{1,2}}{x^2} + \frac{A_{1,3}}{x^3} + \frac{A_{2,1}}{x - 2} + \frac{A_{2,2}}{(x - 2)^2}$$

$$= \frac{A_{1,1}x^2(x - 2)^2 + A_{1,2}x(x - 2)^2 + A_{1,3}(x - 2)^2 + A_{2,1}x^3(x - 2) + A_{2,2}x^3}{x^3(x - 2)^2}$$

$$= \frac{\begin{array}{l} A_{1,1}(x^4 - 4x^3 + 4x^2) + A_{1,2}(x^3 - 4x^2 + 4x) \\ \qquad + A_{1,3}(x^2 - 4x + 4) + A_{2,1}(x^4 - 2x^3) + A_{2,2}x^3 \end{array}}{x^3(x - 2)^2}$$

$$= \frac{(A_{1,1} + A_{2,1})x^4 + (-4A_{1,1} + A_{1,2} - 2A_{2,1} + A_{2,2})x^3 + (4A_{1,1} - 4A_{1,2} + A_{1,3})x^2}{x^3(x - 2)^2}$$

$$+ \frac{(4A_{1,2} - 4A_{1,3})x + 4A_{1,3}}{x^3(x - 2)^2}.$$

Then, equating coefficients of the powers of x, we obtain

$$
\begin{array}{llll}
A_{1,1} & & + A_{2,1} & = 0 \\
-4A_{1,1} + A_{1,2} & & - 2A_{2,1} + A_{2,2} & = 1 \\
4A_{1,1} - 4A_{1,2} & + A_{1,3} & & = 0 \\
4A_{1,2} & - 4A_{1,3} & & = 0 \\
& 4A_{1,3} & & = 2.
\end{array}
$$

Solving this system, we find that $A_{1,1} = \frac{3}{8}, A_{1,2} = \frac{1}{2}, A_{1,3} = \frac{1}{2}, A_{2,1} = -\frac{3}{8}$ and $A_{2,2} = \frac{5}{4}$. Then

$$\int \frac{x^3 + 2}{x^3(x - 2)^2} \, dx = \int \left[\frac{3}{8x} + \frac{1}{2x^2} + \frac{1}{2x^3} - \frac{3}{8(x - 2)} + \frac{5}{4(x - 2)^2} \right] dx$$

$$= \frac{3}{8} \ln |x| - \frac{1}{2x} - \frac{1}{4x^2} - \frac{3}{8} \ln |x - 2| - \frac{5}{4(x - 2)} + C.$$

While the computations in this problem (and others like it) are tedious, it should be kept in mind that the "calculus" problem of integration has been replaced by an algebraic problem.

PROBLEMS 8.6

In Problems 1–24 calculate the given integral by the method of partial fractions.

1. $\int \dfrac{dx}{x(x-1)}$

2. $\int \dfrac{x+1}{x^2+x}\,dx$

3. $\int_2^3 \dfrac{2x-3}{x^2-1}\,dx$

4. $\int_2^3 \dfrac{x^2}{x^2-1}\,dx$

5. $\int_{-1}^0 \dfrac{dx}{(x-1)(x-2)(x-3)}$

6. $\int_{-1}^0 \dfrac{x+2}{(x-1)(x-2)(x-3)}\,dx$

7. $\int \dfrac{x^2+3x+4}{(x-1)(x-2)(x-3)}\,dx$

8. $\int \dfrac{x^3+3x^2+2x+1}{(x-1)(x-2)(x-3)}\,dx$

9. $\int \dfrac{x^5}{x^3-x}\,dx$

10. $\int_0^1 \dfrac{x}{(x-4)^2}\,dx$

11. $\int \dfrac{x}{(x-4)^3}\,dx$

12. $\int \dfrac{x^{10}}{x^2(x+3)}\,dx$

13. $\int \dfrac{dx}{x^2(x+2)^2}$

14. $\int \dfrac{dx}{x^2(x-1)}$

15. $\int \dfrac{x^3+x^2-2}{x^4}\,dx$

16. $\int \dfrac{x^3}{(x-1)^3}\,dx$

17. $\int \dfrac{dx}{x^4-5x^2+4}$

18. $\int \dfrac{x}{x^4-18x^2+81}\,dx$

19. $\int \dfrac{x^4-x^3+x^2-7x+2}{x^3+x^2-14x-24}\,dx$

20. $\int \dfrac{x^2+1}{x^2-1}\,dx$

21. $\int_1^2 \dfrac{x^4}{16x-x^3}\,dx$

22. $\int_0^1 \dfrac{dx}{(x-2)^2(x-3)^2}$

23. $\int \dfrac{x^2+2}{x(x-1)^2(x+1)}\,dx$

24. $\int \dfrac{x^3}{x^4-x^2}\,dx.$

*25. Consider

$$\int \dfrac{t(x)}{q(x)}\,dx = \int \dfrac{x^2+2}{(x-1)^3}\,dx$$

and write

$$\dfrac{x^2+2}{(x-1)^3} = \dfrac{A}{x-1} + \dfrac{B}{(x-1)^2} + \dfrac{C}{(x-1)^3}.$$

    (a) Show that $(x-1)^3 \cdot \dfrac{x^2+2}{(x-1)^3}\Big|_{x=1} = C.$

    (b) Show that $\dfrac{d}{dx}\left[(x-1)^3 \cdot \dfrac{x^2+2}{(x-1)^3}\right]\Big|_{x=1} = B.$

    (c) Show that $\dfrac{d^2}{dx^2}\left[(x-1)^3 \cdot \dfrac{x^2+2}{(x-1)^3}\right]\Big|_{x=1} = 2A.$

    (d) Calculate the integral.

*26. Consider

$$\int \dfrac{t(x)}{q(x)}\,dx = \int \dfrac{x^3+3x+4}{(x-2)^2(x+1)^2}\,dx$$

and write

$$\frac{x^3 + 3x + 4}{(x - 2)^2(x + 1)^2} = \frac{A}{x - 2} + \frac{B}{(x - 2)^2} + \frac{C}{x + 1} + \frac{D}{(x + 1)^2}.$$

(a) Show that $(x - 2)^2 \cdot \dfrac{t(x)}{q(x)} \Big|_{x=2} = B.$

(b) Show that $\dfrac{d}{dx}\left[(x - 2)^2 \dfrac{t(x)}{q(x)}\right]\Big|_{x=2} = A.$

(c) Show that $(x + 1)^2 \dfrac{t(x)}{q(x)} \Big|_{x=-1} = D.$

(d) Show that $\dfrac{d}{dx}\left[(x + 1)^2 \dfrac{t(x)}{q(x)}\right]\Big|_{x=-1} = C.$

(e) Calculate the integral.

In the next two problems we develop a formula for calculating the coefficients of a partial fraction expansion in the case of repeated linear factors.

*27. Write

$$\frac{t(x)}{(x - x_1)^4} = \frac{A}{x - x_1} + \frac{B}{(x - x_1)^2} + \frac{C}{(x - x_1)^3} + \frac{D}{(x - x_1)^4}.$$

(a) Multiply both sides of this equation by $(x - x_1)^4$.
(b) Substitute $x = x_1$ and show that $D = t(x_1)$.
(c) Differentiate both sides of the equation obtained in (a) and substitute $x = x_1$ to show that $C = t'(x_1)$.

(d) Differentiate once more and show that $B = \dfrac{t''(x_1)}{2}$.

(e) Differentiate again to show that $A = \dfrac{t'''(x_1)}{3!}$.

*28. Using the technique suggested in Problem 27, show that in formula (6)

$$A_{i,j} = \frac{1}{(m_i - j)!} \frac{d^{m_i-j}}{dx^{m_i-j}}\left[(x - x_i)^{m_i} \frac{t(x)}{q(x)}\right]\Big|_{x=x_i}$$

where $n! = n(n - 1)(n - 2) \cdots 3 \cdot 2 \cdot 1$.

29. Use the formula in Problem 28 to evaluate

(a) $\displaystyle\int \frac{x^5 + 4x^4 - 3x + 5}{(x - 1)^8}\,dx.$

(b) $\displaystyle\int \frac{x^7 - 3x^5 + 2x^3 + 8x^2 - 20}{x(x + 1)^{10}}\,dx$

(c) $\displaystyle\int \frac{1 + x + x^2 + x^3 + x^4 + x^5 + x^6 + x^7}{(x + 2)^8}\,dx.$

30. Calculate $\displaystyle\int_0^{\pi/2} \frac{\cos x}{\sin^2 x - 5 \sin x + 6}\,dx.$ [*Hint:* Make the substitution $u = \sin x$.]

31. Calculate $\displaystyle\int_0^{\pi} \frac{\sin x}{\cos^2 x + 4 \cos x + 4}\,dx.$

32. Calculate the area bounded by the curve $y = 1/(x^2 + x)$, the x-axis and the lines $x = 1$ and $x = 2$.

33. Calculate the area bounded by the curve $y = (1 + x^2)/(1 + x)^2$, the x- and y-axes and the line $x = 1$.

*8.7 Integration by Partial Fractions—Quadratic Factors

We now discuss the case in which the polynomial $q(x)$ has quadratic factors. To see how this could occur, suppose that $q(x) = x^3 + x = x(x^2 + 1)$. The quadratic $x^2 + 1$ cannot be factored into the linear terms $(x - x_1)(x - x_2)$, where x_1 and x_2 are real numbers since there is no real number x for which $x^2 + 1 = 0$. In this case we say that the quadratic $x^2 + 1$ is *irreducible*.† In general, the quadratic $ax^2 + bx + c$ is irreducible if and only if $b^2 - 4ac < 0$.

EXAMPLE 1. Calculate $\int \dfrac{1}{x^3 + x}\, dx.$

SOLUTION. We write

$$\frac{1}{x^3 + x} = \frac{1}{x(x^2 + 1)} = \frac{A}{x} + \frac{Bx + C}{x^2 + 1} = \frac{A(x^2 + 1) + (Bx + C)x}{x(x^2 + 1)}$$
$$= \frac{(A + B)x^2 + Cx + A}{x(x^2 + 1)}$$

and we obtain

$$A + B = 0, \qquad C = 0, \qquad A = 1.$$

Clearly $B = -1$ and

$$\int \frac{dx}{x(x^2 + 1)} = \int \left[\frac{1}{x} - \frac{x}{x^2 + 1} \right] dx = \ln |x| - \frac{1}{2} \ln |x^2 + 1|$$
$$= \ln \left| \frac{x}{\sqrt{x^2 + 1}} \right| + C.$$

EXAMPLE 2. Calculate $\int \dfrac{x^2 + 2x + 3}{(x^2 + 2x + 2)(x - 2)}\, dx.$

SOLUTION. Since $b^2 - 4ac = 2^2 - 4 \cdot 1 \cdot 2 = -4 < 0$, $x^2 + 2x + 2$ is irreducible and we write

$$\frac{x^2 + 2x + 3}{(x^2 + 2x + 2)(x - 2)} = \frac{A}{x - 2} + \frac{Bx + C}{x^2 + 2x + 2}$$

---

† The student should keep in mind the fact that there is a difference between linear factors of multiplicity two (such as $(x + 1)^2$) and irreducible quadratic factors (such as $x^2 + 1$). To emphasize this point, we calculate

$$\int \frac{dx}{(x + 1)^2} = -\frac{1}{x + 1} \qquad \text{but} \qquad \int \frac{dx}{x^2 + 1} = \tan^{-1} x.$$

$$\frac{x^2 + 2x + 3}{(x^2 + 2x + 2)(x - 2)} = \frac{A(x^2 + 2x + 2) + (Bx + C)(x - 2)}{(x^2 + 2x + 2)(x - 2)}$$

$$= \frac{Ax^2 + 2Ax + 2A + Bx^2 - 2Bx + Cx - 2C}{(x^2 + 2x + 2)(x - 2)}$$

$$= \frac{(A + B)x^2 + (2A - 2B + C)x + (2A - 2C)}{(x^2 + 2x + 2)(x - 2)}$$

so that

$$\begin{aligned} A + B &= 1 \\ 2A - 2B + C &= 2 \\ 2A \qquad - 2C &= 3. \end{aligned}$$

We can solve these equations in a variety of ways. One method is described in Appendix 2. Here we use the first equation to find that $B = 1 - A$ and the third equation to obtain $C = (2A - 3)/2$. Then, from the second equation,

$$2 = 2A - 2B + C = 2A - 2(1 - A) + \frac{2A - 3}{2} = 5A - \frac{7}{2}$$

so that $5A = 2 + \frac{7}{2} = \frac{11}{2}$ and $A = \frac{11}{10}$. Then $B = 1 - A = -\frac{1}{10}$ and

$$C = \frac{2A - 3}{2} = \frac{2(\frac{11}{10}) - 3}{2} = -\frac{4}{10}.$$

Then

$$\int \frac{x^2 + 2x + 3}{(x^2 + 2x + 2)(x - 2)}\, dx = \int \left[\frac{11}{10(x - 2)} - \frac{(x/10) + (4/10)}{(x^2 + 2x + 2)} \right] dx.$$

We must integrate

$$\frac{(x/10) + (4/10)}{x^2 + 2x + 2} = \frac{1}{10}\left(\frac{x + 4}{x^2 + 2x + 2} \right) = \frac{1}{20}\left(\frac{2x + 8}{x^2 + 2x + 2} \right)$$

$$= \frac{1}{20}\left(\frac{2x + 2 + 6}{x^2 + 2x + 2} \right) = \frac{1}{20}\left(\frac{2x + 2}{x^2 + 2x + 2} + \frac{6}{x^2 + 2x + 2} \right).$$

Now,

$$\int \frac{2x + 2}{x^2 + 2x + 2}\, dx = \ln|x^2 + 2x + 2|$$

(that's why we broke the quotient up that way). To integrate $6/(x^2 + 2x + 2)$, we complete the square:

$$\int \frac{6}{x^2 + 2x + 2}\, dx = 6\int \frac{dx}{x^2 + 2x + 1 + 1} = 6\int \frac{dx}{(x + 1)^2 + 1} = 6\tan^{-1}(x + 1).$$

Finally, we obtain

$$\int \frac{x^2 + 2x + 3}{(x^2 + 2x + 2)(x - 2)}\, dx$$

$$= \frac{11}{10}\ln|x - 2| - \frac{1}{20}\{\ln|x^2 + 2x + 2| + 6\tan^{-1}(x + 1)\} + C.$$

These two examples suggest the following 4-step procedure:

CASE 3. Distinct Quadratic Factors. $q(x)$ can be factored into linear terms and quadratic terms of the form $x^2 + ax + b$ where no two quadratic terms are the same.

(i) Treat each linear term as before, and for each quadratic term $x^2 + ax + b$, include the term $(Ax + B)/(x^2 + ax + b)$ in the partial fraction decomposition of $t(x)/q(x)$.

(ii) Write

$$\frac{Ax + B}{x^2 + ax + b} = \frac{A}{2}\left[\frac{2x + \dfrac{2B}{A}}{x^2 + ax + b}\right] = \frac{A}{2}\left[\frac{2x + a + \dfrac{2B}{A} - a}{x^2 + ax + b}\right]$$

$$= \frac{A}{2}\left[\frac{2x + a}{x^2 + ax + b} + \frac{\dfrac{2B}{A} - a}{x^2 + ax + b}\right].$$

(iii) Complete the square of the denominator of the second term in brackets:

$$\frac{Ax + B}{x^2 + ax + b} = \frac{A}{2}\left[\frac{2x + a}{x^2 + ax + b} + \left(\frac{2B}{A} - a\right)\frac{1}{\left(x + \dfrac{a}{2}\right)^2 + b - \dfrac{a^2}{4}}\right].$$

(iv) Integrate:

$$\int \frac{Ax + B}{x^2 + ax + b}\,dx = \frac{A}{2}\int \frac{2x + a}{x^2 + ax + b}\,dx$$

$$+ \frac{A}{2}\left(\frac{2B}{A} - a\right)\int \frac{dx}{\left(x + \dfrac{a}{2}\right)^2 + b - \dfrac{a^2}{4}}$$

$$= \frac{A}{2}\ln|x^2 + ax + b|$$

$$+ \frac{A}{2}\left(\frac{2B}{A} - a\right)\frac{1}{\sqrt{b - \dfrac{a^2}{4}}}\tan^{-1}\frac{x + \dfrac{a}{2}}{\sqrt{b - \dfrac{a^2}{4}}} + C.$$

Note that the last step is justified since the assumption that $x^2 + ax + b$ is irreducible implies that $a^2 - 4b < 0$ or $b - (a^2/4) > 0$ so that we may take its square root. An alternative to step (iv) is to set $x + a/2 = \tan u$ and proceed as in Section 5.

EXAMPLE 3. Calculate $\displaystyle\int \frac{x^3 - 1}{x^3 + 1}\,dx$.

SOLUTION. We first divide since degree($x^3 - 1$) = degree($x^3 + 1$).

$$\frac{x^3 - 1}{x^3 + 1} = 1 - \frac{2}{x^3 + 1}.$$

Now

$$\frac{2}{x^3 + 1} = \frac{2}{(x + 1)(x^2 - x + 1)} = \frac{Ax + B}{x^2 - x + 1} + \frac{C}{x + 1}$$

$$= \frac{(Ax + B)(x + 1) + C(x^2 - x + 1)}{(x + 1)(x^2 - x + 1)}$$

$$= \frac{(A + C)x^2 + (A + B - C)x + (B + C)}{(x + 1)(x^2 - x + 1)}$$

so that

$$A + \qquad C = 0$$
$$A + B - C = 0$$
$$B + C = 2$$

and we find (using a procedure similar to the one used in Example 2) that $A = -\frac{2}{3}$, $B = \frac{4}{3}$, and $C = \frac{2}{3}$. Then

$$\frac{x^3 - 1}{x^3 + 1} = 1 - \frac{-\frac{2}{3}x + \frac{4}{3}}{x^2 - x + 1} - \frac{2}{3(x + 1)} = 1 + \frac{\frac{2}{3}x - \frac{4}{3}}{x^2 - x + 1} - \frac{2}{3(x + 1)}.$$

Now

$$\frac{\frac{2}{3}x - \frac{4}{3}}{x^2 - x + 1} = \frac{1}{3}\left[\frac{2x - 4}{x^2 - x + 1}\right] = \frac{1}{3}\left[\frac{2x - 1}{x^2 - x + 1} - \frac{3}{x^2 - x + 1}\right]$$

$$= \frac{1}{3}\left[\frac{2x - 1}{x^2 - x + 1}\right] - \left[\frac{1}{(x - \frac{1}{2})^2 + \frac{3}{4}}\right].$$

Finally,

$$\int \frac{x^3 - 1}{x^3 + 1}\, dx = \int \left\{1 - \frac{2}{3(x + 1)} + \frac{1}{3}\left(\frac{2x - 1}{x^2 - x + 1}\right) - \left[\frac{1}{(x - \frac{1}{2})^2 + \frac{3}{4}}\right]\right\} dx$$

$$= x - \frac{2}{3}\ln|x + 1| + \frac{1}{3}\ln|x^2 - x + 1| - \frac{1}{\sqrt{\frac{3}{4}}}\tan^{-1}\frac{x - \frac{1}{2}}{\sqrt{\frac{3}{4}}} + C$$

$$= x + \ln\left|\frac{(x^2 - x + 1)^{1/3}}{(x + 1)^{2/3}}\right| - \frac{2}{\sqrt{3}}\tan^{-1}\frac{2x - 1}{\sqrt{3}} + C.$$

We now treat the last case.

EXAMPLE 4. Calculate $\int \frac{x^3 + 3x^2 + 1}{(x^2 + 1)^2}\, dx$.

SOLUTION. We write

$$\frac{x^3 + 3x^2 + 1}{(x^2 + 1)^2} = \frac{Ax + B}{x^2 + 1} + \frac{Cx + D}{(x^2 + 1)^2} = \frac{(Ax + B)(x^2 + 1) + (Cx + D)}{(x^2 + 1)^2}$$

$$= \frac{Ax^3 + Bx^2 + (A + C)x + (B + D)}{(x^2 + 1)^2}$$

and we obtain

$$A = 1$$
$$B = 3$$
$$A + C = 0$$
$$B + D = 1$$

so that $C = -1$, $D = -2$ and

$$\int \frac{x^3 + 3x^2 + 1}{(x^2 + 1)^2} \, dx = \int \left[\frac{x + 3}{x^2 + 1} - \frac{x + 2}{(x^2 + 1)^2} \right] dx$$

$$= \frac{1}{2} \int \left[\frac{2x + 6}{x^2 + 1} - \frac{2x + 4}{(x^2 + 1)^2} \right] dx$$

$$= \frac{1}{2} \int \left[\frac{2x}{x^2 + 1} + \frac{6}{x^2 + 1} - \frac{2x}{(x^2 + 1)^2} - \frac{4}{(x^2 + 1)^2} \right] dx$$

$$= \frac{1}{2} \ln |x^2 + 1| + 3 \tan^{-1} x + \frac{1}{2(x^2 + 1)} - 2 \int \frac{dx}{(x^2 + 1)^2}.$$

To calculate this last integral, we set $x = \tan \theta$. Then

$$-2 \int \frac{dx}{(x^2 + 1)^2} = -2 \int \frac{\sec^2 \theta \, d\theta}{(1 + \tan^2 \theta)^2}$$

$$= -2 \int \frac{\sec^2 \theta}{\sec^4 \theta} \, d\theta$$

$$= -2 \int \cos^2 \theta \, d\theta = - \int (1 + \cos 2\theta) \, d\theta$$

$$= -\theta - \frac{\sin 2\theta}{2} = -\theta - \sin \theta \cos \theta.$$

Now if $x = \tan \theta$, then from Figure 1, $\sin \theta = x / \sqrt{x^2 + 1}$ and $\cos \theta = 1 / \sqrt{x^2 + 1}$ so that

$$-\theta - \sin \theta \cos \theta = - \left(\tan^{-1} x + \frac{x}{x^2 + 1} \right).$$

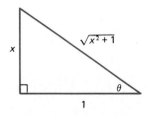

Figure 1

Finally,

$$\int \frac{x^3 + 3x^2 + 1}{(x^2 + 1)^2} \, dx = \ln \sqrt{x^2 + 1} + 2 \tan^{-1} x + \frac{1}{2(x^2 + 1)} - \frac{x}{x^2 + 1} + C.$$

This example suggests the following:

CASE 4. Repeated Quadratic Factors.

(i)   If for $m > 1$ the term $(x^2 + ax + b)^m$ appears in the factoring of $q(x)$, then the partial fraction decomposition of $t(x)/q(x)$ includes terms of the form

$$\frac{A_1x + B_1}{x^2 + ax + b} + \frac{A_2x + B_2}{(x^2 + ax + b)^2} + \cdots + \frac{A_mx + B_m}{(x^2 + ax + b)^m}. \tag{1}$$

(ii)   To integrate the term $(A_kx + B_k)/(x^2 + ax + b)^k$, set it equal to

$$\frac{A_k}{2}\left[\frac{2x + 2(B_k/A_k)}{(x^2 + ax + b)^k}\right] = \frac{A_k}{2}\left[\frac{2x + a}{(x^2 + ax + b)^k}\right]$$

$$+ \frac{A_k}{2}\left(\frac{2B_k}{A_k} - a\right) \cdot \frac{1}{(x^2 + ax + b)^k}$$

(iii)   Complete the square of the denominator in the last term:

$$\frac{1}{(x^2 + ax + b)^k} = \frac{1}{\left[\left(x + \dfrac{a}{2}\right)^2 + b^2 - \dfrac{a^2}{4}\right]^k}.$$

(iv)   Integrate each term. To integrate the term in step (iii), make the substitution

$$x + \frac{a}{2} = \sqrt{b^2 - \frac{a^2}{4}}\tan\theta.$$

EXAMPLE 5.   Calculate $\displaystyle\int \frac{x^2 - 3x + 18}{(x - 1)(x^2 - 2x + 5)^2}\,dx.$

SOLUTION.   We write

$$\frac{x^2 - 3x + 18}{(x - 1)(x^2 - 2x + 5)^2} = \frac{Ax + B}{x^2 - 2x + 5} + \frac{Cx + D}{(x^2 - 2x + 5)^2} + \frac{E}{x - 1}$$

$$= \frac{(Ax + B)(x^2 - 2x + 5)(x - 1) + (Cx + D)(x - 1) + E(x^2 - 2x + 5)^2}{(x^2 - 2x + 5)^2(x - 1)}$$

After a good deal of algebra (which we shall spare you), we obtain $A = -1$, $B = 1$, $C = -3$, $D = 2$ and $E = 1$ so that

$$\int \frac{x^2 - 3x + 18}{(x - 1)(x^2 - 2x + 5)^2}\,dx = \int \left[\frac{-x + 1}{x^2 - 2x + 5} + \frac{-3x + 2}{(x^2 - 2x + 5)^2} + \frac{1}{x - 1}\right]dx$$

$$= \int \left[-\frac{1}{2}\left(\frac{2x - 2}{x^2 - 2x + 5}\right) - \frac{3}{2}\left(\frac{2x - \frac{4}{3}}{(x^2 - 2x + 5)^2}\right) + \frac{1}{x - 1}\right]dx$$

$$= -\frac{1}{2}\ln|x^2 - 2x + 5| + \ln|x - 1| - \frac{3}{2}\int \frac{(2x - 2) + \frac{2}{3}}{(x^2 - 2x + 5)^2}\,dx$$

$$= \ln\left|\frac{x - 1}{\sqrt{x^2 - 2x + 5}}\right| + \frac{3}{2(x^2 - 2x + 5)} - \int \frac{dx}{[(x - 1)^2 + 4]^2}.$$

To obtain the final integral, we substitute $x - 1 = 2 \tan \theta$. Then $dx = 2 \sec^2 \theta$ and

$$-\int \frac{dx}{[(x - 1)^2 + 4]^2} = -\int \frac{2 \sec^2 \theta}{(4 \sec^2 \theta)^2} d\theta = -\frac{1}{8} \int \frac{d\theta}{\sec^2 \theta}$$

$$= -\frac{1}{8} \int \cos^2 \theta \, d\theta = -\frac{1}{8} \int \frac{1 + \cos 2\theta}{2} d\theta$$

$$= -\frac{1}{16} \left(\theta + \frac{\sin 2\theta}{2} \right) = -\frac{1}{16} (\theta + \sin \theta \cos \theta).$$

From Figure 2 and the fact that $\tan \theta = (x - 1)/2$, we find that

$$\sin \theta = \frac{x - 1}{\sqrt{x^2 - 2x + 5}} \qquad \text{and} \qquad \cos \theta = \frac{2}{\sqrt{x^2 - 2x + 5}}$$

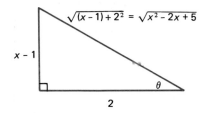

$\sqrt{(x-1)+2^2} = \sqrt{x^2 - 2x + 5}$

$x - 1$

2

θ

Figure 2

so that

$$-\frac{1}{16} (\theta + \sin \theta \cos \theta) = -\frac{1}{16} \left[\tan^{-1} \frac{x - 1}{2} + \frac{2(x - 1)}{x^2 - 2x + 5} \right]$$

$$= -\frac{1}{16} \tan^{-1} \frac{x - 1}{2} - \frac{x - 1}{8(x^2 - 2x + 5)}$$

and, finally,

$$\int \frac{x^3 - 3x + 18}{(x - 1)(x^2 - 2x + 5)} dx = \ln \left| \frac{x - 1}{\sqrt{x^2 - 2x + 5}} \right| - \frac{1}{16} \tan^{-1} \left(\frac{x - 1}{2} \right)$$

$$+ \frac{13 - x}{8(x^2 - 2x + 5)} + C.$$

PROBLEMS 8.7

In Problems 1–16 calculate the given integral.

1. $\int \dfrac{x + 2}{x^2 + x + 1} dx$

2. $\int_0^1 \dfrac{x^3}{x^2 + x + 1} dx$

3. $\int \dfrac{dx}{(x + 2)(x^2 + 1)}$

4. $\int_0^1 \dfrac{4x + 3}{(x^2 + 1)(x^2 + 2)} dx$

5. $\int \dfrac{x^3 + x^2 + 3}{(x^2 + 2)^2} dx$

6. $\int_{-2}^{-1} \dfrac{x + 4}{(x^2 + 4x + 5)^2} dx$

7. $\int \dfrac{x^3}{x^4 - 8x^2 + 16} dx$

8. $\int \dfrac{x}{(x + 1)^2(x^2 + 1)} dx$

9. $\int \dfrac{x}{x^4 - 1} dx$

10. $\int_0^1 \dfrac{x^2 - 2}{(x^2 + 1)^2} dx$

11. $\displaystyle\int_0^1 \frac{(x^3+5)}{(x^2-x+\frac{1}{2})(x^2-6x+10)}\,dx$

12. $\displaystyle\int \frac{dx}{x^4+x^2}$

13. $\displaystyle\int_1^2 \frac{x^4+1}{x^5+x^3}\,dx$

14. $\displaystyle\int \frac{x^2+1}{(x^2+16)^3}\,dx$

15. $\displaystyle\int \frac{x^3+1}{(x^2+4x+13)^2}\,dx$

16. $\displaystyle\int \frac{(2x^3-4x^2+70x-35)\,dx}{(x^2-2x+17)(x^2-10x+26)}$

17. Calculate $\displaystyle\int_0^{\pi/2} \frac{\cos x}{\sin^2 x + 2\sin x + 2}\,dx.$ [*Hint:* See Problem 8.6.30.]

18. Calculate $\displaystyle\int_0^{\pi} \frac{\sin x}{(\cos^2 x - 4\cos x + 13)^2}\,dx.$

19. Calculate the area bounded by the curve $y = x^3/(x^2+1)^3$, the x- and y-axes and the line $x = 1$.

8.8 Other Substitutions

In addition to the integrals discussed in Section 5, there are many other types of expressions that can be integrated by making an appropriate substitution. Unfortunately, many of these substitutions work only for a very small class of integrals. Even when a function can be integrated by substitution, there is usually no obvious way to find the substitution that works, and it is often necessary to resort to trial and error. In this section we provide a few examples of substitutions that do work.

When an integrand contains a term of the form $[f(x)]^{m/n}$, then the substitution

$$u = f(x)^{1/n} \tag{1}$$

will often be useful.

EXAMPLE 1. Calculate $\displaystyle\int \frac{dx}{2+\sqrt{x}}.$

SOLUTION. Let $u = x^{1/2}$. Then $x = u^2$ and $dx = 2u\,du$ so that

$$\int \frac{dx}{2+\sqrt{x}} = \int \frac{2u}{2+u}\,du = \int \left(2 - \frac{4}{u+2}\right)du$$

$$= 2u - 4\ln|u+2| = 2\sqrt{x} - 4\ln(\sqrt{x}+2) + C.$$

If more than one rational power of x appears, we make a substitution of the form $u = x^{1/n}$, where n is chosen to be the least common multiple of the denominators of the fractional powers that appear, so that we end up with a rational function of u which can be integrated by the method of partial fractions.

EXAMPLE 2. Calculate $\displaystyle\int \frac{x-1}{6\sqrt{x}(4+\sqrt[3]{x})}\,dx.$

SOLUTION. The integrand contains terms of the form $x^{1/2}$ and $x^{1/3}$. If we choose $u = x^{1/6}$, then $x^{1/2} = u^3$, $x^{1/3} = u^2$ and since $x = u^6$, $dx = 6u^5$. Thus everything will be expressed in powers of u. We have

$$\int \frac{x-1}{6\sqrt{x}(4 + \sqrt[3]{x})}\, dx = \int \frac{(u^6 - 1)6u^5}{6u^3(4 + u^2)}\, du = \int \frac{u^{11} - u^5}{u^5 + 4u^3}\, du = \int \frac{u^8 - u^2}{u^2 + 4}\, du.$$

Since the degree of the numerator is greater than the degree of the denominator, we divide:

$$\int \frac{u^8 - u^2}{u^2 + 4}\, du = \int \left(u^6 - 4u^4 + 16u^2 - 65 + \frac{260}{u^2 + 4}\right) du$$

$$= \frac{u^7}{7} - \frac{4u^5}{5} + \frac{16u^3}{3} - 65u + 130 \tan^{-1} \frac{u}{2}$$

$$= \frac{x^{7/6}}{7} - \frac{4x^{5/6}}{5} + \frac{16x^{1/2}}{3} - 65x^{1/6} + 130 \tan^{-1}\left(\frac{x^{1/6}}{2}\right) + C.$$

EXAMPLE 3. Calculate $\displaystyle\int_{-1}^{0} x\sqrt[3]{x + 1}\, dx$.

SOLUTION. Let $u = (x + 1)^{1/3}$. Then $u^3 = x + 1$ or $x = u^3 - 1$ and $dx = 3u^2\, du$. When $x = -1$, $u = 0$ and when $x = 0$, $u = 1$. Thus

$$\int_{-1}^{0} x\sqrt[3]{x + 1}\, dx = \int_{0}^{1} (u^3 - 1)u \cdot 3u^2\, du = \int_{0}^{1} (3u^6 - 3u^3)\, du$$

$$= \tfrac{3}{7}u^7 - \tfrac{3}{4}u^4 \Big|_{0}^{1} = -\tfrac{9}{28}.$$

This problem could also be solved by the method of integration by parts.

Certain rational functions of $\sin x$ and $\cos x$ can be integrated with the use of a special trigonometric substitution.

EXAMPLE 4. Calculate $\displaystyle\int \frac{dx}{2 + \cos x}$.

SOLUTION. Let $u = \tan(x/2)$.† Then $x = 2 \tan^{-1} u$ and

$$\boxed{dx = \frac{2}{1 + u^2}\, du.} \qquad\qquad (2)$$

From Figure 1,

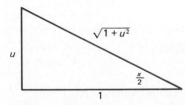

Figure 1

---

† The reason for making this seemingly arbitrary substitution will soon become apparent. It is certainly not obvious that it will work before we try it.

$$\sin \frac{x}{2} = \frac{u}{\sqrt{1 + u^2}} \qquad \text{and} \qquad \cos \frac{x}{2} = \frac{1}{\sqrt{1 + u^2}}.$$

Using the double angle formulas, $\sin 2x = 2 \sin x \cos x$ and $\cos 2x = \cos^2 x - \sin^2 x$, we obtain

$$\boxed{\sin x = 2 \sin \frac{x}{2} \cos \frac{x}{2} = \frac{2u}{1 + u^2}} \qquad\qquad (3)$$

and

$$\boxed{\cos x = \cos^2 \frac{x}{2} - \sin^2 \frac{x}{2} = \frac{1 - u^2}{1 + u^2}.} \qquad\qquad (4)$$

Thus, using (2), (3), and (4)

$$\int \frac{dx}{2 + \cos x} = \int \frac{2/(1 + u^2)}{2 + [(1 - u^2)/(1 + u^2)]} \, du = \int \frac{2}{3 + u^2} \, du$$

$$= \frac{2}{\sqrt{3}} \tan^{-1} \frac{u}{\sqrt{3}}$$

$$= \frac{2}{\sqrt{3}} \tan^{-1} \left[\frac{\tan(x/2)}{\sqrt{3}} \right] + C.$$

EXAMPLE 5. Calculate $\displaystyle\int_0^{\pi/2} \frac{dx}{\sin x + \cos x}$.

SOLUTION. Let $u = \tan x/2$. Then using (2), (3), and (4) we obtain

$$\int_0^{\pi/2} \frac{dx}{\sin x + \cos x} = \int_0^1 \frac{2/(1 + u^2)}{[2u/(1 + u^2)] + [(1 - u^2)/(1 + u^2)]} \, du$$

$$= \int_0^1 \frac{2 \, du}{1 + 2u - u^2} = \int_0^1 \frac{2 \, du}{1 - (u^2 - 2u)}$$

$$= \int_0^1 \frac{2 \, du}{1 + 1 - (u^2 - 2u + 1)} = \int_0^1 \frac{2 \, du}{2 - (u - 1)^2}.$$

Now let $u - 1 = \sqrt{2} \sin \theta$ so that $du = \sqrt{2} \cos \theta \, d\theta$, $\theta = 0$ when $u = 1$ and $\theta = \sin^{-1}(-1/\sqrt{2}) = -\pi/4$ when $u = 0$. Then

$$\int_0^1 \frac{2 \, du}{2 - (u - 1)^2} = \int_{-\pi/4}^0 \frac{2\sqrt{2} \cos \theta}{2 \cos^2 \theta} \, d\theta = \sqrt{2} \int_{-\pi/4}^0 \sec \theta \, d\theta$$

$$= \sqrt{2} \ln |\sec \theta + \tan \theta| \, \Big|_{-\pi/4}^0$$

$$= \sqrt{2}[\ln 1 - \ln(\sqrt{2} - 1)] \approx 1.25.$$

PROBLEMS 8.8

In Problems 1–30 calculate the given integrals by making an appropriate substitution.

1. $\displaystyle\int x\sqrt{1 + x} \, dx$

2. $\displaystyle\int \frac{1 + x}{\sqrt{1 - x}} \, dx$

3. $\displaystyle\int_0^{13} x(1 + 2x)^{2/3} \, dx$

4. $\int_0^1 x^8(1 - x^3)^{1/3}\, dx$

5. $\int \dfrac{dx}{x^3(16 - x^3)^{1/3}}$

6. $\int_0^1 x^8(1 - x^3)^{6/5}\, dx$

7. $\int_0^2 \dfrac{3x^5}{(1 + x^3)^{3/2}}\, dx$

8. $\int \dfrac{x - 4}{\sqrt{x}(1 + \sqrt[3]{x})}\, dx$

9. $\int \dfrac{3x^2 - x}{\sqrt{1 + x}}\, dx$

10. $\int_0^1 \dfrac{dx}{1 + x^{1/5}}$

11. $\int_0^1 \dfrac{dx}{x^{1/3} + x^{1/4}}$

12. $\int \dfrac{x^{1/2}}{x^{1/3} + x^{1/4}}\, dx$

*13. $\int \dfrac{\sqrt{x}}{1 + x^{2/3}}\, dx$ [*Hint:* Factor $u^4 + 1$ into a product of irreducible quadratic factors.]

14. $\int \dfrac{x^3 - x}{\sqrt{x^2 - 2}}\, dx$

*15. $\int \dfrac{\sqrt[3]{1 - x}}{1 + x}\, dx$

16. $\int \dfrac{x}{(x - 1)^4}\, dx$

17. $\int \dfrac{x^{2/3} + 1}{x^{2/3} - 1}\, dx$

*18. $\int \dfrac{x^{3/4} - 16}{x^{3/4} + 16}\, dx$

19. $\int_0^1 x^2\sqrt{2 + 3x}\, dx$

20. $\int \dfrac{x^2}{\sqrt{2 + 3x}}\, dx$

21. $\int \dfrac{1 + x^{1/2}}{x^{5/6} + x^{7/6}}\, dx$

22. $\int \dfrac{dx}{\sqrt{e^{2x} - 1}}$

23. $\int \dfrac{dx}{1 + \sin x}$

24. $\int_0^{\pi/2} \dfrac{dx}{1 + \sin x + \cos x}$

25. $\int_0^{\pi/3} \dfrac{dx}{3 + 2\cos x}$

26. $\int \dfrac{dx}{\sin x + \tan x}$

27. $\int \dfrac{dx}{\sin 2x + \cos 2x}$

28. $\int \dfrac{2 - \sin x}{2 + \sin x}\, dx$

29. $\int \dfrac{\csc x\, dx}{3\csc x + 2\cot x + 2}$

30. $\int \dfrac{dx}{10\sec x - 6}.$

31. Calculate $\int \dfrac{dx}{x\sqrt{x^2 + 4x - 5}}$.   [*Hint:* Try the substitution $u = 1/x$.]

32. Calculate $\int \dfrac{dx}{x^3\sqrt{1 + x + x^2}}$.

☐ 8.9 Numerical Integration

As we have mentioned at several places, there are a great number of continuous functions for which an indefinite integral cannot be expressed in terms of functions we know. In those cases we cannot use the fundamental theorem of calculus to evaluate a definite integral. Nevertheless, it may be very important to approximate the value of such an integral. For that reason many methods have been devised to approximate the value of a definite integral to as many decimal places as are deemed necessary. All these techniques come under the heading of *numerical integration*. We will not discuss this vast subject in great generality here. Rather we will introduce two reasonably effective methods for estimating a definite integral: the *trapezoidal rule* and *Simpson's rule*. For a more complete discussion of numerical integration, the interested reader is referred to a book on numerical analysis.†

Consider the problem of calculating

$$\int_a^b f(x)\, dx. \tag{1}$$

† One reasonably elementary book in this area is the book by S. D. Conte, *Elementary Numerical Analysis*. McGraw-Hill, New York, 1965.

By definition,

$$\int_a^b f(x)\,dx = \lim_{|P| \to 0} [f(x_1^*)\,\Delta x_1 + f(x_2^*)\,\Delta x_2 + \cdots + f(x_n^*)\,\Delta x_n]. \tag{2}$$

In other words, when the lengths of the subintervals in a partition of $[a, b]$ are small, the sum in the right-hand side of (2) gives us a crude approximation to the integral. Here the area is approximated by a sum of areas of rectangles. We saw some examples of this type of approximation in Sections 4.1 and 4.2. We now develop a more efficient way to approximate the integral.

Let f be as in Figure 1 and let us partition the interval $[a, b]$ by the *equally-spaced* points

$$a = x_0 < x_1 < x_2 < \cdots < x_{i-1} < x_i < \cdots < x_n = b, \tag{3}$$

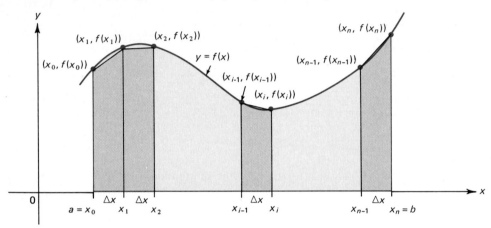

Figure 1

where $x_i - x_{i-1} = \Delta x = (b - a)/n$. In Figure 1, we have indicated that the area under the curve can be approximated by the sum of the areas of n trapezoids. One typical trapezoid is sketched in Figure 2. The area of the trapezoid is the area of the

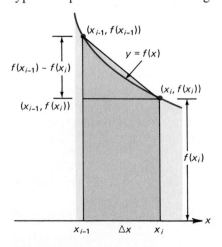

Figure 2

rectangle plus the area of the triangle. But the area of the rectangle is $f(x_i)\,\Delta x$ and the area of the triangle is $\frac{1}{2}[f(x_{i-1}) - f(x_i)]\,\Delta x$ so that

$$\text{area of trapezoid} = f(x_i)\,\Delta x + \tfrac{1}{2}[f(x_{i-1}) - f(x_i)]\,\Delta x$$
$$= \tfrac{1}{2}[f(x_{i-1}) + f(x_i)]\,\Delta x.\dagger \tag{4}$$

Then

$$\int_a^b f(x)\,dx \approx \text{sum of areas of trapezoids}$$

$$= \tfrac{1}{2}[f(x_0) + f(x_1)]\,\Delta x + \tfrac{1}{2}[f(x_1) + f(x_2)]\,\Delta x + \cdots$$
$$+ \tfrac{1}{2}[f(x_{n-2}) + f(x_{n-1})]\,\Delta x + \tfrac{1}{2}[f(x_{n-1}) + f(x_n)]\,\Delta x$$

or

$$\int_a^b f(x)\,dx \approx \tfrac{1}{2}\Delta x[f(x_0) + 2f(x_1) + 2f(x_2) + \cdots + 2f(x_{n-1}) + f(x_n)]. \tag{5}$$

The approximation formula (5) is called the *trapezoidal rule* for numerical integration. Note that since $\Delta x = (b - a)/n$, we can write (5) as

$$\int_a^b f(x)\,dx \approx \frac{b - a}{2n}[f(x_0) + 2f(x_1) + 2f(x_2) + \cdots + 2f(x_{n-1}) + f(x_n)]. \tag{6}$$

$\square$**EXAMPLE 1.** Estimate $\int_1^2 (1/x)\,dx$ by using the trapezoidal rule with $n = 5$ and $n = 10$.

SOLUTION. (i) Here $n = 5$ and

$$\Delta x = \frac{b - a}{b} = \frac{2 - 1}{5} = \frac{1}{5} = 0.2.$$

Then $x_0 = 1$, $x_1 = 1.2$, $x_2 = 1.4$, $x_3 = 1.6$, $x_4 = 1.8$ and $x_5 = 2$. From (5),

$$\int_1^2 \frac{1}{x}\,dx \approx \frac{1}{2}\Delta x[f(x_0) + 2f(x_1) + 2f(x_2) + 2f(x_3) + 2f(x_4) + f(x_5)]$$

$$= \frac{0.2}{2}\left(\frac{1}{1} + \frac{2}{1.2} + \frac{2}{1.4} + \frac{2}{1.6} + \frac{2}{1.8} + \frac{1}{2}\right)$$

$$\approx 0.1(1 + 1.6667 + 1.4286 + 1.25 + 1.1111 + 0.5)$$

$$= 0.1(6.9564) = 0.6956.$$

$\dagger$ Note that this is the same as the average of the area of the rectangle R_{i-1} whose height is $f(x_{i-1})$ (the left-hand endpoint) and the area of the rectangle R_i whose height is $f(x_i)$ (the right-hand endpoint). That is,

$$\frac{1}{2}[(\text{area of } R_{i-1}) + (\text{area of } R_i)] = \frac{1}{2}[f(x_{i-1})\,\Delta x + f(x_i)\,\Delta x] = \frac{1}{2}[f(x_{i-1}) + f(x_i)]\,\Delta x.$$

(ii) Now $n = 10$ and $\Delta x = 1/10 = 0.1$, so that $x_0 = 1$, $x_1 = 1.1, \ldots, x_9 = 1.9$ and $x_{10} = 2$. Thus

$$\int_1^2 \frac{1}{x}\, dx \approx \frac{1}{2}(0.1)\left[1 + \frac{2}{1.1} + \frac{2}{1.2} + \frac{2}{1.3} + \frac{2}{1.4} + \frac{2}{1.5} + \frac{2}{1.6} + \frac{2}{1.7}\right.$$

$$\left. + \frac{2}{1.8} + \frac{2}{1.9} + \frac{1}{2}\right]$$

$$\approx 0.05[1 + 1.8182 + 1.6667 + 1.5385 + 1.4286 + 1.3333$$

$$+ 1.25 + 1.1765 + 1.1111 + 1.0526 + 0.5]$$

$$= 0.05[13.8755] = 0.6938.$$

We can check our calculations by integrating:

$$\int_1^2 \frac{1}{x}\, dx = \ln x \Big|_1^2 = \ln 2 - \ln 1 = \ln 2 \approx 0.6931.$$

We can see that by increasing the number of intervals we increase the accuracy of our answer. This, of course, is not surprising. However we are naturally led to ask what kind of accuracy we can expect by using the trapezoidal rule. In general, there are two kinds of errors encountered when using a numerical method to integrate. The first kind is the kind we have already encountered. This is the error obtained by approximating the curve between the points $(x_{i-1}, f(x_{i-1}))$ and $(x_i, f(x_i))$ by the straight line joining those points. Since we now consider the function at a finite or *discrete* number of points, the error incurred by this approximation is called *discretization error*. However, there is another kind of error which we will always encounter. As we saw in Example 1, we rounded our calculations to four decimal places. Each such "rounding" led to an error in our calculation. The accumulated effect of this rounding is called *round-off error*. Note that as we increase the number of intervals in our calculation, we improve the accuracy of our approximation to the area under the curve. This, evidently, has the effect of reducing the discretization error. On the other hand, an increase in the number of subintervals leads to an increase in the number of computations which, in turn, leads to an increase in the accumulated round-off error. In fact, there is a delicate balance between these two types of errors and often there is an "optimal" number of intervals to be chosen so as to minimize the total error. Round-off error depends on the type of device used for the computations (pencil and paper, slide rule, hand calculator, computer, etc.), and will not be discussed further here. However, we can give a formula for estimating the discretization error incurred in using the trapezoidal rule.

Let the sum in (5) be denoted by T and let ϵ_n^T denote the discretization error:

$$\epsilon_n^T = \int_a^b f(x)\, dx - T$$

when n subintervals are used.

THEOREM 1. Let f be continuous on $[a, b]$ and suppose that f' and f'' exist on $[a, b]$. Then there is a number c in (a, b) such that

$$\epsilon_n^T = -\frac{(b-a)}{12}(\Delta x)^2 f''(c). \tag{7}$$

The proof of this theorem is beyond the scope of this book but it can be found in most standard numerical analysis texts, including the one referenced earlier in this section.

COROLLARY 1. If $|f''(x)| \leq M$ for all x in $[a, b]$, then

$$\left| \epsilon_n^T \right| \leq M \frac{(b-a)^3}{12n^2}. \tag{8}$$

PROOF. From (7)

$$\left| \epsilon_n^T \right| = \frac{b-a}{12} \Delta x^2 |f''(c)| \leq \frac{b-a}{12} \left(\frac{b-a}{n} \right)^2 M = \frac{M(b-a)^3}{12n^2}.$$

EXAMPLE 2. Find a bound on the discretization error incurred when calculating $\int_1^2 (1/x)\, dx$ using the trapezoidal rule with n subintervals.

SOLUTION. $f(x) = 1/x, f'(x) = -1/x^2$ and $f''(x) = 2/x^3$. Hence $f''(x)$ is bounded by 2 for x in $[1, 2]$. Then, from (8),

$$\left| \epsilon_n^T \right| \leq \frac{2(2-1)^3}{12n^2} = \frac{1}{6n^2}.$$

For example, for $n = 5$ we calculated $\int_1^2 (1/x)\, dx \approx 0.6956$. Then, the actual error is

$$\epsilon_n^T \approx 0.6956 - 0.6931 = 0.0025.$$

This compares with a maximum possible error of $1/6n^2 = 1/(6 \cdot 25) = 1/150 \approx 0.0067$. For $n = 10$, the actual error is

$$\epsilon_{10}^T \approx 0.6938 - 0.6931 = 0.0007.$$

This compares with a maximum possible error of $1/6n^2 = 1/600 \approx 0.0017$. Hence we see, in this example at least, that the error bound (8) is a crude estimate of the actual error. Nevertheless, even this crude bound allows us to estimate the accuracy of our calculation in the cases where we can *not* check our answer by integrating. Of course, these are the only cases of interest since we would not use a numerical technique if we could calculate the answer exactly.

☐**EXAMPLE 3.** Use the trapezoidal rule to estimate $\int_0^2 e^{x^2}\, dx$ with a maximum error of 1.

SOLUTION. We must choose n large enough so that $|\epsilon_n^T| \leq 1$. For $f(x) = e^{x^2}$, we have $f'(x) = 2xe^{x^2}$ and $f''(x) = (2 + 4x^2)e^{x^2}$. Since this is an increasing function, its maximum over the interval $[0, 2]$ occurs at 2. Then $M = f''(2) = 18e^4 \approx 983$. Hence, from (8),

$$\left| \epsilon_n^T \right| \leq \frac{M(b-a)^3}{12n^2} \leq \frac{(983)2^3}{12n^2} \approx \frac{655}{n^2}.$$

We need $655/n^2 \leq 1$ or $n^2 \geq 655$ or $n \geq \sqrt{655}$. The smallest n that meets this requirement is $n = 26$. Hence we use the trapezoidal rule with $n = 26$ and $\Delta x = (b-a)/n = 2/26 = 1/13$. We have $x_0 = 0, x_1 = 1/13, x_2 = 2/13, \ldots, x_{25} = 25/13,$

and $x_{26} = 26/13 = 2$. Then

$$\int_0^2 e^{x^2}\, dx \approx \frac{1}{2} \cdot \frac{1}{13} [e^0 + 2e^{(1/13)^2} + 2e^{(2/13)^2} + \cdots + 2e^{(25/13)^2} + e^{(26/13)^2}]$$

$$\approx \frac{1}{26}(1 + 2.012 + 2.048 + 2.109 + 2.199 + 2.319 + 2.475$$

$$+\ 2.673 + 2.921 + 3.230 + 3.614 + 4.092 + 4.689$$

$$+\ 5.437 + 6.378 + 7.572 + 9.097 + 11.059 + 13.603$$

$$+\ 16.933 + 21.328 + 27.184 + 35.060 + 45.756$$

$$+\ 60.427 + 80.751 + 54.598)$$

$$= \frac{1}{26}(430.564) \approx 16.560.$$

This answer is correct to within 1 unit. The next method we discuss will enable us to calculate this integral with greater accuracy and less work.

We now derive this second method for estimating a definite integral. Look at the three sketches in Figure 3. In Figure 3a the area under the curve $y = f(x)$ over the

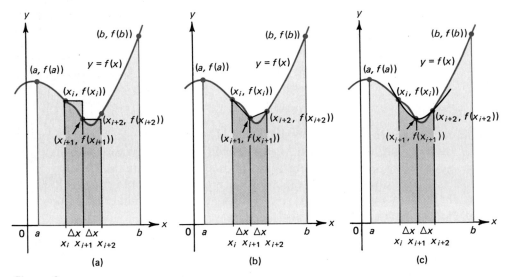

Figure 3

interval $[x_i, x_{i+2}]$ is approximated by rectangles–where the height of each rectangle is the value of the function at an endpoint of an interval. In Figure 3b we have depicted the trapezoidal approximation to this area. The "top" of the first trapezoid is the straight line joining the *two* points $(x_i, f(x_i))$ and $(x_{i+1}, f(x_{i+1}))$ and the "top" of the second trapezoid is given in an analogous manner. In Figure 3c we are approximating the required area by drawing a figure whose "top" is the parabola passing through the *three* points $(x_i, f(x_i))$, $(x_{i+1}, f(x_{i+1}))$ and $(x_{i+2}, f(x_{i+2}))$. As we shall see, this will give us a better approximation to the area under the curve. First, we need to calculate the area depicted in Figure 3c.

THEOREM 2. The area A_{i+2} bounded by the parabola† passing through the points $(x_i, f(x_i))$, $(x_{i+1}, f(x_{i+1}))$ and $(x_{i+2}, f(x_{i+2}))$, the lines $x = x_i$ and $x = x_{i+2}$, and the x-axis (where $x_{i+1} - x_i = x_{i+2} - x_{i+1} = \Delta x$) is given by

$$A_{i+2} = \tfrac{1}{3}\Delta x[f(x_i) + 4f(x_{i+1}) + f(x_{i+2})]. \tag{9}$$

PROOF. First, we find the equation of the parabola. The parabola can be written in the form

$$y = ax^2 + bx + c \tag{10}$$

where a, b, and c are real numbers. Now the area under this parabola is given by

$$A_{i+2} = \int_{x_i}^{x_{i+2}} (ax^2 + bx + c)\, dx = \int_{x_i}^{x_i + 2\Delta x} (ax^2 + bx + c)\, dx$$

$$= \frac{ax^3}{3} + \frac{bx^2}{2} + cx \ \Big|_{x_i}^{x_i + 2\Delta x}$$

$$= \tfrac{1}{3}a(x_i + 2\,\Delta x)^3 + \tfrac{1}{2}b(x_i + 2\,\Delta x)^2 + c(x_i + 2\,\Delta x) - \tfrac{1}{3}ax_i{}^3 - \tfrac{1}{2}bx_i{}^2 - cx$$

$$= \tfrac{1}{3}a(x_i{}^3 + 6x_i{}^2\,\Delta x + 12x_i\,\Delta x^2 + 8\,\Delta x^3) + \tfrac{1}{2}b(x_i{}^2 + 4x_i\,\Delta x + 4\,\Delta x^2)$$
$$\quad + c(x_i + 2\,\Delta x) - \tfrac{1}{3}ax_i{}^3 - \tfrac{1}{2}bx_i{}^2 - cx_i$$

$$= \tfrac{1}{3}a\,\Delta x(6x_i{}^2 + 12x_i\,\Delta x + 8\,\Delta x^2) + \tfrac{1}{2}b\,\Delta x(4x_i + 4\,\Delta x) + 2c\,\Delta x$$

$$= \tfrac{1}{3}a\,\Delta x\,(6x_i{}^2 + 12x_i\,\Delta x + 8\,\Delta x^2) + \underbrace{\tfrac{1}{3}b\,\Delta x\,(6x_i + 6\,\Delta x)}_{\substack{\text{we multiplied and}\\ \text{divided by 3/2}}} + \underbrace{\tfrac{1}{3}(6c\,\Delta x)}_{\substack{\text{we multiplied and}\\ \text{divided by 3}}}$$

$$= \tfrac{1}{3}\Delta x\,[a(6x_i{}^2 + 12x_i\,\Delta x + 8\,\Delta x^2) + b(6x_i + 6\,\Delta x) + 6c]. \tag{11}$$

Now, using the fact that the parabola passes through the points $(x_i, f(x_i))$, $(x_{i+1}, f(x_{i+1}))$ and $(x_{i+2}, f(x_{i+2}))$, we have

$$f(x_i) = ax_i{}^2 + bx_i + c$$
$$f(x_{i+1}) = a(x_i + \Delta x)^2 + b(x_i + \Delta x) + c$$
$$= a(x_i{}^2 + 2x_i\,\Delta x + \Delta x^2) + b(x_i + \Delta x) + c$$
$$f(x_{i+2}) = a(x_i + 2\,\Delta x)^2 + b(x_i + 2\,\Delta x) + c$$
$$= a(x_i{}^2 + 4x_i\,\Delta x + 4\,\Delta x^2) + b(x_i + 2\,\Delta x) + c.$$

Finally,

$$f(x_i) + 4f(x_{i+1}) + f(x_{i+2}) = a(6x_i{}^2 + 12x_i\,\Delta x + 8\,\Delta x^2)$$
$$+ b(6x_i + 6\,\Delta x) + 6c. \tag{12}$$

Comparing (12) with (11) gives us (9) and the proof is complete.

---

†In Figure 3c we assume that f is positive over the interval $[x_i, x_{i+2}]$. The method we are about to develop does *not* require that f be positive. However the method is easier to motivate if we make this assumption.

Now suppose that the interval $[a, b]$ is divided into $2n$ subintervals of equal lengths $\Delta x = (b - a)/2n$. Then, from (9), we have

$$\int_a^b f(x)\, dx \approx A_2 + A_4 + A_6 + \cdots + A_{2n}$$

$$= \tfrac{1}{3}\Delta x[f(x_0) + 4f(x_1) + f(x_2)] + \tfrac{1}{3}\Delta x[f(x_2) + 4f(x_3) + f(x_4)]$$
$$+ \cdots + \tfrac{1}{3}[f(x_{2n-2}) + 4f(x_{2n-1}) + f(x_{2n})]$$

or

$$\int_a^b f(x)\, dx = \tfrac{1}{3}\Delta x\, [f(x_0) + 4f(x_1) + 2f(x_2) + 4f(x_3) + 2f(x_4)$$
$$+ \cdots + 2f(x_{2n-2}) + 4f(x_{2n-1}) + f(x_{2n})]. \tag{13}$$

The approximation (13) is called *Simpson's rule*† (or the *parabolic rule*) for approximating a definite integral. From (13) we see that there is a bit more work needed to estimate an integral using Simpson's rule than by using the trapezoidal rule with the same number of subintervals. However the discretization error in Simpson's rule is a good deal less as suggested in the following theorem, whose proof can be found in the text cited earlier in this section.

THEOREM 3. Let f be continuous on $[a, b]$ and suppose that f', f'', f''' and $f^{(iv)}$ all exist on $[a, b]$. Then the discretization error ϵ_{2n}^S of Simpson's rule (12) using $2n$ equally spaced subintervals of length $\Delta x = (b - a)/2n$ is given by

$$\epsilon_{2n}^S = -\frac{(b - a)}{180}(\Delta x)^4 f^{(iv)}(c), \tag{14}$$

where c is some number in the open interval (a, b).

COROLLARY 2. If $|f^{(iv)}(x)| \le M$ for all x in $[a, b]$, then

$$|\epsilon_{2n}^S| \le \frac{M(b - a)^5}{2880n^4}. \tag{15}$$

PROOF

$$|\epsilon_{2n}^S| = \frac{b - a}{180}(\Delta x)^4\, |f^{(iv)}(c)| \le \left(\frac{b - a}{180}\right)\left(\frac{b - a}{2n}\right)^4 M$$
$$\le \frac{M(b - a)^5}{(180)(16n^4)} = \frac{M(b - a)^5}{2880n^4}.$$

□**EXAMPLE 4.** Use Simpson's rule to estimate $\int_1^2 (1/x)\, dx$ by using 10 subintervals. What is the maximum error in your estimate? Compare this error with the exact answer of $\ln 2$.

†Named after the British mathematician Thomas Simpson (1710–1761) who published the result in his *Mathematical Dissertations on Physical and Analytical Subjects* in 1743.

SOLUTION. Here we have $\Delta x = \frac{1}{10}$ so that, from (13),

$$\int_1^2 \frac{1}{x} dx \approx \frac{1}{30} \left(\frac{1}{1} + \frac{4}{1.1} + \frac{2}{1.2} + \frac{4}{1.3} + \frac{2}{1.4} + \frac{4}{1.5} + \frac{2}{1.6} + \frac{4}{1.7} + \frac{2}{1.8} \right.$$
$$\left. + \frac{4}{1.9} + \frac{1}{2} \right)$$

$$= \frac{1}{30}(1 + 3.636364 + 1.666667 + 3.076923 + 1.428571$$

$$+ 2.666667 + 1.25 + 2.352941 + 1.111111 + 2.105263 + 0.5)$$

$$= \frac{1}{30}(20.794507) \approx 0.693150.$$

To 6 decimal places, $\ln 2 = 0.693147$ so that our answer is very accurate indeed. To calculate the maximum possible error, we first need to calculate $f^{(iv)}$. But $f(x) = 1/x$, $f'(x) = -1/x^2, f''(x) = 2/x^3, f'''(x) = -6/x^4$ and $f^{(iv)}(x) = 24/x^5$. Over the interval $[1, 2]$, $|24/x^4| \le 24$ so that $M = 24$. Then we use formula (15) with $M = 24$ and $n = 5$ (so that $2n = 10$) to obtain

$$|\epsilon_{2n}^S| \le \frac{24}{(2880)5^4} = \frac{24}{(2880)(625)} = \frac{24}{1,800,000} \approx 0.0000133.$$

Our actual error is $0.693150 - 0.693147 = 0.000003$ which is about one-fourth the maximum possible discretization error. Notice that, in this example, Simpson's rule gives a far more accurate anser than the trapezoidal rule using the same number of subintervals (ten).

□**EXAMPLE 5.** Use Simpson's rule to estimate $\int_0^2 e^{x^2} dx$ with a maximum error of 0.1.

SOLUTION. If $f(x) = e^{x^2}$, we have already calculated (in Example 3) that $f''(x) = (2 + 4x^2)e^{x^2}$. Then $f'''(x) = (12x + 8x^3)e^{x^2}$ and $f^{(iv)}(x) = (12 + 48x^2 + 16x^4)e^{x^2}$. This is an increasing function for x in $[0, 2]$ so that $M = f^{(iv)}(2) = 460e^4 \approx 25,115$. Since $(b - a)^5 = 2^5 = 32$, we must choose n such that

$$|\epsilon_{2n}^S| \le \frac{M(b-a)^5}{2880n^4} \approx \frac{(25,115)(32)}{2880n^4} \approx \frac{279}{n^4} < 0.1 = \frac{1}{10}.$$

We then need $n^4 > 2790$ or $n > 2790^{1/4}$. The smallest n which satisfies this inequality is $n = 8$. Thus to obtain the required accuracy we use Simpson's rule with $2n = 16$ subintervals. Then $\Delta x = (b - a)/2n = \frac{2}{16} = \frac{1}{8}$ and we have

$$\int_0^2 e^{x^2} dx \approx \frac{1}{3} \cdot \frac{1}{8}(e^0 + 4e^{(1/8)^2} + 2e^{(2/8)^2} + 2e^{(3/8)^2}$$

$$+ \cdots + 4e^{(14/8)^2} + 2e^{(15/8)^2} + e^{(16/8)^2})$$

$$\approx \frac{1}{24}(1 + 4.0630 + 2.1290 + 4.6040 + 2.5681 + 5.9116$$

$$+ 3.5101 + 8.6014 + 5.4366 + 14.1812 + 9.5415 + 26.4940$$

$$+ 18.9755 + 56.0879 + 42.7619 + 134.5478 + 54.5982)$$

$$= \frac{1}{24}(395.0118) \approx 16.4588.$$

This answer is correct to within one-tenth of a unit. Notice how in calculating $\int_0^2 e^{x^2} dx$, Simpson's rule gives us more accuracy with fewer calculations than the trapezoidal rule.

There are many other methods which can be used to approximate definite integrals. For example, we can find a cubic polynomial that passes through four consecutive points on a curve, find the area under this polynomial, and then develop an approximation method similar to Simpson's rule. A method for finding such a polynomial is discussed in Chapter 13 (Section 13.4). The method we obtain has an error which is proportional to $1/n^6$ when we use $3n$ subintervals. Or, we can go further and find a quartic polynomial which passes through five consecutive points on the curve, and so on. In addition, there are methods in which the points $x_0, x_1, \ldots, x_n$ are *not* equally spaced. One such method is called *Gaussian quadrature*. We shall not discuss this very useful method here except to note that it can be found in any introductory book on numerical analysis. Finally, as we shall discuss in Chapters 13 and 14, techniques using infinite series can be used to approximate certain definite integrals.

We conclude by noting that every definite integral which is known to exist can be evaluated to any number of decimal places of accuracy if one is supplied with the appropriate calculating tool. The problems at the end of this section can all be done reasonably quickly using a scientific hand calculator. For more accuracy, it may be necessary to evaluate a function at hundreds, or even thousands of points. This is a manageable problem only if you have access to a high-speed computer. If, in fact, you do have such access, you should write a computer program to estimate an integral using Simpson's rule and then use it to calculate each of the integrals in the problem set to at least 6 decimal places of accuracy.

□PROBLEMS 8.9

In Problems 1–15 estimate the given definite integral using (i) the trapezoidal rule and (ii) Simpson's rule over the given number of intervals. Then (iii) use the error formula (8) to obtain a bound for the error of the trapezoidal approximation and (iv) use the error formula (15) to obtain a bound for the error of the approximation using Simpson's rule. Then, (v) calculate the integral exactly. Finally, compare the actual errors in your computations with the maximum possible error found in (iii) and (iv).

1. $\int_0^1 x \, dx; \quad n = 4$ 2. $\int_{-2}^2 x \, dx; \quad n = 6$ 3. $\int_0^1 x^2 \, dx; \quad n = 4$

4. $\int_0^1 e^x \, dx; \quad n = 4$ 5. $\int_0^2 e^x \, dx; \quad n = 6$ 6. $\int_1^2 \frac{1}{x^2} \, dx; \quad n = 6$

7. $\int_1^2 \sqrt{x} \, dx; \quad n = 8$ 8. $\int_0^3 \frac{1}{\sqrt{1+x}} \, dx; \quad n = 6$ 9. $\int_0^{\pi/2} \sin x \, dx; \quad n = 4$

10. $\int_0^{\pi/2} \cos x \, dx; \quad n = 4$ 11. $\int_0^{\pi/3} \tan x \, dx; \quad n = 8$

12. $\int_2^5 \frac{x}{\sqrt{x^2 + 1}} \, dx; \quad n = 6$ 13. $\int_1^e \ln x \, dx; \quad n = 6$

14. $\displaystyle\int_0^1 \frac{1}{1+x^2}\,dx; \quad n = 10$

15. $\displaystyle\int_0^1 e^x \sin x\,dx; \quad n = 8.$

In Problems 16–30 approximate the given integral using (i) the trapezoidal rule and (ii) Simpson's rule, with the indicated number of subintervals.

16. $\displaystyle\int_0^1 \sqrt{x + x^2}\,dx; \quad n = 4$

17. $\displaystyle\int_1^2 \frac{\sin x}{x}\,dx; \quad n = 6$

18. $\displaystyle\int_1^3 \frac{x}{\sin x}\,dx; \quad n = 6$

19. $\displaystyle\int_0^1 e^{\sqrt{x}}\,dx; \quad n = 6$

20. $\displaystyle\int_0^1 e^{x^3}\,dx; \quad n = 8$

21. $\displaystyle\int_0^\pi \sin x^2\,dx; \quad n = 8$

22. $\displaystyle\int_1^2 \sqrt{\ln x}\,dx; \quad n = 10$

23. $\displaystyle\int_{-1}^1 e^{-x^2}\,dx; \quad n = 10$

24. $\displaystyle\int_0^1 \sqrt{1 + x^3}\,dx; \quad n = 10$

25. $\displaystyle\int_0^1 \frac{dx}{\sqrt{1 + x^3}}; \quad n = 10$

26. $\displaystyle\int_0^1 xe^{x^3}\,dx; \quad n = 10$

27. $\displaystyle\int_0^1 \ln(1 + e^x)\,dx; \quad n = 8$

28. $\displaystyle\int_1^3 \frac{x^2}{\sqrt[3]{1 + x}}\,dx; \quad n = 10$

29. $\displaystyle\int_0^1 \sinh x^2\,dx; \quad n = 10$

30. $\displaystyle\int_0^{\pi/4} \sin(\tan x)\,dx; \quad n = 10.$

In Problems 31–36 find a bound on the discretization error using the trapezoidal rule and Simpson's rule.

31. The integral of Problem 20.
32. The integral of Problem 21.
33. The integral of Problem 23.
34. The integral of Problem 24.
35. The integral of Problem 26.
36. The integral of Problem 27.

37. The integral $(1/\sqrt{2\pi})\int_{-a}^a e^{-x^2/2}\,dx$ is very important in probability theory. Using Simpson's rule, calculate $(1/\sqrt{2\pi})\int_{-1}^1 e^{-x^2/2}\,dx$ with an error of less than 0.01. [*Hint:* Show that $\int_{-a}^a e^{-x^2}\,dx = 2\int_0^a e^{-x^2/2}\,dx$.]

38. Calculate $(1/\sqrt{2\pi})\int_{-5}^5 e^{-x^2/2}\,dx$ with an error of less than 0.01.

*39. (a) Calculate $(1/\sqrt{2\pi})\int_{-50}^{50} e^{-x^2/2}\,dx$ with an error of less than 0.1.
(b) Can you guess what happens to $(1/\sqrt{2\pi})\int_{-N}^N e^{-x^2/2}\,dx$ as N grows without bound?

40. We know that

$$\int_0^1 \frac{dx}{1+x^2} = \tan^{-1} x \Big|_0^1 = \frac{\pi}{4}.$$

Thus

$$\pi = 4\int_0^1 \frac{dx}{1+x^2}.$$

(i) Using Simpson's rule, how many subintervals does it take to estimate π with an error of less than 0.0001?
(ii) Using this number of subintervals, what is your estimate of π?

41. How many subintervals would it take to estimate $\ln 2$ using the trapezoidal rule applied to the integral $\int_1^2 (1/x)\, dx$ with an error of less than 10^{-10}?

42. How many subintervals would it take to perform the estimate in Problem 41 using Simpson's rule?

*43. The function $J_{1/2}(x) = \sqrt{2/\pi x}\, \sin x$ is called the *Bessel function of order 1/2* and occurs frequently in applications in physics and engineering. Estimate $\int_{1/2}^1 J_{1/2}(x)\, dx$ with an error of less than 0.01.

44. Show that Simpson's rule provides the exact answer for $\int_a^b P_3(x)\, dx$ if P_3 is a polynomial of degree three.

Review Exercises for Chapter Eight

In Exercises 1–50 use one or more of the techniques of this chapter to calculate the given integral.

1. $\displaystyle\int_0^1 \frac{x^2}{\sqrt{4 + x^2}}\, dx$

2. $\displaystyle\int x\sqrt{x + 2}\, dx$

3. $\displaystyle\int \frac{x^2 + 3}{(x^2 + 1)^2}\, dx$

4. $\displaystyle\int \frac{dx}{(x - 1)(x + 3)}$

5. $\displaystyle\int_0^{\pi/2} \cos^3 x \sin^2 x\, dx$

6. $\displaystyle\int_0^1 \frac{\sqrt{x}}{1 + \sqrt{x}}\, dx$

7. $\displaystyle\int xe^{-2x}\, dx$

8. $\displaystyle\int \sin 4x \cos 5x\, dx$

9. $\displaystyle\int \tan^5 x\, dx$

10. $\displaystyle\int_0^{\pi/4} \sec^2 x \tan^3 x\, dx$

11. $\displaystyle\int_0^1 \frac{dx}{1 + \sqrt[3]{x}}$

12. $\displaystyle\int_0^1 (1 + x^2)^{3/2}\, dx$

13. $\displaystyle\int_{\sqrt{2}}^2 \frac{x^2}{(x^2 - 1)^{3/2}}\, dx$

14. $\displaystyle\int \frac{dx}{\cos x - \sin x}$

15. $\displaystyle\int e^{2x} \sin 2x\, dx$

16. $\displaystyle\int_1^2 x^3 \ln x\, dx$

17. $\displaystyle\int \sin^3 x \cos^3 x\, dx$

18. $\displaystyle\int \sec^4 x \tan^2 x\, dx$

19. $\displaystyle\int \csc^3 x \cot^3 x\, dx$

20. $\displaystyle\int \frac{dx}{(x - 1)(x^2 + x + 1)}$

21. $\displaystyle\int \frac{x^5 - 1}{x^3 - x}\, dx$

22. $\displaystyle\int \frac{x^2}{(x + 1)^2(x - 1)^2}\, dx$

23. $\displaystyle\int x \sinh x\, dx$

24. $\displaystyle\int \frac{dx}{x^2 - 4x + 5}$

25. $\displaystyle\int \frac{dx}{x^2 - 4x + 4}$

26. $\displaystyle\int \sin 5x \sin 6x\, dx$

27. $\displaystyle\int \cos 6x \cos 7x\, dx$

28. $\displaystyle\int \frac{1 + x + x^2}{(x^2 + 2x + 3)^2}\, dx$

29. $\displaystyle\int \frac{x^2 + 2}{(x - 3)(x + 4)(x - 5)}\, dx$

30. $\displaystyle\int_0^3 \frac{x^2}{\sqrt{9 + x^2}}\, dx$

31. $\displaystyle\int x^3 \sqrt{x^2 - 4}\, dx$

32. $\displaystyle\int x \cos^{-1} x\, dx$

33. $\displaystyle\int \sec^4 x\, dx$

34. $\displaystyle\int \frac{x^3}{(x + 1)^4}\, dx$

35. $\displaystyle\int \frac{(x + 1)^3}{x^3}\, dx$

36. $\displaystyle\int_0^{\pi/4} \sin x \tan^2 x\, dx$

37. $\displaystyle\int_0^{\pi/2} \sin^6 x\, dx$

38. $\displaystyle\int \sec^4 x \tan^3 x\, dx$

39. $\displaystyle\int_0^1 (1 + x^2)^{-1/2}\, dx$

40. $\displaystyle\int \frac{dx}{2 - \cos x}$

41. $\displaystyle\int \frac{dx}{1 + 5 \sin x}$

42. $\displaystyle\int_0^1 x^3(1 - x)^{2/3}\, dx$

43. $\displaystyle\int \frac{2x^3}{(1 + x^4)^{4/3}}\, dx$

44. $\displaystyle\int \frac{1 - x}{1 + \sqrt{x}}\, dx$

45. $\int \dfrac{x - 2x^3}{\sqrt{2 + 3x}} \, dx$ 　　**46.** $\int \sin(\ln x) \, dx$ 　　**47.** $\int e^{-x} \cos \dfrac{x}{3} \, dx$

48. $\int \sinh x \cosh 2x \, dx$ 　　**49.** $\int \cosh x \cosh 3x \, dx$ 　　**50.** $\int \dfrac{dx}{2 \sec x - 1}$.

In Exercises 51–56 use the trapezoidal rule (T) or Simpson's rule (S) to estimate the given integral with the given number of subintervals.

51. $\displaystyle\int_0^1 e^{x^3} \, dx$; 　T, 　$n = 4$ 　　　　**52.** $\displaystyle\int_0^1 e^{x^3} \, dx$; 　S, 　$n = 4$

53. $\displaystyle\int_0^1 \dfrac{dx}{\sqrt{1 + x^4}}$; 　T, 　$n = 6$ 　　**54.** $\displaystyle\int_0^1 \dfrac{dx}{\sqrt{1 + x^4}}$; 　S, 　$n = 6$

55. $\displaystyle\int_0^{\pi/2} \cos \sqrt{x} \, dx$; 　S, 　$n = 6$ 　　**56.** $\displaystyle\int_{\pi/6}^{\pi/2} \ln(\sin x) \, dx$; 　S, 　$n = 8$.

57. How many subintervals are needed in Exercise 51 to obtain a discretization error less than 0.01?

58. How many subintervals are needed in Exercise 52 to obtain a discretization error less than 0.00001?

59. Use Simpson's rule to estimate $\int_1^2 (1/x^2) \, dx$ with an error of less than 0.0001. Compare your answer with the actual answer that is easily obtained by integration.

60. Answer the questions in Exercise 59 for the integral $\int_1^2 \ln x \, dx$.

NINE

FURTHER APPLICATIONS OF
THE DEFINITE INTEGRAL

In this chapter we continue the discussion of applications of the integral begun in Chapters 4 and 5. The first applications are geometric, and the last contain applications in physics. After completing this chapter the reader should have a sense of the great usefulness of the integral as a tool in many different branches of science.

9.1 Volumes

In this section we show how the definite integral can be used to calculate the volumes of certain solid figures. We begin by deriving the familiar formula for the volume of a right circular cone. Such a cone is depicted in Figure 1. To calculate its volume we place the cone with its vertex at the origin and its central axis along the y-axis (see Figure 2). The line joining the points $(0, 0)$ and (r, h) has the equation $y = (h/r)x$ and

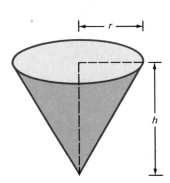

Figure 1

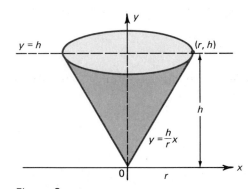

Figure 2

we may think of the cone as the solid obtained by rotating the region bounded by this line, the y-axis, and the line $y = h$ around the y-axis. In this context the cone is called a *solid of revolution*. To calculate its volume, we form a partition P of the interval along the y-axis between $y = 0$ and $y = h$ by using the partition points

$$0 = y_0 < y_1 < y_2 < \cdots < y_{n-1} < y_n = h,$$

where the width of each subinterval given by $\Delta y_i = y_i - y_{i-1}$ is assumed to be small. This has the effect of dividing the cone into "slices" or *disks* parallel to the x-axis. (See

436

Figure 3.) Our problem is to calculate the approximate volume of each disk, and then add up these volumes to obtain an approximation to the total volume of the cone. But it is easy to approximate the volume of a disc. As drawn in Figure 4, each disk is, for Δy_i small, very close to a right circular cylinder with radius $x_i = (r/h)y_i$ and height Δy_i. It is not quite a cylinder because at the top the radius is $(r/h)y_i$ while at the bottom the radius is $(r/h)y_{i-1}$. However if Δy_i is small, this difference will be small.

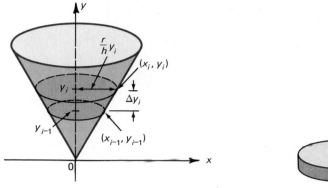

Figure 3   Figure 4

The volume of a cylinder with radius R and height H is $\pi R^2 H$ so that

$$V_i = \text{volume of } i\text{th disk} \approx \pi \left(\frac{r}{h}y_i\right)^2 \Delta y_i = \frac{\pi r^2}{h^2}y_i^2 \, \Delta y_i.$$

Then the approximate volume of the cone is

$$V = V_1 + V_2 + \cdots + V_n \approx \frac{\pi r^2}{h^2}(y_1^2 \, \Delta y_1 + y_2^2 \, \Delta y_2 + \cdots + y_n^2 \, \Delta y_n).$$

This approximation gets better and better as the norm of the partition approaches zero and we have

$$V = \frac{\pi r^2}{h^2} \lim_{|P| \to 0} (y_1^2 \, \Delta y_1 + y_2^2 \, \Delta y_2 + \cdots + y_n^2 \, \Delta y_n)$$

$$= \frac{\pi r^2}{h^2} \int_0^h y^2 \, dy = \frac{\pi r^2}{h^2} \frac{y^3}{3} \Big|_0^h = \frac{1}{3}\pi r^2 h,$$

which is the formula you might have seen earlier for the volume of a cone.

We now calculate the same volume in a different way. We form a partition P of the x-axis between $x = 0$ and $x = r$ by using the partition points

$$0 = x_0 < x_1 < x_2 < \cdots < x_{n-1} < x_n = r,$$

where the length of each subinterval, given by $\Delta x_i = x_i - x_{i-1}$, is assumed to be small. The situation is depicted in Figure 5. This has the effect of dividing the cone into what are called *cylindrical shells* (a cylindrical shell has the shape of an empty tin can with the ends removed). A typical cylindrical shell is drawn in Figure 6. Since the

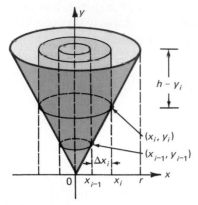

Figure 5   Figure 6

circumference of a circle of radius R is $2\pi R$, the approximate volume of each shell is its circumference $(2\pi x_i)$ times its height times its thickness so that

$$V_i \approx 2\pi x_i (h - y_i)\,\Delta x_i = 2\pi x_i \left(h - \frac{h}{r}x_i\right)\Delta x_i,$$

where V_i now denotes the volume of the ith shell. The volume is approximate because the height of the outside of the shell is $h - y_i$ while the height of the inside of the shell is $h - y_{i-1}$. However this difference is small if Δx_i is small. Proceeding as before, we obtain

$$V = \lim_{|P| \to 0} (V_1 + V_2 + \cdots + V_n)$$

$$= \lim_{|P| \to 0} \left[2\pi x_1 \left(h - \frac{h}{r}x_1\right)\Delta x_1 + 2\pi x_2 \left(h - \frac{h}{r}x_2\right)\Delta x_2 + \cdots \right.$$

$$\left. + 2\pi x_n \left(h - \frac{h}{r}x_n\right)\Delta x_n \right]$$

$$= \int_0^r 2\pi x \left(h - \frac{h}{r}x\right) dx = 2\pi h \int_0^r \left(x - \frac{x^2}{r}\right) dx$$

$$= 2\pi h \left(\frac{x^2}{2} - \frac{x^3}{3r}\right) \bigg|_0^r = 2\pi h \left(\frac{r^2}{2} - \frac{r^2}{3}\right) = 2\pi h \cdot \frac{r^2}{6} = \frac{1}{3}\pi r^2 h,$$

as before.

The two methods used above can be applied in a wide variety of problems. When the curve $y = f(x)$ is revolved about an axis we obtain a *solid of revolution* whose volume can be obtained by the *disk method* or the *cylindrical shell method*.

EXAMPLE   1.   The region bounded by the curve $y = x^2$, the x-axis and the line $x = 2$ is revolved about the x-axis. Calculate the volume of the solid which is generated.

SOLUTION. The solid is depicted in Figure 7. By partitioning the x-axis, we obtain disks. The volume of a typical disk is given by

$$V = \pi r^2 h = \pi y_i^2\,\Delta x_i = \pi x_i^4\,\Delta x_i$$

so that

$$V = \int_0^2 \pi x^4 \, dx = \frac{\pi x^5}{5} \Big|_0^2 = \frac{32\pi}{5}.$$

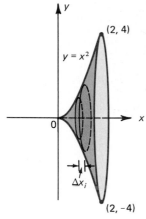

Figure 7

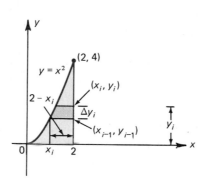

Figure 8

Using cylindrical shells, we have, as in Figure 8,

$$V_i \approx 2\pi y_i (2 - x_i) \Delta y_i = 2\pi y_i (2 - \sqrt{y_i}) \Delta y_i$$

so that

$$V = \int_0^4 2\pi (2 - \sqrt{y}) y \, dy = 2\pi \int_0^4 (2y - y^{3/2}) \, dy = 2\pi \left(y^2 - \frac{2}{5} y^{5/2} \right) \Big|_0^4$$

$$= 2\pi \left(16 - \frac{64}{5} \right) = \frac{32\pi}{5}.$$

It is of course not necessary to use both methods to calculate one volume. Typically one method will be easier to use than the other in a given problem and it will be a matter of judgment in deciding which method to use. The calculation of a given volume follows fairly easily once one of the axes has been partitioned and the volume of a typical disk or shell is calculated, provided that the function involved is easy to integrate. We give below two rules which can be used in the calculation of volumes.

(i)   Let f be continuous and nonnegative on $[a, b]$. (See Figure 9a.) Then the volume of the solid generated by revolving the area below the graph of f about the x-axis between $x = a$ and $x = b$ (see Figure 9b) is given by

$$V = \int_a^b \pi [f(x)]^2 \, dx. \tag{1}$$

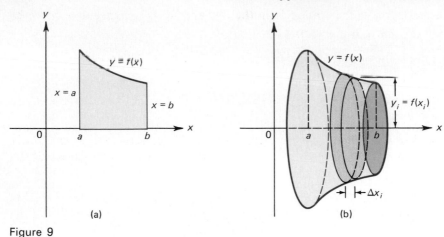

Figure 9

(ii)   If the area in (i) is revolved about the y-axis (see Figure 10) then the volume of the solid generated is given by

$$V = \int_a^b 2\pi x f(x) \, dx.$$ (2)

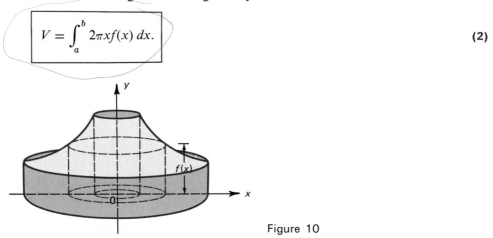

Figure 10

EXAMPLE 2.   Calculate the volume of a sphere of radius r.

SOLUTION. It is not difficult to see that we can generate the sphere of radius r by rotating the semicircle and the points interior to it pictured in Figure 11 around the x-axis. Since $x^2 + y^2 = r^2$ is the equation of the circle of radius r centered at the

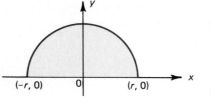

Figure 11

origin, the equation of the semicircle in the first and second quadrants is given by $y = \sqrt{r^2 - x^2}$. Then, by formula (1),

$$V = \pi \int_{-r}^{r} (\sqrt{r^2 - x^2})^2 \, dx = \pi \int_{-r}^{r} (r^2 - x^2) \, dx = \pi \left(r^2 x - \frac{x^3}{3} \right) \Big|_{-r}^{r} = \frac{4}{3}\pi r^3.$$

The same answer is obtained by rotating the semicircle in the first and fourth quadrants around the y-axis and using formula (2).

EXAMPLE 3. The circular disk $(x - 2)^2 + y^2 \leq 1$ is rotated around the y-axis. Find the volume of the "doughnut" shaped region generated. (This solid is called a *torus*.)

SOLUTION. The circular disk $(x - 2)^2 + y^2 \leq 1$ consists of the circle centered at $(2, 0)$ with radius 1 and all points interior to it (see Figure 12a). In the upper part of the

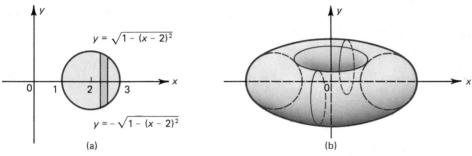

Figure 12

circle (that is, in the first quadrant) $y = \sqrt{1 - (x - 2)^2}$ while in the lower part of the circle, y is negative and is equal to $-\sqrt{1 - (x - 2)^2}$. Thus the height of a typical cylindrical shell is equal to

$$\sqrt{1 - (x_i - 2)^2} - (-\sqrt{1 - (x_i - 2)^2}) = 2\sqrt{1 - (x_i - 2)^2}.$$

Then by formula (2)†, the volume generated is given by

$$V = \int_{1}^{3} 2\pi x (2\sqrt{1 - (x - 2)^2}) \, dx = 4\pi \int_{1}^{3} x\sqrt{1 - (x - 2)^2} \, dx.$$

To integrate this we make the substitution $x - 2 = \sin\theta$. Then

$$dx = \cos\theta \, d\theta, \qquad x = 2 + \sin\theta,$$

and

$$\sqrt{1 - (x - 2)^2} = \sqrt{1 - \sin^2\theta} = \cos\theta.$$

† It is not difficult to see that formula (2) applies here even though the curve being rotated is not nonnegative for all values.

When $x = 1$, $\sin \theta = 1 - 2 = -1$ and $\theta = -\pi/2$. When $x = 3$, $\sin \theta = 1$ and $\theta = \pi/2$. Thus

$$V = 4\pi \int_{-\pi/2}^{\pi/2} (2 + \sin \theta) \cos^2 \theta \, d\theta$$

$$= 8\pi \int_{-\pi/2}^{\pi/2} \cos^2 \theta \, d\theta + 4\pi \int_{-\pi/2}^{\pi/2} \cos^2 \theta \sin \theta \, d\theta$$

$$= 4\pi \int_{-\pi/2}^{\pi/2} (1 + \cos 2\theta) \, d\theta - \frac{4\pi}{3} \cos^3 \theta \Big|_{-\pi/2}^{\pi/2}$$

$$= 4\pi \left(\theta + \frac{\sin 2\theta}{2} \right) \Big|_{-\pi/2}^{\pi/2} - 0 = 4\pi^2.$$

EXAMPLE 4. Find the area generated when the area under one loop of the curve $y = \sin x$ between $x = 0$ and $x = \pi$ is rotated about
(a) the x-axis   (b) the y-axis.

SOLUTION. (a) The area to be rotated is drawn in Figure 13a. From formula (1), the volume of the solid drawn in Figure 13b is given by

$$V = \pi \int_0^\pi \sin^2 x \, dx = \frac{\pi}{2} \int_0^\pi (1 - \cos 2x) \, dx = \frac{\pi}{2} \left(x - \frac{\sin 2x}{2} \right) \Big|_0^{\pi/2} = \frac{\pi^2}{2}.$$

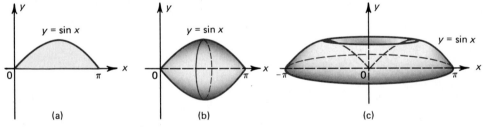

(a)   (b)   (c)

Figure 13

(b) From formula (2), the volume of the solid drawn in Figure 13c is given by

$$V = 2\pi \int_0^\pi x \sin x \, dx.$$

We can integrate this by parts. Let $u = x$ and $dv = \sin x \, dx$. Then $du = dx$, $v = -\cos x$ and

$$V = 2\pi \left(-x \cos x \Big|_0^\pi + \int_0^\pi \cos x \, dx \right) = 2\pi \left(\pi + \sin x \Big|_0^\pi \right) = 2\pi^2.$$

EXAMPLE 5. The area bounded by the curve $y = x^2 + 2$ and the line $y = x + 8$ is rotated around the x-axis. Find the volume of the solid generated.

SOLUTION. The area is depicted in Figure 14a and the solid generated is depicted in Figure 14c. The points of intersection of the two curves are easily found to be $(-2, 6)$ and $(3, 11)$. When the strip depicted in Figure 14a is rotated around the x-axis, it

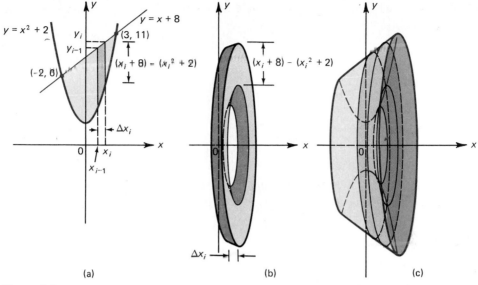

Figure 14

generates a circular ring as in Figure 14b. The volume of this ring is the volume of the large disk (radius $x_i + 8$) minus the volume of the smaller disk (radius $x_i^2 + 2$). Thus

$$V_i = \pi(x_i + 8)^2\, \Delta x_i - \pi(x_i^2 + 2)^2\, \Delta x_i$$

and the total volume is given by

$$V = \pi \int_{-2}^{3} [(x + 8)^2 - (x^2 + 2)^2]\, dx$$

$$= \pi \int_{-2}^{3} (x^2 + 16x + 64 - x^4 - 4x^2 - 4)\, dx$$

$$= \pi \int_{-2}^{3} (-x^4 - 3x^2 + 16x + 60)\, dx$$

$$= \pi \left(-\frac{x^5}{5} - x^3 + 8x^2 + 60x \right)\Big|_{-2}^{3} = 250\pi.$$

The above example can be generalized to obtain the following rule: (See Figure 15)

(iii)   Let f and g be continuous and nonnegative on $[a, b]$ with $f(x) \geq g(x) \geq 0$ for every x in $[a, b]$. Then the volume generated by rotating the region bounded by the curves and lines $y = f(x)$, $y = g(x)$, $x = a$ and $x = b$ around the x-axis is given by

$$V = \pi \int_{a}^{b} \{[f(x)]^2 - [g(x)]^2\}\, dx. \tag{3}$$

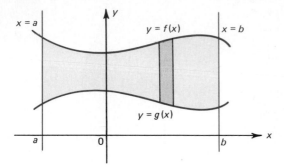

Figure 15

The techniques of this section can also be used to calculate the volumes of certain solids in which parallel cross sections all have the same basic shape. By a *cross section* we mean the figure (in two dimensions) obtained by slicing the solid with a plane. By slicing the figure with parallel planes we obtain a number of disks. (Think of cutting an egg into many thin, parallel slices. Each cross section has the shape of an oval.) Each "slice" has a volume approximately equal to the area of the cross section times its thickness. To calculate the volume of the solid, we simply "add up" the volumes of the separate slices by the now familiar process of integration.

EXAMPLE 6.   The base of a certain solid is the circular disk $x^2 + y^2 \leq 4$ in the xy-plane. Each plane perpendicular to the x-axis cuts the solid in an equilateral triangle. Find the volume of the solid.

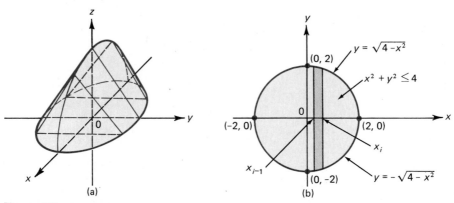

Figure 16

SOLUTION. The solid is drawn in Figure 16a. If the solid is sliced along the x-axis, the volume of a typical slice is $A_i \, \Delta x_i$ where A_i is the area of an equilateral triangle. One side of the triangle has length

$$\sqrt{4 - x_i^2} - (-\sqrt{4 - x_i^2}) = 2\sqrt{4 - x_i^2}$$

(according to Figure 16b). Using Figure 17, we easily calculate that the area of an equilateral triangle with side s is given by $A = (\sqrt{3}/4)s^2$. Thus the area A_i above is

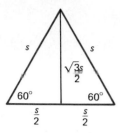

Figure 17

given by

$$A_i = \frac{\sqrt{3}}{4}(4)(4 - x_i^2) = \sqrt{3}(4 - x_i^2) \qquad \text{and} \qquad V_i = \sqrt{3}(4 - x_i^2)\,\Delta x_i$$

so that

$$V = \sqrt{3}\int_{-2}^{2}(4 - x^2)\,dx = \sqrt{3}\left(4x - \frac{x^3}{3}\right)\Big|_{-2}^{2} = \frac{32\sqrt{3}}{3}.$$

PROBLEMS 9.1

In Problems 1–20 find the volume generated when the region bounded by the given curves and lines is rotated about the x-axis.

1. $y = x$, $x = 2$, $y = 0$
2. $y = 2x + 3$, $x = 1$, $x = 4$, $y = 0$
3. $y = 2x^2$, $x = 1$, $y = 0$
4. $y = \sqrt{x + 1}$, $x = 1$, $x = 5$, $y = 0$
5. $y = 4 - x$, $x = 0$, $y = 0$
6. $y = x^3$, $x = 2$, $y = 1$

7. $y = \cos x$, $x = \frac{\pi}{2}$, $y = 0$, $x = 0$
8. $y = \tan x$, $x = \frac{\pi}{4}$, $y = 0$

9. $y = e^x$, $x = -1$, $x = 1$, $y = 0$
10. $y = \sqrt{1 - x^3}$, $x = 0$, $y = 0$
11. $y = (1 - x^2)^{1/4}$, $x = 0$, $y = 0$
*12. $x = y - y^2$, $x = 0$
13. $y = \ln x$, $x = 1$, $x = e$, $y = 0$
14. $y = x^2$, $y = x^3$

15. $y = \sin x$, $y = \cos x$, $0 \le x \le \frac{\pi}{4}$
16. $y = \sec x$, $x = 0$, $x = \frac{\pi}{3}$, $y = 0$

17. $y = xe^x$, $x = 1$, $y = 0$
18. $y = \sin^{-1} x$, $x = 1$, $y = 0$

19. $y = \frac{1}{x}$, $x = 1$, $x = 2$, $y = 0$
20. $y = \sinh^{1/2} x$, $x = 1$, $y = 0$.

In Problems 21–32 find the volumes generated when the region bounded by the given curves and lines is rotated about the y-axis.

21. $x + y = 3$, $x = 0$, $y = 0$
22. $y = \frac{1}{2}x^2$, $x = 1$, $y = 0$

23. $y = x$, $y = 1 - x$, $x = 0$
24. $y = x^{1/3}$, $x = 1$, $y = 0$

25. $y = \ln x$, $x = 1$, $x = 3$, $y = 0$
26. $y = \cos x$, $x = \frac{\pi}{2}$, $y = 0$

27. $y = \sqrt{x}$, $y = x^2$
28. $y = x^2$, $y = x^3$

29. $y = \sin x$, $y = \cos x$, $0 \le x \le \frac{\pi}{4}$
30. $x = y^2 + 1$, $x = 2$

31. $y = e^{x^2}$, $x = 0$, $x = 1$, $y = 0$
32. $y = \cos(x^2)$, $x = 0$, $x = \sqrt{\pi/6}$, $y = 0$.

33. Calculate the volume of the sphere of radius r by rotating the interior of the semicircle $x = \sqrt{r^2 - y^2}$ about the y-axis.

34. The circle centered at $(a, 0)$ with radius r ($r < a$) is rotated about the y-axis. Show that the volume of the torus so generated is equal to $2\pi^2 ar^2$.

35. The equation of an ellipse centered at the origin (see Section 1.9) is given by

$$\frac{x^2}{a^2} + \frac{y^2}{b^2} = 1.$$

Calculate the volume of the "football" shaped solid generated by rotating this ellipse about the x-axis.

36. Find the volume generated by rotating the triangle with vertices at $(1, 1)$, $(2, 3)$, and $(3, 2)$ about the x-axis.

37. Find the volume generated if the triangle of Problem 36 is rotated about the y-axis.

38. What is the volume generated when the area bounded by the line $y = x$ and the curve $y = x^2$ is rotated about the line $x = 2$? About the line $y = 2$?

**39. Let (a_1, b_1), (a_2, b_2) and (a_3, b_3) be three points in the first quadrant. Calculate the volume generated when the triangle with vertices at these points is rotated about the x-axis.

40. Calculate the volume generated by rotating the area bounded by $y = \sqrt{x}$, $x = 1$, and $y = 0$ about the line $x = 4$.

41. Find the volume generated by rotating the area bounded by $y = \sin x$, $x = 0$, $x = \pi/2$, and $y = 0$ about the line $y = 2$.

*42. A water tank in the shape of a sphere has a radius of 10 m. What is the mass of water in the tank if the tank is filled to a depth of 2 m? [*Hint:* The density of water is approximately 1000 kg/m³.]

43. The base of a certain solid is a circle with radius 2 while each parallel cross section of the solid is a square. Find the volume of the solid.

44. Find the volume of the solid in Problem 43 if parallel cross sections are isosceles triangles with height 1.

45. Find the volume of the solid in Problem 43 if parallel cross sections are semicircles.

*46. A right circular cylinder of radius 2 m is cut by two planes. The first is parallel to the base of the cylinder while the second makes an angle of 45° with the base. The two planes intersect at a diameter of the circular cross section of the cylinder, thereby cutting out a wedge. What is the volume of this wedge? [*Hint:* Look at Figure 18. The volume can be calculated two ways by taking (a) cross sections parallel to the base or (b) cross sections perpendicular to it.]

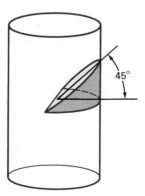

45°

Figure 18

47. Calculate the volume of the wedge in Problem 46 if the second plane intersects the first at an angle of 60°.

48. Calculate the volume of the wedge in Problem 46 if the second plane intersects the first at an angle of α degrees.

49. The base of a certain solid is the region between the line $y = x$ and the curve $y = x^2$. Find the volume of the solid if cross sections perpendicular to the x-axis are
    (a) squares.
    (b) isosceles right triangles with hypotenuses in the xy-plane.
    (c) semicircles.

*50. A water tank is in the shape of the solid generated by rotating the region bounded by $y = x^2$, $x = 0$, and $y = 4$ around the y-axis. If the tank is filled with water, find the work needed to pump all the water to the top of the tank. (Assume that distance is measured in meters.)

*51. A storage tank filled with olive oil is in the shape of the solid generated by rotating the region bounded by $y = \ln x$, $y = 0$, $y = 3$, and $x = 0$ about the y-axis. Find the work needed to pump the oil to the top of the tank. (The density of olive oil is 920 kg/m³.)

9.2 Arc Length

In this section we derive a formula for calculating the length of a curve in the xy-plane (called a *plane curve*). A typical curve is sketched in Figure 1. The problem

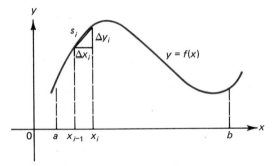

Figure 1

is to calculate the length of the curve from $x = a$ to $x = b$. Such a problem arises in many interesting physical examples. For example, if an object is thrown or shot into the air, then its path as a function of time is a parabola opening downward. To calculate the total length traveled by the object, it is necessary to find the length of the parabola from some initial time t_0 to a final time t_f.

In order to find the length of the curve in Figure 1, we partition the interval $[a, b]$ into n subintervals, find the approximate length s_i of the arc in each subinterval, and add up these lengths. (This technique is, by now, very familiar.) The appearance of the curve near a typical subinterval is sketched in Figure 2. If Δx_i is small, and if

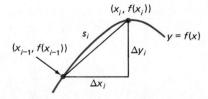

Figure 2

$f(x)$ is continuous in the interval $[x_{i-1}, x_i]$, then the length of the curve between x_{i-1} and x_i, which we shall denote by Δs_i, can be approximated by the straight line segment joining the points $(x_{i-1}, f(x_{i-1}))$ and $(x_i, f(x_i))$. The length of this line segment is, by the Pythagorean theorem, given by

$$\Delta s_i = \sqrt{\Delta x_i^2 + \Delta y_i^2} = \Delta x_i \sqrt{1 + \left(\frac{\Delta y_i}{\Delta x_i}\right)^2}. \tag{1}$$

(The letter s is customarily used for arc length.) Furthermore, if f is differentiable at x_{i-1}, then

$$f'(x_{i-1}) = \lim_{\Delta x_i \to 0} \frac{f(x_{i-1} + \Delta x_i) - f(x_{i-1})}{\Delta x_i} = \lim_{\Delta x_i \to 0} \frac{\Delta y_i}{\Delta x_i}.$$

Thus, for Δx_i small,

$$\frac{\Delta y_i}{\Delta x_i} \approx f'(x_{i-1}). \tag{2}$$

Inserting (2) in (1), we have

$$\Delta s_i \approx \sqrt{1 + [f'(x_{i-1})]^2}\, \Delta x_i$$

and the length of the curve from a to b is approximately

$$s \approx \sqrt{1 + [f'(x_0)]^2}\, \Delta x_1 + \sqrt{1 + [f'(x_1)]^2}\, \Delta x_2 + \cdots + \sqrt{1 + [f'(x_{n-1})]^2}\, \Delta x_n$$

and

$$s = \lim_{|P| \to 0} [\sqrt{1 + f'(x_0)}\, \Delta x_1 + \sqrt{1 + f'(x_1)}\, \Delta x_2 + \cdots + \sqrt{1 + f'(x_{n-1})}\, \Delta x_n]$$

or

$$s = \int_a^b \sqrt{1 + [f'(x)]^2}\, dx. \tag{3}$$

This integral is often written as $\int_0^s ds$ where ds denotes the differential of arc length. We will discuss this notation and its implications in greater detail after our discussion of parametric equations in Chapter 16.

Remark. We can only use formula (3) when $f'(x)$ exists and is continuous in the interval $[a, b]$.

EXAMPLE 1.   Calculate the length of the straight line $y = x$ between $x = 0$ and $x = 1$.

SOLUTION. We easily see that this distance is simply the straight line distance between $(0, 0)$ and $(1, 1)$ and is equal to $\sqrt{2}$. However, we will also calculate it using formula (3), mainly to illustrate for ourselves the formula's validity. Since $f'(x) = 1$, we have

$$s = \int_0^1 \sqrt{1 + 1^2}\, dx = \int_0^1 \sqrt{2}\, dx = \sqrt{2}.$$

EXAMPLE 2.   Calculate the length of the arc of the curve $y = x^2$ between $x = 0$ and $x = 2$.

SOLUTION. Since $f'(x) = 2x$, $[f'(x)]^2 = 4x^2$ and

$$s = \int_0^2 \sqrt{1 + 4x^2}\, dx = 2\int_0^2 \sqrt{x^2 + \tfrac{1}{4}}\, dx.$$

Let $x = \tfrac{1}{2}\tan\theta$. Then

$$s = 2\int_0^{\tan^{-1}4} \sqrt{\tfrac{1}{4}\tan^2\theta + \tfrac{1}{4}} \cdot \tfrac{1}{2}\sec^2\theta\, d\theta = \tfrac{1}{2}\int_0^{\tan^{-1}4} \sec^3\theta\, d\theta$$

$$= \tfrac{1}{4}(\ln|\sec\theta + \tan\theta| + \sec\theta\tan\theta)\Big|_0^{\tan^{-1}4} \qquad \text{(from Example 8.3.10).} \qquad \textbf{(4)}$$

When $\theta = \tan^{-1}4$, $\tan\theta = 4$ and $\sec\theta = \sqrt{17}$ (since $1 + \tan^2\theta = \sec^2\theta$, we have $1 + 4^2 = \sec^2\theta$). Thus from (4) we find that $s = \tfrac{1}{4}[\ln(\sqrt{17} + 4) + 4\sqrt{17}]$.

EXAMPLE 3.   Calculate the length of the arc of the curve $y = 3 + x^{2/3}$ between $x = 1$ and $x = 8$.

SOLUTION. $f'(x) = (2/3)x^{-1/3}$ so that

$$s = \int_1^8 \sqrt{1 + \frac{4}{9x^{2/3}}}\, dx = \int_1^8 \sqrt{\frac{4 + 9x^{2/3}}{9x^{2/3}}}\, dx = \int_1^8 \frac{\sqrt{4 + 9x^{2/3}}}{3x^{1/3}}\, dx.$$

Let $u = \sqrt{4 + 9x^{2/3}}$. Then $u^2 = 4 + 9x^{2/3}$, $9x^{2/3} = u^2 - 4$, $3x^{1/3} = \sqrt{u^2 - 4}$, $x = (u^2 - 4)^{3/2}/27$, and $dx = (1/18)\sqrt{u^2 - 4} \cdot 2u\, du$. When $x = 1$, $u = \sqrt{13}$ and when $x = 8$, $u = \sqrt{40}$. Thus

$$s = \int_{\sqrt{13}}^{\sqrt{40}} \frac{u}{\sqrt{u^2 - 4}} \cdot \frac{1}{18}\sqrt{u^2 - 4} \cdot 2u\, du$$

$$= \frac{1}{9}\int_{\sqrt{13}}^{\sqrt{40}} u^2\, du = \frac{u^3}{27}\Big|_{\sqrt{13}}^{\sqrt{40}} = \frac{1}{27}(40^{3/2} - 13^{3/2}).$$

EXAMPLE 4.   Calculate the length of the loop of the sine function between $x = 0$ and $x = \pi/2$.

SOLUTION. The arc whose length is requested is sketched in Figure 3. Since $f'(x) = \cos x$, we have

$$s = \int_0^{\pi/2} \sqrt{1 + \cos^2 x}\, dx.$$

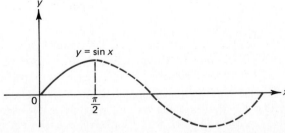

Figure 3

Unfortunately, there is no way to find an indefinite integral for $\sqrt{1 + \cos^2 x}$. We can estimate this integral using either the trapezoidal rule or Simpson's rule. To two decimal places, the value of the integral is 1.93.

EXAMPLE 6. Calculate the length of the arc of the curve $y = \ln x$ between $x = 1$ and $x = \sqrt{3}$.

SOLUTION. Here $f'(x) = 1/x$ so that

$$s = \int_1^{\sqrt{3}} \sqrt{1 + \frac{1}{x^2}}\, dx = \int_1^{\sqrt{3}} \frac{\sqrt{x^2 + 1}}{x}\, dx.$$

Let $u = \tan\theta$. Then

$$s = \int_{\pi/4}^{\pi/3} \frac{\sec\theta \sec^2\theta}{\tan\theta}\, d\theta = \int_{\pi/4}^{\pi/3} \frac{\sec\theta(1 + \tan^2\theta)}{\tan\theta}\, d\theta$$

$$= \int_{\pi/4}^{\pi/3} \left(\frac{\sec\theta}{\tan\theta} + \sec\theta \tan\theta \right) d\theta = \int_{\pi/4}^{\pi/3} \left(\frac{1}{\cos\theta}\frac{\cos\theta}{\sin\theta} + \sec\theta \tan\theta \right) d\theta$$

$$= \int_{\pi/4}^{\pi/3} (\csc\theta + \sec\theta \tan\theta)\, d\theta$$

$$= (-\ln|\csc\theta + \cot\theta| + \sec\theta) \Big|_{\pi/4}^{\pi/3}$$

$$= \left\{ \left[-\ln\left(\frac{2}{\sqrt{3}} + \frac{1}{\sqrt{3}} \right) + 2 \right] - [-\ln(\sqrt{2} + 1) + \sqrt{2}] \right\}$$

$$= \ln(\sqrt{2} + 1) - \ln\sqrt{3} + 2 - \sqrt{2} = \ln\left(\frac{\sqrt{2} + 1}{\sqrt{3}} \right) + 2 - \sqrt{2}.$$

We close this section by noting that the calculation of arc length is usually more difficult than the calculation of volume because it is often difficult or impossible to find an indefinite integral for $\sqrt{1 + [f'(x)]^2}$. This fact illustrates one of the many reasons why numerical techniques are so important.

PROBLEMS 9.2

In Problems 1–10 calculate the length of the arc of the curve $y = f(x)$ over the given interval.

1. $y = 3x + 4$;   [1, 5]
2. $y = 2 + \frac{1}{2}x^2$;   [0, 1]
3. $y = x^{3/2}$;   [0, 1]
4. $y = \frac{1}{3}(x^2 + 2)^{3/2}$;   [0, 3]
5. $y = \frac{1}{6}\left(x^3 + \frac{3}{x}\right)$;   [1, 3]
6. $y = \frac{1}{3}(x^{3/2} - 3\sqrt{x})$;   [1, 4]
7. $y = \frac{x^4}{4} + \frac{1}{8x^2}$;   [1, 2]
*8. $y = 2\ln(1 + x)$;   [0, 1]

9. $y = e^x$;   [1, 2] [*Hint:* Use the substitution $1 + e^{2x} = u^2$.]
10. $y = \cosh x$;   [−1, 1]. [*Hint:* Write $\cosh x$ in exponential form.]

*11. (a) Using the technique of this section, show that the circumference of the unit circle $x^2 + y^2 = 1$ is 2π. [*Hint:* Show that the length of the arc of the circle in the first quadrant between $x = 0$ and $x = 1/\sqrt{2}$ is $\pi/4$.]

(b)   Explain why this calculation cannot be used to find the length of the circumference because it *makes use* of the fact that the length of the circumference of a circle of radius r is $2\pi r$.

*12.   (a)   Show that the circumference of the circle $(x - a)^2 + (y - b)^2 = r^2$ is $2\pi r$.

(b)   Answer part (b) of Problem 11.

*13.   Find the length of the curve $x^{2/3} + y^{2/3} = 1$. [*Hint:* Write y as a function of x and determine the values of x for which the function is defined.]

14.   Find the length of the arc of $y = e^{-x}$ between $x = 0$ and $x = 1$.

15.   Find the length of the arc of the curve $(y + 1)^2 = 4x^3$ between $x = 0$ and $x = 1$.

In Problems 16–24 find a definite integral which is equal to the length of the arc of the given curve over the given interval. Do not try to evaluate the integral.

16. $y = \cos x;$ $\left[0, \dfrac{\pi}{2}\right]$

→ 17. $y^2 = (1 + x^2)^3, y \geq 0;$ $[0, 1]$

18. $y = \tan x;$ $\left[0, \dfrac{\pi}{4}\right]$

19. $y = x \sinh x;$ $[0, 1]$

20. $y = \sin^{-1} x;$ $[0, 1]$

21. $y = e^{x^2};$ $[0, 4]$

22. $y = e^x \sin x;$ $\left[\dfrac{\pi}{6}, \pi\right]$

23. $y = \cosh(1 + x^2);$ $[-1, 1]$

24. $y = \tan^{-1} e^x;$ $[3, 6]$.

9.3 Surface Area

In Section 9.1 we discussed the notion of a solid of revolution obtained by rotating a curve around a given line. In this section we show how to calculate the surface area of such a solid. The technique we use is very similar to the one we used in our calculation of arc length in Section 9.2.

Suppose that the curve in Figure 1a is rotated around the x-axis. We again partition the interval $[a, b]$. When we rotate the "subarc" of the curve $f(x)$ in the interval $[x_{i-1}, x_i]$ around the x-axis, we obtain a thin surface like the one depicted in Figure 1b. If $\Delta x_i = x_i - x_{i-1}$ is small, then the shape of the surface generated by this

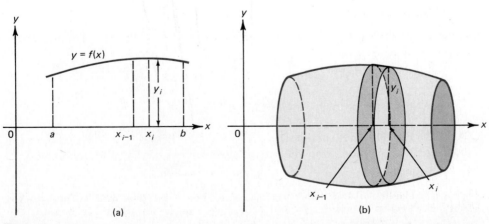

(a)   (b)

Figure 1

arc is (approximately) that of a right circular cylinder. Since area = length × width, the *lateral surface area*† of such a cylinder is equal to its circumference times its width

$$\text{surface area} \approx (2\pi y_i) \times (\text{arc length between } x_{i-1} \text{ and } x_i) \approx 2\pi y_i \, \Delta s_i$$
$$\approx 2\pi y_i \sqrt{1 + [f'(x_i)]^2} \, \Delta x_i \qquad \text{(from Section 9.2)}$$
$$= 2\pi f(x_i) \sqrt{1 + [f'(x_i)]^2} \, \Delta x_i.$$

Note that we must use Δs_i rather than Δx_i in the above formulas because the object whose volume we are calculating is not really a cylinder since its sides slant. The quantity Δs_i is the length of the slanted side.

Then, using our familiar limiting argument, we obtain

$$\text{surface area} = S = \int_a^b 2\pi f(x) \sqrt{1 + [f'(x)]^2} \, dx. \tag{1}$$

Again, we caution that formula (1) can be used only in the case that $f(x)$ possesses a continuous derivative in $[a, b]$.

EXAMPLE 1. Calculate the lateral surface area of a right circular cone with radius r and height h.

SOLUTION. We place the cone so that its vertex is at the origin and its central axis is along the x-axis (see Figure 2). We see that the cone is generated by revolving the

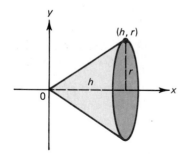

Figure 2

area under the line $y = (r/h)x$ about the x-axis. Then, from formula (1),

$$S = \int_0^h 2\pi \left(\frac{r}{h}x\right) \sqrt{1 + \left(\frac{r}{h}\right)^2} \, dx = 2\pi \frac{r}{h} \sqrt{1 + \left(\frac{r}{h}\right)^2} \int_0^h x \, dx$$
$$= 2\pi \frac{r}{h} \sqrt{\frac{h^2 + r^2}{h^2}} \cdot \frac{h^2}{2} = \pi r \sqrt{r^2 + h^2}.$$

EXAMPLE 2. The area under the curve $y = \frac{1}{2}x^2$ between $x = 0$ and $x = 1$ is rotated about the x-axis. Find the lateral surface area of the solid generated.

---

†The lateral surface area of a solid of revolution is the surface area formed by the rotating curve itself. In the case of a cylinder, the lateral surface area is the surface area of the "sides" of the cylinder—the area of the top and bottom is excluded.

SOLUTION. The solid is pictured in Figure 3. Since $f(x) = \frac{1}{2}x^2$, $f'(x) = x$ and

$$S = \int_0^1 (2\pi)(\tfrac{1}{2}x^2) \sqrt{1 + x^2}\, dx,$$

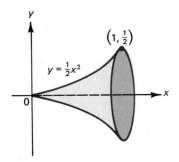

$y = \frac{1}{2}x^2$

$(1, \frac{1}{2})$

Figure 3

Let $x = \tan \theta$. Then

$$S = \pi \int_0^{\pi/4} \tan^2 \theta \sec^3 \theta\, d\theta \qquad \text{(from Example 8.3.13)}$$

$$= \pi \left(-\frac{\ln |\sec \theta + \tan \theta|}{8} + \frac{\sec \theta \tan \theta}{8} + \frac{\sec \theta \tan^3 \theta}{4} \right) \Big|_0^{\pi/4}$$

$$= \pi \left(-\frac{\ln |\sqrt{2} + 1|}{8} + \frac{3\sqrt{2}}{8} \right).$$

EXAMPLE 3.   Calculate the lateral surface area of the solid generated when the area under the curve $y = \sin x$ between $x = 0$ and $x = \pi/2$ is rotated around the x-axis.

SOLUTION. The solid is sketched in Figure 4. Here $f'(x) = \cos x$

$$S = \int_0^{\pi/2} 2\pi \sin x \sqrt{1 + \cos^2 x}\, dx.$$

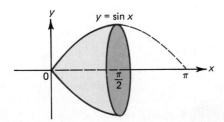

$y = \sin x$

Figure 4

To integrate this, we make the substitution $u = \cos x$. Then $du = -\sin x\, dx$ and

$$S = -2\pi \int_1^0 \sqrt{1 + u^2}\, du = 2\pi \int_0^1 \sqrt{1 + u^2}\, du.$$

We then set $u = \tan \theta$ to obtain

$$S = 2\pi \int_0^{\pi/4} \sec^3 \theta \, d\theta = \pi(\sec \theta \tan \theta + \ln |\sec \theta + \tan \theta|) \Big|_0^{\pi/4}$$

$$= \pi[\sqrt{2} + \ln(\sqrt{2} + 1)].$$

EXAMPLE 4. Calculate the lateral surface area of the solid obtained by revolving the curve $y = x^{1/3}$ about the y-axis between $x = 0$ and $x = 8$.

SOLUTION. The curve is illustrated in Figure 5. By reasoning exactly as in the

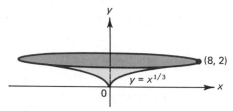

Figure 5

derivation of formula (1), we may assert that when the area between curve $x = g(y)$ and the y-axis is rotated about the y-axis between $y = c$ and $y = d$, the lateral surface area of the resulting solid is given by

$$S = \int_c^d 2\pi g(y) \sqrt{1 + [g'(y)]^2} \, dy. \tag{2}$$

In our example, $y = x^{1/3}$, $x = g(y) = y^3$ and $g'(y) = 3y^2$. If x is in $[1, 8]$, then y is in $[1, 2]$ and we find that

$$S = 2\pi \int_1^2 y^3 \sqrt{1 + 9y^4} \, dy = \frac{\pi}{27}(1 + 9y^4)^{3/2} \Big|_1^2 = \frac{\pi}{27}(145^{3/2} - 10^{3/2}).$$

PROBLEMS 9.3

In Problems 1–12 find the lateral surface area of the solid generated when the area below the given curve over the given interval is rotated about the x-axis.

1. $y = \sqrt{x}$; $[0, 1]$
2. $y = 2\sqrt{x}$; $[0, 3]$
3. $y = 2x$; $[0, 1]$
4. $y = x + 1$; $[1, 4]$
5. $y = \alpha x + \beta$; $[a, b]$
6. $y = e^{-x}$; $[0, 1]$
7. $y = \sqrt{2 - x}$; $[0, 1]$
8. $y = \cos x$; $\left[0, \frac{\pi}{2}\right]$
9. $y = \frac{x^3}{6} + \frac{1}{2x}$; $[1, 2]$
10. $y = \cosh x$; $[0, 1]$
*11. $y = \frac{1}{x}$; $[1, 2]$
12. $y = \frac{3}{2}x^{2/3}$; $[0, 1]$.

13. Show that the surface area of a sphere of radius r is $4\pi r^2$. [*Hint:* Revolve the area in the first quadrant part of the disk $x^2 + y^2 \le r^2$ around the x-axis to obtain the lateral surface area of the hemisphere of radius r, and then multiply by 2.]

14. Calculate the total surface area of the solid obtained by revolving about the y-axis the region bounded by the curve $y = \ln x$, the x-axis, and the line $x = e$.

*15. If the region enclosed by the ellipse $(x^2/a^2) + (y^2/b^2) = 1$ is revolved about the x-axis, find the surface area of the resulting "football-shaped" solid. [Note that there are really two cases: If $a > b$ we indeed get a "football," but if $a < b$ we get something that looks more like a frisbee.]

16. The area below the line $y = 2x + 3$ and above the x-axis between $x = -1$ and $x = 1$ is revolved about the line $y = -1$. Find the resulting surface area. [*Hint:* Be careful.]

17. The arc in the fourth quadrant of the circle $x^2 + (y - 1)^2 = 16$ is revolved about the y-axis. Find the resulting lateral surface area.

*18. When a circle is revolved about a line which does not intersect it, the resulting solid is called a *torus*. Calculate the surface area of the torus obtained by rotating the circle $x^2 + y^2 = r^2$ about the line $x = q$ (where $0 < r < q$).

19. On the earth, a *parallel* is a line around the planet which runs parallel to the equator. A *zone* is any region between two parallels. Show that if we assume that the earth is spherical,† that the area of any zone is given by $S = \pi dh$ where d is the diameter of the earth ($\sim$13,300 km) and h is the distance between the parallels. Note that this area is independent of the position of the zone of the sphere.

9.4 Center of Mass and the First Moment

In the physical applications given in earlier chapters we treated moving objects as though they were particles, having mass but not size. However in the real world objects are made up of many particles. If a moving object is rotating or vibrating, for example, then different particles in the object may exhibit different kinds of behavior. Fortunately we can deal with this kind of situation by representing the motion of the object as the motion of a special point, called the *center of mass* of that object. In general, we can describe the motion of an entire system if we can describe the motion of its center of mass and the motion of the system around that center of mass.

We begin with a system containing two point masses m_1 and m_2, located at distances x_1 and x_2 from a reference point which we label 0. In Figure 1 we have

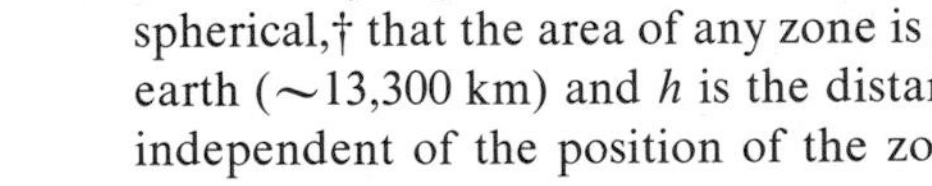

Figure 1

indicated two masses, one of which is located to the left of zero so that its "distance" is treated as negative. We then choose a new point, which we call $\bar{x}$, which has the property that the product of the total mass of the system and the distance between $\bar{x}$ and 0 is equal to the sum of the products of the mass of each particle and the respective distance of each particle to 0. That is,

$$(m_1 + m_2)\bar{x} = m_1 x_1 + m_2 x_2 \tag{1}$$

or

$$\bar{x} = \frac{m_1 x_1 + m_2 x_2}{m_1 + m_2}. \tag{2}$$

The point $\bar{x}$ is called the *center of mass* of the system.

Another way to think of center of mass is given by the motion on a seesaw. We place the two masses on opposite ends of the seesaw as in Figure 2 and label the pivot

† Of course, the earth is not a perfect sphere but is rather flattened at the poles and takes the shape of what is termed an *oblate spheroid*. However, it is very close to a sphere in that the difference between its equatorial diameter and its polar diameter is only about one-third of one percent of the equatorial diameter.

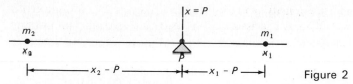

Figure 2

point P. From experience, we know that the farther away a mass is from the pivot point, the more likely the mass will "control" the motion. To put this more precisely, we consider the quantity $m_1(x_1 - P)$, measured in kilogram-meters. By the principle of the lever, the seesaw will exactly balance if

$$m_1(x_1 - P) + m_2(x_2 - P) = 0. \tag{3}$$

The quantity on the left-hand side of (3) is called the *first moment of the system around the line* $x = P$. If $P = 0$, then the first moment is equal to $m_1x_1 + m_2x_2$ and is called the first moment about the *y*-axis (the line $x = 0$). When the first moment is zero, the seesaw will balance and we say that the system is in *equilibrium*. Solving equation (3) for P, we obtain

$$P = \frac{m_1x_1 + m_2x_2}{m_1 + m_2} = \bar{x}.$$

That is, the seesaw will balance if the pivot is at its center of mass. (Of course, in this derivation, we ignored the mass of the seesaw itself.)

If the system contains n particles, where $n > 2$, then we can, in an analogous manner, define the center of mass by

$$\bar{x} = \frac{m_1x_1 + m_2x_2 + \cdots + m_nx_n}{m_1 + m_2 + \cdots + m_n} = \frac{\text{first moment of the system about the } y\text{-axis}}{\text{total mass of the system}}. \tag{4}$$

EXAMPLE 1.   Five particles of masses 2, 1, 4, 3, and 5 kg, respectively, are located on the *x*-axis with *x* values 3, -2, 4, 7, and -3 meters. Calculate the center of mass of the system.

SOLUTION

$$\bar{x} = \frac{m_1x_1 + m_2x_2 + m_3x_3 + m_4x_4 + m_5x_5}{m_1 + m_2 + m_3 + m_4 + m_5}$$

$$= \frac{2(3) + 1(-2) + 4(4) + 3(7) + 5(-3)}{2 + 1 + 4 + 3 + 5} = \frac{26}{15} \text{ m.}$$

Now consider a rigid, horizontal rod with variable density $\rho(x)$ which extends along the *x*-axis from $x = a$ to $x = b$. We can think of the rod as a system of very closely packed particles. In fact, the number of particles (atoms) is so large, and the spacing between them so small, that we can think of the body as having a continuous distribution of mass (that is, having a continuous density function). We first calculate the first moment of the rod and begin by partitioning the interval $[a, b]$:

$$a = x_0 < x_1 < x_2 < \cdots < x_n = b.$$

If $\Delta x_i = x_i - x_{i-1}$ is small, the density $\rho(x)$ (which is assumed to be continuous) is almost constant in $[x_{i-1}, x_i]$. Then the mass m_i of the "subrod" over the subinterval $[x_{i-1}, x_i]$ is given by

$$m_i \approx \rho(x_i)\,\Delta x_i. \tag{5}$$

What we have done here is to approximate the continuous rod by a system of n point masses situated at the n partition points $x_1, x_2, \ldots, x_n$. (See Figure 3.) Combining

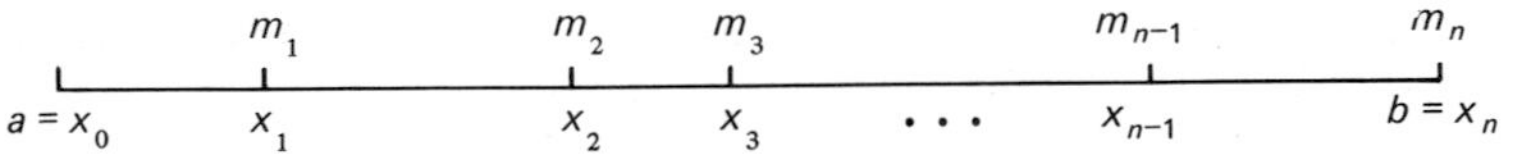

Figure 3

equations (4) and (5), we see that the center of mass of this approximating system is given by

$$\bar{x}_{approx} = \frac{x_1\rho(x_1)\,\Delta x_1 + x_2\rho(x_2)\,\Delta x_2 + \cdots + x_n\rho(x_n)\,\Delta x_n}{\rho(x_1)\,\Delta x_1 + \rho(x_2)\,\Delta x_2 + \cdots + \rho(x_n)\,\Delta x_n}. \tag{6}$$

Now, as the lengths of the individual sections of the partition approach zero, $\bar{x}_{approx}$ approaches the center of mass of the continuous rod. That is

$$\bar{x} = \lim_{|P| \to 0} \bar{x}_{approx} = \frac{\displaystyle\int_a^b x\rho(x)\,dx}{\displaystyle\int_a^b \rho(x)\,dx}. \tag{7}$$

As we saw in Chapter 4, $\int_a^b \rho(x)\,dx = $ the mass of the rod. We define $\int_a^b x\rho(x)\,dx$ to be the *first moment of the rod around the y-axis* (the line $x = 0$). We can then rewrite (7) as

$$\text{coordinate of the center of mass} = \frac{\text{first moment around the } y\text{-axis}}{\text{total mass}}. \tag{8}$$

EXAMPLE 2.   The density ρ of a 4-m nonuniform metal beam is given by $\rho(x) = 2\sqrt{x}$ kg/m, where x is the distance along the beam. Find the center of mass of the beam.

SOLUTION. We place the beam along the x-axis from $x = 0$ to $x = 4$. (See Figure 4.) Then

$$\bar{x} = \frac{2\displaystyle\int_0^4 x\sqrt{x}\,dx}{2\displaystyle\int_0^4 \sqrt{x}\,dx} = \frac{\frac{4}{5}x^{5/2}\big|_0^4}{\frac{4}{3}x^{3/2}\big|_0^4} = \frac{\frac{4}{5}\cdot 32}{\frac{4}{3}\cdot 8} = \frac{12}{5} \text{ m.}$$

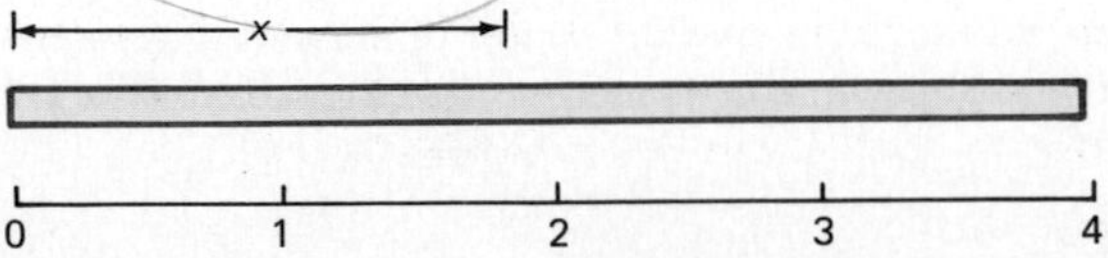

Figure 4

Here the first moment is $128/5$ kg-m and the total mass of the beam is $32/3$ kg.

EXAMPLE 3. The density ρ of a 5-m rod is given by $\rho(x) = \sqrt{25 + x^2}$ kg/m. Find the center of mass of the rod.

SOLUTION

$$\bar{x} = \frac{\displaystyle\int_0^5 x\sqrt{25 + x^2}\, dx}{\displaystyle\int_0^5 \sqrt{25 + x^2}\, dx}.$$

To integrate the denominator, we set $x = 5\tan\theta$. Then

$$\int_0^5 \sqrt{25 + x^2}\, dx = 25 \int_0^{\pi/4} \sec^3\theta\, d\theta$$

$$= \frac{25}{2}(\ln|\sec\theta + \tan\theta| + \sec\theta\tan\theta)\Big|_0^{\pi/4}$$

$$= \frac{25}{2}(\sqrt{2} + \ln(\sqrt{2} + 1)) \quad \text{kg} = \text{the total mass}.$$

Also,

$$\int_0^5 x\sqrt{25 + x^2}\, dx = \frac{1}{3}(25 + x^2)^{3/2}\Big|_0^5 = \frac{1}{3}(50\sqrt{50} - 125)$$

$$= \frac{125}{3}(2\sqrt{2} - 1) \quad \text{kg-m} = \text{the first moment around } x = 0.$$

Thus

$$\bar{x} = \frac{10}{3}\left[\frac{2\sqrt{2} - 1}{\sqrt{2} + \ln(\sqrt{2} + 1)}\right] \quad \text{m} \approx 2.655 \quad \text{m}.$$

(Here the center of mass is slightly more than halfway along the rod.)

Before leaving this section, we note that the first moment is useful in areas other than the study of centers of mass. For example, in Example 5.7.5, we calculated that the work W needed to pump oil to the top of a conical tank was given by

$$W = \int_2^8 9025.2\pi\left(4 - \frac{x}{2}\right)^2 x\, dx.$$

This is the first moment around zero of the "density function" $9025.2\pi[4 - (x/2)]^2$ between $x = 2$ and $x = 8$. The first moment also is useful in probability theory and in many other mathematical applications.

PROBLEMS 9.4

In Problems 1–4 find the first moment around the origin and the center of mass of the system of masses m_i located at the points P_i on the x-axis where m_i is measured in kg and the x-units are meters.

   1. $m_1 = 4$, $m_2 = 6$; $P_1 = (3, 0)$, $P_2 = (-5, 0)$.

2. $m_1 = 3$, $m_2 = 4$, $m_3 = 7$; $P_1 = (2, 0)$, $P_2 = (-3, 0)$, $P_3 = (1, 0)$.

3. $m_1 = 2$, $m_2 = 8$, $m_3 = 5$, $m_4 = 3$; $P_1 = (-6, 0)$, $P_2 = (2, 0)$,
    $P_3 = (20, 0)$, $P_4 = (-1, 0)$.

4. $m_1 = 4$, $m_2 = 1$, $m_3 = 6$, $m_4 = 3$, $m_5 = 10$; $P_1 = (4, 0)$, $P_2 = (10, 0)$,
    $P_3 = (-10, 0)$, $P_4 = (5, 0)$, $P_5 = (-7, 0)$.

In Problems 5–16 a rod of variable density $\rho(x)$ lies along the x-axis in the given interval $[a, b]$.
Find the first moment of the rod about zero, the total mass of the rod, and its center of mass.

5. $\rho(x) = x$; $[0, 1]$

6. $\rho(x) = x^{1/3}$; $[1, 8]$

7. $\rho(x) = x^3$; $[1, 2]$

8. $\rho(x) = \dfrac{1}{x}$; $[1, 3]$

9. $\rho(x) = 2 + \sin x$; $[0, 2\pi]$

10. $\rho(x) = e^x$; $[0, 1]$

11. $\rho(x) = \ln x$; $[1, 4]$

12. $\rho(x) = xe^x$; $[0, 1]$

13. $\rho(x) = x \ln x$; $[1, e]$

14. $\rho(x) = e^x \sin x$; $\left[0, \dfrac{\pi}{2}\right]$

15. $\rho(x) = \dfrac{1}{(x + 1)(x + 2)}$; $[0, 3]$.

16. $\rho(x) = \dfrac{1}{\sqrt{1 - x^2}}$; $\left[0, \dfrac{1}{2}\right]$.

*17. A rod with total mass m of variable density $\rho(x)$ lies along the x-axis between $x = a$ and
    $x = b$. If the rod is split into two pieces at $x = c$, let $\bar{x}_1$ denote the center of mass of the
    first piece over the interval $[a, c]$, let $\bar{x}_2$ denote the center of mass of the second piece
    over the interval $[c, b]$ and let $\bar{x}$ denote the center of mass of the entire (unsplit) rod.
    Show that

$$\bar{x} = \frac{m_1 \bar{x}_1 + m_2 \bar{x}_2}{m_1 + m_2}.$$

18. If the density of a rod is $\rho(x) = x^2$ between $x = 0$ and $x = 3$ and the rod is split at
    $x = 2$, verify the result of Problem 17.

9.5 The Centroid of a Plane Region

We began the last section by considering the center of mass of a finite number of
point masses located along the x-axis. We found that the center of mass $\bar{x}$ was given
by

$$CM, \quad \bar{x} = \frac{\text{first moment of the system around the } y\text{-axis}}{\text{total mass}}. \tag{1}$$

We denote the first moment of the system around the y-axis by M_y and we use the
Greek letter μ to denote the total mass $(m_1 + m_2 + \cdots + m_n)$ of the system. Then
we have

$$\bar{x} = \frac{M_y}{\mu}, \tag{2}$$

Now suppose, instead, that the n masses m_i are located at the points (x_i, y_i) in
the xy-plane. Then we calculate the center of mass by treating separately the x- and
y-coordinates of the points at which the masses are located. For the x-coordinates we

simply use formula (2). We then define the first moment of the system around the x-axis (the line $y = 0$) by

$$M_x = m_1 y_1 + m_2 y_2 + \cdots + m_n y_n, \tag{3}$$

and we have

$$\bar{y} = \frac{M_x}{\mu}. \tag{4}$$

The center of mass of the n particles is then located at the point $(\bar{x}, \bar{y})$. We can think of *the center of mass as a balancing point*. To picture this, imagine that the n masses are placed on a thin, weightless circular plate. Then the plate will balance on a pivot if the pivot is located at the point $(\bar{x}, \bar{y})$. This is illustrated in Figure 1.

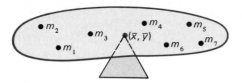

Figure 1

EXAMPLE 1.   Calculate the center of mass of the system of three particles with masses $m_1 = 10$ g, $m_2 = 5$ g, and $m_3 = 8$ g located, respectively, at the points $(2, -1)$, $(3, 2)$, and $(-6, 1)$ (measured in cm).

SOLUTION. We calculate $M_y = 10(2) + 5(3) + 8(-6) = -13$ g-cm, $M_x = 10(-1) + 5(2) + 8(1) = 8$ g-cm and $\mu = 10 + 5 + 8 = 23$ g. Then $\bar{x} = M_y/\mu = -13/23$ cm and $\bar{y} = M_x/\mu = 8/23$ cm and the center of mass is located at the point $(-13/23, 8/23)$.

If the masses in the preceding discussion all have the same mass m, then

$$M_y = mx_1 + mx_2 + \cdots + mx_n = m(x_1 + x_2 + \cdots + x_n)$$
$$M_x = m(y_1 + y_2 + \cdots + y_n)$$

and $\mu = m + m + \cdots + m = nm$ so that

$$\bar{x} = \frac{m(x_1 + x_2 + \cdots + x_n)}{nm} = \frac{x_1 + x_2 + \cdots + x_n}{n} \tag{5}$$

and

$$\bar{y} = \frac{m(y_1 + y_2 + \cdots + y_n)}{nm} = \frac{y_1 + y_2 + \cdots + y_n}{n}. \tag{6}$$

We see that $\bar{x}$ is simply the arithmetic average of the x values, $\bar{y}$ is the arithmetic average of the y values, and the center of mass *does not depend* on the common mass value m. Since, in this setting, the actual value of the mass is irrelevant, we give the center of mass another name: the *centroid*. The centroid is the center of mass in the case of equal masses.

Now let R denote a region in the plane which we can think of as a thin sheet of material with *uniform* (i.e., constant) area density ρ. With this interpretation, the region is called a *plane lamina* and ρ is measured in kg/m², g/cm², lb/ft², or lb/in². (See Figure 2.) If $\rho = 1$, then the total mass of the lamina is equal to the area of the

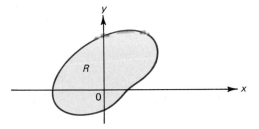

Figure 2

region R. We now calculate the centroid of such a lamina. We can think of the center of mass $(\bar{x}, \bar{y})$ as the pivot point at which the lamina would exactly balance.

Suppose first that R is the region bounded by a continuous, nonnegative function f, the x-axis, and the lines $x = a$ and $x = b$ (see Figure 3). We partition the interval $[a, b]$ in the usual manner:

$$a = x_0 < x_1 < x_2 < \cdots < x_n = b.$$

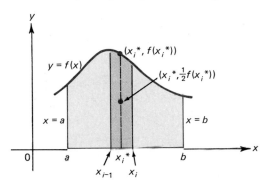

Figure 3

The shaded region in Figure 3 is roughly rectangular in shape. Since density is constant, the center of mass of the rectangular lamina lies in the center of the rectangle which is at the point $(x_i^*, \frac{1}{2}f(x_i^*))$, where $x_i^* = (x_{i-1} + x_i)/2$. The area of the rectangular lamina is, approximately, $f(x_i^*)(x_i - x_{i-1}) = f(x_i^*)\,\Delta x_i$ and the mass of the lamina is

$$\mu_i \approx \rho f(x_i^*)\,\Delta x_i.$$

The first moment of this ith sublamina around the y-axis (mass times average distance to the y-axis) is then given by

$$(M_y)_i \approx \mu_i x_i^* \approx \rho x_i^* f(x_i^*)\,\Delta x_i$$

and the first moment of the entire lamina around the y-axis is defined to be

$$M_y = \lim_{|P| \to 0} [\rho x_1^* f(x_1^*)\,\Delta x_1 + \rho x_2^* f(x_2^*)\,\Delta x_2 + \cdots + \rho x_n^* f(x_n^*)\,\Delta x_n]$$

$$= \rho \int_a^b x f(x)\,dx. \tag{7}$$

Similarly, the first moment of the ith sublamina around the x-axis (mass times average distance to the x-axis) is given by

$$(M_x)_i \approx [\rho f(x_i^*) \, \Delta x_i] \cdot [\tfrac{1}{2} f(x_i^*)] = \frac{\rho}{2} [f(x_i^*)]^2 \, \Delta x_i$$

and the first moment of the entire lamina around the x-axis is defined to be

$$M_x = \lim_{|P| \to 0} \frac{\rho}{2} [f(x_1^*) \, \Delta x_1 + f(x_2^*) \, \Delta x_2 + \cdots + f(x_n^*) \, \Delta x_n]$$

$$= \frac{\rho}{2} \int_a^b [f(x)]^2 \, dx. \tag{8}$$

Since ρ is constant, the mass μ of the lamina is equal to ρ times the area of the region or

$$\mu = \rho \int_a^b f(x) \, dx.$$

Finally, we define the *center of mass* $(\bar{x}, \bar{y})$ of the lamina by

$$\bar{x} = \frac{M_y}{\mu} = \frac{\rho \displaystyle\int_a^b xf(x) \, dx}{\rho \displaystyle\int_a^b f(x) \, dx} = \frac{\displaystyle\int_a^b xf(x) \, dx}{\displaystyle\int_a^b f(x) \, dx} \tag{9}$$

and

$$\bar{y} = \frac{M_x}{\mu} = \frac{(\rho/2) \displaystyle\int_a^b [f(x)]^2 \, dx}{\rho \displaystyle\int_a^b f(x) \, dx} = \frac{\tfrac{1}{2} \displaystyle\int_a^b [f(x)]^2 \, dx}{\displaystyle\int_a^b f(x) \, dx}. \tag{10}$$

Equations (9) and (10) tell us that if the lamina is uniform (i.e., if it has a constant density) then its center of mass depends only on the region and not on the density. As before, we then call the center of mass the *centroid* of the plane region R and we can define the first moments of the *region* R around the x- and y-axes by, respectively,

$$M_y = \int_a^b xf(x) \, dx \qquad \text{and} \qquad M_x = \tfrac{1}{2} \int_a^b [f(x)]^2 \, dx$$

so that (9) and (10) become

$$\bar{x} = \frac{M_y}{A} \qquad \text{and} \qquad \bar{y} = \frac{M_x}{A} \tag{11}$$

where A denotes the area of the region.

EXAMPLE 2. Calculate the centroid of the region in the first quadrant bounded by the curve $y = x^2$, the x-axis and the line $x = 1$.

SOLUTION. We have $M_y = \int_0^1 x^3 \, dx = \frac{1}{4}$, $M_x = \frac{1}{2}\int_0^1 x^4 \, dx = \frac{1}{10}$ and $A = \int_0^1 x^2 \, dx = \frac{1}{3}$. Then

$$\bar{x} = \frac{M_y}{A} = \frac{\frac{1}{4}}{\frac{1}{3}} = \frac{3}{4} \qquad \text{and} \qquad \bar{y} = \frac{\frac{1}{10}}{\frac{1}{3}} = \frac{3}{10}.$$

The centroid is the point $(\frac{3}{4}, \frac{3}{10})$. This is depicted in Figure 4.

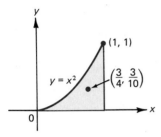

Figure 4

EXAMPLE 3. Calculate the centroid of the region bounded by $y = \sin x$ and the x-axis for x in $[0, \pi/2]$.

SOLUTION. We have

$$M_y = \int_0^{\pi/2} x \sin x \, dx$$

$$= -x \cos x \Big|_0^{\pi/2} + \int_0^{\pi/2} \cos x \, dx = 1 \quad \text{(by integration by parts)},$$

$$M_x = \frac{1}{2}\int_0^{\pi/2} \sin^2 x \, dx$$

$$= \frac{1}{4}\int_0^{\pi/2} (1 - \cos 2x) \, dx = \frac{1}{4}\left(x - \frac{\sin 2x}{2}\right)\Big|_0^{\pi/2} = \frac{\pi}{8},$$

and

$$A = \int_0^{\pi/2} \sin x \, dx = -\cos x \Big|_0^{\pi/2} = 1.$$

Then $(\bar{x}, \bar{y}) = (1/1, (\pi/8)/1) = (1, \pi/8)$. This is depicted in Figure 5.

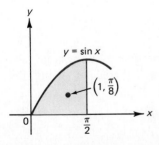

Figure 5

EXAMPLE 4.   Find the centroid of the region bounded by the curve $y = 3x^2 - 6x + 4$, the x- and y-axes, and the line $x = 2$.

SOLUTION. We have

$$M_y = \int_0^2 (3x^3 - 6x^2 + 4x) \, dx = 4,$$

$$M_x = \tfrac{1}{2} \int_0^2 (3x^2 - 6x + 4)^2 \, dx$$

$$= \tfrac{1}{2} \int_0^2 (9x^4 - 36x^3 + 60x^2 - 48x + 16) \, dx = \tfrac{24}{5},$$

and

$$A = \int_0^2 (3x^2 - 6x + 4) \, dx = 4$$

so that $(\bar{x}, \bar{y}) = (1, \tfrac{6}{5})$. This is sketched in Figure 6. Note that, in this case, the centroid lies *outside* the region under consideration.

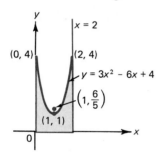

Figure 6

We now generalize these results. Let the region R be bounded by the two curves $f(x)$ and $g(x)$ where both f and g are continuous. Then the situation is as depicted in Figure 7. If x ranges over the interval $[a, b]$, then we again partition the interval and

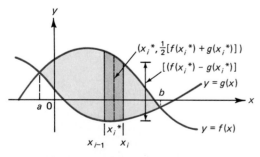

Figure 7

find that the centroid of a typical rectangular shaped subregion is, approximately, $(\bar{x}_i, \bar{y}_i) = (x_i^*, \tfrac{1}{2}(f(x_i^*) + g(x_i^*)))$, where $x_i^* = \tfrac{1}{2}[x_{i-1} + x_i]$. Then, replacing the function f in equation (7) by $f - g$, we find that

$$\boxed{\; \bar{x} = \frac{M_y}{A} \qquad \text{and} \qquad \bar{y} = \frac{M_x}{A} \;}$$

(12)

where

$$M_y = \int_a^b x[f(x) - g(x)] \, dx, \tag{13}$$

$$M_x = \frac{1}{2} \int_a^b [f(x) + g(x)][f(x) - g(x)] \, dx$$

$$M_x = \frac{1}{2} \int_a^b \{[f(x)]^2 - [g(x)]^2\} \, dx, \tag{14}$$

and

$$A = \int_a^b [f(x) - g(x)] \, dx. \tag{15}$$

The expression for M_x follows from the fact that the area of the ith subregion is $\Delta x_i [f(x_i^*) - g(x_i^*)]$.

EXAMPLE 5. Find the centroid of the region bounded by the curves $y = x^3$ and $y = \sqrt{x}$.

SOLUTION. The two curves intersect at $(0, 0)$ and $(1, 1)$. Since $\sqrt{x} \geq x^3$ on $[0, 1]$, we have

$$M_y = \int_0^1 x(\sqrt{x} - x^3) \, dx = \int_0^1 (x^{3/2} - x^4) \, dx = \frac{1}{5}$$

$$M_x = \frac{1}{2} \int_0^1 (x - x^6) \, dx = \frac{5}{28}$$

and

$$A = \int_0^1 (\sqrt{x} - x^3) \, dx = \frac{5}{12}$$

so that $\bar{x} = \frac{12}{25}$ and $\bar{y} = \frac{3}{7}$. This is sketched in Figure 8.

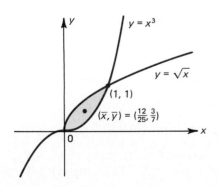

$y = x^3$

$y = \sqrt{x}$

$(1, 1)$

$(\bar{x}, \bar{y}) = (\frac{12}{25}, \frac{3}{7})$

Figure 8

EXAMPLE 6. Calculate the centroid of the region bounded by $y = \sin x$, $y = \cos x$ and the lines $x = \pi/2$ and $x = \pi$.

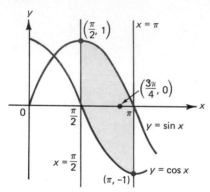

Figure 9

SOLUTION. The region is sketched in Figure 9. We have

$$M_y = \int_{\pi/2}^{\pi} x(\sin x - \cos x)\, dx$$

$$= -x(\cos x + \sin x)\Big|_{\pi/2}^{\pi} + \int_{\pi/2}^{\pi}(\cos x + \sin x)\, dx$$

$$= \frac{3\pi}{2} + (\sin x - \cos x)\Big|_{\pi/2}^{\pi} = \frac{3\pi}{2},$$

$$M_x = \frac{1}{2}\int_{\pi/2}^{\pi}(\sin^2 x - \cos^2 x)\, dx$$

$$= \frac{1}{4}\int_{\pi/2}^{\pi}[(1 - \cos 2x) - (1 + \cos 2x)]\, dx = -\frac{1}{2}\int_{\pi/2}^{\pi}\cos 2x\, dx = 0$$

and

$$A = \int_{\pi/2}^{\pi}(\sin x - \cos x)\, dx = 2$$

so that $(\bar{x}, \bar{y}) = (3\pi/4, 0)$.

The following theorem whose proof is suggested in Problems 39 and 40 can be very useful in the calculation of centroids.

THEOREM 1. (a)   If the plane region R is symmetric about the line $x = c$, then $\bar{x} = c$.
  (b)   If the plane region R is symmetric about the line $y = d$, then $\bar{y} = d$.

EXAMPLE 7.   Calculate the centroid of the semicircle $y = \sqrt{16 - x^2}$.

SOLUTION. The semicircle (half the circle $x^2 + y^2 = 16$) is sketched in Figure 10. Since the region is symmetric about the y-axis (the line $x = 0$), we have $\bar{x} = 0$. We then calculate

$$M_x = \frac{1}{2}\int_0^4 (16 - x^2)\, dx = \frac{64}{3}$$

and since $A = \frac{1}{2}\pi r^2 = 8\pi$, we find that $\bar{y} = 8/3\pi$.

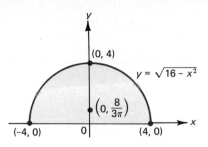

Figure 10

1 - 37

PROBLEMS 9.5

In Problems 1–4 find the center of mass of the system of masses m_i located at the points P_i, where each m_i is measured in grams and x and y units are centimeters.

1. $m_1 = 4$, $m_2 = 6$; $P_1 = (3, 4)$, $P_2 = (-5, 3)$.
2. $m_1 = 3$, $m_2 = 4$, $m_3 = 7$; $P_1 = (2, 3)$, $P_2 = (-3, -4)$, $P_3 = (1, 2)$.
3. $m_1 = 2$, $m_2 = 8$, $m_3 = 5$, $m_4 = 3$; $P_1 = (-6, 2)$, $P_2 = (2, 3)$,
    $P_3 = (20, -5)$, $P_4 = (-1, 8)$.
4. $m_1 = 4$, $m_2 = 1$, $m_3 = 6$, $m_4 = 3$, $m_5 = 10$; $P_1 = (4, 7)$,
    $P_2 = (10, -3)$, $P_3 = (-10, 0)$, $P_4 = (0, 5)$, $P_5 = (6, -6)$.

In Problems 5–20 calculate the centroid of the plane region bounded by the given curve and the x-axis over the indicated interval.

5. $y = 2x + 3$; $[0, 1]$ 6. $y = x^3$; $[1, 2]$ 7. $y = x^{1/3}$; $[0, 1]$

8. $y = x^4$; $[-1, 2]$ 9. $y = \sqrt{10 + x}$; $[-1, 6]$

*10. $y = x\sqrt{x^2 + 9}$; $[0, 3]$ 11. $y = \dfrac{1}{x + 1}$; $[0, 1]$ 12. $y = e^x$; $[0, 1]$

13. $y = \dfrac{1}{(x + 1)(x + 2)}$; $[0, 1]$ 14. $y = \sin 2x$; $\left[0, \dfrac{\pi}{4}\right]$

15. $y = \cos 3x$; $\left[0, \dfrac{\pi}{9}\right]$ 16. $y = \sin^2 x$; $\left[0, \dfrac{\pi}{2}\right]$ 17. $y = \ln x$; $[1, 2]$

18. $y = x^{3/5}$; $[0, 32]$ 19. $y = \dfrac{x}{x - 1}$; $[-1, 0]$ 20. $y = \dfrac{1}{x^2 + 1}$; $[0, 1]$.

21. Find the centroid of the region in the first quadrant bounded by the curve $x = y^2 + 1$ and the line $y = 1$.
*22. Find the centroid of the region in the first quadrant bounded by the curve $x^{2/3} + y^{2/3} = 1$.

In Problems 23–24 find the centroids of the plane regions bounded by the given curves and lines.

                            $x = 0$

23. $y = x^2$, $y = 4x$, $x = 1$ 24. $y = x^2 + 3x + 5$, $y = 2x^2 - x + 8$

25. $y = 2x^5 + 3$, $y = 32x + 3$, $x \ge 0$ 26. $y^2 = 4 + x$, $x + 2y = 4$

27. $y = x^2$, $y = x^3$, $x = 3$, $x \ge 1$ 28. $xy = 1$, $x = 5$, $y = 5$

29. $x + y^2 = 8$, $x + y = 2$ 30. $y = x^{1/3}$, $y = x$

31. $y = \sin x$, $y = \cos x$, $x = 0$, $x = \dfrac{\pi}{4}$ 32. $y = e^{2x}$, $y = e^{-x}$, $x = 2$

33. $y = \ln x$, $y = e^x$; $[1, 2]$ 34. $y = \ln x$, $y = e^{-x}$; $[2, 3]$.

35. Show that the centroid of the ellipse $(x^2/a^2) + (y^2/b^2) = 1$ is the origin.
36. Use Theorem 1 to calculate the centroid of the region bounded by the curve $y = x^8$ and the line $y = 1$.
37. Use Theorem 1 to calculate the centroid of the region bounded by $y = \cos x$ and the x-axis in the interval $[\pi/2, 3\pi/2]$.
38. Find the center of mass of the plane lamina bounded by the curve $y = x^2$, the y-axis and the line $y = 1$, if the area density at any point (x, y) in the lamina is given by $\rho(x, y) = \sqrt{1 - x}\, g/cm^2$. (Assume that units on the x-axis are centimeters.)
*39. Prove Theorem 1(a). [Hint: Divide the interval $[a, b]$ into n subintervals, half of which lie to the left of c and half of which lie to the right. Then treat the line $x = c$ as if it were the y-axis and show that $\bar{x} = 0$.]
*40. Prove Theorem 1(b).
**41. Prove the following special case of the first theorem of Pappus: If the plane region R is revolved about the y-axis and if the y-axis does not intersect R, then the volume generated is equal to the product of the area of R and the length of the circumference of the circle traced by the centroid of R. [Hint: The length of the circumference of the circle traced by the centroid is $2\pi\bar{x}$ so that it is necessary to show that $V = 2\pi\bar{x}A$. To prove this, write down the formula for $\bar{x}$ and A and compare $2\pi\bar{x}A$ with the formula for calculating V by the method of cylindrical shells.]
**42. Prove that the statement of the first theorem of Pappus in Problem 41 holds if R is revolved about any line which does not intersect it.
43. Use the theorem of Pappus to calculate the volume of the torus generated by rotating the circle $(x - a)^2 + y^2 = r^2$ $(r < a)$ about the y-axis. (Compare this with Problem 9.1.34.)
44. Use the first theorem of Pappus to calculate the volume of the solid generated by rotating the triangle with vertices $(-1, 2)$, $(1, 2)$ and $(0, 4)$ about the x-axis.
*45. Use the first theorem of Pappus to calculate the volume of the torus generated by rotating the unit circle about the line $y = 4 - x$.
*46. Use the first theorem of Pappus to calculate the volume of the "elliptical torus" generated by rotating the ellipse $(x^2/a^2) + (y^2/b^2) = 1$ about the line $y = 3a$. Assume that $3a > b > 0$.

*9.6 Moments of Inertia and Kinetic Energy

In Section 4 we defined the first moment around the y-axis (the line $x = 0$) of n masses by

$$M_y = m_1x_1 + m_2x_2 + \cdots + m_nx_n.$$

There, we used the fact that the force exerted by an object on a pivot point is proportional to the mass of the object times its distance from the pivot.

For certain types of problems other kinds of moments are more useful. We define the second moment of the system of n masses around the y-axis by

$$I_y = m_1x_1^2 + m_2x_2^2 + \cdots + m_nx_n^2. \tag{1}$$

EXAMPLE 1. Calculate the second moment I_y of the system of 3 masses $m_1 = 5$ kg, $m_2 = 3$ kg, $m_3 = 4$ kg situated along the x-axis at the points $(3, 0)$, $(-2, 0)$ and $(4, 0)$, respectively, where the units along the x-axis are meters.

SOLUTION. $I_y = 5(3)^2 + 3(-2)^2 + 4(4)^2 = 121$ kg-m^2.

Now consider the motion of a rotating particle of mass m at a distance r from an axis of rotation. (See Figure 1.) The *angular velocity* ω of the particle is measured in

Figure 1

radians per unit time. Thus, for example, if $\omega = 4\pi$ and the unit of time is the second, then the particle makes two complete rotations in one second ($4\pi = 2 \cdot 2\pi$) since there are 2π radians in a circle. The velocity of the particle is then given by

$$v = \omega r \tag{2}$$

since ωr is the distance traveled in one unit of time. Furthermore, the kinetic energy (see Section 5.7) is given by

$$K = \tfrac{1}{2}mv^2 = \tfrac{1}{2}mr^2\omega^2. \tag{3}$$

Suppose that a rotating rigid body contains n particles of masses m_i at distances r_i, respectively, from the axis of rotation. Then the total kinetic energy of the body is given by

$$K = \tfrac{1}{2}(m_1 r_1^2 + m_2 r_2^2 + \cdots + m_n r_n^2)\omega^2. \tag{4}$$

The quantity

$$\boxed{I = m_1 r_1^2 + m_2 r_2^2 + \cdots + m_n r_n^2} \tag{5}$$

is called the *rotational inertia* or *moment of inertia* of the body about the axis of rotation. Note that this quantity is equal to the second moment defined by (1) and depends upon the choice of axis of rotation as well as the shape of the body and the distribution of masses along it.

EXAMPLE 2. A dumbbell consists of two 10-kg weights separated by a 1-m rod. Neglecting the weight of the rod and treating the weights as point masses, calculate the moment of inertia of the dumbbell first about an axis perpendicular to it and passing through the center of the rod; and second, about an axis passing through one of the weights.

SOLUTION. The situation is depicted in Figure 2. We can think of the axis in both cases as a line perpendicular to the plane of the page. In the first case, $m_1 = m_2 = 10$ and $r_1 = r_2 = \tfrac{1}{2}$ so that $I = 10(\tfrac{1}{2})^2 + 10(\tfrac{1}{2})^2 = 5$ kg-m². In the second case, we assume the

Figure 2

axis passes through the mass on the left (the other case leads to the same result). Then $m_1 = m_2 = 10$, $r_1 = 0$ and $r_2 = 1$ so that $I = 10(1)^2 = 10$ kg-m². We see that the rotational inertia is twice as great about an axis through one end as it is about an axis through the center of the dumbbell.

The above ideas can easily be generalized to calculate the moment of inertia of a "continuous" body rotating about a given axis.†

EXAMPLE 3. A rigid 4-m rod spins about its left end 50 times a minute. Its density is given by $\rho(x) = (1 + \sqrt{x})$ kg/m. Find its kinetic energy.

SOLUTION. The rod is sketched in Figure 3. From equation (4) the kinetic energy is

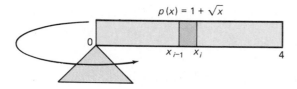

$p(x) = 1 + \sqrt{x}$

Figure 3

given by $K = \frac{1}{2}I\omega^2$. In this problem $\omega = 50(2\pi) = 100\pi$ radians/min. If we partition the interval $[0, 4]$, we may treat each of the n subintervals as point masses. The mass of the ith subinterval is approximately equal to $\rho(x_i)(x_{i-1} - x_i) = (1 + \sqrt{x_i})\,\Delta x_i$. Then the moment of inertia of the ith subinterval is given by

$$I_i \approx m_i r_i{}^2 = [(1 + \sqrt{x_i})\,\Delta x_i]x_i{}^2$$

so that

$$I \approx (1 + \sqrt{x_1})x_1{}^2\,\Delta x_1 + (1 + \sqrt{x_2})x_2{}^2\,\Delta x_2 + \cdots + (1 + \sqrt{x_n})x_n{}^2\,\Delta x_n.$$

Taking the limit as $|P| \to 0$ we obtain

$$I = \int_0^4 (1 + \sqrt{x})x^2\,dx = \int_0^4 (x^2 + x^{5/2})\,dx = \frac{1216}{21}\quad \text{kg-m}^2.$$

Then

$$K = \frac{1}{2}I\omega^2 = \frac{1}{2}\cdot\frac{1216}{21}(100\pi)^2 = \frac{6{,}080{,}000}{21}\pi^2\quad\text{kg-m}^2/\text{min}^2$$

$$\approx 794\quad\text{nt-m} = 794\quad\text{joules.}\ddagger$$

The technique of the above example can be used to calculate moments of inertia for two-dimensional regions.

EXAMPLE 4. The region bounded by the curve $y = \sqrt{x}$, the x-axis and the line $x = 4$ is rotated about the y-axis. Find the moment of inertia if the region has the constant area density of 3 kg/m².

SOLUTION. The region is sketched in Figure 4. We partition the interval $[0, 4]$. We can

†For example, the moment of inertia of the earth about its polar axis is, approximately, $I = 8.08 \times 10^{37}$ kg-m².
‡1 nt = 1 kg-m/sec² = 3600 kg-m/min²; 1 joule = 1 nt-m.

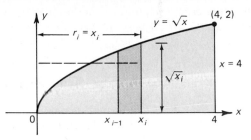

Figure 4

then think of the rotating body as a system of n rotating "rectangularly shaped" masses. The mass m_i of the ith subregion is equal to its area times the density or

$$m_i = 3\sqrt{x_i}\,\Delta x_i.$$

Also, we see from Figure 4 that $r_i \approx x_i$ so that

$$I_i = m_i r_i^2 = 3x_i^{5/2}\,\Delta x_i$$

and, from (4),

$$I \approx 3(x_1^{5/2}\,\Delta x_1 + x_2^{5/2}\,\Delta x_2 + \cdots + x_n^{5/2}\,\Delta x_n)$$

so that

$$I = \lim_{|P| \to 0} 3(x_1^{5/2}\,\Delta x_1 + x_2^{5/2}\,\Delta x_2 + \cdots + x_n^{5/2}\,\Delta x_n)$$

$$= 3\int_0^4 x^{5/2}\,dx = \frac{768}{7}\ \text{kg-m}^2.$$

EXAMPLE 5. Calculate the moment of inertia when the region in Example 4 is rotated about the x-axis.

SOLUTION. The region is the same as before but now we must use horizontal strips to obtain a rotating strip of almost constant radius. This is illustrated in Figure 5. We

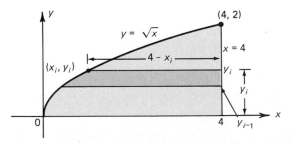

Figure 5

partition the interval $[0, 2]$ along the y-axis to obtain n horizontal rectangular subregions. The mass of each subregion is given by

$$m_i = \text{area} \times \text{density} = 3(4 - x_i)\,\Delta y_i = 3(4 - y_i^2)\,\Delta y_i.$$

Since $r_i = y_i$, we have

$$I_i = m_i r_i^2 = 3(4 - y_i^2)y_i^2\,\Delta y_i$$

and

$$I = 3 \int_0^2 (4 - y^2) y^2 \, dy = \frac{64}{5} \quad \text{kg-m}^2.$$

The above technique can be used in some problems where density is not constant.

EXAMPLE 6. A plane lamina takes the shape of the region bounded by $y = \sin x$, the x- and y-axes and the line $x = \pi/2$. At any point (x, y) on the lamina, the density is given by $\rho(x, y) = x^2 \, \text{gm/cm}^2$. If the lamina is rotated about the y-axis, calculate the moment of inertia if the x- and y-units are centimeters.

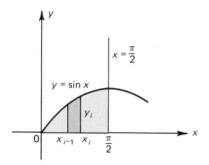

Figure 6

SOLUTION. The lamina is sketched in Figure 6. Here we have

$$m_i = \text{area} \times \text{density} \approx y_i \, \Delta x_i \cdot x_i^2 = x_i^2 \sin x_i \, \Delta x_i$$

and $r_i = x_i$ so that $m_i r_i^2 = x_i^4 \sin x_i \, \Delta x_i$ and

$$I = \int_0^{\pi/2} x^4 \sin x \, dx \qquad \text{(after integrating by parts four times)}$$

$$= \left. (-x^4 \cos x + 4x^3 \sin x + 12x^2 \cos x - 24x \sin x - 24 \cos x) \right|_0^{\pi/2}$$

$$= \left(\frac{\pi^3}{2} - 12\pi + 24 \right) \text{g-cm}^2.$$

EXAMPLE 7. A semicircular lamina of constant density ρ centered at the origin is rotated about the x-axis. Calculate the moment of inertia.

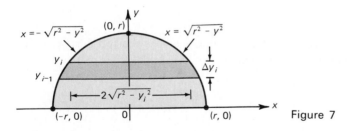

Figure 7

SOUTION. From Figure 7 we see that if $x^2 + y^2 = r^2$ is the equation of the circle, then

$$m_i = (\sqrt{r^2 - y_i^2} - (-\sqrt{r^2 - y_i^2})) \, \Delta y_i \, \rho = 2\rho \sqrt{r^2 - y_i^2} \, \Delta y_i$$

and $r_i = y_i$ so that $m_i r_i^2 = 2\rho y_i^2 \sqrt{r^2 - y_i^2} \, \Delta y_i$ and

$$I = 2\rho \int_0^r y^2 \sqrt{r^2 - y^2} \, dy.$$

To integrate this we set $y = r \sin \theta$ to find that

$$I = 2\rho \int_0^{\pi/2} r^2 \sin^2 \theta \, (r \cos \theta) \, r \cos \theta \, d\theta = 2\rho r^4 \int_0^{\pi/2} \sin^2 \theta \cos^2 \theta \, d\theta$$

$$= \frac{\rho r^4}{2} \int_0^{\pi/2} (1 - \cos 2\theta)(1 + \cos 2\theta) \, d\theta = \frac{\rho r^4}{2} \int_0^{\pi/2} (1 - \cos^2 2\theta) \, d\theta$$

$$= \frac{\rho r^4}{2} \int_0^{\pi/2} \left[1 - \left(\frac{1 + \cos 4\theta}{2} \right) \right] d\theta = \frac{\rho r^4}{2} \int_0^{\pi/2} \left(\frac{1}{2} - \frac{\cos 4\theta}{2} \right) d\theta = \frac{\rho \pi r^4}{8}.$$

EXAMPLE 8. The lamina of Example 7 rotates 20 times per second. Find the kinetic energy.

SOLUTION. Here $\omega = 20(2\pi) = 40\pi$ so that

$$K = \frac{1}{2} I\omega^2 = \frac{1}{2} \left(\frac{\rho \pi r^4}{8} \right) (40\pi)^2 = 100\rho \pi^3 r^4.$$

We conclude this section by defining another interesting physical quantity. If one particle rotates about an axis, then we have seen that $I = mr^2$ or

$$r = \sqrt{\frac{I}{m}}. \tag{6}$$

This quantity is called the *radius of gyration*. If there are more particles, we may define the radius of gyration by using equation (5). If a plane lamina is rotated about an axis and if r denotes its radius of gyration, then in its rotation the lamina exhibits the same behavior as if it were a point mass located r units from the axis.

EXAMPLE 9. Calculate the radius of gyration of the region in Example 4.

SOLUTION. We have $r = \sqrt{I/m}$, where $I = (768/7)$ kg-m². Also, the mass of the region is the product of its density and its area or

$$m = 3 \int_0^4 \sqrt{x} \, dx = 3 \cdot \tfrac{16}{3} = 16 \quad \text{kg.}$$

Then

$$r = \sqrt{\frac{768/7 \text{ kg-m}^2}{16 \text{ kg}}} = \sqrt{\frac{48}{7}} \quad \text{m.}$$

PROBLEMS 9.6

In Problems 1–3 masses are located on the x-axis at the points P_i. Find the moment of inertia about the y-axis.

1. $m_1 = 3$ kg, $m_2 = 2$ kg; $P_1 = (4, 0)$, $P_2 = (-6, 0)$ (measured in m).
2. $m_1 = 4$ g, $m_2 = 1$ g, $m_3 = 5$ g; $P_1 = (-3, 0)$, $P_2 = (-4, 0)$,
   $P_3 = (7, 0)$ (measured in cm).
3. $m_1 = 2$ kg, $m_2 = 3$ kg, $m_3 = 42$ kg, $m_4 = 3$ kg; $P_1 = (4, 0)$,
   $P_2 = (3, 0)$, $P_3 = (-4, 0)$, $P_4 = (-3, 0)$ (measured in m).

4. A dumbbell is composed of two 8-kg masses separated by a bar 1-m long. Ignoring the weight of the bar and treating the masses as point masses, calculate the moment of inertia of the dumbbell about an axis perpendicular to it and passing through
   (a) the center of the rod.
   (b) a point one-third of the way along the rod.

5. Find the kinetic energy in each case of Problem 4 if the dumbbell rotates 25 times per minute.

6. Find the radius of gyration in each case of Problem 4.

7. A dumbbell is unevenly weighted with masses of 5 and 7 kg, respectively, separated by a 1-m bar. Ignoring the weight of the bar and treating the masses as point masses, calculate the moment of inertia of the dumbbell about an axis perpendicular to it and passing through
   (a) the center of the rod. 　　　　　　　(b) the 5-kg mass.
   (c) the 7-kg mass.
   (d) a point one-quarter of the way along the rod, closer to the 5-kg mass.

8. Find the kinetic energy in each case of Problem 7 if the dumbbell rotates 10 times per second.

9. Find the radius of gyration in each case of Problem 7.

10. A 3-m rod with density $\sqrt{16 - x^2}$ kg/m (x being measured from the "left" end) is rotated about its left end. Find the moment of inertia and radius of gyration.

11. Find the kinetic energy if the rod in Problem 10 rotates 20 times per second.

In Problems 12–31 find the moment of inertia and the radius of gyration when the region bounded by the given curves and lines with given constant area density ρ is rotated about the given axis.

12. The rectangle bounded by $x = 0$, $x = 2$, $y = 0$ and $y = 3$; $\rho = 2$; x-axis
13. The region of Problem 12 about the y-axis
14. $y = x^2$, $y = 0$, $x = 1$; $\rho = 3$; x-axis
15. The region of Problem 14 about the y-axis.
16. The region of Problem 14 about the line $x = 2$.
17. The region of Problem 14 about the line $y = -3$
18. $y = \cos x$, $x = 0$, $y = 0$, $x = \pi/2$; $\rho = 2$; x-axis
19. $y = \cos x$, $x = 0$, $x = \pi/2$; $\rho = 2$; x-axis
20. $y = 1/x$, $y = 0$, $x = 1$, $x = 2$; $\rho = 4$; x-axis
21. $y = 1/x$, $y = 0$, $x = 1$, $x = 2$; $\rho = 4$; y-axis
22. $y = e^x$, $y = 0$, $x = 0$, $x = 1$; $\rho = 5$; x-axis
23. $y = e^x$, $y = 0$, $x = 0$, $x = 1$; $\rho = 5$; y-axis
24. $\sqrt{x} + \sqrt{y} = 1$, $y = 0$, $x = 0$; $\rho = 1$; x-axis
25. The circle $(x - 2)^2 + y^2 = 1$; $\rho = 2$; y-axis

26. The upper half of the ellipse $\dfrac{x^2}{a^2} + \dfrac{y^2}{b^2} = 1$; $\rho = 4$; y-axis

27. The circle $(x - 2)^2 + y^2 = 1$; $\rho = 2$; the line $x = -4$
28. The circle $(x - 2)^2 + y^2 = 1$; $\rho = 3$; the line $y = 5$
29. $x = y^2$, $x = 0$, $y = 2$; $\rho = 7$; x-axis
30. $x = y^2$, $y = 0$, $x = 0$, $y = 2$; $\rho = 7$; y-axis
31. The quarter circle $x^2 + y^2 = 1$ in the first quadrant; $\rho = 3$; the line $y = -x$.

32. An aluminum triangle with constant density ρ and vertices at $(0, 0)$, $(0, 4)$, and $(-3, 0)$ is spun with angular velocity ω about the x-axis. Calculate its kinetic energy (assume that the units are kilograms and meters).
33. Calculate the kinetic energy of the triangle in Problem 32 if it is spun about the y-axis.
34. Calculate the kinetic energy in Problems 12, 14, 18, 20, 22, 24, 25, 29, and 31 if the region rotates 8 times per second (assume that the units are kilograms and meters).

*9.7 Fluid Pressure

In this section we show how the integral can be used to calculate fluid pressures. By a *fluid* we mean a substance that can flow. This definition includes both liquids and gases. It is reasonable only to talk about forces applied to the *surface* of a fluid. If the fluid is at rest, then such forces can be applied only perpendicular to the surface (Figure 1a) since if a force were applied tangentially (i.e., from the side as in Figure 1b) then layers of fluid would simply slide over one another, and the fluid would no longer be at rest, as we have assumed.

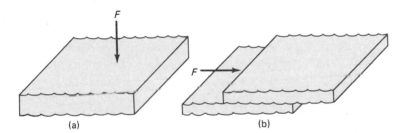

(a) (b) Figure 1

When a force is applied perpendicular to the surface of a fluid it is called a *normal* force. In this case, we define the *pressure P* to be the magnitude of the normal force per unit of area. Commonly used units of pressure are lb/ft^2, $lb/in.^2$, nt/m^2 and nt/cm^2. For example, the air pressure at sea level is $14.7 \, lb/in^2 = 1.013 \times 10^5 \, nt/m^2$.

Now suppose that a thin horizontal lamina is submerged at a certain depth in a fluid and we wish to calculate the total force exerted on the lamina by the column of the fluid on one side of the lamina. The force is given by

$$F = PA, \tag{1}$$

where P is the pressure and A is area. Then

$$P = \frac{F}{A} = \frac{mg}{A}, \tag{2}$$

where g is the force due to gravity. Over a unit of area, the mass of a fluid is given by the product of its density ρ (in $lb/in.^3$, lb/ft^3, kg/m^3 or g/cm^3) times the depth d of the column of fluid.

If the mass is measured in slugs (1 slug = 1 lb/g) then

$$P = \rho d.$$ (3)

If mass is measured in kilograms, then

$$P = \rho dg,$$ (4)

where $g = 9.81$ m/sec². Note that 1 kg-m/sec² = 1 nt.

EXAMPLE 1. A swimming pool with length 5 m, width 8 m and depth 3 m is filled with water. What is the force exerted on the bottom of the pool by the weight of the water?

SOLUTION. The density of water (at 0 °C) is, approximately,

$$\rho = 1000 \quad \text{kg/m}^3.$$

From (4),

$$P = \rho dg = (1000 \text{ kg/m}^3)(3 \text{ m})(9.81 \text{ m/sec}^2) = 29,430 \quad \frac{\text{kg-m/sec}^2}{\text{m}^2}$$

$$= 29,430 \quad \text{nt/m}^2.$$

Then, using (1)

$$F = PA = (29,430 \text{ nt/m}^2)(5 \times 8 \text{ m}^2) = 1,177,200 \quad \text{nt.}$$

EXAMPLE 2. A rectangular sheet of metal 10 ft × 14 ft is submerged horizontally in water to a depth of 12 ft. What is the force exerted on one side of the sheet of metal by the weight of the water?

SOLUTION. The density of water at 0 °C is, approximately,

$$\rho = 62.4 \quad \text{lb/ft}^3.$$

Using (3), we have

$$P = \rho d = (62.4 \text{ lb/ft}^3) \cdot (12 \text{ ft}) = 748.8 \quad \text{lb/ft}^2.$$

Then

$$F = PA = (748.8 \text{ lb/ft}^2)(10 \times 14 \text{ ft}^2) = 104,832 \quad \text{lb.}$$

EXAMPLE 3. *Archimedes' principle*† states that a body immersed in a fluid is buoyed up by a force equal to the weight of the fluid displaced. In the case of a floating object, Archimedes' principle states that a floating body must displace its own weight in water. To demonstrate Archimedes' principle, let a block of width w and area A (see Figure 2) be submerged a distance h in a liquid of density ρ. By equation (4), the force $F_\uparrow$ on the bottom of the block points up and is given by

$$F_\uparrow = PA = \rho g(w + h)A.$$

† It was this discovery that is reputed to have led Archimedes to jump from his bath and run home naked, shouting "Eureka" ("I have found it.").

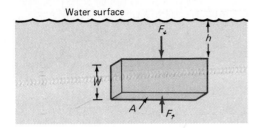

Water surface

Figure 2

Similarly, the force $F_\downarrow$ on the top of the block points down and is given by

$$F_\downarrow = PA = \rho ghA.$$

The net (or resultant) force is given by

$$F_\uparrow - F_\downarrow = \rho g(w + h)A - \rho ghA = \rho gwA = m_l g,$$

where m_l is the total mass of the liquid displaced by the block (since the mass of the block is its volume wA times its density ρ). Thus the block undergoes an upward force equal to the weight of the water it displaces.

In the problems discussed above, we calculated the force of the liquid on a plane surface. That is, water pressure was applied uniformly over a flat surface. However, if a body is submerged in a fluid, then the lateral pressure (pressure on the wall or side of the body) varies with the depth of the fluid so that formulas (3) or (4) cannot be used directly.

EXAMPLE 4. A trough 3 m in length and filled with water has a cross section in the shape of a trapezoid with lower base 2 m, upper base 4 m and a depth of $1\frac{1}{2}$ m. Find the total force due to water pressure on one end of the trough.

SOLUTION. One end of the trough is sketched in Figure 3. To facilitate our calculations, we put in x- and y-coordinates as shown. Then we partition the interval $[0, \frac{3}{2}]$

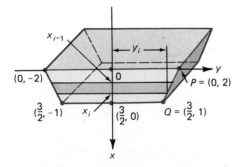

Figure 3

along the "x-axis." The force against the ith strip (shaded) is equal to the pressure along the strip times the area of the strip, and (since pressure is fairly contant on the strip if Δx_i is small) we have

$$F_i = P_i A_i \approx \rho d_i g A_i \qquad \text{(from (4))}.$$

But $\rho = 1000 \text{ kg/m}^3$, $d_i \approx x_i$, $g = 9.81 \text{ m/sec}^2$ and $A_i \approx 2y_i \Delta x_i$. Now we calculate y as a function of x on the edge of the trapezoid. The line PQ on the edge passes through the points $(0, 2)$ and $(\frac{3}{2}, 1)$, leading to the equation

$$\frac{y-2}{x} = \frac{1-2}{\frac{3}{2}} \qquad \text{or} \qquad y = -\frac{2}{3}x + 2.$$

Thus

$$F_i = \rho g x_i (2y_i) \Delta x_i = \rho g x_i \left(-\frac{4}{3}x_i + 4\right) \Delta x_i = 4\rho g \left(x_i - \frac{x_i^2}{3}\right) \Delta x_i.$$

Finally, we let the norm of the partition approach zero and we obtain

$$F = 4\rho g \int_0^{3/2} \left(x - \frac{x^2}{3}\right) dx = 4\rho g \left(\frac{3}{4}\right) = 3\rho g = 3(1000)(9.81) = 29{,}430 \quad \text{nt.}$$

EXAMPLE 5. A cylindrical barrel whose end has a diameter of 4 ft is submerged in seawater (density approximately 64.3 lb/ft³ at 0 °C). Find the total force due to water pressure at one end if the center of the barrel is at a depth of 12 ft.

SOLUTION. One end of the barrel is sketched in Figure 4. We introduce a coordinate

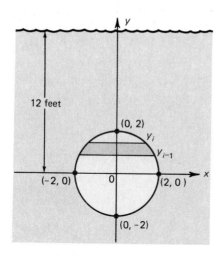

Figure 4

system which places the origin at the center of the barrel. If we now partition the "y-axis" as shown, then the force on a typical horizontal strip is given by

$$F_i = P_i A_i \approx \rho d_i A_i \qquad \text{(from (3))}.$$

But $\rho = 64.3 \text{ lb/ft}^3$, $d_i \approx 12 - y_i$ and $A_i \approx$ (length of strip) $\times (\Delta y_i)$. Since the equation of the circular end is $x^2 + y^2 = 4$, we have $x = \sqrt{4 - y^2}$ to the right of the y-axis and $x = -\sqrt{4 - y^2}$ to the left. Thus the length of the strip is approximately equal to $\sqrt{4 - y_i^2} - (-\sqrt{4 - y_i^2}) = 2\sqrt{4 - y_i^2}$ and we find that $A_i \approx 2\sqrt{4 - y_i^2}\, \Delta y_i$. Then

$$F_i \approx 2\rho(12 - y_i)\sqrt{4 - y_i^2}\, \Delta y_i$$

and

$$F = 2\rho \int_{-2}^{2} (12 - y)\sqrt{4 - y^2}\, dy = 24\rho \int_{-2}^{2} \sqrt{4 - y^2}\, dy - 2\rho \int_{-2}^{2} y\sqrt{4 - y^2}\, dy,$$

The first integral is the area of the semicircle $x = \sqrt{4 - y^2}$ and is equal to $\frac{1}{2}\pi r^2 = \frac{1}{2}\pi(2)^2 = 2\pi$ (or the integral can be calculated by making the substitution $y = \sin\theta$). Thus

$$F = (24\rho)(2\pi) + \frac{2\rho}{3}(4 - y^2)^{3/2}\Big|_{-2}^{2} = (48\pi)\rho + 0$$

$$= (48\pi)(64.3) = 3086.4\pi \quad \text{lb} \approx 9696 \quad \text{lb}.$$

The two examples above illustrate how to go about solving problems involving fluid pressure. We will not cite a general formula since the type of problems encountered can vary so widely. Once a convenient coordinate system is introduced, it is usually not too difficult to obtain the answer.

PROBLEMS 9.7 1~15

1. A circular swimming pool has a radius of 6 m and is 2-m deep. Find the force due to water pressure on the bottom of the pool when the pool is filled with water.
2. Find the force in Problem 1 if the pool is filled with seawater ($\rho \approx 1030$ kg/m³ at 0 °C).
3. A tube of mercury ($\rho = 13.6$ gm/cm³) is inverted over a beaker of mercury as in Figure 5. The pressures P_C and P_D at the points C and D are the same because the points are at the same height. Let P_A denote atmospheric pressure.

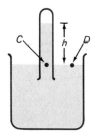

Figure 5

(a) Find a formula giving P_A as a function of h, the height of the column of mercury.
(b) If $h = 76$ cm, show that $P_A = 1.01 \times 10^5$ nt/m². [*Hint:* Use $g = 980$ cm/sec².] This value of P_A is called *one standard atmosphere*.
4. A rectangular swimming pool 5-m wide and 7-m long has a bottom which slopes at a constant rate, being $\frac{1}{2}$-m deep at the shallow end and 4-m deep at the deep end. If the pool is filled with water, find the force due to liquid pressure on the bottom of the pool.
5. A trough filled with water has a cross section in the shape of an equilateral triangle with 5-ft sides (the wide end is up). What is the force due to water pressure on one end of the trough?
6. In Problem 5, find the force if the trough is filled to a depth of (a) 1 ft (b) 2 ft (c) 3 ft.
7. A trough filled with ethyl alcohol ($\rho = 810$ kg/m³) has a cross section in the shape of a semicircle with a radius of 1 m (the open end is up). What is the force due to liquid pressure on one end of the trough?

*8. In Problem 7, find the force if the trough is filled to a depth of (a) 25 cm (b) 50 cm (c) 75 cm.

9. The dam across a certain river has the shape of the parabola $y = x^2/25$. If the river is 25-m across at the top, find the force due to water pressure on the dam.

*10. If the dam across the river in Problem 9 has a face which slopes downward from the water surface at an angle of 45°, find the force due to liquid pressure on it. [*Hint:* First use Problem 9 to find the depth of the water.]

*11. Find the force in Problem 10 if the dam slopes at an angle of 30°.

12. An elliptical plate is submerged in seawater so that the center of the ellipse is at a depth of 6 m. The equation of the ellipse is $(x^2/4) + (y^2/9) = 1$. Find the force due to water pressure on the plate.

13. A cask with sides in the shape of the trapezoid of Example 4 is submerged in seawater so that the top of the trapezoid is level with the surface of the water. Find the force due to liquid pressure on one end of the cask.

14. The gate of a dam is in the shape of an isosceles triangle with a 6-ft base and 4-ft sides. The upper vertex of the triangle is at a depth of 22 ft. Find the force on the gate when the gate is closed.

*15. Find the force on one side of the trough in Example 4.

16. Find the force on the sides of the pool (at the deeper and shallower ends) in Problem 4.

17. Find the force on one side of the trough in Problem 5 if the trough is 10-ft long.

18. The face of a dam is in the shape of one arc of the curve $y = 50 \sin(\pi x/100)$. If the surface of the water is 1-m above the top of the dam, find the force due to water pressure on the face of the dam.

19. A plate is in the shape of the region bounded by $y = e^{-x}$, $x = 0$, $x = 5$, and $y = 0$. If the plate is submerged in seawater so that its bottom edge is at a depth of 10 m, find the force due to water pressure on one side of the plate.

*20. A dam has a circular gate of radius 1 m. If the gate can withstand a force of 20,000 nt, to what height can the dam be filled?

*21. Prove that the total force against one side of a vertical submerged region R is equal to the product of the pressure at the centroid of R times the area of R.

In Problems 22–27 a plate in the shape of the given region is submerged in water with its centroid at the given depth. Use Problem 21 to calculate the force due to liquid pressure.

22. The region bounded by $y = 1/(x + 1)$, $y = 0$, $x = 0$, $x = 1$; $d = 10$ m

23. $y = x^{1/3}$, $y = 0$, $x = 1$; $d = 15$ m

24. $y = e^x$, $y = 0$, $x = 0$, $x = 2$; $d = 25$ ft

25. $y = x^4$, $y = 0$, $x = 2$; $d = 30$ ft

26. $y = 1/(x^2 + 1)$, $y = 0$, $x = 0$, $x = 1$; $d = 8$ m

27. $y = \ln x$, $y = 0$, $x = 1$, $x = 2$; $d = 20$ ft.

Review Exercises for Chapter Nine

In Exercises 1–8 find the volume of the solid generated when the region bounded by the given curves and lines is rotated about the given axis.

1. $y = 3x + 2$; $x = 0$; $y = 0$; $x = 2$; x-axis

2. $y = \sqrt{2x + 1}$; $x = 0$; $y = 0$; $x = 4$; x-axis

3. $y = x^5$; $x = 1$; $y = 0$; x-axis　　　　　　**4.** $y = x^{1/4}$; $x = 1$; $y = 0$; y-axis

5. $y = \sin 2x$; $x = 0$; $x = \pi/4$; x-axis　　　**6.** $y = \cos x$; $x = 0$; $x = \pi/2$; y-axis

7. $(x - 2)^2 + y^2 \le 1$; y-axis　　　　　　　　**8.** $x^2 + (y + 3)^2 \le 1$; x-axis.

9. The base of a certain solid is a circle of radius 3, while each cross section is a square. What is the volume of the solid?

10. Find the volume of the solid in Exercise 9 if each cross section is an equilateral triangle.

In Exercises 11–14 calculate the length of each given arc.

11. $y = 2x + 3$; x in $[2, 7]$

12. $y = x^2$; x in $[0, 2]$

13. $y = \frac{2}{3}x^{3/2}$; x in $[0, 1]$

14. $y = \ln(x^2 - 1)$; x in $[2, 3]$

15. Find a definite integral which is equal to the length of the curve $y = x \sin x$ between $x = 0$ and $x = \pi/2$.

In Exercises 16–19 find the lateral surface area of the solid generated when the area below the given curve over the given interval is rotated about the x-axis.

16. $y = \sqrt{x}$; $[0, 2]$

17. $y = e^x$; $[0, 1]$

18. $y = \cos x$; $[\pi/2, \pi]$

19. $y = -1/x$; $[1, 2]$.

20. If the region enclosed by the ellipse $x^2 + (y^2/4) = 1$ is revolved about the y-axis, find the surface area of the resulting football-shaped solid.

21. Find the surface area if the region of Exercise 20 is revolved about the x-axis.

22. Find the first moment around the origin and the center of mass of the system of masses $m_1 = 2$ kg, $m_2 = 5$ kg, and $m_3 = 8$ kg located along the x-axis at the points $P_1 = (4, 0)$, $P_2 = (-9, 0)$, $P_3 = (2, 0)$.

23. Find the center of mass of a rod lying along the x-axis in the interval $[0, 3]$ with density $\rho(x) = x^2$.

24. Find the center of mass of a rod lying along the x-axis in the interval $[1, 4]$ with density $\rho(x) = 1/x^3$.

25. Find the center of mass of the system of masses $m_1 = 3$ g, $m_2 = 7$ g, $m_3 = 4$ g, $m_4 = 8$ g located at the points $P_1 = (-2, 3)$, $P_2 = (4, 6)$, $P_3 = (3, -7)$, $P_4 = (0, -1)$, where x and y values are measured in centimeters.

26. Calculate the centroid of the plane region bounded by $y = x^2$, $y = 0$, $x = 1$, $x = 3$.

27. Calculate the centroid of the region bounded by $y = \sqrt{1 + x}$, $y = 0$, $x = 0$, $x = 3$.

28. Calculate the centroid of the region bounded by $y = \cos x$, $y = 0$, $x = 0$, $x = \pi/2$.

29. Calculate the centroid of the region bounded by $y = x^2 + 5x + 6$ and $y = x + 3$.

30. Calculate the centroid of the region bounded by $y = e^x$, $y = e^{-x}$, and $x = 1$.

31. Calculate the moment of inertia about the y-axis of the system of masses $m_1 = 3$ kg, $m_2 = 8$ kg, $m_3 = 5$ kg located at the points $P_1 = (4, 0)$, $P_2 = (-6, 0)$, $P_3 = (-2, 0)$.

32. A dumbbell is composed of a 3-kg mass and a 5-kg mass separated by a 1-m bar. Ignoring the weight of the bar and treating the masses as point masses, calculate the moment of inertia of the dumbbell about an axis perpendicular to it and passing through
   (a) the center of the bar.   (b) the 5-kg mass.
   (c) a point one-third of the way along the bar, closer to the 3-kg mass.

33. Calculate the radius of gyration in each case of Exercise 32.

34. If the dumbbell in Problem 32 rotates 15 times per second, calculate the kinetic energy in each of the three cases.

35. Find the moment of inertia and the radius of gyration when the region bounded by $y = x^3$, $y = 0$ and $x = 1$ is rotated about the x-axis, assuming a constant area density of 2 kg/m^2.

36. Find the moment of inertia and the radius of gyration if the region in Exercise 35 is rotated about the y-axis.

37. In Exercise 35 assume that the region is rotated 25 times per second. Calculate the kinetic energy.

38. The region bounded by $y = 1 - x$ and the x- and y-axes is rotated about the x-axis. Assuming a constant area density of 5 g/cm², calculate the moment of inertia and the radius of gyration.

39. Calculate the kinetic energy in Exercise 38 if the region is rotated 15 times per minute.

40. The unit disc $x^2 + y^2 \leq 1$ is rotated about the line $x = 2$. Calculate the moment of inertia and the radius of gyration. Assume that $\rho = 1$.

41. Solve Exercise 40 if the axis of rotation is the line $y = -5$.

42. Show that the second moment of a circle of radius r about an axis b units from its center is $\pi r^2(\frac{1}{4}r^2 + b^2)$.

43. Find the force due to water pressure on the bottom of a rectangular swimming pool 5-m long, 4-m wide, and 2-m deep when it is filled with water.

44. Find the force due to water pressure on the bottom of the pool in Exercise 43 if the bottom slopes downward from a depth of 1 m at one end to a depth of 3 m at the other.

45. A trough filled with water is 5-m long and has a cross section in the shape of a trapezoid 3-m wide at the bottom, $1\frac{1}{2}$-m wide at the top and 2-m deep. What is the force due to water pressure on one end of the trough?

46. Answer the questions in Exercises 44 and 45 if the trough is filled with ethyl alcohol (density $\rho = 810$ kg/m³).

47. In Exercise 45, find the force if the trough is filled with water to a depth of (a) $\frac{1}{2}$m. (b) 1 m. (c) $1\frac{1}{2}$m.

48. A cylindrical barrel whose end has a diameter of 2 m is submerged in seawater (density $\rho = 1030$ kg/m³). Find the total force due to water pressure at one end if the center of the barrel is at a depth of 15 m.

TEN

THE MATHEMATICAL
UNDERPINNINGS OF CALCULUS

In the earlier parts of this text we placed our emphasis on the calculation and applications of limits, derivatives, and integrals. However many of these calculations were based on theoretical results that were left unproved.

In this chapter we fill in the mathematical theory that was omitted earlier. We begin by giving a precise definition to the notion of the limit and then prove the basic limit theorems we relied on so heavily in Chapters 2 and 3. We then discuss continuity, differentiability, and the mean value theorem, which is one of the most important results in mathematical analysis. In the last section we discuss important properties of the integral.

The fact that this theoretical material was omitted from earlier chapters does not imply that it is less interesting or important. All mathematics is based upon precise formulations and proofs, and the results we discussed earlier in this text would be difficult to justify without the material from this chapter.

10.1 The Theory of Limits

In Section 2.2 we gave an informal definition of limits which was intended to appeal to your intuition. The more precise definition we give below states what you already know: namely, that as x gets close to x_0, $f(x)$ gets close to L. Before giving this definition we recall (from Section 1.1) that an *open interval* (a, b) is the set of all points x such that $a < x < b$. A *neighborhood* of the point x_0 is defined as an open interval which contains that point.†

DEFINITION 1. Let $f(x)$ be defined in a neighborhood of the point x_0 (a finite number) except possibly at the point x_0 itself. Then

$$\lim_{x \to x_0} f(x) = L$$

if for every $\epsilon > 0$, there is a $\delta > 0$ such that for all x for which

$$0 < |x - x_0| < \delta, \text{ we have } |f(x) - L| < \epsilon. \tag{1}$$

(We must exclude $|x - x_0| = 0$ since $f(x)$ may not be defined at x_0.)

---

† We can also define a *neighborhood of* ∞ to be an open interval of the form (N, ∞). We will refer to neighborhoods of ∞ in Chapter 12.

This formal definition is difficult to understand at first reading. It is worthwhile at this point to draw a picture of what is happening. We note that $0 < |x - x_0| < \delta$ means (from Section 1.2) that $x_0 - \delta < x < x_0 + \delta$, $x \neq x_0$. Similarly, $|f(x) - L| < \epsilon$ means that $L - \epsilon < f(x) < L + \epsilon$. Figure 1 illustrates that no matter how small ϵ is chosen, δ can be made small enough so that $f(x)$ is within ϵ of L.

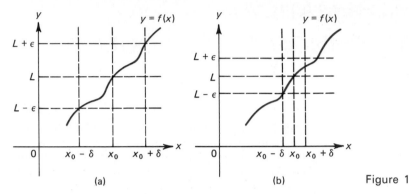

(a)   (b)   Figure 1

The best way to further illustrate this definition is to show by example that our intuitive notion of a limit coincides with the formal definition of a limit.

EXAMPLE 1. Show that $\lim_{x \to 2} 3x = 6$.

SOLUTION. Here $x_0 = 2$, $L = 6$, and $f(x) = 3x$. Let $\epsilon > 0$ be given. Then we must show that there is a $\delta > 0$ such that if $0 < |x - 2| < \delta$, then $|3x - 6| < \epsilon$. Before doing this for any $\epsilon > 0$, we illustrate what is happening by giving ϵ a particular value, say $\epsilon = \frac{1}{10}$. Then we wish to choose δ such that if

$$0 < |x - 2| < \delta, \quad \text{then} \quad |3x - 6| < \tfrac{1}{10}, \quad \text{or} \quad -\tfrac{1}{10} < 3x - 6 < \tfrac{1}{10}.$$

First we add 6 to these last two inequalities:

$$-\tfrac{1}{10} + 6 < 3x < \tfrac{1}{10} + 6.$$

Then we divide by 3:

$$-\tfrac{1}{30} + 2 < x < \tfrac{1}{30} + 2.$$

Next we subtract 2:

$$-\tfrac{1}{30} < x - 2 < \tfrac{1}{30}.$$

This is the same as

$$|x - 2| < \tfrac{1}{30}.$$

Now we make the choice $\delta = \frac{1}{30}$, and pick x such that $0 < |x - 2| < \delta = \frac{1}{30}$. Then we may reverse the process above to show that $|3x - 6| < \epsilon = \frac{1}{10}$ as follows:
If

$$|x - 2| < \tfrac{1}{30},$$

then

$$-\tfrac{1}{30} < x - 2 < \tfrac{1}{30} \quad \text{or} \quad -\tfrac{1}{10} < 3x - 6 < \tfrac{1}{10} \quad \text{or} \quad |3x - 6| < \tfrac{1}{10}.$$

We have therefore shown that for the particular value $\epsilon = \frac{1}{10}$, it is possible to find a $\delta\ (= \frac{1}{30})$ which satisfies the requirements in the definition of the limit (see Figure 2).

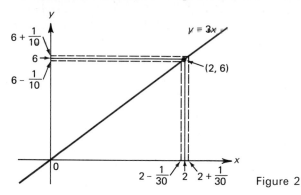

Figure 2

It is important to note that this value of δ is *not* unique. Any value of $\delta < \frac{1}{30}$ will work. This fact is not surprising since it merely states that if we start even closer to 2 than $\frac{1}{30}$ of a unit away, $3x$ will still remain within $\frac{1}{10}$ of a unit from 6.

Using this as an illustration, it is easy to show that for any ϵ, the required δ can be found. Try $\delta = \epsilon/3$ (since $\frac{1}{30} = \frac{1}{10}/3$). Then if

$$|x - 2| < \delta = \epsilon/3,$$

$$-\frac{\epsilon}{3} < x - 2 < \frac{\epsilon}{3} \quad \text{or} \quad -\epsilon < 3x - 6 < \epsilon \quad \text{or} \quad |3x - 6| < \epsilon$$

which is what we had to show. Hence we have proven that

$$\lim_{x \to 2} 3x = 6.$$

Example 1 illustrates the technique for finding a δ that works. We start with the assumption $L - \epsilon < f(x) < L + \epsilon$ and work backwards, using elementary algebraic steps, until we obtain an expression of the form $c < x < d$. Then if δ is chosen so that $x_0 - \delta > c$ and $x_0 + \delta < d$, then $|x - x_0| < \delta$ will imply that $c < x < d$. We can usually show that $|f(x) - L| < \epsilon$ by reversing the steps that led to $c < x < d$.

EXAMPLE 2. Show that $\lim_{x \to -1} (4x + 9) = 5$.

SOLUTION. Let $\epsilon > 0$ be given. Then we must have

$$5 - \epsilon < 4x + 9 < 5 + \epsilon \tag{2}$$

when

$$-1 - \delta < x < -1 + \delta \tag{3}$$

To find an appropriate choice for δ, we subtract 5 from the inequalities (2) to obtain

$$-\epsilon < 4x + 4 < \epsilon$$

or, dividing by 4

$$-\frac{\epsilon}{4} < x + 1 < \frac{\epsilon}{4}$$

and, subtracting 1,

$$-1 - \frac{\epsilon}{4} < x < -1 + \frac{\epsilon}{4}.$$

It is now evident that if $\delta = \epsilon/4$, inequality (2) holds if inequality (3) holds (see Figure 3).

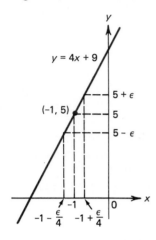

Figure 3

EXAMPLE 3.   Show that $\lim_{x \to 2} x^2 = 4$.

SOLUTION. This example is a bit trickier than the two that preceded it. Let $\epsilon > 0$ be given. We must find a $\delta > 0$ such that if

$$-\delta < x - 2 < \delta \qquad \text{then} \qquad -\epsilon < x^2 - 4 < \epsilon.$$

First we require that δ be less than 1 (for reasons to be made clear shortly). If $-1 < x - 2 < 1$, then $1 < x < 3$, and

$$3 < x + 2 < 5. \tag{4}$$

Now, if

$$-\epsilon < x^2 - 4 < \epsilon, \qquad \text{then} \qquad -\epsilon < (x - 2)(x + 2) < \epsilon$$

and, dividing by $x + 2$, which is positive (since $x + 2 > 3$), we obtain

$$-\frac{\epsilon}{x + 2} < x - 2 < \frac{\epsilon}{x + 2}. \tag{5}$$

From (4), $x + 2 > 3$ and $x + 2 < 5$ so that

$$\frac{1}{x + 2} < \frac{1}{3} \qquad \text{and} \qquad \frac{1}{x + 2} > \frac{1}{5}.$$

(We initially chose $\delta < 1$ so that we could obtain inequalities like these.) In particular, using the second of these inequalities, we have

$$\frac{\epsilon}{x + 2} > \frac{\epsilon}{5} \qquad \text{and} \qquad -\frac{\epsilon}{x + 2} < -\frac{\epsilon}{5} \tag{6}$$

(multiplying an inequality by a negative number changes the direction of the inequality). Thus we may set $\delta = \epsilon/5$ or 1, whichever is smaller. To see that this works, we suppose that

$$-\frac{\epsilon}{5} < x - 2 < \frac{\epsilon}{5}.$$

Then from inequalities (6),

$$-\frac{\epsilon}{x+2} < x - 2 < \frac{\epsilon}{x+2}$$

or, multiplying by $x + 2$ (which is positive)

$$-\epsilon < (x - 2)(x + 2) < \epsilon \qquad \text{and} \qquad -\epsilon < x^2 - 4 < \epsilon$$

which is what we had to show (see Figure 4).

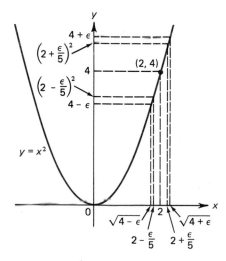

Figure 4

This last example illustrates that the calculation of limits from the definition can be very tedious. Fortunately, as we have seen, there are several theorems which make the calculation of limits (at least, in some cases) a relatively simple process. We will restate and prove the basic limit theorems in the next section. For the remainder of this section we will discuss infinite limits and limits at infinity.

DEFINITION 2. (i) $\lim_{x \to x_0} f(x) = \infty$ if for every $N > 0$ there is a $\delta > 0$ such that $f(x) > N$ for all x such that $0 < |x - x_0| < \delta$.
   (ii) $\lim_{x \to x_0} f(x) = -\infty$ if for every $N > 0$ there is a $\delta > 0$ such that $f(x) < -N$ for all x such that $0 < |x - x_0| < \delta$.

EXAMPLE 4. Show that $\lim_{x \to 0} (1/x^2) = \infty$.

SOLUTION. Let $N > 0$ be chosen, and let $\delta = 1/\sqrt{N}$. If $|x| < 1/\sqrt{N}$, then $-1/\sqrt{N} < x < 1/\sqrt{N}$. If $x > 0$, then $x < 1/\sqrt{N}$ implies that $x^2 < 1/N$. If $x < 0$, then $-1/\sqrt{N} < x$ implies that $x^2 < 1/N$. In either case, $x^2 < 1/N$ so that $1/x^2 > N$. See Figure 5.

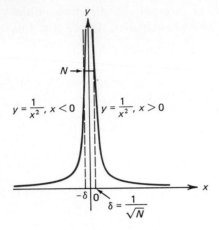

Figure 5

Note. How did we know that choosing $\delta = 1/\sqrt{N}$ would work? The secret is to start with the inequality you want to end up with and work backwards. The inequality $1/x^2 > N$ leads naturally to the inequalities $Nx^2 < 1$, $x^2 < 1/N$ and $x < 1/\sqrt{N}$. Now the choice $\delta = 1/\sqrt{N}$ is obvious.

EXAMPLE 5.   Show that $\lim_{x\to 0}(-1/x^2) = -\infty$.

SOLUTION. Given $N > 0$ we may choose the same δ as in Example 4. Then since $1/x^2 > N$, multiplying by -1 reverses the inequality so that $-1/x^2 < -N$. See Figure 6.

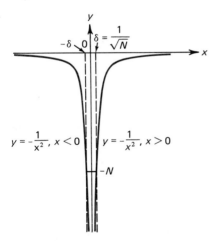

Figure 6

DEFINITION 3. (i) $\lim_{x\to\infty} f(x) = L$ if for every $\epsilon > 0$ there is an $N > 0$ such that if $x > N$, then $|f(x) - L| < \epsilon$.

  (ii) $\lim_{x\to-\infty} f(x) = L$ if for every $\epsilon > 0$, there is an $N > 0$ such that if $x < -N$, then $|f(x) - L| < \epsilon$.

EXAMPLE 6.   Show that $\lim_{x\to\infty} 1/x = 0$.

SOLUTION. Let $\epsilon > 0$ be given. Then for $N = 1/\epsilon$, if $x > N$, then $0 < 1/x < 1/N = \epsilon$. See Figure 7.

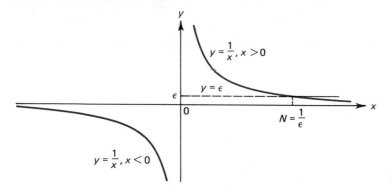

$y = \frac{1}{x}, x > 0$

$y = \epsilon$

ϵ

0

$N = \frac{1}{\epsilon}$

x

$y = \frac{1}{x}, x < 0$

Figure 7

EXAMPLE 7. Show that $\lim_{x \to -\infty} 1/x = 0$.

SOLUTION. Let $\epsilon > 0$ be given. Then for $N = 1/\epsilon$, if $x < -N$, $-\epsilon = -1/N < 1/x < 0$ so that $|1/x| < \epsilon$ which is what we had to show. See Figure 8.

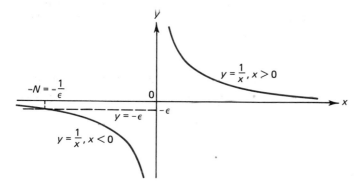

$-N = -\frac{1}{\epsilon}$

$y = \frac{1}{x}, x > 0$

0

$y = -\epsilon$

$-\epsilon$

x

$y = \frac{1}{x}, x < 0$

Figure 8

EXAMPLE 8. Show that $\lim_{x \to \infty} \dfrac{1}{\sqrt{x} + 3} = 0$.

SOLUTION. Let $\epsilon > 0$ be given and let $N = 1/\epsilon^2$. Then, for $x > N, 0 < 1/x < \epsilon^2$, so that

$$\frac{1}{\sqrt{x} + 3} < \frac{1}{\sqrt{x}} < \sqrt{\epsilon^2} = \epsilon.$$

See Figure 9.

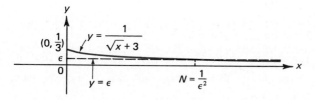

$\left(0, \frac{1}{3}\right)$

$y = \dfrac{1}{\sqrt{x} + 3}$

ϵ

0

$y = \epsilon$

$N = \dfrac{1}{\epsilon^2}$

x

Figure 9

We conclude this section by giving an example of a function which remains bounded, but has no limit.

EXAMPLE 9. Let $f(x)$ be defined by

$$f(x) = \begin{cases} 1, & \text{if } x = 1/2^n \text{ for some integer } n \geq 1 \\ x, & \text{otherwise,} \end{cases}$$

Since $f(x) = x$ for all values of $x \neq 1/2^n$, it may seem that $f(x) \to 0$ as $x \to 0$. We now show that $\lim_{x \to 0} f(x)$ does not exist. First we show that $\lim_{x \to 0} f(x) \neq 0$. To this end we pick $\epsilon > 0$ such that $\epsilon < 1$. For every $\delta > 0$, there is an n such that $1/2^n < \delta$ (since $1/2^n \to 0$ as n gets very large). Hence, no matter how small we choose δ, there will be an x with $0 < x < \delta$ such that $|f(x) - 0| = 1 > \epsilon$. Thus $f(x)$ does not tend to 0 as $x \to 0$. But clearly $f(x)$ cannot have any other limit as $x \to 0$ since $|f(x)| < \epsilon$ if $|x| < \epsilon$ except at the points $1/2^n$. Therefore $f(x)$ has no limit as $x \to 0$. However, if $x_0 > 0$, then it is easy to show that (see Problem 16) $\lim_{x \to x_0} f(x) = x_0$.

At this point you should go back to Sections 2.2, 2.3, and 2.5 and take another look at the limits we calculated in light of the new definitions. You will see that there is a great similarity between the intuitive definitions given in Chapter 2 and the rigorous $\epsilon - \delta$ definitions given here. The reason for providing these more formal definitions is simply that these definitions allow us to *prove* important results like the limit theorems in the next section.

PROBLEMS 10.1

In Problems 1–9 verify the given limits directly from Definition 1. In particular, for $\epsilon = \frac{1}{10}$ and $\epsilon = \frac{1}{100}$, what values of δ will ensure that $|f(x) - L| < \epsilon$ if $0 < |x - x_0| < \delta$?

1. $\lim_{x \to 1} 7x = 7$

2. $\lim_{x \to 4} (4x - 6) = 10$

3. $\lim_{x \to -2} (5x + 1) = -9$

4. $\lim_{x \to -2} x^2 = 4$

5. $\lim_{x \to 3} (x^2 - 6) = 3$

6. $\lim_{x \to 1} (5 - 2x^2) = 3$

7. $\lim_{x \to -1} (1 + x + x^2) = 1$

8. $\lim_{x \to 0} (4 + 2x - 3x^2) = 4$

*9. $\lim_{x \to 2} x^3 = 8$.

10. (a) Prove from Definition 3 that $\lim_{x \to \infty} 1/\sqrt{x/10} = 0$.
     (b) If $\epsilon = 0.01$, how large must N be so that if $x > N$, then $1/\sqrt{x/10} < \epsilon$?
11. (a) Prove from Definition 2 that $\lim_{x \to 0} 1/x^4 = \infty$.
     (b) How small must δ be chosen so that if $|x| < \delta$, then $1/x^4 > 100,000,000$?
12. (a) Prove from Definition 3 that $\lim_{x \to -\infty} |1/(x + 3)| = 0$.
     (b) If $\epsilon = 0.01$, how large must N be chosen so that if $x < -N$, then $1/(x + 3) < \epsilon$?
13. (a) Prove that $\lim_{x \to \infty} (3 + (4/x^3)) = 3$.
     (b) From $\epsilon = 0.001$, how large must N be chosen so that if $x > N$, then $|[3 + (4/x^3)] - 3| < \epsilon$?
14. (a) Show that if α is any positive integer, then $\lim_{x \to \infty} a/x^\alpha = 0$ for any real number a.
     (b) Prove (a) when α is a positive rational number.
     (c) Show that $\lim_{x \to \infty} a/(x^\alpha + b) = 0$ for any real number b.

15.   Prove from Definition 3 that if $|r| < 1$,

$$\lim_{n \to \infty} (1 + r + r^2 + r^3 + \cdots + r^n) = \frac{1}{1 - r}.$$

[*Hint:* See Sections 1.10 and 2.6.]

16.   Let

$$f(x) = \begin{cases} 1, & \text{if } x = 1/2^n \text{ for some integer } n \geq 1 \\ x, & \text{otherwise.} \end{cases}$$

Show that if $x_0 \neq 0$, $\lim_{x \to x_0} f(x) = x_0$.

*17.   Let

$$f(x) = \begin{cases} 1, & x \text{ rational} \\ 0, & x \text{ irrational.} \end{cases}$$

Show that $\lim_{x \to x_0} f(x)$ does not exist for any real number x_0. [*Hint:* Pick any $\epsilon < 1$. Then show that for any $\delta > 0$, there is an x in $(x_0 - \delta, x_0 + \delta)$ such that $|f(x) - f(x_0)| > \epsilon$.]

18.   Let

$$f(x) = \begin{cases} 0, & x < 0 \\ x, & x \geq 0. \end{cases}$$

Prove that $\lim_{x \to 0} f(x) = 0$.

10.2 Proofs of the Limit Theorems

In this section we prove the limit theorems stated in Sections 2.3 and 2.5. In the proofs, we will assume that x_0 and L are real numbers. The cases of infinite limits and limits at infinity will be discussed in the problems.

THEOREM 1 (Theorem 2.3.2).   Let c be any real number and suppose that $\lim_{x \to x_0} f(x)$ exists. Then

$$\lim_{x \to x_0} cf(x) = c \lim_{x \to x_0} f(x).$$

PROOF.   *Case* (*i*): $c = 0$. Then since $\lim_{x \to x_0} 0 = 0$ (see Problem 1), we have

$$\lim_{x \to x_0} cf(x) = \lim_{x \to x_0} 0 = 0 = 0 \cdot \lim_{x \to x_0} f(x).$$

Case (*ii*): $c \neq 0$. Let $\lim_{x \to x_0} f(x) = L$ and let $\epsilon > 0$ be given. Choose $\delta > 0$ so that if $0 < |x - x_0| < \delta$, then $|f(x) - L| < \epsilon/|c|$ (we can always do this since $\lim_{x \to x_0} f(x) = L$). Then

$$|cf(x) - cL| = |c(f(x) - L)| = |c| \, |f(x) - L| < |c| \frac{\epsilon}{|c|} = \epsilon$$

if $0 < |x - x_0| < \delta$ and the theorem is proved.

THEOREM 2 (Theorem 2.3.3).   If $\lim_{x \to x_0} f(x)$ and $\lim_{x \to x_0} g(x)$ both exist (and are finite), then

$$\lim_{x \to x_0} (f(x) + g(x)) = \lim_{x \to x_0} f(x) + \lim_{x \to x_0} g(x).$$

PROOF. Let $\lim_{x \to x_0} f(x) = L_1$ and $\lim_{x \to x_0} g(x) = L_2$. For a given $\epsilon > 0$, choose δ_1 such that if $0 < |x - x_0| < \delta_1$ then $|f(x) - L_1| < \epsilon/2$ and choose δ_2 such that if $0 < |x - x_0| < \delta_2$ then $|g(x) - L_2| < \epsilon/2$. Let δ be the smaller of δ_1 and δ_2 (denoted $\delta = \min\{\delta_1, \delta_2\}$). Then if $0 < |x - x_0| < \delta$,

$$|(f(x) + g(x)) - (L_1 + L_2)| = |(f(x) - L_1) + (g(x) - L_2)|$$

$$\leq |f(x) - L_1| + |g(x) - L_2| = \frac{\epsilon}{2} + \frac{\epsilon}{2} = \epsilon$$

and the theorem is proved. (The last step follows from the triangle inequality; see Section 1.2.)

Remark. It is easy to extend Theorem 2 to a finite sum. We have, if all indicated limits exist,

$$\lim_{x \to x_0} [f_1(x) + f_2(x) + \cdots + f_n(x)] = \lim_{x \to x_0} f_1(x) + \lim_{x \to x_0} f_2(x) + \cdots + \lim_{x \to x_0} f_n(x).$$

THEOREM 3 (Theorem 2.3.4).   If $\lim_{x \to x_0} f(x)$ and $\lim_{x \to x_0} g(x)$ both exist, then

$$\lim_{x \to x_0} f(x) \cdot g(x) = \left[\lim_{x \to x_0} f(x) \right] \left[\lim_{x \to x_0} g(x) \right].$$

PROOF. Let $L_1 = \lim_{x \to x_0} f(x)$ and $L_2 = \lim_{x \to x_0} g(x)$. Let $\epsilon > 0$ be given. We choose four δ's as follows:

(i)   Choose $\delta_1 > 0$ such that if $0 < |x - x_0| < \delta_1$, then $|f(x) - L_1| < 1$ so that $|f(x)| < |L_1| + 1$.

(ii)   Choose δ_2 such that if $0 < |x - x_0| < \delta_2$, then

$$|g(x) - L_2| < \frac{\epsilon}{2} \left(\frac{1}{|L_1| + 1} \right).$$

(iii)   Choose $\delta_3 > 0$ such that if $0 < |x - x_0| < \delta_3$, then

$$|f(x) - L_1| < \frac{\epsilon}{2} \left(\frac{1}{|L_2| + 1} \right).$$

(iv)   Choose $\delta = \min\{\delta_1, \delta_2, \delta_3\}$.

We now show that if $0 < |x - x_0| < \delta$, then $|f(x)g(x) - L_1 L_2| < \epsilon$. This will complete the proof of the theorem. We have, for $0 < |x - x_0| < \delta$,

$$|f(x)g(x) - L_1 L_2| = |f(x)g(x) - f(x)L_2 + f(x)L_2 - L_1 L_2|$$

$$\leq |f(x)g(x) - f(x)L_2| + |f(x)L_2 - L_1 L_2|$$

(by the triangle inequality)

$$\leq |f(x)| |g(x) - L_2| + |L_2| |f(x) - L_1|$$

$$< (|L_1| + 1) \left(\frac{\epsilon}{2} \right) \left(\frac{1}{|L_1| + 1} \right) + |L_2| \left(\frac{\epsilon}{2} \right) \left(\frac{1}{|L_2| + 1} \right)$$

$$< \frac{\epsilon}{2} + \frac{\epsilon}{2} = \epsilon.$$

COROLLARY 1. If $\lim_{x \to x_0} f(x)$ exists and n is a positive integer, then

$$\lim_{x \to x_0} (f(x))^n = \left(\lim_{x \to x_0} f(x) \right)^n.$$

PROOF

$$\lim_{x \to x_0} f^2(x) = \lim_{x \to x_0} f(x) \cdot \lim_{x \to x_0} f(x) = \left[\lim_{x \to x_0} f(x) \right]^2.$$

For $n > 2$, simply use this argument as many times as necessary.

COROLLARY 2 (Theorem 2.3.1). Let $P(x) = c_0 + c_1 x + c_2 x^2 + \cdots + c_n x^n$ be a polynomial where $c_0, c_1, c_2, \ldots, c_n$ are real numbers. Then

$$\lim_{x \to x_0} P(x) = P(x_0) = c_0 + c_1 x_0 + c_2 x_0^2 + \cdots + c_n x_0^n.$$

PROOF. We use the fact that $\lim_{x \to x_0} c_0 = c_0$ and $\lim_{x \to x_0} x = x_0$ (see Problems 1 and 3). Then, using Theorems 1, 2, 3, and Corollary 1 we obtain

$$\lim_{x \to x_0} P(x) = \lim_{x \to x_0} (c_0 + c_1 x + c_2 x^2 + \cdots + c_n x^n)$$

$$= \lim_{x \to x_0} c_0 + c_1 \lim_{x \to x_0} x + c_2 \left[\lim_{x \to x_0} x \right]^2 + \cdots + c_n \left[\lim_{x \to x_0} x \right]^n$$

$$= c_0 + c_1 x_0 + c_2 x_0^2 + c_3 x_0^3 + \cdots + c_n x_0^n = P(x_0).$$

THEOREM 4 (Theorem 2.3.5). If $\lim_{x \to x_0} f(x)$ and $\lim_{x \to x_0} g(x)$ both exist and $\lim_{x \to x_0} g(x) \neq 0$, then

$$\lim_{x \to x_0} \frac{f(x)}{g(x)} = \frac{\lim_{x \to x_0} f(x)}{\lim_{x \to x_0} g(x)}.$$

PROOF. As before, let $L_1 = \lim_{x \to x_0} f(x)$ and $L_2 = \lim_{x \to x_0} g(x)$. Since $L_2 \neq 0$, we show first that $\lim_{x \to x_0} 1/g(x) = 1/L_2$. We must show that for a given $\epsilon > 0$, there is a $\delta > 0$ such that when $|x - x_0| < \delta$,

$$\left| \frac{1}{g(x)} - \frac{1}{L_2} \right| < \epsilon. \tag{1}$$

But the inequality (1) is equivalent to

$$\left| \frac{L_2 - g(x)}{L_2 g(x)} \right| < \epsilon. \tag{2}$$

Select δ_1 such that $0 < |x - x_0| < \delta_1$ implies that $|g(x) - L_2| < |L_2|/2$. Then

$$-\frac{|L_2|}{2} < g(x) - L_2 < \frac{|L_2|}{2}$$

so that

$$L_2 - \frac{|L_2|}{2} < g(x) < L_2 + \frac{|L_2|}{2}.$$

If $L_2 > 0$, then $L_2/2 < g(x) < 3L_2/2$ which implies that $|g(x)| > |L_2|/2$. If $L_2 < 0$, then $3L_2/2 < g(x) < L_2/2$ which implies that $|g(x)| > |L_2|/2$. Then for $0 < |x - x_0| < \delta_1$,

$$|L_2 g(x)| = |L_2|\,|g(x)| > |L_2|\left|\frac{L_2}{2}\right| = \frac{L_2^2}{2}$$

or $1/|L_2 g(x)| < 2/L_2^2$. Similarly, there is a δ_2 such that if $0 < |x - x_0| < \delta_2$, then

$$|L_2 - g(x)| < \frac{L_2^2}{2}\,\epsilon.$$

Now choose $\delta = \min\{\delta_1, \delta_2\}$. Then for $0 < |x - x_0| < \delta$,

$$\left|\frac{L_2 - g(x)}{L_2 g(x)}\right| = \frac{1}{|L_2 g(x)|}\cdot|L_2 - g(x)| < \frac{2}{L_2^2}\cdot\frac{L_2^2}{2}\epsilon = \epsilon,$$

which shows that $\lim_{x\to x_0} 1/g(x) = 1/L_2$. Finally, we obtain from this result and Theorem 3

$$\lim_{x\to x_0}\frac{f(x)}{g(x)} = \lim_{x\to x_0} f(x)\cdot\lim_{x\to x_0}\frac{1}{g(x)} = L_1\cdot\frac{1}{L_2} = \frac{L_1}{L_2}$$

and Theorem 4 is proved.

COROLLARY. Let $r(x) = p(x)/q(x)$ be a rational function (the quotient of two polynomials). Then if $q(x_0) \neq 0$

$$\lim_{x\to x_0} r(x) = r(x_0) = \frac{p(x_0)}{q(x_0)}.$$

PROOF. From Theorem 4 and Corollary 2 of Theorem 3,

$$\lim_{x\to x_0} r(x) = \lim_{x\to x_0}\frac{p(x)}{q(x)} = \frac{\lim_{x\to x_0} p(x)}{\lim_{x\to x_0} q(x)} = \frac{p(x_0)}{q(x_0)} = r(x_0).$$

THEOREM 5 (Theorem 2.5.1—Squeezing Theorem). Suppose that $f(x) \leq g(x) \leq h(x)$ for x in a neighborhood of x_0 and $\lim_{x\to x_0} f(x) = \lim_{x\to x_0} h(x) = L$ where x_0 may be $+\infty$ or $-\infty$. Then

$$\lim_{x\to x_0} g(x) = L.$$

PROOF. We prove this in the case x_0 is finite. For the case $x_0 = +\infty$ or $-\infty$, see Problem 4. Let $\epsilon > 0$ be given. Choose δ_1 such that $|f(x) - L| < \epsilon$ if $0 < |x - x_0| < \delta_1$ and choose δ_2 so that $|h(x) - L| < \epsilon$ if $0 < |x - x_0| < \delta_2$. Let the inequalities $f(x) \leq g(x) \leq h(x)$ hold for x in some open interval (a, b) (recall the definition of a neighborhood). Then since x_0 is not one of the numbers a or b, there exists a δ_3 such that $a < x_0 - \delta_3 < x_0 < x_0 + \delta_3 < b$ and therefore, for $0 < |x - x_0| < \delta_3$, $f(x) \leq g(x) \leq h(x)$. Now let $\delta = \min\{\delta_1, \delta_2, \delta_3\}$. Then if $0 < |x - x_0| < \delta$,

$$g(x) - L \leq h(x) - L < \epsilon.$$

Since $|f(x) - L| < \epsilon$, we have $f(x) - L > -\epsilon$. Then

$$g(x) - L \geq f(x) - L > -\epsilon.$$

Putting (3) and (4) together, we obtain

$$|g(x) - L| < \epsilon$$

and the squeezing theorem is proved.

We have continually talked about *the* limit. We have not yet proven that if a limit exists, then it is unique. We do this now.

THEOREM 6. If $\lim_{x \to x_0} f(x)$ exists, then it is unique.

PROOF. Suppose that $\lim_{x \to x_0} f(x) = L_1$ and $\lim_{x \to x_0} f(x) = L_2$. Let $\epsilon > 0$ be given. There are positive numbers δ_1 and δ_2 such that if $0 < |x - x_0| < \delta = \min\{\delta_1, \delta_2\}$, then $|f(x) - L_1| < \epsilon$ and $|f(x) - L_2| < \epsilon$. Then

$$|L_1 - L_2| = |L_1 - f(x) + f(x) - L_2| \leq |L_1 - f(x)| + |f(x) - L_2|$$
$$= |f(x) - L_1| + |L_2 - f(x)| = \epsilon + \epsilon < 2\epsilon.$$

Thus $|L_1 - L_2| < 2\epsilon$ for *every* positive number ϵ. This can only happen if $L_1 = L_2$ and the theorem is proved.

At this point you should look back at the results in this section and note that each proof depended only on the definition of a limit and the theorems which came before. In the problems and sections to follow, you will find that mathematical proofs are not difficult if you start with basic ideas (usually definitions) and proceed from there.

PROBLEMS 10.2

1. Prove from the definition of a limit that for any finite real number x_0, $\lim_{x \to x_0} c = c$ for any constant c.
2. Prove that $\lim_{x \to \infty} c = c$.
3. Prove that $\lim_{x \to x_0} x = x_0$ for every real number x_0.
4. Prove that if $f(x) \leq g(x) \leq h(x)$ for all x larger than some number N and $\lim_{x \to \infty} f(x) = \lim_{x \to \infty} h(x) = L$, then $\lim_{x \to \infty} g(x) = L$. [*Hint:* Treat the cases $L < \infty$ and $L = \infty$ separately.]
5. Prove that $\lim_{x \to \infty} cf(x) = c \lim_{x \to \infty} f(x)$.
6. Prove that if $\lim_{x \to \infty} f(x)$ and $\lim_{x \to \infty} g(x)$ are finite, then

$$\lim_{x \to \infty} [f(x) + g(x)] = \lim_{x \to \infty} f(x) + \lim_{x \to \infty} g(x).$$

*7. Give an example to show that the result of Problem 6 is false in general if the limits are not assured to be finite. [*Hint:* Find $f(x)$ and $g(x)$ so that $\lim_{x \to \infty} f(x) = \lim_{x \to \infty} g(x) = \infty$ but that $f(x) - g(x)$ is a constant.]
8. Show that if $\lim_{x \to \infty} f(x)$ and $\lim_{x \to \infty} g(x)$ are finite, then

$$\lim_{x \to \infty} f(x)g(x) = \left[\lim_{x \to \infty} f(x)\right]\left[\lim_{x \to \infty} g(x)\right].$$

*9. Give an example to show that the result of Problem 8 is false in general if the limits are not finite. [*Hint:* Find $f(x)$ and $g(x)$ such that $\lim_{x \to \infty} f(x) = \infty$, $\lim_{x \to \infty} g(x) = 0$, and $f(x) \cdot g(x)$ is a constant. Note that the expression $0 \cdot \infty$ is not defined.]

10.   Show that if $\lim_{x\to\infty} f(x)$ is finite and $\lim_{x\to\infty} g(x)$ is finite and nonzero, then

$$\lim_{x\to\infty} \frac{f(x)}{g(x)} = \frac{\lim_{x\to\infty} f(x)}{\lim_{x\to\infty} g(x)}.$$

*11.   Give an example to show that the result of Problem 10 is false in general if the limits are not assumed to be finite.

12.   Prove that if $\lim_{x\to\infty} f(x)$ exists, then it is unique.

10.3 The Theory of One-Sided Limits

In this section we give a precise mathematical definition of the one-sided limits discussed in Section 2.7.

DEFINITION 1.   (i)   Let f be defined on the open interval (x_0, c) for some number $c > x_0$. Suppose that for every $\epsilon > 0$, there is a $\delta > 0$ such that $x_0 < x < x_0 + \delta$ implies that $|f(x) - L| < \epsilon$. Then L is called the *right-hand limit* of $f(x)$ as $x \to x_0$ and is denoted by

$$\lim_{x\to x_0^+} f(x) = L.$$

(ii)   Let f be defined on the open interval (d, x_0) for some number $d < x_0$. If for every $\epsilon > 0$, there is a $\delta > 0$ such that $x_0 - \delta < x < x_0$ implies that $|f(x) - L| < \epsilon$, then L is called the *left-hand limit* of $f(x)$ as $x \to x_0$ and is denoted by

$$\lim_{x\to x_0^-} f(x) = L.$$

EXAMPLE 1.   Prove that $\lim_{x\to 0^+} \sqrt{x} = 0$.

SOLUTION.   Let $\epsilon > 0$ be given. Suppose that $|f(x) - L| = |\sqrt{x} - 0| = \sqrt{x} < \epsilon$. Then $x < \epsilon^2$. This leads to the obvious choice $\delta = \epsilon^2$. Then if $0 < x < \delta$, $0 < \sqrt{x} < \sqrt{\delta} = \epsilon$ and so $\lim_{x\to 0^+} \sqrt{x} = 0$.

EXAMPLE 2. $\lim_{x\to 0^-} \sqrt{x}$ does not exist since $\sqrt{x}$ is not defined if $x < 0$.

EXAMPLE 3.   Let

$$f(x) = \begin{cases} 2x, & x \le 1 \\ x^2, & x > 1 \end{cases}.$$

Show that $\lim_{x\to 1^+} f(x) = 1$ and $\lim_{x\to 1^-} f(x) = 2$.

SOLUTION.   If $1 < x < 1 + \delta$, $f(x) = x^2$. Let $\epsilon > 0$ be given. Then $|f(x) - 1| < \epsilon$ means that $|x^2 - 1| < \epsilon$ or

$$-\epsilon < x^2 - 1 < \epsilon.$$

However, since $x > 1$, we have $x^2 - 1 > 0$ and we need only show that $x^2 - 1 < \epsilon$ for our choice of δ. Let $x < 1 + \delta$. Then $x^2 - 1 < (1 + \delta)^2 - 1 = 2\delta + \delta^2$. If $2\delta + \delta^2 < \epsilon$ and $1 < x < 1 + \delta$, then $x^2 - 1 < \epsilon$. Therefore, we need to choose δ such that $2\delta + \delta^2 < \epsilon$. To do this we first require that $\delta < 1$. Since $2\delta + \delta^2 = \delta(2 + \delta)$, we have $2\delta + \delta^2 = \delta(2 + \delta) < \delta(2 + 1) = 3\delta$. Thus if $\delta < \epsilon/3$, then $2\delta + \delta^2 < 3 \cdot \epsilon/3 = \epsilon$ which shows that $\lim_{x\to 1^+} f(x) = 1$.

Now if $x < 1$, $f(x) = 2x$. Then $|f(x) - 2| < \epsilon$ means that $|2x - 2| < \epsilon$ or $-\epsilon < 2x - 2 < \epsilon$. Since $x < 1$, $2x - 2 < 0$ so we only need to show that $-\epsilon < 2x - 2$. If $\delta = \epsilon/2$, then $1 - \delta < x < 1$ implies that $x > 1 - \delta = 1 - \epsilon/2$ and $2x > 2 - \epsilon$ so that $2x - 2 > -\epsilon$ and, therefore, $\lim_{x \to 1^-} f(x) = 2$.

For infinite limits, the definitions are similar.

DEFINITION 2. (i) $\lim_{x \to x_0^+} f(x) = \infty$ means that for every $N > 0$ there is a $\delta > 0$ such that if $x_0 < x < x_0 + \delta$, then $f(x) > N$.
  (ii) $\lim_{x \to x_0^-} f(x) = \infty$ if for every $N > 0$ there is a $\delta > 0$ such that if $x_0 - \delta < x < x_0$, then $f(x) > N$.
  (iii) $\lim_{x \to x_0^+} f(x) = -\infty$ if for every $N > 0$, there is a $\delta > 0$ such that if $x_0 < x < x_0 + \delta$, then $f(x) < -N$.
  (iv) $\lim_{x \to x_0^-} f(x) = -\infty$ if for every $N > 0$, there is a $\delta > 0$ such that if $x_0 - \delta < x < x_0$, then $f(x) < -N$.

EXAMPLE 4. Prove that $\lim_{x \to 0^+} 1/x = \infty$ and $\lim_{x \to 0^-} 1/x = -\infty$.

SOLUTION. Let $N > 0$ be given. Then if $\delta = 1/N$, $0 < x < \delta$ implies that $x < 1/N$ or $1/x > N$ which shows that $\lim_{x \to 0^+} 1/x = \infty$. Similarly, if $-\delta < x < 0$, then with $\delta = 1/N$, $x > -\delta = -1/N$ implies that $1/x < -N$ and $\lim_{x \to 0^-} 1/x = -\infty$.

We now show that $\lim_{x \to x_0} f(x)$ exists if and only if the right- and left-hand limits exist and are equal. We prove this in the case in which the limits are finite. The infinite case is left as a problem (see Problem 1).

THEOREM 1 (Theorem 2.7.1). $\lim_{x \to x_0} f(x) = L$ if and only if $\lim_{x \to x_0^+} f(x) = L$ and $\lim_{x \to x_0^-} f(x) = L$.

PROOF. First we show that if $\lim_{x \to x_0} f(x) = L$, then $\lim_{x \to x_0^+} f(x) = L$ and $\lim_{x \to x_0^-} f(x) = L$. To show the right-hand limit, let $\epsilon > 0$ be given. Then there is a $\delta > 0$ such that if $0 < |x - x_0| < \delta$, then $|f(x) - L| < \epsilon$. In particular, if $0 < x - x_0 < \delta$, then $0 < |x - x_0| < \delta$ and thus $|f(x) - L| < \epsilon$. Therefore $\lim_{x \to x_0^+} f(x) = L$. The proof for the left-hand limit is similar.

Conversely, suppose that both the right- and left-hand limits exist and are equal to L. For a given $\epsilon > 0$, choose δ_1 such that if $0 < x - x_0 < \delta_1$, then $|f(x) - L| < \epsilon$. Choose δ_2 such that if $-\delta_2 < x - x_0 < 0$, then $|f(x) - L| < \epsilon$. Let $\delta = \min\{\delta_1, \delta_2\}$. Then, if $|x - x_0| < \delta$, we have $0 < x - x_0 < \delta \leq \delta_1$ and $-\delta_2 \leq -\delta < x - x_0 < 0$, so that $|f(x) - L| < \epsilon$ whether $x - x_0$ is positive or negative. Hence $\lim_{x \to x_0} f(x) = L$ and the theorem is proved.

EXAMPLE 5. Let

$$f(x) = \begin{cases} x, & x \leq 1 \\ x^2, & x > 1 \end{cases}.$$

Show that $\lim_{x \to 1} f(x) = 1$.

SOLUTION. If $x > 1$, then $f(x) = x^2$ and, as we showed in Example 3, $\lim_{x \to 1^+} f(x) = \lim_{x \to 1^+} x^2 = 1$. If $x \leq 1$, then $f(x) = x$. Let $\epsilon > 0$ be given and choose $\delta = \epsilon$. If $1 - \delta < x < 1$, then since $f(x) = x$, $1 - \delta < f(x) < 1$. But $\epsilon = \delta$

so that $1 - \epsilon < f(x) < 1$. Thus $\lim_{x \to 1^-} f(x) = \lim_{x \to 1^-} x = 1$. Hence, we have $\lim_{x \to 1^+} f(x) = \lim_{x \to 1^-} f(x) = 1$ which implies, by Theorem 1, that $\lim_{x \to 1} f(x) = 1$.

PROBLEMS 10.3

1. Prove that $\lim_{x \to x_0} f(x) = \infty$ if and only if $\lim_{x \to x_0^+} f(x) = \lim_{x \to x_0^-} f(x) = \infty$.
2. Prove, using the definition, that $\lim_{x \to 4^+} f(x) = 2$ and $\lim_{x \to 4^-} f(x) = 16$ where

$$f(x) = \begin{cases} x^2, & x < 4 \\ \sqrt{x}, & x \geq 4 \end{cases}.$$

3. Show, using the definition and Theorem 1, that if

$$f(x) = \begin{cases} 3x + 2, & x \leq 2 \\ 2x^2, & x > 2 \end{cases},$$

then $\lim_{x \to 2} f(x) = 8$.

4. Find a number c such that if

$$f(x) = \begin{cases} 3x, & x \leq -2 \\ cx^2, & x > -2 \end{cases},$$

then $\lim_{x \to -2} f(x)$ exists.

5. Let $f(x) = [x]$.† Prove that $\lim_{x \to x_0} f(x)$ exists if and only if x_0 is not an integer.

10.4 Continuity

There are two equivalent definitions of continuity at a point x_0. The first of these was given in Section 2.4.

DEFINITION 1 (Continuity). Let $f(x)$ be defined for every x in a neighborhood of x_0. Then f is *continuous* at x_0 if all the following three conditions hold:

    (i) $f(x_0)$ exists.
    (ii) $\lim_{x \to x_0} f(x)$ exists.
    (iii) $\lim_{x \to x_0} f(x) = f(x_0)$.

Using this definition we may immediately prove (using the corollaries to Theorems 10.2.3 and 10.2.4) that every polynomial is continuous at every real number and that if $r(x) = p(x)/q(x)$ with $q(x_0) \neq 0$ (and p and q are polynomials), then r is continuous at x_0.

A second definition of continuity which is often very useful is given below.

DEFINITION 2 (Continuity). The function f is *continuous* at x_0 if f is defined in a neighborhood of x_0 and, for every $\epsilon > 0$, there is a $\delta > 0$ such that if $|x - x_0| < \delta$, then $|f(x) - f(x_0)| < \epsilon$.

Before using these two definitions, we show that they are equivalent. That is, if f is continuous according to Definition 1, then it is continuous according to Definition 2, and vice versa.

THEOREM 1. Definitions 1 and 2 are equivalent.

---

† $[x]$ = the largest integer $\leq x$.

PROOF. (i)   We first show that if a function is continuous according to Definition 1, then it is continuous according to Definition 2. Suppose that f is defined in a neighborhood of x_0 and that $\lim_{x \to x_0} f(x) = f(x_0)$. Let $\epsilon > 0$ be given. Then by the definition of limit, there is a δ such that if $0 < |x - x_0| < \delta$, then $|f(x) - f(x_0)| < \epsilon$. Further, $|f(x_0) - f(x_0)| = 0 < \epsilon$. Thus f is defined in a neighborhood of x_0 and for every $\epsilon > 0$, there is a $\delta > 0$ such that if $|x - x_0| < \delta$, then $|f(x) - f(x_0)| < \epsilon$. Hence f is continuous in the sense of Definition 2.

(ii)   Conversely, if we assume that f is continuous at x_0 in the sense of Definition 2, then f is defined in a neighborhood of x_0 and for every $\epsilon > 0$ there is a $\delta > 0$ such that $|x - x_0| < \delta$ implies $|f(x) - f(x_0)| < \epsilon$. Then by the definition of limit, $\lim_{x \to x_0} f(x) = f(x_0)$ and this is Definition 1.

DEFINITION 3. f is *discontinuous* at x_0 if it is not continuous there.

DEFINITION 4.   The function f is continuous in the closed interval $[a, b]$ if

(i) f is continuous at every x in the open interval (a, b).
(ii) $f(a)$ and $f(b)$ both exist.
(iii) $\lim_{x \to a^+} f(x) = f(a)$ and $\lim_{x \to b^-} f(x) = f(b)$.

Remark. It may be that f is only defined in the interval $[a, b]$. Then f is continuous at every point in $[a, b]$ means the usual thing for x in (a, b). However, by continuity at a, we mean that $\lim_{x \to a^+} f(x) = f(a)$ and by continuity at b we mean that $\lim_{x \to b^-} f(x) = f(b)$. Sometimes this type of continuity at the points a and b is referred to, respectively, as *continuity from the left and continuity from the right*.

We will not cite examples of continuous or discontinuous functions here as there are an ample number given in Sections 2.4 and 2.7.

DEFINITION 5. f has a *jump discontinuity* at x_0 if

(i) $\lim_{x \to x_0^+} f(x)$ and $\lim_{x \to x_0^-} f(x)$ both exist, are finite and are unequal, or
(ii) $f(x_0)$ exists and is finite, both $\lim_{x \to x_0^+} f(x)$ and $\lim_{x \to x_0^-} f(x)$ exist and are equal, but $\lim_{x \to x_0^+} f(x) \neq f(x_0)$.

(For an illustration of the possibility (ii), see Example 2.4.8.)

DEFINITION 6. f is *piecewise continuous* in $[a, b]$ if f is continuous at every point in $[a, b]$ except for a finite number of points at which f has a jump discontinuity.

We now prove a theorem which is extremely useful for the calculation of limits.

THEOREM 2 (Theorem 2.4.2).   If f is continuous at a and if $\lim_{x \to x_0} g(x) = a$, then $\lim_{x \to x_0} f(g(x)) = f(a)$.

PROOF. Let $\epsilon > 0$ be given. We must show that there exists a $\delta > 0$ such that if $|x - x_0| < \delta$, then $|f(g(x)) - f(a)| < \epsilon$. We do this in two steps:

(i)   Since f is continuous at a, by Definition 2 there is a $\delta_1 > 0$ such that if $|x - a| < \delta_1$, then $|f(x) - f(a)| < \epsilon$.
(ii)   Since $\lim_{x \to x_0} g(x) = a$, there is a $\delta > 0$ such that if $|x - x_0| < \delta$, then $|g(x) - a| < \delta_1$. But if $|g(x) - a| < \delta_1$, then, from (i), $|f(g(x)) - f(a)| < \epsilon$ and the theorem is proved.

The trick to using Theorem 2 is to be able to recognize which functions are continuous. We have already seen that all polynomials are continuous and that a rational function is continuous when its denominator is not zero. But we have more than that. Since any differentiable function is continuous (Theorem 2.9.1) we see that almost all the functions we have encountered are continuous.

EXAMPLE 1. Calculate $\lim_{x \to \pi} e^{\sin x}$.

SOLUTION. Since $\lim_{x \to \pi} \sin x = \sin \pi = 0$ and since e^x is continuous, we have $\lim_{x \to \pi} e^{\sin x} = e^0 = 1$.

EXAMPLE 2. Calculate $\lim_{x \to \pi} \sin e^x$.

SOLUTION. $\lim_{x \to \pi} e^x = e^\pi$ and $\sin x$ is continuous so that $\lim_{x \to \pi} \sin e^x = \sin e^\pi \approx -0.91$.

Continuous functions have many other nice properties. As we stated in Chapter 4, continuous functions are always integrable (although, as we have seen, they are not always differentiable—consider $f(x) = |x|$, for example). Two very useful properties of continuous functions are given in the theorems below. The proofs of these theorems are beyond the scope of this text. These proofs and the proofs of related theorems are given in all standard advanced calculus texts.†

THEOREM 3. If f is continuous on the closed, bounded interval $[a, b]$, then f is bounded above and below in that interval. That is, there exist numbers m and M such that

$$m \le f(x) \le M \qquad \text{for every } x \text{ in } [a, b].$$

Moreover, there exist numbers x_1 and x_2 in $[a, b]$ such that $f(x_1) = m$ and $f(x_2) = M$.

> **Remark.** The theorem is not true, in general, on an open interval (a, b). For example $f(x) = 1/x$ is continuous on $(0, 1)$ but as $x \to 0^+$, $f(x) \to \infty$ so that f is not bounded from above on that interval.

Theorem 3 is illustrated in Figure 1 for three different functions.

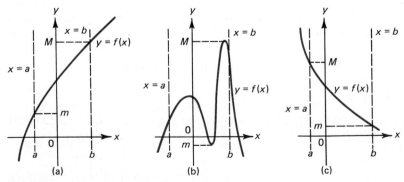

(a)   (b)   (c)

Figure 1

†See, for example, R. C. Buck, *Advanced Calculus,* McGraw-Hill, New York, 1965, Chapter 2.

THEOREM 4 (Intermediate Value Theorem). Let f be continuous on $[a, b]$. Then if c is any number between $f(a)$ and $f(b)$, there is a number $\bar{x}$ in (a, b) such that $f(\bar{x}) = c$.

This theorem has a geometrical interpretation given by Figure 2. The intermediate value theorem could be restated: if the continuous function f takes on the value

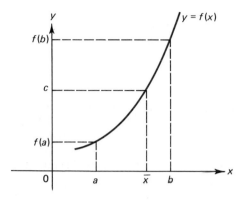

Figure 2

$f(a)$ and $f(b)$, then it takes on every value in between. The intermediate value theorem can be used to prove many interesting results. It is useful, for example, to find roots of a polynomial.

EXAMPLE 3. Show that there is a root of $P(x) = x^3 + x^2 + x - 1$ in the interval $[0, 1]$.

SOLUTION. $P(x)$ is continuous (why?). Since $P(0) = -1$ and $P(1) = 2$, there must be a number $\bar{x}$ in $(0, 1)$ such that $P(\bar{x}) = 0$ (since 0 is between -1 and 2). We can do even better. We have

$$P(\tfrac{1}{2}) = \tfrac{1}{8} + \tfrac{1}{4} + \tfrac{1}{2} - 1 = -\tfrac{1}{8} \quad \text{and} \quad P(\tfrac{3}{4}) = \tfrac{27}{64} + \tfrac{9}{16} + \tfrac{3}{4} - 1 = \tfrac{47}{64}.$$

Therefore there is a root between $\tfrac{1}{2}$ and $\tfrac{3}{4}$. Using a calculator, we can narrow the root down further. Since $P(0.543) = -0.002047993$ and $P(0.544) = 0.000925184$, we have located a root between 0.543 and 0.544. This method of "estimating" a root of a polynomial is very useful when used in conjunction with Newton's method (Section 5.5). We first use the intermediate value theorem to get "close" to the root (within 1 decimal place, say) and then use Newton's method to find the root quickly to as many decimal places of accuracy as needed. Doing this, we find one root of $x^3 + x^2 + x - 1$ is $\bar{x} = 0.5436890127$, correct to 10 decimal places.

PROBLEMS 10.4

1. Let f and g be continuous at x_0 and let c be any constant. Then, using the limit theorems of Section 2 and the first definition of continuity, prove that the following functions are continuous at x_0:
    (a) cf (b) $f + g$
    (c) fg (d) f/g if $g(x_0) \neq 0$.
2. Prove that if $g(x)$ is continuous at x_0 and if $f(x)$ is continuous at $g(x_0)$, then the composite function $f(g(x))$ is continuous at x_0.

In Problems 3–14 use Definition 2 of continuity to show that each function is continuous at the given point.

3. $f(x) = 3x - 2$; $x_0 = 1$

4. $f(x) = x^2 + 1$; $x_0 = -1$

5. $f(x) = 2x + x^2$; $x_0 = 0$

6. $f(x) = \dfrac{1}{x}$; $x_0 = 3$

7. $f(x) = -\dfrac{1}{x^2}$; $x_0 = -5$

8. $f(x) = \dfrac{1}{x+1}$; $x_0 = 0$

9. $f(x) = x^3 + 1$; $x_0 = -1$

10. $f(x) = \begin{cases} 1, & x \le 1 \\ x, & x > 1 \end{cases}$; $x_0 = 1$

11. $f(x) = \begin{cases} x^2, & x \le 2 \\ \frac{1}{2}x^3, & x > 2 \end{cases}$; $x_0 = 2$

12. $f(x) = [x]$; $x_0 = 3.1$

13. $f(x) = |x|$; $x_0 = 0$

14. $f(x) = |x + 2|$; $x_0 = -2$.

15. Prove using Definition 2 that $f(x) = e^x$ is continuous at every real number. [*Hint:* Use the fact that $e^{x+y} - e^x = e^x(e^y - 1)$.]

*16. Prove that $f(x) = \sin x$ is continuous at every real number. [*Hint:* Use the trigonometric identity

$$\sin x - \sin y = 2 \sin \frac{x - y}{2} \cos \frac{x + y}{2}.]$$

17. Let

$$f(x) = \begin{cases} \sin x, & x \ne 0 \\ 0, & x = 0. \end{cases}$$

Show that f is continuous at 0.

*18. Let

$$f(x) = \begin{cases} \sin \dfrac{1}{x}, & x \ne 0 \\ 0, & x = 0. \end{cases}$$

Show that $f(x)$ is discontinuous at 0. This problem provides an example of a function which is bounded ($-1 \le \sin x \le 1$ for every x) but is not continuous.

*19. Show using Definition 2 that $\ln x$ is continuous for every $x > 0$. [*Hint:* Use the fact that $\ln x - \ln y = \ln x/y$.]

In Problems 20–30 use Theorem 2 to calculate the indicated limit.

20. $\displaystyle\lim_{x \to \pi/2} \cos(\sin x)$

21. $\displaystyle\lim_{x \to 3} \sqrt{e^x + 1}$

22. $\displaystyle\lim_{x \to \pi} (e^{\sin x + \cos x})^2$

23. $\displaystyle\lim_{x \to -1} \sec^2 \frac{x}{2}$

24. $\displaystyle\lim_{x \to 0} e^{\tan x}$

25. $\displaystyle\lim_{x \to \pi/2} \sec^2 \frac{x}{2}$

26. $\displaystyle\lim_{x \to 0} \sin^{-1} e^x$

27. $\displaystyle\lim_{x \to \pi/2} \cos^{-1} \sin x$

28. $\displaystyle\lim_{x \to 0} \sin \cosh x$

29. $\displaystyle\lim_{x \to 0} \cosh \sin x$

30. $\displaystyle\lim_{x \to 0} \cos \sinh x$.

31. A car accelerates from 0 to 80 km/hr in 3 seconds. Explain why there is a time between 0 and 30 sec when the car is traveling at exactly 50 km/hr.

*32.   Prove that there is a number whose square is 3. [*Hint:* Let $f(x) = x^2$ and apply the intermediate value theorem.]

In Problems 33–40 show by the definition that the given function is piecewise continuous over the indicated interval. Identify the functions which are continuous.

33. $f(x) = [x] + 1$; $[-2, 2]$

34. $f(x) = \dfrac{x^2 - 1}{x + 1}$; $[-2, 2]$

35. $f(x) = \dfrac{[x]}{x}$; $[1, 3]$

36. $f(x) = \begin{cases} x + x^2, & x \le -1 \\ x^3, & x > -1 \end{cases}$; $[-2, 0]$

37. $f(x) = \begin{cases} x, & x \text{ an integer} \\ \frac{1}{2}, & x \text{ not an integer} \end{cases}$; $[-1, 3]$

38. $f(x) = \begin{cases} 2x^2 + 4, & x \le 0 \\ (x - 2)^2, & x > 0 \end{cases}$; $[-5, 5]$

39. $f(x) = [x]$; $[-2, 2]$

40. $f(x) = |[x]|$; $[-3, 3]$.

41.   Let $f(x) = \begin{cases} x + 3, & x \le 1 \\ 2x + 5, & x > 1 \end{cases}$ and $g(x) = \begin{cases} x^2 + 6, & x \le 1 \\ 5x^3 - 1, & x > 1 \end{cases}$.

   (a)   What is the function $f(x)g(x)$?
   (b)   Show that $\lim_{x \to 1} f(x)$ does not exist.
   (c)   Show that $\lim_{x \to 1} g(x)$ does not exist.
   (d)   Show that $\lim_{x \to 1} f(x)g(x)$ does exist.

*42.   Find two functions $f(x)$ and $g(x)$ such that $\lim_{x \to x_0} f(x)$ and $\lim_{x \to x_0} g(x)$ do not exist but that $\lim_{x \to x_0} f(x)/g(x)$ does exist.

43.   Let

$$f(x) = \begin{cases} p(x), & x \le x_0 \\ q(x), & x > x_0 \end{cases},$$

   where $p(x)$ and $q(x)$ are polynomials. Under what circumstances will $\lim_{x \to x_0} f(x)$ exist?

*44.   Prove that if $f(x_0) > 0$ and f is continuous at x_0, there is a number $\delta > 0$ such that if $x_0 - \delta < x < x_0 + \delta$, then $f(x) > 0$. [*Hint:* Draw a sketch.]

*45.   Prove that if $f(x_0) < 0$ and f is continuous at x_0, then there is a number $\delta > 0$ such that if $x_0 - \delta < x < x_0 + \delta$, then $f(x) < 0$.

46.   Find the maximum M and the minimum m for the function $f(x) = |x + 3|$ on the interval $[-5, 7]$.

*47.   Find M and m for the function

$$f(x) = \frac{x^3 + 3x^2 + 2x + 1}{x + 1}$$

   on the interval $[0, 1]$.

48.   Give an example of a function defined in $[0, 1]$ bounded in $[0, 1]$, continuous for $0 < x \le 1$ but discontinuous at $x = 0$.

*49.   Suppose that f is continuous in $[a, b]$ with $f'(a) < 0$ and $f'(b) > 0$. Using Theorem 3 and Theorem 3.5.2, prove that there is a point c in (a, b) such that either $f'(c) = 0$ or $f'(c)$ does not exist.

50.   Prove that $f(x) = \sqrt{x}$ is continuous in the interval $[0, 1]$.

*51.   Prove that $f(x) = (1 - x^2)^{3/4}$ is continuous in $[-1, 1]$.

52.   Prove that $f(x) = (x^2 - 1)^{3/4}$ is continuous in the intervals $(-\infty, -1]$ and $[1, \infty)$.

53.   Find the largest closed interval over which $f(x) = \sin\sqrt{4 - x^2}$ is continuous.

54.   Show that the function $f(x) = \ln\cos x$ is *not* continuous in any interval of the form $[a, \pi/2]$.

*55.   Show that the function

$$f(x) = \begin{cases} e^{-1/x}, & x \neq 0 \\ 0, & x = 0 \end{cases}$$

is discontinuous at 0.

*56.   Show that the function

$$f(x) = \begin{cases} e^{-1/x^2}, & x \neq 0 \\ 0, & x = 0 \end{cases}$$

is continuous at 0.

10.5 Differentiation and a Proof of the Chain Rule

In Section 3.3 we gave a partial proof of the chain rule (Theorem 3.3.1). In this section we give a complete proof of this important theorem. We begin with an important observation.

By the definition of the derivative,

$$f'(x) = \frac{dy}{dx} = \lim_{\Delta x \to 0} \frac{\Delta y}{\Delta x}. \tag{1}$$

This implies that if Δx is small, then $\Delta y/\Delta x$ is close to dy/dx. For the number x fixed, we define

$$\epsilon(\Delta x) = \frac{dy}{dx} - \frac{\Delta y}{\Delta x} = f'(x) - \frac{\Delta y}{\Delta x}. \tag{2}$$

Then from (1),

$$\lim_{\Delta x \to 0} \epsilon(\Delta x) = \lim_{\Delta x \to 0} \left(\frac{dy}{dx} - \frac{\Delta y}{\Delta x} \right) = 0.$$

Now multiplying both sides of (2) by Δx and rearranging terms, we obtain

$$\Delta y = f'(x)\Delta x - \epsilon(\Delta x) \cdot \Delta x \tag{3}$$

where $\lim_{\Delta x \to 0} \epsilon(\Delta x) = 0$. We have looked at this expression before in a different form. Since Δy is an increment and $f'(x) \Delta x$ is the differential dy, the expression $\epsilon(\Delta x) \cdot \Delta x$ is the difference between the increment and the differential (see Section 3.6).

We are now ready to prove the chain rule.

THEOREM 1 (Chain Rule).   Let g and f be differentiable functions such that for every point x_0 at which $g(x)$ is defined, $f(g(x_0))$ is also defined. Then with $u = g(x)$, the composite function $y = (f \circ g)(x) = f(g(x)) = f(u)$ is a differentiable function of x and

$$\frac{dy}{dx} = \frac{d}{dx}(f \circ g)(x) = \frac{d}{dx}f(g(x)) = f'(g(x))g'(x) = \frac{df}{du}\frac{du}{dx}.$$

PROOF

$$\frac{d}{dx}f(g(x)) = \lim_{\Delta x \to 0} \frac{f(g(x + \Delta x)) - f(g(x))}{\Delta x}$$

if this limit exists. Since f can be written as a function of u, we can rewrite (3) as

$$\Delta f = f'(u)\,\Delta u - \epsilon(\Delta u)\,\Delta u \tag{4}$$

where $\epsilon(\Delta u) \to 0$ as $\Delta u \to 0$, $\Delta u = g(x + \Delta x) - g(x)$, and $f'(u) = f'(g(x))$. Then using (4), we obtain

$$f(g(x + \Delta x)) - f(g(x)) = \Delta f = f'(u)\,\Delta u - \epsilon(\Delta u)\,\Delta u \tag{5}$$

where $\epsilon(\Delta u) \to 0$ as $\Delta u \to 0$. Dividing (5) by Δx, we obtain

$$\frac{f(g(x + \Delta x)) - f(g(x))}{\Delta x} = f'(u)\frac{\Delta u}{\Delta x} - \epsilon(\Delta u)\frac{\Delta u}{\Delta x}.$$

Finally, we take limits as $\Delta x \to 0$. Since $u = g(x)$ and g is differentiable, $\lim_{\Delta x \to 0} \Delta u/\Delta x = du/dx = g'(x)$. Also, $\Delta u = g(x + \Delta x) - g(x)$. Since g is differentiable at x, it is continuous there and $\lim_{\Delta x \to 0} g(x + \Delta x) = g(x)$, or, alternatively, $\lim_{\Delta x \to 0} \Delta u = \lim_{\Delta x \to 0}[g(x + \Delta x) - g(x)] = 0$. Thus, as $\Delta x \to 0$, $\Delta u \to 0$ and $\lim_{\Delta x \to 0} \epsilon(\Delta u) = 0$. Hence

$$\lim_{\Delta x \to 0} \epsilon(\Delta u) \cdot \frac{\Delta u}{\Delta x} = \lim_{\Delta x \to 0} \epsilon(\Delta u) \lim_{\Delta x \to 0} \frac{\Delta u}{\Delta x} = 0 \cdot \frac{du}{dx} = 0.$$

Putting all this together, we obtain

$$\lim_{\Delta x \to 0} \frac{f(g(x + \Delta x)) - f(g(x))}{\Delta x} = \lim_{\Delta x \to 0} f'(u)\frac{\Delta u}{\Delta x} - \lim_{\Delta x \to 0} \epsilon(\Delta u)\frac{\Delta u}{\Delta x}$$

$$= f'(g(x))g'(x) - 0$$

and the chain rule is proved.

10.6 The Mean Value Theorem

In this section we prove one of the most important theorems of mathematics—the mean value theorem. We then use this theorem to give easy proofs of some of the results cited earlier in this text. In Section 10.8 we show how the mean value theorem can be used to prove the fundamental theorem of calculus.

Before stating and proving the mean value theorem, we need one preliminary result.

THEOREM 1 (Rolle's Theorem†). Let f be continuous on the closed interval $[a, b]$ and differentiable on the open interval (a, b). If $f(a) = f(b) = 0$, then there exists at least one number c in (a, b) at which

$$f'(c) = 0. \tag{1}$$

  Remark. The situation described in Rolle's theorem can be depicted as in Figure 1. The proof of the theorem involves describing the situation in the figure.

†Named after the French mathematician Michel Rolle (1652–1719).

PROOF OF ROLLE'S THEOREM. If $f(x) = 0$ for all x in $[a, b]$, then $f'(x) = 0$ in (a, b) and any c in (a, b) will work. If $f(x)$ is not the zero function, then f must become positive or negative in (a, b). Suppose f takes on positive values in (a, b) as in Figure 1.

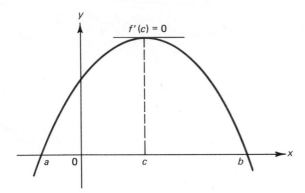

Figure 1

(Negative values can be handled in the same way.) By Theorem 10.4.3, there is a number x_1 in $[a, b]$ such that $f(x_1) = M$ is the maximum value of $f(x)$ on $[a, b]$. x_1 is not equal to a or b since $f(a) = f(b) = 0$ by hypothesis. Therefore, since x_1 is in (a, b), $f'(x_1)$ exists (by hypothesis). Since f has a (local) maximum at x_1, $f'(x_1) = 0$ by Theorem 3.5.2. Taking $c = x_1$, we see that Rolle's theorem is proved.

EXAMPLE 1. Let $f(x) = (x - 1)(x - 2)$. Then since $f(1) = f(2) = 0$, there is a point c in $(1, 2)$ such that $f'(c) = 0$. To find this point it is necessary to differentiate: $f'(x) = (x - 1) + (x - 2) = 2x - 3 = 0$ when $x = \frac{3}{2}$.

EXAMPLE 2. An interesting physical illustration of Rolle's theorem is given by the following situation: Suppose an object (like a ball or rock) is thrown from ground level into the air. Let $s(t)$ denote the height the object reaches at time t. Clearly $s(0) = 0$ and $s(T) = 0$ for some future time T. (The object hits the ground at time T.) We can then conclude, from Rolle's theorem, that there is a time t_1 such that $s'(t_1) = 0$. That is, there is a time at which the velocity of the object is zero—a physical fact confirmed by experience.

We are now ready to state the mean value theorem.

THEOREM 2 (Mean Value Theorem). Let f be continuous on the closed interval $[a, b]$ and differentiable on the open interval (a, b). Then there exists at least one number c in (a, b) such that

$$f'(c) = \frac{f(b) - f(a)}{b - a}. \tag{2}$$

Before proving the mean value theorem, we describe what it says geometrically. Look at Figure 2. We can see that the expression $[f(b) - f(a)]/[b - a]$ represents the slope of the secant line joining the points $(a, f(a))$ and $(b, f(b))$. Then, since $f'(x)$ is the slope of the tangent line to the curve, the mean value theorem states that there is always a number c between a and b such that the slope of the tangent line at the point

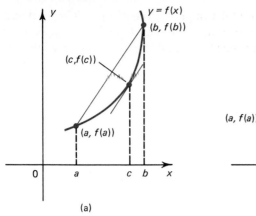

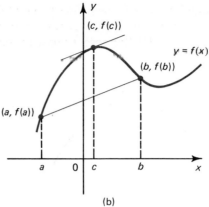

(a)   (b)

Figure 2

$(c, f(c))$ is the same as the slope of the secant line; that is, the secant and tangent lines
are parallel.

PROOF OF THE MEAN VALUE THEOREM. We define the function

$$g(x) = f(x) - \left[\frac{f(b) - f(b)}{b - a}(x - a) + f(a)\right]. \tag{3}$$

The expression

$$y = \frac{f(b) - f(a)}{b - a}(x - a) + f(a)$$

is the equation of a straight line (the secant line discussed above) which passes
through the point $(a, f(a))$ since when $x = a$, $y = f(a)$. See Figure 3. The function

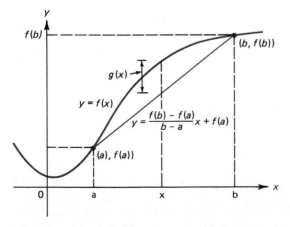

Figure 3

$g(x)$ represents the difference between the curve $f(x)$ and the secant line. We first
observe that since a linear function is everywhere differentiable (being a first degree
polynomial) and since $f(x)$ is continuous on $[a, b]$ and differentiable on (a, b), $g(x)$ is

also continuous on $[a, b]$ and differentiable on (a, b). Moreover,

$$g(a) = f(a) - \left[\frac{f(b) - f(a)}{b - a}(a - a) + f(a)\right] = f(a) - f(a) = 0$$

and

$$g(b) = f(b) - \left[\frac{f(b) - f(a)}{b - a}(b - a) + f(a)\right]$$
$$= f(b) - [f(b) - f(a) + f(a)] = 0.$$

Thus we may use Rolle's theorem and conclude that there is a number c in (a, b) such that

$$g'(c) = 0.$$

But $g'(x) = f'(x) - \{(f(b) - f(a))/(b - a)\}$ so that since $g'(c) = 0$,

$$g'(c) = f'(c) - \left[\frac{f(b) - f(a)}{b - a}\right] = 0$$

and the proof of the mean value theorem is complete.

EXAMPLE 3.   Let $f(x) = x^2$, $a = 2$, $b = 5$. Then $[f(b) - f(a)]/(b - a) = (25 - 4)/3 = 21/3 = 7$. By the mean value theorem there is a number c in $(2, 5)$ such that $f'(c) = 7$. Since $f'(x) = 2x$, we have $c = 7/2$. (See Figure 4.)

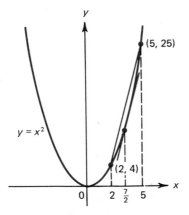

Figure 4

EXAMPLE 4.   Let $f(x) = \sin x$ and let $a = 0$ and $b = \pi/2$. Then

$$\frac{\sin b - \sin a}{b - a} = \frac{1}{\pi/2} = \frac{2}{\pi}.$$

Thus there is a number c in $(0, \pi/2)$ such that $f'(c) = \cos c = 2/\pi \approx 0.6366$ and $c \approx 0.8807$. See Figure 5.

We now prove, using the mean value theorem, two of the theorems cited earlier in the book.

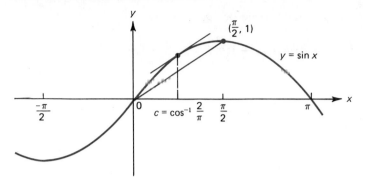

Figure 5

THEOREM 3. If f is differentiable and if $f'(x) = 0$ for every x, then f is a constant function.

PROOF. Pick any two numbers a and b with $a \neq b$. Then there is a number c such that

$$\frac{f(b) - f(a)}{b - a} = f'(c) = 0.$$

Since $b \neq a$, this can only occur if $f(b) - f(a) = 0$ which implies that $f(b) = f(a)$. Since this is true for any numbers a and b, f is a constant function.

THEOREM 4 (Theorem 3.5.1). Let f be differentiable on (a, b) and continuous on $[a, b]$. Then

(i)   if $f'(x) > 0$ for every x in (a, b), f is increasing on $[a, b]$.
(ii)   if $f'(x) < 0$ for every x in (a, b), f is decreasing on $[a, b]$.

(See Definitions 3.5.1 and 3.5.2.)

PROOF. (i)   We must show that if $a \leq x_1 < x_2 \leq b$, then $f(x_2) > f(x_1)$. But

$$\frac{f(x_2) - f(x_1)}{x_2 - x_1} = f'(c) > 0,$$

where $x_1 < c < x_2$. Since $x_2 > x_1$, $x_2 - x_1 > 0$ and we have $f(x_2) - f(x_1) > 0$ which implies that $f(x_2) > f(x_1)$. Part (ii) is proved in the same way.

Notice how much easier (and more complete) this proof is compared to the partial proof given in Section 3.5. The mean value theorem is a powerful tool indeed.

PROBLEMS 10.6

In Problems 1–8 find a number c which satisfies the conclusion of the mean value theorem for the given function and numbers.

1. $f(x) = x^3$; $a = 1$, $b = 2$
2. $f(x) = 1/x$; $a = 1$, $b = 4$
3. $f(x) = \cos x$; $a = 0$, $b = \pi/2$
4. $f(x) = \ln x$; $a = 1$, $b = e$
5. $f(x) = e^x$; $a = 0$, $b = \ln 4$
6. $f(x) = \sqrt{x}$; $a = 1$, $b = 4$
7. $f(x) = 1 + 2x^2$; $a = -1$, $b = 1$
8. $f(x) = \sqrt[3]{x}$; $a = -8$, $b = -1$.

9. Let $f(x) = |x - 1| - 2$.
    (a) Show that $f(3) = f(-1) = 0$.
    (b) Show that there is no number c in $(-1, 3)$ such that $f'(c) = 0$.
    (c) Explain why this does not contradict Rolle's theorem.

10. Show that there is no number c that satisfies the mean value theorem in the interval $[-1, 1]$ when $f(x) = x^{2/3}$.

11. Let $f(x) = 1/x$. Show that there is no c in the interval $(-1, 2)$ such that

$$f'(c) = \frac{f(2) - f(-1)}{2 - (-1)}.$$

Explain why this does *not* contradict the mean value theorem.

12. Prove that if the conditions of the mean value theorem hold and if $f(a) = f(b)$, that there is a point c in (a, b) such that $f'(c) = 0$.

*13. A function is one-to-one (see Section 10.7) if $f(x) = f(y)$ implies that $x = y$. Use the result of Problem 12 to show that if f is differentiable and $f'(x) \neq 0$ for every real x, then f is 1-1.

14. Use the result of Problem 13 to show that a differentiable one-to-one function is always increasing or always decreasing.

15. A function is called a *contraction* on $[a, b]$ if for every x and y in $[a, b]$,

$$|f(x) - f(y)| \leq \lambda |x - y|,$$

where $\lambda < 1$. Show that if f is differentiable and $|f'(x)| \leq \lambda < 1$ for every x in $[a, b]$, then f is a *contraction* on $[a, b]$.

16. Show that $f(x) = e^x$ is a contraction on any interval of the form $(-\infty, -a)$ where $a > 0$.

17. Show that $f(x) = \sin x$ is a contraction on $[0, \pi/4]$.

18. For what values of b is the function $f(x) = x^2 + 1$ a contraction on $[0, b]$?

19. Let $P(x) = c_0 + c_1 x + c_2 x^2 + \cdots + c_n x^n$. Prove that if $P'(x_1) = P'(x_2) = 0$ and $P'(x) \neq 0$ for $x_1 < x < x_2$, then there is at most one number x_3 in (x_1, x_2) such that $P(x_3) = 0$.

20. Use the result of Problem 19 to show that the only root of $x^5 - x = 0$ in the interval $(-1, 1)$ is $x = 0$.

21. If $f''(x)$ exists for every x, show that between any two critical points of x there is a point at which $f''(x) = 0$.

22. Let

$$f(x) = \begin{cases} x^2, & x \leq 1 \\ \alpha x^3, & x > 1 \end{cases}.$$

    (a) For what value of α does the mean value theorem apply in the interval $[0, 2]$?
    (b) Using the value α obtained in part (a), find a number c such that

$$\frac{f(2) - f(0)}{2 - 0} = f'(c).$$

*23. Consider the parabola $y = f(x) = ax^2 + bx + c$. By the mean value theorem, for any two numbers x_1 and x_2, there is a number x_3 in (x_1, x_2) such that the tangent to the parabola at the point $(x_3, f(x_3))$ is parallel to the secant line joining the points $(x_1, f(x_1))$ and $(x_2, f(x_2))$. Show that $x_3 = (x_1 + x_2)/2$ (see Figure 6).

*24. Use Rolle's theorem to show that a quadratic polynomial has at most two real roots. [*Hint:* Show that if there are three real roots, then there are two real roots to the equation $P'(x) = 0$ and one root to the equation $P''(x) = 0$, which is impossible.]

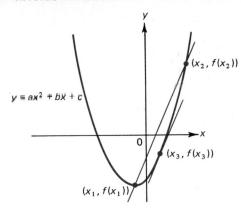

$y = ax^2 + bx + c$

$(x_2, f(x_2))$

$(x_3, f(x_3))$

$(x_1, f(x_1))$

Figure 6

*25. Show that a cubic polynomial has at most 3 real roots. [*Hint:* Follow the steps outlined in Problem 24.]

*26. Use mathematical induction (Appendix 1) to show that an nth degree polynomial has at most n real roots.

27. Use the mean value theorem to show that if f is differentiable and $f'(x) = c$, a constant, then f is a linear function.

28. Use the mean value theorem to show that if f is differentiable and $f'(x)$ is a linear function, then f is a quadratic polynomial.

29. Use the mean value theorem to show that if f is differentiable and $f'(x)$ is a polynomial of degree n, then $f(x)$ is a polynomial of degree $n + 1$. (Do *not* integrate to prove this.)

10.7 Inverse Functions

Let $y = 2x + 3$. Then we can write x as a function of y: $x = (y - 3)/2$. The functions $2x + 3$ and $(x - 3)/2$ are called *inverse functions*. We have seen other examples of inverse functions in Chapters 6 and 7. If $y = \ln x$, then $x = e^y$. The functions e^x and $\ln x$ are inverse functions. Also, if $y = \sin x$, $-\pi/2 \le x \le \pi/2$, then $x = \sin^{-1} y$. The functions $\sin x$ and $\sin^{-1} x$ are inverse functions. In general, we have

DEFINITION 1 (Inverse Functions). The functions f and g are *inverse functions* if

(i) for every x in the domain of g, $g(x)$ is in the domain of f and
$f(g(x)) = x$, and   (1)
(ii) for every x in the domain of f, $f(x)$ is in the domain of g and
$g(f(x)) = x$.   (2)

In this case we write $f(x) = g^{-1}(x)$ or $g(x) = f^{-1}(x)$.   (3)

EXAMPLE 1. If $f(x) = e^x$ and $g(x) = \ln x$, then $f(g(x)) = e^{\ln x} = x$ and $g(f(x)) = \ln e^x = x$ so that e^x and $\ln x$ are indeed inverses.

Note. Example 1 illustrates that $f^{-1}(f(x))$ and $f(f^{-1}(x))$ are *not* the same functions. If $f = e^x$ and $f^{-1} = \ln x$, then $f(f^{-1}(x)) = e^{\ln x}$ and $f^{-1}(f(x)) = \ln e^x$. The first function is defined only when $x > 0$ and the second is defined for *all* x. The two functions have different domains and are therefore different—even though each is equal to x wherever it is defined.

EXAMPLE 2. If $f(x) = \sin x$ and $g(x) = \sin^{-1} x$, then $f(g(x)) = \sin(\sin^{-1} x) = x$ and $\sin^{-1}(\sin x) \neq x$. Note that $D(f \circ g) = [-1, 1]$ and $D(g \circ f) = \mathbb{R}$.

EXAMPLE 3. If $f(x) = 2x + 3$ and $g(x) = (x - 3)/2$, then

$$f(g(x)) = 2\left(\frac{x-3}{2}\right) + 3 = x - 3 + 3 = x$$

and

$$g(f(x)) = \frac{(2x + 3) - 3}{2} = x$$

so that $2x + 3$ and $(x - 3)/2$ are inverse functions.

In general, we find inverse functions by

    (i)   writing y as a function of x,
    (ii) writing x as a function of y, if possible.

The two functions thus obtained are inverses.

We used the phrase "if possible" in step (ii) above since it is not always possible to solve for x as a *function* of y.

EXAMPLE 4. Let $y = x^2$. Then $x = \pm\sqrt{y}$. This means that each value of y comes from two different values of x. See Figure 1. We recall from Section 1.6 that a

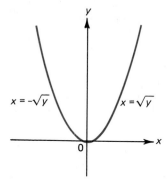

Figure 1

function $y = f(x)$ can be thought of as a rule that assigns to each x in its domain a *unique* value of y. Suppose we try to define the inverse of x^2 by taking the positive square root. Then we would, hopefully, be able to show that $f = x^2$ and $g = \sqrt{x}$ are inverses. Let $x = -2$. Then $g(f(x)) = \sqrt{(-2)^2} = \sqrt{4} = 2 \neq -2 = x$ so that $g(f(x)) \neq x$.

To avoid problems like the one just encountered, we make the following definition.

DEFINITION 2. The function $y = f(x)$ is *one-to-one* on the interval $[a, b]$, written 1-1, if for every pair of numbers x_1 and x_2 in $[a, b]$, $f(x_1) = f(x_2)$ only when $x_1 = x_2$. That is, each value of $f(x)$ comes from only one value of x.

THEOREM 1. If f is an increasing or decreasing function on $[a, b]$, then f is 1-1 on $[a, b]$.

Remark. This theorem is illustrated in Figure 2.

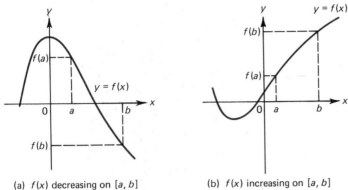

(a) $f(x)$ decreasing on $[a, b]$ (b) $f(x)$ increasing on $[a, b]$ Figure 2

PROOF. Suppose f is increasing and suppose that $f(x) = f(y)$. If $x > y$, then $f(x) > f(y)$. If $y > x$, then $f(y) > f(x)$. Since both these cases are ruled out, we have $x = y$. The proof in the case f is decreasing is identical.

THEOREM 2. If f is continuous on $[a, b]$ and has a continuous derivative on (a, b) satisfying $f'(x) \neq 0$ for every x in (a, b), then f is 1-1 on $[a, b]$.

PROOF. Suppose that there are points x_1 and x_2 in (a, b) such that $f'(x_1) < 0$ and $f'(x_2) > 0$. Then since f' is continuous, the intermediate value theorem (Theorem 10.4.4) implies that there is an x_3 between x_1 and x_2 such that $f'(x_3) = 0$. But this is impossible since $f'(x) \neq 0$ on (a, b). Therefore on (a, b) $f'(x) > 0$ or $f'(x) < 0$. But if $f'(x) > 0$, then f is increasing (by Theorem 10.6.4) and if $f'(x) < 0$, f is decreasing. In either case, Theorem 1 tells us that f is 1-1 on $[a, b]$.

> *Remark.* There are 1-1 functions for which the condition $f'(x) \neq 0$ does not hold. For example, the function x^3 is 1-1 in $[-1, 1]$ but $f'(0) = 0$.

Why are we interested in 1-1 functions? Because if $y = f(x)$ is 1-1 then every value of y comes from a unique value of x so that we can write x as a function of y, and this new function is the inverse of y. That is, *every 1-1 function has an inverse*.

EXAMPLE 5. Let $f(x) = \sqrt{x}$. Then $f' = 1/2\sqrt{x} \neq 0$ for $x > 0$. Thus for $x > 0$, f has an inverse which we find by setting $y = \sqrt{x}$ so that $x = y^2$ and, therefore, $f^{-1} = x^2$.

EXAMPLE 6. Let $f(x) = e^x$. Then $f' = e^x > 0$ so that f has an inverse which we have already seen to be $\ln x$.

EXAMPLE 7. Let $f(x) = y = \sin x$. Then $f' = \cos x \neq 0$ in $(-\pi/2, \pi/2)$ so that f is 1-1 in $[-\pi/2, \pi/2]$. That is why we write $y = \sin^{-1} x$, $-\pi/2 \leq y \leq \pi/2$.

It is not always possible to explicitly calculate an inverse function even when such a function is known to exist, as the following example illustrates.

EXAMPLE 8.   Let $f(x) = x^5 + x^3 + x$. Then $f'(x) = 5x^4 + 3x^2 + 1$ which is always positive so that f^{-1} exists. However, there is no way to explicitly solve the equation $y = x^5 + x^3 + x$ for x. This does not contradict the existence of f^{-1}. It simply illustrates the limitations of our algebraic techniques.

EXAMPLE 9.   Let $f(x) = x^3 + 4$. Then $f'(x) = 3x^2$ which is zero when $x = 0$. Nevertheless $f(x)$ is increasing and 1-1 as a glance at Figure 3 illustrates. Thus f^{-1} exists and solving $y = x^3 + 4$ for x yields $f^{-1}(x) = \sqrt[3]{x} - 4$.

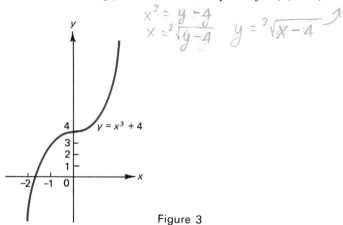

$$x^3 = y - 4$$
$$x = \sqrt[3]{y-4} \qquad y = \sqrt[3]{x-4}$$

Figure 3

EXAMPLE 10.   We have seen that $y = f(x) = x^2$ does not have an inverse. However if we redefine f by $f(x) = x^2$, $x > 0$, then f does have an inverse given by $f^{-1} = \sqrt{x}$.

In Section 3.7 we suggested the relationship $dy/dx = 1/(dx/dy)$. For example if $y = e^x$, then $x = \ln y$, $dy/dx = e^x$, and $dx/dy = 1/y = 1/e^x$ so that $dy/dx = 1/(dx/dy)$. That this fact is true in some generality is given by the following theorem.

THEOREM 3.   Let $y = f(x)$ be differentiable in some neighborhood of a point x_0 and suppose that f is 1-1 in a neighborhood of x_0 and that $f'(x_0) \neq 0$. If $x = g(y) = f^{-1}(y)$ is an inverse function of $f(x)$ which is continuous in a neighborhood of $y_0 = f(x_0)$, then $f^{-1}(y) = g(y)$ is differentiable at y_0 and

$$g'(y_0) = \frac{1}{f'(x_0)} \qquad \text{or, equivalently,} \qquad \frac{dx}{dy} = \frac{1}{dy/dx}. \qquad (4)$$

PROOF: Since $\lim\limits_{\Delta x \to 0} \dfrac{f(x_0 + \Delta x) - f(x_0)}{\Delta x} = f'(x_0) \neq 0$, we have, by Theorem 10.2.4,

$$\lim_{\Delta x \to 0} \frac{\Delta x}{f(x_0 + x) - f(x_0)} = \frac{1}{f'(x_0)}. \qquad (5)$$

Now $x_0 = g(y_0)$ and since $\Delta y = f(x_0 + \Delta x) - f(x_0) = f(x_0 + \Delta x) - y_0$, we have $y_0 + \Delta y = f(x_0 + \Delta x)$ or $x_0 + \Delta x = g(y_0 + \Delta y)$ and $\Delta x = g(y_0 + \Delta y) - g(y_0)$. Thus

$$\lim_{\Delta y \to 0} \Delta x = \lim_{\Delta y \to 0} [g(y_0 + \Delta y) - g(y_0)] = 0 \qquad (6)$$

by the continuity of g. Finally, since $y_0 = f(x_0)$, we find that

$$g'(y_0) = \lim_{\Delta y \to 0} \frac{g(y_0 + \Delta y) - g(y_0)}{\Delta y} = \lim \frac{g(f(x_0 + \Delta x)) - g(f(x_0))}{f(x_0 + \Delta x) - f(x_0)}$$

$$= \lim_{\Delta y \to 0} \frac{(x_0 + \Delta x) - x_0}{f(x_0 + \Delta x) - f(x_0)} = \lim_{\Delta y \to 0} \frac{\Delta x}{f(x_0 + \Delta x) - f(x_0)}$$

$$= \text{(from (6))} \lim_{\Delta x \to 0} \frac{\Delta x}{f(x_0 + \Delta x) - f(x_0)} = \frac{1}{f'(x_0)}.$$

This completes the proof of the theorem.

Note. It can be shown (see Problem 31) that if $f'(x_0) \neq 0$, then $f(x)$ is 1-1 in a neighborhood of x_0; thus the assumption that $f(x)$ is 1-1 is not really necessary in the statement of the theorem.

EXAMPLE 11. Let $y = \sqrt{x}$. Then $dy/dx = 1/2\sqrt{x}$. The inverse function is $x = g(y) = y^2$. Then

$$\frac{dx}{dy} = 2y = 2\sqrt{x} = \frac{1}{dy/dx} = \frac{1}{1/2\sqrt{x}}.$$

EXAMPLE 12. Let $y = f(x) = x^2 + 4x + 3$. To find the inverse function, we note that

$$x^2 + 4x + (3 - y) = 0,$$

or

$$x = \frac{-4 \pm \sqrt{16 - 4(3 - y)}}{2} = -2 \pm \frac{\sqrt{4 + 4y}}{2} = -2 \pm \sqrt{1 + y}.$$

The graph of $y = f(x)$ is given in Figure 4. It is apparent that for every value of

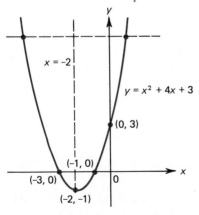

$x = -2$

$y = x^2 + 4x + 3$

$(0, 3)$

$(-1, 0)$

$(-3, 0)$

$(-2, -1)$

Figure 4

$y > -1$, there are *two* different values of x. Thus f is not 1-1 and f^{-1} does not exist. However, we may define

$$f_1(x) = x^2 + 4x + 3, \quad x \geq -2 \quad \text{and} \quad f_2(x) = x^2 + 4x + 3, \quad x \leq -2$$

(that is, we have restricted the domain of f in two different ways to obtain two different functions). Then f_1 and f_2 are both 1-1 and we have the inverse functions

$$g_1(y) = f_1^{-1}(y) = -2 + \sqrt{1+y} \qquad \text{and} \qquad g_2(y) = f_2^{-1}(y) = -2 - \sqrt{1+y}.$$

Note that both $f_1^{-1}(y)$ and $f_2^{-1}(y)$ are not defined for $y < -1$. To verify (4), we have, for $x \geq -2$,

$$\frac{dx}{dy} = \frac{dg_1}{dy} = \frac{1}{2\sqrt{1+y}}.$$

But since $x = -2 + \sqrt{1+y}$, $\sqrt{1+y} = x + 2$. Therefore

$$\frac{1}{2\sqrt{1+y}} = \frac{1}{2x+4} = \frac{1}{dy/dx}.$$

We obtain a similar result for $x \leq -2$ $(x = g_2(y))$.

EXAMPLE 13. If $y = x^5 + x^3 + x$, then as we saw in Example 8, f^{-1} exists but cannot be explicitly calculated. Nevertheless, we can calculate

$$\frac{df^{-1}}{dy} = \frac{dx}{dy} = \frac{1}{dy/dx} = \frac{1}{5x^4 + 3x^2 + 1}.$$

This answer is not exactly what we want since we would like to calculate df^{-1}/dy as a function of y, not of x. Nevertheless it *is* quite useful since, for example, if we know a point (x_0, y_0) on the curve $x = f^{-1}(y)$, we could calculate the derivative of f^{-1} at that point.

PROBLEMS 10.7

In Problems 1–30 determine intervals over which the given function is 1-1 and determine the inverse function over each such interval, if possible. Then use Theorem 3 to calculate the derivatives of each inverse function.

1. $f(x) = 3x + 5$

2. $f(x) = 17 - 2x$

3. $f(x) = \dfrac{1}{x}$

4. $f(x) = \dfrac{1}{x-2}$

5. $f(x) = \sqrt{4x+3}$

6. $f(x) = \sqrt[3]{x+1}$

7. $f(x) = 1 - x^3$

8. $f(x) = (x+1)^2$

9. $f(x) = e^{x-2}$

10. $f(x) = \sqrt{e^x}$

11. $f(x) = \ln(2x+5)$

12. $y = \ln(x^2 + 1)$

13. $y = \tan 3x$

14. $y = \cos 2x$

15. $y = \csc x$

16. $y = \cot x$

17. $y = \cosh 2x$

18. $y = \sinh(x/2)$

19. $y = \sqrt{\tanh x}$

20. $y = \dfrac{x}{x+1}$

21. $y = xe^x$

22. $y = \dfrac{\ln x}{x}$

23. $y = e^{\sin x}$

24. $y = x^2 + 2x + 2$

25. $y = x^2 - 5x + 6$

*26. $y = x^4 + 4x^2 + 4$ [*Hint:* Use the quadratic formula.]

27. $y = x^3 + x$

28. $y = x^9 + x^7 + x^3 + x + 1$

29. $y = e^{x^2}$

30. $y = \sin x^2$.

*31. Prove that if $f(x)$ is differentiable in a neighborhood of the point x_0 and if $f'(x_0) \neq 0$, then $f(x)$ is 1-1 in some neighborhood of x_0. [*Hint:* Suppose that $f'(x_0) > 0$. Then in some neighborhood of x_0, $f'(x) > 0$. If x_1 and x_2 are two points in this neighborhood and if $x_1 > x_2$, then $f(x_1) > f(x_2)$. A similar result holds if $f'(x_0) < 0$.]

*32. Suppose that f is continuous and increasing on $[a, b]$. Show that f^{-1} is increasing on $[f(a), f(b)]$.

*33. Suppose that f is continuous and increasing on $[a, b]$. Show that f^{-1} is continuous on $[f(a), f(b)]$.

34. Prove that

$$\frac{d}{dx} x^{1/n} = \frac{1}{n} x^{(1/n)-1}$$

if n is a positive integer. [*Hint:* Let $f(x) = x^n$. Then find f^{-1} and use Theorem 3.]

10.8 Integration

In Chapter 4 we introduced the definite and indefinite integrals. We showed there and in Chapter 8 how certain integrals could be calculated. The basic theorem underlying these calculations is that if f is piecewise continuous in $[a, b]$, then $\int_a^b f(x)\, dx$ exists. The proof of this theorem requires tools from advanced calculus that cannot be developed in an elementary calculus text. Nevertheless, there are interesting results that can be derived using facts about continuous and differentiable functions that we already know. In this section we will obtain some of these results, assuming only the basic integrability theorem cited above.

We begin with the theorem which was used in Section 4.5 to prove the fundamental theorem of calculus.

THEOREM 1 (Theorem 4.5.1). If f is continuous on $[a, b]$ then $G(x) = \int_a^x f(t)\, dt$ is continuous on $[a, b]$, differentiable on (a, b), and for every x in (a, b),

$$G'(x) = f(x). \tag{1}$$

PROOF. We show first that $G'(x) = f(x)$ for x in (a, b). Let x_0 be in (a, b). We have

$$G(x_0 + \Delta x) = \int_a^{x_0 + \Delta x} f(t)\, dt = \int_a^{x_0} f(t)\, dt + \int_{x_0}^{x_0 + \Delta x} f(t)\, dt. \tag{2}$$

Then

$$G(x_0 + \Delta x) - G(x_0) = \int_a^{x_0 + \Delta x} f(t)\, dt - \int_a^{x_0} f(t)\, dt = \int_{x_0}^{x_0 + \Delta x} f(t)\, dt. \tag{3}$$

Let $\epsilon > 0$ be chosen. Because f is continuous, there is a $\delta > 0$ such that if $|x - x_0| < \delta$, then $|f(x) - f(x_0)| < \epsilon$. If $\Delta x < \delta$, then for any x in the interval $[x_0, x_0 + \Delta x]$, $|f(x) - f(x_0)| < \epsilon$ or

$$f(x_0) - \epsilon < f(x) < f(x_0) + \epsilon.$$

Then, by Theorem 4.3.8, with $m = f(x_0) - \epsilon$ and $M = f(x_0) + \epsilon$, we have

$$[f(x_0) - \epsilon]\, \Delta x < \int_{x_0}^{x_0 + \Delta x} f(t)\, dt < [f(x_0) + \epsilon]\, \Delta x.$$

Using (3) in (1), we obtain

$$[f(x_0) - \epsilon]\, \Delta x < G(x_0 + \Delta x) - G(x_0) < [f(x_0) + \epsilon]\, \Delta x$$

and, dividing by Δx (we assume that $\Delta x > 0$; the case $\Delta x < 0$ can be dealt with in a similar fashion),

$$f(x_0) - \epsilon < \frac{G(x_0 + \Delta x) - G(x_0)}{\Delta x} < f(x_0) + \epsilon.$$

If we let $\Delta x \to 0$, we can assume that $\epsilon \to 0$ since $\lim_{\Delta x \to 0} f(x_0 + \Delta x) = f(x_0)$. Thus

$$G'(x_0) = \lim_{\Delta x \to 0} \frac{G(x_0 + \Delta x) - G(x_0)}{\Delta x} = f(x_0)$$

which shows that if x_0 is in (a, b), the theorem is true. The only things left to worry about are the endpoints a and b. To show continuity at the endpoints, we must show that $\lim_{x \to a^+} G(x) = G(a)$ and $\lim_{x \to b^-} G(x) = G(b)$, or alternatively, we must show that

$$\lim_{\Delta x \to 0^+} \int_a^{a+\Delta x} f(t)\, dt = 0 = \int_a^a f(t)\, dt = G(a) \qquad (4)$$

and

$$\lim_{\Delta x \to 0^+} \int_a^{b-\Delta x} f(t)\, dt = \int_a^b f(t)\, dt = G(b). \qquad (5)$$

To prove (4), suppose that $\epsilon > 0$ is given. Then there is a δ such that if x is in $[a, b]$ and $x - a < \delta$, $|f(x) - f(a)| < \epsilon$. If $\Delta x < \delta$, we have $(a + \Delta x) - a < \delta$ and in the interval $[a, a + \Delta x]$, then $f(a) - \epsilon < f(x) < f(a) + \epsilon$ so that

$$\int_a^{a+\Delta x} (f(a) - \epsilon)\, dx < \int_a^{a+\Delta x} f(x)\, dx < \int_a^{a+\Delta x} (f(a) + \epsilon)\, dx$$

or

$$(f(a) - \epsilon)\, \Delta x < \int_a^{a+\Delta x} f(x)\, dx < (f(a) + \epsilon)\, \Delta x$$

by Theorem 4.3.8. Since as $\epsilon \to 0$, we can assume that $\Delta x \to 0$, we let $\epsilon \to 0$ in the above expression to prove (4). To prove (5), we observe that

$$\int_a^b f(t)\, dt - \int_a^{b-\Delta x} f(t)\, dt = \int_{b-\Delta x}^b f(t)\, dt$$

and the same argument used to prove (4) shows that

$$\lim_{\Delta x \to 0^+} \int_{b-\Delta x}^b f(t)\, dt = 0 \quad \text{so that} \quad \lim_{\Delta x \to 0^+} \int_a^{b-\Delta x} f(t)\, dt = \int_a^b f(t)\, dt.$$

Theorem 1 can be used to show that if f is continuous and $\int_a^b f(t)\, dt$ exists, then an indefinite integral $\int f(t)\, dt$ also exists. This is obvious since, by the definition of the indefinite integral, $G(x)$ defined above *is* an indefinite integral.

We now prove that any two definite integrals differ by a constant.

THEOREM 2 (Theorem 4.4.1). If F and G are differentiable and if $F'(x) = G'(x)$, then $F(x) = G(x) + C$.

PROOF. Let $H(x) = F(x) - G(x)$. Then $H'(x) = F'(x) - G'(x) = 0$. Therefore, by

Theorem 10.6.3, $H(x) = C$ and the theorem is proved.

In Section 4.5 we used Theorem 1 to prove the fundamental theorem of calculus. We now show that this central theorem can be proved without invoking Theorem 1 by using the mean value theorem.

THEOREM 3 (Fundamental Theorem of Calculus). Let f be continuous on $[a, b]$. If F is some indefinite integral of f on $[a, b]$, then $\int_a^b f(t)\, dt = F(b) - F(a)$.

PROOF. Let P be a partition of $[a, b]$:

$$a = x_0 < x_1 < x_2 < \cdots < x_{n-1} < x_n = b.$$

Then

$$
\begin{aligned}
F(b) - F(a) &= F(x_n) - F(x_0) \\
&= [F(x_n) - F(x_{n-1})] + [F(x_{n-1}) - F(x_{n-2})] \\
&\quad + \cdots + [F(x_2) - F(x_1)] + [F(x_1) - F(x_0)].
\end{aligned}
$$

(Notice how all terms except the first and last cancel off.) Since F is continuous in $[a, b]$ and differentiable on (a, b), we may use the mean value theorem in each subinterval to find a number $x_i{}^*$ such that

$$F(x_i) - F(x_{i-1}) = F'(x_i{}^*)(x_i - x_{i-1}) = f(x_i{}^*)\,\Delta x_i$$

for $i = 1, 2, \ldots, n$, where $x_{i-1} < x_i{}^* < x_i$. Thus for every partition P, there exist points $x_i{}^*$ in each subinterval such that

$$f(x_1{}^*)\,\Delta x_1 + f(x_2{}^*)\,\Delta x_2 + \cdots + f(x_n{}^*)\,\Delta x_n = F(b) - F(a). \tag{6}$$

Since $\int_a^b f(t)\, dt$ exists (f is continuous),

$$\int_a^b f(t)\, dt = \lim_{|P| \to 0} [f(x_1{}^*)\,\Delta x_1 + f(x_2{}^*)\,\Delta x_2 + \cdots + f(x_n{}^*)\,\Delta x_n]$$

exists and is independent of the way the $x_i{}^*$'s are chosen. Thus, we can see that for every partition P, there are points $x_i{}^*$ such that (6) holds. Therefore

$$\lim_{|P| \to 0} [f(x_1{}^*)\,\Delta x_1 + f(x_2{}^*)\,\Delta x_2 + \cdots + f(x_n{}^*)\,\Delta x_n] = \lim_{|P| \to 0} [F(b) - F(a)]$$

$$= F(b) - F(a)$$

and the theorem is proved.

We close this section with a theorem that is sometimes very useful in applications. It is one of several mean value theorems for integrals.

THEOREM 4 (Mean Value Theorem for Integrals). Suppose that f and g are continuous on $[a, b]$ and that $g(x)$ is never zero on (a, b). Then there exists a number c in $[a, b]$ such that

$$\int_a^b f(x)g(x)\, dx = f(c) \int_a^b g(x)\, dx. \tag{7}$$

PROOF. Since f is continuous on $[a, b]$, there exist numbers m and M such that

$$m \leq f(x) \leq M \tag{8}$$

and there are numbers x_1 and x_2 in $[a, b]$ such that $f(x_1) = m$ and $f(x_2) = M$ (from Theorem 10.4.3). Since $g(x) \neq 0$ and g is continuous, we see that either $g(x) > 0$ or $g(x) < 0$ on (a, b). (Otherwise the intermediate value theorem would be contradicted.) Suppose that $g > 0$. The case $g < 0$ is left as a problem (see Problem 1). Then we multiply (8) by $g(x)$ to obtain

$$mg(x) \leq f(x)g(x) \leq Mg(x)$$

which implies by the comparison theorem (Theorem 4.3.7) that

$$m \int_a^b g(x)\, dx \leq \int_a^b f(x)g(x)\, dx \leq M \int_a^b g(x)\, dx. \tag{9}$$

Since $g(x) > 0$, $\int_a^b g(x)\, dx > 0$ and dividing (9) by this integral yields

$$m \leq \frac{\displaystyle\int_a^b f(x)g(x)\, dx}{\displaystyle\int_a^b g(x)\, dx} \leq M.$$

Since $m = f(x_1)$ and $M = f(x_2)$ are values for $f(x)$, $[\int_a^b f(x)g(x)\, dx]/\int_a^b g(x)\, dx$ is a number between $f(x_1)$ and $f(x_2)$ so that, by the intermediate value theorem, there is a c such that

$$f(c) = \frac{\displaystyle\int_a^b f(x)g(x)\, dx}{\displaystyle\int_a^b g(x)\, dx}$$

and the theorem is proved.

PROBLEMS 10.8

1. Show that the mean value theorem for integrals holds in the case $g < 0$ on (a, b).
2. Prove that for any continuous function $f \geq 0$

$$\int_0^\pi f(t) \sin t\, dt \leq \int_0^\pi f(t)\, dt.$$

3. Prove that if f is continuous and positive and if $a > 0$, then

$$\int_0^a e^{-t^2} f(t)\, dt \leq \int_0^a f(t)\, dt.$$

4. Let $G(u) = \int_a^u f(t)\, dt$, where f is continuous. Use the chain rule and Theorem 1 to prove that

$$\frac{dG(u)}{dx} = f(u)\frac{du}{dx}.$$

5. Use the results of Problem 4 to prove that if $b(x)$ is differentiable and f is continuous,

$$\frac{d}{dx} \int_a^{b(x)} f(t)\, dt = f(b(x))b'(x).$$

6. Use the result of Problem 5 to show that if $a(x)$ is differentiable,

$$\frac{d}{dx} \int_{a(x)}^b f(t)\, dt = -f(a(x))a'(x).$$

7. Using Problems 5 and 6, show that

$$\frac{d}{dx} \int_{a(x)}^{b(x)} f(t)\, dt = f(b(x))b'(x) - f(a(x))a'(x).$$

[*Hint:* Write $\int_{a(x)}^{b(x)} f = \int_{a(x)}^c f + \int_c^{b(x)} f.$]

In Problems 8–12 use Problem 7 to differentiate the given function.

8. $G(x) = \int_x^{x^2} \sin t^2\, dt$ 9. $G(x) = \int_{\sin x}^{\cos x} e^{\sqrt{t}}\, dt$ 10. $G(x) = \int_{\sqrt{x}}^{\sqrt[3]{x}} t^2\, dt$

11. $G(x) = \int_{\ln x}^{e^x} \dfrac{\cos t}{t}\, dt$ 12. $G(x) = \int_{\sinh x}^{\cosh x} \tanh^2 t\, dt.$

*13. Prove another form of the mean value theorem for integrals: Let f be continuous on $[a, b]$. Then there exists a number c in (a, b) such that $\int_a^b f(t)\, dt = f(c) \cdot (b - a)$. [*Hint:* Apply the mean value theorem for derivatives to the function $G(x)$ defined in Theorem 1.]

14. Using the result of Problem 13, prove that

$$\int_0^{2\pi} \sin t^3\, dt \le 2\pi.$$

15. Prove that $\displaystyle\int_0^1 e^{t^7}\, dt \le e$,   16. Prove that $\displaystyle\int_0^{e-1} \ln \sqrt{t + 1} \le e - 1.$

17. Prove that $\displaystyle\int_0^{\pi/2} (\sin^8 t \sqrt[3]{\cos t})^{t^5 - 3}\, dt \le \pi/2.$

Review Exercises for Chapter Ten

In Exercises 1–5 use the ϵ–δ definition of the limit to verify the given limits.

1. $\displaystyle\lim_{x \to -1} 4x = -4$ 2. $\displaystyle\lim_{x \to 3} x^2 + 2 = 11$ 3. $\displaystyle\lim_{x \to \infty} \left(\frac{1}{x} + \frac{1}{x^2}\right) = 0$

4. $\displaystyle\lim_{x \to \infty} \frac{1}{\sqrt[3]{x} + 3} = 0$ 5. $\displaystyle\lim_{x \to -3} |x + 4| = 1.$

6. Let

$$f(x) = \begin{cases} 1, & x \text{ an integer} \\ 2, & x \text{ not an integer.} \end{cases}$$

Show that $\lim_{x \to \infty} f(x)$ does not exist.

7. Prove that if $\lim_{x \to x_0} f(x) = L_1$ and $\lim_{x \to x_0} g(x) = L_2$ (where x_0, L_1, and L_2 are finite), then

$$\lim_{x \to x_0} \{[1 + f(x)][1 + g(x)]\} = 1 + L_1 + L_2 + L_1 L_2.$$

8. Give an example of two functions $f(x)$ and $g(x)$ such that $\lim_{x \to x_0} \{[1 + f(x)][1 + g(x)]\}$ exists but that neither $\lim_{x \to x_0} f(x)$ nor $\lim_{x \to x_0} g(x)$ exist.

9. Show that if

$$f(x) = \begin{cases} 2x + 3, & x \le -2 \\ x^2 - 5, & x > -2, \end{cases}$$

then $\lim_{x \to -2} f(x)$ exists. Calculate that limit.

10. Let

$$f(x) = \begin{cases} x^2 + 3, & x \le 3 \\ x^3 + \alpha, & x > 3. \end{cases}$$

For what value of α does $\lim_{x \to 3} f(x)$ exist? Calculate this limit.

11. Show that $f(x) = [4x]$ is continuous at $x = \frac{1}{3}$ and discontinuous at $x = \frac{1}{2}$.

12. Show that $f(x) = -1/(x^5 - 2)$ is continuous at $x = 2$.

13. Show that

$$f(x) = \begin{cases} e^{-1/x^4}, & x \ne 0 \\ 0, & x = 0 \end{cases}$$

is continuous at 0.

14. Show that

$$f(x) = \begin{cases} e^{-1/x^3}, & x \ne 0 \\ 0, & x = 0 \end{cases}$$

is discontinuous at 0.

15. Calculate $\lim_{x \to 3} e^{\cos(x-3)}$.   16. Calculate $\lim_{x \to \pi/2} \ln \sin x$.

17. Prove, using the intermediate value theorem, that there is a number whose square is 7.

18. Show that $f(x) = [x] - \frac{5}{2}$ is piecewise continuous in $[0, 8]$.

19. Show that

$$f(x) = \begin{cases} x^3 - 7x + 2, & x \le 1 \\ x^5 - \sqrt[3]{x}, & x > 1 \end{cases}$$

is piecewise continuous in $[-10, 10]$.

20. Find the largest closed interval over which $f(x) = e^{\sqrt{9-x^2}}$ is continuous.

21. Show that the function $f(x) = \ln \sin x$ is not continuous over any interval of the form $[0, a]$.

In Exercises 22–25 find the c which satisfies the conclusion of the mean value theorem for the given function and numbers.

22. $f(x) = x^5$; $a = 0$, $b = 1$ 23. $f(x) = \ln x$; $a = 1$, $b = e^3$
24. $f(x) = \sqrt[5]{x}$; $a = -32$, $b = -1$ 25. $f(x) = \sin x$; $a = 0$, $b = \pi/2$.
26. Show that $f(x) = \ln x$ is a contraction on the interval $[2, 100]$.
27. Let

$$f(x) = \begin{cases} x^3, & x \le 1 \\ \alpha x^4, & x > 1. \end{cases}$$

(a) For what value of α does the mean value theorem apply in the interval $[0, 2]$?

(b) Using the value of α obtained in part (a), find a number c such that

$$[f(2) - f(0)]/(2 - 0) = f'(c).$$

28. Use Rolle's theorem to show that any fourth degree polynomial has at most four real roots.

In Exercises 29–36 determine intervals over which the given function is 1–1 and determine the inverse function over each such interval, if possible. Then calculate the derivative of each inverse function.

29. $f(x) = \sqrt{2x - 5}$ 　　　30. $f(x) = \sqrt[3]{x - 2}$ 　　　31. $f(x) = e^{x+4}$

32. $f(x) = \ln(4x - 3)$ 　　　33. $f(x) = \operatorname{sech} x$ 　　　34. $f(x) = \sqrt[3]{e^x + 2}$

35. $y = x^2 + 5x + 6$ 　　　　36. $y = x^4 + 3x^2 + 2.$

37. Prove that for any continuous function $f \geq 0$

$$\int_{-\pi/2}^{\pi/2} f(t) \cos t\, dt \leq \int_{-\pi/2}^{\pi/2} f(t)\, dt.$$

38. Calculate $\dfrac{d}{dx} \displaystyle\int_{x^2}^{x^3} e^{t^3}\, dt.$

39. Calculate $\dfrac{d}{dx} \displaystyle\int_{\sin x}^{\cos x} \dfrac{1}{\ln t}\, dt.$

40. Prove that $\displaystyle\int_0^{\pi/4} \tan^{17/3} t\, dt \leq \pi/4.$

41. Prove that $\displaystyle\int_0^1 [(e^{x-1})\sqrt[3]{\sin(x^2 + 3)}]^5\, dx \leq 1.$

ELEVEN

POLAR COORDINATES

11.1 The Polar Coordinate System

In Section 1.3 we introduced the Cartesian plane (the xy-plane) and up to this point we have represented points in the plane by their x- and y-coordinates. There are many other ways to represent points in the plane, the most important of which is called the *polar coordinate system*.

We begin by choosing a fixed point, which we label 0, and a ray (half-line) which extends in one direction from 0 which we label $0A$. The fixed point is called the *pole* (or origin) and the line $0A$ is called the *polar axis*. In Figure 1 the polar axis is drawn as a horizontal line which extends indefinitely to the right (just like the positive x-axis). This is done as a matter of convention although it would be correct to have the polar axis extend in any direction.

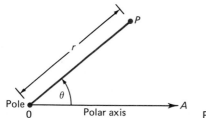

Figure 1

If $P \neq 0$ is any other point in the plane, let r denote the distance between 0 and P and let θ represent the angle (in radians or degrees) between $0A$ and $0P$, measured counterclockwise from $0A$ to $0P$. With this representation we initially use the convention that $0 \leq \theta < 2\pi$ and $r > 0$ (except at the pole). Then every point in the plane, except the pole, can be represented by a pair of numbers (r, θ) where $r > 0$ and $0 \leq \theta < 2\pi$. The pole can be represented as $(0, \theta)$ for any number θ. The representation $P = (r, \theta)$ is called *the polar representation* of the point and r and θ are called the *polar coordinates* of P. Some typical points are sketched in Figure 2.

If we are given two numbers r and θ with $r \geq 0$ and $0 \leq \theta < 2\pi$, then we can draw the point $P = (r, \theta)$ in the plane. But what do we do if either $r < 0$ or $\theta > 2\pi$? To avoid this difficulty we simply extend our definition. Let (r, θ) be any pair of real numbers. To locate the point $P = (r, \theta)$, we first rotate the polar axis through an angle of θ in the counterclockwise direction if θ is positive and in the clockwise direction if θ is negative. Let us call this new ray $0B$. Then if $r > 0$, the point P is placed on the

524

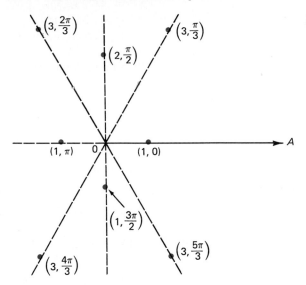

Figure 2

ray $0B$, r units from the pole. If $r < 0$, the ray $0B$ is extended backwards through the pole and the point P is then located $|r|$ units from the pole along this extended ray.

EXAMPLE 1. Plot the following points:

(a) $\left(2, \frac{8\pi}{3}\right)$ (b) $\left(-1, \frac{\pi}{4}\right)$ (c) $\left(3, -22\frac{1}{2}\pi\right)$ (d) $\left(-2, -\frac{\pi}{2}\right)$.

SOLUTION. The four points are plotted in Figure 3.

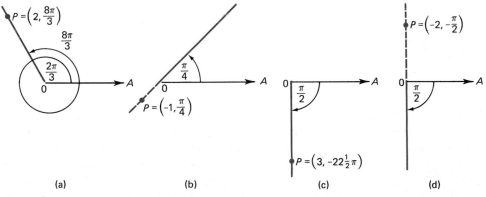

(a)   (b)   (c)   (d)

Figure 3

In (a), since $8\pi/3 = 2\pi + (2\pi/3)$, the line $0B$ makes an angle of $2\pi/3$ with $0A$ and the point P lies 2 units along this line.

In (b), we rotate $\pi/4$ units and then extend $0B$ backwards since $r < 0$. The point P is located 1 unit along this extended line.

In (c), we rotate in a clockwise direction and find that the line $0B$ makes an angle of $\pi/2$ with the polar axis since $22\frac{1}{2}\pi = 11(2\pi) + (\pi/2)$. The point P is located 3 units along $0B$.

In (d), we rotate $0A$ in a clockwise direction $\pi/2$ radians and then extend $0B$ backward (and therefore upward) since $r < 0$. The point P is then located 2 units along this line.

We immediately notice that if we do not restrict r and θ, then there are many (in fact an infinite number of) representations for each point in the plane. For example, in Figure 3d, we see that $(-2, -\pi/2) = (2, \pi/2) = (2, (\pi/2) + 2n\pi)$ for any integer n. In general,

$$P = (r, \theta) = (-r, \theta + \pi) = (r, \theta + 2n\pi) \qquad \text{for} \quad n = \pm 1, \pm 2, \pm 3, \dots.$$

(see Problem 36). Actually, these ordered pairs are not equal. The equal sign indicates that the points they represent are the same. Therefore, with this understanding we will write $(r_1, \theta_1) = (r_2, \theta_2)$ in the rest of this chapter when the points they represent are the same, even though r_1 may not be equal to r_2, and θ_1 may not be equal to θ_2. However to avoid difficulties, if a point $P \neq 0$ is given, then we can write $P = (r, \theta)$ in polar coordinates in a unique way if we specify that $r > 0$ and $0 \leq \theta < 2\pi$. If P is the pole 0, then $P = (0, \theta)$ for any real number θ and there is no unique representation unless we specify a value for θ.

What is the relationship between polar and rectangular (Cartesian) coordinates? To find it, we place the pole at the origin in the xy-plane with the polar axis along the x-axis as in Figure 4. Let $P = (x, y) = (r, \theta)$ be a point in the plane. Then, as is evident from Figure 4, we have

$$\cos \theta = \frac{x}{r} \qquad \text{and} \qquad \sin \theta = \frac{y}{r}$$

or

$$\boxed{x = r \cos \theta \qquad \text{and} \qquad y = r \sin \theta.} \tag{1}$$

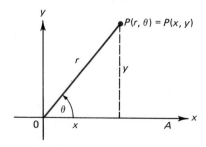

Figure 4

Similarly, from the Pythagorean theorem we see that $x^2 + y^2 = r^2$ so that if we specify that $r > 0$, we have

$$\boxed{r = \sqrt{x^2 + y^2}.} \tag{2}$$

Finally, we can calculate θ from x and y by the relation

$$\tan \theta = \frac{y}{x}, \quad \text{if} \quad x \neq 0. \tag{3}$$

Before citing examples, we note that the above conversion formulas have only been illustrated for $r > 0$ and $0 \leq \theta < \pi/2$. The fact that $0 \leq \theta < \pi/2$ in Figure 4 is irrelevant since, using the circular definition of $\sin x$ and $\cos x$ in Section 7.2, we see that $x/r = \cos \theta$ and $y/r = \sin \theta$ for any real number θ.

To deal with the case $r < 0$, we note the identity

$$(-r, \theta) = (r, \theta + \pi). \tag{4}$$

The proof is left as an exercise (see Problem 36).

With these formulas we can always convert from polar to rectangular coordinates in a unique way. To convert from rectangular to polar coordinates in a unique way it is necessary to specify that $r > 0$ and $0 \leq \theta < 2\pi$.

EXAMPLE 2. Convert from polar to rectangular coordinates:

(a) $\left(3, \frac{\pi}{6}\right)$ (b) $\left(4, \frac{2\pi}{3}\right)$ (c) $\left(-6, \frac{\pi}{4}\right)$ (d) $(2, 0)$ (e) $(1, -\pi)$.

SOLUTION. (a) $x = 3 \cos \pi/6$ and $y = 3 \sin \pi/6$ so that $(x, y) = (3\sqrt{3}/2, 3/2)$.

(b) $x = 4 \cos 2\pi/3 = -4 \cos \pi/3 = -4/2 = -2$ and $y = 4 \sin 2\pi/3 = 4 \sin \pi/3 = 4\sqrt{3}/2 = 2\sqrt{3}$ so that $(x, y) = (-2, 2\sqrt{3})$.

(c) $(-6, \pi/4) = (6, (\pi/4) + \pi) = (6, 5\pi/4)$ (from   (4))   and   we   have $x = 6 \cos 5\pi/4 = 6(-\sqrt{2}/2) = -3\sqrt{2}$, $y = 6 \sin 5\pi/4 = -3\sqrt{2}$ and $(x, y) = (-3\sqrt{2}, -3\sqrt{2})$.

(d) $x = 2 \cos 0 = 2$ and $y = 2 \sin 0 = 0$ so that $(x, y) = (2, 0)$.

(e) Since $(1, -\pi) = (1, -\pi + 2\pi) = (1, \pi)$, we have $x = \cos \pi = -1$, $y = \sin \pi = 0$ and $(x, y) = (-1, 0)$.

These 5 points are sketched in Figure 5.

EXAMPLE 3. Convert from rectangular to polar coordinates (with $r > 0$ and $0 \leq \theta < 2\pi$):

(a) $(1, \sqrt{3})$.   (b) $(2, -2)$.   (c) $(-4\sqrt{3}, -4)$.

SOLUTION.   (a) $r = \sqrt{x^2 + y^2} = \sqrt{1^2 + \sqrt{3}^2} = \sqrt{1 + 3} = 2$ and $\tan \theta = \sqrt{3}/1 = \sqrt{3}$. Since $(1, \sqrt{3})$ is in the first quadrant, we have $\theta = \pi/3$ so that $(r, \theta) = (2, \pi/3)$.

(b) $r = \sqrt{(2)^2 + (-2)^2} = \sqrt{8} = 2\sqrt{2}$ and $\tan \theta = -2/2 = -1$. Since $(2, -2)$ is in the fourth quadrant, $\theta = 7\pi/4$ and $(r, \theta) = (2\sqrt{2}, 7\pi/4)$.

(c) $r = \sqrt{(-4\sqrt{3})^2 + (-4)^2} = \sqrt{48 + 16} = 8$ and

$\tan \theta = -4/(-4\sqrt{3}) = 1/\sqrt{3}$.

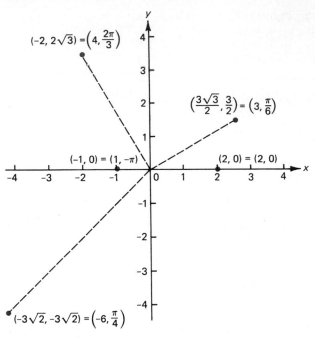

Figure 5

Since $(-4\sqrt{3}, -4)$ is in the third quadrant, we have $\theta = 7\pi/6$ and $(r, \theta) = (8, 7\pi/6)$. These three points are sketched in Figure 6.

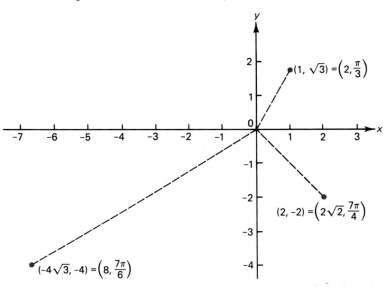

Figure 6

PROBLEMS 11.1

In Problems 1–18 a point is given in polar coordinates. Write the point in rectangular coordinates and plot it in the xy-plane showing both representations.

1. $(3, 0)$ 2. $(4, \pi/4)$ 3. $(-5, 0)$
4. $(-4, \pi/4)$ 5. $(6, 7\pi/6)$ 6. $(-3, -\pi/6)$

7.	$(5, \pi)$	8.	$(5, -\pi)$	9.	$(-5, -\pi)$
10.	$(2, 3\pi/4)$	11.	$(-2, 3\pi/4)$	12.	$(-2, 5\pi/4)$
13.	$(2, 7\pi/4)$	14.	$(-2, -\pi/4)$	15.	$(1, 3\pi/2)$
16.	$(1, -\pi/2)$	17.	$(-1, \pi/2)$	18.	$(-1, -3\pi/2)$.

In Problems 19–35, a point is given in rectangular coordinates. Write the point in polar coordinates with $r > 0$ and $0 \le \theta < 2\pi$. Then plot the point with both representations in the *xy*-plane.

19.	$(3, 0)$	20.	$(7, 0)$	21.	$(-3, 0)$
22.	$(1, 1)$	23.	$(-1, -1)$	24.	$(1, -1)$
25.	$(-1, 1)$	26.	$(0, 1)$ [*Hint:* Draw a sketch first.]		
27.	$(0, -1)$	28.	$(2, 2\sqrt{3})$	29.	$(-2, 2\sqrt{3})$
30.	$(2, -2\sqrt{3})$	31.	$(-2, -2\sqrt{3})$	32.	$(2\sqrt{3}, 2)$
33.	$(2\sqrt{3}, -2)$	34.	$(-2\sqrt{3}, 2)$	35.	$(-2\sqrt{3}, -2)$.

36. Show that $(-r, \theta) = (r, \theta + \pi)$. [*Hint:* Draw a sketch.]

11.2 Graphing in Polar Coordinates

In rectangular coordinates we define the graph of the equation $y = f(x)$, or, more generally, $F(x, y) = 0$, as the set of points (x, y) whose coordinates satisfy the equation. In polar coordinates, however, we must be careful since each point in the plane has an infinite number of representations.

DEFINITION 1. The graph of an equation written in polar coordinates r and θ consists of those points P having at least one representation $P = (r, \theta)$ whose coordinates satisfy the equation.

EXAMPLE 1. Find the graph of the polar equation $r = 1$.

SOLUTION. The set of points for which $r = 1$ is simply the set of points one unit from the pole, which is the definition of the unit circle. Thus the graph of $r = 1$ is the unit circle. To see this in another way, note that $r = \sqrt{x^2 + y^2} = 1$ implies that $x^2 + y^2 = 1$. This is sketched in Figure 1, together with the curves $r = 2$ (the circle of radius 2) and $r = 3$.

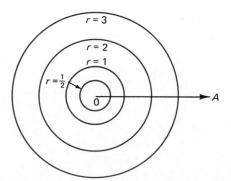

Figure 1

Remark. Every point on the circle $r = 1$ has a representations for which $r \neq 1$ since the point $(-1, \theta)$ is on the circle for any real number θ. This presents no problem since $(-1, \theta) = (1, \theta + \pi)$.

EXAMPLE 2. Sketch the curve $\theta = \pi/4$.

SOLUTION. The curve $\theta = \pi/4$ is the straight line passing through the pole making an angle of $\pi/4$ with the polar axis (see Figure 2). It extends in both directions because it

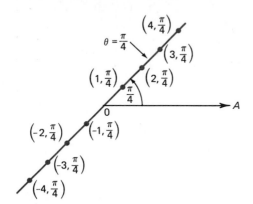

Figure 2

contains points of the form $(-r, \pi/4) = (r, 5\pi/4)$. To see in another way that this is a straight line, note that $y/x = \tan \theta = \tan \pi/4 = 1$ so that $y = x$ in the rectangular representation of the line.

Examples 1 and 2 give us some very useful information. Previously, we sketched graphs by representing a typical point $P = (x_0, y_0)$ as the intersection of the line $x = x_0$ with the line $y = y_0$ (Figure 3a). In polar coordinates we can represent the point $P = (r_0, \theta_0)$ as the intersection of the circle $r = r_0$ with the ray $\theta = \theta_0$ $(r \geq 0)$ (Figure 3b).

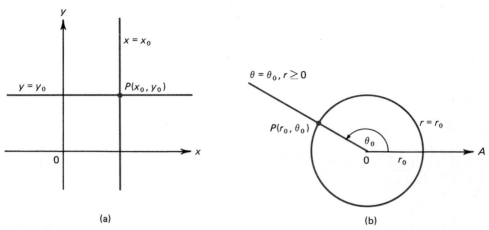

(a)   (b)

Figure 3

We now consider the graphs of some more general curves in polar coordinates. To aid us in obtaining sketches of these curves, we cite three rules which are often useful:

(i)   If in a polar equation θ can be replaced by $-\theta$ without changing the equation, then the polar graph is symmetric about the polar axis (see Figure 4a).

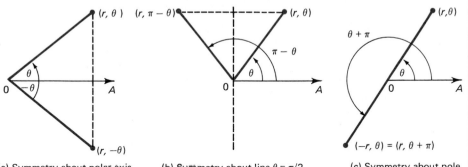

(a) Symmetry about polar axis   (b) Symmetry about line $\theta = \pi/2$ (c) Symmetry about pole

Figure 4

(ii)   If θ can be replaced by $\pi - \theta$ without changing the equation, then the polar graph is symmetric about the line $\theta = \pi/2$ (see Figure 4b).

(iii)   If r can be replaced by $-r$ or, equivalently, if θ can be replaced by $\theta + \pi$ without changing the equation, then the polar graph is symmetric about the pole (see Figure 4c).

EXAMPLE 3.   Sketch the graph of the curve $r = \cos\theta$.

SOLUTION. Since $\cos(-\theta) = \cos\theta$, rule (i) implies that the graph is symmetric about the polar axis so we need only consider values of θ between 0 and π. In Table 1, we tabulate r for some typical values of θ in $[0, \pi]$. From this table we can plot the

TABLE 1

θ	0	$\dfrac{\pi}{6}$	$\dfrac{\pi}{4}$	$\dfrac{\pi}{3}$	$\dfrac{\pi}{2}$	$\dfrac{2\pi}{3}$	$\dfrac{3\pi}{4}$	$\dfrac{5\pi}{6}$	π
r	1	$\dfrac{\sqrt{3}}{2}$	$\dfrac{\sqrt{2}}{2}$	$\dfrac{1}{2}$	0	$-\dfrac{1}{2}$	$-\dfrac{\sqrt{2}}{2}$	$-\dfrac{\sqrt{3}}{2}$	-1

curve for $0 \le \theta < \pi$. This is done in Figure 5. Here we have used the fact that

$$\left(-\frac{1}{2}, \frac{2\pi}{3}\right) = \left(\frac{1}{2}, \frac{5\pi}{3}\right), \quad \left(-\frac{\sqrt{2}}{2}, \frac{3\pi}{4}\right) = \left(\frac{\sqrt{2}}{2}, \frac{7\pi}{4}\right)$$

and

$$\left(-\frac{\sqrt{3}}{2}, \frac{5\pi}{6}\right) = \left(\frac{\sqrt{3}}{2}, \frac{11\pi}{6}\right).$$

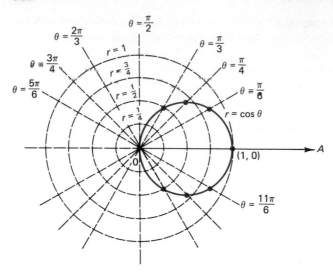

Figure 5

From the sketch in Figure 5, it appears that the graph is a circle. To see this analytically, we convert to rectangular coordinates. Since $r = \cos\theta$, we have, for $r \neq 0$,

$$r^2 = r\cos\theta$$

($r = 0$ is just the origin which is already known to be on the graph) or

$$x^2 + y^2 = x.$$

This can be written

$$x^2 + y^2 - x = 0.$$

Then, completing the square, $(x - \frac{1}{2})^2 + y^2 = \frac{1}{4}$ which is the equation of a circle centered at $(\frac{1}{2}, 0)$ with radius $\frac{1}{2}$.

In general, the graph of the equation $r = a\cos\theta + b\sin\theta$ is a circle for any real numbers a and b. (See Problems 66, 67, and 68.)

EXAMPLE 4.   Sketch the curve $r = 1 + \sin\theta$.

SOLUTION. Since $\sin(\pi - \theta) = \sin\theta$, the curve is symmetric about the line $\theta = \pi/2$. We therefore need only consider values of θ in $[0, \pi/2]$ and $[3\pi/2, 2\pi]$. Typical values of the function are given in Table 2. We use these values in Figure 6 and then use the

TABLE 2

θ	0	$\frac{\pi}{6}$	$\frac{\pi}{4}$	$\frac{\pi}{3}$	$\frac{\pi}{2}$	$\frac{3\pi}{2}$	$\frac{5\pi}{3}$	$\frac{7\pi}{4}$	$\frac{11\pi}{6}$
$r = 1 + \sin\theta$	1	$\frac{3}{2}$	$1 + \frac{\sqrt{2}}{2}$	$1 + \frac{\sqrt{3}}{2}$	2	0	$1 - \frac{\sqrt{3}}{2}$	$1 - \frac{\sqrt{2}}{2}$	$\frac{1}{2}$

symmetry to reflect the curve about the line $\theta = \pi/2$. The heart-shaped curve we have sketched is called a *cardioid,* from the Greek *kardia,* meaning heart.

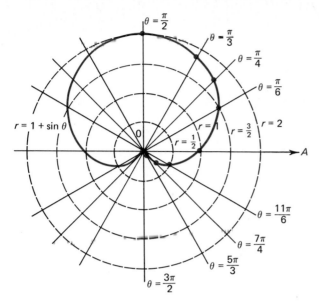

Figure 6

EXAMPLE 5.   Sketch the curve $r = 3 - 2 \cos \theta$.

SOLUTION. The curve is symmetric about the polar axis and we therefore calculate values of r for θ in $[0, \pi]$ (Table 3). We then obtain the graph sketched in Figure 7. The curve sketched in Figure 7 is called a *limaçon.*†

TABLE 3

θ	0	$\dfrac{\pi}{6}$	$\dfrac{\pi}{4}$	$\dfrac{\pi}{3}$	$\dfrac{\pi}{2}$	$\dfrac{2\pi}{3}$	$\dfrac{3\pi}{4}$	$\dfrac{5\pi}{6}$	π
$r = 3 - 2 \cos \theta$	1	$3 - \sqrt{3}$	$3 - \sqrt{2}$	2	3	4	$3 + \sqrt{2}$	$3 + \sqrt{3}$	5

EXAMPLE 6.   Sketch the curve $r = 1 + 2 \cos \theta$.

SOLUTION. This curve is also symmetric about the polar axis and we therefore tabulate the function for θ in $[0, \pi]$ (Table 4). Using the facts that

$$\left(1 - \sqrt{2}, \frac{3\pi}{4}\right) = \left(\sqrt{2} - 1, \frac{7\pi}{4}\right), \quad \left(1 - \sqrt{3}, \frac{5\pi}{6}\right) = \left(\sqrt{3} - 1, \frac{11\pi}{6}\right)$$

and

$$(-1, \pi) = (1, 2\pi) = (1, 0),$$

†From the Latin word *limax* meaning snail. The curve is also referred to as *Pascal's limaçon* since it was discovered by Etienne Pascal (1588–1640), father of the famous French mathematician Blaise Pascal (1623–1662).

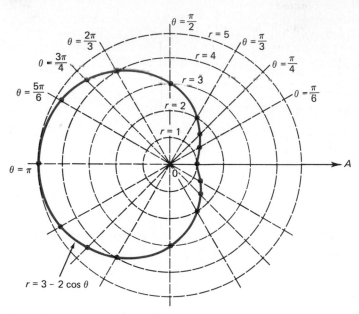

Figure 7

together with the symmetry around the polar axis, we obtain the graph in Figure 8. The curve sketched above is also called a limaçon.

TABLE 4

θ	0	$\dfrac{\pi}{6}$	$\dfrac{\pi}{4}$	$\dfrac{\pi}{3}$	$\dfrac{\pi}{2}$	$\dfrac{2\pi}{3}$	$\dfrac{3\pi}{4}$	$\dfrac{5\pi}{6}$	π
$r = 1 + 2\cos\theta$	3	$1 + \sqrt{3}$	$1 + \sqrt{2}$	2	1	0	$1 - \sqrt{2}$	$1 - \sqrt{3}$	-1

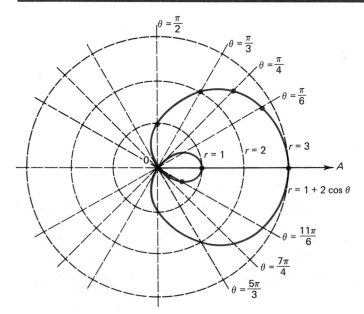

Figure 8

We can generalize the results of the last three examples.

The graph of the equation

$$r = a \pm b \cos \theta \quad \text{or} \quad r = a \pm b \sin \theta$$

is a limaçon which can take one of three possible shapes:

    (i) If $a = b$, we obtain a cardioid as in Figure 6.
    (ii) If $a > b$, we obtain a limaçon having the appearance of the curve in Figure 7.
    (iii) If $b > a$, we obtain a limaçon with a loop as in Figure 8.

EXAMPLE 7. Sketch the curve $r = 3 \cos 2\theta$.

SOLUTION. We first test for symmetry. Replacing θ by $-\theta$ we have $3 \cos 2(-\theta) = 3 \cos 2\theta$ implying that the curve is symmetric about the polar axis. Similarly, replacing θ by $\pi - \theta$ yields the equations

$$3 \cos 2(\pi - \theta) = 3 \cos(2\pi - 2\theta) = 3 \cos(-2\theta) = 3 \cos 2\theta$$

so that the curve is symmetric about the line $\theta = \pi/2$. Finally, replacing θ by $\pi + \theta$ we obtain

$$3 \cos 2(\pi + \theta) = 3 \cos(2\pi + 2\theta) = 3 \cos 2\theta$$

so that the curve is also symmetric about the pole. Thus it is only necessary to sketch the curve for $0 \le \theta \le \pi/2$ and then reflect about the polar axis, the line $x = \pi/2$ and the pole. Values for r and θ are given in Table 5. Now

$$\left(-\frac{3}{2}, \frac{\pi}{3} \right) = \left(\frac{3}{2}, \frac{4\pi}{3} \right), \quad \left(\frac{-3\sqrt{3}}{2}, \frac{5\pi}{12} \right) = \left(\frac{3\sqrt{3}}{2}, \frac{17\pi}{12} \right)$$

and

$$\left(-3, \frac{\pi}{2} \right) = \left(3, \frac{3\pi}{2} \right).$$

TABLE 5

θ	0	$\dfrac{\pi}{12}$	$\dfrac{\pi}{8}$	$\dfrac{\pi}{6}$	$\dfrac{\pi}{4}$	$\dfrac{\pi}{3}$	$\dfrac{5\pi}{12}$	$\dfrac{\pi}{2}$
$r = 3 \cos 2\theta$	3	$\dfrac{3\sqrt{3}}{2}$	$\dfrac{3\sqrt{2}}{2}$	$\dfrac{3}{2}$	0	$-\dfrac{3}{2}$	$-\dfrac{3\sqrt{3}}{2}$	-3

Then, using symmetry, we obtain the curve sketched in Figure 9. This curve is called a *four-leafed rose*.

In general, any curve of the form

$$r = a \cos 2\theta \quad \text{or} \quad r = a \sin 2\theta$$

is a *four-leafed rose*.

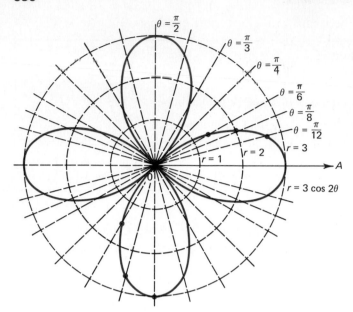

$$\theta = \frac{\pi}{2}$$

$$\theta = \frac{\pi}{3}$$

$$\theta = \frac{\pi}{4}$$

$$\theta = \frac{\pi}{6}$$

$$\theta = \frac{\pi}{8}$$

$$\theta = \frac{\pi}{12}$$

$r = 1$ $r = 2$ $r = 3$

A

$r = 3 \cos 2\theta$

Figure 9

EXAMPLE 8. Sketch the curve $r^2 = 4 \sin 2\theta$.

SOLUTION. Two things should be immediately evident from the equation. First, since $(-r)^2 = r^2$, the graph is symmetric about the pole. Second, since $r^2 \geq 0$, the function is only defined for values of θ such that $\sin 2\theta \geq 0$. If θ is restricted to the interval $[0, 2\pi]$, then $\sin 2\theta \geq 0$ if and only if θ is in $[0, \pi/2]$ or $[\pi, 3\pi/2]$. Then using Table 6 and the symmetry about the pole, we obtain the graph sketched in Figure 10. This curve is called a *lemniscate.†*

TABLE 6

θ	0	$\frac{\pi}{12}$	$\frac{\pi}{8}$	$\frac{\pi}{6}$	$\frac{\pi}{4}$	$\frac{\pi}{3}$	$\frac{5\pi}{12}$	$\frac{\pi}{2}$
$r = \sqrt{4 \sin 2\theta}$	0	$\sqrt{2}$	$2^{3/4}$	$\sqrt{2} \cdot 3^{1/4}$	2	$\sqrt{2} \cdot 3^{1/4}$	$\sqrt{2}$	0

> In general, any curve with the equation
>
> $$r^2 = a \sin b\theta \qquad \text{or} \qquad r^2 = a \cos b\theta$$
>
> is a *lemniscate.*

EXAMPLE 9. Sketch the curve $r = \theta$, $\theta \geq 0$.

SOLUTION. This curve has no symmetry. To sketch its graph, we note that as we move around the pole, θ increases as r increases. The graph is sketched in Figure 11 and the

†From the Greek *lemniskos* and the Latin *lemniscus* meaning knotted ribbon. Actually, this curve was originally called the *lemniscate of Bernoulli,* named after the Swiss mathematician Jacques Bernoulli (1654–1705) who described the curve in the *Acta Eruditorum* published in 1694.

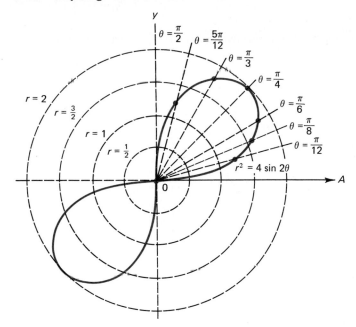

Figure 10

curve depicted is called the *spiral of Archimedes*.† It intersects the polar axis when
$\theta = 2n\pi$, n an integer, and it intersects the line $\theta = \pi/2$ when $\theta = [(2n + 1)/2]\pi$.

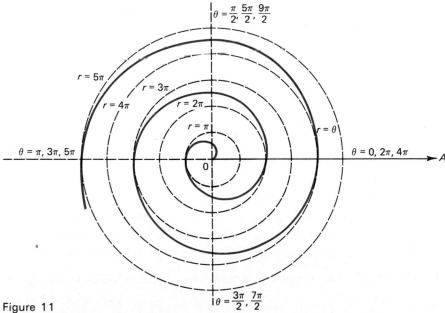

Figure 11

†Archimedes devoted a great deal of study to the spiral that can be described by the polar equation $r = a\theta$.
He used it initially in his attempt to solve the ancient problem of trisecting the angle. Later he calculated
part of its area by his method of exhaustion. He described his work on this subject in one of his most
important works entitled *On Spirals*.

PROBLEMS 11.2

In Problems 1–65, sketch the graph of the given equation, indicating any symmetry about the polar axis, the line $\theta = \pi/2$, and/or the pole.

1. $r = 5$ **2.** $r = 7$ **3.** $r = -4$

4. $\theta = 3\pi/8$ **5.** $\theta = -\pi/6$ **6.** $\theta = 13\pi/5$

7. $r = 5 \sin \theta$ **8.** $r = 5 \cos \theta$ **9.** $r = -5 \sin \theta$

10. $r = -5 \cos \theta$ **11.** $r = 5 \cos \theta + 5 \sin \theta$

12. $r = -5 \cos \theta + 5 \sin \theta$ **13.** $r = 5 \cos \theta - 5 \sin \theta$

14. $r = -5 \cos \theta - 5 \sin \theta$ **15.** $r = 2 + 2 \sin \theta$

16. $r = 2 + 2 \cos \theta$ **17.** $r = 2 - 2 \sin \theta$

18. $r = 2 - 2 \cos \theta$ **19.** $r = -2 + 2 \sin \theta$

20. $r = -2 + 2 \cos \theta$ **21.** $r = -2 - 2 \sin \theta$

22. $r = -2 - 2 \cos \theta$ **23.** $r = 1 + 3 \sin \theta$

24. $r = 3 + \sin \theta$ **25.** $r = -2 + 4 \cos \theta$

26. $r = -4 + 2 \cos \theta$ **27.** $r = -3 - 4 \cos \theta$

28. $r = -4 - 3 \cos \theta$ **29.** $r = -3 - 4 \sin \theta$

30. $r = -4 - 3 \sin \theta$ **31.** $r = 4 - 3 \cos \theta$

32. $r = 3 - 4 \cos \theta$ **33.** $r = 4 + 3 \sin \theta$

34. $r = 3 + 4 \sin \theta$ **35.** $r = 3 \sin 2\theta$

36. $r = -3 \cos 2\theta$ **37.** $r = -3 \sin 2\theta$

38. $r = 3 \cos 2\theta$

39. $r = 5 \sin 3\theta$ (this is called a *three-leafed rose*)

40. $r = 5 \cos 3\theta$ **41.** $r = -5 \sin 3\theta$

42. $r = -5 \cos 3\theta$

43. $r = 2 \cos 4\theta$ (this is called an *eight-leafed rose*)

44. $r = 2 \sin 4\theta$ **45.** $r = -2 \cos 4\theta$

46. $r = 3\theta, \quad \theta \geq 0$ (46–48 are all spirals of Archimedes)

47. $r = -5\theta, \quad \theta \geq 0$ **48.** $r = \theta/2, \quad \theta \geq 0$

49. $r = e^{\theta}$ (this is called a *logarithmic spiral* since $\ln r = \theta$)

50. $r = e^{\theta/2}$ **51.** $r = e^{3\theta}$

52. $r^2 = \cos 2\theta$ **53.** $r^2 = \sin 2\theta$

54. $r^2 = -\cos 2\theta$ **55.** $r^2 = -\sin 2\theta$

56. $r^2 = 4 \sin 2\theta$ **57.** $r^2 = -25 \cos 2\theta$

58. $r^2 = -25 \sin 2\theta$

59. $r = \sin \theta \tan \theta$ (this is called a *cissoid*)

60. $r = 2 - 3 \sec \theta$ (this is called a *conchoid*)

61. $r = 4 + 3 \csc \theta$

62. $r^2 = \theta$ (this is called a *parabolic spiral*)

63. $(r + 1)^2 = 3\theta$ **64.** $r = |\sin \theta|$ **65.** $r = |\cos \theta|$.

66. By translating into rectangular coordinates and completing the square, show that for any real number a, the graph of the equation $r = a \cos \theta$ is a circle in the xy-plane with center at $(a/2, 0)$ and radius $a/2$.

67. Show that the graph of the equation $r = b \sin \theta$ in the xy-plane is a circle with radius $a/2$ centered at $(0, b/2)$.

68. Show that the graph of the equation $r = a \cos \theta + b \sin \theta$ in the xy-plane is a circle centered at $(a/2, b/2)$ with radius $\sqrt{a^2 + b^2}/2$.

69. Find the polar equation of the circle centered at $(-2, \frac{3}{2})$ with radius $\frac{5}{2}$.

70. Show that the polar equation $r \cos \theta = a$ is the equation of a vertical line for any real number $a \neq 0$.

71. Show that the polar equation $r \sin \theta = a$ is the equation of a horizontal line for any real number $a \neq 0$.

72. Show that the polar equation $r(a \cos \theta + b \sin \theta) = c$ with a, b, c real numbers is the equation of a straight line.

73. Sketch the graphs of
    (a) $r \sin \theta = 3$ (b) $r \cos \theta = -2$ (c) $3r \cos \theta = 8$
    (d) $r(2 \sin \theta - 3 \cos \theta) = 4$ (e) $r(-5 \sin \theta + 10 \cos \theta) = 20.$

11.3 Points of Intersection of Graphs of Polar Equations

Suppose that we have two functions given in rectangular coordinates: $y = f(x)$ and $y = g(x)$. To find the points of intersection of the graphs of these equations, we simply find the roots of the equation $f(x) = g(x)$. This method works because every point in the xy-plane has a unique representation. However, as we have already seen, the same is not true in polar coordinates. Therefore to find the points of intersection of the graphs of $r = f(\theta)$ and $r = g(\theta)$, the "analytic" method described above will usually not locate all such points, since the graphs can intersect for different representations of the points of intersection.

EXAMPLE 1. Find all points of intersection of the curves $r = \sin \theta$ and $r = \cos \theta$.

SOLUTION. We first try our "old" method of setting $\sin \theta = \cos \theta$. Then $\tan \theta = 1$ or $\theta = (\pi/4) + n\pi$ for any integer n, and $(r, \theta) = (1/\sqrt{2}, \pi/4)$ is one point of intersection. Next we draw the graphs of these two curves. From the discussion of last section, we know that $r = \sin \theta$ and $r = \cos \theta$ are the equations of circles whose graphs are sketched in Figure 1. Evidently the pole is also a point of intersection. How did we

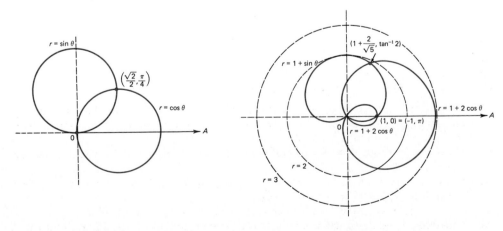

Figure 1   Figure 2

miss it? The answer is simply that the pole can be represented as $(0, \theta)$ for any real number θ. On the curve $r = \sin \theta, r = 0$ when $\theta = n\pi$ and on $r = \cos \theta, r = 0$ when $\theta = (n + \frac{1}{2})\pi$. Since these two representations of the pole are not the same, our old method will not locate this point of intersection.

In general, to find all points of intersection of two polar equations, it is almost always essential to sketch their graphs.

EXAMPLE 2.   Find the points of intersection of the graphs of the equations $r = 1 + \sin \theta$ and $r = 1 + 2 \cos \theta$.

SOLUTION. We first set $1 + \sin \theta = 1 + 2 \cos \theta$ to obtain $\tan \theta = 2$ or $\theta = \tan^{-1} 2$. When $\theta = \tan^{-1} 2$, $\sin \theta = 2/\sqrt{5}$ and $\cos \theta = 1/\sqrt{5}$ (draw a triangle to verify this). To check for other points of intersection, we sketch the two limaçons (see Examples 11.2.4 and 11.2.6) in Figure 2. Again we find that the pole is a point of intersection. In addition, the curves intersect at the point $(1, 0)$ since, when $\theta = \pi$, $r = 1 + 2 \cos \theta = 1 - 2 = -1$ and $(-1, \pi) = (1, 2\pi) = (1, 0)$. Thus three points of intersection are at $(1 + (2/\sqrt{5}), \tan^{-1} 2), (1, 0)$ and 0. But, surprisingly, these are not all. There are two additional points of intersection that fall within the width of the green lines (and so are not visible on our sketch). To find them, we first note that $\theta = \pi + \tan^{-1} 2$ (in the third quadrant) satisfies $\tan \theta = 2$ since the tangent function is periodic of period π. Then $\sin (\pi + \tan^{-1} 2) = -\sin (\tan^{-1} 2) = -2/\sqrt{5}$ and $\cos (\pi + \tan^{-1} 2) = -\cos (\tan^{-1} 2) = -1/\sqrt{5}$. Thus $r = 1 + \sin \theta = 1 + 2 \cos \theta = 1 - 2/\sqrt{5}$ and we see that $(1 - (2/\sqrt{5}), \pi + \tan^{-1} 2)$ is a fourth point of intersection. It is more difficult to find the fifth point of intersection. We begin with the identity

$$(r, \theta) = (-r, \theta + \pi).$$

This suggests that we consider the possibility that the two curves intersect at a point (r, θ) for which $r = 1 + \sin \theta$ and $-r = 1 + 2 \cos (\theta + \pi)$. (Explain why!). But then $r = -1 - 2 \cos (\theta + \pi) = -1 + 2 \cos \theta$ and we are led to the equation

$$1 + \sin \theta = -1 + 2 \cos \theta \tag{1}$$

or

$$2 + \sin \theta = 2 \cos \theta$$

and, squaring,

$$4 + 4 \sin \theta + \sin^2 \theta =$$
$$4 \cos^2 \theta = 4 - 4 \sin^2 \theta$$

or

$$5 \sin^2 \theta + 4 \sin \theta = 0$$

with solutions

$\sin \theta = 0$ and $\sin \theta = \dfrac{-4}{5}$. But if $\sin \theta = 0$, we again obtain the point $(1,0)$. If $\sin \theta = -\dfrac{4}{5}$, then θ is in the third or fourth quadrant. To find out which, we again start with equation (1) to obtain

$$\sin^2 \theta = (-2 + 2 \cos \theta)^2 = 4 - 8 \cos \theta + 4 \cos^2 \theta$$

or, using $\sin^2 \theta = 1 - \cos^2 \theta$,

$$5 \cos^2 \theta - 8 \cos \theta + 3 = 0$$

with solutions $\cos \theta = -1$ and $\cos \theta = -3/5$. Hence $\cos \theta = -3/5, \theta$ is in the fourth quadrant and $\theta = -\sin^{-1}(4/5) = -\cos^{-1}(3/5)$, giving us the last point of intersection $(1/5, -\sin^{-1}(4/5))$. (Note that this is the same point as $(-1/5, \pi - \cos^{-1}(3/5))$.)

From the last example it is evident that finding all points of intersection can be very tricky. A method for determining all points of intersection without sketching the graphs is outlined in Problems 22 and 23.

PROBLEMS 11.3

In Problems 1–20 find all points of intersection of the graphs of the two polar equations.

1. $r = 1, r = \sin \theta$
2. $r = \frac{1}{2}, r = \cos \theta$
3. $r = 2(1 - \sin \theta), r = 2(1 - \cos \theta)$
4. $r = 1 - \sin \theta, r = 1 + \cos \theta$
5. $r = 3, \theta = \pi/4$ (Be careful!)
6. $r = -2, \theta = \pi/3$
7. $r = -2 + 2 \cos \theta, r = -4 + 2 \sin \theta$
8. $r = \sqrt{2} \sin \theta, r = \cos \theta$
9. $r = \sqrt{3} \sin \theta, r = \cos \theta$
10. $r^2 = \sin 2\theta, r^2 = \cos 2\theta$
11. $r = \frac{3}{2}, r = 3 \cos 2\theta$
12. $r = 2, r = 4 \sin 2\theta$
13. $r^2 = \cos \theta, r = \cos \theta$
14. $r^2 = \sin \theta, r = \sin \theta$
15. $r = 0, r^2 = 0, r, \theta \geq 0, r \leq 2$
16. $r = 5 \sin 3\theta, r = 5 \cos 3\theta$
17. $r = \sin 2\theta, r = \cos 2\theta$
18. $r = \pi/3, r = 2\theta$
19. $r = \sin \theta, r = \sec \theta$
20. $r = |\sin \theta|, r = |\cos \theta|$.

21. Show that all points of intersection of the two curves $r \cos \theta = a$ and $r^2 \sin 2\theta = b$ lie on a straight line parallel to the polar axis.

*22. If the graphs of the equations $r = f(\theta)$ and $r = g(\theta)$ intersect at a point $(r, \theta) \neq 0$, explain why one of the following three equations must hold for r and θ:
   (i) $r = f(\theta), r = g(\theta)$.
   (ii) $r = f(\theta), r = g(\theta + 2n\pi)$, for some integer n.
   (iii) $r = f(\theta), -r = g(\theta + (2n + 1)\pi)$.

23. Show that if $r = f(\theta)$ and $r = g(\theta)$ intersect at the pole, then there exist numbers θ_1 and θ_2 such that $f(\theta_1) = g(\theta_2) = 0$.

Using the results of Problems 22 and 23 we can find all points of intersection of the graphs of $r = f(\theta)$ and $r = g(\theta)$ by checking the pole and solving equations (i), (ii), and (iii). In Problems 24–34 use this method to find all points of intersection.

24. $r = 1 - \sin \theta, r = 1 + \sqrt{3} \cos \theta$
25. $r = \tan 2\theta, r = 1$

26. $r = 3 \tan \theta \sin \theta, r = 3 \cos \theta$ **27.** $r = 8(1 + \sin \theta), r = \dfrac{6}{1 - \sin \theta}$

28. $r = 4(1 + \cos \theta), r = \dfrac{3}{1 - \cos \theta}$ **29.** $r = \dfrac{1}{2} \tan \theta, r = \dfrac{1}{\sqrt{3}} \sin \theta$

30. $r^2 = \sin \theta, r = \csc \theta$ **31.** $r^2 = -\cos \theta, r = \sec \theta$

32. $r = \csc \theta, r = \cot \theta$ **33.** $r = \theta, r = 2\theta, \theta \geq 0$

34. $r = 2\theta, r = 3\theta, \theta \geq 0.$

11.4 Derivatives and Tangent Lines

As we have seen repeatedly in this text, the derivative of a given function tells us a great deal about the behavior of that function. If $r = f(\theta)$, then $dr/d\theta$, which we can easily calculate, tells us how r changes when θ changes. However other applications of the derivative cannot be immediately extended to polar coordinates.

Suppose we wish to find the slope of the tangent line to a given curve given in polar coordinates. We can immediately see that this is *not* given by $dr/d\theta$. Let $r = c$, a constant. Then $dr/d\theta = 0$, but this is not the slope of the tangent to the curve $r = c$ since $r = c$ is the equation of a circle. To calculate the slope of the tangent line, it is necessary to translate into rectangular coordinates and then calculate dy/dx.

From Section 11.1 we have

$$x = r \cos \theta \qquad \text{and} \qquad y = r \sin \theta.$$

Since we are assuming that r is a function of θ, we may also consider x and y to be functions of θ. Then, using the product rule and differentiating with respect to θ, we obtain

$$\frac{dx}{d\theta} = \frac{dr}{d\theta} \cos \theta - r \sin \theta \tag{1}$$

and

$$\frac{dy}{d\theta} = \frac{dr}{d\theta} \sin \theta + r \cos \theta. \tag{2}$$

By the chain rule, $dy/dx = (dy/d\theta)/(dx/d\theta)$, so that

$$\boxed{\frac{dy}{dx} = \frac{(dr/d\theta) \sin \theta + r \cos \theta}{(dr/d\theta) \cos \theta - r \sin \theta}.} \tag{3}$$

EXAMPLE 1. Calculate the slope of the line tangent to the limaçon $r = 1 + 2 \cos \theta$ at the point $(2, \pi/3)$ (see Example 11.2.6).

SOLUTION. Here $dr/d\theta = -2 \sin \theta = -\sqrt{3}$ when $\theta = \pi/3$. Then substituting $\theta = \pi/3$ and $r = 2$ into (3), we obtain

$$\frac{dy}{dx} = \frac{-\sqrt{3} \cdot (\sqrt{3}/2) + 2 \cdot \frac{1}{2}}{-\sqrt{3} \cdot \frac{1}{2} - 2 \cdot (\sqrt{3}/2)} = \frac{1}{3\sqrt{3}}.$$

EXAMPLE 2. Find the equation of the line tangent to the four-leafed rose $r = 3 \cos 2\theta$ at the point $(3/\sqrt{2}, \pi/8)$ (see Example 11.2.7).

SOLUTION. Since $dr/d\theta = -6 \sin 2\theta = -6/\sqrt{2}$ when $\theta = \pi/8$, substitution of $r = 3/\sqrt{2}$ and of $\theta = \pi/8$ into equation (3) yields

$$\frac{dy}{dx} = \frac{-(6/\sqrt{2})\cdot(1/\sqrt{2}) + (3/\sqrt{2})\cdot(1/\sqrt{2})}{-(6/\sqrt{2})\cdot(1/\sqrt{2}) - (3/\sqrt{2})\cdot(1/\sqrt{2})} = \frac{1}{3}.$$

At $(3/\sqrt{2}, \pi/8)$, $x = r \cos \theta = (3/\sqrt{2})\cdot \cos \pi/8$ and $y = r \sin \theta = (3/\sqrt{2}) \sin \pi/8$. Then the equation of the tangent line is given by

$$\frac{y - (3/\sqrt{2}) \sin \pi/8}{x - (3/\sqrt{2}) \cos \pi/8} = \frac{1}{3} \quad \text{or} \quad 3y - x = \frac{9}{\sqrt{2}} \sin \frac{\pi}{8} - \frac{3}{\sqrt{2}} \cos \frac{\pi}{8}.$$

Note. We can calculate $\sin \pi/8$ and $\cos \pi/8$ by the half-angle formula. We have

$$\sin \frac{\pi}{8} = \sqrt{\frac{1 - \cos \pi/4}{2}} = \sqrt{\frac{1 - (1/\sqrt{2})}{2}} = \sqrt{\frac{\sqrt{2} - 1}{2\sqrt{2}}}$$

and

$$\cos \frac{\pi}{8} = \sqrt{\frac{1 + \cos \pi/4}{2}} = \sqrt{\frac{\sqrt{2} + 1}{2\sqrt{2}}}.$$

As the two examples above illustrate, the calculation of dy/dx can be tedious. Using some elementary geometry, we can simplify our work. From Section 7.4, we see that dy/dx can be thought of as the tangent of the angle α that the tangent line makes with the positive x-axis, or, in polar coordinates, the tangent of the angle it makes with the polar axis (see Figure 1). Then equation (3) can be written

$$\tan \alpha = \frac{(dr/d\theta) \sin \theta + r \cos \theta}{(dr/d\theta) \cos \theta - r \sin \theta}. \tag{4}$$

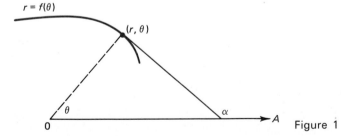

$r = f(\theta)$

(r, θ)

Figure 1

If $\cos \theta \neq 0$, then we can divide numerator and denominator of (4) by $\cos \theta$ to obtain

$$\tan \alpha = \frac{(dr/d\theta) \tan \theta + r}{(dr/d\theta) - r \tan \theta}. \tag{5}$$

Now let β denote the angle formed by the intersection of the line $\theta = \theta_0$ and the tangent line (see Figure 2). From elementary geometry, we have $\pi = \theta + \beta + (\pi - \alpha)$ or

$$\beta = \alpha - \theta. \tag{6}$$

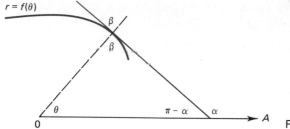

Figure 2

Recalling the formula (see Problem 7.3.22)

$$\tan(A - B) = \frac{\tan A - \tan B}{1 + \tan A \tan B},$$ (7)

we have

$$\tan \beta = \tan(\alpha - \theta) = \frac{\tan \alpha - \tan \theta}{1 + \tan \alpha \tan \theta}.$$ (8)

Substituting (5) into (8) we obtain

$$\tan \beta = \frac{\dfrac{(dr/d\theta)\tan\theta + r}{(dr/d\theta) - r\tan\theta} - \tan\theta}{1 + \dfrac{(dr/d\theta)\tan\theta + r}{(dr/d\theta) - r\tan\theta}\cdot\tan\theta}.$$

Then multiplying numerator and denominator by $(dr/d\theta) - r\tan\theta$ we have,

$$\tan \beta = \frac{(dr/d\theta)\tan\theta + r - (dr/d\theta)\tan\theta + r\tan^2\theta}{(dr/d\theta) - r\tan\theta + (dr/d\theta)\tan^2\theta + r\tan\theta}$$

$$= \frac{r(1 + \tan^2\theta)}{(dr/d\theta)(1 + \tan^2\theta)} = \frac{r}{dr/d\theta}$$

and, finally,

$$\boxed{\tan \beta = \tan(\alpha - \theta) = \frac{r}{dr/d\theta}.}$$ (9)

Often it will be more convenient to calculate $\tan \beta$ in a given problem. The next example illustrates how the slope of the tangent line may be calculated by concentrating on the angle β.

EXAMPLE 3. Calculate the slope of the line tangent to the limaçon $r = 1 + 2 \sin \theta$ at the point $(2, \pi/6)$.

SOLUTION. We have $dr/d\theta = 2 \cos \theta = \sqrt{3}$ when $\theta = \pi/6$. Then, from (9),

$$\tan \beta = \frac{r}{dr/d\theta} = \frac{2}{\sqrt{3}}.$$

But we can rearrange equation (8) and solve for $\tan\alpha$ to obtain

$$\tan\alpha = \frac{\tan\theta + \tan\beta}{1 - \tan\theta\tan\beta}. \tag{10}$$

Then substituting $\tan\theta = \tan\pi/6 = 1/\sqrt{3}$ and $\tan\beta = 2/\sqrt{3}$ into (10), we have

$$\frac{dy}{dx} = \tan\alpha = \frac{(1/\sqrt{3}) + (2/\sqrt{3})}{1 - (1/\sqrt{3})\cdot(2/\sqrt{3})} = 3\sqrt{3}.$$

EXAMPLE 4. Calculate the slope of the line tangent to the curve $r = \theta$ at the point $(\pi/4, \pi/4)$.

SOLUTION. Since $dr/d\theta = 1$, $\tan\beta = r/(dr/d\theta) = \pi/4$. Then, from (10),

$$\frac{dy}{dx} = \tan\alpha = \frac{\tan(\pi/4) + \pi/4}{1 - \tan(\pi/4)\cdot\pi/4} = \frac{1 + (\pi/4)}{1 - (\pi/4)} = \frac{4 + \pi}{4 - \pi}.$$

As a further application of the usefulness of the angle β, suppose that $r = f(\theta)$ and that θ^* denotes a number for which $f(\theta^*) = 0$. Then the curve passes through the pole when $\theta = \theta^*$. Thus the angle between the line $\theta = \theta^*$ and the line tangent to the curve at $(0, \theta^*)$ is 0 or, put another way, the line $\theta = \theta^*$ *is* a tangent line to the curve $r = f(\theta)$ at the pole (see Figure 3).

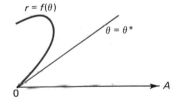

$r = f(\theta)$

$\theta = \theta^*$

A

0

Figure 3

In general, the following rule is useful for sketching the graphs of equations in polar coordinates.

If $\theta_1, \theta_2, \ldots, \theta_n$ are roots of $f(\theta) = 0$, then there is a branch of the graph of $r = f(\theta)$ which is tangent to the line $\theta = \theta_i$ at the pole for each i, $i = 1, 2, \ldots, n$.

EXAMPLE 5. In Example 11.2.4 we sketched the graph of the cardioid $r = 1 + \sin\theta$. We can improve that sketch by noting that since $r = 0$ when $\theta = 3\pi/2$ (so that $\sin\theta = -1$), the cardioid is tangent to the line $\theta = 3\pi/2$ at the pole. This is sketched in Figure 4.

EXAMPLE 6. In Example 11.2.7 we sketched the graph of the four-leafed rose $r = 3\cos 2\theta$. We have $r = 0$ when $\theta = \pi/4$ and $3\pi/4$ (so that $\cos 2\theta = 0$). Therefore the four-leafed rose has branches which are tangent to the lines $\theta = \pi/4$ and $\theta = 3\pi/4$ at the pole. This is sketched in Figure 5.

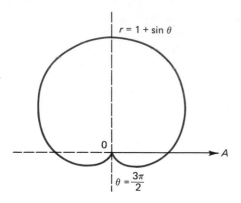

Figure 4

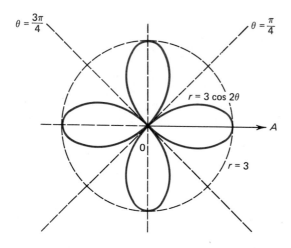

Figure 5

PROBLEMS 11.4

In Problems 1–10 calculate the slope of the line tangent to the given curve at the given point by using equation (3) or (4).

1. $r = 1$; $(1, \pi/2)$
2. $r = 3$; $(3, 0)$
3. $r = \sin \theta$; $(1/\sqrt{2}, \pi/4)$
4. $r = 2 \cos \theta$; $(1, \pi/3)$
5. $r = 2 - 3 \sin \theta$; $(1/2, \pi/6)$
6. $r = 2 + 2 \cos \theta$; $(3, \pi/3)$
7. $r = 2\theta$; $(\pi, \pi/2)$
8. $r = 3 \cos 2\theta$; $(0, \pi/4)$
9. $r = -4 \sin 2\theta$; $(2\sqrt{3}, 5\pi/3)$
10. $r = e^\theta$; $(e^{\pi/2}, \pi/2)$.

In Problems 11–24 calculate the tangent of the angle β between the radial line $\theta = \theta_0$ and the tangent line at the given point and then use that calculation to find the slope of the tangent line.

11. $r = 3 \sin \theta$; $(3, \pi/2)$
12. $r = -2 \cos \theta$; $(\sqrt{3}, 5\pi/6)$
13. $r = 4 - 3 \cos \theta$; $(5/2, \pi/3)$
14. $r = 5 \sin 3\theta$; $(5/\sqrt{2}, \pi/12)$
15. $r = e^{\theta/2}$; $(e^{\pi/4}, \pi/2)$
16. $r = 3\theta$; $(3\pi/2, \pi/2)$
17. $r = \sin \theta \tan \theta$; $(1/\sqrt{2}, \pi/4)$
18. $r = 2 \cos 4\theta$; $(1, \pi/12)$
19. $r^2 = \cos 2\theta$; $(1/\sqrt{2}, \pi/6)$
20. $r = 2 + 3 \sec \theta$; $(8, \pi/3)$
21. $r = 9 - 3 \csc \theta$; $(3, \pi/6)$
22. $r^2 = \theta$; $(\sqrt{\pi}/2, \pi/4)$
23. $r = 7 \cos 7\theta$; $(7/\sqrt{2}, \pi/28)$
24. $r = 2 \tan \theta$; $(2, \pi/4)$.

25. The curve $r = ae^{b\theta}$, where a and b are nonzero real numbers, is called a *logarithmic spiral*. Show that the angle β between the tangent line and the radial line joining the pole and the point of tangency is a constant. Use this information to sketch the curve.

*26. Using the half angle formula, show that for any real number $a \neq 0$, $\tan \beta = \tan \theta/2$ at any point on the cardioid $r = a(1 - \cos \theta)$.

*27. By calculating dy/dx, find all relative maxima and minima of $r = 1 + 2 \cos \theta$.

28. Find all relative maxima and minima of $r = 3 \cos 2\theta$.

*29. Find all relative maxima and minima of $r = 1 - \sin \theta$.

*30. Let $r = f(\theta)$ and $r = g(\theta)$ be the equations of two curves which intersect at the point (r_0, θ_0). Let β_1 and β_2 denote, respectively, the angles formed by the intersections of the line $\theta = \theta_0$ and the tangent lines to the two curves at the point (θ_0, r_0). If we define the angle γ between two curves at a point as the positive angle formed by the intersection of their tangent lines at that point, show that at (r_0, θ_0),

$$\gamma = \beta_1 - \beta_2.$$

31. Show that, in Problem 30,

$$\tan \gamma = \frac{\tan \beta_1 - \tan \beta_2}{1 + \tan \beta_1 \tan \beta_2}.$$

In Problems 32–39 use the results of Problems 30 and 31 to calculate (to the nearest degree) the angle between the given curves at all nonpolar points of intersection. (Use the tangent values in Table A.4.)

32. $r = \sin \theta$; $r = \cos \theta$
33. $r = 5 \sin 2\theta$; $r = 5 \cos 2\theta$
34. $r = 1 + \sin \theta$; $r = 1 - \cos \theta$
35. $r = 2$; $r = 4 \cos \theta$
36. $r = 1 + \sin \theta$; $r = 1 + 2 \cos \theta$
37. $r = \sin \theta$; $r = \sec \theta$
38. $r = 1 - \sin \theta$; $r = 1 + \cos \theta$
39. $r^2 = \cos \theta$; $r = \cos \theta$.

40. Show that the curves $r = a \cos \theta$ and $r = b \cos \theta$ intersect at right angles for any nonzero real numbers a and b.

41. Show that the curves $r^2 = a^2 \sec 2\theta$ and $r^2 = b^2 \csc 2\theta$ intersect at right angles.

42. Show that, except at the pole, the curves $r = a(1 - \cos \theta)$ and $r = a(1 + \cos \theta)$ intersect at right angles.

In Problems 43–56 find all values θ_i for which the given curve has a branch which is tangent to the line $\theta = \theta_i$ at the pole. Then sketch the curve.

43. $r = \sin \theta$
44. $r = \cos \theta$
45. $r = \sin 2\theta$
46. $r = \cos 2\theta$
47. $r = a \sin 2\theta, a > 0$
48. $r = a \cos 2\theta, a > 0$
49. $r^2 = \sin 2\theta$
50. $r^2 = \cos 2\theta$
51. $r^2 = a \sin 2\theta$
52. $r^2 = a \cos 2\theta$; $a > 0$.
53. $r = 4 \sin 3\theta$
54. $r = a(1 - \sin \theta)$
55. $r = (1 + 2 \sin \theta)$
56. $r = 7(\cos 2\theta - \sqrt{3} \sin 2\theta)$.

11.5 Areas in Polar Coordinates

In this section we derive a method for calculating the area of a region whose boundary is given in polar coordinates. We begin by recalling the formula for the area of a sector. Consider the sector bounded by the lines $\theta = \theta_1$ and $\theta = \theta_2$ and the circle $r = r_0$ (the shaded region in Figure 1). Let φ denote the angle (given in radians) between these two lines ($\varphi = \theta_2 - \theta_1$) and let A denote the area of the sector. Then,

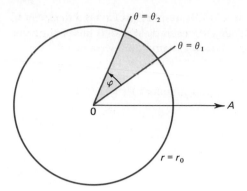

Figure 1

since the area of the circle is πr_0^2, we have the proportion

$$\frac{\varphi}{2\pi} = \frac{A}{\pi r_0^2} \tag{1}$$

or

$$A = \tfrac{1}{2} r_0^2 \varphi. \tag{2}$$

Now consider the plane region bounded by the polar curve $r = f(\theta)$ and the lines $\theta = \alpha$ and $\theta = \beta$ (see Figure 2). To calculate the area of this region, we divide

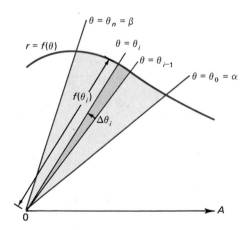

Figure 2

the region into n subregions by the rays $\alpha = \theta_0 < \theta_1 < \theta_2 < \cdots < \theta_{n-1} < \theta_n = \beta$. We use the notation $\Delta\theta_i$ to denote the angle between two successive lines:

$$\Delta\theta_i = \theta_i - \theta_{i-1}.$$

If $\Delta\theta_i$ is small, then the ith subregion is very close in shape to a sector and, by (2), its area is given by

$$A_i \approx \tfrac{1}{2} r_i^2 \, \Delta\theta_i = \tfrac{1}{2} [\, f(\theta_i)]^2 \, \Delta\theta_i. \tag{3}$$

Equation (3) is approximate because the "radius" of the subregion varies. However, if $\Delta\theta_i$ is small, then the variation is small. If we let the notation $|P| \to 0$ mean that the maximum $\Delta\theta_i$ approaches 0, we have, using a familiar argument

$$A = \lim_{|P|\to 0} (A_1 + A_2 + \cdots + A_n)$$

$$= \lim_{|P|\to 0} \{\tfrac{1}{2}[f(\theta_1)]^2\, \Delta\theta_1 + \tfrac{1}{2}[f(\theta_2)]^2\, \Delta\theta_2 + \cdots + \tfrac{1}{2}[f(\theta_n)]^2\, \Delta\theta_n\}$$

or

$$A = \tfrac{1}{2}\int_\alpha^\beta [f(\theta)]^2\, d\theta.$$

EXAMPLE 1.   Calculate the area of the region enclosed by the graph of the cardioid $r = 1 + \sin\theta$.

SOLUTION. The curve is symmetric about the line $\theta = \pi/2$ and is sketched in Figure 3.

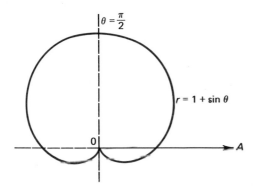

$r = 1 + \sin\theta$

Figure 3

(See Example 11.2.4.) Thus we need only calculate the area for θ in $[-\pi/2, \pi/2]$ and then multiply it by 2. We have

$$\text{half the area} = \frac{1}{2}\int_{-\pi/2}^{\pi/2}(1 + \sin\theta)^2\, d\theta = \frac{1}{2}\int_{-\pi/2}^{\pi/2}(1 + 2\sin\theta + \sin^2\theta)\, d\theta$$

$$= \frac{1}{2}\theta\,\Big|_{-\pi/2}^{\pi/2} + \int_{-\pi/2}^{\pi/2}\sin\theta\, d\theta + \frac{1}{4}\int_{-\pi/2}^{\pi/2}(1 - \cos 2\theta)\, d\theta$$

$$= \frac{\pi}{2} + 0 + \frac{\pi}{4} = \frac{3\pi}{4}$$

and the total area $A = 2 \cdot 3\pi/4 = 3\pi/2$.

EXAMPLE 2.   Find the area enclosed by the smaller loop of the limaçon $r = 1 + 2\cos\theta$.

SOLUTION. The curve is sketched in Figure 4 (see Example 11.2.6). The curve is symmetric about the polar axis and the inner loop is determined by the interval

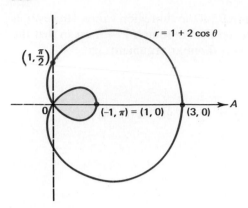

Figure 4

$(2\pi/3, 4\pi/3)$ (see Table 11.2.4). Then

$$A = \frac{1}{2} \int_{2\pi/3}^{4\pi/3} (1 + 2 \cos \theta)^2 \, d\theta = \frac{1}{2} \int_{2\pi/3}^{4\pi/3} (1 + 4 \cos \theta + 4 \cos^2 \theta) \, d\theta$$

$$= \frac{1}{2} \left[\theta + 4 \sin \theta + 2 \left(\theta + \frac{\sin 2\theta}{2} \right) \right] \Big|_{2\pi/3}^{4\pi/3} = \frac{2\pi - 3\sqrt{3}}{2}.$$

EXAMPLE 3. Find the area inside the circle $r = 5 \sin \theta$ and outside the limaçon $r = 2 + \sin \theta$.

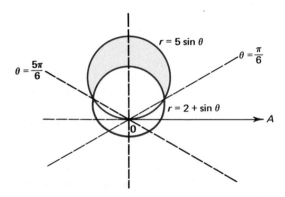

Figure 5

SOLUTION. The region is sketched in Figure 5. The two curves intersect whenever $2 + \sin \theta = 5 \sin \theta$, or $\sin \theta = \frac{1}{2}$, so that $\theta = \pi/6$ and $\theta = 5\pi/6$. The area of the region can be calculated by calculating the area of the limaçon between $\pi/6$ and $5\pi/6$ and subtracting it from the area of the circle for θ in that interval. By symmetry, we need only calculate the area between $\theta = \pi/6$ and $\theta = \pi/2$ and then multiply it by 2. We therefore have

$$A = \int_{\pi/6}^{\pi/2} (5 \sin \theta)^2 \, d\theta - \int_{\pi/6}^{\pi/2} (2 + \sin \theta)^2 \, d\theta$$

(we have already multiplied by 2)

$$= \int_{\pi/6}^{\pi/2} (25 \sin^2 \theta - 4 - 4 \sin \theta - \sin^2 \theta) \, d\theta$$

$$= \int_{\pi/6}^{\pi/2} [12(1 - \cos 2\theta) - 4 - 4 \sin \theta]\, d\theta$$

$$= (8\theta - 6 \sin 2\theta + 4 \cos \theta) \Big|_{\pi/6}^{\pi/2} = \frac{8\pi}{3} + \sqrt{3}.$$

PROBLEMS 11.5

In Problems 1–10 calculate the area enclosed by the curve and the two lines.

1. $r = \theta; \theta = 0; \theta = \pi/2; 0 \leq r \leq \pi/2$
2. $r = 3\theta; \theta = 0; \theta = \pi$
3. $r = 2 \cos \theta; \theta = -\pi/2; \theta = \pi/2$
4. $r = a \sin \theta; \theta = 0; \theta = \pi$
5. $r = a \cos 5\theta; \theta = -\pi/10; \theta = \pi/10$
6. $r = a \sin 3\theta; \theta = 0; \theta = \pi/6$
7. $r = 2\theta^2; \theta = 0; \theta = \pi/2$
8. $r = e^\theta; \theta = 0; \theta = 2\pi/3$
9. $r = 5\theta^4; \theta = 0; \theta = \pi$
10. $r = 1/\theta; \theta = \pi/6; \theta = \pi/2.$

In Problems 11–26 sketch the given curve and find the entire area enclosed by it. Where appropriate, assume that $a > 0$.

11. $r = a \sin \theta$
12. $r = a \cos \theta$
13. $r = a(\cos \theta + \sin \theta)$
14. $r = a(\cos \theta - \sin \theta)$
15. $r = a(\sin \theta - \cos \theta)$
16. $r = 3 + 2 \sin \theta$
17. $r = 3 + 2 \cos \theta$
18. $r = 3 - 2 \sin \theta$
19. $r = 3 - 2 \cos \theta$
20. $r = a \cos 2\theta$
21. $r = a \sin 2\theta$
22. $r = a \sin 3\theta$
23. $r = a \cos 5\theta$
24. $r^2 = a^2 \sin 2\theta$
25. $r^2 = a \cos 2\theta$
26. $r = \sqrt{\cos \theta}.$

In Problems 27–40 sketch the given region and find its area.

27. The smaller loop of the limaçon $r = 2 - 3 \sin \theta$.
28. The smaller loop of the limaçon $r = 2 + 3 \cos \theta$.
29. The smaller loop of the limaçon $r = a + b \sin \theta, b > a > 0$.
30. The area outside the smaller loop and inside the larger loop of the limaçon in Problem 27.
31. The area outside the smaller loop and inside the larger loop of the limaçon of Problem 28.
32. The area outside the smaller loop and inside the larger loop of the limaçon of Problem 29.
33. The region bounded by the graphs of $r = 2 \sin \theta, r = 2 + 2 \sin \theta, \theta = 0,$ and $\theta = \pi.$
34. The region inside the cardioid $r = a(1 + \sin \theta)$ and outside the circle $r = a, a > 0.$
*35. The region outside the circle $r = a \cos \theta$ and inside the lemniscate $r^2 = a^2 \sin 2\theta.$
36. The region inside the two circles $r = 3 \sin \theta$ and $r = 3 \cos \theta.$
37. The region enclosed by the graphs of $r = 4 \sin 2\theta$ and $r = 4 \cos 2\theta.$
* 38. The region inside the circle $r = \cos \theta$ and inside the four-leafed rose $r = \cos 2\theta.$
39. The region enclosed by one "petal" of the rose $r = a \cos n\theta, a > 0.$
* 40. The region enclosed by both the circles $r = a \sin \theta$ and $r = b \cos \theta, a > 0$ and $b > 0.$

41. A roller coaster encloses a region bounded by the graph of the equation $r^2 = 9 \sin 2\theta.$ If r is measured in meters, what is the area of this region?

Review Exercises for Chapter Eleven

In Exercises 1–6 convert from polar to rectangular coordinates.

1. $(2, 0)$
2. $(3, \pi/6)$
3. $(-7, \pi/2)$
4. $(3, 23\pi/3)$
5. $(-1, -\pi/2)$
6. $(2, 11\pi/12).$

In Exercises 7–12 convert from rectangular to polar coordinates with $r > 0$ and $0 \le \theta < 2\pi$.

7. $(2, 0)$ 8. $(1, \sqrt{3})$ 9. $(\sqrt{3}, -1)$

10. $(6, 6)$ 11. $(-6, -6)$ 12. $(-6, 6)$.

In Exercises 13–26 sketch the given equation, indicating any symmetry about the polar axis, the line $\theta = \pi/2$, and/or the pole.

13. $r = 8$ 14. $\theta = \pi/3$ 15. $r = 2 \cos \theta$

16. $r = 3 - 3 \sin \theta$ 17. $r = 3 - 2 \sin \theta$ 18. $r = 2 - 3 \sin \theta$

19. $r = 5 \cos 2\theta$ 20. $r^2 = 4 \sin 2\theta$ 21. $r = 3 \sin 4\theta$

22. $r = 3\theta, \theta \ge 0$ 23. $r = e^{2\theta}$ 24. $r^2 = 4\theta, r \ge 0$

25. $r \sin \theta = 4$ 26. $r \cos \theta = -2$

27.   Find, in rectangular coordinates, the equation of the circle

$$r = 3 \sin \theta + 4 \cos \theta.$$

28.   Find the polar equation of the circle centered at $(\frac{5}{2}, -6)$ with radius $\frac{13}{2}$.

In Exercises 29–36 find all points of intersection of the graphs of the two polar equations.

29. $r = 1; r = \cos \theta$ 30. $r = 2 + \cos \theta; r = 2 - \sin \theta$

31. $r^2 = \sin \theta; r = \sin \theta$ 32. $r^2 = 4 \cos 2\theta; r^2 = 4 \sin 2\theta$

33. $r = \theta; r^2 = 4\theta; \theta, r \ge 0$ 34. $r = \sin \theta; r = \sqrt{3} \cos \theta$

35. $r = \sqrt{3} \tan 2\theta; r = 1$ 36. $r = 2 \cot \theta; r = 8 \cos \theta$.

In Exercises 37–46 calculate the slope of the line tangent to the given curve at the given point by two methods: directly from formula (11.4.3) and indirectly by first calculating $\tan \beta$.

37. $r = 1; (1, 3\pi/2)$ 38. $r = \sin \theta; (1/2, \pi/6)$

39. $r = 4\theta; (\pi, \pi/4)$ 40. $r = 3 \sin 2\theta; (3/\sqrt{2}, \pi/8)$

41. $r = 3 + 3 \sin \theta; (9/2, \pi/6)$ 42. $r = e^{\theta/3}; (e^{\pi/6}, \pi/2)$

43. $r^2 = \sin 2\theta; (1, \pi/4)$ 44. $r = 4 \sin 3\theta; (2\sqrt{3}, \pi/9)$

45. $r = 10 \sin 10\theta; (10, \pi/20)$ 46. $r = 3 \tan \theta; (3, \pi/4)$.

47.   Calculate to the nearest degree the angle formed by the curves $r = 2 \sin \theta$ and $r = 2 \cos \theta$ at each of their nonpolar points of intersection.

48.   Do the same for the curves $r = 3 - 3 \cos \theta$ and $r = 3 + 3 \sin \theta$.

49.   Find all numbers θ_i for which the curve $r = 2 \sin 2\theta$ has a branch which is tangent to the line $\theta = \theta_i$ at the pole. Then sketch the curve.

50.   Do the same for the curve $r = -4(1 - 2 \cos \theta)$.

In Exercises 51–61 find the area of the given region.

51.   The region bounded by $r = \theta, \theta = \pi/6$ and $\theta = \pi/2, \pi/6 \le r \le \pi/2$.

52.   The region bounded by $r = a \cos 4\theta, \theta = 0$, and $\theta = \pi/12$.

53.   The region enclosed by the circle $r = a(\sin \theta + \cos \theta)$.

54.   The region enclosed by the limaçon $r = 4 + 3 \cos \theta$.

55.   The region enclosed by the cardioid $r = -2 - 2 \sin \theta$.

56.   The region enclosed by the smaller loop of the limaçon $r = 3 - 4 \cos \theta$.

57.   The region outside the smaller loop and inside the larger loop of the limaçon in Exercise 56.

58.   The region inside the cardioid $r = 4(1 + \cos \theta)$ and outside the circle $r = 4$.

59.   The region inside the two circles $r = 2 \sin \theta$ and $r = 2 \cos \theta$.

60.   The region enclosed by the lemniscate $r^2 = 36 \cos 2\theta$.

61.   The region enclosed by one petal of the rose $r = 4 \sin 10\theta$.

TWELVE

INDETERMINATE FORMS AND IMPROPER INTEGRALS

12.1 The Indeterminate Form 0/0 and L'Hôpital's Rule

In earlier chapters we encountered quotients of the form $f(x)/g(x)$. In Theorem 10.2.4 we showed that if the limits $\lim_{x \to x_0} f(x)$ and $\lim_{x \to x_0} g(x)$ both exist, then

$$\lim_{x \to x_0} \frac{f(x)}{g(x)} = \frac{\lim_{x \to x_0} f(x)}{\lim_{x \to x_0} g(x)} \tag{1}$$

provided that $\lim_{x \to x_0} g(x) \neq 0$.

However this last condition is frequently an obstacle in important applications. For example, in Section 7.5 we showed that $\lim_{x \to 0}[(\sin x)/x] = 1$. We did this despite the fact that $\lim_{x \to 0} x = 0$ so that the rule (1) did not apply. As another example, we saw (in Problem 2.2.4) that

$$\lim_{x \to 1} \frac{(x - 1)(x - 2)}{x - 1} = -1,$$

again despite the fact that $\lim_{x \to 1}(x - 1) = 0$.

In general, we have the following definition.

DEFINITION 1. Let f and g be two functions having the property that $\lim_{x \to x_0} f(x) = 0$ and $\lim_{x \to x_0} g(x) = 0$. Then the function f/g has *the indeterminate form 0/0 at x_0*.

We now give a rule for finding $\lim_{x \to x_0}[f(x)/g(x)]$ in certain instances in the case f/g has the indeterminate form 0/0 at x_0. The proof of this result will be given in Section 12.2 and other indeterminate forms will be considered in Section 12.3.

THEOREM 1 (L'Hôpital's† Rule for the Indeterminate Form 0/0). Let x_0 be a real number, $+\infty$, or $-\infty$, and let f and g be two functions that satisfy

---

†This theorem is named after the French mathematician Marquis de L'Hôpital (1661–1704). L'Hôpital included this theorem in a book, considered to be the first calculus textbook ever written, published in 1696. Actually the theorem was first proven by the great Swiss mathematician Jean Bernoulli (1667–1748) who was one of L'Hôpital's tutors and who sent L'Hôpital the proof in a letter in 1694.

(i) f and g are differentiable at every point in a neighborhood† of x_0, except possibly at x_0 itself,

(ii) $\lim_{x \to x_0} f(x) = 0$ and $\lim_{x \to x_0} g(x) = 0$, and

(iii) $\lim_{x \to x_0} [f'(x)/g'(x)] = L$, where L is a real number, $+\infty$, or $-\infty$.

Then

$$\lim_{x \to x_0} \frac{f(x)}{g(x)} = L.$$

That is, under the conditions of the theorem, the limit of the quotient of the two functions is equal to the limit of the quotient of their derivatives.‡ Furthermore, Theorem 1 is also true for right- and left-hand limits.

EXAMPLE 1. Calculate $\lim_{x \to 0} [(\sin x)/x]$.

SOLUTION. Since $f(x) = \sin x$ and $g(x) = x$ are everywhere differentiable and have the limiting form $0/0$ at $x = 0$, we can apply L'Hôpital's rule to find that

$$\lim_{x \to 0} \frac{\sin x}{x} = \lim_{x \to 0} \frac{\frac{d}{dx} \sin x}{\frac{d}{dx} x} = \lim_{x \to 0} \frac{\cos x}{1} = \lim_{x \to 0} \cos x = 1.$$

EXAMPLE 2. Calculate $\displaystyle\lim_{x \to 1} \frac{x^3 - 5x^2 + 6x - 2}{x^5 - 4x^4 + 7x^2 - 9x + 5}$.

SOLUTION. We first note that

$$\lim_{x \to 1} (x^3 - 5x^2 + 6x - 2) = 0 \quad \text{and} \quad \lim_{x \to 1} (x^5 - 4x^4 + 7x^2 - 9x + 5) = 0,$$

so that the indicated limit has the form $0/0$. Then, applying L'Hôpital's rule,

$$\lim_{x \to 1} \frac{x^3 - 5x^2 + 6x - 2}{x^5 - 4x^4 + 7x^2 - 9x + 5} = \lim_{x \to 1} \frac{3x^2 - 10x + 6}{5x^4 - 16x^3 + 14x - 9}$$

$$= \frac{-1}{-6} = \frac{1}{6}.$$

EXAMPLE 3. Calculate $\lim_{x \to 0^+} [\sqrt{x}/(\sin \sqrt{x})]$.

SOLUTION. Since $\sqrt{x}$ is only defined for $x \geq 0$ and since we are only considering a right-hand limit, a neighborhood of 0 is an interval of the form $(0, b)$. In this context the hypotheses of Theorem 1 hold and we have

$$\lim_{x \to 0^+} \frac{\sqrt{x}}{\sin \sqrt{x}} = \lim_{x \to 0^+} \frac{1/2\sqrt{x}}{(\cos \sqrt{x}) \cdot (1/2\sqrt{x})} = \lim_{x \to 0^+} \frac{1}{\cos \sqrt{x}} = 1.$$

† Recall from Section 10.1 that by a neighborhood of ∞ we mean an open interval of the form (a, ∞) for some real number a.

‡ We emphasize that we are taking the quotient of the two derivatives (f'/g'), *not* the derivative of the quotient (($f/g)'$).

EXAMPLE 4.   Calculate $\lim_{x \to 0}[(\sin x)/x^3]$.

SOLUTION

$$\lim_{x \to 0} \frac{\sin x}{x^3} = \lim_{x \to 0} \frac{\cos x}{3x^2} = +\infty,$$

EXAMPLE 5.   Calculate $\lim_{x \to 0}(e^x - 1)/x$.

SOLUTION

$$\lim_{x \to 0} \frac{e^x - 1}{x} = \lim_{x \to 0} \frac{e^x}{1} = \lim_{x \to 0} e^x = 1.$$

EXAMPLE 6.   Calculate $\lim_{x \to 1}(\ln x)/(x - 1)$.

SOLUTION

$$\lim_{x \to 1} \frac{\ln x}{x - 1} = \lim_{x \to 1} \frac{1/x}{1} = \lim_{x \to 1} \frac{1}{x} = 1.$$

EXAMPLE 7.   Calculate $\lim_{x \to \infty}[1 + (1/x)]^x$.

SOLUTION. First note that the expression $\lim_{x \to \infty}[1 + (1/x)]^x$ was used in Section 6.1 to define the number e. We cannot apply L'Hôpital's rule directly since the indicated limit does not have the form of a quotient. Let us define $y = [1 + (1/x)]^x$. Then

$$\ln y = x \ln \left(1 + \frac{1}{x}\right) = \frac{\ln[1 + (1/x)]}{1/x}.$$

Since $\lim_{x \to \infty} \ln[1 + (1/x)] = 0$ and $\lim_{x \to \infty} 1/x = 0$, we can now apply L'Hôpital's rule to obtain

$$\lim_{x \to \infty} \ln y = \lim_{x \to \infty} \frac{\ln[1 + (1/x)]}{1/x} = \lim_{x \to \infty} \frac{\frac{1}{1 + (1/x)} \cdot \frac{-1}{x^2}}{-1/x^2}$$

$$= \lim_{x \to \infty} \frac{1}{1 + (1/x)} = 1.$$

Hence $\ln y \to 1$ as $x \to \infty$ so that $y \to e^1 = e$ as $x \to \infty$ (which is, of course, what we expected).

EXAMPLE 8.   Calculate $\lim_{x \to -2} \dfrac{3x^3 + 16x^2 + 28x + 16}{x^5 + 4x^4 + 4x^3 + 3x^2 + 12x + 12}$.

SOLUTION. First we note that

$$\lim_{x \to -2}(3x^3 + 16x^2 + 28x + 16) = 0$$

and

$$\lim_{x \to -2}(x^5 + 4x^4 + 4x^3 + 3x^2 + 12x + 12) = 0$$

so that

$$\lim_{x \to -2} \frac{3x^3 + 16x^2 + 28x + 16}{x^5 + 4x^4 + 4x^3 + 3x^2 + 12x + 12}$$

$$= \lim_{x \to -2} \frac{9x^2 + 32x + 28}{5x^4 + 16x^3 + 12x^2 + 6x + 12}.$$

But

$$\lim_{x \to -2} (9x^2 + 32x + 28) = 0$$

and

$$\lim_{x \to -2} (5x^4 + 16x^3 + 12x^2 + 6x + 12) = 0$$

so we simply apply L'Hôpital's rule again:

$$= \lim_{x \to -2} \frac{18x + 32}{20x^3 + 48x^2 + 24x + 6} = \frac{-4}{-10} = \frac{2}{5}.$$

Warning: Do not try to apply L'Hôpital's rule when either the numerator or denominator of f/g has a finite, nonzero limit at x_0. For example, we easily see that

$$\lim_{x \to 0} \frac{x}{1 + \sin x} = \frac{\lim_{x \to 0} x}{\lim_{x \to 0}(1 + \sin x)} = \frac{0}{1} = 0.$$

But if we try to apply L'Hôpital's rule we obtain

$$\lim_{x \to 0} \frac{x}{1 + \sin x} = \lim_{x \to 0} \frac{1}{\cos x} = 1,$$

which is an incorrect result.

PROBLEMS 12.1

In Problems 1–26, calculate the given limit.

1. $\lim_{x \to 0} \dfrac{1 - \cos x}{x}$

2. $\lim_{x \to \pi/2} \dfrac{x - (\pi/2)}{\cos x}$

3. $\lim_{x \to 0^+} \dfrac{\sin x}{\sqrt{x}}$

4. $\lim_{x \to 0} \dfrac{\sin x}{x^5}$

5. $\lim_{x \to 1} \dfrac{x^4 - x^3 + x^2 - 1}{x^3 - x^2 + x - 1}$

6. $\lim_{x \to \infty} \dfrac{1/x}{\ln[1 + (1/x)]}$

7. $\lim_{x \to 0} \dfrac{e^x - 1}{x(3 + x)}$

8. $\lim_{x \to 0} \dfrac{x + \sin 5x}{x - \sin 5x}$

9. $\lim_{x \to 0} \dfrac{x + \sin ax}{x - \sin ax}, a \neq 1$

10. $\lim_{x \to 0} \dfrac{x - \sin x}{x^3}$

11. $\displaystyle\lim_{x\to1^-} \frac{x^2 - 1}{\sqrt{1 - x}}$

12. $\displaystyle\lim_{x\to0} \frac{3 + x - 3e^x}{x(2 + 5e^x)}$

*13. $\displaystyle\lim_{x\to1^-} \frac{\sqrt{1 - x^3}}{\sqrt{1 - x^4}}$

*14. $\displaystyle\lim_{x\to0} \left(\frac{2}{\sin x} - \frac{2}{x}\right)$

15. $\displaystyle\lim_{x\to0} \frac{4^x - 3^x}{\sqrt{x}}$

16. $\displaystyle\lim_{x\to\pi/2} \frac{\cos x}{\sin 2x}$

17. $\displaystyle\lim_{x\to0} (1 + x)^{2/x}$

18. $\displaystyle\lim_{x\to0} \frac{\tan^{-1} x}{x}$

19. $\displaystyle\lim_{x\to0} \frac{x^2}{\sin^{-1} x}$

20. $\displaystyle\lim_{x\to0} \frac{\sin^{-1} x - x}{5x^2}$

*21. $\displaystyle\lim_{x\to0} \frac{\sinh x - \sin x}{\sin^3 x}$

22. $\displaystyle\lim_{x\to0} (x + e^x)^{1/x}$

23. $\displaystyle\lim_{x\to4\pi} \frac{(x - 4\pi)^2}{\ln \cos x}$

24. $\displaystyle\lim_{x\to-\infty} \frac{e^x}{2^x}$

25. $\displaystyle\lim_{x\to x_0} \frac{\sqrt{x} - \sqrt{x_0}}{x - x_0}$

26. $\displaystyle\lim_{x\to0} \frac{\tanh ax}{\tanh bx}$.

27. Show that if a and b are real numbers and if $x_0 > 0$, then

$$\lim_{x\to x_0} \frac{x^a - x_0{}^a}{x^b - x_0{}^b} = \frac{a}{b} x_0^{a-b}.$$

*28. Calculate $\displaystyle\lim_{x\to0^+} \frac{\displaystyle\int_0^x \cos t^2 \, dt}{\displaystyle\int_0^x e^{t^2} \, dt}$. [*Hint:* See Theorem 10.8.1.]

29. Calculate $\displaystyle\lim_{x\to0^+} \frac{\displaystyle\int_0^x \sin t^3 \, dt}{x^4}$.

30. In Section 1.10 we showed that the sum of the geometric progression

$$S_n = 1 + a + a^2 + \cdots + a^n = \frac{1 - a^{n+1}}{1 - a}.$$

Calculate $\lim_{a\to1} S_n$ and compare this with the sum of the first $n + 1$ terms of the geometric progression with $a = 1$.

31. Show that $\lim_{x\to0}(1 + x)^{a/x} = e^a$ for any real number $a \neq 0$.
32. Calculate $\lim_{x\to\infty} x \sin(1/x)$.
33. Show that $\lim_{x\to\infty} x^a \sin(1/x) = 0$ if $0 < a < 1$.
34. Show that $\lim_{x\to\infty} x^a \sin(1/x) = +\infty$ if $a > 1$.
35. In Problem 6.4.17 we discussed the simple electric circuit shown in Figure 1 and found that the current at any time t is given by $I(t) = (E/R) + [I(0) - (E/R)]e^{-Rt/L}$. Assuming that $I(0) = 0$ find, for fixed t, E, and L, $\lim_{R\to0^+} I(t)$.

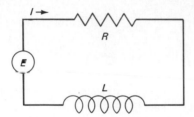

Figure 1

*12.2 Proof of L'Hôpital's Rule

In this section we will prove L'Hôpital's rule for the indeterminate form $0/0$. Before doing this, however, we need to prove another kind of mean value theorem which will be useful here and in Chapter 13.

THEOREM 1 (Cauchy† Mean Value Theorem). Suppose that the two functions f and g are continuous in the closed interval $[a, b]$ and differentiable in the open interval (a, b). Suppose further that $g'(x) \neq 0$ for x in (a, b). Then there exists at least one number c in (a, b) such that

$$\frac{f(b) - f(a)}{g(b) - g(a)} = \frac{f'(c)}{g'(c)}. \tag{1}$$

PROOF. Let

$$h(x) = [g(b) - g(a)][f(x) - f(a)] - [g(x) - g(a)][f(b) - f(a)].$$

Then clearly $h(a) = h(b) = 0$. Since h is continuous in $[a, b]$ and differentiable in (a, b), we can apply Rolle's theorem (Theorem 10.6.1) which tells us that there exists at least one number c in (a, b) such that $h'(c) = 0$. But

$$h'(c) = [g(b) - g(a)][f'(c)] - [g'(c)][f(b) - f(a)] = 0. \tag{2}$$

Then $f'(c)[g(b) - g(a)] = g'(c)[f(b) - f(a)]$. Since $g'(c) \neq 0$ by assumption, we can divide by it and by $[g(b) - g(a)]$ to obtain equation (1) and the proof is complete.

> **Note.** It is *not* necessary to assume that $g(b) - g(a) \neq 0$ since this fact is implied by the hypotheses (see Problem 1).

We can now prove L'Hôpital's rule for the indeterminate form $0/0$.

THEOREM 2. Let x_0 be a real number, $+\infty$, or $-\infty$ and let f and g be two functions which satisfy

    (i) f and g are differentiable at every point in a neighborhood of x_0, except possibly at x_0 itself.
    (ii) $\lim_{x \to x_0} f(x) = 0$ and $\lim_{x \to x_0} g(x) = 0$.
    (iii) $\lim_{x \to x_0} [f'(x)/g'(x)] = L$, where L is a real number, $+\infty$, or $-\infty$.

† Augustin-Louis Cauchy (1789–1867) was a great French mathematician who made significant contributions to many branches of mathematics. He is generally credited with the modern definition of the definite integral in terms of the limit of certain sums.

Then $\lim_{x\to x_0}[f(x)/g(x)] = L$. Furthermore the theorem is true for right- and left-hand limits as well.

PROOF. We prove the theorem first in the case $x \to x_0^+$ where x_0 is a real number. By (i) we can choose a positive number $\Delta_1 x$ such that f and g are differentiable in $(x_0, x_0 + \Delta_1 x)$ and continuous in $(x_0, x_0 + \Delta_1 x]$. Moreover, since by (iii) $\lim_{x\to x_0^+}[f'(x)/g'(x)]$ exists, we can choose a $\Delta_2 x$ so that $f'(x)/g'(x)$ exists in $(x_0, x_0 + \Delta_2 x)$. This follows from the definition of a limit. Moreover, if $[f'(x)/g'(x)]$ exists in $(x_0, x_0 + \Delta_2 x)$, then $g'(x) \neq 0$ for x in $(x_0, x_0 + \Delta_2 x)$. Now let $\Delta x = \min(\Delta_1 x, \Delta_2 x)$. Since $\lim_{x\to x_0} f(x) = 0$, either $f(x_0) = 0$ or f is not defined at x_0. In the latter case we simply extend the f by defining it to be zero at x_0. Similarly, either $g(x_0) = 0$ or g can be defined to be zero at x_0. Then, by hypothesis (ii), both functions are continuous in the closed interval $[x_0, x_0 + \Delta x]$. Since $g'(x) \neq 0$ in $(x_0, x_0 + \Delta x)$, we can apply the Cauchy mean value theorem which tells us that there is a number $c_{\Delta x}$ in $(x_0, x_0 + \Delta x)$† such that

$$\frac{f'(c_{\Delta x})}{g'(c_{\Delta x})} = \frac{f(x_0 + \Delta x) - f(x_0)}{g(x_0 + \Delta x) - g(x_0)} = \frac{f(x_0 + \Delta x)}{g(x_0 + \Delta x)}.$$

Now $c_{\Delta x} \to x_0^+$ as $\Delta x \to 0^+$ so

$$\lim_{x\to x_0^+} \frac{f(x)}{g(x)} = \lim_{\Delta x \to 0^+} \frac{f(x_0 + \Delta x)}{g(x_0 + \Delta x)} = \lim_{\Delta x \to 0^+} \frac{f'(c_{\Delta x})}{g'(c_{\Delta x})}$$

$$= \lim_{c_{\Delta x} \to x_0^+} \frac{f'(c_{\Delta x})}{g'(c_{\Delta x})} = \lim_{x\to x_0^+} \frac{f'(x)}{g'(x)} = L$$

If $x \to x_0^-$, then we obtain the same proof by considering the interval $(x_0 - \Delta x, x_0)$. If $x \to x_0^-$ and $x \to x_0^+$, then $x \to x_0$. Finally, for the case $x \to \infty$,

$$\lim_{x\to\infty} \frac{f'(x)}{g'(x)} = \lim_{y\to 0^+} \frac{f'(1/y)}{g'(1/y)} = \lim_{y\to 0^+} \frac{f'(1/y)\cdot(-1/y^2)}{g'(1/y)\cdot(-1/y^2)}$$

$$= \lim_{y\to 0^+} \frac{f(1/y)}{g(1/y)} = \lim_{x\to\infty} \frac{f(x)}{g(x)} \qquad \text{(by the case proven above)}$$

since

$$\left[f\left(\frac{1}{y}\right)\right]' = f'\left(\frac{1}{y}\right)\frac{d}{dy}\left(\frac{1}{y}\right) = f'\left(\frac{1}{y}\right)\left(-\frac{1}{y^2}\right)$$

and the case $x \to \infty$ is proved. The case $x \to -\infty$ is handled in similar manner.

PROBLEM 12.2

1. Use the mean value theorem to show that under the hypotheses of Theorem 1, $g(b) \neq g(a)$.

---

†The notation $c_{\Delta x}$ implies that the number depends on the choice of Δx.

12.3 Other Indeterminate Forms

There are other situations in which $\lim_{x \to x_0}[f(x)/g(x)]$ cannot be directly evaluated. One important case is defined below.

DEFINITION 1. Let f and g be two functions having the property that $\lim_{x \to x_0} f(x) = \pm\infty$ and $\lim_{x \to x_0} g(x) = \pm\infty$, where x_0 is real, $+\infty$ or $-\infty$. Then the function f/g has the *indeterminate form* ∞/∞ at x_0.

THEOREM 1 (L'Hôpital's Rule for the Indeterminate Form ∞/∞). Let x_0 be a real number, $+\infty$, or $-\infty$ and let f and g be two functions that satisfy

　(i)　f and g are differentiable at every point in a neighborhood of x_0, except possibly at x_0 itself,
　(ii)　$\lim_{x \to x_0} f(x) = \pm\infty$ and $\lim_{x \to x_0} g(x) = \pm\infty$, and
　(iii)　$\lim_{x \to x_0}[f'(x)/g'(x)] = L$, where L is a real number, $+\infty$, or $-\infty$.

Then

$$\lim_{x \to x_0} \frac{f(x)}{g(x)} = L. \tag{1}$$

The result is also true for right- and left-hand limits.

　The proof of this theorem is more complicated than the proof of the earlier version of L'Hôpital's rule and is omitted.†

EXAMPLE 1. Calculate $\lim_{x \to \infty}(e^x/x)$.

SOLUTION. Since $e^x \to \infty$ as $x \to \infty$, L'Hôpital's rule applies and we have

$$\lim_{x \to \infty} \frac{e^x}{x} = \lim_{x \to \infty} \frac{e^x}{1} = \infty.$$

EXAMPLE 2. Calculate $\lim_{x \to 0^+}(x \ln x)$.

SOLUTION. We write $x \ln x = (\ln x)/(1/x)$. Then, since $\lim_{x \to 0^+} \ln x = -\infty$ and $\lim_{x \to 0^+}(1/x) = \infty$, we have

$$\lim_{x \to 0^+} (x \ln x) = \lim_{x \to 0^+} \frac{\ln x}{1/x} = \lim_{x \to 0^+} \frac{1/x}{-1/x^2} = \lim_{x \to 0^+} -x = 0.$$

EXAMPLE 3. Calculate $\lim_{x \to 0^+} x^{ax}$, $a > 0$.

SOLUTION. If $y = x^{ax}$, then $\ln y = ax \ln x$ and

$$\lim_{x \to 0^+} \ln y = \lim_{x \to 0^+}(ax \ln x) = \lim_{x \to 0^+} \frac{\ln x}{1/ax} = \lim_{x \to 0^+} \frac{1/x}{-1/ax^2} = \lim_{x \to 0^+} (-ax) = 0.$$

Since $\ln y \to 0$, we have $y = x^{ax} \to 1$ as $x \to 0^+$.

†For a proof consult R. C. Buck, *Advanced Calculus,* McGraw-Hill, New York, 1965, p. 92.

EXAMPLE 4. Calculate $\lim_{x\to\infty} x^a e^{-bx}$ for any real number a and any positive real number b.

SOLUTION. If $a = 0$, then $\lim_{x\to\infty}(x^0 e^{-bx}) = \lim_{x\to\infty}(1/e^{bx}) = 0$. If $a < 0$, then $-a > 0$ and $\lim_{x\to\infty}(x^a e^{-bx}) = \lim_{x\to\infty}(e^{-bx}/x^{-a})$. Since $-a > 0$, $x^{-a} \to \infty$ as $x \to \infty$ and since $e^{-bx} \to 0$, we have, evidently, $\lim_{x\to\infty}(e^{-bx}/x^{-a}) = 0$. The case $a > 0$ is the most interesting. We can write $\lim_{x\to\infty}(x^a e^{-bx}) = \lim_{x\to\infty}(x^a/e^{bx})$. Here both numerator and denominator approach ∞ as $x \to \infty$ so we can apply L'Hôpital's rule as often as necessary to obtain

$$\lim_{x\to\infty} \frac{x^a}{e^{bx}} = \lim_{x\to\infty} \frac{ax^{a-1}}{be^{bx}} = \lim_{x\to\infty} \frac{a(a-1)x^{a-2}}{b^2 e^{bx}} = \cdots$$

$$= \lim_{x\to\infty} \frac{a(a-1)\cdots(a-k+1)x^{a-k}}{b^k e^{bx}}.$$

We have applied L'Hôpital's rule k times, where k is the smallest integer such that $a - k \le 0$. Then we use the fact that $a - k \le 0$ to conclude that

$$\boxed{\lim_{x\to\infty} x^a e^{-bx} = 0 \qquad \text{if } b > 0.} \tag{2}$$

The above result is very interesting and useful. It tells us that the exponential function e^x grows much faster than any power function. In particular, we can conclude from (2) that if $P_n(x)$ is a polynomial of degree n, then we have

THEOREM 2

$$\lim_{x\to\infty}[P_n(x)e^{-x}] = 0, \qquad \text{for all values of } n. \tag{3}$$

We will make use of this fact in later chapters.

There are other indeterminate forms that can be dealt with by applying L'Hôpital's rule. For example, in Example 2 we calculated $\lim_{x\to 0^+}(x \ln x)$. Since $\lim_{x\to 0^+} x = 0$ and $\lim_{x\to 0^+} \ln x = -\infty$, this indeterminate expression is really of the form $0 \cdot \infty$. An indeterminate form of this type can often be treated by putting it into one of the forms $0/0$ or ∞/∞.

Another indeterminate expression is one in the form $\infty - \infty$. Again, it is necessary to put it into one of the forms $0/0$ or ∞/∞.

EXAMPLE 5. Calculate $\lim_{x\to 0^+}(\csc x - \cot x)$.

SOLUTION. We first note that

$$\lim_{x\to 0^+} \csc x = \lim_{x\to 0^+} \frac{1}{\sin x} = \infty$$

and

$$\lim_{x\to 0^+} \cot x = \lim_{x\to 0^+} \frac{\cos x}{\sin x} = \infty.$$

We now write

$$\lim_{x \to 0^+}(\csc x - \cot x) = \lim_{x \to 0^+} \frac{1}{\sin x} - \frac{\cos x}{\sin x}$$

$$= \lim_{x \to 0^+} \frac{1 - \cos x}{\sin x}$$

$$= \lim_{x \to 0^+} \frac{\sin x}{\cos x} = 0.$$

In the next to the last line we used L'Hôpital's rule for the form $0/0$.

Finally, we may encounter indeterminate expressions of the form 0^0, ∞^0, and 1^∞. In Example 12.1.7, we calculated $\lim_{x \to \infty}[1 + (1/x)]^x$ which is of the form 1^∞ since $\lim_{x \to \infty}[1 + (1/x)] = 1$ and $\lim_{x \to \infty} x = \infty$. In Example 3 we calculated $\lim_{x \to 0^+} x^{ax}$ which is of the form 0^0 for obvious reasons. In both cases we dealt with the problem by taking logarithms.

EXAMPLE 6. Calculate $\lim_{x \to \infty} x^{1/x}$.

SOLUTION. This is of the form ∞^0. We set $y = x^{1/x}$. Then $\ln y = (1/x) \ln x$ and

$$\lim_{x \to \infty} \ln y = \lim_{x \to \infty} \frac{\ln x}{x} = \lim_{x \to \infty} \frac{1/x}{1} = 0$$

so that $\ln y \to 0$ and $y = x^{1/x} \to 1$.

In all these problems we emphasize that in order to calculate an indeterminate expression by using L'Hôpital's rule, it is necessary to write it either as a $0/0$ form or as an ∞/∞ form.

PROBLEMS 12.3

In Problems 1–32 evaluate the given limit.

1. $\lim_{x \to \infty} \dfrac{x^3 + 3x + 4}{2x^3 - 4x + 2}$

2. $\lim_{x \to \infty} \dfrac{4x^{5/2} + 3\sqrt{x} - 10}{3x^{5/2} - 8x^{3/2} + 45x^2}$

3. $\lim_{x \to \infty} \dfrac{\ln x}{\sqrt{x}}$

4. $\lim_{x \to \pi/2^-} \dfrac{\sec x}{\tan x}$

5. $\lim_{x \to \infty} \dfrac{\ln x}{x^a}, \; a > 0$

6. $\lim_{x \to 0^+} \dfrac{\ln(\sin x)}{\ln(\tan x)}$

*7. $\lim_{x \to 0+} \dfrac{\ln x}{\cot x}$

8. $\lim_{x \to \pi/2^-} \dfrac{\tan 2x}{\tan x}$

9. $\lim_{x \to \infty} \left(x \sin \dfrac{1}{x}\right)$

10. $\lim_{x \to 0^+} x^{-1/x}$

*11. $\lim\limits_{x \to \infty} \dfrac{x}{e^{\sqrt{x}}}$

*12. $\lim\limits_{x \to \infty} (xe^{-x^a}),\ 0 < a < 1$

13. $\lim\limits_{x \to 0^+} x^{\sin x}$

14. $\lim\limits_{x \to \pi/2^+} \left(x - \dfrac{\pi}{2} \right)^{\cos x}$

15. $\lim\limits_{x \to 0^+} (x \ln \sin x)$

16. $\lim\limits_{x \to 0^+} (\sin x)^x$

17. $\lim\limits_{x \to \infty} \left(1 + \dfrac{1}{2x} \right)^{x^2}$

*18. $\lim\limits_{x \to \infty} \left(1 + \dfrac{1}{ax} \right)^{x^a},\ a > 0$

19. $\lim\limits_{x \to 0^+} (1 + \sinh x)^{1/x}$

20. $\lim\limits_{x \to 0^+} (1 + \sinh x)^{a/x},\ a > 0$

21. $\lim\limits_{x \to 0^+} (\csc 2x \sin 3x)$

22. $\lim\limits_{x \to (\pi/2)^-} (\sec x \cos 3x)$

23. $\lim\limits_{x \to 0^+} (\sin x)^{(\sin x) - x}$

24. $\lim\limits_{x \to (\pi/2)^-} (\cos x)^{\sec x}$

25. $\lim\limits_{x \to 2} x^{1/(2-x)}$

26. $\lim\limits_{x \to \infty} (1 + 5x)^{e^{-x}}$

27. $\lim\limits_{x \to \infty} \left(\cos \dfrac{1}{x} \right)^x$

28. $\lim\limits_{x \to 3} \left[\dfrac{1}{x - 3} - \dfrac{1}{(x - 3)^2} \right]$

29. $\lim\limits_{x \to 1^-} \left(\dfrac{1}{1 - x} - \dfrac{1}{\ln x} \right)$

30. $\lim\limits_{x \to \infty} (\sqrt{x^2 + x} - x)$

31. $\lim\limits_{x \to 2^+} \left[\dfrac{x}{x - 2} - \dfrac{1}{\ln(x - 1)} \right]$

32. $\lim\limits_{x \to \infty} (x^3 - \sqrt{x^6 - 5x^3 + 3}).$

*33. Since $\lim_{x \to \infty} \sin x$ does not exist (but oscillates continuously between -1 and $+1$) it is evident that $\lim_{x \to \infty} e^{-\sin x}$ does not exist. But

$$\lim\limits_{x \to \infty} e^{-\sin x} = \lim\limits_{x \to \infty} \dfrac{2x + \sin 2x}{(2x + \sin 2x)e^{\sin x}}.$$

Now show successively that
(a) $2 + 2 \cos 2x = 4 \cos^2 x.$ [*Hint:* $\cos 2x = \cos^2 x - \sin^2 x.$]
(b) $\lim\limits_{x \to \infty} \dfrac{4 \cos x}{(2x + 4 \cos x + \sin 2x)e^{\sin x}} = 0.$

Then apply L'Hôpital's rule (using (a) and (b)) to show that

(c) $\lim\limits_{x \to \infty} \dfrac{2x + \sin 2x}{(2x + \sin 2x)e^{\sin x}} = 0.$

Thus $\lim_{x \to \infty} e^{-\sin x} = 0.$ Where have we been led astray?

34. Show that $\lim_{x \to 0}(e^{-1/x^2}/x) = 0.$ [*Hint:* Write $\lim_{x \to 0}(e^{-1/x^2}/x) = \lim_{x \to \infty}((1/x)/e^{1/x^2}).$]
35. Show that for every integer n $\lim_{x \to 0}(e^{-1/x^2}/x^n) = 0.$
36. Use the result of Problem 35 to show that the function

$$f(x) = \begin{cases} e^{-1/x^2}, & x \neq 0 \\ 0, & x = 0 \end{cases}$$

has continuous derivatives of all orders at every real number x.

12.4 Improper Integrals

In our introduction to the definite integral in Chapter 4, we cited Theorem 4.3.1 which states that $\int_a^b f(x)\,dx$ exists if $f(x)$ is piecewise continuous in the closed interval $[a, b]$. However, in many interesting applications, one of two situations occurs: either (1) a or b is infinite, or (2) f becomes infinite at one or more values in the interval $[a, b]$. If one of these cases occurs, we say that the integral in question is an *improper integral*. In this section we will learn how to evaluate improper integrals.

CASE 1. $b = +\infty$, $a = -\infty$, or both.

Before dealing with the general case, we will begin with an example.

EXAMPLE 1.   Calculate the area in the first quadrant under the curve $y = e^{-x}$.

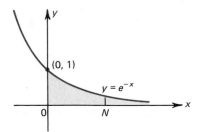

Figure 1

SOLUTION. The region in question is sketched in Figure 1. For the first time we are asked to calculate the area of a region which stretches an infinite distance. To deal with this situation, we calculate the area from 0 to N where N is some "large" number and then we see what happens as $N \to \infty$. We have

$$\text{Area from 0 to } N = A_0^N = \int_0^N e^{-x}\,dx = -e^{-x}\Big|_0^N = 1 - e^{-N}.$$

Then

$$\text{Total area} = \lim_{N\to\infty} A_0^N = \lim_{N\to\infty}(1 - e^{-N}) = 1.$$

Thus the total area in the first quadrant under the curve $y = e^{-x}$ is 1.

This example leads to the following general definitions.

DEFINITION 1.   (i)   Let a be a real number and let f be a function having the property that $\int_a^N f(x)\,dx$ exists for every real number $N \geq a$. Then we define the *improper integral*

$$\boxed{\int_a^\infty f(x)\,dx = \lim_{N\to\infty}\int_a^N f(x)\,dx} \tag{1}$$

provided that this limit exists.

If $\int_a^\infty f(x)\,dx$ exists and is finite, we say that the improper integral is *convergent*.

If the limit in (1) does not exist, or if it exists and is infinite, then we say that the improper integral is *divergent*.

(ii)   If $\int_{-N}^{b} f(x)\, dx$ exists for every real N such that $-N \le b$, we define

$$\int_{-\infty}^{b} f(x)\, dx = \lim_{N \to \infty} \int_{-N}^{b} f(x)\, dx \qquad (2)$$

whenever the limit exists. We define the terms convergent and divergent as in (i).

(iii)   If $\int_{-M}^{0} f(x)\, dx$ and $\int_{0}^{N} f(x)\, dx$ exist for every real N and M, then we define

$$\int_{-\infty}^{\infty} f(x)\, dx = \lim_{N \to \infty} \int_{0}^{N} f(x)\, dx + \lim_{M \to \infty} \int_{-M}^{0} f(x)\, dx \qquad (3)$$

whenever both these limits exist.

EXAMPLE 2.   Evaluate $\displaystyle\int_{1}^{\infty} \frac{1}{x}\, dx$.

SOLUTION. $\int_{1}^{N} (1/x)\, dx = \ln x\,|_{1}^{N} = \ln N - \ln 1 = \ln N$.   But $\lim_{N \to \infty} \ln N = \infty$ so that the improper integral is divergent.

EXAMPLE 3.   Evaluate $\displaystyle\int_{0}^{\infty} e^{x}\, dx$.

SOLUTION. $\int_{0}^{N} e^{x}\, dx = e^{N} - 1$ which approaches ∞ as $N \to \infty$ so that this improper integral is also divergent.

EXAMPLE 4.   Evaluate $\displaystyle\int_{-\infty}^{0} e^{x}\, dx$.

SOLUTION. $\int_{-N}^{0} e^{x}\, dx = 1 - e^{-N}$ which $\to 1$ as $N \to \infty$ so that $\int_{-\infty}^{0} e^{x}\, dx$ is convergent and is equal to 1.

EXAMPLE 5.   Evaluate $\displaystyle\int_{0}^{\infty} \cos x\, dx$.

SOLUTION. $\int_{0}^{N} \cos x\, dx = \sin x\,|_{0}^{N} = \sin N$ which has no limit as $N \to \infty$ so that the improper integral $\int_{0}^{\infty} \cos x\, dx$ diverges. This makes sense graphically. In Figure 2 the

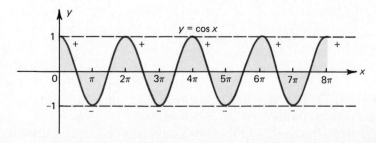

Figure 2

integral in question is the area of the shaded region. This area has no finite value because we must keep on adding and subtracting areas of equal size.

EXAMPLE 6. Evaluate $\int_{-\infty}^{\infty} xe^{-x^2}\,dx$.

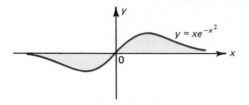

$y = xe^{-x^2}$

Figure 3

SOLUTION. The function xe^{-x^2} is sketched in Figure 3. $\int_0^N xe^{-x^2}\,dx = -\tfrac{1}{2}e^{-x^2}\big|_0^N = \tfrac{1}{2}(1 - e^{-N^2}) \to \tfrac{1}{2}$ as $N \to \infty$ and $\int_{-M}^0 xe^{-x^2}\,dx = -\tfrac{1}{2}e^{-x^2}\big|_{-M}^0 = \tfrac{1}{2}(e^{-M^2} - 1) \to -\tfrac{1}{2}$ as $M \to \infty$. Thus, since both limits exist,

$$\int_{-\infty}^{\infty} xe^{-x^2}\,dx = \int_0^{\infty} xe^{-x^2}\,dx + \int_{-\infty}^0 xe^{-x}\,dx = \frac{1}{2} - \frac{1}{2} = 0.$$

EXAMPLE 7. Calculate $\int_{-\infty}^{\infty} x^3\,dx$.

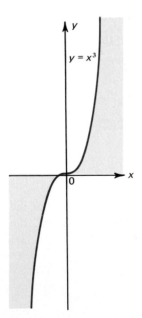

$y = x^3$

Figure 4

SOLUTION. The area to be calculated is sketched in Figure 4. Since $\lim_{N\to\infty} \int_0^N x^3\,dx = \lim_{N\to\infty}(N^4/4) = \infty$, the integral diverges. Note, however, that

$$\lim_{N\to\infty} \int_{-N}^{N} x^3\, dx = \lim_{N\to\infty} \frac{x^4}{4}\bigg|_{-N}^{N} = \lim_{N\to\infty}\left(\frac{N^4}{4} - \frac{N^4}{4}\right) = \lim_{N\to\infty} 0 = 0$$

which explains why we *do not* define $\int_{-\infty}^{\infty} f(x)\, dx$ as $\lim_{N\to\infty}\int_{-N}^{N} f(x)\, dx$. The problem here is that $\int_0^\infty x^3\, dx = \infty$ and $\int_{-\infty}^0 x^3\, dx = -\infty$. We simply cannot "cancel off" infinite terms. The expression $\infty - \infty$ is not defined.

EXAMPLE 8. Evaluate $\displaystyle\int_1^\infty (1/x^k)\, dx,\ k > 0,\ k \neq 1$.

SOLUTION

$$\int_1^\infty \frac{1}{x^k}\, dx = \lim_{N\to\infty} \int_1^N x^{-k}\, dx = \frac{x^{-k+1}}{-k+1}\bigg|_1^N = \frac{1}{k-1}(1 - N^{1-k}).$$

If $0 < k < 1$, then $N^{1-k} \to \infty$ as $n \to \infty$ and the integral diverges. If $k > 1$, then $N^{1-k} = 1/N^{k-1} \to 0$ as $N \to \infty$ and the integral converges. Combining this with Example 2 we have

$$\int_1^\infty \frac{1}{x^k}\, dx \quad \begin{cases} \text{diverges if } 0 < k \leq 1 \\ \text{converges to } \dfrac{1}{k-1} \text{ if } k > 1. \end{cases}$$

EXAMPLE 9. Evaluate $\displaystyle\int_{-\infty}^{\infty} \frac{dx}{x^2 - 4x + 9}$.

SOLUTION

$$\int_0^N \frac{dx}{x^2 - 4x + 9} = \int_0^N \frac{dx}{(x-2)^2 + 5} = \frac{1}{\sqrt{5}} \tan^{-1}\frac{x-2}{\sqrt{5}}\bigg|_0^N$$

$$= \frac{1}{\sqrt{5}}\left(\tan^{-1}\frac{N-2}{\sqrt{5}} - \tan^{-1}\left(-\frac{2}{\sqrt{5}}\right)\right)$$

and $\lim_{N\to\infty} \tan^{-1}[(N-2)/\sqrt{5}] = \pi/2$. Similarly,

$$\int_{-M}^0 \frac{dx}{x^2 - 4x + 9} = \frac{1}{\sqrt{5}} \tan^{-1}\frac{x-2}{\sqrt{5}}\bigg|_{-M}^0$$

$$= \frac{1}{\sqrt{5}}\left[\tan^{-1}\left(-\frac{2}{\sqrt{5}}\right) - \tan^{-1}\frac{-M-2}{\sqrt{5}}\right].$$

Thus

$$\int_0^\infty \frac{dx}{x^2 - 4x + 5} = \frac{1}{\sqrt{5}}\left[\frac{\pi}{2} - \tan^{-1}\left(-\frac{2}{\sqrt{5}}\right)\right],$$

$$\int_{-\infty}^0 \frac{dx}{x^2 - 4x + 5} = \frac{1}{\sqrt{5}}\left[\tan^{-1}\left(-\frac{2}{\sqrt{5}}\right) - \left(-\frac{\pi}{2}\right)\right],$$

and

$$\int_{-\infty}^{\infty} \frac{dx}{x^2 - 4x + 5} = \int_{0}^{\infty} \frac{dx}{x^2 - 4x + 5} + \int_{-\infty}^{0} \frac{dx}{x^2 - 4x + 5}$$

$$= \frac{1}{\sqrt{5}} \left(\frac{\pi}{2} + \frac{\pi}{2} \right) = \frac{\pi}{\sqrt{5}}.$$

We now turn to the second type of improper integral.

CASE 2. *a* and *b* are finite and *f* becomes infinite at some number in the closed interval [*a*, *b*].

As before, we begin with an example.

EXAMPLE 10. Calculate the area in the first quadrant under the curve $y = 1/\sqrt{x}$ between $x = 0$ and $x = 1$.

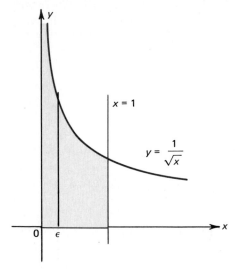

Figure 5

SOLUTION. The region in question is sketched in Figure 5. As in Example 1, the region is infinite. However, if ϵ is a small, positive real number, then we can calculate $\int_{\epsilon}^{1} (1/\sqrt{x}) \, dx$ and see what happens as $\epsilon \to 0^+$. We have

$$A_{\epsilon}^{1} = \int_{\epsilon}^{1} \frac{1}{\sqrt{x}} dx = 2\sqrt{x} \mid_{\epsilon}^{1} = 2(1 - \sqrt{\epsilon}).$$

Then

$$\text{total area} = \lim_{\epsilon \to 0^+} A_{\epsilon}^{1} = \lim_{\epsilon \to 0^+} 2(1 - \sqrt{\epsilon}) = 2.$$

Before citing the general definition, we remark that if *f* is continuous on (*a*, *b*] but $f(x) \to +\infty$ or $-\infty$ as $x \to a^+$, then the graph of *f* has a *vertical asymptote* at $x = a$. That is, as $x \to a^+$, the sketch of $y = f(x)$ approaches the vertical line $x = a$. In Example 10 we see this illustrated for $a = 0$.

DEFINITION 2. Let a and b be finite numbers.

(i) If $\int_{a+\epsilon}^{b} f(x)\, dx$ exists for every ϵ in $(0, b - a]$ and if f has a vertical asymptote at $x = a$, then

$$\int_{a}^{b} f(x)\, dx = \lim_{\epsilon \to 0^+} \int_{a+\epsilon}^{b} f(x)\, dx \tag{4}$$

provided that the limit exists.

(ii) If $\int_{a}^{b-\epsilon} f(x)\, dx$ exists for every ϵ in $(0, b - a]$ and if f has a vertical asymptote at $x = b$, then

$$\int_{a}^{b} f(x)\, dx = \lim_{\epsilon \to 0^+} \int_{a}^{b-\epsilon} f(x)\, dx \tag{5}$$

provided that the limit exists.

(iii) If for c in (a, b) f has a vertical asymptote at $x = c$ and if the integrals $\int_{a}^{c-\epsilon_1} f(x)\, dx$ and $\int_{c+\epsilon_2}^{b} f(x)\, dx$ exist for ϵ_1 in $(0, c - a]$ and ϵ_2 in $(0, b - c]$, then

$$\int_{a}^{b} f(x)\, dx = \lim_{\epsilon_1 \to 0^+} \int_{a}^{c-\epsilon_1} f(x)\, dx + \lim_{\epsilon_2 \to 0^+} \int_{c+\epsilon_2}^{b} f(x)\, dx \tag{6}$$

provided that both these integrals exist.

For each of the cases (i), (ii), and (iii), the improper integral is *convergent* if the appropriate limit or limits exist and are finite. Otherwise it is *divergent*.

The three cases of Definition 2 are illustrated in Figure 6. Of course, we could have just as well drawn the curves approaching $-\infty$ instead of $+\infty$, but the basic idea is the same.

In Example 10, we found that $\int_{0}^{1} (1/\sqrt{x})\, dx$ is a converging improper integral.

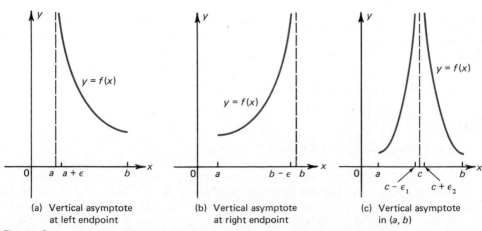

(a) Vertical asymptote at left endpoint

(b) Vertical asymptote at right endpoint

(c) Vertical asymptote in (a, b)

Figure 6

EXAMPLE 11. Evaluate $\int_0^1 (1/x)\,dx$.

SOLUTION. $\int_\epsilon^1 (1/x)\,dx = -\ln \epsilon$ which $\to \infty$ as $\epsilon \to 0^+$ so that $\int_0^1 (1/x)\,dx$ diverges.

EXAMPLE 12. Calculate $\displaystyle\int_0^2 \frac{dx}{(x-2)^{4/5}}$.

SOLUTION. The integral is not defined at $x = 2$. We have

$$\int_0^2 \frac{dx}{(x-2)^{4/5}} = \lim_{\epsilon \to 0^+} \int_0^{2-\epsilon} \frac{dx}{(x-2)^{4/5}} = \lim_{\epsilon \to 0^+} 5(x-2)^{1/5} \Big|_0^{2-\epsilon}$$

$$= \lim_{\epsilon \to 0^+} 5[(-\epsilon)^{1/5} - (-2)^{1/5}] = 5 \cdot 2^{1/5}.$$

EXAMPLE 13. Calculate $\displaystyle\int_0^3 \frac{dx}{(x-1)^3}$.

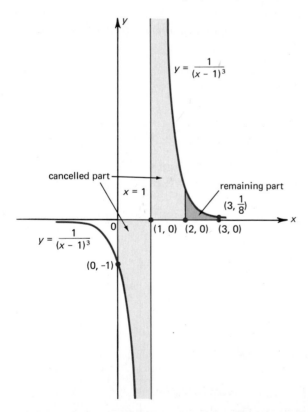

cancelled part

$y = \dfrac{1}{(x-1)^3}$

$x = 1$

remaining part

$\left(3, \dfrac{1}{8}\right)$

$y = \dfrac{1}{(x-1)^3}$

$(1, 0)$ $(2, 0)$ $(3, 0)$

$(0, -1)$

Figure 7

SOLUTION. See Figure 7. In this problem the integrand has a vertical asymptote at $x = 1$ so that we must consider two improper integrals:

$$\int_0^3 \frac{dx}{(x-1)^3} = \int_0^1 \frac{dx}{(x-1)^3} + \int_1^3 \frac{dx}{(x-1)^3}.$$

We have

$$\int_0^{1-\epsilon} \frac{dx}{(x-1)^3} = -\frac{1}{2(x-1)^2}\Big|_0^{1-\epsilon} = \frac{1}{2}\left[1 - \frac{1}{(-\epsilon)^2}\right]$$

which $\to -\infty$ as $\epsilon \to 0^+$ so that $\int_0^1[1/(x-1)^3]\,dx$ diverges, implying that $\int_0^3[1/(x-1)^3]\,dx$ also diverges.

> **Note.** If we had not noticed the discontinuity at $x = 1$ we could have blindly integrated to obtain
>
> $$\int_0^3 \frac{dx}{(x-1)^3} = -\frac{1}{2(x-1)^2}\Big|_0^3 = \frac{1}{2}\left(1 - \frac{1}{4}\right) = \frac{3}{8}$$
>
> which is, of course, wrong. The reason that we obtain this finite answer is that two infinite areas have been illegitimately canceled off. (This is illustrated in Figure 7.)

EXAMPLE 14. Calculate $\displaystyle\int_{-1}^1 \frac{dx}{\sqrt{1-x^2}}$.

SOLUTION. Here we have vertical asymptotes at both endpoints. Before integrating we make the substitution $x = \sin\theta$. Then

$$\int_{-1}^1 \frac{dx}{\sqrt{1-x^2}} = \int_{-\pi/2}^{\pi/2} \frac{\cos x\,dx}{\cos x} = \pi.$$

Here we have transformed an improper integral into an ordinary one. This is legitimate as long as the vertical asymptotes occur at the endpoints.

EXAMPLE 15. Calculate $\displaystyle\int_{-1}^8 (1/\sqrt[3]{x})\,dx$.

SOLUTION. Since we have a vertical asymptote at $x = 0$, we write this as the sum of two integrals:

$$\int_{-1}^8 \frac{1}{\sqrt[3]{x}}\,dx = \int_{-1}^0 \frac{1}{\sqrt[3]{x}}\,dx + \int_0^8 \frac{1}{\sqrt[3]{x}}\,dx$$

$$= \lim_{\epsilon_1\to 0^+}\int_{-1}^{-\epsilon_1} x^{-1/3}\,dx + \lim_{\epsilon_2\to 0^+}\int_{\epsilon_2}^8 x^{-1/3}\,dx$$

$$= \lim_{\epsilon_1\to 0^+} \tfrac{3}{2}x^{2/3}\Big|_{-1}^{-\epsilon_1} + \lim_{\epsilon_2\to 0^+} \tfrac{3}{2}x^{2/3}\Big|_{\epsilon_2}^8$$

$$= \lim_{\epsilon_1\to 0^+} \tfrac{3}{2}[(-\epsilon_1)^{2/3} - 1] + \lim_{\epsilon_2\to 0^+} \tfrac{3}{2}[4 - \epsilon_2^{3/2}]$$

$$= -\tfrac{3}{2} + 6 = \tfrac{9}{2}.$$

EXAMPLE 16. Calculate $\displaystyle\int_{-3}^3 (1/x)\,dx$.

SOLUTION. We shall show that this integral is divergent. Since there is a vertical asymptote at 0, we have

$$\int_{-3}^{3} \frac{dx}{x} = \int_{-3}^{0} \frac{dx}{x} + \int_{0}^{3} \frac{dx}{x} = \lim_{\epsilon \to 0^+} \int_{-3}^{-\epsilon} \frac{dx}{x} + \lim_{\epsilon \to 0^+} \int_{\epsilon}^{3} \frac{dx}{x}.$$

Here we can go astray. By the limit theorem 10.2.2,

$$\lim_{x \to x_0} [f(x) + g(x)] = \lim_{x \to x_0} f(x) + \lim_{x \to x_0} g(x)$$

if *both these limits exist and are finite*. However,

$$\lim_{\epsilon \to 0^+} \int_{-3}^{-\epsilon} \frac{dx}{x} = \lim_{\epsilon \to 0^+} \ln |x| \Big|_{-3}^{-\epsilon} = \lim_{\epsilon \to 0^+} (\ln \epsilon - \ln 3) = -\infty$$

so that $\int_{-3}^{3} (1/x) \, dx$ diverges. Note that if we blindly use the limit theorem cited above, we have

$$\lim_{\epsilon \to 0^+} \int_{-3}^{-\epsilon} \frac{dx}{x} + \lim_{\epsilon \to 0^+} \int_{\epsilon}^{3} \frac{dx}{x} = \lim_{\epsilon \to 0^+} \left(\int_{-3}^{-\epsilon} \frac{dx}{x} + \int_{\epsilon}^{3} \frac{dx}{x} \right)$$

$$= \lim_{\epsilon \to 0^+} (\ln \epsilon - \ln 3 + \ln 3 - \ln \epsilon)$$

$$= \lim_{\epsilon \to 0^+} 0 = 0,$$

which is wrong. Here again we illegitimately canceled off two infinite areas.

As we see in Problems 64–72 and in the chapters to come, improper integrals arise frequently in applications. One interesting application is in the area of economics.

In Section 6.5, we discussed the notion of present value, which is defined as the current worth of an investment paying a certain sum of money at some time in the future. If $a(t)$ denotes a stream of income, then the present value of the money received up to the time T years in the future is given by (see Equation 6.5.11)

$$\text{present value} = \int_{0}^{T} a(t)e^{-it} \, dt, \tag{7}$$

where i denotes the annual interest rate. If the income is to be received indefinitely, then we would be interested in calculating the present value of the investment for an indefinitely long time in the future. This leads to the equation

$$\boxed{\text{present value} = \int_{0}^{\infty} a(t)e^{-it} \, dt.} \tag{8}$$

EXAMPLE 17. A corporation makes an investment from which it expects to receive a monthly return of $200 + 1.5t$ dollars, where t is the time (in years) since the investment was made. What is the present value of this investment assuming an annual interest rate of 7%?

SOLUTION. We have the annual return $a(t) = 12(200 + 1.5t) = 2400 + 18t$. Since $i = 0.07$, we have

$$\text{present value} = \int_0^\infty (2400 + 18t)e^{-0.07t} \, dt$$

$$= \lim_{N \to \infty} \int_0^N (2400 + 18t)e^{-0.07t} \, dt$$

then integrating by parts we have

$$= \lim_{N \to \infty} \left\{ -\frac{1}{0.07}(2400 + 18t)e^{-0.07t} \Big|_0^N + \frac{18}{0.07} \int_0^N e^{-0.07t} \, dt \right\}$$

$$= \lim_{N \to \infty} \left\{ \frac{1}{0.07}[2400 - (2400 + 18N)e^{-0.07N}] - \frac{18}{(0.07)^2} e^{-0.07t} \Big|_0^N \right\}$$

$$= \frac{2400}{0.07} + \frac{18}{(0.07)^2} \approx \$37,959.$$

In this calculation we used the fact that $\lim_{N \to \infty} Ne^{-0.07N} = 0$ (see Example 12.3.4).

PROBLEMS 12.4

In Problems 1–45 determine whether the given improper integral converges or diverges. If it converges, calculate its value.

1. $\int_0^\infty e^{-2x} \, dx$

2. $\int_{-\infty}^4 e^{3x} \, dx$

3. $\int_{-\infty}^\infty e^{-0.01x} \, dx$

4. $\int_0^\infty xe^{-2x} \, dx$

5. $\int_{-\infty}^\infty x^3 e^{-x^4} \, dx$

6. $\int_{-\infty}^\infty x^2 e^{-x^3} \, dx$

7. $\int_{-\infty}^1 \frac{dx}{\sqrt{4-x}}$

8. $\int_{-\infty}^1 \frac{dx}{(4-x)^{3/2}}$

9. $\int_{10}^\infty \frac{dx}{x^2 - 1}$

10. $\int_0^\infty \frac{dx}{1+x^2}$

11. $\int_a^\infty \frac{dx}{b^2 + x^2}, \ a > 0$

12. $\int_0^\infty \frac{2x}{x^2 + 1} \, dx$

13. $\int_0^\infty \frac{1}{x^3} \, dx$

14. $\int_0^\infty \frac{1}{\sqrt{x}} \, dx$

15. $\int_0^1 \frac{dx}{x^{3/2}}$

16. $\int_0^{\pi/2} \tan x \, dx$

17. $\int_2^3 \frac{dx}{\sqrt{x-2}}$

18. $\int_2^3 \frac{dx}{(x-2)^{1/3}}$

19. $\int_2^3 \frac{dx}{(x-2)^{4/3}}$

20. $\int_0^{\pi/4} \csc x \, dx$

21. $\int_0^{\pi/2} \sec x \, dx$

22. $\int_0^2 \frac{dx}{(x-1)^{1/3}}$

23. $\int_0^2 \frac{x \, dx}{(x^2 - 1)^{1/3}}$

24. $\int_{-2}^2 \frac{dx}{x^2 - 1}$

25. $\int_1^\infty \frac{dx}{x \ln x}$

26. $\int_0^\infty e^{-x} \sin x \, dx$

27. $\int_{-\infty}^0 e^x \cos x \, dx$

28. $\int_0^\infty e^{-ax} \sin bx \, dx, \ a, \ b > 0$

29. $\int_0^5 \frac{dx}{\sqrt{25 - x^2}}$

30. $\displaystyle\int_0^\infty \frac{x^5}{e^{x^6}}\,dx$

31. $\displaystyle\int_0^2 \frac{dx}{x-1}$

32. $\displaystyle\int_{-2}^0 \frac{dx}{(x+2)^5}$

33. $\displaystyle\int_{-2}^0 \frac{dx}{(x+2)^{1/5}}$

34. $\displaystyle\int_0^\pi \tan x\,dx$

35. $\displaystyle\int_2^4 \frac{dx}{(x-3)^7}$

36. $\displaystyle\int_2^4 \frac{dx}{(x-3)^{1/7}}$

*37. $\displaystyle\int_0^{\pi/2} \frac{dx}{1-\sin x}$

*38. $\displaystyle\int_0^\pi \frac{dx}{1-\cos x}$

39. $\displaystyle\int_{-\infty}^\infty \frac{x^2}{x^2+1}\,dx$

40. $\displaystyle\int_5^\infty \frac{dx}{x^2-6x+8}$

41. $\displaystyle\int_1^\infty \frac{dx}{x^2-6x+8}$

42. $\displaystyle\int_{-\infty}^0 \frac{x}{(x^2+1)^{5/2}}\,dx$

43. $\displaystyle\int_5^\infty \frac{dx}{x(\ln x)^2}$

44. $\displaystyle\int_0^1 \ln x\,dx$

45. $\displaystyle\int_0^\infty \frac{e^{-2\sqrt{x}}}{\sqrt{x}}\,dx.$

46. In Example 2 we showed that the area under the curve $y = 1/x$ for $x \geq 1$ is infinite. Show that if this curve is revolved about the x-axis, then the volume of the resulting solid of revolution is finite and calculate this volume.

47. Find the volume of the solid generated when the curve $y = e^{-x}$ for $x \geq 0$ is rotated about the x-axis.

48. Find the volume generated when the curve in Problem 47 is rotated about the y-axis.

49. Find the lateral surface area of the solid in Problem 47.

50. Find the centroid of the region bounded by $y = e^{-x}$ and the x- and y-axes.

51. What is the present value of an annuity if the stream of income is $a(t) = 10,000(1 + e^{0.05t})$, paid annually, over an indefinite period of time, assuming an annual interest rate of 8%?

52. What is the present value of an annuity if the stream of income is $a(t) = 20 + 2t + \frac{1}{2}t^2$, paid weekly, over an indefinite period of time if the annual interest rate is 6%?

53. An investment broker has a choice between investing $100,000 in a growing business or putting the money into bonds yielding a 12% annual return, compounded continuously for an indefinite period of time. The annual income from the business is projected to be $4000(1 + e^{t/10})$ dollars after t years. Where should he put his money? [*Hint:* See Example 6.5.16.]

54. Show that $\int_a^b [1/(x-b)^r]\,dx$ converges if $0 < r < 1$ and diverges if $r \geq 1$.

*55. Show by mathematical induction (Appendix 1) and integration by parts that for any integer $n > 0$, $\int_0^\infty x^n e^{-x}\,dx = n!$.

*56. Calculate $\int_0^\infty x^n e^{-ax}\,dx$ for $a > 0$.

*57. Let f and g be two continuous functions such that $f(x) \geq g(x) \geq 0$ for $x \geq a$. Show that
   (a) If $\int_a^\infty f(x)\,dx$ converges, then $\int_a^\infty g(x)\,dx$ converges.
   (b) If $\int_a^\infty g(x)\,dx$ diverges, then $\int_a^\infty f(x)\,dx$ diverges.
   These two results are called *comparison theorems*. [*Hint:* For (i) show that for N large $\int_a^N g(x)\,dx \leq \int_a^N f(x)\,dx \leq L$ where $L = \int_a^\infty f(x)\,dx$. For (ii) show that $\int_a^N f(x)\,dx \geq \int_a^N g(x)\,dx$ which $\to \infty$ as $N \to \infty$.]

58. Use the result of Problem 57 to show that $\int_4^\infty (1/\sqrt{x^3-1+\sin x})\,dx$ converges.

59. Show that $\int_1^\infty (\sqrt{1+x^{1/8}}/x^{3/4})\,dx$ diverges.

60. Show that $\int_1^\infty [1/\ln(1+x)]\,dx$ diverges.

61. Show that $\int_{-\infty}^\infty e^{-x^2}\,dx$ converges.

62. Show that $\int_1^\infty [(\sin^2 x)/x^2]\,dx$ converges.

*63. Show that $\int_0^1 (1/\sqrt{\sin x})\,dx$ converges. [*Hint:* Show that $\sin x \geq kx$ for some constant k so that $1/\sqrt{\sin x} \leq 1/\sqrt{kx}$ which has a convergent integral on $[0, 1]$.]

A *probability density function* is a function f whose domain is the set of all real numbers such that

(1) $f(x) \geq 0$ for all x in $\mathbb{R}$.

(2) $\int_{-\infty}^{\infty} f(x)\, dx = 1$.

This function is very useful in probability theory.

64. Show that

$$f(x) = \begin{cases} 0, & x < a \\ \dfrac{1}{b-a}, & a \leq x \leq b \\ 0, & x \geq b \end{cases}$$

is a probability density function. This is called a *uniform* density function.

65. Show that

$$f(x) = \begin{cases} \dfrac{1}{a} e^{-x/a}, & x \geq 0 \\ 0, & x < 0 \end{cases}$$

for $a > 0$ is a probability density function. This is called the *exponential* density function.

66. Show that $\int_{-\infty}^{\infty} (1/\sqrt{2\pi})\, e^{-t^2/2}\, dt$ converges. The integrand here is called the *unit normal* density function. Tables of $\int_{-\infty}^{\infty} (1/\sqrt{2\pi})\, e^{-t^2/2}\, dt$ are available in most books on probability theory.

The *expected value* or *mean* μ associated with the density function f is defined by

$$\mu = \int_{-\infty}^{\infty} x f(x)\, dx.$$

That is, the expected value is the first moment of the density function. We can think of the expected value as a weighted average of various probabilities.

67. Calculate the expected value associated with the probability density function given in Problem 64.

68. Calculate the expected value associated with the probability density function given in Problem 65.

69. Calculate the expected value associated with the probability density function given in Problem 66.

The *variance* σ^2 associated with the probability density function f is defined by

$$\sigma^2 = \int_{-\infty}^{\infty} (x - \mu)^2 f(x)\, dx.$$

We can think of the variance as the "average" square of the deviation from the mean. We define the *standard deviation* σ associated with the function f to be the square root of the variance σ^2.

70. Calculate the variance and standard deviation associated with the probability density function given in Problem 64.

71. Calculate the variance and standard deviation associated with the probability density function given in Problem 65.

72. Show that the variance and standard deviation associated with the probability density function given in Problem 66 are finite.

Review Exercises for Chapter Twelve

In Exercises 1–20 evaluate the given limit.

1. $\lim\limits_{x \to 0} \dfrac{\sin 3x}{2x}$

2. $\lim\limits_{x \to \infty} \dfrac{\ln x}{x^2}$

3. $\lim\limits_{x \to 0} \dfrac{2x^3 + 3x + 4}{x - x^3}$

4. $\lim\limits_{x \to \pi/4} \dfrac{\cos 2x}{(\pi/4) - x}$

5. $\lim\limits_{x \to \infty} 3xe^{-x}$

6. $\lim\limits_{x \to 0} \dfrac{1 - e^x}{x^2 + x}$

7. $\lim\limits_{x \to 0^+} \dfrac{\sqrt{x}}{\sin x}$

8. $\lim\limits_{x \to 0^+} (\sin x)^{2x}$

9. $\lim\limits_{x \to \pi/4^-} [(\pi/4) - x]^{\cos 2x}$

10. $\lim\limits_{x \to 0^+} x^{3x}$

11. $\lim\limits_{x \to 0} \dfrac{\sin x}{x^3}$

12. $\lim\limits_{x \to 0} \dfrac{x - \sin x}{2x^3}$

13. $\lim\limits_{x \to \pi/2^-} \dfrac{\tan x}{\sec x}$

14. $\lim\limits_{x \to 0^+} \dfrac{\ln(\tan x)}{\ln(\sin x)}$

15. $\lim\limits_{x \to \infty} (x - \sqrt{x^2 + x})$

16. $\lim\limits_{x \to \infty} (x^4 - \sqrt{x^8 + x^4 + 1})$

17. $\lim\limits_{x \to \infty} \dfrac{e^x}{2^x}$

18. $\lim\limits_{x \to \infty} \dfrac{e^x}{4^x}$

19. $\lim\limits_{x \to 0} x^{10} e^{-1/x^2}$

20. $\lim\limits_{x \to 0^+} \dfrac{\int_0^x e^{t^3}\, dt}{\int_0^x e^{t^2}\, dt}$.

In Exercises 21–40 determine whether the given improper integral converges or diverges. If it converges, calculate its value.

21. $\displaystyle\int_0^\infty e^{-3x}\, dx$

22. $\displaystyle\int_{-\infty}^2 e^{50x}\, dx$

23. $\displaystyle\int_0^\infty x^3 e^{-7x}\, dx$

24. $\displaystyle\int_{-\infty}^\infty x^2 e^{-x^3}\, dx$

25. $\displaystyle\int_{50}^\infty \dfrac{dx}{x^2 - 4}$

26. $\displaystyle\int_0^1 \dfrac{dx}{\sqrt[4]{x}}$

27. $\displaystyle\int_0^1 \dfrac{dx}{x^4}$

28. $\displaystyle\int_{-\infty}^\infty \dfrac{dx}{x^3}$

29. $\displaystyle\int_{-\infty}^0 \dfrac{dx}{(1 - x)^{5/2}}$

30. $\displaystyle\int_3^4 \dfrac{dx}{(x - 3)^{6/5}}$

31. $\displaystyle\int_0^{\pi/2} \csc x\, dx$

32. $\displaystyle\int_0^4 \dfrac{dx}{x - 2}$

33. $\displaystyle\int_0^4 \dfrac{dx}{\sqrt{x - 2}}$

34. $\displaystyle\int_3^5 \dfrac{dx}{(x - 4)^{10}}$

35. $\displaystyle\int_0^\infty \frac{x^3}{x^3+1}\, dx$

36. $\displaystyle\int_0^e \ln x \, dx$

37. $\displaystyle\int_2^\infty \frac{dx}{(x-1)^{1/100}}$

38. $\displaystyle\int_0^\infty e^{-x} \sin 2x \, dx$

39. $\displaystyle\int_0^\infty \frac{dx}{x^2-3x+2}$

40. $\displaystyle\int_{-\infty}^0 e^{2x} \cos 3x \, dx.$

41. Find the volume generated when the region in the first quadrant bounded by $y = 1/(x+1)^2$ is rotated about the x-axis.

42. Find the volume generated if the region in Exercise 41 is rotated about the y-axis.

43. Find the centroid of the region in the first quadrant bounded by $y = 1/(1+x)^3$.

44. What is the present value of an annuity if the annual stream of income is $5000(1 + e^{0.04t})$, paid annually over an indefinite period of time, assuming an annual interest rate of 7% compounded continuously?

45. Show that $\int_2^\infty (\sqrt{x^{4/5} - 1}/x^{21/5})\, dx$ converges.

46. Show that $\int_0^{\pi/2} (1/\sqrt{\cos x})\, dx$ converges.

47. Show that $\int_{10}^\infty (1/\ln\sqrt{2+x})\, dx$ diverges.

48. Show that the function

$$f(x) = \begin{cases} \frac{1}{5}e^{-x/5}, & x \geq 0 \\ 0, & x < 0 \end{cases}$$

is a probability density function.

49. Calculate the expected value associated with the density function given in Exercise 48.

50. Calculate the variance and standard deviation associated with the density function given in Exercise 48.

THIRTEEN

TAYLOR POLYNOMIALS, APPROXIMATION, AND INTERPOLATION

It is a fact that many functions arising in applications are difficult to deal with. A continuous function, for example, may take a complicated form; or it may take a simple form which, nevertheless, cannot be integrated.

For this reason mathematicians and physicists have developed methods for approximating certain functions by other functions which are much easier to handle. Some of the easiest functions to deal with are the polynomials, since, in addition to having other useful properties, they can be differentiated and integrated any number of times and still remain polynomials. In this chapter, we will show two ways that certain continuous functions can be approximated by polynomials.

13.1 Taylor's Theorem and Taylor Polynomials

In this section we show how a function can be approximated as closely as desired by a polynomial, provided that the function possesses a sufficient number of derivatives.

We begin by reminding you of the factorial notation defined for all positive integers n:

$$n! = n(n - 1)(n - 2) \cdots 3 \cdot 2 \cdot 1. \tag{1}$$

That is, $n!$ is the product of the first n integers. For example, $3! = 3 \cdot 2 \cdot 1 = 6$ and $5! = 5 \cdot 4 \cdot 3 \cdot 2 \cdot 1 = 120$. By convention, we define 0! to be equal to 1. We can now state the principal result of this chapter.

THEOREM 1 (Taylor's† Formula with Remainder). Let the function f and its first $n + 1$ derivatives be continuous on the closed interval $[a, b]$. Then

$$\boxed{\begin{aligned} f(b) = f(a) &+ \frac{f'(a)}{1!}(b - a) + \frac{f''(a)}{2!}(b - a)^2 \\ &+ \cdots + \frac{f^{(n)}(a)}{n!}(b - a)^n + R_n(b), \end{aligned}} \tag{2}$$

remainder

---

†Named after the English mathematician Brook Taylor (1685–1731) who published this theorem in *Methodus Incrementorum* in 1715. There was a considerable controversy over whether Taylor's discovery was, in fact, a plagiarism of an earlier result of the Swiss mathematician Jean Bernoulli (1667–1748).

578

where the *remainder term* $R_n(b)$ is given by

$$R_n(b) = \int_a^b \frac{(b-t)^n}{n!} f^{(n+1)}(t)\, dt, \tag{3}$$

We will defer the proof of this theorem until Section 13.2. We immediately note, however, that the remainder term is in a very complicated form. Fortunately, there is an alternative way to write $R_n(b)$ which is given in the theorem below. The proof of this theorem will also be given in Section 13.2.

THEOREM 2.† Under the assumptions of Theorem 1, there is a number c in the interval (a, b) such that

$$R_n(b) = \frac{f^{(n+1)}(c)}{(n+1)!}(b-a)^{n+1}. \tag{4}$$

If the hypotheses of Theorem 1 hold over the interval $[a, b]$ and if $a < x < b$, then they certainly hold over the smaller interval $[a, x]$. Then, substituting x for b in (2) and using (4), we have, for any x in (a, b),

$$f(x) = f(a) + \frac{f'(a)}{1!}(x-a) + \frac{f''(a)}{2!}(x-a)^2 + \cdots$$
$$+ \frac{f^{(n)}(a)}{n!}(x-a)^n + \frac{f^{(n+1)}(c)}{(n+1)!}(x-a)^{n+1} \tag{5}$$

for some c in the interval (a, x).

In Equation (5) the nth degree polynomial

$$P_n(x) = f(a) + \frac{f'(a)}{1!}(x-a) + \frac{f''(a)}{2!}(x-a)^2$$
$$+ \cdots + \frac{f^{(n)}(a)}{n!}(x-a)^n \tag{6}$$

is called the *nth degree Taylor polynomial* of the function f at a. We then can write

$$f(x) = P_n(x) + R_n(x), \tag{7}$$

where

$$R_n(x) = \frac{f^{(n+1)}(c)}{(n+1)!}(x-a)^{n+1} \tag{8}$$

for some number c in the interval (a, x).

---

†This form of the remainder is called the *Lagrange form,* named after the French mathematician Joseph Lagrange (1736–1813).

Before continuing our discussion, we have to answer the question "What is the good of all this?" In the next section we will show that if we make some very reasonable assumptions about the function f, then we can show that the remainder term $R_n(x)$ given by (8) actually approaches 0 as $n \to \infty$. Thus, as n increases, $P_n(x)$ is an increasingly useful approximation to the function f over the interval $[a, b]$.

For the remainder of this section, we will calculate some Taylor polynomials. In Section 13.3 we will use the fact that $R_n(x)$ becomes small to find some very interesting approximations.

EXAMPLE 1. Calculate the fifth degree Taylor polynomial of $f(x) = \sin x$ at 0.

SOLUTION. We have $f(x) = \sin x$, $f'(x) = \cos x$, $f''(x) = -\sin x$, $f'''(x) = -\cos x$, $f^{(iv)}(x) = \sin x$ and $f^{(v)}(x) = \cos x$. Then $f(0) = 0$, $f'(0) = 1$, $f''(0) = 0$, $f'''(0) = -1$, $f^{(iv)}(0) = 0$, $f^{(v)}(0) = 1$ and we obtain

$$P_5(x) = f(0) + \frac{f'(0)}{1!}(x) + \frac{f''(0)}{2!}x^2 + \frac{f'''(0)}{3!}x^3 + \frac{f^{(iv)}(0)}{4!}x^4 + \frac{f^{(v)}(0)}{5!}x^5$$

$$= x - \frac{x^3}{3!} + \frac{x^5}{5!} = x - \frac{x^3}{6} + \frac{x^5}{120}.$$

EXAMPLE 2. Calculate the fifth degree Taylor polynomial of $f(x) = \sin x$ at $\pi/6$.

SOLUTION. Using the derivatives found in Example 1 we have $f(\pi/6) = 1/2$, $f'(\pi/6) = \sqrt{3}/2$, $f''(\pi/6) = -1/2$, $f'''(\pi/6) = -\sqrt{3}/2$, $f^{(iv)}(\pi/6) = 1/2$ and $f^{(v)}(\pi/6) = \sqrt{3}/2$ so that, in this case,

$$P_5(x) = \frac{1}{2} + \frac{\sqrt{3}}{2}\left(x - \frac{\pi}{6}\right) - \frac{1}{2}\frac{[x - (\pi/6)]^2}{2!} - \frac{\sqrt{3}}{2}\frac{[x - (\pi/6)]^3}{3!}$$

$$+ \frac{1}{2}\frac{[x - (\pi/6)]^4}{4!} + \frac{\sqrt{3}}{2}\frac{[x - (\pi/6)]^5}{5!}$$

$$= \frac{1}{2} + \frac{\sqrt{3}}{2}\left(x - \frac{\pi}{6}\right) - \frac{1}{4}\left(x - \frac{\pi}{6}\right)^2 - \frac{\sqrt{3}}{12}\left(x - \frac{\pi}{6}\right)^3$$

$$+ \frac{1}{48}\left(x - \frac{\pi}{6}\right)^4 + \frac{\sqrt{3}}{240}\left(x - \frac{\pi}{6}\right)^5.$$

The two examples above illustrate that, in many cases, it is easiest to calculate the Taylor polynomial at 0. In this situation we have

$$P_n(x) = f(0) + f'(0)x + \frac{f''(0)}{2!}x^2 + \cdots + \frac{f^{(n)}(0)}{n!}x^n. \tag{9}$$

EXAMPLE 3. Find the eighth degree Taylor polynomial of $f(x) = e^x$ at 0.

SOLUTION. Here $f(x) = f'(x) = f''(x) = \cdots = f^{(viii)}(x) = e^x$ and $e^0 = 1$ so that

$$P_8(x) = 1 + x + \frac{x^2}{2!} + \frac{x^3}{3!} + \frac{x^4}{4!} + \frac{x^5}{5!} + \frac{x^6}{6!} + \frac{x^7}{7!} + \frac{x^8}{8!}.$$

EXAMPLE 4. Find the fifth degree Taylor polynomial of $f(x) = 1/(1 - x)$ at 0.

SOLUTION. Here

$$f(x) = \frac{1}{1 - x}, \qquad f'(x) = \frac{1}{(1 - x)^2}, \qquad f''(x) = \frac{2}{(1 - x)^3},$$

$$f'''(x) = \frac{6}{(1 - x)^4}, \qquad f^{(iv)}(x) = \frac{24}{(1 - x)^5}, \quad \text{and} \quad f^{(v)}(x) = \frac{120}{(1 - x)^6}.$$

Thus $f(0) = 1$, $f'(0) = 1$, $f''(0) = 2$, $f'''(0) = 6$, $f^{(iv)}(0) = 24$, $f^{(v)}(x) = 120$, and

$$P_5(x) = 1 + x + \frac{2x^2}{2!} + \frac{6x^3}{3!} + \frac{24x^4}{4!} + \frac{120x^5}{5!}$$

$$= 1 + x + x^2 + x^3 + x^4 + x^5.$$

This result should not be surprising. In Section 2.6, we showed that if $|r| < 1$, then

$$1 + r + r^2 + r^3 + \cdots + r^n + \cdots = \frac{1}{1 - r}$$

(see Equation 2.6.5). Thus we see that for $|x| < 1$, a reasonable approximation to $1/(1 - x)$ is given by $1 + x + x^2 + \cdots + x^n$ for a large number n. We will return to this problem in our discussion of infinite series in Chapter 14.

Note that in Examples 1, 2, and 3 the given function had continuous derivatives of all orders defined for all real numbers. In Example 4 $f(x)$ is defined over intervals of the form $[0, b]$ where $b < 1$. Thus Taylor's theorem does *not* apply for $x \geq 1$. *It is always necessary to check whether the hypotheses of Taylor's theorem hold over a given interval.*

PROBLEMS 13.1

In Problems 1–26 find the Taylor polynomial of the given degree n for the function f at the number a.

1. $f(x) = \cos x$; $a = \pi/4$; $n = 6$ 2. $f(x) = \sqrt{x}$; $a = 1$; $n = 4$
3. $f(x) = \ln x$; $a = e$; $n = 5$ 4. $f(x) = \ln(1 + x)$; $a = 0$; $n = 5$

5. $f(x) = \dfrac{1}{x}$; $a = 1$; $n = 4$ 6. $f(x) = \dfrac{1}{1 + x}$; $a = 0$; $n = 5$

7. $f(x) = \tan x$; $a = 0$; $n = 4$ 8. $f(x) = \tan^{-1} x$; $a = 0$; $n = 6$

9. $f(x) = \tan x$; $a = \pi$; $n = 4$ 10. $f(x) = \tan^{-1} x$; $a = \pi/4$; $n = 4$

11. $f(x) = \dfrac{1}{1 + x^2}$; $a = 0$; $n = 4$ 12. $f(x) = \dfrac{1}{\sqrt{x}}$; $a = 4$; $n = 3$

13. $f(x) = \sinh x$; $a = 0$; $n = 4$ 14. $f(x) = \cosh x$; $a = 0$; $n = 4$
15. $f(x) = \ln \sin x$; $a = \pi/2$, $n = 3$ 16. $f(x) = \ln \cos x$; $a = 0$; $n = 3$

17. $f(x) = \dfrac{1}{\sqrt{4 - x}}$; $a = 0$; $n = 4$ 18. $f(x) = \dfrac{1}{\sqrt{4 - x}}$; $a = 3$; $n = 4$

19. $f(x) = e^{\alpha x}$; $a = 0$; $n = 6$; α real   20. $f(x) = \sin \alpha x$; $a = 0$; $n = 6$; α real
21. $f(x) = \sin^{-1} x$; $a = 0$; $n = 3$ 22. $f(x) = 1 + x + x^2$; $a = 0$, $n = 10$
23. $f(x) = a_0 + a_1 x + a_2 x^2 + a_3 x^3$, $a = 1$, $n = 10$

24. $f(x) = e^{x^2}$; $a = 0$; $n = 4$ **25.** $f(x) = \sin x^2$; $a = 0$; $n = 4$
26. $f(x) = \cos x^2$; $a = 0$; $n = 4$.

27. Show that the nth degree Taylor polynomial of $f(x) = 1/(1 - x)$ at 0 is given by

$$P_n(x) = 1 + x + x^2 + \cdots + x^n.$$

*13.2 A Proof of Taylor's Theorem and Estimates on the Remainder Term

In this section we give proofs of the two theorems stated in Section 13.1. We then show how it is possible to estimate the maximum size of the remainder term $R_n(x)$ given by equation (13.1.8).

  There are at least two essentially different proofs of Taylor's theorem. The one we give here makes use of the now familiar technique of integration by parts. A less direct proof is suggested in the problem set (see Problems 1–5).

PROOF OF TAYLOR'S THEOREM. We must show that

$$f(b) = f(a) + \frac{f'(a)}{1!}(b - a) + \frac{f''(a)}{2!}(b - a)^2 + \cdots$$

$$+ \frac{f^{(n)}(a)}{n!}(b - a)^n + R_n(b), \tag{1}$$

where

$$R_n(b) = \int_a^b \frac{(b - t)^n}{n!} f^{(n+1)}(t)\, dt. \tag{2}$$

Remember that we are assuming that the functions $f, f', f'', \ldots, f^{(n)}, f^{(n+1)}$ are continuous on $[a, b]$. To prove (1) and (2), we begin with the familiar formula

$$f(b) - f(a) = \int_a^b f'(t)\, dt \tag{3}$$

which is a statement of the fundamental theorem of calculus. Set $u = f'(t)$ and $dv = dt$ in (3). Then $du = f''(t)\, dt$ and instead of using the obvious indefinite integral $v = t$, we use the indefinite integral $v = t - b = -(b - t)$ (the reason for this choice will be evident momentarily). Then

$$\int_a^b f'(t)\, dt = uv \Big|_a^b - \int_a^b v\, du = -f'(t)(b - t)\Big|_a^b + \int_a^b (b - t)f''(t)\, dt \tag{4}$$

and, combining (3) and (4),

$$f(b) - f(a) = f'(a)(b - a) + \int_a^b (b - t)f''(t)\, dt \tag{5}$$

which is Taylor's formula in the case $n = 1$. Now, setting $u = f''(t)$ and $dv = (b - t)\, dt$, we obtain $du = f'''(t)\, dt$, $v = -(b - t)^2/2$ and

$$\int_a^b (b - t) f''(t)\, dt = -f''(t) \frac{(b - t)^2}{2} \Big|_a^b + \int_a^b \frac{(b - t)^2}{2} f'''(t)\, dt$$

$$= f''(a) \frac{(b - a)^2}{2} + \int_a^b \frac{(b - t)^2}{2} f'''(t)\, dt \qquad (6)$$

so that substituting (6) into (5) gives us

$$f(b) - f(a) = f'(a)(b - a) + f''(a) \frac{(b - a)^2}{2} + \int_a^b \frac{(b - t)^2}{2} f'''(t)\, dt \qquad (7)$$

which is Taylor's formula in the case $n = 2$. It is clear that we can continue in this manner. To complete the proof, we will show how to proceed from the $(n - 1)$st to the nth step. Suppose that we have obtained

$$f(b) - f(a) = f'(a)(b - a) + f''(a) \frac{(b - a)^2}{2!} + \cdots$$

$$+ f^{(n-1)}(a) \frac{(b - a)^{n-1}}{(n - 1)!} + \int_a^b \frac{(b - t)^{n-1}}{(n - 1)!} f^{(n)}(t)\, dt \qquad (8)$$

Then, setting $u = f^{(n)}(t)\, dt$ and $dv = [(b - t)^{n-1}/(n - 1)!]\, dt$, we have

$$du = f^{(n+1)}(t)\, dt, \qquad v = -\frac{(b - t)^n}{n(n - 1)!} = -\frac{(b - t)^n}{n!}$$

and

$$\int_a^b \frac{(b - t)^{n-1}}{(n - 1)!} f^{(n)}(t)\, dt = -f^{(n)}(t) \frac{(b - t)^n}{n!} \Big|_a^b + \int_a^b \frac{(b - t)^n}{n!} f^{(n+1)}(t)\, dt$$

$$= f^{(n)}(a) \frac{(b - a)^n}{n!} + \int_a^b \frac{(b - t)^n}{n!} f^{(n+1)}(t)\, dt. \qquad (9)$$

Substituting (9) into (8) then yields Taylor's formula.

The proof we have given above is really a proof by *mathematical induction*. The reader is referred to Appendix 1 for a discussion of this important method of mathematical proof.

We now derive the simplified form of the remainder theorem.

PROOF OF THEOREM 13.1.2. We must show that there is a number c in (a, b) such that

$$R_n(b) = \int_a^b \frac{(b - t)^n}{n!} f^{(n+1)}(t)\, dt = \frac{f^{(n+1)}(c)}{(n + 1)!} (b - a)^{n+1}. \qquad (10)$$

Since $f^{(n+1)}$ is continuous on $[a, b]$ (by assumption), there exist (by Theorem 10.4.3) numbers t_1 and t_2 in $[a, b]$ such that

$$f^{(n+1)}(t_1) \le f^{(n+1)}(t) \le f^{(n+1)}(t_2)$$

for every t in $[a, b]$. Then

$$\frac{(b - t)^n}{n!} f^{(n+1)}(t_1) \le \frac{(b - t)^n}{n!} f^{(n+1)}(t) \le \frac{(b - t)^n}{n!} f^{(n+1)}(t_2) \tag{11}$$

for every t in $[a, b]$. Next, we integrate (11) and use the comparison theorem (Theorem 4.3.7) to obtain

$$\int_a^b \frac{(b - t)^n}{n!} f^{(n+1)}(t_1) \, dt \le \int_a^b \frac{(b - t)^n}{n!} f^{(n+1)}(t) \, dt$$

$$\le \int_a^b \frac{(b - t)^n}{n!} f^{(n+1)}(t_2) \, dt. \tag{12}$$

Since $f^{(n+1)}(t_1)$ and $f^{(n+1)}(t_2)$ are constants, we have, from (12),

$$f^{(n+1)}(t_1) \int_a^b \frac{(b - t)^n}{n!} \, dt \le R_n(b) \le f^{(n+1)}(t_2) \int_a^b \frac{(b - t)^n}{n!} \, dt.$$

But $\int_a^b [(b - t)^n / n!] \, dt = [(b - a)^{n+1} / (n + 1)!]$ so that

$$f^{(n+1)}(t_1) \frac{(b - a)^{n+1}}{(n + 1)!} \le R_n(b) \le f^{(n+1)}(t_2) \frac{(b - a)^{n+1}}{(n + 1)!}. \tag{13}$$

Since $f^{(n+1)}(t) [(b - a)^{n+1}/(n + 1)!]$ is continuous, by the intermediate value theorem (Theorem 10.4.4), it takes on all values between $f^{(n+1)}(t_1) [(b - a)^{n+1}/(n + 1)!]$ and $f^{(n+1)}(t_2) [(b - a)^{n+1}/(n + 1)!]$. Since by (13) $R_n(b)$ is between these values, there exists a number c between t_1 and t_2 (and therefore in (a, b)) such that

$$R_n(b) = f^{(n+1)}(c) \frac{(b - a)^{n+1}}{(n + 1)!}$$

and the theorem is proved.

The following result is central for applications.

THEOREM 1.   Under the assumptions of Taylor's theorem, there exists a number M_n such that

$$|R_n(x)| \le M_n \frac{|x - a|^{n+1}}{(n + 1)!} \tag{14}$$

for every x in $[a, b]$ and every positive integer n.

PROOF. Since $f^{(n+1)}$ is continuous on $[a, b]$, it is bounded above and below on that interval (from Theorem 10.4.3). That is, there is a number M_n such that $|f^{(n+1)}(x)| \le M_n$ for every x in $[a, b]$. Since, by Theorem 13.1.2,

$$R_n(x) = f^{(n+1)}(c) \frac{(x - a)^{n+1}}{(n + 1)!}$$

with c in (a, x),

$$|R_n(x)| = |f^{(n+1)}(c)| \frac{(x-a)^{n+1}}{(n+1)!} \leq M_n \frac{(x-a)^{n+1}}{(n+1)!}$$

and the theorem is proved.

In the next section we will show how this bound on the magnitude of the remainder term can be used to find some very interesting approximations. Since many of these examples use the Taylor polynomial at 0, we restate the results of this section for that special case: If $f, f', \ldots, f^{(n+1)}$ are continuous in $[0, b]$, then for any x in $[0, b]$

$$f(x) = f(0) + f'(0) + f''(0)\frac{x^2}{2!} + \cdots + f^{(n)}(0)\frac{x^n}{n!} + R_n(x) \tag{15}$$

and there is a number M_n such that

$$|R_n(x)| \leq M_n \frac{|x|^{n+1}}{(n+1)!}.$$

We close this section by showing that if $|M_n| \leq k$ for some constant k and for all $n \geq 1$, then the remainder terms actually approach zero as n increases.

THEOREM 2.   Let x be any real number. Then $x^n/n! \to 0$ as $n \to \infty$.

PROOF. Let N be an integer such that $N > 2|x|$. Then $|x| < N/2$ and for $n > N$ and $k > 0$,

$$\frac{|x|}{N+k} < \frac{N}{2(N+k)} = \frac{1}{2}\left(\frac{N}{N+k}\right) < \frac{1}{2}.$$

Then

$$\frac{x^n}{n!} \leq \frac{|x|^n}{n!} = \frac{\overbrace{|x| \cdot |x| \cdots |x|}^{N \text{ times}}}{1 \cdot 2 \cdot 3 \cdots N} \cdot \underbrace{\frac{|x|}{N+1} \cdot \frac{|x|}{N+2} \cdots \frac{|x|}{n}}_{\substack{n-N \text{ terms} \\ \text{each of which is } < \frac{1}{2}}} < \frac{|x|^N}{N!}\left(\frac{1}{2}\right)^{n-N}$$

$$= \frac{|x|^N}{N!} \cdot 2^N \cdot \left(\frac{1}{2}\right)^n.$$

Since x and N are fixed, we define $M = (|x|^N/N!) \cdot 2^N$ so that for each $n > N$, $|x|^n/n! < M(\frac{1}{2})^n$ which obviously approaches zero as $n \to \infty$.

THEOREM 3.   If $|M_n| \leq K$ for all $n \geq 1$, and if the hypotheses of Taylor's theorem hold, then

$$|R_n(x)| \to 0 \qquad \text{as} \quad n \to \infty. \tag{16}$$

PROOF. Using Theorems 1 and 2, we have

$$|R_n(x)| \leq M_n \frac{|x-a|^{n+1}}{(n+1)!} \leq K \left|\frac{(x-a)^{n+1}}{(n+1)!}\right| \to 0$$

PROBLEMS 13.2

1. Let $A(x) = f(b) - f(x) - f'(x)(b - x) - [f''(x)/2!](b - x)^2 - \cdots - [f^{(n)}(x)/n!](b - x)^n$ and let $B(x) = (b - x)^{n+1}/(n + 1)!$. Show that $A(b) = B(b) = 0$.

2. Show that $A'(x) = -[f^{(n+1)}(x)/n!](b - x)^n$ and that $B'(x) = -(1/n!)(b - x)^n$.

3. Prove that there is a number c in (a, b) such that $A(a) = [A'(c)/B'(c)]B(c)$. [*Hint:* Show that A and B are continuous and differentiable on $[a, b]$ and that for every x in (a, b), $B'(x) \neq 0$. Then apply the Cauchy mean value theorem (Theorem 12.2.1) and use the result of Problem 1.]

4. Using Problems 1, 2, and 3, show that

$$A'(a) = \frac{f^{(n+1)}(c)}{(n + 1)!}(b - a)^{n+1}.$$

5. Prove Taylor's theorem with $R_n(x)$ in the form

$$\frac{f^{(n+1)}(c)}{(n + 1)!}(b - a)^{n+1}.$$

☐ 13.3 Approximation Using Taylor Polynomials

In this section we show how Taylor's formula can be used as a tool for making approximations. In many of the examples that follow, results have been obtained by making use of a hand calculator. If a hand calculator is available, we suggest that you use it to check the computations.

For convenience, we summarize the results of the preceding two sections:

Let $f, f', f'', \ldots, f^{(n+1)}$ be continuous on $[a, b]$. Then for any x in $[a, b]$,

$$f(x) = f(a) + f'(a)(x - a) + f''(a)\frac{(x - a)^2}{2!} + \cdots$$

$$+ f^{(n)}(a)\frac{(x - a)^n}{n!} + R_n(x), \tag{1}$$

where

$$R_n(x) = \int_a^x \frac{(x - t)^n}{n!} f^{(n+1)}(t) \, dt = f^{(n+1)}(c)\frac{(x - a)^{n+1}}{n + 1!} \tag{2}$$

for some c (which depends on x) in the interval (a, x). Moreover, there exists a number M_n such that for every x in $[a, x]$,

$$|R_n(x)| \leq M_n \frac{(x - a)^{n+1}}{(n + 1)!}. \tag{3}$$

☐ **EXAMPLE 1.** In Example 13.1.1, we found that the fifth degree Taylor polynomial of $f(x) = \sin x$ at 0 is $P_5(x) = x - (x^3/3!) + (x^5/5!)$. We then have

$$\sin x = x - \frac{x^3}{3!} + \frac{x^5}{5!} + R_5(x), \tag{4}$$

where $R_5(x) = \int_0^x [(x - t)^5/5!] \sin^{(vi)}(t)\, dt$. But $\sin^{(vi)}(t) = -\sin t$ and $|\sin t| \leq 1$. Then from (3), if x is in $[0, 1]$,

$$|R_5(x)| \leq \frac{1(x - 0)^6}{6!} = \frac{x^6}{720},$$

For example, suppose we wish to calculate $\sin \pi/10$. From (4)

$$\sin \frac{\pi}{10} = \frac{\pi}{10} - \frac{\pi^3}{3!10^3} + \frac{\pi^5}{5!10^5} + R_n(x)$$

with

$$|R_n(x)| \leq \frac{(\pi/10)^6}{720} \approx 0.0000013.$$

We find that

$$\sin \frac{\pi}{10} \approx \frac{\pi}{10} - \frac{1}{3!} \frac{\pi^3}{10^3} + \frac{1}{5!} \frac{\pi^5}{10^5} \approx 0.3141593 - 0.0051677 + 0.0000255$$

$$= 0.3090171.$$

The actual value of $\sin \pi/10 = \sin 18° = 0.3090170$, correct to 7 decimal places so our actual error is 0.0000001 which is quite a bit less than 0.0000013. This illustrates the fact that the actual error (the value of the remainder term) is often quite a bit smaller than the theoretical upper bound on the error given by formula (3).

□**EXAMPLE 2.** Use a Taylor polynomial to estimate $e^{0.3}$ with an error of less than 0.0001.

SOLUTION. For convenience, choose the interval $[0, 1]$. Then on $[0, 1]$, e^x and all its derivatives have a maximum value of $e^1 = e$. Then, for any n, if we use a Taylor polynomial at 0 of degree n, we have

$$|R_n(0.3)| \leq e \cdot \frac{(0.3)^{n+1}}{(n + 1)!}.$$

We must choose n so that $e \cdot (0.3)^{n+1}/(n + 1)! < 0.0001$. For $n = 3$, $e \cdot (0.3)^4/4! \approx 0.0009$ while for $n = 4$, $e \cdot (0.3)^5/5! \approx 0.00006$ so we choose a fourth degree Taylor polynomial for our approximation:

$$P_5(x) = 1 + x + \frac{x^2}{2!} + \frac{x^3}{3!} + \frac{x^4}{4!}$$

(see Example 13.1.3). Then

$$P_5(0.3) = 1 + 0.3 + \frac{(0.3)^2}{2!} + \frac{(0.3)^3}{3!} + \frac{(0.3)^4}{4!}$$

$$\approx 1 + 0.3 + 0.045 + 0.0045 + 0.00034 = 1.34984.$$

The actual value of $e^{0.3}$ is 1.34986, correct to 5 decimal places. Thus the error in our calculation is, approximately, 0.00002, one-third the calculated upper bound on the error of 0.00006.

☐**EXAMPLE 3.** Compute $\cos[(\pi/4) + 0.1]$ with an error of less than 0.00001.

SOLUTION. In this problem it is clearly convenient to use the Taylor expansion at $\pi/4$. Since all derivatives of $\cos x$ are bounded by 1, we have, for any n,

$$|R_n(x)| \leq \frac{[x - (\pi/4)]^{n+1}}{(n + 1)!}$$

so that $R_n[(\pi/4) + 0.1] \leq (0.1)^{n+1}/(n + 1)!$. If $n = 2$, then $(0.1)^3/3! = 0.00017$ and for $n = 3$, $(0.1)^4/4! = 0.000004 < 0.00001$. Thus we need to calculate $P_3(x)$ at $\pi/4$. But

$$P_3(x) = \cos\frac{\pi}{4} - \left(\sin\frac{\pi}{4}\right)\left(x - \frac{\pi}{4}\right) - \left(\cos\frac{\pi}{4}\right)\frac{[x - (\pi/4)]^2}{2}$$

$$+ \left(\sin\frac{\pi}{4}\right)\frac{[x - (\pi/4)]^3}{6}$$

$$= \frac{1}{\sqrt{2}}\left\{1 - \left(x - \frac{\pi}{4}\right) - \frac{[x - (\pi/4)]^2}{2} + \frac{[x - (\pi/4)]^3}{6}\right\}$$

and for $x = (\pi/4) + 0.1$,

$$\cos\left(\frac{\pi}{4} + 0.1\right) \approx P_4\left(\frac{\pi}{4} + 0.1\right) = \frac{1}{\sqrt{2}}\left[1 - 0.1 - \frac{(0.1)^2}{2} + \frac{(0.1)^3}{6}\right]$$

$$\approx 0.63298.$$

This is correct to five decimal places.

We now consider a more general example.

THE LOGARITHM

In Section 1.10 we showed that for any number $x \neq 1$, the formula for the sum of a geometric progression is given by

$$1 + u + u^2 + \cdots + u^n = \frac{1 - u^{n+1}}{1 - u} = \frac{1}{1 - u} - \frac{u^{n+1}}{1 - u}. \tag{5}$$

This leads to the expression

$$\boxed{\frac{1}{1 - u} = 1 + u + u^2 + \cdots + u^n + \frac{u^{n+1}}{1 - u}.} \tag{6}$$

If $u < 1$, then equation (6) is just the Taylor expansion of the function $1/(1 - u)$ around 0 (see Example 13.1.4 and Problem 13.1.27), with $R_n(u) = u^{n+1}/(1 - u)$. Setting $u = -t$ in (6) leads to the expression

$$\boxed{\frac{1}{1 + t} = 1 - t + t^2 - t^3 + \cdots + (-1)^n t^n + (-1)^{n+1}\frac{t^{n+1}}{1 + t}.} \tag{7}$$

Integration of both sides of (7) from 0 to x yields

$$\ln(1 + x) = x - \frac{x^2}{2} + \frac{x^3}{3} - \cdots + (-1)^n \frac{x^{n+1}}{n+1}$$

$$+ (-1)^{n+1} \int_0^x \frac{t^{n+1}}{t+1} \, dt \qquad\qquad (8)$$

which is valid if $x > -1$ (since $\ln(1 + x)$ is not defined for $x \leq -1$). Formula (8) gives us the Taylor expansion of $\ln(1 + x)$ about 0 for $x > -1$. We now show that if $-1 < x \leq 1$,

$$R_n(x) = (-1)^{n+1} \int_0^x \frac{t^{n+1}}{(t+1)} \, dt \to 0 \qquad \text{as} \quad n \to \infty.$$

CASE 1. $0 \leq x \leq 1$. Then for t in $[0, 1]$, $1/(1 + t) \leq 1$ so that

$$|R_n(x)| = \int_0^x \frac{t^{n+1}}{t+1} \, dt \leq \int_0^x t^{n+1} \, dt = \frac{x^{n+2}}{n+2} \qquad\qquad (9)$$

which approaches 0 as $n \to \infty$.

CASE 2. $-1 < x < 0$. First, we note that

$$(-1)^{n+1} \int_0^x \frac{t^{n+1}}{t+1} \, dt = (-1)^n \int_x^0 \frac{t^{n+1}}{t+1} \, dt.$$

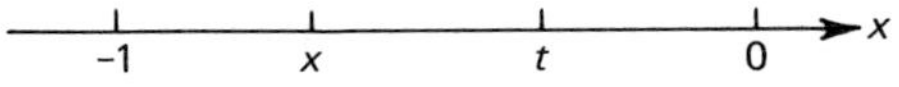

Figure 1

In this latter integral t is in the interval $[x, 0]$ where $-1 < x < 0$ (see Figure 1). Then we have $1 \geq 1 + t \geq 1 + x$, so that

$$1 \leq \frac{1}{1+t} \leq \frac{1}{1+x}$$

and

$$|R_n(x)| \leq \int_x^0 \frac{|t|^{n+1}}{|t+1|} \, dt \leq \int_x^0 \frac{(-t)^{n+1}}{(x+1)} \, dt \qquad \text{(since } |t| = -t \text{ in } [x, 0])$$

$$= \frac{1}{x+1} \int_x^0 (-t)^{n+1} \, dt = -\frac{1}{x+1} \frac{(-t)^{n+2}}{(n+2)} \Big|_x^0 = \frac{(-x)^{n+2}}{(1+x)(n+2)}$$

so that

$$|R_n(x)| \leq \frac{(-x)^{n+2}}{(1+x)(n+2)} \qquad\qquad (10)$$

and since $-1 < x < 0$ implies that $0 < -x < 1$, we have $|R_n(x)| \to 0$ as $n \to \infty$.

□**EXAMPLE 4.** Calculate $\ln 1.4$ with an error of less than 0.001.

SOLUTION.   Here $x = 0.4$ and, from (9), we need to find an n such that $(0.4)^{n+2}/(n + 2) < 0.001$. We have $(0.4)^5/5 \approx 0.00205$ and $(0.4)^6/6 \approx 0.00068$ so that choosing $n = 4$ ($n + 2 = 6$), we obtain the Taylor polynomial (from (8))

$$P_4(x) = x - \frac{x^2}{2} + \frac{x^3}{3} - \frac{x^4}{4} + \frac{x^5}{5}$$

(remember, from (8) the last term in $P_n(x)$ is $(-1)^n x^{n+1}/(n + 1)$) and

$$\ln 1.4 \approx P_4(0.4) = 0.4 - \frac{(0.4)^2}{2} + \frac{(0.4)^3}{3} - \frac{(0.4)^4}{4} + \frac{(0.4)^5}{5}$$

$$\approx 0.4 - 0.08 + 0.02133 - 0.0064 + 0.00205 = 0.33698.$$

The actual value of $\ln 1.4$ is 0.33647, correct to five decimal places, so that the error is $0.33698 - 0.33647 = 0.00051$. This is slightly less than the maximum possible error of 0.00068 and so, in this case, our error bound is fairly sharp.

□**EXAMPLE 5.**   Calculate $\ln 0.9$ with an error of less than 0.001.

SOLUTION.   Here we need to find $\ln(1 + x)$ for $x = -0.1$. Using (10), we have

$$|R_n(x)| \le \frac{(-x)^{n+2}}{(1 + x)(n + 2)} = \frac{(0.1)^{n+2}}{0.9(n + 2)}.$$

For $n = 1$, $(0.1)^3/(0.9)(3) = 0.00037$ so that we need only to evaluate $P_1(-0.1) = (-0.1) - [(-0.1)^2/2] = -0.10500$. The value of $\ln 0.9$ correct to five decimal places is -0.10536 so that our actual error of 0.00036 and our maximum error of 0.00037 almost coincide.

THE ARC TANGENT

If in equation (6) we substitute $u = -t^2$, then for any number t we obtain

$$\frac{1}{1 + t^2} = 1 - t^2 + t^4 - t^6 + \cdots + (-1)^n t^{2n} + (-1)^{n+1} \frac{t^{2(n+1)}}{1 + t^2}. \tag{11}$$

Integration of both sides of (11) from 0 to x yields

$$\tan^{-1} x = x - \frac{x^3}{3} + \frac{x^5}{5} - \frac{x^7}{7} + \cdots + \frac{(-1)^n x^{2n+1}}{2n + 1}$$

$$+ (-1)^{n+1} \int_0^x \frac{t^{2(n+1)}}{1 + t^2} \, dt. \tag{12}$$

This is the Taylor expansion for $\tan^{-1} x$ and is valid for any real number $x \ge 0$. Since $1/(1 + t^2) \le 1$,

$$|R_n(x)| = \int_0^x \frac{t^{2(n+1)}}{1 + t^2} \le \int_0^x t^{2(n+1)} \, dt = \frac{x^{2n+3}}{2n + 3}, \tag{13}$$

which approaches 0 as $n \to \infty$ if $|x| \le 1$.

Equation (12) gives us a formula for calculating π.† Setting $x = 1$, we have

$$\frac{\pi}{4} = \tan^{-1} 1 = 1 - \frac{1}{3} + \frac{1}{5} - \frac{1}{7} + \frac{1}{9} - \frac{1}{11} + \cdots.$$

The error here (given by (13)) is $1/(2n + 3)$ which approaches zero very slowly. To get an error < 0.001 we would need $1/(2n + 3) < 1/1000$ or $2n + 3 > 1000$ and $n \geq 499$. For example $1 - \frac{1}{3} + \frac{1}{5} - \frac{1}{7} + \frac{1}{9} - \frac{1}{11} + \frac{1}{13} - \frac{1}{15} = 0.75427$ while $\pi/4$ is 0.78540. A better way to approximate π is suggested in Problems 31 and 32.

PROBLEMS 13.3

In Problems 1–10 find a bound for $|R_n(x)|$ for x in the given interval.

1. $f(x) = \sin x$; $a = \pi/4$; $n = 6$; $x \in [0, \pi/2]$
2. $f(x) = \sqrt{x}$; $a = 1$; $n = 4$; $x \in [\frac{1}{4}, 4]$
3. $f(x) = 1/x$; $a = 1$; $n = 4$; $x \in [\frac{1}{2}, 2]$
4. $f(x) = \tan x$; $a = 0$; $n = 4$; $x \in [0, \pi/4]$
5. $f(x) = 1/\sqrt{x}$; $a = 5$; $n = 5$; $x \in [19/4, 21/4]$
6. $f(x) = \sinh x$; $a = 0$; $n = 4$; $x \in [0, 1]$
7. $f(x) = \ln \cos x$; $a = 0$; $n = 3$; $x \in [0, \pi/6]$
8. $f(x) = e^{\alpha x}$; $a = 0$; $n = 4$; $x \in [0, 1]$
9. $f(x) = e^{x^2}$; $a = 0$; $n = 4$; $x \in [0, \frac{1}{3}]$
10. $f(x) = \sin x^2$; $a = 0$; $n = 4$; $x \in [0, \pi/4]$.

In Problems 11–26 use a Taylor polynomial to estimate the given number with the given degree of accuracy.

11. $\sin\left(\frac{\pi}{6} + 0.2\right)$; error < 0.001

12. $\sin 33°$; error < 0.001 [*Hint:* Convert to radians.]

13. $\tan\left(\frac{\pi}{4} + 0.1\right)$; error < 0.01

14. e; error < 0.0001 [*Hint:* You may assume that $2 < e < 3$]

15. e^{-1}; error < 0.001 16. $\ln 2$; error < 0.1

17. $\ln 1.5$; error < 0.001 18. $\ln 0.5$; error < 0.0001

19. e^3; error < 0.01 [*Hint:* See Problem 14.]
20. $\tan^{-1} 0.5$; error < 0.001

*21. $\sinh \frac{1}{2}$; error < 0.01 *22. $\cosh \frac{1}{2}$; error < 0.01
23. $\sin 100°$; error < 0.001 24. $\cos 195°$; error < 0.001
25. $1/\sqrt{1.1}$; error < 0.001 *26. $\ln \cos 0.3$; error < 0.01.

27. Use the result of Problem 9 to estimate $\int_0^{1/3} e^{x^2} \, dx$. What is the maximum error of your estimate?

28. Use the result of Problem 7 to estimate $\int_0^{\pi/6} \ln \cos x \, dx$. What is the maximum error of your estimate?

*29. Use the formula $\tan(A \pm B) = (\tan A \pm \tan B)/(1 \mp \tan A \tan B)$ to prove that

---

† This formula was discovered by the Scottish mathematician James Gregory (1638–1675) and was first published in 1712.

$\tan(4 \tan^{-1} 1/5 - \tan^{-1} 1/239) = 1$. [*Hint:* First calculate $\tan 2(\tan^{-1} 1/5)$ and $\tan 4(\tan^{-1} 1/5)$.] This implies that $4 \tan^{-1} 1/5 - \tan^{-1} 1/239 = \pi/4$.

*30. Use the result of Problem 29 to calculate π to 5 decimal places.

31. Let $f(x) = (1 + x)^n$, where n is an integer.
(a) Show that $f^{(n+1)}(x) - 0$.
(b) Show that

$$f(x) = 1 + nx + \frac{n(n-1)}{2!}x^2 + \frac{n(n-1)(n-2)}{3!}x^3 + \cdots + x^n.$$

This result is called the *binomial theorem*. Also see Appendix 4.

32. Use the binomial theorem to calculate
(a) $(1.2)^4$ (b) $(0.8)^5$ (c) $(1.03)^4$.

33. Let $f(x) = (1 + x)^\alpha$ where α is any real number.
(a) Show that

$$f(x) = 1 + \alpha x + \frac{\alpha(\alpha-1)}{2!}x^2 + \frac{\alpha(\alpha-1)(\alpha-2)}{3!}x^3 + \cdots$$

$$+ \frac{\alpha(\alpha-1)(\alpha-2)\cdots(\alpha-n+1)}{n!}x^n + R_n(x).$$

(b) Show that if $|x| < 1$, then $R_n(x) \to 0$ as $n \to \infty$. This result is called the *general binomial expansion*.

34. Use the result of Problem 33 to calculate the following numbers to four decimal places of accuracy.
(a) $\sqrt{1.2}$ (b) $(0.9)^{3/4}$ (c) $(1.8)^{1/4}$
(d) $\dfrac{1}{\sqrt[3]{1.01}}$ (e) $2^{5/3}$ (f) $(0.4)^{1.6}$.

*13.4 Polynomial Interpolation

In Sections 13.1–13.3 we considered the problem of approximating a given differentiable function by a polynomial. In some applications it is more important to find a polynomial which is *equal to* the function f at a certain number of specified points in some interval over which f is defined. Such a polynomial is said to *interpolate* the function f at these points. For example, in Figure 1 the fourth degree polynomial P_4 interpolates the function f at the points x_1, x_2, x_3, and x_4. The existence of an interpolating polynomial is given in the theorem below.

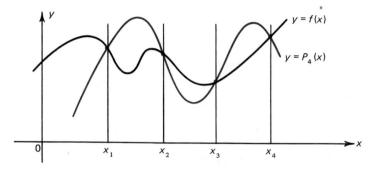

Figure 1

THEOREM 1. Let the function f be defined on some interval $[a, b]$ and let $x_0, x_1,$ $x_2, \ldots, x_n$ be $n + 1$ *distinct points* of $[a, b]$. Then there exists a *unique* polynomial $q(x)$ of degree less than or equal to n such that

$$q(x_k) = f(x_k) \qquad \text{for} \quad k = 0, 1, 2, \ldots, n, \tag{1}$$

This polynomial is called the *Lagrange† interpolating polynomial.*

PROOF. Define the polynomials $r_k(x)$, $k = 0, 1, \ldots, n$, by

$$r_k(x) = \frac{(x - x_0)(x - x_1) \cdots (x - x_{k-1})(x - x_{k+1}) \cdots (x - x_n)}{(x_k - x_0)(x_k - x_1) \cdots (x_k - x_{k-1})(x_k - x_{k+1}) \cdots (x_k - x_n)}. \tag{2}$$

We note the following obvious facts about each r_k:

    (i) r_k is a polynomial of degree n (the denominator of the quotient in equation (2) is a constant),
    (ii) $r_k(x_k) = 1$, and
    (iii) $r_k(x_j) = 0$ if $j \neq k$.

We now define

$$\boxed{q(x) = f(x_0)r_0(x) + f(x_1)r_1(x) + \cdots + f(x_n)r_n(x).} \tag{3}$$

Since $q(x)$ is a sum of polynomials of degree n, $q(x)$ is of degree $\leq n$. Moreover, using facts (ii) and (iii) above, we easily see that $q(x_k) = f(x_k)$ for each k, $k = 0, 1, 2, \ldots, n$, and this proves that $q(x)$ *is* an interpolating polynomial. We still must prove that this polynomial is unique. Suppose that $q(x)$ and $s(x)$ are two interpolating polynomials. Let $u(x) = q(x) - s(x)$. Then $u(x)$ is a polynomial of degree $\leq n$ and $u(x_k) = q(x_k) - s(x_k) = f(x_k) - f(x_k) = 0$ for $k = 0, 1, 2, \ldots, n$. But any nonzero polynomial of degree n has at most n zeros (see Problem 10.6.26). Since $u(x)$ has the $n + 1$ distinct zeros $x_0, x_1, \ldots, x_n$, we must conclude that $u(x) = 0$ for every x so that $q(x) = s(x)$, thereby proving the uniqueness of the interpolating polynomial.

EXAMPLE 1. Find the polynomial which interpolates the function $y = 1/x$ at the values 1, 2, 3, and 4.

TABLE 1

k	0	1	2	3
x_k	1	2	3	4
$f(x_k)$	1	$\frac{1}{2}$	$\frac{1}{3}$	$\frac{1}{4}$

SOLUTION. We refer to Table 1. Using (2) we have

$$r_0(x) = \frac{(x - x_1)(x - x_2)(x - x_3)}{(x_0 - x_1)(x_0 - x_2)(x_0 - x_3)} = \frac{(x - 2)(x - 3)(x - 4)}{(1 - 2)(1 - 3)(1 - 4)}$$

$$= -\frac{1}{6}(x - 2)(x - 3)(x - 4)$$

†Named after the same Lagrange referred to in Section 13.2. We will refer to him again in later chapters.

$$r_1(x) = \frac{(x - x_0)(x - x_2)(x - x_3)}{(x_1 - x_0)(x_1 - x_2)(x_1 - x_3)} = \frac{(x - 1)(x - 3)(x - 4)}{(2 - 1)(2 - 3)(2 - 4)}$$

$$= \frac{1}{2}(x - 1)(x - 3)(x - 4)$$

$$r_2(x) = \frac{(x - x_0)(x - x_1)(x - x_3)}{(x_2 - x_0)(x_2 - x_1)(x_2 - x_3)} = \frac{(x - 1)(x - 2)(x - 4)}{(3 - 1)(3 - 2)(3 - 4)}$$

$$= -\frac{1}{2}(x - 1)(x - 2)(x - 4)$$

$$r_3(x) = \frac{(x - x_0)(x - x_1)(x - x_2)}{(x_3 - x_0)(x_3 - x_1)(x_3 - x_2)} = \frac{(x - 1)(x - 2)(x - 3)}{(4 - 1)(4 - 2)(4 - 3)}$$

$$= \frac{1}{6}(x - 1)(x - 2)(x - 3).$$

Then the interpolating polynomial is of third degree and is given by

$$\begin{aligned}
q(x) &= r_0(x)f(x_0) + r_1(x)f(x_1) + r_2(x)f(x_2) + r_3(x)f(x_3) \\
&= -\tfrac{1}{6}(x - 2)(x - 3)(x - 4) + \tfrac{1}{4}(x - 1)(x - 3)(x - 4) \\
&\quad - \tfrac{1}{6}(x - 1)(x - 2)(x - 4) + \tfrac{1}{24}(x - 1)(x - 2)(x - 3) \\
&= -\tfrac{1}{24}(x^3 - 10x^2 + 35x - 50).
\end{aligned}$$

(We've spared you the missing algebraic steps.) This is illustrated in Figure 2.

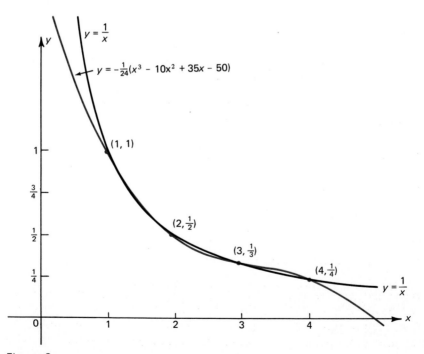

Figure 2

When $n = 1$ in equations (2) and (3) we obtain the formula for *linear inter-polation*. In this case we interpolate at the two points x_0 and x_1. Then

$$r_0(x) = \frac{x - x_1}{x_0 - x_1}, \qquad r_1(x) = \frac{x - x_0}{x_1 - x_0}$$

and

$$q(x) = f(x_0)\left(\frac{x - x_1}{x_0 - x_1}\right) + f(x_1)\left(\frac{x - x_0}{x_1 - x_0}\right)$$

$$= \frac{(x_1 - x)f(x_0) + (x - x_0)f(x_1)}{x_1 - x_0}. \tag{4}$$

This is easily seen to be the equation of the straight line passing through the points $(x_0, f(x_0))$ and $(x_1, f(x_1))$. The situation is illustrated in Figure 3.

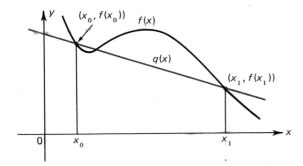

Figure 3

☐**EXAMPLE 2.** Using the values $\sin 35° = 0.5736$ and $\sin 40° = 0.6428$, use linear interpolation to find an approximate value for $\sin 37°$.

SOLUTION. The equation of the straight line joining the points $(x_0, f(x_0)) = (35°, 0.5736)$ and $(x_1, f(x_1)) = (40°; 0.6428)$ is given by

$$q(x) = \frac{(40 - x)(0.5736) + (x - 35)(0.6428)}{40 - 35}$$

$$= (0.01384)x + 0.0892.$$

Substituting $x = 37°$ yields

$$q(37°) = (0.01384)(37) + (0.0892) \approx 0.6013.$$

The correct value of $\sin 37°$ is 0.6018.

The preceding examples lead to the obvious question—how good an approximation is given by the interpolating polynomial? Intuitively, it seems that the farther we are from an interpolating point (a point of the form $(x_k, f(x_k))$), the greater the distance between the interpolating polynomial and the original function. A precise expression for the error is given by the following theorem.

THEOREM 2. Suppose that $f, f', \ldots, f^{(n+1)}$ are continuous over the interval $[a, b]$. Let x be any number in $[a, b]$ and let I_x denote the smallest closed interval containing the numbers $x, x_0, x_1, \ldots, x_n$. Then there is a number x^* in I_x such that

$$f(x) - q(x) = \frac{Q(x)}{(n+1)!} f^{(n+1)}(x^*), \tag{5}$$

where $Q(x)$ is defined as $Q(x) = (x - x_0)(x - x_1)(x - x_2) \cdots (x - x_n)$ and $q(x)$ is the interpolating polynomial for $f(x)$ which interpolates at the points $(x_0, f(x_0))$, $(x_1, f(x_1)), \ldots, (x_n, f(x_n))$.

> **Remark.** We cannot use formula (5) to calculate the actual "error" $f(x) - q(x)$ since the number x^* is not known. However, as with Taylor's formula, it is often possible to use the formula to obtain a bound on the maximum difference between a function and its interpolating polynomial.

PROOF OF THEOREM 2. If $x = x_k$ for some k, then there is nothing to prove as both sides of (5) are equal to zero. Therefore suppose that $x \neq x_k$ and define $\alpha = [f(x) - q(x)]/Q(x)$ (since x is fixed, α is a constant). Define the function $S(t)$ by

$$S(t) = f(t) - q(t) - \alpha Q(t). \tag{6}$$

If $t = x_k$ for some k, then

$$S(x_k) = f(x_k) - q(x_k) - \alpha Q(x_k) = 0.$$

Similarly,

$$S(x) = f(x) - q(x) - \left[\frac{f(x) - q(x)}{Q(x)}\right] Q(x) = 0.$$

Recall that Rolle's theorem (Theorem 10.6.1) states that if f and f' are continuous on $[a, b]$ and if $f(a) = f(b) = 0$, then there is a number c in (a, b) such that $f'(c) = 0$. Now, the function S is zero for at least $n + 2$ numbers in the interval I_x. Rolle's theorem implies that S' has at least $n + 1$ zeros in that interval (one zero of S' between every two consecutive zeros of S). Then S'' has at least n zeros, S''' has at least $n - 1$ zeros and, finally, $S^{(n+1)}$ has at least one zero at a number, which we shall call x^*, in the interval (a, b). If we differentiate both sides of (6) $n + 1$ times, we obtain

$$S^{(n+1)}(t) = f^{(n+1)}(t) - q^{(n+1)}(t) - \alpha Q^{(n+1)}(t). \tag{7}$$

But since $Q(t) = t^{n+1} + \cdots$, we have $Q^{(n+1)}(t) = (n + 1)!$ (see Problems 3.8.16 and 3.8.17). Thus substituting $t = x^*$ in (7), we have

$$S^{(n+1)}(x^*) = 0 = f^{(n+1)}(x^*) - q^{(n+1)}(x^*) - \alpha(n + 1)!.$$

But $q^{(n+1)}(x) = 0$ since $q(x)$ is a polynomial of degree at most n (see Problem 3.8.17) and we obtain

$$f^{(n+1)}(x^*) = \alpha(n + 1)! = \frac{f(x) - q(x)}{Q(x)}(n + 1)!$$

Finally, after rearranging terms, we find that

$$f(x) - q(x) = \frac{Q(x)}{(n+1)!} f^{(n+1)}(x^*),$$

which is what we want to prove.

EXAMPLE 3. If there is only one interpolating point $(x_0, f(x_0))$, then $n = 0$, $q(x) = f(x_0)$, $Q(x) = (x - x_0)$ and, from (5), we obtain

$$f(x) - f(x_0) = (x - x_0)f'(x^*), \tag{8}$$

where x^* is in the smallest interval containing x and x_0. This is one of the intervals $[x_0, x]$ or $[x, x_0]$. Does this look familiar? It should, for it is simply the mean value theorem of differential calculus.

EXAMPLE 4. If there are two interpolating points we obtain the linear interpolating polynomial given by (4). Then the error is given by

$$f(x) - q(x) = \frac{(x - x_0)(x - x_1)}{2} f''(x^*).$$

Suppose that $|f''(x)| \le M$ for some constant M. Define the function h by

$$h(x) = \frac{Q(x)}{2} = \frac{(x - x_0)(x - x_1)}{2} = \frac{x^2 - (x_0 + x_1)x + x_0 x_1}{2}$$

and we see that $h'(x) = x - [(x_0 + x_1)/2]$ which is zero when $x = (x_0 + x_1)/2$. In the interval $[x_0, x_1]$, $h(x) \le 0$. Since $h''(x) = 1 > 0$, $h(x)$ has a minimum at $x = (x_0 + x_1)/2$. Since $h(x)$ is negative, this means that $|h(x)|$ has a maximum at $(x_0 + x_1)/2$ in the interval $[x_0, x_1]$. If $x = (x_0 + x_1)/2$, then $h(x) = -(x_1 - x_0)^2/8$ and we obtain the bound

$$\boxed{|f(x) - q(x)| \le \frac{(x_1 - x_0)^2}{8} M.} \tag{9}$$

□**EXAMPLE 5.** In Example 2, we have $x_0 = 35° \approx 0.61087$ rad and $x_1 = 40° \approx 0.69813$ rad. Since $f(x) = \sin x$, $f''(x) = -\sin x$ and $|f''(x)| \le 1$. Then from (9) we obtain

$$|f(x) - q(x)| \le \frac{(0.69813 - 0.61087)^2}{8} \approx 0.00095.$$

In our calculation the actual error was $0.6018 - 0.6013 = 0.0005$.

EXAMPLE 6. In Example 1 we approximated the function $f(x) = 1/x$ by a cubic polynomial $q(x) = -(1/24)(x^3 - 10x^2 + 35x - 50)$. How well does $q(x)$ approximate $f(x)$ over the interval $[1, 4]$?

SOLUTION. We will derive a formula for an upper bound on the maximum error of cubic interpolation. Let the four numbers x_0, x_1, x_2, and x_3 be equally spaced as in

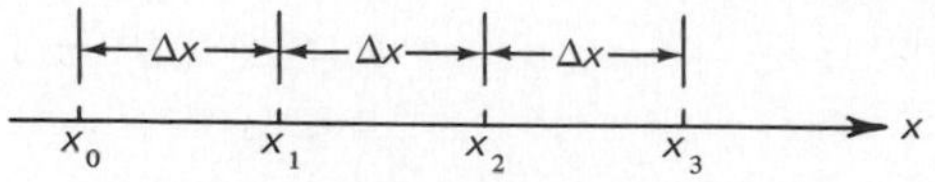

Figure 4

Figure 4. Here $Q(x) = (x - x_0)(x - x_1)(x - x_2)(x - x_3)$. It is not difficult to show that the maximum of $Q(x)$ over the interval $[x_0, x_3]$ occurs at $x = (x_1 + x_2)/2$ and that $Q((x_1 + x_2)/2) = 9(\Delta x)^4/16$ (see Problem 14). Then if M denotes the maximum value of $f^{(iv)}(x)$ on $[x_0, x_3]$, we obtain

$$|f(x) - q(x)| \le \frac{1}{4!}\frac{9}{16}(\Delta x)^4 M = \frac{3}{128}(\Delta x)^4 M. \tag{10}$$

Returning to our example, since $f(x) = 1/x$, we have $f'(x) = -(1/x^2), f''(x) = 2/x^3$, $f'''(x) = -(6/x^4)$ and $f^{(iv)}(x) = 24/x^5$ which is bounded by $M = 24$ on the interval $[1, 4]$. Since $\Delta x = 1$ here, we see that

$$|f(x) - q(x)| \le \frac{3 \cdot 24}{128} = \frac{9}{16} = 0.5625.$$

For example, if $x = \frac{3}{2}$, then $f(\frac{3}{2}) = \frac{2}{3} = 0.6667$ and $q(\frac{3}{2}) = 0.6927$ with an error of 0.026. If $x = \frac{5}{2}$, then $f(x) = \frac{2}{5} = 0.4$ and $q(\frac{5}{2}) = 0.3906$ with an error of 0.0094. Here the actual error is much less than the maximum error as calculated by the result of Theorem 2. Actually, the error bound obtained here is unusually large. If we had instead chosen $\Delta x = 0.1$, we would have obtained the error bound

$$|f(x) - q(x)| \le \frac{9}{16}(0.1)^4 = 0.00005625.$$

Thus the cubic polynomial which interpolates the function $f(x) = 1/x$ at the numbers 1, 1.1, 1.2, and 1.3 gives values of $1/x$ correct to at least 4 decimal places for any x in the interval $[1, 1.3]$.

□**EXAMPLE 7.** In order to see how interpolation improves as we continue to increase the degree of the interpolating polynomial, let us interpolate the function $f(x) = \cos x$ at two, three, and four evenly spaced points in the interval $[0, \pi/2]$.

SOLUTION. (a)   Two points: We find the line passing through the points $(0, 1)$ and $(\pi/2, 0)$. This is given by

$$q_1(x) = -\frac{2}{\pi}x + 1 \approx -0.637x + 1, \qquad 0 \le x \le \frac{\pi}{2}.$$

Since

$$\left|\frac{d^2}{dx^2}\cos x\right| = |\cos x| \le 1,$$

we have $M = 1$ and, by (9), the maximum error of interpolation is

$$|\cos x - q_1(x)| \le \frac{(\pi/2)^2}{8}M = \frac{\pi^2}{32} \approx 0.318.$$

(b)   Three points: Here we interpolate at the points $(0, 1)$, $(\pi/4, 1/\sqrt{2})$ and $(\pi/2, 0)$. Then we find that

$$r_0(x) = \frac{\left(x - \frac{\pi}{4}\right)\left(x - \frac{\pi}{2}\right)}{\left(0 - \frac{\pi}{4}\right)\left(0 - \frac{\pi}{2}\right)} = \frac{x^2 - \frac{3\pi}{4}x + \frac{\pi^2}{8}}{\frac{\pi^2}{8}} = \frac{8x^2}{\pi^2} - \frac{6x}{\pi} + 1$$

$$r_1(x) = \frac{x\left(x - \frac{\pi}{2}\right)}{\left(\frac{\pi}{4} - 0\right)\left(\frac{\pi}{4} - \frac{\pi}{2}\right)} = \frac{x^2 - \frac{\pi}{2}x}{-\frac{\pi^2}{16}} = -\frac{16x^2}{\pi^2} + \frac{8}{\pi}x$$

$$r_2(x) = \frac{x\left(x - \frac{\pi}{4}\right)}{\left(\frac{\pi}{2} - 0\right)\left(\frac{\pi}{2} - \frac{\pi}{4}\right)} = \frac{x^2 - \frac{\pi}{4}x}{\frac{\pi^2}{8}} = \frac{8x^2}{\pi^2} - \frac{2x}{\pi}$$

and

$$q_2(x) = 1\left(\frac{8x^2}{\pi^2} - \frac{6x}{\pi} + 1\right) + \frac{1}{\sqrt{2}}\left(-\frac{16x^2}{\pi^2} + \frac{8x}{\pi}\right) + 0\left(\frac{8x^2}{\pi^2} - \frac{2x}{\pi}\right)$$

or

$$q_2(x) = \frac{8(1 - \sqrt{2})x^2}{\pi^2} + \frac{(4\sqrt{2} - 6)x}{\pi} + 1$$

$$\approx -0.336x^2 - 0.109x + 1, \qquad 0 \leq x \leq \frac{\pi}{2}.$$

To find the maximum error, we first note that, as before, $M = 1$ (since the third derivative of $\cos x = \sin x$). Then

$$Q(x) = x\left(x - \frac{\pi}{4}\right)\left(x - \frac{\pi}{2}\right) \qquad \text{and} \qquad Q'(x) = 3x^2 - \frac{3\pi}{2}x + \frac{\pi^2}{8}$$

which is zero when $x = (\pi/4)[1 \pm (\sqrt{3}/3)]$. (This follows from the quadratic formula.) We then can easily see that

$$Q''\left(\frac{\pi}{4}\left(1 - \frac{\sqrt{3}}{3}\right)\right) < 0 \qquad \text{and} \qquad Q\left(\frac{\pi}{4}\left(1 - \frac{\sqrt{3}}{3}\right)\right) \approx 0.186.$$

Hence we find, from (5), that

$$|\cos x - q_2(x)| \leq \frac{0.186}{3!} \cdot 1 \approx 0.031.$$

Note that in going from linear to quadratic interpolation, the maximum error has been reduced by a factor of ten.

(c)   Four points: Now we seek the cubic interpolating polynomial. If we interpolate at $(0, 1)$, $(\pi/6, \sqrt{3}/2)$, $(\pi/3, 1/2)$, $(\pi/2, 0)$, we find that (we have left out a lot of algebra)

$$r_0(x) = \frac{\left(x - \frac{\pi}{6}\right)\left(x - \frac{\pi}{3}\right)\left(x - \frac{\pi}{2}\right)}{\left(0 - \frac{\pi}{6}\right)\left(0 - \frac{\pi}{3}\right)\left(0 - \frac{\pi}{2}\right)} = -\frac{36x^3}{\pi^3} + \frac{36x^2}{\pi^2} - \frac{11x}{\pi} + 1,$$

$$r_1(x) = \frac{x\left(x - \frac{\pi}{3}\right)\left(x - \frac{\pi}{2}\right)}{\frac{\pi}{6}\left(\frac{\pi}{6} - \frac{\pi}{3}\right)\left(\frac{\pi}{6} - \frac{\pi}{2}\right)} = \frac{108x^3}{\pi^3} - \frac{90x^2}{\pi^2} + \frac{18x}{\pi},$$

$$r_2(x) = \frac{x\left(x - \frac{\pi}{6}\right)\left(x - \frac{\pi}{2}\right)}{\frac{\pi}{3}\left(\frac{\pi}{3} - \frac{\pi}{6}\right)\left(\frac{\pi}{3} - \frac{\pi}{2}\right)} = -\frac{108x^3}{\pi^3} + \frac{72x^2}{\pi^2} - \frac{9x}{\pi},$$

$$r_3(x) = \frac{x\left(x - \frac{\pi}{6}\right)\left(x - \frac{\pi}{3}\right)}{\frac{\pi}{2}\left(\frac{\pi}{2} - \frac{\pi}{6}\right)\left(\frac{\pi}{2} - \frac{\pi}{3}\right)} = \frac{36x^3}{\pi^3} - \frac{18x^2}{\pi^2} + \frac{2x}{\pi},$$

and

$$q_3(x) = 1\left(\frac{-36x^3}{\pi^3} + \frac{36x^2}{\pi^2} - \frac{11x}{\pi} + 1\right) + \frac{\sqrt{3}}{2}\left(\frac{108x^3}{\pi^3} - \frac{90x^2}{\pi^2} + \frac{18x}{\pi}\right)$$
$$+ \frac{1}{2}\left(-\frac{108x^3}{\pi^3} + \frac{72x^2}{\pi^2} - \frac{9x}{\pi}\right)$$
$$= \left(\frac{54\sqrt{3} - 90}{\pi^3}\right)x^3 + \left(\frac{72 - 45\sqrt{3}}{\pi^2}\right)x^2 + \left(\frac{9\sqrt{3} - (31/2)}{\pi}\right)x + 1$$

or

$$q_3(x) \approx 0.114x^3 - 0.602x^2 + 0.028x + 1, \qquad 0 \le x \le \frac{\pi}{2}.$$

Since $\Delta x = \pi/6$ and $M = 1$ here, we can use formula (10) to obtain the error bound

$$|\cos x - q_3(x)| \le \frac{1}{24} \cdot \frac{9}{16}\left(\frac{\pi}{6}\right)^4 \cdot 1 \approx 0.0018.$$

We now see that the added work of using cubic interpolation instead of quadratic interpolation has decreased the maximum error by a factor of 17.

To actually test the accuracy of the three interpolating polynomials we have obtained, we calculate a typical value, say $\cos\frac{1}{2}$. We then obtain

$$q_1\left(\frac{1}{2}\right) = -\frac{0.637}{2} + 1 = 0.6815,$$

$$q_2\left(\frac{1}{2}\right) = -\frac{0.336}{4} - \frac{0.109}{2} + 1 = 0.8615$$

and

$$q_3\left(\frac{1}{2}\right) = \frac{0.114}{8} - \frac{0.602}{4} + \frac{0.028}{2} + 1 \approx 0.8778.$$

Since $\cos\frac{1}{2} \approx \cos 28.6° \approx .8776$, we obtain the actual errors

$$|\cos\tfrac{1}{2} - q_1(\tfrac{1}{2})| = 0.1961, \quad |\cos\tfrac{1}{2} - q_2(\tfrac{1}{2})| = 0.0161,$$

and

$$|\cos\tfrac{1}{2} - q_3(\tfrac{1}{2})| = 0.0002.$$

The three interpolating polynomials are sketched in Figure 5.

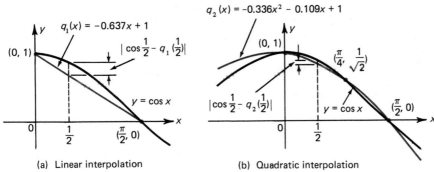

(a) Linear interpolation   (b) Quadratic interpolation

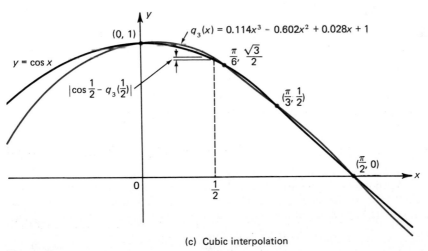

(c) Cubic interpolation

Figure 5

PROBLEMS 13.4

In Problems 1–10 find the Lagrange polynomial which interpolates the given function at the given points. Graph the function and its interpolating polynomial over the interval of interpolation.

1. $f(x) = \sin x$; $x = 0$; $x = \pi/2$

2. $f(x) = \ln x$; $x = 1$; $x = 3$

3. $f(x) = \sqrt{1 + x}$; $x = 0$; $x = 3$; $x = 8$ 4. $f(x) = 1/x^2$; $x = 1$; $x = 2$; $x = 3$
5. $f(x) = e^x$; $x = -1$; $x = 0$; $x = 1$
6. $f(x) = \sin x$; $x = 0$; $x = \pi/6$; $x = \pi/3$; $x = \pi/2$
7. $f(x) = \tan x$; $x = 0$; $x = \pi/6$; $x = \pi/4$ 8. $f(x) = x^4$; $x = 0$; $x = 1$
9. $f(x) = x^4$; $x = 0$; $x = \frac{1}{2}$; $x = 1$
10. $f(x) = x^4$; $x = 0$; $x = \frac{1}{3}$; $x = \frac{2}{3}$; $x = 1$.

☐11. Given that $\ln 2 = 0.693147$ and that $\ln 3 = 1.09861$, use linear interpolation to estimate $\ln 2.3$. What is the maximum error of your estimate?

☐12. In addition to the information in Problem 11, assume that $\ln 2.5 = 0.916291$. Estimate $\ln 2.3$ using quadratic interpolation. What is the maximum error of your estimate?

☐13. Given that $\ln \frac{7}{3} = 0.847298$ and $\ln \frac{8}{3} = 0.980829$, and using the information in Problem 11, find the cubic polynomial which interpolates $\ln x$ at the numbers 2, $2\frac{1}{3}$, $2\frac{2}{3}$, and 3. Use this to estimate $\ln 2.3$. What is the maximum error now?

14. Show that the maximum value of $(x - x_0)(x - x_1)(x - x_2)(x - x_3)$ occurs at $x = (x_1 + x_2)/2$ and is equal to $\frac{9}{16}(\Delta x)^4$ assuming that $x_1 = x_0 + \Delta x$, $x_2 = x_0 + 2\Delta x$ and $x_3 = x_0 + 3\Delta x$.

*15. *Quartic interpolation.* Consider the problem of interpolating the function f at the five equally spaced values $x_k = x_0 + k\Delta x$, $k = 0, 1, 2, 3, 4$.
(a) Find the maximum value of $|Q(x)| = |(x - x_0)(x - x_1)(x - x_2)(x - x_3)$
$(x - x_4)|$ over the interval $[x_0, x_4]$.
(b) If $|f^{(v)}(x)| \leq M$ on $[x_0, x_4]$, find a bound on the interpolation error $|f(x) - q(x)|$ on $[x_0, x_4]$ where $P(x)$ is the fourth-degree Lagrange interpolating polynomial.

☐16. Given that $\sqrt{\frac{5}{4}} = 1.11803$, $\sqrt{\frac{3}{2}} = 1.22474$, $\sqrt{\frac{7}{4}} = 1.32288$, and $\sqrt{2} = 1.41421$, find the quartic (fourth-degree) polynomial that interpolates $f(x) = \sqrt{x}$ over the interval $[1, 2]$ at the five equally spaced points $1, \frac{5}{4}, \frac{3}{2}, \frac{7}{4}, 2$.

17. Use the polynomial obtained in Problem 16 to estimate $\sqrt{1.2}$ and $\sqrt{1.7}$. What is an upper bound on the error in each of these estimates?

*18. Let $q(x)$ be the polynomial that interpolates $f(x)$ at the $n + 1$ points $x_0, x_1, \ldots, x_n$. Show that if $x \neq x_k$, $q(x)$ can be written

$$q(x) = Q(x)\left[\frac{f(x_0)}{(x - x_0)Q'(x_0)} + \frac{f(x_1)}{(x - x_1)Q'(x_1)} + \cdots + \frac{f(x_n)}{(x - x_n)Q'(x_n)}\right].$$

19. Use the result of Problem 18 and L'Hôpital's rule to show that for each k, $\lim_{x \to x_k} q(x) = f(x_k)$.

20. Show that if f is a polynomial of degree n, then the Lagrange polynomial which interpolates f at $n + 1$ points is f itself. This proves that a polynomial of degree n is completely determined if we specify its values at $n + 1$ points.

21. Find the interpolating polynomials which interpolate $f(x) = \sin x$ at 2, 3, and 4 equally spaced points in the interval $[0; \pi/2]$. Calculate an error bound for each of these polynomials. Then find actual errors by using the polynomials to estimate $\sin \frac{1}{2}$ and $\sin 1$.

Review Exercises for Chapter Thirteen

In Exercises 1–8 find the Taylor polynomial of the given degree at the number a.

1. $f(x) = e^x$; $a = 0$; $n = 3$ 2. $f(x) = \ln x$; $a = 1$; $n = 4$
3. $f(x) = \sin x$; $a = \pi/6$; $n = 3$ 4. $f(x) = \cos x$; $a = \pi/2$; $n = 5$

5. $f(x) = \cot x$; $a = \pi/2$; $n = 4$ **6.** $f(x) = \sinh x$; $a = 0$; $n = 3$

7. $f(x) = x^3 - x^2 + 2x + 3$; $a = 0$; $n = 8$ **8.** $f(x) = e^{-x^2}$; $a = 0$; $n = 5$.

In Exercises 9–13 find a bound for $|R_n(x)|$ for x in the given interval.

9. $f(x) = \cos x$; $a = \pi/6$; $n = 5$; $x \in [0, \pi/2]$

10. $f(x) = \sqrt[9]{x}$; $a = 1$; $n = 4$; $x \in [\frac{7}{8}, \frac{9}{8}]$

11. $f(x) = e^x$; $a = 0$; $n = 6$; $x \in [-\ln e, \ln e]$

12. $f(x) = \cot x$; $a = \pi/2$; $n = 2$; $x \in [\pi/4, 3\pi/4]$

13. $f(x) = e^{-x^2}$; $a = 0$; $n = 4$; $x \in [-1, 1]$.

In Exercises 14–20 use a Taylor polynomial to estimate the given number with the given degree of accuracy.

14. $\cos[(\pi/3) + 0.1]$; error < 0.001

15. $\cos 43°$; error < 0.001

16. $\cot[(\pi/4) + 0.1]$; error < 0.001

*17. $\ln 2$; error < 0.0001 [*Hint:* Look at $\ln[(1 + x)/(1 - x)]$]

18. e^2; error < 0.0001

19. $\tan^{-1} 0.3$; error < 0.0001

20. $\ln \sin[(\pi/2) + 0.2]$; error < 0.01.

21. Use a Taylor polynomial of degree 4 to estimate $\int_0^{1/2} \cos x^2 \, dx$. What is the maximum error of your estimate?

22. Use a Taylor polynomial of degree 4 to estimate $\int_0^1 e^{-x^3} \, dx$. What is the maximum error of your estimate?

In Exercises 23–28 find the Lagrange polynomial which interpolates the given function at the given points. Graph the function and its interpolating polynomial over the interval of interpolation.

23. $f(x) = \sin x$; $x = 0$; $x = \pi/4$; $x = \pi/2$

24. $f(x) = e^x$; $x = 0$; $x = 1$; $x = 2$

25. $f(x) = 1/(1 + x^2)$; $x = 0$; $x = 1$; $x = 2$

26. $f(x) = x^5$; $x = 0$; $x = 1$; $x = -1$

27. $f(x) = x^5$; $x = 0$; $x = 1$; $x = 2$; $x = 3$

28. $f(x) = x^5$; $x = \frac{1}{2}$; $x = 2.7$; $x = 5.3$; $x = 7.8$; $x = 10.4$; $x = 17.6$.

29. Given that $\sqrt[3]{2} = 1.259921$, use linear interpolation to estimate $\sqrt[3]{1.7}$. What is the maximum error of your estimate?

30. In Exercise 29, use the fact that $\sqrt[3]{1.5} = 1.144714$ together with quadratic interpolation to estimate $\sqrt[3]{1.7}$. What is the maximum error of your estimate?

FOURTEEN

SEQUENCES AND SERIES

In Chapter 1 (Section 1.10) we encountered the sum of the geometric progression

$$S_n = 1 + r + r^2 + \cdots + r^n$$

and showed that for $r \neq 1$, this sum is equal to $(1 - r^{n+1})/(1 - r)$. In Chapter 2 (Section 2.6) we showed that if $|r| < 1$, then the "infinite" sum

$$S = 1 + r + r^2 + r^3 + \cdots$$

is equal to $1/(1 - r)$ and used this fact to explain Zeno's paradox.

There are many other kinds of infinite sums that can and do arise in applications. For example, we shall see that many continuous functions can be written as "infinite" Taylor polynomials.

An infinite sum of the type we have been discussing is called an *infinite series*. In this chapter we will discuss the theory of infinite series and will show how it can give us a great deal of information about a wide variety of functions.

14.1 Sequences of Real Numbers

Consider the infinite set of numbers

$$A = \left\{ \frac{1}{2}, \frac{1}{4}, \frac{1}{8}, \ldots, \frac{1}{2^n}, \ldots \right\}. \tag{1}$$

That is, the set A consists of all numbers of the form $1/2^n$ for each positive integer n. There is another way to describe this set. We define the function f by the rule $f(n) = 1/2^n$, where the domain of f is the set of positive integers. Then the set A is precisely the set of values taken by the function f.

In general, we have the following definition:

DEFINITION 1. A *sequence* of real numbers is a function whose domain is the set of positive integers. The values taken by the function are called *terms* of the sequence.

> *Notation.* We will often denote the terms of a sequence by a_n. Thus if the function given in Definition 1 is f, then $a_n = f(n)$. With this notation, we can denote the set of values taken by the sequence by $\{a_n\}$.

EXAMPLE 1. The following are sequences of real numbers:

(a) $\{a_n\} = \left\{ \dfrac{1}{4^n} \right\}$ (b) $\{a_n\} = \{ \sqrt{n} \}$ (c) $\{a_n\} = \left\{ \dfrac{1}{n!} \right\}$

(d) $\{a_n\} = \{\sin n\}$ (e) $\{a_n\} = \left\{\dfrac{e^n}{n!}\right\}$ (f) $\{a_n\} = \left\{\dfrac{n^3 + 1}{n^2 + 2}\right\}.$

We sometimes denote a sequence by writing out the values $\{a_1, a_2, a_3, \ldots\}$.

EXAMPLE 2.   We write out the values of the sequences in Example 1:

(a) $\left\{\dfrac{1}{4}, \dfrac{1}{16}, \dfrac{1}{64}, \dfrac{1}{256}, \ldots, \dfrac{1}{4^n}, \ldots\right\}.$

(b) $\{1, \sqrt{2}, \sqrt{3}, \sqrt{4}, \ldots, \sqrt{n}, \ldots\}.$

(c) $\left\{1, \dfrac{1}{2}, \dfrac{1}{6}, \dfrac{1}{24}, \ldots, \dfrac{1}{n!}, \ldots\right\}.$

(d) $\{\sin 1, \sin 2, \sin 3, \sin 4, \ldots, \sin n, \ldots\}.$

(e) $\left\{e, \dfrac{e^2}{2}, \dfrac{e^3}{6}, \dfrac{e^4}{24}, \ldots, \dfrac{e^n}{n!}, \ldots\right\}.$

(f) $\left\{\dfrac{2}{3}, \dfrac{9}{6}, \dfrac{28}{11}, \dfrac{65}{18}, \ldots, \dfrac{n^3 + 1}{n^2 + 2}, \ldots\right\}.$

EXAMPLE 3.   Find the general term a_n of the sequence

$$\{-1, 1, -1, 1, -1, 1, -1, \ldots\}.$$

SOLUTION.   We see that $a_1 = -1$, $a_2 = 1$, $a_3 = -1$, $a_4 = 1, \ldots.$ Hence

$$a_n = \begin{cases} -1, & \text{if } n \text{ is odd} \\ 1, & \text{if } n \text{ is even} \end{cases}.$$

A more concise way to write this is

$$a_n = (-1)^n.$$

In Section 2.6, we indicated that if $|r| > 1$, then as $n \to \infty$, $1/r^n \to 0$. For the remainder of this section, we will be concerned with calculating the limit of a sequence as $n \to \infty$. Since a sequence is a special type of function, our definition of the limit of a sequence is going to be very similar to the definition of $\lim_{x \to \infty} f(x)$.

DEFINITION 2.   A sequence $\{a_n\}$† has the limit L if for every $\epsilon > 0$ there exists an integer $N > 0$ such that if $n \geq N$, then $|a_n - L| \leq \epsilon$. We write

$$\lim_{n \to \infty} a_n = L. \tag{2}$$

Intuitively, this definition states that $a_n \to L$ if as n increases without bound, a_n gets arbitrarily close to L (see Definitions 2.5.3 and 10.1.3).

DEFINITION 3.   If the limit in (2) exists and if L is finite, we say that the sequence *converges* or is *convergent*. Otherwise we say that the sequence *diverges* or is *divergent*.

---

†To be precise $\{a_n\}$ denotes the set of values taken by the sequence. There is a difference between the sequence, which is a function f, and the values $a_n = f(n)$ taken by this function. However, because it is more convenient to write down the values the sequence takes, we will, from now on, use the notation $\{a_n\}$ to denote a sequence.

EXAMPLE 4. The sequence $\{1/2^n\}$ is convergent since, as we have seen, $\lim_{n\to\infty} 1/2^n = 0$.

EXAMPLE 5. The sequence $\{r^n\}$ is divergent for $r > 1$ since $\lim_{n\to\infty} r^n = \infty$ if $r > 1$.

EXAMPLE 6. The sequence $\{(-1)^n\}$ is divergent since the values a_n alternate between -1 and $+1$ but do not stay close to any fixed number as n becomes large.

Since we have a large body of theory and experience behind us in the calculation of ordinary limits, we would like to make use of that experience to calculate limits of sequences. The following theorem, whose proof is left as a problem (see Problem 34), is extremely useful.

THEOREM 1. If $\lim_{x\to\infty} f(x) = L$, and if f is defined for every positive integer, then the limit of the sequence $\{a_n\} = \{f(n)\}$ is also equal to L. That is, $\lim_{x\to\infty} f(x) = \lim_{n\to\infty} a_n = L$.

EXAMPLE 7. Calculate $\lim_{n\to\infty} 1/n^2$.

SOLUTION. Since $\lim_{x\to\infty} 1/x^2 = 0$, we have $\lim_{n\to\infty} 1/n^2 = 0$ (by Theorem 1).

EXAMPLE 8. Does the sequence $\{e^n/n\}$ converge or diverge?

SOLUTION. Since $\lim_{x\to\infty} e^x/x = \lim_{x\to\infty} e^x/1$ (by L'Hôpital's rule) $= \infty$, we find that the sequence diverges.

> **Remark.** It should be emphasized that Theorem 1 does *not* say that if $\lim_{x\to\infty} f(x)$ does not exist, then $\{a_n\} = \{f(n)\}$ diverges. For example, let
>
> $$f(x) = \begin{cases} x^2, & x \neq n, \text{ a positive integer} \\ 1/n, & x = n, \text{ a positive integer.} \end{cases}$$
>
> Then $\lim_{x\to\infty} f(x)$ does not exist but $\lim_{n\to\infty} f(n) = \lim_{n\to\infty} 1/n = 0$. In Example 8, we can conclude that $\{e^n/n\}$ diverges only from the fact that $\lim_{x\to\infty}(e^x/x) = \infty$ also implies that, by continuity, $\lim_{n\to\infty}(e^n/n) = \infty$. This fact does not follow from Theorem 1.

EXAMPLE 9. Let $\{a_n\} = \{[1 + (1/n)]^n\}$. Does this sequence converge or diverge?

SOLUTION. Since $\lim_{x\to\infty}[1 + (1/x)]^x = e$, we see that a_n converges to the limit e.

EXAMPLE 10. Does the sequence $\{3^n/n^3\}$ converge or diverge?

SOLUTION. By L'Hôpital's rule

$$\lim_{x\to\infty} \frac{3^x}{x^3} = \lim_{x\to\infty} \frac{3^x \ln 3}{3x^2} = \lim_{x\to\infty} \frac{3^x(\ln 3)^2}{6x} = \lim_{x\to\infty} \frac{3^x(\ln 3)^3}{6}$$

$$= \frac{(\ln 3)^3}{6} \lim_{x\to\infty} 3^x = \infty$$

Thus it is clear that $\lim_{n\to\infty}(3^n/n^3) = \infty$ so that the sequence $\{a_n\}$ diverges.

EXAMPLE 11.   Let $p(x) = c_0 + c_1 x + \cdots + c_m x^m$ and $q(x) = d_0 + d_1 x + \cdots + d_r x^r$. In Problem 2.5.33 we showed that if the rational function $r(x) = p(x)/q(x)$, then

$$\lim_{x \to \infty} \frac{p(x)}{q(x)} = \begin{cases} 0, & \text{if } m < r \\ \pm\infty, & \text{if } m > r. \\ c_m/d_r, & \text{if } m = r \end{cases}$$

Thus the sequence $\{p(n)/q(n)\}$ converges to 0 if $m < r$, converges to c_m/d_r if $m = r$, and diverges if $m > r$.

EXAMPLE 12. Does the sequence $\{(5n^3 + 2n^2 + 1)/(2n^3 + 3n + 4)\}$ converge or diverge?

SOLUTION. Here $m = r = 3$ so that by the result of Example 11, the sequence converges to $c_3/d_3 = \frac{5}{2}$.

EXAMPLE 13.   Does the sequence $\{n^{1/n}\}$ converge or diverge?

SOLUTION. Since $\lim_{x \to \infty} x^{1/x} = 1$ (see Example 12.3.6), the sequence converges to 1.

EXAMPLE 14.   Let α be any real number. Determine the convergence or divergence of the sequence $\{[1 + (\alpha/n)]^n\}$.

SOLUTION. Let $y = [1 + (\alpha/x)]^x$. Then

$$\ln y = x \ln\left(1 + \frac{\alpha}{x}\right) = \frac{\ln[1 + (\alpha/x)]}{1/x}.$$

Since the numerator and denominator of this expression approach zero as $x \to \infty$, we apply L'Hôpital's rule to find that

$$\lim_{x \to \infty} \frac{\ln\left(1 + \dfrac{\alpha}{x}\right)}{1/x} = \lim_{x \to \infty} \frac{\dfrac{1}{1 + (\alpha/x)} \cdot \dfrac{-\alpha}{x^2}}{-1/x^2} = \alpha$$

so that $\ln y \to \alpha$ and $y \to e^\alpha$. Therefore the sequence $\{[1 + (\alpha/n)]^n\}$ converges to e^α.

EXAMPLE 15.   Determine the convergence or divergence of the sequence $\{1 + (-1)^n\}$.

SOLUTION. We see that $a_1 = 1 + (-1)^1 = 1 - 1 = 0, a_2 = 1 + (-1)^2 = 1 + 1 = 2$, $a_3 = 0, a_4 = 2$, and so on. In fact, we have

$$a_n = \begin{cases} 0, & \text{if } n \text{ is odd} \\ 2, & \text{if } n \text{ is even.} \end{cases}$$

Thus a_n does not stay close to any one finite number as $n \to \infty$ and the sequence is divergent.

EXAMPLE 16.   Determine the convergence or divergence of the sequence $\{(\ln n)/n\}$.

SOLUTION. $\lim_{x\to\infty}[(\ln x)/x] = \lim_{x\to\infty}[(1/x)/1] = 0$ so that the sequence converges to 0.

EXAMPLE 17. Determine the convergence or divergence of the sequence $\{(\sin \alpha n)/n^{\beta}\}$, where α is a real number and $\beta > 0$.

SOLUTION. Since $-1 \leq \sin \alpha x \leq 1$, we see that

$$-\frac{1}{x^{\beta}} \leq \frac{\sin \alpha x}{x^{\beta}} \leq \frac{1}{x^{\beta}} \qquad \text{for any } x > 0.$$

But $\pm\lim_{x\to\infty} 1/x^{\beta} = 0$ and so, by the squeezing theorem (Theorem 2.5.1), $\lim_{x\to\infty}[(\sin \alpha x)/x^{\beta}] = 0$. Therefore the sequence $\{(\sin \alpha n)/n^{\beta}\}$ converges to 0.

As in Example 17, the squeezing theorem can often be used to calculate the limit of a sequence.

PROBLEMS 14.1

In Problems 1–9 find the first five terms of the given sequence.

1. $\left\{\dfrac{1}{3^n}\right\}$ 2. $\left\{\dfrac{n+1}{n}\right\}$ 3. $\left\{1 - \dfrac{1}{4^n}\right\}$

4. $\{\sqrt[3]{n}\}$ 5. $\{e^{1/n}\}$ 6. $\{n \cos n\}$

7. $\{\sin n\pi\}$ 8. $\{\cos n\pi\}$ 9. $\left\{\sin \dfrac{n\pi}{2}\right\}.$

In Problems 10–27 determine whether the given sequence is convergent or divergent. If it is convergent, find its limit.

10. $\left\{\dfrac{3}{n}\right\}$ 11. $\left\{\dfrac{1}{\sqrt{n}}\right\}$ 12. $\left\{\dfrac{n+1}{n^{5/2}}\right\}$

13. $\{\sin n\}$ 14. $\{\sin n\pi\}$ 15. $\left\{\cos\left(n + \dfrac{\pi}{2}\right)\right\}$

16. $\left\{\dfrac{n^5 + 3n^2 + 1}{n^6 + 4n}\right\}$ 17. $\left\{\dfrac{4n^5 - 3}{7n^5 + n^2 + 2}\right\}$ 18. $\left\{\left(1 + \dfrac{4}{n}\right)^n\right\}$

19. $\left\{\left(1 + \dfrac{1}{4n}\right)^n\right\}$ 20. $\left\{\dfrac{\sqrt{n}}{\ln n}\right\}$

21. $\{\sqrt{n+3} - n\}$ [*Hint:* multiply and divide by $\sqrt{n+3} + n$.]

22. $\left\{\dfrac{2^n}{n!}\right\}$ 23. $\left\{\dfrac{\alpha^n}{n!}\right\}$ (α real)   24. $\left\{\dfrac{4}{\sqrt{n^2 + 3} - n}\right\}$

25. $\left\{\dfrac{(-1)^n n^3}{n^3 + 1}\right\}$ 26. $\{(-1)^n \cos n\pi\}.$ 27. $\left\{\dfrac{(-1)^n}{\sqrt{n}}\right\}.$

In Problems 28–33 find the general term a_n of the given sequence.

28. $\{1, -2, 3, -4, 5, -6, \dots\}$ 29. $\{1, 2\cdot 5, 3\cdot 5^2, 4\cdot 5^3, 5\cdot 5^4, \dots\}$

30. $\{\frac{1}{2}, \frac{2}{3}, \frac{3}{4}, \frac{4}{5}, \frac{5}{6}, \dots\}$ 31. $\{\frac{1}{2}, \frac{3}{4}, \frac{7}{8}, \frac{15}{16}, \frac{31}{32}, \dots\}$

*32. $\{\frac{1}{3}, \frac{2}{5}, \frac{3}{7}, \frac{4}{9}, \frac{5}{11}, \dots\}$ 33. $\{1, -\frac{1}{3}, \frac{1}{9}, -\frac{1}{27}, \dots\}.$

*34. Prove Theorem 1. [*Hint:* Use Definition 10.1.3.]

35. Show that if $\{a_n\}$ and $\{b_n\}$ are two sequences such that $|a_n| \leq |b_n|$ for each n, then if $|b_n|$ converges to 0, $|a_n|$ also converges to 0. [*Hint:* Use the squeezing theorem.]

36. Use the result of Problem 35 to show that the sequence $\{(a \sin bn + c \cos dn)/n^{p^2}\}$ converges to 0 for any real numbers a, b, c, d, and p.

37. Prove that if $|r| < 1$, then the sequence $\{nr^n\}$ converges to 0.

*38. Show that if $\{a_n\}$ converges, then the $\lim_{n\to\infty} a_n$ is unique. [*Hint:* Assume that $\lim_{n\to\infty} a_n = L$, $\lim_{n\to\infty} a_n = M$ and $L \neq M$. Then choose $\epsilon = \frac{1}{2}|L - M|$ to show that Definition 2 is violated.]

14.2 Bounded and Monotonic Sequences

There are certain kinds of sequences which have special properties worthy of mention.

DEFINITION 1. (i) The sequence $\{a_n\}$ is *bounded above* if there is a number M_1 such that

$$a_n \leq M_1 \tag{1}$$

for every positive integer n.

(ii) It is *bounded below* if there is a number M_2 such that

$$M_2 \leq a_n \tag{2}$$

for every positive integer n.

(iii) It is *bounded* if there is a number $M > 0$ such that

$$|a_n| \leq M$$

for every positive integer n.

The numbers M_1, M_2 and M are called, respectively, an *upper bound*, a *lower bound*, and a *bound* for $\{a_n\}$.

(iv) If the sequence is not bounded, it is called *unbounded*.

Remark. If $\{a_n\}$ is bounded above and below, then it is bounded. Simply set $M = \max\ \{|M_1|, |M_2|\}$.

EXAMPLE 1. The sequence $\{\sin n\}$ has the upper bound of 1, the lower bound of -1, and the bound of 1 since $-1 \leq \sin n \leq 1$ for every n.

EXAMPLE 2. The sequence $\{(-1)^n\}$ has the upper bound 1, the lower bound -1 and the bound 1.

EXAMPLE 3. The sequence $\{2^n\}$ is bounded below by 2 but has no upper bound, and so is unbounded.

EXAMPLE 4. The sequence $\{(-1)^n 2^n\}$ is bounded neither below nor above.

It turns out that

> Every convergent sequence is bounded.

THEOREM 1. If the sequence $\{a_n\}$ is convergent, then it is bounded.

PROOF. Before giving the technical details, we remark that the idea behind the proof is easy. For if $\lim_{n \to \infty} a_n = L$, then a_n is close to the finite number L if n is large. Thus, for example, $|a_n| \leq 2|L|$ if N is large enough. Since a_n is a real number for every n, the first few terms of the sequence are bounded and these two facts give us a bound for the entire sequence.

Now to the details: Let $\epsilon = 1$. Then there is an $N > 0$ such that (according to Definition 14.1.2)

$$|a_n - L| \leq 1 \qquad \text{if} \quad n \geq N. \tag{3}$$

Let

$$K = \max\{|a_1|, |a_2|, \ldots, |a_N|\}. \tag{4}$$

Since each a_n is finite, K, being the maximum of a finite number of terms, is also finite. Now let

$$M = \max\{|L| + 1, K\}. \tag{5}$$

It follows from (4) that if $n \leq N$, then $|a_n| \leq K$. If $n \geq N$, then, from (3), $|a_n| \leq |L| + 1$ so, in either case, $|a_n| \leq M$ and the theorem is proved.

Sometimes it is difficult to find a bound for a convergent sequence.

□**EXAMPLE 5.** Find an M such that $5^n/n! \leq M$.

SOLUTION. We know from Theorem 13.2.2 than $\lim_{n \to \infty} x^n/n! = 0$ for every real number x. In particular $\{5^n/n!\}$ is convergent and therefore must be bounded. Perhaps the easiest way to find the bound is to tabulate a few values as in Table 1.

TABLE 1

n	$\dfrac{5^n}{n!}$	n	$\dfrac{5^n}{n!}$
1	5	7	15.5
2	12.5	8	9.69
3	20.83	9	5.38
4	26.04	10	2.69
5	26.04	20	0.000039
6	21.7		

It is clear from the table that the maximum value of a_n occurs at $n = 4$ or $n = 5$ and is equal to 26.04. Of course, any number larger than 26.04 is also a bound for the sequence.

Theorem 1 is useful in another way. Since every convergent sequence is bounded, it follows that

> Every unbounded sequence is divergent.

EXAMPLE 6. The following sequences are divergent since they are clearly unbounded:

(a) $\{\ln \ln n\}$ (b) $\{n \sin n\}$ (c) $\{(-\sqrt{2})^n\}$.

The converse of Theorem 1 is *not* true. That is, it is not true that every bounded sequence is convergent. For example, the sequences $\{(-1)^n\}$ and $\{\sin n\}$ are both bounded *and* divergent. Since boundedness alone does not ensure convergence, we need some other property. We investigate this now.

DEFINITION 2.   (i)   The sequence $\{a_n\}$ is *increasing* if $a_n \le a_{n+1}$ for every $n \ge 1$.
　　　(ii)   The sequence $\{a_n\}$ is *decreasing* if $a_n \ge a_{n+1}$ for every $n \ge 1$.
　　　(iii)   The sequence $\{a_n\}$ is *monotonic* if it is either increasing or decreasing.

DEFINITION 3.   (i)   The sequence $\{a_n\}$ is *strictly increasing* if $a_n < a_{n+1}$ for every $n \ge 1$.
　　　(ii)   The sequence $\{a_n\}$ is *strictly decreasing* if $a_n > a_{n+1}$ for every $n \ge 1$.
　　　(iii)   The sequence $\{a_n\}$ is *strictly monotonic* if it is either strictly increasing or strictly decreasing.

EXAMPLE 7. The sequence $\{1/2^n\}$ is strictly decreasing since $1/2^n > 1/2^{n+1}$ for every n.

EXAMPLE 8. Determine whether the sequence $\{2n/(3n + 2)\}$ is increasing, decreasing or not monotonic.

SOLUTION. If we write out the first few terms of the sequence, we find that $\{2n/(3n + 2)\} = \{\frac{2}{5}, \frac{4}{8}, \frac{6}{11}, \frac{8}{14}, \frac{10}{17}, \frac{12}{20}, \ldots\}$. Since these terms are strictly increasing, we suspect that $\{2n/(3n + 2)\}$ is an increasing sequence. To check this, we try to verify that $a_n < a_{n+1}$. But

$$a_{n+1} = \frac{2(n + 1)}{3(n + 1) + 2} = \frac{2n + 2}{3n + 5}.$$

Then $a_n < a_{n+1}$ implies that

$$\frac{2n}{3n + 2} < \frac{2n + 2}{3n + 5}.$$

Multiplying both sides of this inequality by $(3n + 2)(3n + 5)$ we obtain

$$(2n)(3n + 5) < (2n + 2)(3n + 2) \quad \text{or} \quad 6n^2 + 10n < 6n^2 + 10n + 4.$$

Since this last inequality is obviously true for all $n \ge 1$, we can reverse our steps to conclude that $a_n < a_{n+1}$ and the sequence is strictly increasing.

EXAMPLE 9. Determine whether the sequence $\{-[n/(n + 1)]\}$ is increasing, decreasing or not monotonic.

SOLUTION. The first five terms are $-\frac{1}{2}$, $-\frac{2}{3}$, $-\frac{3}{4}$, $-\frac{4}{5}$, and $-\frac{5}{6}$. It appears that the sequence is strictly decreasing. To verify this we note that

$$a_{n+1} = -\left[\frac{n + 1}{(n + 1) + 1}\right] = -\left(\frac{n + 1}{n + 2}\right).$$

Then

$$-\frac{n}{n+1} > -\left(\frac{n+1}{n+2}\right)$$

implies that

$$-n(n+2) > -(n+1)^2$$

or, multiplying by -1,

$$n(n+2) < (n+1)^2 \qquad \text{and} \qquad n^2 + 2n < n^2 + 2n + 1$$

which is indeed true for all $n \geq 1$ and the result follows as before (that is, by reversing the steps).

EXAMPLE 10. The sequence $\{[n/4]\}$ is increasing, but not strictly increasing. Here $[x]$ is the "greatest integer" function (see Example 2.4.7). The first ten terms are 0, 0, 0, 1, 1, 1, 1, 2, 2, 2, 2, 3. For example, $a_9 = [\frac{9}{4}] = 2$.

EXAMPLE 11. The sequence $\{(-1)^n\}$ is not monotonic since successive terms oscillate between $+1$ and -1.

The importance of monotonic functions is given in the theorem below.

THEOREM 2. A bounded monotonic sequence is convergent.

PROOF. We will prove this theorem for the case in which the sequence $\{a_n\}$ is increasing. The proof of the other case is similar. Since $\{a_n\}$ is bounded, there is a number M such that $a_n \leq M$ for every n. Let L be the smallest such upper bound.† Now let $\epsilon > 0$ be given. Then there is a number $N > 0$ such that $a_N > L - \epsilon$. If this were not true, then we would have $a_n \leq L - \epsilon$ for all $n \geq 1$. Then $L - \epsilon$ would be an upper bound for $\{a_n\}$ and, since $L - \epsilon < L$, this would contradict the choice of L as the smallest such upper bound. Since $\{a_n\}$ is increasing, we have, for $n \geq N$

$$L - \epsilon < a_N \leq a_n \leq L < L + \epsilon. \tag{6}$$

But the inequalities (6) imply that $|a_n - L| \leq \epsilon$ for $n \geq N$ which proves, according to the definition of convergence (Definition 14.1.2) that $\lim_{n \to \infty} a_n = L$.

> **Remark.** We have actually proved a stronger result. Namely, that if the sequence $\{a_n\}$ is bounded above and increasing, then it converges to its least upper bound. Similarly if $\{a_n\}$ is bounded below and decreasing, then it converges to its greatest lower bound.

EXAMPLE 12. In Example 8 we saw that the sequence $\{2n/(3n+2)\}$ is strictly increasing. Also, since $2n/(3n+2) < 3n/(3n+2) < 3n/3n = 1$, we see that $\{a_n\}$ is also bounded so that by Theorem 2, $\{a_n\}$ is convergent. We easily find that $\lim_{n \to \infty} 2n/(3n+2) = \frac{2}{3}$.

EXAMPLE 13. Show that the sequence $\{(1 + 2^n)^{1/n}\}$ converges.

---

†The number L is called the *least upper bound* for the sequence $\{a_n\}$. It is an axiom of the real number system that every set of real numbers that is bounded above has a least upper bound and that every set of real numbers that is bounded below has a *greatest lower bound*. This axiom is called the *completeness axiom* and is of paramount importance in theoretical mathematical analysis.

SOLUTION. Since $1 \leq 1 + 2^n \leq 2^n + 2^n = 2 \cdot 2^n$, we find that

$$1 = 1^{1/n} \leq (1 + 2^n)^{1/n} \leq (2 \cdot 2^n)^{1/n} = 2^{1/n} \cdot (2^n)^{1/n} = 2^{1/n} \cdot 2 \leq 4.$$

Thus $1 \leq a_n \leq 4$ and $\{a_n\}$ is bounded. We now show that $a_n > a_{n+1}$. We do this with a string of inequalities:

$$(1 + 2^n)^{(n+1)/n} = (1 + 2^n)^{1/n}(1 + 2^n)^{n/n} = (1 + 2^n)^{1/n}(1 + 2^n)$$
$$> (2^n)^{1/n}(1 + 2^n) = 2(1 + 2^n) = 2 + 2^{n+1} > 1 + 2^{n+1}.$$

Thus $(1 + 2^n)^{(n+1)/n} > 1 + 2^{n+1}$ and, taking the $(n + 1)$st root of each side, we obtain

$$(1 + 2^n)^{1/n} > (1 + 2^{n+1})^{1/(n+1)}.$$

Hence, $a_n > a_{n+1}$ which implies that $\{a_n\}$ is monotonic, and, by Theorem 2, is convergent.

PROBLEMS 14.2

In Problems 1–12 determine whether the given sequence is bounded or unbounded. If it is bounded, find the smallest bound for $|a_n|$.

1. $\left\{\dfrac{1}{n + 1}\right\}$.

2. $\{\sin n\pi\}$

3. $\{\cos n\pi\}$

4. $\{\sqrt{n} \sin n\}$

5. $\left\{\dfrac{2^n}{1 + 2^n}\right\}$

6. $\left\{\dfrac{2^n + 1}{2^n}\right\}$

7. $\left\{\dfrac{1}{n!}\right\}$

□8. $\left\{\dfrac{3^n}{n!}\right\}$

9. $\left\{\dfrac{n^2}{n!}\right\}$

10. $\left\{\dfrac{2n}{2^n}\right\}$

11. $\left\{\dfrac{\ln n}{n}\right\}$

*12. $\{ne^{-n}\}$.

13. Show that for $n > 2^{10}$, $n^{10}/n! > (n + 1)^{10}/(n + 1)!$ and use this to conclude that $\{n^{10}/n!\}$ is bounded.

In Problems 14–28 determine whether the given sequence is increasing, strictly increasing, decreasing, strictly decreasing, or not monotonic.

14. $\{\sin n\pi\}$

15. $\left\{\dfrac{3^n}{2 + 3^n}\right\}$

16. $\left\{\left[\left(\dfrac{n}{25}\right)^{1/3}\right]\right\}$

17. $\{n + (-1)^n \sqrt{n}\}$

18. $\left\{\dfrac{\sqrt{n + 1}}{n}\right\}$

19. $\left\{\dfrac{n!}{n^n}\right\}$

20. $\left\{\dfrac{n^n}{n!}\right\}$

21. $\left\{\dfrac{2n!}{1 \cdot 3 \cdot 5 \cdot 7 \cdot \ldots \cdot (2n - 1)}\right\}$

*22. $\{n + \cos n\}$

23. $\left\{\dfrac{2^{2n}}{n!}\right\}$

24. $\left\{\dfrac{\sqrt{n} - 1}{n}\right\}$

25. $\left\{\dfrac{n - 1}{n + 1}\right\}$

26. $\left\{\ln\left(\dfrac{3n}{n + 1}\right)\right\}$

27. $\{\ln n - \ln(n + 2)\}$

28. $\left\{\left(1 + \dfrac{3}{n}\right)^{1/n}\right\}$.

*29. Show that the sequence $\{(2^n + 3^n)^{1/n}\}$ is convergent. [*Hint:* First show that $2 < a_n < 6$. Then show that $(2^n + 3^n)^{(n+1)/n} > 2^{n+1} + 3^{n+1}$ and take $(n + 1)$st roots to conclude that $a_n > a_{n+1}$.]

*30. Show that $\{(a^n + b^n)^{1/n}\}$ is convergent for any positive real numbers a and b. [*Hint:* First do Problem 29.]

*31. Show that the sequence $\{n!/n^n\}$ is bounded. [*Hint:* Show that $n!/n^n < (n + 1)!/(n + 1)^n$ for sufficiently large n. Use the fact that $[1 + (1/n)]^n \to e$ as $n \to \infty$.]

32. Prove that the sequence $\{n!/n^n\}$ converges. [*Hint:* Use the result and hint of Problem 31.]

33. Use Theorem 2 to show that $\{\ln n - \ln(n + 4)\}$ converges.

34. Show that the sequence of Problem 21 is convergent.

35. Show that the sequence $\{2 \cdot 5 \cdot 8 \cdot 11 \cdot \cdots \cdot (3n - 1)/3^n n!\}$ is convergent.

14.3 The $\sum$ Notation

In this section we introduce a notation for writing sums which will be very useful in the rest of the chapter. If

$$S_n = a_1 + a_2 + a_3 + \cdots + a_n \tag{1}$$

then we write†

$$S_n = \sum_{k=1}^{n} a_k \tag{2}$$

which is read "S_n is the sum of the terms a_k as k goes from 1 to n in increments of one."

EXAMPLE 1. Calculate $\sum_{k=1}^{4} k$.

SOLUTION. Here $a_k = k$ so that

$$\sum_{k=1}^{4} k = 1 + 2 + 3 + 4 = 10.$$

EXAMPLE 2. Calculate $\sum_{k=1}^{5} k^2$.

SOLUTION. Here $a_k = k^2$ and

$$\sum_{k=1}^{5} k^2 = 1^2 + 2^2 + 3^2 + 4^2 + 5^2 = 55.$$

EXAMPLE 3. Calculate $\sum_{k=0}^{n} r^k$ for $r \neq 1$.

---

†The Greek letter Σ (sigma) was first used to denote a sum by the great Swiss mathematician Leonhard Euler (1707–1783). In this context Σ is called the *summation sign*.

SOLUTION. $\sum_{k=0}^{n} r^n = r^0 + r^1 + r^2 + \cdots + r^n = (1 - r^{n+1})/(1 - r)$ as this is simply the sum of a geometric progression.

EXAMPLE 4.   Write the sum $S_8 = 1 - 2 + 3 - 4 + 5 - 6 + 7 - 8$ using the summation sign.

SOLUTION. Since $1 = (-1)^2$, $-2 = (-1)^3 \cdot 2$, $3 = (-1)^4 \cdot 3, \ldots$, we have

$$S_8 = \sum_{k=1}^{8} (-1)^{k+1} k.$$

EXAMPLE 5.   Write the sum

$$S = a^4 + a^6 + a^8 + \cdots + a^{28}$$

using the summation sign.

SOLUTION. Since $a^4 = a^{2 \cdot 2}$, $a^6 = a^{2 \cdot 3}$, $a^8 = a^{2 \cdot 4}, \ldots$, we have

$$S = \sum_{k=2}^{14} a^{2k}.$$

The summation sign could have been used in the definition of the definite integral. Since

$$\int_a^b f(x)\, dx = \lim_{|P| \to 0} \{ f(x_1^*)\, \Delta x_1 + f(x_2^*)\, \Delta x_2 + \cdots + f(x_n^*)\, \Delta x_n \}$$

(see Definition 4.3.1), we can write

$$\int_a^b f(x)\, dx = \lim_{|P| \to 0} \sum_{k=1}^{n} f(x_k^*)\, \Delta x_k.$$

The Σ sign is used to simplify notation. After a little practice you should become comfortable with its use. We remark that the letter k in the Σ notation can be replaced by any other symbol without changing anything. It is a "dummy" variable. Thus $\sum_{k=1}^{n} f(k) = \sum_{j=1}^{n} f(j) = \sum_{i=1}^{n} f(i)$ just as $\int_a^b f(x)\, dx = \int_a^b f(t)\, dt = \int_a^b f(\mu)\, d\mu$.

PROBLEMS 14.3

In Problems 1–6 evaluate the given sums.

1. $\displaystyle\sum_{k=1}^{4} 2k$

2. $\displaystyle\sum_{i=1}^{3} i^3$

3. $\displaystyle\sum_{k=0}^{6} 1$

4. $\displaystyle\sum_{k=1}^{8} 3^k$

5. $\displaystyle\sum_{i=2}^{5} \frac{i}{i+1}$

6. $\displaystyle\sum_{j=5}^{7} \frac{2j+3}{j-2}$.

In Problems 7–16 write each sum using the $\sum$ notation.

7. $1 + 2 + 4 + 8 + 16$

8. $1 - 3 + 9 - 27 + 81 - 243$

9. $\dfrac{2}{3} + \dfrac{3}{4} + \dfrac{4}{5} + \dfrac{5}{6} + \dfrac{6}{7} + \dfrac{7}{8} + \cdots + \dfrac{n}{n+1}$

10. $1 - \dfrac{1}{2!} + \dfrac{1}{3!} - \dfrac{1}{4!} + \dfrac{1}{5!} - \dfrac{1}{6!} + \dfrac{1}{7!}$

11. $1 + 2^{1/2} + 3^{1/3} + 4^{1/4} + 5^{1/5} + \cdots + n^{1/n}$

12. $1 + x^3 + x^6 + x^9 + x^{12} + x^{15} + x^{18} + x^{21}$

13. $x^5 - x^{10} + x^{15} - x^{20} + x^{25} - x^{30} + x^{35} - x^{40}$

14. $-1 + \dfrac{1}{a} - \dfrac{1}{a^2} + \dfrac{1}{a^3} - \dfrac{1}{a^4} + \dfrac{1}{a^5} - \dfrac{1}{a^6} + \dfrac{1}{a^7} - \dfrac{1}{a^8} + \dfrac{1}{a^9}$

*15. $1 \cdot 3 + 3 \cdot 5 + 5 \cdot 7 + 7 \cdot 9 + 9 \cdot 11 + 11 \cdot 13 + 13 \cdot 15 + 15 \cdot 17$

*16. $2^2 \cdot 4 + 3^2 \cdot 6 + 4^2 \cdot 8 + 5^2 \cdot 10 + 6^2 \cdot 12 + 7^2 \cdot 14.$

14.4 Infinite Series

In our discussion of geometric progressions we obtained a sum of the form $\sum_{k=0}^{n} r^k$. In Section 2.6 we indicated that as $n \to \infty$, this sum approaches the value $1/(1-r)$ if $|r| < 1$. We could write this limit as

$$\lim_{n \to \infty} \sum_{k=0}^{n} r^k = \sum_{k=0}^{\infty} r^k = \frac{1}{1-r}. \tag{1}$$

The above infinite sum is called the *geometric series* and, since this sum is finite, we say that the geometric series *converges* if $|r| < 1$.

Let us look at this from a different point of view. Let S_n denote the sum of the first $n + 1$ terms of the geometric series. Then

$$S_n = 1 + r + r^2 + \cdots + r^n = \frac{1 - r^{n+1}}{1-r}, \qquad r \neq 1.$$

For each n we obtain the number S_n and, therefore, we can define a new sequence $\{S_n\}$ to be the sequence of *partial sums* of the geometric series. If $|r| < 1$, then

$$\lim_{n \to \infty} S_n = \lim_{n \to \infty} \frac{1 - r^{n+1}}{1-r} = \frac{1}{1-r}.$$

That is, the convergence of the geometric series is implied by the convergence of the sequence of partial sums $\{S_n\}$.

We now give a more general definition of these concepts.

DEFINITION 1. Let $\{a_n\}$ be a sequence. Then the infinite sum

$$\sum_{k=1}^{\infty} a_k = a_1 + a_2 + a_3 + \cdots + a_n + \cdots \tag{2}$$

is called an *infinite series* (or, simply, *series*). The *partial sums* of the series are given by

$$S_n = \sum_{k=1}^{n} a_k.$$

The term S_n is called the *nth partial sum* of the series. If the sequence of partial sums $\{S_n\}$ converges to L, then the infinite series $\sum_{k=1}^{\infty} a_k$ *converges* to L and we write

$$\sum_{k=1}^{\infty} a_k = L. \tag{3}$$

Otherwise we say that the series $\sum_{k=1}^{\infty} a_k$ *diverges*.

> **Remark.** Occasionally a series will be written with the first term other than a_1. For example, $\sum_{k=0}^{\infty} (\frac{1}{2})^k$ and $\sum_{k=2}^{\infty} 1/(\ln k)$ are both examples of infinite series. In the second case we must start with $k = 2$ since $1/(\ln 1)$ is not defined.

EXAMPLE 1. We can write the number $\frac{1}{3}$ as

$$\frac{1}{3} = 0.33333\ldots = \frac{3}{10} + \frac{3}{100} + \frac{3}{1000} + \cdots + \frac{3}{10^n} + \cdots. \tag{4}$$

This is an infinite series. Here $a_n = 3/10^n$ and

$$S_n = \frac{3}{10} + \frac{3}{100} + \cdots + \frac{3}{10^n} = \overbrace{0.333\ldots 3}^{n \text{ places}}.$$

We can formally prove that this converges by noting that

$$S_n = \frac{3}{10}\left(1 + \frac{1}{10} + \cdots + \frac{1}{10^{n-1}}\right) = \frac{3}{10}\sum_{k=0}^{n-1}\left(\frac{1}{10}\right)^k$$

$$= \frac{3}{10}\left[\frac{1 - \left(\frac{1}{10}\right)^n}{1 - \frac{1}{10}}\right] = \frac{3}{10}\left[\frac{1 - \left(\frac{1}{10}\right)^n}{\frac{9}{10}}\right] = \frac{1}{3}\left[1 - \left(\frac{1}{10}\right)^n\right] \to \frac{1}{3}$$

as $n \to \infty$.

As a matter of fact, any decimal number x can be thought of as a convergent infinite series, for if $x = 0.\,a_1a_2a_3\ldots a_n\ldots$, then

$$x = \frac{a_1}{10} + \frac{a_2}{100} + \frac{a_3}{1000} + \cdots + \frac{a_n}{10^n} + \cdots = \sum_{k=1}^{\infty}\frac{a_k}{10^k}.$$

EXAMPLE 2. Express the *repeating decimal* $0.123123123\ldots$ as a rational number (the quotient of two integers).

SOLUTION

$$0.123123123\ldots = 0.123 + 0.000123 + 0.000000123 + \cdots$$

$$= \frac{123}{10^3} + \frac{123}{10^6} + \frac{123}{10^9} + \cdots = \frac{123}{10^3}\left(1 + \frac{1}{10^3} + \frac{1}{(10^3)^2} + \cdots\right)$$

$$= \frac{123}{1000}\left[\frac{1}{1 - (1/1000)}\right] = \frac{123}{1000}\cdot\frac{1000}{999} = \frac{123}{999} = \frac{41}{333}.$$

In general, we can use the geometric series to write any repeating decimal in the form of a fraction by using the technique of Examples 1 or 2. In fact, the rational numbers are exactly those real numbers which can be written as repeating decimals.

EXAMPLE 3. Consider the infinite series $\sum_{k=1}^{\infty} 1/k(k+1)$. We write the first three partial sums:

$$S_1 = \sum_{k=1}^{1} \frac{1}{k(k+1)} = \frac{1}{1 \cdot 2} = \frac{1}{2}$$

$$S_2 = \sum_{k=1}^{2} \frac{1}{k(k+1)} = \frac{1}{1 \cdot 2} + \frac{1}{2 \cdot 3} = \frac{1}{2} + \frac{1}{6} = \frac{2}{3}$$

$$S_3 = \sum_{k=1}^{3} \frac{1}{k(k+1)} = \frac{1}{1 \cdot 2} + \frac{1}{2 \cdot 3} + \frac{1}{3 \cdot 4} = \frac{1}{2} + \frac{1}{6} + \frac{1}{12} = \frac{3}{4}.$$

In general, we can use partial fractions to see that

$$a_k = \frac{1}{k(k+1)} = \frac{1}{k} - \frac{1}{k+1}$$

so that

$$S_n = \left(\frac{1}{1} - \frac{1}{2}\right) + \left(\frac{1}{2} - \frac{1}{3}\right) + \left(\frac{1}{3} - \frac{1}{4}\right) + \cdots + \left(\frac{1}{n-1} - \frac{1}{n}\right)$$

$$+ \left(\frac{1}{n} - \frac{1}{n+1}\right)$$

$$= 1 - \frac{1}{n+1},$$

since all other terms cancel. Since $\lim_{n \to \infty} S_n = \lim_{n \to \infty} \{1 - [1/(n+1)]\} = 1$, we see that

$$\sum_{k=1}^{\infty} \frac{1}{k(k+1)} = 1.$$

Remark. Often, it is *not* possible to calculate the exact sum of an infinite series, even if it can be shown that the series converges.

EXAMPLE 4. Consider the series

$$\sum_{k=1}^{\infty} \frac{1}{k} = 1 + \frac{1}{2} + \frac{1}{3} + \frac{1}{4} + \cdots + \frac{1}{n} + \cdots. \tag{5}$$

This series is called the *harmonic series*. Although $a_n = 1/n \to 0$ as $n \to \infty$, it is not difficult to show that the harmonic series diverges.† To see this, we write

$$\sum_{k=1}^{\infty} \frac{1}{k} = 1 + \frac{1}{2} + \underbrace{\left(\frac{1}{3} + \frac{1}{4}\right)}_{>\frac{1}{2}} + \underbrace{\left(\frac{1}{5} + \frac{1}{6} + \frac{1}{7} + \frac{1}{8}\right)}_{>\frac{1}{2}} + \underbrace{\left(\frac{1}{9} + \cdots + \frac{1}{16}\right)}_{>\frac{1}{2}} + \cdots$$

$$\overset{\text{2 terms}}{\qquad\qquad} \overset{\text{4 terms}}{\qquad\qquad} \overset{\text{8 terms}}{\qquad\qquad}$$

Here we have written the terms in groups containing 2^n numbers. Note that $\frac{1}{3} + \frac{1}{4} > \frac{2}{4} = \frac{1}{2}$, $\frac{1}{5} + \frac{1}{6} + \frac{1}{7} + \frac{1}{8} > \frac{1}{8} + \frac{1}{8} + \frac{1}{8} + \frac{1}{8} = \frac{1}{2}$ and so on. Thus $\sum_{k=1}^{\infty} 1/k > 1 + \frac{1}{2} + \frac{1}{2} + \cdots$ and the series diverges.

EXAMPLE 5. Calculate $\displaystyle\sum_{k=3}^{\infty} \left(\frac{3}{4}\right)^k$.

SOLUTION. We know that $\sum_{k=0}^{\infty} \left(\frac{3}{4}\right)^k = 1/(1 - \frac{3}{4}) = 4$. But

$$\sum_{k=0}^{\infty} \left(\tfrac{3}{4}\right)^k = 1 + \tfrac{3}{4} + \left(\tfrac{3}{4}\right)^2 + \left(\tfrac{3}{4}\right)^3 + \cdots = 1 + \tfrac{3}{4} + \left(\tfrac{3}{4}\right)^2 + \sum_{k=3}^{\infty} \left(\tfrac{3}{4}\right)^k.$$

Thus

$$\sum_{k=3}^{\infty} \left(\tfrac{3}{4}\right)^k = 4 - \{1 + \tfrac{3}{4} + \left(\tfrac{3}{4}\right)^2\} = 4 - \tfrac{37}{16} = \tfrac{27}{16}.$$

It is often difficult to determine whether a series converges or diverges. For that reason a number of techniques have been developed to make it easier to do so. We will present some easy facts here, and then will develop additional techniques in the two sections that follow.

THEOREM 1. Let c be a constant. Suppose that $\sum_{k=1}^{\infty} a_k$ and $\sum_{k=1}^{\infty} b_k$ both converge. Then $\sum_{k=1}^{\infty}(a_k + b_k)$ and $\sum_{k=1}^{\infty} ca_k$ converge, and

(i) $$\sum_{k=1}^{\infty} (a_k + b_k) = \sum_{k=1}^{\infty} a_k + \sum_{k=1}^{\infty} b_k, \tag{6}$$

(ii) $$\sum_{k=1}^{\infty} ca_k = c \sum_{k=1}^{\infty} a_k. \tag{7}$$

This theorem should not be surprising. Since the sum in a series is the limit of a sequence (the sequence of partial sums), the first part, for example, simply restates the fact that the limit of the sum is the sum of the limits.

---

†This example clearly shows that even though the sequence $\{a_n\}$ converges to 0, the series Σa_n may, in fact, diverge.

PROOF. (i)   Let $S = \sum_{k=1}^{\infty} a_k$ and $T = \sum_{k=1}^{\infty} b_k$. The partial sums are given by $S_n = \sum_{k=1}^{n} a_k$ and $T_n = \sum_{k=1}^{n} b_k$. Then

$$\sum_{k=1}^{\infty} (a_k + b_k) = \lim_{n \to \infty} \sum_{k=1}^{n} (a_k + b_k)$$

$$= \lim_{n \to \infty} \left(\sum_{k=1}^{n} a_k + \sum_{k=1}^{n} b_k \right) = \lim_{n \to \infty} (S_n + T_n)$$

$$= \lim_{n \to \infty} S_n + \lim_{n \to \infty} T_n = S + T = \sum_{k=1}^{\infty} a_k + \sum_{k=1}^{\infty} b_k.$$

(ii) $$\sum_{k=1}^{\infty} ca_k = \lim_{n \to \infty} \sum_{k=1}^{n} ca_k = \lim_{n \to \infty} c \sum_{k=1}^{n} a_k = \lim_{n \to \infty} cS_n$$

$$= c \lim_{n \to \infty} S_n = cS = c \sum_{k=1}^{\infty} a_k.$$

EXAMPLE 6.   Show that $\sum_{k=1}^{\infty} \{ [1/k(k + 1)] + (\frac{5}{6})^k \}$ converges.

SOLUTION. This follows since $\sum_{k=1}^{\infty} 1/k(k + 1)$ converges (Example 3) and $\sum_{k=1}^{\infty} (\frac{5}{6})^k$ converges since $\sum_{k=1}^{\infty} (\frac{5}{6})^k = \sum_{k=0}^{\infty} (\frac{5}{6})^k - (\frac{5}{6})^0$ (we added and subtracted the term $(\frac{5}{6})^0 = 1) = 1/(1 - \frac{5}{6}) - 1 = 5$.

EXAMPLE 7.   Does $\sum_{k=1}^{\infty} 1/50k$ converge or diverge?

SOLUTION. We show that the series diverges by assuming that it converges to obtain a contradiction. If $\sum_{k=1}^{\infty} 1/50k$ did converge, then $50 \sum_{k=1}^{\infty} 1/50k$ would also converge by Theorem 1. But then $50 \sum_{k=1}^{\infty} 1/50k = \sum_{k=1}^{\infty} 50 \cdot 1/50k = \sum_{k=1}^{\infty} 1/k$ and this is the harmonic series which we know diverges. Hence $\sum_{k=1}^{\infty} 1/50k$ diverges.

Another useful test is given by the following.

THEOREM 2.   If $\sum_{k=1}^{\infty} a_k$ converges, then $\lim_{n \to \infty} a_n = 0$.

PROOF. Let $S = \sum_{k=1}^{\infty} a_k$. Then the partial sums S_n and S_{n-1} are given by

$$S_n = \sum_{k=1}^{n} a_k = a_1 + a_2 + \cdots + a_{n-1} + a_n$$

and

$$S_{n-1} = \sum_{k=1}^{n-1} a_k = a_1 + a_2 + \cdots + a_{n-1}$$

so that

$$S_n - S_{n-1} = a_n.$$

Then

$$\lim_{n\to\infty} a_n = \lim_{n\to\infty} (S_n - S_{n-1}) = \lim_{n\to\infty} S_n - \lim_{n\to\infty} S_{n-1} = S - S = 0.$$

Remark. We have already seen that the converse of this theorem is false. The convergence of $\{a_n\}$ to 0 does *not* imply that $\sum_{k=1}^{\infty} a_k$ converges. For example, the harmonic series does not converge but the sequence $\{1/n\}$ does converge to zero.

COROLLARY. If $\{a_n\}$ does not converge to 0, then $\sum_{k=1}^{\infty} a_n$ diverges.

EXAMPLE 8. $\displaystyle\sum_{k=1}^{\infty} (-1)^k$ diverges since the sequence $\{(-1)^k\}$ does not converge to zero.

EXAMPLE 9. $\displaystyle\sum_{k=1}^{\infty} \frac{k}{k+100}$ diverges since $\displaystyle\lim_{n\to\infty} a_n = \lim_{n\to\infty} \frac{n}{n+100} = 1 \neq 0.$

PROBLEMS 14.4

In Problems 1–15 a convergent infinite series is given. Find its sum.

1. $\displaystyle\sum_{k=0}^{\infty} \frac{1}{4^k}$

2. $\displaystyle\sum_{k=0}^{\infty} \left(-\frac{2}{3}\right)^k$

3. $\displaystyle\sum_{k=2}^{\infty} \frac{1}{2^k}$

4. $\displaystyle\sum_{k=1}^{\infty} \frac{1}{2^{k-1}}$

5. $\displaystyle\sum_{k=-3}^{\infty} \frac{1}{2^{k+3}}$

6. $\displaystyle\sum_{k=3}^{\infty} \left(\frac{2}{3}\right)^k$

7. $\displaystyle\sum_{k=0}^{\infty} \frac{100}{5^k}$

8. $\displaystyle\sum_{k=0}^{\infty} \frac{5}{100^k}$

9. $\displaystyle\sum_{k=2}^{\infty} \frac{1}{k(k+1)}$

10. $\displaystyle\sum_{k=3}^{\infty} \frac{1}{k(k-1)}$

11. $\displaystyle\sum_{k=0}^{\infty} \frac{1}{(k+1)(k+2)}$

12. $\displaystyle\sum_{k=-1}^{\infty} \frac{1}{(k+3)(k+4)}$

13. $\displaystyle\sum_{k=2}^{\infty} \frac{2^{k+3}}{3^k}$

14. $\displaystyle\sum_{k=2}^{\infty} \frac{2^{k+4}}{3^{k-1}}$

15. $\displaystyle\sum_{k=4}^{\infty} \frac{5^{k-2}}{6^{k+1}}.$

In Problems 16–24 write the repeating decimals as rational numbers.

16. $0.666\ldots$

17. $0.353535\ldots$

18. $0.282828\ldots$

19. $0.717171\ldots$

20. $0.214214214\ldots$

21. $0.501501501\ldots$

22. $0.124242424\ldots$

23. $0.11362362362\ldots$

24. $0.513651365136\ldots.$

25. Prove that the geometric series diverges if $|r| \geq 1$. [*Hint:* Use the corollary to Theorem 2.]

In Problems 26–30 use Theorem 1 to calculate the sum of the convergent series.

26. $\displaystyle\sum_{k=0}^{\infty} \left[\frac{1}{2^k} + \frac{1}{5^k}\right]$

27. $\displaystyle\sum_{k=1}^{\infty} \left[\frac{1}{k(k+1)} + \frac{1}{(k+1)(k+2)}\right]$

28. $\displaystyle\sum_{k=0}^{\infty} \left[\frac{3}{5^k} - \frac{7}{4^k}\right]$

29. $\displaystyle\sum_{k=1}^{\infty} \left[\frac{8}{5^k} - \frac{7}{(k+3)(k+4)}\right]$

30. $\displaystyle\sum_{k=3}^{\infty} \left[\frac{12 \cdot 2^{k+1}}{3^{k-2}} - \frac{15 \cdot 3^{k+1}}{4^{k+2}} \right].$

31. Show that for any nonzero real numbers a and b, $\sum_{k=1}^{\infty} a/bk$ diverges.

32. Show that if the sequences $\{a_k\}$ and $\{b_k\}$ differ only for a finite number of terms, then $\sum_{k=1}^{\infty} a_k$ and $\sum_{k=1}^{\infty} b_k$ either both converge or both diverge.

33. Use the result of Problem 32 to show that $\sum_{k=1}^{\infty} 1/(k+6)$ diverges.

*34. Show that if $\sum_{k=1}^{\infty} a_k$ converges and $\sum_{k=1}^{\infty} b_k$ diverges, then $\sum_{k=1}^{\infty} (a_k + b_k)$ diverges. [*Hint:* Assume that $\sum_{k=1}^{\infty} (a_k + b_k)$ converges and then show that this leads to a contradiction of Theorem 1.]

35. Use the result of Problem 34 to show that $\sum_{k=1}^{\infty} (3/2^k + 2 \cdot 5^k)$ diverges.

36. Give an example in which $\sum_{k=1}^{\infty} a_k$ and $\sum_{k=1}^{\infty} b_k$ both diverge but $\sum_{k=1}^{\infty} (a_k + b_k)$ converges.

37. Use the geometric series to show that

$$\frac{1}{1+x} = \sum_{k=0}^{\infty} (-1)^k x^k$$

for any real number x with $|x| < 1$.

38. Show that $1/(1 + x^2) = \sum_{k=0}^{\infty} (-1)^k x^{2k}$ if $|x| < 1$.

*39. At what time between 1 P.M. and 2 P.M. is the minute hand of a clock exactly over the hour hand? [*Hint:* The minute hand moves 12 times as fast as the hour hand. Start at 1:00 P.M. When the minute hand has reached 1, the hour hand points to $1 + \frac{1}{12}$; when the minute hand has reached $1 + \frac{1}{12}$, the hour hand has reached $1 + \frac{1}{12} + \frac{1}{12} \cdot \frac{1}{12}$; etc. Now add up the geometric series.]

*40. At what time between 7 A.M. and 8 A.M. is the minute hand exactly over the hour hand?

41. A ball is dropped from a height of 8 m. Each time it hits the ground it rebounds to a height of two-thirds the height from which it fell. Find the total distance traveled by the ball until it comes to rest (that is, until it stops bouncing).

14.5 Series with Nonnegative Terms

In this section we consider series of the form $\sum_{k=1}^{\infty} a_k$ where each a_k is nonnegative. Such series are often easier to handle than others. One fact is easy to prove. The sequence $\{S_n\}$ of partial sums is an increasing sequence since $S_{n+1} = S_n + a_{n+1}$ and $a_{n+1} \geq 0$ for every n. Then if $\{S_n\}$ is bounded, it is convergent by Theorem 14.2.2 and we have

THEOREM 1. An infinite series of nonnegative terms is convergent if and only if its sequence of partial sums is bounded.

EXAMPLE 1. Show that $\displaystyle\sum_{k=1}^{\infty} \frac{1}{k^2}$ is convergent.

SOLUTION. We group the terms as follows:

$$\sum_{k=1}^{\infty} \frac{1}{k^2} = \frac{1}{1^2} + \frac{1}{2^2} + \frac{1}{3^2} + \frac{1}{4^2} + \frac{1}{5^2} + \frac{1}{6^2} + \frac{1}{7^2} + \frac{1}{8^2} + \cdots + \frac{1}{15^2} + \cdots$$

$$\overbrace{\qquad}^{\text{2 terms}} \qquad \overbrace{\qquad\qquad}^{\text{4 terms}} \qquad \overbrace{\qquad\qquad\qquad}^{\text{8 terms}}$$

$$\le 1 + \frac{1}{2^2} + \frac{1}{2^2} + \frac{1}{4^2} + \frac{1}{4^2} + \frac{1}{4^2} + \frac{1}{4^2} + \frac{1}{8^2} + \cdots + \frac{1}{8^2} + \cdots$$

$$= 1 + \frac{2}{2^2} + \frac{4}{4^2} + \frac{8}{8^2} + \cdots = 1 + \frac{1}{2} + \frac{1}{4} + \frac{1}{8} + \cdots$$

$$= \sum_{k=0}^{\infty} \frac{1}{2^k} = 2.$$

Thus the sequence of partial sums is bounded by 2 and is therefore convergent.

With Theorem 1, the convergence or divergence of a series of nonnegative terms depends on whether or not its partial sums are bounded. There are four basic tests that can be used to determine this. We will deal with these one at a time.

THEOREM 2 (Comparison Test).   Let $\Sigma_{k=1}^{\infty} a_k$ be a series with $a_k \ge 0$ for every k.

---

(1)   If there exists a convergent series $\Sigma_{k=1}^{\infty} b_k$ and a number N such that $a_k \le b_k$ for every $k \ge N$, then $\Sigma_{k=1}^{\infty} a_k$ converges.
(ii)   If there exists a divergent series $\Sigma_{k=1}^{\infty} c_k$ and a number N such that $a_k \ge c_k$ for every $k \ge N$, then $\Sigma_{k=1}^{\infty} a_k$ diverges.

PROOF. In either case the sum of the first N terms is finite so we need only consider the series $\Sigma_{k=N+1}^{\infty} a_k$ since if this is convergent or divergent, then the addition of a finite number of terms does not affect the convergence or divergence.

(i) $\Sigma_{k=N+1}^{\infty} b_k$ is a nonnegative series (since $b_k \ge a_k \ge 0$ for $k > N$) and is convergent. Thus the partial sums $T_n = \Sigma_{k=N+1}^{n} b_k$ are bounded. If $S_n = \Sigma_{k=N+1}^{n} a_k$, then $S_n \le T_n$ and so the partial sums of $\Sigma_{k=N+1}^{\infty} a_k$ are also bounded, implying that $\Sigma_{k=N+1}^{\infty} a_k$ is convergent.

(ii)   Let $U_n = \Sigma_{k=N+1}^{n} c_k$. By Theorem 1 these partial sums are unbounded since $\Sigma_{k=N+1}^{\infty} c_k$ diverges. Since, in this case, $S_n \ge U_n$, the partial sums of $\Sigma_{k=N+1}^{\infty} a_k$ are also unbounded and the series $\Sigma_{k=N+1}^{\infty} a_k$ diverges.

EXAMPLE 2.   Determine whether $\displaystyle\sum_{k=1}^{\infty} \frac{1}{\sqrt{k}}$ converges or diverges.

SOLUTION. Since $1/\sqrt{k} \ge 1/k$ for $k \ge 1$, and since $\Sigma_{k=1}^{\infty} 1/k$ diverges, we see that, by the comparison test, $\Sigma_{k=1}^{\infty} 1/\sqrt{k}$ diverges.

EXAMPLE 3.   Determine whether $\displaystyle\sum_{k=1}^{\infty} \frac{1}{k!}$ converges or diverges.

SOLUTION. If $k \ge 4$, $k! \ge 2^k$. To see this, note that $4! = 24$ and $2^4 = 16$. Then $5! = 5 \cdot 24$ and $2^5 = 2 \cdot 16$ and since $5 > 2$, $5! > 2^5$, and so on. Then, since $\Sigma_{k=1}^{\infty} 1/2^k$ converges, we see that $\Sigma_{k=1}^{\infty} 1/k!$ converges. In fact, as we shall show in Section 14.8, it converges to e. That is,

$$e = 1 + 1 + \frac{1}{2!} + \frac{1}{3!} + \frac{1}{4!} + \cdots. \tag{1}$$

THEOREM 3 (The Integral Test). Let f be a function which is continuous, positive, and decreasing for all $x \geq 1$. Then the series

$$\sum_{k=1}^{\infty} f(k) = f(1) + f(2) + f(3) + \cdots + f(k) + \cdots \tag{2}$$

converges if $\int_1^{\infty} f(x)\, dx$ converges, and diverges if $\int_1^n f(x)\, dx \to \infty$ as $n \to \infty$.

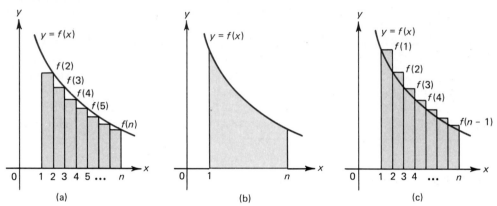

(a)   (b)   (c)

Figure 1

PROOF. The idea behind this proof is fairly easy. Take a look at Figure 1. Comparing areas, we immediately see that

$$f(2) + f(3) + \cdots + f(n) \leq \int_1^n f(x)\, dx \leq f(1) + f(2) + \cdots + f(n-1).$$

If $\lim_{n \to \infty} \int_1^n f(x)\, dx$ is finite, then the partial sums $[f(2) + f(3) + \cdots + f(n)]$ are bounded and the series converges. On the other hand, if $\lim_{n \to \infty} \int_1^n f(x)\, dx = \infty$, then the partial sums $[f(1) + f(2) + \cdots + f(n-1)]$ are unbounded and the series diverges.

EXAMPLE 4. Consider the series $\sum_{k=1}^{\infty} 1/k^{\alpha}$ with $\alpha > 0$. We have already seen that this series diverges for $\alpha = 1$ (the harmonic series) and converges for $\alpha = 2$ (Example 1). Now let $f(x) = 1/x^{\alpha}$. Then for $\alpha \neq 1$

$$\int_1^n f(x)\, dx = \int_1^n \frac{1}{x^{\alpha}}\, dx = \frac{x^{1-\alpha}}{1-\alpha} \bigg|_1^n = \frac{1}{1-\alpha} (n^{1-\alpha} - 1).$$

This last expression converges to $1/(\alpha - 1)$ if $\alpha > 1$ and diverges if $\alpha < 1$. For $\alpha = 1$

$$\int_1^n f(x)\, dx = \int_1^n \frac{1}{x}\, dx = \ln x \bigg|_1^n = \ln n$$

which diverges. (This is another proof that the harmonic series diverges.) Hence

$$\sum_{k=1}^{\infty} \frac{1}{k^{\alpha}} \quad \begin{cases} \text{diverges if } \alpha \leq 1 \\ \text{converges if } \alpha > 1. \end{cases}$$

EXAMPLE 5. Determine whether $\displaystyle\sum_{k=1}^{\infty} \frac{\ln k}{k^2}$ converges or diverges.

SOLUTION. We easily see, using L'Hôpital's rule, that

$$\lim_{x \to \infty} \frac{\ln x}{\sqrt{x}} = \lim_{x \to \infty} \frac{1/x}{-1/2\sqrt{x}} = -2 \lim_{x \to \infty} \frac{\sqrt{x}}{x} = 0$$

so that for k sufficiently large, $\ln k \leq \sqrt{k}$. Thus

$$\frac{\ln k}{k^2} \leq \frac{\sqrt{k}}{k^2} = \frac{1}{k^{3/2}}.$$

But $\sum_{k=1}^{\infty} 1/k^{3/2}$ converges by the result of Example 4 and, therefore, by the comparison test, $\sum_{k=1}^{\infty} (\ln k)/k^2$ also converges.

EXAMPLE 6. Determine whether $\displaystyle\sum_{k=1}^{\infty} \frac{1}{k \ln(k + 5)}$ converges or diverges.

SOLUTION. First we note that $1/[k \ln(k + 5)] > 1/[(k + 5) \ln(k + 5)]$. Also,

$$\int_1^n \frac{dx}{(x + 5) \ln(x + 5)} = \ln \ln(x + 5) \Big|_1^n = \ln \ln(n + 5) - \ln \ln 6$$

which diverges so that $\sum_{k=1}^{\infty} 1/[k \ln(k + 5)]$ also diverges.

THEOREM 4 (The Ratio Test). Let $\displaystyle\sum_{k=1}^{\infty} a_k$ be a series with $a_k > 0$ for every k and suppose that

$$\lim_{n \to \infty} \frac{a_{n+1}}{a_n} = L. \tag{3}$$

Then

(i)   If $L < 1$, $\sum_{k=1}^{\infty} a_k$ converges;
(ii) if $L > 1$, $\sum_{k=1}^{\infty} a_k$ diverges;
(iii) If $L = 1$, $\sum_{k=1}^{\infty} a_k$ may converge or diverge and the ratio test is inconclusive, and some other test must be used.

PROOF. (i)   Pick $\epsilon > 0$ such that $L + \epsilon < 1$. By the definition of the limit in (3), there is a number $N > 0$ such that if $n \geq N$, we have

$$\frac{a_{n+1}}{a_n} < L + \epsilon.$$

Then

$$a_{n+1} < a_n(L + \epsilon), \qquad a_{n+2} < a_{n+1}(L + \epsilon) < a_n(L + \epsilon)^2$$

and

$$a_{n+k} < a_n(L + \epsilon)^k \tag{4}$$

for each $k \geq 1$ and each $n \geq N$. In particular, for $k \geq N$, we use (4) to obtain

$$a_k = a_{(k-N)+N} \leq a_N(L + \epsilon)^{k-N}.$$

Then

$$S_n = \sum_{k=N}^{n} a_k \leq \sum_{k=N}^{n} a_N(L + \epsilon)^{k-N} = \frac{a_N}{(L + \epsilon)^N} \sum_{k=N}^{n} (L + \epsilon)^k.$$

But since $L + \epsilon < 1$, $\sum_{k=0}^{\infty}(L + \epsilon)^k = 1/[1 - (L + \epsilon)]$ (since this last sum is the sum of a geometric series). Thus

$$S_n \leq \frac{a_N}{(L + \epsilon)^N} \cdot \frac{1}{1 - (L + \epsilon)}$$

and so the partial sums of $\sum_{k=N}^{\infty} a_k$ are bounded, implying that $\sum_{k=N}^{\infty} a_k$ converges. Thus $\sum_{k=1}^{\infty} a_k = \sum_{k=1}^{N-1} a_k + \sum_{k=N}^{\infty} a_k$ also converges.

(ii)   If $L > 1$, pick ϵ such that $L - \epsilon > 1$. Then for $n \geq N$, the same proof as before (with the inequalities reversed) shows that

$$a_k \geq a_N(L - \epsilon)^{k-N}$$

and that

$$S_n = \sum_{k=N}^{n} a_k > \frac{a_N}{(L - \epsilon)^N} \sum_{k=N}^{n} (L - \epsilon)^k.$$

But since $L - \epsilon > 1$, $\sum_{k=N}^{\infty}(L - \epsilon)^k$ diverges so that the partial sums S_n are unbounded and $\sum_{k=N}^{\infty} a_k$ diverges.

(iii)   To illustrate (iii), we show that $L = 1$ can occur for a converging or diverging series.

(a)   The harmonic series $\sum_{k=1}^{\infty} 1/k$ diverges. But

$$\lim_{n \to \infty} \frac{a_{n+1}}{a_n} = \lim_{n \to \infty} \frac{1/(n + 1)}{1/n} = \lim_{n \to \infty} \frac{n}{n + 1} = 1.$$

(b) The series $\sum_{k=1}^{\infty} 1/k^2$ converges. Here

$$\lim_{n \to \infty} \frac{a_{n+1}}{a_n} = \lim_{n \to \infty} \frac{1/(n+1)^2}{1/n^2} = \lim_{n \to \infty} \left(\frac{n}{n+1}\right)^2 = 1.$$

Remark. The ratio test is very useful. But in those cases where $L = 1$, we must try another test to determine whether the series converges or diverges.

EXAMPLE 7. We have used the comparison test to show that $\sum_{k=1}^{\infty} 1/k!$ converges. Using the ratio test, we find that

$$\lim_{n \to \infty} \frac{a_{n+1}}{a_n} = \lim_{n \to \infty} \frac{1/(n+1)!}{1/n!} = \lim_{n \to \infty} \frac{n!}{(n+1)!} = \lim_{n \to \infty} \frac{1}{n+1} = 0 < 1$$

so that the series converges.

EXAMPLE 8. Determine whether the series $\sum_{k=0}^{\infty} (100)^k/k!$ converges or diverges.

SOLUTION. Here

$$\lim_{n \to \infty} \frac{a_{n+1}}{a_n} = \lim_{n \to \infty} \frac{(100)^{n+1}/(n+1)!}{(100)^n/n!} = \lim_{n \to \infty} \frac{100}{n+1} = 0$$

so that the series converges.

EXAMPLE 9. Determine whether the series $\sum_{k=1}^{\infty} k^k/k!$ converges or diverges.

SOLUTION

$$\lim_{n \to \infty} \frac{a_{n+1}}{a_n} = \lim_{n \to \infty} \frac{[(n+1)^{n+1}/(n+1)!]}{n^n/n!} = \lim_{n \to \infty} \frac{(n+1)^{n+1}}{(n+1)n^n}$$

$$= \lim_{n \to \infty} \left(\frac{n+1}{n}\right)^n = \lim_{n \to \infty} \left(1 + \frac{1}{n}\right)^n = e > 1$$

so that the series diverges.

EXAMPLE 10. Determine whether the series $\sum_{k=1}^{\infty} (k+1)/k(k+2)$ converges or diverges.

SOLUTION. Here

$$\lim_{n \to \infty} \frac{a_{n+1}}{a_n} = \lim_{n \to \infty} \frac{(n+2)/(n+1)(n+3)}{(n+1)/n(n+2)} = \lim_{n \to \infty} \frac{n(n+2)^2}{(n+1)^2(n+3)}$$

$$= \lim_{n \to \infty} \frac{n^3 + 4n^2 + 4n}{n^3 + 5n^2 + 6n + 3} = 1.$$

Thus the ratio test fails. However, for $k \geq 1$, $(k + 1)/(k + 2) > \frac{1}{2}$ (explain why) so that $(k + 1)/k(k + 2) > 1/2k$ and $\sum_{k=1}^{\infty} 1/2k = \frac{1}{2}\sum_{k=1}^{\infty} 1/k$ which diverges so that, by the comparison test, $\sum_{k=1}^{\infty}(k + 1)/k(k + 2)$ also diverges.

THEOREM 5 (The Root Test). Let $\sum_{k=1}^{\infty} a_k$ be a series with $a_k > 0$ and suppose that $\lim_{n\to\infty} (a_n)^{1/n} = R$. Then

---

(i)   If $R < 1$, $\displaystyle\sum_{k=1}^{\infty} a_k$ converges.

(ii)   If $R > 1$, $\displaystyle\sum_{k=1}^{\infty} a_k$ diverges.

(iii)   If $R = 1$, the series either converges or diverges, and no conclusion can be drawn from this test.

The proof of this theorem is similar to the proof of the ratio test and is left as an exercise (see Problems 47–49).

EXAMPLE 11. Determine whether $\displaystyle\sum_{k=2}^{\infty} 1/(\ln k)^k$ converges or diverges.

SOLUTION. Note first that we start at $k = 2$ since $1/(\ln 1)^1$ is not defined.

$$\lim_{n\to\infty} \left[\frac{1}{(\ln n)^n}\right]^{1/n} = \lim_{n\to\infty} \frac{1}{\ln n} = 0$$

so that the series converges. This is a good example of the use of the root test as no other test could be easily applied in this problem.

EXAMPLE 12. Determine whether the series $\displaystyle\sum_{k=1}^{\infty} 3^k/k^5$ converges or diverges.

SOLUTION

$$\lim_{n\to\infty} (a_n)^{1/n} = \lim_{n\to\infty} \left(\frac{3^n}{n^5}\right)^{1/n} = \lim_{n\to\infty} \frac{3}{n^{5/n}}$$

$$= 3 \lim_{n\to\infty} \left(\frac{1}{n^{1/n}}\right)^5 = 3 \lim_{n\to\infty} (n^{-1/n})^5.$$

But if $y = x^{-1/x}$, then

$$\ln y = -(1/x)\ln x \qquad \text{and} \qquad \lim_{x\to\infty} [-(\ln x)/x] = \lim_{x\to\infty} [(-1/x)/1] = 0$$

so that $\ln y \to 0$ and $y = x^{-1/x} \to 1$ as $x \to \infty$. Hence $\lim_{n\to\infty} n^{-1/n} = 1$ and

$$3 \lim_{n\to\infty} (n^{-1/n})^5 = 3 \left(\lim_{n\to\infty} n^{-1/n}\right)^5 = 3 \cdot 1^5 = 3 > 1$$

so that the series diverges.

PROBLEMS 14.5

In Problems 1–45 determine the convergence or divergence of the given series.

1. $\displaystyle\sum_{k=1}^{\infty} \frac{1}{k^2 + 1}$

2. $\displaystyle\sum_{k=10}^{\infty} \frac{1}{k(k - 3)}$

3. $\displaystyle\sum_{k=4}^{\infty} \frac{1}{5k + 50}$

4. $\displaystyle\sum_{k=1}^{\infty} \frac{1}{\sqrt{k^2 + 2k}}$

5. $\displaystyle\sum_{k=1}^{\infty} \frac{1}{\sqrt{k^3 + 1}}$

6. $\displaystyle\sum_{k=1}^{\infty} \frac{\ln k}{k^3}$

7. $\displaystyle\sum_{k=2}^{\infty} \frac{1}{k^2 + 1}$

8. $\displaystyle\sum_{k=2}^{\infty} \frac{4}{k \ln k}$

9. $\displaystyle\sum_{k=0}^{\infty} ke^{-k}$

10. $\displaystyle\sum_{k=3}^{\infty} k^3 e^{-k^4}$

11. $\displaystyle\sum_{k=5}^{\infty} \frac{1}{k(\ln k)^3}$

12. $\displaystyle\sum_{k=4}^{\infty} \frac{1}{k^2 \sqrt{\ln k}}$

13. $\displaystyle\sum_{k=1}^{\infty} \frac{1}{(3k - 1)^{3/2}}$

14. $\displaystyle\sum_{k=1}^{\infty} \frac{1}{\sqrt{k^2 + 3}}$

15. $\displaystyle\sum_{k=1}^{\infty} \frac{2^k}{k^2}$

16. $\displaystyle\sum_{k=1}^{\infty} \frac{5^k}{k^5}$

17. $\displaystyle\sum_{k=1}^{\infty} \frac{r^k}{k^r}, \, 0 < r < 1$

18. $\displaystyle\sum_{k=1}^{\infty} \frac{r^k}{k^r}, \, r > 1$

19. $\displaystyle\sum_{k=1}^{\infty} \frac{1}{50 + \sqrt{k}}$

20. $\displaystyle\sum_{k=3}^{\infty} \left(\frac{k}{k + 1}\right)^k$

21. $\displaystyle\sum_{k=1}^{\infty} \left(\frac{k}{k + 1}\right)^{1/k}$

22. $\displaystyle\sum_{k=2}^{\infty} \frac{k!}{k^k}$

23. $\displaystyle\sum_{k=1}^{\infty} \frac{k^k}{(2k)!}$

24. $\displaystyle\sum_{k=2}^{\infty} \frac{1}{k\sqrt{\ln k}}$

25. $\displaystyle\sum_{k=1}^{\infty} \frac{k^2}{50k}$

26. $\displaystyle\sum_{k=2}^{\infty} \frac{k}{(\ln k)^k}$

27. $\displaystyle\sum_{k=1}^{\infty} \frac{4^k}{k^3}$

28. $\displaystyle\sum_{k=1}^{\infty} \frac{k^{2/3}}{10^k}$

29. $\displaystyle\sum_{k=2}^{\infty} \left(1 + \frac{1}{k}\right)^k$

30. $\displaystyle\sum_{k=1}^{\infty} \frac{\sqrt{k} \ln k}{k^3 + 1}$

31. $\displaystyle\sum_{k=1}^{\infty} \frac{\sqrt{k}}{3k^2 + 2k + 20}$

32. $\displaystyle\sum_{k=1}^{\infty} \frac{e^k}{k^5}$

33. $\displaystyle\sum_{k=1}^{\infty} \frac{e^k}{k!}$

34. $\displaystyle\sum_{k=4}^{\infty} \frac{1}{k \ln \ln k}$

35. $\displaystyle\sum_{k=1}^{\infty} \sin \frac{1}{k}$

36. $\displaystyle\sum_{k=1}^{\infty} \frac{3}{(k + 2)\sqrt{\ln(k + 2)}}$

37. $\displaystyle\sum_{k=1}^{\infty} \frac{e^{1/k}}{k^2}$

38. $\displaystyle\sum_{k=1}^{\infty} \frac{\tan^{-1} k}{1 + k^2}$

39. $\displaystyle\sum_{k=1}^{\infty} \text{sech } k$

40. $\displaystyle\sum_{k=10}^{\infty} \frac{1}{k(\ln k)(\ln \ln k)}$

41. $\displaystyle\sum_{k=1}^{\infty} \frac{1}{\cosh^2 k}$

42. $\displaystyle\sum_{k=1}^{\infty} \frac{3^k + k}{k! + 2}$

43. $\displaystyle\sum_{k=1}^{\infty} \frac{k(k + 2)}{(k + 1)(k + 3)(k + 5)}$

44. $\displaystyle\sum_{k=1}^{\infty} \tan^{-1} k$

45. $\displaystyle\sum_{k=1}^{\infty} \frac{1}{\sqrt{k} \ln^{10} k}.$

46. Show that $\Sigma_{k=0}^{\infty} x^k/k!$ converges for every real number x.

*47. Prove part (i) of the root test (Theorem 5). [*Hint:* If $k < 1$, choose $\epsilon > 0$ so that $R + \epsilon < 1$. Show that there is an N such that if $n \geq N$, then $a_n < (R + \epsilon)^n$. Then complete the proof by comparing Σa_k with the sum of a geometric series.]

48. Prove part (ii) of the root test. [*Hint:* Follow the steps of Problem 47.]

49. Show that if $a_n^{1/n} \to 1$, then $\Sigma_{k=1}^{\infty} a_k$ may converge or diverge. [*Hint:* Consider $\Sigma \, 1/k$ and $\Sigma \, 1/k^2$.]

50. Prove that $k!/k^k \to 0$ as $k \to \infty$.

*51. Let $S_n = \Sigma_{k=1}^{n} 1/k$. Show that

$$\ln(n + 1) < S_n < \ln n + 1.$$

[*Hint:* Use the inequality $1/(k + 1) \leq 1/x \leq 1/k$ if $0 < k \leq x \leq k + 1$, integrate, and add as in the proof of the integral test.]

52. Let $S_n = \ln n! = \ln 2 + \ln 3 + \cdots + \ln n$. By calculating $\int_1^n \ln x \, dx$ and comparing areas as in the proof of the integral test, show that for $n \geq 2$

$$\ln(n - 1)! < n \ln n - n + 1 < \ln n!$$

and that

$$(n - 1)! < n^n e^{-(n-1)} < n!.$$

53. Let S_n denote the nth partial sum of the harmonic series. Show that $S_n > \ln(n + 1)$ for $n > 1$. [*Hint:* Consider Figure 2. Show that S_n is the sum of the areas of the first n rectangles in the figure.]

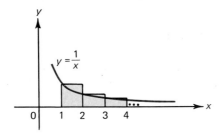

Figure 2

54. (a) Show that the sequence $\{S_n - \ln(n + 1)\}$ is increasing.

    *(b) Show that this sequence is bounded by 1. [*Hint:* $S_n - \ln(n + 1) =$ the sum of the areas of "triangular" shaped regions in the region. Move those "triangles" over so that they all fit into the first rectangle.]

    (c) Show that $\lim_{n \to \infty}[S_n - \ln(n + 1)]$ exists. This limit is denoted by γ which is called the *Euler constant*

$$\gamma = \lim_{n \to \infty}\left[1 + \frac{1}{2} + \frac{1}{3} + \cdots + \frac{1}{n} - \ln(n + 1)\right]$$

$$= \lim_{n \to \infty}\left[1 + \frac{1}{2} + \frac{1}{3} + \cdots + \frac{1}{n} - \ln n\right].$$

It is the third transcendental number you have seen and it arises in physical applications. To seven decimal places

$$\gamma = 0.5772157.$$

14.6 Alternating Series: Absolute and Conditional Convergence

In Section 14.5 all the series we dealt with had positive terms. In this section we consider special types of series which have positive and negative terms.

DEFINITION 1. The series $\sum_{k=1}^{\infty} a_k$ is said to *converge absolutely* if the series $\sum_{k=1}^{\infty} |a_k|$ converges.

EXAMPLE 1. The series

$$\sum_{k=1}^{\infty} \frac{(-1)^{k+1}}{k^2} = \frac{1}{1^2} - \frac{1}{2^2} + \frac{1}{3^2} - \frac{1}{4^2} + \cdots$$

converges absolutely because $\sum_{k=1}^{\infty} |(-1)^{k+1}/k^2| = \sum_{k=1}^{\infty} 1/k^2$ converges.

EXAMPLE 2. The series

$$\sum_{k=1}^{\infty} \frac{(-1)^{k+1}}{k} = \frac{1}{1} - \frac{1}{2} + \frac{1}{3} - \frac{1}{4} + \frac{1}{5} + \cdots$$

does not converge absolutely because $\sum_{k=1}^{\infty} 1/k$ diverges.

The importance of absolute convergence is given in the theorem below.

THEOREM 1. If $\sum_{k=1}^{\infty} |a_k|$ converges, then $\sum_{k=1}^{\infty} a_k$ also converges. That is,

> Absolute convergence implies convergence.

Remark. The converse of this theorem is false. That is, there are series which are convergent but not absolutely convergent. We will see examples of this shortly.

PROOF. Since $a_k \leq |a_k|$, we have

$$0 \leq a_k + |a_k| \leq 2|a_k|.$$

Since $\sum_{k=1}^{\infty} |a_k|$ converges, we see that $\sum_{k=1}^{\infty} (a_k + |a_k|)$ converges by the comparison test. Then as $a_k = (a_k + |a_k|) - |a_k|$, $\sum_{k=1}^{\infty} a_k$ converges because it is the sum of two convergent series.

EXAMPLE 3. The series $\sum_{k=1}^{\infty} (-1)^{k+1}/k^2$ considered in Example 1 converges since it converges absolutely.

DEFINITION 2. A series in which successive terms have opposite signs is called an *alternating series*.

EXAMPLE 4. The series

$$\sum_{k=1}^{\infty} \frac{(-1)^{k+1}}{k^2} = 1 - \frac{1}{2^2} + \frac{1}{3^2} - \frac{1}{4^2} + \cdots$$

is an alternating series.

EXAMPLE 5.   The series

$$\sum_{k=1}^{\infty} \frac{(-1)^{k+1}}{k} = 1 - \frac{1}{2} + \frac{1}{3} - \frac{1}{4} + \frac{1}{5} - \frac{1}{6} + \cdots$$

is an alternating series.

EXAMPLE 6.   The series $1 + \frac{1}{2} - \frac{1}{3} - \frac{1}{4} + \frac{1}{5} + \frac{1}{6} - \cdots$ is not an alternating series because two successive terms have the same sign.

Let us consider the series of Example 5:

$$S = 1 - \tfrac{1}{2} + \tfrac{1}{3} - \tfrac{1}{4} + \tfrac{1}{5} - \tfrac{1}{6} + \cdots.$$

Calculating successive partial sums, we find that

$$S_1 = 1, \quad S_2 = \tfrac{1}{2}, \quad S_3 = \tfrac{5}{6}, \quad S_4 = \tfrac{7}{12}, \quad S_5 = \tfrac{47}{60}, \quad \ldots.$$

It is clear that this series is not diverging to infinity and that the partial sums are getting "narrowed down." At this point it is reasonable to suspect that the series converges. But it does *not* converge absolutely (since the series of absolute values is the harmonic series) and we cannot use any of the tests of the previous section since the terms are not nonnegative. The result we need is given in the theorem below.

THEOREM 2 (Alternating Series Test).   Let $\{a_k\}$ be a decreasing sequence of positive numbers such that $\lim_{k\to\infty} a_k = 0$. Then the alternating series $\sum_{k=1}^{\infty}(-1)^{k+1}a_k = a_1 - a_2 + a_3 - a_4 + \cdots$ converges.

PROOF.   Looking at the odd numbered partial sums of this series, we find that

$$S_{2n+1} = (a_1 - a_2) + (a_3 - a_4) + (a_5 - a_6) + \cdots + (a_{2n-1} - a_{2n}) + a_{2n+1}.$$

Since $\{a_k\}$ is decreasing, all the terms in parentheses are nonnegative so that $S_{2n+1} \geq 0$ for every n. Moreover,

$$S_{2n+3} = S_{2n+1} - a_{2n+2} + a_{2n+3} = S_{2n+1} - (a_{2n+2} - a_{2n+3}),$$

and since $a_{2n+2} - a_{2n+3} \geq 0$, we have

$$S_{2n+3} \leq S_{2n+1}.$$

Hence the sequence of odd numbered partial sums is bounded below by 0 and decreasing and is therefore convergent by Theorem 14.2.2. Thus $S_{2n+1} \to L$. Now let us consider the sequence of even numbered partial sums. We find that $S_{2n+2} = S_{2n+1} - a_{2n+2}$ and since $a_{2n+2} \to 0$

$$\lim_{n\to\infty} S_{2n+2} = \lim_{n\to\infty} S_{2n+1} - \lim_{n\to\infty} a_{2n+2} = L - 0 = L$$

so that the even partial sums also converge to L. Since both the odd and even sums converge to L, we see that the partial sums converge to L and the proof is complete.

EXAMPLE 7.   The following alternating series are convergent by the alternating series test:

(a) $1 - \dfrac{1}{2} + \dfrac{1}{3} - \dfrac{1}{4} + \dfrac{1}{5} - \dfrac{1}{6} + \cdots$

(b) $1 - \dfrac{1}{\sqrt{2}} + \dfrac{1}{\sqrt{3}} - \dfrac{1}{\sqrt{4}} + \dfrac{1}{\sqrt{5}} - \dfrac{1}{\sqrt{6}} + \dfrac{1}{\sqrt{7}} - \cdots$

(c) $\dfrac{1}{\ln 2} - \dfrac{1}{\ln 3} + \dfrac{1}{\ln 4} - \dfrac{1}{\ln 5} + \dfrac{1}{\ln 6} - \cdots$

(d) $1 - \dfrac{1}{2} + \dfrac{1}{2^2} - \dfrac{1}{2^3} + \dfrac{1}{2^4} - \dfrac{1}{2^5} + \dfrac{1}{2^6} - \dfrac{1}{2^7} + \cdots$

DEFINITION 3. An alternating series is said to be *conditionally convergent* if it is convergent but not absolutely convergent.

EXAMPLE 8. In Example 7, all the series are conditionally convergent except the last one.

It is not difficult to estimate the sum of a convergent alternating series. We again consider the series

$$S = 1 - \tfrac{1}{2} + \tfrac{1}{3} - \tfrac{1}{4} + \tfrac{1}{5} - \cdots.$$

Suppose we wish to approximate S by its nth partial sum S_n. Then

$$S - S_n = \pm \left(\dfrac{1}{n+1} - \dfrac{1}{n+2} + \dfrac{1}{n+3} - \dfrac{1}{n+4} + \cdots \right) = R_n.$$

But we can estimate the remainder term R_n:

$$|R_n| = \left| \left(\dfrac{1}{n+1} - \left(\dfrac{1}{n+2} - \dfrac{1}{n+3} \right) - \left(\dfrac{1}{n+4} - \dfrac{1}{n+5} \right) - \cdots \right) \right| \leq \dfrac{1}{n+1}.$$

That is, the error is less than the first term which we left out! For example, $|S - S_{20}| \leq 1/21 \approx 0.0476$.

In general, we have the following result whose proof is left as an exercise (see Problem 31).

THEOREM 3. If $S = \Sigma_{k=1}^{\infty} a_k$ is a convergent alternating series with monotone decreasing terms, then for any n,

$$|S - S_n| \leq |a_{n+1}|. \tag{1}$$

▢**EXAMPLE 9.** The series

$$\sum_{k=1}^{\infty} \dfrac{(-1)^{k+1}}{\ln(k+1)} = \dfrac{1}{\ln 2} - \dfrac{1}{\ln 3} + \dfrac{1}{\ln 4} - \dfrac{1}{\ln 5} + \cdots$$

can be approximated by S_n with an error of less than $1/\ln(n+2)$. For example, with $n = 10$, $1/\ln(n+2) = 1/\ln 12 \approx 0.4$. Hence the sum

$$\sum_{k=1}^{\infty} \dfrac{(-1)^{k+1}}{\ln(k+1)} = \dfrac{1}{\ln 2} - \dfrac{1}{\ln 3} + \cdots$$

can be approximated by

$$S_{10} = \frac{1}{\ln 2} - \frac{1}{\ln 3} + \frac{1}{\ln 4} - \frac{1}{\ln 5} + \frac{1}{\ln 6} - \frac{1}{\ln 7}$$

$$+ \frac{1}{\ln 8} - \frac{1}{\ln 9} + \frac{1}{\ln 10} - \frac{1}{\ln 11} \approx 0.7197$$

with an error of less than 0.4.

There is one fascinating fact about an alternating series which is conditionally but not absolutely convergent:

> By reordering the terms of a conditionally convergent series, the new series of rearranged terms can be made to add up to any real number.

Let us illustrate this fact with the series

$$S = 1 - \tfrac{1}{2} + \tfrac{1}{3} - \tfrac{1}{4} + \tfrac{1}{5} - \tfrac{1}{6} + \cdots.$$

The odd-numbered terms sum to a divergent series:

$$1 + \tfrac{1}{3} + \tfrac{1}{5} + \tfrac{1}{7} + \cdots. \tag{2}$$

The even-numbered terms are likewise a divergent series:

$$-\tfrac{1}{2} - \tfrac{1}{4} - \tfrac{1}{6} - \cdots. \tag{3}$$

If either of these series converged, then the other one would too (by the comparison test), and then the entire series would be absolutely convergent (which we know to be false). Now choose any real number, say 1.5. Then:

(i)   Choose enough terms from the series (2) so that the sum exceeds 1.5. We can do this since the series diverges.

$$1 + \tfrac{1}{3} + \tfrac{1}{5} = 1.53333 \ldots.$$

(ii)   Add enough negative terms from (3) so that the sum is now just under 1.5.

$$1 + \tfrac{1}{3} + \tfrac{1}{5} - \tfrac{1}{2} = 1.0333 \ldots.$$

(iii)   Add more terms from (2) until 1.5 is exceeded.

$$1 + \tfrac{1}{3} + \tfrac{1}{5} - \tfrac{1}{2} + \tfrac{1}{7} + \tfrac{1}{9} + \tfrac{1}{11} + \tfrac{1}{13} + \tfrac{1}{15} = 1.5218.$$

(iv)   Again subtract terms from (3) until the sum is again under 1.5.

$$1 + \tfrac{1}{3} + \tfrac{1}{5} - \tfrac{1}{2} + \tfrac{1}{7} + \tfrac{1}{9} + \tfrac{1}{11} + \tfrac{1}{13} + \tfrac{1}{15} - \tfrac{1}{4} = 1.2718.$$

We continue the process to "converge" to 1.5. Since the terms in each series are decreasing to 0, the amount above or below 1.5 will approach 0 and the partial sums converge.

We will show in Section 14.9 that, without rearranging,

$$\sum_{k=1}^{\infty} \frac{(-1)^{k+1}}{k} = 1 - \tfrac{1}{2} + \tfrac{1}{3} - \tfrac{1}{4} + \tfrac{1}{5} - \tfrac{1}{6} + \cdots = \ln 2 \approx 0.693147.$$

Remark. It should be noted that *any* rearrangement of the terms of an *absolutely converging* series converges to the same number.

PROBLEMS 14.6

In Problems 1–30 determine whether the given series is absolutely convergent, conditionally convergent, or divergent.

1. $\displaystyle\sum_{k=1}^{\infty} (-1)^k$

2. $\displaystyle\sum_{k=1}^{\infty} \frac{(-1)^{k+1}}{2k}$

3. $\displaystyle\sum_{k=2}^{\infty} \frac{(-1)^k}{k \ln k}$

4. $\displaystyle\sum_{k=1}^{\infty} \frac{(-1)^k}{k^{3/2}}$

5. $\displaystyle\sum_{k=2}^{\infty} \frac{(-1)^k k}{\ln k}$

6. $\displaystyle\sum_{k=1}^{\infty} \frac{(-1)^k \ln k}{k}$

7. $\displaystyle\sum_{k=1}^{\infty} \frac{(-1)^{k+1}}{5k - 4}$

8. $\displaystyle\sum_{k=1}^{\infty} \sin \frac{k\pi}{2}$

9. $\displaystyle\sum_{k=0}^{\infty} \cos \frac{k\pi}{2}$

10. $\displaystyle\sum_{k=1}^{\infty} \frac{(-3)^k}{k!}$

11. $\displaystyle\sum_{k=1}^{\infty} \frac{k!}{(-3)^k}$

12. $\displaystyle\sum_{k=1}^{\infty} \frac{(-2)^k}{k^2}$

13. $\displaystyle\sum_{k=1}^{\infty} \frac{k^2}{(-2)^k}$

14. $\displaystyle\sum_{k=2}^{\infty} \frac{(-1)^{k+1}}{\sqrt{k(k-1)}}$

15. $\displaystyle\sum_{k=2}^{\infty} \frac{(-1)^k k^2}{k^3 + 1}$

16. $\displaystyle\sum_{k=1}^{\infty} \frac{\cos \frac{k\pi}{6}}{k^2}$

17. $\displaystyle\sum_{k=3}^{\infty} \frac{\sin \frac{k\pi}{7}}{k^3}$

18. $\displaystyle\sum_{k=1}^{\infty} \frac{(-1)^k(k+2)}{k(k+1)}$

19. $\displaystyle\sum_{k=2}^{\infty} \frac{(-1)^k k(k+1)}{(k+2)^3}$

20. $\displaystyle\sum_{k=2}^{\infty} \frac{(-1)^k k(k+1)}{(k+2)^4}$

21. $\displaystyle\sum_{k=1}^{\infty} \frac{(-1)^k 2^k}{k}$

22. $\displaystyle\sum_{k=1}^{\infty} \frac{(-1)^{k+1}}{k!}$

23. $\displaystyle\sum_{k=1}^{\infty} \frac{(-1)^k k^k}{k!}$

24. $\displaystyle\sum_{k=1}^{\infty} \frac{(-1)^k \sqrt{k}}{k+3}$

25. $\displaystyle\sum_{k=2}^{\infty} \frac{(-1)^k(k^2+3)}{k^3+4}$

26. $\displaystyle\sum_{k=2}^{\infty} \frac{(-1)^k}{\sqrt[3]{\ln k}}$

27. $\displaystyle\sum_{k=1}^{\infty} \frac{(-1)^k k^2}{4+k^2}$

28. $\displaystyle\sum_{k=1}^{\infty} (-1)^k \left(1 + \frac{1}{k}\right)^k$

29. $\displaystyle\sum_{k=2}^{\infty} \frac{(-1)^k}{k \sqrt{\ln k}}$

30. $\displaystyle\sum_{k=2}^{\infty} \frac{(-1)^k k^3}{k^3 + 2k^2 + k - 1}$.

*31. Prove Theorem 3. [*Hint:* Assume that the odd-numbered terms are positive. Show that the sequence $\{S_{2n}\}$ is increasing and that $S_{2n} < S_{2n+2} < S$ for all $n \geq 1$. Then show that the sequence of odd-numbered partial sums is decreasing and that $S < S_{2n+1} < S_{2n-1}$ for all $n \geq 1$. Conclude that (a) $0 < S - S_{2n} < a_{2n+1}$ for all $n \geq 1$ and that (b) $0 < S_{2n-1} - S < -a_{2n}$. Use inequalities (a) and (b) to prove the theorem.]

In Problems 32–37 use the result of Theorem 3 to estimate the given sum to within the indicated accuracy.

32. $\displaystyle\sum_{k=1}^{\infty} \frac{(-1)^{k+1}}{k!}$; error < 0.001

33. $\displaystyle\sum_{k=1}^{\infty} \frac{(-1)^{k+1}}{k^2}$; error < 0.01

34. $\displaystyle\sum_{k=1}^{\infty} \frac{(-1)^{k+1}}{k^4}$; error < 0.0001

□35. $\displaystyle\sum_{k=2}^{\infty} \frac{(-1)^{k+1}}{k \ln k}$; error < 0.1

36. $\displaystyle\sum_{k=1}^{\infty} \frac{(-1)^{k+1}}{k^k}$; error < 0.0001 □**37.** $\displaystyle\sum_{k=1}^{\infty} \frac{(-1)^{k+1}}{\sqrt{k}}$; error < 0.1.

38. Find the first ten terms of a rearrangement of the series $\sum_{k=1}^{\infty}(-1)^{k+1}/k$ which converges to 0.

39. Find the first ten terms of a rearrangement of the series $\sum_{k=1}^{\infty}(-1)^{k+1}/k$ which converges to 0.3.

40. Explain why there is no rearrangement of the series $\sum_{k=1}^{\infty}(-1)^k/k^2$ which converges to -1.

41. Prove that if $\sum_{k=1}^{\infty} a_k$ is a convergent series of nonzero terms, then $\sum_{k=1}^{\infty} 1/|a_k|$ diverges.

42. Show that if $\sum_{k=1}^{\infty} a_k$ is absolutely convergent, then $\sum_{k=1}^{\infty} a_k^{p}$ is convergent for any number $p \geq 1$.

43. Given an example of a sequence $\{a_k\}$ such that $\sum_{k=1}^{\infty} a_k^2$ converges but that $\sum_{k=1}^{\infty} a_k$ diverges.

14.7 Power Series

In previous sections we discussed infinite series of real numbers. Here we discuss series of functions.

DEFINITION 1

> (i)   A *power series in x* is a series of the form
>
> $$\sum_{k=0}^{\infty} a_k x^k = a_0 + a_1 x + a_2 x^2 + \cdots + a_n x^n + \cdots. \qquad (1)$$
>
> (ii)   A power series in $x - x_0$ is a series of the form
>
> $$\sum_{k=0}^{\infty} (x - x_0)^k = a_0 + a_1(x - x_0) + a_2(x - x_0)^2$$
>
> $$+ \cdots + a_n(x - x_0)^n + \cdots, \qquad (2)$$
>
> where x_0 is a real number.

Remark. A power series in $x - x_0$ can be converted to a power series in u by the change of variables $u = x - x_0$. Then $\sum_{k=0}^{\infty} a_k(x - x_0)^k = \sum_{k=0}^{\infty} a_k u^k$. Thus, in our discussion, we will only consider power series of the form $\sum_{k=0}^{\infty} a_k x^k$ since these include series of type (2) as well.

DEFINITION 2.   (i)   A power series is said to *converge* at x if the series of real numbers $\sum_{k=0}^{\infty} a_k x^k$ converges. Otherwise it is said to *diverge*.

(ii)   A power series is said to converge in a set D of real numbers if it converges for every real number x in D.

EXAMPLE 1.   For what real numbers does the power series

$$\sum_{k=0}^{\infty} \frac{x^k}{3^k} = 1 + \frac{x}{3} + \frac{x^2}{3^2} + \frac{x^3}{3^3} + \cdots$$

converge?

SOLUTION. The nth term in this series is $x^n/3^n$. Using the ratio test, we find that

$$\lim_{n\to\infty}\frac{|a_{n+1}|}{|a_n|} = \lim_{n\to\infty}\frac{|x^{n+1}|/3^{n+1}}{|x^n/3^n|} = \lim_{n\to\infty}\left|\frac{x}{3}\right| = \left|\frac{x}{3}\right|.^{\dagger}$$

Thus the power series converges absolutely if $|x/3| < 1$ or $|x| < 3$ and diverges if $|x| > 3$. The case $|x| = 3$ has to be treated separately. For $x = 3$

$$\sum_{k=0}^{\infty}\frac{x^k}{3^k} = \sum_{k=0}^{\infty}\frac{3^k}{3^k} = \sum_{k=0}^{\infty}1^k$$

which diverges. For $x = -3$

$$\sum_{k=0}^{\infty}\frac{x^k}{3^k} = \sum_{k=0}^{\infty}(-1)^k$$

which also diverges. Thus the series converges in the open interval $(-3, 3)$. We will show in Theorem 1 that since the series diverges for $x = 3$, it diverges for $|x| > 3$ so that the conditional convergence at any x for which $|x| > 3$ is ruled out.

EXAMPLE 2. For what values of x does the series $\sum_{k=0}^{\infty}x^k/(k+1)$ converge?

SOLUTION. Here $a_n = x^n/(n+1)$ so that

$$\lim_{n\to\infty}\frac{|a_{n+1}|}{|a_n|} = \lim_{n\to\infty}\left|\frac{x^{n+1}/(n+2)}{x^n/(n+1)}\right| = |x|\lim_{n\to\infty}\frac{n+1}{n+2} = |x|,$$

since $\lim_{n\to\infty}(n+1)/(n+2) = 1$. Thus the series converges absolutely for $|x| < 1$ and diverges for $|x| > 1$. If $x = 1$, then

$$\sum_{k=0}^{\infty}\frac{x^k}{k+1} = \sum_{k=0}^{\infty}\frac{1}{k+1}$$

which diverges since this is the harmonic series. If $x = -1$, then

$$\sum_{k=0}^{\infty}\frac{x^k}{k+1} = \sum_{k=0}^{\infty}\frac{(-1)^k}{k+1} = 1 - \frac{1}{2} + \frac{1}{3} - \frac{1}{4} + \cdots$$

which converges conditionally by the alternating series test. In sum, the power series $\sum_{k=0}^{\infty}x^k/(k+1)$ converges in the half-open interval $[-1, 1)$.

The following theorem is of great importance in determining the range of values over which a power series converges.

THEOREM 1. (i) If $\sum_{k=0}^{\infty}a_k x^k$ converges at x_0, $x_0 \neq 0$, then it converges absolutely at all x such that $|x| < |x_0|$.

(ii) If $\sum_{k=0}^{\infty}a_k x^k$ diverges at x_0, then it diverges at all x such that $|x| > |x_0|$.

$\dagger$ We put in the absolute value bars since the ratio test only applies to a series of *positive terms*. However, as this example shows, we can use the ratio test to test for the absolute convergence of any series of nonzero terms by inserting absolute value bars, thereby making all the terms positive.

PROOF. (i)   Since $\Sigma_{k=0}^{\infty} a_k x_0{}^k$ converges, $a_k x_0{}^k \to 0$ as $k \to \infty$ by Theorem 14.4.2. This implies that for all k sufficiently large, $|a_k x_0{}^k| < 1$. Then, if $|x| < |x_0|$ and if k is sufficiently large,

$$|a_k x^k| = \left| a_k \frac{x_0{}^k x^k}{x_0{}^k} \right| = |a_k x_0{}^k| \left| \frac{x}{x_0} \right|^k < \left| \frac{x}{x_0} \right|^k.$$

Since $|x| < |x_0|$, $|x/x_0| < 1$ and the geometric series $\Sigma_{k=0}^{\infty} |x/x_0|^k$ converges. Thus $\Sigma_{k=0}^{\infty} |a_k x^k|$ converges by the comparison test.

(ii)   Suppose $|x| > |x_0|$ and $\Sigma_{k=0}^{\infty} a_k x_0{}^k$ diverges. If $\Sigma_{k=0}^{\infty} a_k x_0{}^k$ did converge, then, by part (i), $\Sigma_{k=0}^{\infty} a_k x_0{}^k$ would also converge. This contradiction completes the proof of the theorem.

Theorem 1 is very useful for it enables us to place all power series in one of three categories:

Category 1: $\Sigma_{k=0}^{\infty} a_k x^k$ converges only at 0.

Category 2: $\Sigma_{k=0}^{\infty} a_k x^k$ converges for all real numbers.

Category 3: There exists a nonzero real number R, called the *radius of convergence* of the power series, such that $\Sigma_{k=0}^{\infty} a_k x^k$ converges if $|x| < R$ and diverges if $|x| > R$. At $x = R$ and at $x = -R$, the series may converge or diverge.

We can extend the notion of radius of convergence to categories 1 and 2:

1: In Category 1 we say that the radius of convergence is 0.
2: In Category 2 we say that the radius of convergence is ∞.

Note. The series in Examples 1 and 2 both fall into Category 3. In Example 1, $R = 3$, and in Example 2, $R = 1$.

EXAMPLE 3.   For what values of x does the series $\displaystyle\sum_{k=0}^{\infty} k! x^k$ converge?

SOLUTION. Here

$$\lim_{n \to \infty} \left| \frac{a_{n+1}}{a_n} \right| = \lim_{n \to \infty} \left| \frac{(n+1)! x^{n+1}}{n! x^n} \right| = |x| \lim_{n \to \infty} (n+1) = \infty$$

so that if $x \neq 0$, the series diverges. Thus $R = 0$ and the series falls into Category 1.

EXAMPLE 4.   For what values of x does the series $\displaystyle\sum_{k=0}^{\infty} x^k/k!$ converge?

SOLUTION. Here

$$\lim_{n \to \infty} \left| \frac{a_{n+1}}{a_n} \right| = \lim_{n \to \infty} \left| \frac{x^{n+1}/(n+1)!}{x^n/n!} \right| = |x| \lim_{n \to \infty} \frac{1}{n+1} = 0$$

so that the series converges for every real number x. Here $R = \infty$ and the series falls into Category 2.

In going through these examples, we find that there is an easy way to calculate the radius of convergence. The proof of the following theorem is easy and is left as an exercise (see Problem 36).

THEOREM 2. Consider the power series $\sum_{k=0}^{\infty} a_k x^k$ and suppose that $\lim_{n\to\infty} |a_{n+1}/a_n|$ exists and is equal to L.

(i) If $L = \infty$, then $R = 0$ and the series falls into Category 1.
(ii) If $L = 0$, then $R = \infty$ and the series falls into Category 2.
(iii) If $0 < L < \infty$, then $R = 1/L$ and the series falls into Category 3.

With this theorem in mind, we calculate the *interval of convergence*† of a power series in one or two steps.

(i) Calculate R. If $R = 0$ the series converges only at 0, and if $R = \infty$ the interval of convergence is $(-\infty, \infty)$.

(ii) If $0 < R < \infty$ check the values $x = -R$ and $x = R$. Then the interval of convergence is $(-R, R)$, $[-R, R)$, $(-R, R]$, or $[-R, R]$ depending on the convergence or divergence of the series at $x = R$ and $x = -R$.

Note. In Example 1, the interval of convergence is $(-3, 3)$ and in Example 2 the interval of convergence is $[-1, 1)$.

EXAMPLE 5. Find the radius of convergence and interval of convergence of the power series $\sum_{k=0}^{\infty} 2^k x^k /\ln(k + 2)$.

SOLUTION. Here $a_n = 2^n/\ln(n + 2)$ and

$$L = \lim_{n\to\infty} \left| \frac{a_{n+1}}{a_n} \right| = \lim_{n\to\infty} \left| \frac{2^{n+1}/\ln(n + 3)}{2^n/\ln(n + 2)} \right|$$

$$= 2 \lim_{n\to\infty} \frac{\ln(n + 2)}{\ln(n + 3)} = 2.$$

Thus $R = 1/L = \frac{1}{2}$. For $x = \frac{1}{2}$, $\sum_{k=0}^{\infty} 2^k x^k/\ln(k + 2) = \sum_{k=0}^{\infty} 1/\ln(k + 2)$ which diverges by comparison with the harmonic series since $1/\ln(k + 2) > 1/(k + 2)$ if $k \geq 1$. If $x = -\frac{1}{2}$ then $\sum_{k=0}^{\infty} 2^k x^k/\ln(k + 2) = \sum_{k=0}^{\infty} (-1)^k/\ln(k + 2)$ which converges by the alternating series test. Thus the interval of convergence is $[-\frac{1}{2}, \frac{1}{2})$.

EXAMPLE 6. Find the radius of convergence and interval of convergence of the power series $\sum_{k=0}^{\infty} (-1)^k (x - 3)^k/(k + 1)^2$.

SOLUTION. We make the substitution $u = x - 3$. The series then becomes $\sum_{k=0}^{\infty} (-1)^k u^k/(k + 1)^2$. This implies that

$$L = \lim_{n\to\infty} \left| \frac{a_{n+1}}{a_n} \right| = \lim_{n\to\infty} \frac{(n + 1)^2}{(n + 2)^2} = 1$$

so that $R = 1$. If $u = -1$, $\sum_{k=0}^{\infty} (-1)^k u^k/(k + 1)^2 = \sum_{k=0}^{\infty} 1/(k + 1)^2$ which converges. If $u = 1$, $\sum_{k=0}^{\infty} (-1)^k u^k/(k + 1)^2 = \sum_{k=0}^{\infty} (-1)^k/(k + 1)^2$ which also con-

† This is the interval over which the series converges.

verges. Thus the interval of convergence of the series $\sum_{k=0}^{\infty} (-1)^k u^k / (k + 1)^2$ is $[-1, 1]$. Since $u = x - 3$, the original series converges for $-1 \le x - 3 \le 1$ or $2 \le x \le 4$. Hence the interval of convergence is $[2, 4]$.

PROBLEMS 14.7

In Problems 1–33 find the radius of convergence and interval of convergence of the given power series.

1. $\sum_{k=0}^{\infty} \dfrac{x^k}{6^k}$

2. $\sum_{k=0}^{\infty} \dfrac{(-1)^k x^k}{8^k}$

3. $\sum_{k=0}^{\infty} \dfrac{(x + 1)^k}{3^k}$

4. $\sum_{k=0}^{\infty} \dfrac{(-1)^k (x - 3)^k}{4^k}$

5. $\sum_{k=0}^{\infty} (3x)^k$

6. $\sum_{k=0}^{\infty} \dfrac{x^k}{k^2 + 1}$

7. $\sum_{k=0}^{\infty} \dfrac{(x - 1)^k}{k^3 + 3}$

8. $\sum_{k=2}^{\infty} \dfrac{x^k}{(\ln k)^2}$

9. $\sum_{k=0}^{\infty} \dfrac{(x + 17)^k}{k!}$

10. $\sum_{k=2}^{\infty} \dfrac{x^k}{k \ln k}$

11. $\sum_{k=0}^{\infty} x^{2k}$

12. $\sum_{k=1}^{\infty} \dfrac{x^{2k}}{k}$

13. $\sum_{k=1}^{\infty} \dfrac{(-1)^k x^{2k}}{k^k}$

14. $\sum_{k=1}^{\infty} \dfrac{k x^k}{\ln(k + 1)}$

15. $\sum_{k=0}^{\infty} \dfrac{(-1)^k k x^k}{\sqrt{k + 1}}$

16. $\sum_{k=1}^{\infty} \dfrac{x^k}{k^k}$

*17. $\sum_{k=2}^{\infty} \dfrac{x^k}{(\ln k)^k}$ [*Hint:* Use the root test.]

*18. $\sum_{k=1}^{\infty} \dfrac{3^k x^k}{k^5}$ [*Hint:* See Example 14.5.12.]

19. $\sum_{k=1}^{\infty} \dfrac{(-2x)^k}{k^4}$

20. $\sum_{k=0}^{\infty} \dfrac{(2x + 3)^k}{k!}$

21. $\sum_{k=0}^{\infty} \dfrac{(2x + 3)^k}{5^k}$

22. $\sum_{k=0}^{\infty} \dfrac{(3x - 5)^k}{3^{2k}}$

23. $\sum_{k=0}^{\infty} \left(\dfrac{k}{15}\right)^k x^k$

24. $\sum_{k=0}^{\infty} (-1)^k x^{2k}$

25. $\sum_{k=0}^{\infty} (-1)^k x^{2k+1}$

26. $\sum_{k=1}^{\infty} \dfrac{(\ln k)(x + 3)^k}{k + 1}$

*27. $\sum_{k=1}^{\infty} k^k (x + 1)^k$

*28. $\sum_{k=1}^{\infty} \dfrac{k^k}{k!} x^k$ [*Hint:* See Example 14.5.9.]

29. $\sum_{k=0}^{\infty} \dfrac{(x + 10)^k}{(k + 1)3^k}$

*30. $\sum_{k=1}^{\infty} \dfrac{k!}{k^k} x^k$

31. $\sum_{k=0}^{\infty} [1 + (-1)^k] x^k$

32. $\sum_{k=0}^{\infty} \dfrac{[1 + (-1)^k]}{k!} x^k$

33. $\sum_{k=1}^{\infty} \dfrac{[1 + (-1)^k]}{k} x^k$.

34. Show that the interval of convergence of the power series $\sum_{k=0}^{\infty} (ax + b)^k / c^k$ with $a > 0$ is $((-c - b)/a, (c - b)/a)$.

35. Prove that if the interval of convergence of a power series is $(a, b]$, then the power series is conditionally convergent at b.

36.   Prove Theorem 2. [*Hint:* Show that if $|x| < 1/L$, then the series converges absolutely by applying the ratio test. Then show that if $|x| > 1/L$, the series diverges.]

37.   Show that if $\lim_{n\to\infty} |a_n|^{1/n} = L$, then the radius of convergence of $\Sigma_{k=0}^{\infty} a_k x^k$ is $1/L$.

38.   Show that if the radius of convergence of $\Sigma_{k=0}^{\infty} a_k x^k$ is R, and if $m > 0$ is an integer, then the radius of convergence of the power series $\Sigma_{k=0}^{\infty} a_k x^{mk}$ is $R^{1/m}$.

14.8 Differentiation and Integration of Power Series

Consider the power series

$$\sum_{k=0}^{\infty} a_k(x - x_0)^k = a_0 + a_1(x - x_0) + a_2(x - x_0)^2 + \cdots \tag{1}$$

with a given interval of convergence *I*. For each *x* in *I*, we may define a new function $f(x)$ by

$$f(x) = \sum_{k=0}^{\infty} a_k(x - x_0)^k. \tag{2}$$

As we shall see in this section and in Section 14.9, many familiar functions can be written as power series. In this section we shall discuss some properties of a function given in the form of equation (2).

EXAMPLE 1.   We know that

$$\sum_{k=0}^{\infty} x^k = 1 + x + x^2 + \cdots = \frac{1}{1 - x} \qquad \text{if} \quad |x| < 1. \tag{3}$$

Thus the function $1/(1 - x)$ can be defined by

$$f(x) = \frac{1}{1 - x} = \sum_{k=0}^{\infty} x^k \qquad \text{if} \quad |x| < 1.$$

EXAMPLE 2.   Substituting x^4 for *x* in (3) leads to the equality

$$f(x) = \frac{1}{1 - x^4} = \sum_{k=0}^{\infty} x^{4k} = 1 + x^4 + x^8 + x^{12} + \cdots \qquad \text{if} \quad |x| < 1.$$

EXAMPLE 3.   Substituting $-x$ for *x* in (3) leads to the equality

$$f(x) = \frac{1}{1 - (-x)} = \frac{1}{1 + x} = \sum_{k=0}^{\infty} (-1)^k x^k$$

$$= 1 - x + x^2 - x^3 + x^4 - \cdots \qquad \text{if} \quad |x| < 1.$$

Once we see that certain functions can be written as power series, the next question that naturally arises is whether such functions can be differentiated and integrated. The remarkable theorem given below ensures that every function represented as a power series can be differentiated and integrated at any x such that $|x| < R$, the radius of convergence. Moreover, we see how the derivative and integral can be calculated. The proof of this theorem is long (but not conceptually difficult) and so is omitted.†

THEOREM 1. Let the power series $\sum_{k=0}^{\infty} a_k x^k$ have the radius of convergence $R > 0$. Let

$$f(x) = \sum_{k=0}^{\infty} a_k x^k = a_0 + a_1 x + a_2 x^2 + \cdots \qquad \text{for} \quad |x| < R.$$

Then, for $|x| < R$,

(i) $f(x)$ is continuous,
(ii) the derivative $f'(x)$ exists, and

$$f'(x) = \frac{d}{dx} a_0 + \frac{d}{dx} a_1 x + \frac{d}{dx} a_2 x^2 + \cdots$$

$$= a_1 + 2a_2 x + 3a_3 x^2 + \cdots = \sum_{k=1}^{\infty} k a_k x^{k-1}. \qquad (4)$$

(iii) The indefinite integral $\int f(x)\, dx$ exists and

$$\int f(x)\, dx = \int a_0\, dx + \int a_1 x\, dx + \int a_2 x^2\, dx + \cdots$$

$$= a_0 x + a_1 \frac{x^2}{2} + a_2 \frac{x^3}{3} + \cdots = \sum_{k=0}^{\infty} a_k \frac{x^{k+1}}{k+1} + C. \qquad (5)$$

Moreover, the two series $\sum_{k=1}^{\infty} k a_k x^{k-1}$ and $\sum_{k=0}^{\infty} a_k x^{k+1}/(k+1)$ both have the radius of convergence R.

> **Remark 1.** Simply put, this theorem says that the derivative of a converging power series is the series of derivatives of its terms and that the integral of a converging power series is the series of integrals of its terms.

> **Remark 2.** A more concise statement of Theorem 1 is: *A power series may be differentiated and integrated term-by-term within its radius of convergence.*

EXAMPLE 4. From Example 1 we see that

$$\frac{1}{1-x} = 1 + x + x^2 + \cdots = \sum_{k=0}^{\infty} x^k \qquad \text{for} \quad |x| < 1.$$

Then differentiating and integrating yields

$$\frac{1}{(1-x)^2} = \frac{d}{dx}\left(\frac{1}{1-x}\right) = 1 + 2x + 3x^2 + \cdots$$

$$= \sum_{k=1}^{\infty} k x^{k-1} \qquad \text{for} \quad |x| < 1, \qquad (6)$$

†For a proof see R. C. Buck, *Advanced Calculus,* McGraw-Hill, New York, 1965, p. 198.

and

$$-\ln(1-x) = \int \frac{dx}{1-x} = x + \frac{x^2}{2} + \frac{x^3}{3} + \cdots = \sum_{k=0}^{\infty} \frac{x^{k+1}}{k+1} = \sum_{k=1}^{\infty} \frac{x^k}{k} \quad (7)$$

for $|x| < 1$. It is easy to verify that the radii of convergence of series (6) and (7) are both 1.

EXAMPLE 5. From Example 3, we have

$$\frac{1}{1+x} = 1 - x + x^2 - x^3 + \cdots = \sum_{k=0}^{\infty} (-1)^k x^k. \quad (8)$$

Substituting $u = x + 1$, we have $x = u - 1$ and $1/(1+x) = 1/u$. If $-1 < x < 1$, then $0 < u < 2$ and we obtain

$$\frac{1}{u} = 1 - (u-1) + (u-1)^2 - (u-1)^3 + \cdots = \sum_{k=0}^{\infty} (-1)^k (u-1)^k$$

for $0 < u < 2$. Integration then yields

$$\ln u = \int \frac{du}{u} = u - \frac{(u-1)^2}{2} + \frac{(u-1)^3}{3} - \cdots + C.$$

Since $\ln 1 = 0$, we immediately find that $C = -1$ so that

$$\ln u = \sum_{k=0}^{\infty} (-1)^k \frac{(u-1)^{k+1}}{k+1} \quad (9)$$

for $0 < u < 2$. Here we have expressed the logarithmic function as a power series.

EXAMPLE 6. The series

$$f(x) = 1 + x + \frac{x^2}{2!} + \frac{x^3}{3!} + \cdots = \sum_{k=0}^{\infty} \frac{x^k}{k!} \quad (10)$$

converges for every real number x (i.e., $R = \infty$—see Example 14.7.4). But

$$f'(x) = \frac{d}{dx} 1 + \frac{d}{dx} x + \frac{d}{dx} \frac{x^2}{2!} + \cdots = 1 + x + \frac{x^2}{2!} + \cdots = f(x).$$

Thus f satisfies the differential equation

$$f' = f$$

and so, from the discussion in Section 6.4, we find that

$$f(x) = ce^x \quad (11)$$

for some constant c. Then substituting $x = 0$ into equations (10) and (11) yields

$$f(0) = 1 = ce^0 = c$$

so that $f(x) = e^x$ and we have obtained the important expansion which is valid for any real number x:

$$e^x = 1 + x + \frac{x^2}{2!} + \frac{x^3}{3!} + \cdots = \sum_{k=0}^{\infty} \frac{x^k}{k!}. \tag{12}$$

For example, if we substitute the value $x = 1$ into (12), we obtain partial sum approximations for $e = 1 + 1 + 1/2! + 1/3! + \cdots$. (See Table 1). The last value $(\sum_{k=0}^{8} 1/k!)$ is correct to five decimal places.

TABLE 1

n	0	1	2	3	4	5	6	7	8
$S_n = \sum_{k=0}^{n} \frac{1}{k!}$	1	2	2.5	2.66667	2.70833	2.71667	2.71806	2.71825	2.71828

□**EXAMPLE 7.** Substituting $x = -x$ in (12) we obtain

$$e^{-x} = 1 - x + \frac{x^2}{2!} - \frac{x^3}{3!} + \cdots = \sum_{k=0}^{\infty} (-1)^k \frac{x^k}{k!}. \tag{13}$$

Since this is an alternating series if $x > 0$, Theorem 14.6.3 tells us that the error $|S - S_n|$ in approximating e^{-x} for $x > 0$ is bounded by $|a_{n+1}| = x^{n+1}/(n + 1)!$.† For example, to calculate e^{-1} with an error of less than 0.0001, we must have $|S - S_n| \le 1/(n + 1)! < 0.0001 = 1/10{,}000$ or $(n + 1)! > 10{,}000$. If $n = 7$, $(n + 1)! = 8! = 40{,}320$ so that $\sum_{k=0}^{7} (-1)^k/k!$ will approximate e^{-1} correct to four decimal places. We obtain

$$e^{-1} \approx 1 - 1 + \frac{1}{2!} - \frac{1}{3!} + \frac{1}{4!} - \frac{1}{5!} + \frac{1}{6!} - \frac{1}{7!}$$

$$= \frac{1}{2} - \frac{1}{6} + \frac{1}{24} - \frac{1}{120} + \frac{1}{720} - \frac{1}{5{,}040}$$

$$= 0.5 - 0.16667 + 0.04167 - 0.00833 + 0.00139 - 0.0002$$

$$= 0.36786$$

and $e^{-1} \approx 0.3679$ correct to four decimal places.

EXAMPLE 8. Consider the series

$$f(x) = 1 - \frac{x^2}{2!} + \frac{x^4}{4!} - \frac{x^6}{6!} + \cdots = \sum_{k=0}^{\infty} (-1)^k \frac{x^{2k}}{(2k)!}. \tag{14}$$

It is easy to see that $R = \infty$ since the series is absolutely convergent for every x by comparison with the series (12) for e^x. (The series (12) is larger than the series (14)

---

† We can apply Theorem 14.6.3 here because the terms in the sequence $\{1/n!\}$ are monotone decreasing.

since it contains the terms $x^n/n!$ for n both even and odd, not just for n even as in (14).) Differentiating, we obtain

$$f'(x) = -x + \frac{x^3}{3!} - \frac{x^5}{5!} + \frac{x^7}{7!} - \cdots = \sum_{k=0}^{\infty} (-1)^{k+1} \frac{x^{2k+1}}{(2k+1)!} \tag{15}$$

Since this series has a radius of convergence $R = \infty$, we can differentiate once more to obtain

$$f''(x) = -1 + \frac{x^2}{2!} - \frac{x^4}{4!} + \frac{x^6}{6!} - \cdots = \sum_{k=0}^{\infty} \frac{(-1)^{k+1} x^{2k}}{(2k)!} = -f(x).$$

Thus we see that f satisfies the differential equation

$$f'' + f = 0. \tag{16}$$

Moreover, from equations (14) and (15) we see that

$$f(0) = 1 \quad \text{and} \quad f'(0) = 0. \tag{17}$$

We discussed this differential equation in Section 7.7. It is easily seen that the function $f(x) = \cos x$ satisfies equation (16) together with the conditions (17). In fact, although we do not prove this here, it is the only function which does this.† Thus we have

$$\boxed{\cos x = 1 - \frac{x^2}{2!} + \frac{x^4}{4!} - \frac{x^6}{6!} + \cdots = \sum_{k=0}^{\infty} (-1)^k \frac{x^{2k}}{(2k)!}.} \tag{18}$$

Since

$$\frac{d}{dx} \cos x = -\sin x,$$

we obtain from (15) the series (after multiplying both sides by -1)

$$\boxed{\sin x = x - \frac{x^3}{3!} + \frac{x^5}{5!} - \frac{x^7}{7!} + \cdots = \sum_{k=0}^{\infty} (-1)^k \frac{x^{2k+1}}{(2k+1)!}.} \tag{19}$$

Power series expansions can be very useful for integration.

□**EXAMPLE 9.** Calculate $\int_0^1 e^{-t^2} \, dt$ with an error of < 0.0001.

SOLUTION. Substituting t^2 for x in (13), we find that

$$e^{-t^2} = 1 - t^2 + \frac{t^4}{2!} - \frac{t^6}{3!} + \cdots = \sum_{k=0}^{\infty} (-1)^k \frac{t^{2k}}{k!}. \tag{20}$$

† This follows from the uniqueness of solutions to linear differential equations, a fact discussed in Chapter 20.

Then we integrate to find that

$$\int_0^x e^{-t^2}\, dt = x - \frac{x^3}{3} + \frac{x^5}{5 \cdot 2!} - \frac{x^7}{7 \cdot 3!} + \cdots = \sum_{k=0}^{\infty} \frac{(-1)^k x^{2k+1}}{(2k+1)k!}. \tag{21}$$

The error $|S - S_n|$ is bounded by

$$|a_{n+1}| = \left| \frac{x^{2(n+1)+1}}{[2(n+1)+1](n+1)!} \right|.$$

In our example $x = 1$ so we need to choose n so that

$$\frac{1}{(2n+3)(n+1)!} < 0.0001 \qquad \text{or} \qquad (2n+3)(n+1)! > 10{,}000.$$

If $n = 6$, then $(2n+3)(n+1)! = (15)(7!) = 75{,}600$. With this choice of n ($n = 5$ is too small), we obtain

$$\int_0^1 e^{-t^2}\, dt \approx 1 - \frac{1}{3} + \frac{1}{5 \cdot 2!} - \frac{1}{7 \cdot 3!} + \frac{1}{9 \cdot 4!} - \frac{1}{11 \cdot 5!} + \frac{1}{13 \cdot 6!}$$

$$\approx 1 - 0.33333 + 0.1 - 0.02381 + 0.00463 - 0.00076 + 0.00011$$

$$= 0.74684$$

and, to four decimal places,

$$\int_0^1 e^{-t^2}\, dt = 0.7468.$$

Note that in Section 8.9, we discussed some more general ways to calculate definite integrands which do not require the existence of a series expansion of the function being integrated. However, as seen in Example 9 above, a power series provides an easy method of numerical integration when the power series representation of a function is readily obtainable.

PROBLEMS 14.8

1. By substituting x^2 for x in (8), find a series expansion for $1/(1 + x^2)$ which is valid for $|x| < 1$.
2. Integrate the series obtained in Problem 1 to obtain a series expansion for $\tan^{-1} x$.
3. Use the result of Problem 2 to obtain an estimate of π which is correct to 2 decimal places.
4. Use the series expansion for $\ln x$ to calculate to two decimal places of accuracy:
   (a) $\ln 0.5$; (b) $\ln 1.6$.

In Problems 5–13 estimate the given integral to within the given accuracy.

5. $\displaystyle\int_0^1 e^{-t^2}\, dt$; error < 0.01
6. $\displaystyle\int_0^1 e^{-t^3}\, dt$; error < 0.001
7. $\displaystyle\int_0^{1/2} \cos t^2\, dt$; error < 0.001
8. $\displaystyle\int_0^{1/2} \sin t^2\, dt$; error < 0.0001

9. $\int_0^1 t^2 e^{-t^2} \, dt$; error < 0.01 [*Hint:* The series expansion of $t^2 e^{-t^2}$ is obtained by multiplying each term of the series expansion of e^{-t^2} by t^2.]

10. $\int_0^{1/4} t^5 e^{-t^5} \, dt$; error < 0.0001 11. $\int_0^1 \cos \sqrt{t} \, dt$; error < 0.01

12. $\int_0^1 t \sin \sqrt{t} \, dt$; error < 0.001 13. $\int_0^{1/2} \dfrac{dt}{1 + t^8}$; error < 0.0001.

14. Find a series expansion for xe^x that is valid for all real values of x.
15. Use the result of Problem 14 to find a series expansion for $\int_0^x t e^t \, dt$.
16. Use the result of Problem 15 to show that $\sum_{k=0}^{\infty} 1/(k+2)k! = 1$.
17. Find a power series expansion for $\int_0^x [\ln(1 + t)/t] \, dt$.
18. Expand $1/x$ as a power series of the form $\sum_{k=0}^{\infty} a_k(x - 1)^k$. What is the interval of convergence of this series?
*19. Define the function $J(x)$ by $J(x) = \sum_{k=0}^{\infty} [(-1)^k/(k!)^2](x/2)^{2k}$.
   (a) What is the interval of convergence of this series?
   (b) Show that $J(x)$ satisfies the differential equation

$$x^2 J''(x) + x J'(x) + x^2 J(x) = 0.$$

The function $J(x)$ is called a *Bessel function* of order zero.†

14.9 Taylor and Maclaurin Series

In the last two sections we used the fact that, within its interval of convergence, the function

$$f(x) = \sum_{k=0}^{\infty} a_k(x - x_0)^k$$

is a function which is differentiable and integrable. In this section we look more closely at the coefficients a_k and show that they can be represented in terms of derivatives of the function f.

We begin with the case $x_0 = 0$ and assume that $R > 0$ so that the theorem on power series differentiation applies. We have

$$f(x) = \sum_{k=0}^{\infty} a_k x^k = a_0 + a_1 x + a_2 x^2 + \cdots + a_n x^n + \cdots \tag{1}$$

and, clearly,

$$f(0) = a_0 + 0 + 0 + \cdots + 0 + \cdots = a_0. \tag{2}$$

If we differentiate (1), we obtain

$$f'(x) = \sum_{k=1}^{\infty} k a_k x^{k-1} = a_1 + 2a_2 x + 3a_3 x^2 + \cdots + n a_n x^{n-1} + \cdots \tag{3}$$

---

†Named after the German physicist and mathematician Wilhelm Bessel (1784–1846) who used the function in his study of planetary motion.

and

$$f'(0) = a_1. \tag{4}$$

Continuing to differentiate, we obtain

$$f''(x) = \sum_{k=2}^{\infty} k(k-1)a_k x^{k-2} = 2a_2 + 3a_3 x + 4 \cdot 3a_4 x^2$$
$$+ \cdots + n(n-1)a_n x^{n-2} + \cdots$$

and

$$f''(0) = 2a_2$$

or

$$a_2 = \frac{f''(0)}{2} = \frac{f''(0)}{2!}. \tag{5}$$

Similarly

$$f'''(x) = \sum_{k=3}^{\infty} k(k-1)(k-2)a_k x^{k-3}$$
$$= 3 \cdot 2a_3 + 4 \cdot 3 \cdot 2a_4 x + 5 \cdot 4 \cdot 3a_5 x^2 + \cdots$$
$$+ n(n-1)(n-2)a_n x^{n-3} + \cdots$$

and

$$f'''(0) = 3 \cdot 2a_3$$

or

$$a_3 = \frac{f'''(0)}{3 \cdot 2} = \frac{f'''(0)}{3!}. \tag{6}$$

It is not difficult to see that this pattern continues and that, for every positive integer n,

$$a_n = \frac{f^{(n)}(0)}{n!}. \tag{7}$$

For $n = 0$, we use the convention $0! = 1$ and $f^{(0)}(x) = f(x)$. Then formula (7) holds for every n and we have

$$\boxed{\begin{aligned} f(x) &= \sum_{k=0}^{\infty} a_k x^k = \sum_{k=0}^{\infty} \frac{f^{(k)}(0)}{k!} x^k \\ &= f(0) + f'(0)x + f''(0)\frac{x^2}{2!} + \cdots + f^{(n)}(0)\frac{x^n}{n!} + \cdots. \end{aligned}} \tag{8}$$

In the general case, if

$$f(x) = \sum_{k=0}^{\infty} a_k(x - x_0)^k = a_0 + a_1(x - x_0) + a_2(x - x_0)^2$$

$$+ \cdots + a_n(x - x_0)^n + \cdots. \tag{9}$$

Then

$$f(x_0) = a_0$$

and, differentiating as before, we find that

$$a_n = \frac{f^{(n)}(x_0)}{n!} \tag{10}$$

so that the series in (9) can be written as

$$f(x) = \sum_{k=0}^{\infty} \frac{f^{(k)}(x_0)}{k!}(x - x_0)^k$$

$$= f(x_0) + f'(x_0)(x - x_0) + f''(x_0)\frac{(x - x_0)^2}{2!} + \cdots$$

$$+ f^{(n)}(x)\frac{(x - x_0)^n}{n!} + \cdots. \tag{11}$$

The series in (11) is called the *Taylor series*† of the function f at x_0. The special case $x_0 = 0$ in (8) is called a *Maclaurin series*.‡

Warning. We have shown here that *if* $f(x) = \sum_{k=0}^{\infty} a_k(x - x_0)^k$, then f is infinitely differentiable (i.e., f has derivatives of all orders) and that the series for f is the Taylor series (or Maclaurin series if $x_0 = 0$) of f. What we have *not* shown is that if f is infinitely differentiable at x_0, then f has a Taylor series expansion at x_0. In general, this last statement is false as we shall see in Example 3 below.

EXAMPLE 1. Find the Maclaurin series for e^x.

SOLUTION. If $f(x) = e^x$, then $f(0) = f'(0) = \cdots = f^{(k)}(0) = 1$ and

$$e^x = \sum_{k=0}^{\infty} \frac{x^k}{k!} = 1 + x + \frac{x^2}{2!} + \frac{x^3}{3!} + \cdots + \frac{x^n}{n!} + \cdots. \tag{12}$$

---

† We see that the first n terms of the Taylor series of a function are simply the Taylor polynomial described in Section 13.1. The history of Taylor series is somewhat muddied. It has been claimed that the basis for its development was found in India before 1550! (Taylor published the result in 1715.) For an interesting discussion of this, see the paper by C. T. Rajagopal and T. V. Vedamurthi, "On the Hindu Proof of Gregory's Series," Scripta Mathematica, **17**, 65–74 (1951).

‡ Named after Colin Maclaurin (1698–1746) who was probably the outstanding British mathematician in the generation after Newton. He was a child prodigy who became a professor of mathematics at the University of Aberdeen, Scotland, at the age of 19.

This is the series we obtained in Example 14.8.6. It is important to note here that what this example shows is that *if* e^x has a Maclaurin series expansion, then the series must be the series (12). It does not show that e^x actually does have such a series expansion. To prove that the series in (12) is really equal to e^x, we differentiate as in Example 14.8.6, and use the fact that the only continuous function which satisfies

$$f'(x) = f(x), \qquad f(0) = 1$$

is the function e^x.

EXAMPLE 2. Assuming that the function $f(x) = \cos x$ can be written as a Maclaurin series, find that series.

SOLUTION. If $f(x) = \cos x$, then $f(0) = 1$, $f'(0) = 0$, $f''(0) = -1$, $f'''(0) = 0$, $f^{(iv)}(0) = 1$, and so on so that if

$$\cos x = \sum_{k=0}^{\infty} a_k x^k,$$

then

$$\cos x = f(0) + f'(0) + \frac{f''(0)x^2}{2!} + \frac{f'''(0)x^3}{3!} + \frac{f^{(iv)}(0)x^4}{4!} + \cdots$$

or

$$\boxed{\cos x = 1 - \frac{x^2}{2!} + \frac{x^4}{4!} - \frac{x^6}{6!} + \cdots = \sum_{k=0}^{\infty} \frac{(-1)^k x^{2k}}{(2k)!}.} \tag{13}$$

This is the series found in Example 14.8.8.

> **Note.** Again, this does not prove that the equality (13) is correct. It only shows that *if* $\cos x$ has a Maclaurin expansion, then the expansion must be given by (13).

EXAMPLE 3. Let

$$f(x) = \begin{cases} e^{-1/x^2} & \text{if } x \neq 0 \\ 0 & \text{if } x = 0 \end{cases}.$$

Find a Maclaurin expansion for f if one exists.

SOLUTION. First we note that since $\lim_{x \to 0} e^{-1/x^2} = 0$, f is continuous. Now $\lim_{x \to 0} e^{-1/x^2}/x^n = 0$ for every positive integer n (see Problem 12.3.35). For example,

$$\lim_{x \to 0} \frac{e^{-1/x^2}}{x} = \lim_{x \to 0} \frac{1/x}{e^{1/x^2}} = \lim_{x \to 0} \frac{-1/x^2}{-(2/x^3)e^{1/x^2}} = \lim_{x \to 0} \frac{x}{2e^{1/x^2}} = 0$$

and

$$\lim_{x \to 0} \frac{e^{-1/x^2}}{x^3} = \lim_{x \to 0} \frac{1/x^3}{e^{1/x^2}} = \lim_{x \to 0} \frac{-3/x^4}{-(2/x^3)e^{1/x^2}} = \frac{3}{2} \lim_{x \to 0} \frac{e^{-1/x^2}}{x} = 0.$$

Now for $x \neq 0$, $f'(x) = (2/x^3)e^{-1/x^2} \to 0$ as $x \to 0$ so that f' is continuous at 0. Similarly, $f''(x) = [(4/x^6) - (6/x^4)]e^{-1/x^2}$ which also $\to 0$ as $x \to 0$ by the above limit result. In fact, *every* derivative of f is continuous and $f^{(n)}(0) = 0$ for every n. Thus f is infinitely differentiable and, *if* it had a Maclaurin series, then we would have

$$f(x) = f(0) + f'(0)x + f''(0)\frac{x^2}{2!} + \cdots.$$

But $f(0) = f'(0) = f''(0) = \cdots = 0$ so that the Maclaurin series would be the zero series. But since f is obviously not the zero function, we can only conclude that f does *not* have a Maclaurin series.

This last example illustrates that infinite differentiability is not sufficient to guarantee that a given function can be represented by its Taylor series. Something more is needed.

DEFINITION 1.   We say that a function f is *analytic* at x_0 if f can be represented by a Taylor series in some neighborhood of x_0.

We see that the functions e^x and $\cos x$ are analytic at 0 while the function

$$f(x) = \begin{cases} e^{-1/x^2}, & x \neq 0 \\ 0, & x = 0 \end{cases}$$

is not. A condition that guarantees analyticity of an infinitely differentiable function is given below.

THEOREM 1.   Suppose that the function f has continuous derivatives of all orders in a neighborhood N of the number x_0. Then

---

f is analytic at x_0 if and only if

$$\lim_{n \to \infty} R_n(x) = \lim_{n \to \infty} \frac{f^{(n+1)}(c_n)}{(n+1)!}(x - x_0)^{n+1} = 0 \tag{14}$$

for every x in N where c_n is between x_0 and x.

Remark.   The expression on the right-hand side of (14) is simply the remainder term given by Taylor's theorem (Theorems 13.1.1 and 13.1.2).

PROOF.   The hypotheses of Taylor's theorem apply so that we can write, for any n,

$$f(x) = P_n(x) + R_n(x), \tag{15}$$

where $P_n(x)$ is the nth Taylor polynomial for f. To show that f is analytic, it is necessary to show that

$$\lim_{n \to \infty} P_n(x) = f(x) \tag{16}$$

for every x in N. But, if x is in N, we obtain, from (14) and (15),

$$\lim_{n \to \infty} P_n(x) = \lim_{n \to \infty} [f(x) - R_n(x)] = f(x) - \lim_{n \to \infty} R_n(x) = f(x) - 0 = f(x)$$

and the theorem is proved.

EXAMPLE 4. If $f(x) = e^x$, then $f^{(n)}(x) = e^x$ and

$$\lim_{n \to \infty} \frac{f^{(n+1)}(c_n)}{(n+1)!}(x-0)^{n+1} = \lim_{n \to \infty} \frac{e^{c_n}x^{n+1}}{(n+1)!} = \quad \le e^{|x|} \lim_{n \to \infty} \frac{|x|^{n+1}}{(n+1)!} \to 0$$

since $|x|^{n+1}/(n+1)!$ is the $(n+2)$nd term in the converging power series $\sum_{k=0}^{\infty} |x|^k/k!$ and the terms in a converging power series $\to 0$ by Theorem 14.4.2. Since this is true for any $x \in \mathbb{R}$, we may take $N = (-\infty, \infty)$ to conclude that the series (12) is valid for every real number x.

EXAMPLE 5. Let $f(x) = \cos x$. Since all derivatives of $\cos x$ are equal to $\pm\sin x$ or $\pm\cos x$, we see that $|f^{(n+1)}(c_n)| \le 1$. Then for $x_0 = 0$, $|R_n(x)| \le |x|^{n+1}/(n+1)!$ which $\to 0$ as $n \to \infty$ so that the series (13) is also valid for every real number x.

EXAMPLE 6. It can be shown (although it is extremely difficult) that for the function in Example 3, $R_n(x) \not\to 0$. This occurs because the derivatives $f^{(n)}(x)$ have larger and larger maxima for x in $(-\infty, \infty)$ and so are *not* bounded by some fixed constant M. Then $\lim_{n \to \infty} f^{(n+1)}(c_n)$ is not finite and

$$\lim_{n \to \infty} f^{(n+1)}(c_n) \frac{x^{n+1}}{(n+1)!}$$

is not 0 if $x \ne 0$.

EXAMPLE 7. Find the Taylor expansion for $f(x) = \ln x$ at $x = 1$.

SOLUTION. Since $f'(x) = 1/x$, $f''(x) = -1/x^2$, $f'''(x) = 2/x^3$, $f^{(iv)}(x) = -6/x^4, \ldots, f^{(n)}(x) = (-1)^{n+1}(n-1)!/x^n$, we find that $f(1) = 0$, $f'(1) = 1$, $f''(1) = -1$, $f'''(1) = 2$, $f^{(iv)}(1) = -6, \ldots, f^{(n)}(1) = (-1)^{n+1}(n-1)!$. Then, wherever valid,

$$\ln x = \sum_{k=0}^{\infty} f^{(k)}(1) \frac{(x-1)^k}{k!}$$

$$= 0 + (x-1) - \frac{(x-1)^2}{2} + \frac{2(x-1)^3}{3!} - \frac{3!(x-1)^4}{4!} + \frac{4!(x-1)^5}{5!} + \cdots$$

or

$$\ln x = (x-1) - \frac{(x-1)^2}{2} + \frac{(x-1)^3}{3} - \frac{(x-1)^4}{4} + \cdots$$

$$= \sum_{k=1}^{\infty} \frac{(-1)^{k+1}(x-1)^k}{k}. \tag{17}$$

Now define the neighborhood $N = (\alpha, \infty)$ for $0 < \alpha < 1$. We have, for $x \in N$, $|f^{(n+1)}(x)| = n!/x^{n+1} < n!/\alpha^{n+1}$ since $x > \alpha$. Thus

$$|R_n(x)| = \left| f^{(n+1)}(c_n) \frac{(x-1)^{n+1}}{(n+1)!} \right| \le \frac{1}{\alpha^{n+1}} \left| \frac{(x-1)^{n+1}}{n+1} \right|$$

$$= \left| \frac{[(x-1)/\alpha]^{n+1}}{n+1} \right|.$$

This converges to 0 if $|(x - 1)/\alpha| < 1$ or if $1 - \alpha < x < 1 + \alpha$. Since α can be any number in $(0, 1)$, we see that the series (17) is equal to $\ln x$ if $0 < x < 2$. If $x > 2$, then the series (17) obviously diverges. If $x = 0$, then the series (17) is the negative of the harmonic series which diverges to $-\infty$. Note that $\lim_{x \to 0^+} \ln x = -\infty$. Finally, if $x = 2$, the series (17) is the alternating series $1 - \frac{1}{2} + \frac{1}{3} - \frac{1}{4} + \cdots$ which converges. Thus the series (17) represents $\ln x$ for $0 < x \le 2$ and, as a corollary, we see that

$$\ln 2 = 1 - \frac{1}{2} + \frac{1}{3} - \frac{1}{4} + \frac{1}{5} - \cdots = \sum_{k=1}^{\infty} \frac{(-1)^{k+1}}{k}. \tag{18}$$

EXAMPLE 8. Find the Maclaurin series for $f(x) = \sinh x$.

SOLUTION. Here $f'(x) = \cosh x, f''(x) = \sinh x$, and so on so that $f(0) = 0, f'(0) = 1, f''(0) = 0, f'''(0) = 1, \ldots$, and

$$\sinh x = x + \frac{x^3}{3!} + \frac{x^5}{5!} + \frac{x^7}{7!} + \cdots. \tag{19}$$

Let x be any real number. If $|x| < c$, choose $N = [-2c, 2c]$. Since $\sinh x$ and $\cosh x$ are continuous in N, they are bounded by some constant M for x in N. Then

$$|R_n(x)| = \left| \frac{f^{(n+1)}(c_n) x^{n+1}}{(n + 1)!} \right| < \frac{M|x^{n+1}|}{(n + 1)!} \to 0 \quad \text{as} \quad n \to \infty.$$

Thus the series (19) is valid for every real number x.

EXAMPLE 9. Find a Taylor series for $f(x) = \sin x$ at $x = \pi/3$.

SOLUTION. Here $f(\pi/3) = \sqrt{3}/2$, $f'(\pi/3) = 1/2$, $f''(\pi/3) = -\sqrt{3}/2$, $f'''(\pi/3) = -1/2$, and so on so that

$$\sin x = \frac{\sqrt{3}}{2} + \frac{1}{2}\left(x - \frac{\pi}{3}\right) - \frac{\sqrt{3}}{2}\frac{[x - (\pi/3)]^2}{2!} - \frac{1}{2}\frac{[x - (\pi/3)]^3}{3!}$$
$$+ \frac{\sqrt{3}}{2}\frac{[x - (\pi/3)]^4}{4!} + \cdots.$$

The proof that this series is valid for every real number x is similar to the proof in Example 5 and is therefore omitted.

We conclude this section with a list of useful Maclaurin series:

$$e^x = \sum_{k=0}^{\infty} \frac{x^k}{k!} = 1 + x + \frac{x^2}{2!} + \frac{x^3}{3!} + \cdots \tag{20}$$

$$\cos x = \sum_{k=0}^{\infty} \frac{(-1)^k x^{2k}}{(2k)!} = 1 - \frac{x^2}{2!} + \frac{x^4}{4!} - \frac{x^6}{6!} + \cdots \tag{21}$$

$$\sin x = \sum_{k=0}^{\infty} \frac{(-1)^k x^{2k+1}}{(2k + 1)!} = x - \frac{x^3}{3!} + \frac{x^5}{5!} - \frac{x^7}{7!} + \cdots \tag{22}$$

$$\cosh x = \sum_{k=0}^{\infty} \frac{x^{2k}}{(2k)!} = 1 + \frac{x^2}{2!} + \frac{x^4}{4!} + \frac{x^6}{6!} + \cdots \tag{23}$$

$$\sinh x = \sum_{k=0}^{\infty} \frac{x^{2k+1}}{(2k+1)!} = x + \frac{x^3}{3!} + \frac{x^5}{5!} + \frac{x^7}{7!} + \cdots \tag{24}$$

You are asked to prove, in Problems 1 and 2, that the series (22) and (23) are valid for every real number x.

PROBLEMS 14.9

1. Prove that the series (22) represents $\sin x$ for all real x.
2. Prove that the series (23) represents $\cosh x$ for all real x.
3. Find the Taylor series for e^x at 1.
4. Find the Maclaurin series for e^{-x}.
5. Find the Taylor series for $\cos x$ at $\pi/4$.
6. Find the Taylor series for $\sinh x$ at $\ln 2$.
7. Find the Maclaurin series for $e^{\alpha x}$, α real.
8. Find the Maclaurin series for xe^x.
9. Find the Maclaurin series for $x^2 e^{-x^2}$.
10. Find the Maclaurin series for $(\sin x)/x$.
11. Find the Taylor series for e^x at $x = -1$.
12. Find the Maclaurin series for $\sin^2 x$. [*Hint:* $\sin^2 x = (1 - \cos 2x)/2$.]
13. Find the Taylor series for $(x - 1) \ln x$ at 1. Over what interval is this representation valid?
14. Find the first four terms of the Maclaurin series for $\tan x$. What is its interval of convergence?
15. Find the first four terms of the Taylor series for $\csc x$ at $\pi/2$. What is its interval of convergence?
16. Find the first five terms of the Maclaurin series for $\ln \cos x$. What is its interval of convergence? [*Hint:* $\tan x \, dx = -\ln |\cos x|$.]
17. Find the Taylor series of $\sqrt{x}$ at $x = 4$. What is its radius of convergence?
18. Find the Maclaurin series of $\sin^{-1} x$. What is its radius of convergence? [*Hint:* Expand $1/\sqrt{1 - x^2}$ and integrate.]
19. Use the Maclaurin series for $\sin x$ to obtain the Maclaurin series for $\sin x^2$.
20. Find the Maclaurin series for $\cos x^2$.
21. Differentiate the Maclaurin series for $\sin x$ and show that it is equal to the Maclaurin series for $\cos x$.
22. Differentiate the Maclaurin series for $\sinh x$ and show that it is equal to the Maclaurin series for $\cosh x$.
23. Using the fact that if f has a Taylor series at x_0 then the Taylor series is given by (11), show that $1/(1 - x) = 1 + x + x^2 + \cdots$ is the Taylor series for $1/(1 - x)$ when $|x| < 1$.
24. Find the Maclaurin series for $\tan^{-1} x$. What is its interval of convergence? [*Hint:* Integrate the series for $1/(1 + x^2)$.]
25. Prove that for any real number r

$$(1 + x)^r = 1 + rx + \frac{r(r-1)}{2!}x^2 + \frac{r(r-1)(r-2)}{3!}x^3 + \cdots$$

$$+ \frac{r(r-1)\cdots(r-n+1)}{n!}x^n + \cdots$$

for $|x| < 1$. This series is called the *binomial series*.

26. Show that for any real number r

$$1 + \frac{r}{2} + \frac{r(r-1)}{2^2 2!} + \cdots + \frac{r(r-1)\cdots(r-n+1)}{2^n n!} = \left(\frac{3}{2}\right)^r.$$

□27. The *Error function* (which arises in mathematical statistics) is defined by

$$\text{erf}(x) = \frac{2}{\sqrt{\pi}} \int_0^x e^{-t^2} \, dt.$$

(a) Find a Maclaurin series for $\text{erf}(x)$ by integrating the Maclaurin series for e^{-x^2}.
(b) Use the series obtained in (a) to estimate, with an error of < 0.0001, $\text{erf}(1)$ and $\text{erf}(\frac{1}{2})$.

□28. The *complementary error function* is defined by

$$\text{erfc}(x) = 1 - \text{erf}(x) = 1 - \frac{2}{\sqrt{\pi}} \int_0^x e^{-t^2} \, dt.$$

Find a Maclaurin series for $\text{erfc}(x)$ and use it to estimate $\text{erfc}(1)$ and $\text{erfc}(\frac{1}{2})$ with a maximum error of 0.0001.

*29. The *sine integral* is defined by

$$\text{Si}(x) = \int_0^x \frac{\sin t}{t} \, dt.$$

(a) Show that $\text{Si}(x)$ is defined and continuous for all real x.
(b) Find a Maclaurin series expansion for $\text{Si}(x)$.
(c) Estimate $\text{Si}(1)$ and $\text{Si}(\frac{1}{2})$ with a maximum error of 0.0001.

Review Exercises for Chapter Fourteen

1. Find the first five terms of the sequence $\{(n-2)/n\}$.
2. Find the first seven terms of the sequence $\{n^2 \sin n\}$.

In Exercises 3–8 determine whether the given sequence is convergent or divergent. If it is convergent, find its limit.

3. $\left\{\dfrac{-7}{n}\right\}$ **4.** $\{\cos \pi n\}$ **5.** $\left\{\dfrac{\ln n}{\sqrt{n}}\right\}$

6. $\left\{\dfrac{7^n}{n!}\right\}$ **7.** $\left\{\left(1 - \dfrac{2}{n}\right)^n\right\}$ **8.** $\left\{\dfrac{3}{\sqrt{n^2 + 8} - n}\right\}$.

9. Find the general term of the sequence

$$\frac{1}{8}, \frac{3}{16}, \frac{5}{32}, \frac{7}{64}, \ldots.$$

10. Find the general term of the sequence

$$1, -\frac{1}{5}, \frac{1}{25}, -\frac{1}{125}, \frac{1}{625}, \cdots.$$

In Exercises 11–19 determine whether the given sequence is bounded or unbounded and increasing, decreasing, or not monotonic.

11. $\{\sqrt{n}\cos n\}$

12. $\left\{\dfrac{3}{n+2}\right\}$

13. $\left\{\dfrac{2^n}{1+2^n}\right\}$

14. $\left\{\dfrac{n^n}{n!}\right\}$

15. $\left\{\dfrac{n!}{n^n}\right\}$

16. $\left\{\dfrac{\sqrt{n}+1}{n}\right\}$

17. $\left\{\left(1-\dfrac{1}{n}\right)^{1/n}\right\}$

18. $\left\{\dfrac{n-7}{n+4}\right\}$

19. $\{(3^n+5^n)^{1/n}\}.$

20. Write using the $\sum$ notation:

$$1 - 2^{1/2} + 3^{1/3} - 4^{1/4} + 5^{1/5} - \cdots + n^{1/n}.$$

21. Evaluate $\displaystyle\sum_{k=2}^{10} 4^k.$

22. Evaluate $\displaystyle\sum_{k=1}^{\infty} \frac{1}{3^k}.$

23. Evaluate $\displaystyle\sum_{k=3}^{\infty} [(\tfrac{3}{4})^k - (\tfrac{2}{5})^k].$

24. Evaluate $\displaystyle\sum_{k=2}^{\infty} \frac{1}{k(k-1)}.$

25. Write $0.79797979\ldots$ as a rational number.
26. Write $0.142314231423\ldots$ as a rational number.
27. At what time between 9 P.M. and 10 P.M. is the minute hand of a clock exactly over the hour hand?

In Exercises 28–40 determine the convergence or divergence of the given series.

28. $\displaystyle\sum_{k=1}^{\infty} \frac{1}{k^3 - 5}$

29. $\displaystyle\sum_{k=5}^{\infty} \frac{1}{k(k+6)}$

30. $\displaystyle\sum_{k=1}^{\infty} \frac{1}{\sqrt{k^3+4}}$

31. $\displaystyle\sum_{k=2}^{\infty} \frac{3}{\ln k}$

32. $\displaystyle\sum_{k=4}^{\infty} \frac{1}{\sqrt[3]{k^3+50}}$

33. $\displaystyle\sum_{k=1}^{\infty} \frac{r^k}{k^r}, \; 0 < r < 1$

34. $\displaystyle\sum_{k=1}^{\infty} \frac{k!}{k^k}$

35. $\displaystyle\sum_{k=2}^{\infty} \frac{10^k}{k^5}$

36. $\displaystyle\sum_{k=1}^{\infty} \frac{k^{6/5}}{8^k}$

37. $\displaystyle\sum_{k=1}^{\infty} \frac{\sqrt{k}\ln(k+3)}{k^2+2}$

38. $\displaystyle\sum_{k=1}^{\infty} \operatorname{csch} k$

39. $\displaystyle\sum_{k=2}^{\infty} \frac{e^{1/k}}{k^{3/2}}$

40. $\displaystyle\sum_{k=1}^{\infty} \frac{k(k+6)}{(k+1)(k+3)(k+5)}.$

In Exercises 41–52 determine whether the given alternating series is absolutely convergent, conditionally convergent, or divergent.

41. $\displaystyle\sum_{k=1}^{\infty} \frac{(-1)^{k+1}}{50k}$

42. $\displaystyle\sum_{k=2}^{\infty} \frac{(-1)^k \sqrt{k}}{\ln k}$

43. $\displaystyle\sum_{k=2}^{\infty} \frac{(-1)^{k+1}}{\sqrt{k}(k-1)}$

44. $\displaystyle\sum_{k=2}^{\infty} \frac{(-1)^k k^2}{k^3 + 1}$

45. $\displaystyle\sum_{k=2}^{\infty} \frac{(-1)^k k^2}{k^4 + 1}$

46. $\displaystyle\sum_{k=2}^{\infty} \frac{(-1)^k k^3}{k^3 + 1}$

47. $\displaystyle\sum_{k=3}^{\infty} \frac{(-1)^k (k+2)(k+3)}{(k+1)^3}$

48. $\displaystyle\sum_{k=2}^{\infty} \frac{(-1)^k 3^k}{3k}$

49. $\displaystyle\sum_{k=1}^{\infty} \frac{(-1)^k k^k}{k!}$

50. $\displaystyle\sum_{k=1}^{\infty} \frac{(-1)^k k^4}{k^4 + 20k^3 + 17k + 2}$

51. $\displaystyle\sum_{k=1}^{\infty} (-1)^k \left(1 + \frac{1}{k}\right)^k$

52. $\displaystyle\sum_{k=1}^{\infty} \frac{(-1)^k k!}{k^k}$.

53. Calculate $\sum_{k=1}^{\infty} (-1)^{k+1}/k^3$ with an error of less than 0.001.
54. Calculate $\sum_{k=0}^{\infty} (-1)^k/k!$ with an error of less than 0.0001.
55. Find the first ten terms of a rearrangement of the series $\sum_{k=1}^{\infty} (-1)^{k+1}/k$ which converges to $\frac{1}{2}$.

In Exercises 56–67 find the radius of convergence and interval of convergence of the given power series.

56. $\displaystyle\sum_{k=0}^{\infty} \frac{x^k}{3^k}$

57. $\displaystyle\sum_{k=0}^{\infty} \frac{(-1)^k x^k}{3^k}$

58. $\displaystyle\sum_{k=0}^{\infty} \frac{x^k}{k^2 + 2}$

59. $\displaystyle\sum_{k=1}^{\infty} \frac{x^k}{k^k}$

60. $\displaystyle\sum_{k=2}^{\infty} \frac{x^k}{(2 \ln k)^k}$

61. $\displaystyle\sum_{k=0}^{\infty} \frac{(3x + 5)^k}{k!}$

62. $\displaystyle\sum_{k=0}^{\infty} \frac{(3x - 5)^k}{3^k}$

63. $\displaystyle\sum_{k=0}^{\infty} \left(\frac{k}{6}\right)^k x^k$

64. $\displaystyle\sum_{k=0}^{\infty} (-1)^k x^{3k}$

65. $\displaystyle\sum_{k=1}^{\infty} \frac{\ln k (x - 1)^k}{k + 2}$

66. $\displaystyle\sum_{k=1}^{\infty} \frac{[1 - (-1)^{k+1}]}{k} x^k$

67. $\displaystyle\sum_{k=0}^{\infty} \frac{(x + 8)^k}{(k + 1)3^k}$.

68. Estimate $\int_0^{1/2} e^{-t^4} \, dt$ with an error of less than 0.00001.
69. Estimate $\int_0^{1/2} \sin t^2 \, dt$ with an error of less than 0.001.
70. Estimate $\int_0^{1/2} t^3 e^{-t^3} \, dt$ with an error of less than 0.001.
71. Estimate $\int_0^{1/2} [1/(1 + t^4)] \, dt$ with an error of less than 0.00001.
72. Find the Maclaurin series for $x^2 e^x$.
73. Find the Taylor series for e^x at $\ln 2$.
74. Find the Maclaurin series for $\cos^2 x$. [*Hint:* $\cos^2 x = (1 + \cos 2x)/2$.]
75. Find the Maclaurin series for $\sin \alpha x$, α real.

FIFTEEN

VECTORS IN THE PLANE

With this chapter we begin a new subject in the study of calculus–the study of vectors and vector functions. This will lead us, in Chapter 18, to the study of functions of two and more variables, and it is a sharp departure from the first 14 chapters of this book.

The study of vectors began essentially with the work of the great Irish mathematician Sir William Rowan Hamilton (1805–1865). Hamilton was a genius who, by the age of twelve, had mastered not only the languages of continental Europe but also Greek, Latin, Sanskrit, Hebrew, Chinese, Persian, Arabic, Malay, Hindi, Bengali, and several others as well. In his twenties, Hamilton turned to science, and his treatises on mechanics and optics provide the basis for much of modern physics.

In his thirties this remarkable man began his research in mathematics. His desire to find a way to represent certain objects in the plane and in space led to the discovery of what he called *quaternions,* which were the precursors of the modern theory of vectors. Throughout Hamilton's life and for the remainder of the nineteenth century, there was considerable debate over the usefulness of quaternions and vectors. At the end of the century, the great British physicist Lord Kelvin wrote that quaternions ". . . although beautifully ingenious, have been an unmixed evil to those who have touched them in any way . . . . vectors . . . have never been of the slightest use to any creature."

But Kelvin was wrong. Today nearly all branches of classical and modern physics are represented using the language of vectors. Vectors are also used with increasing frequency in the social and biological sciences.

In this chapter we will explore properties of vectors in the plane. When going through this material the reader should keep in mind that, like most important discoveries, vectors have been a source of great controversy—a controversy that was not resolved until well into the twentieth century.†

15.1 Vectors and Vector Operations

In many applications of mathematics to the physical and biological sciences and engineering, scientists are concerned with entities that have both magnitude (length) and direction. Examples include the notions of force, velocity, acceleration, and momentum. It is frequently useful to express these quantities geometrically.

†For interesting discussions of the development of modern vector analysis, consult the book by M. J. Crowe, *A History of Vector Analysis,* Univ. of Notre Dame Press, Notre Dame, 1967, or Morris Kline's excellent book *Mathematical Thought from Ancient to Modern Times,* Oxford Univ. Press, New York, 1972, Chapter 32.

Let P and Q be two points in the plane. Then the *directed line segment* from P to Q, denoted by $\overrightarrow{PQ}$, is the straight line segment that extends from P to Q (see Figure 1a). Note that the directed line segments $\overrightarrow{PQ}$ and $\overrightarrow{QP}$ are different since they point in opposite directions (Figure 1b).

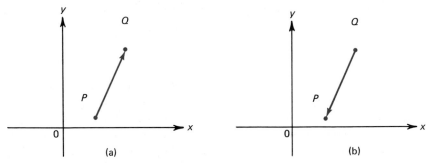

Figure 1

The point P in the directed line segment $\overrightarrow{PQ}$ is called the *initial point* of the segment and the point Q is called the *terminal point*. The two important properties of a directed line segment are its magnitude (length) and its direction. If two directed line segments $\overrightarrow{PQ}$ and $\overrightarrow{RS}$ have the same magnitude and direction, we say that they are *equivalent* no matter where they are located with respect to the origin. The directed line segments in Figure 2 are all equivalent.

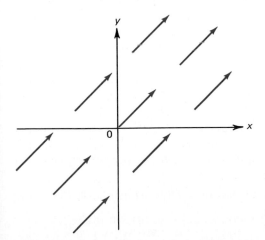

Figure 2

DEFINITION 1 (Geometric Definition of a Vector).   The set of all directed line segments equivalent to a given directed line segment is called a *vector*. Any directed line segment in that set is called a *representation* of the vector.

> **Remark.** The directed line segments in Figure 2 are all representations of the same vector.

> **Notation.** We will denote vectors by lowercase boldface letters such as **v**, **w**, **a**, **b**, etc.

From Definition 1 we see that a given vector **v** can be represented in many different ways. In fact, let $\overrightarrow{PQ}$ be a representation of **v**. Then, without changing magnitude or direction, we can move $\overrightarrow{PQ}$ in a parallel way so that its initial point is shifted to the origin. We then obtain the directed line segment $\overrightarrow{OR}$ which is another representation of the vector **v** (see Figure 3). Now suppose that R has the Cartesian

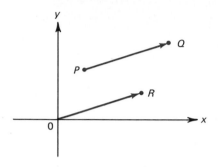

Figure 3

coordinates (a, b). Then we can describe the directed line segment $\overrightarrow{OR}$ by the coordinates (a, b). That is, $\overrightarrow{OR}$ is the directed line segment with initial point $(0, 0)$ and terminal point (a, b). Since one representation of a vector is as good as another, we can write the vector **v** as (a, b).

DEFINITION 2 (Algebraic Definition of a Vector).   A *vector* **v** in the xy-plane is an ordered pair of real numbers (a, b). The numbers a and b are called the *components* of the vector **v**. The *zero vector* is the vector $(0, 0)$.

> *Remark 1.* With this definition, a point in the xy-plane can be thought of as a vector pointing from the origin to that point.

> *Remark 2.* The zero vector has a magnitude of 0. Therefore, since the initial and terminal points coincide, we say that the zero vector has no direction.

> *Remark 3.* We emphasize that Definitions 1 and 2 describe precisely the same objects. Each point of view (geometric and algebraic) has its advantages.

Since we will often have to differentiate between real numbers and vectors (which are pairs of real numbers), we shall use the term *scalar*† to denote a real number.

Since a vector is really a set of equivalent directed line segments, we define the

---

† The term *scalar* originated with Hamilton. His definition of the quaternion included what he called a *real part* and an *imaginary part*. In his paper, "On Quaternions, or on a New System of Imaginaries in Algebra," in *Philosophical Magazine*, 3rd Ser., **25**, 26–27 (1844), he wrote

> The algebraically *real* part may receive . . . all values contained on the one *scale* of progression of numbers from negative to positive infinity; we shall call it therefore the *scalar part,* or simply the *scalar* of the quaternion . . . .

Moreover, in the same paper, Hamilton went on to define the imaginary part of his quaternion as the *vector* part. Although this was not the first usage of the word *vector*, it was the first time it was used in the context of Definitions 1 and 2. In fact, it is fair to say that the paper from which the above quotation was taken marks the beginning of modern vector analysis.

magnitude or *length* of a vector as the length of any one of its representations and its *direction* as the direction of any one of its representations. Using the representation $\overrightarrow{OR}$, and writing the vector $\mathbf{v} = (a, b)$, we find that

$$|\mathbf{v}| = \text{magnitude of } \mathbf{v} = \sqrt{a^2 + b^2}. \qquad (1)$$

This follows from the Pythagorean theorem (see Figure 4). We have used the notation $|\mathbf{v}|$ to denote the magnitude of **v**. Note that $|\mathbf{v}|$ *is a scalar.*

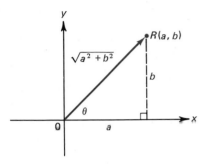

Figure 4

EXAMPLE 1. Calculate the magnitudes of the vectors (a) $(2, 2)$; (b) $(2, 2\sqrt{3})$; (c) $(-2\sqrt{3}, 2)$; (d) $(-3, -3)$; (e) $(6, -6)$.

SOLUTION

(a) $|\mathbf{v}| = \sqrt{2^2 + 2^2} = \sqrt{8} = 2\sqrt{2}.$ (b) $|\mathbf{v}| = \sqrt{2^2 + (2\sqrt{3})^2} = 4.$
(c) $|\mathbf{v}| = \sqrt{(-2\sqrt{3})^2 + 2^2} = 4.$
(d) $|\mathbf{v}| = \sqrt{(-3)^2 + (-3)^2} = \sqrt{18} = 3\sqrt{2}.$
(e) $|\mathbf{v}| = \sqrt{6^2 + (-6)^2} = \sqrt{72} = 6\sqrt{2},$

We now define the *direction* of the vector $\mathbf{v} = (a, b)$ to be the angle θ, measured in radians, that the vector makes with the positive x-axis. By convention, we choose θ such that $0 \le \theta < 2\pi$. It follows from Figure 4 that if $a \ne 0$, then

$$\tan \theta = \frac{b}{a} \qquad (2)$$

EXAMPLE 2. Calculate the directions of the vectors in Example 1.

SOLUTION. We depict these five vectors in Figure 5.

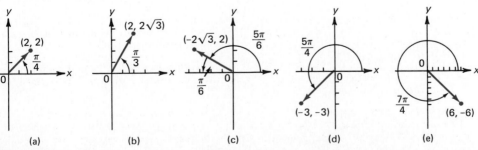

Figure 5

(a)   Here **v** is in the first quadrant and since $\tan \theta = 2/2 = 1$, $\theta = \pi/4$.

(b)   Here $\theta = \tan^{-1} 2\sqrt{3}/2 = \tan^{-1} \sqrt{3} = \pi/3$.

(c)   We see that **v** is in the second quadrant and, since $\tan^{-1} 2/(2\sqrt{3}) = \tan^{-1} 1/\sqrt{3} = \pi/6$, we see from the figure that $\theta = \pi - (\pi/6) = 5\pi/6$.

(d)   Here **v** is in the third quadrant and, since $\tan^{-1} 1 = \pi/4$, $\theta = \pi + (\pi/4) = 5\pi/4$.

(e)   Since **v** is in the fourth quadrant and since $\tan^{-1}(-1) = -\pi/4$, $\theta = 2\pi - (\pi/4) = 7\pi/4$.

Let $\mathbf{v} = (a, b)$. Then, as we have seen, **v** can be represented in many different ways. For example, the representation of **v** with the initial point (c, d) has the terminal point $(c + a, d + b)$. This is depicted in Figure 6. It is easy to show that the directed line segment $\overrightarrow{PQ}$ in Figure 6 has the same magnitude and direction as the vector **v** (see Problems 48 and 49).

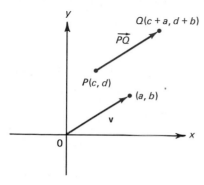

Figure 6

EXAMPLE 3.   Find a representation of the vector $(2, -1)$ whose initial point is the point $P = (5, -8)$.

SOLUTION. Let $Q = (5 + 2, -8 - 1) = (7, -9)$. Then $\overrightarrow{PQ}$ is a representation of the vector $(2, -1)$. This is illustrated in Figure 7.

We now turn to the question of adding vectors and multiplying them by scalars.

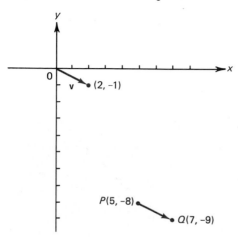

Figure 7

DEFINITION 3. Let $\mathbf{u} = (a_1, b_1)$ and $\mathbf{v} = (a_2, b_2)$ be two vectors in the plane and let α be a scalar. Then we define

(i) $\mathbf{u} + \mathbf{v} = (a_1 + a_2, b_1 + b_2)$, and
(ii) $\alpha\mathbf{u} = (\alpha a_1, \alpha b_1)$.

That is,

> *to add two vectors, we add their corresponding components and to multiply a vector by a scalar, we multiply each of its components by that scalar.*

EXAMPLE 4. Let $\mathbf{u} = (1, 3)$ and $\mathbf{v} = (-2, 4)$. Calculate (a) $\mathbf{u} + \mathbf{v}$; (b) $3\mathbf{u}$; (c) $-\mathbf{v}$; (d) $\mathbf{u} - \mathbf{v}$; and (e) $-3\mathbf{u} + 5\mathbf{v}$.

SOLUTION

(a) $\mathbf{u} + \mathbf{v} = (1 + (-2), 3 + 4) = (-1, 7)$; (b) $3\mathbf{u} = 3(1, 3) = (3, 9)$;
(c) $-\mathbf{v} = (-1)(-2, 4) = (2, -4)$;
(d) $\mathbf{u} - \mathbf{v} = \mathbf{u} + -\mathbf{v} = (1 + 2, 3 - 4) = (3, -1)$;
(e) $-3\mathbf{u} + 5\mathbf{v} = (-3, -9) + (-10, 20) = (-13, 11)$.

There are interesting geometric interpretations of vector addition and scalar multiplication. First, let $\mathbf{v} = (a, b)$ and let α be any scalar. Then

$$|\alpha\mathbf{v}| = |(\alpha a, \alpha b)| = \sqrt{\alpha^2 a^2 + \alpha^2 b^2} = |\alpha|\sqrt{a^2 + b^2} = |\alpha|\,|\mathbf{v}|.$$

> *That is multiplying a vector by a scalar has the effect of multiplying the length of the vector by the absolute value of that scalar.*

Moreover, if $\alpha > 0$, then $\alpha\mathbf{v}$ is in the same quadrant as $\mathbf{v}$ and, since $\tan^{-1}(\alpha b/\alpha a) = \tan^{-1}(b/a)$, the direction of $\alpha\mathbf{v}$ is the same as the direction of $\mathbf{v}$. If $\alpha < 0$, then the direction of $\alpha\mathbf{v}$ is equal to the direction of $\mathbf{v}$ plus π (which is the direction of $-\mathbf{v}$.)

EXAMPLE 5.   Let $\mathbf{v} = (1, 1)$.   Then $|\mathbf{v}| = \sqrt{1 + 1} = \sqrt{2}$ and $|2\mathbf{v}| = |(2, 2)| = \sqrt{2^2 + 2^2} = \sqrt{8} = 2\sqrt{2} = 2|\mathbf{v}|$.   Also, $|-2\mathbf{v}| = \sqrt{(-2)^2 + (-2)^2} = 2\sqrt{2} = 2|\mathbf{v}|$. Moreover the direction of $2\mathbf{v}$ is $\pi/4$ while the direction of $-2\mathbf{v}$ is $5\pi/4$. This is illustrated in Figure 8.

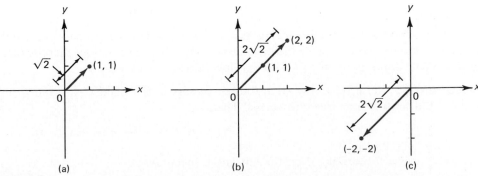

(a)   (b)   (c)

Figure 8

Now suppose we add the vectors $\mathbf{u} = (a_1, b_1)$ and $\mathbf{v} = (a_2, b_2)$ as in Figure 9. From the figure we see that the vector $\mathbf{u} + \mathbf{v} = (a_1 + a_2, b_1 + b_2)$ can be obtained by shifting the representation of the vector $\mathbf{v}$ so that its initial point coincides with the terminal point (a_1, b_1) of the vector $\mathbf{v}$. We can therefore obtain the vector $\mathbf{u} + \mathbf{v}$ by drawing a parallelogram with one vertex at the origin and sides $\mathbf{u}$ and $\mathbf{v}$. Then $\mathbf{u} + \mathbf{v}$ is the vector which points from the origin along the diagonal of the parallelogram.

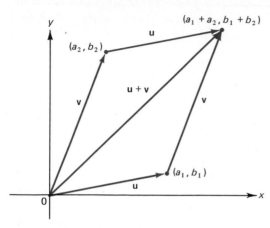

Figure 9

Note. Since a straight line is the shortest distance between two points, it immediately follows from Figure 9 that

$$|\mathbf{u} + \mathbf{v}| \leq |\mathbf{u}| + |\mathbf{v}|. \tag{3}$$

For obvious reasons, the inequality (3) is called the *triangle inequality*.

We can also use Figure 9 to obtain a geometric representation of the vector $\mathbf{u} - \mathbf{v}$. Since $\mathbf{u} = \mathbf{u} - \mathbf{v} + \mathbf{v}$, the vector $\mathbf{u} - \mathbf{v}$ is the vector that must be added to $\mathbf{v}$ to obtain $\mathbf{u}$. This is illustrated in Figure 10a.

(a)   (b)   Figure 10

The following theorem lists several properties that hold for any vectors $\mathbf{u}$, $\mathbf{v}$, and $\mathbf{w}$ and any scalars α and β. Since the proof is very easy, we leave it as an exercise (see Problem 58). Some parts of this theorem have already been proven.

THEOREM 1. Let $\mathbf{u}$, $\mathbf{v}$, and $\mathbf{w}$ be any three vectors in the plane, let α and β be scalars, and let $\mathbf{0}$ denote the zero vector. Then

    (i) $\mathbf{u} + \mathbf{v} = \mathbf{v} + \mathbf{u}$.   (ii) $\mathbf{u} + (\mathbf{v} + \mathbf{w}) = (\mathbf{u} + \mathbf{v}) + \mathbf{w}$.
    (iii) $\mathbf{v} + \mathbf{0} = \mathbf{v}$.
    (iv) $0\mathbf{v} = \mathbf{0}$ (here the 0 on the left is the scalar zero).

(v) $\alpha\mathbf{0} = \mathbf{0}.$

(vi) $(\alpha\beta)\mathbf{v} = \alpha(\beta\mathbf{v}).$

(vii) $\mathbf{v} + (-\mathbf{v}) = \mathbf{0}.$

(viii) $1 \cdot \mathbf{v} = \mathbf{v}.$

(ix) $(\alpha + \beta)\mathbf{v} = \alpha\mathbf{v} + \beta\mathbf{v}.$

(x) $|\alpha\mathbf{v}| = |\alpha|\,|\mathbf{v}|.$

(xi) $\alpha(\mathbf{u} + \mathbf{v}) = \alpha\mathbf{u} + \alpha\mathbf{v}.$

(xii) $|\mathbf{u} + \mathbf{v}| \leq |\mathbf{u}| + |\mathbf{v}|.$

Many of the above properties can be illustrated geometrically. For example, the rule (i), which is called the *commutative law for vector addition,* is illustrated in Figure 11. Similarly, the rule (ii), which is called the *associative law for vector addition,* is illustrated in Figure 12.

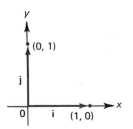

Figure 11

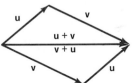

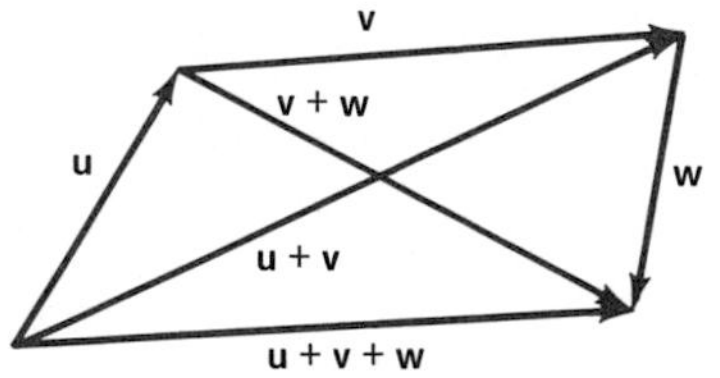

Figure 12

When a set of vectors together with a set of scalars and the operations of addition and scalar multiplication have the properties given in Theorem 1, we say that the vectors comprise a *vector space.* The set of vectors of the form (a, b) where a and b are real numbers is denoted $\mathbb{R}^2$. We shall not discuss properties of abstract vector spaces here, except to say that many abstract vector spaces have properties very similar to the properties of the vector space $\mathbb{R}^2$.

There are two special vectors in $\mathbb{R}^2$ which allow us to represent other vectors in $\mathbb{R}^2$ in a convenient way. We shall denote the vector $(1, 0)$ by the vector symbol $\mathbf{i}$ and

Figure 13

the vector $(0, 1)$ by the vector symbol $\mathbf{j}$ (see Figure 13).† If (a, b) denotes any other vector in $\mathbb{R}^2$, then, since $(a, b) = a(1, 0) + b(0, 1)$, we may write

$$\mathbf{v} = (a, b) = a\mathbf{i} + b\mathbf{j}.$$

Moreover, any vector in $\mathbb{R}^2$ can be represented in a unique way in the form $a\mathbf{i} + b\mathbf{j}$ since the representation of (a, b) as a point in the plane is unique. (Put another way, a point in the xy-plane has one and only one x-coordinate and one and only one y-coordinate.) Thus Theorem 1 holds with this new representation as well.

† The symbols i and j were first used by Hamilton. He defined his quarternion as a quantity of the form $a + b\mathbf{i} + c\mathbf{j} + d\mathbf{k}$ where a was the "scalar part" and $b\mathbf{i} + c\mathbf{j} + d\mathbf{k}$ the "vector part." In Section 17.2 we will write vectors in space in the form $b\mathbf{i} + c\mathbf{j} + d\mathbf{k}$.

When the vector **v** is written in the form $\mathbf{v} = a\mathbf{i} + b\mathbf{j}$, we say that **v** *is resolved into its horizontal and vertical components* since, obviously, a is the horizontal component of **v** while b is its vertical component. The vectors **i** and **j** are called *basis vectors* for the vector space $\mathbb{R}^2$.

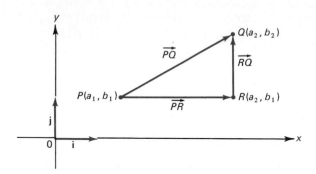

Figure 14

Now suppose that a vector **v** can be represented by the directed line segment $\overrightarrow{PQ}$ where $P = (a_1, b_1)$ and $Q = (a_2, b_2)$. (See Figure 14.) If we label the point (a_2, b_1) as R, then we immediately see that

$$\mathbf{v} = \overrightarrow{PQ} = \overrightarrow{PR} + \overrightarrow{RQ}. \tag{4}$$

But the length of $\overrightarrow{PR}$ is $a_2 - a_1$ and since $\overrightarrow{PR}$ has the same direction as **i**, we can write

$$\overrightarrow{PQ} = (a_2 - a_1)\mathbf{i}. \tag{5}$$

Similarly,

$$\overrightarrow{RQ} = (b_2 - b_1)\mathbf{j} \tag{6}$$

and we may write (using (4), (5), and (6))

$$\mathbf{v} = (a_2 - a_1)\mathbf{i} + (b_2 - b_1)\mathbf{j}. \tag{7}$$

EXAMPLE 6. Resolve the vector represented by the directed line segment from $(-2, 3)$ to $(1, 5)$ into its vertical and horizontal components.

SOLUTION. Using (7), we have

$$\mathbf{v} = (a_2 - a_1)\mathbf{i} + (b_2 - b_1)\mathbf{j} = (1 - (-2))\mathbf{i} + (5 - 3)\mathbf{j} = 3\mathbf{i} + 2\mathbf{j}.$$

We conclude this section by defining a kind of vector that is very useful in certain types of applications.

DEFINITION 4. A *unit vector* **u** is a vector which has length 1.

EXAMPLE 7. The vector $\mathbf{u} = (1/2)\mathbf{i} + (\sqrt{3}/2)\mathbf{j}$ is a unit vector since

$$|\mathbf{u}| = \sqrt{\left(\frac{1}{2}\right)^2 + \left(\frac{\sqrt{3}}{2}\right)^2} = \sqrt{\frac{1}{4} + \frac{3}{4}} = 1.$$

Let $\mathbf{u} = a\mathbf{i} + b\mathbf{j}$ be a unit vector. Then $|\mathbf{u}| = \sqrt{a^2 + b^2} = 1$, so $a^2 + b^2 = 1$ and $\mathbf{u}$ is a point on the unit circle (see Figure 15). If θ is the direction of $\mathbf{u}$, then we

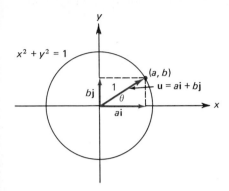

Figure 15

immediately see that $a = \cos\theta$ and $b = \sin\theta$. Thus any unit vector $\mathbf{u}$ can be written in the form

$$\mathbf{u} = (\cos\theta)\mathbf{i} + (\sin\theta)\mathbf{j} \tag{8}$$

where θ is the direction of $\mathbf{u}$.

EXAMPLE 8. The unit vector $\mathbf{u} = (1/2)\mathbf{i} + (\sqrt{3}/2)\mathbf{j}$ of Example 7 can be written in the form (8) with $\theta = \cos^{-1} 1/2 = \pi/3$.

Finally,

let $\mathbf{v}$ be any nonzero vector. Then $\mathbf{u} = \mathbf{v}/|\mathbf{v}|$ is a unit vector having the same direction as $\mathbf{v}$

(see Problem 31).

EXAMPLE 9. Find a unit vector having the same direction as $\mathbf{v} = 2\mathbf{i} - 3\mathbf{j}$.

SOLUTION. Here $|\mathbf{v}| = \sqrt{4 + 9} = \sqrt{13}$ so that $\mathbf{u} = \mathbf{v}/|\mathbf{v}| = (2/\sqrt{13})\mathbf{i} - (3/\sqrt{13})\mathbf{j}$ is the required unit vector.

EXAMPLE 10. Find a vector $\mathbf{v}$ whose direction is $5\pi/4$ and whose magnitude is 7.

SOLUTION. A unit vector $\mathbf{u}$ with direction $5\pi/4$ is given by

$$\mathbf{u} = \left(\cos\frac{5\pi}{4}\right)\mathbf{i} + \left(\sin\frac{5\pi}{4}\right)\mathbf{j} = -\frac{1}{\sqrt{2}}\mathbf{i} - \frac{1}{\sqrt{2}}\mathbf{j}.$$

Then $\mathbf{v} = 7\mathbf{u} = -(7/\sqrt{2})\mathbf{i} - (7/\sqrt{2})\mathbf{j}$. This is sketched in Figure 16a. In Figure 16b we have translated $\mathbf{v}$ so that it points toward the origin. This representation of $\mathbf{v}$ will be useful in Section 3.

We conclude this section with a summary of properties of vectors, given in Table 1.

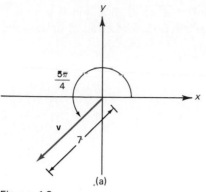

(a)

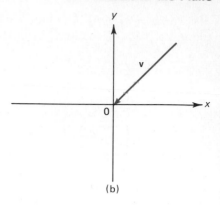

(b)

Figure 16

TABLE 1

Object	Intuitive definition	Expression in terms of components if $\mathbf{u} = u_1\mathbf{i} + u_2\mathbf{j}$, $\mathbf{v} = v_1\mathbf{i} + v_2\mathbf{j}$ and $\mathbf{u} = (u_1, u_2)$, $\mathbf{v} = (v_1, v_2)$
Vector $\mathbf{v}$	Magnitude and direction $\uparrow\mathbf{v}$	$v_1\mathbf{i} + v_2\mathbf{j}$ or (v_1, v_2)
$|\mathbf{v}|$	Magnitude or length of $\mathbf{v}$	$\sqrt{v_1{}^2 + v_2{}^2}$
$\alpha\mathbf{v}$	$\uparrow\mathbf{v}$ $\uparrow\alpha\mathbf{v}$ (Here $\alpha = 2$)	$\alpha v_1\mathbf{i} + \alpha v_2\mathbf{j}$ or $(\alpha v_1, \alpha v_2)$
$-\mathbf{v}$	$\uparrow\mathbf{v}$ $\downarrow-\mathbf{v}$	$-v_1\mathbf{i} - v_2\mathbf{j}$ or $(-v_1, -v_2)$ or $-(v_1, v_2)$
$\mathbf{u} + \mathbf{v}$	$\mathbf{u}+\mathbf{v}$ $\mathbf{v}$ $\mathbf{u}$	$(u_1 + v_1)\mathbf{i} + (u_2 + v_2)\mathbf{j}$ or $(u_1 + v_1, u_2 + v_2)$
$\mathbf{u} - \mathbf{v}$	$\mathbf{v}$ $\mathbf{u} - \mathbf{v}$ $\mathbf{u}$	$(u_1 - v_1)\mathbf{i} + (u_2 - v_2)\mathbf{j}$ or $(u_1 - v_1, u_2 - v_2)$

PROBLEMS 15.1

In Problems 1–6 a vector $\mathbf{v}$ and a point P are given. Find a point Q such that the directed line segment $\overrightarrow{PQ}$ is a representation of $\mathbf{v}$. Sketch $\mathbf{v}$ and $\overrightarrow{PQ}$.

1. $\mathbf{v} = (2, 5)$; $P = (1, -2)$

2. $\mathbf{v} = (5, 8)$; $P = (3, 8)$

3. $\mathbf{v} = (-3, 7)$; $P = (7, -3)$

4. $\mathbf{v} = -\mathbf{i} - 7\mathbf{j}$; $P = (0, 1)$

5. $\mathbf{v} = 5\mathbf{i} - 3\mathbf{j}$; $P = (-7, -2)$

6. $\mathbf{v} = e\mathbf{i} + \pi\mathbf{j}$; $P = (\pi, \sqrt{2})$.

In Problems 7–18 find the magnitude and direction of the given vector.

7. $\mathbf{v} = (4, 4)$

8. $\mathbf{v} = (-4, 4)$

9. $\mathbf{v} = (4, -4)$

10. $\mathbf{v} = (-4, -4)$

11. $\mathbf{v} = (\sqrt{3}, 1)$

12. $\mathbf{v} = (1, \sqrt{3})$

13. $\mathbf{v} = (-1, \sqrt{3})$

14. $\mathbf{v} = (1, -\sqrt{3})$

15. $\mathbf{v} = (-1, -\sqrt{3})$

16. $\mathbf{v} = (1, 2)$

17. $\mathbf{v} = (-5, 8)$

18. $\mathbf{v} = (11, -14)$.

In Problems 19–26 write in the form $a\mathbf{i} + b\mathbf{j}$ the vector $\mathbf{v}$ that is represented by $\overrightarrow{PQ}$. Sketch $\overrightarrow{PQ}$ and $\mathbf{v}$.

19. $P = (1, 2)$; $Q = (1, 3)$

20. $P = (2, 4)$; $Q = (-7, 4)$

21. $P = (5, 2); \; Q = (-1, 3)$
22. $P = (8, -2); \; Q = (-3, -3)$
23. $P = (7, -1); \; Q = (-2, 4)$
24. $P = (3, -6); \; Q = (8, 0)$
25. $P = (-3, -8); \; Q = (-8, -3)$
26. $P = (2, 4); \; Q = (-4, -2).$

27. Let $u = (2, 3)$ and $v = (-5, 4)$. Find
    (a) $3u$ (b) $u + v$ (c) $v - u$ (d) $2u - 7v$.
    Sketch these vectors.

28. Let $u = 2i - 3j$ and $v = -4i + 6j$. Find
    (a) $u + v$ (b) $u - v$ (c) $3u$
    (d) $-7v$ (e) $8u - 3v$ (f) $4v - 6u$.
    Sketch.

29. Show that the vectors i and j are unit vectors.
30. Show that the vector $(1/\sqrt{2})i + (1/\sqrt{2})j$ is a unit vector.
31. Show that if $v = ai + bj$, then $v = (a/\sqrt{a^2 + b^2})i + (b/\sqrt{a^2 + b^2})j$ is a unit vector having the same direction as v.

In Problems 32–37 find a unit vector having the same direction as the given vector.

32. $v = 2i + 3j$ 33. $v = i - j$ 34. $v = 3i + 4j$
35. $v = 3i - 4j$ 36. $v = -3i + 4j$ 37. $v = ai + aj$.

38. If $v = ai + bj$, show that $a/\sqrt{a^2 + b^2} = \cos\theta$ and $b/\sqrt{a^2 + b^2} = \sin\theta$ where θ is the direction of v.
39. If $v = 2i - 3j$, find $\sin\theta$ and $\cos\theta$.
40. If $v = -3i + 8j$, find $\sin\theta$ and $\cos\theta$.

A vector v has a direction opposite to that of a vector u if direction $v =$ direction $u + \pi$. In Problems 41–46 find a unit vector v which has a direction opposite the direction of the given vector u.

41. $u = i + j$ 42. $u = 2i - 3j$
43. $u = -3i + 4j$ 44. $u = -2i + 3j$
45. $u = -3i - 4j$ 46. $u = 8i - 3j$.

47. Let $u = 2i - 3j$ and $v = -i + 2j$. Find a unit vector having the same direction as
    (a) $u + v$ (b) $2u - 3v$ (c) $3u + 8v$.
48. Let $P = (c, d)$ and $Q = (c + a, d + b)$. Show that the magnitude of $\overrightarrow{PQ}$ is $\sqrt{a^2 + b^2}$.
49. Show that the direction of $\overrightarrow{PQ}$ in Problem 48 is the same as the direction of the vector (a, b). [*Hint:* If $R = (a, b)$, show that the line passing through the points P and Q is parallel to the line passing through the point 0 and R.]

In Problems 50–57 find a vector v having the given magnitude and direction. [*Hint:* See Example 10.]

50. $|v| = 3; \; \theta = \pi/6$ 51. $|v| = 8; \; \theta = \pi/3$ 52. $|v| = 7; \; \theta = \pi$
53. $|v| = 4; \; \theta = \pi/2$ 54. $|v| = 1; \; \theta = \pi/4$ 55. $|v| = 6; \; \theta = 2\pi/3$
56. $|v| = 8; \; \theta = 3\pi/2$ 57. $|v| = 6; \; \theta = 11\pi/6$.

58. Prove Theorem 1. [*Hint:* Use the definitions of addition and scalar multiplication of vectors.]
59. Show algebraically (i.e., strictly from the definitions of vector addition and magnitude) that for any two vectors u and v, $|u + v| \le |u| + |v|$.
60. Show that if neither u nor v is the zero vector, then $|u + v| = |u| + |v|$ if and only if u is a scalar multiple of v.

15.2 The Dot Product

In Section 15.1 we showed how a vector could be multiplied by a scalar, but not how two vectors can be multiplied. Actually, there are several ways to define the product of two vectors, and in this section we shall discuss one of them. We shall discuss a second product operation in Section 17.4.

DEFINITION 1. Let $u = (a_1, b_1) = a_1 i + b_1 j$ and $v = (a_2, b_2) = a_2 i + b_2 j$. Then the *dot product* of u and v, denoted $u \cdot v$, is defined by

$$u \cdot v = a_1 a_2 + b_1 b_2. \tag{1}$$

> **Remark.** The dot product of two vectors is a *scalar*. For this reason the dot product is often called the *scalar product*.

EXAMPLE 1. If $u = (1, 3)$, and $v = (4, -7)$, then $u \cdot v = 1(4) + 3(-7) = 4 - 21 = -17$.

THEOREM 1. For any vectors u, v, w and scalar α,

(i) $u \cdot v = v \cdot u$.

(ii) $(u + v) \cdot w = u \cdot w + v \cdot w$.

(iii) $(\alpha u) \cdot v = \alpha(u \cdot v)$.

(iv) $|u| = \sqrt{u \cdot u}$.

PROOF. Let $u = (a_1, b_1)$, $v = (a_2, b_2)$ and $w = (a_3, b_3)$.

(i) $u \cdot v = a_1 a_2 + b_1 b_2 = a_2 a_1 + b_2 b_1 = v \cdot u$.

(ii) $(u + v) \cdot w = (a_1 + a_2, b_1 + b_2) \cdot (a_3, b_3) = (a_1 + a_2)a_3 + (b_1 + b_2)b_3$
$= a_1 a_3 + b_1 b_3 + a_2 a_3 + b_2 b_3 = u \cdot w + v \cdot w$.

(iii) $(\alpha u) \cdot v = (\alpha a_1, \alpha b_1) \cdot (a_2, b_2) = \alpha a_1 a_2 + \alpha b_1 b_2$
$= \alpha(a_1 a_2 + b_1 b_2) = \alpha(u \cdot v)$.

(iv) $\sqrt{u \cdot u} = \sqrt{(a_1, b_1) \cdot (a_1, b_1)} = \sqrt{a_1^2 + b_1^2} = |u|$.

The dot product is useful in a wide variety of applications. A simple one is given below.

DEFINITION 2. Let u and v be two vectors. Then the angle φ between u and v is defined to be the smallest angle† between the representations of u and v that have the origin as their initial points. If $u = \alpha v$ for some scalar α, then we define $\varphi = 0$ if $\alpha > 0$ and $\varphi = \pi$ if $\alpha < 0$.

THEOREM 2. Let u and v be two nonzero vectors. Then if φ is the angle between them,

$$\cos \varphi = \frac{u \cdot v}{|u| \, |v|}. \tag{2}$$

PROOF. The law of cosines (see Problem 7.4.14) states that in the triangle of Figure 1,

$$c^2 = a^2 + b^2 - 2ab \cos C. \tag{3}$$

---

† The smallest angle will be in the interval $[0, \pi]$.

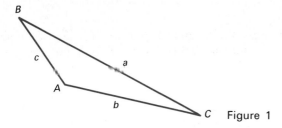

Figure 1

We now place the representations of **u** and **v** with initial points at the origin so that **u** = (a_1, b_1) and **v** = (a_2, b_2) (see Figure 2). Then, from the law of cosines,

$$|v - u|^2 = |v|^2 + |u|^2 - 2|u| |v| \cos \varphi.$$

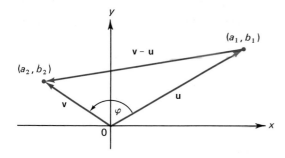

Figure 2

But

$$|v - u|^2 = (v - u) \cdot (v - u) = v \cdot v - 2u \cdot v + u \cdot u$$
$$= |v|^2 - 2u \cdot v + |u|^2.$$

Thus, after simplification, we obtain

$$-2u \cdot v = -2|u| |v| \cos \varphi$$

from which the theorem follows.

Remark. Using Theorem 2 we could define the dot product **u** · **v** by

$$u \cdot v = |u| |v| \cos \varphi. \tag{4}$$

EXAMPLE 2. Find the cosine of the angle between the vectors **u** = $2i + 3j$ and **v** = $-7i + j$.

SOLUTION. $u \cdot v = -14 + 3 = -11,$ $|u| = \sqrt{2^2 + 3^2} = \sqrt{13}$ and $|v| = \sqrt{(-7)^2 + 1^2} = \sqrt{50}$ so that

$$\cos \varphi = \frac{u \cdot v}{|u| |v|} = \frac{-11}{\sqrt{13} \sqrt{50}} = \frac{-11}{\sqrt{650}} \approx -0.4315.$$

DEFINITION 3. Two nonzero vectors **u** and **v** are *parallel* if the angle between them is 0 or π.

EXAMPLE 3. Show that the vectors **u** = $(2, -3)$ and **v** = $(-4, 6)$ are parallel.

SOLUTION

$$\cos \varphi = \frac{\mathbf{u \cdot v}}{|\mathbf{u}|\,|\mathbf{v}|} = \frac{-8-18}{\sqrt{13}\sqrt{52}} = \frac{-26}{\sqrt{13}(2\sqrt{13})} = \frac{-26}{2(13)} = -1$$

so that $\varphi = \pi$.

THEOREM 3. If $\mathbf{u} \neq \mathbf{0}$, then $\mathbf{v} = \alpha\mathbf{u}$ for some nonzero constant α if and only if $\mathbf{u}$ and $\mathbf{v}$ are parallel.

PROOF. The proof is easy and is left as an exercise (see Problem 43).

DEFINITION 4. The nonzero vectors $\mathbf{u}$ and $\mathbf{v}$ are called *orthogonal* (or *perpendicular*) if the angle between them is $\pi/2$.

EXAMPLE 4. Show that the vectors $\mathbf{u} = 3\mathbf{i} - 4\mathbf{j}$ and $\mathbf{v} = 4\mathbf{i} + 3\mathbf{j}$ are orthogonal.

SOLUTION. $\mathbf{u \cdot v} = 3 \cdot 4 - 4 \cdot 3 = 0$. This implies that $\cos \varphi = (\mathbf{u \cdot v})/(|\mathbf{u}|\,|\mathbf{v}|) = 0$. Since φ is in the interval $[0, \pi]$, $\varphi = \pi/2$.

THEOREM 4. The nonzero vectors $\mathbf{u}$ and $\mathbf{v}$ are orthogonal if and only if $\mathbf{u \cdot v} = 0$.

PROOF. This proof is also easy and is left as an exercise (see Problem 44).

EXAMPLE 5. Using vectors, show that the diagonals of a rectangle are orthogonal if and only if the rectangle is a square.

SOLUTION. We place the rectangle so that two of its sides lie along the x- and y-axes (see Figure 3). One of the diagonals is the vector $\mathbf{u} = (a, b)$. The other is the vector

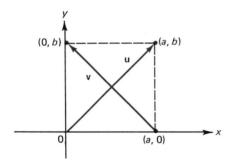

Figure 3

$\mathbf{v} = (0 - a, b - 0) = (-a, b)$ (see Equation 15.1.7). Then $\mathbf{u \cdot v} = -a^2 + b^2$. This is zero if and only if $a = b$. But if $a = b$, the rectangle is a square.

We shall see a number of other "geometric" applications of vectors in Section 3.

EXAMPLE 6. Let $\mathbf{u} = \mathbf{i} + 4\mathbf{j}$ and $\mathbf{v} = 3\mathbf{i} + \alpha\mathbf{j}$. Determine α such that (a) $\mathbf{u}$ and $\mathbf{v}$ are orthogonal; (b) $\mathbf{u}$ and $\mathbf{v}$ are parallel.

SOLUTION. (a) We have $\mathbf{u \cdot v} = 3 + 4\alpha$. In order that $\mathbf{u}$ and $\mathbf{v}$ be orthogonal, we must have $\mathbf{u \cdot v} = 0$. This implies that $3 + 4\alpha = 0$ or $\alpha = -\frac{3}{4}$.

(b)   Here we must have $\varphi = 0$ or π so that $\cos \varphi = \pm 1$. Then

$$\cos \varphi = \frac{\mathbf{u \cdot v}}{|\mathbf{u}|\,|\mathbf{v}|} = \frac{3 + 4\alpha}{\sqrt{17}\sqrt{9 + \alpha^2}} = \pm 1.$$

Squaring both sides of this last equation, we obtain

$$9 + 24\alpha + 16\alpha^2 = 17(9 + \alpha^2) = 153 + 17\alpha^2.$$

This leads to the quadratic equation

$$\alpha^2 - 24\alpha + 144 = 0 = (\alpha - 12)^2,$$

with the single solution $\alpha = 12$.

A number of interesting problems involve the notion of the projection of one vector along another. Before defining this, we prove the following theorem.

THEOREM 5. Let **v** be a nonzero vector. Then for any other vector **u**, the vector $\mathbf{w} = \mathbf{u} - [(\mathbf{u}\cdot\mathbf{v})\mathbf{v}/|\mathbf{v}|^2]$ is orthogonal to **v**.

PROOF

$$\mathbf{w}\cdot\mathbf{v} = \left[\mathbf{u} - \frac{(\mathbf{u}\cdot\mathbf{v})\mathbf{v}}{|\mathbf{v}|^2}\right]\cdot\mathbf{v} = \mathbf{u}\cdot\mathbf{v} - \frac{(\mathbf{u}\cdot\mathbf{v})(\mathbf{v}\cdot\mathbf{v})}{|\mathbf{v}|^2}$$

$$= \mathbf{u}\cdot\mathbf{v} - \frac{(\mathbf{u}\cdot\mathbf{v})|\mathbf{v}|^2}{|\mathbf{v}|^2} = \mathbf{u}\cdot\mathbf{v} - \mathbf{u}\cdot\mathbf{v} = 0.$$

The vectors **u**, **v**, and **w** are illustrated in Figure 4.

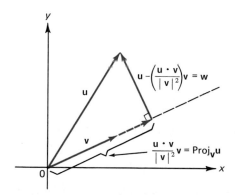

Figure 4

DEFINITION 5. Let **u** and **v** be nonzero vectors. Then the *projection of* **u** *on* **v** is a vector, denoted $\mathrm{Proj}_\mathbf{v}\,\mathbf{u}$, which is defined by

$$\mathrm{Proj}_\mathbf{v}\,\mathbf{u} = \frac{\mathbf{u}\cdot\mathbf{v}}{|\mathbf{v}|^2}\,\mathbf{v}. \qquad (5)$$

The component of **u** *in the direction* **v** *is* $\mathbf{u}\cdot\mathbf{v}/|\mathbf{v}|$.

Note that $\mathbf{v}/|\mathbf{v}|$ is a unit vector in the direction of **v**.

Remark 1. From Figure 4 and the fact that $\cos \varphi = (\mathbf{u} \cdot \mathbf{v})/(|\mathbf{u}|\,|\mathbf{v}|)$, we find that

> $\mathbf{v}$ and $\text{Proj}_\mathbf{v}\,\mathbf{u}$ have (i) the same direction if $\mathbf{u} \cdot \mathbf{v} > 0$ and (ii) opposite directions if $\mathbf{u} \cdot \mathbf{v} < 0$.

Remark 2. $\text{Proj}_\mathbf{v}\,\mathbf{u}$ can be thought of as the "v-component" of the vector $\mathbf{u}$. We shall see an illustration of this in our discussion of force in the next section.

Remark 3. If $\mathbf{u}$ and $\mathbf{v}$ are orthogonal, then $\mathbf{u} \cdot \mathbf{v} = 0$ so that $\text{Proj}_\mathbf{v}\,\mathbf{u} = \mathbf{0}$.

Remark 4. An alternative definition of projection is: If $\mathbf{u}$ and $\mathbf{v}$ are nonzero vectors, then $\text{Proj}_\mathbf{v}\,\mathbf{u}$ is the unique vector having the properties

> (i) $\text{Proj}_\mathbf{v}\,\mathbf{u}$ is parallel to $\mathbf{v}$ and (ii) $\mathbf{u} - \text{Proj}_\mathbf{v}\,\mathbf{u}$ is orthogonal to $\mathbf{v}$.

EXAMPLE 7.   Let $\mathbf{u} = 2\mathbf{i} + 3\mathbf{j}$ and $\mathbf{v} = \mathbf{i} + \mathbf{j}$. Calculate $\text{Proj}_\mathbf{v}\,\mathbf{u}$.

SOLUTION. $\text{Proj}_\mathbf{v}\,\mathbf{u} = (\mathbf{u} \cdot \mathbf{v})\mathbf{v}/|\mathbf{v}|^2 = [5/(\sqrt{2})^2]\mathbf{v} = (5/2)\mathbf{i} + (5/2)\mathbf{j}$. This is illustrated in Figure 5.

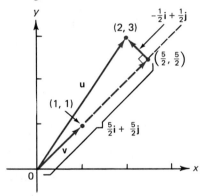

Figure 5

EXAMPLE 8.   Let $\mathbf{u} = 2\mathbf{i} - 3\mathbf{j}$ and $\mathbf{v} = \mathbf{i} + \mathbf{j}$. Find $\text{Proj}_\mathbf{v}\,\mathbf{u}$.

SOLUTION. Here $(\mathbf{u} \cdot \mathbf{v})/|\mathbf{v}|^2 = -\frac{1}{2}$ so that $\text{Proj}_\mathbf{v}\,\mathbf{u} = -\frac{1}{2}\mathbf{i} - \frac{1}{2}\mathbf{j}$. This is illustrated in Figure 6.

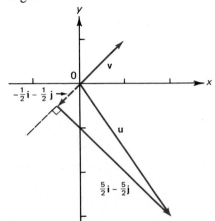

Figure 6

PROBLEMS 15.2

In Problems 1–10 calculate both the dot product of the two vectors and the cosine of the angle between them.

1. $u = i + j$; $v = i - j$
2. $u = 3i$; $v = -7j$
3. $u = -5i$; $v = 18j$
4. $u = \alpha i$; $v = \beta j$; α, β real
5. $u = 2i + 5j$; $v = 5i + 2j$
6. $u = 2i + 5j$; $v = 5i - 2j$
7. $u = -3i + 4j$; $v = -2i - 7j$
8. $u = 4i + 5j$; $v = 5i - 4j$
9. $u = 11i - 8j$; $v = 4i - 7j$
10. $u = -13i + 8j$; $v = 8i + 13j$.

11. Show that for any real numbers α and β, the vectors $u = \alpha i + \beta j$ and $v = \beta i - \alpha j$ are orthogonal.

12. Let u, v, and w denote three arbitrary vectors. Explain why the product $u \cdot v \cdot w$ is *not defined*.

In Problems 13–20 determine whether the given vectors are orthogonal, parallel, or neither. Then sketch each pair.

13. $u = 3i + 5j$; $v = -6i - 10j$
14. $u = 2i + 3j$; $v = 6i - 4j$
15. $u = 2i + 3j$; $v = 6i + 4j$
16. $u = 2i + 3j$; $v = -6i + 4j$
17. $u = 7i$; $v = -23j$
18. $u = 2i - 6j$; $v = -i + 3j$
19. $u = i + j$; $v = \alpha i + \alpha j$; α real
20. $u = -2i + 3j$; $v = -i + 2j$.

21. Let $u = 3i + 4j$ and $v = i + \alpha j$. Determine α such that
    (a) u and v are orthogonal
    (b) u and v are parallel
    (c) the angle between u and v is $\pi/4$
    (d) the angle between u and v is $\pi/3$.

22. Let $u = -2i + 5j$ and $v = \alpha i - 2j$. Determine α such that
    (a) u and v are orthogonal.
    (b) u and v are parallel.
    (c) the angle between u and v is $2\pi/3$.
    (d) the angle between u and v is $\pi/3$.

23. In Problem 21 show that there is no value of α for which u and v have opposite directions.

24. In Problem 22 show that there is no value of α for which u and v have the same direction.

In Problems 25–38 calculate $\text{Proj}_v\, u$.

25. $u = 3i$; $v = i + j$
26. $u = -5j$; $v = i + j$
27. $u = 2i + j$; $v = i - 2j$
28. $u = 2i + 3j$; $v = 4i + j$
29. $u = i + j$; $v = 2i - 3j$
30. $u = i + j$; $v = 2i + 3j$
31. $u = 4i + 5j$; $v = 2i + 4j$
32. $u = 4i + 5j$; $v = 2i - 4j$
33. $u = -4i + 5j$; $v = 2i - 4j$
34. $u = -4i - 5j$; $v = -2i - 4j$
35. $u = \alpha i + \beta j$; $v = i + j$; α, β real and positive
36. $u = i + j$; $v = \alpha i + \beta j$; α, β real and positive
37. $u = \alpha i - \beta j$; $v = i + j$; α, β real and positive with $\alpha > \beta$
38. $u = \alpha i - \beta j$; $v = i + j$; α, β real and positive with $\alpha < \beta$.

39. Let $u = a_1 i + b_1 j$ and $v = a_2 i + b_2 j$. Give a condition on a_1, b_1, a_2, and b_2 which will ensure that v and $\text{Proj}_v\, u$ have the same direction.

40. In Problem 39, give a condition which will ensure that v and $\text{Proj}_v\, u$ have opposite directions.

41. Let $P = (2, 3)$, $Q = (5, 7)$, $R = (2, -3)$ and $S = (1, 2)$. Calculate $\text{Proj}_{\overrightarrow{PQ}}\, \overrightarrow{RS}$ and $\text{Proj}_{\overrightarrow{RS}}\, \overrightarrow{PQ}$.

42. Let $P = (-1, 3)$, $Q = (2, 4)$, $R = (-6, -2)$ and $S = (3, 0)$. Calculate $\text{Proj}_{\overrightarrow{PQ}} \overrightarrow{RS}$ and $\text{Proj}_{\overrightarrow{RS}} \overrightarrow{PQ}$.

43. Prove that $\mathbf{u}$ and $\mathbf{v}$ are parallel if and only if $\mathbf{v} = \alpha\mathbf{u}$ for some constant α. [*Hint:* Show that $\cos \varphi = \pm 1$ if and only if $\mathbf{v} = \alpha\mathbf{u}$.]

44. Prove that $\mathbf{u}$ and $\mathbf{v}$ are orthogonal if and only if $\mathbf{u} \cdot \mathbf{v} = 0$.

45. Show that the vector $\mathbf{v} = a\mathbf{i} + b\mathbf{j}$ is orthogonal to the line $ax + by + c = 0$.

46. Show that the vector $\mathbf{u} = b\mathbf{i} - a\mathbf{j}$ is parallel to the line $ax + by + c = 0$.

47. A triangle has vertices $(1, 3)$, $(4, -2)$ and $(-3, 6)$. Find the cosine of each of its angles.

48. A triangle has vertices (a_1, b_1), (a_2, b_2) and (a_3, b_3). Find a formula for the cosines of each of its angles.

*49. The *Cauchy–Schwartz inequality* states that for any real numbers a_1, a_2, b_1, and b_2,

$$\left| \sum_{k=1}^{2} a_k b_k \right| \leq \left(\sum_{k=1}^{2} a_k^2 \right)^{1/2} \left(\sum_{k=1}^{2} b_k^2 \right)^{1/2}.$$

Use the dot product to prove this formula. Under what circumstances can the inequality be replaced by an equality?

50. Prove that the shortest distance between a point and a line is measured along a line through the point and perpendicular to the line.

51. Find the distance between $P = (2, 3)$ and the line through the points $Q = (-1, 7)$ and $R = (3, 5)$. [*Hint:* Draw a picture and use the Pythagorean theorem.]

52. Find the distance between $(3, 7)$ and the line along the vector $\mathbf{v} = 2\mathbf{i} - 3\mathbf{j}$.

15.3 Some Applications of Vectors

In this section we discuss some elementary applications of vectors. Further examples of the use of vectors will be given in the next chapter after we discuss the differentiation of vector functions.

We begin by showing how vectors can be used to prove some basic geometric rules. First we derive a simple rule that is very useful.

Suppose that P and Q are two points in the plane. Let R be a point on the straight line segment joining P and Q and suppose that $|\overrightarrow{PR}| = a$ and $|\overrightarrow{RQ}| = b$. Then we seek a formula for the vector $\mathbf{w} = \overrightarrow{OR}$ in terms of the vectors $\mathbf{u} = \overrightarrow{OP}$ and $\mathbf{v} = \overrightarrow{OQ}$ (see Figure 1).

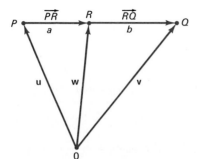

Figure 1

THEOREM 1. In the above context

$$\mathbf{w} = \frac{b}{a+b}\mathbf{u} + \frac{a}{a+b}\mathbf{v}. \tag{1}$$

PROOF. From Figure 1, we see that $\overrightarrow{PQ} = \mathbf{v} - \mathbf{u}$. Also, $\overrightarrow{PR}$, $\overrightarrow{RQ}$, and $\overrightarrow{PQ}$ have the same direction so that $\overrightarrow{PR} = [a/(a+b)]\overrightarrow{PQ}$ and $\overrightarrow{RQ} = [b/(a+b)]\overrightarrow{PQ}$. From Figure 1, we have

$$\mathbf{w} = \mathbf{u} + \overrightarrow{PR} = \mathbf{u} + \left(\frac{a}{a+b}\right)\overrightarrow{PQ} = \mathbf{u} + \frac{a}{a+b}(\mathbf{v} - \mathbf{u})$$

$$= \left(1 - \frac{a}{a+b}\right)\mathbf{u} + \frac{a}{a+b}\mathbf{v} = \frac{b}{a+b}\mathbf{u} + \frac{a}{a+b}\mathbf{v}$$

and the theorem is proved.

EXAMPLE 1. Show that the line segment joining the midpoints of two sides of a triangle has the length of half the third side and is parallel to it.

SOLUTION. We refer to Figure 2. If D and E are the midpoints of sides AB and BC, respectively, then, from (1)

$$\overrightarrow{OD} = \tfrac{1}{2}\overrightarrow{OA} + \tfrac{1}{2}\overrightarrow{OB} \qquad \text{and} \qquad \overrightarrow{OE} = \tfrac{1}{2}\overrightarrow{OC} + \tfrac{1}{2}\overrightarrow{OB}.$$

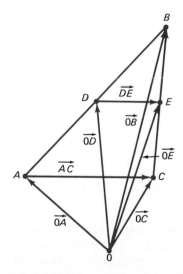

Figure 2

We note that

$$\overrightarrow{DE} = \overrightarrow{OE} - \overrightarrow{OD} \qquad \text{and} \qquad \overrightarrow{AC} = \overrightarrow{OC} - \overrightarrow{OA}.$$

Then

$$\overrightarrow{DE} = \overrightarrow{OE} - \overrightarrow{OD} = (\tfrac{1}{2}\overrightarrow{OC} + \tfrac{1}{2}\overrightarrow{OB}) - (\tfrac{1}{2}\overrightarrow{OA} + \tfrac{1}{2}\overrightarrow{OB})$$
$$= \tfrac{1}{2}(\overrightarrow{OC} - \overrightarrow{OA}) = \tfrac{1}{2}\overrightarrow{AC}.$$

Thus $\overrightarrow{DE}$ and $\overrightarrow{AC}$ have the same direction and

$$|\overrightarrow{DE}| = \tfrac{1}{2}|\overrightarrow{AC}|$$

which is what we wanted to show.

EXAMPLE 2. Show that the midpoints of the sides of a quadrilateral are the vertices of a parallelogram.

SOLUTION. The situation is sketched in Figure 3. For simplicity, we let **A** denote the

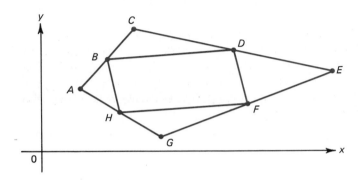

Figure 3

vector $\overrightarrow{OA}$, **B** the vector $\overrightarrow{OB}$, etc. Since B, D, F, and H are midpoints, we have, from (1)

$$\mathbf{B} = \tfrac{1}{2}\mathbf{A} + \tfrac{1}{2}\mathbf{C}, \qquad \mathbf{D} = \tfrac{1}{2}\mathbf{C} + \tfrac{1}{2}\mathbf{E}, \qquad \mathbf{F} = \tfrac{1}{2}\mathbf{E} + \tfrac{1}{2}\mathbf{G}, \qquad \mathbf{H} = \tfrac{1}{2}\mathbf{A} + \tfrac{1}{2}\mathbf{G}.$$

Then

$$\mathbf{B} - \mathbf{D} = \tfrac{1}{2}\mathbf{A} + \tfrac{1}{2}\mathbf{C} - (\tfrac{1}{2}\mathbf{C} + \tfrac{1}{2}\mathbf{E}) = \tfrac{1}{2}\mathbf{A} - \tfrac{1}{2}\mathbf{E}$$

and

$$\mathbf{H} - \mathbf{F} = \tfrac{1}{2}\mathbf{A} + \tfrac{1}{2}\mathbf{G} - (\tfrac{1}{2}\mathbf{E} + \tfrac{1}{2}\mathbf{G}) = \tfrac{1}{2}\mathbf{A} - \tfrac{1}{2}\mathbf{E}.$$

Thus the vectors $\mathbf{B} - \mathbf{D}$ and $\mathbf{H} - \mathbf{F}$ are the same. Similarly, we have $\mathbf{D} - \mathbf{F} = \mathbf{B} - \mathbf{H}$. Thus opposite sides in the quadrilateral $BDFH$ are parallel and have equal magnitudes which implies that $BDFH$ is a parallelogram.

    In earlier chapters we solved problems involving the notion of force. We started such problems with a statement like "a force of x nt is applied to. . . ." Implicit in such a problem is the idea that a force with a certain magnitude (measured in pounds or newtons) is exerted in a certain direction. In this context, force can (and should) be thought of as a vector.

    If more than one force is applied to an object, then we define the *resultant* of the forces applied to the object to be the *vector sum* of these forces. We can think of the resultant as the *net* applied force.

EXAMPLE 3. A force of 3 nt is applied to the left side of an object, one of 4 nt is applied from the bottom, and a force of 7 nt is applied from an angle of $\pi/4$ to the horizontal. What is the resultant of forces applied to the object?

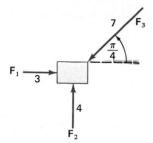

Figure 4

SOLUTION. The forces are indicated in Figure 4. We write each force as a magnitude times a unit vector in the indicated direction. For convenience, we can think of the center of the object as being at the origin. Then $F_1 = 3i$, $F_2 = 4j$, and $F_3 = -(7/\sqrt{2})(i + j)$. This last follows from the fact that the vector $-(1/\sqrt{2})(i + j)$ is a unit vector pointed toward the origin making an angle of $\pi/4$ with the x-axis (see Example 15.1.10). Then the resultant is given by

$$F = F_1 + F_2 + F_3 = \left(3 - \frac{7}{\sqrt{2}}\right)i + \left(4 - \frac{7}{\sqrt{2}}\right)j.$$

The magnitude of F is

$$|F| = \sqrt{\left(3 - \frac{7}{\sqrt{2}}\right)^2 + \left(4 - \frac{7}{\sqrt{2}}\right)^2} = \sqrt{74 - \frac{98}{\sqrt{2}}} \approx 2.17 \quad nt.$$

The direction θ can be calculated by first finding the unit vector in the direction of F:

$$\frac{F}{|F|} = \frac{3 - (7/\sqrt{2})}{\sqrt{74 - (98/\sqrt{2})}}i + \frac{4 - (7/\sqrt{2})}{\sqrt{74 - (98/\sqrt{2})}}j$$

$$= (\cos\theta)i + (\sin\theta)j.$$

Then

$$\cos\theta = \frac{3 - (7/\sqrt{2})}{\sqrt{74 - (98/\sqrt{2})}} \approx -0.8990$$

and $\theta \approx 2.6883 \approx 206°$ (or $-154°$). This is illustrated in Figure 5.

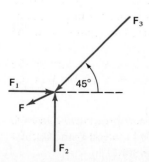

Figure 5

EXAMPLE 4. In Example 3, what additional force must be applied in order that the object remain at rest?

SOLUTION. We must apply a force F_4 so that the resultant of the four forces is **0**. If $F_4 = a\mathbf{i} + b\mathbf{j}$, then the new resultant is given by

$$F^* = F_1 + F_2 + F_3 + F_4 = \left(3 - \frac{7}{\sqrt{2}} + a\right)\mathbf{i} + \left(4 - \frac{7}{\sqrt{2}} + b\right)\mathbf{j}.$$

In order that this be **0**, we must have

$$a = \frac{7}{\sqrt{2}} - 3 \quad \text{and} \quad b = \frac{7}{\sqrt{2}} - 4$$

so that $\mathbf{F_4} = -\mathbf{F}$ where $\mathbf{F}$ is the resultant of F_1, F_2, and F_3. This is sketched in Figure 6.

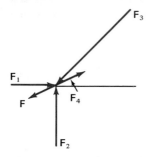

Figure 6

Recall from Section 5.7 that the work done by a force F in moving an object a distance d is given by

$$W = Fd \tag{2}$$

where units of work are nt-m or ft-lb. In Formula (2) it is assumed that the force is applied in the same direction as the direction of motion. However, this is not always the case. In general, we may define

$$W = (\text{component of } \mathbf{F} \text{ in the direction of motion}) \times (\text{distance moved}). \tag{3}$$

If the object moves from P to Q, then the distance moved is $|\overrightarrow{PQ}|$. The vector **d**, one of whose representations is $\overrightarrow{PQ}$, is called a *displacement vector*. Then from Definition 15.2.5,

$$\text{component of } \mathbf{F} \text{ in direction of motion} = \frac{\mathbf{F} \cdot \mathbf{d}}{|\mathbf{d}|}. \tag{4}$$

Finally, combining (3) and (4), we obtain

$$W = \frac{\mathbf{F} \cdot \mathbf{d}}{|\mathbf{d}|} |\mathbf{d}| = \mathbf{F} \cdot \mathbf{d}. \tag{5}$$

That is, *the work done is the dot product of the force* **F** *and the displacement vector* **d**. Note that if **F** acts in the direction **d** and if φ denotes the angle between **F** and **d**, then $\mathbf{F} \cdot \mathbf{d} = |\mathbf{F}|\,|\mathbf{d}| \cos \varphi = |\mathbf{F}|\,|\mathbf{d}| \cos 0 = |\mathbf{F}|\,|\mathbf{d}|$ which is the formula (2).

EXAMPLE 5. A force of 4 nt has the direction of $\pi/3$. What is the work done in moving an object from the point $(1, 2)$ to the point $(5, 4)$ where distances are measured in meters?

SOLUTION. A unit vector with direction $\pi/3$ is given by $\mathbf{u} = (\cos \pi/3)\mathbf{i} + (\sin \pi/3)\mathbf{j}$ $= (1/2)\mathbf{i} + (\sqrt{3}/2)\mathbf{j}$. Thus $\mathbf{F} = 4\mathbf{u} = 2\mathbf{i} + 2\sqrt{3}\mathbf{j}$. The displacement vector $\mathbf{d}$ is given by $(5 - 1)\mathbf{i} + (4 - 2)\mathbf{j} = 4\mathbf{i} + 2\mathbf{j}$. Thus

$$W = \mathbf{F} \cdot \mathbf{d} = (2\mathbf{i} + 2\sqrt{3}\mathbf{j}) \cdot (4\mathbf{i} + 2\mathbf{j}) = (8 + 4\sqrt{3}) \approx 14.93 \quad \text{nt-m.}$$

The component of $\mathbf{F}$ in the direction of motion is sketched in Figure 7.

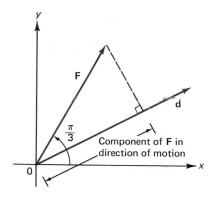

Figure 7

PROBLEMS 15.3

In Problems 1–3 the letter $\mathbf{P}$ represents the vector $\overrightarrow{OP}$, the letter $\mathbf{Q}$ represents the vector $\overrightarrow{OQ}$, and the letter $\mathbf{R}$ represents the vector $\overrightarrow{OR}$.

1. The point S is on the line PQ, two-thirds the distance from P to Q. If R is a fourth point, express the vector $\overrightarrow{RS}$ in terms of $\mathbf{P}$, $\mathbf{Q}$, and $\mathbf{R}$.

2. Answer the question in Problem 1 if the point S is eight-elevenths the distance from P to Q.

3. If S is on the line PQ and one-quarter the distance from P to Q, and T is on the line PR three-fifths the distance from P to R, express the vector $\overrightarrow{ST}$ in terms of $\mathbf{P}$, $\mathbf{Q}$, and $\mathbf{R}$.

*4. Prove that the medians of a triangle intersect at a fixed point which is two-thirds the distance from each vertex to the opposite side.

5. Use the result of Problem 4 to find the point of intersection of the medians of the triangle with vertices at $(3, 4)$, $(2, -1)$ and $(-3, 2)$.

6. Show that the diagonals of a parallelogram bisect each other.

7. Show that the diagonals of a rhombus are orthogonal.

8. Show that in a trapezoid, the line segments which join the midpoints of the two sides that are not parallel is parallel to the parallel sides and has a length equal to the average of the lengths of the parallel sides.

9. Use vector methods to show that the angles opposite the equal sides in an isosceles triangle are equal.

In Problems 10–18 find the resultant of the forces acting on an object. Then find the force that must be applied so that the object will remain at rest.

10. 2 nt (from right), 5 nt (from above)

11.   2 nt (from left), 5 nt (from below)

12.   3 nt (from left), 5 nt (from right), 3 nt (from above)

13.   10 lb (from right), 8 lb (from below)

14.   5 lb (from above), 4 lb (from direction $\pi/6$)

15.   6 nt (from left), 4 nt (from direction $\pi/4$), 2 nt (from direction $\pi/3$)

16.   2 nt (from above), 3 nt (from direction $3\pi/4$)

17.   5 nt (from direction $\pi/3$), 5 nt (from direction $2\pi/3$)

18.   7 nt (from direction $\pi/6$), 7 nt (from direction $\pi/3$), 14 nt (from direction $5\pi/4$).

In Problems 19–27 find the work done when the force with given magnitude and direction moves an object from P to Q. All distances are measured in meters. (Note that work can be negative.)

19. $|\mathbf{F}| = 3$ nt; $\theta = 0$; $P = (2, 3)$; $Q = (1, 7)$

20. $|\mathbf{F}| = 2$ nt; $\theta = \pi/2$; $P = (5, 7)$; $Q = (1, 1)$

21. $|\mathbf{F}| = 6$ nt; $\theta = \pi/4$; $P = (2, 3)$; $Q = (-1, 4)$

22. $|\mathbf{F}| = 4$ nt; $\theta = \pi/6$; $P = (-1, 2)$; $Q = (3, 4)$

23. $|\mathbf{F}| = 7$ nt; $\theta = 2\pi/3$; $P = (4, -3)$; $Q = (1, 0)$

24. $|\mathbf{F}| = 3$ nt; $\theta = 3\pi/4$; $P = (2, 1)$; $Q = (1, 2)$

25. $|\mathbf{F}| = 6$ nt; $\theta = \pi$; $P = (3, -8)$; $Q = (5, 10)$

26. $|\mathbf{F}| = 4$ nt; θ is direction of $2\mathbf{i} + 3\mathbf{j}$; $P = (2, 0)$; $Q = (-1, 3)$

27. $|\mathbf{F}| = 5$ nt; θ is direction of $-3\mathbf{i} + 2\mathbf{j}$; $P = (1, 3)$; $Q = (4, -6)$.

Review Exercises for Chapter Fifteen

In Exercises 1–6 find the magnitude and direction of the given vector.

1. $\mathbf{v} = (3, 3)$ 2. $\mathbf{v} = -3\mathbf{i} + 3\mathbf{j}$ 3. $\mathbf{v} = (2, -2\sqrt{3})$

4. $\mathbf{v} = (\sqrt{3}, 1)$ 5. $\mathbf{v} = -12\mathbf{i} - 12\mathbf{j}$ 6. $\mathbf{v} = \mathbf{i} + 4\mathbf{j}$.

In Exercises 7–10 write in the form $a\mathbf{i} + b\mathbf{j}$ the vector $\mathbf{v}$ that is represented by $\overrightarrow{PQ}$. Sketch $\overrightarrow{PQ}$ and $\mathbf{v}$.

7. $P = (2, 3)$; $Q = (4, 5)$ 8. $P = (1, -2)$; $Q = (7, 12)$

9. $P = (-1, -6)$; $Q = (3, -4)$ 10. $P = (-1, 3)$; $Q = (3, -1)$.

11.   Let $\mathbf{u} = (2, 1)$ and $\mathbf{v} = (-3, 4)$. Find
        (a) $5\mathbf{u}$ (b) $\mathbf{u} - \mathbf{v}$ (c) $-8\mathbf{u} + 5\mathbf{v}$.

12.   Let $\mathbf{u} = -4\mathbf{i} + \mathbf{j}$ and $\mathbf{v} = -3\mathbf{i} - 4\mathbf{j}$. Find
        (a) $-3\mathbf{v}$ (b) $\mathbf{u} + \mathbf{v}$ (c) $3\mathbf{u} - 6\mathbf{v}$.

In Exercises 13–19 find a unit vector having the same direction as the given vector.

13. $\mathbf{v} = \mathbf{i} + \mathbf{j}$ 14. $\mathbf{v} = -\mathbf{i} + \mathbf{j}$ 15. $\mathbf{v} = 2\mathbf{i} + 5\mathbf{j}$

16. $\mathbf{v} = -7\mathbf{i} + 3\mathbf{j}$ 17. $\mathbf{v} = 3\mathbf{i} + 4\mathbf{j}$ 18. $\mathbf{v} = -2\mathbf{i} - 2\mathbf{j}$

19. $\mathbf{v} = a\mathbf{i} - a\mathbf{j}$.

20.   If $\mathbf{v} = 4\mathbf{i} - 7\mathbf{j}$, find $\sin \theta$ and $\cos \theta$, where θ is the direction of $\mathbf{v}$.

21.   Find a unit vector with direction opposite to that of $\mathbf{v} = 5\mathbf{i} + 2\mathbf{j}$.

22.   Find two unit vectors orthogonal to $\mathbf{v} = \mathbf{i} - \mathbf{j}$.

23.   Find a unit vector with direction opposite to that of $\mathbf{v} = 10\mathbf{i} - 7\mathbf{j}$.

In Exercises 24–27 find a vector $\mathbf{v}$ having the given magnitude and direction.

24. $|\mathbf{v}| = 2$; $\theta = \pi/3$ 25. $|\mathbf{v}| = 1$; $\theta = \pi/2$

26. $|\mathbf{v}| = 4$; $\theta = \pi$ 27. $|\mathbf{v}| = 7$; $\theta = 5\pi/6$.

In Exercises 28–31 calculate the dot product of the two vectors and the cosine of the angle between them.

28. $u = i - j; v = i + 2j$

29. $u = -4i; v = 11j$

30. $u = 4i - 7j; v = 8i + 6j$

31. $u = -i - 2j; v = 4i + 5j$.

In Exercises 32–37 determine whether the given vectors are orthogonal, parallel, or neither. Then sketch each pair.

32. $u = 2i - 6j; v = -i + 3j$

33. $u = 4i - 5j; v = 5i - 4j$

34. $u = 4i - 5j; v = -5i + 4j$

35. $u = -7i - 7j; v = i + j$

36. $u = -7i - 7j; v = -i + j$

37. $u = -7i - 7j; v = -i - j$.

38. Let $u = 2i + 3j$ and $v = 4i + \alpha j$. Determine α such that

    (a) u and v are orthogonal

    (b) u and v are parallel

    (c) the angle between u and v is $\pi/4$

    (d) the angle between u and v is $\pi/6$.

In Exercises 39–44 calculate $\text{Proj}_v\, u$.

39. $u = 14i; v = i + j$

40. $u = 14i; v = i - j$

41. $u = 3i - 2j; v = 3i + 2j$

42. $u = 3i + 2j; v = i - 3j$

43. $u = 2i - 5j; v = -3i - 7j$

44. $u = 4i - 5j; v = -3i - j$.

45. Let $P = (3, -2)$, $Q = (4, 7)$, $R = (-1, 3)$ and $S = (2, -1)$. Calculate $\text{Proj}_{\overrightarrow{PQ}}\ \overrightarrow{RS}$ and $\text{Proj}_{\overrightarrow{RS}}\ \overrightarrow{PQ}$.

In Exercises 46–48 calculate the resultant of the forces acting on an object.

46. 3 nt (from left), 2 nt (from below)

47. 5 nt (from right), 2 nt (from left), 3 nt (from above)

48. 2 nt (from direction $\pi/4$), 5 nt (from left), 4 nt (from direction $2\pi/3$), 6 nt (from direction $3\pi/4$).

In Exercises 49–52 find the work done when the force with given magnitude and direction moves an object from P to Q. All distances are measured in meters.

49. $|F| = 2$ nt; $\theta = \pi/4$; $P = (1, 6)$; $Q = (2, 4)$

50. $|F| = 3$ nt; $\theta = \pi/2$; $P = (3, -5)$; $Q = (2, 7)$

51. $|F| = 11$ nt; $\theta = \pi/6$; $P = (-1, -2)$; $Q = (-7, -4)$

52. $|F| = 8$ nt; $\theta = 2\pi/3$; $P = (-1, 4)$; $Q = (5, -6)$.

SIXTEEN

VECTOR FUNCTIONS, VECTOR DIFFERENTIATION, AND PARAMETRIC EQUATIONS

16.1 Vector Functions and Parametric Equations

In Chapter 15 we considered vectors (in the plane) which could be written as

$$\mathbf{v} = (a, b) = a\mathbf{i} + b\mathbf{j}. \tag{1}$$

In this chapter we see what happens when the numbers a and b in (1) are replaced by functions $f_1(t)$ and $f_2(t)$.

DEFINITION 1. Let f_1 and f_2 be functions of the real variable t. Then for all values of t for which $f_1(t)$ and $f_2(t)$ are defined, we define the *vector-valued function* $\mathbf{f}$ by

$$\boxed{\mathbf{f}(t) = (f_1(t), f_2(t)) = f_1(t)\mathbf{i} + f_2(t)\mathbf{j}.} \tag{2}$$

The *domain* of $\mathbf{f}$ is the intersection of the domains of f_1 and f_2.

> *Remark.* For simplicity, we will refer to vector-valued functions as *vector functions*.

EXAMPLE 1. Let $\mathbf{f}(t) = f_1(t)\mathbf{i} + f_2(t)\mathbf{j} = (1/t)\mathbf{i} + \sqrt{t + 1}\mathbf{j}$. Find the domain of $\mathbf{f}$.

SOLUTION. The domain of $\mathbf{f}$ is the set of all t for which f_1 and f_2 are defined. Since $f_1(t)$ is defined for $t \neq 0$ and $f_2(t)$ is defined for $t \geq -1$, we see that the domain of $\mathbf{f}$ is the set $\{t : t \geq -1 \text{ and } t \neq 0\}$.

Let $\mathbf{f}$ be a vector function. Then for each t in the domain of $\mathbf{f}$, the endpoint of the vector $f_1(t)\mathbf{i} + f_2(t)\mathbf{j}$ is a point (x, y) in the xy-plane where

$$x = f_1(t) \qquad \text{and} \qquad y = f_2(t). \tag{3}$$

The set of all such points form a *curve C* in the xy-plane. Equation (2) is called the *vector equation* of C while equations (3) are called the *parametric equations* or *parametric representation* of C. In this context, the variable t is called a *parameter*.

EXAMPLE 2. Describe the curve given by the vector equation

$$\mathbf{f}(t) = (\cos t)\mathbf{i} + (\sin t)\mathbf{j}. \tag{4}$$

SOLUTION. We first see that for every t, $|\mathbf{f}(t)| = 1$ since $|\mathbf{f}(t)| = \sqrt{\cos^2 t + \sin^2 t} = 1$. Moreover, if we write the curve in its parametric representation, we find that

$$x = \cos t, \qquad y = \sin t \tag{5}$$

and, since $\cos^2 t + \sin^2 t = 1$, we have

$$x^2 + y^2 = 1$$

which is, of course, the equation of the unit circle. This is sketched in Figure 1. Note that, in the sketch, the parameter t does not appear. The representation $x^2 + y^2 = 1$ is called the *Cartesian equation* of the curve given by (5).

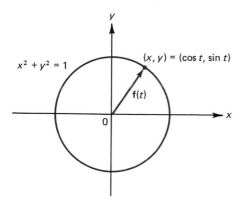

Figure 1

EXAMPLE 3. Describe and sketch the curve given parametrically by $x = t + 3$, $y = t^2 - t + 2$.

SOLUTION. With problems of this type, the easiest thing to do is to write, if possible, t as a function of x or y. Since $x = t + 3$, we immediately see that $t = x - 3$ and $y = t^2 - t + 2 = (x - 3)^2 - (x - 3) + 2 = x^2 - 7x + 14$. This is the Cartesian equation of the curve and is the equation of a parabola. It is sketched in Figure 2.

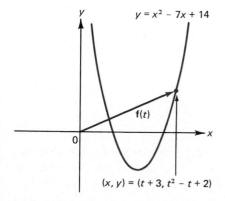

Figure 2

EXAMPLE 4. Describe and sketch the curve given parametrically by $x = \cosh t$ and $y = \sinh t$.

SOLUTION. We start with formula (7.9.6)

$$\cosh^2 t - \sinh^2 t = 1.$$

Thus, $x^2 - y^2 = 1$ which is the equation of a hyperbola (see Section 1.9). This is sketched in Figure 3. Note that the graph of the hyperbola $x^2 - y^2 = 1$ includes the dotted curve in the figure. This must be excluded from the graph of (cosh t, sinh t) since $x = \cosh t \geq 1$.

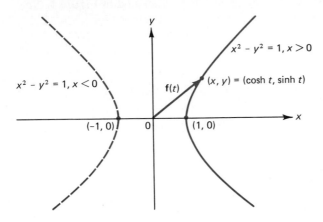

$x^2 - y^2 = 1, x < 0$

$x^2 - y^2 = 1, x > 0$

$(x, y) = (\cosh t, \sinh t)$

$f(t)$

$(-1, 0)$ 0 $(1, 0)$

Figure 3

EXAMPLE 5. Describe and sketch the curve given by the vector equation

$$\mathbf{f}(r) = (1 - r^4)\mathbf{i} + r^2\mathbf{j}. \tag{6}$$

SOLUTION. First we note that the parameter in this problem is r instead of t. This makes absolutely no difference since, as in the case of the variable of integration, the parameter is a "dummy" variable. Now, to get a feeling for the shape of this curve, we display, in Table 1, values of x and y for various values of r. Plotting some of

TABLE 1

r	0	$\frac{1}{2}$	1	$\frac{3}{2}$	2
$x = 1 - r^4$	1	$\frac{15}{16}$	0	$-\frac{65}{16}$	-15
$y = r^2$	0	$\frac{1}{4}$	1	$\frac{9}{4}$	4

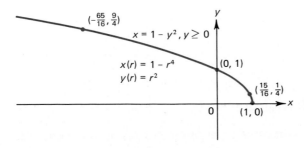

$\left(-\frac{65}{16}, \frac{9}{4}\right)$

$x = 1 - y^2, y \geq 0$

$x(r) = 1 - r^4$
$y(r) = r^2$

$(0, 1)$

$\left(\frac{15}{16}, \frac{1}{4}\right)$

$(1, 0)$

Figure 4

these points leads to the sketch in Figure 4. To write the Cartesian equation for this curve, we square both sides of the equation $y = r^2$ to obtain $y^2 = r^4$ and $x = 1 - r^4 = 1 - y^2$ which is the equation of the parabola sketched in Figure 5. Note that this is *not* the same as the curve sketched in Figure 4 since the parametric representation $y = r^2$ requires that y be positive. Thus the curve described by (6) is only *half* the parabola described by the equation $x = 1 - y^2$.

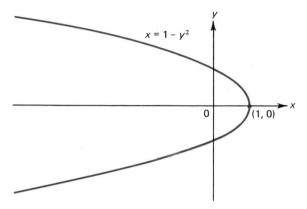

Figure 5

We now obtain the parametric representation of a line L. Let $P = (x_1, y_1)$ and $Q = (x_2, y_2)$ be two points on the line and let R be a third point on the line (see

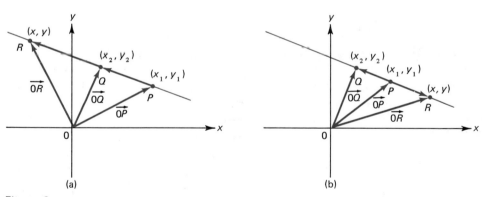

(a)   (b)

Figure 6

Figure 6). Since the vectors $\overrightarrow{PQ}$ and $\overrightarrow{PR}$ have the same or opposite directions, they are parallel and, by Theorem 15.2.3, we have

$$\overrightarrow{PR} = t\,\overrightarrow{PQ} \tag{7}$$

for some real number t (t is positive if $\overrightarrow{PQ}$ and $\overrightarrow{PR}$ have the same direction and t is negative if $\overrightarrow{PQ}$ and $\overrightarrow{PR}$ have opposite directions). Then, from Figure 6, we see that

$$\overrightarrow{OR} = \overrightarrow{OP} + \overrightarrow{PR} = \overrightarrow{OP} + t\,\overrightarrow{PQ}. \tag{8}$$

But $\overrightarrow{PQ} = (x_2 - x_1)\mathbf{i} + (y_2 - y_1)\mathbf{j}$ and $\overrightarrow{OP} = x_1\mathbf{i} + y_1\mathbf{j}$ so that, from (8), if $R = (x, y)$ we obtain

$$x\mathbf{i} + y\mathbf{j} = x_1\mathbf{i} + y_1\mathbf{j} + t(x_2 - x_1)\mathbf{i} + t(y_2 - y_1)\mathbf{j}$$

or

$$x\mathbf{i} + y\mathbf{j} = [x_1 + t(x_2 - x_1)]\mathbf{i} + [y_1 + t(y_2 - y_1)]\mathbf{j}.$$

Thus a parametric equation of the line passing through the points (x_1, y_1) and (x_2, y_2) is

$$x = x_1 + t(x_2 - x_1) \qquad \text{and} \qquad y = y_1 + t(y_2 - y_1). \tag{9}$$

EXAMPLE 6. Find a parametric representation of the line passing through the points $P = (2, 5)$ and $Q = (-3, 9)$.

SOLUTION. Using (9), we immediately find that

$$x = 2 - 5t \qquad \text{and} \qquad y = 5 + 4t. \tag{10}$$

Note that the problem asked for *a* parametric representation of the line. This is because there are several such representations. For example, reversing the roles of P and Q, we obtain the representation (from the new $P = (-3, 9)$ and $Q = (2, 5)$)

$$x = -3 + 5t \qquad \text{and} \qquad y = 9 - 4t. \tag{11}$$

Although these representations look different, they do represent the same line as is seen by inserting a few sample values for t. For example, if $t = 1$, from (10) we obtain the point $(-3, 9)$. This point is obtained from (11) by setting $t = 0$.

To obtain another parametric representation, we first calculate that the slope-intercept form of the line passing through the points $(2, 5)$ and $(-3, 9)$ is given by

$$\frac{y - 5}{x - 2} = \frac{9 - 5}{-3 - 2} = -\frac{4}{5} \qquad \text{or} \qquad y = -\frac{4}{5}x + \frac{33}{5}.$$

Then, setting $t = x$, we obtain the "slope–intercept" parametrization

$$x = t, \qquad y = -\frac{4}{5}t + \frac{33}{5}.$$

The following example shows how vector analysis can be useful in a simple physical problem.

EXAMPLE 7. A cannonball shot from a cannon has an initial velocity of 600 m/sec. The muzzle of the cannon is inclined at an angle of 30°. Ignoring air resistance, determine the path of the cannonball.

SOLUTION. We place the x- and y-axes so that the mouth of the cannon is at the origin (see Figure 7). The velocity vector $\mathbf{v}$ can be resolved into its vertical and horizontal components

$$\mathbf{v} = v_x\mathbf{i} + v_y\mathbf{j}.$$

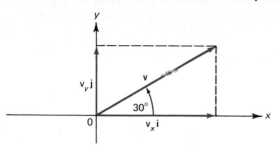

Figure 7

A unit vector in the direction of **v** is

$$\mathbf{u} = (\cos 30°)\mathbf{i} + (\sin 30°)\mathbf{j} = \frac{\sqrt{3}}{2}\mathbf{i} + \frac{1}{2}\mathbf{j}$$

so that, initially,

$$\mathbf{v} = |\mathbf{v}|\mathbf{u} = 600\mathbf{u} = 300\sqrt{3}\mathbf{i} + 300\mathbf{j}.$$

The scalar $|\mathbf{v}|$ is called the *speed* of the cannonball. Thus $v_x = 300\sqrt{3}$ m/sec (the initial speed in the horizontal direction) and $v_y = 300$ m/sec (the initial speed in the vertical direction). Now the vertical acceleration (due to gravity) is

$$a_y = -9.81 \text{ m/sec}^2 \qquad \text{and} \qquad v_y = \int a_y\, dt = -9.81t + C.$$

Since, initially, $v_y(0) = 300$, we have $C = 300$ and

$$v_y = -9.81t + 300.$$

Then

$$y(t) = \int v_y\, dt = -\frac{9.81t^2}{2} + 300t + C_1$$

and, since $y(0) = 0$ (we start at the origin), we find that

$$y(t) = -\frac{9.81t^2}{2} + 300t.$$

To calculate the x-component of the position vector, we note that, ignoring air resistance, the velocity in the horizontal direction is constant†; that is

$$v_x = 300\sqrt{3} \quad \text{m/sec.}$$

Then $x(t) = \int v_x\, dt = 300\sqrt{3}t + C_2$ and, since $x(0) = 0$, we obtain

$$x = 300\sqrt{3}t.$$

Thus the position vector describing the location of the cannonball is

$$\mathbf{s}(t) = x(t)\mathbf{i} + y(t)\mathbf{j} = 300\sqrt{3}t\mathbf{i} + \left(300t - \frac{9.81t^2}{2}\right)\mathbf{j}.$$

† There are no forces acting on the ball in the horizontal direction.

To obtain the Cartesian equation of this curve, we start with

$$x = 300\sqrt{3}t$$

so that

$$t = \frac{x}{300\sqrt{3}}$$

and

$$y = 300\left(\frac{x}{300\sqrt{3}}\right) - \frac{9.81}{2}\frac{x^2}{(300\sqrt{3})^2} = \frac{x}{\sqrt{3}} - \frac{9.81}{540{,}000}x^2.$$

This parabola is sketched in Figure 8.

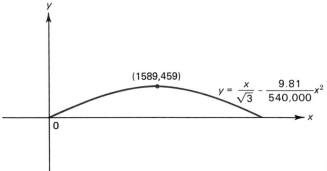

Figure 8

We close this section by describing parametrically a curve which, historically, has been of great interest. Suppose that a wheel of radius r is rolling in a straight line without slipping. Let P be a fixed point on the wheel a distance s from the center. As the wheel moves, the curve traced by the point P is called a *trochoid*† and if $s = r$ (that is, if P is on the circumference of the wheel) then the curve traced is called a cycloid.‡

To simplify our computations, we place the wheel so that it rolls in a clockwise fashion on the x-axis (see Figure 9a). Suppose that P moves through an angle of α radians to reach its position in Figure 9b. Then the new position vector $\overrightarrow{OP}$ is easily seen to be given by

$$\overrightarrow{OP} = \overrightarrow{OR} + \overrightarrow{RC} + \overrightarrow{CP}. \tag{12}$$

†From the Greek word meaning wheel.
‡From the Greek word meaning circle. The cycloid was a source of great controversy in the seventeenth century. Many of its properties were discovered by the French mathematician Gilles Persone de Roberval (1602–1675) although the curve was first discussed by Galileo. Unfortunately Roberval, for unknown reasons, did not publish his discoveries concerning the cycloid, which meant that he lost credit for most of them. The ensuing arguments over who discovered what were so bitter that the cycloid became known as the "Helen of geometers" (after Helen of Troy—the source of the intense jealousy that led to the Trojan war).

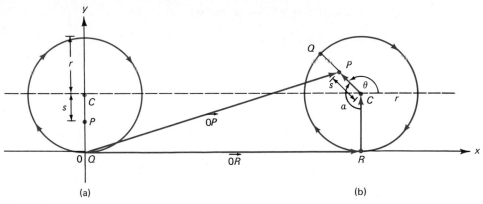

Figure 9

The length of $\overrightarrow{OR}$ is the length of the arc of the circle that is generated by moving through an angle of α radians and is equal to α times the radius of the circle $= \alpha r$. Thus

$$\overrightarrow{OR} = \alpha r\mathbf{i}.$$

Clearly $\overrightarrow{RC} = r\mathbf{j}$, and if θ denotes the angle the vector $\overrightarrow{CP}$ makes with the positive x-axis, then

$$\overrightarrow{CP} = s[(\cos \theta)\mathbf{i} + (\sin \theta)\mathbf{j}].$$

Hence, using (12),

$$\overrightarrow{OP} = \alpha r\mathbf{i} + r\mathbf{j} + s(\cos \theta)\mathbf{i} + s(\sin \theta)\mathbf{j}$$
$$= [\alpha r + s(\cos \theta)]\mathbf{i} + [r + s(\sin \theta)]\mathbf{j}. \tag{13}$$

But, since $\alpha + \theta = 3\pi/2$, we have $\theta = 3\pi/2 - \alpha$ so that

$$\cos \theta = \cos\left(\frac{3\pi}{2} - \alpha\right) = -\sin \alpha, \qquad \sin \theta = \sin\left(\frac{3\pi}{2} - \alpha\right) = -\cos \alpha$$

and the equation of the trochoid becomes

$$x\mathbf{i} + y\mathbf{j} = \overrightarrow{OP} = (r\alpha - s \sin \alpha)\mathbf{i} + (r - s \cos \alpha)\mathbf{j} \tag{14}$$

or

$$x = r\alpha - s \sin \alpha, \qquad y = r - s \cos \alpha. \tag{15}$$

When $r = s$, P is on the circumference of the wheel and we obtain the cycloid given parametrically by

$$x = r(\alpha - \sin \alpha), \qquad y = r(1 - \cos \alpha). \tag{16}$$

Note that $y \geq 0$ in (16) and $y = 0$ when $\cos \alpha = 1$ which occurs when α is a multiple of 2π. For these values of α, $x = r\alpha = 2\pi n r$. Moreover, y takes on its maximum value $2r$ when $\alpha = \pi, 3\pi, 5\pi, \ldots$. With this information, we can sketch the cycloid as in Figure 10.

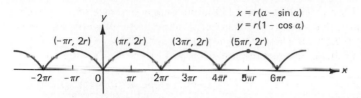

Figure 10

Remark. Here is one instance in which it is much easier to represent a curve parametrically. If you doubt this, try to write the Cartesian equation of the trochoid (equation (14)) or the simpler cycloid (see Problem 49). You will quickly learn to appreciate the advantages of parametric representation. In fact, while it is possible to find the Cartesian equation of the trochoid, the parametric representation is much easier to use.

PROBLEMS 16.1

In Problems 1–8 find the domain of each vector-valued function.

1. $f(t) = \dfrac{1}{t}i + \dfrac{1}{t-1}j$

2. $f(t) = \sqrt{t}i + \dfrac{1}{t}j$

3. $f(s) = \dfrac{1}{s^2-1}i + (s^2-1)j$

4. $f(u) = e^{1/u}i + e^{-1/(u+1)}j$

5. $f(s) = (\ln s)i + \ln(1-s)j$

6. $f(r) = (\sin r)i + rj$

7. $f(t) = (\sec t)i + (\csc t)j$

8. $f(w) = (\tan w)i + (\cot w)j.$

In Problems 9–24 find the Cartesian equation of each curve and then sketch the curve in the *xy*-plane.

9. $f(t) = t^2i + 2tj$

10. $f(t) = (2t-3)i + t^2j$

11. $f(t) = t^2i + t^3j$

12. $f(t) = 3(\sin t)i + 3(\cos t)j$

13. $f(t) = (2t-1)i + (4t+3)j$

14. $f(t) = 2(\cosh t)i + 2(\sinh t)j$

15. $f(t) = (t^4 + t^2 + 1)i + t^2j$

16. $f(t) = t^2i + t^8j$

17. $f(t) = t^3i + (t^9 - 1)j$

18. $f(t) = ti + e^tj$

19. $f(t) = e^ti + t^2j$

20. $f(t) = (t^2 + t - 3)i + \sqrt{t}j$

*21. $f(t) = e^t(\sin t)i + e^t(\cos t)j$ [*Hint:* Show that $|f(t)| = e^t$.]

22. $f(t) = i + (\tan t)j$

23. $f(t) = e^ti + e^{2t}j$

24. $f(t) = (\tan t)i + (\sec t)j.$

In Problems 25–30 find two parametric representations of the straight line that passes through the given points.

25. $(2, 4)$; $(1, 6)$

26. $(-3, 2)$; $(0, 4)$

27. $(3, 5)$; $(-1, -7)$

28. $(4, 6)$; $(7, 9)$

29. $(-2, 3)$; $(4, 7)$

30. $(-4, 0)$; $(3, -2)$.

31. The equation $(x^2/a^2) + (y^2/b^2) = 1$ is the equation of an ellipse (Section 1.9). Show that the curve given by vector equation $f(t) = a(\cos t)i + b(\sin t)j$ is an ellipse.

32. A cannonball is shot upward from ground level at an angle of $45°$ with an initial speed of 1300 ft/sec. Find a parametric representation of the path of the cannonball. Then find the Cartesian equation of this path.

33. How many feet (horizontally) does the cannonball in Problem 32 travel before it hits the ground?

34. How many meters (horizontally) does the cannonball in Example 7 travel before it hits the ground?

35. An object is thrown down from the top of a 150-m building at an angle of 30° (below the horizontal) with an initial velocity of 100 m/sec. Determine a parametric representation of the path of the object. [*Hint:* Draw a picture.]

36. When the object in Problem 35 hits the ground, how far is it from the base of the building?

37. A point is located 25 cm from the center of a wheel 1 m in diameter. Find the parametric representation of the curve traced by that point as the wheel turns.

38. Answer the question in Problem 37 if the point is located on the circumference of the wheel.

*39. A *hypocycloid*† is a curve generated by the motion of a point P on the circumference of a circle which rolls internally, without slipping, on a larger circle (see Figure 11). Assume that the radius of the large circle is a, while that of the smaller circle is b. If θ is the angle in Figure 11, show that a parametric representation of the hypocycloid is

$$x = (a - b) \cos \theta + b \cos \left(\frac{a - b}{b} \right) \theta \quad \text{and} \quad y = (a - b) \sin \theta - b \sin \left(\frac{a - b}{b} \right) \theta.$$

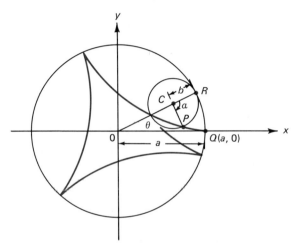

Figure 11

[*Hint:* First show that $a\theta = b\alpha$.]

40. If $a = 4b$ in Problem 39, then the curve generated is called a *hypocycloid of four cusps*. Sketch this curve.

*41. Show that the Cartesian equation of the hypocyloid of four cusps is

$$x^{2/3} + y^{2/3} = a^{2/3}.$$

[*Hint:* Use the identities $\cos 3\theta = 4 \cos^3 \theta - 3 \cos \theta$ and $\sin 3\theta = 3 \sin \theta - 4 \sin^3 \theta$ and the fact that $a - b = 3b$ and $(a - b)/b = 3$.]

42. If $a = nb$ in Problem 39, then the curve generated is called a *hypocycloid of n cusps*. Sketch the hypocycloid of seven cusps.

43. Show that the hypocycloid of two cusps is a straight line segment.

44. Calculate the area bounded by the hypocycloid of four cusps. [*Hint:* Use the result of Problem 41.]

*45. An *epicycloid*‡ is a curve generated by the motion of a point on the circumference of a circle that rolls externally, without slipping, on a fixed circle. Show that the parametric

†The "hypo" comes from the Greek *hupo* meaning "under."
‡From the Greek *epi* meaning "upon" or "over."

representation of an epicycloid is given by

$$x = (a + b) \cos \theta - b \cos \left(\frac{a + b}{b} \right) \theta \quad \text{and} \quad y = (a + b) \sin \theta - b \sin \left(\frac{a + b}{b} \right) \theta$$

where a and b are as in Problem 39.

46. If $a = 4b$, show that the epicycloid generated has the sketch given in Figure 12.

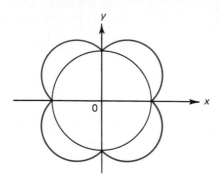

Figure 12

47. Sketch the epicycloid in the case $a = 5b$.
48. Sketch the epicycloid in the case $a = 8b$.
*49. Show that the Cartesian equation of the cycloid is

$$x = r \left[\cos^{-1} \left(\frac{r - y}{y} \right) - \sqrt{2ry - y^2} \right].$$

50. Show that a parametric representation of the hyperbola $(x^2/a^2) - (y^2/b^2) = 1$ is given by

$$x = a \sec \theta, \quad y = b \tan \theta.$$

16.2 The Equation of the Tangent Line to a Parametric Curve

Suppose that $x = f_1(t)$ and $y = f_2(t)$ is the parametric representation of a curve C. We would like to be able to calculate the equation of the line tangent to the curve without having to determine the Cartesian equation of the curve. However, there are complications that might occur since the curve could intersect itself. This happens if there are two numbers $t_1 \neq t_2$ such that $f_1(t_1) = f_1(t_2)$ and $f_2(t_1) = f_2(t_2)$. Thus there are three possibilities. At a given point, the curve could have

(i)   a unique tangent,
(ii) no tangent, or
(iii) two or more tangents.

This is illustrated in Figure 1. A condition that ensures that there is at least one tangent line at each point is that

$$\boxed{f_1' \text{ and } f_2' \text{ are continuous and } [f_1'(t)]^2 + [f_2'(t)]^2 \neq 0.} \tag{1}$$

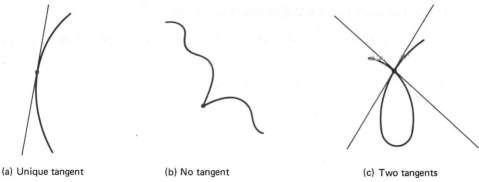

(a) Unique tangent   (b) No tangent   (c) Two tangents

Figure 1

THEOREM 1. Let (x_0, y_0) be on the curve C given by $x = f_1(t)$ and $y = f_2(t)$ and suppose that $f_1'(t_0) \neq 0$. If the curve passes through (x_0, y_0) when $t = t_0$,† then the slope m of the line tangent to C at (x_0, y_0) is given by

$$m = \frac{f_2'(t_0)}{f_1'(t_0)}. \tag{2}$$

PROOF. We refer to Figure 2. From the figure we see that

$$\frac{dy}{dx} = \lim_{\Delta t \to 0} \frac{f_2(t_0 + \Delta t) - f_2(t_0)}{f_1(t_0 + \Delta t) - f_1(t_0)}$$

$$= \lim_{\Delta t \to 0} \frac{[f_2(t_0 + \Delta t) - f_2(t_0)]/\Delta t}{[f_1(t_0 + \Delta t) - f_1(t_0)]/\Delta t}$$

$$= \frac{f_2'(t_0)}{f_1'(t_0)}.$$

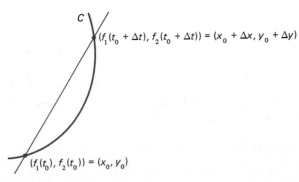

C

$(f_1(t_0 + \Delta t), f_2(t_0 + \Delta t)) = (x_0 + \Delta x, y_0 + \Delta y)$

$(f_1(t_0), f_2(t_0)) = (x_0, y_0)$

Figure 2

Here we have used the fact that $f_1'(t_0) \neq 0$ so that $f_1(t_0 + \Delta t) - f_1(t_0) \neq 0$ if Δt is sufficiently small.

---

†The curve may pass through the point (x_0, y_0) for other values of t as well, and therefore may have other tangent lines at that point as in Figure 1c.

Using Theorem 1, we find that the tangent line to C at t_0 is given by

$$\frac{y - y_0}{x - x_0} = \frac{f_2'(t_0)}{f_1'(t_0)}$$

or, after simplification,

$$\boxed{f_2'(t_0)(x - x_0) - f_1'(t_0)(y - y_0) = 0.} \tag{3}$$

Note that if $f_1'(t_0) = 0$, then equation (3) is still valid and we obtain

$$f_2'(t_0)(x - x_0) = 0. \tag{4}$$

Since, by assumption, $[f_1'(t_0)]^2 + [f_2'(t_0)]^2 \neq 0$, the assumption $f_1'(t_0) = 0$ implies that $f_2'(t_0) \neq 0$. Thus we can divide by $f_2'(t_0)$ in (4) to obtain the equation

$$x - x_0 = 0 \quad \text{or} \quad x = x_0. \tag{5}$$

In this case the tangent line is vertical. If $f_2'(t_0) = 0$, then C has a horizontal tangent at t_0. In sum, if condition (1) holds

> (i) C has a vertical tangent line at t_0 if
>
> $$\left.\frac{dx}{dt}\right|_{t=t_0} = f_1'(t_0) = 0.$$
>
> (ii) C has a horizontal tangent line at t_0 if
>
> $$\left.\frac{dy}{dt}\right|_{t=t_0} = f_2'(t_0) = 0.$$

EXAMPLE 1.   Find the equation of a line (or lines) tangent to the curve $x = e^t$, $y = e^{-t}$ at the point $(1, 1)$.

SOLUTION. The point $(1, 1)$ is reached only when $t = 0$. Then

$$\frac{dy}{dx} = \frac{d(e^{-t})/dt}{d(e^t)/dt} = -\left.\frac{e^{-t}}{e^t}\right|_{t=0} = -1$$

and the equation of the tangent line is

$$\frac{y - 1}{x - 1} = -1 \quad \text{or} \quad y = -x + 2.$$

EXAMPLE 2.   Find the equation of the line tangent to the curve

$$x = 2t^3 - 15t^2 + 24t + 7, \quad y = t^2 + t + 1 \quad \text{at} \quad t = 2.$$

SOLUTION. Here $f_1'(2) = 6t^2 - 30t + 24|_{t=2} = -12$ and $f_2'(2) = 2t + 1|_{t=2} = 5$. When $t = 2$, $x = 11$, and $y = 7$. Then, using (3), we obtain

$$5(x - 11) + 12(y - 7) = 0 \quad \text{or} \quad 5x + 12y = 139.$$

EXAMPLE 3.   Find all horizontal and vertical tangents to the curve of Example 2.

SOLUTION. There are vertical tangents when

$$6t^2 - 30t + 24 = 0 = t^2 - 5t + 4 = (t - 4)(t - 1)$$

or when $t = 1$ and $t = 4$. When $t = 1$, $x = 18$ and when $t = 4$, $x = -9$. Thus the vertical tangents are the lines $x = 18$ and $x = -9$. There is a horizontal tangent when $2t + 1 = 0$ or $t = -\frac{1}{2}$. When $t = -\frac{1}{2}$, $y = \frac{3}{4}$ and the line $y = \frac{3}{4}$ is a horizontal tangent.

EXAMPLE 4. Find all points at which the curve $x = 2 + 7 \cos \theta$, $y = 8 + 3 \sin \theta$ has a vertical or horizontal tangent.

SOLUTION. We find that $f_1'(\theta) = x'(\theta) = -7 \sin \theta$ which is 0 when $\theta = n\pi$ for some integer n. If $\theta = n\pi$, then

$$x = \begin{cases} 9, & n \text{ even} \\ -5, & n \text{ odd} \end{cases} \quad \text{and} \quad y = 8$$

so that there are vertical tangents at $(9, 8)$ and $(-5, 8)$. Similarly, $f_2'(\theta) = y'(\theta) = 3 \cos \theta$ which is 0 when $\theta = (n + \frac{1}{2})\pi$ for some integer n. If $\theta = (n + \frac{1}{2})\pi$, then

$$x = 2 \quad \text{and} \quad y = \begin{cases} 11, & n \text{ even} \\ 5, & n \text{ odd} \end{cases}$$

so that the curve has horizontal tangents at $(2, 11)$ and $(2, 5)$.

EXAMPLE 5. Find the tangent or tangents to the curve

$$x = t^2 - 3t + 5, \qquad y = t^3 + t^2 - 10t + 9$$

at the point $(3, 1)$.

SOLUTION. It is easy to verify that $x = 3$ and $y = 1$ when $t = 1$ and $t = 2$. Hence there are *two* tangents to the curve at the point $(3, 1)$. When $t = 1$, we have

$$\frac{dy}{dx} = \frac{3t^2 + 2t - 10}{2t - 3}\bigg|_{t=1} = \frac{-5}{-1} = 5$$

and when $t = 2$

$$\frac{dy}{dx} = \frac{3t^2 + 2t - 10}{2t - 3}\bigg|_{t=2} = \frac{6}{1} = 6.$$

Thus the two tangent lines have the equations

$$\frac{y - 1}{x - 3} = 5 \quad \text{or} \quad y = 5x - 14$$

and

$$\frac{y - 1}{x - 3} = 6 \quad \text{or} \quad y = 6x - 17.$$

PROBLEMS 16.2

In Problems 1–16 find the slope of the line tangent to the given curve for the given value of the parameter.

1. $x = t^2 - 2$; $y = 4t$; $t = 3$
2. $x = t^3$; $y = t^4 - 5$; $t = -1$
3. $x = t^2$; $y = \sqrt{1 - t}$; $t = \frac{1}{2}$
4. $x = t + 4$; $y = t^3 - t + 4$; $t = 2$

5. $x = e^{2t}$; $y = e^{-2t}$; $t = 1$
6. $x = \cos \theta$; $y = \sin \theta$; $\theta = \frac{\pi}{4}$

7. $x = \cos 2\theta$; $y = \sin 2\theta$; $\theta = \frac{\pi}{4}$
8. $x = \tan \theta$; $y = \sec \theta$; $\theta = \frac{\pi}{4}$

9. $x = \sec \theta$; $y = \tan \theta$; $\theta = \frac{\pi}{4}$
10. $x = \cosh t$; $y = \sinh t$; $t = 0$

11. $x = \sqrt{1 - \sin \theta}$; $y = \sqrt{1 + \cos \theta}$; $\theta = 0$
12. $x = 8 \cos \theta$; $y = -3 \sin \theta$; $\theta = \frac{2\pi}{3}$

13. $x = \cos^3 \theta$; $y = \sin^3 \theta$; $\theta = \frac{\pi}{6}$
14. $x = \cos^3 \theta$; $y = \sin^3 \theta$; $\theta = \frac{\pi}{2}$

15. $x = \theta$; $y = \frac{1}{\theta}$; $\theta = \frac{\pi}{4}$
16. $x = t \cosh t$; $y = \frac{(\tanh t)}{(1 + t)}$; $t = 0$.

In Problems 17–22 find the equation of the tangent line.

17. The curve of Problem 1
18. The curve of Problem 4
19. The curve of Problem 5
20. The curve of Problem 7
21. The curve of Problem 11
22. The curve of Problem 13.

In Problems 23–34 find the points (in the form (x, y)) at which the given curves have vertical and horizontal tangents.

23. $x = t^2 - 1$; $y = t^2 - 4$
24. $x = 2 \cos \theta$; $y = 3 \sin \theta$
25. $x = \sin^3 \theta$; $y = \cos^3 \theta$
26. $x = \sin^n \theta$; $y = \cos^n \theta$ (n a positive integer)

27. $x = \sin 3\theta$; $y = \cos 5\theta$
28. $x = \sin \theta + \cos \theta$; $y = \sin \theta - \cos \theta$
29. $x = e^{3t}$; $y = e^{-5t}$
30. $x = \sinh t$; $y = \cosh t$
31. $x = \cosh t$; $y = \sinh t$
*32. $x = \theta \sin \theta$; $y = \theta \cos \theta$
33. $x = 1/\sqrt{1 - t^2}$; $y = \sqrt{1 - t^2}$
34. $x = \ln(1 + t^2)$; $y = \ln(1 + t^3)$.

35. Suppose a curve is given in polar coordinates by the equation $r = f(\theta)$. Show that the curve can then be given parametrically by

$$x = f(\theta) \cos \theta \quad \text{and} \quad y = f(\theta) \sin \theta.$$

In Problems 36–43 use the result of Problem 35 to determine the slope of the tangent line to the given curve for the given value of θ.

36. $r = 5 \sin \theta$; $\theta = \pi/6$
37. $r = 5 \cos \theta + 5 \sin \theta$; $\theta = \pi/4$
38. $r = -4 + 2 \cos \theta$; $\theta = \pi/3$
39. $r = 4 + 3 \sin \theta$; $\theta = 2\pi/3$
40. $r = 3 \sin 2\theta$; $\theta = \pi/6$
41. $r = 5 \sin 3\theta$; $\theta = \pi/4$
42. $r = e^{\theta/2}$; $\theta = 0$
43. $r^2 = \cos 2\theta$; $\theta = \pi/6$.

44. Find the equations of the two tangents to the curve $x = t^3 - 2t^2 - 3t + 11$, $y = t^2 - 2t - 5$ at the point $(11, -2)$.

*45. Calculate the equations of the three lines that are tangent to the curve $x = t^3 - 2t^2 - t + 3$, $y = 2t - t^2 - (2/t)$ at the point $(1, -1)$.

46. Calculate the slope of the tangent to the cycloid given in Equations 16.1.16 and show that the tangent is vertical when α is a multiple of 2π.

47. For what values of θ does the trochoid given by equations (16.1.15) have a vertical tangent?

16.3 The Differentiation and Integration
of a Vector Function

In Section 16.2 we showed how the derivative dy/dx could be calculated when x and y were given parametrically in terms of t. In this section we shall show how to calculate the derivative of a vector function.

Since a derivative is a special kind of limit, we first need to define the limit of a vector function. This definition is what you might expect.

DEFINITION 1. Let $\mathbf{f}(t) = f_1(t)\mathbf{i} + f_2(t)\mathbf{j}$. Let t_0 be any real number, $+\infty$, or $-\infty$. Then if $\lim_{t \to t_0} f_1(t)$ and $\lim_{t \to t_0} f_2(t)$ both exist, we define

$$\lim_{t \to t_0} \mathbf{f}(t) = \left[\lim_{t \to t_0} f_1(t)\right]\mathbf{i} + \left[\lim_{t \to t_0} f_2(t)\right]\mathbf{j}. \tag{1}$$

Thus, in order to calculate the limit of a vector function, it is only necessary to calculate two ordinary limits.

EXAMPLE 1. Let $\mathbf{f}(t) = [(\sin t)/t]\mathbf{i} + [\ln(3 + t)]\mathbf{j}$. Calculate $\lim_{t \to 0} \mathbf{f}(t)$.

SOLUTION

$$\lim_{t \to 0} \mathbf{f}(t) = \left[\lim_{t \to 0} \frac{\sin t}{t}\right]\mathbf{i} + \left[\lim_{t \to 0} \ln(3 + t)\right]\mathbf{j} = \mathbf{i} + (\ln 3)\mathbf{j}.$$

DEFINITION 2. $\mathbf{f}$ is *continuous* at t_0 if the component functions f_1 and f_2 are continuous at t_0. This means that

(i) $\mathbf{f}$ is defined at t_0 (ii) $\lim_{t \to t_0} \mathbf{f}(t)$ exists   (iii) $\lim_{t \to t_0} \mathbf{f}(t) = \mathbf{f}(t_0)$.

DEFINITION 3. Let f be defined at t. Then $\mathbf{f}$ is *differentiable* at t if

$$\lim_{\Delta t \to 0} \frac{\mathbf{f}(t + \Delta t) - \mathbf{f}(t)}{\Delta t} \tag{2}$$

exists and is finite. The vector function $\mathbf{f}'$ defined by

$$\mathbf{f}'(t) = \frac{d\mathbf{f}}{dt} = \lim_{\Delta t \to 0} \frac{\mathbf{f}(t + \Delta t) - \mathbf{f}(t)}{\Delta t} \tag{3}$$

is called the *derivative* of $\mathbf{f}$ and the domain of $\mathbf{f}'$ is the set of all t such that the limit in (2) exists.

DEFINITION 4. The vector function $\mathbf{f}$ is differentiable on the interval I if $\mathbf{f}'(t)$ exists for every t in I.

Before giving examples of the calculation of derivatives, we prove a theorem which makes this calculation no more difficult than the calculation of "ordinary" derivatives.

THEOREM 1. If $\mathbf{f}(t) = f_1(t)\mathbf{i} + f_2(t)\mathbf{J}$, then at any value t for which $f_1'(t)$ and $f_2'(t)$ exist,

$$\boxed{\mathbf{f}'(t) = f_1'(t)\mathbf{i} + f_2'(t)\mathbf{j}.} \tag{4}$$

PROOF

$$\mathbf{f}'(t) = \lim_{\Delta t \to 0} \frac{\mathbf{f}(t + \Delta t) - \mathbf{f}(t)}{\Delta t}$$

$$= \lim_{\Delta t \to 0} \frac{[f_1(t + \Delta t)\mathbf{i} + f_2(t + \Delta t)\mathbf{j}] - [f_1(t)\mathbf{i} + f_2(t)\mathbf{j}]}{\Delta t}$$

$$= \lim_{\Delta t \to 0} \frac{[f_1(t + \Delta t) - f_1(t)]\mathbf{i} + [f_2(t + \Delta t) - f_2(t)]\mathbf{j}}{\Delta t}$$

$$= \lim_{\Delta t \to 0} \left[\frac{f_1(t + \Delta t) - f_1(t)}{\Delta t}\right]\mathbf{i} + \lim_{\Delta t \to 0} \left[\frac{f_2(t + \Delta t) - f_2(t)}{\Delta t}\right]\mathbf{j}$$

$$= f_1'(t)\mathbf{i} + f_2'(t)\mathbf{j}.$$

EXAMPLE 2. Let $\mathbf{f}(t) = (\cos t)\mathbf{i} + e^{2t}\mathbf{j}$. Calculate $\mathbf{f}'(t)$.

SOLUTION

$$\mathbf{f}'(t) = \frac{d}{dt}(\cos t)\mathbf{i} + \frac{d}{dt}e^{2t}\mathbf{j} = -(\sin t)\mathbf{i} + 2e^{2t}\mathbf{j}.$$

Once we know how to calculate the first derivative of $\mathbf{f}$, we can calculate higher derivatives as well.

DEFINITION 5. If the function $\mathbf{f}'$ is differentiable at t, we define the *second derivative of* $\mathbf{f}$ to be the derivative of $\mathbf{f}'$. That is

$$\boxed{\mathbf{f}'' = (\mathbf{f}')'.} \tag{5}$$

EXAMPLE 3. Let $\mathbf{f}(t) = (\ln t)\mathbf{i} + (1/t)\mathbf{j}$. Calculate $\mathbf{f}''(t)$.

SOLUTION. $\mathbf{f}'(t) = (1/t)\mathbf{i} - (1/t^2)\mathbf{j}$ so that

$$\mathbf{f}''(t) = -\frac{1}{t^2}\mathbf{i} + \frac{2}{t^3}\mathbf{j}.$$

Note that $\mathbf{f}'$ and $\mathbf{f}''$ are defined for all $t > 0$. ($\mathbf{f}'$ and $\mathbf{f}''$ are not defined for $t \leq 0$ because $\ln t$ is not defined for $t \leq 0$.)

We now seek a geometric interpretation for $\mathbf{f}'$. As we have seen, the set of vectors $\mathbf{f}(t) = f_1(t)\mathbf{i} + f_2(t)\mathbf{j}$ for t in the domain of $\mathbf{f}$ form a curve C in the plane (see

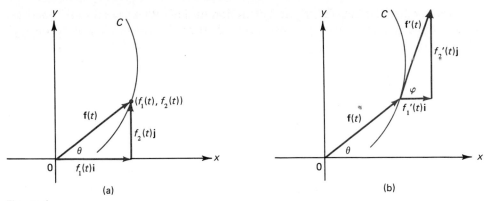

Figure 1

Figure 1a). For fixed t, $\mathbf{f}(t)$ is the vector drawn in this figure. If φ denotes the direction of $\mathbf{f}'$, then (see Eq. 15.1.2)

$$\tan \varphi = \frac{f_2'(t)}{f_1'(t)} = \frac{dy/dt}{dx/dt} = \frac{dy}{dx}. \tag{6}$$

This last step is justified by the chain rule. This implies that $\mathbf{f}'(t)$ is tangent to the curve $\mathbf{f}$ at the point $(f_1(t), f_2(t))$ (see Figure 1b). This is a natural extension of the usual notion of a tangent line to the case of a vector function.

To see this in another way, we see in Figure 2 that the vector $\mathbf{f}(t + \Delta t) - \mathbf{f}(t)$ is a secant vector whose direction approaches that of the tangent vector as $\Delta t \to 0$.

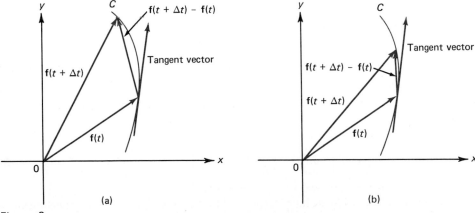

Figure 2

It is sometimes useful to calculate a *unit tangent vector* to a curve C. This is a tangent vector with a magnitude of 1. The unit tangent vector is usually denoted $\mathbf{T}$ and can be calculated by the formula

$$\mathbf{T} = \frac{\mathbf{f}'(t)}{|\mathbf{f}'(t)|} \tag{7}$$

for any number t so long as $\mathbf{f}'(t) \neq 0$. This follows since $\mathbf{f}'(t)$ is a tangent vector and $\mathbf{f}'/|\mathbf{f}'|$ is a unit vector.

EXAMPLE 4. Find the unit tangent vector to the curve $\mathbf{f} = (\ln t)\mathbf{i} + (1/t)\mathbf{j}$ at $t = 1$.

SOLUTION. $\mathbf{f}'(t) = (1/t)\mathbf{i} - (1/t^2)\mathbf{j} = \mathbf{i} - \mathbf{j}$ when $t = 1$. Since $|\mathbf{f}'| = |1 - \mathbf{J}| = \sqrt{2}$, we find that

$$\mathbf{T} = \frac{1}{\sqrt{2}}\mathbf{i} - \frac{1}{\sqrt{2}}\mathbf{j}.$$

EXAMPLE 5. Find a unit tangent vector to the curve $\mathbf{f} = (\cos t)\mathbf{i} + e^t\mathbf{j}$ at $t = \pi/6$.

SOLUTION. $\mathbf{f}'(t) = -(\sin t)\mathbf{i} + e^t\mathbf{j} = -\frac{1}{2}\mathbf{i} + e^{\pi/6}\mathbf{j}$ at $t = \pi/6$. In addition

$$\left| -\frac{1}{2}\mathbf{i} + e^{\pi/6}\mathbf{j} \right| = \sqrt{\frac{1}{4} + e^{\pi/3}} = \frac{\sqrt{1 + 4e^{\pi/3}}}{2}$$

so that

$$\mathbf{T} = \frac{2}{\sqrt{1 + 4e^{\pi/3}}}\left(-\frac{1}{2}\mathbf{i} + e^{\pi/6}\mathbf{j} \right).$$

Note. A unit tangent vector can be calculated so long as $\mathbf{f}'(t) \neq 0$ so that $\mathbf{f}'(t)/|\mathbf{f}'(t)|$ is defined.

We now turn to the integration of vector functions.

DEFINITION 6. (i) Let $\mathbf{f}(t) = f_1(t)\mathbf{i} + f_2(t)\mathbf{j}$ and suppose that the component functions f_1 and f_2 have indefinite integrals. Then we define the *indefinite integral of* $\mathbf{f}$ by

$$\int \mathbf{f}(t)\,dt = \left\{ \int f_1(t)\,dt \right\}\mathbf{i} + \left\{ \int f_2(t)\,dt \right\}\mathbf{j} + \mathbf{C} \tag{8}$$

where $\mathbf{C}$ is a constant vector of integration.

   (ii) If f_1 and f_2 are integrable over the interval $[a, b]$, then we define the *definite integral of* $\mathbf{f}$ by

$$\int_a^b \mathbf{f}(t)\,dt = \left(\int_a^b f_1(t)\,dt \right)\mathbf{i} + \left(\int_a^b f_2(t)\,dt \right)\mathbf{j}. \tag{9}$$

Remark 1. The indefinite integral (or antiderivative) of $\mathbf{f}$ is a new vector function and is *not unique* since $\int f_1(t)\,dt$ and $\int f_2(t)\,dt$ are not unique.

Remark 2. The definite integral of $\mathbf{f}$ is a constant vector since $\int_a^b f_1(t)\,dt$ and $\int_a^b f_2(t)\,dt$ are constants.

EXAMPLE 6. Let $\mathbf{f}(t) = (\cos t)\mathbf{i} + (\sin t)\mathbf{j}$. Calculate

   (a) $\int \mathbf{f}(t)\,dt$ and (b) $\int_0^{\pi/2} \mathbf{f}(t)\,dt$.

SOLUTION

(a) $\int \mathbf{f}(t)\,dt = \left(\int \cos t\,dt\right)\mathbf{i} + \left(\int \sin t\,dt\right)\mathbf{j}$

$\qquad\qquad = (\sin t + C_1)\mathbf{i} + (-\cos t + C_2)\mathbf{j}$

$\qquad\qquad = (\sin t)\mathbf{i} - (\cos t)\mathbf{j} + \mathbf{C}$

where $\mathbf{C} = C_1\mathbf{i} + C_2\mathbf{j}$ is a constant vector.

(b) $\displaystyle\int_0^{\pi/2} \mathbf{f}(t)\,dt = \left(\sin t\,\bigg|_0^{\pi/2}\right)\mathbf{i} + \left(-\cos t\,\bigg|_0^{\pi/2}\right)\mathbf{j} = \mathbf{i} + \mathbf{j}.$

EXAMPLE 7. Let $\mathbf{f}(t) = (1/t)\mathbf{i} + \sqrt{t+1}\,\mathbf{j}$. Calculate $\displaystyle\int_1^3 \mathbf{f}(t)\,dt$.

SOLUTION

$\displaystyle\int_1^3 \mathbf{f}(t)\,dt = \left(\int_1^3 \frac{1}{t}\,dt\right)\mathbf{i} + \left(\int_1^3 \sqrt{t+1}\,dt\right)\mathbf{j} = \ln t\,\bigg|_1^3\,\mathbf{i} + \frac{2}{3}(t+1)^{3/2}\,\bigg|_1^3\,\mathbf{j}$

$\qquad\qquad = (\ln 3)\mathbf{i} + \frac{2}{3}(8 - 2\sqrt{2})\mathbf{j}.$

EXAMPLE 8. Let $\mathbf{f}(t) = e^{-t}\mathbf{i} + \dfrac{1}{1+t^2}\mathbf{j}$. Calculate $\displaystyle\int_0^\infty \mathbf{f}(t)\,dt$.

SOLUTION. Although this is now an improper integral, the method of calculation remains unchanged. We have

$\displaystyle\int_0^\infty \mathbf{f}(t)\,dt = \left(\int_0^\infty e^{-t}\,dt\right)\mathbf{i} + \left(\int_0^\infty \frac{dt}{1+t^2}\right)\mathbf{j}$

$\qquad\qquad = \left(\lim_{N\to\infty}\int_0^N e^{-t}\,dt\right)\mathbf{i} + \left(\lim_{N\to\infty}\int_0^N \frac{dt}{1+t^2}\right)\mathbf{j}$

$\qquad\qquad = \lim_{N\to\infty}\left(-e^{-t}\,\bigg|_0^N\,\mathbf{i} + \tan^{-1} t\,\bigg|_0^N\,\mathbf{j}\right)$

$\qquad\qquad = \lim_{N\to\infty}[(1 - e^{-N})\mathbf{i} + (\tan^{-1} N)\mathbf{j}] = \mathbf{i} + \frac{\pi}{2}\mathbf{j}.$

EXAMPLE 9. If $\mathbf{f}'(t) = t\mathbf{i} + t^3\mathbf{j}$ and $\mathbf{f}(0) = \mathbf{i} - 2\mathbf{j}$, calculate $\mathbf{f}(t)$.

SOLUTION. We first find that

$\mathbf{f}(t) = \int \mathbf{f}'(t)\,dt = \left(\frac{t^2}{2} + C_1\right)\mathbf{i} + \left(\frac{t^4}{4} + C_2\right)\mathbf{j}$

$\qquad = \frac{t^2}{2}\mathbf{i} + \frac{t^4}{4}\mathbf{j} + \mathbf{C}.$　　　　　　　　　　　　　　　　(10)

To evaluate **C**, we substitute the value $t = 0$ into (10) to find that $\mathbf{f}(0) = \mathbf{C} = \mathbf{i} - 2\mathbf{j}$
so that

$$\mathbf{f}(t) = \frac{t^2}{2}\mathbf{i} + \frac{t^4}{4}\mathbf{j} + \mathbf{i} - 2\mathbf{j} = \left(\frac{t^2}{2} + 1\right)\mathbf{i} + \left(\frac{t^4}{4} - 2\right)\mathbf{j}.$$

PROBLEMS 16.3

In Problems 1–10 calculate the first and second derivatives of the given vector function.

1. $\mathbf{f}(t) = t\mathbf{i} - t^5\mathbf{j}$

2. $\mathbf{f}(t) = (1 + t^2)\mathbf{i} + \frac{2}{t}\mathbf{j}$

3. $\mathbf{f}(t) = (\sin 2t)\mathbf{i} + (\cos 3t)\mathbf{j}$

4. $\mathbf{f}(t) = \frac{t}{1 + t}\mathbf{i} - \frac{1}{\sqrt{t}}\mathbf{j}$

5. $\mathbf{f}(t) = (\ln t)\mathbf{i} + e^{3t}\mathbf{j}$

6. $\mathbf{f}(t) = e^t(\sin t)\mathbf{i} + e^t(\cos t)\mathbf{j}$

7. $\mathbf{f}(t) = (\tan t)\mathbf{i} + (\sec t)\mathbf{j}$

8. $\mathbf{f}(t) = (\tan^{-1} t)\mathbf{i} + (\sin^{-1} t)\mathbf{j}$

9. $\mathbf{f}(t) = (\ln \cos t)\mathbf{i} + (\ln \sin t)\mathbf{j}$

10. $\mathbf{f}(t) = (\cosh t)\mathbf{i} + (\sinh t)\mathbf{j}.$

In Problems 11–20 find the unit tangent vector to the given curve for the given value of t.

11. $\mathbf{f}(t) = t^2\mathbf{i} + t^3\mathbf{j}; t = 1$

12. $\mathbf{f}(t) = t\mathbf{i} + \frac{1}{t}\mathbf{j}; t = 1$

13. $\mathbf{f}(t) = (\cos t)\mathbf{i} + (\sin t)\mathbf{j}; t = 0$

14. $\mathbf{f}(t) = (\cos t)\mathbf{i} + (\sin t)\mathbf{j}; t = \frac{\pi}{2}$

15. $\mathbf{f}(t) = (\cos t)\mathbf{i} + (\sin t)\mathbf{j}; t = \frac{\pi}{4}$

16. $\mathbf{f}(t) = (\cos t)\mathbf{i} + (\sin t)\mathbf{j}; t = \frac{3\pi}{4}$

17. $\mathbf{f}(t) = (\tan t)\mathbf{i} + (\sec t)\mathbf{j}; t = 0$

18. $\mathbf{f}(t) = (\ln t)\mathbf{i} + e^{2t}\mathbf{j}; t = 1$

19. $\mathbf{f}(t) = \frac{t}{t + 1}\mathbf{i} + \frac{t + 1}{t}\mathbf{j}; t = 2$

20. $\mathbf{f}(t) = \frac{t + 1}{t}\mathbf{i} + \frac{t}{t + 1}\mathbf{j}; t = 2.$

In Problems 21–28 calculate the indicated integral.

21. $\displaystyle\int_0^2 (t^2\mathbf{i} + t^3\mathbf{j})\, dt$

22. $\displaystyle\int [(\sin 2t)\mathbf{i} + e^t\mathbf{j}]\, dt$

23. $\displaystyle\int (t^{-1/2}\mathbf{i} + t^{1/2}\mathbf{j})\, dt$

24. $\displaystyle\int_0^{\pi/4} [(\cos 2t)\mathbf{i} - (\sin 2t)\mathbf{j}]\, dt$

25. $\displaystyle\int_0^1 [(\sinh t)\mathbf{i} - (\cosh t)\mathbf{j}]\, dt$

26. $\displaystyle\int_1^e [(1/t)\mathbf{i} - (3/t)\mathbf{j}]\, dt$

27. $\displaystyle\int [(\ln t)\mathbf{i} + te^t\mathbf{j}]\, dt$

28. $\displaystyle\int [(\tan t)\mathbf{i} + (\sec t)\mathbf{j}]\, dt.$

29. If $\mathbf{f}'(t) = t^3\mathbf{i} - t^5\mathbf{j}$ and if $\mathbf{f}(0) = 2\mathbf{i} + 5\mathbf{j}$, find $\mathbf{f}(t)$.
30. If $\mathbf{f}'(t) = (1/\sqrt{t})\mathbf{i} + \sqrt{t}\mathbf{j}$ and if $\mathbf{f}(1) = -2\mathbf{i} + \mathbf{j}$, find $\mathbf{f}(t)$.
31. If $\mathbf{f}'(t) = (\cos t)\mathbf{i} + (\sin t)\mathbf{j}$ and if $\mathbf{f}(\pi/2) = \mathbf{i}$, find $\mathbf{f}(t)$.
32. Find $\mathbf{f}(t)$ in Problem 31 if $\mathbf{f}(\pi/2) = \mathbf{j}$.
33. The ellipse $(x^2/a^2) + (y^2/b^2) = 1$ can be written parametrically as $x = a \cos \theta$, $y = b \sin \theta$. Find the unit tangent vector to the ellipse at $\theta = \pi/4$.
34. Find a unit tangent vector to the cycloid $\mathbf{f}(t) = r(\alpha - \sin \alpha)\mathbf{i} + r(1 - \cos \alpha)\mathbf{j}$ for $\alpha = 0$.
35. Find a unit tangent vector to the cycloid of Problem 34 for $\alpha = \pi/2$.
36. Find a unit tangent vector to the cycloid of Problem 34 for $\alpha = \pi/3$.
37. Find, for $\theta = \pi/6$, the unit tangent vector to the hypocycloid of Problem 16.1.39 assuming that $a = 5$ and $b = 2$.

16.4 Some Differentiation Formulas

In Section 3.1 we gave a number of rules for differentiation. Many of these rules carry over to the differentiation of a vector function as we see in the theorem below.

THEOREM 1. Let **f** and **g** be vector functions which are differentiable in an interval *I*. Let the scalar function *h* be differentiable in *I*. Finally, let α be a scalar and let **v** be a constant vector. Then

    (i) **f** + **g** is differentiable and

$$\frac{d}{dt}(\mathbf{f} + \mathbf{g}) = \frac{d\mathbf{f}}{dt} + \frac{d\mathbf{g}}{dt} = \mathbf{f}' + \mathbf{g}'. \tag{1}$$

    (ii) $\alpha\mathbf{f}$ is differentiable and

$$\frac{d}{dt}\alpha\mathbf{f} = \alpha\frac{d\mathbf{f}}{dt} = \alpha\mathbf{f}'. \tag{2}$$

    (iii) $\mathbf{v}\cdot\mathbf{f}$ is differentiable and

$$\frac{d}{dt}\mathbf{v}\cdot\mathbf{f} = \mathbf{v}\cdot\frac{d\mathbf{f}}{dt} = \mathbf{v}\cdot\mathbf{f}'. \tag{3}$$

    (iv) $h\mathbf{f}$ is differentiable and

$$\frac{d}{dt}h\mathbf{f} = h\frac{d\mathbf{f}}{dt} + \frac{dh}{dt}\mathbf{f} = h\mathbf{f}' + h'\mathbf{f}. \tag{4}$$

    (v) $\mathbf{f}\cdot\mathbf{g}$ is differentiable and

$$\frac{d}{dt}\mathbf{f}\cdot\mathbf{g} = \mathbf{f}\cdot\frac{d\mathbf{g}}{dt} + \frac{d\mathbf{f}}{dt}\cdot\mathbf{g} = \mathbf{f}\cdot\mathbf{g}' + \mathbf{f}'\cdot\mathbf{g}. \tag{5}$$

PROOF

    (i) $\dfrac{d}{dt}(\mathbf{f} + \mathbf{g}) = \dfrac{d}{dt}(f_1\mathbf{i} + f_2\mathbf{j} + g_1\mathbf{i} + g_2\mathbf{j})$

$$= \frac{d}{dt}[(f_1 + g_1)\mathbf{i} + (f_2 + g_2)\mathbf{j}] = (f_1 + g_1)'\mathbf{i} + (f_2 + g_2)'\mathbf{j}$$

$$= (f_1' + g_1')\mathbf{i} + (f_2' + g_2')\mathbf{j}$$

$$= (f_1'\mathbf{i} + f_2'\mathbf{j}) + (g_1'\mathbf{i} + g_2'\mathbf{j}) = \mathbf{f}' + \mathbf{g}'.$$

    (ii) $\dfrac{d}{dt}(\alpha\mathbf{f}) = \dfrac{d}{dt}(\alpha f_1\mathbf{i} + \alpha f_2\mathbf{j}) = \alpha f_1'\mathbf{i} + \alpha f_2'\mathbf{j}$

$$= \alpha(f_1'\mathbf{i} + f_2'\mathbf{j}) = \alpha\mathbf{f}'.$$

(iii)   Let $\mathbf{v} = v_1\mathbf{i} + v_2\mathbf{j}$ where v_1 and v_2 are constants. Then

$$\mathbf{v}\cdot\mathbf{f} = v_1 f_1 + v_2 f_2 \quad \text{and} \quad (\mathbf{v}\cdot\mathbf{f})' = v_1 f_1' + v_2 f_2' = \mathbf{v}\cdot\mathbf{f}'.$$

(iv) $\dfrac{d}{dt} h\mathbf{f} = \dfrac{d}{dt}(hf_1\mathbf{i} + hf_2\mathbf{j}) = (hf_1)'\mathbf{i} + (hf_2)'\mathbf{j}$

$$= (hf_1' + h'f_1)\mathbf{i} + (hf_2' + h'f_2)\mathbf{j}$$
$$= h(f_1'\mathbf{i} + f_2'\mathbf{j}) + h'(f_1\mathbf{i} + f_2\mathbf{j}) = h\mathbf{f}' + h'\mathbf{f}.$$

(v) $\mathbf{f}\cdot\mathbf{g} = (f_1\mathbf{i} + f_2\mathbf{j})\cdot(g_1\mathbf{i} + g_2\mathbf{j}) = f_1 g_1 + f_2 g_2$ so that

$$\frac{d}{dt}(\mathbf{f}\cdot\mathbf{g}) = f_1' g_1 + f_1 g_1' + f_2 g_2' + f_2' g_2$$

$$= f_1 g_1' + f_2 g_2' + f_1' g_1 + f_2' g_2$$
$$= (f_1\mathbf{i} + f_2\mathbf{j})\cdot(g_1'\mathbf{i} + g_2'\mathbf{j}) + (f_1'\mathbf{i} + f_2'\mathbf{j})\cdot(g_1\mathbf{i} + g_2\mathbf{j})$$
$$= \mathbf{f}\cdot\mathbf{g}' + \mathbf{f}'\cdot\mathbf{g}.$$

EXAMPLE 1.   Let $\mathbf{f}(t) = t\mathbf{i} + t^3\mathbf{j}$, $\mathbf{g}(t) = (\cos t)\mathbf{i} + (\sin t)\mathbf{j}$ and $\mathbf{v} = 2\mathbf{i} - 3\mathbf{j}$. Calculate (a) $(\mathbf{f} + \mathbf{g})'$, (b) $(\mathbf{v}\cdot\mathbf{f})$, and (c) $(\mathbf{f}\cdot\mathbf{g})'$.

SOLUTION

(a) $(\mathbf{f} + \mathbf{g})' = \mathbf{f}' + \mathbf{g}' = (\mathbf{i} + 3t^2\mathbf{j}) + [-(\sin t)\mathbf{i} + (\cos t)\mathbf{j}]$
$$= (1 - \sin t)\mathbf{i} + (3t^2 + \cos t)\mathbf{j}.$$

(b) $(\mathbf{v}\cdot\mathbf{f})' = \mathbf{v}\cdot\mathbf{f}' = (2\mathbf{i} - 3\mathbf{j})\cdot(\mathbf{i} + 3t^2\mathbf{j}) = 2 - 9t^2.$

(c) $(\mathbf{f}\cdot\mathbf{g})' = \mathbf{f}\cdot\mathbf{g}' + \mathbf{f}'\cdot\mathbf{g}$
$$= (t\mathbf{i} + t^3\mathbf{j})\cdot[-(\sin t)\mathbf{i} + (\cos t)\mathbf{j}]$$
$$+ (\mathbf{i} + 3t^2\mathbf{j})\cdot[(\cos t)\mathbf{i} + (\sin t)\mathbf{j}]$$
$$= -t \sin t + t^3 \cos t + \cos t + 3t^2 \sin t$$
$$= (\cos t)(t^3 + 1) + (\sin t)(3t^2 - t).$$

EXAMPLE 2.   Let $\mathbf{f}(t) = \sqrt{t}\,\mathbf{i} + t^2\mathbf{j}$, $\mathbf{g}(t) = e^t\mathbf{i} + (\ln t)\mathbf{j}$ and $h(t) = \cosh t$. Calculate (a) $(\mathbf{f}\cdot\mathbf{g})'$ and (b) $(h\mathbf{f})'$.

SOLUTION

(a) $(\mathbf{f}\cdot\mathbf{g})' = \mathbf{f}\cdot\mathbf{g}' + \mathbf{f}'\cdot\mathbf{g}$
$$= (\sqrt{t}\,\mathbf{i} + t^2\mathbf{j})\cdot\left(e^t\mathbf{i} + \frac{1}{t}\mathbf{j}\right) + \left(\frac{1}{2\sqrt{t}}\mathbf{i} + 2t\mathbf{j}\right)\cdot[e^t\mathbf{i} + (\ln t)\mathbf{j}]$$

$$= \sqrt{t}\,e^t + t + \frac{e^t}{2\sqrt{t}} + 2t \ln t.$$

(b) $(h\mathbf{f})' = h\mathbf{f}' + h'\mathbf{f} = \cosh t \left(\frac{1}{2\sqrt{t}}\mathbf{i} + 2t\mathbf{j}\right) + \sinh t(\sqrt{t}\,\mathbf{i} + t^2\mathbf{j})$

$$= \left(\frac{\cosh t}{2\sqrt{t}} + \sqrt{t}\sinh t\right)\mathbf{i} + (2t\cosh t + t^2\sinh t)\mathbf{j}.$$

EXAMPLE 3. Let $f(t) = (\cos t)\mathbf{i} + (\sin t)\mathbf{j}$. Calculate $\mathbf{f} \cdot \mathbf{f}'$.

SOLUTION

$$\mathbf{f} \cdot \mathbf{f}' = [(\cos t)\mathbf{i} + (\sin t)\mathbf{j}] \cdot [-(\sin t)\mathbf{i} + (\cos t)\mathbf{j}]$$
$$= -\cos t \sin t + \sin t \cos t = 0.$$

Example 3 can be generalized to the following interesting result:

THEOREM 2. Let $\mathbf{f}(t) = f_1\mathbf{i} + f_2\mathbf{j}$ be a differentiable vector function such that $|\mathbf{f}(t)| = \sqrt{f_1^2(t) + f_2^2(t)}$ is constant. Then

$$\mathbf{f} \cdot \mathbf{f}' = 0. \tag{6}$$

PROOF. Suppose $|\mathbf{f}(t)| = C$, a constant. Then

$$\mathbf{f} \cdot \mathbf{f} = f_1^2 + f_2^2 = |\mathbf{f}|^2 = C^2$$

so that

$$\frac{d}{dt}(\mathbf{f} \cdot \mathbf{f}) = \frac{d}{dt} C^2 = 0.$$

But

$$\frac{d}{dt}(\mathbf{f} \cdot \mathbf{f}) = \mathbf{f} \cdot \mathbf{f}' + \mathbf{f}' \cdot \mathbf{f} = 2\mathbf{f} \cdot \mathbf{f}' = 0$$

so that

$$\mathbf{f} \cdot \mathbf{f}' = 0.$$

Note. In Example 3, $|\mathbf{f}(t)| = \sqrt{\cos^2 t + \sin^2 t} = 1$ for every t.

There is an interesting geometric application of Theorem 2. Let $\mathbf{f}(t)$ be a differentiable vector function. For all t for which $\mathbf{f}'(t) \neq 0$, we let $\mathbf{T}(t)$ denote the unit tangent vector to the curve $\mathbf{f}(t)$. Then, since $|\mathbf{T}(t)| = 1$ by the definition of a *unit tangent vector*, we have, from Theorem 2 (assuming that $\mathbf{T}'(t)$ exists)

$$\boxed{\mathbf{T}(t) \cdot \mathbf{T}'(t) = 0.} \tag{7}$$

That is, $\mathbf{T}'(t)$ is *orthogonal* to $\mathbf{T}(t)$. Recalling from Section 3.1 that a line perpendicular to a tangent line is called a *normal line,* we call the vector $\mathbf{T}'$ a *normal vector* to the curve $\mathbf{f}$. Finally, whenever $\mathbf{T}'(t) \neq 0$, we can define the *unit normal vector* to the curve $\mathbf{f}$ at t as

$$\boxed{\mathbf{n}(t) = \frac{\mathbf{T}'(t)}{|\mathbf{T}'(t)|}.} \tag{8}$$

EXAMPLE 4. Calculate a unit normal vector to the curve $\mathbf{f}(t) = (\cos t)\mathbf{i} + (\sin t)\mathbf{j}$ at $t = \pi/4$.

SOLUTION. First, we calculate $\mathbf{f}'(t) = -(\sin t)\mathbf{i} + (\cos t)\mathbf{j}$ and, since $|\mathbf{f}'(t)| = 1$, we find that

$$\mathbf{T}(t) = \frac{\mathbf{f}'(t)}{|\mathbf{f}'(t)|} = -(\sin t)\mathbf{i} + (\cos t)\mathbf{j}.$$

Then

$$\mathbf{T}'(t) = -(\cos t)\mathbf{i} - (\sin t)\mathbf{j} = \mathbf{n}(t)$$

since $|\mathbf{T}'(t)| = 1$ so that, at $t = \pi/4$,

$$\mathbf{n} = -\frac{1}{\sqrt{2}}\mathbf{i} - \frac{1}{\sqrt{2}}\mathbf{j}.$$

This is sketched in Figure 1 (remember that $\mathbf{f}(t) = (\cos t)\mathbf{i} + (\sin t)\mathbf{j}$ is the parametric equation of the unit circle). The reason the vector $\mathbf{n}$ points inward is that it is the negative of the position vector at $t = \pi/4$.

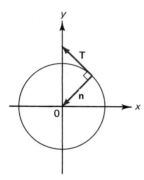

Figure 1

EXAMPLE 5. Calculate a unit normal vector to the curve $\mathbf{f}(t) = [(t^3/3) - t]\mathbf{i} + t^2\mathbf{j}$ at $t = 3$.

SOLUTION. Here $\mathbf{f}'(t) = (t^2 - 1)\mathbf{i} + 2t\mathbf{j}$ and

$$|\mathbf{f}'(t)| = \sqrt{(t^2 - 1)^2 + 4t^2} = \sqrt{t^4 - 2t^2 + 1 + 4t^2} = \sqrt{t^4 + 2t^2 + 1} = t^2 + 1.$$

Thus

$$\mathbf{T}(t) = \frac{\mathbf{f}'(t)}{|\mathbf{f}'(t)|} = \frac{t^2 - 1}{t^2 + 1}\mathbf{i} + \frac{2t}{t^2 + 1}\mathbf{j}.$$

Then

$$\mathbf{T}'(t) = \frac{d}{dt}\left(\frac{t^2 - 1}{t^2 + 1}\right)\mathbf{i} + \frac{d}{dt}\left(\frac{2t}{t^2 + 1}\right)\mathbf{j} = \frac{4t}{(t^2 + 1)^2}\mathbf{i} + \frac{2 - 2t^2}{(t^2 + 1)^2}\mathbf{j}.$$

Finally

$$|\mathbf{T}'(t)| = \left\{\left[\frac{4t}{(t^2 + 1)^2}\right]^2 + \left[\frac{2 - 2t^2}{(t^2 + 1)^2}\right]^2\right\}^{1/2}$$

$$= \frac{1}{(t^2 + 1)^2}(16t^2 + 4 - 8t^2 + 4t^4)^{1/2}$$

$$= \frac{1}{(t^2 + 1)^2} \sqrt{4t^4 + 8t^2 + 4} = \frac{2}{(t^2 + 1)^2} \sqrt{t^4 + 2t^2 + 1}$$

$$= \frac{2(t^2 + 1)}{(t^2 + 1)^2} = \frac{2}{t^2 + 1}$$

so that

$$\mathbf{n}(t) = \frac{\mathbf{T}'(t)}{|\mathbf{T}'(t)|} = \frac{t^2 + 1}{2} \left\{ \frac{4t}{(t^2 + 1)^2} \mathbf{i} + \frac{2 - 2t^2}{(t^2 + 1)^2} \mathbf{j} \right\} = \frac{2t}{t^2 + 1} \mathbf{i} + \frac{1 - t^2}{t^2 + 1} \mathbf{j}.$$

At $t = 3$, $\mathbf{T}(t) = \frac{4}{5}\mathbf{i} + \frac{3}{5}\mathbf{j}$ and $\mathbf{n} = \frac{3}{5}\mathbf{i} - \frac{4}{5}\mathbf{j}$. Note that $\mathbf{T}(3) \cdot \mathbf{n}(3) = 0$. This is sketched in Figure 2.

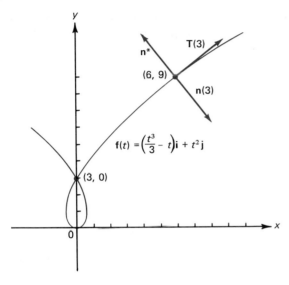

Figure 2

In Figure 2 we see that there are two vectors perpendicular to $\mathbf{T}$ at the point P: one pointing "inward" ($\mathbf{n}$) and one pointing "outward" ($\mathbf{n}^*$). In general, the situation is as in Figure 3. How do we know which of the two vectors is $\mathbf{n}$? If φ is the direction

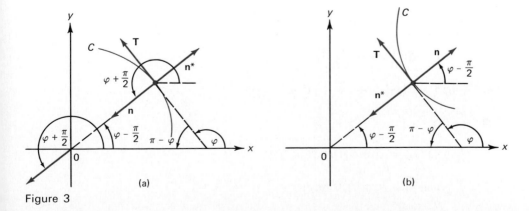

(a)   (b)

Figure 3

of **T**, then, since **T** is a unit vector,

$$\mathbf{T} = (\cos \varphi)\mathbf{i} + (\sin \varphi)\mathbf{j}$$

so that

$$\frac{d\mathbf{T}}{dt} = \frac{d\mathbf{T}}{d\varphi} \cdot \frac{d\varphi}{dt}$$

and

$$\left|\frac{d\mathbf{T}}{dt}\right| = \left|\frac{d\mathbf{T}}{d\varphi}\right| \left|\frac{d\varphi}{dt}\right| = \left|\frac{d\varphi}{dt}\right| \qquad \left(\text{since } \left|\frac{d\mathbf{T}}{d\varphi}\right| = 1\right).$$

Then

$$\mathbf{n}(t) = \frac{d\mathbf{T}/dt}{|d\mathbf{T}/dt|} = \frac{d\mathbf{T}}{d\varphi} \cdot \frac{d\varphi/dt}{|d\varphi/dt|} = \frac{d\varphi/dt}{|d\varphi/dt|}[-(\sin \varphi)\mathbf{i} + (\cos \varphi)\mathbf{j}].$$

Since $d\varphi/dt$ is a scalar, $(d\varphi/dt)/|d\varphi/dt| = \pm 1$. It is $+1$ in Figure 3a since φ increases as t increases so that $d\varphi/dt > 0$, while it is -1 in Figure 3b since φ decreases as t increases. In Figure 3a, then,

$$\mathbf{n}(t) = -(\sin \varphi)\mathbf{i} + (\cos \varphi)\mathbf{j} = \cos[\varphi + (\pi/2)]\mathbf{i} + \sin[\varphi + (\pi/2)]\mathbf{j}.$$

That is, the direction of **n** is $\varphi + (\pi/2)$ and **n** must point inward (on the concave side of C) as in Figure 3a. In the other case, $(d\varphi/dt)/|d\varphi/dt| = -1$ so that $\mathbf{n}(t) = (\sin \varphi)\mathbf{i} - (\cos \varphi)\mathbf{j} = \cos[\varphi - (\pi/2)]\mathbf{i} + \sin[\varphi - (\pi/2)]\mathbf{j}$. That is, the direction of **n** is $\varphi - (\pi/2)$ and so **n** points as in Figure 3b. In either case,

> the unit normal vector **n** *always points in the direction of the concave side of the curve.*

PROBLEMS 16.4

In Problems 1–15 calculate the indicated derivative.

1. $\dfrac{d}{dt}\{[2t\mathbf{i} + (\cos t)\mathbf{j}] + [(\tan t)\mathbf{i} - (\sec t)\mathbf{j}]\}$

2. $\dfrac{d}{dt}\{e^t[(\cos t)\mathbf{i} + (\sin t)\mathbf{j}]\}$

3. $\dfrac{d}{dt}\left\{(3\mathbf{i} - 4\mathbf{j}) \cdot \left(\sqrt{t}\,\mathbf{i} - \dfrac{1}{\sqrt{t}}\,\mathbf{j}\right)\right\}$

4. $\dfrac{d}{dt}\left\{(t^3\mathbf{i} - t^5\mathbf{j}) \cdot \left(-\dfrac{1}{t^3}\,\mathbf{i} + \dfrac{1}{t^5}\,\mathbf{j}\right)\right\}$

5. $\dfrac{d}{dt}\{(t^3\mathbf{i} - t^5\mathbf{j}) + (-t^3\mathbf{i} + t^5\mathbf{j})\}$

6. $\dfrac{d}{dt}\{[(\sinh t)\mathbf{i} + (\cosh t)\mathbf{j}] \cdot (-2\mathbf{i} + 7\mathbf{j})\}$

7. $\dfrac{d}{dt}\{[(\sinh t)\mathbf{i} + (\cosh t)\mathbf{j}]3 \sin t\}$

8. $\dfrac{d}{dt}\{[(\sinh t)\mathbf{i} + (\cosh t)\mathbf{j}]\cdot[(\sin t)\mathbf{i} + (\cos t)\mathbf{j}]\}$

9. $\dfrac{d}{dt}\left\{\left(\dfrac{t+1}{t}\mathbf{i} + \dfrac{t}{t+1}\mathbf{j}\right)\cdot\left(t^{10}\mathbf{i} - \dfrac{1}{t^{10}}\mathbf{j}\right)\right\}$

10. $\dfrac{d}{dt}\{[e^{2t}(\cos t)\mathbf{i} + e^{2t}(\sin t)\mathbf{j}]\cdot(t\mathbf{i} + t\mathbf{j})\}$

11. $\dfrac{d}{dt}\{[(\ln t)\mathbf{i} + (\ln t^{3})\mathbf{j}]\cdot[(\tanh t)\mathbf{i} + (\operatorname{sech} t)\mathbf{j}]\}$

12. $\dfrac{d}{dt}\{[(\sin^{-1} t)\mathbf{i} + (\cos^{-1} t)\mathbf{j}]\cdot(2\mathbf{i} - 10\mathbf{j})\}$

13. $\dfrac{d}{dt}\{[(\sin^{-1} t)\mathbf{i} + (\cos^{-1} t)\mathbf{j}] + [(\cos t)\mathbf{i} + (\sin t)\mathbf{j}]\}$

14. $\dfrac{d}{dt}\{[(\sin^{-1} t)\mathbf{i} + (\cos^{-1} t)\mathbf{j}](\cos t)\}$

15. $\dfrac{d}{dt}\{[(\sin^{-1} t)\mathbf{i} + (\cos^{-1} t)\mathbf{j}]\cdot[(\tan^{-1} t)\mathbf{i} + (\cos t)\mathbf{j}]\}$.

In Problems 16–29 find $\mathbf{T}(t)$, $\mathbf{n}(t)$, and the particular vectors $\mathbf{T}$ and $\mathbf{n}$ when $t = t_0$. Then sketch the curve near $t = t_0$ and include the vectors $\mathbf{T}$ and $\mathbf{n}$ in your sketch.

16. $\mathbf{f} = (\cos 3t)\mathbf{i} + (\sin 3t)\mathbf{j}$; $t = 0$
17. $\mathbf{f} = (\cos 5t)\mathbf{i} + (\sin 5t)\mathbf{j}$; $t = \pi/2$
18. $\mathbf{f} = 2(\cos 4t)\mathbf{i} + 2(\sin 4t)\mathbf{j}$; $t = \pi/4$
19. $\mathbf{f} = -3(\cos 10t)\mathbf{i} - 3(\sin 10t)\mathbf{j}$; $t = \pi$
20. $\mathbf{f} = 8(\cos t)\mathbf{i} + 8(\sin t)\mathbf{j}$; $t = \pi/4$
21. $\mathbf{f} = 4t\mathbf{i} + 2t^{2}\mathbf{j}$; $t = 1$
*22. $\mathbf{f} = (2 + 3t)\mathbf{i} + (8 - 5t)\mathbf{j}$; $t = 3$
*23. $\mathbf{f} = (4 - 7t)\mathbf{i} + (-3 + 5t)\mathbf{j}$; $t = -5$
*24. $\mathbf{f} = (a + bt)\mathbf{i} + (c + dt)\mathbf{j}$, a, b, c, d real; $t = t_0$
25. $\mathbf{f} = 2t\mathbf{i} + (e^{-t} + e^{t})\mathbf{j}$; $t = 0$
26. $\mathbf{f} = (t - \cos t)\mathbf{i} + (1 - \sin t)\mathbf{j}$; $t = \pi/2$
27. $\mathbf{f} = (t - \cos t)\mathbf{i} + (1 - \sin t)\mathbf{j}$; $t = \pi$
28. $\mathbf{f} = (t - \cos t)\mathbf{i} + (1 - \sin t)\mathbf{j}$; $t = \pi/4$
29. $\mathbf{f} = (\ln \sin t)\mathbf{i} + (\ln \cos t)\mathbf{j}$; $t = \pi/6$.

16.5 Arc Length Revisited

In Section 9.2 we derived a formula for the length of an arc of the curve $y = f(x)$ between $x = a$ and $x = b$ (equation (9.2.3)). We now derive a formula for arc length in a more general setting. Let the curve C be given parametrically by

$$x = f_1(t), \qquad y = f_2(t). \tag{1}$$

We shall assume in this section that f_1' and f_2' exist. Let t_0 be a fixed number. This fixes a point $P_0 = (x_0, y_0) = (f_1(t_0), f_2(t_0))$ on the curve (see Figure 1). The arrows in Figure 1 indicate the direction a point moves along the curve as t increases. We define the function $s(t)$ by

$$\boxed{s(t) = \text{length of path of curve } C \text{ from } t_0 \text{ to } t.}$$

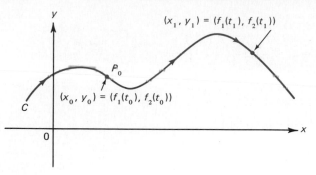

Figure 1

The following theorem allows us to calculate the length of a curve given parametrically.

THEOREM 1. Suppose that f_1' and f_2' are continuous in the interval $[t_0, t_1]$. Then $s(t)$ is a differentiable function of t for $t \in [t_0, t_1]$ and

$$\frac{ds}{dt} = \sqrt{(f_1'(t))^2 + (f_2'(t))^2} = \sqrt{\left(\frac{dx}{dt}\right)^2 + \left(\frac{dy}{dt}\right)^2}. \tag{2}$$

SKETCH OF PROOF. The proof of this theorem is quite difficult. However, it is possible to give an intuitive idea of what is going on. Consider an arc of the curve between t and $t + \Delta t$ (see Figure 2). First we note that as $\Delta t \to 0$, the length of the secant line L

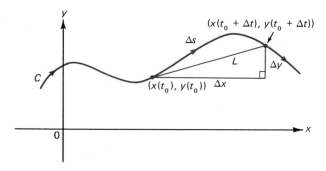

Figure 2

approaches the length of the arc Δs between the points $(x(t_0), y(t_0))$ and $(x(t_0 + \Delta t), y(t_0 + \Delta t))$. Thus, we have

$$\lim_{\Delta t \to 0} \frac{\Delta s}{L} = 1. \tag{3}$$

Then, since $L = \sqrt{\Delta x^2 + \Delta y^2}$, we have

$$\frac{\Delta s}{\Delta t} = \left(\frac{\Delta s}{L}\right)\left(\frac{L}{\Delta t}\right) = \left(\frac{\Delta s}{L}\right)\left(\frac{\sqrt{\Delta x^2 + \Delta y^2}}{\Delta t}\right) = \left(\frac{\Delta s}{L}\right)\left(\sqrt{\left(\frac{\Delta x}{\Delta t}\right)^2 + \left(\frac{\Delta y}{\Delta t}\right)^2}\right)$$

and

$$\frac{ds}{dt} = \lim_{\Delta x \to 0} \frac{\Delta s}{\Delta t} = \lim_{\Delta t \to 0} \frac{\Delta s}{L} \cdot \lim_{\Delta t \to 0} \sqrt{\left(\frac{\Delta x}{\Delta t}\right)^2 + \left(\frac{\Delta y}{\Delta t}\right)^2} = \sqrt{\left(\frac{dx}{dt}\right)^2 + \left(\frac{dy}{dt}\right)^2}.$$

The last step can be justified under the assumption that dx/dt and dy/dt are continuous, but its proof is beyond the scope of this book.†

Using equation (2), we can immediately calculate the length of the arc from t_0 to t_1. We have

---

$s(t_1)$ = length of arc from t_0 to t_1

$$= \int_{t_0}^{t_1} \left(\frac{ds}{dt}\right) dt = \int_{t_0}^{t_1} \sqrt{\left(\frac{dx}{dt}\right)^2 + \left(\frac{dy}{dt}\right)^2}\, dt. \qquad (4)$$

Using the chain rule (applied to differentials), we have

$$\frac{ds}{dt}\, dt = ds \qquad (5)$$

and so (4) becomes

$$s(t_1) = \int_{t_0}^{t_1} ds \qquad (6)$$

where

$$ds = \sqrt{\left(\frac{dx}{dt}\right)^2 + \left(\frac{dy}{dt}\right)^2}\, dt. \qquad (7)$$

In this context, the variable s is called the *parameter of arc length*. We will discuss this further in the next section. Note, however, that while x measures distance along the horizontal axis and y measures distance along the vertical axis, s measures distance *along the curve* given parametrically by equations (1).

Before giving examples, we note that formula (4) is a generalization of the formula for arc length we derived in Section 9.2. For if $y = f(x)$, we have the parametric representation

$$x = t, \qquad y = f(t)$$

and

$$\int_{t_0}^{t_1} \sqrt{\left(\frac{dx}{dt}\right)^2 + \left(\frac{dy}{dt}\right)^2}\, dt = \int_{t_0}^{t_1} \sqrt{1 + [f'(t)]^2}\, dt$$

which is the formula for arc length given in equation (9.2.3).

---

† See, for example, R. C. Buck, *Advanced Calculus*, McGraw-Hill, New York, 1965, p. 321.

EXAMPLE 1. We use formula (4) to calculate the length of the circumference of a circle given parametrically by $x = \cos t$, $y = \sin t$. Then

$$ds = \sqrt{\left(\frac{dx}{dt}\right)^2 + \left(\frac{dy}{dt}\right)^2}\, dt = \sqrt{(-\sin t)^2 + (\cos t)^2}\, dt = dt$$

so that

$$s = \int_0^{2\pi} ds = \int_0^{2\pi} dt = 2\pi.$$

EXAMPLE 2. Let the curve C be given by $x = t^2$ and $y = t^3$. Calculate the length of the arc from $t = 0$ to $t = 3$.

SOLUTION. Here

$$\sqrt{\left(\frac{dx}{dt}\right)^2 + \left(\frac{dy}{dt}\right)^2} = \sqrt{(2t)^2 + (3t^2)^2} = \sqrt{4t^2 + 9t^4} = t\sqrt{4 + 9t^2}$$

so that

$$s = \int_0^3 t\sqrt{4 + 9t^2}\, dt = \frac{1}{27}(4 + 9t^2)^{3/2}\Big|_0^3 = \frac{85^{3/2} - 8}{27}.$$

Suppose now that a curve is given in polar coordinates by $r = f(\theta)$. Since $x = r\cos\theta$ and $y = r\sin\theta$ (equation (11.1.1)), we have

$$x = f(\theta)\cos\theta \qquad\text{and}\qquad y = f(\theta)\sin\theta$$

so that

$$\frac{dx}{d\theta} = f'(\theta)\cos\theta - f(\theta)\sin\theta, \qquad \frac{dy}{d\theta} = f'(\theta)\sin\theta + f(\theta)\cos\theta$$

and

$$\left(\frac{dx}{d\theta}\right)^2 + \left(\frac{dy}{d\theta}\right)^2 = [f'(\theta)]^2\cos^2\theta - 2f'(\theta)f(\theta)\sin\theta\cos\theta + [f(\theta)]^2\sin^2\theta$$

$$+ [f'(\theta)]^2\sin^2\theta + 2f'(\theta)f(\theta)\sin\theta\cos\theta + [f(\theta)]^2\cos^2\theta$$

$$= [f(\theta)]^2 + [f'(\theta)]^2 = r^2 + \left(\frac{dr}{d\theta}\right)^2.$$

Thus the formula for arc length in polar coordinates becomes

$$s = \int_{\theta=\theta_0}^{\theta=\theta_1} \sqrt{r^2 + \left(\frac{dr}{d\theta}\right)^2}\, d\theta. \tag{8}$$

EXAMPLE 3. Calculate the arc length of the cardioid $r = a(1 + \cos\theta)$.

SOLUTION. The curve is sketched in Figure 3. Since the curve is symmetric about the polar axis, we calculate the arc length between $\theta = 0$ and $\theta = \pi$ and then multiply the

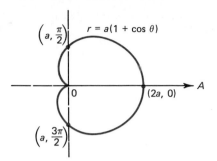

$r = a(1 + \cos \theta)$

$\left(a, \dfrac{\pi}{2}\right)$

0

$(2a, 0)$

A

$\left(a, \dfrac{3\pi}{2}\right)$

Figure 3

result by 2. We have

$$s = \int_0^{\pi} \sqrt{r^2 + \left(\frac{dr}{d\theta}\right)^2}\, d\theta = \int_0^{\pi} a\sqrt{(1 + \cos \theta)^2 + \sin^2 \theta}\, d\theta$$

$$= a \int_0^{\pi} \sqrt{1 + 2 \cos \theta + \cos^2 \theta + \sin^2 \theta}\, d\theta$$

$$= a \int_0^{\pi} \sqrt{(2 + 2 \cos \theta)}\, d\theta = a \int_0^{\pi} \sqrt{\frac{4 + 4 \cos \theta}{2}}\, d\theta$$

$$= 2a \int_0^{\pi} \sqrt{\frac{1 + \cos \theta}{2}}\, d\theta = 2a \int_0^{\pi} \cos \frac{\theta}{2}\, d\theta = 4a \sin \frac{\theta}{2}\,\Big|_0^{\pi} = 4a.$$

The total arc length is therefore $8a$.

Remark. If we ignored symmetry and tried to integrate from 0 to 2π, we would obtain

$$s = 2a \int_0^{2\pi} \sqrt{\frac{1 + \cos \theta}{2}}\, d\theta = 2a \int_0^{2\pi} \cos \frac{\theta}{2}\, d\theta = 4a \sin \frac{\theta}{2}\,\Big|_0^{2\pi} = 0.$$

The problem is that $\cos \theta/2 \geq 0$ for $0 \leq \theta \leq \pi$ and $\cos \theta/2 \leq 0$ for $\pi \leq \theta \leq 2\pi$. We see that $\sqrt{(1 + \cos \theta)/2} = \cos(\theta/2)$ for $0 \leq \theta \leq \pi$ and $\sqrt{(1 + \cos \theta)/2} = -\cos(\theta/2)$ for $\pi \leq \theta \leq 2\pi$. Thus, it would be necessary to write

$$s = 2a \left(\int_0^{\pi} \cos \frac{\theta}{2}\, d\theta - \int_{\pi}^{2\pi} \cos \frac{\theta}{2}\, d\theta \right)$$

$$= 4a \left(\sin \frac{\theta}{2}\,\Big|_0^{\pi} - \sin \frac{\theta}{2}\,\Big|_{\pi}^{2\pi} \right) = 4a[1 - (-1)] = 8a$$

as before. In other problems, the reader is cautioned to pay attention to signs when taking square roots so that "positive" and "negative" arc lengths are not inadvertantly "canceled off."

EXAMPLE 4. Calculate the length of the spiral of Archimedes† $r = a\theta$ from $\theta = 0$ to $\theta = 2\pi$ (see Example 11.2.9).

SOLUTION. The curve is sketched in Figure 4. From (8), we have, since $dr/d\theta = a$

$$s = \int_0^{2\pi} \sqrt{a^2\theta^2 + a^2} = a \int_0^{2\pi} \sqrt{1 + \theta^2}\, d\theta.$$

---

† While Archimedes was able to approximate the area enclosed by this spiral, he was unable to calculate its length. As we shall see, even with the advantage of calculus, this calculation is fairly complicated.

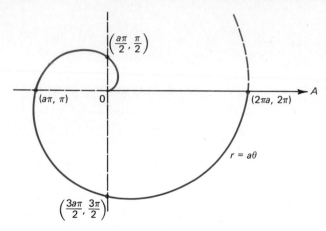

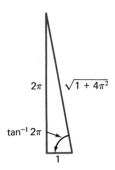

Figure 4

So far, this problem is easy, but the calculation of this integral is tedious. We first make the substitution $\theta = \tan t$. Then $d\theta = \sec^2 t \, dt$, $\sqrt{1 + \theta^2} = \sqrt{1 + \tan^2 t} = \sec t$, and

$$a \int_0^{2\pi} \sqrt{1 + \theta^2} \, d\theta = a \int_0^{\tan^{-1} 2\pi} \sec^3 \theta \, d\theta$$

$$= \frac{a}{2} (\ln|\sec t + \tan t| + \sec t \tan t)\Big|_0^{\tan^{-1} 2\pi}$$

($\int \sec^3 x \, dx$ is calculated in Example 8.3.10). From the triangle in Figure 5, we find that $\sec(\tan^{-1} 2\pi) = \sqrt{1 + 4\pi^2}$ and $\tan(\tan^{-1} 2\pi) = 2\pi$ so that

$$s = \frac{a}{2} [\ln(\sqrt{1 + 4\pi^2} + 2\pi) + 2\pi \sqrt{1 + 4\pi^2}].$$

2π $\sqrt{1 + 4\pi^2}$

$\tan^{-1} 2\pi$

1   Figure 5

For example, if $a = 1$, then $s \approx 21.256$.

There is a more concise way to write our formula for arc length using vector notation. Let the curve C be given by

$$\mathbf{f}(t) = f_1(t)\mathbf{i} + f_2(t)\mathbf{j}. \tag{9}$$

Then

$$\mathbf{f}'(t) = f_1'(t)\mathbf{i} + f_2'(t)\mathbf{j}$$

and

$$|\mathbf{f}'(t)| = \sqrt{[f_1'(t)]^2 + [f_2'(t)]^2} = \frac{ds}{dt} \tag{10}$$

so that the length of the arc between t_0 and t_1 is given by

$$\boxed{s = \int_{t_0}^{t_1} |\mathbf{f}'(t)| \, dt.} \tag{11}$$

EXAMPLE 5. Calculate the length of the arc of the curve $\mathbf{f}(t) = (2t - t^2)\mathbf{i} + \frac{8}{3}t^{3/2}\mathbf{j}$ between $t = 1$ and $t = 3$.

SOLUTION. $\mathbf{f}'(t) = (2 - 2t)\mathbf{i} + 4\sqrt{t}\mathbf{j}$ and

$$|\mathbf{f}'(t)| = \sqrt{4 - 8t + 4t^2 + 16t} = 2\sqrt{t^2 + 2t + 1} = 2(t + 1)$$

so that

$$s = \int_1^3 2(t + 1) \, dt = (t^2 + 2t)\Big|_1^3 = 12.$$

PROBLEMS 16.5

In Problems 1–20 find the length of the arc over the given interval or the length of the closed curve.

1. $x = t^3$; $y = t^2$; $1 \le t \le 4$
2. $x = \cos 2\theta$; $y = \sin 2\theta$; $0 \le \theta \le \pi/2$
3. $x = t^3 + 1$; $y = 3t^2 + 2$; $0 \le t \le 2$
4. $x = 1 + t$; $y = (1 + t)^{3/2}$; $0 \le t \le 1$
5. $x = \dfrac{1}{\sqrt{t + 1}}$; $y = \dfrac{t}{[2(t + 1)]}$; $0 \le t \le 4$
6. $x = e^t \cos t$; $y = e^t \sin t$; $0 \le t \le \pi/2$
7. $x = \sin^2 t$; $y = \cos^2 t$; $0 \le t \le \pi/2$
8. The hypocycloid of four cusps $x = a \cos^3 \theta$, $y = a \sin^3 \theta$, $a > 0$ [Hint: Calculate the length in the first quadrant and multiply by 4.]
*9. The cardioid $r = a(1 + \sin \theta)$ [Hint:

$$\int \sqrt{1 + \sin \theta} \, d\theta = \int \sqrt{1 + \sin \theta} \cdot \frac{\sqrt{1 - \sin \theta}}{\sqrt{1 - \sin \theta}} \, d\theta.$$

Pay attention to signs.]
10. One arc of the cycloid $x = a(\theta - \sin \theta)$, $y = a(1 - \cos \theta)$, $a > 0$
11. $x = t^3$; $y = t^2$; $-1 \le t \le 1$ [Hint: $\sqrt{t^2} = -t$ for $t < 0$.]
12. $r = a \sin \theta$; $0 \le \theta \le \pi/2$ 13. $r = a \cos \theta$; $0 \le \theta \le \pi$
14. $r = a\theta$; $0 \le \theta \le 6\pi$ 15. $r = e^\theta$; $0 \le \theta \le 3$
16. $r = \theta^2$; $0 \le \theta \le \pi$ 17. $r = 6 \cos^2(\theta/2)$; $0 \le \theta \le \pi/2$
18. $r = \sin^3(\theta/3)$; $0 \le \theta \le \pi/2$ [Hint: $\sin^2(\theta/3) = \frac{1}{2}(1 - \cos(2\theta/3))$.]
19. $\mathbf{f}(t) = e^t(\sin t)\mathbf{i} + e^t(\cos t)\mathbf{j}$; $0 \le t \le \pi/2$
20. $\mathbf{f}(t) = 3(\cos \theta)\mathbf{i} + 3(\sin \theta)\mathbf{j}$; $0 \le \theta \le 2\pi$

21.   The parametric representation of the ellipse $(x^2/a^2) + (y^2/b^2) = 1$ is given by $x = a \cos \theta$, $y = b \sin \theta$. Find an integral which represents the length of the circumference of an ellipse but do not try to evaluate it. The integral you obtain is called an *elliptic integral* and it arises in a variety of physical applications. It cannot be integrated (except numerically) unless $a = b$.

*22.   A tack is stuck in the front tire of a bicycle wheel with a diameter of 1 meter. What is the total distance traveled by the tack if the bicycle moves a total of 30π meters?

16.6 Arc Length as a Parameter

In many problems it is convenient to use the arc length s as a parameter. We can think of the vector function **f** as the *position vector* of a particle moving in the xy-plane. Then if $P_0 = (x_0, y_0)$ is a fixed point on the curve C described by the vector function **f**, we may write

$$\mathbf{f}(s) = x(s)\mathbf{i} + y(s)\mathbf{j}. \tag{1}$$

where s is the distance along the curve measured from P_0. In this way we can determine the x and y components of the position vector as we move s units *along the curve*.

EXAMPLE 1.   Write the vector $\mathbf{f}(t) = (\cos t)\mathbf{i} + (\sin t)\mathbf{j}$ (which describes the unit circle) with arc length as a parameter.

SOLUTION. We have $ds/dt = 1$ (from Example 16.5.1) so that

$$s = \int_0^t ds = \int_0^t \frac{ds}{dt} dt = \int_0^t dt = t.$$

Thus, we may write

$$\mathbf{f}(s) = (\cos s)\mathbf{i} + (\sin s)\mathbf{j}.$$

For example, if we begin at the point $(1, 0)$ and move π units along the unit circle (which is half the unit circle) then we move to the point $(\cos \pi, \sin \pi) = (-1, 0)$. This is what we would expect. See Figure 1.

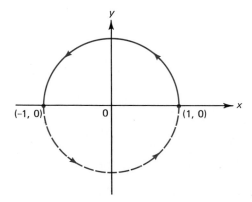

Figure 1

EXAMPLE 2. Let $\mathbf{f}(t) = (2t - t^2)\mathbf{i} + \frac{8}{3}t^{3/2}\mathbf{j}$. Write this curve with arc length as a parameter.

SOLUTION. Suppose that the fixed point is $P_0 = (0,0)$ when $t = 0$. Then, from Example 16.5.5, we have

$$\frac{ds}{dt} = 2(t + 1),$$

so that

$$s = \int_0^t 2(t + 1) \, dt = t^2 + 2t.$$

This leads to the equation

$$t^2 + 2t - s = 0 \quad \text{and} \quad t = \frac{-2 + \sqrt{4 + 4s}}{2} = \sqrt{1 + s} - 1.$$

We took the positive square root here since it is assumed that t starts at 0 and increases. Then

$$x = 2t - t^2 = 4\sqrt{1 + s} - 4 - s$$
$$y = \frac{8}{3}t^{3/2} = \frac{8}{3}(\sqrt{1 + s} - 1)^{3/2}$$

and we obtain

$$\mathbf{f}(s) = (4\sqrt{1 + s} - 4 - s)\mathbf{i} + \frac{8}{3}(\sqrt{1 + s} - 1)^{3/2}\mathbf{j}.$$

As Example 2 illustrates, writing $\mathbf{f}$ with arc length as a parameter can be tedious.

There is an interesting and important relationship between position vectors, tangent vectors, and normal vectors that becomes apparent when we use s as a parameter.

THEOREM 1. If the curve C is parametrized by $\mathbf{f}(s) = x(s)\mathbf{i} + y(s)\mathbf{j}$ where s is arc length and x and y have continuous derivatives, then the unit tangent vector $\mathbf{T}$ is given by

$$\boxed{\mathbf{T} = \frac{d\mathbf{f}}{ds}.} \qquad (2)$$

PROOF. With *any* parametrization of C, the unit tangent vector is given by (see equation (16.3.7))

$$\mathbf{T} = \frac{\mathbf{f}'(t)}{|\mathbf{f}'(t)|}.$$

So choosing $t = s$ yields

$$\mathbf{T} = \frac{d\mathbf{f}/ds}{|d\mathbf{f}/ds|}.$$

But, from equation (16.5.10), $|\mathbf{f}'(t)| = ds/dt$ so that

$$\left|\frac{d\mathbf{f}}{ds}\right| = \left|\frac{d\mathbf{f}/dt}{ds/dt}\right| = \left|\frac{\mathbf{f}'(t)}{ds/dt}\right| = 1$$

and the proof is complete.

Theorem 1 is quite useful in that it provides a check of our calculation of the parametrization in terms of arc length. For if

$$\mathbf{f}(s) = x(s)\mathbf{i} + y(s)\mathbf{j}$$

then

$$\mathbf{T} = \frac{d\mathbf{f}}{ds} = \frac{dx}{ds}\mathbf{i} + \frac{dy}{ds}\mathbf{j}.$$

But $|\mathbf{T}| = 1$ so that $|\mathbf{T}|^2 = 1$ which implies that

$$\left(\frac{dx}{ds}\right)^2 + \left(\frac{dy}{ds}\right)^2 = 1. \tag{3}$$

We can apply this in Example 2. We have

$$x(s) = 4\sqrt{1+s} - 4 - s \quad\text{and}\quad y(s) = \tfrac{8}{3}(\sqrt{1+s} - 1)^{3/2}$$

so that

$$\frac{dx}{ds} = \frac{2}{\sqrt{1+s}} - 1, \quad \frac{dy}{ds} = \frac{2}{\sqrt{1+s}}(\sqrt{1+s} - 1)^{1/2}$$

and

$$\left(\frac{dx}{ds}\right)^2 + \left(\frac{dy}{ds}\right)^2 = \frac{4}{1+s} - \frac{4}{\sqrt{1+s}} + 1 + \frac{4}{1+s}(\sqrt{1+s} - 1) = 1,$$

as expected.

PROBLEMS 16.6

In the following problems find parametric equations in terms of the arc length s measured from the point reached when $t = 0$. Verify your solution by using formula (3).

1. $\mathbf{f} = 3t^2\mathbf{i} + 2t^3\mathbf{j}$
2. $\mathbf{f} = t^3\mathbf{i} + t^2\mathbf{j}$
3. $\mathbf{f} = (t^3 + 1)\mathbf{i} + (t^2 - 1)\mathbf{j}$
4. $\mathbf{f} = (3t^2 + a)\mathbf{i} + (2t^3 + b)\mathbf{j}$
5. $\mathbf{f} = 3(\cos\theta)\mathbf{i} + 3(\sin\theta)\mathbf{j}$
6. $\mathbf{f} = a(\sin\theta)\mathbf{i} + a(\cos\theta)\mathbf{j}$
7. $\mathbf{f} = a(\cos\theta)\mathbf{i} + a(\sin\theta)\mathbf{j}$
8. $\mathbf{f} = 3(\cos t + t\sin t)\mathbf{i} + 3(\sin t - t\cos t)\mathbf{j}$
9. $\mathbf{f} = (a + b\cos\theta)\mathbf{i} + (c + b\sin\theta)\mathbf{j}$
10. $\mathbf{f} = ae^t(\cos t)\mathbf{i} + ae^t(\sin t)\mathbf{j}$
11. One cusp of the hypocycloid of four cusps

$$\mathbf{f} = a(\cos^3\theta)\mathbf{i} + a(\sin^3\theta)\mathbf{j}, \quad 0 \le \theta \le \tfrac{\pi}{2}, \quad a > 0$$

12. The cycloid $x = a(\theta - \sin\theta)$, $y = a(1 - \cos\theta)$, $a > 0$.

16.7 Velocity, Acceleration, Force, and Momentum

Suppose that an object is moving in the plane. Then we can describe its motion parametrically by the vector function

$$\mathbf{f}(t) = f_1(t)\mathbf{i} + f_2(t)\mathbf{j}. \tag{1}$$

In this context $\mathbf{f}$ is called the *position vector* of the object and the curve described by $\mathbf{f}$ is called the *trajectory* of the object. We then have the following definition.

DEFINITION 1.   If $\mathbf{f}'$ and $\mathbf{f}''$ exist, then

$$\text{(i)} \quad \mathbf{v}(t) = \mathbf{f}'(t) = f_1'(t)\mathbf{i} + f_2'(t)\mathbf{j} \tag{2}$$

is called the *velocity vector* of the moving object at time t.

$$\text{(ii)} \quad \mathbf{a}(t) = \frac{d\mathbf{v}}{dt} = \mathbf{f}''(t) = f_1''(t)\mathbf{i} + f_2''(t)\mathbf{j} \tag{3}$$

is called the *acceleration vector* of the object.

This definition is, of course, not surprising. It simply extends to the vector case our notion of velocity as the derivative of position and acceleration as the derivative of velocity.

DEFINITION 2.   (i)   The *speed* $v(t)$ of a moving object is the magnitude of the velocity vector.

(ii)   The *acceleration scalar* $a(t)$ is the magnitude of the acceleration vector.

> **Remark 1.** Since we have already shown that $|\mathbf{f}'(t)| = ds/dt$ (equation (16.5.10)), we have, since $v(t) = |\mathbf{v}(t)| = |\mathbf{f}'(t)|$,
>
> $$v(t) = \frac{ds}{dt}. \tag{4}$$

> **Remark 2.** Although $\mathbf{a}(t)$ is the derivative of $\mathbf{v}(t)$, it is *not true* in general that $a(t)$ is the derivative of the speed $v(t)$. For example, consider the motion along the unit circle given by
>
> $$\mathbf{f}(t) = (\cos t)\mathbf{i} + (\sin t)\mathbf{j}.$$
>
> Then
>
> $$\mathbf{v}(t) = -(\sin t)\mathbf{i} + (\cos t)\mathbf{j} \quad \text{and} \quad \mathbf{a}(t) = -(\cos t)\mathbf{i} - (\sin t)\mathbf{j}.$$
>
> But $v(t) = |\mathbf{v}(t)| = 1$ so that $dv/dt = 0$. But $a(t) = |\mathbf{a}(t)| = 1$ which is, evidently, not equal to dv/dt.

EXAMPLE 1.   A particle is moving along the circle with the position vector $\mathbf{f} = 3(\cos 2t)\mathbf{i} + 3(\sin 2t)\mathbf{j}$. Calculate $\mathbf{v}(t)$, $\mathbf{a}(t)$, $v(t)$, $a(t)$ and find the velocity and acceleration vectors when $t = \pi/6$.

SOLUTION. Here $\mathbf{v}(t) = \mathbf{f}'(t) = -6(\sin 2t)\mathbf{i} + 6(\cos 2t)\mathbf{j}$ and $\mathbf{a}(t) = -12(\cos 2t)\mathbf{i} - 12(\sin 2t)\mathbf{j}$. Then $v(t) = |\mathbf{v}(t)| = \sqrt{36 \sin^2 2t + 36 \cos^2 2t} = 6$ m/sec   and $a(t) = |\mathbf{a}(t)| = 12$ m/sec^2. Finally, $\mathbf{v}(\pi/6) = -3\sqrt{3}\mathbf{i} + 3\mathbf{j}$ and $\mathbf{a}(\pi/6) = -6\mathbf{i} - 6\sqrt{3}\mathbf{j}$. We sketch these vectors in Figure 1. Note that $\mathbf{a}(t) \cdot \mathbf{v}(t) = 0$. This follows from Theorem 16.4.2 and the fact that $|\mathbf{v}(t)|$ is constant.

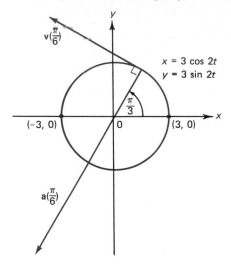

$$x = 3 \cos 2t$$
$$y = 3 \sin 2t$$

Figure 1

EXAMPLE 2. An object is moving in the plane with position vector $\mathbf{f} = [(t^3/3) - t]\mathbf{i} + t^2\mathbf{j}$. Calculate $\mathbf{v}(t)$, $\mathbf{a}(t)$, $v(t)$, and $a(t)$ when $t = 3$ (see Example 16.4.5). All distances are measured in centimeters.

SOLUTION. $\mathbf{v}(t) = (t^2 - 1)\mathbf{i} + 2t\mathbf{j}$ so that $v(t) = \sqrt{(t^2 - 1)^2 + 4t^2} = (t^2 + 1)$ cm/sec. Then $\mathbf{a}(t) = 2t\mathbf{i} + 2\mathbf{j}$ and $a(t) = \sqrt{4t^2 + 4} = 2\sqrt{t^2 + 1}$ cm/sec^2. When $t = 3$, we find that

$$\mathbf{v}(3) = 8\mathbf{i} + 6\mathbf{j}, \qquad \text{and} \qquad \mathbf{a}(3) = 6\mathbf{i} + 2\mathbf{j}$$
$$v(3) = 10 \text{ cm/sec}, \qquad \qquad a(3) = 2\sqrt{10} \text{ cm/sec}^2.$$

This is sketched in Figure 2. Note that, in this case, $v(t)$ is not constant and that $\mathbf{v}(t) \cdot \mathbf{a}(t) \neq 0$.

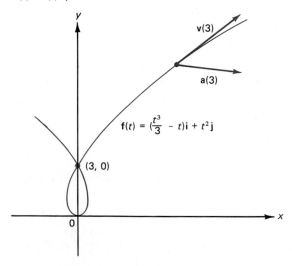

$$f(t) = \left(\frac{t^3}{3} - t\right)\mathbf{i} + t^2\mathbf{j}$$

Figure 2

We now calculate the vectors which describe the motion of an object in the plane. First, we make the simplifying assumption that the only force acting on the particle is the force of gravity. We ignore, for example, the frictional force due to air resistance. The force of gravity is directed vertically downward. There is no force acting in the horizontal direction. Thus the acceleration vector is given by

$$\mathbf{a} = -g\mathbf{j}, \tag{5}$$

where the constant $g = 9.81 \text{ m/sec}^2 = 32.2 \text{ ft/sec}^2$. In addition

$$\mathbf{F} = m\mathbf{a} = -mg\mathbf{j}$$

where m is the mass of the object. To obtain the velocity vector $\mathbf{v}(t)$, we simply integrate (5) to obtain

$$\mathbf{v}(t) = -gt\mathbf{j} + \mathbf{C} \tag{6}$$

where $\mathbf{C}$ is a constant vector. But, from (6), we find that

$$\mathbf{v}(0) = \mathbf{C}$$

so that $\mathbf{C}$ is the initial velocity vector $\mathbf{v}_0$ (the velocity at $t = 0$) and (6) becomes

$$\mathbf{v}(t) = -gt\mathbf{j} + \mathbf{v}_0. \tag{7}$$

Integrating (7), we obtain the position vector

$$\mathbf{f}(t) = -\tfrac{1}{2}gt^2\mathbf{j} + \mathbf{v}_0 t + \mathbf{D} \tag{8}$$

where $\mathbf{D}$ is another constant vector. Evaluating (8) at $t = 0$ yields $\mathbf{D} = \mathbf{f}(0) =$ the initial position vector $\mathbf{f}_0$ so that (8) becomes

$$\boxed{\mathbf{f}(t) = -\tfrac{1}{2}gt^2\mathbf{j} + \mathbf{v}_0 t + \mathbf{f}_0.} \tag{9}$$

EXAMPLE 3.   A cannon whose muzzle is tilted upward at an angle of 30° shoots a ball at an initial velocity of 600 m/sec. (a) Find the position vector at all times $t \geq 0$. (b) How much time does the ball spend in the air? (c) How far does the cannonball travel? (d) How high does the ball get? (e) How far from the cannon does the ball land? (f) What is the velocity and speed of the ball at the time of impact with the earth? Assume that the mouth of the cannon is at ground level.

SOLUTION. (a)   From Example 16.1.7 we find, resolving the initial velocity vector into its vertical and horizontal components, that $\mathbf{v}_0 = 300\sqrt{3}\,\mathbf{i} + 300\mathbf{j}$. Moreover, if we place the origin so that it coincides with the mouth of the cannon, then the initial position vector $\mathbf{f}_0$ is $(0, 0) = 0\mathbf{i} + 0\mathbf{j}$. From (9),

$$\mathbf{f}(t) = -\tfrac{1}{2}gt^2\mathbf{j} + 300\sqrt{3}\,t\mathbf{i} + 300t\mathbf{j} + \mathbf{0}$$
$$= 300\sqrt{3}\,t\mathbf{i} + (300t - \tfrac{1}{2}gt^2)\mathbf{j}.$$

   (b)   The ball hits the ground when the vertical component is zero; that is, when $300t - \tfrac{1}{2}gt^2 = 0$ which occurs at $t = 600/g$ sec.
   (c)   The total distance traveled is

$$s = \int_0^{600/g} \left(\frac{ds}{dt}\right) dt \quad \text{sec.}$$

But

$$\frac{ds}{dt} = v(t) = |\mathbf{v}(t)| = |\mathbf{f}'(t)| = |300\sqrt{3}\mathbf{i} + (300 - gt)\mathbf{j}|$$

$$- \sqrt{270,000 + (300 - gt)^2}.$$

Thus using formula (16.5.4),

$$s = \int_0^{600/g} \left(\frac{ds}{dt}\right) dt = \int_0^{600/g} \sqrt{270,000 + (300 - gt)^2} \, dt \quad \text{sec}.$$

This integral could be evaluated by making the substitution $300 - gt = \sqrt{270,000}\tan\theta$ or it could be evaluated numerically using one of the techniques of Section 8.9. However we will stop here and simply leave the answer in terms of the integral.

(d)   The maximum height is achieved when $dy/dt = 0$. That is, when $300 - gt = 0$ or when $t = 300/g$. For that value of t, $y_{\max} = 45,000/g \approx 4587.2 \text{ m} = 4.587 \text{ km}$.

(e)   In $600/g$ sec, the x-component of $\mathbf{f}$ increases from 0 to $(300\sqrt{3})\cdot(600/g) \text{ m} \approx 31,780.7 \text{ m} \approx 31.78 \text{ km}$.

(f)   Since the speed $= v(t) = \sqrt{270,000 + (300 - gt)^2}$ m/sec, we find that upon impact, $t = 600/g$ sec so that

$$v\left(\frac{600}{g}\right) = \sqrt{270,000 + \left(300 - g\cdot\frac{600}{g}\right)^2} = \sqrt{270,000 + 90,000}$$

$$= \sqrt{360,000} = 600 \text{ m/sec}.$$

Also, $\mathbf{a}(t) = -gt\mathbf{j}$ so that the acceleration scalar $a(t) = gt = g(600/g) = 600 \text{ m/sec}^2$ upon impact. This is all illustrated in Figure 3.

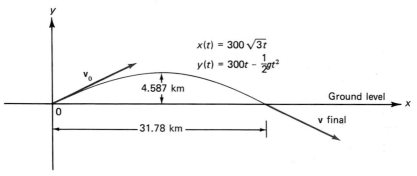

Figure 3

There is another convenient application of vectors applied to the motion of an object.

DEFINITION 3.   The *momentum* **P** of a particle at any time t is a vector defined as the product of the mass m of the particle and its velocity **v**. That is,

$$\mathbf{P} = m\mathbf{v}. \tag{10}$$

Newton's second law of motion states that the *rate of change of momentum of a moving object is proportional to the resultant force and is in the direction of that force.* That is,

$$\mathbf{F} = \frac{d\mathbf{P}}{dt}. \tag{11}$$

If the mass of the object is constant, then, using (10) and (11), we have the familiar law

$$\mathbf{F} = \frac{d}{dt}m\mathbf{v} = m\frac{d\mathbf{v}}{dt} = m\mathbf{a}. \tag{12}$$

If mass is not constant, then (12) becomes

$$\mathbf{F} = \frac{d}{dt}m\mathbf{v} = m\frac{d\mathbf{v}}{dt} + \mathbf{v}\frac{dm}{dt}. \tag{13}$$

EXAMPLE 4. The cannonball of Example 3 has a mass of 8 kg. Find the force acting on the ball at any time t.

SOLUTION. Here m is constant so that $\mathbf{F} = m\mathbf{a} = (8 \text{ kg})(-g\mathbf{j}) = -8g\mathbf{j}$ which is a force with a magnitude of $8\ gt$ nt acting vertically downward. There is no force in the horizontal direction.

EXAMPLE 5. An object of mass m moves in the elliptical orbit given by $\mathbf{f}(t) = a(\cos \alpha t)\mathbf{i} + b(\sin \alpha t)\mathbf{j}$. Find the force acting on the object at any time t.

SOLUTION. We easily find that $\mathbf{a}(t) = \mathbf{f}''(t) = -\alpha^2 a(\cos \alpha t)\mathbf{i} - \alpha^2 b(\sin \alpha t)\mathbf{j} = -\alpha^2\mathbf{f}(t)$. Since m is constant, $\mathbf{F} = m\mathbf{a} = -m\alpha^2\mathbf{f}(t)$. Thus the force always acts in the direction opposite to the direction of the position vector (thereby pointing *toward* the origin) and has a magnitude proportional to the distance of the object from the origin. Such a force is called a *central force* (see Figure 4).

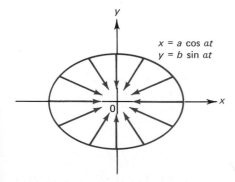

$x = a \cos at$
$y = b \sin at$

Figure 4

If the vector sum of forces acting on a system is zero, then, since $\mathbf{0} = \mathbf{F} = \mathbf{P}'(t)$, we find that the *total momentum of the system is constant.* This fact is called the *principle of the conservation of linear momentum.* We now see how that principle can be applied to the motion of a rocket.

Let the mass of a rocket be denoted by m and suppose that the rocket is traveling with velocity $\mathbf{v}(t)$. The rocket is propelled by gas emissions. We assume that the exhaust velocity of the gases ejected from the rocket, relative to the rocket, is $\mathbf{u}(t)$. Then the total velocity of the gas (relative to the earth, say) is $\mathbf{u} + \mathbf{v}$. The rate of change of momentum on the rocket is

$$\mathbf{P}'(t) = \frac{d}{dt}(m\mathbf{v}) = m\frac{d\mathbf{v}}{dt} + \mathbf{v}\frac{dm}{dt}.$$

As the rocket loses mass, the exhaust gains it, so that the rate of change of momentum of the gases is given by

$$-(\mathbf{u} + \mathbf{v})\frac{dm}{dt}.$$

Here dm/dt is the rate at which exhaust gas accumulates. The total force acting on the rocket–gas system is the sum of the rates of change of momentum for each of the two components of the system; that is,

$$\mathbf{F} = m\frac{d\mathbf{v}}{dt} + \mathbf{v}\frac{dm}{dt} - \mathbf{u}\frac{dm}{dt} - \mathbf{v}\frac{dm}{dt} = m\frac{d\mathbf{v}}{dt} - \mathbf{u}\frac{dm}{dt}$$

or

$$\mathbf{F} + \mathbf{u}\frac{dm}{dt} = m\frac{d\mathbf{v}}{dt}. \tag{14}$$

The term $\mathbf{u}\,dm/dt$ is called the *thrust* of the rocket. Note that $dm/dt < 0$ (since the rocket is losing mass) so that the thrust has the direction *opposite* to that of the velocity $\mathbf{u}$ of the exhaust gases.

Now, if the rocket is free of any gravitational field, then there are no external forces acting on the system so that $\mathbf{F} = 0$ and (14) becomes

$$\mathbf{u}\frac{dm}{dt} = m\frac{d\mathbf{v}}{dt}. \tag{15}$$

If $\mathbf{u}$ is constant, then (15) becomes

$$\frac{d\mathbf{v}}{dt} = \frac{\mathbf{u}}{m}\frac{dm}{dt}$$

and integration yields

$$\mathbf{v}(t) - \mathbf{v}_0 = \mathbf{u}\int_{m_0}^{m}\frac{dm}{m} = \mathbf{u}\ln m\,\Big|_{m_0}^{m}$$

or

$$\mathbf{v} - \mathbf{v}_0 = -\mathbf{u}\ln\left(\frac{m_0}{m}\right). \tag{16}$$

Here $\mathbf{v}_0$ is the initial velocity vector and m_0 is the initial mass of the rocket and the fuel combined. Thus, once the rocket has escaped all gravitational fields, we can calculate its velocity at any time if we know the velocity of the escaping gases and the proportion of gases that have been expelled.

EXAMPLE 6. A 1000-kg rocket carrying 2000 kg of fuel is motionless in space. If its engine starts and if its exhaust velocity is 0.7 km/sec, what is its speed when all its fuel is consumed?

SOLUTION. Here $v_0 = 0$, Then

$$|\mathbf{v}| = |\mathbf{u}| \ln \frac{m_0}{m} = 0.7 \ln \frac{3000}{1000} = 0.7 \ln 3 \approx 0.769 \quad \text{km/sec}$$

$$\approx 2768 \quad \text{km/hr} \ (\approx 1720 \quad \text{mi/hr}).$$

EXAMPLE 7. The above calculation was made in the absence of gravitational forces. Let us now assume that the rocket is fired upward from the earth and assume that gas is ejected at the constant rate of 100 kg/sec. Then the force propelling the rocket upward is equal to thrust and is given by

$$F_{\text{thrust}} = \left| \mathbf{u} \frac{dm}{dt} \right| = |(700 \text{ m/sec})(-100 \text{ kg/sec})| = 70,000 \quad \text{nt}.$$

The thrust is opposed by the force of gravity.

$$F_{\text{grav}} = -mg = (-3000 \text{ kg})(9.81 \text{ m/sec}^2) = -29,430 \quad \text{nt}.$$

(Note that we have treated $\mathbf{u}$ and $\mathbf{F}$ as scalars since we are assuming only motion in one direction: straight up.) We see that the initial net upward force is given by

$$F_{\text{thrust}} + F_{\text{grav}} = 70,000 - 29,430 = 40,570 \quad \text{nt}.$$

Finally, just before all the fuel is used up, the force due to gravity is

$$F_{\text{grav}} = (-1000 \text{ kg})(9.81 \text{ m/sec}^2) = -9810 \quad \text{nt}$$

so that the net force is $70,000 - 9810 = 60,190$ nt. Of course, when the fuel burns out, the rocket will fall back to earth.

> ***Remark.*** In this example we made the assumption that the rocket did not get so high so as to partially escape the effect of the earth's gravitational field; that is, so that the force of the earth's gravitational attraction could not be represented by $-mg$.

PROBLEMS 16.7

In Problems 1–14 the position vector of a moving particle is given. For the indicated value of t, calculate the velocity vector, the acceleration vector, the speed, and the acceleration scalar. Then sketch the portion of the trajectory showing the velocity and acceleration vectors.

1. $\mathbf{f} = (\cos 3t)\mathbf{i} + (\sin 3t)\mathbf{j}; \ t = 0$
2. $\mathbf{f} = (\cos 5t)\mathbf{i} + (\sin 5t)\mathbf{j}; \ t = \pi/2$
3. $\mathbf{f} = 2(\cos 4t)\mathbf{i} + 2(\sin 4t)\mathbf{j}; \ t = \pi/6$
4. $\mathbf{f} = -3(\cos 10t)\mathbf{i} - 3(\sin 10t)\mathbf{j}; \ t = \pi$
5. $\mathbf{f} = 4t\mathbf{i} + 2t^2\mathbf{j}; \ t = 1$
6. $\mathbf{f} = (2 + 3t)\mathbf{i} + (8 - 5t)\mathbf{j}; \ t = 3$
7. $\mathbf{f} = (4 - 7t)\mathbf{i} + (-3 + 5t)\mathbf{j}; \ t = -5$
8. $\mathbf{f} = (a + bt)\mathbf{i} + (c + dt)\mathbf{j}; \ t = t_0, \ a, \ b, \ c, \ d \ \text{real}$
9. $\mathbf{f} = 2t\mathbf{i} + (e^{-t} + e^t)\mathbf{j}; \ t = 0.$
10. $\mathbf{f} = \cosh t\mathbf{i} + 2t\mathbf{j}; \ t = 2$
11. $\mathbf{f} = (t - \cos t)\mathbf{i} + (1 - \sin t)\mathbf{j}; \ t = \pi/2$

12. $\mathbf{f} = (t - \cos t)\mathbf{i} + (1 - \sin t)\mathbf{j}; \; t = \pi$

13. $\mathbf{f} = (t - \cos t)\mathbf{i} + (1 - \sin t)\mathbf{j}; \; t = \pi/4$

14. $\mathbf{f} = (\ln \sin t)\mathbf{i} + (\ln \cos t)\mathbf{j}; \; t = \pi/6.$

15. A bullet is shot from a gun with an initial velocity of 1200 m/sec. The gun is inclined at an angle of 45°. Find (a) the total distance traveled by the bullet, (b) the horizontal distance traveled by the bullet, (c) the maximum height it reaches, and (d) the speed of the bullet at impact. Assume that the gun is held at ground level.

16. Answer the questions of Problem 15 if the angle of inclination of the gun is 60° and the initial velocity is 3000 ft/sec.

17. A man is standing at the top of a 200-m building. He throws a ball horizontally with an initial speed of 20 m/sec.

    (a) How far does the ball travel in its path?

    (b) How far from the base of the building does the ball hit the ground?

    (c) At what angle with the horizontal does the ball hit the ground, and

    (d) with what speed?

18. An airplane, flying horizontally at a height of 1500 m and with a speed 450 km/hr releases a bomb.

    (a) How long does it take the bomb to hit the ground?

    (b) How far does it travel in the horizontal direction?

    (c) What is its speed of impact?

19. A girl is standing 20 ft from a tall building. She throws a ball at an angle of 60° toward the building with an initial speed of 60 ft/sec. If her hand is 4 ft above the ground when she releases the ball, how high up the building does the ball hit?

*20. A man can throw a football a maximum distance of 60 yd. What is the maximum speed, in ft/sec, that he can throw the football?

21. A 2000-kg rocket carrying 3000 kg of fuel is initially motionless in space. Its exhaust velocity is 1 km/sec. What is its velocity when all its fuel is consumed?

22. If the rocket in Problem 21 is launched from the earth and if gas is emitted at a constant rate of 75 kg/sec, find

    (a) the net initial force acting on the rocket.

    (b) the net force just before all the fuel is expended.

23. In Problem 22, what is the minimum thrust needed to get the rocket off the ground?

24. Suppose that two masses, m_1 and m_2, sitting on a frictionless table are connected by a spring (see Figure 5). The two blocks are then pulled apart and released. Show that if $\mathbf{v}_1$ and $\mathbf{v}_2$ are the velocities of the two masses, that

$$\mathbf{v}_1 = -\frac{m_2}{m_1}\mathbf{v}_2.$$

Figure 5

[*Hint:* Use the principle of conservation of linear momentum to show that the initial momentum is equal to the final momentum. Then show that the final momentum is given by $m_1v_1 + m_2v_2$.]

*16.8 Curvature and the Acceleration Vector

The derivative dy/dx of a curve $y = f(x)$ measures the rate of change of the vertical component of the curve with respect to the horizontal component. As we have seen, the derivative ds/dt represents the change in the length of the arc traced out by the vector $\mathbf{f} = x(t)\mathbf{i} + y(t)\mathbf{j}$ as t increases. Another quantity of interest is the rate of change of the direction of the curve with respect to the length of the curve. That is, how much does the direction change for every one unit change in the arc length? We are thus led to the following definition.

DEFINITION 1. (i) Let the curve C be given by the differentiable vector function $\mathbf{f}(t) = f_1(t)\mathbf{i} + f_2(t)\mathbf{j}$. Let $\varphi(t)$ denote the direction of $\mathbf{f}'(t)$. Then the *curvature* of C, denoted by $\kappa(t)$, is the absolute value of the rate of change of direction with respect to arc length; that is,

$$\kappa(t) = \left| \frac{d\varphi}{ds} \right|. \tag{1}$$

(ii)   The *radius of curvature* $\rho(t)$ is defined by

$$\rho(t) = \frac{1}{\kappa(t)}. \tag{2}$$

Remark 1. The curvature is a measure of *how fast* the curve turns as we move along it.

Remark 2. If $\kappa(t) = 0$, we say that the radius of curvature is *infinite*. To understand this, note that if $\kappa(t) = 0$, then the "curve" does not bend and, so, is a straight line (see Example 2). A straight line can be thought of as an arc of a circle with *infinite* radius.

In Chapter 17 we will need a definition of curvature that does not depend on the angle φ.

THEOREM 1. If $\mathbf{T}(t)$ denotes the unit tangent vector to $\mathbf{f}$, then

$$\kappa(t) = \left| \frac{d\mathbf{T}}{ds} \right|. \tag{3}$$

PROOF. By the chain rule (which applies just as well to vector-valued functions)

$$\frac{d\mathbf{T}}{ds} = \frac{d\mathbf{T}}{d\varphi} \frac{d\varphi}{ds}.$$

But, since φ is the direction of $\mathbf{f}'$, and therefore also the direction of $\mathbf{T}$, we have

$$\mathbf{T} = (\cos \varphi)\mathbf{i} + (\sin \varphi)\mathbf{j}$$

so that

$$\frac{d\mathbf{T}}{d\varphi} = -(\sin \varphi)\mathbf{i} + (\cos \varphi)\mathbf{j} \quad \text{and} \quad \left| \frac{d\mathbf{T}}{d\varphi} \right| = \sqrt{\sin^2 \varphi + \cos^2 \varphi} = 1.$$

Thus

$$\left| \frac{d\mathbf{T}}{ds} \right| = \left| \frac{d\mathbf{T}}{d\varphi} \right| \left| \frac{d\varphi}{ds} \right| = 1 \left| \frac{d\varphi}{ds} \right| = \kappa(t)$$

and the theorem is proved.

We now derive an easier way to calculate $\kappa(t)$.

THEOREM 2. With the curve C given as in Definition 1, the curvature of C is given by the formula

$$\kappa(t) = \frac{|(dx/dt)(d^2y/dt^2) - (dy/dt)(d^2x/dt^2)|}{[(dx/dt)^2 + (dy/dt)^2]^{3/2}} \qquad (4)$$

where $x(t) = f_1(t)$ and $y(t) = f_2(t)$.

PROOF. By the chain rule,

$$\frac{d\varphi}{ds} = \frac{d\varphi}{dt}\frac{dt}{ds} = \frac{d\varphi/dt}{ds/dt}. \qquad (5)$$

From equation (16.5.10),

$$\frac{ds}{dt} = \sqrt{\left(\frac{dx}{dt}\right)^2 + \left(\frac{dy}{dt}\right)^2} \qquad (6)$$

and, from equation (16.3.6), we obtain

$$\tan \varphi = \frac{dy/dt}{dx/dt} \qquad (7)$$

or

$$\varphi = \tan^{-1}\frac{dy/dt}{dx/dt}. \qquad (8)$$

Differentiating both sides of (8) with respect to t, we obtain

$$\frac{d\varphi}{dt} = \frac{1}{1 + \left(\frac{dy/dt}{dx/dt}\right)^2}\frac{d}{dt}\left(\frac{\frac{dy}{dt}}{\frac{dx}{dt}}\right) = \frac{\left(\frac{dx}{dt}\right)^2}{\left(\frac{dx}{dt}\right)^2 + \left(\frac{dy}{dt}\right)^2} \cdot \frac{\frac{dx}{dt}\frac{d^2y}{dt^2} - \frac{dy}{dt}\frac{d^2x}{dt^2}}{\left(\frac{dx}{dt}\right)^2}. \qquad (9)$$

Substitution of (6) and (9) into (5) completes the proof of the theorem.

EXAMPLE 1. We would certainly expect that the curvature of a circle would be constant and that its radius of curvature would be its radius. Show that this is true.

SOLUTION. The circle of radius r centered at the origin is given parametrically by

$$x = r \cos t, \qquad y = r \sin t.$$

Then $dx/dt = -r \sin t$, $d^2x/dt^2 = -r \cos t$, $dy/dt = r \cos t$, and $d^2y/dt^2 = -r \sin t$ so that, from (4),

$$\kappa(t) = \frac{|r^2 \sin^2 t + r^2 \cos^2 t|}{[r^2 \sin^2 t + r^2 \cos^2 t]^{3/2}} = \frac{r^2}{r^3} = \frac{1}{r}$$

and

$$\rho(t) = \frac{1}{\kappa(t)} = r,$$

as expected.

EXAMPLE 2. Show that for a straight line, $\kappa(t) = 0$.

SOLUTION. A line can be represented parametrically (see equation (16.1.9)) by $x = x_1 + t(x_2 - x_1)$ and $y = y_1 + t(y_2 - y_1)$. Then $dx/dt = x_2 - x_1$, $d^2x/dt^2 = 0$, $dy/dt = (y_2 - y_1)$ and $d^2y/dt^2 = 0$. Substitution of these values into (4) immediately yields $\kappa(t) = 0$.

EXAMPLE 3. Find the curvature and radius of curvature of the curve given parametrically by $x = (t^3/3) - t$ and $y = t^2$ (see Example 16.4.5) at $t = 2\sqrt{2}$.

SOLUTION. We have $dx/dt = t^2 - 1$, $d^2x/dt^2 = 2t$, $dy/dt = 2t$, and $d^2y/dt^2 = 2$ so that

$$\kappa(t) = \frac{|2(t^2 - 1) - 2t(2t)|}{[(t^2 - 1)^2 + (2t)^2]^{3/2}} = \frac{2(t^2 + 1)}{(t^4 + 2t^2 + 1)^{3/2}} = \frac{2(t^2 + 1)}{(t^2 + 1)^3} = \frac{2}{(t^2 + 1)^2}.$$

Then $\kappa(2\sqrt{2}) = 2/81$ and the radius of curvature $\rho(2\sqrt{2}) = 1/\kappa(2\sqrt{2}) = 81/2$. A portion of this curve near $t = 2\sqrt{2}$ is sketched in Figure 1. For reference purposes, the unit tangent vector T at $t = 2\sqrt{2}$ is included.

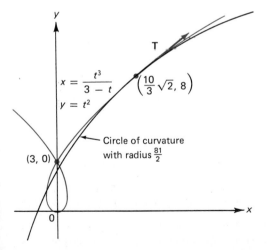

Figure 1

If we are given the Cartesian equation of a curve, $y = f(x)$, then the formula for the curvature, now denoted by $\kappa(x)$, is simpler. Proceeding as before, we have

$$\frac{d\varphi}{dx} = \frac{d\varphi}{ds}\frac{ds}{dx} \qquad \text{or} \qquad \frac{d\varphi}{ds} = \frac{d\varphi/dx}{ds/dx}.$$

From equation (9.2.3) we have

$$\frac{ds}{dx} = \sqrt{1 + \left(\frac{dy}{dx}\right)^2}.$$

Also, since $\tan \varphi = dy/dx$, we have

$$\varphi = \tan^{-1}\frac{dy}{dx} \quad \text{and} \quad \frac{d\varphi}{dx} = \frac{1}{1 + (dy/dx)^2} \cdot \frac{d^2y}{dx^2}$$

so that

$$\kappa(x) = \left|\frac{d\varphi}{ds}\right| = \frac{|d^2y/dx^2|}{[1 + (dy/dx)^2]^{3/2}}. \tag{10}$$

EXAMPLE 4. Calculate the curvature and radius of curvature of the curve $y = 3/x$ at the point $(1, 3)$.

SOLUTION. Here $dy/dx = -3/x^2$ and $d^2y/dx^2 = 6/x^3$ and

$$\kappa(x) = \frac{|6/x^3|}{[1 + (9/x^4)]^{3/2}} = \frac{|6/x^3|}{[(9 + x^4)/x^4]^{3/2}} = \frac{|6x^3|}{(9 + x^4)^{3/2}}.$$

When $x = 1$, $\kappa(1) = 6/10\sqrt{10} = 3\sqrt{10}/50 \approx 0.19$. Then $\rho(1) = 50/3\sqrt{10} \approx 5.27$. The curve is sketched in Figure 2.

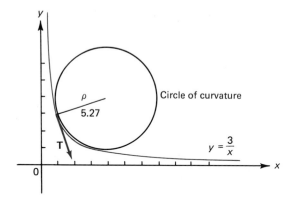

Circle of curvature

$y = \dfrac{3}{x}$

ρ

5.27

T

Figure 2

Note 1. In Figures 1 and 2, the circle with radius of curvature $\rho(t)$ which lies on the concave side of C is called the *circle of curvature*.

Note 2. The curve $y = 3/x$ is concave up for $x > 0$. Graphically, it seems as if the curve is "flattening out" as x gets larger. This is easy to prove since $\lim_{x\to\infty} \kappa(x) = 0$. In fact, it can be shown (but not every easily) that if any curve is concave up (or down) for x sufficiently large, then $\lim_{x\to\infty} \kappa(x) = 0$.

There is an interesting relationship between curvature and acceleration vectors. If a particle is moving along the curve C with position vector

$$\mathbf{f}(t) = x(t)\mathbf{i} + y(t)\mathbf{j},$$

then, from the last section, the acceleration vector is given by

$$\mathbf{a}(t) = \frac{d^2x}{dt^2}\mathbf{i} + \frac{d^2y}{dt^2}\mathbf{j}. \tag{11}$$

The representation (11) resolves $\mathbf{a}$ into its horizontal and vertical components. However there is another representation that is often more useful. Imagine yourself driving on the highway. If the car in which you are riding accelerates forward, you are pressed to the back of your seat. If it turns sharply to one side, you are thrown to the other. Both motions are due to acceleration. The second force is related to the rate at which the car turns which is, of course, related to the curvature of the road. Thus, we would like to express the acceleration vector as a component in the direction of motion and a component somehow related to the curvature of the path. How do we measure this? A glance at either Figure 1 or Figure 2 reveals the answer. The radial line of the circle of curvature is perpendicular to the unit tangent vector $\mathbf{T}$ since $\mathbf{T}$ is tangent to the circle of curvature and, in a circle, tangent and radial lines at a point are orthogonal. (See Example 16.4.3.) But a vector which is perpendicular to $\mathbf{T}$ has the direction of the unit normal vector $\mathbf{n}$. Thus the component of acceleration in the direction $\mathbf{n}$ will be a measure of the acceleration due to turning. We would like to write

$$\mathbf{a} = a_{\mathbf{T}}\mathbf{T} + a_{\mathbf{n}}\mathbf{n} \tag{12}$$

where $a_{\mathbf{T}}$ and $a_{\mathbf{n}}$ are, respectively, the components of $\mathbf{a}$ in the tangential and normal directions. Before calculating $a_{\mathbf{T}}$ and $a_{\mathbf{n}}$, we need the following result.

THEOREM 3. For a curve C, let $\mathbf{T}$ and $\mathbf{n}$ be the unit tangent and unit normal vectors, respectively, and let φ be as before. Then

(i) $\dfrac{d\mathbf{T}}{d\varphi} = \pm\mathbf{n}$ $\tag{13}$

and

(ii) $\dfrac{d\mathbf{T}}{ds} = \kappa\mathbf{n},$ $\tag{14}$

where κ is the curvature of C. In (13), the plus sign is taken if φ increases as s increases and the minus sign is taken if φ decreases as s increases.

PROOF. (i)   By the definition of $\mathbf{T}$,

$$\mathbf{T} = (\cos\varphi)\mathbf{i} + (\sin\varphi)\mathbf{j}.$$

Then

$$\frac{d\mathbf{T}}{d\varphi} = -(\sin\varphi)\mathbf{i} + (\cos\varphi)\mathbf{j} = \pm\mathbf{n}\dagger$$

since $\mathbf{T}\cdot(d\mathbf{T}/d\varphi) = 0$ and $|(-(\sin\varphi)\mathbf{i} + (\cos\varphi)\mathbf{j}| = 1$.

† To see why $d\mathbf{T}/d\varphi = -\mathbf{n}$ if $d\varphi/ds < 0$, look again at the discussion at the end of Section 16.3. There we showed that if $d\varphi/ds < 0$ (Figure 16.3.3(b)), then

$$\mathbf{n} = \sin\varphi\mathbf{i} - \cos\varphi\mathbf{j} = -[-\sin\varphi\mathbf{i} + \cos\varphi\mathbf{j}] = -\frac{d\mathbf{T}}{d\varphi}.$$

If $d\varphi/ds > 0$, then

(ii) $\dfrac{d\mathbf{T}}{ds} = \dfrac{d\mathbf{T}}{d\varphi}\dfrac{d\varphi}{ds} = \mathbf{n}\kappa$

from the definition of κ. If $d\varphi/ds < 0$, then

$$\dfrac{d\mathbf{T}}{ds} = \dfrac{d\mathbf{T}}{d\varphi}\dfrac{d\varphi}{ds} = (-\mathbf{n})(-\kappa) = \kappa\mathbf{n}$$

We can now show how to write $\mathbf{a}(t)$ in its tangential and normal components.

THEOREM 4.   Let the curve C be given by the vector function $\mathbf{f}(t)$. Then the acceleration vector $\mathbf{a}(t) = \mathbf{f}''(t)$ can be written as

$$\boxed{\ \mathbf{a} = \dfrac{d^2s}{dt^2}\,\mathbf{T} + \left(\dfrac{ds}{dt}\right)^2 \kappa\mathbf{n}\ } \tag{15}$$

where $\mathbf{T}$ and $\mathbf{n}$ are the unit tangent and unit normal vectors to the curve, s is the arc length measured from the point reached when $t = 0$, and κ is the curvature.

Note.   Since $v = ds/dt$ is the speed, we can write (15) as

$$\mathbf{a} = \dfrac{dv}{dt}\,\mathbf{T} + v^2\kappa\mathbf{n} = \dfrac{dv}{dt}\,\mathbf{T} + \dfrac{v^2}{\rho}\,\mathbf{n}. \tag{16}$$

Thus the tangential component of acceleration $a_T = dv/dt$ and the normal component of acceleration is $a_n = v^2/\rho$.

PROOF OF THEOREM 4. We have

$$\mathbf{v} = \dfrac{d\mathbf{f}}{dt} = \dfrac{d\mathbf{f}}{ds}\dfrac{ds}{dt} = \mathbf{T}\dfrac{ds}{dt}$$

(from Theorem 16.6.1 $\mathbf{T} = d\mathbf{f}/ds$). Then

$$\mathbf{a} = \dfrac{d\mathbf{v}}{dt} = \dfrac{d}{dt}\left(\mathbf{T}\dfrac{ds}{dt}\right) = \dfrac{d^2s}{dt^2}\,\mathbf{T} + \dfrac{ds}{dt}\dfrac{d\mathbf{T}}{dt}.$$

But

$$\dfrac{ds}{dt}\dfrac{d\mathbf{T}}{dt} = \dfrac{ds}{dt}\left(\dfrac{d\mathbf{T}}{ds}\dfrac{ds}{dt}\right) = \left(\dfrac{ds}{dt}\right)^2\dfrac{d\mathbf{T}}{ds} = \left(\dfrac{ds}{dt}\right)^2 \kappa\mathbf{n}.$$

The last step follows from Theorem 3 and the theorem is proved.

The result given by (15) or (16) is very important in physics. If an object of constant mass m is traveling along a trajectory, then the force acting to keep the object on that trajectory is given by

$$\mathbf{F} = m\mathbf{a} = m\dfrac{dv}{dt}\,\mathbf{T} + mv^2\kappa\mathbf{n}. \tag{17}$$

The term $mv^2\kappa$ is the magnitude of the force necessary to keep the object from "moving off" the trajectory because of the force exerted on the object caused by turning.

EXAMPLE 5. A 1500-kg race car is driven at a speed of 150 km/hr on a circular race track of radius 50 m. What frictional force must be exerted by the tires on the road surface to keep the car from skidding?

SOLUTION. The friction of the tires must be equal to the normal component with respect to the circular race track of the force due to acceleration. That is,

$$F = mv^2\kappa = \frac{mv^2}{\rho} = (1500 \text{ kg}) \frac{(150,000 \text{ m})^2}{(3600 \text{ sec})^2} \cdot \frac{1}{50 \text{ m}}$$

$$= 52,083\tfrac{1}{3} \frac{\text{kg-m}}{\text{sec}^2} = 52,083\tfrac{1}{3} \text{ nt.}$$

EXAMPLE 6. Let the car of Example 5 have the *coefficient of friction* μ. This means that the maximum frictional force that can be exerted by the car on the road surface is *μmg* where *mg* is the *normal force* of the car on the road; that is, the force of the car on the road due to gravity. What is the minimum value μ can take in order that the car not slide off the road?

SOLUTION. We must have $\mu mg \geq 52,083\tfrac{1}{3}$ nt. But $\mu mg = \mu(9.81)(1500)$ so that we obtain

$$\mu \geq \frac{52,083\tfrac{1}{3}}{(9.81)(1500)} \approx 3.54.$$

PROBLEMS 16.8

In Problems 1–20 find the curvature and radius of curvature for each of the following. Sketch the unit tangent vector and the circle of curvature.

1. $\mathbf{f} = 2\cos t\mathbf{i} + 2\sin t\mathbf{j}; \; t = \pi/4$
2. $\mathbf{f} = 2\cos t\mathbf{i} + 2\sin t\mathbf{j}; \; t = \pi/27$
3. $\mathbf{f} = t\mathbf{i} + t^2\mathbf{j}; \; t = 1$
4. $\mathbf{f} = 3\sin t\mathbf{i} + 4\cos t\mathbf{j}; \; t = 0$
5. $\mathbf{f} = 3\sin t\mathbf{i} + 4\cos t\mathbf{j}; \; t = \pi/2$
6. $\mathbf{f} = 3\sin t\mathbf{i} + 4\cos t\mathbf{j}; \; t = \pi/4$
7. $\mathbf{f} = (\cos t + t\sin t)\mathbf{i} + (\sin t - t\cos t)\mathbf{j}; \; t = \pi/6$
8. $y = x^2; \; (0, 0)$
9. $y = x^2; \; (1, 1)$
10. $xy = 1; \; (1, 1)$
11. $y = e^x; \; (0, 1)$
12. $y = e^x; \; (1, e)$
13. $y = \ln x; \; (1, 0)$
14. $y = \cos x; \; (\pi/3, \tfrac{1}{2})$
15. $y = ax^2 + bx + c; \; (0, c), a, b, c$ real
16. $y = \ln\cos x; \; (\pi/4, \ln 1/\sqrt{2})$
17. $y = \sqrt{1 - x^2}; \; (0, 1)$
18. $y = \sin^{-1} x; \; (1, \pi/2)$
19. $x = \cos y; \; (0, \pi/2)$
20. $x = y^3; \; (1, 1).$

21. At what point on the parabola $y = ax^2$ is the curvature a maximum?
22. At what point on the curve $y = \ln x$ is the curvature a maximum?
23. For what value of t in the interval $[0, 2\pi]$ is the curvature of the curve $\mathbf{f}(t) = a(\cos^3 t)\mathbf{i} + a(\sin^3 t)\mathbf{j}$ a minimum? For what value is it a maximum?
*24. Let $r = f(\theta)$ be the equation of a curve in polar coordinates. Show that

$$\kappa(\theta) = \frac{|r^2 + 2(dr/d\theta)^2 - r\, d^2r/d\theta^2|}{[r^2 + (dr/d\theta)^2]^{3/2}}.$$

[*Hint:* Use $x = r\cos\theta$ and $y = r\sin\theta$.]

In Problems 25–30 use the result of Problem 24 to calculate the curvature. In all cases assume that $a > 0$.

25. $r = a \sin 2\theta; \ \theta = \pi/8$

26. $r = a\theta; \ \theta = 1$

27. $r = 2a \cos \theta; \ \theta = \pi/3$

28. $r = a(1 + \sin \theta); \ \theta = \pi/2$

29. $r = a(1 - \cos \theta); \ \theta = \pi$

30. $r = e^{a\theta}; \ \theta = 1$.

In Problems 31–38 find the tangential and normal components of acceleration for each of the given position vectors.

31. $\mathbf{f} = (\cos 2t)\mathbf{i} + (\sin 2t)\mathbf{j}$

32. $\mathbf{f} = 2(\cos t)\mathbf{i} + 3(\sin t)\mathbf{j}$

33. $\mathbf{f} = t\mathbf{i} + t^2\mathbf{j}$

34. $\mathbf{f} = t\mathbf{i} + (\cos t)\mathbf{j}$

35. $\mathbf{f} = (t^3 - 3t)\mathbf{i} + (t^2 - 1)\mathbf{j}$

36. $\mathbf{f} = e^{-t}\mathbf{i} + e^t\mathbf{j}$

37. $\mathbf{f} = t^2\mathbf{i} + t^3\mathbf{j}$

38. $\mathbf{f} = (\sin t^2)\mathbf{i} + (\cos t^2)\mathbf{j}$.

39.   Show that if a particle is moving in a circle at constant velocity, then the tangential component of acceleration is zero.

40.   Suppose that the driver of the car of Examples 5 and 6 reduces his speed by a factor of M. Show that the frictional force needed to keep the car from skidding is reduced by a factor of M^2.

41.   A locomotive traveling at 80 km/hr and weighing 100,000 kg is moving on a curved stretch of track. The equation of the curved section is the parabola $y = x^2 - x$ m. What is the frictional force exerted by the wheels of the locomotive on the track at the "point" $(0, 0)$?

42.   If the coefficient of friction for the locomotive in Problem 41 is 2.5, what is the maximum speed it can achieve at the point $(0, 0)$ without derailing?

43.   If the race car of Example 5 is placed on a track with half the radius of the original one, how much slower would it have to be driven so as not to increase the normal component of acceleration?

*44.   A man swings a rope attached to a bucket containing 3 kg of water. The pail rotates in the vertical plane in a circular path with a radius of 1 m. What is the smallest number of revolutions that must be made every minute in order that the water stay in the pail? [*Hint:* Calculate the pressure of the water on the bottom of the pail. This can be determined by first calculating the normal force of the motion. Then the water will stay in the bucket if this normal force exceeds the force due to gravity.]

Review Exercises for Chapter Sixteen

In Exercises 1–8 find the Cartesian equation of each curve and then sketch the curve in the xy-plane.

1. $\mathbf{f}(t) = t\mathbf{i} + 2t\mathbf{j}$.

2. $\mathbf{f}(t) = (2t - 6)\mathbf{i} + t^2\mathbf{j}$.

3. $\mathbf{f}(t) = t^2\mathbf{i} + (2t - 6)\mathbf{j}$.

4. $\mathbf{f}(t) = t^2\mathbf{i} + t^4\mathbf{j}$.

5. $\mathbf{f}(t) = (\cos 4t)\mathbf{i} + (\sin 4t)\mathbf{j}$.

6. $\mathbf{f}(t) = 4(\sin t)\mathbf{i} + 9(\sin t)\mathbf{j}$.

7. $\mathbf{f}(t) = t^6\mathbf{i} + t^2\mathbf{j}$.

8. $\mathbf{f}(t) = e^t(\cos t)\mathbf{i} + e^t(\sin t)\mathbf{j}$.

In Exercises 9–16 find the slope of the line tangent to the given curve for the given value of the parameter, and then find all points at which the curve has vertical and horizontal tangents.

9. $x = t^3; \ y = 6t; \ t = 1$

10. $x = t^7; \ y = t^8 - 5; \ t = 2$

11. $x = \sin 5\theta; \ y = \cos 5\theta; \ \theta = \pi/3$

12. $x = \cos^2 \theta; \ y = -3\theta; \ \theta = \pi/4$

13. $x = \cosh t; \ y = \sinh t; \ t = 0$

14. $x = 2/\theta; \ y = -3\theta; \ \theta = 10\pi$

15. $x = 3 \cos \theta; \ y = 4 \sin \theta; \ \theta = \pi/3$

16. $x = 3 \cos \theta; \ y = -4 \sin \theta; \ \theta = 2\pi/3$.

17. Find the slope of the line tangent to the polar curve $r = -3 \cos \theta$ for $\theta = \pi/3$.
18. Find the slope of the line tangent to the polar curve $r = \sin 2\theta$ for $\theta = \pi/8$.
19. Find the slope of the line tangent to the spiral of Archimedes $r = \theta$ for $\theta = \pi$.
20. Calculate the equation of the line that is tangent to the curve $x = t^2 - t - 2$, $y = t^2 - t + 1$ at the point $(4, 7)$.

In Exercises 21–24 find the first and second derivatives of the given vector functions.

21. $\mathbf{f}(t) = 2t\mathbf{i} - t^3\mathbf{j}$
22. $\mathbf{f}(t) = (1/t^2)\mathbf{i} + \sqrt[3]{t}\mathbf{j}$.
23. $\mathbf{f}(t) = (\cos 5t)\mathbf{i} + 2(\sin 4t)\mathbf{j}$
24. $\mathbf{f}(t) = (\tan t)\mathbf{i} + (\cot t)\mathbf{j}$.

In Exercises 25–30 find the unit tangent and unit normal vectors to the given curve for the given value of t.

25. $\mathbf{f}(t) = t^4\mathbf{i} + t^5\mathbf{j}$; $t = 1$
26. $\mathbf{f}(t) = (\cos 2t)\mathbf{i} + (\sin 2t)\mathbf{j}$; $t = \pi/6$.
27. $\mathbf{f}(t) = (\sin 5t)\mathbf{i} + (\cos 5t)\mathbf{j}$; $t = \pi/30$
28. $\mathbf{f}(t) = (\cosh 2t)\mathbf{i} + (\sinh 2t)\mathbf{j}$; $t = 0$
29. $\mathbf{f}(t) = (\ln t)\mathbf{i} + \sqrt{t}\mathbf{j}$; $t = 1$
30. $\mathbf{f}(t) = (\tan t)\mathbf{i} + (\cot t)\mathbf{j}$; $t = \pi/3$.

In Exercises 31–34 calculate the integral.

31. $\displaystyle\int_0^3 (t^3\mathbf{i} + t^5\mathbf{j}) \, dt$

32. $\displaystyle\int [(\cos 3t)\mathbf{i} + (\sin 3t)\mathbf{j}] \, dt$

33. $\displaystyle\int (\sqrt{t}\mathbf{i} + \sqrt[3]{t}\mathbf{j}) \, dt$

34. $\displaystyle\int_0^{\pi/3} [(\cos t)\mathbf{i} + (\tan t)\mathbf{j}] \, dt$.

35. If $\mathbf{f}'(t) = t^7\mathbf{i} - t^6\mathbf{j}$ and if $\mathbf{f}(0) = -\mathbf{i} + 3\mathbf{j}$, find $\mathbf{f}(t)$.

In Exercises 36–40 calculate the derivative.

36. $\dfrac{d}{dt} [(2t\mathbf{i} + \sqrt{t}\mathbf{j}) \cdot (4\mathbf{i} - 3\mathbf{j})]$

37. $\dfrac{d}{dt} \{[(\cos t)\mathbf{i} + (\sin t)\mathbf{j}] \cdot [(\cosh t)\mathbf{i} - (\sinh t)\mathbf{j}]\}$

38. $\dfrac{d}{dt} [(e^t)(t^2\mathbf{i} - t^{3/2}\mathbf{j})]$

39. $\dfrac{d}{dt} \left[\left(\dfrac{t}{1+t}\mathbf{i} - \dfrac{t+1}{t}\mathbf{j}\right) \cdot (e^t\mathbf{i} + e^{-t}\mathbf{j})\right]$

40. $\dfrac{d}{dt} \{(2\mathbf{i} - 11\mathbf{j}) \cdot [-(\tan t)\mathbf{i} + (\sec t)\mathbf{j}]\}$.

In Exercises 41–46 find the length of the arc over the given interval or the length of the closed curve.

41. $x = \cos 4\theta$; $y = \sin 4\theta$; $0 \le \theta \le \pi/12$
42. $x = e^t \sin t$; $y = e^t \cos t$; $0 \le t \le \pi/2$
43. $r = 2(1 + \cos \theta)$
44. $r = 5 \sin \theta$
45. $r = \theta^2$; $1 \le \theta \le 5$
46. $r = 2\theta$; $0 \le \theta \le 20\pi$.

In Exercises 47–50 find parametric equations in terms of the arc length s measured from the point reached when $t = 0$. Verify your answer by using formula (16.6.3).

47. $\mathbf{f} = 3t\mathbf{i} + 4t^{3/2}\mathbf{j}$
48. $\mathbf{f} = \frac{2}{9}t^{9/2}\mathbf{i} + \frac{1}{3}t^3\mathbf{j}$
49. $\mathbf{f} = 2(\cos 3t)\mathbf{i} + 2(\sin 3t)\mathbf{j}$
50. $\mathbf{f} = e^t(\sin t)\mathbf{i} + e^t(\cos t)\mathbf{j}$.

In Exercises 51–56 the position vector of a moving particle is given. For the indicated value of t calculate the velocity vector, the acceleration vector, the speed, and the acceleration scalar. Then sketch a portion of the trajectory showing the velocity and acceleration vectors.

51. $\mathbf{f} = (\cos 2t)\mathbf{i} + (\sin 2t)\mathbf{j};\ t = \pi/6$
52. $\mathbf{f} = 6t\mathbf{i} + 2t^3\mathbf{j};\ t = 1$
53. $\mathbf{f}(t) = (2^t + e^{-t})\mathbf{i} + 2t\mathbf{j};\ t = 0$
54. $\mathbf{f} = (1 - \cos t)\mathbf{i} + (t - \sin t)\mathbf{j};\ t = \pi/2$
55. $\mathbf{f} = 2(\sinh t)\mathbf{i} + 4t\mathbf{j};\ t = 1$
56. $\mathbf{f} = (3 + 5t)\mathbf{i} + (2 + 8t)\mathbf{j};\ t = 6.$

57. A boy throws a ball into the air with an initial speed of 30 m/sec. The ball is thrown from shoulder level (1.5 m above the ground) at an angle of 30° with the horizontal. Ignoring air resistance, find
(a) the total distance traveled by the ball,
(b) the horizontal distance traveled by the ball,
(c) the maximum height the ball reaches,
(d) the speed of the ball at impact with the ground, and
(e) the angle at which the path of the ball meets the ground at the point of impact.
58. A 1500-kg rocket carrying 1000 kg of fuel starts from a position motionless in space. Its exhaust velocity is 1.5 km/sec. What is its velocity when all its fuel is consumed?
59. If the rocket in Exercise 58 is launched from the earth and if gas is emitted at a constant rate of 40 kg/sec, find
(a) the net initial force acting on the rocket, and
(b) the net force just before all the fuel is expended.
60. In Exercise 59, what is the minimum thrust needed to get the rocket off the ground?

In Exercises 61–70 find the curvature and radius of curvature for each of the following. Sketch the unit tangent vector and the circle of curvature.

61. $\mathbf{f} = (\cos 2t)\mathbf{i} + (\sin 2t)\mathbf{j};\ t = \pi/3$
62. $\mathbf{f} = t^2\mathbf{i} + 2t\mathbf{j};\ t = 2$
63. $\mathbf{f} = 4(\cos t)\mathbf{i} + 9(\sin t)\mathbf{j};\ t = \pi/4$
64. $y = 2x^2;\ (0, 0)$
65. $xy = 1;\ (2, \frac{1}{2})$
66. $y = e^{-x};\ (1, 1/e)$
67. $y = \sqrt{x};\ (4, 2)$
68. $r = 3\cos 2\theta;\ \theta = \pi/6$
69. $r = 1 + \sin \theta;\ \theta = \pi/2$
70. $r = 3\theta;\ \theta = \pi.$

In Exercises 71–74 find the tangential and normal components of acceleration for each of the given position vectors.

71. $\mathbf{f} = 2(\sin t)\mathbf{i} + 2(\cos t)\mathbf{j}$
72. $\mathbf{f} = 4(\cos t)\mathbf{i} + 9(\sin t)\mathbf{j}$
73. $\mathbf{f} = 3t^2\mathbf{i} + 2t^3\mathbf{j}$
74. $\mathbf{f} = (\cos t^2)\mathbf{i} + (\sin t^2)\mathbf{j}.$

75. A 1300-kg race car is driven at a speed of 175 kg/hr on a circular race track of radius 65 m. What frictional force must be exerted by the tires on the road surface to keep the car from skidding?

SEVENTEEN

VECTORS IN SPACE

In Chapters 15 and 16 we developed a theory of vectors in the plane. In this chapter we shall extend that development to vectors in space and shall find that there are great similarities between the two notions.

17.1 The Rectangular Coordinate System in Space

In Section 1.3 we showed how any point in a plane can be represented as an ordered pair of real numbers. It is not surprising, then, that any point in space can be represented by an *ordered triple* of real numbers:

$$(a, b, c) \tag{1}$$

where a, b, and c are real numbers.

DEFINITION 1. The set of ordered triples of the form (1) is called *real three-dimensional space* and is denoted $\mathbb{R}^3$.

There are many ways to represent a point in $\mathbb{R}^3$. Two other methods will be given in Section 17.9. However, the most common representation (given in Definition 1) is very similar to the representation of a point in the plane by its x- and y-coordinates. We begin, as before, by choosing a point in $\mathbb{R}^3$ and calling it the *origin,* denoted by 0. Then we draw three mutually perpendicular axes, called the *coordinate axes,* which we label the *x-axis,* the *y-axis,* and the *z-axis*. These axes can be selected in a variety of ways, but the most common selection has the x- and y-axes drawn horizontally with the z-axis vertical. On each axis, we choose a positive direction and measure distance along each axis as the number of units in this positive direction measured from the origin.

The two basic systems of drawing these axes are depicted in Figure 1. If the axes are placed as in Figure 1a, then the system is called a *right-handed system;* if they are placed as in Figure 1b, the system is a *left-handed system*. In the figures, the arrows indicate the positive directions on the axes. The reason for this choice of terms is as follows: In a right-handed system, if you place your right hand so that your index finger points in the positive direction of the x-axis while your middle finger points in the positive direction of the y-axis, then your thumb will point in the positive direction of the z-axis. This is illustrated in Figure 2. For a left-handed system, the same rule will work for your left hand. For the remainder of this text, we will follow common practice and depict the coordinate axes using a right-handed system.

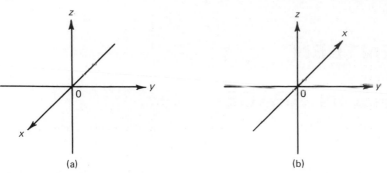

(a)   (b)   Figure 1

If you still have trouble visualizing the placement of these axes, do the follow-ing. Face any uncluttered corner (on the floor) of the room in which you are sitting. Call the corner the origin. Then the x-axis lies along the floor, along the wall, and to your left; the y-axis lies along the floor, along the wall, and to your right; and the z-axis lies along the vertical intersection of the two perpendicular walls. This is illustrated in Figure 3.

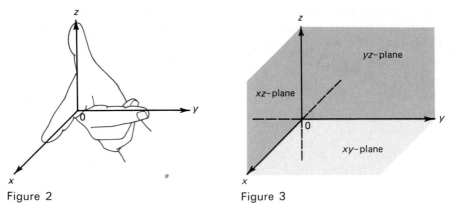

Figure 2   Figure 3

The three axes in our system determine three *coordinate planes* which we will call the xy-plane, the xz-plane, and the yz-plane. The xy-plane contains the x- and y-axes and is simply the plane with which we have been dealing in most of this book. The xz- and yz-planes can be thought of in a similar way.

Having built our structure of coordinate axes and planes, we can describe any point P in $\mathbb{R}^3$ in a unique way

$$P = (x, y, z), \tag{2}$$

where the first coordinate x is the distance from the yz-plane to P (measured in the positive direction of the x-axis), the second coordinate y is the distance from the xz-plane to P (measured in the positive direction of the y-axis), and the third coordinate z is the distance from the xy-plane to P (measured in the positive direction of the z-axis). Thus, for example, any point in the xy-plane has z-coordinate 0; any point in the xz-plane has y-coordinate 0; and any point in the yz-plane has x-coordinate 0. Some representative points are sketched in Figure 4.

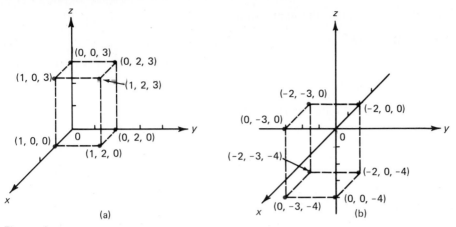

Figure 4

In this system, the three coordinate planes divide $\mathbb{R}^3$ into eight *octants*, just as in $\mathbb{R}^2$ the two coordinate axes divide the plane into four quadrants. The first octant is always chosen to be the one in which the three coordinates are positive.

The coordinate system we have just established is often referred to as the *rectangular coordinate system* or the *Cartesian coordinate system*. Once we are comfortable with the notion of depicting a point in this system, then we can generalize many of our ideas from the plane.

THEOREM 1. Let $P = (x_1, y_1, z_1)$ and $Q = (x_2, y_2, z_2)$ be two points in space. Then the distance $\overline{PQ}$ between P and Q is given by

$$\boxed{\overline{PQ} = \sqrt{(x_1 - x_2)^2 + (y_1 - y_2)^2 + (z_1 - z_2)^2}.} \tag{3}$$

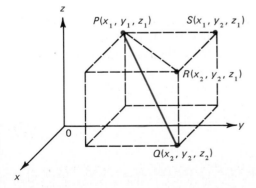

Figure 5

PROOF. The two points are sketched in Figure 5. From the Pythagorean theorem, since the line segments PR and RQ are perpendicular, the triangle PQR is a right triangle and

$$\overline{PQ}^2 = \overline{PR}^2 + \overline{RQ}^2. \tag{4}$$

But, using the Pythagorean theorem again,

$$\overline{PR}^2 = \overline{PS}^2 + \overline{SR}^2 \tag{5}$$

so, combining (4) and (5)

$$\overline{PQ}^2 = \overline{PS}^2 + \overline{SR}^2 + \overline{RQ}^2. \tag{6}$$

Since the x and z coordinates of P and S are equal,

$$\overline{PS}^2 = (y_2 - y_1)^2. \tag{7}$$

Similarly,

$$\overline{RS}^2 = (x_2 - x_1)^2 \tag{8}$$

and

$$\overline{RQ}^2 = (z_2 - z_1)^2. \tag{9}$$

Thus, using (7), (8), and (9) in (6) yields

$$\overline{PQ}^2 = (x_2 - x_1)^2 + (y_2 - y_1)^2 + (z_2 - z_1)^2$$

and the proof is complete.

EXAMPLE 1. Calculate the distance between the points $(3, -1, 6)$ and $(-2, 3, 5)$.

SOLUTION. $\overline{PQ} = \sqrt{[3 - (-2)]^2 + [-1 - 3]^2 + (6 - 5)^2} = \sqrt{42}$.

DEFINITION 2. The *graph* of an equation in $\mathbb{R}^3$ is the set of all points in $\mathbb{R}^3$ whose coordinates satisfy the equation.

One of our first examples of a graph in $\mathbb{R}^2$ was the graph of the unit circle: $x^2 + y^2 = 1$. This can easily be generalized.

DEFINITION 3. A *sphere* is the set of points in space equidistant from a given point. The common point is called the *center* of the sphere and the common distance is called the *radius* of the sphere.

EXAMPLE 2. Suppose that the center of a sphere is the origin $(0, 0, 0)$ and the radius of the sphere is 1. Let (x, y, z) be a point on the sphere. Then, from (3),

$$1 = \sqrt{(x - 0)^2 + (y - 0)^2 + (z - 0)^2}.$$

Simplifying and squaring, we obtain

$$x^2 + y^2 + z^2 = 1 \tag{10}$$

which is the equation of the *unit sphere*.

In general, if the center of a sphere is (a, b, c), the radius is r, and if (x, y, z) is a point on the sphere, we obtain

$$r = \sqrt{(x - a)^2 + (y - b)^2 + (z - c)^2}$$

or

$$(x - a)^2 + (y - b)^2 + (z - c)^2 = r^2. \tag{11}$$

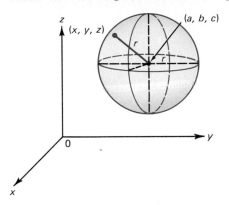

Figure 6

This is sketched in Figure 6.

EXAMPLE 3. Find the equation of the sphere with center at $(1, -3, 2)$ and radius 5.

SOLUTION. From (11) we obtain

$$(x - 1)^2 + (y + 3)^2 + (z - 2)^2 = 25.$$

EXAMPLE 4. Show that

$$x^2 - 6x + y^2 + 2y + z^2 + 10z + 5 = 0 \tag{12}$$

is the equation of a sphere and find its center and radius.

SOLUTION. Completing the squares (see Section 1.5), we obtain

$$x^2 - 6x = (x - 3)^2 - 9$$
$$y^2 + 2y = (y + 1)^2 - 1$$
$$z^2 + 10z = (z + 5)^2 - 25$$

so that

$$0 = x^2 - 6x + y^2 + 2y + z^2 + 10z + 5$$
$$= (x - 3)^2 - 9 + (y + 1)^2 - 1 + (z + 5)^2 - 25 + 5$$

and

$$(x - 3)^2 + (y + 1)^2 + (z + 5)^2 = 30$$

which is the equation of a sphere with center $(3, -1, -5)$ and radius $\sqrt{30}$.

> **Warning.** Not every *second degree* equation in a form similar to (12) is the equation of a sphere. For example, if the number 5 in (12) is replaced by 40, we obtain
>
> $$(x - 3)^2 + (y + 1)^2 + (z + 5)^2 = -5.$$
>
> Clearly, the sum of squares cannot be negative so that there are *no* points in $\mathbb{R}^3$ which satisfy this equation. On the other hand, if we replaced the 5 by 35, we would obtain
>
> $$(x - 3)^2 + (y + 1)^2 + (z + 5)^2 = 0.$$
>
> This equation can hold only when $x = 3$, $y = -1$, and $z = -5$. In this case the graph of the equation contains the single point $(3, -1, 5)$.

PROBLEMS 17.1

In Problems 1–15 sketch the given point in $\mathbb{R}^3$.

1. $(1, 4, 2)$ **2.** $(3, -2, 1)$ **3.** $(-1, 5, 7)$
4. $(8, -2, 3)$ **5.** $(-2, 1, -2)$ **6.** $(1, -2, 1)$
7. $(3, 2, -5)$ **8.** $(-2, -3, -8)$ **9.** $(2, 0, 4)$
10. $(-3, -8, 0)$ **11.** $(0, 4, 7)$ **12.** $(1, 3, 0)$
13. $(3, 0, 0)$ **14.** $(0, 8, 0)$ **15.** $(0, 0, -7)$.

In Problems 16–25 find the distance between the two points.

16. $(8, 1, 6)$; $(8, 1, 4)$ **17.** $(3, -4, 3)$; $(3, 2, 5)$
18. $(3, -4, 7)$; $(3, -4, 9)$ **19.** $(-2, 1, 3)$; $(4, 1, 3)$
20. $(2, -7, 5)$; $(8, -7, -1)$ **21.** $(1, 3, -2)$; $(4, 7, -2)$
22. $(3, 1, 2)$; $(1, 2, 3)$ **23.** $(5, -6, 4)$; $(3, 11, -2)$
24. $(-1, -7, -2)$; $(-4, 3, -5)$ **25.** $(8, -2, -3)$; $(-7, -5, 1)$.

26. Find the equation of the sphere with center at $(2, -1, 4)$ and radius 2.

27. Find the equation of the sphere with center at $(-1, 8, -3)$ and radius $\sqrt{5}$.

28. Show that the equation $x^2 + y^2 + z^2 - 4x - 4y + 8z + 8 = 0$ is the equation of a sphere and find its center and radius.

29. Do the same for the equation $x^2 + y^2 + z^2 + 3x - y + 2z - 1 = 0$.

30. Find a number α such that the equation $x^2 + y^2 + z^2 - 2x + 8y - 5z + \alpha = 0$ has exactly one solution.

31. In Problem 30 give a condition on α in order that the equation have no solution.

32. The equation $x^2 + y^2 + z^2 + ax + by + cz + d = 0$ is called a *second-degree equation in three variables*. Let $e = d - (a/2)^2 - (b/2)^2 - (c/2)^2$. Show that the equation
(a) is the equation of a sphere if $e < 0$,
(b) has exactly one solution if $e = 0$, and
(c) has no solutions if $e > 0$.

33. Three points P, Q, and R are *collinear* if they lie on the same straight line. Show that, in $\mathbb{R}^2$, P, Q, and R are collinear if $\overline{PR} = \overline{PQ} + \overline{QR}$ or $\overline{PQ} = \overline{PR} + \overline{QR}$. Use this last fact in $\mathbb{R}^3$ to show that the points $(-1, -1, -1)$, $(5, 8, 2)$ and $(-3, -4, -2)$ are collinear.

34. Show that the points $(3, 0, 1)$, $(0, -4, 0)$ and $(6, 4, 2)$ are collinear.

*35. Let $P = (x_1, y_1, z_1)$ and $Q = (x_2, y_2, z_2)$. Show that the midpoint of PQ is the point $R = ((x_1 + x_2)/2, (y_1 + y_2)/2, (z_1 + z_2)/2)$. [*Hint:* Show that P, Q, and R are collinear and that $\overline{PR} = \overline{RQ}$.]

36. Find the midpoint of the line joining the points $(2, -1, 4)$ and $(5, 7, -3)$.

37. Find the equation of the sphere which has a diameter with endpoints $(3, 1, -2)$ and $(4, 1, 6)$. [*Hint:* Find the center and radius of the sphere by using the result of Problem 35.]

38. One sphere is said to be *inscribed* in a second sphere if it has a smaller radius and the same center as the second sphere. Find the equation of a sphere of radius 1 inscribed in the sphere given by $x^2 + y^2 + z^2 - 2x - 4y + z - 2 = 0$.

39. Find the volumes of the spheres of Problems 37 and 38.

40. Find the surface areas of the spheres of Problems 37 and 38.

17.2 Vectors in $\mathbb{R}^3$

In Chapter 15 we developed properties of vectors in the plane, $\mathbb{R}^2$. Given the similarity between the coordinate systems in $\mathbb{R}^2$ and $\mathbb{R}^3$, it should come as no surprise to learn that vectors in R^2 and R^3 have very similar structures. In this section we will develop the notion of a vector in space. The development will closely follow the development in Sections 15.1 and 15.2 and, therefore, some of the details will be omitted.

Let P and Q be two distinct points in R^3. Then the *directed line segment* $\overrightarrow{PQ}$ is the straight line segment that extends from P to Q. Two directed line segments are *equivalent* if they have the same magnitude and direction. A *vector* in R^3 is the set of all directed line segments equivalent to a given directed line segment and any directed line segment $\overrightarrow{PQ}$ in that set is called a *representation* of the vector.

So far, our definitions are identical. For convenience, we will choose P to be the origin so that the vector $v = \overrightarrow{OQ}$ can be described by the coordinates (x, y, z) of the point Q. Then the magnitude of $v = |v| = \sqrt{x^2 + y^2 + z^2}$ (from Theorem 17.1.1).

EXAMPLE 1. Let $v = (1, 3, -2)$. Find $|v|$.

SOLUTION. $|v| = \sqrt{1^2 + 3^2 + (-2)^2} = \sqrt{14}$.

Let $u = (x_1, y_1, z_1)$ and $v = (x_2, y_2, z_2)$ be two vectors and let α be a real number (scalar). Then we define

$$u + v = (x_1 + x_2, y_1 + y_2, z_1 + z_2)$$

and

$$\alpha u = (\alpha x_1, \alpha y_1, \alpha z_1).$$

This is the same definition of vector addition and scalar multiplication we had before and is illustrated in Figure 1.

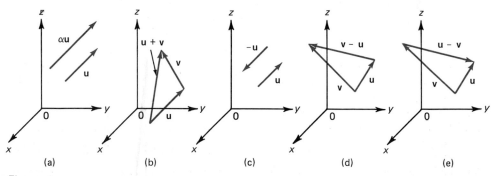

(a)   (b)   (c)   (d)   (e)

Figure 1

EXAMPLE 2. Let $u = (2, 3, -1)$ and $v = (-6, 2, 4)$. Find (a) $u + v$, (b) $3v$, (c) $-u$, and (d) $4u - 3v$.

SOLUTION. (a) $u + v = (2 - 6, 3 + 2, -1 + 4) = (-4, 5, 3)$
(b) $3v = (-18, 6, 12)$
(c) $-u = (-2, -3, 1)$
(d) $4u - 3v = (8, 12, -4) - (-18, 6, 12) = (26, 6, -16)$.

The following theorem extends to three dimensions the results of Theorem 15.1.1. Its proof is easy and is left as an exercise (Problem 54).

THEOREM 1. Let **u**, **v**, and **w** be any three vectors in space, let α and β be scalars, and let **0** denote the zero vector $(0, 0, 0)$. Then

(i) $\mathbf{u} + \mathbf{v} = \mathbf{v} + \mathbf{u}$ (ii) $\mathbf{u} + (\mathbf{v} + \mathbf{w}) = (\mathbf{u} + \mathbf{v}) + \mathbf{w}$
(iii) $\mathbf{v} + \mathbf{0} = \mathbf{v}$ (iv) $0\mathbf{v} = \mathbf{0}$
(v) $\alpha\mathbf{0} = \mathbf{0}$ (vi) $(\alpha\beta)\mathbf{v} = \alpha(\beta\mathbf{v})$
(vii) $\mathbf{v} + (-\mathbf{v}) = \mathbf{0}$ (viii) $1 \cdot \mathbf{v} = \mathbf{v}$
(ix) $(\alpha + \beta)\mathbf{v} = \alpha\mathbf{v} + \beta\mathbf{v}$ (x) $|\alpha\mathbf{v}| = |\alpha|\,|\mathbf{v}|$
(xi) $\alpha(\mathbf{u} + \mathbf{v}) = \alpha\mathbf{u} + \alpha\mathbf{v}$ (xii) $|\mathbf{u} + \mathbf{v}| \le |\mathbf{u}| + |\mathbf{v}|$

A unit vector **u** is a vector with magnitude 1. If **v** is any nonzero vector, then $\mathbf{u} = \mathbf{v}/|\mathbf{v}|$ is a unit vector having the same direction as **v**.

EXAMPLE 3. Find a unit vector having the same direction as $\mathbf{v} = (2, 4, -3)$.

SOLUTION. Since $|\mathbf{v}| = \sqrt{2^2 + 4^2 + (-3)^2} = \sqrt{29}$, we have

$$\mathbf{u} = (2/\sqrt{29}, 4/\sqrt{29}, -3/\sqrt{29}).$$

We can now formally define the direction of a vector in $\mathbb{R}^3$. We cannot define it to be the angle θ the vector makes with the positive x-axis, since, for example, if $0 < \theta < \pi/2$, then there are an *infinite number* of vectors making the angle θ with the positive x-axis, and these together form a cone (see Figure 2).

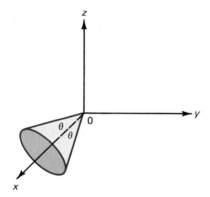

Figure 2

DEFINITION 1. The *direction* of a nonzero vector **v** in $\mathbb{R}^3$ is defined to be the direction† of the unit vector $\mathbf{u} = \mathbf{v}/|\mathbf{v}|$.

> *Remark.* We could have defined the direction of a vector **v** in $\mathbb{R}^2$ in this way. For if $\mathbf{u} = \mathbf{v}/|\mathbf{v}|$, then $\mathbf{u} = (\cos\theta, \sin\theta)$ where θ is the direction of **v**.

It would still be nice to define the direction of a vector in terms of some angles. Let **v** be the vector $\overrightarrow{OP}$ depicted in Figure 3. We define α to be the angle between **v**

† At this point, we have only an intuitive feeling for the "direction" of a unit vector so that this definition is not mathematically precise. However, as we shall shortly see, the direction of a unit vector is determined by the angles the vector makes with the coordinate axes.

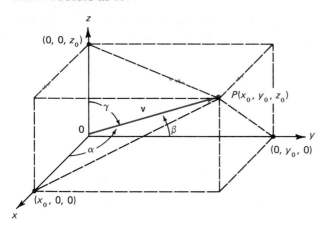

Figure 3

and the positive x-axis, β the angle between $\mathbf{v}$ and the positive y-axis, and γ the angle between $\mathbf{v}$ and the positive z-axis. The angles α, β, and γ are called the *direction angles* of the vector $\mathbf{v}$. Then, from Figure 3,

$$\cos \alpha = \frac{x_0}{|\mathbf{v}|}, \qquad \cos \beta = \frac{y_0}{|\mathbf{v}|}, \qquad \text{and} \qquad \cos \gamma = \frac{z_0}{|\mathbf{v}|}. \tag{1}$$

If $\mathbf{v}$ is a unit vector, then $|\mathbf{v}| = 1$ and

$$\cos \alpha = x_0, \qquad \cos \beta = y_0, \qquad \cos \gamma = z_0. \tag{2}$$

By definition, each of these three angles lie between 0 and π. The cosines of these angles are called the *direction cosines* of the vector $\mathbf{v}$. Note, from (1), that

$$\cos^2 \alpha + \cos^2 \beta + \cos^2 \gamma = \frac{x_0^2 + y_0^2 + z_0^2}{|\mathbf{v}|^2} = \frac{x_0^2 + y_0^2 + z_0^2}{x_0^2 + y_0^2 + z_0^2} = 1. \tag{3}$$

If α, β, and γ are any three numbers between 0 and π such that condition (3) is satisfied, then they uniquely determine a unit vector given by $\mathbf{u} = (\cos \alpha, \cos \beta, \cos \gamma)$.

Remark. If $\mathbf{v} = (a, b, c)$ and $|\mathbf{v}| \neq 0$, then the numbers a, b, and c are called *direction numbers* of the vector $\mathbf{v}$.

EXAMPLE 4.   Find the direction cosines of the vector $v = (4, -1, 6)$.

SOLUTION. The direction of $\mathbf{v}$ is $\mathbf{v}/|\mathbf{v}| = \mathbf{v}/\sqrt{53} = (4/\sqrt{53}, -1/\sqrt{53}, 6/\sqrt{53})$. Then $\cos \alpha = 4/\sqrt{53} \approx 0.5494$, $\cos \beta = -1/\sqrt{53} \approx -0.1374$ and $\cos \gamma = 6/\sqrt{53} \approx 0.8242$. From these, we use Table A.4 to obtain $\alpha \approx 56.7° \approx 0.989$ rad, $\beta \approx 97.9° \approx 1.71$ rad, and $\gamma = 34.5° \approx 0.602$ rad. The vector, along with the angles α, β, and γ, is sketched in Figure 4.

EXAMPLE 5.   Find a vector $\mathbf{v}$ of magnitude 7 whose direction cosines are $1/\sqrt{6}$, $1/\sqrt{3}$, and $1/\sqrt{2}$.

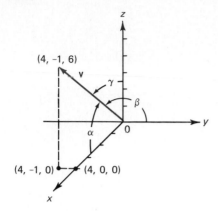

Figure 4

SOLUTION. Let $\mathbf{u} = (1/\sqrt{6}, 1/\sqrt{3}, 1/\sqrt{2})$. Then $\mathbf{u}$ is a unit vector since $|\mathbf{u}| = 1$. Thus the direction of $\mathbf{v}$ is given by $\mathbf{u}$ and so

$$\mathbf{v} = |\mathbf{v}|\mathbf{u} = 7\mathbf{u} = \left(\frac{7}{\sqrt{6}}, \frac{7}{\sqrt{3}}, \frac{7}{\sqrt{2}}\right).$$

Note. We can solve this problem because $(1/\sqrt{6})^2 + (1/\sqrt{3})^2 + (1/\sqrt{2})^2 = 1$.

It is interesting to note that if $\mathbf{v}$ in $\mathbb{R}^2$ is written

$$\mathbf{v} = (\cos\theta)\mathbf{i} + (\sin\theta)\mathbf{j},$$

where θ is the direction of $\mathbf{v}$, then $\cos\theta$ and $\sin\theta$ are the direction cosines of $\mathbf{v}$. Here $\alpha = \theta$ and we define β to be the angle that $\mathbf{v}$ makes with the y-axis (see Figure 5).

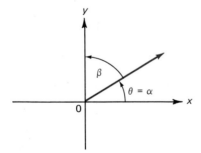

Figure 5

Then $\beta = (\pi/2) - \alpha$ so that

$$\cos\beta = \cos\left(\frac{\pi}{2} - \alpha\right) = \sin\alpha$$

and $\mathbf{v}$ can be written in the "direction cosine" form

$$\mathbf{v} = (\cos\alpha)\mathbf{i} + (\cos\beta)\mathbf{j}.$$

In Chapter 15 we showed how any vector in the plane can be written in terms of the basis vectors $\mathbf{i}$ and $\mathbf{j}$. To extend this idea to $\mathbb{R}^3$ we define

$$\boxed{\mathbf{i} = (1,0,0), \quad \mathbf{j} = (0,1,0), \quad \text{and} \quad \mathbf{k} = (0,0,1).} \tag{4}$$

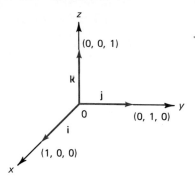

Figure 6

Here **i, j,** and **k** are unit vectors. The vector **i** lies along the x-axis, **j** along the y-axis, and **k** along the z-axis. These are sketched in Figure 6. If $\mathbf{v} = (x, y, z)$ is any vector in $\mathbb{R}^3$, then

$$\mathbf{v} = (x, y, z) = (x, 0, 0) + (0, y, 0) + (0, 0, z) = x\mathbf{i} + y\mathbf{j} + z\mathbf{k}. \tag{5}$$

That is, *any vector* **v** *in* $\mathbb{R}^3$ *can be written in a unique way in terms of the vectors* **i, j,** *and* **k.**

Let $P = (a_1, b_1, c_1)$ and $Q = (a_2, b_2, c_2)$. Then as in Section 15.1, the vector $\mathbf{v} = \overrightarrow{PQ}$ can be written

$$\mathbf{v} = (a_2 - a_1)\mathbf{i} + (b_2 - b_1)\mathbf{j} + (c_2 - c_1)\mathbf{k} \tag{6}$$

(see Problem 55 for details).

EXAMPLE 6. Find a vector in space that can be represented by the directed line segment from $(2, -1, 4)$ to $(5, 1, -3)$.

SOLUTION. $\mathbf{v} = (5 - 2)\mathbf{i} + (1 - (-1))\mathbf{j} + (-3 - 4)\mathbf{k} = 3\mathbf{i} + 2\mathbf{j} - 7\mathbf{k}.$

We now turn to the notion of dot product (or scalar product) in $\mathbb{R}^3$.

DEFINITION 2. If $\mathbf{u} = x_1\mathbf{i} + y_1\mathbf{j} + z_1\mathbf{k}$ and $\mathbf{v} = x_2\mathbf{i} + y_2\mathbf{j} + z_2\mathbf{k}$, then we define

$$\mathbf{u} \cdot \mathbf{v} = x_1 x_2 + y_1 y_2 + z_1 z_2. \tag{7}$$

As before the dot product of two vectors is a *scalar*. Note that $\mathbf{i} \cdot \mathbf{i} = 1$, $\mathbf{j} \cdot \mathbf{j} = 1$, $\mathbf{k} \cdot \mathbf{k} = 1$, $\mathbf{i} \cdot \mathbf{j} = 0$, $\mathbf{j} \cdot \mathbf{k} = 0$, and $\mathbf{i} \cdot \mathbf{k} = 0$.

EXAMPLE 7. If $\mathbf{u} = 2\mathbf{i} - 3\mathbf{j} - 4\mathbf{k}$ and $\mathbf{v} = -3\mathbf{i} + \mathbf{j} - 2\mathbf{k}$, calculate $\mathbf{u} \cdot \mathbf{v}$.

SOLUTION. $\mathbf{u} \cdot \mathbf{v} = 2(-3) + (-3)(1) + (-4)(-2) = -1.$

THEOREM 2. For any vectors **u, v,** and **w** in space, and for any scalar α

(i) $\mathbf{u} \cdot \mathbf{v} = \mathbf{v} \cdot \mathbf{u}.$
(ii) $(\mathbf{u} + \mathbf{v}) \cdot \mathbf{w} = \mathbf{u} \cdot \mathbf{w} + \mathbf{v} \cdot \mathbf{w}.$
(iii) $(\alpha \mathbf{u}) \cdot \mathbf{v} = \alpha(\mathbf{u} \cdot \mathbf{v}).$
(iv) $|\mathbf{u}| = \sqrt{\mathbf{u} \cdot \mathbf{u}}.$

PROOF. The proof is almost identical to the proof of Theorem 15.2.1 and is left as an exercise (see Problem 56).

THEOREM 3. If φ denotes the angle between two nonzero vectors **u** and **v**, we have

$$\cos \varphi = \frac{\mathbf{u} \cdot \mathbf{v}}{|\mathbf{u}| \, |\mathbf{v}|}. \tag{8}$$

PROOF. The proof is almost identical to the proof of Theorem 15.2.2 and is left as an exercise (see Problem 57).

EXAMPLE 8. Calculate the cosine of the angle between $\mathbf{u} = 3\mathbf{i} - \mathbf{j} + 2\mathbf{k}$ and $\mathbf{v} = 4\mathbf{i} + 3\mathbf{j} - \mathbf{k}$.

SOLUTION. $\mathbf{u} \cdot \mathbf{v} = 7$, $|\mathbf{u}| = \sqrt{14}$, and $|\mathbf{v}| = \sqrt{26}$ so that $\cos \varphi = 7/\sqrt{(14)(26)} = 7/\sqrt{364} \approx 0.3669$ and $\varphi \approx 68.5° \approx 1.2$ rad.

DEFINITION 3. Two nonzero vectors **u** and **v** are
  (i)   *parallel* if the angle between them is 0 or π, and
  (ii) *orthogonal* (or *perpendicular*) if the angle between them is $\pi/2$.

THEOREM 4. (i)   If $\mathbf{u} \neq 0$, then **u** and **v** are parallel if and only if $\mathbf{v} = \alpha \mathbf{u}$ for some constant α.
  (ii)   If **u** and **v** are nonzero, then **u** and **v** are orthogonal if and only if $\mathbf{u} \cdot \mathbf{v} = 0$.

PROOF. Again the proof is easy and is left as an exercise (see Problem 58).

EXAMPLE 9. Show that the vectors $\mathbf{u} = \mathbf{i} + 3\mathbf{j} - 4\mathbf{k}$ and $\mathbf{v} = -2\mathbf{i} - 6\mathbf{j} + 8\mathbf{k}$ are parallel.

SOLUTION. Here $(\mathbf{u} \cdot \mathbf{v})/(|\mathbf{u}| \, |\mathbf{v}|) = -52/(\sqrt{26} \sqrt{104}) = -52/(\sqrt{26} \cdot 2 \cdot \sqrt{26}) = -1$ so that $\cos \theta = -1$, $\theta = \pi$, and **u** and **v** are parallel (but have opposite directions). Another way to see this is to note that $\mathbf{v} = -2\mathbf{u}$ so that, by Theorem 4, **u** and **v** are parallel. This is sketched in Figure 7.

EXAMPLE 10. Find a number α such that $\mathbf{u} = 8\mathbf{i} - 2\mathbf{j} + 4\mathbf{k}$ and $\mathbf{v} = 2\mathbf{i} + 3\mathbf{j} + \alpha\mathbf{k}$ are orthogonal.

SOLUTION. We must have $0 = \mathbf{u} \cdot \mathbf{v} = 10 + 4\alpha$ so that $\alpha = -5/2$. The vectors **u** and **v** are sketched in Figure 8.

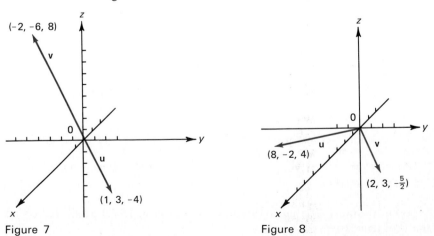

Figure 7   Figure 8

We now turn to the definition of the projection of one vector on another. First, we state the theorem which is the analog of Theorem 15.2.5 (and which has an identical proof).

THEOREM 5. Let **v** be a nonzero vector. Then for any other vector **u,**

$$\mathbf{w} = \mathbf{u} - \frac{\mathbf{u \cdot v}}{|\mathbf{v}|^2}\mathbf{v}$$

is orthogonal to **v.**

DEFINITION 4. Let **u** and **v** be nonzero vectors. Then the *projection of* **u** *on* **v,** denoted Proj$_\mathbf{v}$ **u,** is defined by

$$\text{Proj}_\mathbf{v}\,\mathbf{u} = \frac{\mathbf{u \cdot v}}{|\mathbf{v}|^2}\,\mathbf{v}. \tag{9}$$

The *component* of **u** in the direction **v** is given by $(\mathbf{u \cdot v})/|\mathbf{v}|$.

EXAMPLE 11. Let **u** $= 2\mathbf{i} + 3\mathbf{j} + \mathbf{k}$ and **v** $= \mathbf{i} + 2\mathbf{j} - 6\mathbf{k}$. Find Proj$_\mathbf{v}$ **u.**

SOLUTION. Here $(\mathbf{u \cdot v})/|\mathbf{v}|^2 = 2/41$ so that

$$\text{Proj}_\mathbf{v}\,\mathbf{u} = \frac{2}{41}\mathbf{i} + \frac{4}{41}\mathbf{j} - \frac{12}{41}\mathbf{k}.$$

The component of **u** in the direction **v** is $(\mathbf{u \cdot v})/|\mathbf{v}| = 2/\sqrt{41}$.

Note that, as in the planar case, Proj$_\mathbf{v}$ **u** is a vector which has the same direction as **v** if $\mathbf{u \cdot v} > 0$ and the direction opposite to that of **v** if $\mathbf{u \cdot v} < 0$.

PROBLEMS 17.2

In Problems 1–20 find the magnitude and the direction cosines of the given vector.

1. $\mathbf{v} = 3\mathbf{j}$	**2.** $\mathbf{v} = -3\mathbf{i}$	**3.** $\mathbf{v} = 14\mathbf{k}$
4. $\mathbf{v} = -8\mathbf{j}$	**5.** $\mathbf{v} = 4\mathbf{i} - \mathbf{j}$	**6.** $\mathbf{v} = \mathbf{i} + 2\mathbf{k}$
7. $\mathbf{v} = -2\mathbf{i} + 3\mathbf{j}$	**8.** $\mathbf{v} = \mathbf{i} + \mathbf{j} + \mathbf{k}$	**9.** $\mathbf{v} = \mathbf{i} - \mathbf{j} + \mathbf{k}$
10. $\mathbf{v} = \mathbf{i} + \mathbf{j} - \mathbf{k}$	**11.** $\mathbf{v} = -\mathbf{i} + \mathbf{j} + \mathbf{k}$	**12.** $\mathbf{v} = \mathbf{i} - \mathbf{j} - \mathbf{k}$
13. $\mathbf{v} = -\mathbf{i} + \mathbf{j} - \mathbf{k}$	**14.** $\mathbf{v} = -\mathbf{i} - \mathbf{j} + \mathbf{k}$	**15.** $\mathbf{v} = -\mathbf{i} - \mathbf{j} - \mathbf{k}$
16. $\mathbf{v} = 2\mathbf{i} + 5\mathbf{j} - 7\mathbf{k}$	**17.** $\mathbf{v} = -7\mathbf{i} + 2\mathbf{j} - 13\mathbf{k}$	**18.** $\mathbf{v} = \mathbf{i} + 7\mathbf{j} - 7\mathbf{k}$
19. $\mathbf{v} = -3\mathbf{i} - 3\mathbf{j} + 8\mathbf{k}$	**20.** $\mathbf{v} = -2\mathbf{i} - 3\mathbf{j} - 4\mathbf{k}$.	

21. The three direction angles of a certain unit vector are the same and are between 0 and $\pi/2$. What is the vector?

22. Find a vector of magnitude 12 which has the same direction as the vector of Problem 21.

23. Show that there is no unit vector whose direction angles are $\pi/6$, $\pi/3$, and $\pi/4$.

24. Let $P = (2, 1, 4)$ and $Q = (3, -2, 8)$. Find a unit vector in the direction of $\overrightarrow{PQ}$.

25. Let $P = (-3, 1, 7)$ and $Q = (8, 1, 7)$. Find a unit vector whose direction is opposite that of $\overrightarrow{PQ}$.

26. In Problem 25 find all points R such that $\overrightarrow{PR} \perp \overrightarrow{PQ}$.

*27. Show that the set of points which satisfy the condition of Problem 26 and the condition $|\overrightarrow{PR}| = 1$ form a circle.

In Problems 28–46 let $u = 2i - 3j + 4k$, $v = -2i - 3j + 5k$, $w = i - 7j + 3k$, and $t = 3i + 4j + 5k$.

28. Calculate $u + v$
29. Calculate $2u - 3v$
30. Calculate $-18u$
31. Calculate $w - u - v$
32. Calculate $t + 3w - v$
33. Calculate $2u - 7w + 5v$
34. Calculate $2v + 7t - w$
35. Calculate $u \cdot v$
36. Calculate $|w|$
37. Calculate $u \cdot w - w \cdot t$
38. Calculate the angle between u and w
39. Calculate the angle between t and w
40. Calculate the angle between v and t
41. Calculate $\text{Proj}_v u$
42. Calculate $\text{Proj}_u v$
43. Calculate $\text{Proj}_t w$
44. Calculate $\text{Proj}_w t$
45. Calculate $\text{Proj}_w u$
46. Calculate $\text{Proj}_t v$.

47. Find the distance between the point $P = (2, 1, 3)$ and the line passing through the points $Q = (-1, 1, 2)$ and $R = (6, 0, 1)$. [*Hint:* See Problem 15.2.50.]
48. Find the distance from the point $P = (1, 0, 1)$ to the line passing through the points $Q = (2, 3, -1)$ and $R = (6, 1, -3)$.
49. Show that the points $P = (3, 5, 6)$, $Q = (1, 2, 7)$ and $R = (6, 1, 0)$ are the vertices of a right triangle.
50. Show that the points $P = (3, 2, -1)$, $Q = (4, 1, 6)$, $R = (7, -2, 3)$ and $S = (8, -3, 10)$ are the vertices of a parallelogram.
*51. A solid figure in space with exactly four vertices is called a *tetrahedron* (see Figure 9).

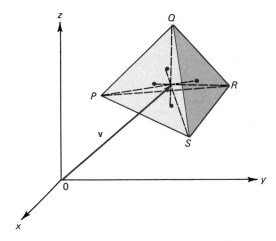

Figure 9

Let P represent the vector $\overrightarrow{OP}$, Q the vector $\overrightarrow{OQ}$, and so on. A line is drawn from each vertex to the centroid of the opposite side. Show that these four lines meet at the endpoint of the vector

$$v = \frac{P + Q + R + S}{4}.$$

52. A force of 3 nt acts in the direction of the vector with direction cosines $(1/\sqrt{6}, 1/\sqrt{3}, 1/\sqrt{2})$. Find the work done in moving the object from the point $(1, 2, 3)$ to the point $(2, 8, 11)$ where distance is measured in meters. [*Hint:* See Example 15.3.5.]
53. Find the work done when a force of 3 nt acting in the direction of the vector

$\mathbf{v} = \mathbf{i} + \mathbf{j} - \mathbf{k}$ moves an object from $(-1, 3, 4)$ to $(3, 7, -2)$. Again distance is measured in meters.

54. Prove Theorem 1.
55. Prove that formula (6) is correct. [*Hint:* Follow the steps leading to formula (15.1.7).]
56. Prove Theorem 2.
57. Prove Theorem 3.
58. Prove Theorem 4.
*59. Let PQR be a triangle in $\mathbb{R}^3$. Show that if a force of N nt moves an object around the triangle, then the work done by that force is zero.
*60. Find the angle between the diagonal of a cube and the diagonal of one of its faces.

17.3 Lines in $\mathbb{R}^3$

In Section 1.4 we derived the equation of a line in the plane. In that section we showed that the equation of the line could be determined if we knew either (i) two points on the line, or (ii) one point on the line and the direction (slope) of the line. Moreover, in Section 16.1 we showed how the equation of a line could be written parametrically. In $\mathbb{R}^3$, our intuition tells us that the basic ideas are the same. Since two points determine a line, we should be able to calculate the equation of a line in space if we know two points on it. Alternatively, if we know one point and the direction of a line, we should also be able to find its equation.

We begin with two points $P = (x_1, y_1, z_1)$ and $Q = (x_2, y_2, z_2)$ on a line L. A vector parallel to L is a vector with representation $\overrightarrow{PQ}$ or (from formula 17.2.6)

$$\mathbf{v} = (x_2 - x_1)\mathbf{i} + (y_2 - y_1)\mathbf{j} + (z_2 - z_1)\mathbf{k} \tag{1}$$

is a vector parallel to L. Now let $R = (x, y, z)$ be another point on the line. Then $\overrightarrow{PR}$ is parallel to $\overrightarrow{PQ}$ which is parallel to $\mathbf{v}$ so that, by Theorem 17.2.4,

$$\overrightarrow{PR} = t\mathbf{v} \tag{2}$$

for some real number t. Now look at Figure 1. From the figure, we have (in each of the three possible cases)

$$\overrightarrow{OR} = \overrightarrow{OP} + \overrightarrow{PR} \tag{3}$$

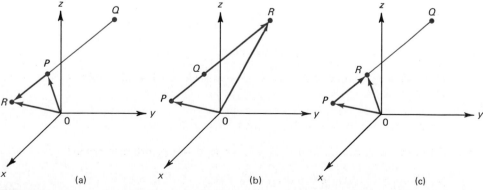

(a) (b) (c)

Figure 1

and, combining (2) and (3)

$$\overrightarrow{PR} = \overrightarrow{OR} - \overrightarrow{OP} = t\mathbf{v}$$

or,

$$\overrightarrow{OR} = \overrightarrow{OP} + t\mathbf{v}. \tag{4}$$

Equation (4) is called the *vector equation* of the line L. For if R is on L, then (4) is satisfied for some real number t. Conversely, if (4) is satisfied, then, reversing our steps, we see that $\overrightarrow{PR}$ is parallel to $\mathbf{v}$ which means that R is on $\mathbf{v}$.

If we write out the components of equation (4), we obtain

$$x\mathbf{i} + y\mathbf{j} + z\mathbf{k} = x_1\mathbf{i} + x_2\mathbf{j} + x_3\mathbf{k} + t(x_2 - x_1)\mathbf{i} + t(y_2 - y_1)\mathbf{j} + t(z_2 - z_1)\mathbf{k}$$

or

$$\boxed{\begin{aligned} x &= x_1 + t(x_2 - x_1), \\ y &= y_1 + t(y_2 - y_1), \\ z &= z_1 + t(z_2 - z_1). \end{aligned}} \tag{5}$$

The equations (5) are called the *parametric equations* of a line.

Finally, solving for t in (5), and defining $x_2 - x_1 = a$, $y_2 - y_1 = b$, and $z_2 - z_1 = c$, we find that

$$\boxed{\frac{x - x_1}{a} = \frac{y - y_1}{b} = \frac{z - z_1}{c}.} \tag{6}$$

The equations (6) are called the *symmetric equations* of the line.

Here a, b, and c are direction numbers of the vector $\mathbf{v}$. Of course, equations (6) are valid only if a, b, and c are nonzero.

EXAMPLE 1. Find a vector equation, parametric equations, and symmetric equations of a line L passing through the points $P = (2, -1, 6)$ and $Q = (3, 1, -2)$.

SOLUTION. First we calculate

$$\mathbf{v} = (3 - 2)\mathbf{i} + (1 - (-1))\mathbf{j} + (-2 - 6)\mathbf{k} = \mathbf{i} + 2\mathbf{j} - 8\mathbf{k}.$$

Then, from (4), if $R = (x, y, z)$ is on the line,

$$\overrightarrow{OR} = x\mathbf{i} + y\mathbf{j} + z\mathbf{k} = \overrightarrow{OP} + t\mathbf{v} = 2\mathbf{i} - \mathbf{j} + 6\mathbf{k} + t(\mathbf{i} + 2\mathbf{j} - 8\mathbf{k})$$

or

$$x = 2 + t, \qquad y = -1 + 2t, \qquad \text{and} \qquad z = 6 - 8t.$$

Finally, since $a = 1$, $b = 2$, and $c = -8$, we find the symmetric equations

$$\frac{x - 2}{1} = \frac{y + 1}{2} = \frac{z - 6}{-8}. \tag{7}$$

To check this, we verify that $(2, -1, 6)$ and $(3, 1, -2)$ are indeed on the line. We have

$$\frac{2-2}{1} = \frac{-1+1}{2} = \frac{6-6}{-8} = 0$$

and

$$\frac{3-2}{1} = \frac{1+1}{2} = \frac{-2-6}{-8} = 1.$$

Other points on the line can be found. For example, if $t = 3$, we obtain

$$3 = \frac{x-2}{1} = \frac{y+1}{2} = \frac{z-6}{-8}$$

which yields the point $(5, 5, -18)$.

EXAMPLE 2.   Find the symmetric equation of the line passing through the point $(1, -2, 4)$ and parallel to the vector $v = i + j - k$.

SOLUTION. We simply choose $\overrightarrow{OP} = i - 2j + 4k$ and v as above. Then $a = 1, b = 1$, $c = -1$ and we obtain

$$\frac{x-1}{1} = \frac{y+2}{1} = \frac{z-4}{-1}.$$

What happens if one of the direction numbers a, b, or c is zero?

EXAMPLE 3.   Find the symmetric equation of the lines containing the points $P = (3, 4, -1)$ and $Q = (-2, 4, 6)$.

SOLUTION. Here $v = -5i + 7k$ and $a = -5, b = 0, c = 7$. Then a parametric representation of the line is

$$x = 3 - 5t, \qquad y = 4, \qquad \text{and} \qquad z = -1 + 7t.$$

Solving for t, we find that

$$\frac{x-3}{-5} = \frac{z+1}{7} \qquad \text{and} \qquad y = 4.$$

The equation $y = 4$ is the equation of a plane parallel to the xz-plane so that we have obtained an equation of a line in that plane.

EXAMPLE 4.   Find symmetric equations of the line in the xy-plane which passes through the points $(x_1, y_1, 0)$ and $(x_2, y_2, 0)$.

SOLUTION. Here $v = (x_2 - x_1)i + (y_2 - y_1)j$ and we obtain

$$\frac{x - x_1}{x_2 - x_1} = \frac{y - y_1}{y_2 - y_1} \qquad \text{and} \qquad z = 0.$$

We can rewrite this as

$$y - y_1 = \left(\frac{y_2 - y_1}{x_2 - x_1}\right)(x - x_1).$$

Here $(y_2 - y_1)/(x_2 - x_1) = m$, the slope of the line and when $x = 0$, $y = y_1 - [(y_2 - y_1)/(x_2 - x_1)]x_1 = b$, the y-intercept of the line. That is, $y = mx + b$, which is the slope–intercept form of a line in the xy-plane. Thus we see that the symmetric equations of a line in space are really a generalization of the equation of a line in the plane.

Now, what happens if two of the direction numbers are 0?

EXAMPLE 5. Find the symmetric equations of the line passing through the points $P = (2, 3, -2)$ and $Q = (2, -1, -2)$.

SOLUTION. Here $\mathbf{v} = -4\mathbf{j}$ so that $a = 0$, $b = -4$, and $c = 0$. A parametric representation of the line is, by equation (4), given by

$$x = 2, \qquad y = 3 - 4t, \qquad z = -2.$$

Now $x = 2$ is the equation of a plane parallel to the yz-plane while $z = -2$ is the equation of a plane parallel to the xy-plane. Their intersection is the line $x = 2$, $z = -2$, which is parallel to the y-axis. In fact, the equation $y = 3 - 4t$ says, essentially, that y can take on any value (while x and z remain fixed).

The results of Examples 1, 3, and 5 are summarized in Theorem 1.

THEOREM 1. Let L be a line passing through the point (x_1, y_1, z_1) and parallel to the line $\mathbf{v} = a\mathbf{i} + b\mathbf{j} + c\mathbf{k}$. Then symmetric equations of the line are

(i) $(x - x_1)/a = (y - y_1)/b = (z - z_1)/c$,

if a, b, and c are all nonzero.

(ii) $x = x_1, (y - y_1)/b = (z - z_1)/c$,

if $a = 0$. Then the line is parallel to the yz-plane. If either b or $c = 0$, but $a \neq 0$, similar results hold.

(iii) $x = x_1, y = y_1, z_1 + ct$

if a and b are 0. Then the line is parallel to the z-axis. If a and c or b and c are 0, similar results hold.

> *Warning.* The parametric or symmetric equations of a line are *not* unique. To see this, simply choose two other points on the line.

EXAMPLE 6. In Example 1 the line contains the point $(5, 5, -18)$. Choose $P = (5, 5, -18)$ and $Q = (3, 1, -2)$. We find that $\mathbf{v} = -2\mathbf{i} - 4\mathbf{j} + 16\mathbf{k}$ so that

$$x = 5 - 2t, \qquad y = 5 - 4t, \qquad \text{and} \qquad z = -18 + 16t.$$

(Note that if $t = \frac{3}{2}$, we obtain $(x, y, z) = (2, -1, 6)$.) The symmetric equations are now

$$\frac{x - 5}{-2} = \frac{y - 5}{-4} = \frac{z + 18}{16}.$$

In the plane, two lines which are not parallel have exactly one point of intersection. In $\mathbb{R}^3$, this is not the case. For example, the line L_1: $x = 2$, $y = 3$ (parallel to the z-axis) and L_2: $x = 1$, $z = 3$ (parallel to the y-axis) are not parallel and have no points in common. It takes a bit of work to determine whether two lines in $\mathbb{R}^3$ do have a point in common (they usually do not).

EXAMPLE 7. Determine whether the lines

$$L_1: \quad x = 1 + t, \quad y = -3 + 2t, \quad z = -2 - t$$

and

$$L_2: \quad x = 17 + 3s, \quad y = 4 + s, \quad z = -8 - s$$

have a point of intersection.

SOLUTION. To avoid confusion we have written L_2 with s instead of t as the parameter. If (x, y, z) is a point common to both lines, then we must have

$$1 + t = 17 + 3s, \qquad -3 + 2t = 4 + s, \qquad -2 - t = -8 - s.$$

With a little bit of algebra, we find that the numbers $t = 1$ and $s = -5$ satisfy all three of these equations. Thus, when $t = 1$ (or $s = -5$) we obtain the unique point of intersection $(2, -1, -3)$.

EXAMPLE 8. Determine whether the lines

$$x = 2 - t, \quad y = 1 + t, \quad z = -2 - t$$

and

$$x = 1 + s, \quad y = -2s, \quad z = 3 + 2s$$

have a point in common.

SOLUTION. We must have

$$2 - t = 1 + s, \qquad 1 + t = -2s, \qquad -2 - t = 3 + 2s.$$

The first of these equations implies that $t = 1 - s$. Substituting this in the second equation we find that

$$1 + (1 - s) = -2s$$

or $s = -2$ and then $t = 1 - s = 3$. But if $s = -2$ and $t = 3$, the third equation implies that

$$-2 - 3 = 3 - 4 \qquad \text{or} \qquad -5 = -1$$

which is, of course, false. Thus there are *no* values of s and t which satisfy the three equations and, therefore, the two lines have no points of intersection.

In general, in order to have a point of intersection, we must find a solution to three equations in the two variables s and t. Since each equation can be thought of as a line in the st-plane, there will be a solution only if the three lines intersect at the *same point*. Since three lines in a plane rarely intersect at the same point, we can justify our earlier statement that two lines in $\mathbb{R}^3$ rarely have a point of intersection.

PROBLEMS 17.3

In Problems 1–22 find a vector equation, parametric equations, and symmetric equations of the indicated line.

1. Containing $(2, 1, 3)$ and $(1, 2, -1)$
2. Containing $(1, -1, 1)$ and $(-1, 1, -1)$
3. Containing $(1, 3, 2)$ and $(2, 4, -2)$
4. Containing $(-2, 4, 5)$ and $(3, 7, 2)$
5. Containing $(-4, 1, 3)$ and $(-4, 0, 1)$
6. Containing $(2, 3, -4)$ and $(2, 0, -4)$
7. Containing $(1, 2, 3)$ and $(3, 2, 1)$
8. Containing $(7, 1, 3)$ and $(-1, -2, 3)$
9. Containing $(1, 2, 4)$ and $(1, 2, 7)$
10. Containing $(-3, -1, -6)$ and $(-3, 1, 6)$
11. Containing $(2, 2, 1)$ and parallel to $2\mathbf{i} - \mathbf{j} - \mathbf{k}$
12. Containing $(-1, -6, 2)$ and parallel to $4\mathbf{i} + \mathbf{j} - 3\mathbf{k}$
13. Containing $(1, 0, 3)$ and parallel to $\mathbf{i} - \mathbf{j}$
14. Containing $(2, 1, -4)$ and parallel to $\mathbf{i} + 4\mathbf{k}$
15. Containing $(-1, -2, 5)$ and parallel to $-3\mathbf{j} + 7\mathbf{k}$
16. Containing $(-2, 3, -2)$ and parallel to $4\mathbf{k}$
17. Containing $(-1, -3, 1)$ and parallel to $-7\mathbf{j}$
18. Containing $(2, 1, 5)$ and parallel to $3\mathbf{i}$
19. Containing (a, b, c) and parallel to $d\mathbf{i} + e\mathbf{j}$
20. Containing (a, b, c) and parallel to $d\mathbf{k}$
21. Containing $(4, 1, -6)$ and parallel to $(x - 2)/3 = (y + 1)/6 = (z - 5)/2$
22. Containing $(3, 1, -2)$ and parallel to $(x + 1)/3 = (y + 3)/2 = (z - 2)/(-4)$.

23. Let L_1 be given by

$$\frac{x - x_1}{a_1} = \frac{y - y_1}{b_1} = \frac{z - z_1}{c_1}$$

and L_2 be given by

$$\frac{x - x_1}{a_2} = \frac{y - y_1}{b_2} = \frac{z - z_1}{c_2}.$$

Show that L_1 is orthogonal to L_2 if and only if $a_1 a_2 + b_1 b_2 + c_1 c_2 = 0$.

24. Show that the lines

$$L_1: \quad \frac{x - 3}{2} = \frac{y + 1}{4} = \frac{z - 2}{-1} \quad \text{and} \quad L_2: \quad \frac{x - 3}{5} = \frac{y + 1}{-2} = \frac{z - 3}{2}$$

are orthogonal.

25. Show that the lines

$$L_1: \quad \frac{x - 1}{1} = \frac{y + 3}{2} = \frac{z + 3}{3} \quad \text{and} \quad L_2: \quad \frac{x - 3}{3} = \frac{y - 1}{6} = \frac{z - 8}{9}$$

are equations of the same straight line.

In Problems 26–31 determine whether the given pair of lines has a point of intersection. If so, find it.

26. $L_1: x = 2 + t,\ y = -1 + 2t,\ z = 3 + 4t;\ L_2: x = 9 + s,\ y = -2 - s,\ z = 1 - 2s$
27. $L_1: x = 3 + 2t,\ y = 2 - t,\ z = 1 + t;\ L_2: x = 4 - s,\ y = -2 + 3s,\ z = 2 + 2s$

28. L_1: $x = 4 - 3t$, $y = 1 + 7t$, $z = -2 - 8t$; L_2: $x = 5 + s$, $y = 3 - s$, $z = 1 + 2s$
29. L_1: $x = 2 - 5t$, $y = 1 + t$, $z = 3 + 4t$; L_2: $x = -3 + 4s$, $y = 2 - s$, $z = 7 + 6s$
30. L_1: $x = 4 - t$, $y = 7 + 5t$, $z = 2 - 3t$; L_2: $x = 1 + 2s$, $y = 6 - 2s$, $z = 10 + 3s$
31. L_1: $x = 1 + t$, $y = 2 - t$, $z = 3t$; L_2: $x = 3s$, $y = 2 - s$, $z = 2 + s$.
32. Let L be given in its vector form $\overrightarrow{OR} = \overrightarrow{OP} + t\mathbf{v}$. Find a number t such that $\overrightarrow{OR}$ is perpendicular to $\mathbf{v}$.
33. Use the result of Problem 32 to find the distance between the line L (containing P and parallel to $\mathbf{v}$) and the origin when
     (a) $P = (2, 1, -4)$; $\mathbf{v} = \mathbf{i} + \mathbf{j} + \mathbf{k}$,
     (b) $P = (1, 2, -3)$; $\mathbf{v} = 3\mathbf{i} - \mathbf{j} - \mathbf{k}$,
     (c) $P = (-1, -4, 2)$; $\mathbf{v} = -\mathbf{i} + \mathbf{j} + 2\mathbf{k}$.

17.4 The Cross Product of Two Vectors

To this point the only product of vectors which we have considered has been the dot or scalar product. We now define a new product, called the *cross product†* (or *vector product*) which is defined only in $\mathbb{R}^3$.

DEFINITION 1. Let $\mathbf{u} = a_1\mathbf{i} + b_1\mathbf{j} + c_1\mathbf{k}$ and $\mathbf{v} = a_2\mathbf{i} + b_2\mathbf{j} + c_2\mathbf{k}$. Then the *cross product (vector product)* of $\mathbf{u}$ and $\mathbf{v}$, denoted $\mathbf{u} \times \mathbf{v}$, is a new vector defined by

$$\mathbf{u} \times \mathbf{v} = (b_1c_2 - c_1b_2)\mathbf{i} + (c_1a_2 - a_1c_2)\mathbf{j} + (a_1b_2 - b_1a_2)\mathbf{k}. \qquad (1)$$

Note that the result of the cross product is a vector while the result of the dot product is a scalar.

Here the cross product seems to have been defined somewhat arbitrarily. There are obviously many ways to define a vector product. Why was this definition chosen? We will answer that question in this section by demonstrating some of the properties of the cross product and illustrating some of its uses.

EXAMPLE 1. Let $\mathbf{u} = \mathbf{i} - \mathbf{j} + 2\mathbf{k}$ and $\mathbf{v} = 2\mathbf{i} + 3\mathbf{j} - 4\mathbf{k}$. Calculate $\mathbf{w} = \mathbf{u} \times \mathbf{v}$.

SOLUTION. Using formula (1),

$$\mathbf{w} = [(-1)(-4) - (2)(3)]\mathbf{i} + [(2)(2) - (1)(-4)]\mathbf{j} + [(1)(3) - (-1)(2)]\mathbf{k}$$
$$= -2\mathbf{i} + 8\mathbf{j} + 5\mathbf{k}.$$

Note. In this example, $\mathbf{u} \cdot \mathbf{w} = \mathbf{v} \cdot \mathbf{w} = 0$. That is, $\mathbf{u} \times \mathbf{v}$ is orthogonal to both $\mathbf{u}$ and $\mathbf{v}$. As we shall shortly see, the cross product of $\mathbf{u}$ and $\mathbf{v}$ is always orthogonal to $\mathbf{u}$ and $\mathbf{v}$.

Before continuing our discussion of the uses of the cross product, we remark that there is an easy way to calculate $\mathbf{u} \times \mathbf{v}$ if you are familiar with the elementary properties of 3×3 determinants. If you are not, we suggest that you turn to Appendix 2 where these properties are discussed.

† The cross product was defined by Hamilton in one of a series of papers discussing his quaternions which were published in *Philosophical Magazine* between the years 1844 and 1850.

THEOREM 1

$$\mathbf{u} \times \mathbf{v} = \begin{vmatrix} \mathbf{i} & \mathbf{j} & \mathbf{k} \\ a_1 & b_1 & c_1 \\ a_2 & b_2 & c_2 \end{vmatrix}.$$

PROOF

$$\begin{vmatrix} \mathbf{i} & \mathbf{j} & \mathbf{k} \\ a_1 & b_1 & c_1 \\ a_2 & b_2 & c_2 \end{vmatrix} = \mathbf{i}\begin{vmatrix} b_1 & c_1 \\ b_2 & c_2 \end{vmatrix} - \mathbf{j}\begin{vmatrix} a_1 & c_1 \\ a_2 & c_2 \end{vmatrix} + \mathbf{k}\begin{vmatrix} a_1 & b_1 \\ a_2 & b_2 \end{vmatrix}$$

$$= (b_1 c_2 - c_1 b_2)\mathbf{i} + (a_2 c_1 - a_1 c_2)\mathbf{j} + (a_1 b_2 - b_1 a_2)\mathbf{k}$$

which is equal to $\mathbf{u} \times \mathbf{v}$ according to Definition 1.

Remark. The above determinant makes sense only if it is expanded by the first row.

EXAMPLE 2.   Calculate $\mathbf{u} \times \mathbf{v}$ where $\mathbf{u} = 2\mathbf{i} + 4\mathbf{j} - 5\mathbf{k}$ and $\mathbf{v} = -3\mathbf{i} - 2\mathbf{j} + \mathbf{k}$.

SOLUTION

$$\mathbf{u} \times \mathbf{v} = \begin{vmatrix} \mathbf{i} & \mathbf{j} & \mathbf{k} \\ 2 & 4 & -5 \\ -3 & -2 & 1 \end{vmatrix} = (4 - 10)\mathbf{i} - (2 - 15)\mathbf{j} + (-4 + 12)\mathbf{k}$$

$$= -6\mathbf{i} + 13\mathbf{j} + 8\mathbf{k}.$$

The following theorem summarizes some properties of the cross product.

THEOREM 2.   Let $\mathbf{u}$, $\mathbf{v}$, and $\mathbf{w}$ be vectors in $\mathbb{R}^3$, and let α be a scalar. Then

   (i) $\mathbf{u} \times \mathbf{0} = \mathbf{0} \times \mathbf{u} = \mathbf{0}.$
   (ii) $\mathbf{u} \times \mathbf{v} = -(\mathbf{v} \times \mathbf{u}).$
   (iii) $(\alpha\mathbf{u} \times \mathbf{v}) = \alpha(\mathbf{u} \times \mathbf{v}).$
   (iv) $\mathbf{u} \times (\mathbf{v} + \mathbf{w}) = (\mathbf{u} \times \mathbf{v}) + (\mathbf{u} \times \mathbf{w}).$
   (v) $(\mathbf{u} \times \mathbf{v}) \cdot \mathbf{w} = \mathbf{u} \cdot (\mathbf{v} \times \mathbf{w}).$ (This is called the *scalar triple product* of $\mathbf{u}$, $\mathbf{v}$
and $\mathbf{w}$.)
   (vi) $\mathbf{u} \cdot (\mathbf{u} \times \mathbf{v}) = \mathbf{v} \cdot (\mathbf{u} \times \mathbf{v}) = 0.$ (That is, $\mathbf{u} \times \mathbf{v}$ is orthogonal to both $\mathbf{u}$ and $\mathbf{v}$.)
   (vii)   If $\mathbf{u}$ and $\mathbf{v}$ are parallel, then $\mathbf{u} \times \mathbf{v} = \mathbf{0}.$

PROOF. (i)   Let $\mathbf{u} = a_1\mathbf{i} + b_1\mathbf{j} + c_1\mathbf{k}$. Then

$$\mathbf{u} \times \mathbf{0} = \begin{vmatrix} \mathbf{i} & \mathbf{j} & \mathbf{k} \\ a_1 & b_1 & c_1 \\ 0 & 0 & 0 \end{vmatrix} = 0\mathbf{i} + 0\mathbf{j} + 0\mathbf{k} = \mathbf{0}.$$

Similarly $\mathbf{0} \times \mathbf{u} = \mathbf{0}$.

   (ii)   Let $\mathbf{v} = a_2\mathbf{i} + b_2\mathbf{j} + c_2\mathbf{k}$. Then

$$\mathbf{u} \times \mathbf{v} = \begin{vmatrix} \mathbf{i} & \mathbf{j} & \mathbf{k} \\ a_1 & b_1 & c_1 \\ a_2 & b_2 & c_2 \end{vmatrix} = - \begin{vmatrix} \mathbf{i} & \mathbf{j} & \mathbf{k} \\ a_2 & b_2 & c_2 \\ a_1 & b_1 & c_1 \end{vmatrix} = -(\mathbf{v} \times \mathbf{u}),$$

since interchanging the rows of a determinant has the effect of multiplying that determinant by -1 (see Theorem 5(iii) in Appendix 2).

$$\text{(iii)} \quad \alpha\mathbf{u} \times \mathbf{v} = \begin{vmatrix} \mathbf{i} & \mathbf{j} & \mathbf{k} \\ \alpha a_1 & \alpha b_1 & \alpha c_1 \\ a_2 & b_2 & c_2 \end{vmatrix} = \alpha \begin{vmatrix} \mathbf{i} & \mathbf{j} & \mathbf{k} \\ a_1 & b_1 & c_1 \\ a_2 & b_2 & c_2 \end{vmatrix} = \alpha(\mathbf{u} \times \mathbf{v}).$$

The second equality follows from Theorem 5(iv) in Appendix 2.

   (iv)   Let $\mathbf{w} = a_3\mathbf{i} + b_3\mathbf{j} + c_3\mathbf{k}$. Then

$$\mathbf{u} \times (\mathbf{v} + \mathbf{w}) = \begin{vmatrix} \mathbf{i} & \mathbf{j} & \mathbf{k} \\ a_1 & b_1 & c_1 \\ a_2 + a_3 & b_2 + b_3 & c_2 + c_3 \end{vmatrix} = \begin{vmatrix} \mathbf{i} & \mathbf{j} & \mathbf{k} \\ a_1 & b_1 & c_1 \\ a_2 & b_2 & c_2 \end{vmatrix} + \begin{vmatrix} \mathbf{i} & \mathbf{j} & \mathbf{k} \\ a_1 & b_1 & c_1 \\ a_3 & b_3 & c_3 \end{vmatrix}$$

$$= (\mathbf{u} \times \mathbf{v}) + (\mathbf{u} \times \mathbf{w}).$$

The second equality is easily verified by direct calculation.

$$\text{(v)} \quad (\mathbf{u} \times \mathbf{v}) \cdot \mathbf{w} = [(b_1 c_2 - c_1 b_2)\mathbf{i} + (c_1 a_2 - a_1 c_2)\mathbf{j} + (a_1 b_2 - b_1 a_2)\mathbf{k}]$$
$$\cdot (a_3\mathbf{i} + b_3\mathbf{j} + c_3\mathbf{k})$$
$$= b_1 c_2 a_3 - c_1 b_2 a_3 + c_1 a_2 b_3 - a_1 c_2 b_3 + a_1 b_2 c_3 - b_1 a_2 c_3.$$

We can easily show that $\mathbf{u} \cdot (\mathbf{v} \times \mathbf{w})$ is equal to the same expression.

   Note. For an interesting geometric interpretation of the scalar triple product, see Problem 41.

   (vi)   We know that $\mathbf{u} \cdot (\mathbf{u} \times \mathbf{v}) = (\mathbf{u} \times \mathbf{v}) \cdot \mathbf{u}$ (since the dot product is commutative—see Theorem 17.2.2(i)). But, from parts (ii) and (v) of this theorem,

$$(\mathbf{u} \times \mathbf{v}) \cdot \mathbf{u} = \mathbf{u} \cdot (\mathbf{v} \times \mathbf{u}) = \mathbf{u} \cdot (-\mathbf{u} \times \mathbf{v}) = -\mathbf{u} \cdot (\mathbf{u} \times \mathbf{v}).$$

Thus $\mathbf{u} \cdot (\mathbf{u} \times \mathbf{v}) = -\mathbf{u} \cdot (\mathbf{u} \times \mathbf{v})$ which can only occur if $\mathbf{u} \cdot (\mathbf{u} \times \mathbf{v}) = 0$. A similar computation shows that $\mathbf{v} \cdot (\mathbf{u} \times \mathbf{v}) = 0$.

   (vii)   If $\mathbf{u}$ and $\mathbf{v}$ are parallel, then $\mathbf{v} = \alpha\mathbf{u}$ for some scalar α (from Theorem 17.2.4) so that

$$\mathbf{u} \times \mathbf{v} = \begin{vmatrix} \mathbf{i} & \mathbf{j} & \mathbf{k} \\ a_1 & b_1 & c_1 \\ \alpha a_1 & \alpha b_1 & \alpha c_1 \end{vmatrix} = \mathbf{0}$$

since a determinant with two proportional rows is zero (see Theorem 5(ii) in Appendix 2).

Note. We could have proved this theorem without using determinants, but the proof would have involved many more computations.

What happens when we take cross products of the basis vectors **i, j, k**? It is easy to verify the following:

$$\mathbf{i} \times \mathbf{i} = \mathbf{j} \times \mathbf{j} = \mathbf{k} \times \mathbf{k} = 0 \qquad (2)$$

$$\mathbf{i} \times \mathbf{j} = \mathbf{k}, \qquad \mathbf{k} \times \mathbf{i} = \mathbf{j}, \qquad \mathbf{j} \times \mathbf{k} = \mathbf{i} \qquad (3)$$

$$\mathbf{j} \times \mathbf{i} = -\mathbf{k}, \qquad \mathbf{i} \times \mathbf{k} = -\mathbf{j}, \qquad \mathbf{k} \times \mathbf{j} = -\mathbf{i} \qquad (4)$$

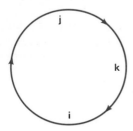

Figure 1

To remember these, consider the circle in Figure 1. The cross product of two consecutive vectors in the clockwise direction is positive, while the cross product of two consecutive vectors in the counterclockwise direction is negative. Note that the above formulas show that the cross product is *not* associative since, for example, $\mathbf{i} \times (\mathbf{i} \times \mathbf{j}) = \mathbf{i} \times \mathbf{k} = \mathbf{j}$ while $(\mathbf{i} \times \mathbf{i}) \times \mathbf{j} = 0 \times \mathbf{j} = 0$ so that

$$\mathbf{i} \times (\mathbf{i} \times \mathbf{j}) \neq (\mathbf{i} \times \mathbf{i}) \times \mathbf{j}.$$

In general,

$$\mathbf{u} \times (\mathbf{v} \times \mathbf{w}) \neq (\mathbf{u} \times \mathbf{v}) \times \mathbf{w}. \qquad (5)$$

EXAMPLE 3. Calculate $(3\mathbf{i} + 4\mathbf{k}) \times (2\mathbf{i} - 3\mathbf{j})$.

SOLUTION. This is a good example of the usefulness of Theorem 2 and formulas (2), (3), and (4). We have

$$\begin{aligned}(3\mathbf{i} + 4\mathbf{k}) \times (2\mathbf{i} - 3\mathbf{j}) &= (3 \cdot 2)(\mathbf{i} \times \mathbf{i}) + (4 \cdot 2)(\mathbf{k} \times \mathbf{i}) - (3 \cdot 3)(\mathbf{i} \times \mathbf{j}) \\ &\quad + 4(-3)(\mathbf{k} \times \mathbf{j}) \\ &= 0 + 8\mathbf{j} - 9\mathbf{k} + 12\mathbf{i} = 12\mathbf{i} + 8\mathbf{j} - 9\mathbf{k}.\end{aligned}$$

EXAMPLE 4. Find a line orthogonal to the lines $(x - 1)/3 = (y + 6)/4 = (z - 2)/-2$ and $(x + 2)/-3 = (y - 3)/4 = (z + 1)/1$ and passing through the point $(2, -1, 1)$.

SOLUTION. The directions of these lines are

$$\mathbf{v}_1 = 3\mathbf{i} + 4\mathbf{j} - 2\mathbf{k} \qquad \text{and} \qquad \mathbf{v}_2 = -3\mathbf{i} + 4\mathbf{j} + \mathbf{k}.$$

A vector orthogonal to these vectors is

$$\mathbf{w} = \mathbf{v}_1 \times \mathbf{v}_2 = \begin{vmatrix} \mathbf{i} & \mathbf{j} & \mathbf{k} \\ 3 & 4 & -2 \\ -3 & 4 & 1 \end{vmatrix} = 12\mathbf{i} + 3\mathbf{j} + 24\mathbf{k}.$$

Then a line satisfying the requested conditions is given by

$$L_1: \quad \frac{x-2}{12} = \frac{y+1}{3} = \frac{z-1}{24}.$$

Note. $\mathbf{w}_1 = \mathbf{v}_2 \times \mathbf{v}_1 = -(\mathbf{v}_1 \times \mathbf{v}_2)$ is also orthogonal to $\mathbf{v}_1$ and $\mathbf{v}_2$ so another line is given by

$$L_2: \quad \frac{x-2}{-12} = \frac{y+1}{-3} = \frac{z-1}{-24}.$$

However L_1 and L_2 are really the same line. (Explain why.)

The preceding example leads to a basic question. We know that $\mathbf{u} \times \mathbf{v}$ is a vector orthogonal to $\mathbf{u}$ and $\mathbf{v}$. But there are always *two* unit vectors orthogonal to $\mathbf{u}$ and $\mathbf{v}$ (see Figure 2). The vectors $\mathbf{n}$ and $-\mathbf{n}$ (**n** stands for normal, of course) are both

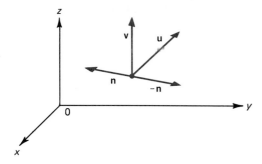

Figure 2

orthogonal to $\mathbf{u}$ and $\mathbf{v}$. Which one is in the direction of $\mathbf{u} \times \mathbf{v}$? The answer is given by the *right-hand rule*. If the right hand is placed so that the index finger points in the direction of $\mathbf{u}$ and the middle finger points in the direction of $\mathbf{v}$, then the thumb points in the direction of $\mathbf{u} \times \mathbf{v}$ (see Figure 3).

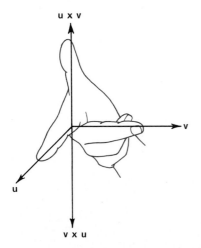

Figure 3

Having discussed the direction of the vector $\mathbf{u} \times \mathbf{v}$ we now turn to a discussion of its magnitude.

THEOREM 3. If φ is the angle between **u** and **v**, then

$$|\mathbf{u} \times \mathbf{v}| = |\mathbf{u}|\,|\mathbf{v}|\sin\varphi. \tag{6}$$

PROOF. It is easy to show (by comparing components) that

$$|\mathbf{u} \times \mathbf{v}|^2 = |\mathbf{u}|^2|\mathbf{v}|^2 - (\mathbf{u}\cdot\mathbf{v})^2 \tag{7}$$

(see Problem 36). Then since $(\mathbf{u}\cdot\mathbf{v})^2 = |\mathbf{u}|^2|\mathbf{v}|^2\cos^2\varphi$ (from Theorem 17.2.3),

$$|\mathbf{u} \times \mathbf{v}|^2 = |\mathbf{u}|^2|\mathbf{v}|^2 - |\mathbf{u}|^2|\mathbf{v}|^2\cos^2\varphi = |\mathbf{u}|^2|\mathbf{v}|^2(1 - \cos^2\varphi)$$
$$= |\mathbf{u}|^2|\mathbf{v}|^2\sin^2\varphi$$

and the theorem follows after taking square roots of both sides.

There is an interesting geometric interpretation of Theorem 3. The vectors **u** and **v** are sketched in Figure 4 and can be thought of as two adjacent sides of a parallelogram. Then, from elementary geometry, we see that

$$\text{area of the parallelogram} = |\mathbf{u}|\,|\mathbf{v}|\sin\varphi = |\mathbf{u} \times \mathbf{v}|. \tag{8}$$

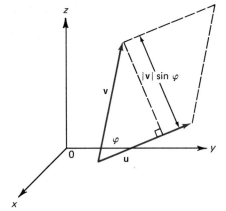

Figure 4

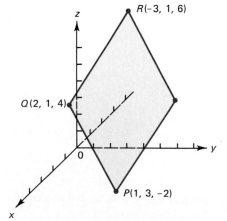

Figure 5

EXAMPLE 5. Find the area of the parallelogram with consecutive vertices at $\mathbf{P} = (1, 3, -2)$, $\mathbf{Q} = (2, 1, 4)$, and $\mathbf{R} = (-3, 1, 6)$.

SOLUTION. The parallelogram is sketched in Figure 5. We have

$$\text{area} = |\overrightarrow{PQ} \times \overrightarrow{QR}| = |(\mathbf{i} - 2\mathbf{j} + 6\mathbf{k}) \times (-5\mathbf{I} + 2\mathbf{k})|$$

$$= \begin{vmatrix} \mathbf{i} & \mathbf{j} & \mathbf{k} \\ 1 & -2 & 6 \\ -5 & 0 & 2 \end{vmatrix} = |-4\mathbf{i} - 32\mathbf{j} - 10\mathbf{k}| = \sqrt{1140} \quad \text{square units.}$$

PROBLEMS 17.4

In Problems 1–20 find the cross product $\mathbf{u} \times \mathbf{v}$.

1. $\mathbf{u} = \mathbf{i} - 2\mathbf{j}; \mathbf{v} = 3\mathbf{k}$
2. $\mathbf{u} = 3\mathbf{i} - 7\mathbf{j}; \mathbf{v} = \mathbf{i} + \mathbf{k}$
3. $\mathbf{u} = \mathbf{i} - \mathbf{j}; \mathbf{v} = \mathbf{j} + \mathbf{k}$
4. $\mathbf{u} = -7\mathbf{k}; \mathbf{v} = \mathbf{j} + 2\mathbf{k}$
5. $\mathbf{u} = -2\mathbf{i} + 3\mathbf{j}; \mathbf{v} = 7\mathbf{i} + 4\mathbf{k}$
6. $\mathbf{u} = a\mathbf{i} + b\mathbf{j}; \mathbf{v} = c\mathbf{i} + d\mathbf{j}$
7. $\mathbf{u} = a\mathbf{i} + b\mathbf{k}; \mathbf{v} = c\mathbf{i} + d\mathbf{k}$
8. $\mathbf{u} = a\mathbf{j} + b\mathbf{k}; \mathbf{v} = c\mathbf{i} + d\mathbf{k}$
9. $\mathbf{u} = 2\mathbf{i} - 3\mathbf{j} + \mathbf{k}; \mathbf{v} = \mathbf{i} + 2\mathbf{j} + \mathbf{k}$
10. $\mathbf{u} = 3\mathbf{i} - 4\mathbf{j} + 2\mathbf{k}; \mathbf{v} = 6\mathbf{I} - 3\mathbf{j} + 5\mathbf{k}$
11. $\mathbf{u} = -3\mathbf{i} - 2\mathbf{j} + \mathbf{k}; \mathbf{v} = 6\mathbf{i} + 4\mathbf{j} - 2\mathbf{k}$
12. $\mathbf{u} = \mathbf{i} + 7\mathbf{j} - 3\mathbf{k}; \mathbf{v} = -\mathbf{i} - 7\mathbf{j} + 3\mathbf{k}$
13. $\mathbf{u} = \mathbf{i} - 7\mathbf{j} - 3\mathbf{k}; \mathbf{v} = -\mathbf{i} + 7\mathbf{j} - 3\mathbf{k}$
14. $\mathbf{u} = 2\mathbf{i} - 3\mathbf{j} + 5\mathbf{k}; \mathbf{v} = 3\mathbf{i} - \mathbf{j} - \mathbf{k}$
15. $\mathbf{u} = 10\mathbf{i} + 7\mathbf{j} - 3\mathbf{k}; \mathbf{v} = -3\mathbf{i} + 4\mathbf{j} - 3\mathbf{k}$
16. $\mathbf{u} = 2\mathbf{i} + 4\mathbf{j} - 6\mathbf{k}; \mathbf{v} = -\mathbf{i} - \mathbf{j} + 3\mathbf{k}$
17. $\mathbf{u} = 2\mathbf{i} - \mathbf{j} + \mathbf{k}; \mathbf{v} = 4\mathbf{i} + 2\mathbf{j} + 2\mathbf{k}$
18. $\mathbf{u} = 3\mathbf{i} - \mathbf{j} + 8\mathbf{k}; \mathbf{v} = \mathbf{i} + \mathbf{j} - 4\mathbf{k}$
19. $\mathbf{u} = a\mathbf{i} + a\mathbf{j} + a\mathbf{k}; \mathbf{v} = b\mathbf{i} + b\mathbf{j} + b\mathbf{k}$
20. $\mathbf{u} = a\mathbf{i} + b\mathbf{j} + c\mathbf{k}; \mathbf{v} = a\mathbf{i} + b\mathbf{j} - c\mathbf{k}$.

21. Find two unit vectors orthogonal to both $\mathbf{u} = 2\mathbf{i} - 3\mathbf{j}$ and $\mathbf{v} = 4\mathbf{j} + 3\mathbf{k}$.
22. Find two unit vectors orthogonal to both $\mathbf{u} = \mathbf{i} + \mathbf{j} + \mathbf{k}$ and $\mathbf{v} = \mathbf{i} - \mathbf{j} - \mathbf{k}$.
23. Use the cross product to find the sine of the angle φ between the vectors $\mathbf{u} = 2\mathbf{i} + \mathbf{j} - \mathbf{k}$ and $\mathbf{v} = -3\mathbf{i} - 2\mathbf{j} + 4\mathbf{k}$.
24. Use the dot product to calculate the cosine of the angle between the vectors of Problem 23. Then show that for the values you have calculated, $\sin^2 \varphi + \cos^2 \varphi = 1$.

In Problems 25–28 find a line orthogonal to the two given lines and passing through the given point.

25. $\dfrac{x + 2}{-3} = \dfrac{y - 1}{4} = \dfrac{z}{-5}; \dfrac{x - 3}{7} = \dfrac{y + 2}{-2} = \dfrac{z - 8}{3}; (1, -3, 2)$

26. $\dfrac{x - 2}{-4} = \dfrac{y + 3}{-7} = \dfrac{z + 1}{3}; \dfrac{x + 2}{3} = \dfrac{y - 5}{-4} = \dfrac{z + 3}{-2}; (-4, 7, 3)$

27. $x = 3 - 2t, \quad y = 4 + 3t, \quad z = -7 + 5t; \quad x = -2 + 4s, \quad y = 3 - 2s, \quad z = 3 + s;$ $(-2, 3, 4)$

28. $x = 4 + 10t, y = -4 - 8t, z = 3 + 7t; x = -2t, y = 1 + 4t, z = -7 - 3t; (4, 6, 0)$.

In Problems 29–34 find the area of the parallelogram with the given adjacent vertices.

29. $(1, -2, 3); (2, 0, 1); (0, 4, 0)$
30. $(-2, 1, 1); (2, 2, 3); (-1, -2, 4)$
31. $(-2, 1, 0); (1, 4, 2); (-3, 1, 5)$
32. $(7, -2, -3); (-4, 1, 6); (5, -2, 3)$

33. $(a, 0, 0)$; $(0, b, 0)$; $(0, 0, c)$ 34. $(a, b, 0)$; $(a, 0, b)$; $(0, a, b)$.

35. Show that if $\mathbf{u} = (a_1, b_1, c_1)$, $\mathbf{v} = (a_2, b_2, c_2)$,   and $\mathbf{w} = (a_3, b_3, c_3)$

$$\mathbf{u} \cdot (\mathbf{v} \times \mathbf{w}) = \begin{vmatrix} a_1 & b_1 & c_1 \\ a_2 & b_2 & c_2 \\ a_3 & b_3 & c_3 \end{vmatrix}.$$

36. Show that $|\mathbf{u} \times \mathbf{v}|^2 = |\mathbf{u}|^2 |\mathbf{v}|^2 - (\mathbf{u} \cdot \mathbf{v})^2$. [*Hint:* Write out in terms of components.]

37. Show that the area of the triangle PQR is given by $A = \frac{1}{2}|\overrightarrow{PQ} \times \overrightarrow{QR}|$.

38. Use the result of Problem 37 to calculate the area of the triangle with vertices at $(2, 1, -4)$, $(1, 7, 2)$, $(3, -2, 3)$.

39. Calculate the area of a triangle with vertices at $(3, 1, 7)$, $(2, -3, 4)$, $(7, -2, 4)$.

40. Calculate the area of the triangle with vertices at $(1, 0, 0)$, $(0, 1, 0)$, and $(0, 0, 1)$. Sketch this triangle.

*41. Let $\mathbf{u}, \mathbf{v}$, and $\mathbf{w}$ be three vectors that are not in the same plane. Then they form the sides of a parallelepiped in space (see Figure 6). Prove that the volume of the parallelepiped is given by $V = |(\mathbf{u} \times \mathbf{v}) \cdot \mathbf{w}|$. [*Hint:* The area of the base is $|\mathbf{u} \times \mathbf{v}|$.]

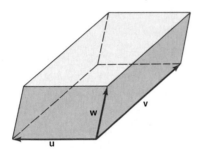

Figure 6

42. Calculate the volume of the parallelepiped determined by the vectors $\mathbf{u} = 2\mathbf{i} - \mathbf{j} + \mathbf{k}$, $\mathbf{v} = 3\mathbf{i} + 2\mathbf{j} - 2\mathbf{k}$, $\mathbf{w} = 3\mathbf{i} + 2\mathbf{j}$.

43. Calculate the volume of the parallelepiped determined by the vectors $\mathbf{i} - \mathbf{j}$, $3\mathbf{i} + 2\mathbf{k}$, $-7\mathbf{j} + 3\mathbf{k}$.

44. Calculate the volume of the parallelepiped determined by the vectors $\overrightarrow{PQ}$, $\overrightarrow{PR}$, and $\overrightarrow{PS}$ where $P = (2, 1, -1)$, $Q = (-3, 1, 4)$, $R = (-1, 0, 2)$, and $S = (-3, -1, 5)$.

*45. Calculate the distance between the lines

$$L_1: \quad \frac{x - 2}{3} = \frac{y - 5}{2} = \frac{z - 1}{-1} \quad \text{and} \quad L_2: \quad \frac{x - 4}{-4} = \frac{y - 5}{4} = \frac{z + 2}{1}.$$

[*Hint:* The distance is measured along a vector $\mathbf{v}$ which is perpendicular to both L_1 and L_2. Let P be a point on L_1 and Q a point on L_2. Then the length of the projection of $\overrightarrow{PQ}$ on $\mathbf{v}$ is the distance between the lines, measured along a vector which is perpendicular to them both. See Figure 7.]

46. Find the distance between the lines

$$L_1: \quad \frac{x + 2}{3} = \frac{y - 7}{-4} = \frac{z - 2}{2} \quad \text{and} \quad L_2: \quad \frac{x - 1}{-3} = \frac{y + 2}{4} = \frac{z + 1}{1}.$$

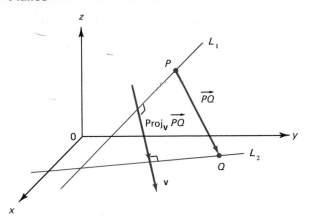

Figure 7

47. Find the distance between the lines $x = 2 - 3t$, $y = 1 + 2t$, $z = -2 - t$, and $x = 1 + 4s$, $y = -2 - s$, $z = 3 + s$.

48. Find the distance between the lines $x = -2 - 5t$, $y = -3 - 2t$, $z = 1 + 4t$, and $x = 2 + 3s$, $y = -1 + s$, $z = 3s$.

17.5 Planes

In Section 17.3 we derived the equation of a line in space by specifying a point on the line and a vector *parallel* to this line. We can derive the equation of a plane in space by specifying a point in the plane and a vector orthogonal to every vector in the plane. This orthogonal vector is called a *normal vector* and is denoted by **N**. (See Figure 1.)

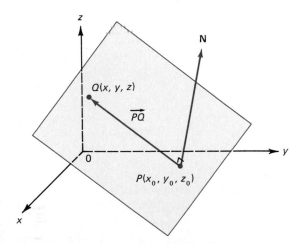

Figure 1

DEFINITION 1. Let P be a point in space and let **N** be a given vector. Then the set of all points Q for which $\overrightarrow{PQ}$ and **N** are orthogonal comprises a *plane* in $\mathbb{R}^3$.

Notation. We will usually denote a plane by the symbol π.

If we have $P = (x_0, y_0, z_0)$ and $N = a\mathbf{i} + b\mathbf{j} + c\mathbf{k}$, then if $Q = (x, y, z)$, $\overrightarrow{PQ} = (x - x_0)\mathbf{i} + (y - y_0)\mathbf{j} + (z - z_0)\mathbf{k}$ and if $\overrightarrow{PQ} \perp N$ then $\overrightarrow{PQ} \cdot N = 0$. But this implies that

$$\boxed{a(x - x_0) + b(y - y_0) + c(z - z_0) = 0.} \qquad (1)$$

A more common way to write the equation of a plane is easily derived from (1):

$$ax + by + cz = d,$$

where

$$d = ax_0 + by_0 + cz_0 = \overrightarrow{OP} \cdot N. \qquad (2)$$

EXAMPLE 1. Find the plane π passing through the point $(2, 5, 1)$ and normal to the vector $N = \mathbf{i} - 2\mathbf{j} + 3\mathbf{k}$.

SOLUTION. From (1), we immediately obtain

$$(x - 2) - 2(y - 5) + 3(z - 1) = 0$$

or

$$x - 2y + 3z = -5. \qquad (3)$$

This plane is sketched in Figure 2.

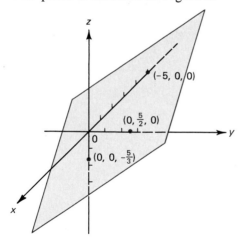

Figure 2

Important Remark. The plane is easily sketched by setting $x = y = 0$ in (3) to obtain $(0, 0, -\frac{5}{3})$, $x = z = 0$ to obtain $(0, \frac{5}{2}, 0)$ and $y = z = 0$ to obtain $(-5, 0, 0)$. These three points all lie on the plane.

The three coordinate planes are easily represented as follows.

(i)   *The xy-plane.* This plane passes through the origin $(0, 0, 0)$ and any vector lying along the z-axis is normal to it. The simplest such vector is $\mathbf{k}$. Thus, from (1), we obtain

$$0(x - 0) + 0(y - 0) + 1(z - 0) = 0$$

which yields

$$z = 0 \tag{4}$$

as the equation of the xy-plane. (This should not be very surprising.)

(ii) *The xz-plane* has the equation

$$y = 0. \tag{5}$$

(iii) *The yz-plane* has the equation

$$z = 0. \tag{6}$$

Three points which are not collinear determine a plane since they determine two nonparallel vectors which intersect at a point.

EXAMPLE 2. Find the equation of the plane π passing through the points $P = (1, 2, 1)$, $Q = (-2, 3, -1)$, $R = (1, 0, 4)$.

SOLUTION. The vectors $\overrightarrow{PQ} = -3\mathbf{i} + \mathbf{j} - 2\mathbf{k}$ and $\overrightarrow{QR} = 3\mathbf{i} - 3\mathbf{j} + 5\mathbf{k}$ lie on the plane and are therefore orthogonal to the normal vector so that

$$\mathbf{N} = \overrightarrow{PQ} \times \overrightarrow{QR} = \begin{vmatrix} \mathbf{i} & \mathbf{j} & \mathbf{k} \\ -3 & 1 & -2 \\ 3 & -3 & 5 \end{vmatrix} = -\mathbf{i} + 9\mathbf{j} + 6\mathbf{k}$$

and we obtain

$$\pi: \quad -(x - 1) + 9(y - 2) + 6(z - 1) = 0$$

or

$$-x + 9y + 6z = 23.$$

Note that if we choose another point, say Q, we get the equation

$$-(x + 2) + 9(y - 3) + 6(z + 1) = 0$$

which reduces to

$$-x + 9y + 6z = 23.$$

This is sketched in Figure 3.

DEFINITION 2. Two planes are *parallel* if their normal vectors are parallel; that is, if the cross product of their normal vectors is zero.

Two parallel planes are drawn in Figure 4.

EXAMPLE 3. The planes π_1: $2x + 3y - z = 3$ and π_2: $-4x - 6y + 2z = 8$ are parallel since $\mathbf{N}_1 = 2\mathbf{i} + 3\mathbf{j} - \mathbf{k}$, $\mathbf{N}_2 = -4\mathbf{i} - 6\mathbf{j} + 2\mathbf{k}$ and $\mathbf{N}_2 = -2\mathbf{N}_1$ (and $\mathbf{N}_1 \times \mathbf{N}_2 = 0$).

If two planes are not parallel, then they intersect in a straight line.

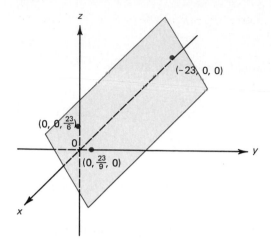

Figure 3

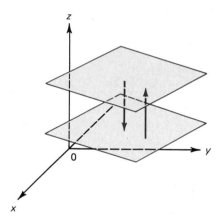

Figure 4

EXAMPLE 4. Find all points of intersection of the planes $2x - y - z = 3$ and $x + 2y + 3z = 7$.

SOLUTION. When the planes intersect, we have

$$2x - y - z = 3$$

and

$$x + 2y + 3z = 7.$$

Multiplying the first equation by 2 and adding it to the second, we obtain

$$\begin{array}{r} 4x - 2y - 2z = 6 \\ \underline{x + 2y + 3z = 7} \\ 5x + z = 13 \end{array}$$

or $z = -5x + 13$. Then, from the first equation,

$$y = 2x - z - 3 = 2x - (-5x + 13) - 3 = 7x - 16.$$

Then, setting $x = t$, we obtain the parametric representation of the line of intersection:

$$x = t, \qquad y = -16 + 7t, \qquad z = 13 - 5t.$$

This is sketched in Figure 5. Note that this line is orthogonal to both normal vectors.

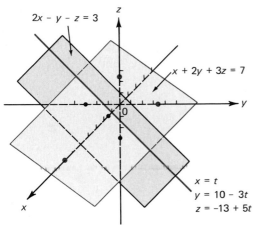

Figure 5

We conclude this section by indicating how the distance from a plane to a point can be calculated. Look at Figure 6. If Q is the point, then the required distance is the distance measured along a line orthogonal to π. That is, the shortest distance is

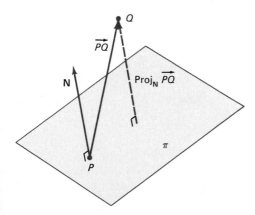

Figure 6

obtained by "dropping a perpendicular" from the point to the plane. This is done by calculating (for any point P on the plane)

$$D = |\text{Proj}_N \overrightarrow{PQ}| = \frac{|\overrightarrow{PQ} \cdot \mathbf{N}|}{|\mathbf{N}|}. \tag{7}$$

EXAMPLE 5.   Find the distance d between the plane $2x - y + 3z = 6$ and the point $Q = (3, 5, -7)$.

SOLUTION. One point on the plane is $P = (3, 0, 0)$ and $\mathbf{N} = 2\mathbf{i} - \mathbf{j} + 3\mathbf{k}$. Then $\overrightarrow{PQ} = 5\mathbf{j} - 7\mathbf{k}$, $|\overrightarrow{PQ} \cdot \mathbf{N}| = 26$ and $|\mathbf{N}| = \sqrt{14}$ so that

$$D = \frac{26}{\sqrt{14}}.$$

EXAMPLE 6. Find the distance from the plane $ax + by + cz = d$ to the origin.

SOLUTION. If $a \neq 0$, one point on the plane is $P = (d/a, 0, 0)$. (If $a = 0$ but $b \neq 0$, a point on the plane is $(0, d/b, 0)$ leading to the same result.) Then $\mathbf{N} = a\mathbf{i} + b\mathbf{j} + c\mathbf{k}$ and $\overrightarrow{PO} = (d/a)\mathbf{i}$ so that $|\overrightarrow{OP} \cdot \mathbf{N}| = |d|$, $\mathbf{N} = \sqrt{a^2 + b^2 + c^2}$ and

$$D = \frac{|d|}{\sqrt{a^2 + b^2 + c^2}}.$$

PROBLEMS 17.5

In Problems 1–16 find the equation of the plane and sketch it.

1. $P = (0, 0, 0)$; $\mathbf{N} = \mathbf{i}$
2. $P = (0, 0, 0)$; $\mathbf{N} = \mathbf{j}$
3. $P = (0, 0, 0)$; $\mathbf{N} = \mathbf{k}$
4. $P = (1, 2, 3)$; $\mathbf{N} = \mathbf{i} + \mathbf{j}$
5. $P = (1, 2, 3)$; $\mathbf{N} = \mathbf{i} + \mathbf{k}$
6. $P = (1, 2, 3)$; $\mathbf{N} = \mathbf{j} + \mathbf{k}$
7. $P = (2, -1, 6)$; $\mathbf{N} = 3\mathbf{i} - \mathbf{j} + 2\mathbf{k}$
8. $P = (-4, -7, 5)$; $\mathbf{N} = -3\mathbf{i} - 4\mathbf{j} + \mathbf{k}$
9. $P = (-3, 11, 2)$; $\mathbf{N} = 4\mathbf{i} + \mathbf{j} - 7\mathbf{k}$
10. $P = (3, -2, 5)$; $\mathbf{N} = 2\mathbf{i} - 7\mathbf{j} - 8\mathbf{k}$
11. $P = (4, -7, -3)$; $\mathbf{N} = -\mathbf{i} - \mathbf{j} - \mathbf{k}$
12. $P = (8, 1, 0)$; $\mathbf{N} = -7\mathbf{i} + \mathbf{j} + 2\mathbf{k}$
13. Containing $(1, 2, -4)$, $(2, 3, 7)$, and $(4, -1, 3)$
14. Containing $(-7, 1, 0)$, $(2, -1, 3)$ and $(4, 1, 6)$
15. Containing $(1, 0, 0)$, $(0, 1, 0)$, and $(0, 0, 1)$
16. Containing $(2, 3, -2)$, $(4, -1, -1)$, $(3, 1, 2)$.

Two planes are *orthogonal* if their normal vectors are orthogonal. In Problems 17–23 determine whether the given planes are parallel, orthogonal, coincident (i.e., the same), or none of these.

17. π_1: $x + y + z = 2$; π_2: $2x + 2y + 2z = 4$
18. π_1: $x - y + z = 3$; π_2: $-3x + 3y - 3z = -9$
19. π_1: $2x - y + z = 3$; π_2: $x + y - z = 7$
20. π_1: $2x - y + z = 3$; π_2: $x + y + z = 3$
21. π_1: $3x - 2y + 7z = 4$; π_2: $-2x + 4y + 2x = 16$
22. π_1: $-4x + 4y - 6z = 7$; π_2: $2x - 2y + 3z = -3$
23. π_1: $-4x + 4y - 6z = 6$; π_2: $2x - 2y + 3z = -3$.

In Problems 24–26 find the equation of all points of intersection of the two planes.

24. π_1: $x - y + z = 2$; π_2: $2x - 3y + 4z = 7$
25. π_1: $3x - y + 4z = 3$; π_2: $-4x - 2y + 7z = 8$
26. π_1: $-2x - y + 17z = 4$; π_2: $2x - y - z = -7$.

In Problems 27–30 find the distance from the given point to the given plane.

27. $(2, -1, 4)$; $3x - y + 7z = 2$
28. $(4, 0, 1)$; $2x - y + 8z = 3$
29. $(-7, -2, -1)$; $-2x + 8z = -5$
30. $(-3, 0, 2)$; $-3x + y + 5z = 0$.

31. Prove that the distance between the plane $ax + by + cz = d$ and the point (x_0, y_0, z_0) is given by

$$D = \frac{|ax_0 + by_0 + cz_0 - d|}{\sqrt{a^2 + b^2 + c^2}}.$$

The *angle between two planes* is defined to be the angle between their normal vectors. In Problems 32–34 find the angle between the two planes

32. of Problem 24 **33.** of Problem 25 **34.** of Problem 26.

*35. Let **u** and **v** be two nonparallel vectors in a plane π. Show that if **w** is any other vector in π, then there exist scalars α and β such that

w $= \alpha$**u** $+ \beta$**v**.

This is called the *parametric representation* of the plane π. [*Hint:* Draw a parallelogram in which α**u** and β**v** form adjacent sides and the diagonal vector is **w**.]

36. Three vectors **u**, **v**, and **w** are called *coplanar* if they all lie in the same plane π. Show that if **u**, **v**, and **w** all pass through the origin, then they are coplanar if and only if the scalar triple product equals zero:

u $\cdot$ (**v** $\times$ **w**) $= 0.$

In Problems 37–41 determine whether the three given position vectors (i.e., one endpoint at the origin) are coplanar. If they are coplanar, find the equation of the plane containing them.

37. **u** $= 2$**i** $- 3$**j** $+ 4$**k**; **v** $= 7$**i** $- 2$**j** $+ 3$**k**; **w** $= 9$**i** $- 5$**j** $+ 7$**k**
38. **u** $= -3$**i** $+ $**j** $+ 8$**k**; **v** $= -2$**i** $- 3$**j** $+ 5$**k**; **w** $= 2$**i** $+ 14$**j** $- 4$**k**
39. **u** $= 2$**i** $+ $**j** $- 2$**k**; **v** $= 2$**i** $- $**j** $- 2$**k**; **w** $= 2$**i** $- $**j** $+ 2$**k**
40. **u** $= 3$**i** $- 2$**j** $+ $**k**; **v** $= $**i** $+ $**j** $- 5$**k**; **w** $= -$**i** $+ 5$**j** $- 16$**k**
41. **u** $= 2$**i** $- $**j** $- $**k**; **v** $= 4$**i** $+ 3$**j** $+ 2$**k**; **w** $= 6$**i** $+ 7$**j** $+ 5$**k**.

*17.6 Quadric Surfaces

A *surface* in space is defined as the set of points satisfying the equation $F(x, y, z) = 0$. For example, the equation

$$F(x, y, z) = x^2 + y^2 + z^2 - 1 = 0 \tag{1}$$

is the equation of the unit sphere, as we saw in Section 1. In this section we shall take a brief look at some of the most commonly encountered surfaces in $\mathbb{R}^3$. We will take a more detailed look at general surfaces in $\mathbb{R}^3$ in Chapter 18.

Having already discussed the sphere, we turn our attention to the cylinder.

DEFINITION 1. Let a line L and a plane curve C be given. A *cylinder* is the surface generated when a line parallel to L moves around C, remaining parallel to L. The line L is called the *generatrix* of the cylinder and the curve C is called its *directrix*.

EXAMPLE 1. Let L be the z-axis and C the circle $x^2 + y^2 = a^2$ in the xy-plane. Sketch the cylinder.

SOLUTION. As we move a line along the circle $x^2 + y^2 = a^2$ and parallel to the z-axis, we obtain the *right circular cylinder* $x^2 + y^2 = a^2$ sketched in Figure 1.

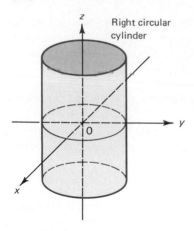

Right circular
cylinder

Figure 1

EXAMPLE 2. If L is the x-axis and C is given by $y = z^2$, sketch the resulting cylinder.

SOLUTION. The curve $y = z^2$ is a parabola in the yz-plane. As we move along it, parallel to the x-axis, we obtain the *parabolic cylinder* sketched in Figure 2.

EXAMPLE 3. If L is the x-axis and C is given by $y^2 + (z^2/4) = 1$, sketch the resulting cylinder.

SOLUTION. $y^2 + (z^2/4) = 1$ is an ellipse in the yz-plane. As we move along it parallel to the x-axis, we obtain the *elliptic cylinder* sketched in Figure 3.

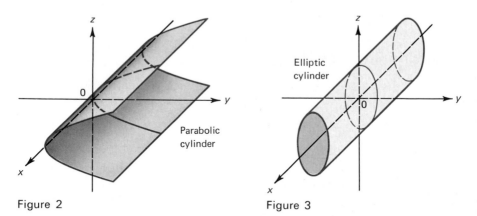

Parabolic
cylinder

Elliptic
cylinder

Figure 2   Figure 3

As we see from the discussion in Section 1.9 and Appendix 3, a second-degree equation in the variables x and y forms a circle, parabola, ellipse, or hyperbola (or a degenerate form of one of these such as a single point or a straight line) in the plane. In $\mathbb{R}^3$ we have the following definition.

DEFINITION 2. A *quadric surface* in $\mathbb{R}^3$ is the graph of a second-degree equation in the variables x, y, and z. Such an equation takes the form

$$Ax^2 + By^2 + Cz^2 + Dxy + Exz + Fyz + Gx + Hy + Jz + K = 0.$$

We have already seen sketches of several quadric surfaces. We list below the standard forms† of eleven types of nondegenerate‡ quadric surfaces. Although we shall not prove this here, any quadric surface can be written in one of these forms by a translation or rotation of the coordinate axes. (See Appendix 3 for a discussion of translation and rotation of axes.) Here is the list:

1. Sphere
2. Right circular cylinder
3. Parabolic cylinder
4. Elliptic cylinder
5. Hyperbolic cylinder
6. Ellipsoid
7. Hyperboloid of one sheet
8. Hyperboloid of two sheets
9. Elliptic paraboloid
10. Hyperbolic paraboloid
11. Elliptic cone

We have already seen sketches of surfaces 1, 2, 3 and 4.

5. *The hyperbolic cylinder:* $(y^2/b^2) - (x^2/a^2) = 1$. This is the equation of a hyperbola in the xy-plane. See Figure 4.

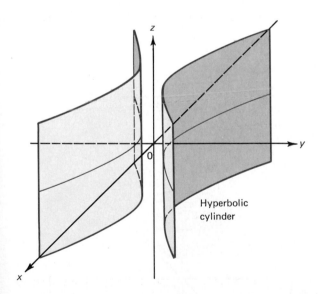

Hyperbolic cylinder

Figure 4

We can best describe the remaining six surfaces by looking at *cross sections* parallel to a given coordinate plane. For example, in the unit sphere $x^2 + y^2 + z^2 = 1$, cross sections parallel to the xy-plane are circles. To see this let

† A standard form is one in which the surface has its center or vertex at the origin.
‡ By degenerate we again mean a surface whose graph contains a finite number of points (or none). We saw examples of degenerate spheres (0 or 1 point) in Section 1.

$z = c$ where $-1 < c < 1$ (this is a plane parallel to the xy-plane). Then

$$x^2 + y^2 = 1 - c^2$$

which is the equation of a circle in the xy-plane.

6. *The ellipsoid:* $(x^2/a^2) + (y^2/b^2) + (z^2/c^2) = 1$. This is sketched in Figure 5. It is a closed "watermelon-shaped" surface. Cross sections parallel to the xy-plane, the xz-plane and the yz-plane are all ellipses.

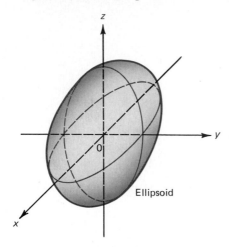

Ellipsoid

Figure 5

Note. If the ellipse $(x^2/a^2) + (y^2/b^2) = 1$ is revolved about the x-axis, the resulting surface is the ellipsoid $(x^2/a^2) + (y^2/b^2) + (z^2/b^2) = 1$. What do you get if the ellipse is rotated around the y-axis? (See Problem 35.)

7. *The hyperboloid of one sheet:* $(x^2/a^2) + (y^2/b^2) - (z^2/c^2) = 1$. This is sketched in Figure 6. Cross sections parallel to the xy-plane are ellipses. Cross sections parallel to the xz-plane and the yz-plane are hyperbolas.

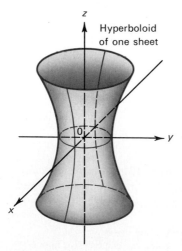

Hyperboloid of one sheet

Figure 6

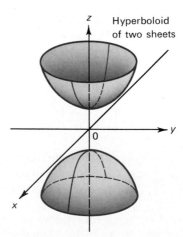

Hyperboloid of two sheets

Figure 7

8. *The hyperboloid of two sheets:* $(z^2/c^2) - (x^2/a^2) - (y^2/b^2) = 1$. This is sketched in Figure 7. Cross sections are as for the hyperboloid of one sheet. Note that the equation implies that $|z| \geq |c|$ (explain why).

9. *The elliptic paraboloid:* $z = (x^2/a^2) + (y^2/b^2)$, This is sketched in Figure 8. For each fixed z, $(x^2/a^2) + (y^2/b^2) = z$ is the equation of an ellipse. Hence cross sections parallel to the xy-plane are ellipses. If x or y is fixed, then we obtain parabolas. Hence cross sections parallel to the xz- or yz-plane are parabolas.

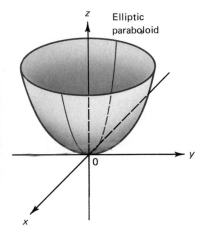

Figure 8

10. *The hyperbolic paraboloid:* $z = (x^2/a^2) - (y^2/b^2)$. This is sketched in Figure 9. For each fixed z, we obtain a hyperbola parallel to the xy-plane. Hence cross sections parallel to the xy-plane are hyperbolas, If x or y is fixed, we obtain parabolas. Thus cross sections parallel to the xz- and yz-planes are parabolas. The shape of the graph suggests why the hyperbolic paraboloid is often called a *saddle surface*.

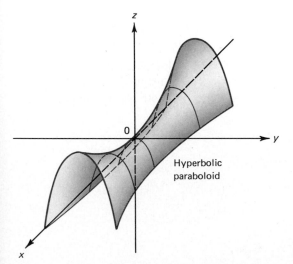

Figure 9

11. *The elliptic cone:* $(x^2/a^2) + (y^2/b^2) = z^2$. This is sketched in Figure 10. We get one cone for $z > 0$ and another for $z < 0$. Cross sections cut by planes not passing through the origin are either parabolas, ellipses, or hyperbolas. When $a = b$ the cross sections are the *conic sections* discussed in Appendix 3. If $a = b$, we obtain the equation of a *circular cone*.

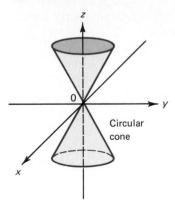

Circular
cone

Figure 10

EXAMPLE 4. Describe the surface given by

$$x^2 - 4y^2 - 9z^2 = 25.$$

SOLUTION. Dividing by 25, we obtain

$$\frac{x^2}{25} - \frac{4y^2}{25} - \frac{9z^2}{25} = 1 \quad \text{or} \quad \frac{x^2}{25} - \frac{y^2}{(5/2)^2} - \frac{z^2}{(5/3)^2} = 1.$$

This is the equation of a hyperboloid of two sheets. Note that here cross sections parallel to the yz-plane are ellipses.

EXAMPLE 5. Describe the surface given by $4x - 4y^2 - z^2 = 0$.

SOLUTION. Dividing by 4 we obtain

$$x = y^2 + \frac{z^2}{4}.$$

This is the equation of an elliptic paraboloid. The only difference between this and the one described under surface (9) is that now cross sections parallel to the yz-plane are ellipses while cross sections parallel to the other coordinate planes are parabolas. This is sketched in Figure 11.

EXAMPLE 6. Describe the surface $x^2 + 8x - 2y^2 + 8y + z^2 = 0$.

SOLUTION. Completing the squares we obtain

$$(x + 4)^2 - 2(y - 2)^2 + z^2 = 8$$

or, dividing by 8 and rearranging,

$$\frac{(x + 4)^2}{8} - \frac{(y - 2)^2}{4} + \frac{z^2}{8} = 1.$$

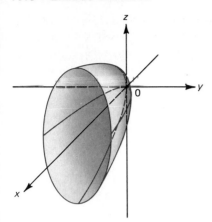

Figure 11

This is a hyperboloid of one sheet which, however, is centered at $(-4, 2, 0)$ instead of the origin. Moreover, cross sections parallel to the xz-plane are ellipses while those parallel to the other coordinate planes are hyperbolas.

As a rule, to identify a quadric surface with x, y, and z terms, completing the square will generally resolve the problem. If there are xy, xz, or yz terms present, then it is necessary to rotate the axes, a technique that we will not discuss here.

Warning. Watch out for degenerate surfaces. For example, there are clearly no points which satisfy

$$x^2 + \frac{y^2}{4} + \frac{z^2}{9} + 1 = 0.$$

PROBLEMS 17.6

In Problems 1–7 draw a sketch of the given cylinder. Here the directrix C is given. The generatrix L is the axis of the variable missing in the equation.

1. $y = \sin x$	**2.** $z = \sin y$	**3.** $y = \cos z$
4. $y = \cosh x$	**5.** $z = x^3$	**6.** $z = \lvert y \rvert$

7. $x^2 + y^2 + 2y = 0$. [*Hint:* Complete the square.]

In Problems 8–34 identify the quadric surface and sketch it.

8. $x^2 + y^2 = 4$ **9.** $y^2 + z^2 = 4$

10. $x^2 + z^2 = 4$ **11.** $\dfrac{x^2}{4} - \dfrac{y^2}{9} = 1$

12. $x^2 + 4z^2 = 1$ **13.** $y^2 + 2y + 4z^2 = 1$
14. $x^2 - z^2 = 1$ **15.** $3x^2 - 4y^2 = 4$
16. $x^2 + y^2 + z^2 = 1$ **17.** $x^2 + y^2 - z^2 = 1$
18. $y^2 + 2y - z^2 + x^2 = 1$ **19.** $x^2 + 2y^2 + 3z^2 = 4$
20. $x + 2y^2 + 3z^2 = 4$ **21.** $x^2 - 2y - 3z^2 = 4$
22. $x^2 + 4x + y^2 + 6y - z^2 - 8z = 2$ **23.** $4x - x^2 + y^2 + z^2 = 0$
24. $5x^2 + 7y^2 + 8z^2 = 8z$ **25.** $x^2 - y^2 - z^2 = 1$
26. $x^2 - 2y^2 - 3z^2 = 4$ **27.** $z^2 - x^2 - y^2 = 2$
28. $y^2 - 3x^2 - 3z^2 = 27$ **29.** $x^2 + y^2 + z^2 = 2(x + y + z)$

30. $4x^2 + 4y^2 + 16z^2 = 16$ **31.** $4y^2 - 4x^2 + 8z^2 = 16$
32. $z + x^2 - y^2 = 0$ **33.** $-x^2 + 2x + y^2 - 6y = z$
34. $x + y + z = x^2 + z^2$.

35. Identify and sketch the surface generated when the ellipse $(x^2/a^2) + (y^2/b^2) = 1$ is revolved about the y-axis.

17.7 Vector Functions and Parametric Equations in $\mathbb{R}^3$

Having seen how vectors in $\mathbb{R}^2$ generalize to vectors in $\mathbb{R}^3$, it is easy to imagine properties of vector functions in $\mathbb{R}^3$. We will describe these briefly as almost all their properties have already been proved in Chapter 16.

DEFINITION 1. Let $f_1, f_2,$ and f_3 be functions of the real variable t. Then for all values of t for which $f_1(t), f_2(t)$ and $f_3(t)$ are defined, we define the *vector-valued function* **f** by

$$\mathbf{f}(t) = f_1(t)\mathbf{i} + f_2(t)\mathbf{j} + f_3(t)\mathbf{k}. \tag{1}$$

The set of points traced out by the end of the vector **f** is called a *curve* in space.

EXAMPLE 1. Sketch the curve $\mathbf{f}(t) = (\cos t)\mathbf{i} + 4(\sin t)\mathbf{j} + t\mathbf{k}$.

SOLUTION. Here $x = \cos t$ and $y = 4 \sin t$ so that, eliminating t, we obtain $x^2 + (y^2/16) = 1$, which is the equation of an ellipse in the xy-plane. Since $z = t$ increases as t increases, the curve is a spiral which climbs up the side of an elliptical cylinder as sketched in Figure 1. This curve is called an *elliptical helix*.

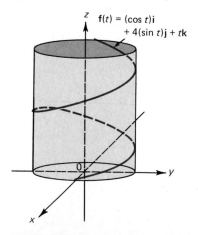

z $\mathbf{f}(t) = (\cos t)\mathbf{i} + 4(\sin t)\mathbf{j} + t\mathbf{k}$

Figure 1

DEFINITION 2. If $\mathbf{f}(t) = f_1(t)\mathbf{i} + f_2(t)\mathbf{j} + f_3(t)\mathbf{k}$, and if $\lim_{t \to t_0} f_1(t), \lim_{t \to t_0} f_2(t)$, and $\lim_{t \to t_0} f_3(t)$ exist, then we define

$$\lim_{t \to t_0} \mathbf{f}(t) = \left[\lim_{t \to t_0} f_1(t)\right]\mathbf{i} + \left[\lim_{t \to t_0} f_2(t)\right]\mathbf{j} + \left[\lim_{t \to t_0} f_3(t)\right]\mathbf{k}. \tag{2}$$

DEFINITION 3

$$\frac{d\mathbf{f}(t)}{dt} = \mathbf{f}'(t) = \lim_{\Delta t \to 0} \frac{\mathbf{f}(t + \Delta t) - \mathbf{f}(t)}{\Delta t}. \tag{3}$$

THEOREM 1. $\mathbf{f}$ is differentiable if and only if its component functions f_1, f_2, and f_3 are differentiable and

$$\boxed{\mathbf{f}'(t) = f_1'(t)\mathbf{i} + f_2'(t)\mathbf{j} + f_3'(t)\mathbf{k}.} \tag{4}$$

EXAMPLE 2. Find the derivative of

$$\mathbf{f}(t) = (\cos t)\mathbf{i} + (\sin t)\mathbf{j} + t\mathbf{k}.$$

SOLUTION. $\mathbf{f}'(t) = -(\sin t)\mathbf{i} + (\cos t)\mathbf{j} + \mathbf{k}.$

The following theorem is new.

THEOREM 2. If $\mathbf{f}$ and $\mathbf{g}$ are differentiable, then $\mathbf{f} \times \mathbf{g}$ is differentiable and

$$\boxed{(\mathbf{f} \times \mathbf{g})' = (\mathbf{f}' \times \mathbf{g}) + (\mathbf{f} \times \mathbf{g}').} \tag{5}$$

PROOF. The proof is an easy application of the product rule of differentiation and is left as an exercise (see Problem 25).

EXAMPLE 3. Let $\mathbf{f}(t) = (\cos t)\mathbf{i} + (\sin t)\mathbf{j} + t\mathbf{k}$ and $\mathbf{g}(t) = t^2\mathbf{i} + t^3\mathbf{j} + t^4\mathbf{k}$. Calculate

$$\frac{d}{dt}(\mathbf{f} \times \mathbf{g}).$$

SOLUTION

$$
\begin{aligned}
(\mathbf{f} \times \mathbf{g})' &= (\mathbf{f}' \times \mathbf{g}) + (\mathbf{f} \times \mathbf{g}') \\
&= [-(\sin t)\mathbf{i} + (\cos t)\mathbf{j} + \mathbf{k}] \times (t^2\mathbf{i} + t^3\mathbf{j} + t^4\mathbf{k}) \\
&\quad + [(\cos t)\mathbf{i} + (\sin t)\mathbf{j} + t\mathbf{k}] \times (2t\mathbf{i} + 3t^2\mathbf{j} + 4t^3\mathbf{k}) \\
&= \begin{vmatrix} \mathbf{i} & \mathbf{j} & \mathbf{k} \\ -\sin t & \cos t & 1 \\ t^2 & t^3 & t^4 \end{vmatrix} + \begin{vmatrix} \mathbf{i} & \mathbf{j} & \mathbf{k} \\ \cos t & \sin t & t \\ 2t & 3t^2 & 4t^3 \end{vmatrix} \\
&= (t^4 \cos t - t^3)\mathbf{i} + (t^2 + t^4 \sin t)\mathbf{j} - (t^3 \sin t + t^2 \cos t)\mathbf{k} \\
&\quad + (4t^3 \sin t - 3t^3)\mathbf{i} + (2t^2 - 4t^3 \cos t)\mathbf{j} + (3t^2 \cos t - 2t \sin t)\mathbf{k} \\
&= (4t^3 \sin t + t^4 \cos t - 4t^3)\mathbf{i} + (t^4 \sin t - 4t^3 \cos t + 3t^2)\mathbf{j} \\
&\quad + [2t^2 \cos t - (t^3 + 2t) \sin t]\mathbf{k}.
\end{aligned}
$$

Facts about arc length and tangent vectors hold exactly as in $\mathbb{R}^3$. Proofs of the parts of the following theorem are identical to the proofs given in Chapter 16.

THEOREM 3. Let **f** have a continuous derivative and let $\mathbf{T}(t)$ and $s(t)$ denote the unit tangent vector and arc length, respectively. Then

    (i)   if $\mathbf{f}'(t) \neq 0$, $\mathbf{T}(t) = \mathbf{f}'(t)/|\mathbf{f}'(t)|$ (see equation 16.3.7);
    (ii) if $|\mathbf{f}(t)|$ is constant, then $\mathbf{f} \cdot \mathbf{f}' = 0$ (see Theorem 16.4.2);
    (iii) $ds/dt = |\mathbf{f}'(t)|$ (see Theorem 16.5.1 and equation (16.5.10));
    (iv) $s(t_1) = \int_{t_0}^{t_1} |\mathbf{f}'(t)|\, dt$ (see Equation (16.5.6)), where $s(t)$ measures arc length from a reference point $P_0 = (x(t_0), y(t_0), z(t_0))$; and
    (v) $\mathbf{T} = d\mathbf{f}/ds$ (see Theorem 16.6.1).

EXAMPLE 4. Let $\mathbf{f}(t) = (\cos t)\mathbf{i} + (\sin t)\mathbf{j} + t\mathbf{k}$. (This is called a *circular helix*.)

    (a) Calculate $\mathbf{T}(t)$ at $t = \pi/3$.
    (b) Find the length of the arc from $t = 0$ to $t = 4$.

SOLUTION

    (a) $\mathbf{f}'(t) = -(\sin t)\mathbf{i} + (\cos t)\mathbf{j} + \mathbf{k}$ and

$$|\mathbf{f}'(t)| = \sqrt{\sin^2 t + \cos^2 t + 1} = \sqrt{2}$$

so that

$$\mathbf{T}(t) = -\frac{\sin t}{\sqrt{2}}\mathbf{i} + \frac{\cos t}{\sqrt{2}}\mathbf{j} + \frac{1}{\sqrt{2}}\mathbf{k}$$

and at $t = \pi/3$,

$$\mathbf{T}(t) = -\frac{\sqrt{3}}{2\sqrt{2}}\mathbf{i} + \frac{1}{2\sqrt{2}}\mathbf{j} + \frac{1}{\sqrt{2}}\mathbf{k}.$$

    (b) Since $ds/dt = \sqrt{2}$

$$s(4) = \int_0^4 \sqrt{2}\, dt = 4\sqrt{2}.$$

The curvature $\kappa(t)$ and the unit normal vector $\mathbf{n}(t)$ are defined differently in $\mathbb{R}^3$. They would have to be. For one thing, there is a whole "plane-full" of unit vectors orthogonal to **T**. Also, curvature cannot be defined as $|d\varphi/ds|$ since there are now three angles which define the direction of C. Therefore, we use an alternative definition of curvature (see Theorem 16.8.1).

DEFINITION 4. If **f** has a continuous derivative, the *curvature* of **f** is given by

$$\boxed{\kappa(t) = \left|\frac{d\mathbf{T}}{ds}\right|.} \qquad (6)$$

Remark. By Theorem 16.8.1, equation (6) is equivalent in $\mathbb{R}^2$ to the definition $\kappa(t) = |d\varphi/ds|$ which was our $\mathbb{R}^2$ definition.

DEFINITION 5. The *principal unit normal vector* **n** is defined by

$$\boxed{\mathbf{n}(t) = \frac{1}{\kappa(t)}\frac{d\mathbf{T}}{ds}.} \qquad (7)$$

Remark. Having defined a unit normal vector, we must show that it is orthogonal to the unit tangent vector.

THEOREM 4. $\mathbf{n}(t) \perp \mathbf{T}(t)$.

PROOF. Since $1 = \mathbf{T} \cdot \mathbf{T}$, we differentiate both sides with respect to s to obtain

$$0 = \mathbf{T} \cdot \frac{d\mathbf{T}}{ds} + \frac{d\mathbf{T}}{ds} \cdot \mathbf{T} = 2\mathbf{T} \cdot \frac{d\mathbf{T}}{ds} = (2\kappa)\mathbf{T} \cdot \mathbf{n}.$$

EXAMPLE 5. Find the curvature and principal unit normal vector, $\kappa(t)$ and $\mathbf{n}(t)$, for the curve of Example 4 at $t = \pi/3$.

SOLUTION. Since $ds/dt = \sqrt{2}$, $s = \sqrt{2}t$ and $t = s/\sqrt{2}$ so that

$$\mathbf{T}(s) = -\frac{\sin(s/\sqrt{2})}{\sqrt{2}}\mathbf{i} + \frac{\cos(s/\sqrt{2})}{\sqrt{2}}\mathbf{j} + \frac{1}{\sqrt{2}}\mathbf{k}$$

and

$$\frac{d\mathbf{T}}{ds} = -\frac{\cos(s/\sqrt{2})}{2}\mathbf{i} - \frac{\sin(s/\sqrt{2})}{2}\mathbf{j}$$

so that

$$\kappa(t) = \sqrt{\frac{\cos^2(s/\sqrt{2})}{4} + \frac{\sin^2(s/\sqrt{2})}{4}} = \frac{1}{2}$$

and

$$\mathbf{n}(t) = \frac{1}{\kappa(t)}\frac{d\mathbf{T}}{ds} = -(\cos t)\mathbf{i} - (\sin t)\mathbf{j}.$$

At $t = \pi/3$,

$$\mathbf{n}\left(\frac{\pi}{3}\right) = -\frac{1}{2}\mathbf{i} - \frac{\sqrt{3}}{2}\mathbf{j}.$$

Remark. When it is inconvenient to solve for t in terms of s, we can calculate $d\mathbf{T}/ds$ directly by the relation

$$\frac{d\mathbf{T}}{ds} = \frac{d\mathbf{T}/dt}{ds/dt}. \tag{8}$$

There is a third vector which is often useful in applications.

DEFINITION 6. The *binormal vector* $\mathbf{B}$ to the curve $\mathbf{f}$ is defined by

$$\mathbf{B} = \mathbf{T} \times \mathbf{n}. \tag{9}$$

From this definition we see that $\mathbf{B}$ is orthogonal to both $\mathbf{T}$ and $\mathbf{n}$. Moreover, since $\mathbf{T} \perp \mathbf{n}$, the angle θ between $\mathbf{T}$ and $\mathbf{n}$ is $\pi/2$ and

$$|\mathbf{B}| = |\mathbf{T} \times \mathbf{n}| = |\mathbf{T}|\,|\mathbf{n}|\sin\theta = 1$$

so that $\mathbf{B}$ is a unit vector.

EXAMPLE 6. Find the binormal vector to the curve of Example 5 at $t = \pi/3$.

SOLUTION

$$\mathbf{B} = \mathbf{T} \times \mathbf{n} = \begin{vmatrix} \mathbf{i} & \mathbf{j} & \mathbf{k} \\ -\dfrac{\sin t}{\sqrt{2}} & \dfrac{\cos t}{\sqrt{2}} & \dfrac{1}{\sqrt{2}} \\ -\cos t & -\sin t & 0 \end{vmatrix} = \dfrac{\sin t}{\sqrt{2}}\mathbf{i} - \dfrac{\cos t}{\sqrt{2}}\mathbf{j} + \dfrac{1}{\sqrt{2}}\mathbf{k}.$$

When $t = \pi/3$, $\mathbf{B} = (\sqrt{3}/2\sqrt{2})\mathbf{i} - (1/2\sqrt{2})\mathbf{j} + (1/\sqrt{2})\mathbf{k}$. The vectors **T**, **n**, and **B** are sketched in Figure 2.

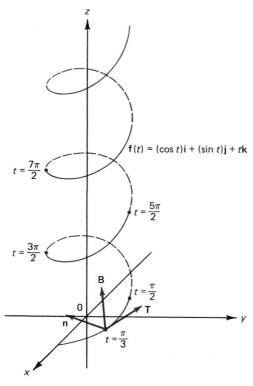

Figure 2

Our definitions of velocity and acceleration vectors remain unchanged in $\mathbb{R}^3$. Suppose that an object is moving in space. Then we can describe its motion parametrically by the vector function

$$\mathbf{f}(t) = f_1(t)\mathbf{i} + f_2(t)\mathbf{j} + f_3(t)\mathbf{k}. \tag{10}$$

In this context **f** is again called the *position vector* of the object and the curve described by **f** is called the *trajectory* of the point.

DEFINITION 7. If $\mathbf{f}'$ and $\mathbf{f}''$ exist, then

$$\text{(i)} \quad \mathbf{v}(t) = \mathbf{f}'(t) = f_1'(t)\mathbf{i} + f_2'(t)\mathbf{j} + f_3'(t)\mathbf{k} \tag{11}$$

is called the *velocity vector,* and

(ii) $\mathbf{a}(t) = \dfrac{d\mathbf{v}}{dt} = \mathbf{f}''(t) = f_1''(t)\mathbf{i} + f_2''(t)\mathbf{j} + f_3''(t)\mathbf{k}$ (12)

is called the *acceleration vector.*

(iii)   The *speed* of the object $v(t)$ is the magnitude of the velocity vector:

$v(t) = |\mathbf{v}(t)|.$ (13)

(iv)   The *acceleration scalar* $a(t)$ is the magnitude of the acceleration vector:

$a(t) = |\mathbf{a}(t)|.$ (14)

EXAMPLE 7. Find the velocity, acceleration, speed, and acceleration scalar of an object whose position vector is given by

$\mathbf{f}(t) = (\cos t)\mathbf{i} + (\sin t)\mathbf{j} + t^3\mathbf{k}.$

SOLUTION

$$\mathbf{v}(t) = -(\sin t)\mathbf{i} + (\cos t)\mathbf{j} + 3t^2\mathbf{k}$$

$$\mathbf{a}(t) = -(\cos t)\mathbf{i} - (\sin t)\mathbf{j} + 6t\mathbf{k}$$

$$v(t) = \sqrt{\sin^2 t + \cos^2 t + (3t^2)^2} = \sqrt{1 + 9t^4}$$

$$a(t) = \sqrt{\cos^2 t + \sin^2 t + (6t)^2} = \sqrt{1 + 36t^2}.$$

For example, at $t = 1$, $v(t) = \sqrt{10}$ and $a(t) = \sqrt{37}$.

As in the $\mathbb{R}^2$ case, the acceleration vector $\mathbf{a}(t)$ can be resolved into its tangential and normal components.

THEOREM 5

$$\mathbf{a}(t) = \frac{d^2s}{dt^2}\mathbf{T} + \left(\frac{ds}{dt}\right)^2\kappa\mathbf{n}.$$

PROOF. $\mathbf{v}(t) = \dfrac{d\mathbf{f}}{dt} = \dfrac{d\mathbf{f}}{ds}\dfrac{ds}{dt} = \mathbf{T}\dfrac{ds}{dt}$ (from Theorem 3(v)). Then

$$\mathbf{a}(t) = \frac{d\mathbf{v}}{dt} = \mathbf{T}\frac{d^2s}{dt^2} + \frac{d\mathbf{T}}{dt}\frac{ds}{dt} = \mathbf{T}\frac{d^2s}{dt^2} + \left(\frac{d\mathbf{T}}{ds}\frac{ds}{dt}\right)\left(\frac{ds}{dt}\right)$$

$$= \frac{d^2s}{dt^2}\mathbf{T} + \kappa\mathbf{n}\left(\frac{ds}{dt}\right)^2$$

and the theorem is proved.

Using this result, we can prove

THEOREM 6

$$\kappa = \frac{|\mathbf{f}' \times \mathbf{f}''|}{|\mathbf{f}'|^3}.$$

PROOF. $\mathbf{f}' \times \mathbf{f}'' = (\mathbf{T}\, ds/dt) \times [(d^2s/dt^2)\mathbf{T} + \kappa\mathbf{n}(ds/dt)^2]$. Now $\mathbf{T} \times \mathbf{T} = 0$ and $\mathbf{T} \times \mathbf{n} = \mathbf{B}$. Thus using Theorem 17.4.2(vii),

$$\mathbf{f}' \times \mathbf{f}'' = \left(\frac{ds}{dt}\right)^3 \kappa\mathbf{B}.$$

Then, taking absolute values and using the fact that $|ds/dt| = |\mathbf{f}'|$,

$$|\mathbf{f}' \times \mathbf{f}''| = \kappa|\mathbf{f}'|^3$$

from which the result follows.

Theorem 6 is useful for calculating curvature when it is not easy to write $\mathbf{f}$ in terms of the arc length parameter s.

EXAMPLE 8. Calculate the curvature of $\mathbf{f} = t^2\mathbf{i} + t^3\mathbf{j} + t^4\mathbf{k}$ at $t = 1$.

SOLUTION. Here

$$\mathbf{f}' = 2t\mathbf{i} + 3t^2\mathbf{j} + 4t^3\mathbf{k}$$
$$|\mathbf{f}'| = \sqrt{4t^2 + 9t^4 + 16t^6} = t\sqrt{4 + 9t^2 + 16t^4}$$
$$\mathbf{f}'' = 2\mathbf{i} + 6t\mathbf{j} + 12t^2\mathbf{k}$$

$$\mathbf{f}' \times \mathbf{f}'' = \begin{vmatrix} \mathbf{i} & \mathbf{j} & \mathbf{k} \\ 2t & 3t^2 & 4t^3 \\ 2 & 6t & 12t^2 \end{vmatrix} = 12t^4\mathbf{i} - 16t^3\mathbf{j} + 6t^2\mathbf{k}$$

$$|\mathbf{f}' \times \mathbf{f}''| = \sqrt{144t^8 + 256t^6 + 36t^4} = t^2\sqrt{144t^4 + 256t^2 + 36}.$$

Thus

$$\kappa = \frac{\sqrt{144t^4 + 256t^2 + 36}}{t(4 + 9t^2 + 16t^4)^{3/2}}.$$

At $t = 1$, $\kappa = \sqrt{436}/29^{3/2} \approx 0.1337$.

PROBLEMS 17.7

In Problems 1–6 find the unit tangent vector $\mathbf{T}$ for the given value of t.

1. $\mathbf{f}(t) = t\mathbf{i} + t^2\mathbf{j} + t^3\mathbf{k}$; $t = 1$
2. $\mathbf{f}(t) = t^3\mathbf{i} + t^5\mathbf{j} + t^7\mathbf{k}$; $t = 1$
3. $\mathbf{f}(t) = t\mathbf{i} + e^t\mathbf{j} + e^{-t}\mathbf{k}$; $t = 0$
4. $\mathbf{f}(t) = t^2\mathbf{i} + t^2\mathbf{j} + t^{5/2}\mathbf{k}$; $t = 4$
5. $\mathbf{f}(t) = 4(\cos 2t)\mathbf{i} + 9(\sin 2t)\mathbf{j} + t\mathbf{k}$; $t = \pi/4$
6. $\mathbf{f}(t) = (\cosh t)\mathbf{i} + (\sinh t)\mathbf{j} + t^2\mathbf{k}$; $t = 0$.

7. Find the arc length of the curve $\mathbf{f} = 2(\cos 3t)\mathbf{i} + 2(\sin 3t)\mathbf{j} + t^2\mathbf{k}$ between $t = 0$ and $t = 10$.
8. Find the arc length of the curve $\mathbf{f} = e^t(\cos 2t)\mathbf{i} + e^t(\sin 2t)\mathbf{j} + e^t\mathbf{k}$ between $t = 1$ and $t = 4$.
9. Find the arc length of the curve $\mathbf{f} = \frac{2}{3}t^3\mathbf{i} + (1 + t^{9/2})\mathbf{j} + (1 - t^{9/2})\mathbf{k}$ between $t = 0$ and $t = 2$.

In Problems 10–15 find the velocity vector, the speed, the acceleration vector and the acceleration scalar for the given value of t.

10. $\mathbf{f} = t^2\mathbf{i} + t^3\mathbf{j} + t^4\mathbf{k}; \ t = 2$

11. $\mathbf{f} = (\cos t)\mathbf{i} + (\sin t)\mathbf{j} + t^4\mathbf{k}; \ t = 1$

12. $\mathbf{f} = (\ln t)\mathbf{i} + \dfrac{1}{t}\mathbf{j} + \dfrac{1}{t^2}\mathbf{k}; \ t = 1$

13. $\mathbf{f} = (\cosh t)\mathbf{i} + (\sinh t)\mathbf{j} + t\mathbf{k}; \ t = 0$

14. $\mathbf{f} = e^t\mathbf{i} + e^{-t}\mathbf{j} + \sqrt{t}\mathbf{k}; \ t = 4$

15. $\mathbf{f} = \sqrt{t}\mathbf{i} + \sqrt[3]{t}\mathbf{j} + \sqrt[4]{t}\mathbf{k}; \ t = 1.$

In Problems 16–23 find the unit tangent vector **T**, the unit normal vector **n**, the binormal vector **B** and the curvature κ at the given value of t. Check that $\mathbf{n} \cdot \mathbf{T} = 0$. [*Hint:* Use equation (8).]

16. $\mathbf{f} = a(\sin t)\mathbf{i} + a(\cos t)\mathbf{j} + t\mathbf{k}; \ t = \pi/4; \ a > 0$

17. $\mathbf{f} = a(\sin t)\mathbf{i} + a(\cos t)\mathbf{j} + t\mathbf{k}; \ t = \pi/6; \ a > 0$

18. $\mathbf{f} = a(\cos t)\mathbf{i} + b(\sin t)\mathbf{j} + t\mathbf{k}; \ t = 0; \ a > 0, \ b > 0, \ a \neq b$

19. $\mathbf{f} = a(\cos t)\mathbf{i} + b(\sin t)\mathbf{j} + t\mathbf{k}; \ t = \pi/2; \ a > 0, \ b > 0, \ a \neq b$

20. $\mathbf{f} = t\mathbf{i} + t^2\mathbf{j} + t^3\mathbf{k}; \ t = 0$

21. $\mathbf{f} = e^t(\cos 2t)\mathbf{i} + e^t(\sin 2t)\mathbf{j} + e^t\mathbf{k}; \ t = 0$

22. $\mathbf{f} = e^t(\cos 2t)\mathbf{i} + e^t(\sin 2t)\mathbf{j} + e^t\mathbf{k}; \ t = 0$

23. $\mathbf{f} = \mathbf{i} + t\mathbf{j} + t^2\mathbf{k}; \ t = 1.$

24.   Let $\mathbf{f} = a(\cos t)\mathbf{i} + a(\sin t)\mathbf{j} + t\mathbf{k}$ (a circular helix). Show that the angle between the unit tangent vector **T** and the z-axis is constant.

25.   Show by writing out the component functions that $(\mathbf{f} \times \mathbf{g})' = (\mathbf{f}' \times \mathbf{g}) + (\mathbf{f} \times \mathbf{g}')$.

26.   Show that the curvature of a straight line in space is zero.

*27.   Let **B** be the binormal vector to a curve **f**, and let s denote arc length. Show that if $d\mathbf{B}/ds \neq 0$, then $d\mathbf{B}/ds \perp \mathbf{B}$ and $d\mathbf{B}/ds \perp \mathbf{T}$. [*Hint:* Differentiate $\mathbf{B} \cdot \mathbf{T} = 0$ and $\mathbf{B} \cdot \mathbf{B} = 1$.]

*28.   Since, from Problem 27, $d\mathbf{B}/ds$ is orthogonal to **B** and **T**, it must be parallel to **n**. Hence

$$\frac{d\mathbf{B}}{ds} = -\tau\mathbf{n}$$

for some number τ. This number is called the *torsion* of the curve C and is a measure of how much the curve twists. Show that

$$\frac{d\mathbf{n}}{ds} = -\kappa\mathbf{T} + \tau\mathbf{B}.$$

**29.   Show that

$$\mathbf{f}'''(t) = \left(\frac{d^3s}{dt^3} - \left(\frac{ds}{dt}\right)^3\kappa^2\right)\mathbf{T} + \left(3\left(\frac{d^2s}{dt^2}\right)\frac{ds}{dt}\kappa + \left(\frac{ds}{dt}\right)^2\frac{d\kappa}{dt}\right)\mathbf{n} + \left(\frac{ds}{dt}\right)^3\kappa\tau\mathbf{B}.$$

**30.   Use the result of Problem 29 to show that

$$\tau = \frac{\mathbf{f}''' \cdot (\mathbf{f}' \times \mathbf{f}'')}{\kappa^2(ds/dt)^6}.$$

*31.   Calculate the torsion of $\mathbf{f} = (\cos t)\mathbf{i} + (\sin t)\mathbf{j} + t\mathbf{k}$ at $t = \pi/6$.

*32.   Calculate the torsion of $\mathbf{f} = e^t(\cos t)\mathbf{i} + e^t(\sin t)\mathbf{j} + e^t\mathbf{k}$ at $t = 0$.

17.8 Cylindrical and Spherical Coordinates

In this chapter, so far, we have represented points using rectangular (Cartesian) coordinates. However, there are many ways to represent points in space, some of which are quite bizarre. In this section we will briefly introduce two common ways to represent points in space. The first is the generalization of the polar coordinate system in the plane.

In the *cylindrical coordinate system* a point P is given by

$$P = (r, \theta, z) \tag{1}$$

where $r > 0, 0 \le \theta < 2\pi$, r and θ are polar coordinates of the projection of P onto the xy-plane, called the *polar plane,* and z is the distance (measured in the positive direction) of this plane from P (see Figure 1). In this figure, $\overrightarrow{OQ}$ is the projection of $\overrightarrow{OP}$ on the xy-plane.

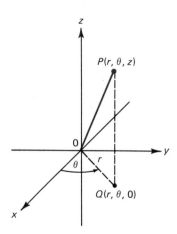

Figure 1

EXAMPLE 1. Discuss the graphs of the equations in cylindrical coordinates

(a) $r = c, c > 0$ (b) $\theta = c$ (c) $z = c$
(d) $r_1 \le r \le r_2, \theta_1 \le \theta \le \theta_2, z_1 \le z \le z_2$.

SOLUTION. (a) If $r = c$, a constant, then θ and z can vary freely and we obtain a right circular cylinder with radius c. (See Figure 2a.) This is the generalization of the circle whose equation in polar coordinates is $r = c$ and is the reason why the system is called the cylindrical coordinate system.

(b) If $\theta = c$, we obtain a plane through the z-axis (see Figure 2b).

(c) If $z = c$, we obtain a plane parallel to the polar plane. This is sketched in Figure 2c.

(d) $r_1 \le r \le r_2$ gives the region between the cylinders $r = r_1$ and $r = r_2$. $\theta_1 \le \theta \le \theta_2$ is the "wedge-shaped" region between the planes $\theta = \theta_1$ and $\theta = \theta_2$ (there are two such regions). Finally, $z_1 \le z \le z_2$ is the "slice" of space between the planes $z = z_1$ and $z = z_2$. Putting this together we get the rectangularly shaped solid in Figure 3.

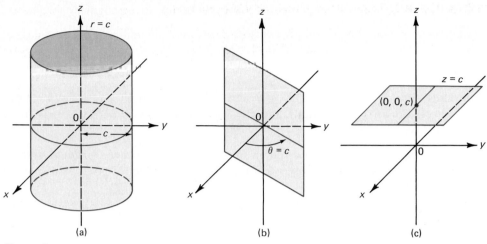

(a)   (b)   (c)

Figure 2

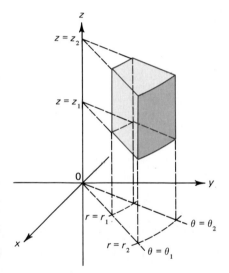

Figure 3

To translate from cylindrical to rectangular coordinates and back again, we have the formulas

$$x = r \cos \theta, \qquad y = r \sin \theta, \qquad z = z \tag{2}$$

and

$$r = \sqrt{x^2 + y^2}, \qquad \tan \theta = \frac{y}{x}, \qquad z = z. \tag{3}$$

These need no proof as they follow from the formulas for polar coordinates in Chapter 11.

EXAMPLE 2. Convert $P = (2, \pi/3, -5)$ from cylindrical to rectangular coordinates.

SOLUTION. $x = r \cos \theta = 2 \cos \pi/3 = 1, y = r \sin \theta = 2 \sin \pi/3 = \sqrt{3}$, and $z = -5$. Thus, in rectangular coordinates $P = (1, \sqrt{3}, -5)$.

EXAMPLE 3. Convert $(-3, 3, 7)$ from rectangular to cylindrical coordinates.

SOLUTION. $r = \sqrt{3^2 + 3^2} = 3\sqrt{2}$, $\theta = \tan^{-1}(3/-3) = \tan^{-1}(-1) = 3\pi/4$ (since $(-3, 3)$ is in the second quadrant), and $z = 7$ so that $(-3, 3, 7) = (3\sqrt{2}, 3\pi/4, 7)$ in cylindrical coordinates.

The second new coordinate system in space is the *spherical coordinate system*. This is not a generalization of any system in the plane. A typical point P in space is represented as

$$P = (\rho, \theta, \varphi) \tag{4}$$

where $\rho \geq 0, 0 \leq \theta < 2\pi, 0 \leq \varphi \leq \pi$, and

ρ is the (positive) distance between the point and the origin,
θ is the same as in cylindrical coordinates, and
φ is the angle between $\overrightarrow{OP}$ and the positive z-axis.

This is illustrated in Figure 4.

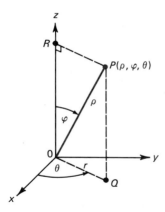

Figure 4

EXAMPLE 4. Discuss the graphs of the equations in spherical coordinates

(a) $\rho = c, c > 0$ (b) $\theta = c$ (c) $\varphi = c$
(d) $\rho_1 \leq \rho \leq \rho_2, \theta_1 \leq \theta \leq \theta_2, \varphi_1 \leq \varphi \leq \varphi_2$.

SOLUTION. (a)   The set of points for which $\rho = c$ is the set of points c units from the origin. This constitutes a sphere centered at the origin with radius c and is what gives the coordinate system its name. This is sketched in Figure 5a.

(b)   As with cylindrical coordinates, the graph is a plane containing the z-axis. See Figure 5b.

(c)   If $0 < c < \pi/2$, then the graph of $\varphi = c$ is obtained by rotating the vector

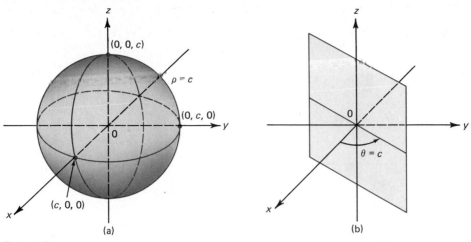

Figure 5

$\overrightarrow{OP}$ around the z-axis. This yields the circular cone sketched in Figure 6a. If $\pi/2 < c < \pi$, then we obtain the cone of Figure 6b. Finally, if $c = 0$ we obtain the positive z-axis and if $c = \pi$ we obtain the negative z-axis.

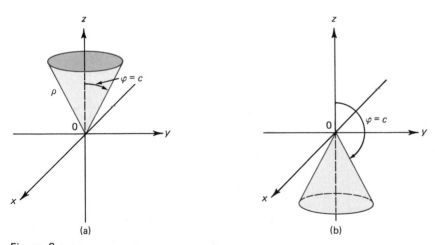

Figure 6

(d) $\rho_1 \leq \rho \leq \rho_2$ gives the region between the sphere $\rho = \rho_1$ and the sphere $\rho = \rho_2$. $\theta_1 \leq \theta \leq \theta_2$ consists, as before, of the two wedge-shaped regions between the planes $\theta = \theta_1$ and $\theta = \theta_2$. Finally, $\varphi_1 \leq \varphi \leq \varphi_2$ yields the region between the cones $\varphi = \varphi_1$ and $\varphi = \varphi_2$. Putting this together, we obtain the solid (called a *spherical wedge*) half of which is sketched in Figure 7.

To convert from spherical to rectangular coordinates, we have, from Figure 4

$$x = \overline{OQ} \cos \theta \qquad \text{and} \qquad y = \overline{OQ} \sin \theta.$$

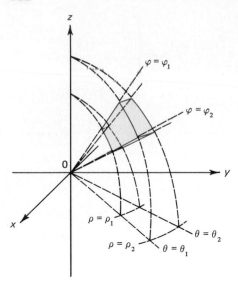

Figure 7

But, from Figure 4,

$$\overline{OQ} = \overline{PR} = \rho \sin \varphi \tag{5}$$

so that

$$\boxed{\begin{aligned} x &= \rho \sin \varphi \cos \theta \\ &\text{and} \\ y &= \rho \sin \varphi \sin \theta. \end{aligned}} \tag{6}$$
$$\tag{7}$$

Finally, from triangle OPR we have

$$\boxed{z = \rho \cos \varphi.} \tag{8}$$

To convert to spherical coordinates from rectangular coordinates, we have

$$\boxed{\overline{OP} = \sqrt{x^2 + y^2 + z^2} = \rho} \tag{9}$$

so that, from (8),

$$\boxed{\cos \varphi = \frac{z}{\rho} = \frac{z}{\sqrt{x^2 + y^2 + z^2}}}$$

Knowing ρ and φ, we can then calculate θ from equation (6) or (7).

EXAMPLE 5. Convert $\left(4, \dfrac{\pi}{6}, \dfrac{\pi}{4}\right)$ from spherical to rectangular coordinates.

SOLUTION

$$x = \rho \sin \varphi \cos \theta = 4 \sin \frac{\pi}{4} \cos \frac{\pi}{6} = 4 \frac{\sqrt{2}}{2} \frac{\sqrt{3}}{2} = \sqrt{6},$$

$$y = \rho \sin \varphi \sin \theta = 4 \sin \frac{\pi}{4} \sin \frac{\pi}{6} = 4 \frac{\sqrt{2}}{2} \frac{1}{2} = \sqrt{2},$$

and

$$z = \rho \cos \varphi = 4 \cos \frac{\pi}{4} = 2\sqrt{2}.$$

Therefore, in rectangular coordinates, $(4, \pi/6, \pi/4)$ becomes $(\sqrt{6}, \sqrt{2}, 2\sqrt{2})$.

EXAMPLE 6.   Convert $(1, \sqrt{3}, -2)$ from rectangular to spherical coordinates.

SOLUTION. $\rho = \sqrt{x^2 + y^2 + z^2} = \sqrt{1 + 3 + 4} = \sqrt{8} = 2\sqrt{2}$. Then, from (7),

$$\cos \varphi = \frac{z}{\rho} = \frac{-2}{2\sqrt{2}} = -\frac{1}{\sqrt{2}}$$

so that $\varphi = 3\pi/4$. Finally, from (5),

$$\cos \theta = \frac{x}{\rho \sin \varphi} = \frac{1}{(2\sqrt{2})(1/\sqrt{2})} = \frac{1}{2}$$

so that $\theta = \pi/3$, and we find that $(1, \sqrt{3}, -2)$ is $(2\sqrt{2}, \pi/3, 3\pi/4)$ in spherical coordinates.

We will not discuss the graphs of different kinds of surfaces given in cylindrical and spherical coordinates as that would take us too far afield. Both coordinate systems are useful in a wide variety of physical applications. For example, points on the earth and its interior are much more easily described using a spherical coordinate system than a rectangular system.

We will return to cylindrical and spherical coordinates in Chapter 19 in our discussion of multiple integration. For convenience, we reproduce the formulas for changing from one coordinate system to another that we derived in this section.

(i)   From cylindrical to rectangular coordinates

$$\begin{array}{l} x = r \cos \theta \\ y = r \sin \theta \\ z = z. \end{array}$$

(ii)   From rectangular to cylindrical coordinates

$$\begin{array}{l} r = \sqrt{x^2 + y^2} \\ \tan \theta = \dfrac{y}{x}, \quad 0 \le \theta < 2\pi \\ z = z. \end{array}$$

(iii)   From spherical to rectangular coordinates

$$x = \rho \sin \varphi \cos \theta$$
$$y = \rho \sin \varphi \sin \theta$$
$$z = \rho \cos \varphi.$$

(iv)   From rectangular to spherical coordinates

$$\rho = \sqrt{x^2 + y^2 + z^2}$$

$$\varphi = \cos^{-1}\frac{z}{\rho}, \qquad 0 \le \varphi \le \pi$$

$$\cos \theta = \frac{x}{\rho \sin \varphi} \quad \text{or} \quad \sin \theta = \frac{y}{\rho \sin \varphi}, \qquad 0 \le \theta < 2\pi.$$

PROBLEMS 17.8

In Problems 1–9 convert from cylindrical to rectangular coordinates.

1. $(2, \pi/3, 5)$ **2.** $(1, 0, -3)$ **3.** $(8, 2\pi/3, 1)$
4. $(4, \pi/2, -7)$ **5.** $(3, 3\pi/4, 2)$ **6.** $(10, 5\pi/3, -3)$
7. $(10, \pi, -3)$ **8.** $(13, 3\pi/2, 4)$ **9.** $(7, 5\pi/4, 2)$.

In Problems 10–18 convert from rectangular to cylindrical coordinates.

10. $(1, 0, 0)$ **11.** $(0, 1, 0)$ **12.** $(0, 0, 1)$
13. $(1, 1, 2)$ **14.** $(-1, 1, 4)$ **15.** $(2, 2\sqrt{3}, -5)$
16. $(2\sqrt{3}, -2, 8)$ **17.** $(2, -2\sqrt{3}, 4)$ **18.** $(-2\sqrt{3}, -2, 1)$.

In Problems 19–27 convert from spherical to rectangular coordinates.

19. $(2, 0, \pi/3)$ **20.** $(4, \pi/4, \pi/4)$ **21.** $(6, \pi/2, \pi/3)$
22. $(3, \pi/6, 5\pi/6)$ **23.** $(7, 7\pi/4, 3\pi/4)$ **24.** $(3, \pi/2, \pi/2)$
25. $(4, \pi/3, 2\pi/3)$ **26.** $(4, 2\pi/3, \pi/3)$ **27.** $(5, 11\pi/6, 5\pi/6)$.

In Problems 28–36 convert from rectangular to spherical coordinates.

28. $(1, 1, 0)$ **29.** $(1, 1, \sqrt{2})$ **30.** $(1, -1, \sqrt{2})$
31. $(-1, -1, \sqrt{2})$ **32.** $(1, -\sqrt{3}, 2)$ **33.** $(-\sqrt{3}, -1, 2)$
34. $(2, \sqrt{3}, 4)$ **35.** $(-2, \sqrt{3}, -4)$ **36.** $(-\sqrt{3}, -2, -4)$.

37. Write the equation $x^2 + y^2 + z^2 = 25$ in cylindrical and spherical coordinates.
38. Write the plane $ax + by + cz = d$ in cylindrical and spherical coordinates.
39. Write the equation $r = 9 \sin \theta$ in rectangular coordinates.
40. Write the equation $r^2 \sin 2\theta = z^3$ in rectangular coordinates.
41. Write the surface $x^2 + y^2 - z^2 = 1$ in cylindrical and spherical coordinates.
42. Write the surface $x^2 + 4y^2 + 4z^2 = 1$ in cylindrical and rectangular coordinates.
43. Write the equation $z = r^2$ in rectangular coordinates.
44. Write the equation $\rho \cos \varphi = 1$ in rectangular coordinates.
45. Write the equation $\rho^2 \sin \varphi \cos \varphi = 1$ in rectangular coordinates.
46. Write the equation $\rho = \sin \theta \cos \varphi$ in rectangular coordinates.
47. Write the equation $z = r^2 \sin 2\theta$ in rectangular coordinates.
48. Write the equation $\rho = 2 \cot \theta$ in rectangular coordinates.

49. Write the equation $x^2 + (y - 3)^2 + z^2 = 9$ in spherical coordinates.
50. Using the dot product, show that $\cos \varphi = z/\rho$.

Review Exercises for Chapter Seventeen

In Exercises 1–4 sketch the two given points and then find the distance between them.

1. $(3, 1, 2); (-1, -3, -4)$ **2.** $(4, -1, 7); (-5, 1, 3)$
3. $(-2, 4, -8); (0, 0, 6)$ **4.** $(2, -7, 0); (0, 5, -8)$.
5. Find the equation of the sphere centered at $(-1, 4, 2)$ with radius 3.
6. Show that $x^2 + y^2 + z^2 - 4x + 8y + 10z = 12$ is the equation of a sphere and find its center and radius.
7. Show that the points $(1, 3, 0)$, $(3, -1, -2)$, and $(-1, 7, 2)$ are collinear.

In Exercises 8–11 find the magnitude and the direction cosines of the given vector.

8. $\mathbf{v} = 2\mathbf{i} - \mathbf{k}$ **9.** $\mathbf{v} = 3\mathbf{j} + 11\mathbf{k}$
10. $\mathbf{v} = \mathbf{i} - 2\mathbf{j} - 3\mathbf{k}$ **11.** $\mathbf{v} = -4\mathbf{i} + \mathbf{j} + 6\mathbf{k}$.

12. Find a unit vector in the direction of $\overrightarrow{PQ}$, where $P = (3, -1, 2)$ and $Q = (-4, 1, 7)$.
13. Find a unit vector whose direction is opposite that of $\overrightarrow{PQ}$ where $P = (1, -3, 0)$ and $Q = (-7, 1, -4)$.

In Exercises 14–21 let $\mathbf{u} = \mathbf{i} - 2\mathbf{j} + 3\mathbf{k}$, $\mathbf{v} = -3\mathbf{i} + 2\mathbf{j} + 5\mathbf{k}$, and $\mathbf{w} = 2\mathbf{i} - 4\mathbf{j} + \mathbf{k}$. Calculate

14. $\mathbf{u} - \mathbf{v}$ **15.** $3\mathbf{v} + 5\mathbf{w}$ **16.** $\text{Proj}_{\mathbf{v}} \mathbf{w}$
17. $\text{Proj}_{\mathbf{w}} \mathbf{u}$ **18.** $2\mathbf{u} - 4\mathbf{v} + 7\mathbf{w}$ **19.** $\mathbf{u} \cdot \mathbf{w} - \mathbf{w} \cdot \mathbf{v}$
20. the angle between $\mathbf{u}$ and $\mathbf{v}$ **21.** the angle between $\mathbf{v}$ and $\mathbf{w}$.

22. Find the distance from the point $P = (3, -1, 2)$ to the line passing through the points $Q = (-2, -1, 6)$ and $R = (0, 1, -8)$.
23. Find the work done when a force of 4 nt acting in the direction of the vector $\mathbf{v} = -\mathbf{i} + \mathbf{j} + \mathbf{k}$ moves an object from $(2, 1, -6)$ to $(3, 5, 8)$ (distance is in meters).

In Exercises 24–27 find a vector equation, parametric equations, and symmetric equations of the given line.

24. Containing $(3, -1, 4)$ and $(-1, 6, 2)$
25. Containing $(-4, 1, 0)$ and $(3, 0, 7)$
26. Containing $(3, 1, 2)$ and parallel to $3\mathbf{i} - \mathbf{j} - \mathbf{k}$
27. Containing $(1, -2, -3)$ and parallel to $\dfrac{x + 1}{5} = \dfrac{y - 2}{-3} = \dfrac{z - 4}{2}$.

28. Show that the lines L_1: $x = 3 - 2t$, $y = 4 + t$, $z = -2 + 7t$ and L_2: $x = -3 + s$, $y = 2 - 4s$, $z = 1 + 6s$ have no points of intersection.
29. Find the distance from the origin to the line passing through the point $(3, 1, 5)$ and having the direction $\mathbf{v} = 2\mathbf{i} - \mathbf{j} + \mathbf{k}$.

In Exercises 30–33 find the cross product $\mathbf{u} \times \mathbf{v}$.

30. $\mathbf{u} = 3\mathbf{i} - \mathbf{j}$; $\mathbf{v} = 2\mathbf{i} + 4\mathbf{k}$
31. $\mathbf{u} = 7\mathbf{j}$; $\mathbf{v} = \mathbf{i} - \mathbf{k}$
32. $\mathbf{u} = 4\mathbf{i} - \mathbf{j} + 7\mathbf{k}$; $\mathbf{v} = -7\mathbf{i} + \mathbf{j} - 2\mathbf{k}$
33. $\mathbf{u} = -2\mathbf{i} + 3\mathbf{j} - 4\mathbf{k}$; $\mathbf{v} = -3\mathbf{i} + \mathbf{j} - 10\mathbf{k}$.

34. Find two unit vectors orthogonal to both $\mathbf{u} = \mathbf{i} - \mathbf{j} + 3\mathbf{k}$ and $\mathbf{v} = -2\mathbf{i} - 3\mathbf{j} + 4\mathbf{k}$.

35. Find the equation of the line passing through $(-1, 2, 4)$ and orthogonal to L_1: $(x-1)/4 = (y+6)/3 = z/-2$ and L_2: $(x+3)/5 = (y-1)/1 = (z+3)/4$.

36. Calculate the area of the parallelogram with the adjacent vertices $(1, 4, -2)$, $(-3, 1, 6)$ and $(1, -2, 3)$.

37. Calculate the area of the triangle with vertices at $(2, 1, 3)$, $(-4, 1, 7)$ and $(-1, -1, 3)$.

38. Calculate the volume of the parallelepiped determined by the vectors $\mathbf{i} + \mathbf{j}$, $2\mathbf{i} - 3\mathbf{k}$, and $2\mathbf{j} + 7\mathbf{k}$.

In Exercises 39–41 find the equation of the plane containing the given point and orthogonal to the given normal vector.

39. $P = (1, 3, -2)$; $\mathbf{N} = \mathbf{i} + \mathbf{k}$ 40. $P = (1, -4, 6)$; $\mathbf{N} = 2\mathbf{j} - 3\mathbf{k}$

41. $P = (-4, 1, 6)$; $\mathbf{N} = 2\mathbf{i} - 3\mathbf{j} + 5\mathbf{k}$.

42. Find the equation of the plane containing the points $(-2, 4, 1)$, $(3, -7, 5)$, and $(-1, -2, -1)$.

43. Find all points of intersection of the planes
    π_1: $-x + y + z = 3$,   and π_2: $-4x + 2y - 7z = 5$.

44. Find all points of intersection of the planes
    π_1: $-4x + 6y + 8z = 12$,   and π_2: $2x - 3y - 4z = 5$.

45. Find all points of intersection of the planes
    π_1: $3x - y + 4z = 8$,   and π_2: $-3x - y - 11z = 0$.

46. Find the distance from $(1, -2, 3)$ to the plane $2x - y - z = 6$.

47. Find the angle between the planes of Exercise 43.

48. Show that the position vectors $\mathbf{u} = \mathbf{i} - 2\mathbf{j} + \mathbf{k}$, $\mathbf{v} = 3\mathbf{i} + 2\mathbf{j} - 3\mathbf{k}$, and $\mathbf{w} = 9\mathbf{i} - 2\mathbf{j} - 3\mathbf{k}$ are coplanar and find the equation of the plane containing them.

In Exercises 49–51 draw a sketch of the given cylinder. The directrix C is given. The generatrix L is the axis of the variable missing in the equation.

49. $x = \cos y$ 50. $y = z^2$ 51. $z = \sqrt[3]{x}$.

In Exercises 52–58 identify the quadric surface and sketch it.

52. $x^2 + y^2 = 9$ 53. $x^2 - \dfrac{y^2}{4} + \dfrac{z^2}{9} = 1$

54. $x^2 - \dfrac{y^2}{4} - \dfrac{z^2}{9} = 1$ 55. $-9x^2 + 16y^2 - 9z^2 = 25$

56. $4x^2 + y^2 + 4z^2 - 8y = 0$ 57. $x = \dfrac{y^2}{4} - \dfrac{z^2}{9}$

58. $y^2 + z^2 = x^2$.

In Exercises 59–61 find the unit tangent vector $\mathbf{T}$ for the given value for t.

59. $\mathbf{f}(t) = t^3\mathbf{i} - \sqrt{t}\mathbf{j} + 2t\mathbf{k}$; $t = 1$

60. $\mathbf{f}(t) = e^t\mathbf{i} + e^{-t}\mathbf{j} + t^3\mathbf{k}$; $t = 0$

61. $\mathbf{f}(t) = (\cos 3t)\mathbf{i} - (\sin 3t)\mathbf{j} + t\mathbf{k}$; $t = \pi/9$.

62. Find the arc length of the curve $\mathbf{f} = 3(\sin 2t)\mathbf{i} + 3(\cos 2t)\mathbf{j} + t^2\mathbf{k}$ between $t = 0$ and $t = 5$.

63. If the position vector of a moving particle is given by $\mathbf{f} = t^3\mathbf{i} + (\cos t)\mathbf{j} - (\sin t)\mathbf{k}$, find the velocity, the speed, the acceleration vector and the acceleration scalar when $t = \pi/4$.

64. Answer the questions in Exercise 63 for $\mathbf{f} = e^t\mathbf{i} + (\ln t)\mathbf{j} - t^{3/2}\mathbf{k}$ at $t = 1$.

In Exercises 65–67 find the unit tangent vector **T**, the unit normal vector **n**, the binormal vector **B**, and the curvature κ for the given value of t.

65. $\mathbf{f} = 2(\cos t)\mathbf{i} + 2(\sin t)\mathbf{j} + t\mathbf{k}; \ t = \pi/6$

66. $\mathbf{f} = t^2\mathbf{i} + t^3\mathbf{j} - t\mathbf{k}; \ t = 0$

67. $\mathbf{f} = e^t\mathbf{i} + e^t(\cos t)\mathbf{j} + e^t(\sin t)\mathbf{k}; \ t = 0.$

68. Find the torsion of the function in Exercise 66 at $t = 0$.

In Exercises 69–76 convert from one coordinate system to another as indicated.

69. $(3, \pi/6, -1)$, cylindrical to rectangular

70. $(2, 2, -4)$, rectangular to cylindrical

71. $(2, 2\pi/3, 4)$, cylindrical to rectangular

72. $(-2, 2\sqrt{3}, -4)$, rectangular to cylindrical

73. $(3, \pi/3, \pi/4)$, spherical to rectangular

74. $(2, 7\pi/3, 2\pi/3)$, spherical to rectangular

75. $(-1, 1, -\sqrt{2})$, rectangular to spherical

76. $(2, -\sqrt{3}, 4)$, rectangular to spherical.

77. Write the equation $x^2 + y^2 + z^2 = 25$ in cylindrical and spherical coordinates.

78. Write the equation $r^2 \cos 2\theta = z^3$ in rectangular coordinates.

79. Write the equation $x^2 - y^2 + z^2 = 1$ in cylindrical and spherical coordinates.

80. Write the equation $z = r^3$ in rectangular coordinates.

81. Write the equation $\rho \sin \varphi = 1$ in rectangular coordinates.

EIGHTEEN

DIFFERENTIATION OF FUNCTIONS OF TWO OR MORE VARIABLES

18.1 Introduction

For most of the functions we have so far encountered in this book, we have been able to write $y = f(x)$. This means that we could write the variable y explicitly in terms of the single variable x. However, in a great variety of applications, it is necessary to write the quantity of interest in terms of two or more variables. We have already encountered this. For example, the volume of a right circular cylinder is given by

$$V = \pi r^2 h,$$

where r is the radius of the cylinder and h is its height. That is, V is a function of the *two* variables r and h.

As a second example, we discussed (see equation (6.2.11)) the *ideal gas law* which relates pressure, volume, and temperature for an ideal gas: We have

$$PV = nRT,$$

where P is the pressure of the gas, V is the volume, T is the absolute temperature (i.e., in degrees Kelvin), n is the number of moles of the gas and R is a constant. Solving for P, we find that

$$P = \frac{nRT}{V}.$$

That is, we can write P as a function of the *three* variables n, T, and V.

As a third example, we saw in Problem 3.2.39 that according to Poiseuille's law, the resistance R of a blood vessel of length l and radius r is given by

$$R = \frac{\alpha l}{r^4},$$

where α is a constant of proportionality. If l and r are allowed to vary, then R is written as a function of the *two* variables l and r.

As a final example, let the vector **v** be given by

$$\mathbf{v} = x\mathbf{i} + y\mathbf{j} + z\mathbf{k}.$$

798

Then the magnitude of **v**, |**v**|, is given by

$$|\mathbf{v}| = \sqrt{x^2 + y^2 + z^2}.$$

That is, the magnitude of **v** is written as a function of the *three* variables x, y, and z.

The four examples we just cited represent only the tip of the iceberg. It is probably fair to say that very few physical, biological, or economic quantities can be properly expressed in terms of one variable alone. Often, we write these quantities in terms of one variable simply because functions of only one variable are the easiest functions to handle.

In this chapter we shall see how many of the operations we have studied in our discussion of the "one-variable" calculus can be extended to functions of several variables. We shall begin by discussing the basic notions of limits and continuity, and shall go on to differentiation and applications of differentiation. In Chapter 19 we shall discuss the integration of functions of several variables.

18.2 Functions of Several Variables

In Chapters 16 and 17 we discussed the notion of vector-valued functions. For example, if $\mathbf{f}(t) = f_1(t)\mathbf{i} + f_2(t)\mathbf{j}$, then for every t in the domain of **f** we obtain a vector $\mathbf{f}(t)$. In defining functions of two or more variables, the situation is somewhat reversed. For example, in the formula for the volume of a cylinder discussed in Section 18.1, we have

$$V = \pi r^2 h. \tag{1}$$

Here the volume V is written as a function of the two variables r and h. Put another way, for every ordered pair of positive real numbers (r, h) there is a unique positive number V such that $V = \pi r^2 h$. To indicate the dependence of V on the variables r and h, we write $V(r, h)$. That is, V can be thought of as a function which assigns a positive real number to every ordered pair of real numbers. Thus we see what we meant when we said that the situation is "reversed." Instead of having a vector function of one variable (a scalar) we have a scalar function of an ordered pair of variables (which is a vector in $\mathbb{R}^2$). We now give a definition of a function of two variables.

DEFINITION 1 (Function of Two Variables). Let D be a subset of $\mathbb{R}^2$. Then a *function of two variables f* is a rule which assigns to each point (ordered pair) (x, y) in D a unique real number which we denote $f(x, y)$. The set D is called the *domain* of f. The set $\{ f(x, y): (x, y) \in D \}$ which is the set of values the function f takes on is called the *range* of f.

EXAMPLE 1. Let $D = \mathbb{R}^2$. For each point $(x, y) \in \mathbb{R}^2$ we assign the number $f(x, y) = x^2 + y^4$. Since $x^2 + y^4 \geq 0$ for every pair of real numbers (x, y), we see that the range of f is $\mathbb{R}^+$, the set of nonnegative real numbers.

When the domain D is not given, we shall take the domain of f to be the largest subset of $\mathbb{R}^2$ for which the expression $f(x, y)$ makes sense.

EXAMPLE 2.   Find the domain and range of the function f given by $f(x, y) = \sqrt{4 - x^2 - y^2}$ and find $f(0, 1)$ and $f(1, -1)$.

SOLUTION. Clearly f is defined when the expression under the square root sign is nonnegative. Thus $D = \{(x, y): x^2 + y^2 \leq 4\}$. This is the disk† centered at the origin with radius 2. This is sketched in Figure 1. Since x^2 and y^2 are nonnegative,

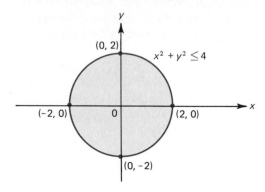

Figure 1

$4 - x^2 - y^2$ is largest when $x = y = 0$. This means that the largest value of $\sqrt{4 - x^2 - y^2} = \sqrt{4} = 2$. Since $x^2 + y^2 \leq 4$, the smallest value of $4 - x^2 - y^2$ is 0, taken when $x^2 + y^2 = 4$ (at all points on the circle $x^2 + y^2 = 4$). Thus the range of f is the closed interval $[0, 2]$. Finally,

$$f(0, 1) = \sqrt{4 - 0^2 - 1^2} = \sqrt{3} \quad \text{and} \quad f(-1, 1) = \sqrt{4 - (-1)^2 - 1^2} = \sqrt{2}.$$

Remark. We emphasize that the domain of f is a subset of $\mathbb{R}^2$ while the range is a subset of $\mathbb{R}$, the real numbers.

In Chapter 1 we wrote $y = f(x)$. That is, we used the letter y to denote the value of a function of one variable. Here, we can use the letter z (or any other letter for that matter) to denote the value taken by f, which is now a function of two variables. We then write

$$z = f(x, y). \tag{2}$$

EXAMPLE 3.   Let $z = f(x, y) = \ln(2x - y + 1)$. Find the domain and range of f and calculate $f(3, 2)$ and $f(5, -7)$.

SOLUTION. $\ln(2x - y + 1)$ is defined only when $2x - y + 1 > 0$ which occurs when $y < 2x + 1$. This is the equation of the half plane "below" (but not including) the line $y = 2x + 1$. It is sketched in Figure 2. Since $2x - y + 1$ can take any positive value, $\ln(2x - y + 1)$ can take any real value. Thus the range of f is $\mathbb{R}$. Finally, $f(3, 2) = \ln(6 - 2 + 1) = \ln 5 \approx 1.61$ and $f(5, -7) = \ln[10 - (-7) + 1] = \ln 18 \approx 2.89$.

Caution. In Figures 1 and 2 we sketched the *domains* of two different functions. We did *not* sketch their graphs. The graph of a function of two or more variables is more complicated and will be discussed shortly.

---

†The *disk* of radius r is the circle of radius r together with all points interior to that circle.

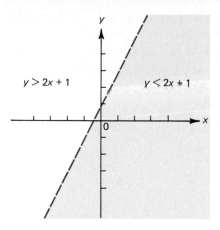

$y > 2x + 1$

$y < 2x + 1$

Figure 2

EXAMPLE 4. Find the domain and range of the function

$$z = f(x, y) = \frac{\sqrt{x^2 + y^2 - 9}}{x - y}.$$

SOLUTION. $\sqrt{x^2 + y^2 - 9}$ is defined only when $x^2 + y^2 \geq 9$. This is the circle $x^2 + y^2 = 9$ with radius 3 and its exterior. However $f(x, y)$ is also not defined when $x - y = 0$ or when $y = x$. Thus all points on the line $y = x$ are excluded from the domain of f. We have domain of $f = \{(x, y): x^2 + y^2 \geq 9 \text{ and } x \neq y\}$. This is sketched in Figure 3. The points on the dotted line are excluded.

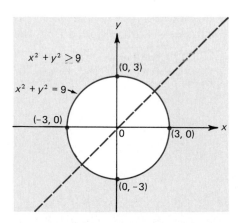

$x^2 + y^2 \geq 9$

$x^2 + y^2 = 9$

$(0, 3)$

$(-3, 0)$

$(3, 0)$

$(0, -3)$

Figure 3

We now calculate the range of f. We first note that $\sqrt{x^2 + y^2 - 9}$ can take any nonnegative value. Since $x - y$ can be positive or negative, we see that $\sqrt{x^2 + y^2 - 9}/(x - y)$ can take on any real value so that the range of f is $\mathbb{R}$. Note that $f(x, y) = 0$ at any point on the circle except at the points $(\sqrt{3/2}, \sqrt{3/2})$ and $(-\sqrt{3/2}, -\sqrt{3/2})$. The function is not defined at these last two points.

EXAMPLE 5. Let $z = f(x, y) = 3/\sqrt{x^2 - y}$. Find the domain and range of f.

SOLUTION. $f(x, y)$ is defined if $x^2 - y > 0$ or $y < x^2$. This is the region in the plane "outside" the parabola $y = x^2$ and is sketched in Figure 4. Note that the parabola $y = x^2$ is *not* in the domain of f. The range of f is the set of positive real numbers since when $x^2 - y > 0$, $3/\sqrt{x^2 - y}$ can take on any positive real number.

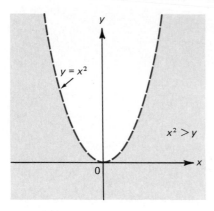

Figure 4

EXAMPLE 6. Find the domain and range of $f(x, y) = \tan^{-1}(y/x)$.

SOLUTION. $\tan^{-1}(y/x)$ is defined so long as $x \neq 0$. Hence domain of $f = \{(x, y): x \neq 0\}$. From the definition of $\tan^{-1} x$ in Section 7.8, we find that the range of f is the open interval $(-\pi/2, \pi/2)$.

EXAMPLE 7. Find the domain and range of $f(x, y) = \ln(x^2 - y^2)$.

SOLUTION. We must have $x^2 - y^2 > 0$ or $x^2 > y^2$. The region $\{(x, y): x^2 > y^2\}$ is sketched in Figure 5. Note that the lines $y = x$ and $y = -x$ are excluded from the domain.

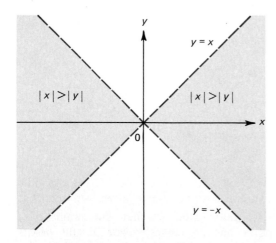

Figure 5

We now turn to the definition of a function of three variables.

DEFINITION 2 (Function of Three Variables). Let D be a subset of $\mathbb{R}^3$. Then a *function of three variables f* is a rule which assigns to each point (ordered triple)

(x, y, z) in D a unique real number which we denote $f(x, y, z)$. The set D is called the *domain* of f and the set $\{ f(x, y, z): (x, y, z) \in D \}$, which is the set of values the function f takes on, is called the *range* of f.

Note. We will often use the letter w to denote the values that a function of three variables takes. We then have

$$w = f(x, y, z). \tag{3}$$

EXAMPLE 8. Let $w = f(x, y, z) = \sqrt{1 - x^2 - (y^2/4) - (z^2/9)}$. Find the domain and range of f. Calculate $f(0, 1, 1)$.

SOLUTION. $f(x, y, z)$ is defined if $1 - x^2 - (y^2/4) - (z^2/9) \geq 0$ which occurs if $x^2 + (y^2/4) + (z^2/9) \leq 1$. From Section 17.6 we see that the equation $x^2 + (y^2/4) + (z^2/9) = 1$ is the equation of an ellipsoid. Thus the domain of f is the set of points (x, y, z) in $\mathbb{R}^3$ which are on and interior to this ellipsoid (sketched in Figure 6). The range of f is the closed interval $[0, 1]$. This follows because

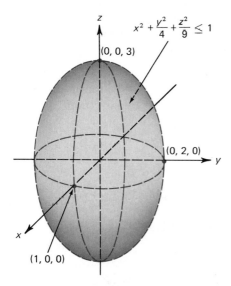

$$x^2 + \frac{y^2}{4} + \frac{z^2}{9} \leq 1$$

$(0, 0, 3)$

$(0, 2, 0)$

$(1, 0, 0)$

Figure 6

$x^2 + (y^2/4) + (z^2/4) \geq 0$ so that $1 - x^2 - (y^2/4) - (z^2/4)$ is maximized when $x = y = z = 0$. Finally,

$$f(0, 1, 1) = \sqrt{1 - 0^2 - \frac{1^2}{4} - \frac{1^2}{9}} = \sqrt{1 - \frac{1}{4} - \frac{1}{9}} = \sqrt{1 - \frac{13}{36}} = \frac{\sqrt{23}}{6}.$$

EXAMPLE 9. Let $w = f(x, y, z) = \ln(x + 2y + z - 3)$. Find the domain and range of f. Calculate $f(1, 2, 3)$.

SOLUTION. $f(x, y, z)$ is defined when $x + 2y + z - 3 > 0$ which occurs when $x + 2y + z > 3$. The equation $x + 2y + z = 3$ is the equation of a plane. Thus the domain of f is the *half-space* above (but not including) this plane. The plane is sketched in Figure 7. As in Example 3, the range of f is $\mathbb{R}$. Finally, $f(1, 2, 3) = \ln(1 + 4 + 3 - 3) = \ln 5 \approx 1.61$.

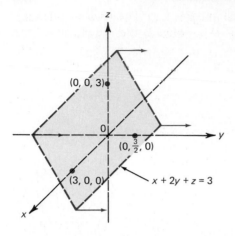

Figure 7

We will not here define a function of more than three variables as such a definition would be very similar to Definitions 1 and 2.

We now turn to a discussion of the graph of a function. Recall from Section 1.6 that the graph of a function of one variable f is the set of all points (x, y) in the plane such that $y = f(x)$. Using this as our model, we have

DEFINITION 3.   The *graph* of a function of two variables f is the set of all points (x, y, z) in $\mathbb{R}^3$ such that $z = f(x, y)$. The graph of a function of two variables is called a *surface* in $\mathbb{R}^3$.

EXAMPLE 10.   Sketch the graph of the function

$$z = f(x, y) = \sqrt{1 - x^2 - y^2}. \tag{4}$$

SOLUTION. We first note that $z \geq 0$. Then squaring both sides of (4) we have $z^2 = 1 - x^2 - y^2$ or $x^2 + y^2 + z^2 = 1$. This is, of course, the equation of the unit sphere. However, since $z \geq 0$, the graph of f is the hemisphere sketched in Figure 8.

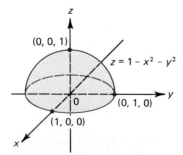

Figure 8

It is often very difficult to sketch the graph of a function $z = f(x, y)$ since, except for the quadric surfaces we discussed in Section 17.6, we really do not have a vast "catalog" of surfaces to which to refer. Moreover, the techniques of curve plotting in three dimensions are tedious, to say the least, and plotting points in space will, except in the most trivial of cases, not get us very far. The situation is even worse when we try to sketch the graph of a function of three variables $w = f(x, y, z)$. We would need *four dimensions* to sketch such a surface. Being mortal, we are limited to three dimensions and we see that we have reached the point where our comfortable three-dimensional geometry fails us. We are *not* saying that curves and surfaces in four-dimensional space do not exist. They do. Only we are not able to sketch them.

We will not spend much time in this text discussing spaces of dimension higher than three. To give you a taste of the subject, we can define four-dimensional space $\mathbb{R}^4$ as the set of all "points" or *four-tuples* (x, y, z, w) where x, y, z and w are real numbers. Then the graph of a function $w = f(x, y, z)$ is the set of points (x, y, z, w) in $\mathbb{R}^4$ with $w = f(x, y, z)$. It is important to note that the only difference between a graph of a function of two or three variables is our ability to sketch one (sometimes) but not the other. If we do not insist on being able to sketch things, then there is really no essential difference between $\mathbb{R}^3$ or $\mathbb{R}^4$ (or $\mathbb{R}^5, \mathbb{R}^6, \ldots,$ for that matter). For the ramainder of this chapter we will focus on $\mathbb{R}^2$ and $\mathbb{R}^3$, keeping in mind that all our definitions apply to higher dimensions as well.

We have stated that it is usually very difficult to sketch the graph of a function of two variables. Fortunately, there is a way to describe the graph of such a function in two dimensions. The idea for what we are about to do comes from cartographers (map makers). Cartographers have the problems of indicating three-dimensional features (such as mountains and valleys) on a two-dimensional surface. They solve the problem by drawing a *relief map* (*topographical map*). This is a map in which points of constant elevation are joined to form curves, called *contour curves*. The closer together these contour curves are drawn, the steeper the terrain. A portion of a typical contour map is sketched in Figure 9.

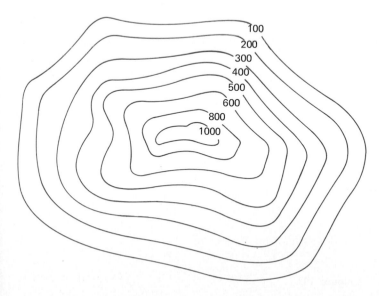

Figure 9

We can use the same idea to depict the function $z = f(x, y)$ graphically. If z is fixed, then the equation $f(x, y) = z$ is the equation of a curve in the plane called a *level curve*. Each value of z gives us such a curve. In other words, a level curve is the intersection of the curve $z = f(x, y)$ with the plane $z = c$. This idea is best illustrated with some examples.

EXAMPLE 11. Sketch the level curves of $z = x^2 + y^2$.

SOLUTION. If $z > 0$, then $z = a^2$ for some positive number $a > 0$. Hence all level curves are circles of the form $x^2 + y^2 = a^2$. The number a^2 can be thought of as the "elevation" of points on a level curve. Some of these curves are sketched in Figure 10.

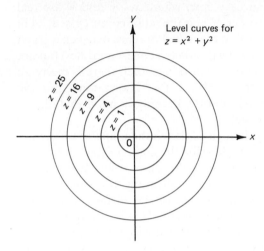

Level curves for
$z = x^2 + y^2$

$z = 25$
$z = 16$
$z = 9$
$z = 4$
$z = 1$

Figure 10

Each level curve encloses a "slice" of the actual graph of the function in three dimensions. In this case, each circle is the projection onto the xy-plane of a part of the surface in space. Actually, this example is especially simple because we can, without much difficulty, sketch the graph in space. From Section 17.6 the equation $z = x^2 + y^2$ is the equation of an elliptic paraboloid (actually a circular paraboloid). It is sketched in Figure 11. In this easy case we can see that if we slice this figure

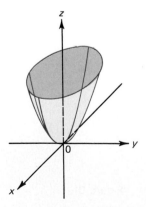

Figure 11

parallel to the xy-plane, we obtain the discs enclosed by the level curves. In most cases, of course, we will not be able to easily sketch the graph in space so that we will have to rely on our "relief map" sketch in $\mathbb{R}^2$.

EXAMPLE 12. Let $z = f(x, y) = 2x - y + 4$. Describe the level curves.

SOLUTION. The level curves are the lines $2x - y + 4 = c$ or $y = 2x + (4 - c)$. Some of these are sketched in Figure 12. Note that here the graph of the function in space is the plane $2x - y - z + 4 = 0$.

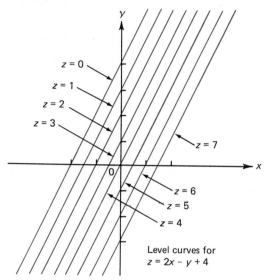

Figure 12

EXAMPLE 13. Let $z = \ln(x + y)$. Describe the level curves.

SOLUTION. The level curves take the form $\ln(x + y) = c$ or $x + y = e^c$ or $y = -x + e^c$. These are straight lines, some of which are sketched in Figure 13. In this example it would be more difficult to sketch the surface $z = f(x, y) = \ln(x + y)$.

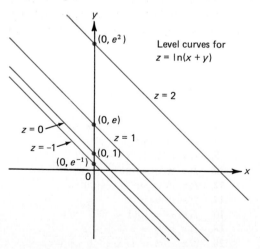

Figure 13

EXAMPLE 14. Let $z = 1/xy$. Describe the level curves.

SOLUTION. If $z = c \neq 0$, then $1/xy = c$ or $xy = 1/c$. These are the equations of hyperbolas, some of which are sketched in Figure 14.

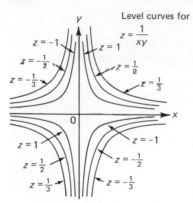

Level curves for

$$z = \frac{1}{xy}$$

Figure 14

There are some interesting applications of level curves in the sciences and economics. Three are given below.

(i)   Let $T(x, y)$ denote the temperature at a point (x, y) in the xy-plane. The level curves $T(x, y) = c$ are called *isothermal curves*. All points on such a curve have the same temperature.

(ii)   Let $V(x, y)$ denote the voltage (or potential) at a point in the xy-plane. The level curves $V(x, y) = c$ are called *equipotential curves*. All points on such a curve have the same voltage.

(iii)   A manufacturer makes two products. Let x and y denote the number of units of the first and second products produced during a given year. If $P(x, y)$ denotes the profit the manufacturer receives each year, then the level curves $P(x, y) = c$ are *constant profit* curves. All points on such a curve yield the same profit.

We conclude this section by noting that the idea behind level curves can be used to describe functions of three variables of the form $w = f(x, y, z)$. For each fixed w, the equation $f(x, y, z) = w$ is the equation of a surface in space, called a *level surface*. Since surfaces are so difficult to draw, we will not pursue the notion of level surfaces here.

PROBLEMS 18.2

In Problems 1–35 find the domain and range of the indicated function.

1. $f(x, y) = \sqrt{x^2 + y^2}$
2. $f(x, y) = \sqrt{1 + x + y}$
3. $f(x, y) = x/y$
4. $f(x, y) = \sqrt{1 - x^2 - 4y^2}$
5. $f(x, y) = \sqrt{1 - x^2 + 4y^2}$
6. $f(x, y) = \sin(x + y)$
7. $f(x, y) = e^x + e^y$
8. $f(x, y) = 1/(x^2 - y^2)^{3/2}$
9. $f(x, y) = \tan(x - y)$
10. $f(x, y) = \sqrt{(x + y)/(x - y)}$
11. $f(x, y) = \sqrt{(x - y)/(x + y)}$
12. $f(x, y) = \sin^{-1}(x + y)$
13. $f(x, y) = \cos^{-1}(x - y)$
14. $f(x, y) = y/|x|$
15. $f(x, y) = (x^2 - y^2)/(x + y)$
16. $f(x, y) = \ln(1 + x^2 - y^2)$
*17. $f(x, y) = (x/2y) + (2y/x)$
18. $f(x, y, z) = x + y + z$
19. $f(x, y, z) = \sqrt{x + y + z}$
20. $f(x, y, z) = 1/\sqrt{x^2 + y^2 + z^2}$
21. $f(x, y, z) = 1/\sqrt{x^2 - y^2 + z^2}$
22. $f(x, y, z) = 1/\sqrt{x^2 - y^2 - z^2}$
23. $f(x, y, z) = \sqrt{-x^2 - y^2 - z^2}$
24. $f(x, y, z) = \ln(x - 2y - 3z + 4)$
25. $f(x, y, z) = xy/z$
26. $f(x, y, z) = \sin(x + y - z)$
27. $f(x, y, z) = \sin^{-1}(x + y - z)$
28. $f(x, y, z) = \ln(x + y - z)$

29. $f(x, y, z) = \tan^{-1}[(x + y)/z]$

30. $f(x, y, z) = e^{xy+z}$

31. $f(x, y, z) = (e^x + e^y)/e^z$

32. $f(x, y, z) = xyz$

33. $f(x, y, z) = 1/xyz$

34. $f(x, y, z) = x/(y + z)$

35. $f(x, y, z) = \sin x + \cos y + \sin z.$

In Problems 36–42 sketch the graph of the given function.

36. $z = 4x^2 + 4y^2$

37. $y = x^2 + 4z^2$

38. $x = 4z^2 - 4y^2$

39. $z = x^2 - 4y^2$

40. $z = \sqrt{x^2 + 4y^2 + 4}$

41. $x = \sqrt{x^2 - 4z^2 + 4}$

42. $y = \sqrt{4 - z^2 - 4y^2}.$

In Problems 43–50 describe the level curves of the given function and sketch these curves for the given values of z.

43. $z = \sqrt{1 + x + y}$; $z = 0, 1, 5, 10$

44. $z = x/y$; $z = 1, 3, 5, -1, -3$

45. $z = \sqrt{1 - x^2 - 4y^2}$; $z = 0, \frac{1}{4}, \frac{1}{2}, 1$

46. $z = \sqrt{1 + x^2 - y}$; $z = 0, 1, 2, 5$

47. $z = \cos^{-1}(x - y)$; $z = 0, \pi/6, \pi/3, \pi/2$

48. $z = \sqrt{\dfrac{x + y}{x - y}}$; $z = 0, 1, 2, 5$

49. $z = \tan(x + y)$; $z = 0, 1, -1, \sqrt{3}$

50. $z = \tan^{-1}(x - y^2)$; $z = 0, \pi/6, \pi/4.$

51. The temperature T at any point on an object in the plane is given by $T(x, y) = 20 + x^2 + 4y^2$. Sketch the isothermal curves for $T = 50$, $T = 60$ and $T = 70$ degrees.

52. The voltage at a point (x, y) on a metal plate placed in the xy-plane is given by $V(x, y) = \sqrt{1 - 4x^2 - 9y^2}$. Sketch the equipotential curves for $V = 1.0$ V, $V = 0.5$ V, and $V = 0.25$ V.

53. A manufacturer earns $P(x, y) = 100 + 2x^2 + 3y^2$ dollars each year for producing x and y units, respectively, of two products. Sketch the constant profit curves for $P = \$100$, $P = \$200$, and $P = \$1000$.

18.3 Limits and Continuity

In this section we discuss the fundamental concepts of limits and continuity for functions of two and three variables. Recall that in defining a limit in Chapter 2 and Chapter 10, fundamental to our discussion were the notions of an open and a closed interval. We could say, for example, that x was close to x_0 if $|x - x_0|$ was sufficiently small, or, equivalently, if x was contained in a small open interval (neighborhood) centered at x_0. It is interesting to see how these ideas extend to $\mathbb{R}^2$ or $\mathbb{R}^3$.

Let (x_0, y_0) be a point in $\mathbb{R}^2$. What do we mean by the equation

$$|(x, y) - (x_0, y_0)| = r? \tag{1}$$

Since (x, y) and (x_0, y_0) are vectors in $\mathbb{R}^2$,

$$|(x, y) - (x_0, y_0)| = |(x - x_0, y - y_0)| = \sqrt{(x - x_0)^2 + (y - y_0)^2} \tag{2}$$

(see equation (15.1.1)). Inserting (2) in (1) and squaring both sides yields

$$(x - x_0)^2 + (y - y_0)^2 = r^2 \tag{3}$$

which is the equation of the circle of radius r centered at (x_0, y_0). Then the set of points whose coordinates (x, y) satisfy the inequality

$$|(x, y) - (x_0, y_0)| < r \tag{4}$$

is the set of all points in $\mathbb{R}^2$ interior to the circle given by (3). This is sketched in Figure 1a. Similarly, the inequality

$$|(x, y) - (x_0, y_0)| \le r \tag{5}$$

describes the set of all points interior to and on the circle given by (3). This is sketched in Figure 1b. This leads to the following definition.

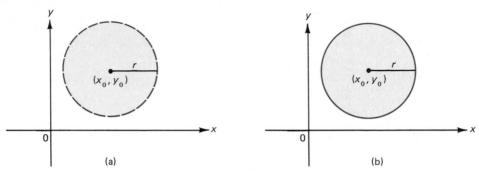

Figure 1

DEFINITION 1. (i) The *open disk D_r* centered at (x_0, y_0) with radius r is the subset of $\mathbb{R}^2$ given by

$$\{(x, y): |(x, y) - (x_0, y_0)| < r\}.$$

(ii) The *closed disk* centered at (x_0, y_0) with radius r is the subset of $\mathbb{R}^2$ given by

$$\{(x, y): |(x, y) - (x_0, y_0)| \le r\}.$$

(iii)   The *boundary* of the open or closed disk defined in (i) or (ii) is the circle

$$\{(x, y): |(x, y) - (x_0, y_0)| = r\}.$$

(iv)   A *neighborhood* of a point (x_0, y_0) in $\mathbb{R}^2$ is an open disk centered at (x_0, y_0).

> **Remark.** In this definition the words *open* and *closed* have meanings very similar to their meanings in the terms "open interval" and "closed interval." An open interval does not contain its endpoints. An open disk does not contain any point on its boundary. Similarly, a closed interval contains all its boundary points as does a closed disk.

With these definitions it is easy to define a limit of a function of two variables. Intuitively we say that $f(x, y)$ approaches the limit L as (x, y) approaches (x_0, y_0) if $f(x, y)$ gets arbitrarily "close" to L as (x, y) gets "close" to (x_0, y_0) from any direction. We define this notion precisely as follows.

DEFINITION 2 (Definition of a Limit). Let $f(x, y)$ be defined in a neighborhood of (x_0, y_0) but not necessarily at (x_0, y_0) itself. Then the *limit of $f(x, y)$ as (x, y) ap-*

proaches (x_0, y_0) *is* L, written

$$\lim_{(x,y)\to(x_0,y_0)} f(x, y) = L \tag{6}$$

if for every number $\epsilon > 0$, there is a number $\delta > 0$ such that $|f(x, y) - L| < \epsilon$ for every (x, y) (not including (x_0, y_0)) in the open disk centered at (x_0, y_0) with radius δ.

Remark. This definition is illustrated in Figure 2.

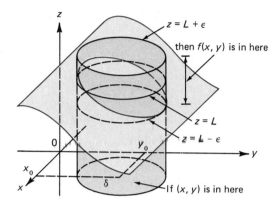

Figure 2

EXAMPLE 1. Show, using the definition of a limit, that $\lim_{(x,y)\to(1,2)} (3x + 2y) = 7$.

SOLUTION. Let $\epsilon > 0$ be given. We need to choose a $\delta > 0$ such that $|3x + 2y - 7| < \epsilon$ if $0 < \sqrt{(x-1)^2 + (y-2)^2} < \delta$. We start with the inequality $|3x + 2y - 7| < \epsilon$ and work backwards to find a suitable value for δ. We use the triangle inequality (see equation (1.2.9)):

$$|3x + 2y - 7| = |3x - 3 + 2y - 4| \le |3x - 3| + |2y - 4|$$
$$= 3|x - 1| + 2|y - 2|.$$

Now, $|x - 1| \le \sqrt{(x-1)^2 + (y-2)^2}$ (explain why) and $|y - 2| \le \sqrt{(x-1)^2 + (y-2)^2}$. Thus

$$3|x - 1| + 2|y - 2| \le 5\sqrt{(x-1)^2 + (y-2)^2}.$$

Now let $\delta = \epsilon/5$. Then, if $\sqrt{(x-1)^2 + (y-2)^2} < \delta$, we have

$$|3x + 2y - 7| \le 3|x - 1| + 2|y - 2| \le 5\sqrt{(x-1)^2 + (y-2)^2}$$
$$< 5\delta = 5\frac{\epsilon}{5} = \epsilon$$

and the requested limit is shown.

In the intuitive definition of a limit which preceded Definition 2, we used the phrase "if $f(x, y)$ gets arbitrarily close to L as (x, y) gets close to (x_0, y_0) *from any direction.*" Recall that $\lim_{x \to x_0} f(x) = L$ only if $\lim_{x \to x_0^+} f(x) = \lim_{x \to x_0^-} f(x) = L$. That is, the limit exists only if we get the same value from either side of x_0. In $\mathbb{R}^2$ the situation is more complicated because (x, y) can approach (x_0, y_0) not just from two,

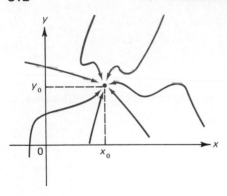

Figure 3

but from an *infinite number* of directions. Some of these are illustrated in Figure 3. Thus the only way we can verify a limit is by making use of Definition 2 or some appropriate limit theorem that can be proven directly from Definition 2. We illustrate the kinds of problems we can encounter in the next two examples.

EXAMPLE 2. Let $f(x, y) = (y^2 - x^2)/(y^2 + x^2)$ for $(x, y) \neq (0, 0)$. We shall show that $\lim_{(x,y)\to(0,0)} f(x, y)$ does not exist. There are an infinite number of approaches to the origin. For example, if we approach along the x-axis, then $y = 0$ and

$$\lim_{(x,y)\to(0,0)} \frac{y^2 - x^2}{y^2 + x^2} = \lim_{(x,y)\to(0,0)} \frac{-x^2}{x^2} = \lim_{(x,y)\to(0,0)} -1 = -1.$$

On the other hand, if we approach along the y-axis, then $x = 0$ and

$$\lim_{(x,y)\to(0,0)} \frac{y^2 - x^2}{y^2 + x^2} = \lim_{(x,y)\to(0,0)} \frac{y^2}{y^2} = \lim_{(x,y)\to(0,0)} 1 = 1.$$

Thus we get different answers depending on how we approach the origin. To prove that the limit cannot exist, we note that we have shown that in any open disk centered at the origin, there are points at which f takes on the values $+1$ and -1. Hence f cannot have a limit as $(x, y) \to (0, 0)$.

This example leads to the general rule:

If we get two or more different values for $\lim_{(x,y)\to(x_0,y_0)} f(x, y)$ *as we approach* (x_0, y_0) *along different paths, then* $\lim_{(x,y)\to(x_0,y_0)} f(x, y)$ *does not exist.*

EXAMPLE 3. Let $f(x, y) = xy^2/(x^2 + y^4)$. We shall show that $\lim_{(x,y)\to(0,0)} f(x, y)$ does not exist. First, let us approach zero along a straight line passing through the origin which is not the y-axis. Then $y = mx$ and

$$f(x, y) = \frac{x(m^2 x^2)}{x^2 + m^4 x^4} = \frac{m^2 x}{1 + m^2 x^2}$$

which $\to 0$ as $x \to 0$. Thus, along every straight line, $f(x, y) \to 0$ as $(x, y) \to (0, 0)$. But if we approach $(0, 0)$ along the parabola $x = y^2$, then

$$f(x, y) = \frac{y^2(y^2)}{y^4 + y^4} = \frac{1}{2}$$

so that, along this parabola, $f(x, y) \to \frac{1}{2}$ as $(x, y) \to (0, 0)$ and the limit does not exist.

EXAMPLE 4. Prove that $\lim_{(x,y)\to(0,0)}[xy^2/(x^2 + y^2)] = 0$. It is easy to show that this limit is zero along any straight line passing through the origin. But, as we have seen, this is not enough. We must rely on our definition. Let $\epsilon > 0$ be given; then we must show that there is a $\delta > 0$ such that if $0 < \sqrt{x^2 + y^2} < \delta$, then

$$|f(x, y)| = \left| \frac{xy^2}{x^2 + y^2} \right| < \epsilon.$$

But $|x| \le \sqrt{x^2 + y^2}$ and $y^2 \le x^2 + y^2$ so that

$$\left| \frac{xy^2}{x^2 + y^2} \right| \le \frac{(\sqrt{x^2 + y^2})(x^2 + y^2)}{x^2 + y^2} = \sqrt{x^2 + y^2}.$$

Hence if we choose $\delta = \epsilon$, we will have $|f(x, y)| < \epsilon$ if $\sqrt{x^2 + y^2} < \delta$.

Remark. This example illustrates the fact that while the existence of different limits for different paths implies that no limit exists, only the definition or an appropriate limit theorem can be used to prove that a limit does exist.

As Examples 1 and 4 illustrate, it is often tedious to calculate limits from the definition. Fortunately, just as in the case of functions of a single variable, there are a number of theorems that greatly facilitate the calculation of limits. As you will recall, one of our major results was that if f is continuous, then $\lim_{x \to x_0} f(x) = f(x_0)$. We now define continuity of a function of two variables.

DEFINITION 3. (i)   Let $f(x, y)$ be defined at every point (x, y) in a neighborhood of (x_0, y_0). Then f is *continuous* at (x_0, y_0) if all of the following conditions hold:

(a) $f(x_0, y_0)$ exists (that is, (x_0, y_0) is in the domain of f).
(b) $\lim_{(x,y)\to(x_0,y_0)} f(x, y)$ exists.
(c) $\lim_{(x,y)\to(x_0,y_0)} f(x, y) = f(x_0, y_0)$.

(ii)   If one or more of these three conditions fails to hold, then f is said to be *discontinuous* at (x_0, y_0).
(iii) f is *continuous in a subset* S of $\mathbb{R}^2$ if f is continuous at every point (x, y) in S.

Remark. Condition (c) tells us that if a function f is continuous at (x_0, y_0), then we can calculate $\lim_{(x,y)\to(x_0,y_0)} f(x, y)$ by evaluation.

EXAMPLE 5. Let $f(x, y) = xy^2/(x^2 + y^2)$. f is discontinuous at $(0, 0)$ because $f(0, 0)$ is not defined.

EXAMPLE 6. Let

$$f(x, y) = \begin{cases} \dfrac{xy^2}{x^2 + y^4}, & (x, y) \neq (0, 0) \\ 0, & (x, y) = (0, 0). \end{cases}$$

Here f is defined at $(0, 0)$ but it is still discontinuous there because $\lim_{(x,y) \to (0,0)} f(x, y)$ does not exist (see Example 3).

EXAMPLE 7. Let

$$f(x, y) = \begin{cases} \dfrac{xy^2}{x^2 + y^2}, & (x, y) \neq (0, 0) \\ 0, & (x, y) = (0, 0). \end{cases}$$

Here f is continuous at $(0, 0)$ according to Example 4.

EXAMPLE 8. Let

$$f(x, y) = \begin{cases} \dfrac{xy^2}{x^2 + y^2}, & (x, y) \neq (0, 0) \\ 1, & (x, y) = (0, 0). \end{cases}$$

Then f is discontinuous at $(0, 0)$ because $\lim_{(x,y) \to (0,0)} f(x, y) = 0 \neq f(0, 0) = 1$ so that condition (c) is violated.

Naturally, we would like to know what functions are continuous. We can answer this question if we look at the continuous functions of one variable. We start with polynomials.

DEFINITION 4. (i) A *polynomial* $p(x, y)$ in the two variables x and y is a finite sum of terms of the form Ax^my^n, where m and n are positive integers and A is a real number.
　(ii)　A *rational function* $r(x, y)$ in the two variables x and y is a function which can be written as the quotient of two polynomials: $r(x, y) = p(x, y)/q(x, y)$.

EXAMPLE 9. $p(x, y) = 5x^5y^2 + 12xy^9 - 37x^{82}y^5 + x + 4y - 6$ is a polynomial.

EXAMPLE 10

$$r(x, y) = \frac{8x^3y^7 - 7x^2y^4 + xy - 2y}{1 - 3y^3 + 7x^2y^2 + 18yx^7}$$

is a rational function.

The limit theorems we proved in Chapter 10 can be extended, with minor modifications, to functions of two variables. We will not state them here. However, by using them it is not difficult to prove the following theorem about continuous functions.

THEOREM 1. (i)　Any polynomial p is continuous at any point in $\mathbb{R}^2$.
　(ii)　Any rational function $r = p/q$ is continuous at any point (x, y) for which $q(x, y) \neq 0$.

(iii)   If f and g are continuous at (x_0, y_0), then $f + g$, $f - g$, and $f \cdot g$ are continuous at (x_0, y_0).†

(iv)   If f and g are continuous at (x_0, y_0) and if $g(x_0, y_0) \neq 0$, then f/g is continuous at (x_0, y_0).

(v)   If f is continuous at (x_0, y_0) and if h is a function of one variable which is continuous at $f(x_0, y_0)$, then the composite function $h \circ f$, defined by $(h \circ f)(x, y) = h(f(x, y))$, is continuous at (x_0, y_0).

EXAMPLE 11. Calculate $\displaystyle \lim_{(x,y) \to (4,1)} \frac{x^3 y^2 - 4xy}{x + 6xy^3}$.

SOLUTION. $(x^3 y^2 - 4xy)/(x + 6xy^3)$ is continuous at $(4, 1)$ so we can calculate the limit by evaluation. We have

$$\lim_{(x,y) \to (4,1)} \frac{x^3 y^2 - 4xy}{x + 6xy^3} = \left. \frac{x^3 y^2 - 4xy}{x + 6xy^3} \right|_{(4,1)} = \frac{64 \cdot 1 - 4 \cdot 4 \cdot 1}{4 + 6 \cdot 4 \cdot 1} = \frac{48}{28} = \frac{12}{7}.$$

EXAMPLE 12. Calculate $\displaystyle \lim_{(x,y) \to (\pi/6, \pi/3)} \sin(x + y)$.

SOLUTION. $f(x, y) = x + y$ is continuous as is $h(x) = \sin x$. Then, by (v), $(h \circ f)(x, y) = \sin(x + y)$ is continuous and, therefore,

$$\lim_{(x,y) \to (\pi/6, \pi/3)} \sin(x + y) = \sin\left(\frac{\pi}{6} + \frac{\pi}{3}\right) = \sin\left(\frac{\pi}{2}\right) = 1.$$

EXAMPLE 13. For what values of (x, y) is the function

$$f(x, y) = \frac{x^3 + 7xy^5 - 8y^6}{x^2 - y^2}$$

continuous?

SOLUTION. Since f is a rational function in x and y, f is continuous so long as the denominator is not zero; that is, when $y \neq \pm x$. f is continuous at every point (x, y) in $\mathbb{R}^2$ not on one of the lines $y = x$ and $y = -x$.

EXAMPLE 14. What is the largest subset of $\mathbb{R}^2$ on which $f(x, y) = \ln(2x + y)$ is continuous?

SOLUTION. Since $\ln x$ is defined and continuous only when $x > 0$, $\ln(2x + y)$ is continuous when $2x + y > 0$. The set $\{(x, y): 2x + y > 0\}$ is the half plane which lies to the right of the line $2x + y = 0$ as illustrated in Figure 4.

All the ideas in this section can be generalized to functions of three variables. A *neighborhood* of radius r of the point (x_0, y_0, z_0) in $\mathbb{R}^3$ consists of all points in $\mathbb{R}^3$ interior to the sphere

$$(x - x_0)^2 + (y - y_0)^2 + (z - z_0)^2 = r^2.$$

†$(f + g)(x, y) = f(x, y) + g(x, y)$, just as in the case of functions of one variable. Similarly, $(f \cdot g)(x, y) = [f(x, y)][g(x, y)]$ and $(f/g)(x, y) = f(x, y)/g(x, y)$.

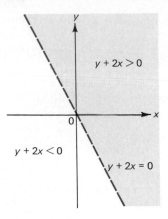

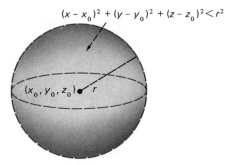

Figure 4

That is, the neighborhood (also called an *open ball*) is given by

$$\{(x, y, z): (x - x_0)^2 + (y - y_0)^2 + (z - z_0)^2 < r^2\}.$$

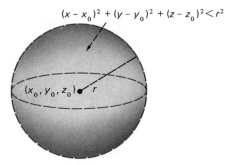

Figure 5

A typical neighborhood is sketched in Figure 5. The *closed ball* is given by

$$\{(x, y, z): (x - x_0)^2 + (y - y_0)^2 + (z - z_0)^2 \le r^2\}.$$

With this definition, the definition of the limit $\lim_{(x,y,z) \to (x_0, y_0, z_0)} f(x, y, z) = L$ is analogous to that given in Definition 2. All other definitions, remarks, and theorems given in this section apply to functions of three (or more) variables.

PROBLEMS 18.3

In Problems 1–5 sketch the indicated region.

  1. The open disk centered at $(3, 0)$ with radius 2
  2. The open disk centered at $(3, 2)$ with radius 3
  3. The closed disk centered at $(-1, 1)$ with radius 1
  4. The open ball centered at $(1, 0, 0)$ with radius 1
  5. The closed ball centered at $(0, 1, 1)$ with radius 2.

In Problems 6–11 use Definition 2 to verify the indicated limit.

  6. $\displaystyle\lim_{(x,y)\to(1,2)} (3x + y) = 5$
  7. $\displaystyle\lim_{(x,y)\to(3,-1)} (x - 7y) = 10$

8. $\displaystyle\lim_{(x,y)\to(5,-2)} (ax + by) = 5a - 2b$

*9. $\displaystyle\lim_{(x,y)\to(1,1)} \frac{x}{y} = 1$

10. $\displaystyle\lim_{(x,y)\to(0,0)} \frac{2x^2 y}{x^2 + y^2} = 0$

11. $\displaystyle\lim_{(x,y)\to(4,1)} (x^2 + 3y^2) = 19.$

In Problems 12–21 show that the given limit does not exist.

12. $\displaystyle\lim_{(x,y)\to(0,0)} \frac{x + y}{x - y}$

13. $\displaystyle\lim_{(x,y)\to(0,0)} \frac{xy}{x^2 - y^2}$

14. $\displaystyle\lim_{(x,y)\to(1,1)} \frac{xy}{x^2 + y^2}$

15. $\displaystyle\lim_{(x,y)\to(0,0)} \frac{xy^3}{x^4 + y^4}$

16. $\displaystyle\lim_{(x,y)\to(0,0)} \frac{xy}{x^3 + y^3}$

17. $\displaystyle\lim_{(x,y)\to(0,0)} \frac{(x^2 + y^2)^2}{x^4 + y^4}$

18. $\displaystyle\lim_{(x,y)\to(0,0)} \frac{x^2 - 2y}{y^2 + 2x}$

19. $\displaystyle\lim_{(x,y)\to(0,0)} \frac{ax^2 + by}{cy^2 + dx},\ a, b, c, d > 0$

20. $\displaystyle\lim_{(x,y,z)\to(0,0,0)} \frac{xy + 2xz + 3yz}{x^2 + y^2 + z^2}$

21. $\displaystyle\lim_{(x,y,z)\to(0,0,0)} \frac{xyz}{x^3 + y^3 + z^3}.$

In Problems 22–25 show that the indicated limit exists and calculate it.

22. $\displaystyle\lim_{(x,y)\to(0,0)} \frac{3xy}{\sqrt{x^2 + y^2}}$

23. $\displaystyle\lim_{(x,y)\to(0,0)} \frac{5x^2 y^2}{x^4 + y^2}$

24. $\displaystyle\lim_{(x,y)\to(0,0)} \frac{x^3 + y^3}{x^2 + y^2}$

25. $\displaystyle\lim_{(x,y,z)\to(0,0,0)} \frac{yx^2 + z^3}{x^2 + y^2 + z^2}.$

In Problems 26–35 calculate the indicated limit.

26. $\displaystyle\lim_{(x,y)\to(-1,2)} (xy + 4y^2 x^3)$

27. $\displaystyle\lim_{(x,y)\to(-1,2)} \frac{4x^3 y^2 - 2xy^5 + 7y - 1}{3xy - y^4 + 3x^3}$

28. $\displaystyle\lim_{(x,y)\to(-4,3)} \frac{1 + xy}{1 - xy}$

29. $\displaystyle\lim_{(x,y)\to(\pi,\pi/3)} \ln(x + y)$

30. $\displaystyle\lim_{(x,y)\to(1,2)} \ln(1 + e^{x+y})$

31. $\displaystyle\lim_{(x,y)\to(2,5)} \sinh\left(\frac{x + 1}{y - 2}\right)$

32. $\displaystyle\lim_{(x,y)\to(1,1)} \frac{x - y}{x^2 - y^2}$ [Hint: Divide.]

33. $\displaystyle\lim_{(x,y)\to(2,2)} \frac{x^3 - 2xy + 3x^2 - 2y}{x^2 y + 4y^2 - 6x^2 + 24y}$

34. $\displaystyle\lim_{(x,y,z)\to(1,1,3)} \frac{xy^2 - 4xz^2 + 5yz}{3z^2 - 8z^3 y^7 x^4 + 7x - y + 2}$

35. $\displaystyle\lim_{(x,y,z)\to(4,1,3)} \ln(x - yz + 4x^3 y^5 z).$

In Problems 36–49 describe the maximum region over which the given function is continuous.

36. $f(x, y) = \sqrt{x - y}$

37. $f(x, y) = \dfrac{x^3 + 4xy^6 - 7x^4}{x^2 + y^2}$

38. $f(x, y) = \dfrac{x^3 + 4xy^6 - 7x^4}{x^3 - y^3}$

*39. $f(x, y) = \dfrac{xy^3 - 17x^2y^5 + 8x^3y}{xy + 3y - 4x - 12}$

40. $f(x, y) = \ln(3x + 2y + 6)$

41. $f(x, y) = \tan^{-1}(x - y)$

42. $f(x, y) = e^{xy+2}$

43. $f(x, y) = \dfrac{x^3 - 1 + 3y^5x^2}{1 - xy}$

44. $f(x, y) = \dfrac{x}{\sqrt{1 - (x^2/4) - y^2}}$

45. $f(x, y, z) = e^{(xy+yz-\sqrt{x})}$

46. $f(x, y, z) = \dfrac{xyz^2 + yzx^2 - 3x^3yz^5}{x - y + 2z + 4}$

47. $f(x, y, z) = y \ln(xz)$

48. $f(x, y) = \cos^{-1}(x^2 - y)$

49. $f(x, y, z) = \dfrac{1}{\sqrt{1 - x^2 - y^2 - z^2}}.$

50. Find a function $g(x)$ such that the function

$$f(x, y) = \begin{cases} \dfrac{x^2 - y^2}{x - y}, & x \neq y \\ g(x), & x = y \end{cases}$$

is continuous at every point in $\mathbb{R}^2$.

51. Find a number c such that the function

$$f(x, y) = \begin{cases} \dfrac{3xy}{\sqrt{x^2 + y^2}}, & (x, y) \neq (0, 0) \\ c, & (x, y) = (0, 0) \end{cases}$$

is continuous at the origin.

52. Find a number c such that

$$f(x, y) = \begin{cases} \dfrac{xy}{|x| + |y|}, & \text{if } (x, y) \neq (0, 0) \\ c, & \text{if } (x, y) = (0, 0) \end{cases}$$

is continuous at the origin.

*53. Discuss the continuity at the origin of

$$f(x, y, z) = \begin{cases} \dfrac{yz - x^2}{x^2 + y^2 + z^2}, & \text{if } (x, y, z) \neq (0, 0, 0) \\ 0, & \text{if } (x, y, z) = (0, 0, 0). \end{cases}$$

18.4 Partial Derivatives

In this section we show one of the ways a function of several variables can be differentiated. The idea is simple. Let $z = f(x, y)$. If we keep one of the variables, say y, fixed, then f can be treated as a function of x only and we can calculate the derivative (if it exists) of f with respect to x. This new function is called the *partial derivative of f with respect to x* and is denoted $\partial f/\partial x$.† Before giving a more formal definition, we give an example.

EXAMPLE 1.   Let $z = f(x, y) = x^2 y + \sin xy^2$. Calculate $\partial f/\partial x$.

SOLUTION. Treating y as if it were constant, we have

$$\frac{\partial f}{\partial x} = \frac{\partial}{\partial x}(x^2 y + \sin xy^2) = \frac{\partial}{\partial x} x^2 y + \frac{\partial}{\partial x} \sin xy^2 = 2xy + y^2 \cos xy^2.$$

DEFINITION 1.   Let $z = f(x, y)$. Then

(i)   the *partial derivative of f with respect to x* is the function

$$\frac{\partial z}{\partial x} = \frac{\partial f}{\partial x} = \lim_{\Delta x \to 0} \frac{f(x + \Delta x, y) - f(x, y)}{\Delta x}. \tag{1}$$

$\partial f/\partial x$ is defined at every point (x, y) in the domain of f such that the limit (1) exists.

(ii)   the *partial derivative of f with respect to y* is the function

$$\frac{\partial z}{\partial y} = \frac{\partial f}{\partial y} = \lim_{\Delta y \to 0} \frac{f(x, y + \Delta y) - f(x, y)}{\Delta y}. \tag{2}$$

$\partial f/\partial y$ is defined at every point (x, y) in the domain of f such that the limit (2) exists.

> **Remark 1.** This definition allows us to calculate partial derivatives in the same way we calculate ordinary derivatives by allowing only one of the variables to vary.

> **Remark 2.** The partial derivatives $\partial f/\partial x$ and $\partial f/\partial y$ gives us the rate of change of f as each of the variables x and y change with the other one held fixed. They do *not* tell us how f changes when x and y change simultaneously. We will discuss this different topic in Section 6.

> **Remark 3.** It should be emphasized that while the functions $\partial f/\partial x$ and $\partial f/\partial y$ are computed with one of the variables held constant, each is a function of both variables.

EXAMPLE 2.   Let $f(x, y) = \sqrt{x + y^2}$. Calculate $\partial f/\partial x$ and $\partial f/\partial y$.

SOLUTION

$$\frac{\partial f}{\partial x} = \frac{1}{2\sqrt{x + y^2}} \frac{\partial}{\partial x}(x + y^2) = \frac{1}{2\sqrt{x + y^2}}(1 + 0) = \frac{1}{2\sqrt{x + y^2}}$$

since we are treating y as a constant.

$$\frac{\partial f}{\partial y} = \frac{1}{2\sqrt{x + y^2}} \frac{\partial}{\partial y}(x + y^2) = \frac{1}{2\sqrt{x + y^2}}(0 + 2y) = \frac{y}{\sqrt{x + y^2}}.$$

---

† The symbol ∂ of partial derivatives is not a letter from any alphabet, but an invented mathematical symbol which may be read "partial." Historically, the difference between an ordinary and partial derivative was not recognized at first, and the same symbol d was used for both. The symbol ∂ was introduced in the eighteenth century by the mathematicians Alexis Fontaine des Bertins (1705–1771), Leonhard Euler (1707–1783), Alexis-Claude Clairaut (1713–1765) and Jean Le Rond d'Alembert (1717–1783) and was used in their development of the theory of partial differentiation.

EXAMPLE 3. Let $z = (x/y) \sin(x^2 y^3)$. Calculate $\partial z / \partial x$ and $\partial z / \partial y$.

SOLUTION. We use the product rule of differentiation.

$$\frac{\partial z}{\partial x} = \frac{x}{y} \frac{\partial}{\partial x} [\sin(x^2 y^3)] + \left[\frac{\partial}{\partial x} \left(\frac{x}{y} \right) \right] \sin(x^2 y^3)$$

$$= \frac{x}{y} [\cos(x^2 y^3)][2xy^3] + \frac{1}{y} \sin(x^2 y^3)$$

$$= 2x^2 y^2 \cos(x^2 y^3) + \frac{1}{y} \sin(x^2 y^3).$$

$$\frac{\partial z}{\partial y} = \frac{x}{y} \frac{\partial}{\partial y} [\sin(x^2 y^3)] + \left[\frac{\partial}{\partial y} \left(\frac{x}{y} \right) \right] \sin(x^2 y^3)$$

$$= \frac{x}{y} [\cos(x^2 y^3)][3x^2 y^2] - \frac{x}{y^2} \sin(x^2 y^3)$$

$$= 3x^3 y \cos(x^2 y^3) - \frac{x}{y^2} \sin(x^2 y^3).$$

EXAMPLE 4. Let $f(x, y) = (1 + x^2 + y^5)^{4/3}$. Calculate $\partial f / \partial x$ and $\partial f / \partial y$ at the point $(3, 1)$.

SOLUTION

$$\frac{\partial f}{\partial x} = \frac{4}{3} (1 + x^2 + y^5)^{1/3} \frac{\partial}{\partial x} (1 + x^2 + y^5)$$

$$= \frac{4}{3} (1 + x^2 + y^5)^{1/3} \cdot 2x = \frac{8x}{3} (1 + x^2 + y^5)^{1/3}.$$

At $(3, 1)$,

$$\frac{\partial f}{\partial x} = \frac{(8)(3)}{3} (1 + 3^2 + 1^5)^{1/3} = 8 \sqrt[3]{11}.$$

$$\frac{\partial f}{\partial y} = \frac{4}{3} (1 + x^2 + y^5)^{1/3} \frac{\partial}{\partial y} (1 + x^2 + y^5)$$

$$= \frac{4}{3} (1 + x^2 + y^5)^{1/3} \cdot 5y^4 = \frac{20y^4}{3} (1 + x^2 + y^5)^{1/3}.$$

At $(3, 1)$,

$$\frac{\partial f}{\partial y} = \frac{20}{3} \sqrt[3]{11}.$$

    We now obtain a geometric interpretation of the partial derivative. Let $z = f(x, y)$. As we saw in Section 18.2, this is the equation of a surface in $\mathbb{R}^3$. To obtain $\partial z / \partial x$, we hold y fixed at some constant value y_0. The equation $y = y_0$ is a plane in space parallel to the xz-plane (whose equation is $y = 0$). Thus, if y is constant, $\partial z / \partial x$ is the rate of change of f with respect to x as x changes along the curve C which is at the intersection of the surface $z = f(x, y)$ and the plane $y = y_0$. This is indicated in Figure 1. To be more precise, if (x_0, y_0, z_0) is a point on the

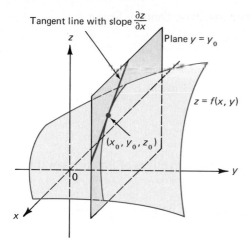

Figure 1

surface $z = f(x, y)$, then $\partial z/\partial x$ evaluated at (x_0, y_0) is the slope of the line tangent to the surface at the point (x_0, y_0, z_0) which lies in the plane $y = y_0$. Analogously, $\partial z/\partial y$ evaluated at (x_0, y_0) is the slope of the line tangent to the surface at the point (x_0, y_0, z_0) which lies in the plane $x = x_0$ (since x is held fixed in order to calculate $\partial z/\partial y$). This is illustrated in Figure 2.

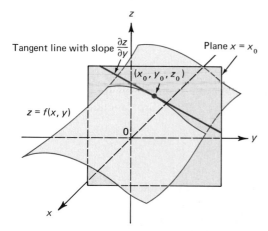

Figure 2

There are other ways to denote partial derivatives. We will often write

$$f_x = \frac{\partial f}{\partial x} \quad \text{and} \quad f_y = \frac{\partial f}{\partial y}. \tag{3}$$

If f is a function of other variables, say s and t, then we may write $\partial f/\partial s = f_s$ and $\partial f/\partial t = f_t$.

EXAMPLE 5. As we calculated in Section 9.1, the volume of a cone of radius r and height h is given by

$$V = \tfrac{1}{3}\pi r^2 h.$$

Then the rate of change of V with respect to r (with h fixed) is given by

$$\frac{\partial V}{\partial r} = V_r = \frac{\partial}{\partial r}(\pi r^2 h) = 2\pi r h$$

and the rate of change of V with respect to h (with r fixed) is given by

$$\frac{\partial V}{\partial h} = V_h = \frac{\partial}{\partial h}(\pi r^2 h) = \pi r^2.$$

We now turn to the question of finding partial derivatives of functions of three variables.

DEFINITION 2.   Let $w = f(x, y, z)$ be defined in a neighborhood of the point (x, y, z). Then

(i)   *the partial derivative of f with respect to x is the function*

$$\frac{\partial w}{\partial x} = \frac{\partial f}{\partial x} = f_x = \lim_{\Delta x \to 0} \frac{f(x + \Delta x, y, z) - f(x, y, z)}{\Delta x}. \tag{4}$$

f_x is defined at every point (x, y, z) in the domain of f at which the limit in (4) exists.

(ii)   *the partial derivative of f with respect to y is the function*

$$\frac{\partial w}{\partial y} = \frac{\partial f}{\partial y} = f_y = \lim_{\Delta y \to 0} \frac{f(x, y + \Delta y, z) - f(x, y, z)}{\Delta y}. \tag{5}$$

f_y is defined at every point (x, y, z) in the domain of f at which the limit in (5) exists.

(iii)   *the partial derivative of f with respect to z is the function*

$$\frac{\partial w}{\partial z} = \frac{\partial f}{\partial z} = f_z = \lim_{\Delta z \to 0} \frac{f(x, y, z + \Delta z) - f(x, y, z)}{\Delta z}. \tag{6}$$

f_z is defined at any point (x, y, z) in the domain of f at which the limit in (6) exists.

Remark. As can be seen from Definition 2, to calculate $\partial f/\partial x$, we simply treat y and z as constants and then calculate an ordinary derivative.

EXAMPLE 6.   Let $w = f(x, y, z) = xz + e^{y^2 z} + \sqrt{xy^2 z^3}$. Calculate $\partial w/\partial x$, $\partial w/\partial y$, and $\partial w/\partial z$.

SOLUTION.   To calculate $\partial w/\partial x$, we keep y and z fixed. Then

$$\frac{\partial w}{\partial x} = \frac{\partial f}{\partial x} = f_x = z + \frac{\partial}{\partial x} e^{y^2 z} + \frac{1}{2\sqrt{xy^2 z^3}} \frac{\partial}{\partial x}(xy^2 z^3)$$

$$= z + 0 + \frac{y^2 z^3}{2\sqrt{xy^2 z^3}} = z + \frac{y^2 z^3}{2\sqrt{xy^2 z^3}}.$$

To calculate $\partial w / \partial y$, we keep x and z fixed. Then

$$\frac{\partial w}{\partial y} = \frac{\partial f}{\partial y} = f_y = \frac{\partial}{\partial y} xz + e^{y^2 z} \frac{\partial}{\partial y}(y^2 z) + \frac{1}{2\sqrt{xy^2 z^3}} \frac{\partial}{\partial y}(xy^2 z^3)$$

$$= 0 + 2yze^{y^2 z} + \frac{2xyz^3}{2\sqrt{xy^2 z^3}} = 2yze^{y^2 z} + \frac{xyz^3}{\sqrt{xy^2 z^3}}.$$

To calculate $\partial w / \partial z$, we keep x and y fixed. Then

$$\frac{\partial w}{\partial z} = \frac{\partial f}{\partial z} = f_z = x + e^{y^2 z} \frac{\partial}{\partial z} y^2 z + \frac{1}{2\sqrt{xy^2 z^3}} \frac{\partial}{\partial z}(xy^2 z^3)$$

$$= x + y^2 e^{y^2 z} + \frac{3xy^2 z^2}{2\sqrt{xy^2 z^3}}.$$

EXAMPLE 7. Let C denote the oxygen consumption of a fur-bearing animal, let T_b denote its internal body temperature (in °C), let T_f denote the outside temperature of its fur, and let w denote its weight (in kg). It has been experimentally determined that a reasonable model for the oxygen consumption of the animal is given by

$$C = \frac{5(T_b - T_f)}{2w^{2/3}}.$$

Calculate (i) C_{T_b}, (ii) C_{T_f}, and (iii) C_w.

SOLUTION

$$\text{(i)} \quad C_{T_b} = \frac{\partial C}{\partial T_b} = \frac{\partial}{\partial T_b} \left(\frac{5T_b}{2w^{2/3}} - \frac{5T_f}{2w^{2/3}} \right) = \frac{5}{2w^{2/3}};$$

$$\text{(ii)} \quad C_{T_f} = -\frac{5}{2w^{2/3}};$$

$$\text{(iii)} \quad C_w = \frac{\partial}{\partial w} \frac{5}{2}(T_b - T_f)w^{-2/3} = -\frac{5}{3}(T_b - T_f)w^{-5/3}.$$

Note that (i) and (ii) imply that, since $w > 0$, an increase in internal body temperature leads to an increase in oxygen consumption (if T_f and w do not change) while an increase in fur temperature leads to a decrease in oxygen consumption (with T_b and w held constant). Does this make sense intuitively? Furthermore, if T_b and T_f are held constant, then, assuming that $T_b > T_f$, an increase in the animal's weight will lead to a decrease in its oxygen consumption.

We can use Definitions 1 and 2 to develop a definition of partial derivatives of functions of four or more variables. Although we shall not give a formal definition, the general idea is to hold all the variables but one constant, and then perform an "ordinary" differentiation.

In Chapter 2 we showed that a differentiable function is continuous. The situation is more complicated for a function of two variables. In fact there are functions which are discontinuous at a point but for which all partial derivatives exist at that point.

EXAMPLE 8.   Let

$$f(x, y) = \begin{cases} \dfrac{xy}{x^2 + y^2}, & (x, y) \neq (0, 0) \\ 0, & (x, y) = (0, 0). \end{cases}$$

Show that f_x and f_y exist at $(0, 0)$ but that f is not continuous there.

SOLUTION. We first show that $\lim_{(x,y)\to(0,0)} f(x, y)$ does not exist so that f cannot be continuous at $(0, 0)$. To show that we first let $(x, y) \to (0, 0)$ along the line $y = x$. Then

$$\frac{xy}{x^2 + y^2} = \frac{x^2}{x^2 + x^2} = \frac{1}{2}$$

so that if $\lim_{(x,y)\to(0,0)} f(x, y)$ existed, it would have to equal $\frac{1}{2}$. But if we now let $(x, y) \to (0, 0)$ along the line $y = -x$, we have

$$\frac{xy}{x^2 + y^2} = \frac{-x^2}{x^2 + x^2} = -\frac{1}{2}$$

so that along this line the limit is $-\frac{1}{2}$. Hence the limit does not exist. On the other hand we have

$$f_x(0, 0) = \lim_{\Delta x \to 0} \frac{f(0 + \Delta x, 0) - f(0, 0)}{\Delta x} = \lim_{\Delta x \to 0} \frac{\dfrac{(0 + \Delta x) \cdot 0}{\Delta x^2 + 0^2} - 0}{\Delta x^2}$$

$$= \lim_{\Delta x \to 0} \frac{0}{\Delta x^2} = \lim_{\Delta x \to 0} 0 = 0.$$

Similarly, $f_y(0, 0) = 0$. Hence both f_x and f_y exist at $(0, 0)$ (and are equal to zero) even though f is not continuous there.

     In the next section we will show a relationship between continuity and partial derivatives and in Section 18.6 we shall show that a certain kind of differentiability does imply continuity.

PROBLEMS 18.4

In Problems 1–12 calculate $\partial z/\partial x$ and $\partial z/\partial y$.

1. $z = x^2 y$
2. $z = xy^2$
3. $z = 3e^{xy^3}$
4. $z = \sin(x^2 + y^3)$
5. $z = 4x/y^5$
6. $z = e^y \tan x$
7. $z = \ln(x^3 y^5 - 2)$
8. $z = \sqrt{xy + 2y^3}$
9. $z = (x + 5y \sin x)^{4/3}$
10. $z = \sinh(2x - y)$

*11. $z = \sin^{-1}\left(\dfrac{x^2 - y^2}{x^2 + y^2}\right)$

12. $z = \sec xy$.

In Problems 13–20 calculate the value of the given partial derivative at the given point.

13. $f(x, y) = x^3 - y^4$; $f_x(1, -1)$
14. $f(x, y) = \ln(x^2 + y^4)$; $f_y(3, 1)$
15. $f(x, y) = \sin(x + y)$; $f_x(\pi/6, \pi/3)$
16. $f(x, y) = e^{\sqrt{x^2 + y}}$; $f_y(0, 4)$

17. $f(x, y) = \sinh(x - y)$; $f_x(3, 3)$

18. $f(x, y) = \sqrt[3]{x^2 y - y^2 x^5}$; $f_y(-2, 4)$

19. $f(x, y) = \dfrac{x^2 - y^2}{x^2 + y^2}$; $f_y(2, -3)$

20. $f(x, y) = \tan^{-1} \dfrac{y}{x}$; $f_x(4, 4)$.

In Problems 21–29 calculate $\partial w / \partial x$, $\partial w / \partial y$, and $\partial w / \partial z$.

21. $w = xyz$

22. $w = \sqrt{x + y + z}$

23. $w = \dfrac{x + y}{z}$

24. $w = \dfrac{x^2 - y^2 + z^2}{x^2 + y^2 + z^2}$

25. $w = \ln(x^3 + y^2 + z)$

26. $w = e^{x + 2y + 3z}$

27. $w = \sin(xyz)$

28. $w = \tan^{-1} xz/y$

29. $w = \cosh \sqrt{x + 2y + 5z}$.

In Problems 30–39 calculate the value of the given partial derivative at the given point.

30. $f(x, y, z) = xyz$; $f_x(2, 3, 4)$

31. $f(x, y, z) = \sqrt{x + 2y + 3z}$; $f_y(2, -1, 3)$

32. $f(x, y, z) = \dfrac{x - y}{z}$; $f_z(-3, -1, 2)$

33. $f(x, y, z) = \sin(z^2 - y^2 + x)$; $f_y(0, 1, 0)$

34. $f(x, y, z) = \ln(x + 2y + 3z)$; $f_z(2, 2, 5)$

35. $f(x, y, z) = \tan^{-1} xy/z$; $f_x(1, 2, -2)$

36. $f(x, y, z) = \dfrac{y^3 - z^5}{x^2 y + z}$; $f_y(4, 0, 1)$

37. $f(x, y, z) = \sqrt{\dfrac{x + y - z}{x + y + z}}$; $f_z(1, 1, 1)$

38. $f(x, y, z) = e^{xy}(\cosh z - \sinh z)$; $f_z(2, 3, 0)$

39. $f(x, y, z) = \sqrt{x^2 + y^2 + z^2}$; $f_x(a, b, c)$.

40. Let $f(x, y) = \begin{cases} \dfrac{x + y}{x - y}, & (x, y) \neq (0, 0) \\ 0, & (x, y) = (0, 0). \end{cases}$

(a) Show that f is not continuous at $(0, 0)$. [*Hint:* Show that $\lim_{(x,y) \to (0,0)} f(x, y)$ does not exist.]

(b) Do $f_x(0, 0)$ and $f_y(0, 0)$ exist?

41. A *partial differential equation* is an equation involving partial derivatives. Show that the function $z = f(x, y) = e^{(x + \sqrt{3}y)/4} - 4x - 2y - 4 - 2\sqrt{3}$ satisfies the partial differential equation $\partial z / \partial x + \sqrt{3}\, \partial z / \partial y - z = 4x + 2y$.

42. Show that the function $f(x, y) = z = e^{x + (5/4)y} - \frac{7}{2} e^{y/2} + \frac{7}{2}$ is a solution to the partial differential equation $3f_x - 4f_y + 2f = 7$ which also satisfies $f(x, 0) = e^x$.

43. Find the equation of the line tangent to the surface $z = x^3 - 4y^3$ at the point $(1, -1, 5)$ which

(a) lies in the plane $x = 1$.   (b) lies in the plane $y = -1$.

44. Find the equation of the line tangent to the surface $z = \tan^{-1}(y/x)$ at the point $(\sqrt{3}, 1, \pi/6)$ which

(a) lies in the plane $x = \sqrt{3}$.   (b) lies in the plane $y = 1$.

45. Find the equation of the line tangent to the surface $x^2 + 4y^2 + 4z^2 = 9$ which lies on the plane $y = 1$ at the point $(1, 1, 1)$.

46. The ideal gas law states that $P = nRT/V$ (see Section 18.1). Assume that the number of moles n of an ideal gas and the temperature T of the gas are held constant at the values 10 and 20 °C ($= 293$ °K), respectively. What is the rate of change of the pressure, as a function of the volume, when the volume of the gas is 2 liters?

47. For any ideal gas show that $\dfrac{\partial V}{\partial T}\dfrac{\partial T}{\partial P}\dfrac{\partial P}{\partial V} = -1$.

48. In Section 6.5 we showed that the present value of an annuity for which B dollars are to be paid every year for t years is given by

$$A_0 = \frac{B}{i}\left[1 - \left(\frac{1}{1+i}\right)^t\right] \qquad \text{(see equation (6.5.10)).}$$

(a) If t is fixed, how does the present value of annuity change as the rate of interest changes?
(b) How is the present value changing with respect to i if $B = \$500$ and $i = 6\%$?
(c) If i is fixed, how does the present value of the annuity change as the number of years during which the payments are made is increased?

49. A fur-bearing animal weighing 10 kg has a constant internal body temperature of 23 °C. Using the model of Example 7, if the outside temperature is dropping, how is the oxygen consumption of the animal changing when the outside temperature of its fur is 5 °C?

50. The cost to a manufacturer of producing x units of product A and y units of product B is given (in dollars) by

$$C(x, y) = \frac{50}{2 + x} + \frac{125}{(3 + y)^2}.$$

Calculate the marginal cost of each of the two products.

51. The revenue received from the manufacturer of Problem 50 is given by

$$R(x, y) = \ln(1 + 50x + 75y) + \sqrt{1 + 40x + 125y}.$$

Calculate the marginal revenue from each of the two products.

52. If a particle is falling in a fluid, then according to *Stoke's law* the velocity of the particle is given by

$$V = \frac{2g}{9}(\rho_P - \rho_f)\frac{r^2}{\nu},$$

where g is the acceleration due to gravity, ρ_P is the density of the particle, ρ_f is the density of the fluid, r is the radius of the particle (in cm) and ν is the absolute viscosity of the liquid. Calculate V_{ρ_P}, V_{ρ_f}, V_r, and V_ν.

53. Let $f(x, y) = g(x)h(y)$ where g and h are differentiable functions of a single variable. Show that the partial derivatives $\partial f/\partial x$ and $\partial f/\partial y$ exist and that

$$\frac{\partial f}{\partial x} = g'(x)h(y) \qquad \text{and} \qquad \frac{\partial f}{\partial y} = g(x)h'(y).$$

54. Let g, h, and k be differentiable and let $f(x, y, z) = g(x)h(y)k(z)$. Show that $\partial f/\partial x$, $\partial f/\partial y$, and $\partial f/\partial z$ all exist and

$$\frac{\partial f}{\partial x} = g'(x)h(y)k(z), \quad \frac{\partial f}{\partial y} = g(x)h'(y)k(z), \quad \text{and} \quad \frac{\partial f}{\partial z} = g(x)h(y)k'(z).$$

18.5 Higher Order Partial Derivatives

We have seen that if $y = f(x)$, then

$$y' = \frac{df}{dx} \quad \text{and} \quad y'' = \frac{d^2 f}{dx^2} = \frac{d}{dx}\left(\frac{df}{dx}\right).$$

That is the second derivative of f is the derivative of the first derivative of f. Analogously, if $z = f(x, y)$, then we can differentiate each of the two "first" partial derivatives $\partial f/\partial x$ and $\partial f/\partial y$ with respect to both x and y to obtain four *second partial derivatives* as follows:

(i) Differentiate twice with respect to x:

$$\boxed{\frac{\partial^2 z}{\partial x^2} = \frac{\partial^2 f}{\partial x^2} = f_{xx} = \frac{\partial}{\partial x}\left(\frac{\partial f}{\partial x}\right).} \tag{1}$$

(ii) Differentiate first with respect to x and then with respect to y:

$$\boxed{\frac{\partial^2 z}{\partial y\, \partial x} = \frac{\partial^2 f}{\partial y\, \partial x} = f_{xy} = \frac{\partial}{\partial y}\left(\frac{\partial f}{\partial x}\right).} \tag{2}$$

(iii) Differentiate first with respect to y and then with respect to x:

$$\boxed{\frac{\partial^2 z}{\partial x\, \partial y} = \frac{\partial^2 f}{\partial x\, \partial y} = f_{yx} = \frac{\partial}{\partial x}\left(\frac{\partial f}{\partial y}\right).} \tag{3}$$

(iv) Differentiate twice with respect to y:

$$\boxed{\frac{\partial^2 z}{\partial y^2} = \frac{\partial^2 f}{\partial y^2} = f_{yy} = \frac{\partial}{\partial y}\left(\frac{\partial f}{\partial y}\right).} \tag{4}$$

Remark 1. The derivatives $\partial^2 f/\partial x\, \partial y$ and $\partial^2 f/\partial y\, \partial x$ are called the *mixed second* partials.

Remark 2. It is much easier to denote the second partials by $f_{xx}, f_{xy}, f_{yx},$ and f_{yy}. We will therefore use this notation for the remainder of this section. Note that the symbol f_{xy} indicates that we differentiate first with respect to x and then with respect to y.

EXAMPLE 1. Let $z = f(x, y) = x^3 y^2 - xy^5$. Calculate the four second partial derivatives.

SOLUTION. We have $f_x = 3x^2 y^2 - y^5$ and $f_y = 2x^3 y - 5xy^4$. Then

(i) $f_{xx} = \dfrac{\partial}{\partial x}(f_x) = 6xy^2;$

(ii) $f_{xy} = \dfrac{\partial}{\partial y}(f_x) = 6x^2 y - 5y^4;$

(iii) $f_{yx} = \dfrac{\partial}{\partial x}(f_y) = 6x^2y - 5y^4$; and

(iv) $f_{yy} = \dfrac{\partial}{\partial y}(f_y) = 2x^3 - 20xy^3$.

EXAMPLE 2. Let $z = f(x, y) = \sin xy^3$. Calculate the four second partial derivatives of f.

SOLUTION. We have $f_x = y^3 \cos xy^3$ and $f_y = 3xy^2 \cos xy^3$. Then

$$f_{xx} = -y^6 \sin xy^3$$
$$f_{xy} = y^3(-3xy^2 \sin xy^3) + 3y^2 \cos xy^3 = -3xy^5 \sin xy^3 + 3y^2 \cos xy^3$$
$$f_{yx} = (3xy^2)(-y^3 \sin xy^3) + 3y^2 \cos xy^3 = -3xy^5 \sin xy^3 + 3y^2 \cos xy^3$$
$$f_{yy} = 3xy^2(-3xy^2 \sin xy^3) + 6xy \cos xy^3 = -9x^2y^4 \sin xy^3 + 6xy \cos xy^3.$$

In the last two problems we saw that $f_{xy} = f_{yx}$. This is no accident as we see by the following theorem whose proof can be found in any advanced calculus text.†

THEOREM 1. Suppose that f, f_x, f_y, f_{xy}, and f_{yx} are all continuous at (x_0, y_0). Then

$$\boxed{f_{xy}(x_0, y_0) = f_{yx}(x_0, y_0).}\tag{5}$$

This result is often referred to as the *equality of mixed partials.*‡

The definition of second partial derivatives and the theorem on the equality of mixed partials is easily extended to functions of three variables. If $w = f(x, y, z)$, then we have the nine second partial derivatives (assuming that they exist)

$$\dfrac{\partial^2 f}{\partial x^2} = f_{xx}, \qquad \dfrac{\partial^2 f}{\partial x\,\partial y} = f_{yx}, \qquad \dfrac{\partial^2 f}{\partial x\,\partial z} = f_{zx},$$

$$\dfrac{\partial^2 f}{\partial y^2} = f_{yy}, \qquad \dfrac{\partial^2 f}{\partial y\,\partial x} = f_{xy}, \qquad \dfrac{\partial^2 f}{\partial y\,\partial z} = f_{zy},$$

$$\dfrac{\partial^2 f}{\partial z^2} = f_{zz}, \qquad \dfrac{\partial^2 f}{\partial z\,\partial x} = f_{xz}, \qquad \dfrac{\partial^2 f}{\partial z\,\partial y} = f_{yz}.$$

THEOREM 2. If f, f_x, f_y, f_z, and all six mixed partials are continuous at a point (x_0, y_0, z_0), then at that point

$$\boxed{\begin{aligned} f_{xy} &= f_{yx}, \\ f_{xz} &= f_{zx}, \\ f_{yz} &= f_{zy}. \end{aligned}}$$

EXAMPLE 3. Let $f(x, y, z) = xy^3 - zx^5 + x^2yz$. Calculate all nine second partial derivatives and show that all three pairs of mixed partials are equal.

---

† See, for example, R. C. Buck, *Advanced Calculus,* McGraw-Hill, New York, 1965, p. 248.
‡ This theorem was first proved by Euler in a 1734 paper devoted to a problem in hydrodynamics.

SOLUTION. We have

$$f_x = y^3 - 5zx^4 + 2xyz,$$
$$f_y = 3xy^2 + x^2z,$$

and

$$f_z = -x^5 + x^2y.$$

Then

$$f_{xx} = -20zx^3 + 2yz, \qquad f_{yy} = 6xy, \qquad f_{zz} = 0,$$

$$f_{xy} = \frac{\partial}{\partial y}(y^3 - 5zx^4 + 2xyz) = 3y^2 + 2xz,$$

$$f_{yx} = \frac{\partial}{\partial x}(3xy^2 + x^2z) = 3y^2 + 2xz,$$

$$f_{xz} = \frac{\partial}{\partial z}(y^3 - 5zx^4 + 2xyz) = -5x^4 + 2xy,$$

$$f_{zx} = \frac{\partial}{\partial x}(-x^5 + x^2y) = -5x^4 + 2xy,$$

$$f_{yz} = \frac{\partial}{\partial z}(3xy^2 + x^2z) = x^2,$$

$$f_{zy} = \frac{\partial}{\partial y}(-x^5 + x^2y) = x^2.$$

We conclude this section by pointing out that we can easily define partial derivatives of orders higher than two. For example,

$$f_{zyx} = \frac{\partial^3 f}{\partial x\, \partial y\, \partial z} = \frac{\partial}{\partial x}\left(\frac{\partial^2 f}{\partial y\, \partial z}\right) = \frac{\partial}{\partial x}(f_{zy}).$$

EXAMPLE 4. Calculate f_{xxx}, f_{xzy}, f_{yxz}, and f_{yxzx} for the function of Example 3.

SOLUTION. We easily obtain the three third partial derivatives:

$$f_{xxx} = \frac{\partial}{\partial x}(f_{xx}) = \frac{\partial}{\partial x}(-20zx^3 + 2yz) = -60zx^2$$

$$f_{xzy} = \frac{\partial}{\partial y}(f_{xz}) = \frac{\partial}{\partial y}(-5x^4 + 2xy) = 2x$$

$$f_{yxz} = \frac{\partial}{\partial z}(f_{yx}) = \frac{\partial}{\partial z}(3y^2 + 2xz) = 2x.$$

Note that $f_{xzy} = f_{yxz}$. This again is no accident and follows from the generalization of Theorem 2 to mixed third partial derivatives. Finally, the fourth partial derivative f_{yxzx} is given by

$$f_{yxzx} = \frac{\partial}{\partial x}(f_{yxz}) = \frac{\partial}{\partial x}(2x) = 2.$$

PROBLEMS 18.5

In Problems 1–12 calculate the four second partial derivatives and show that the mixed partials are equal.

1. $f(x, y) = x^2 y$
2. $f(x, y) = xy^2$
3. $f(x, y) = 3e^{xy^3}$
4. $f(x, y) = \sin(x^2 + y)$
5. $f(x, y) = 4x/y^5$
6. $f(x, y) = e^y \tan x$
7. $f(x, y) = \ln(x^3 y^5 - 2)$
8. $f(x, y) = \sqrt{xy + 2y^3}$
9. $f(x, y) = (x + 5y \sin x)^{4/3}$
10. $f(x, y) = \sinh(2x - y)$

*11. $f(x, y) = \sin^{-1}\left(\dfrac{x^2 - y^2}{x^2 + y^2}\right)$

12. $f(x, y) = \sec xy$.

In Problems 13–21 calculate the nine second partial derivatives and show that the three pairs of mixed partials are equal.

13. $f(x, y, z) = xyz$
14. $f(x, y, z) = x^2 y^3 z^4$
15. $f(x, y, z) = (x + y)/z$
16. $f(x, y, z) = \sin(x + 2y + z^2)$
17. $f(x, y, z) = \tan^{-1} xz/y$
18. $f(x, y, z) = \cos xyz$
19. $f(x, y, z) = e^{3xy} \cos z$
20. $f(x, y, z) = \ln(xy + z)$
21. $f(x, y, z) = \cosh \sqrt{x + yz}$.

22. How many third partial derivatives are there for a function of
    (a) two variables?   (b) three variables?
23. How many fourth partial derivatives are there for a function of
    (a) two variables?   (b) three variables?
*24. How many nth partial derivatives are there for a function of
    (a) two variables?   (b) three variables?

In Problems 25–30 calculate the given partial derivative.

25. $f(x, y) = x^2 y^3 + 2y;\ f_{xyx}$
26. $f(x, y) = \sin(2xy^4);\ f_{xyy}$
27. $f(x, y) = \ln(3x - 2y);\ f_{yxy}$
28. $f(x, y, z) = x^2 y + y^2 z - 3\sqrt{xz};\ f_{xyz}$
29. $f(x, y, z) = \cos(x + 2y + 3z);\ f_{zzx}$
30. $f(x, y, z) = e^{xy} \sin z;\ f_{zxyx}$.

31. Consider a string which is tightly stretched between two fixed points 0 and L on the x-axis. The string is pulled back and released at a time $t = 0$, causing it to vibrate. One position of the string is sketched in Figure 1. Let $y(x, t)$ denote the height of the string at

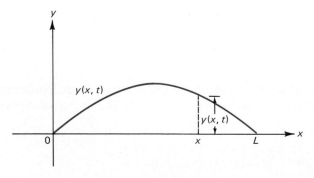

Figure 1

any time $t \geq 0$ and at any point x in the interval $[0, L]$. It can be shown that $y(x, t)$ satisfies the partial differential equation†

$$\frac{\partial^2 y}{\partial t^2} = c^2 \frac{\partial^2 y}{\partial x^2},$$

where c is a constant. This equation is called the one-dimensional *wave equation*. Show that

$$y(x, t) = \tfrac{1}{2}[(x - ct)^2 + (x + ct)^2]$$

is a solution to the wave equation, where c is any constant.

32. Consider a cylindrical rod composed of a uniform heat-conducting material of length l and radius r and assume that heat can enter and leave the rod only through its ends. Let $T(x, t)$ denote the absolute temperature at time t at a point x units along the rod (see Figure 2). Then it can be shown‡ that T satisfies the partial differential equation

$$\frac{\partial T}{\partial t} = \delta \frac{\partial^2 T}{\partial x^2},$$

where δ is a positive constant called the *diffusivity* of the rod. This equation is called the *heat equation* (or *diffusion equation*). Show that the function

$$T(x, t) = e^{-\alpha^2 \delta t} \sin \alpha x$$

satisfies the heat equation for any constant α.

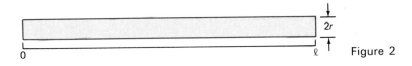

Figure 2

33. Show that $T(x, y) = (1/\sqrt{t})e^{-x^2/4\delta t}$ satisfies the heat equation.
34. Find constants α and β such that $T(x, t) = e^{\alpha x + \beta t}$ satisfies the heat equation.
35. One of the most important partial differential equations of mathematical physics is *Laplace's equation in* $\mathbb{R}^2$ given by

$$\frac{\partial^2 f}{\partial x^2} + \frac{\partial^2 f}{\partial y^2} = 0.$$

Show that the function $f(x, y) = x^2 - y^2$ satisfies Laplace's equation.
36. Show that $f(x, y) = \tan^{-1}(y/x)$ satisfies Laplace's equation.
37. Show that $f(x, y) = \ln(x^2 + y^2)$ satisfies Laplace's equation.
38. Show that $f(x, y) = \sin x \sinh y$ satisfies Laplace's equation.
39. Laplace's equation in $\mathbb{R}^3$ is given by

$$\frac{\partial^2 f}{\partial x^2} + \frac{\partial^2 f}{\partial y^2} + \frac{\partial^2 f}{\partial z^2} = 0.$$

Show that $f(x, y, z) = x^2 + y^2 - 2z^2$ satisfies Laplace's equation in $\mathbb{R}^3$.

---

†For a derivation of this equation, see W. Derrick and S. Grossman, *Elementary Differential Equations with Applications*, Addison-Wesley, Reading, Mass., 1976, p. 430.
‡See Derrick and Grossman, *Elementary Differential Equations*, p. 7.

18.6 Differentiability and the Gradient

In this section we discuss the notion of the differentiability of a function of several variables. There are several ways to introduce this subject and the way we have chosen is designed to illustrate the great similarities between differentiation of functions of one variable and differentiation of functions of several variables.

We begin with a function of one variable,

$$y = f(x).$$

If f is differentiable, then

$$f'(x) = \frac{dy}{dx} = \lim_{\Delta x \to 0} \frac{\Delta y}{\Delta x}. \tag{1}$$

Then, if we define the new function $\epsilon(\Delta x)$ by†

$$\epsilon(\Delta x) = f'(x) - \frac{\Delta y}{\Delta x} \tag{2}$$

we have

$$\lim_{\Delta x \to 0} \epsilon(\Delta x) = \lim_{\Delta x \to 0} \left(f'(x) - \frac{\Delta y}{\Delta x} \right) = f'(x) - \lim_{\Delta x \to 0} \frac{\Delta y}{\Delta x}$$

$$= f'(x) - f'(x) = 0. \tag{3}$$

Multiplying both sides of (2) by Δx and rearranging terms, we obtain

$$\Delta y = f'(x)\,\Delta x + \epsilon(\Delta x)\,\Delta x$$

Note that here Δy depends on both Δx and x. Finally, since $\Delta y = f(x + \Delta x) - f(x)$, we obtain

$$f(x + \Delta x) - f(x) = f'(x)\,\Delta x + \epsilon(\Delta x)\,\Delta x. \tag{4}$$

Why did we do all this? We did so in order to be able to state the following "alternative" definition of differentiability of a function f of one variable.

DEFINITION 1 (Alternate Definition of Differentiability). Let f be a function of one variable. Then f is differentiable at a number x if there is a function $L(x)$ and a function $g(\Delta x)$ such that

$$f(x + \Delta x) - f(x) = L(x)\,\Delta x + g(\Delta x) \tag{5}$$

where $\lim_{\Delta x \to 0}[g(\Delta x)/\Delta x] = 0$.

> **Note.** If f is differentiable at x, then from (4) and (5) we see that $L(x) = f'(x)$ and $g(\Delta x) = \epsilon(\Delta x)\,\Delta x$.

We now show how this new, one-variable definition of differentiability can be extended to functions of several variables. First, we introduce some condensed

---

†We used this function in our proof of the chain rule in Section 10.5.

notation. Since a point (x, y) is a vector in $\mathbb{R}^2$, we shall write (as we have done before) $\mathbf{x} = (x, y)$. Then, if $z = f(x, y)$, we can simply write

$$z = f(\mathbf{x}). \tag{6}$$

Similarly, if $w = f(x, y, z)$, we may write

$$w = f(\mathbf{x}) \tag{7}$$

where $\mathbf{x}$ is the vector (x, y, z). With this notation, we may use the symbol $\Delta\mathbf{x}$ to denote the vector $(\Delta x, \Delta y)$ in $\mathbb{R}^2$ or $(\Delta x, \Delta y, \Delta z)$ in $\mathbb{R}^3$.

Let us try to use our "usual" definition of a derivative. We might be tempted to write

$$f'(\mathbf{x}) = \lim_{\Delta\mathbf{x} \to 0} \frac{f(\mathbf{x} + \Delta\mathbf{x}) - f(\mathbf{x})}{\Delta\mathbf{x}}.$$

However, this makes no sense because we are unable to divide by the vector $\Delta\mathbf{x}$. We shall instead give a definition of differentiability which is analogous to Definition 1 above.

DEFINITION 2. Let f be a function of two variables which is defined in a neighborhood of a point $\mathbf{x} = (x, y)$. Let $\Delta\mathbf{x} = (\Delta x, \Delta y)$. Then f is *differentiable* at $\mathbf{x}$ if there are functions $\mathbf{L}$ and g such

$$\boxed{f(\mathbf{x} + \Delta\mathbf{x}) - f(\mathbf{x}) = \mathbf{L}(\mathbf{x}) \cdot \Delta\mathbf{x} + g(\Delta\mathbf{x})} \tag{8}$$

where

$$\boxed{\lim_{\Delta\mathbf{x} \to 0} \frac{g(\Delta\mathbf{x})}{|\Delta\mathbf{x}|} = 0.} \tag{9}$$

The function $\mathbf{L}(\mathbf{x})$ is called the *gradient* of f.

Remark 1. Although formulas (5) and (8) look very similar, there are two fundamental differences. First, $f' = L$ is a scalar while $\mathbf{L}$ is a vector. Second, $L(x)\,\Delta x = f'(x)\,\Delta x$ is an ordinary product of real numbers while $\mathbf{L}(\mathbf{x}) \cdot \Delta\mathbf{x}$ is a dot product of vectors. We will derive an expression for the gradient vector $\mathbf{L}(\mathbf{x})$ shortly.

Remark 2. If $z = f(x, y) = f(\mathbf{x})$, then the change or increment in z, denoted Δz or Δf, in moving from (x, y) to $(x + \Delta x, y + \Delta y)$ is given by

$$\Delta z = \Delta f = f(x + \Delta x, y + \Delta y) - f(x, y).$$

Using this notation (8) becomes

$$\Delta f(\mathbf{x}) = \Delta z = \mathbf{L}(\mathbf{x}) \cdot \Delta\mathbf{x} + g(\Delta\mathbf{x}). \tag{10}$$

Remark 3. The gradient of f is usually denoted ∇f, which is read "del" f. This symbol, an inverted Greek delta, was first used in the 1850s, although the name *del* first appeared in print only in 1901. The symbol ∇ is also called *nabla*. This is because someone once suggested to the Scottish mathematician Peter Guthrie Tait (1831–1901) that ∇ looks like an Assyrian harp, the Assyrian name of which is nabla.† Tait, incidentally, was one of the mathematicians who helped carry on Hamilton's development of the theory of quaternions and vectors in the nineteenth century. Using the symbol ∇f, equation (8) becomes

$$f(\mathbf{x} + \Delta \mathbf{x}) - f(\mathbf{x}) = \nabla f(\mathbf{x}) \cdot \Delta \mathbf{x} + g(\Delta \mathbf{x}). \tag{11}$$

Remark 4. We can write out (11) without using vector notation, although this is more cumbersome. Let

$$\nabla f(\mathbf{x}) = f_1(x, y)\mathbf{i} + f_2(x, y)\mathbf{j}.$$

Then (11) may be written

$$f(x + \Delta x, y + \Delta y) - f(x, y) = f_1(x, y)\,\Delta x + f_2(x, y)\,\Delta y + g(\Delta x, \Delta y). \tag{12}$$

In fact, it can be shown (see equations (31), (33), and (34)) that if (12) holds with

$$\lim_{(\Delta x, \Delta y) \to (0,0)} \frac{g(\Delta x, \Delta y)}{\sqrt{\Delta x^2 + \Delta y^2}} = 0,$$

then there exist functions ϵ_1 and ϵ_2 such that

$$g(\Delta x, \Delta y) = \epsilon_1(\Delta x, \Delta y)\,\Delta x + \epsilon_2(\Delta x, \Delta y)\,\Delta y \tag{13}$$

with

$$\lim_{(\Delta x, \Delta y) \to (0,0)} \epsilon_1(\Delta x, \Delta y) = 0 \quad \text{and} \quad \lim_{(\Delta x, \Delta y) \to (0,0)} \epsilon_2(\Delta x, \Delta y) = 0. \tag{14}$$

As we shall see in subsequent sections, the gradient of f is important for a great variety of reasons. We shall give a geometric interpretation of the gradient in Section 18.8. Before giving a simple method for calculating gradients, we calculate some gradients directly from the definition.

EXAMPLE 1. Let $f(x, y) = 3x - 4y + 2$. Show that f is differentiable and calculate ∇f.

SOLUTION

$$\begin{aligned}
f(x + \Delta x, y + \Delta y) - f(x, y) &= [3(x + \Delta x) - 4(y + \Delta y) + 2] - [3x - 4y + 2] \\
&= 3x + 3\,\Delta x - 4y - 4\,\Delta y + 2 - 3x + 4y - 2 \\
&= 3\,\Delta x - 4\,\Delta y.
\end{aligned}$$

Thus, in (12), $f_1 = 3$, $f_2 = -4$ and $g(\Delta x, \Delta y) = 0$ so that $\nabla f = 3\mathbf{i} - 4\mathbf{j}$.

EXAMPLE 2. Let $f(x, y) = xy$. Show that f is differentiable and calculate ∇f.

SOLUTION

$$\begin{aligned}
f(x + \Delta x, y + \Delta y) - f(x, y) &= (x + \Delta x)(y + \Delta y) - xy \\
&= xy + y\,\Delta x + x\,\Delta y + \Delta x\,\Delta y - xy \\
&= y\,\Delta x + x\,\Delta y + \Delta x\,\Delta y.
\end{aligned}$$

†Fortunately, most (but certainly not all) of the mathematical terms currently in use have more to do with the objects they describe. We might further point out that ∇ is also called atled, which is delta spelled backwards (no kidding).

Thus, in (12), we may take $f_1(x, y) = y, f_2(x, y) = x$, and $g(\Delta x, \Delta y) = \Delta x \, \Delta y$. To show that f is differentiable, we must show that

$$\lim_{|\Delta \mathbf{x}| \to 0} \frac{g(\Delta x, \Delta y)}{|\Delta \mathbf{x}|} = \lim_{|\Delta \mathbf{x}| \to 0} \frac{\Delta x \, \Delta y}{|\Delta \mathbf{x}|} = 0.$$

But $|\Delta \mathbf{x}| = \sqrt{\Delta x^2 + \Delta y^2}$ so we must show that

$$\lim_{(\Delta x, \Delta y) \to (0,0)} \frac{\Delta x \, \Delta y}{\sqrt{\Delta x^2 + \Delta y^2}} = 0.$$

This follows from the fact that $|\Delta x| \le \sqrt{\Delta x^2 + \Delta y^2}$ and $|\Delta y| \le \sqrt{\Delta x^2 + \Delta y^2}$ so that

$$\frac{|\Delta x \, \Delta y|}{\sqrt{\Delta x^2 + \Delta y^2}} \le \frac{\sqrt{\Delta x^2 + \Delta y^2} \sqrt{\Delta x^2 + \Delta y^2}}{\sqrt{\Delta x^2 + \Delta y^2}} = \sqrt{\Delta x^2 + \Delta y^2}$$

which approaches 0 as $(\Delta x, \Delta y) \to (0, 0)$. Thus f is differentiable and $\nabla f(x, y) = y\mathbf{i} + x\mathbf{j}$.

The next theorem shows us how we can greatly simplify our calculations of gradients.

THEOREM 1. Let f, f_x, and f_y be defined and continuous in a neighborhood of $\mathbf{x} = (x, y)$. Then f is differentiable at $\mathbf{x}$ and

$$\boxed{\nabla f(\mathbf{x}) = f_x(\mathbf{x})\mathbf{i} + f_y(\mathbf{x})\mathbf{j}.} \tag{15}$$

Remark 1. It is essential that f_x and f_y be continuous in this theorem. In Example 18.4.8, we saw an example of a function for which f_x and f_y exist at $(0, 0)$ but f itself is not even defined there.

Remark 2. If the hypotheses of Theorem 1 are satisfied, then f is said to be *continuously differentiable* at $\mathbf{x}$.

As the proof of Theorem 1 is very difficult, we shall postpone it until the end of the section. The proof depends upon the mean value theorem discussed in Section 10.6.

EXAMPLE 3. Let $z = f(x, y) = \sin xy^2 + e^{x^2 y^3}$. Show that f is differentiable and calculate ∇f. Find $\nabla f(1, 1)$.

SOLUTION. $\partial f / \partial x = y^2 \cos xy^2 + 2xy^3 e^{x^2 y^3}$ and $\partial f / \partial y = 2xy \cos xy^2 + 3x^2 y^2 e^{x^2 y^3}$. Since $\partial f / \partial x$ and $\partial f / \partial y$ are continuous, f is differentiable and

$$\nabla f(x, y) = (y^2 \cos xy^2 + 2xy^3 e^{x^2 y^3})\mathbf{i} + (2xy \cos xy^2 + 3x^2 y^2 e^{x^2 y^3})\mathbf{j}.$$

At $(1, 1)$, $\nabla f(1, 1) = (\cos 1 + 2e)\mathbf{i} + (2 \cos 1 + 3e)\mathbf{j}$.

In Section 18.4 we showed that the existence of all of its partial derivatives at a point does *not* ensure that a function is continuous at that point. However, differentiability (according to Definition 2) does ensure continuity.

THEOREM 2. If f is differentiable at $x_0 = (x_0, y_0)$, then f is continuous at x_0.

PROOF. We must show that $\lim_{(x,y)\to(x_0,y_0)} f(x, y) = f(x_0, y_0)$ or equivalently, $\lim_{\mathbf{x}\to\mathbf{x}_0} f(\mathbf{x}) = f(\mathbf{x}_0)$. But if we define $\Delta\mathbf{x}$ by $\Delta\mathbf{x} = \mathbf{x} - \mathbf{x}_0$, this is the same as showing that

$$\lim_{\Delta\mathbf{x}\to 0} f(\mathbf{x}_0 + \Delta\mathbf{x}) = f(\mathbf{x}_0). \tag{16}$$

Since f is differentiable at $\mathbf{x}_0$,

$$f(\mathbf{x}_0 + \Delta\mathbf{x}) - f(\mathbf{x}_0) = \nabla f(\mathbf{x}_0) \cdot \Delta\mathbf{x} + g(\Delta\mathbf{x}). \tag{17}$$

But as $\Delta\mathbf{x} \to 0$, both terms on the right-hand side of (17) approach zero so that

$$\lim_{\Delta\mathbf{x}\to 0} [f(\mathbf{x}_0 + \Delta\mathbf{x}) - f(\mathbf{x}_0)] = 0$$

which means that (16) holds and the theorem is proved.

The converse to this theorem is false as it is in one-variable calculus. That is, there are functions which are continuous, but not differentiable, at a given point. For example, the function

$$f(x, y) = \sqrt[3]{x} + \sqrt[3]{y}$$

is continuous at any point (x, y) in $\mathbb{R}^2$. But

$$\nabla f(x, y) = \frac{1}{3x^{2/3}}\mathbf{i} + \frac{1}{3y^{2/3}}\mathbf{j}$$

which is not defined at any point (x, y) for which either x or y is zero. That is, $\nabla f(x, y)$ is not defined on the x- and y-axes. Hence f is not differentiable along these axes.

All the definitions and theorems in this section hold for functions of three or more variables. We give the equivalent results for functions of three variables below.

DEFINITION 2′. Let f be a function of three variables which is defined in a neighborhood of $\mathbf{x} = (x, y, z)$ and let $\Delta\mathbf{x} = (\Delta x, \Delta y, \Delta z)$. Then f is *differentiable* at $\mathbf{x}$ if there are functions ∇f and g such that

$$\boxed{f(\mathbf{x} + \Delta\mathbf{x}) - f(\mathbf{x}) = \nabla f \cdot \Delta\mathbf{x} + g(\Delta\mathbf{x}),} \tag{18}$$

where

$$\boxed{\lim_{|\Delta\mathbf{x}|\to 0} \frac{g(\Delta\mathbf{x})}{|\Delta\mathbf{x}|} = 0.} \tag{19}$$

Equivalently, we can write

$$f(x + \Delta x, y + \Delta y, z + \Delta z) - f(x, y, z) = f_1(x, y, z)\,\Delta x + f_2(x, y, z)\,\Delta y$$
$$+ f_3(x, y, z)\,\Delta z + g(\Delta x, \Delta y, \Delta z), \tag{20}$$

where

$$\lim_{(\Delta x, \Delta y, \Delta z) \to (0,0,0)} \frac{g(\Delta x, \Delta y, \Delta z)}{\sqrt{\Delta x^2 + \Delta y^2 + \Delta z^2}} = 0. \tag{21}$$

THEOREM 1'. If f, $\partial f/\partial x$, $\partial f/\partial y$, and $\partial f/\partial z$ exist and are continuous in a neighborhood of $\mathbf{x} = (x, y, z)$, then f is differentiable at $\mathbf{x}$ and

$$\boxed{\nabla f(\mathbf{x}) = f_x(\mathbf{x})\mathbf{i} + f_y(\mathbf{x})\mathbf{j} + f_z(\mathbf{x})\mathbf{k}.} \tag{22}$$

THEOREM 2'. Let f be a function of three variables which is differentiable at $\mathbf{x}_0$. Then f is continuous at $\mathbf{x}_0$.

EXAMPLE 4. Let $f(x, y, z) = xy^2z^3$. Show that f is differentiable at any point $\mathbf{x}_0$, calculate ∇f and find $\nabla f(3, -1, 2)$.

SOLUTION. $\partial f/\partial x = y^2z^3$, $\partial f/\partial y = 2xyz^3$ and $\partial f/\partial z = 3xy^2z^2$. Since f, $\partial f/\partial x$, $\partial f/\partial y$, and $\partial f/\partial z$ are all continuous, we know that f is differentiable and that

$$\nabla f = y^2z^3\mathbf{i} + 2xyz^3\mathbf{j} + 3xy^2z^2\mathbf{k}$$

and

$$\nabla f(3, -1, 2) = 8\mathbf{i} - 48\mathbf{j} + 36\mathbf{k}.$$

We conclude this section with a proof of Theorem 1. The proof of Theorem 1′ is similar. This proof should be omitted if you are not familiar with the material in Section 10.6.

PROOF OF THEOREM 1. We begin by restating the mean value theorem of a function f of one variable.

MEAN VALUE THEOREM. Let f be continuous on $[a, b]$ and differentiable on (a, b). Then there is a number c in (a, b) such that

$$f(b) - f(a) = f'(c)(b - a). \tag{23}$$

Now we have assumed that f, f_x, and f_y are all continuous in a neighborhood N of $\mathbf{x} = (x, y)$. Choose $\Delta \mathbf{x}$ so small such that $\mathbf{x} + \Delta \mathbf{x}$ is in N. Then

$$\Delta f(\mathbf{x}) = f(x + \Delta x, y + \Delta y) - f(x, y)$$

$$\overset{\text{this term was added and subtracted}}{= [f(x + \Delta x, y + \Delta y) - \overbrace{f(x + \Delta x, y)}] + [f(x + \Delta x, y) - f(x, y)].} \tag{24}$$

If $x + \Delta x$ is fixed, then $f(x + \Delta x, y)$ is a function of y which is continuous and differentiable in the interval $[y, y + \Delta y]$. Hence by the mean value theorem, there is a number c_2 between y and $y + \Delta y$ such that

$$f(x + \Delta x, y + \Delta y) - f(x + \Delta x, y) = f_y(x + \Delta x, c_2)[(y + \Delta y) - y]$$
$$= f_y(x + \Delta x, c_2)\,\Delta y. \tag{25}$$

Similarly, with y fixed, $f(x, y)$ is a function of x only and we obtain

$$f(x + \Delta x, y) - f(x, y) = f_x(c_1, y) \, \Delta x \tag{26}$$

where c_1 is between x and $x + \Delta x$. Thus using (25) and (26) in (24), we have

$$\Delta f(\mathbf{x}) = f_x(c_1, y) \, \Delta x + f_y(x + \Delta x, c_2) \, \Delta y. \tag{27}$$

Now both f_x and f_y are continuous at $\mathbf{x} = (x, y)$ so that since c_1 is between x and $x + \Delta x$ and c_2 is between y and $y + \Delta y$, we obtain

$$\lim_{\Delta \mathbf{x} \to 0} f_x(c_1, y) = f_x(x, y) = f_x(\mathbf{x}) \tag{28}$$

and

$$\lim_{\Delta \mathbf{x} \to 0} f_y(x + \Delta x, c_2) = f_y(x, y) = f_y(\mathbf{x}). \tag{29}$$

Now let

$$\epsilon_1(\Delta \mathbf{x}) = f_x(c_1, y) - f_x(x, y). \tag{30}$$

From (28) it follows that

$$\lim_{|\Delta \mathbf{x}| \to 0} \epsilon_1(\Delta \mathbf{x}) = 0. \tag{31}$$

Similarly, if

$$\epsilon_2(\Delta \mathbf{x}) = f_y(x + \Delta x, c_2) - f_y(x, y) \tag{32}$$

then

$$\lim_{|\Delta \mathbf{x}| \to 0} \epsilon_2(\Delta \mathbf{x}) = 0. \tag{33}$$

Now define

$$g(\Delta \mathbf{x}) = \epsilon_1(\Delta \mathbf{x}) \, \Delta x + \epsilon_2(\Delta \mathbf{x}) \, \Delta y. \tag{34}$$

From (31) and (33), it follows that

$$\lim_{|\Delta \mathbf{x}| \to 0} \frac{g(\Delta \mathbf{x})}{|\Delta \mathbf{x}|} = 0. \tag{35}$$

Finally, since

$$f_x(c_1, y) = f_x(x, y) + \epsilon_1(\Delta \mathbf{x}) \qquad \text{(from (30))} \tag{36}$$

and

$$f_y(x + \Delta x, c_2) = f_y(x, y) + \epsilon_2(\Delta \mathbf{x}) \qquad \text{(from (32))} \tag{37}$$

we may substitute (36) and (37) into (27) to obtain

$$\Delta f(\mathbf{x}) = f(\mathbf{x} + \Delta \mathbf{x}) - f(\mathbf{x}) = [f_x(\mathbf{x}) + \epsilon_1(\Delta \mathbf{x})] \, \Delta x + [f_y(\mathbf{x}) + \epsilon_2(\Delta \mathbf{x})] \, \Delta y$$
$$= f_x(\mathbf{x}) \, \Delta x + f_y(\mathbf{x}) \, \Delta y + g(\Delta \mathbf{x}) = (f_x \mathbf{i} + f_y \mathbf{j}) \cdot (\Delta \mathbf{x}) + g(\Delta \mathbf{x})$$

where $\lim_{\Delta \mathbf{x} \to 0} [g(\Delta \mathbf{x}) / |\Delta \mathbf{x}|] \to 0$ and the proof is (at last) complete.

PROBLEMS 18.6

1. Let $f(x, y) = x^2 + y^2$. Show by using Definition 2 that f is differentiable at any point in $\mathbb{R}^2$.
2. Let $f(x, y) = x^2 y^2$. Show by using Definition 2 that f is differentiable at any point in $\mathbb{R}^2$.
3. Let $f(x, y)$ be any polynomial in the variables x and y. Show that f is continuous.

In Problems 4–24 calculate the gradient of the given function. If a point is also given, evaluate the gradient at that point.

4. $f(x, y) = (x + y)^2$
5. $f(x, y) = e^{\sqrt{xy}}$; $(1, 1)$
6. $f(x, y) = \cos(x - y)$; $(\pi/2, \pi/4)$
7. $f(x, y) = \ln(2x - y + 1)$
8. $f(x, y) = \sqrt{x^2 + y^3}$
9. $f(x, y) = \tan^{-1}(y/x)$, $(3, 3)$
10. $f(x, y) = y \tan(y - x)$
11. $f(x, y) = x^2 \sinh y$

12. $f(x, y) = \sec(x + 3y)$; $(0, 1)$
13. $f(x, y) = \dfrac{x - y}{x + y}$; $(3, 1)$

14. $f(x, y) = \dfrac{x^2 - y^2}{x^2 + y^2}$
15. $f(x, y) = \dfrac{e^{x^2} - e^{-y^2}}{3y}$

16. $f(x, y, z) = xyz$; $(1, 2, 3)$
17. $f(x, y, z) = \sin x \cos y \tan z$; $(\pi/6, \pi/4, \pi/3)$
18. $f(x, y, z) = \dfrac{x^2 - y^2 + z^2}{3xy}$; $(1, 2, 0)$

19. $f(x, y, z) = x \ln y - z \ln x$
20. $f(x, y, z) = xy^2 + y^2 z^3$; $(2, 3, -1)$
21. $f(x, y, z) = (y - z)e^{x+2y+3z}$; $(-4, -1, 3)$
22. $f(x, y, z) = x \sin y \ln z$; $(1, 0, 1)$

23. $f(x, y, z) = \dfrac{x - z}{\sqrt{1 - y^2 + x^2}}$; $(0, 0, 1)$

24. $f(x, y, z) = x \cosh z - y \sin x$.

25. Show that if f and g are differentiable functions of three variables, that
$$\nabla(f + g) = \nabla f + \nabla g.$$

26. Show that if f and g are differentiable functions of three variables, then fg is differentiable and
$$\nabla(fg) = f(\nabla g) + g(\nabla f).$$

*27. Show that $\nabla f = 0$ if and only if f is constant.
*28. Show that if $\nabla f = \nabla g$, then there is a constant c for which $f(x, y) = g(x, y) = c$. [*Hint:* Use the result of Problem 27.]
*29. What is the most general function f such that $\nabla f(\mathbf{x}) = \mathbf{x}$ for every $\mathbf{x}$ in $\mathbb{R}^2$?

*30. Let $f(x, y) = \begin{cases} (x^2 + y^2) \sin \dfrac{1}{\sqrt{x^2 + y^2}}, & (x, y) \neq (0, 0) \\ 0, & (x, y) = (0, 0). \end{cases}$

(a) Calculate $f_x(0, 0)$ and $f_y(0, 0)$.
(b) Show that f is *not* differentiable at $(0, 0)$.
(c) Explain why f_x and f_y are *not* continuous at $(0, 0)$.

18.7 The Chain Rule

In this section we derive the chain rule for functions of two and three variables. Let us recall the chain rule for the composition of two functions of one variable:

Let $y = f(u)$ and $u = g(x)$ and assume that f and g are differentiable. Then

$$\frac{dy}{dx} = \frac{dy}{du}\frac{du}{dx} = f'(g(x))g'(x). \tag{1}$$

If $z = f(x, y)$ is a function of two variables, then there are two versions of the chain rule.

THEOREM 1. Let $z = f(x, y)$ be differentiable and suppose that $x = x(t)$ and $y = y(t)$. Assume further that dx/dt and dy/dt exist and are continuous. Then z can be written as a function of the parameter t and

$$\boxed{\frac{dz}{dt} = \frac{\partial z}{\partial x}\frac{dx}{dt} + \frac{\partial z}{\partial y}\frac{dy}{dt} = f_x\frac{dx}{dt} + f_y\frac{dy}{dt}} \tag{2}$$

We can write this result using our gradient notation. If $\mathbf{g}(t) = x(t)\mathbf{i} + y(t)\mathbf{j}$, then $\mathbf{g}'(t) = (dx/dt)\mathbf{i} + (dy/dt)\mathbf{j}$ and (2) can be written

$$\boxed{\frac{d}{dt}f(x(t), y(t)) = (f \circ \mathbf{g})'(t) = f(\mathbf{g}(t))' = \nabla f \cdot \mathbf{g}'(t).} \tag{3}$$

We will not use this notation in this section, but shall refer to it in Section 18.8.

THEOREM 2. Let $z = f(x, y)$ be differentiable and suppose that x and y are functions of the two variables r and s. That is, $x = x(r, s)$ and $y = y(r, s)$. Suppose further that $\partial x/\partial r$, $\partial x/\partial s$, $\partial y/\partial r$, and $\partial y/\partial s$ all exist and are continuous. Then z can be written as a function of r and s and

$$\boxed{\frac{\partial z}{\partial r} = \frac{\partial z}{\partial x}\frac{\partial x}{\partial r} + \frac{\partial z}{\partial y}\frac{\partial y}{\partial r}} \tag{4}$$

$$\boxed{\frac{\partial z}{\partial s} = \frac{\partial z}{\partial x}\frac{\partial x}{\partial s} + \frac{\partial z}{\partial y}\frac{\partial y}{\partial s}.} \tag{5}$$

We will leave the proofs of these theorems until the end of this section.

EXAMPLE 1. Let $z = f(x, y) = xy^2$. Let $x = \cos t$ and $y = \sin t$. Calculate dz/dt.

SOLUTION

$$\frac{dz}{dt} = \frac{\partial z}{\partial x}\frac{dx}{dt} + \frac{\partial z}{\partial y}\frac{dy}{dt} = y^2(-\sin t) + 2xy(\cos t) =$$

$$= (\sin^2 t)(-\sin t) + 2(\cos t)(\sin t)(\cos t)$$

$$= 2\sin t \cos^2 t - \sin^3 t.$$

We can calculate this another way. Since $z = xy^2$, we have $z = (\cos t)(\sin^2 t)$. Then

$$\frac{dz}{dt} = (\cos t)(2 \sin t \cos t) + (\sin^2 t)(-\sin t)$$

$$= 2 \sin t \cos^2 t - \sin^3 t.$$

EXAMPLE 2. Let $z = f(x, y) = \sin xy^2$. Suppose that $x = r/s$ and $y = e^{r-s}$. Calculate $\partial z/\partial r$ and $\partial z/\partial s$.

SOLUTION

$$\frac{\partial z}{\partial r} = \frac{\partial z}{\partial x}\frac{\partial x}{\partial r} + \frac{\partial z}{\partial y}\frac{\partial y}{\partial r} = (y^2 \cos xy^2)\frac{1}{s} + (2xy \cos xy^2)e^{r-s}$$

$$= \frac{e^{2(r-s)} \cos\left[\frac{r}{s}e^{2(r-s)}\right]}{s} + \frac{2r}{s}\left\{\cos\left[\frac{r}{s}e^{2(r-s)}\right]\right\}e^{2(r-s)}$$

and

$$\frac{\partial z}{\partial s} = \frac{\partial z}{\partial x}\frac{\partial x}{\partial s} + \frac{\partial z}{\partial y}\frac{\partial y}{\partial s} = (y^2 \cos xy^2)\frac{-r}{s^2} + (2xy \cos xy^2)(-e^{r-s})$$

$$= \frac{-re^{2(r-s)} \cos\left[\frac{r}{s}e^{2(r-s)}\right]}{s^2} - \frac{2r}{s}\left\{\cos\left[\frac{r}{s}e^{2(r-s)}\right]\right\}e^{2(r-s)}$$

EXAMPLE 3. Let $z = f(x, y) = y/x$, $x = rs$ and $y = r/s$. Calculate $\partial z/\partial r$.

SOLUTION

$$\frac{\partial z}{\partial r} = \frac{\partial z}{\partial x}\frac{\partial x}{\partial r} + \frac{\partial z}{\partial y}\frac{\partial y}{\partial r} = \frac{-y}{x^2}s + \frac{1}{x}\frac{1}{s}$$

$$= -\frac{(r/s) \cdot s}{r^2s^2} + \frac{1}{rs^2} = -\frac{1}{rs^2} + \frac{1}{rs^2} = 0.$$

This seems surprising. Or does it? We have $z = y/x = r/s(rs) = 1/s^2$ which is a function of s only. Thus f is constant with respect to r and we must have $\partial z/\partial r = 0$.

The chain rules given in Theorem 1 and Theorem 2 can easily be extended to functions of three or more variables.

THEOREM 1'. Let $w = f(x, y, z)$ be a differentiable function. If $x = x(t)$, $y = y(t)$, $z = z(t)$, and if dx/dt, dy/dt, and dz/dt exist and are continuous, then

$$\boxed{\frac{dw}{dt} = \frac{\partial w}{\partial x}\frac{dx}{dt} + \frac{\partial w}{\partial y}\frac{dy}{dt} + \frac{\partial w}{\partial z}\frac{dz}{dt}.} \tag{6}$$

THEOREM 2'. Let $w = f(x, y, z)$ be a differentiable function and let $x = x(r, s)$, $y = y(r, s)$ and $z = z(r, s)$. Then if all indicated partial derivatives exist and are continuous, we have

$$\frac{\partial w}{\partial r} = \frac{\partial w}{\partial x}\frac{\partial x}{\partial r} + \frac{\partial w}{\partial y}\frac{\partial y}{\partial r} + \frac{\partial w}{\partial z}\frac{\partial z}{\partial r} \qquad (7)$$

and

$$\frac{\partial w}{\partial s} = \frac{\partial w}{\partial x}\frac{\partial x}{\partial s} + \frac{\partial w}{\partial y}\frac{\partial y}{\partial s} + \frac{\partial w}{\partial z}\frac{\partial z}{\partial s}. \qquad (8)$$

THEOREM 3. Let $w = f(x, y, z)$ be a differentiable function and let $x = x(r, s, t)$, $y = y(r, s, t)$ and $z = z(r, s, t)$. Then if all indicated partial derivatives exist and are continuous

$$\frac{\partial w}{\partial r} = \frac{\partial w}{\partial x}\frac{\partial x}{\partial r} + \frac{\partial w}{\partial y}\frac{\partial y}{\partial r} + \frac{\partial w}{\partial z}\frac{\partial z}{\partial r},$$

$$\frac{\partial w}{\partial s} = \frac{\partial w}{\partial x}\frac{\partial x}{\partial s} + \frac{\partial w}{\partial y}\frac{\partial y}{\partial s} + \frac{\partial w}{\partial z}\frac{\partial z}{\partial s}$$

and

$$\frac{\partial w}{\partial t} = \frac{\partial w}{\partial x}\frac{\partial x}{\partial t} + \frac{\partial w}{\partial y}\frac{\partial y}{\partial t} + \frac{\partial w}{\partial z}\frac{\partial z}{\partial t}. \qquad (9)$$

EXAMPLE 4.   Let $w = yz^2 - x^3$, and let $x = e^{r-t}$, $y = \ln(r + 2s + 3t)$, and $z = \sqrt{rs + t}$. Calculate $\partial w/\partial r$, $\partial w/\partial s$, and $\partial w/\partial t$.

SOLUTION

$$\frac{\partial w}{\partial r} = \frac{\partial w}{\partial x}\frac{\partial x}{\partial r} + \frac{\partial w}{\partial y}\frac{\partial y}{\partial r} + \frac{\partial w}{\partial z}\frac{\partial z}{\partial r} = -3x^2 e^{r-t} + z^2\left(\frac{1}{r + 2s + 3t}\right)$$

$$+ 2yz\left(\frac{s}{2\sqrt{rs + t}}\right) = -3e^{3(r-t)} + \frac{rs + t}{r + 2s + 3t} + s\ln(r + 2s + 3t).$$

$$\frac{\partial w}{\partial s} = \frac{\partial w}{\partial x}\frac{\partial x}{\partial s} + \frac{\partial w}{\partial y}\frac{\partial y}{\partial s} + \frac{\partial w}{\partial z}\frac{\partial z}{\partial s}$$

$$= -3x^2 \cdot 0 + z^2\frac{2}{r + 2s + 3t} + 2yz\left(\frac{r}{2\sqrt{rs + t}}\right)$$

$$= \frac{2(rs + t)}{r + 2s + 3t} + r\ln(r + 2s + 3t).$$

$$\frac{\partial w}{\partial t} = \frac{\partial w}{\partial x}\frac{\partial x}{\partial t} + \frac{\partial w}{\partial y}\frac{\partial y}{\partial t} + \frac{\partial w}{\partial z}\frac{\partial z}{\partial t}$$

$$= -3x^2(-e^{r-t}) + z^2 \left(\frac{3}{r + 2s + 3t} \right) + 2yz \left(\frac{1}{2\sqrt{rs + t}} \right)$$

$$= 3e^{3(r+t)} + \frac{3(rs + t)}{r + 2s + 3t} + \ln(r + 2s + 3t).$$

We saw in Section 5.1 that the chain rule is useful to solve "related rates" type problems. The chain rules for functions of two or more variables are useful in a similar way.

EXAMPLE 5. The radius of a right circular cone is increasing at a rate of 3 cm/sec while its height is increasing at a rate of 5 cm/sec. How fast is the volume increasing when $r = 15$ cm and $h = 25$ cm?

SOLUTION. We are asked to find dV/dt where $V = \frac{1}{3}\pi r^2 h$. But

$$\frac{dV}{dt} = \frac{\partial V}{\partial r} \frac{dr}{dt} + \frac{\partial V}{\partial h} \frac{dh}{dt} = \frac{2}{3} \pi r h(3) + \frac{1}{3} \pi r^2(5)$$

$$= 2\pi r h + \frac{5}{3}\pi r^2 = \pi \left(2 \cdot 15 \cdot 25 + \frac{5}{3}(15)^2 \right)$$

$$= 1125\pi \quad \text{cm}^3/\text{sec.}$$

EXAMPLE 6. According to the ideal gas law (see Section 18.1), the pressure, volume, and absolute temperature of n moles of an ideal gas are related by

$$PV = nRT,$$

where R is a constant. Suppose that the volume of an ideal gas is increasing at a rate of 10 cm³/min and the pressure is decreasing at a rate of 0.3 nt/cm²/min (see Section 9.7 for a discussion of pressure). How is the temperature of the gas changing when the volume of 5 moles of a gas is 100 cm³, the pressure is 2 nt/cm² and the temperature is 293 °K ($= 20$ °C)?

SOLUTION. We have $T = PV/nR$, where $n = 5$. Then

$$\frac{dT}{dt} = \frac{\partial T}{\partial P} \frac{dP}{dt} + \frac{\partial T}{\partial V} \frac{dV}{dt} = \frac{V}{nR}(-0.3) + \frac{P}{nR}(10) = \frac{100}{5R}(-0.3) + \frac{2}{5R}(10)$$

$$= -2/R \quad \text{°K/min.}$$

We conclude this section with a proof of Theorem 2. Theorem 1 follows easily from Theorem 2. (Explain why.)

PROOF OF THEOREM 2. We will show that

$$\frac{\partial z}{\partial r} = \frac{\partial z}{\partial x} \frac{\partial x}{\partial r} + \frac{\partial z}{\partial y} \frac{\partial y}{\partial r}.$$

Equation (5) follows in an identical manner. Since $x = x(r, s)$, a change Δr in r will cause a change Δx in x and a change Δy in y. We may therefore write

$$\Delta x = x(r + \Delta r, s) - x(r, s)$$

and

$$\Delta y = y(r + \Delta r, s) - y(r, s).$$

Since z is differentiable, we may write

$$\Delta z = f(x + \Delta x, y + \Delta y) - f(x, y) = \frac{\partial z}{\partial x} \Delta x + \frac{\partial z}{\partial y} \Delta y + g(\Delta x, \Delta y). \tag{10}$$

Using the equivalent form given in equation (18.6.13), we may write

$$\Delta z = \frac{\partial z}{\partial x} \Delta x + \frac{\partial z}{\partial y} \Delta y + \epsilon_1(\Delta x, \Delta y) \Delta x + \epsilon_2(\Delta x, \Delta y) \Delta y \tag{11}$$

where

$$\lim_{(\Delta x, \Delta y) \to (0,0)} \epsilon_1(\Delta x, \Delta y) = \lim_{(\Delta x, \Delta y) \to (0,0)} \epsilon_2(\Delta x, \Delta y) = 0. \tag{12}$$

Then dividing both sides of (11) by Δr and taking limits, we obtain

$$\lim_{\Delta r \to 0} \frac{\Delta z}{\Delta r} = \lim_{\Delta r \to 0} \left\{ \frac{\partial z}{\partial x} \frac{\Delta x}{\Delta r} + \frac{\partial z}{\partial y} \frac{\Delta y}{\Delta r} + \epsilon_1(\Delta x, \Delta y) \frac{\Delta x}{\Delta r} + \epsilon_2(\Delta x, \Delta y) \frac{\Delta y}{\Delta r} \right\}. \tag{13}$$

Since x and y are continuous functions of r, we have

$$\lim_{\Delta r \to 0} \Delta x = 0 \quad \text{and} \quad \lim_{\Delta r \to 0} \Delta y = 0$$

so that

$$\lim_{\Delta r \to 0} \epsilon_1(\Delta x, \Delta y) = \lim_{(\Delta x, \Delta y) \to (0,0)} \epsilon_1(\Delta x, \Delta y) = 0$$

and

$$\lim_{\Delta r \to 0} \epsilon_2(\Delta x, \Delta y) = \lim_{(\Delta x, \Delta y) \to (0,0)} \epsilon_2(\Delta x, \Delta y) = 0.$$

Thus the limits in (13) become

$$\frac{\partial z}{\partial r} = \frac{\partial z}{\partial x} \frac{\partial x}{\partial r} + \frac{\partial z}{\partial y} \frac{\partial y}{\partial r} + 0 \cdot \frac{\partial x}{\partial r} + 0 \cdot \frac{\partial y}{\partial r}$$

and the theorem is proved.

PROBLEMS 18.7

In Problems 1–11 use the chain rule to calculate dz/dt or dw/dt. Check your answer by first writing z or w as a function of t and then differentiating.

1. $z = xy$, $x = e^t$, $y = e^{2t}$
2. $z = x^2 + y^2$, $x = \cos t$, $y = \sin t$
3. $z = y/x$, $x = t^2$, $y = t^3$
4. $z = e^x \sin y$, $x = \sqrt{t}$, $y = \sqrt[3]{t}$
5. $z = \tan^{-1}(y/x)$, $x = \cos 3t$, $y = \sin 5t$
6. $z = \sinh(x - 2y)$, $x = 2t^4$, $y = t^2 + 1$

7. $w = x^2 + y^2 + z^2$, $x = \cos t$, $y = \sin t$, $z = t$
8. $w = xy - yz + zx$, $x = e^t$, $y = e^{2t}$, $z = e^{3t}$
9. $w = (x + y)/z$, $x = t$, $y = t^2$, $z = t^3$
10. $w = \sin(x + 2y + 3z)$, $x = \tan t$, $y = \sec t$, $z = t^5$
11. $w = \ln(2x - 3y + 4z)$, $x = e^t$, $y = \ln t$, $z = \cosh t$.

In Problems 12–26 use the chain rule to calculate the indicated partial derivatives.

12. $z = xy$; $x = r + s$; $y = r - s$; $\partial z/\partial r$ and $\partial z/\partial s$
13. $z = x^2 + y^2$; $x = \cos(r + s)$; $y = \sin(r - s)$; $\partial z/\partial r$ and $\partial z/\partial s$
14. $z = y/x$; $x = e^r$; $y = e^s$; $\partial z/\partial r$ and $\partial z/\partial s$
15. $z = \sin(y/x)$; $x = r/s$; $y = s/r$; $\partial z/\partial r$ and $\partial z/\partial s$
16. $z = e^{x+y}/e^{x-y}$; $x = \ln rs$; $y = \ln(r/s)$; $\partial z/\partial r$ and $\partial z/\partial s$
17. $z = x^2 y^3$; $x = r - s^2$; $y = 2s + r$; $\partial z/\partial r$ and $\partial z/\partial s$
18. $w = x + y + z$; $x = rs$; $y = r + s$; $z = r - s$; $\partial w/\partial r$ and $\partial w/\partial s$
19. $w = xy/z$; $x = r$, $y = s$, $z = t$; $\partial w/\partial r$, $\partial w/\partial s$, and $\partial w/\partial t$
20. $w = xy/z$; $x = r + s$, $y = t - r$, $z = s + 2t$; $\partial w/\partial r$, $\partial w/\partial s$, and $\partial w/\partial t$
21. $w = \sin xyz$; $x = s^2 r$, $y = r^2 s$, $z = r - s$; $\partial w/\partial r$ and $\partial w/\partial s$
22. $w = \sinh(x + 2y + 3z)$; $x = \sqrt{r + s}$, $y = \sqrt[3]{s - t}$, $z = 1/(r + t)$; $\partial w/\partial r$, $\partial w/\partial s$, and $\partial w/\partial t$
23. $w = x^2 y + yz^2$; $x = rst$, $y = rs/t$, $z = 1/rst$; $\partial w/\partial r$, $\partial w/\partial s$, and $\partial w/\partial t$
24. $w = \ln(x + 2y + 3z)$; $x = rt^3 + s$; $y = t - s^5$, $z = e^{r+s}$; $\partial w/\partial r$, $\partial w/\partial s$, and $\partial w/\partial t$
25. $w = e^{xy/z}$; $x = r^2 + t^2$, $y = s^2 - t^2$, $z = r^2 + s^2$; $\partial w/\partial r$, $\partial w/\partial s$ and $\partial w/\partial t$
*26. $u = xy + w^2 - z^3$; $x = t + r - q$, $y = q^2 + s^2 - t + r$, $z = (qr + st)/r^4$, $w = (r - s)/(t + q)$; $\partial u/\partial r$, $\partial u/\partial s$, $\partial u/\partial t$, and $\partial u/\partial q$.

27. The radius of a right circular cone is increasing at a rate of 7 in./min while its height is decreasing at a rate of 20 in./min. How fast is the volume changing when $r = 45$ in. and $h = 100$ in.? Is the volume increasing or decreasing?

28. The radius of a right circular cylinder is decreasing at a rate of 12 cm/sec while its height is increasing at a rate of 25 cm/sec. How is the volume changing when $r = 180$ cm and $h = 500$ cm? Is the volume increasing or decreasing?

29. The volume of 10 moles of an ideal gas is decreasing at a rate of 25 cm³/min and its temperature is increasing at a rate of 1 °C/min. How fast is the pressure changing when $V = 1000$ cm³, $P = 3$ nt/cm² and $T = 303$ °K ($= 30$ °C)? Is the pressure increasing or decreasing? Leave your answer in terms of R.

30. The pressure of 8 moles of an ideal gas is decreasing at a rate of 0.4 nt/cm²/min while the temperature is decreasing at a rate of 0.5 °K/min. How fast is the volume of the gas changing when $V = 1000$ cm³, $P = 3$ nt/cm² and $T = 303$ °K? Is the volume increasing or decreasing?

31. The angle A of a triangle ABC is increasing at a rate of 3 degrees/sec, the side AB is increasing at a rate of 1 cm/sec and the side AC is decreasing at a rate of 2 cm/sec. How fast is the side BC changing when $A = 30$ degrees, $AB = 10$ cm and $AC = 24$ cm? Is the length of BC increasing or decreasing? [*Hint:* Use the law of cosines.]

32. The wave equation (see Problem 18.5.31) is the partial differential equation

$$\frac{\partial^2 y}{\partial t^2} = c^2 \frac{\partial^2 y}{\partial x^2}.$$

Show that if f is any differentiable function, then

$$y(x, t) = \tfrac{1}{2}[f(x - ct) + f(x + ct)]$$

is a solution to this equation.

33. Let $z = f(x, y)$ be differentiable. Write the expression $(\partial z/\partial x)^2 + (\partial z/\partial y)^2$ in terms of polar coordinates r and θ.

34. Let $w = f(x, y, z)$ be differentiable and let $x = r \cos \theta$, $y = r \sin \theta$ and $z = t$. Calculate $\partial w/\partial r$, $\partial w/\partial \theta$, and $\partial w/\partial t$. (These are cylindrical coordinates.)

35. Let $w = f(x, y, z)$ be differentiable and let $x = \rho \sin \varphi \sin \theta$, $y = \rho \sin \varphi \cos \theta$, and $z = \rho \cos \varphi$. Calculate $\partial w/\partial \rho$, $\partial w/\partial \varphi$, and $\partial w/\partial \theta$. (These are spherical coordinates.)

*36. Let $z = f(x, y)$ and let $x = x(r, s)$, $y = y(r, s)$. Show that

$$\frac{\partial^2 z}{\partial r^2} = \frac{\partial}{\partial x}\left(\frac{\partial z}{\partial x}\frac{\partial x}{\partial r} + \frac{\partial z}{\partial y}\frac{\partial y}{\partial r}\right)\left(\frac{\partial x}{\partial r}\right) + \frac{\partial}{\partial y}\left(\frac{\partial z}{\partial x}\frac{\partial x}{\partial r} + \frac{\partial z}{\partial y}\frac{\partial y}{\partial r}\right)\frac{\partial y}{\partial r}.$$

**37. Laplace's equation (see Problem 18.5.35) is the partial differential equation

$$\frac{\partial^2 f}{\partial x^2} + \frac{\partial^2 f}{\partial y^2} = 0.$$

If we write (x, y) in polar coordinates ($x = r \cos \theta$, $y = r \sin \theta$), show that Laplace's equation becomes

$$\frac{\partial^2 f}{\partial r^2} + \frac{1}{r^2}\frac{\partial^2 f}{\partial \theta^2} + \frac{1}{r}\frac{\partial f}{\partial r} = 0.$$

[*Hint:* Write $\partial^2 f/\partial x^2$ and $\partial^2 f/\partial y^2$ in terms of r and θ using the result of Problem 36.]

18.8 Tangent Planes, Normal Lines, and Gradients

Let $z = f(x, y)$ be a function of two variables. As we have seen, the graph of f is a surface in $\mathbb{R}^3$. More generally, the graph of the equation $F(x, y, z) = 0$ is a surface in $\mathbb{R}^3$. The surface $F(x, y, z) = 0$ is called *differentiable* at a point (x_0, y_0, z_0) if $\partial F/\partial x$, $\partial F/\partial y$, and $\partial F/\partial z$ all exist and are continuous at (x_0, y_0, z_0). Just as a differentiable curve in $\mathbb{R}^2$ has a unique tangent line at each point, a differentiable surface in $\mathbb{R}^3$ has a unique tangent plane at each point. We shall formally define what we mean by a tangent plane to a surface after a bit, although it should be easy enough to visualize (see Figure 1). We note here that not every surface has a tangent plane at every point. For example, the cone $z = \sqrt{x^2 + y^2}$ clearly has no tangent plane at the origin (see Figure 2).

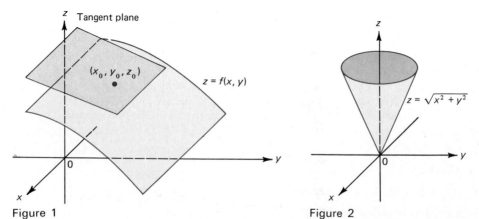

Figure 1

Figure 2

Assume that the surface S given by $F(x, y, z) = 0$ is differentiable. Let C be any curve lying on S. That is, C can be given parametrically by $\mathbf{g}(t) = x(t)\mathbf{i} + y(t)\mathbf{j} + z(t)\mathbf{k}$. (Recall, from Definition 17.7.1, the definition of a curve in $\mathbb{R}^3$.) Then for points on the curve, $F(x, y, z)$ can be written as a function of t and, from the vector form of the chain rule (Equation 18.7.3), we have

$$F'(t) = \nabla F \cdot \mathbf{g}'(t). \tag{1}$$

But since $F(x(t), y(t), z(t)) = 0$ for all t (since $(x(t), y(t), z(t))$ is on S), we see that $F'(t) = 0$ for all t. But $\mathbf{g}'(t)$ is tangent to the curve C for every number t. Thus (1) implies that

> *The gradient of F at a point $\mathbf{x}_0 = (x_0, y_0, z_0)$ in S is orthogonal to the tangent vector to any curve C remaining on S and passing through $\mathbf{x}_0$.*

This is illustrated in Figure 3.

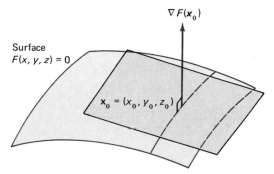

$\nabla F(\mathbf{x}_0)$

Surface
$F(x, y, z) = 0$

$\mathbf{x}_0 = (x_0, y_0, z_0)$

Figure 3

Thus if we think of all the vectors tangent to a surface at a point $\mathbf{x}_0$ as comprising a plane, then $\nabla F(\mathbf{x}_0)$ is a *normal* vector to that plane. This motivates the following definition.

DEFINITION 1.   Let F be differentiable at $\mathbf{x}_0 = (x_0, y_0, z_0)$ and let the surface S be defined by $F(x, y, z) = 0$.

(i)   Then the *tangent plane* to S is the plane passing through the point (x_0, y_0, z_0) with normal vector $\nabla F(\mathbf{x}_0)$.

(ii)   The *normal line* to S at $\mathbf{x}_0$ is the line passing through $\mathbf{x}_0$ having the same direction as $\nabla F(\mathbf{x}_0)$.

EXAMPLE 1.   Find the equation of the tangent plane and symmetric equations of the normal line to the ellipsoid $x^2 + (y^2/4) + (z^2/9) = 3$ at the point $(1, 2, 3)$.

SOLUTION.   Since $F(x, y, z) = x^2 + (y^2/4) + (z^2/9) - 3 = 0$, we have

$$\nabla F = \frac{\partial F}{\partial x}\mathbf{i} + \frac{\partial F}{\partial y}\mathbf{j} + \frac{\partial F}{\partial z}\mathbf{k} = 2x\mathbf{i} + \frac{y}{2}\mathbf{j} + \frac{2z}{9}\mathbf{k}.$$

Then $\nabla F(1, 2, 3) = 2\mathbf{i} + \mathbf{j} + \frac{2}{3}\mathbf{k}$ and the equation of the tangent plane is

$$2(x - 1) + (y - 2) + \tfrac{2}{3}(z - 3) = 0$$

or

$$2x + y + \tfrac{2}{3}z = 6.$$

The normal line is given by

$$\frac{x-1}{2} = y - 2 = \tfrac{3}{2}(z - 3).$$

The situation is even simpler if we can write the surface in the form $z = f(x, y)$. Then $F(x, y, z) = f(x, y) - z = 0$ so that

$$F_x = f_x, \qquad F_y = f_y, \qquad \text{and} \qquad F_z = -1$$

and the normal vector $\mathbf{N}$ to the tangent plane is

$$\mathbf{N} = f_x(x_0, y_0)\mathbf{i} + f_y(x_0, y_0)\mathbf{j} - \mathbf{k}. \tag{2}$$

Remark. One interesting consequence of this fact is that if $z = f(x, y)$ and if $\nabla f(x_0, y_0) = \mathbf{0}$, then *the tangent plane to the surface at* $(x_0, y_0, f(x_0, y_0))$ *is parallel to the xy-plane* (i.e., it is horizontal). This occurs because at $(x_0, y_0, f(x_0, y_0))$, $\mathbf{N} = (\partial f/\partial x)\mathbf{i} + (\partial f/\partial y)\mathbf{j} - \mathbf{k} = \nabla f - \mathbf{k} = -\mathbf{k}$. Thus the z-axis is normal to the tangent plane.

EXAMPLE 2. Find the tangent plane and normal line to the surface $z = x^3y^5$ at the point (2, 1, 8).

SOLUTION. $\mathbf{N} = (\partial z/\partial x)\mathbf{i} + (\partial z/\partial y)\mathbf{j} - \mathbf{k} = 3x^2y^5\mathbf{i} + 5x^3y^4\mathbf{j} - \mathbf{k} = 12\mathbf{i} + 40\mathbf{j} - \mathbf{k}$ at (2, 1, 8). Then the tangent plane is given by

$$12(x - 2) + 40(y - 1) - (z - 8) = 0$$

or

$$12x + 40y - z = 56.$$

The normal line is

$$\frac{x-2}{12} = \frac{y-1}{40} = \frac{z-8}{-1}.$$

We can write the equation of the tangent plane to a surface $z = f(x, y)$ so that it looks like the equation of the tangent line to a curve in $\mathbb{R}^2$. This will further illustrate the connection between the derivative of a function of one variable and the gradient. Recall from Section 17.5 that if P is a point on a plane and $\mathbf{N}$ is a normal vector, then if Q denotes any other point on the plane, the equation of the plane can be written

$$\overrightarrow{PQ} \cdot \mathbf{N} = 0. \tag{3}$$

In this case, since $z = f(x, y)$, a point on the surface takes the form $(x, y, z) = (x, y, f(x, y))$. Then, since $\mathbf{N} = f_x\mathbf{i} + f_y\mathbf{j} - \mathbf{k}$, the equation of the tangent plane at $(x_0, y_0, f(x_0, y_0))$ becomes, using (3),

$$\begin{aligned}
0 &= \{(x, y, z) - (x_0, y_0, z_0)\} \cdot (f_x, f_y, -1) \\
&= (x - x_0, y - y_0, z - z_0) \cdot (f_x, f_y, -1) \\
&= (x - x_0)f_x + (y - y_0)f_y - (z - z_0).
\end{aligned} \tag{4}$$

We can rewrite (4) as

$$z = f(x_0, y_0) + (x - x_0)f_x + (y - y_0)f_y. \tag{5}$$

Denote (x_0, y_0) by $\mathbf{x}_0$ and (x, y) by $\mathbf{x}$. Then (5) can be written

$$z = f(\mathbf{x}_0) + (\mathbf{x} - \mathbf{x}_0) \cdot \nabla f(\mathbf{x}_0). \tag{6}$$

Recall that if $y = f(x)$ is differentiable at x_0, then the equation of the tangent line to the curve at the point $(x_0, f(x_0))$ is given by

$$\frac{y - f(x_0)}{x - x_0} = f'(x_0)$$

or

$$y = f(x_0) + (x - x_0)f'(x_0). \tag{7}$$

This similarity between (6) and (7) illustrates quite vividly the importance of the gradient vector of a function of several variables as the generalization of the derivative of a function of one variable.

PROBLEMS 18.8

In Problems 1–16 find the equation of the tangent plane and symmetric equations of the normal line to the given surface at the given point.

1. $x^2 + y^2 + z^2 = 1$; $(1, 0, 0)$
2. $x^2 + y^2 + z^2 = 1$; $(0, 1, 0)$
3. $x^2 + y^2 + z^2 = 1$; $(0, 0, 1)$
4. $x^2 - y^2 + z^2 = 1$; $(1, 1, 1)$
5. $\dfrac{x^2}{a^2} + \dfrac{y^2}{b^2} + \dfrac{z^2}{c^2} = 3$; (a, b, c)
6. $\dfrac{x^2}{a^2} + \dfrac{y^2}{b^2} + \dfrac{z^2}{c^2} = 3$; $(-a, b, -c)$
7. $x^{1/2} + y^{1/2} + z^{1/2} = 6$; $(4, 1, 9)$
8. $ax + by + cz = d$; $\left(\dfrac{1}{a}, \dfrac{1}{b}, \dfrac{d-2}{c}\right)$
9. $xyz = 4$; $(1, 2, 2)$
10. $xy^2 - yz^2 + zx^2 = 1$; $(1, 1, 1)$
11. $4x^2 - y^2 - 5z^2 = 15$; $(3, 1, -2)$
12. $xe^y - ye^z = 1$; $(1, 0, 0)$
13. $\sin xy - 2\cos yz = 0$; $\left(\dfrac{\pi}{2}, 1, \dfrac{\pi}{3}\right)$
14. $x^2 + y^2 + 4x + 2y + 8z = 7$; $(2, -3, -1)$
15. $e^{xyz} = 5$; $(1, 1, \ln 5)$
16. $\sqrt{\dfrac{x+y}{z-1}} = 1$; $(1, 1, 3)$.

In Problems 17–24 write the equation of the tangent plane in the form (6) and find the symmetric equations of the normal line to the given surface.

17. $z = xy^2$; $(1, 1, 1)$
18. $z = \ln(x - 2y)$; $(3, 1, 0)$
19. $z = \sin(2x + 5y)$; $\left(\dfrac{\pi}{8}, \dfrac{\pi}{20}, 1\right)$
20. $z = \sqrt{\dfrac{x+y}{x-y}}$; $(5, 4, 3)$
21. $z = \tan^{-1}\dfrac{y}{x}$; $\left(-2, 2, \dfrac{-\pi}{4}\right)$
22. $z = \sinh xy^2$; $(0, 3, 0)$
23. $z = \sec(x - y)$; $\left(\dfrac{\pi}{2}, \dfrac{\pi}{6}, 2\right)$
24. $z = e^x \cos y + e^y \cos x$; $\left(\dfrac{\pi}{2}, 0, e^{\pi/2}\right)$.

*25. Find the two points of intersection of the surface $z = x^2 + y^2$ and the line
$$(x - 3)/1 = (y + 1)/(-1) = (z + 2)/(-2).$$

26. At each of the points of intersection found in Problem 25, find the cosine of the angle between the given line and the normal line to the surface.
*27. Show that every line normal to the surface of a sphere passes through the center of the sphere.
28. Show that every line normal to the cone $z^2 = ax^9 + ay^2$ intersects the z-axis.
29. Let f be a differentiable function of one variable and let $z = yf(y/x)$. Show that all tangent planes to the surface defined by this equation have a point in common.
30. The angle between two surfaces at a point of intersection is defined to be the angle between their normal lines. Show that if two surfaces $F(x, y, z) = 0$ and $G(x, y, z) = 0$ intersect at right angles at a point $\mathbf{x}_0$, then

$$\nabla F(\mathbf{x}_0) \cdot \nabla G(\mathbf{x}_0) = 0.$$

31. Show that the sum of the squares of the x-, y- and z-intercepts of any plane tangent to the surface $x^{2/3} + y^{2/3} + z^{2/3} = a^{2/3}$ is constant.
32. The equation $F(x, y) = 0$ defines a curve in $\mathbb{R}^2$. Show that if F is differentiable then $\nabla F(x, y)$ is normal to the curve at every point.
33. Use the result of Problem 32 to find the equation of the tangent line to a curve $F(x, y) = 0$ at a point (x_0, y_0).

In Problems 34–39 use the results of Problems 32 and 33 to find a normal vector and the equation of the tangent line to the curve at the given point.

34. $xy = 5$; $(1, 5)$

35. $x^2 + xy + y^2 + 3x - 5y = 6$; $(1, -2)$

36. $\dfrac{x + y}{x - y} = 7$; $(4, 3)$

37. $xe^{xy} = 1$; $(1, 0)$

38. $\dfrac{x^2}{4} + \dfrac{y^2}{16} = 1$; $(\sqrt{2}, 2\sqrt{2})$

39. $\tan(x + y) = 1$; $\left(\dfrac{\pi}{4}, 0\right)$.

40. Show that at any point (x, y), $y \neq 0$, the curve $x/y = a$ is orthogonal to the curve $x^2 + y^2 = r^2$ for any constants a and r.

18.9 Directional Derivatives and Gradients

Let us take another look at the partial derivatives $\partial f/\partial x$ and $\partial f/\partial y$ of the function $z = f(x, y)$. We have

$$\frac{\partial f}{\partial x} = \lim_{\Delta x \to 0} \frac{f(x + \Delta x, y) - f(x, y)}{\Delta x}$$

which we can write as

$$\frac{\partial f}{\partial x} = \lim_{h \to 0} \frac{f(x + h, y) - f(x, y)}{h}. \tag{1}$$

Similarly,

$$\frac{\partial f}{\partial y} = \lim_{h \to 0} \frac{f(x, y + h) - f(x, y)}{h}. \tag{2}$$

Writing

$$\mathbf{x} = x\mathbf{i} + y\mathbf{j},$$

we see that

$$(x + h, y) = (x + h)\mathbf{i} + y\mathbf{j} = x\mathbf{i} + y\mathbf{j} + h\mathbf{i} = \mathbf{x} + h\mathbf{i}$$

and

$$(x, y + h) = \mathbf{x} + h\mathbf{j}.$$

Thus we can rewrite (1) and (2) as

$$\frac{\partial f}{\partial x} = \lim_{h \to 0} \frac{f(\mathbf{x} + h\mathbf{i}) - f(\mathbf{x})}{h} \tag{3}$$

and

$$\frac{\partial f}{\partial y} = \lim_{h \to 0} \frac{f(\mathbf{x} + h\mathbf{j}) - f(\mathbf{x})}{h}. \tag{4}$$

With this notation we can say that $\partial f/\partial x$ measures the rate of change in f as we move in the direction $\mathbf{i}$ and $\partial f/\partial y$ measures the rate of change of f as we move in the direction $\mathbf{j}$. More generally, we may wish to calculate the rate of change of f as we move in any direction $\mathbf{u}$, where $\mathbf{u}$ is a unit vector.

DEFINITION 1. Let f be defined in a neighborhood of a point $\mathbf{x}$ in $\mathbb{R}^2$ and let $\mathbf{u}$ be a unit vector. Then the *directional derivative of f in the direction* $\mathbf{u}$, denoted $f_{\mathbf{u}}'(\mathbf{x})$, is given by

$$f_{\mathbf{u}}'(\mathbf{x}) = \lim_{h \to 0} \frac{f(\mathbf{x} + h\mathbf{u}) - f(\mathbf{x})}{h}. \tag{5}$$

Remark 1. Note that if $\mathbf{u} = \mathbf{i}$, then $f_{\mathbf{i}}'(\mathbf{x}) = f_x(\mathbf{x})$ and if $\mathbf{u} = \mathbf{j}$, then $f_{\mathbf{j}}'(\mathbf{x}) = f_y(\mathbf{x})$ since in these cases equation (5) becomes (3) or (4).

Remark 2. Definition 1 makes sense if f is a function of three variables. Then, of course, $\mathbf{u}$ represents a unit vector in $\mathbb{R}^3$.

Before we try to compute any directional derivatives, we prove a theorem which makes their calculation very easy.

THEOREM 1. If f is differentiable at $\mathbf{x}$, then f has a directional derivative in *every* direction and

$$f_{\mathbf{u}}'(\mathbf{x}) = \nabla f(\mathbf{x}) \cdot \mathbf{u}. \tag{6}$$

PROOF. Since f is differentiable at $\mathbf{x}$, we have (see equation (18.6.8))

$$f(\mathbf{x} + h\mathbf{u}) - f(\mathbf{x}) = \nabla f(\mathbf{x}) \cdot h\mathbf{u} + g(h\mathbf{u}). \tag{7}$$

Now, from the definition of differentiability,

$$\lim_{|h\mathbf{u}|\to 0} \frac{g(h\mathbf{u})}{|h\mathbf{u}|} = 0.$$

But since $\mathbf{u}$ is a unit vector, $|h\mathbf{u}| = |h|\,|\mathbf{u}| = |h|$. Thus we find that

$$\lim_{h\to 0} \frac{g(h\mathbf{u})}{h} = 0.$$

Then, dividing both sides of (7) by h and taking limits, we obtain

$$\lim_{h\to 0} \frac{f(\mathbf{x} + h\mathbf{u}) - f(\mathbf{x})}{h} = \lim_{h\to 0} \frac{\nabla f(\mathbf{x}) \cdot h\mathbf{u}}{h} + \lim_{h\to 0} \frac{g(h\mathbf{u})}{h}$$

$$= \lim_{h\to 0} \frac{h(\nabla f(\mathbf{x}) \cdot \mathbf{u})}{h} + \lim_{h\to 0} \frac{g(h\mathbf{u})}{h} = \nabla f(\mathbf{x}) \cdot \mathbf{u}$$

and the proof is complete. Note that this proof works equally well if f is a function of three (or more) variables.

EXAMPLE 1. Let $z = f(x, y) = xy^2$. Calculate the directional derivative of f in the direction of the vector $\mathbf{v} = 2\mathbf{i} + 3\mathbf{j}$ at the point $(4, -1)$.

SOLUTION. A unit vector in the direction $\mathbf{v}$ is $\mathbf{u} = (2/\sqrt{13})\mathbf{i} + (3/\sqrt{13})\mathbf{j}$. Also, $\nabla f = y^2\mathbf{i} + 2xy\mathbf{j}$. Thus

$$f_{\mathbf{u}}'(x, y) = \nabla f(\mathbf{x}) \cdot \mathbf{u} = \frac{2y^2}{\sqrt{13}} + \frac{6xy}{\sqrt{13}} = \frac{2y^2 + 6xy}{\sqrt{13}}.$$

At $(4, -1)$, $f_{\mathbf{u}}'(4, -1) = -22/\sqrt{13}$.

EXAMPLE 2. Let $f(x, y, z) = x \ln y - e^{xz^3}$. Calculate the directional derivative of f in the direction of the vector $\mathbf{v} = \mathbf{i} - \mathbf{j} + 3\mathbf{k}$. Evaluate this derivative at the point $(-5, 1, -2)$.

SOLUTION. A unit vector in the direction of $\mathbf{v}$ is $\mathbf{u} = (1/\sqrt{11})\mathbf{i} - (1/\sqrt{11})\mathbf{j} + (3/\sqrt{11})\mathbf{k}$ and

$$\nabla f = (\ln y - z^3 e^{xz^3})\mathbf{i} + \frac{x}{y}\mathbf{j} - 3xz^2 e^{xz^3}\mathbf{k}.$$

Thus

$$f_{\mathbf{u}}'(\mathbf{x}) = \nabla f(\mathbf{x}) \cdot \mathbf{u} = \frac{\ln y - z^3 e^{xz^3} - (x/y) - 9xz^2 e^{xz^3}}{\sqrt{11}}$$

and at $(-5, 1, -2)$

$$f_{\mathbf{u}}'(-5, 1, -2) = \frac{5 + 188e^{40}}{\sqrt{11}}.$$

There is an interesting geometric interpretation of the directional derivative. By Definitions 15.2.5 and 17.2.4, the projection of ∇f on $\mathbf{u}$ is given by

$$\text{Proj}_{\mathbf{u}}\, \nabla f = \frac{\nabla f \cdot \mathbf{u}}{|\mathbf{u}|^2}\, \mathbf{u}$$

and the component of ∇f in the direction $\mathbf{u}$ is given by

$$\frac{\nabla f \cdot \mathbf{u}}{|\mathbf{u}|} = \nabla f \cdot \mathbf{u}.$$

Thus the *directional derivative of f in the direction* $\mathbf{u}$ *is the component of the gradient of f in the direction* $\mathbf{u}$. This is illustrated in Figure 1.

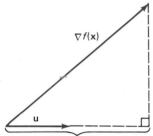

$\nabla f(\mathbf{x})$

$\mathbf{u}$

Component of $\nabla f(\mathbf{x})$ in direction $\mathbf{u}$ Figure 1

We now derive another remarkable property of the gradient. Recall that $\mathbf{u} \cdot \mathbf{v} = |\mathbf{u}|\, |\mathbf{v}| \cos \theta$, where θ is the smallest angle between the vectors $\mathbf{u}$ and $\mathbf{v}$. Thus, the directional derivative of f in the direction $\mathbf{u}$ can be written, using Theorem 1, as

$$f_{\mathbf{u}}'(\mathbf{x}) = \nabla f(\mathbf{x}) \cdot \mathbf{u} = |\nabla f(\mathbf{x})|\, |\mathbf{u}| \cos \theta$$

or, since $\mathbf{u}$ is a unit vector,

$$f_{\mathbf{u}}'(\mathbf{x}) = |\nabla f(\mathbf{x})| \cos \theta. \tag{8}$$

Now $\cos \theta = 1$ when $\theta = 0$. This occurs when $\mathbf{u}$ has the direction of ∇f. Similarly $\cos \theta = -1$ when $\theta = \pi$ which occurs when $\mathbf{u}$ has the direction of $-\nabla f$. Thus, since $-1 \le \cos \theta \le 1$, equation (8) implies the following important result:

THEOREM 2.   Let f be differentiable. Then f increases most rapidly in the direction of its gradient and decreases most rapidly in the direction opposite to that of its gradient.

EXAMPLE 3.   Consider the sphere $x^2 + y^2 + z^2 = 1$. We can write the upper half of this sphere (i.e., the upper hemisphere) as $z = f(x, y) = \sqrt{1 - x^2 - y^2}$. Then

$$\nabla f = \frac{-x}{\sqrt{1 - x^2 - y^2}}\, \mathbf{i} + \frac{-y}{\sqrt{1 - x^2 - y^2}}\, \mathbf{j}.$$

The direction of ∇f is

$$\tan^{-1}\left(\frac{-y/\sqrt{1 - x^2 - y^2}}{-x/\sqrt{1 - x^2 - y^2}} \right) = \tan^{-1}\left(\frac{-y}{-x} \right).$$

But $\tan^{-1}(-y/-x)$ is the direction of the vector $-x\mathbf{i} - y\mathbf{j}$ which points from (x, y) to $(0, 0)$. Thus if we start at a point (x, y, z) on the sphere, the path of steepest ascent (increase) is a great circle passing through the point $(0, 0, 1)$ (called the *north pole* of the sphere).

EXAMPLE 4. The distribution of voltage on a metal plate is given by

$$V = 50 - x^2 - 4y^2.$$

(a) At the point $(1, -2)$, in what direction does the voltage increase most rapidly?

(b) In what direction does it decrease most rapidly?

(c) What is the magnitude of this increase or decrease?

SOLUTION. $\nabla V = v_x\mathbf{i} + v_y\mathbf{j} = -2x\mathbf{i} - 8y\mathbf{j}$. At $(1, -2)$, $\nabla V = -2\mathbf{i} + 16\mathbf{j}$. Thus

(a) the voltage increases most rapidly as we move in the direction $-2\mathbf{i} + 16\mathbf{j}$.

(b) it decreases most rapidly in the direction $2\mathbf{i} - 16\mathbf{j}$.

(c) the magnitude of the increase or decrease is $\sqrt{2^2 + 16^2} = \sqrt{260}$.

EXAMPLE 5. In Example 4, describe the path of a particle that starts at the point $(1, -2)$ and moves in the direction of greatest voltage increase.

SOLUTION. The path of the particle will be that of the gradient. If the particle follows the path $\mathbf{f}(t) = x(t)\mathbf{i} + y(t)\mathbf{j}$, then since the direction of the path is $\mathbf{f}'(t) = x'(t)\mathbf{i} + y'(t)\mathbf{j}$ and since this direction is also given by $\nabla V = -2x\mathbf{i} - 8y\mathbf{j}$, we must have

$$x'(t) = -2x(t) \qquad \text{and} \qquad y'(t) = -8y(t).$$

From Section 6.4, the solutions to these differential equations are

$$x(t) = c_1 e^{-2t} \qquad \text{and} \qquad y(t) = c_2 e^{-8t}.$$

But $x(0) = 1$ and $y(0) = -2$ so that

$$x(t) = e^{-2t} \qquad \text{and} \qquad y(t) = -2e^{-8t}.$$

Then, since $e^{-8t} = (e^{-2t})^4$, we see that the particle moves along the path

$$y = -2x^4.$$

One other fact about directional derivatives and gradients should be mentioned here. Since the gradient vector is normal to the curve $f(x, y) = 0$, we say that the directional derivative of f in the direction of the gradient is the *normal derivative* of f and is denoted df/dn. We then have, from (8),

$$\text{normal derivative} = \frac{df}{dn} = |\nabla f|. \tag{9}$$

EXAMPLE 6. Let $f(x, y) = xy^2$. Calculate the normal derivative. Evaluate df/dn at the point $(3, -2)$.

SOLUTION. $\nabla f = y^2\mathbf{i} + 2xy\mathbf{j}$. Then $df/dn = |\nabla f| = \sqrt{y^4 + 4x^2y^2}$. At $(3, -2)$, $df/dn = \sqrt{16 + 144} = \sqrt{160}$.

PROBLEMS 18.9

In Problems 1–15 calculate the directional derivative of the given function at the given point in the direction of the given vector **v**.

1. $f(x, y) = xy$ at $(2, 3)$; $\mathbf{v} = \mathbf{i} + 3\mathbf{j}$
2. $f(x, y) = 2x^2 - 3y^2$ at $(1, -1)$; $\mathbf{v} = -\mathbf{i} + 2\mathbf{j}$
3. $f(x, y) = \ln(x + 3y)$ at $(2, 4)$; $\mathbf{v} = \mathbf{i} + \mathbf{j}$
4. $f(x, y) = ax^2 + by^2$ at (c, d); $\mathbf{v} = \alpha\mathbf{i} + \beta\mathbf{j}$
5. $f(x, y) = \tan^{-1}(y/x)$ at $(2, 2)$; $\mathbf{v} = 3\mathbf{i} - 2\mathbf{j}$
6. $f(x, y) = (x - y)/(x + y)$ at $(4, 3)$; $\mathbf{v} = -\mathbf{i} - 2\mathbf{j}$
7. $f(x, y) = xe^y + ye^x$ at $(1, 2)$; $\mathbf{v} = \mathbf{i} + \mathbf{j}$
8. $f(x, y) = \sin(2x + 3y)$ at $(\pi/12, \pi/9)$; $\mathbf{v} = -2\mathbf{i} + 3\mathbf{j}$
9. $f(x, y, z) = xy + yz + xz$ at $(1, 1, 1)$; $\mathbf{v} = \mathbf{i} + \mathbf{j} + \mathbf{k}$
10. $f(x, y, z) = xy^3z^5$ at $(-3, -1, 2)$; $\mathbf{v} = -\mathbf{i} - 2\mathbf{j} + \mathbf{k}$
11. $f(x, y, z) = \ln(x + 2y + 3z)$ at $(1, 2, 0)$; $\mathbf{v} = 2\mathbf{i} + \mathbf{j} - \mathbf{k}$
12. $f(x, y, z) = xe^{yz}$ at $(2, 0, -4)$; $\mathbf{v} = -\mathbf{i} + 2\mathbf{j} + 5\mathbf{k}$
13. $f(x, y, z) = x^2y^3 + z\sqrt{x}$ at $(1, -2, 3)$; $\mathbf{v} = 5\mathbf{j} + \mathbf{k}$
14. $f(x, y, z) = e^{-(x^2+y^2+z^2)}$ at $(1, 1, 1)$; $\mathbf{v} = \mathbf{i} + 3\mathbf{j} - 5\mathbf{k}$
15. $f(x, y, z) = \dfrac{1}{\sqrt{x^2 + y^2 + z^2}}$ at $(-1, 2, 3)$; $\mathbf{v} = \mathbf{i} - \mathbf{j} + \mathbf{k}$.

16. The voltage (potential) at any point on a metal structure is given by

$$v(x, y, z) = \frac{1}{0.02 + \sqrt{x^2 + y^2 + z^2}}.$$

At the point $(1, -1, 2)$, in what direction does the voltage increase most rapidly?

17. The temperature at any point in a solid metal ball centered at the origin is given by

$$T(x, y, z) = 100e^{-(x^2+y^2+z^2)}.$$

(a) Where is the ball hottest?
(b) Show that at any point (x, y, z) on the ball, the direction of greatest increase in temperature is a vector pointing toward the origin.

18. The temperature distribution of a ball centered at the origin is given by

$$T(x, y, z) = \frac{100}{x^2 + y^2 + z^2 + 1}.$$

(a) Where is the ball hottest?
(b) Find the direction of greatest decrease of heat at the point $(3, -1, 2)$.
(c) Find the direction of greatest increase in heat. Does this vector point toward the origin?

19. The temperature distribution on a plate is given by

$$T(x, y) = 1 - \frac{x^2}{a^2} - \frac{y^2}{b^2}.$$

Find the path of a heat-seeking particle (i.e., a particle that always moves in the direction of greatest increase in heat) if it starts at the point (a, b).

20. Find the path of the particle in Problem 19 if it starts at the point $(-a, b)$.
21. The height of a mountain is given by

$$h(x, y) = 3000 - 2x^2 - y^2,$$

where the x-axis points north and the y-axis points east and all distances are measured in meters. Suppose that a mountain climber is at the point $(30, -20, 800)$.
(a)   If the climber moves in the southwest direction, will he (or she) ascend or descend?
(b)   In what direction should the climber move so as to ascend most rapidly?

* **22.**   Prove that if $w = f(x, y, z)$ is differentiable at x, then all the first partials exist at x and

$$\nabla f(\mathbf{x}) = f_x(\mathbf{x})\mathbf{i} + f_y(\mathbf{x})\mathbf{j} + f_z(\mathbf{x})\mathbf{k}.$$

(This is the converse to Theorem 18.6.1.)

In Problems 23–26 calculate the normal derivative at the given point.

23. $f(x, y) = x + 2y$ at $(1, 4)$ 　　　　　　　　**24.** $f(x, y) = e^{x+3y}$ at $(1, 0)$

25. $f(x, y) = \tan^{-1}\dfrac{y}{x}$ at $(-1, -1)$ 　　　**26.** $f(x, y) = \sqrt{\dfrac{x - y}{x + y}}$ at $(3, 1)$.

*18.10 Conservative Vector Fields and the Gradient

In this brief section we show how the gradient can arise in a fundamental way in physics. Suppose that for every $\mathbf{x}$ in $\mathbb{R}^2$ (or $\mathbb{R}^3$), $\mathbf{f}(\mathbf{x})$ is another vector in $\mathbb{R}^2$ (or $\mathbb{R}^3$). Then $\mathbf{f}$ is really a vector function of a vector. We call such a function a *vector field*.† An important question that can arise is: does there exist a function $F(\mathbf{x})$ such that $\nabla F(\mathbf{x}) = \mathbf{f}(\mathbf{x})$? That is, is $\mathbf{f}$ the gradient of some function F? In Section 18.12 we will prove that the answer to this question is yes, provided that $\mathbf{f}$ is what we call an *exact differential*. However in this section we shall assume that there exists a function F such that

$$\mathbf{f} = \nabla F. \tag{1}$$

If $\mathbf{f} = -\nabla F$ for some function F, then $\mathbf{f}$ is said to be a *conservative vector field* and F is called a *potential function* for $\mathbf{f}$. The reason for this terminology will be made clear shortly. (If $\mathbf{f} = \nabla F$, then $\mathbf{f} = -\nabla(-F)$ so that the introduction of the minus sign does not cause any problem.)

Now let $\mathbf{g}(t) = x(t)\mathbf{i} + y(t)\mathbf{j}$ be a differentiable curve and suppose that a particle of mass m moves along it. Suppose further that the force acting on the particle at any time t is given by $\mathbf{f}(\mathbf{x}(t))$ where $\mathbf{f}$ is assumed to be a conservative vector field. By Newton's second law,

$$\mathbf{f}(\mathbf{x}(t)) = m\mathbf{a}(t) = m\mathbf{x}''(t). \tag{2}$$

But since $\mathbf{f}$ is conservative, (1) implies that $\mathbf{f}(\mathbf{x}) = -\nabla F(\mathbf{x})$ for some differentiable function F. Then we have

$$-\nabla F(\mathbf{x}(t)) = m\mathbf{x}''(t)$$

or

$$m\mathbf{x}'' + \nabla F(\mathbf{x}) = 0. \tag{3}$$

†We will discuss vector fields in more detail in Chapter 19.

We now take the dot product of both sides of (3) with $\mathbf{x}'$ to obtain

$$m\mathbf{x}' \cdot \mathbf{x}'' + \nabla F(\mathbf{x}) \cdot \mathbf{x}' = 0. \tag{4}$$

But, by the product rule,

$$\frac{d}{dt}|\mathbf{x}'(t)|^2 = \frac{d}{dt}(\mathbf{x}'(t) \cdot \mathbf{x}'(t)) = \mathbf{x}'(t) \cdot \mathbf{x}''(t) + \mathbf{x}''(t) \cdot \mathbf{x}'(t)$$

$$= 2\mathbf{x}'(t) \cdot \mathbf{x}''(t) \tag{5}$$

and, by equation (18.7.3).

$$\frac{d}{dt}F(\mathbf{x}(t)) = \nabla F(\mathbf{x}(t)) \cdot \mathbf{x}'(t). \tag{6}$$

Using (5) and (6) in (4), we obtain

$$\frac{d}{dt}\left[\frac{1}{2}m|\mathbf{x}'|^2 + F(\mathbf{x}(t))\right] = 0$$

which implies that

$$\frac{1}{2}m|\mathbf{x}'|^2 + F(\mathbf{x}(t)) = C, \tag{7}$$

where C is a constant. This is one of the versions of the *law of conservation of energy*. The term $\frac{1}{2}m|\mathbf{x}'|^2 = \frac{1}{2}m|\mathbf{v}|^2$ is called the *kinetic energy* of the particle and the term $F(\mathbf{x}(t))$ is called the *potential energy* of the particle. Equation (7) tells us simply that if the force function $\mathbf{f}$ is conservative, then the total energy of the system is constant and, moreover, the potential function F of $\mathbf{f}$ represents the potential energy of the system.

The *principle of the conservation of energy* states that energy may be transformed from one kind to another but cannot be created or destroyed; that is, the total energy is constant. Thus it seems reasonable that force fields in classical physics are conservative (although, since the work of Einstein, it has been found that energy can be transformed into mass and vice versa so that there are forces which are not conservative). One example of a conservative force is given by the force of gravitational attraction. Let m_1 represent the mass of a (relatively) fixed object in space and let m_2 denote the mass of an object moving near the fixed object. Then the magnitude of the gravitational force between the objects is given by (see Example 3.2.8)

$$|\mathbf{f}| = G\frac{m_1m_2}{r^2}, \tag{8}$$

where r is the distance between the objects and G is a universal constant. If we assume that the first object is at the origin, then we may denote the position of the second object by $\mathbf{x}(t)$ and then, since $r = |\mathbf{x}|$, (8) can be written

$$|\mathbf{f}| = \frac{Gm_1m_2}{|\mathbf{x}|^2}. \tag{9}$$

Also, the force acts toward the origin; that is, in the direction opposite to that of the positive vector $\mathbf{x}$. Therefore,

$$\text{direction of } \mathbf{f} = -\frac{\mathbf{x}}{|\mathbf{x}|} \tag{10}$$

so that, from (9) and (10),

$$\mathbf{f}(t) = \frac{\alpha \mathbf{x}(t)}{|\mathbf{x}(t)|^3}, \tag{11}$$

where $\alpha = -Gm_1m_2$. We now prove that $\mathbf{f}$ is conservative. To show this we must come up with a function F such that $\mathbf{f} = -\nabla F$. Here we will pull F "out of a hat" but, in Section 18.12, we will show you how to construct such an F (if one exists). Let

$$F(\mathbf{x}) = \frac{\alpha}{|\mathbf{x}|}.$$

Then $F(\mathbf{x}) = \alpha/\sqrt{x^2 + y^2 + z^2}$ and

$$-\nabla F(\mathbf{x}) = -\alpha \left(\frac{-x}{(x^2 + y^2 + z^2)^{3/2}}\mathbf{i} + \frac{-y}{(x^2 + y^2 + z^2)^{3/2}}\mathbf{j} + \frac{-z}{(x^2 + y^2 + z^2)^{3/2}}\mathbf{k} \right)$$

$$= \frac{\alpha}{|\mathbf{x}|^3}(x\mathbf{i} + y\mathbf{j} + z\mathbf{k}) = \frac{\alpha \mathbf{x}}{|\mathbf{x}|^3} = \mathbf{f}.$$

Thus $\mathbf{f}$ is conservative so that, with respect to gravitational forces, the law of conservation of energy holds.

PROBLEMS 18.10

1. Show that the force given by $\mathbf{f}(\mathbf{x}) = -\alpha \mathbf{x}/|\mathbf{x}|^2$ is conservative by finding a potential function for it.
2. Do the same for the force given by $\mathbf{f}(\mathbf{x}) = -\alpha \mathbf{x}/|\mathbf{x}|^4$.
3. If $k \geq 1$ is a positive integer and α is a constant, show that the force $\mathbf{f}(\mathbf{x}) = -\alpha \mathbf{x}/|\mathbf{x}|^k$ is conservative.
4. If $\mathbf{x} = x\mathbf{i} + y\mathbf{j}$, show that the force $\mathbf{f}(x, y) = y\mathbf{i} + x\mathbf{j}$ is conservative.
5. Show that if $\mathbf{f}$ and $\mathbf{g}$ are conservative, then $\alpha \mathbf{f} + \beta \mathbf{g}$ is also conservative for any constants α and β.
*6. Show that the force $\mathbf{f}(x, y) = y\mathbf{i} - x\mathbf{j}$ is *not* conservative. [*Hint:* Assume that $\mathbf{f} = -\nabla F$ so that $-y = \partial F/\partial x$ and $x = \partial F/\partial y$. Then integrate to obtain a contradiction.]

18.11 The Total Differential
   and Approximation

In Section 3.6 we used the notions of increments and differentials to approximate a function. We used the fact that if Δx was small, then

$$f(x + \Delta x) - f(x) = \Delta y \approx f'(x)\,\Delta x. \tag{1}$$

We also defined the differential dy by

$$dy = f'(x)\,dx = f'(x)\,\Delta x \tag{2}$$

(since dx is defined to be equal to Δx). Note that in (2), it is not required that Δx be small.

We now extend these ideas to functions of two or three variables.

DEFINITION 1.   Let $f = f(\mathbf{x})$ be a function of two or three variables and let $\Delta \mathbf{x} = (\Delta x, \Delta y)$ or $(\Delta x, \Delta y, \Delta z)$.

   (i)   The *increment of f,* denoted Δf, is defined by

$$\Delta f = f(\mathbf{x} + \Delta \mathbf{x}) - f(\mathbf{x}). \tag{3}$$

   (ii)   The *total differential of f,* denoted df, is given by

$$df = \nabla f(\mathbf{x}) \cdot \Delta \mathbf{x}. \tag{4}$$

Note that equation (4) is very similar in form to equation (2).

   Remark 1. If f is a function of two variables, then (3) and (4) become:

$$\Delta f = f(x + \Delta x, y + \Delta y) - f(x, y) \tag{5}$$

and the total differential

$$df = f_x(x, y)\, \Delta x + f_y(x, y)\, \Delta y. \tag{6}$$

   Remark 2. If f is a function of three variables, then (3) and (4) become:

$$\Delta f = f(x + \Delta x, y + \Delta y, z + \Delta z) - f(x, y, z) \tag{7}$$

and

$$df = f_x(x, y, z)\, \Delta x + f_y(x, y, z)\, \Delta y + f_z(x, y, z)\, \Delta z. \tag{8}$$

   Remark 3. Note that in the definition of the total differential, it is *not* required that $|\Delta \mathbf{x}|$ be small.

From Theorems 18.6.1 and 18.6.1′ and the definition of differentiability, we see that *if* $|\Delta \mathbf{x}|$ is small and if f is differentiable, then

$$\Delta f \approx df. \tag{9}$$

We can use the relation (9) to approximate functions of several variables in much the same way we used the relation (1) to approximate the values of functions of one variable.

EXAMPLE 1.   The sides of a rectangle are 5.02 cm and 2.97 cm. Use the total differential to estimate the area of the rectangle.

SOLUTION.   The area of the rectangle is $A(x, y) = xy$. We need to estimate $A(5.02, 2.97)$. We know that $A(5, 3) = 5 \cdot 3 = 15$. Also, $\nabla A = A_x \mathbf{i} + A_y \mathbf{j} = y\mathbf{i} + x\mathbf{j} = 3\mathbf{i} + 5\mathbf{j}$ at $(5, 3)$. Then, since $\Delta x = 0.02$ and $\Delta y = -0.03$, we have

$$\Delta A = A(5 + 0.02, 3 - 0.03) - A(5, 3) \approx dA = 3\, \Delta x + 5\, \Delta y$$
$$= 3(0.02) + 5(-0.03) = -0.09$$

and we obtain

$$A(5 + 0.02, 3 - 0.03) = A(5, 3) + \Delta A \approx A(5, 3) + dA = 15 - 0.09 = 14.91$$

The exact area is $(5.02)(2.97) = 14.9094$ so that $\Delta A = 14.9094 - 15 = -0.0906$. The difference $\Delta A - dA = (-0.0906) - (-0.09) = -0.0006$. This problem provides a good example of the use of the total differential. In this problem

$$A(x + \Delta x, y + \Delta y) - A(x, y) = (x + \Delta x)(y + \Delta y) - xy = y\,\Delta x + x\,\Delta y + \Delta x\,\Delta y$$
$$= \nabla A \cdot \Delta \mathbf{x} + \Delta x\,\Delta y = dA + \Delta x\,\Delta y.$$

$\Delta x\,\Delta y$ is the "error in approximation" between ΔA and dA. We verify that $\Delta x\,\Delta y = (0.02)(-0.03) = -0.0006$, as calculated above.

EXAMPLE 2.   Use the total differential to estimate $\sqrt{(2.98)^2 + (4.03)^2}$.

SOLUTION. Let $f(x, y) = \sqrt{x^2 + y^2}$. Then we are asked to calculate $f(2.98, 4.03)$. We know that $f(3, 4) = \sqrt{3^2 + 4^2} = 5$. Thus we need to calculate $f(3 - 0.02, 4 + 0.03)$. Now

$$\nabla f(\mathbf{x}) = \frac{x}{\sqrt{x^2 + y^2}}\mathbf{i} + \frac{y}{\sqrt{x^2 + y^2}}\mathbf{j} = \frac{3}{5}\mathbf{i} + \frac{4}{5}\mathbf{j} \qquad \text{at} \qquad (3, 4).$$

Then, using (6),

$$df = \frac{3}{5}\Delta x + \frac{4}{5}\Delta y = (0.6)(-0.02) + (0.8)(0.03) = 0.012.$$

Hence

$$f(3 - 0.02, 4 + 0.03) - f(3, 4) = \Delta f \approx df = 0.012$$

so that

$$f(2.98, 4.03) \approx f(3, 4) + 0.012 = 5.012.$$

The exact value of $\sqrt{(2.98)^2 + (4.03)^2}$ is $\sqrt{8.8804 + 16.2409} = \sqrt{25.1213} \approx 5.012115$ so that $\Delta y \approx 0.012115$ and our approximation is very good indeed.

EXAMPLE 3.   Use the total differential to estimate $(1.95)^4(3.04)^3(0.97)^5$.

SOLUTION. Let $f(x, y, z) = x^4y^3z^5$. Then we need to calculate $f(1.95, 3.04, 0.97) = f(2 - 0.05, 3 + 0.04, 1 - 0.03)$. We have $f(2, 3, 1) = 2^43^31^5 = 432$. Also

$$\nabla f = 4x^3y^3z^5\mathbf{i} + 3x^4y^2z^5\mathbf{j} + 5x^4y^3z^4\mathbf{k} = 864\mathbf{i} + 432\mathbf{j} + 2160\mathbf{k}.$$

Then, from (8)

$$\Delta f \approx df = (864)(-0.05) + 432(0.04) + (2160)(-0.03)$$
$$= -43.2 + 17.28 - 64.8 = -90.72.$$

Thus

$$f(x + \Delta x, y + \Delta y, z + \Delta z) \approx f(x, y, z) + df$$

so that

$$f(1.95, 3.04, 0.97) \approx 432 - 90.72 = 341.28.$$

Finally, $(1.95)^4(3.04)^3(0.97)^5 \approx 348.833$.

EXAMPLE 4. The radius of a cone is measured to be 15 cm and the height of the cone is measured to be 25 cm. There is a maximum error of ± 0.02 cm in the measurement of the radius and ± 0.05 cm in the measurement of the height. (a) What is the approximate volume of the cone? (b) What is the maximum error in the calculation of the volume?

SOLUTION. (a) $V = \frac{1}{3}\pi r^2 h \approx \frac{1}{3}\pi (15)^2 25 = 1875\pi$ cm$^3 \approx 5890.5$ cm^3.

(b) $\nabla V = V_r \mathbf{i} + V_h \mathbf{j} = \frac{2}{3}\pi rh \mathbf{i} + \frac{1}{3}\pi r^2 \mathbf{j} = \pi(250\mathbf{i} + 75\mathbf{j})$. Then choosing $\Delta x = 0.02$ and $\Delta y = 0.05$ to find the maximum error, we have

$$\Delta V \approx dV = \nabla V \cdot \Delta x = \pi(250(0.02) + 75(0.05)) = \pi(5 + 3.75)$$
$$= 8.75\pi \approx 27.5 \quad \text{cm}^3.$$

Thus the maximum error in the calculation is, approximately, 27.5 cm^3 which means that

$$5890.5 - 27.5 < V < 5890.5 + 27.5$$

or

$$5863 \text{ cm}^3 < V < 5918 \text{ cm}^3.$$

Note that an error of 27.5 cm^3 is only a *relative error* of $27.5/5890.5 \approx 0.0047$, which is a very small relative error (see Section 3.6 for a discussion of relative error.)

EXAMPLE 5. A cylindrical tin can has an inside radius of 5 cm and a height of 12 cm. The thickness of the tin is 0.2 cm. Estimate the amount of tin needed to construct the can (excluding its ends).

SOLUTION. We need to estimate the difference between the "outer" and "inner" volumes of the can. We have $V = \pi r^2 h$. The inner volume is $\pi(5^2)(12) = 300\pi$ cm^3 and the outer volume is $\pi(5.2)^2(12.2)$. The difference is

$$\Delta V = \pi(5.2)^2(12.2) - 300\pi \approx dV.$$

Since $\nabla V = 2\pi rh \mathbf{i} + \pi r^2 \mathbf{j} = \pi(120\mathbf{i} + 25\mathbf{j})$, we have

$$dV = \pi(120(0.2) + 25(0.2)) = 29\pi.$$

Thus the amount of tin needed is, approximately, 29π cm$^3 \approx 91.1$ cm^3.

PROBLEMS 18.11

In Problems 1–12 calculate the total differential df.

1. $f(x, y) = xy^3$

2. $f(x, y) = \tan^{-1}\frac{y}{x}$

3. $f(x, y) = \sqrt{\dfrac{x - y}{x + y}}$

4. $f(x, y) = xe^y$

5. $f(x, y) = \ln(2x + 3y)$

6. $f(x, y) = \sin(x - 4y)$

7. $f(x, y, z) = xy^2 z^5$

8. $f(x, y, z) = \dfrac{xy}{z}$

9. $f(x, y, z) = \ln(x + 2y + 3z)$

10. $f(x, y, z) = \sec xy - \tan z$

11. $f(x, y, z) = \cosh(xy - z)$

12. $f(x, y, z) = \dfrac{x - z}{y + 3x}$.

13. Let $f(x, y) = xy^2$.
   (a)   Calculate explicitly the difference $\Delta f - df$.
   (b)   Verify your answer by calculating $\Delta f \to df$ at the point $(1, 2)$, where $\Delta x = -0.01$ and $\Delta y = 0.03$.

*14. Repeat the steps of Problem 13 for the function $f(x, y) = x^3 y^2$.

In Problems 15–23, use the total differential to estimate the given number

15. $3.01/5.99$

16. $19.8\sqrt{65}$

17. $\sqrt{35.6}\,\sqrt[3]{64.08}$

18. $(2.01)^4(3.04)^7 - (2.01)(3.04)^9$

19. $\sqrt{(5.02 - 3.96)/(5.02 + 3.96)}$

20. $((4.95)^2 + (7.02))^{1/5}$

21. $(3.02)(1.97)/\sqrt{8.95}$

22. $\sin(11\pi/24)\cos(13\pi/36)$

23. $(7.92)\sqrt{5.01} - (0.98)^2$.

24. The radius and height of a cylinder are, approximately, 10 cm and 20 cm, respectively. The maximum error in approximations are ± 0.03 cm and ± 0.07 cm.
   (a)   What is the approximate volume of the cylinder?
   (b)   What is the approximate maximum error in this calculation?

25. Two sides of a triangular shaped piece of land were measured to be 50 m and 110 m and with an error of at most ± 0.3 m in each measurement. The angle between the sides was measured to be 60 degrees with an error of at most ± 1 degree.
   (a)   Using the law of cosines, find the approximate length of the third side of the triangle.
   (b)   What is the approximate maximum error of your measurement? [*Hint:* It will be necessary to convert 1 degree to radians.]

26. How much wood is contained in the sides of a rectangular box with sides of inside measurements 1 m, 1.2 m, and 1.6 m if the thickness of the wood making up the sides is 5 cm ($= 0.05$ m)?

27. When three resistors are connected in parallel, the total resistance R (measured in ohms) is given by

$$\frac{1}{R} = \frac{1}{r_1} + \frac{1}{r_2} + \frac{1}{r_3},$$

where r_1, r_2, and r_3 are the three separate resistances. If $r_1 = 6 \pm 0.1$ ohms, $r_2 = 8 \pm 0.03$ ohms, and $r_3 = 12 \pm 0.15$ ohms,
   (a)   estimate R, and
   (b)   find an approximate value for the maximum error in your estimate.

28. The volume of 10 moles of an ideal gas was calculated to be 500 cm^3 at a temperature of 40 °C ($= 313$ °K). If the maximum error in each measurement n, V and T was $\frac{1}{2}\%$,
   (a)   calculate the approximate pressure of the gas (in nt/cm^2), and
   (b)   find the approximate error in your computation. [*Hint:* Recall that according to the ideal gas law, $PV = nRT$.]

18.12 Exact Differentials or How to Obtain a Function from Its Gradient

If $z = F(x, y)$ is differentiable, then the gradient of F exists and

$$\nabla F = F_x(x, y)\mathbf{i} + F_y(x, y)\mathbf{j}. \tag{1}$$

We can write (1) as

$$\nabla F = P(x, y)\mathbf{i} + Q(x, y)\mathbf{j}. \tag{2}$$

In this section we shall try to reverse this process. This corresponds roughly to finding an indefinite integral of a function of one variable. We shall ask the following question: Give a *vector field* of the form

$$\mathbf{f}(x, y) = P(x, y)\mathbf{i} + Q(x, y)\mathbf{j}, \tag{3}$$

does there exist a differentiable function F such that $\nabla F = \mathbf{f}$? (If the answer is yes, then, as we saw in Section 18.10, the vector field $\mathbf{f}$ is called *conservative* and the function $-F$ is called a *potential function* for $\mathbf{f}$.) Another way to look at this problem is to write the expression

$$P(x, y)\, \Delta x + Q(x, y)\, \Delta y \tag{4}$$

and ask the question: Does there exist a differentiable function F such that $dF = P(x, y)\, \Delta x + Q(x, y)\, \Delta y$? If the answer is yes, then $P(x, y)\, \Delta x + Q(x, y)\, \Delta y$ is called an *exact differential*.

Before answering these two questions in general, we shall look at two examples which illustrate what can happen.

EXAMPLE 1.   Let $\mathbf{f} = 9x^2y^2\mathbf{i} + 6x^3y\mathbf{j}$. Show that there is a function F such that $\nabla F = \mathbf{f}$.

SOLUTION. If there is such an F, then

$$\frac{\partial F}{\partial x} = 9x^2y^2 \quad \text{and} \quad \frac{\partial F}{\partial y} = 6x^3y. \tag{5}$$

Integrating both sides of the first of these equations *with respect to x*, we obtain

$$F(x, y) = 3x^3y^2 + g(y). \tag{6}$$

The term $g(y)$ is a *constant of integration* since

$$\frac{\partial}{\partial x} g(y) = 0.$$

Next we differentiate (6) *with respect to y* and use the second equation of (5):

$$\frac{\partial F}{\partial y} = 6x^3y + g'(y) = 6x^3y.$$

Thus $g'(y) = 0$ and $g(y) = C$, a constant. Hence, for any constant C,

$$F(x, y) = 3x^3y^2 + C$$

is a function which satisfies $\nabla F = \mathbf{f}$. (This should be checked.)

> **Remark.** When we integrate a function of x and y with respect to x, the "constant" of integration is always a function of y. Analogously, when we integrate a function of x and y with respect to y, the "constant" of integration is always a function of x.

EXAMPLE 2. Show that there is *no* function F which satisfies

$$\nabla F = y\mathbf{i} - x\mathbf{j}.$$

SOLUTION. Suppose that $\nabla F = y\mathbf{i} - x\mathbf{j}$. Then $\partial F/\partial x = y$ and $\partial F/\partial y = -x$. Integration of the first of these equations with respect to x yields

$$F(x, y) = yx + g(y)$$

so that

$$\frac{\partial F}{\partial y} = x + g'(y) = -x.$$

This implies that $g'(y) = -2x$ which is impossible since g and g' are functions of y only. Hence there is no function F whose gradient is $y\mathbf{i} - x\mathbf{j}$.

We now state a general result which tells us whether $\mathbf{f}$ is a gradient of some function F (that is, whether $\mathbf{f}$ is conservative). Its proof is difficult and is omitted.†

THEOREM 1. Let $\mathbf{f}(x, y) = P(x, y)\mathbf{i} + Q(x, y)\mathbf{j}$ and suppose that P, Q, $\partial P/\partial y$, and $\partial Q/\partial x$ are continuous. Then $\mathbf{f}$ is the gradient of a function F if and only if

$$\boxed{\frac{\partial P}{\partial y} = \frac{\partial Q}{\partial x}.} \qquad (7)$$

Notation. We will say that the vector field $P(x, y)\mathbf{i} + Q(x, y)\mathbf{j}$ is *exact* if condition (7) holds; that is, if $\partial P/\partial y = \partial Q/\partial x$.

EXAMPLE 3. Show that the vector field

$$\mathbf{f}(x, y) = \left(4x^3y^3 + \frac{1}{x}\right)\mathbf{i} + \left(3x^4y^2 - \frac{1}{y}\right)\mathbf{j}$$

is exact and find all functions F for which $\nabla F = \mathbf{f}$.

SOLUTION. $P(x, y) = 4x^3y^3 + (1/x)$ and $Q(x, y) = 3x^4y^2 - (1/y)$ so that

$$\frac{\partial P}{\partial y} = 12x^3y^2 = \frac{\partial Q}{\partial x}.$$

If $\nabla F = \mathbf{f}$, then $\partial F/\partial x = P$ so that

$$F(x, y) = \int \left(4x^3y^3 + \frac{1}{x}\right)dx = x^4y^3 + \ln|x| + g(y).$$

Differentiating with respect to y, we have

$$Q = \frac{\partial F}{\partial y} = 3x^4y^2 + g'(y) = 3x^4y^2 - \frac{1}{y}$$

†For a proof see R. C. Buck, *Advanced Calculus*, McGraw-Hill, New York, 1963, p. 424.

so that $g'(y) = -1/y$ and $g(y) = -\ln|y| + C$ and, finally,

$$F(x, y) = x^4y^3 + \ln|x| - \ln|y| + C = x^4y^3 + \ln\left|\frac{x}{y}\right| + C.$$

This answer should be checked. Note that the presence of the constant C indicates that the "potential function" F is not unique.

We close this section by pointing out that it is now easy to check whether a force field **f** is conservative. It is necessary only to check whether or not **f** is exact.

PROBLEMS 18.12

In Problems 1–14 test for exactness. If the given vector field is exact, find all functions F for which $\nabla F = \mathbf{f}$.

1. $\mathbf{f}(x, y) = 2xy\mathbf{i} + (x^2 + 1)\mathbf{j}$

2. $\mathbf{f}(x, y) = (4x^3 - ye^{xy})\mathbf{i} + (\tan y - xe^{xy})\mathbf{j}$

3. $\mathbf{f}(x, y) = (4x^2 - 4y^2)\mathbf{i} + (8xy - \ln y)\mathbf{j}$

4. $\mathbf{f}(x, y) = [x\cos(x + y) + \sin(x + y)]\mathbf{i} + x\cos(x + y)\mathbf{j}$

5. $\mathbf{f}(x, y) = 2x(\cos y)\mathbf{i} + x^2(\sin y)\mathbf{j}$

6. $\mathbf{f}(x, y) = \left[\dfrac{\ln(\ln y)}{x} + \dfrac{2}{3}xy^3\right]\mathbf{i} + \left[\dfrac{\ln x}{y\ln y} + x^2y^2\right]\mathbf{j}$

7. $\mathbf{f}(x, y) = (x - y\cos x)\mathbf{i} - (\sin x)\mathbf{j}$

8. $\mathbf{f}(x, y) = e^{x^2y}\mathbf{i} + e^{x^2y}\mathbf{j}$

9. $\mathbf{f}(x, y) = (3x\ln x + x^5 - y)\mathbf{i} - x\mathbf{j}$

10. $\mathbf{f}(x, y) = [(1/x^2) + y^2]\mathbf{i} + (2xy)\mathbf{j}$

11. $\mathbf{f}(x, y) = \left(\tan^2 x - \dfrac{y}{x^2 + y^2}\right)\mathbf{i} + \left(\dfrac{x}{x^2 + y^2} - e^y\right)\mathbf{j}$

12. $\mathbf{f}(x, y) = [\cos(x + 3y) - x^4]\mathbf{i} + [3\cos(x + 3y) - (2/y)]\mathbf{j}$

13. $\mathbf{f}(x, y) = (x^2 + y^2 + 1)\mathbf{i} - (xy + y)\mathbf{j}$

14. $\mathbf{f}(x, y) = [-(1/x^3) + 4x^3y]\mathbf{i} + (\sin y + \sqrt{y} + x^4)\mathbf{j}$.

Consider the vector field

$$\mathbf{f}(x, y, z) = P(x, y, z)\mathbf{i} + Q(x, y, z)\mathbf{j} + R(x, y, z)\mathbf{k}.$$

f is exact (i.e., conservative) if there is a differentiable function F such that

$$\nabla F(x, y, z) = \mathbf{f}(x, y, z).$$

If P, Q, R, $\partial P/\partial y$, $\partial P/\partial z$, $\partial Q/\partial x$, $\partial Q/\partial z$, $\partial R/\partial x$, and $\partial R/\partial y$ are continuous, then, analogously to Theorem 1, **f** is exact if and only if

$$\frac{\partial P}{\partial y} = \frac{\partial Q}{\partial x}, \qquad \frac{\partial R}{\partial x} = \frac{\partial P}{\partial z}, \qquad \text{and} \qquad \frac{\partial Q}{\partial z} = \frac{\partial R}{\partial y}.$$

In Problems 15–20 use the above result to test for exactness. If **f** is exact, find all functions F for which $\nabla F = \mathbf{f}$.

15. $\mathbf{f} = \mathbf{i} + \mathbf{j} + \mathbf{k}$

16. $\mathbf{f} = yz\mathbf{i} + xz\mathbf{j} + xy\mathbf{k}$

17. $\mathbf{f} = [(y/z) + x^2]\mathbf{i} + [(x/z) - \sin y]\mathbf{j} + [\cos z - (xy/z^2)]\mathbf{k}$

18. $\mathbf{f} = \left(\dfrac{1}{x + 2y + 3z} - 3x\right)\mathbf{i} + \left(\dfrac{2}{x + 2y + 3z} + y^2\right)\mathbf{j} + \left(\dfrac{3}{x + 2y + 3z} - \tan^{-1}z\right)\mathbf{k}$

19. $\mathbf{f} = (e^{yz} + y)\mathbf{i} + (xze^{yz} - x)\mathbf{j} + (xye^{yz} + 2z)\mathbf{k}$
20. $\mathbf{f} = (2xy^3 + x + z)\mathbf{i} + (3x^2y^2 - y)\mathbf{j} + (x + \sin z)\mathbf{k}.$

*21.   A particle is moving in space with position vector $\mathbf{x}(t)$ and is subjected to a force $\mathbf{f}$ of magnitude $\alpha/|\mathbf{x}(t)|^3$ (α a constant) directed towards the origin. Show that the force is conservative and find all potential functions for $\mathbf{f}$.

*22.   Answer the questions in Problem 21 if the magnitude of the force is $\alpha/|\mathbf{x}(t)|^k$ for some positive integer k.

*23.   Suppose that a moving particle is subjected to a force of constant magnitude which always points toward the origin. Show that the force is conservative. [*Hint:* If the path is given by $\mathbf{x}(t) = x(t)\mathbf{i} + y(t)\mathbf{j}$, find a unit vector which points towards the origin.]

18.13 Maxima and Minima for a Function of Two Variables

In Sections 3.5 and 5.2 we discussed methods for obtaining maximum and minimum values for a function of one variable. The first basic fact we used was that if f is continuous on a closed bounded interval $[a, b]$, then f takes on a maximum and minimum value in $[a, b]$. We then defined a critical point to be a number x at which $f'(x) = 0$ or for which $f(x)$ exists but $f'(x)$ does not. Finally, we defined local maxima and minima (which occur at critical points) and gave conditions on second derivatives which ensured that a critical point was a local maximum or minimum.

The theory for functions of two variables is more complicated, but some of the basic ideas are the same. The first thing we need to know is that a function *has* a maximum or minimum. The proof of the following theorem can be found in any advanced calculus book.†

THEOREM 1.   Let f be a function of two variables which is continuous on the closed disk D. Then there exist points $\mathbf{x}_1$ and $\mathbf{x}_2$ in D such that

$$m = f(\mathbf{x}_1) \leq f(\mathbf{x}) \leq f(\mathbf{x}_2) = M \tag{1}$$

for any other point $\mathbf{x}$ in D. The numbers m and M are called, respectively, the *global minimum* and the *global maximum* for the function f over the disk D.

We now define what we mean by local minima and maxima.

DEFINITION 1 (Definition of Local Maxima and Minima).   Let f be defined in a neighborhood of a point $\mathbf{x}_0 = (x_0, y_0)$. Then f has a

(i)   *local maximum* at $\mathbf{x}_0$ if there is a neighborhood N_1 of $\mathbf{x}_0$ such that for every point $\mathbf{x}$ in N_1,

$$f(\mathbf{x}) \leq f(\mathbf{x}_0). \tag{2}$$

(ii)   *local minimum* at $\mathbf{x}_0$ if there is a neighborhood N_2 of $\mathbf{x}_0$ such that for every point $\mathbf{x}$ in N_2,

$$f(\mathbf{x}) \geq f(\mathbf{x}_0). \tag{3}$$

† See, for example, Buck, *Advanced Calculus*, p. 73.

(iii) If f has a local maximum or minimum at $\mathbf{x}_0$, then $\mathbf{x}_0$ is called an *extreme point* of f.

Note. Compare this definition with Definition 3.5.4.

A rough sketch of a function with several local maxima is given in Figure 1.

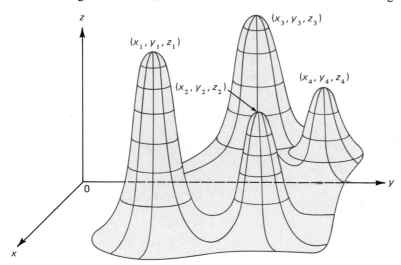

Figure 1

The following theorem is a natural generalization of Theorem 3.5.2.

THEOREM 2. Let f be differentiable at $\mathbf{x}_0$ and suppose that $\mathbf{x}_0$ is an extreme point of f. Then

$$\nabla f(\mathbf{x}_0) = \mathbf{0}. \tag{4}$$

PROOF. We must show that $f_x(x_0, y_0)$ and $f_y(x_0, y_0) = 0$. To prove that $f_x(x_0, y_0) = 0$, define a function h by

$$h(x) = f(x, y_0).$$

Then $h(x_0) = f(x_0, y_0)$ and $h'(x_0) = f_x(x_0, y_0)$. But since f has a local maximum (or minimum) at (x_0, y_0), h has a local maximum (or minimum) at x_0 so that, by Theorem 3.5.2, $h'(x_0) = 0$. Thus $f_x(x_0, y_0) = 0$. In a very similar way, we can show that $f_y(x_0, y_0) = 0$ by defining $h(y) = f(x_0, y)$, and the theorem is proved.

DEFINITION 2. $\mathbf{x}_0$ is a critical point of f if f is differentiable at $\mathbf{x}_0$ and $\nabla f(\mathbf{x}_0) = \mathbf{0}$.

Remark 1. Definitions 1 and 2 and Theorems 1 and 2 hold for functions of three variables with very little change.

Remark 2. By the remark preceding Example 18.8.2, we see that *if* $\mathbf{x}_0$ *is a critical point of* f, *then the tangent plane to the surface* $z = f(x, y)$ *is horizontal* (i.e., parallel to the xy-plane).

Remark 3. Note that this differs from our definition of a critical point of a function of one variable in that, for a function of one variable, x_0 is a critical point if $f(x_0)$ is defined but $f'(x_0)$ does not exist.

As in the case of a function of one variable, the fact that $\mathbf{x}_0$ is a critical point of f does not guarantee that $\mathbf{x}_0$ is an extreme point of f, as Example 3 below illustrates.

EXAMPLE 1. Let $f(x, y) = 1 + x^2 + 3y^2$. Then $\nabla f = 2x\mathbf{i} + 6y\mathbf{j}$ which is zero only when $(x, y) = (0, 0)$. Thus $(0, 0)$ is the only critical point and it is, clearly, a local (and global) minimum. This is illustrated in Figure 2.

EXAMPLE 2. Let $f(x, y) = 1 - x^2 - 3y^2$. Then $\nabla f = -2x\mathbf{i} - 6y\mathbf{j}$ which is zero only at the origin. In this case $(0, 0)$ is a local (and global) maximum for f. See Figure 3.

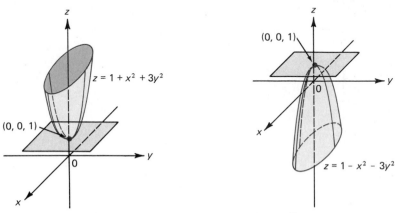

Figure 2   Figure 3

EXAMPLE 3. Let $f(x, y) = x^2 - y^2$. Then $\nabla f = 2x\mathbf{i} - 2y\mathbf{j}$ which again is zero only at $(0, 0)$. But $(0, 0)$ is *neither* a local maximum nor a local minimum for f. To see this we simply note that f can take positive and negative values in any neighborhood of $(0, 0)$ since $f(x, y) > 0$ if $|x| > |y|$ and $f(x, y) < 0$ if $|x| < |y|$. This is illustrated by Figure 4. The figure sketched in Figure 4 is called a *hyperbolic paraboloid* or *saddle surface* (see Figure 17.6.9) and we have the following definition:

DEFINITION 3. If $\mathbf{x}_0$ is a critical point of f but is not an extreme point of f, then $\mathbf{x}_0$ is called a *saddle point* of f.

The three examples above illustrate that a critical point of f may be a local maximum, a local minimum, or a saddle point. There is another possibility.

EXAMPLE 4. Let $f(x, y) = \sqrt{x^2 + y^2}$. Then it is obvious that f has a local (and global) minimum at $(0, 0)$. But

$$\nabla f = \frac{x}{\sqrt{x^2 + y^2}}\mathbf{i} + \frac{y}{\sqrt{x^2 + y^2}}\mathbf{j}$$

so that $(0, 0)$ is *not* a critical point since $\nabla f(0, 0)$ *does not exist*. We see this clearly in Figure 5. The graph of f (which is a cone) does not have a tangent plane at $(0, 0)$.

Examples 1, 2, and 3 above indicate that more is needed to determine whether a critical point is an extreme point (in most cases, it will not be at all obvious). You might suspect that the answer has something to do with the signs of the second partial

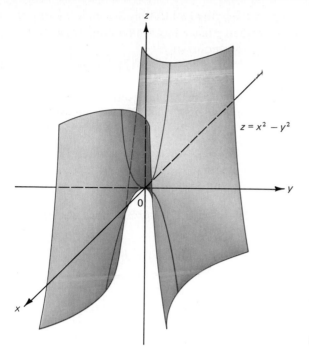

$z = x^2 - y^2$

Figure 4

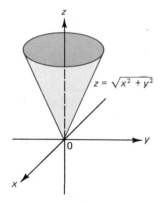

$z = \sqrt{x^2 + y^2}$

Figure 5

derivatives of f, as in the case of functions of one variable. The following theorem gives the answer. Although we could give its proof using results we already know, the proof is very tedious so we shall omit it. Its proof can be found (as usual) in any advanced calculus text.†

THEOREM 3 (Second Derivatives Test). Let f and all its first and second partial derivatives be continuous in a neighborhood of the critical point (x_0, y_0). Let

$$D(x, y) = f_{xx}(x, y)f_{yy}(x, y) - [f_{xy}(x, y)]^2 \tag{5}$$

and let D denote $D(x_0, y_0)$.

---

†See, for example, J. M. H. Olmsted, *Advanced Calculus,* Appleton-Century Crofts, New York, 1961, p. 283.

   (i)   If $D > 0$ and $f_{xx}(x_0, y_0) > 0$, then f has a local minimum at (x_0, y_0).
   (ii) If $D > 0$ and $f_{xx}(x_0, y_0) < 0$, then f has a local maximum at (x_0, y_0).
  (iii) If $D < 0$ then (x_0, y_0) is a saddle point of f.
  (iv) If $D = 0$ then any of the preceding alternatives is possible.

Remark. This theorem does not provide the full answer to the question of what are the extrema of f in $\mathbb{R}^2$. It is still necessary to check points at which ∇f does not exist (as in Example 4).

EXAMPLE 5. Let $f(x, y) = 1 + x^2 + 3y^2$. Then, as we saw in Example 1, $(0, 0)$ is the only critical point of f. But $f_{xx} = 2, f_{yy} = 6$, and $f_{xy} = 0$ so that $D(0, 0) = 12$ and $f_{xx} > 0$ which *proves* that f has a local maximum at $(0, 0)$.

EXAMPLE 6. Let $f(x, y) = -x^2 - y^2 + 2x + 4y + 5$. Determine the nature of the critical points of f.

SOLUTION. $\nabla f = (-2x + 2)\mathbf{i} + (-2y + 4)\mathbf{j}$. We see that $\nabla f = \mathbf{0}$ when

$$-2x + 2 = 0 \quad \text{and} \quad -2y + 4 = 0$$

which occurs only at the point $(1, 2)$. Also, $f_{xx} = -2, f_{yy} = -2$, and $f_{xy} = 0$ so that $D = 4$ and $f_{xx} < 0$ which implies that there is a local maximum at $(1, 2)$. At $(1, 2)$, $f = 10$.

EXAMPLE 7. Let $f(x, y) = 2x^3 - 24xy + 16y^3$. Determine the nature of the critical points of f.

SOLUTION. We have $\nabla f(x, y) = (6x^2 - 24y)\mathbf{i} + (-24x + 48y^2)\mathbf{j}$. If $\nabla f = \mathbf{0}$, we have

$$6(x^2 - 4y) = 0 \quad \text{and} \quad -24(x - 2y^2) = 0.$$

Clearly one critical point is $(0, 0)$. To obtain another, we must solve the simultaneous equations

$$x^2 - 4y = 0,$$
$$x - 2y^2 = 0.$$

The second equation tells us that $x = 2y^2$. Substituting this into the first equation yields

$$4y^4 - 4y = 0$$

with solutions $y = 0$ and $y = 1$, giving us the critical points $(0, 0)$ and $(2, 1)$. Now $f_{xx} = 12x, f_{yy} = 96y$, and $f_{xy} = -24$ so that

$$D(x, y) = (12x)(96y) - 24^2 = 1152xy - 576.$$

Then $D(0, 0) = -576$ and $D(2, 1) = 1728$. Since $f_{xx}(2, 1) > 0$, we find that $(0, 0)$ is a saddle point and $(2, 1)$ is a local minimum.

EXAMPLE 8. Let $f(x, y) = 4x^2 - 4xy + y^2 + 5$. Determine the nature of the critical points of f.

SOLUTION. $\nabla f = (8x - 4y)\mathbf{i} + (-4x + 2y)\mathbf{j}$. We see that $\nabla f = \mathbf{0}$ when

$$8x - 4y = 0$$

and

$$-4x + 2y = 0.$$

This occurs if (x, y) is any point on the line $y = 2x$. Thus, in this case, there are an *infinite number* of critical points. Now $f_{xx} = 8$, $f_{yy} = 2$ and $f_{xy} = -4$ so that $D = 8 \cdot 2 - 16 = 0$. It is impossible to determine the nature of the critical points of x by the second derivatives test. However, since we can find that

$$f(x, y) = (2x - y)^2 + 5,$$

we see that $f(x, y) \geq 5$ and $f(x, y) = 5$ when $y = 2x$. Thus *every point* on the line $y = 2x$ is a local (and global) minimum.

EXAMPLE 9. If $f(x, y) = -(4x^2 - 4xy + y^2) + 5$, then following the reasoning of Example 8 it is easy to see that *every point* on the line $y = 2x$ is a local maximum, but that $D(x, y) = 0$ at any such point.

EXAMPLE 10. Let $f(x, y) = x^3 - y^3$. Then $\nabla f = 3x^2\mathbf{i} - 3y^2\mathbf{j}$ and the only critical point is $(0, 0)$. But $f_{xx} = 6x$, $f_{yy} = -6y$ and $f_{xy} = 0$ so that $D(x, y) = -36xy$ and $D(0, 0) = 0$. In this case it is clear that, since f can take on positive and negative values in any neighborhood of $(0, 0)$, $(0, 0)$ is a saddle point of f.

Remark. Examples 8, 9, and 10 illustrate that any of the three cases can occur when $D = 0$.

EXAMPLE 11. A rectangular wooden box with an open top is to contain α cm³, where α is a given positive number. Ignoring the thickness of the wood, how is the box to be constructed so as to use the smallest amount of wood?

SOLUTION. If the dimensions of the box are x, y, and z, then

$$V = xyz = \alpha.$$

We must minimize the area of the sides of the box, where the area is given by

$$A = xy + 2xz + 2yz$$

(see Figure 6). But $z = \alpha/xy$ so that the problem becomes one of minimizing $A(x, y)$ where

$$A(x, y) = xy + 2x\frac{\alpha}{xy} + 2y\frac{\alpha}{xy} = xy + \frac{2\alpha}{y} + \frac{2\alpha}{x}.$$

Now

$$\nabla A(x, y) = \left(y - \frac{2\alpha}{x^2}\right)\mathbf{i} + \left(x - \frac{2\alpha}{y^2}\right)\mathbf{j}.$$

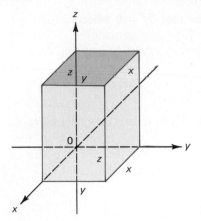

Figure 6

This is zero when

$$y = \frac{2\alpha}{x^2} \quad \text{and} \quad x = \frac{2\alpha}{y^2}$$

or

$$yx^2 = 2\alpha = xy^2.$$

One obvious critical point is $(0, 0)$. If $xy \neq 0$, then we can divide the last expression by xy to obtain

$$x = y = \frac{2\alpha}{xy}$$

so that

$$x = \frac{2\alpha}{xy} = \frac{2\alpha}{x^2}, \quad x^3 = 2\alpha$$

and

$$x = y = \sqrt[3]{2\alpha}.$$

Thus $(\sqrt[3]{2\alpha}, \sqrt[3]{2\alpha})$ is a second critical point. We must throw out $(0, 0)$ because at $(0, 0)$ $V = 0$, not α. It is easy to verify that the point $(\sqrt[3]{2\alpha}, \sqrt[3]{2\alpha})$ does indeed give us a local minimum since $D(\sqrt[3]{2\alpha}, \sqrt[3]{2\alpha}) > 0$ and

$$A_{xx}(\sqrt[3]{2\alpha}, \sqrt[3]{2\alpha}) = \frac{4\alpha}{x^3} \bigg|_{\sqrt[3]{2\alpha}} = \frac{4\alpha}{2\alpha} = 2$$

which is greater than 0. Finally, we obtain

$$z = \frac{\alpha}{xy} = \frac{\alpha}{x^2} = \frac{\alpha x}{x^3} = \frac{\alpha x}{2\alpha} = \frac{x}{2}$$

so that we can minimize the amount of wood needed by building a box with a square base and a height equal to half its length (or width).

There is an interesting way that we can use the theory of this section to derive a result which is very useful for statistical analysis and, in fact, any analysis involving the use of a great deal of data. Suppose n data points $(x_1, y_1), (x_2, y_2), \ldots, (x_n, y_n)$ are collected. For example, the x's may represent average tree growth and the y's average daily temperature in a given year in a certain forest. Or x may represent a week's sales and y a week's profit for a certain business. The question arises as to whether we can "fit" these data points to a straight line. That is, is there a straight line which runs "more or less" through the points. If so, then we can write y as a linear function of x, with obvious computational advantages.

The problem is to find the "best" straight line $y = mx + b$ passing through or near these points. Look at Figure 7. If (x_i, y_i) is one of our n points, then, on the line

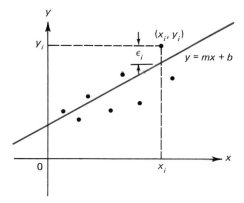

Figure 7

$y = mx + b$, corresponding to x_i we obtain $y_i = mx_i + b$. The "error" ϵ_i between the y value of our actual point and the "approximating" value on the line is given by

$$\epsilon_i = y - mx_i - b. \tag{6}$$

One way to choose the approximating line is to use the line which minimizes the squares of the errors. This is called the *least-squares* criterion for choosing the line. That is, we want to choose m and b such that the function

$$f(m, b) = \epsilon_1{}^2 + \epsilon_2{}^2 + \cdots + \epsilon_n{}^2$$

$$= \sum_{i=1}^{n} \epsilon_i{}^2 = \sum_{i=1}^{n} (y_i - mx_i - b)^2 \tag{7}$$

is a minimum. To do this, we calculate

$$\frac{\partial}{\partial m}(y_i - mx_i - b)^2 = -2x_i(y_i - mx_i - b)$$

and

$$\frac{\partial}{\partial b}(y_i - mx_i - b) = -2(y_i - mx_i - b).$$

Hence

$$\frac{\partial f}{\partial m} = -2 \sum_{i=1}^{n} x_i (y_i - mx_i - b)$$

and

$$\frac{\partial f}{\partial b} = -2 \sum_{i=1}^{n} (y_i - mx_i - b).$$

Setting $\partial f/\partial m = 0$ and $\partial f/\partial b = 0$ and rearranging terms, we obtain

$$\sum_{i=1}^{n} (x_i y_i - mx_i^2 - bx_i) = 0$$

and

$$\sum_{i=1}^{n} (y_i - mx_i - b) = 0.$$

This leads to the system of two equations in the "unknowns" m and b

$$\left\{ \sum_{i=1}^{n} x_i^2 \right\} m + \left\{ \sum_{i=1}^{n} x_i \right\} b = \sum_{i=1}^{n} x_i y_i \tag{8}$$

and

$$\left\{ \sum_{i=1}^{n} x_i \right\} m + nb = \sum_{i=1}^{n} y_i. \tag{9}$$

Here we have used the fact that $\sum_{i=1}^{n} b = nb$. The system (8) and (9) is not hard to solve for m and b. To do so, we multiply both sides of (8) by n and both sides of (9) by $\sum_{i=1}^{n} x_i$ and then subtract to finally obtain

$$m = \frac{n \sum_{i=1}^{n} x_i y_i - \{\sum_{i=1}^{n} x_i\}\{\sum_{i=1}^{n} y_i\}}{n \sum_{i=1}^{n} x_i^2 - \{\sum_{i=1}^{n} x_i\}^2} \tag{10}$$

and

$$b = \frac{\{\sum_{i=1}^{n} x_i^2\}\{\sum_{i=1}^{n} y_i\} - \{\sum_{i=1}^{n} x_i\}\{\sum_{i=1}^{n} x_i y_i\}}{n \sum_{i=1}^{n} x_i^2 - \{\sum_{i=1}^{n} x_i\}^2}. \tag{11}$$

Of course, this procedure only works if

$$n \sum_{i=1}^{n} x_i^2 - \left(\sum_{i=1}^{n} x_i \right)^2 \neq 0.$$

The line $y = mx + b$ given by (10) and (11) is called the *regression line* for the n points.

EXAMPLE 12. Find the regression line for the points $(1, 2)$, $(2, 4)$, and $(5, 5)$.

TABLE 1

l	x_i	y_i	x_i^2	$x_i y_i$
1	1	2	1	2
2	2	4	4	8
3	5	5	25	25
$\Sigma_{i=1}^3$	8	11	30	35

SOLUTION. We tabulate some appropriate values in Table 1. Then, from (10),

$$m = \frac{3(35) - (8)(11)}{3(30) - 8^2} = \frac{17}{26} \approx 0.654$$

and

$$b = \frac{(30)(11) - 8(35)}{26} = \frac{50}{26} \approx 1.923.$$

Thus the regression line is

$$y = 0.654x + 1.923.$$

This is all illustrated in Figure 8.

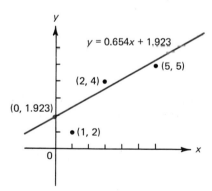

Figure 8

PROBLEMS 18.13

In Problems 1–18 determine the nature of the critical points of the given function.

1. $f(x, y) = 7x^2 - 8xy + 3y^2 + 1$

2. $f(x, y) = x^2 + y^3 - 3xy$

3. $f(x, y) = x^2 + 3y^2 + 4x - 6y + 3$

4. $f(x, y) = x^2 + y^2 + 4xy + 6y - 3$

5. $f(x, y) = x^2 + y^2 + 4x - 2y + 3$

6. $f(x, y) = xy^2 + x^2y - 3xy$

7. $f(x, y) = x^3 + 3xy^2 + 3y^2 - 15x + 2$

8. $f(x, y) = x^3 + y^3 - 3xy$

9. $f(x, y) = \dfrac{1}{y} - \dfrac{1}{x} - 4x + y$

10. $f(x, y) = \dfrac{1}{x} + \dfrac{2}{y} + 2x + y + 1$

11. $f(x, y) = x^2 - xy + y^2 + 2x + 2y$

12. $f(x, y) = xy + \dfrac{8}{x} + \dfrac{1}{y}$

13. $f(x, y) = (4 - x - y)xy$
*14. $f(x, y) = \sin x + \sin y + \sin(x + y)$
15. $f(x, y) = 2x^2 + y^2 + 2/x^2y$
16. $f(x, y) = 4x^2 + 12xy + 9y^2 + 25$
17. $f(x, y) = x^{25} - y^{25}$
18. $f(x, y) = \tan xy$.

19. Find three numbers whose sum is 50 and such that their product is a maximum.
20. Find three numbers x, y, and z whose sum is 50 such that the product xy^2z^3 is a maximum.
21. Find three numbers whose sum is 50 and the sum of whose squares is a minimum.
22. Use the methods of this section to find the minimum distance from the point $(1, -1, 2)$ to the plane $x + 2y - z = 4$. [*Hint:* Express in terms of x and y the distance between $(1, -1, 2)$ and a point (x, y, z) on the plane.]
23. Find the minimum distance between the point $(2, 0, 1)$ and the plane $-x + y + z = 8$.
24. What is the maximum volume of an open-top rectangular box that can be built from a square meters of wood?
*25. Find the dimensions of the rectangular box of maximum volume that can be inscribed in the ellipsoid

$$\frac{x^2}{a^2} + \frac{y^2}{b^2} + \frac{z^2}{c^2} = 1$$

whose faces are parallel to the coordinate planes.
26. Show that the rectangular box inscribed in a sphere which encloses the greatest volume has the dimensions of a cube.
27. A company uses two types of raw materials, I and II, for its product. If it uses x units of I and y units of II it can produce U units of the finished item where

$$U(x, y) = 8xy + 32x + 40y - 4x^2 - 6y^2.$$

Each unit of I costs \$10 and each unit of II costs \$4. Each unit of the product can be sold for \$40. How can the company maximize its profits?
28. Find the regression line for the points $(1, 1)$, $(2, 3)$, and $(3, 6)$. Sketch the line and the points.
29. Find the regression line for the points $(-1, 3)$, $(1, 2)$, $(2, 0)$, and $(4, -2)$. Sketch the line and the points.
30. Find the regression line for the points $(1, 4)$, $(3, -2)$, $(5, 8)$, and $(7, 3)$. Sketch the line and the points.
31. Show that if n points lie on the same straight line L, then the regression line for the n points is L.
*32. Show that the function $e^{(x^2+y^2)/y} + (11/x) + (5/y) + \sin x^2y$ has a global minimum in the first quadrant.

18.14 Constrained Maxima and Minima—Lagrange Multipliers

In the last section we saw how to find the maximum and minimum of a function of two variables by taking gradients and applying a second derivative test. It often happens that there are side conditions (or *constraints*) attached to a problem. For example, we have been asked to find the shortest distance from a point (x_0, y_0) to a line $y = mx + b$. We could write this problem as

minimize the function: $z = \sqrt{(x - x_0)^2 + (y - y_0)^2}$

subject to the constraint: $y - mx - b = 0$.

As another example, suppose that a sphere is heated and a function $w = T(x, y, z)$ gives the temperature of every point on the surface of the sphere. Then, if the sphere is given by $x^2 + y^2 + z^2 = r^2$, and if we wish to find the hottest point on the sphere, we have the problem

maximize: $w = T(x, y, z)$

subject to the constraint: $x^2 + y^2 + z^2 - r^2 = 0.$

We now generalize these two examples. Let f and g be functions of two variables. Then we can formulate a *constrained maximization (or minimization)* problem as follows:

maximize (or minimize): $z = f(x, y)$ (1)

subject to the constraint: $g(x, y) = 0.$ (2)

If f and g are functions of three variables, we have the problem

maximize (or minimize): $w = f(x, y, z)$ (3)

subject to the constraint: $g(x, y, z) = 0.$ (4)

We now develop a method for dealing with problems of the type (1), (2) or (3), (4). Let C be a curve in $\mathbb{R}^2$ or $\mathbb{R}^3$ given parametrically by the differentiable function $\mathbf{F}(t)$. That is, C is given by

$$\mathbf{F}(t) = x(t)\mathbf{i} + y(t)\mathbf{j} \qquad (\text{in } \mathbb{R}^2)$$

or

$$\mathbf{F}(t) = x(t)\mathbf{i} + y(t)\mathbf{j} + z(t)\mathbf{k}.$$

Let $f(\mathbf{x})$ denote the function (of two or three variables) that is to be maximized.

THEOREM 1. Suppose that f is differentiable at a point $\mathbf{x}_0$ and that among all points on a curve C, f takes its maximum (or minimum) value at $\mathbf{x}_0$. Then $\nabla f(\mathbf{x}_0)$ is orthogonal to C at $\mathbf{x}_0$. That is, since $\mathbf{F}'(t)$ is tangent to C, if $\mathbf{x}_0 = \mathbf{F}(t_0)$, then

$$\boxed{\nabla f(\mathbf{x}_0) \cdot \mathbf{F}'(t_0) = 0.}$$ (5)

PROOF. For $\mathbf{x}$ on C, $\mathbf{x} = \mathbf{F}(t)$ so that the composite function $f(\mathbf{F}(t))$ has a maximum (or minimum) at t_0. Therefore its derivative at t_0 is 0. By the chain rule,

$$\frac{d}{dt} f(\mathbf{F}(t)) = \nabla f(\mathbf{F}(t)) \cdot \mathbf{F}'(t)$$

and, at t_0,

$$0 = \nabla f(\mathbf{F}(t_0)) \cdot \mathbf{F}'(t_0) = \nabla f(\mathbf{x}_0) \cdot \mathbf{F}'(t_0)$$

and the theorem is proved.

For f as a function of two variables, the result of Theorem 1 is illustrated in Figure 1.

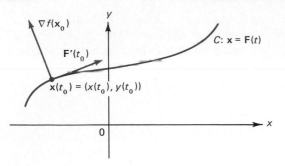

Figure 1

We can use Theorem 1 to make an interesting geometric observation.† Suppose that, subject to the constraint $g(x, y) = 0$, f takes its maximum (or minimum) at the point $x_0 = (x_0, y_0)$. The equation $g(x, y) = 0$ determines a curve C in the xy-plane and, by Theorem 1, $\nabla f(x_0, y_0)$ is orthogonal to C at (x_0, y_0). But from Section 18.8, $\nabla g(x_0, y_0)$ is also orthogonal to C at (x_0, y_0). Thus we see that

$$\boxed{\nabla g(x_0, y_0) \text{ and } \nabla f(x_0, y_0) \text{ are parallel.}}$$

Hence there is a number λ such that

$$\boxed{\nabla f(x_0, y_0) = \lambda \, \nabla g(x_0, y_0).}$$

We can extend this observation to the following rule which applies equally well to functions of three or more variables:

> If, subject to the constraint $g(\mathbf{x}) = 0$, f takes its maximum (or minimum) value at a point $\mathbf{x}_0$, then there is a number λ such that
>
> $$\nabla f(\mathbf{x}_0) = \lambda \, \nabla g(\mathbf{x}_0).$$ **(6)**

The number λ is called a *Lagrange multiplier*. We shall illustrate the Lagrange multiplier technique with a number of examples.

EXAMPLE 1.   Find the maximum and minimum values of $f(x, y) = xy^2$ subject to the condition $x^2 + y^2 = 1$.

SOLUTION. We have $g(x, y) = x^2 + y^2 - 1 = 0$. Then

$$\nabla f = y^2\mathbf{i} + 2xy\mathbf{j} \qquad \text{and} \qquad \nabla g = 2x\mathbf{i} + 2y\mathbf{j}.$$

At a maximizing or minimizing point, we have

$$\nabla f = \lambda \nabla g$$

---

†This observation was first made by the great French mathematician Joseph-Louis Lagrange (1736–1813) who is considered by many to have been the best mathematician of the eighteenth century, only Euler having been a close rival. Lagrange made many contributions to mathematics, the best known of which is his development of the calculus of variations, a subject dealing generally with problems of optimization.

or

$$y^2\mathbf{i} + 2xy\mathbf{j} = 2x\lambda\mathbf{i} + 2y\lambda\mathbf{j}$$

which leads to the equations

$$y^2 = 2x\lambda$$
$$2xy = 2y\lambda.$$

Multiplying the first equation by y and the second by x, we obtain

$$y^3 = 2xy\lambda,$$
$$2x^2y = 2xy\lambda$$

or

$$y^3 = 2x^2y.$$

But $x^2 = 1 - y^2$ so that

$$y^3 = 2(1 - y^2)y = 2y - 2y^3$$

or

$$3y^3 = 2y.$$

The solutions to this last equation are $y = 0$ and $y = \pm\sqrt{\frac{2}{3}}$. This leads to the six points

$$(1, 0),\ (-1, 0),\ \left(\frac{1}{\sqrt{3}}, \sqrt{\frac{2}{3}}\right),\ \left(-\frac{1}{\sqrt{3}}, \sqrt{\frac{2}{3}}\right),\ \left(\frac{1}{\sqrt{3}}, -\sqrt{\frac{2}{3}}\right),$$

$$\left(-\frac{1}{\sqrt{3}}, -\sqrt{\frac{2}{3}}\right).$$

Evaluating $f(x, y)$ at xy^2 at these points, we have

$$f(1, 0) = f(-1, 0) = 0,\ f\left(\frac{1}{\sqrt{3}}, \sqrt{\frac{2}{3}}\right) = f\left(\frac{1}{\sqrt{3}}, -\sqrt{\frac{2}{3}}\right) = \frac{2}{3\sqrt{3}},$$

and

$$f\left(-\frac{1}{\sqrt{3}}, \sqrt{\frac{2}{3}}\right) = f\left(-\frac{1}{\sqrt{3}}, -\sqrt{\frac{2}{3}}\right) = -\frac{2}{3\sqrt{3}}.$$

Thus the maximum value of f is $2/3\sqrt{3}$ and the minimum value of f is $-2/3\sqrt{3}$. Note that there is neither a maximum nor a minimum at $(1, 0)$ and $(-1, 0)$ even though $\nabla f(1, 0) = 0\nabla g(1, 0)$ and $\nabla f(-1, 0) = 0\nabla g(-1, 0) = 0$.

EXAMPLE 2. Find the point on the sphere $x^2 + y^2 + z^2 = 1$ closest to and farthest from the point $(1, 2, 3)$.

SOLUTION. We wish to minimize and maximize $f(x, y, z) = (x - 1)^2 + (y - 2)^2 + (z - 3)^2$ subject to $g(x, y, z) = x^2 + y^2 + z^2 - 1 = 0$. We have $\nabla f(x, y, z) =$

$2(x-1)\mathbf{i} + 2(y-2)\mathbf{j} + 2(z-3)\mathbf{k}$, and $\nabla g(x, y, z) = 2x\mathbf{i} + 2y\mathbf{j} + 2z\mathbf{k}$. Condition (6) implies that at a maximizing point, $\nabla f = \lambda \nabla g$ so that

$$2(x-1) = 2x\lambda,$$
$$2(y-2) = 2y\lambda,$$
$$2(z-3) = 2z\lambda.$$

If $\lambda \neq 1$, then we find that

$$x - 1 = x\lambda \quad \text{or} \quad x - x\lambda = 1 \quad \text{or} \quad x(1-\lambda) = 1$$

and

$$x = \frac{1}{1-\lambda}.$$

Similarly, we obtain

$$y = \frac{2}{1-\lambda} \quad \text{and} \quad z = \frac{3}{1-\lambda}.$$

Then

$$1 = x^2 + y^2 + z^2 = \frac{1}{(1-\lambda)^2}(1^2 + 2^2 + 3^2) = \frac{14}{(1-\lambda)^2}$$

so that

$$(1-\lambda)^2 = 14, \quad (1-\lambda) = \pm\sqrt{14}$$

and

$$\lambda = 1 \pm \sqrt{14}.$$

If $\lambda = 1 + \sqrt{14}$, then $(x, y, z) = (-1/\sqrt{14}, -2/\sqrt{14}, -3/\sqrt{14})$. If $\lambda = 1 - \sqrt{14}$, then $(x, y, z) = (1/\sqrt{14}, 2/\sqrt{14}, 3/\sqrt{14})$. Finally, evaluation shows us that f is maximized at $(-1/\sqrt{14}, -2/\sqrt{14}, -3/\sqrt{14})$ and that f is minimized at $(1/\sqrt{14}, 2/\sqrt{14}, 3/\sqrt{14})$.

EXAMPLE 3. In Example 18.13.11 we found a way to minimize the amount of wood needed to construct an open-top rectangular box with a fixed volume α. The problem was to minimize

$$A(x, y, z) = xy + 2xz + 2yz$$

subject to the constraint $xyz = \alpha$ or

$$g(x, y, z) = xyz - \alpha = 0.$$

Solve this problem using Lagrange multipliers.

SOLUTION. $\nabla A = (y + 2z)\mathbf{i} + (x + 2z)\mathbf{j} + (2x + 2y)\mathbf{k}$ and $\nabla g = yz\mathbf{i} + xz\mathbf{j} + xy\mathbf{k}$. Then the condition $\nabla A = \lambda \nabla g$ leads to the equations

$$y + 2z = \lambda yz,$$
$$x + 2z = \lambda xz,$$
$$2x + 2y = \lambda xy.$$

To solve, we multiply the three equations by x, y, and z, respectively, to obtain

$$xy + 2xz = \lambda xyz,$$
$$xy + 2yz = \lambda xyz,$$
$$2xz + 2yz = \lambda xyz.$$

Thus

$$xy + 2xz = xy + 2yz = 2xz + 2yz.$$

The first equality indicates that

$$2xz = 2yz$$

and, since $z \neq 0$ (since otherwise xyz would equal 0), we have

$$x = y.$$

Then, the second equality tells us that

$$xy = 2xz$$

or

$$x^2 = 2xz$$

and, since $x \neq 0$,

$$x = 2z.$$

Thus we have $x = y = 2z$ and, since $xyz = \alpha$,

$$xyz = (2z)(2z)z = \alpha \qquad \text{and} \qquad 4z^3 = \alpha$$

so that

$$z = \sqrt[3]{\frac{\alpha}{4}} = \sqrt[3]{\frac{2\alpha}{8}} = \frac{\sqrt[3]{2\alpha}}{2}$$

and

$$x = y = \sqrt[3]{2\alpha}.$$

This is the answer we obtained earlier.

We conclude this section by noting that we can use Lagrange multipliers in $\mathbb{R}^3$ if there are two or more constraint equations. Suppose, for example, we wish to maximize (or minimize) $w = f(x, y, z)$ subject to the constraints

$$g(x, y, z) = 0 \tag{7}$$

and

$$h(x, y, z) = 0. \tag{8}$$

Each of the equations (7) and (8) represents a surface in $\mathbb{R}^3$ and their intersection forms a curve in $\mathbb{R}^3$. By an argument very similar to the one we used earlier (but applied in $\mathbb{R}^3$ instead of $\mathbb{R}^2$), we find that if f is maximized (or minimized) at

(x_0, y_0, z_0), then $\nabla f(x_0, y_0, z_0)$ is in the plane determined by $\nabla g(x_0, y_0, z_0)$ and $\nabla h(x_0, y_0, z_0)$. This means that there are numbers λ and μ such that

$$\nabla f(x_0, y_0, z_0) = \lambda \nabla g(x_0, y_0, z_0) + \mu \nabla h(x_0, y_0, z_0) \qquad (9)$$

(see Problem 17.5.35). Formula (9) is the generalization of formula (6) in the case of two constraints.

EXAMPLE 4. Find the maximum value of $w = xyz$ among all points (x, y, z) lying on the planes $x + y + z = 30$ and $x + y - z = 0$.

SOLUTION. Setting $f(x, y, z) = xyz$, $g(x, y, z) = x + y + z - 30$ and $h(x, y, z) = x + y - z$, we obtain

$$\nabla f = yz\mathbf{i} + xz\mathbf{j} + xy\mathbf{k}$$
$$\nabla g = \mathbf{i} + \mathbf{j} + \mathbf{k}$$
$$\nabla h = \mathbf{i} + \mathbf{j} - \mathbf{k}$$

and, using equation (9) to obtain the maximum, we obtain the equations

$$yz = \lambda + \mu,$$
$$xz = \lambda + \mu,$$
$$xy = \lambda - \mu.$$

Multiplying the three equations by x, y, and z, respectively, we find that

$$xyz = (\lambda + \mu)x,$$
$$xyz = (\lambda + \mu)y,$$
$$xyz = (\lambda - \mu)z.$$

If $\lambda + \mu = 0$, then $yz = 0$ and $xyz = 0$, which clearly is not a maximum value. Thus we can divide the first two equations by $\lambda + \mu$ to find that

$$x = y.$$

Since $x + y - z = 0$, we have $2x - z = 0$ or $z = 2x$. But, then,

$$30 = x + y + z = x + x + 2x = 4x$$

or

$$x = \frac{15}{2}.$$

Then

$$y = \frac{15}{2}, \qquad z = 15$$

and the maximum value of xyz occurs at $(15/2, 15/2, 15)$ and is equal to $(15/2)(15/2)15 = 843\frac{3}{4}$.

PROBLEMS 18.14

1. Use Lagrange multipliers to find the minimum distance from the point $(1, 2)$ to the line $2x + 3y = 5$,

2. Use Lagrange multipliers to find the minimum distance from the point $(3, -2)$ to the line $y = 2 - x$.

3. Use Lagrange multipliers to find the minimum distance from the point $(1, -1, 2)$ to the plane $x + y - z = 3$.

4. Use Lagrange multipliers to find the minimum distance from the point $(3, 0, 1)$ to the plane $2x - y + 4z = 5$.

5. Use Lagrange multipliers to find the minimum distance from the plane $ax + by + cz = d$ to the origin.

*6. Find the maximum and minimum values of $x^2 + y^2$ subject to the condition $x^3 + y^3 = 6xy$.

7. Find the maximum and minimum values of $2x^2 + xy + y^2 - 2y$ subject to the condition $y = 2x - 1$.

8. Find the maximum and minimum values of $x^2 + y^2 + z^2$ subject to the condition $z^2 = x^2 - 1$.

9. Find the maximum and minimum values of $x^3 + y^3 + z^3$ if (x, y, z) lies on the sphere $x^2 + y^2 + z^2 = 4$.

10. Find the maximum and minimum values of $x + y + z$ if (x, y, z) lies on the sphere $x^2 + y^2 + z^2 = 1$.

11. Find the maximum and minimum values of xyz if (x, y, z) is on the ellipsoid $x^2 + (y^2/4) + (z^2/9) = 1$.

12. Solve Problem 11 if (x, y, z) is on the ellipsoid $(x^2/a^2) + (y^2/b^2) + (z^2/c^2) = 1$.

*13. Minimize the function $x^2 + y^2 + z^2$ for (x, y, z) on the planes $3x - y + z = 6$ and $x + 2y + 2z = 2$.

14. Find the maximum value of $x^3 + y^3 + z^3$ for (x, y, z) on the planes $x + y + z = 2$ and $x + y - z = 3$.

15. Find the maximum and minimum distances from the origin to a point on the ellipse $(x^2/a^2) + (y^2/b^2) = 1$.

16. Find the maximum and minimum distances from the origin to a point on the ellipsoid $(x^2/a^2) + (y^2/b^2) + (z^2/c^2) = 1$.

*17. Find the maximum value of $x_1 + x_2 + x_3 + x_4$ subject to $x_1^2 + x_2^2 + x_3^2 + x_4^2 = 1$. [*Hint:* Use the obvious generalization of Lagrange multipliers to functions of four variables.]

*18. Find the maximum value of $x_1 + x_2 + \cdots + x_n$ subject to $x_1^2 + x_2^2 + \cdots + x_n^2 = 1$.

19. Using Lagrange multipliers show that among all rectangles with the same perimeter the square encloses the greatest area.

*20. Show that among all triangles having the same perimeter, the equilateral triangle has the greatest area.

21. Find the maximum and minimum values of xyz subject to $x^2 + z^2 = 1$ and $x = y$.

22. Find the volume of the largest rectangular parallelepiped that can be inscribed in the ellipsoid $x^2 + 4y^2 + 9z^2 = 9$.

23. A silo is in the shape of a cylinder topped with a cone (see Figure 2). If the radius of each is 6 m, and the total surface area is 200 m² (excluding the base), what are the heights of the cylinder and cone that maximize the volume enclosed by the silo?

24. Show that among all triangles inscribed in a circle, the equilateral triangle has the greatest perimeter.

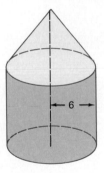

Figure 2

25. The base of an open-top rectangular box costs $3 per square meter to construct, while the sides cost only $1 per square meter. Find the dimensions of the box of greatest volume that can be constructed for $36.

26. A manufacturing company has three plants I, II, and III, which produce x, y, and z units, respectively, of a certain product. The annual revenue from this production is given by

    $$R(x, y, z) = 6xyz^2 - 400,000x - 400,000y - 400,000z.$$

    If the company is to produce 1000 units annually, how should it allocate production so as to maximize profits?

27. A firm has $250,000 to spend on labor and raw materials. The output of the firm is αxy where α is a constant and x and y are, respectively, the quantity of labor and raw materials consumed. If the unit price of hiring labor is $5000 and the unit price of raw materials is $2500, find the ratio of x to y that maximizes output.

28. The temperature of a point (x, y, z) on the unit sphere is given by $T(x, y, z) = xy + yz$. What is the hottest point on the sphere?

Review Exercises for Chapter Eighteen

In Exercises 1–6 find the domain and range of the indicated function.

1. $f(x, y) = \sqrt{x^2 - y^2}$

2. $f(x, y) = \dfrac{1}{\sqrt{x^2 + y^2}}$

3. $f(x, y) = \cos(x + 3y)$

4. $f(x, y, z) = \sqrt{1 - x^2 - y^2 - z^2}$

5. $f(x, y, z) = \dfrac{1}{\sqrt{x^2 + y^2 + z^2 - 1}}$

6. $f(x, y) = \ln(x - y + 4z - 3)$.

In Exercises 7–10 describe the level curves of the given function and sketch these curves for the given values of z.

7. $z = \sqrt{1 - x - y}$; $z = 0, 1, 3, 8$

8. $z = \sqrt{1 - y^2 + x}$; $z = 0, 2, 4, 7$

9. $z = \ln(x - 3y)$; $z = 0, 1, 2, 3$

10. $z = \dfrac{x^2 + y^2}{x^2 - y^2}$; $z = 1, 3, 6$.

11. Sketch the open disk centered at $(-1, 2)$ with radius 4.

12. Sketch the closed ball centered at $(1, 2, 3)$ with radius 2.

13. Show that $\displaystyle \lim_{(x,y)\to(0,0)} \dfrac{xy}{y^2 - x^2}$ does not exist.

14. Show that $\displaystyle\lim_{(x,y)\to(0,0)} \frac{y^2 - 2x}{y^2 + 2x}$ does not exist.

15. Show that $\displaystyle\lim_{(x,y)\to(0,0)} \frac{4xy^3}{x^2 + y^4} = 0$.

16. Show that $\displaystyle\lim_{(x,y,z)\to(0,0,0)} \frac{zy^2 + x^3}{x^2 + y^2 + z^2} = 0$.

17. Calculate $\displaystyle\lim_{(x,y)\to(1,-2)} \frac{1 + x^2y}{2 - y}$.

18. Calculate $\displaystyle\lim_{(x,y)\to(1,4)} \ln(1 + x + \sqrt{y})$.

19. Calculate $\displaystyle\lim_{(x,y,z)\to(2,-1,1)} \frac{x^2 - yz^3}{1 + xyz - 2y^5}$.

20. Find the maximum region over which the function $f(x, y) = \ln(1 - 2x + 3y)$ is continuous.

21. Find the maximum region over which the function $f(x, y, z) = \ln(x - y - z + 4)$ is continuous.

22. Find the maximum region over which the function

$$f(x, y, z) = \frac{1}{\sqrt{1 - x^2 + y^2 - z^2}}$$

is continuous.

23. Find a number c such that the function

$$f(x, y) = \begin{cases} \dfrac{-2xy}{\sqrt{x^2 + y^2}}, & (x, y) \neq (0, 0) \\ c, & (x, y) = (0, 0) \end{cases}$$

is continuous at the origin.

In Exercises 24–35 calculate all first partial derivatives.

24. $f(x, y) = \dfrac{y}{x}$

25. $f(x, y) = \cos(x - 3y)$

26. $f(x, y) = \dfrac{1}{\sqrt{x^2 - y^3}}$

27. $f(x, y) = \tan^{-1}\dfrac{y}{1 + x}$

28. $f(x, y, z) = \ln(x - y + 4z)$

29. $f(x, y, z) = \dfrac{1}{\sqrt{x^2 + y^2 + z^2}}$

30. $f(x, y) = \cosh\dfrac{y}{x^2}$

31. $f(x, y, z) = \sec\dfrac{x - y}{z}$

32. $f(x, y, z) = (x^2y - y^3z^5 + x\sqrt{z})^{2/3}$

33. $f(x, y, z) = \dfrac{x^2 - y^3}{y^3 + z^4}$

34. $f(x, y, z, w) = \dfrac{x - z + w}{y + 2w - x}$

35. $f(x, y, z, w) = e^{(x-w)/(y+z)}$

In Exercises 36–41 calculate all second partial derivatives and show that all pairs of mixed partials are equal.

36. $f(x, y) = xy^3$

37. $f(x, y) = \tan^{-1}\dfrac{y}{x}$

38. $f(x, y) = \sqrt{x^2 - y^2}$

39. $f(x, y) = \dfrac{x + y}{x - y}$

40. $f(x, y, z) = \ln(2 - 3x + 4y - 7z)$

41. $f(x, y, z) = \dfrac{1}{\sqrt{1 - x^2 - y^2 - z^2}}.$

42. If $f(x, y, z) = x^2 y^3 - zx^5$, calculate f_{yzx}.

43. If $f(x, y, z) = \dfrac{x - y}{z}$, calculate f_{zxxyz}.

In Exercises 44–51 calculate the gradient of the given function at the given point.

44. $f(x, y) = x^2 - y^3$; $(1, 2)$

45. $f(x, y) = \tan^{-1}\dfrac{y}{x}$; $(-1, -1)$

46. $f(x, y) = \dfrac{x - y}{x + y}$; $(3, 2)$

47. $f(x, y) = \cos(x - 2y)$; $\left(\dfrac{\pi}{2}, \dfrac{\pi}{6}\right)$

48. $f(x, y, z) = xy + yz^3$; $(1, 2, -1)$

49. $f(x, y, z) = \dfrac{x - y}{3z}$; $(2, 1, 4)$

50. $f(x, y, z) = \dfrac{1}{\sqrt{x^2 + y^2 + z^2}}$; (a, b, c)

51. $f(x, y, z) = e^{-(x^2 + y^3 + z^4)}$; $(0, -1, 1).$

In exercises 52–60 use the chain rule to calculate the indicated derivative.

52. $z = 2xy$; $x = \cos t$, $y = \sin t$; dz/dt
53. $z = \sin^{-1}(y/x)$; $x = 1 + t$; $y = t^2$; dz/dt
54. $w = \ln(1 - x - 2y + 3z)$; $x = e^t \sin t$, $y = e^t \cos t$, $z = t^2$; dw/dt
55. $z = y/x$; $x = r - s$; $y = r + s$; $\partial z/\partial s$
56. $z = xy^3$; $x = r/s$; $y = s^2/r$; $\partial z/\partial r$
57. $z = \sin(x - y)$; $x = e^{r+s}$, $y = e^{r-s}$; $\partial z/\partial s$
58. $w = xyz$; $x = rs$, $y = r/s$, $z = s^2 r^3$; $\partial w/\partial r$ and $\partial w/\partial s$
59. $w = x^3 y + y^3 z$; $x = rst$; $y = rs/t$, $z = rt/s$; $\partial w/\partial s$ and $\partial w/\partial t$
60. $w = \ln(x + 2y + 5z)$; $x = e^{r+s+t}$, $y = \sqrt{rst^2}$, $z = 1/\sqrt{r + s + t}$; $\partial w/\partial r$ and $\partial w/\partial t.$

In Exercises 61–66 find the equation of the tangent plane and symmetric equations of the normal line to the given surface at the given point.

61. $x^2 + y^2 + z^2 = 3$; $(1, 1, 1)$

62. $x^{1/2} + y^{3/2} + z^{1/2} = 3$; $(1, 0, 4)$

63. $3x - y + 5z = 15$; $(-1, 2, 4)$

64. $xy^2 - yz^3 = 0$, $(1, 1, 1)$

65. $xyz = 6$; $(-2, 1, -3)$

66. $\sqrt{\dfrac{x - y}{y + z}} = \dfrac{1}{2}$; $(2, 1, 3).$

In Exercises 67–72 calculate the directional derivative of the given function at the given point, in the direction of the given vector **v**.

67. $f(x, y) = y/x$, at $(1, 2)$; $\mathbf{v} = \mathbf{i} - \mathbf{j}$
68. $f(x, y) = 3x^2 - 4xy$ at $(3, -1)$; $\mathbf{v} - 2\mathbf{i} + 5\mathbf{j}$
69. $f(x, y) = \tan^{-1} y/x$ at $(1, -1)$; $\mathbf{v} = -3\mathbf{i} + 2\mathbf{j}$
70. $f(x, y, z) = xy^2 - zy^3$ at $(1, 2, 3)$; $\mathbf{v} = \mathbf{i} - \mathbf{j} + 2\mathbf{k}$
71. $f(x, y, z) = \dfrac{1}{\sqrt{x^2 + y^2 + z^2}}$ at $(1, -1, 2)$; $\mathbf{v} = -2\mathbf{i} + \mathbf{j} - 3\mathbf{k}$
72. $f(x, y, z) = e^{-(x + y^2 - xz)}$ at $(1, 0, -1)$; $\mathbf{v} = 2\mathbf{i} + 5\mathbf{j} + \mathbf{k}.$

73. Show that the force $\mathbf{f}(\mathbf{x}) = -3\mathbf{x}/|\mathbf{x}|^5$ is conservative and find a potential function for **f**.
74. Show that the force $\mathbf{f}(x, y) = y^2 \mathbf{i} - x^2 \mathbf{j}$ is not conservative.

In Exercises 75–80 calculate the total differential df.

75. $f(x, y) = x^3 y^2$

76. $f(x, y) = \cos^{-1}\left(\dfrac{y}{x}\right)$

77. $f(x, y) = \sqrt{\dfrac{x+1}{y-1}}$

78. $f(x, y, z) = xy^5 z^3$

79. $f(x, y, z) = \ln(x - y + 4z)$

80. $f(x, y) = \left(\dfrac{x-y}{y+z}\right)^{1/3}$.

In Exercises 81–83 estimate the given number using the total differential.

81. $4.03/6.97$

82. $\sqrt{(4.97)^2 + (12.02)^2}$

83. $\sqrt{3.97}(10.05 - 1.03)^{3/2}$.

84. How much wood is contained in the sides of a rectangular box with sides of inside measurements 1.5 m, 1.3 m, and 2 m if the thickness of the wood making up the sides is 3 cm $(= 0.03$ m)?

In Exercises 85–90 test for exactness. If the vector field is exact, find all functions F for which $\nabla F = \mathbf{f}$.

85. $\mathbf{f}(x, y) = (y^2 + 1)\mathbf{i} + (2xy)\mathbf{j}$

86. $\mathbf{f}(x, y) = -(\sin y)\mathbf{i} + (y - x \cos y)\mathbf{j}$

87. $\mathbf{f}(x, y) = (2xy)\mathbf{i} + [(1/y^2) - x^2]\mathbf{j}$

88. $\mathbf{f}(x, y) = (\sin x + \sqrt{x} + y^4)\mathbf{i} + [(-1/y^3) + 4y^3 x]\mathbf{j}$

89. $\mathbf{f}(x, y, z) = (y^3 z^5 + 3x)\mathbf{i} + (-4\sqrt{y} + 3xy^2 z^5)\mathbf{j} + (\sin z + 5xy^3 z^4)\mathbf{k}$

90. $\mathbf{f}(x, y, z) = [x + (1/z)]\mathbf{i} + [y - (1/z)]\mathbf{j} + [z + (y - x)/z^2]\mathbf{k}$.

In Exercises 91–96 determine the nature of the critical points of the given function.

91. $f(x, y) = 6x^2 + 14y^2 - 16xy + 2$

92. $f(x, y) = x^5 - y^5$

93. $f(xy) = \dfrac{1}{y} + \dfrac{2}{x} + 2y + x + 4$

94. $f(x, y) = 49 - 16x^2 + 24xy - 9y^2$

95. $f(x, y) = x^2 + y^2 + \dfrac{2}{xy^2}$

96. $f(x, y) = \cot xy$.

97. Find the minimum distance from the point $(2, -1, 4)$ to the plane $x - y + 3z = 7$.

98. Solve Exercise 97 using Lagrange multipliers.

99. What is the maximum volume of an open-top rectangular box that can be built from 10 square meters of wood?

100. What is the smallest amount of wood needed to build an open-top rectangular box enclosing a volume of 25 cubic meters?

101. What are the dimensions of the rectangular parallelepiped of maximum volume that can be inscribed in the ellipsoid $x^2 + 9y^2 + 4z^2 = 36$?

102. Solve Exercise 101 using Lagrange multipliers.

103. Use Lagrange multipliers to find the minimum distance from the plane $x + 3y + 5z = 2$ to the origin.

104. Find the maximum and minimum values of $xy^3 z^2$ if (x, y, z) lies on the plane $x - y + 2z = 2$.

105. Minimize the function $x^2 + y^2 + z^2$ for (x, y, z) on the planes $2x + y + z = 2$ and $x - y - 3z = 4$.

106. Find the regression line through the points $(1, 4), (-2, 3), (1, 1)$, and $(2, 6)$. Sketch the line and the points.

NINETEEN

MULTIPLE INTEGRATION

In Chapter 4 we introduced the concept of the definite integral of a function of one variable. In this chapter we show how the notion of the definite integral can be extended to functions of several variables. In particular, we shall discuss the double integral of a function of two variables and the triple integral of a function of three variables.

Multiple integrals were developed essentially by Euler (whom we have referred to earlier). Euler had a clear conception of the double integral over a bounded region enclosed by arcs, and in a paper published in 1769,† he gave a procedure for evaluating double integrals by repeated integration. We shall discuss Euler's technique in Section 19.2. Multiple integrals were first used by Newton in his work (which appeared in the *Principia*) on the gravitational attraction exerted by spheres on particles. It should be mentioned, however, that Newton had only a geometric interpretation of a multiple integral. The more precise analytical definitions had to await the work of Euler and, later, Lagrange.

We shall begin our discussion of multiple integration with an introduction to the double integral. Our development will closely parallel the development of the definite integral in Chapter 4.

19.1 Volume under a Surface and the Double Integral

In our development of the definite integral in Chapter 4 we began by calculating the area under a curve $y = f(x)$ (and above the x-axis) for x in the interval $[a, b]$. We initially assumed that, on $[a, b], f(x) \geq 0$. We can carry out a similar development by obtaining an expression which represents a volume in $\mathbb{R}^3$.

We begin by considering an especially simple case. Let R denote the rectangle in $\mathbb{R}^2$ given by

$$R = \{(x, y): a \leq x \leq b \text{ and } c \leq y \leq d\}. \tag{1}$$

This rectangle is sketched in Figure 1. Let $z = f(x, y)$ be a continuous function which is nonnegative over R. That is, $f(x, y) \geq 0$ for every (x, y) in R. We now ask: What is the volume "under" the surface $z = f(x, y)$ and "over" the rectangle R? The volume requested is sketched in Figure 2.

---

† *Novi Comm. Acad. Sci. Petrop.*, **14**, 1769, pages 72–103.

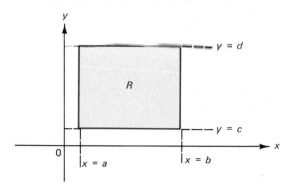

Figure 1

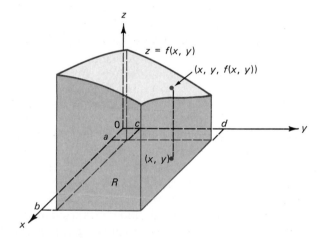

Figure 2

We shall calculate this volume in much the same way we calculated the area under a curve in Chapter 4. We begin by "partitioning" the rectangle. This is accomplished by partitioning the intervals $[a, b]$ and $[c, d]$ in the same manner as in Chapter 4:

$$a = x_0 < x_1 < x_2 < \cdots < x_{n-1} < x_n = b \tag{2}$$

$$c = y_0 < y_1 < y_2 < \cdots < y_{m-1} < y_m = d. \tag{3}$$

We then define

$$\Delta x_i = x_i - x_{i-1}, \tag{4}$$

$$\Delta y_j = y_j - y_{j-1} \tag{5}$$

and define the subrectangles R_{ij} by

$$R_{ij} = \{(x, y): x_{i-1} \leq x \leq x_i \text{ and } y_{j-1} \leq y \leq y_j\} \tag{6}$$

for $i = 1, 2, \ldots, n$ and $j = 1, 2, \ldots, m$. This is sketched in Figure 3. Note that there are nm subrectangles R_{ij} covering the rectangle R. We next define the *norm* of the partition of R, denoted $|P|$, to be the largest area among the nm subrectangles R_{ij}.

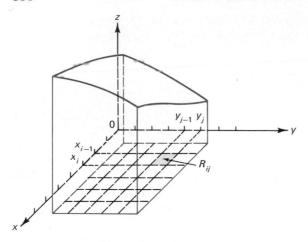

Figure 3

We can estimate the volume we are trying to calculate by estimating the volume under the surface and over each of the subrectangles, and then adding up these volumes. Let $(x_i{}^*, y_j{}^*)$ be a point in R_{ij}. Then the approximate volume V_{ij} under the surface and over R_{ij} is given by

$$V_{ij} \approx f(x_i{}^*, y_j{}^*)\, \Delta x_i\, \Delta y_j. \tag{7}$$

The expression on the right-hand side of (7) is simply the volume of the parallele-piped (three-dimensional box) with base R_{ij} and height $f(x_i{}^*, y_j{}^*)$. This corresponds to the approximate area $A_i \approx f(x_i{}^*)\, \Delta x_i$ that we used in Section 4.2. Unless $f(x, y)$ is a plane parallel to the xy-plane, the expression $f(x_i{}^*, y_j{}^*)\, \Delta x_i\, \Delta y_j$ will not in general be equal to the volume under the surface S. But if $\Delta x_i\, \Delta y_j$ is small, the approximation will be a good one. The difference between the actual V_{ij} and the approximate volume given in (7) is illustrated in Figure 4. We now add up these approximate

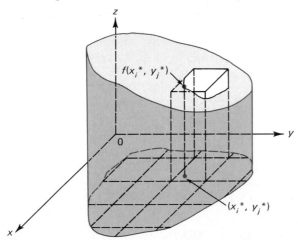

Figure 4

volumes to obtain an approximation to the total volume sought. The total volume is

$$V = V_{11} + V_{12} + \cdots + V_{1m} + V_{21} + V_{22} + \cdots + V_{2m}$$
$$+ \cdots + V_{n1} + V_{n2} + \cdots + V_{nm}. \tag{8}$$

To simplify notation, we use the summation sign Σ introduced in Chapter 14. Since we are summing over two variables i and j, we need two such signs:

$$V = \sum_{i=1}^{n} \sum_{j=1}^{m} V_{ij}. \tag{9}$$

The expression in (9) is called a *double sum*. If we "write out"† the expression in (9), we obtain the expression in (8). Then combining (7) and (9), we have

$$V \approx \sum_{i=1}^{n} \sum_{j=1}^{m} f(x_i^*, y_j^*) \, \Delta x_i \, \Delta y_j. \tag{10}$$

As the norm of the partition approaches zero, the number of subrectangles R_{ij} increases without bound and the area of each R_{ij} approaches zero. This implies that the volume approximation given by (7) is getting closer and closer to the "true" volume over R_{ij}. Thus the approximation (10) gets better and better as $|P| \to 0$ which enables us to write

$$V = \lim_{|P| \to 0} \sum_{i=1}^{n} \sum_{j=1}^{m} f(x_i^*, y_j^*) \, \Delta x_i \, \Delta y_j. \tag{11}$$

The limit in (11) is equal to the volume we are seeking if (i) the limit exists and (ii) is the same no matter how R is partitioned and no matter which points (x_i^*, y_j^*) are chosen in the subrectangles R_{ij}.

EXAMPLE 1.   Calculate the volume under the plane $z = x + 2y$ and over the rectangle $R = \{(x, y): 1 \le x \le 2 \text{ and } 3 \le y \le 5\}$.

SOLUTION. The solid whose volume we wish to calculate is sketched in Figure 5. For simplicity, we partition each of the intervals $[1, 2]$ and $[3, 5]$ into n subintervals of equal length (i.e., $m = n$):

$$1 = x_0 < x_1 < \cdots < x_n = 2,$$
$$3 = y_0 < y_1 < \cdots < y_n = 5,$$

where

$$x_i = 1 + \frac{i}{n}, \qquad \Delta x_i = \frac{1}{n}$$

and

$$y_j = 3 + \frac{2j}{n}, \qquad \Delta y_j = \frac{2}{n}.$$

---

†This is done by summing over j first and then over i. For example,

$$\sum_{i=1}^{3} \sum_{j=1}^{4} a_{ij} = \sum_{i=1}^{3} (a_{i1} + a_{i2} + a_{i3} + a_{i4})$$

$$= a_{11} + a_{12} + a_{13} + a_{14} + a_{21} + a_{22} + a_{23} + a_{24} + a_{31} + a_{32} + a_{33} + a_{34}.$$

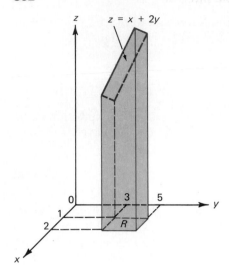

Figure 5

Then choosing $x_i^* = x_i$ and $y_i^* = y_i$ we obtain

$$V_{ij} \approx f(x_i^*, y_j^*)\,\Delta x_i\,\Delta y_j = (x_i + 2y_j)\,\Delta x_i\,\Delta y_j$$

$$= \left\{ \left(1 + \frac{i}{n}\right) + 2\left(3 + \frac{2j}{n}\right) \right\} \frac{1}{n} \cdot \frac{2}{n}$$

$$= \left(7 + \frac{i}{n} + \frac{4j}{n}\right) \frac{2}{n^2}.$$

This implies that

$$\sum_{i=1}^{n} \sum_{j=1}^{n} V_{ij} \approx \sum_{i=1}^{n} \sum_{j=1}^{n} \left(\frac{14}{n^2} + \frac{2i}{n^3} + \frac{8j}{n^3}\right)$$

$$= \underbrace{\sum_{i=1}^{n} \sum_{j=1}^{n} \frac{14}{n^2}}_{①} + \underbrace{\sum_{i=1}^{n} \sum_{j=1}^{n} \frac{2i}{n^3}}_{②} + \underbrace{\sum_{i=1}^{n} \sum_{j=1}^{n} \frac{8j}{n^3}}_{③}.$$

It is not difficult to evaluate each of these double sums. There are n^2 terms in each sum. Since $14/n^2$ does not depend on i or j, we evaluate the sum ① by simply adding up the term $14/n^2$ a total of n^2 times. Thus

$$\sum_{i=1}^{n} \sum_{j=1}^{n} \frac{14}{n^2} = n^2 \left(\frac{14}{n^2}\right) = 14.$$

Next, if we set $i = 1$ in ③, then we have $\sum_{j=1}^{n} 8j/n^3$. Similarly, setting $i = 2, 3, 4, \ldots, n$ in ③ yields $\sum_{j=1}^{n} 8j/n^3$. Thus in ③, we obtain the term $(\sum_{j=1}^{n} 8j/n^3)$ n times.

But

$$\sum_{j=1}^{n} \frac{8j}{n^3} = \frac{8}{n^3} \sum_{j=1}^{n} j = \frac{8}{n^3} (1 + 2 + \cdots + n)$$

$$= \frac{8}{n^3} \left(\frac{n(n+1)}{2} \right) \qquad \text{(from equation (4.2.10))}$$

$$= \frac{4(n+1)}{n^2}.$$

Thus

$$\sum_{i=1}^{n} \sum_{j=1}^{n} \frac{8j}{n^3} = n \left\{ \sum_{j=1}^{n} \frac{8j}{n^3} \right\} = n \left(\frac{4(n+1)}{n^2} \right) = \frac{4(n+1)}{n}.$$

To calculate ②, we use the same argument as in ③:

$$\sum_{j=1}^{n} \frac{2i}{n^3} = n \left(\frac{2i}{n^3} \right) = \frac{2i}{n^2},$$

so that

$$\sum_{i=1}^{n} \sum_{j=1}^{n} \frac{2i}{n^3} = \sum_{i=1}^{n} \frac{2i}{n^2} = \frac{2}{n^2} \sum_{i=1}^{n} i = \frac{2}{n^2} \left[\frac{n(n+1)}{2} \right] = \frac{n+1}{n}.$$

Finally, we have

$$\sum_{i=1}^{n} \sum_{j=1}^{n} V_{ij} \approx 14 + \frac{4(n+1)}{n} + \frac{n+1}{n}.$$

But, if $|P| \to 0$, then the number of subrectangles R_{ij} increases without bound. Hence

$$V = \lim_{|P| \to 0} \sum_{i=1}^{n} \sum_{j=1}^{n} f(x_i^*, y_j^*) \, \Delta x_i \, \Delta y_j = \lim_{n \to \infty} \sum_{i=1}^{n} \sum_{j=1}^{n} f(x_i^*, y_j^*) \, \Delta x_i \, \Delta y_j$$

$$= \lim_{n \to \infty} \left[14 + 4 \left(\frac{n+1}{n} \right) + \frac{n+1}{n} \right] = 14 + 4 + 1 = 19.$$

The calculation we just made was very tedious. Instead of making other calculations like this one, we shall instead define the double integral and shall, in Section 19.2, show how double integrals can be easily calculated.

DEFINITION 1 (The Double Integral). Let $z = f(x, y)$ and let the rectangle R be given by (1). Suppose that

$$\lim_{|P| \to 0} \sum_{i=1}^{n} \sum_{j=1}^{m} f(x_i^*, y_j^*) \, \Delta x_i \, \Delta y_j$$

exists and is independent of the way in which the rectangle R is partitioned and the way in which the points (x_i^*, y_j^*) are chosen. Then the *double integral of f over R,* written $\iint_R f(x, y)\, dx\, dy$, is defined by

$$\iint_R f(x, y)\, dx\, dy = \lim_{|P| \to 0} \sum_{i=1}^{n} \sum_{j=1}^{m} f(x_i^*, y_j^*)\, \Delta x_i\, \Delta y_j. \tag{12}$$

If the limit in (12) exists, then the function f is said to be *integrable* over R.

Remark 1. $\iint_R f(x, y)\, dx\, dy$ is a number, not a function. This is analogous to the fact that the definite integral $\int_a^b f(x)\, dx$ is a number. We will not encounter indefinite double integrals in this book.

Remark 2. The above definition says nothing about volumes (just as the definition of the definite integral in Section 4.3 says nothing about areas). For example, if $f(x, y)$ takes on negative values in R, then the limit in (12) will not represent the volume under the surface $z = f(x, y)$. However, the limit in (12) may still exist and, in that case, f will be integrable over R.

Remark 3. The words "and is independent of the way in which the rectangle R is partitioned and the way in which the points (x_i^*, y_j^*) are chosen" cannot be omitted. We will not give an example of why this is so, but an example could be found along the lines of Problem 4.3.46.

Remark 4. The letters x and y represent, as in the case of the definite integral of a function of one variable, "dummy" variables. They could be replaced by any other letters without changing a thing. Thus, for example, we could write

$$\iint_R f(x, y)\, dx\, dy = \iint_R f(s, t)\, ds\, dt = \iint_R f(u, v)\, du\, dv.$$

As we already stated, we will not calculate any other double integrals in this section but shall wait until Section 19.2 to see how these calculations can be made simple. We should note, however, that the result of Example 1 can now be restated as

$$\iint_R (x + 2y)\, dx\, dy = 19,$$

where R is the rectangle $\{(x, y): 1 \le x \le 2 \text{ and } 3 \le y \le 5\}$.

What functions are integrable over a rectangle R? The following theorem is the double integral analog of Theorem 4.3.1.

THEOREM 1. If f is continuous on R, then f is integrable over R.

We will not give proofs of the theorems stated in this section regarding double integrals. The proofs are similar to, but more complicated than, the analogous proofs for theorems on definite integrals stated in Section 4.3. The proofs of all these theorems can be found in any standard advanced calculus text.†

We now turn to the question of defining double integrals over regions in $\mathbb{R}^2$ which are not rectangular. We will denote a region in $\mathbb{R}^2$ by Ω. The two types of

† See, for example, the book by R. C. Buck, *Advanced Calculus,* McGraw-Hill, New York, 1965.

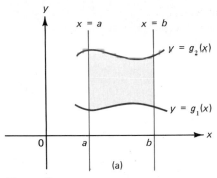
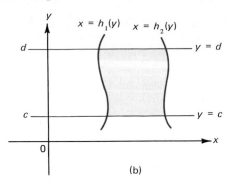

(a)   (b)

Figure 6

regions in which we will be most interested are illustrated in Figure 6. In this figure, g_1 and g_2, h_1 and h_2 denote continuous functions. A more general region Ω is sketched in Figure 7. We assume that the region is bounded. That means that there is a number M such that for every (x, y) in Ω, $|(x, y)| = \sqrt{x^2 + y^2} \leq M$. Since Ω is bounded, we can draw a rectangle R around it. Let f be defined over Ω. We then define a new function F by

$$F(x, y) = \begin{cases} f(x, y), & \text{for } (x, y) \text{ in } \Omega \\ 0, & \text{for } (x, y) \text{ in } R \text{ but not in } \Omega. \end{cases} \tag{13}$$

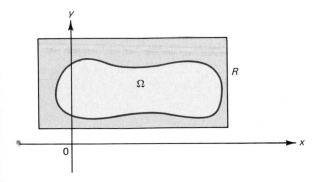

Figure 7

DEFINITION 2. Let f be defined for (x, y) in Ω and let F be defined by (13). Then we write

$$\iint_\Omega f(x, y)\, dx\, dy = \iint_R F(x, y)\, dx\, dy \tag{14}$$

if the integral on the right exists. In this case we say that f is *integrable* over Ω.

> **Remark.** If we divide R into nm subrectangles as in Figure 8, then we can see what is happening. For each subrectangle R_{ij} which lies entirely in Ω, $F = f$ so that the volume of the "parallelepiped" above R_{ij} is given by
>
> $$V_{ij} \approx f(x_i^*, y_j^*)\, \Delta x_i\, \Delta y_j = F(x_i^*, y_j^*)\, \Delta x_i\, \Delta y_j.$$

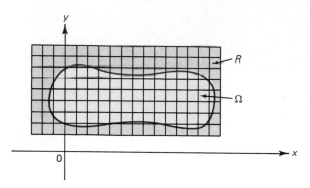

Figure 8

However, if R_{ij} is in R but not in Ω, then $F = 0$ so that

$$V_{ij} \approx F(x_i{}^*, y_j{}^*)\, \Delta x_i\, \Delta y_j = 0.$$

Finally, if R_{ij} is partly in Ω and partly outside of Ω, then there is no real problem since, as $|P| \to 0$, the sum of the volumes above these rectangles (along the boundary of Ω) will approach zero—unless the boundary of Ω is very complicated indeed. Thus we see that the limit of the sum of the volumes of the "parallelepipeds" above R is the same as the limit of the sum of the volumes of the "parallelepipeds" above Ω. This should help explain the "reasonableness" of the expression (14).

THEOREM 2. Let Ω be one of the regions depicted in Figure 6, where the functions g_1 and g_2 or h_1 and h_2 are continuous. Let F be defined by (13). Then if f is continuous over Ω, f is integrable over Ω and its integral is given by (14).

> **Remark 1.** There are some regions Ω which are so complicated that there are functions continuous but not integrable over Ω. We will not concern ourselves with such regions in this book.
>
> **Remark 2.** *If f is nonnegative and integrable over Ω, then*
>
> $$\iint_\Omega f(x, y)\, dx\, dy$$
>
> *is the volume under the surface $z = f(x, y)$ and over the region Ω.*
>
> **Remark 3.** *If the function $f(x, y) = 1$ is integrable over Ω, then*
>
> $$\iint_\Omega 1\, dx\, dy = \iint_\Omega dx\, dy \tag{15}$$
>
> *is equal to the area of the region Ω.* To see this, note that each
>
> $$V_{ij} \approx f(x_i{}^*, y_j{}^*)\, \Delta x_i\, \Delta y_j = \Delta x_i\, \Delta y_j$$
>
> so that the double integral (15) is the limit of the sum of areas of rectangles in Ω.

We close this section by stating five theorems about double integrals. Each one is analogous to a theorem about definite integrals.

THEOREM 3. If f is integrable over Ω, then for any constant c, cf is integrable over Ω and

$$\iint_{\Omega} cf(x, y)\, dx\, dy = c \iint_{\Omega} f(x, y)\, dx\, dy \qquad (16)$$

(see Theorem 4.3.4).

THEOREM 4. If f and g are integrable over Ω, then $f + g$ is integrable over Ω and

$$\iint_{\Omega} [f(x, y) + g(x, y)]\, dx\, dy = \iint_{\Omega} f(x, y)\, dx\, dy + \iint_{\Omega} g(x, y)\, dx\, dy \qquad (17)$$

(see Theorem 4.3.5).

THEOREM 5. If f is integrable over Ω and $\Omega = \Omega_1 \cup \Omega_2$, where Ω_1 and Ω_2 have no points in common except perhaps those of their common boundary, then f is integrable over Ω_1 and Ω_2 and

$$\iint_{\Omega} f(x, y)\, dx\, dy = \iint_{\Omega_1} f(x, y)\, dx\, dy + \iint_{\Omega_2} f(x, y)\, dx\, dy$$

(see Theorem 4.3.3). A typical region Ω is depicted in Figure 9.

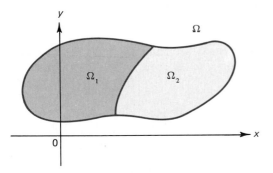

Figure 9

THEOREM 6. If f and g are integrable over Ω and $f(x, y) \le g(x, y)$ for every (x, y) in Ω, then

$$\iint_{\Omega} f(x, y)\, dx\, dy \le \iint_{\Omega} g(x, y)\, dx\, dy \qquad (18)$$

(see Theorem 4.3.7).

THEOREM 7. Let f be integrable over Ω. Suppose that there exist constants m and M such that

$$m \le f(x, y) \le M \qquad (19)$$

for every (x, y) in Ω. Then if A_Ω denotes the area of Ω,

$$mA_\Omega \le \iint_\Omega f(x, y)\, dx\, dy \le MA_\Omega \tag{20}$$

(see Theorem 4.3.8).

Theorem 7 can be useful for estimating double integrals.

EXAMPLE 2. Let Ω be the rectangle $\{(x, y): a \le x \le b \text{ and } c \le y \le d\}$. Find upper and lower bounds for

$$\iint_\Omega \sin(x - 3y^3)\, dx\, dy.$$

SOLUTION. Since $-1 \le \sin(x - 3y^3) \le 1$, and since $A_\Omega = (b - a)(d - c)$, we have, using (20),

$$-(b - a)(d - c) \le \iint_\Omega \sin(x - 3y^3)\, dx\, dy \le (b - a)(d - c).$$

EXAMPLE 3. Let Ω be the disk $\{(x, y): x^2 + y^2 \le 1\}$. Find upper and lower bounds for

$$\iint_\Omega \frac{1}{1 + x^2 + y^2}\, dx\, dy.$$

SOLUTION. Since $0 \le x^2 + y^2 \le 1$ in Ω, we easily see that

$$\frac{1}{2} \le \frac{1}{1 + x^2 + y^2} \le 1.$$

Since the area of the disk is π, we have

$$\frac{\pi}{2} \le \iint_\Omega \frac{1}{1 + x^2 + y^2}\, dx\, dy \le \pi.$$

PROBLEMS 19.1

In Problems 1–8 let Ω denote the rectangle $\{(x, y): 0 \le x \le 3 \text{ and } 1 \le y \le 2\}$. Use the technique employed in Example 1 to calculate the given double integral. Use Theorem 3 and/or 4 where appropriate.

1. $\iint_\Omega (2x + 3y)\, dx\, dy$
2. $\iint_\Omega (x - y)\, dx\, dy$
3. $\iint_\Omega (y - x)\, dx\, dy$
4. $\iint_\Omega (ax + by + c)\, dx\, dy$
5. $\iint_\Omega (x^2 + y^2)\, dx\, dy$
   [*Hint:* Use formula 4.2.7.]
6. $\iint_\Omega (x^2 - y^2)\, dx\, dy$
7. $\iint_\Omega (2x^2 + 3y^2)\, dx\, dy$
8. $\iint_\Omega (ax^2 + by^2 + cx + ey + f)\, dx\, dy.$

In Problems 9–14 let Ω denote the rectangle $\{(x, y): -1 \le x \le 0 \text{ and } -2 \le y \le 3\}$. Calculate the double integral.

9. $\iint_\Omega (x + y) \, dx \, dy$
10. $\iint_\Omega (3x - y) \, dx \, dy$
11. $\iint_\Omega (y - 2x) \, dx \, dy$
12. $\iint_\Omega (x^2 + 2y^2) \, dx \, dy$
13. $\iint_\Omega (y^2 - x^2) \, dx \, dy$
14. $\iint_\Omega (3x^2 - 5y^2) \, dx \, dy$.

In Problems 15–19 use Theorem 7 to obtain upper and lower bounds for the given integral.

15. $\iint_\Omega (x^5 y^2 + xy) \, dx \, dy$, where Ω is the rectangle $\{(x, y): 0 \le x \le 1 \text{ and } 1 \le y \le 2\}$
16. $\iint_\Omega e^{-(x^2 + y^2)} \, dx \, dy$, where Ω is the disk $x^2 + y^2 \le 4$
*17. $\iint_\Omega [(x - y)/(4 - x^2 - y^2)] \, dx \, dy$, where Ω is the disk $x^2 + y^2 \le 1$
18. $\iint_\Omega \cos(\sqrt{|x|} - \sqrt{|y|}) \, dx \, dy$, where Ω is the region of Problem 17
19. $\iint_\Omega \ln(1 + x + y) \, dx \, dy$, where Ω is the region bounded by the lines $y = x$, $y = 1 - x$ and the x-axis.

*20. Let Ω be the region of Problem 19. Which is greater:

$$\iint_\Omega e^{(x^2 + y^2)} \, dx \, dy \quad \text{or} \quad \iint_\Omega (x^2 + y^2) \, dx \, dy?$$

19.2 The Calculation of Double Integrals

In this section we derive an easy method for calculating $\iint_\Omega f(x, y) \, dx \, dy$, where Ω is one of the regions depicted in Figure 18.1.6.

We begin, as in Section 1, by considering

$$\iint_R f(x, y) \, dx \, dy, \tag{1}$$

where R is the rectangle

$$R = \{(x, y): a \le x \le b \text{ and } c \le y \le d\}. \tag{2}$$

If $z = f(x, y) \ge 0$ for (x, y) in R, then the double integral in (1) is the volume under the surface $z = f(x, y)$ and over the rectangle R in the xy-plane. We now calculate this volume by partitioning the x-axis taking "slices" parallel to the yz-plane. This is illustrated in Figure 1. We can approximate the volume by adding up the volumes of

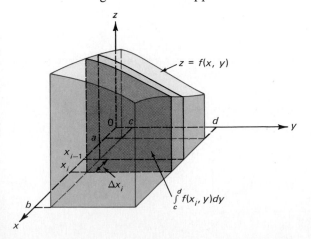

Figure 1

the various "slices." The face of each "slice" lies in the plane $x = x_i$ and the volume of the ith slice is the area of its face times its thickness, Δx_i. What is the area of the face? If x is fixed, then $z = f(x, y)$ can be thought of as a curve lying in the plane $x = x_i$. Thus the area of the ith face is the area bounded by this curve, the y-axis, and

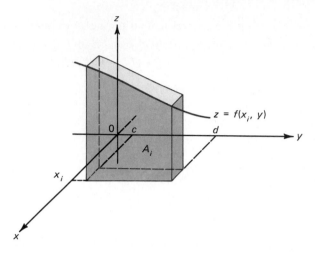

Figure 2

the lines $y = c$ and $y = d$. This is sketched in Figure 2. If $f(x_i, y)$ is a continuous function of y, then the area of the ith face, denoted A_i, is given by

$$A_i = \int_c^d f(x_i, y) \, dy.$$

Then the volume of the ith slice is given by

$$V_i = \left\{ \int_c^d f(x_i, y) \, dy \right\} \Delta x_i$$

so that, adding up these "subvolumes" and taking the limit as the norm of the partition of the x-axis approaches zero, we obtain

$$V = \int_a^b \left\{ \int_c^d f(x, y) \, dy \right\} dx. \tag{3}$$

The expression in (3) is called a *repeated integral* or *iterated integral*. Since we also have

$$V = \iint_R f(x, y) \, dx \, dy$$

we obtain

$$\boxed{\iint_R f(x, y) \, dx \, dy = \int_a^b \left\{ \int_c^d f(x, y) \, dy \right\} dx.} \tag{4}$$

Remark 1. Usually we will write equation (4) without braces. We then have

$$\iint_R f(x, y)\, dx\, dy = \int_a^b \int_c^d f(x, y)\, dy\, dx. \tag{5}$$

Remark 2. We should emphasize that the first integration in $\int_a^b \int_c^d f(x, y)\, dy\, dx$ is performed by treating x as a constant.

Similarly, if we instead begin by partitioning the y-axis, we find that the area of the face of a "slice" lying in the plane $y = y_i$ is given by

$$A_i = \int_a^b f(x, y_i)\, dx$$

so that

$$V = \int_c^d \left\{ \int_a^b f(x, y)\, dx \right\} dy \tag{6}$$

and

$$\iint_R f(x, y)\, dx\, dy = \int_c^d \int_a^b f(x, y)\, dx\, dy. \tag{7}$$

EXAMPLE 1.   Calculate the volume under the plane $z = x + 2y$ and over the rectangle

$$R = \{(x, y)\colon 1 \le x \le 2 \text{ and } 3 \le y \le 5\}.$$

SOLUTION. We calculated this volume in Example 19.1.1. Using equation (5), we have

$$V = \iint_R (x + 2y)\, dx\, dy = \int_1^2 \left\{ \int_3^5 (x + 2y)\, dy \right\} dx$$

$$= \int_1^2 \left\{ xy + y^2 \Big|_3^5 \right\} dx = \int_1^2 [(5x + 25) - (3x + 9)]\, dx$$

$$= \int_1^2 (2x + 16)\, dx = (x^2 + 16x) \Big|_1^2 = 19.$$

Similarly, using equation (7),

$$V = \int_3^5 \left\{ \int_1^2 (x + 2y)\, dx \right\} dy = \int_3^5 \left\{ \frac{x^2}{2} + 2yx \Big|_1^2 \right\} dy$$

$$= \int_3^5 \left[(2 + 4y) - \left(\frac{1}{2} + 2y \right) \right] dy = \int_3^5 \left(2y + \frac{3}{2} \right) dy$$

$$= \left(y^2 + \frac{3}{2} y \right) \Big|_3^5 = 19.$$

EXAMPLE 2. Calculate the volume of the region beneath the surface $z = xy^2 + y^3$ and over the rectangle $R = \{(x, y): 0 \le x \le 2 \text{ and } 1 \le y \le 3\}$.

SOLUTION. Using equation (5),

$$V = \int_0^2 \int_1^3 (xy^2 + y^3)\, dy\, dx = \int_0^2 \left(\frac{xy^3}{3} + \frac{y^4}{4} \Big|_1^3 \right) dx$$

$$= \int_0^2 \left[\left(9x + \frac{81}{4} \right) - \left(\frac{x}{3} + \frac{1}{4} \right) \right] dx = \int_0^2 \left(\frac{26}{3}x + 20 \right) dx$$

$$= \left(\frac{13x^2}{3} + 20x \right) \Big|_0^2 = \frac{52}{3} + 40 = \frac{172}{3}.$$

We calculate the same volume using equation (7). This is not necessary, but it does provide us with a method for checking our answer.

$$V = \int_1^3 \int_0^2 (xy^2 + y^3)\, dx\, dy = \int_1^3 \left\{ \frac{x^2}{2} y^2 + xy^3 \Big|_0^2 \right\} dy$$

$$= \int_1^3 (2y^2 + 2y^3)\, dy = \left(\frac{2y^3}{3} + \frac{y^4}{2} \right) \Big|_1^3 = \left\{ \left(18 + \frac{81}{2} \right) - \left(\frac{2}{3} + \frac{1}{2} \right) \right\} = \frac{172}{3}.$$

We now extend our results to more general regions. Let

$$\Omega = \{(x, y): a \le x \le b \text{ and } g_1(x) \le y \le g_2(x)\}. \tag{8}$$

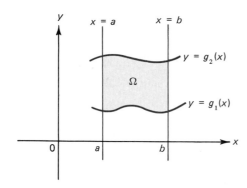

Figure 3

This region is sketched in Figure 3. We assume that for every x in $[a, b]$,

$$g_1(x) \le g_2(x). \tag{9}$$

If we partition the x-axis as before, then we obtain slices lying in the planes $x = x_i$, a typical one of which is sketched in Figure 4. Then

$$A_i = \int_{g_1(x_i)}^{g_2(x_i)} f(x_i, y)\, dy, \qquad V_i = \left\{ \int_{g_1(x_i)}^{g_2(x_i)} f(x_i, y)\, dy \right\} \Delta x_i$$

and

$$V = \iint_\Omega f(x, y)\, dx\, dy = \int_a^b \int_{g_1(x)}^{g_2(x)} f(x, y)\, dy\, dx. \tag{10}$$

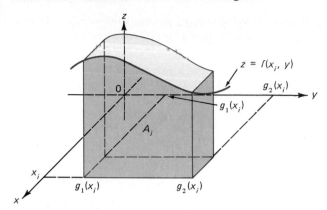

Figure 4

Similarly, let

$$\Omega = \{(x, y): g_1(y) \le x \le g_2(y) \text{ and } c \le y \le d\} \tag{11}$$

(see Figure 5). Then

$$V = \int_c^d \int_{g_1(y)}^{g_2(y)} f(x, y) \, dx \, dy. \tag{12}$$

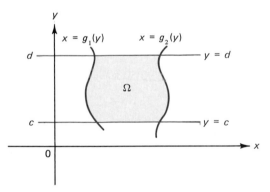

Figure 5

We summarize these results in the following theorem:

THEOREM 1.   Let f be continuous over a region Ω given by equation (8) or (11).

(i)   If Ω is of the form (8) where g_1 and g_2 are continuous, then

$$\iint_\Omega f(x, y) \, dx \, dy = \int_a^b \int_{g_1(x)}^{g_2(x)} f(x, y) \, dy \, dx.$$

(ii)   If Ω is of the form (11), where g_1 and g_2 are continuous, then

$$\iint_\Omega f(x, y) \, dx \, dy = \int_c^d \int_{g_1(y)}^{g_2(y)} f(x, y) \, dx \, dy.$$

Remark 1. We have not actually proved this theorem here, but have merely indicated why it should be so. A rigorous proof can be found in any advanced calculus text.

Remark 2. Note that this theorem says nothing about volume. It can be used to calculate any double integral if the hypotheses of the theorem are satisfied and if each function being integrated has an indefinite integral which can be written in terms of elementary functions.

EXAMPLE 3. Find the volume of the solid under the surface $z = x^2 + y^2$ and lying above the region

$$\Omega = \{(x, y): 0 \le x \le 1 \text{ and } x^2 \le y \le \sqrt{x}\}.$$

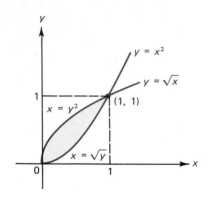

Figure 6

SOLUTION. Ω is sketched in Figure 6. We see that $0 \le x \le 1$ and $x^2 \le y \le \sqrt{x}$. Then using (10),

$$V = \int_0^1 \int_{x^2}^{\sqrt{x}} (x^2 + y^2) \, dy \, dx = \int_0^1 \left\{ x^2 y + \frac{y^3}{3} \Big|_{x^2}^{\sqrt{x}} \right\} dx$$

$$= \int_0^1 \left\{ x^2 \sqrt{x} + \frac{(\sqrt{x})^3}{3} - \left(x^2 \cdot x^2 + \frac{(x^2)^3}{3} \right) \right\} dx$$

$$= \int_0^1 \left(x^{5/2} + \frac{x^{3/2}}{3} - x^4 - \frac{x^6}{3} \right) dx$$

$$= \frac{2x^{7/2}}{7} + \frac{2x^{5/2}}{15} - \frac{x^5}{5} - \frac{x^7}{21} \Big|_0^1 = \frac{2}{7} + \frac{2}{15} - \frac{1}{5} - \frac{1}{7} = \frac{8}{105}.$$

We can calculate this integral in another way. We note that x varies between the curves $x = y^2$ and $x = \sqrt{y}$. Then, using (12), since $0 \le y \le 1$ and $y^2 \le x \le \sqrt{y}$,

$$V = \int_0^1 \int_{y^2}^{\sqrt{y}} (x^2 + y^2) \, dx \, dy$$

which is easily seen to be equal to 8/105.

EXAMPLE 4. Let $f(x, y) = x^2 y$. Calculate the integral of f over the region bounded by the x-axis and the semicircle $x^2 + y^2 = 4$, $y \ge 0$.

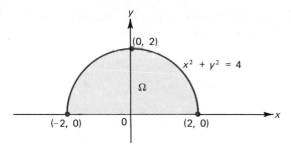

Figure 7

SOLUTION. The region of integration is sketched in Figure 7. Using equation (8), we see that $0 \le y \le \sqrt{4 - x^2}$, $-2 \le x \le 2$, so that, integrating first with respect to y, we obtain

$$\iint_\Omega x^2y \, dx \, dy = \int_{-2}^{2} \int_{0}^{\sqrt{4-x^2}} x^2y \, dy \, dx$$

$$= \int_{-2}^{2} \left\{ \frac{x^2y^2}{2} \, \Big|_{0}^{\sqrt{4-x^2}} \right\} dx = \int_{-2}^{2} \frac{x^2(4 - x^2)}{2} \, dx$$

$$= \int_{-2}^{2} \left(2x^2 - \frac{x^4}{2} \right) dx = \left(\frac{2x^3}{3} - \frac{x^5}{10} \right) \Big|_{-2}^{2} = \frac{64}{15}.$$

We can also use equation (11) and integrate first with respect to x. Then $-\sqrt{4 - y^2} \le x \le \sqrt{4 - y^2}$, $0 \le y \le 2$, and

$$V = \int_{0}^{2} \int_{-\sqrt{4-x^2}}^{\sqrt{4-y^2}} x^2y \, dx \, dy = \int_{0}^{2} \left(\frac{x^3y}{3} \, \Big|_{-\sqrt{4-y_2}}^{\sqrt{4-y^2}} \right) dy$$

$$= \int_{0}^{2} \frac{2}{3} (4 - y^2)^{3/2}y \, dy = \frac{-2}{15} (4 - y^2)^{5/2} \, \Big|_{0}^{2} = \frac{64}{15}.$$

EXAMPLE 5. Evaluate $\displaystyle\int_{1}^{2} \int_{1}^{x^2} \frac{x}{y} \, dy \, dx$.

SOLUTION

$$\int_{1}^{2} \int_{1}^{x^2} \frac{x}{y} \, dy \, dx = \int_{1}^{2} \left\{ x \ln y \, \Big|_{1}^{x^2} \right\} dx = \int_{1}^{2} x \ln x^2 \, dx = \int_{1}^{2} 2x \ln x \, dx.$$

It is necessary to use integration by parts to complete the problem. Setting $u = \ln x$ and $dv = 2x \, dx$, we have $du = (1/x) \, dx$, $v = x^2$ and

$$\int_{1}^{2} 2x \ln x \, dx = x^2 \ln x \, \Big|_{1}^{2} - \int_{1}^{2} x \, dx = 4 \ln 2 - \frac{x^2}{2} \, \Big|_{1}^{2} = 4 \ln 2 - \frac{3}{2}.$$

There is an easier way to calculate the double integral. We simply *reverse the order of integration*. The region of integration is sketched in Figure 8. If we want to integrate first with respect to x, we note that we can describe the region by

$$\Omega = \{(x, y) : \sqrt{y} \le x \le 2 \text{ and } 1 \le y \le 4\}.$$

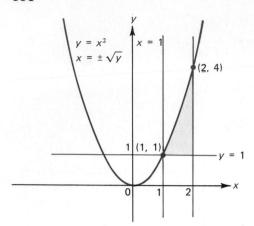

Figure 8

Then

$$\int_1^2 \int_1^{x^2} \frac{x}{y}\, dy\, dx = \iint_\Omega \frac{x}{y}\, dx\, dy = \int_1^4 \int_{\sqrt{y}}^2 \frac{x}{y}\, dx\, dy = \int_1^4 \left\{ \frac{x^2}{2y} \Big|_{\sqrt{y}}^2 \right\} dx$$

$$= \int_1^4 \left(\frac{2}{y} - \frac{1}{2} \right) dy = \left(2\ln y - \frac{y}{2} \right) \Big|_1^4 = 2\ln 4 - \frac{3}{2}$$

$$= 4\ln 2 - \frac{3}{2}.$$

Note that, in this case, it is much easier to integrate first with respect to x. •

EXAMPLE 6. Use double integration to find the area in the first quadrant bounded by the curves $y = \sin x$, $y = \cos x$ and the y-axis.

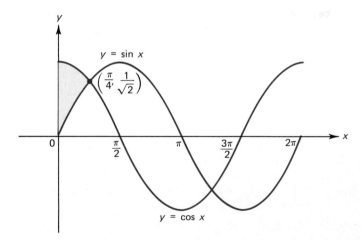

Figure 9

SOLUTION. The required area is sketched in Figure 9. From formula (19.1.15),

$$A = \iint_\Omega dx\, dy.$$

But $\Omega = \{(x, y): 0 \leq x \leq \pi/4 \text{ and } \sin x \leq y \leq \cos x\}$. Thus

$$A = \int_0^{\pi/4} \int_{\sin x}^{\cos x} dy\, dx = \int_0^{\pi/4} \left\{ y \Big|_{\sin x}^{\cos x} \right\} dx = \int_0^{\pi/4} (\cos x - \sin x)\, dx$$

$$= \sin x + \cos x \Big|_0^{\pi/4} = \sqrt{2} - 1.$$

EXAMPLE 7. Find the volume in the first octant bounded by the three coordinate planes and the surface $z = 1/(1 + x + 3y)^3$.

SOLUTION. The solid here extends over the infinite region $\{(x, y): 0 \leq x \leq \infty$ and $0 \leq y \leq \infty\}$. Thus

$$V = \int_0^\infty \int_0^\infty \frac{1}{(1 + x + 3y)^3}\, dx\, dy = \int_0^\infty \lim_{N \to \infty} \left\{ -\frac{1}{2(1 + x + 3y)^2} \Big|_0^N \right\} dy$$

$$= \int_0^\infty \frac{1}{2(1 + 3y)^2}\, dy = \lim_{N \to \infty} \left[-\frac{1}{6(1 + 3y)} \Big|_0^N \right] = \frac{1}{6}.$$

Note that improper double integrals can be treated in the same way that we treat improper "single" integrals.

EXAMPLE 8. Find the volume of the solid bounded by the circular paraboloid $x = y^2 + z^2$ and the plane $x = 1$.

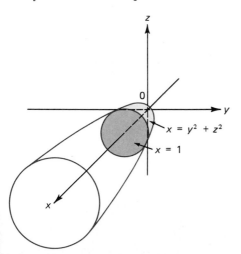

Figure 10

SOLUTION. The solid is sketched in Figure 10. When $x = 1$, we obtain the circle $\Omega: y^2 + z^2 = 1$. Thus we need to integrate the function $x = y^2 + z^2$ over this circle. We have

$$V = \iint_\Omega (y^2 + z^2)\, dy\, dz = \int_{-1}^1 \int_{-\sqrt{1-z^2}}^{\sqrt{1-z^2}} (y^2 + z^2)\, dy\, dz$$

$$= \int_{-1}^1 \left\{ \frac{y^3}{3} + z^2 y \Big|_{-\sqrt{1-z^2}}^{\sqrt{1-z^2}} \right\} dz = \int_{-1}^1 \left\{ \frac{2}{3}(1 - z^2)^{3/2} + 2z^2 \sqrt{1 - z^2} \right\} dz.$$

To integrate this we set $z = \sin \theta$. This yields

$$V = \int_{-\pi/2}^{\pi/2} \left\{ \frac{2}{3} \cos^3 \theta + 2 \sin^2 \theta \cos \theta \right\} \cos \theta \, d\theta$$

$$= \int_{-\pi/2}^{\pi/2} \left\{ \frac{2}{3} \cos^4 \theta + 2 \sin^2 \theta \cos^2 \theta \right\} d\theta$$

$$= \int_{-\pi/2}^{\pi/2} \left\{ \frac{1}{6} (1 + \cos 2\theta)^2 + \frac{1}{2} (1 - \cos 2\theta)(1 + \cos 2\theta) \right\} d\theta$$

$$= \int_{-\pi/2}^{\pi/2} \left(\frac{1}{6} + \frac{1}{3} \cos 2\theta + \frac{1}{6} \cos^2 2\theta + \frac{1}{2} - \frac{1}{2} \cos^2 2\theta \right) d\theta$$

$$= \int_{-\pi/2}^{\pi/2} \left\{ \frac{2}{3} + \frac{1}{3} \cos 2\theta - \frac{1}{6} (1 + \cos 4\theta) \right\} d\theta$$

$$= \int_{-\pi/2}^{\pi/2} \left(\frac{1}{2} + \frac{1}{3} \cos 2\theta - \frac{1}{6} \cos 4\theta \right) d\theta = \frac{\pi}{2}.$$

EXAMPLE 9.   Find the volume of the solid bounded by the coordinate planes and the plane $2x + y + z = 2$.

SOLUTION.   We have $z = 2 - 2x - y$ and this must be integrated over the region in the xy-plane bounded by the line $2x + y = 2$ (obtained when $z = 0$) and the x- and y-axes. See Figure 11. We therefore have

$$V = \int_0^1 \int_0^{2-2x} (2 - 2x - y) \, dy \, dx = \int_0^1 \left\{ 2y - 2xy - \frac{y^2}{2} \Big|_0^{2-2x} \right\} dx$$

$$= \int_0^1 \left\{ 2(2 - 2x) - 2x(2 - 2x) - \frac{(2 - 2x)^2}{2} \right\} dx$$

$$= \int_0^1 (2x^2 - 4x + 2) \, dx = \frac{2x^3}{3} - 2x^2 + 2x \Big|_0^1 = \frac{2}{3}.$$

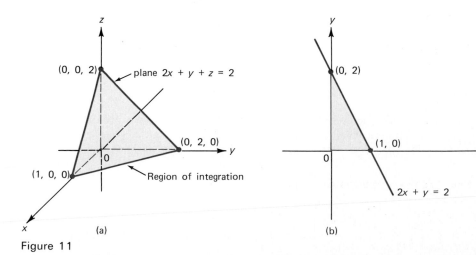

(0, 0, 2)   plane $2x + y + z = 2$

(0, 2, 0)

(1, 0, 0)   Region of integration

(a)

(0, 2)

(1, 0)

$2x + y = 2$

(b)

Figure 11

PROBLEMS 19.2

In Problems 1–23 evaluate the given double integral.

1. $\int_0^1 \int_0^9 xy^2 \, dx \, dy$

2. $\int_{-1}^3 \int_2^4 (x^2 - y^3) \, dy \, dx$

3. $\int_2^5 \int_0^4 e^{(x-y)} \, dx \, dy$

4. $\int_0^1 \int_{x^2}^x x^3 y \, dy \, dx$

5. $\int_2^4 \int_{1+y}^{2+3y} (x - y^2) \, dx \, dy$

6. $\int_{\pi/4}^{\pi/3} \int_{\sin x}^{\cos x} (x + 2y) \, dy \, dx$

7. $\int_0^3 \int_{-\sqrt{9-y^2}}^{\sqrt{9-y^2}} x^2 y \, dy \, dx$

8. $\int_1^2 \int_{y^5}^{3y^5} \frac{1}{x} \, dx \, dy.$

9. $\iint_\Omega (x^2 + y^2) \, dx \, dy$, where $\Omega = \{(x, y): 1 \le x \le 2 \text{ and } -1 \le y \le 1\}$.

10. $\iint_\Omega 2xy \, dx \, dy$, where $\Omega = \{(x, y): 0 \le x \le 4 \text{ and } 1 \le y \le 3\}$.

11. $\iint_\Omega (x - y)^2 \, dx \, dy$, where $\Omega = \{(x, y): -2 \le x \le 2 \text{ and } 0 \le y \le 1\}$.

12. $\iint_\Omega \sin(2x + 3y) \, dx \, dy$, where $\Omega = \{(x, y): 0 \le x \le \pi/6 \text{ and } 0 \le y \le \pi/18\}$.

13. $\iint_\Omega x e^{(x^2+y)} \, dx \, dy$, where Ω is the region of Problem 10.

14. $\iint_\Omega (x - y^2) \, dx \, dy$, where Ω is the region in the first quadrant bounded by the x-axis, the y-axis, and the unit circle.

15. $\iint_\Omega (x^2 + y) \, dx \, dy$, where Ω is the region of Problem 14.

16. $\iint_\Omega (x^3 - y^3) \, dx \, dy$, where Ω is the region of Problem 14.

17. $\iint_\Omega (x + 2y) \, dx \, dy$, where Ω is the triangular region bounded by the lines $y = x$, $y = 1 - x$, and the x-axis.

18. $\iint_\Omega e^{x+2y} \, dx \, dy$, where Ω is the region of Problem 17.

19. $\iint_\Omega (x^2 + y) \, dx \, dy$, where Ω is the region in the first quadrant between the parabolas $y = x^2$ and $y = 1 - x^2$.

20. $\iint_\Omega (1/\sqrt{y}) \, dx \, dy$, where Ω is the region of Problem 19.

21. $\iint_\Omega (y/\sqrt{x^2 + y^2}) \, dx \, dy$, where $\Omega = \{(x, y): 1 \le x \le y \text{ and } 1 \le y \le 2\}$.

22. $\iint_\Omega [e^{-y}/(1 + x^2)] \, dx \, dy$, where Ω is the first quadrant.

23. $\iint_\Omega (x + y) e^{-(x+y)} \, dy \, dx$, where Ω is the first quadrant.

In Problems 24–33 (a) change the order of integration. Then (b) evaluate the given integral, and (c) sketch the region over which the integral is taken.

24. $\int_0^2 \int_{-1}^3 dx \, dy$

25. $\int_0^4 \int_{-5}^8 (x + y) \, dy \, dx$

26. $\int_2^4 \int_1^y (y^3/x^3) \, dx \, dy$

27. $\int_0^1 \int_0^x dy \, dx$

28. $\int_0^1 \int_x^1 dy \, dx$

29. $\int_0^{\pi/2} \int_0^{\cos y} y \, dx \, dy$

30. $\int_0^2 \int_0^{\sqrt{4-y^2}} (4 - x^2)^{3/2} \, dx \, dy$

31. $\int_0^1 \int_{\sqrt{x}}^{\sqrt[3]{x}} (1 + y^6) \, dy \, dx$

32. $\int_0^1 \int_{\sqrt{y}}^1 \sqrt{3 - x^3} \, dx \, dy$

33. $\int_0^\infty \int_x^\infty [1/(1 + y^2)^{7/5}] \, dy \, dx.$

34. Show that if both integrals exist then

$$\int_0^\infty \int_0^x f(x, y) \, dy \, dx = \int_0^\infty \int_y^\infty f(x, y) \, dx \, dy.$$

[*Hint:* Draw a picture.]

In Problems 35–44 find the volume of the given solid.

35. The solid bounded by the plane $x + y + z = 3$ and the three coordinate planes
*36. The solid bounded by the planes $x = 0$, $z = 0$, $x + 2y + z = 6$ and $x - 2y + z = 6$
37. The solid bounded by the cylinders $x^2 + y^2 = 4$ and $y^2 + z^2 = 4$
38. The solid bounded by the cylinder $x^2 + z^2 = 1$ and the planes $y = 0$ and $y = 2$
39. The ellipsoid $x^2 + 4y^2 + 9z^2 = 36$
40. The solid bounded above by the sphere $x^2 + y^2 + z^2 = 9$ and below by the plane $z = \sqrt{5}$
41. The solid bounded by the planes $y = 0$, $y = x$, and the cylinder $x + z^2 = 2$
*42. The solid bounded by the parabolic cylinder $x = z^2$ and the planes $y = 1, y = 5, z = 1$ and $x = 0$.
*43. The solid bounded by the paraboloid $y = x^2 + z^2$ and the plane $x + y = 3$
44. The solid bounded by the surface $z = e^{-(x+y)}$ and the three coordinate planes.

*45. Use a double integral to find the area bounded by the x-axis and the curves $y = x^3 + 1$ and $y = 3 - x^2$.
46. Use a double integral to find the area in the first quadrant bounded by the curves $y = x^{1/m}$ and $y = x^{1/n}$, where m and n are positive and $n > m$.
47. Let $f(x, y) = g(x)h(y)$, where g and h are continuous. Let Ω be the rectangle $\{(x, y): a \le x \le b$ and $c \le y \le d\}$. Show that

$$\iint_\Omega f(x, y)\, dx\, dy = \left\{ \int_a^b g(x)\, dx \right\} \cdot \left\{ \int_c^d h(y)\, dy \right\}.$$

48. Sketch the solid whose volume is given by

$$V = \int_1^3 \int_0^2 (x + 3y)\, dx\, dy.$$

*49. Sketch the solid whose volume is given by

$$V = \int_0^1 \int_{x^2}^{\sqrt{x}} \sqrt{x^2 + y^2}\, dy\, dx.$$

*19.3 Density, Mass, and Center of Mass

Let $\rho(x, y)$ denote the density of a plane object (like a thin lamina, for example). Suppose that the object occupies a region Ω in the xy-plane. Then the mass of a small rectangle of sides Δx and Δy centered at the point (x, y) is approximated by

$$\rho(x, y)\, \Delta x\, \Delta y \tag{1}$$

and the total mass of the object is

$$\mu = \iint_\Omega \rho(x, y)\, dx\, dy. \tag{2}$$

Compare this with the formula for the mass of an object lying along the x-axis with density $\rho(x)$ (see equation (4.4.10)).

In Sections 9.4 and 9.5 we showed how to calculate the first moment and center of mass of an object around the x- and y-axes. For example, we defined

$$M_y = \int_a^b x\rho(x)\, dx \tag{3}$$

to be the first moment about the y-axis when we had a system of masses distributed along the x-axis. Similarly, we calculated the x-coordinate of the center of mass of the object to be

$$\bar{x} = \frac{\displaystyle\int_a^b x\rho(x)\,dx}{\displaystyle\int_a^b \rho(x)\,dx} = \frac{\text{first moment about } y\text{-axis}}{\text{mass}} = \frac{M_y}{\mu} \tag{4}$$

(see equation (9.4.9)).

When, in Section 9.5, we calculated the centroid of a plane region, we found that it was necessary to assume that the region had a constant area density ρ. However, by using double integrals we can get away from this restriction. Consider the plane region whose mass is given by (2). Then we define

$$M_y = \text{first moment around } y\text{-axis} = \iint_\Omega x\rho(x,y)\,dx\,dy. \tag{5}$$

These formulas are derived in essentially the same way that we derived formulas (9.5.9) and (9.5.10). Look at Figure 1. The first moment about the y-axis of a small

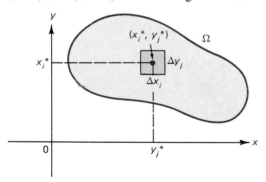

Figure 1

rectangle centered at (x, y) is given by

$$x_i \rho(x_i^*, y_i^*)\,\Delta x_i\,\Delta y_i \tag{6}$$

and if we add up these moments for all such "subrectangles" and take a limit, we arrive at equation (5). Finally, we define the center of mass of the plane region to be the point $(\bar{x}, \bar{y})$ where

$$\bar{x} = \frac{M_y}{\mu} = \frac{\iint_\Omega x\rho(x,y)\,dx\,dy}{\iint_\Omega \rho(x,y)\,dx\,dy} \tag{7}$$

and

$$\bar{y} = \frac{M_x}{\mu} = \frac{\iint_\Omega y\rho(x,y)\,dx\,dy}{\iint_\Omega \rho(x,y)\,dx\,dy}. \tag{8}$$

As before, it ρ is constant, the center of mass of a region is called the *centroid* of that region.

EXAMPLE 1. A plano lamina has the shape of the triangle bounded by the lines $y = x$, $y = 2 - x$ and the x-axis. Its density function is given by $\rho(x, y) = 1 + 2x + y$. If distance is measured in meters and mass is measured in kilograms, find the mass and center of mass of the lamina.

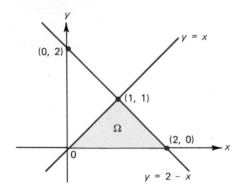

Figure 2

SOLUTION. The region is sketched in Figure 2. The mass is given by

$$\mu = \iint_{\Omega} \rho(x, y)\, dx\, dy = \int_0^1 \int_y^{2-y} (1 + 2x + y)\, dx\, dy$$

$$= \int_0^1 \left\{ x + x^2 + xy \Big|_y^{2-y} \right\} dy = \int_0^1 (6 - 4y - 2y^2)\, dy$$

$$= \left(6y - 2y^2 - \frac{2y^3}{3} \right) \Big|_0^1 = \frac{10}{3} \quad \text{kg.}$$

Then

$$M_y = \int_0^1 \int_y^{2-y} x(1 + 2x + y)\, dx\, dy$$

$$= \int_0^1 \int_y^{2-y} (x + 2x^2 + xy)\, dx\, dy = \int_0^1 \left\{ \frac{x^2}{2} + \frac{2x^3}{3} + \frac{x^2 y}{2} \Big|_y^{2-y} \right\} dy$$

$$= \int_0^1 \left(\frac{22}{3} - 8y + 2y^2 - \frac{4y^3}{3} \right) dy$$

$$= \frac{22}{3} y - 4y^2 + \frac{2y^3}{3} - \frac{y^4}{3} \Big|_0^1 = \frac{11}{3} \quad \text{kg-m.}$$

$$M_x = \int_0^1 \int_y^{2-y} y\,(1 + 2x + y)\, dx\, dy = \int_0^1 \left\{ xy + x^2 y + xy^2 \Big|_y^{2-y} \right\} dy$$

$$= \int_0^1 (6y - 4y^2 - 2y^3)\, dy = 3y^2 - \frac{4y^3}{3} - \frac{y^4}{2} \Big|_0^1 = \frac{7}{6} \quad \text{kg-m.}$$

Thus

$$\bar{x} = \frac{M_y}{\mu} = \frac{11/3}{10/3} = \frac{11}{10} \text{ m} \quad \text{and} \quad \bar{y} = \frac{M_x}{\mu} = \frac{7/6}{10/3} = \frac{7}{20} \text{ m}.$$

EXAMPLE 2. The density at any point on a semicircular plane lamina is proportional to the square of the distance from the point to the center of the circle. Find the center of mass of the lamina.

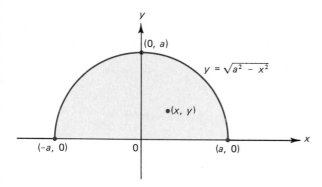

Figure 3

SOLUTION. We place the lamina with its center at the origin as in Figure 3 and denote its radius by a. We have $\rho(x, y) = \alpha(x^2 + y^2)$, where α is a constant of proportionality. Then

$$\mu = \int_{-a}^{a} \int_{0}^{\sqrt{a^2-x^2}} \alpha(x^2 + y^2)\, dy\, dx = \alpha \int_{-a}^{a} \left[x^2 y + \frac{y^3}{3} \Big|_{0}^{\sqrt{a^2-x^2}} \right] dx$$

$$= \alpha \int_{-a}^{a} \left[x^2 \sqrt{a^2 - x^2} + \frac{(a^2 - x^2)^{3/2}}{3} \right] dx.$$

Making the substitution $x = a \sin \theta$, we obtain

$$\mu = \alpha a^4 \int_{-\pi/2}^{\pi/2} \left(\sin^2 \theta \cos \theta + \frac{\cos^3 \theta}{3} \right) \cos \theta\, d\theta$$

$$= \alpha a^4 \int_{-\pi/2}^{\pi/2} \left(\sin^2 \theta \cos^2 \theta + \frac{\cos^4 \theta}{3} \right) d\theta$$

$$= \alpha a^4 \int_{-\pi/2}^{\pi/2} \left[\frac{(1 - \cos 2\theta)(1 + \cos 2\theta)}{4} + \frac{(1 + \cos 2\theta)^2}{12} \right] d\theta$$

$$= \frac{\alpha a^4}{4} \int_{-\pi/2}^{\pi/2} \left[(1 - \cos^2 2\theta) + \frac{(1 + 2 \cos 2\theta + \cos^2 2\theta)}{3} \right] d\theta$$

$$= \frac{\alpha a^4}{4} \int_{-\pi/2}^{\pi/2} \left(\frac{4}{3} + \frac{2 \cos 2\theta}{3} - \frac{2 \cos^2 2\theta}{3} \right) d\theta$$

$$= \frac{\alpha a^4}{4} \int_{-\pi/2}^{\pi/2} \left[\frac{4}{3} + \frac{2 \cos 2\theta}{3} - \left(\frac{1 + \cos 4\theta}{3} \right) \right] d\theta$$

$$= \frac{\alpha a^4}{4} \int_{-\pi/2}^{\pi/2} \left(1 + \frac{2 \cos 2\theta}{3} - \frac{\cos 4\theta}{3} \right) d\theta = \frac{\alpha \pi a^4}{4}.$$

Similarly,

$$M_y = \int_{-a}^{a} \int_{0}^{\sqrt{a^2-x^2}} \alpha x(x^2 + y^2) \, dy \, dx = \alpha \int_{-a}^{a} \left\{ x^3 y + \frac{xy^3}{3} \Big|_{0}^{\sqrt{a^2-x^2}} \right\} dx$$

$$= \alpha \int_{-a}^{a} \left[x^3 \sqrt{a^2 - x^2} + \frac{x(a^2 - x^2)^{3/2}}{3} \right] dx$$

$$= \alpha a^5 \int_{-\pi/2}^{\pi/2} \left(\sin^3 \theta \cos^2 \theta + \frac{\cos^4 \theta \sin \theta}{3} \right) d\theta$$

$$= \alpha a^5 \int_{-\pi/2}^{\pi/2} \left[\sin \theta (1 - \cos^2 \theta) \cos^2 \theta + \frac{\cos^4 \theta \sin \theta}{3} \right] d\theta$$

$$= \alpha a^5 \int_{-\pi/2}^{\pi/2} \left(\cos^2 \theta \sin \theta - \frac{2 \cos^4 \theta \sin \theta}{3} \right) d\theta$$

$$= \alpha a^5 \left\{ \frac{\cos^3 \theta}{3} - \frac{2 \cos^5 \theta}{15} \right\} \Big|_{-\pi/2}^{\pi/2} = 0.$$

Because of symmetry, this result should not be at all surprising. Finally, we calculate

$$M_x = \alpha \int_{-a}^{a} \int_{0}^{\sqrt{a^2-x^2}} y(x^2 + y^2) \, dy \, dx = \alpha \int_{-a}^{a} \left\{ \frac{x^2 y^2}{2} + \frac{y^4}{4} \Big|_{0}^{\sqrt{a^2-x^2}} \right\} dx$$

$$= \alpha \int_{-a}^{a} \left[\frac{x^2}{2} (a^2 - x^2) + \frac{(a^2 - x^2)^2}{4} \right] dx$$

$$= \frac{\alpha}{4} \int_{-a}^{a} (a^4 - x^4) \, dx = \frac{\alpha}{4} \left\{ a^4 x - \frac{x^5}{5} \Big|_{-a}^{a} \right\} = \frac{2\alpha a^5}{5}.$$

Thus

$$\bar{x} = \frac{M_y}{\mu} = 0 \qquad \text{and} \qquad \bar{y} = \frac{M_x}{\mu} = \frac{(2/5)\alpha a^5}{\alpha a^4 \pi/4} = \frac{8a}{5\pi}.$$

PROBLEMS 19.3

In the following problems find the mass and center of the mass of an object which lies in the given region with the given area density function.

1. $\Omega = \{(x, y): 1 \le x \le 2, -1 \le y \le 1\}$; $\rho(x, y) = x^2 + y^2$
2. $\Omega = \{(x, y): 0 \le x \le 4, 1 \le y \le 3\}$; $\rho(x, y) = 2xy$
3. $\Omega = \{(x, y): 0 \le x \le \pi/6, 0 \le y \le \pi/18\}$; $\rho(x, y) = \sin(2x + 3y)$
4. $\Omega = \{(x, y): -2 \le x \le 2, 0 \le y \le 1\}$; $\rho(x, y) = (x - y)^2$
5. Ω is the region of Problem 2; $\rho(x, y) = xe^{x-y}$
6. Ω is the quarter of the unit circle lying in the first quadrant; $\rho(x, y) = x + y^2$
7. Ω is the region of Problem 6; $\rho(x, y) = x^2 + y$
8. Ω is the region of Problem 6; $\rho(x, y) = x^3 + y^3$
9. Ω is the triangular region bounded by the lines $y = x$, $y = 1 - x$, and the x-axis; $\rho(x, y) = x + 2y$
10. Ω is the region of Problem 9; $\rho(x, y) = e^{x+2y}$
11. Ω is the first quadrant; $\rho(x, y) = e^{-y}/(1 + x)^3$
12. Ω is the first quadrant; $\rho(x, y) = (x + y)e^{-(x+y)}$.

19.4 Double Integrals in Polar Coordinates

In this section we shall see how to evaluate double integrals of functions in the form $z = f(r, \theta)$, where r and θ denote the polar coordinates of a point in the plane. Let $z = f(r, \theta)$ and let Ω denote the "polar rectangle"

$$\theta_1 \le \theta \le \theta_2, \qquad r_1 \le r \le r_2. \tag{1}$$

This region is sketched in Figure 1. We shall calculate the volume of the solid between the surface $z = f(r, \theta)$ and the region Ω. We partition Ω by small "polar rectangles"

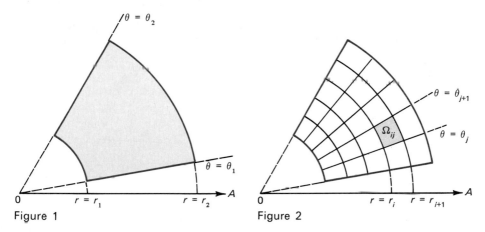

Figure 1   Figure 2

and calculate the volume over such a region (see Figure 2). The volume of the part of the solid over the region Ω_{ij} is given, approximately, by

$$f(r_i, \theta_j) A_{ij}, \tag{2}$$

where A_{ij} is the area of Ω_{ij}. Recall from Section 11.5 that if $r = f(\theta)$, then the area bounded by the lines $\theta = \alpha$, $\theta = \beta$ and the curve $r = f(\theta)$ is given by

$$A = \int_{\alpha}^{\beta} \tfrac{1}{2}[f(\theta)]^2 \, d\theta. \tag{3}$$

Thus

$$A_{ij} = \tfrac{1}{2} \int_{\theta_j}^{\theta_{j+1}} (r_{i+1}{}^2 - r_i{}^2) \, d\theta$$

$$= \tfrac{1}{2}(r_{i+1}{}^2 - r_i{}^2)\theta \Big|_{\theta_j}^{\theta_{j+1}} = \tfrac{1}{2}(r_{i+1}{}^2 - r_i{}^2)(\theta_{j+1} - \theta_j)$$

$$= \tfrac{1}{2}(r_{i+1} + r_i)(r_{i+1} - r_i)(\theta_{j+1} - \theta_j)$$

$$= \tfrac{1}{2}(r_{i+1} + r_i) \, \Delta r_i \, \Delta \theta_j.$$

But if Δr_i is small, then $r_{i+1} \approx r_i$ and we have

$$A_{ij} \approx \tfrac{1}{2}(2r_i) \, \Delta r_i \, \Delta \theta_j = r_i \, \Delta r_i \, \Delta \theta_j.$$

Then

$$V_{ij} \approx f(r_i, \theta_j)A_{ij} \approx f(r_i, \theta_j)r_i \, \Delta r_i \, \Delta \theta_j$$

so that, adding up the individual volumes and taking a limit, we obtain

$$V = \iint_\Omega f(r, \theta) r \, dr \, d\theta \tag{4}$$

Note. Do not forget the extra r in the above formula.

EXAMPLE 1. Find the volume enclosed by the sphere $x^2 + y^2 + z^2 = a^2$.

SOLUTION. We shall calculate the volume enclosed by the hemisphere $z = \sqrt{a^2 - x^2 - y^2}$ and then multiply by two. To do this, we first note that this is the volume of the solid under the hemisphere and above the circle $x^2 + y^2 = a^2$. We use polar coordinates since, in polar coordinates, $x^2 + y^2 = (r \cos \theta)^2 + (r \sin \theta^2) = r^2$. On the circle, $0 \le r \le a$ and $0 \le \theta \le 2\pi$ (see Figure 3). Then $z = \sqrt{a^2 - (x^2 + y^2)} = \sqrt{a^2 - r^2}$ so that, by (4),

$$V = \int_0^{2\pi} \int_0^a \sqrt{a^2 - r^2} \, r \, dr \, d\theta = \int_0^{2\pi} \left\{ -\frac{1}{3}(a^2 - r^2)^{3/2} \Big|_0^a \right\} d\theta$$

$$= \int_0^{2\pi} \frac{1}{3} a^3 \, d\theta = \frac{2\pi a^3}{3}.$$

Thus the volume of the sphere is $2(2\pi a^3/3) = (4/3)\pi a^3$.

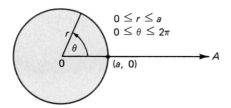

$$0 \le r \le a$$
$$0 \le \theta \le 2\pi$$

Figure 3

EXAMPLE 2. Find the volume of the solid bounded above by the surface $z = 3 + r$ and below by the cardioid $r = 1 + \sin \theta$.

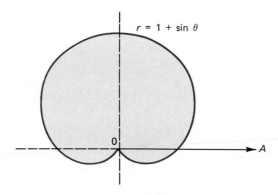

$r = 1 + \sin \theta$

Figure 4

SOLUTION. The cardioid is sketched in Figure 4 and can be described by

$$\Omega = \{(r, \theta): 0 \le \theta \le 2\pi \text{ and } 0 \le r \le 1 + \sin \theta\}.$$

Then, from (4),

$$V = \int_0^{2\pi} \int_0^{1+\sin\theta} (3 + r)r \, dr \, d\theta = \int_0^{2\pi} \left\{ \frac{3r^2}{2} + \frac{r^3}{3} \Big|_0^{1+\sin\theta} \right\} d\theta$$

$$= \int_0^{2\pi} \left\{ \frac{3}{2}(1 + 2\sin\theta + \sin^2\theta) + \frac{1}{3}(1 + 3\sin\theta + 3\sin^2\theta + \sin^3\theta) \right\} d\theta$$

$$= \int_0^{2\pi} \left(\frac{11}{6} + 4\sin\theta + \frac{5}{2}\sin^2\theta + \frac{\sin^3\theta}{3} \right) d\theta$$

$$= \int_0^{2\pi} \left(\frac{11}{6} + 4\sin\theta + \frac{5}{4}(1 - \cos 2\theta) + \frac{\sin\theta}{3}(1 - \cos^2\theta) \right) d\theta$$

$$= \left(\frac{11}{6}\theta - 4\cos\theta + \frac{5}{4}\theta - \frac{5}{8}\sin 2\theta - \frac{\cos\theta}{3} + \frac{\cos^3\theta}{9} \right) \Big|_0^{2\pi} = \frac{37}{6}\pi.$$

Remark. We can also use this technique to calculate areas. If $f(r, \theta) = 1$, then

$$\iint_\Omega r \, dr \, d\theta = \text{area of } \Omega. \tag{5}$$

[See Remark 3 following Theorem 19.1.2.]

EXAMPLE 3. Calculate the area enclosed by the cardioid $r = 1 + \sin\theta$.

SOLUTION

$$A = \int_0^{2\pi} \int_0^{1+\sin\theta} r \, dr \, d\theta = \int_0^{2\pi} \left\{ \frac{r^2}{2} \Big|_0^{1+\sin\theta} \right\} d\theta$$

$$= \frac{1}{2} \int_0^{2\pi} (1 + 2\sin\theta + \sin^2\theta) \, d\theta = \frac{1}{2} \int_0^{2\pi} \left(1 + 2\sin\theta + \frac{1 - \cos 2\theta}{2} \right) d\theta$$

$$= \frac{1}{2} \left(\frac{3\theta}{2} - 2\cos\theta - \frac{\sin 2\theta}{4} \right) \Big|_0^{2\pi} = \frac{3\pi}{2}.$$

As we shall see, it is often very useful to write a double integral in terms of polar coordinates. Let $z = f(x, y)$ be a function defined over a region Ω. Then, using polar coordinates, we can write

$$z = f(r \cos\theta, r \sin\theta) \tag{6}$$

and we can also describe Ω in terms of polar coordinates. The volume of the solid under f and over Ω is the same whether we use rectangular or polar coordinates. Thus, writing the volume in both rectangular and polar coordinates, we obtain the useful *change of variables formula*

$$\iint_\Omega f(x, y) \, dx \, dy = \iint_\Omega f(r \cos\theta, r \sin\theta) r \, dr \, d\theta. \tag{7}$$

EXAMPLE 4. The density at any point on a semicircular plane lamina is proportional to the square of the distance from the point to the center of the circle. Find the mass of the lamina.

SOLUTION. We solved this problem in Example 19.3.2 but we shall show how much easier are our calculations using polar coordinates. We have $\rho(x, y) = \alpha(x^2 + y^2) = \alpha r^2$ and $\Omega = \{(r, \theta): 0 \leq r \leq a \text{ and } 0 \leq \theta \leq \pi\}$, where a is the radius of the circle. Then

$$\mu = \int_0^\pi \int_0^a (\alpha r^2) r \, dr \, d\theta = \int_0^\pi \left\{\frac{\alpha r^4}{4}\Big|_0^a\right\} d\theta = \frac{\alpha a^4}{4} \int_0^\pi d\theta = \frac{\alpha \pi a^4}{4}.$$

Compare this with the calculation in Example 19.3.2.

EXAMPLE 5. Find the volume of the solid bounded by the xy-plane, the cylinder $x^2 + y^2 = 4$, and the paraboloid $z = 2(x^2 + y^2)$.

SOLUTION. The volume requested is the volume under the surface $z = 2(x^2 + y^2) = 2r^2$ and above the circle $x^2 + y^2 = 4$. Thus

$$V = \int_0^{2\pi} \int_0^2 2r^2 \cdot r \, dr \, d\theta = \int_0^{2\pi} \left\{\frac{r^4}{2}\Big|_0^2\right\} d\theta = 16\pi.$$

EXAMPLE 6. In probability theory one of the most important integrals that is encountered is the integral

$$\int_{-\infty}^\infty e^{-x^2} dx.$$

We now show how a combination of double integrals and polar coordinates can be used to evaluate it. Let

$$I = \int_0^\infty e^{-x^2} dx.$$

Then, by symmetry, $\int_{-\infty}^\infty e^{-x^2} dx = 2I$. Thus we need only to evaluate I. But, since any dummy variable can be used in a definite integral, we also have

$$I = \int_0^\infty e^{-y^2} dy.$$

Thus

$$I^2 = \left(\int_0^\infty e^{-x^2} dx\right)\left(\int_0^\infty e^{-y^2} dy\right)$$

or, from the result of Problem 19.2.47,

$$I^2 = \int_0^\infty \int_0^\infty e^{-x^2} e^{-y^2} dx \, dy = \int_0^\infty \int_0^\infty e^{-(x^2+y^2)} dx \, dy = \iint_\Omega e^{-(x^2+y^2)} dx \, dy,$$

where Ω denotes the first quadrant. In polar coordinates, the first quadrant can be written

$$\Omega = \left\{ (r, \theta) : 0 \le r < \infty \text{ and } 0 \le \theta \le \frac{\pi}{2} \right\}.$$

Thus, since $x^2 + y^2 = r^2$, we obtain

$$I^2 = \int_0^{\pi/2} \int_0^{\infty} e^{-r^2} r \, dr \, d\theta = \int_0^{\pi/2} \left(\lim_{N \to \infty} \int_0^{N} e^{-r^2} r \, dr \right) d\theta$$

$$= \int_0^{\pi/2} \left(\lim_{N \to \infty} -\frac{1}{2} e^{-r^2} \Big|_0^N \right) d\theta = \frac{1}{2} \int_0^{\pi/2} d\theta = \frac{\pi}{4}.$$

Hence $I^2 = \pi/4$ so that $I = \sqrt{\pi}/2$ and

$$\int_{-\infty}^{\infty} e^{-x^2} dx = 2I = \sqrt{\pi}.$$

By making the substitution $u = x/\sqrt{2}$, it is easy to show that

$$\int_{-\infty}^{\infty} e^{-x^2/2} dx = \sqrt{2\pi}.$$

The function $\rho(x) = (1/\sqrt{2\pi}) e^{-x^2/2}$ is called the *density function for the unit normal distribution*. We have just shown that $\int_{-\infty}^{\infty} \rho(x) \, dx = 1$.

EXAMPLE 7. Evaluate $\int\int_\Omega [1/(1 + x^2 + y^2)^{5/2}] \, dx \, dy$, where Ω is the wedge-shaped region bounded by the line $y = x$, the circle $x^2 + y^2 = 4$, and the x-axis.

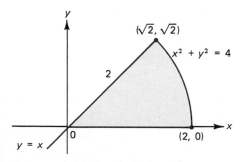

Figure 5

SOLUTION. The region is sketched in Figure 5. Using polar coordinates we have, since $1 + x^2 + y^2 = 1 + r^2$,

$$V = \int_0^{\pi/4} \int_0^2 \frac{r \, dr}{(1 + r^2)^{5/2}} \, d\theta$$

$$= \int_0^{\pi/4} \left[-\frac{1}{3} (1 + r^2)^{-3/2} \Big|_0^2 \right] d\theta = \frac{1}{3} \int_0^{\pi/4} \left(1 - \frac{1}{5^{3/2}} \right) d\theta$$

$$= \frac{1}{3} \left(1 - \frac{1}{5^{3/2}} \right) \frac{\pi}{4} \approx 0.24.$$

PROBLEMS 19.4

In Problems 1–5 calculate the volume under the given surface which lies over the given region Ω.

1. $z = r$; Ω is the circle of radius a
2. $z = r^n$, n a positive integer; Ω is the circle of radius a
3. $z = 3 - r$; Ω is the circle $r = 2 \cos \theta$
4. $z = r^2$; Ω is the cardioid $r = 4(1 - \cos \theta)$
5. $z = r^3$; Ω is the region enclosed by the spiral of Archimedes $r = a\theta$ and the polar axis for θ between 0 and 2π.

In Problems 6–15 calculate the area of the region enclosed by the given curve or curves.

6. $r = 1 - \cos \theta$ 7. $r = 4(1 + \cos \theta)$
8. $r = 1 + 2 \cos \theta$ 9. $r = 3 - 2 \sin \theta$
10. $r^2 = \cos 2\theta$ 11. $r^2 = 4 \sin 2\theta$
12. $r = a + b \sin \theta$, $a > b > 0$ 13. $r = \tan \theta$ and the line $\theta = \pi/4$
14. Outside the circle $r = 6$ and inside the cardioid $r = 4(1 + \sin \theta)$
15. Inside the cardioid $r = 2(1 + \cos \theta)$ but outside the circle $r = 2$.

16. Find the volume of the solid bounded above by the sphere $x^2 + y^2 + z^2 = 4a^2$, below by the xy-plane, and on the sides by the cylinder $x^2 + y^2 = a^2$.
17. Find the area of the region interior to the curve $(x^2 + y^2)^3 = 9y^2$.
18. Find the volume of the solid bounded by the cone $x^2 + y^2 = z^2$ and the cylinder $x^2 + y^2 = 4y$.
19. Find the volume of the solid bounded by the cone $z^2 = x^2 + y^2$ and the paraboloid $2z = x^2 + y^2$.
20. Find the volume of the solid bounded by the cylinder $x^2 + y^2 = 9$ and the hyperboloid $x^2 + y^2 - z^2 = 1$.
21. Find the volume of the solid centered at the origin which is bounded above by the surface $z = e^{-(x^2+y^2)}$ and below by the unit circle.
22. Find the centroid of the region bounded by $r = \cos \theta + 2 \sin \theta$.
23. Find the centroid of the region bounded by the limaçon $r = 3 + \sin \theta$.
24. Find the centroid of the region bounded by the limaçon $r = a + b \cos \theta$, $a > b > 0$.

19.5 Surface Area

In Section 9.2 we found a formula for the length of a plane curve (given by $y = f(x)$) by showing that the length of a small "piece" of the curve was approximately equal to $\sqrt{1 + [f'(x)]^2} \, \Delta x$. We now establish the area of a surface $z = f(x, y)$ which lies over a region Ω in the xy-plane.

We begin by calculating the surface area $\Delta \sigma$ over a rectangle ΔR with sides Δx and Δy. We have used the Greek letter σ to denote surface area. The situation is depicted in Figure 1 in which it is assumed that $f(x, y) > 0$ for (x, y) in Ω. We assume that f is differentiable over R. If Δx and Δy are small, then the region $PQSR$ in space has, approximately, the shape of a parallelogram. Thus, by equation (17.4.8),

$$\Delta \sigma \approx \text{area of parallelogram} = |\overrightarrow{PQ} \times \overrightarrow{PR}|. \qquad (1)$$

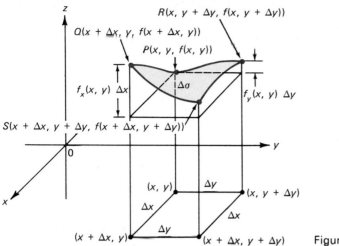

Figure 1

Now

$$\vec{PQ} = (x + \Delta x, y, f(x + \Delta x, y)) - (x, y, f(x, y))$$
$$= (\Delta x, 0, f(x + \Delta x, y) - f(x, y)).$$

But if Δx is small, then

$$\frac{f(x + \Delta x, y) - f(x, y)}{\Delta x} \approx f_x(x, y)$$

so that

$$f(x + \Delta x, y) - f(x, y) \approx f_x(x, y)\,\Delta x$$

and

$$\vec{PQ} \approx (\Delta x, 0, f_x(x, y)\,\Delta x). \tag{2}$$

Similarly,

$$\vec{PR} = (x, y + \Delta y, f(x, y + \Delta y)) - (x, y, f(x, y))$$
$$= (0, \Delta y, f(x, y + \Delta y) - f(x, y))$$

or, if Δy is small,

$$\vec{PR} \approx (0, \Delta y, f_y(x, y)\,\Delta y). \tag{3}$$

Thus, from (2) and (3),

$$\vec{PQ} \times \vec{PR} \approx \begin{vmatrix} \mathbf{i} & \mathbf{j} & \mathbf{k} \\ \Delta x & 0 & f_x(x, y)\,\Delta x \\ 0 & \Delta y & f_y(x, y)\,\Delta y \end{vmatrix}$$
$$= -f_x(x, y)\,\Delta x\,\Delta y\,\mathbf{i} - f_y(x, y)\,\Delta x\,\Delta y\,\mathbf{j} + \Delta x\,\Delta y\,\mathbf{k}$$
$$= (-f_x(x, y)\mathbf{i} - f_y(x, y)\mathbf{j} + \mathbf{k})\,\Delta x\,\Delta y$$

so that, from (1),

$$\Delta\sigma \approx \sqrt{f_x^2(x, y) + f_y^2(x, y) + 1}\; \Delta x\, \Delta y. \tag{4}$$

Finally, adding up the surface area over rectangles which partition Ω and taking a limit, yields

$$\boxed{\sigma = \iint\limits_{\Omega} \sqrt{1 + f_x^2(x, y) + f_y^2(x, y)}\; dx\, dy.} \tag{5}$$

Remark. We caution that equation (5) has *not* been proved. However if f is differentiable, it certainly is plausible.

EXAMPLE 1. Calculate the surface area cut from the cylinder $y^2 + z^2 = 9$ by the planes $x = 0$, $x = 1$, $y = 0$, and $y = 2$.

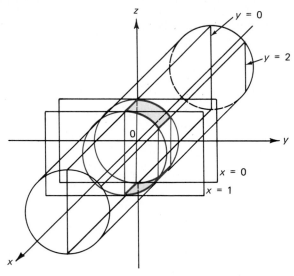

Figure 2

SOLUTION. The required area is sketched in Figure 2. It consists of two equal parts: one for $z > 0$ and one for $z < 0$. We will calculate the surface area for $z > 0$ and multiply the result by 2. For $z > 0$, we have $z = \sqrt{9 - y^2}$. Ω is the rectangle $\{(x, y): 0 \le x \le 1 \text{ and } 0 \le y \le 2\}$. Then

$$f_x = 0, \qquad f_y = -\frac{y}{\sqrt{9 - y^2}}$$

so that

$$\sigma = \int_0^2 \int_0^1 \sqrt{0 + \left(\frac{-y}{\sqrt{9 - y^2}}\right)^2 + 1}\; dx\, dy = \int_0^2 \int_0^1 \sqrt{1 + \frac{y^2}{9 - y^2}}\; dx\, dy$$

$$= 3\int_0^2 \int_0^1 \frac{1}{\sqrt{9 - y^2}}\; dx\, dy = 3\int_0^2 \left(\frac{x}{\sqrt{9 - y^2}}\Big|_0^1\right) dy = 3\int_0^2 \frac{dy}{\sqrt{9 - y^2}}$$

$$= 3\sin^{-1}\frac{y}{3}\Big|_0^2 = 3\sin^{-1}\frac{2}{3}.$$

Thus the total surface area is $6\sin^{-1}\frac{2}{3} \approx 4.38$.

EXAMPLE 2. Find the area of the part of the surface $z = \frac{2}{3}(x^{3/2} + y^{3/2})$ cut by the planes $x = 0$, $x = 4$, $y = 1$, and $y = 3$.

SOLUTION. $f_x = \sqrt{x}$ and $f_y = \sqrt{y}$ so that

$$\sigma = \int_0^4 \int_1^3 \sqrt{1 + x + y}\, dy\, dx = \int_0^4 \left[\frac{2}{3}(1 + x + y)^{3/2}\Big|_1^3\right] dx$$

$$= \frac{2}{3} \int_0^4 [(4 + x)^{3/2} - (2 + x)^{3/2}]\, dx$$

$$= \frac{4}{15}[(4 + x)^{5/2} - (2 + x)^{5/2}]\Big|_0^4 = \frac{4}{15}[8^{5/2} - 6^{5/2} + 4^{5/2} - 2^{5/2}] \approx 119.$$

EXAMPLE 3. Find the surface area of the circular paraboloid $z = x^2 + y^2$ between the xy-plane and the plane $z = 9$.

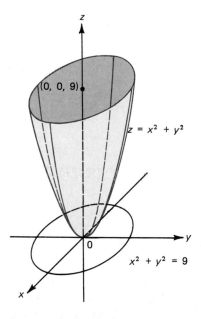

(0, 0, 9)

$z = x^2 + y^2$

$x^2 + y^2 = 9$

Figure 3

SOLUTION. The surface area requested is sketched in Figure 3. The region Ω is the disk $x^2 + y^2 \leq 9$. We have

$$f_x = 2x, \qquad f_y = 2y$$

so that

$$\sigma = \iint_\Omega \sqrt{1 + 4x^2 + 4y^2}\, dy\, dx.$$

Clearly this problem calls for the use of polar coordinates. We have

$$\sigma = \int_0^{2\pi} \int_0^3 \sqrt{1 + 4r^2}\, r\, dr\, d\theta = \int_0^{2\pi} \left[\frac{1}{12}(1 + 4r^2)^{3/2}\Big|_0^3\right] d\theta$$

$$= \frac{\pi}{6}(37^{3/2} - 1) \approx 117.3.$$

EXAMPLE 4. Calculate the area of the part of the surface $z = x^3 + y^4$ which lies over the square $\{(x, y): 0 \le x \le 1, 0 \le y \le 1\}$.

SOLUTION. $f_x = 3x^2$ and $f_y = 4y^3$ so that

$$\sigma = \int_0^1 \int_0^1 \sqrt{1 + 9x^4 + 16y^6} \, dx \, dy.$$

However this is as far as we can go unless we resort to numerical techniques to approximate this double integral. As with ordinary definite integrals, many double integrals cannot be integrated in terms of functions that we know. However there are a great number of techniques for approximating a double integral numerically which parallel the techniques discussed in Section 8.9.

PROBLEMS 19.5

In Problems 1–9 find the area of the part of the surface that lies over the given region.

1. $z = x + 2y$; $\Omega = \{(x, y): 0 \le x \le y, 0 \le y \le 2\}$
2. $z = 4x + 7y$; $\Omega = $ region between $y = x^2$ and $y = x^5$
3. $z = ax + by$; $\Omega = $ upper half of unit circle
4. $z = y^2$; $\Omega = \{(x, y): 0 \le x \le 2, 0 \le y \le 4\}$
*5. $z = 3 + x^{2/3}$; $\Omega = \{(x, y): -1 \le x \le 1, 1 \le y \le 2\}$
6. $z = (x^4/4) + (1/8x^2)$; $\Omega = \{(x, y): 1 \le x \le 2, 0 \le y \le 5\}$
7. $z = \frac{1}{3}(y^2 + 2)^{3/2}$; $\Omega = \{(x, y): -4 \le x \le 7, 0 \le y \le 3\}$
8. $z = 2 \ln(1 + y)$; $\Omega = \{(x, y): 0 \le x \le 2, 0 \le y \le 1\}$
*9. $(z + 1)^2 = 4x^3$; $\Omega = \{(x, y): 0 \le x \le 1, 0 \le y \le 2\}$.

*10. Calculate the lateral surface area of the cylinder $y^{2/3} + z^{2/3} = 1$ for x in the interval $[0, 2]$.
11. Find the surface area of the hemisphere $x^2 + y^2 + z^2 = a^2$, $z \ge 0$.
*12. Find the surface area of the part of the sphere $x^2 + y^2 + z^2 = a^2$ that is also inside the cylinder $x^2 + y^2 = ay$.
13. Find the area of the surface in the first octant cut from the cylinder $x^2 + y^2 = 16$ by the plane $y = z$.
14. Find the area of the portion of the sphere $x^2 + y^2 + z^2 = 16z$ lying within the circular paraboloid $z = x^2 + y^2$.
15. Find the area of the surface cut from the hyperbolic paraboloid $4z = x^2 - y^2$ by the cylinder $x^2 + y^2 = 16$.
*16. Let $z = f(x, y)$ be the equation of a plane (i.e., $z = ax + by + c$). Show that over the region Ω, the area of the plane is given by

$$\sigma = \iint\limits_{\Omega} \sec \gamma \, dx \, dy,$$

where γ is the angle between the normal vector $\mathbf{N}$ to the plane and the positive z-axis. [*Hint:* Show, using the dot product, that

$$\cos \gamma = \frac{\mathbf{N} \cdot \mathbf{k}}{|\mathbf{N}|} = \frac{1}{\sqrt{1 + a^2 + b^2}}.]$$

In Problems 17–20 find a double integral which represents the area of the given surface over the given region. Do *not* try to evaluate the integral.

17. $z = x^3 + y^3$; Ω is the unit circle

18. $z = \ln(x + 2y)$; $\Omega = \{(x, y): 0 \leq x \leq 1, 0 \leq y \leq 4\}$
19. $z = \sqrt{1 + x + y}$; Ω is the triangle bounded by $y = x$, $y = 4 - x$, and the y-axis
20. $z = e^{x-y}$; Ω is the ellipse $4x^2 + 9y^2 = 36$.

***21.** Find a double integral which represents the volume of the ellipsoid $(x^2/a^2) + (y^2/b^2) + (z^2/c^2) = 1$.

19.6 The Triple Integral

In this section we discuss the idea behind the triple integral of a function of three variables $f(x, y, z)$ over a region S in $\mathbb{R}^3$. This is really a simple extension of the double integral. For that reason we shall omit a number of technical details.

We start with a parallelepiped π in $\mathbb{R}^3$. This can be written

$$\pi = \{(x, y, z): a_1 \leq x \leq a_2, b_1 \leq y \leq b_2, c_1 \leq z \leq c_2\} \qquad (1)$$

and is sketched in Figure 1. We then partition the three intervals, $[a_1, a_2]$, $[b_1, b_2]$, and $[c_1, c_2]$ by

$$a_1 = x_0 < x_1 < \cdots < x_n = a_2$$
$$b_1 = y_0 < y_1 < \cdots < y_m = b_2$$
$$c_1 = z_0 < z_1 < \cdots < z_p = c_2$$

to obtain nmp "boxes." The volume of a typical box B_{ijk} is given by

$$V_{ijk} = \Delta x_i \, \Delta y_j \, \Delta z_k \qquad (2)$$

when $\Delta x_i = x_i - x_{i-1}$, $\Delta y_j = y_j - y_{j-1}$, and $\Delta z_k = z_k - z_{k-1}$. We then form the sum

$$\sum_{i=1}^{n} \sum_{j=1}^{m} \sum_{k=1}^{p} f(x_i^*, y_j^*, z_k^*) \, \Delta x_i \, \Delta y_j \, \Delta z_k, \qquad (3)$$

where (x_i^*, y_j^*, z_k^*) is in B_{ijk}. By the norm $|P|$ of the partition P of π we mean the largest volume of the boxes contained in the partition of π. We then take the limit as $|P| \to 0$.

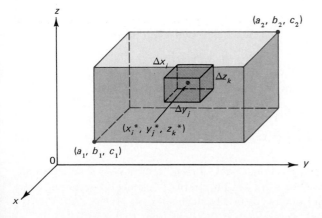

Figure 1

DEFINITION 1. Let $w = f(x, y, z)$ and let the parallelepiped π be given by (1). Suppose that

$$\lim_{|P| \to 0} \sum_{i=1}^{n} \sum_{j=1}^{m} \sum_{k=1}^{p} f(x_i^*, y_j^*, z_k^*) \, \Delta x_i \, \Delta y_j \, \Delta z_k$$

exists and is independent of the way in which π is partitioned and the way in which the points (x_i^*, y_j^*, z_k^*) are chosen. Then the *triple integral of f over* π, written $\iiint_\pi f(x, y, z) \, dx \, dy \, dz$, is defined by

$$\iiint_\pi f(x, y, z) \, dx \, dy \, dz = \lim_{|P| \to 0} \sum_{i=1}^{n} \sum_{j=1}^{m} \sum_{k=1}^{p} f(x_i^*, y_j^*, z_k^*) \, \Delta x_i \, \Delta y_j \, \Delta z_k. \qquad (4)$$

As with double integrals, we can write triple integrals as iterated (or repeated) integrals. If π is defined by (1), we have

$$\iiint_\pi f(x, y, z) \, dx \, dy \, dz = \int_{a_1}^{a_2} \int_{b_1}^{b_2} \int_{c_1}^{c_2} f(x, y, z) \, dz \, dy \, dx. \qquad (5)$$

EXAMPLE 1. Evaluate $\iiint_\pi xy \cos yz \, dx \, dy \, dz$, where π is the parallelepiped

$$\left\{ (x, y, z): 0 \le x \le 1, 0 \le y \le 1, 0 \le z \le \frac{\pi}{2} \right\}.$$

SOLUTION

$$\iiint_\pi xy \cos yz \, dx \, dy \, dz = \int_0^1 \int_0^1 \int_0^{\pi/2} xy \cos yz \, dz \, dy \, dx$$

$$= \int_0^1 \int_0^1 \left\{ xy \cdot \frac{1}{y} \sin yz \, \Big|_0^{\pi/2} \right\} dy \, dx = \int_0^1 \int_0^1 x \sin \frac{\pi}{2} y \, dy \, dx$$

$$= \int_0^1 \left\{ -\frac{2}{\pi} x \cos \frac{\pi}{2} y \, \Big|_0^1 \right\} dx = \int_0^1 \frac{2}{\pi} x \, dx = \frac{x^2}{\pi} \Big|_0^1 = \frac{1}{\pi}.$$

We now define the triple integral over a more general region S. We assume that S is bounded. Then we can enclose S in a parallelepiped π and define a new function F by

$$F(x, y, z) = \begin{cases} f(x, y, z), & \text{if } (x, y, z) \text{ is in } S \\ 0. & \text{if } (x, y, z) \text{ is in } \pi, \text{ but not in } S. \end{cases}$$

We then define

$$\iiint_S f(x, y, z) \, dx \, dy \, dz = \iiint_\pi F(x, y, z) \, dx \, dy \, dz. \qquad (6)$$

Remark 1. If f is continuous over S and if S is a region of the type we shall discuss below, then $\iiint_S f(x, y, z) \, dx \, dy \, dz$ will exist. The proof of this fact is beyond the scope of this text but can be found in any advanced calculus text.

Remark 2. If $f \geq 0$ on S, then the triple integral $\iiint_S f(x, y, z)\, dx\, dy\, dz$ represents the "volume" in four-dimensional space $\mathbb{R}^4$ of the region bounded above by f and below by S. We cannot, of course, draw this volume but otherwise the theory of volumes carries over to four (and more) dimensions.

Now let S take the form

$$S = \{(x, y, z): a_1 \leq x \leq a_2,\ g_1(x) \leq y \leq g_2(x),\ h_1(x, y) \leq z \leq h_2(x, y)\}. \tag{7}$$

What does such a solid look like? We first note that the equations $z = h_1(x, y)$ and $z = h_2(x, y)$ are the equations of surfaces in $\mathbb{R}^3$. The equations $y = g_1(x)$ and $y = g_2(x)$ are equations of cylinders in $\mathbb{R}^3$ and the equations $x = a_1$ and $x = a_2$ are

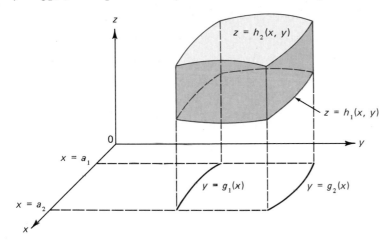

Figure 2

equations of planes in $\mathbb{R}^3$. The solid S is sketched in Figure 2. We assume that g_1, g_2, h_1, and h_2 are continuous. If f is continuous, then $\iiint_S f(x, y, z)\, dx\, dy\, dz$ will exist and

$$\iiint_S f(x, y, z)\, dx\, dy\, dz = \int_{a_1}^{a_2} \int_{g_1(x)}^{g_2(x)} \int_{h_1(x, y)}^{h_2(x, y)} f(x, y, z)\, dz\, dy\, dx. \tag{8}$$

Note the similarity between equations (8) and (19.2.10).

EXAMPLE 2.   Evaluate $\iiint_S 2x^3 y^2 z\, dx\, dy\, dz$, where S is the region

$$\{(x, y, z): 0 \leq x \leq 1,\ x^2 \leq y \leq x,\ x - y \leq z \leq x + y\}.$$

SOLUTION

$$\iiint_S 2x^3 y^2 z\, dx\, dy\, dz = \int_0^1 \int_{x^2}^{x} \int_{x-y}^{x+y} 2x^3 y^2 z\, dz\, dy\, dx$$

$$= \int_0^1 \int_{x^2}^{x} \left\{ x^3 y^2 z^2 \Big|_{x-y}^{x+y} \right\} dy\, dx$$

$$= \int_0^1 \int_{x^2}^{x} x^3 y^2 [(x + y)^2 - (x - y)^2]\, dy\, dx$$

$$= \int_0^1 \int_{x^2}^x 4x^4 y^3 \, dy \, dx = \int_0^1 \left\{ x^4 y^4 \Big|_{x^2}^x \right\} dx$$

$$= \int_0^1 (x^8 - x^{12}) \, dx = \frac{1}{9} - \frac{1}{13} = \frac{4}{117}.$$

Many of the applications we saw for the double integral can be extended to the triple integral. We present three of them below.

I. VOLUME

Let the region S be defined by (7). Then, since $\Delta x_i \, \Delta y_j \, \Delta z_k$ represents the volume of a "box" in S, when we add up the volumes of these boxes and take a limit we obtain the total volume of S. That is,

$$\text{volume of } S = \iiint_S dx \, dy \, dz. \qquad (9)$$

EXAMPLE 3. Calculate the volume of the region of Example 2.

SOLUTION

$$V = \int_0^1 \int_{x^2}^x \int_{x-y}^{x+y} dz \, dy \, dx = \int_0^1 \int_{x^2}^x \left\{ z \Big|_{x-y}^{x+y} \right\} dy \, dx$$

$$= \int_0^1 \int_{x^2}^x 2y \, dy \, dx = \int_0^1 \left\{ y^2 \Big|_{x^2}^x \right\} dx = \int_0^1 (x^2 - x^4) \, dx$$

$$= \frac{1}{3} - \frac{1}{5} = \frac{2}{15}.$$

EXAMPLE 4. Find the volume of the tetrahedron formed by the planes $x = 0$, $y = 0$, $z = 0$ and $x + (y/2) + (z/4) = 1$.

SOLUTION. The tetrahedron is sketched in Figure 3. We see that z ranges from 0 to the plane $x + (y/2) + (z/4) = 1$ or $z = 4[1 - x - (y/2)]$. This last plane intersects the xy-plane in a line whose equation (obtained by setting $z = 0$) is given by $0 = 1 - x - (y/2)$ or $y = 2(1 - x)$ so that y ranges from 0 to $2(1 - x)$ (see Figure 4). Finally, this line intersects the x-axis at the point $(1, 0, 0)$ so that x ranges from 0 to 1 and we have

$$V = \int_0^1 \int_0^{2(1-x)} \int_0^{4(1-x-y/2)} dz \, dy \, dx = \int_0^1 \int_0^{2(1-x)} 4\left(1 - x - \frac{y}{2}\right) dy \, dx$$

$$= 4 \int_0^1 \left\{ y(1 - x) - \frac{y^2}{4} \Big|_0^{2(1-x)} \right\} dx = 4 \int_0^1 [2(1 - x)^2 - (1 - x)^2] \, dx$$

$$= -\frac{4}{3}(1 - x)^3 \Big|_0^1 = \frac{4}{3}.$$

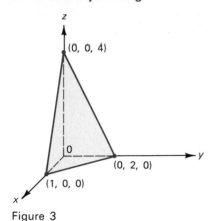

Figure 3

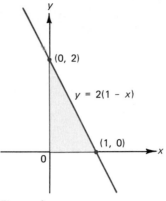

Figure 4

Remark. It was not necessary to integrate in the order z, then y, then x. We could have written, for example,

$$0 \le x \le 1 - \frac{y}{2} - \frac{z}{4}.$$

The intersection of this plane with the yz-plane is the line $0 = 1 - (y/2) - (z/4)$ or $z = 4[1 - (y/2)]$. The intersection of this line with the y-axis occurs at the point $(0, 2, 0)$. Thus

$$V = \int_0^2 \int_0^{4(1-y/2)} \int_0^{1-y/2-z/4} dx \, dz \, dy$$

$$= \int_0^2 \int_0^{4(1-y/2)} \left(1 - \frac{y}{2} - \frac{z}{4}\right) dz \, dy = \int_0^2 \left\{ \left(1 - \frac{y}{2}\right) z - \frac{z^2}{8} \Big|_0^{4(1-y/2)} \right\} dy$$

$$= \int_0^2 \left\{ 4\left(1 - \frac{y}{2}\right)^2 - \frac{16[1 - (y/2)]^2}{8} \right\} dy = 2 \int_0^2 \left(1 - \frac{y}{2}\right)^2 dy$$

$$= -\frac{4}{3}\left(1 - \frac{y}{2}\right)^3 \Big|_0^2 = \frac{4}{3}.$$

We could also integrate in any of four other orders (xyz, yzx, yxz, and zxy) to obtain the same result.

II. DENSITY AND MASS

Let the function $\rho(x, y, z)$ denote the density (in kg/m³, say) of a solid S in $\mathbb{R}^3$. Then on a "box" of sides Δx_i, Δy_j and Δz_k, the approximate mass of the box will be equal to $\rho(x_i^*, y_j^*, z_k^*) \, \Delta x_i \, \Delta y_j \, \Delta z_k$ if Δx_i, Δy_j, and Δz_k are small. We then obtain

$$\text{total mass of } S = \mu(S) = \iiint_S \rho(x, y, z) \, dx \, dy \, dz. \tag{10}$$

EXAMPLE 5. The density of the solid of Example 2 is given by $\rho(x, y, z) = x + 2y + 4z$ kg/m³. Calculate the total mass of the solid.

SOLUTION

$$\mu(S) = \int_0^1 \int_{x^2}^x \int_{x-y}^{x+y} (x + 2y + 4z)\, dz\, dy\, dx$$

$$= \int_0^1 \int_{x^2}^x \left\{ (x + 2y)z + 2z^2 \Big|_{x-y}^{x+y} \right\} dy\, dx$$

$$= \int_0^1 \int_{x^2}^x (10xy + 4y^2)\, dy\, dx = \int_0^1 \left\{ 5xy^2 + \frac{4y^3}{3} \Big|_{x^2}^x \right\} dx$$

$$= \int_0^1 \left(5x^3 - 5x^5 + \frac{4}{3}x^3 - \frac{4}{3}x^6 \right) dx = \int_0^1 \left(\frac{19}{3}x^3 - 5x^5 - \frac{4}{3}x^6 \right) dx$$

$$= \frac{19}{12} - \frac{5}{6} - \frac{4}{21} = \frac{47}{84} \quad \text{kg.}$$

III. FIRST MOMENTS AND CENTER OF MASS

In $\mathbb{R}^3$, we use the symbol M_{yz} to denote the first moment with respect to the yz-plane. Similarly M_{xz} denotes the first moment with respect to the xz-plane, and M_{xy} denotes the first moment with respect to the xy-plane. Since the distance from a point (x, y, z) to the yz-plane is x, and so on, we may use familiar reasoning to obtain

$$M_{yz} = \iiint_S x\rho(x, y, z)\, dx\, dy\, dz \tag{11}$$

$$M_{xz} = \iiint_S y\rho(x, y, z)\, dx\, dy\, dz \tag{12}$$

and

$$M_{xy} = \iiint_S z\rho(x, y, z)\, dx\, dy\, dz. \tag{13}$$

The *center of mass* of S is then given by

$$(\bar{x}, \bar{y}, \bar{z}) = \left(\frac{M_{yz}}{\mu}, \frac{M_{xz}}{\mu}, \frac{M_{xy}}{\mu} \right), \tag{14}$$

where μ denotes the mass of S. If $\rho(x, y, z)$ is constant, then as in $\mathbb{R}^1$ and $\mathbb{R}^2$, the center of mass is called the *centroid*.

EXAMPLE 6. Find the center of mass of the solid in Example 5.

SOLUTION. We have already found that $\mu = 47/84$. We next calculate (some of the details are omitted)

$$M_{yz} = \int_0^1 \int_{x^2}^x \int_{x-y}^{x+y} x(x + 2y + 4z)\, dz\, dy\, dx$$

$$= \int_0^1 \int_{x^2}^x \left\{ (x^2 + 2xy)z + 2xz^2 \Big|_{x-y}^{x+y} \right\} dy\, dx$$

$$= \int_0^1 \int_{x^2}^x (10x^2 y + 4xy^2)\, dy\, dx = \int_0^1 \left(5x^4 - 5x^6 + \frac{4}{3}x^4 - \frac{4}{3}x^7 \right) dx$$

$$= 1 - \frac{5}{7} + \frac{4}{15} - \frac{1}{6} = \frac{27}{70}.$$

$$M_{xz} = \int_0^1 \int_{x^2}^x \int_{x-y}^{x+y} y(x + 2y + 4z)\, dz\, dy\, dx$$

$$= \int_0^1 \int_{x^2}^x \left\{ (xy + 2y^2)z + 2yz^2 \Big|_{x-y}^{x+y} \right\} dy\, dx$$

$$= \int_0^1 \int_{x^2}^x (10xy^2 + 4y^3)\, dy\, dx = \int_0^1 \left\{ \frac{10}{3} xy^3 + y^4 \Big|_{x^2}^x \right\} dx$$

$$= \int_0^1 \left(\frac{10}{3}x^4 - \frac{10}{3}x^7 + x^4 - x^8 \right) dx = \frac{2}{3} - \frac{5}{12} + \frac{1}{5} - \frac{1}{9} = \frac{61}{180}.$$

$$M_{xy} = \int_0^1 \int_{x^2}^x \int_{x-y}^{x+y} z(x + 2y + 4z)\, dz\, dy\, dx$$

$$= \int_0^1 \int_{x^2}^x \left\{ (x + 2y)\frac{z^2}{2} + \frac{4z^3}{3} \Big|_{x-y}^{x+y} \right\} dy\, dz$$

$$= \int_0^1 \int_{x^2}^x \left[(x + 2y)(2xy) + \frac{4}{3}(6x^2 y + 2y^3) \right] dy\, dx$$

$$= \int_0^1 \int_{x^2}^x \left(10x^2 y + 4xy^2 + \frac{8}{3}y^3 \right) dy\, dx$$

$$= \int_0^1 \left\{ 5x^2 y^2 + \frac{4}{3}xy^3 + \frac{2}{3}y^4 \Big|_{x^2}^x \right\} dx$$

$$= \int_0^1 \left(5x^4 - 5x^6 + \frac{4}{3}x^4 - \frac{4}{3}x^7 + \frac{2}{3}x^4 - \frac{2}{3}x^8 \right) dx$$

$$= \int_0^1 \left(7x^4 - 5x^6 - \frac{4}{3}x^7 - \frac{2}{3}x^8 \right) dx = \frac{7}{5} - \frac{5}{7} - \frac{1}{6} - \frac{2}{27} = \frac{841}{1890}.$$

Thus

$$(\bar{x}, \bar{y}, \bar{z}) = \left(\frac{M_{yz}}{\mu}, \frac{M_{xz}}{\mu}, \frac{M_{xy}}{\mu} \right) = \left(\frac{27/70}{47/84}, \frac{61/180}{47/84}, \frac{841/1890}{47/84} \right)$$

$$= \left(\frac{162}{235}, \frac{427}{705}, \frac{1682}{2115} \right) \approx (0.689, 0.606, 0.795).$$

Here distances are measured in meters.

EXAMPLE 7. A solid S is bounded above by the plane $z = y$, below by the xy-plane, and lies inside the parabolic cylinder $y = 1 - x^2$. The density at any point in S is constant. Find the centroid of S.

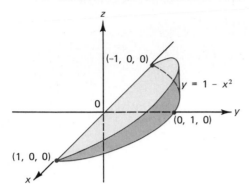

Figure 5

SOLUTION. The solid is sketched in Figure 5. Let the density be given by the constant ρ. Then

$$\mu(S) = \iiint_S \rho\, dz\, dy\, dx.$$

But S is given by $\{(x, y, z): -1 \le x \le 1, 0 \le y \le 1 - x^2, 0 \le z \le y\}$. Thus

$$\mu(S) = \int_{-1}^{1} \int_{0}^{1-x^2} \int_{0}^{y} \rho\, dz\, dy\, dx = \rho \int_{-1}^{1} \int_{0}^{1-x^2} y\, dy\, dx$$

$$= \rho \int_{-1}^{1} \left\{ \frac{y^2}{2} \Big|_{0}^{1-x^2} \right\} dx = \frac{\rho}{2} \int_{-1}^{1} (1 - 2x^2 + x^4)\, dx = \frac{8\rho}{15}.$$

Then

$$M_{yz} = \rho \int_{-1}^{1} \int_{0}^{1-x^2} \int_{0}^{y} x\, dz\, dy\, dx = \rho \int_{-1}^{1} \int_{0}^{1-x^2} xy\, dy\, dx$$

$$= \frac{\rho}{2} \int_{-1}^{1} (x - 2x^3 + x^5)\, dx = 0.$$

(This tells us that $\bar{x} = 0$ which is no surprise since the solid is symmetric about the yz-plane.)

$$M_{xz} = \rho \int_{-1}^{1} \int_{0}^{1-x^2} \int_{0}^{y} y\, dz\, dy\, dx = \rho \int_{-1}^{1} \int_{0}^{1-x^2} y^2\, dy\, dx$$

$$= \rho \int_{-1}^{1} \frac{(1 - x^2)^3}{3}\, dx = \frac{\rho}{3} \int_{-1}^{1} (1 - 3x^2 + 3x^4 - x^6)\, dx = \frac{32\rho}{105}.$$

Finally, from the calculation of M_{xz},

$$M_{xy} = \rho \int_{-1}^{1} \int_{0}^{1-x^2} \int_{0}^{y} z\, dz\, dy\, dx = \rho \int_{-1}^{1} \int_{0}^{1-x^2} \frac{y^2}{2}\, dy\, dx$$

$$= \frac{1}{2} M_{xz} = \frac{16\rho}{105}.$$

Thus the centroid is given by

$$(\bar{x}, \bar{y}, \bar{z}) = \left(\frac{0}{8\rho/15}, \frac{32\rho/105}{8\rho/15}, \frac{16\rho/105}{8\rho/15}\right) = \left(0, \frac{4}{7}, \frac{2}{7}\right).$$

PROBLEMS 19.6

In Problems 1–7 evaluate the repeated triple integral.

1. $\displaystyle\int_0^1 \int_0^y \int_0^x y \, dz \, dx \, dy$

2. $\displaystyle\int_0^2 \int_{-z}^z \int_{y-z}^{y+z} 2xz \, dx \, dy \, dz$

3. $\displaystyle\int_{a_1}^{a_2} \int_{b_1}^{b_2} \int_{c_1}^{c_2} dy \, dx \, dz$

4. $\displaystyle\int_0^{\pi/2} \int_0^{\pi/2} \int_0^z \sin(x/z) \, dx \, dz \, dy$

5. $\displaystyle\int_1^2 \int_{1-y}^{1+y} \int_0^{yz} 6xyz \, dx \, dz \, dy$

6. $\displaystyle\int_0^1 \int_0^{\sqrt{1-x^2}} \int_0^x yz \, dz \, dy \, dx$

7. $\displaystyle\int_0^1 \int_{-\sqrt{1-x^2}}^{\sqrt{1-x^2}} \int_{-\sqrt{1-x^2-y^2}}^{\sqrt{1-x^2-y^2}} z^2 \, dz \, dy \, dx.$ [*Hint:* Use polar coordinates.]

8. Change the order of integration in Problem 1 and write the integral in the form
(a) $\int_?^? \int_?^? \int_?^? y \, dx \, dy \, dz$ (b) $\int_?^? \int_?^? \int_?^? y \, dy \, dz \, dx.$

9. Write the integral of Problem 6 in the form
(a) $\int_?^? \int_?^? \int_?^? yz \, dx \, dy \, dz$ (b) $\int_?^? \int_?^? \int_?^? yz \, dy \, dz \, dx.$

10. Write the integral of Problem 7 in the form $\int_?^? \int_?^? \int_?^? z^2 \, dx \, dz \, dy.$

In Problems 11–18 find the volume of the given solid.

11. The tetrahedron with vertices at the points $(0, 0, 0)$, $(1, 0, 0)$, $(0, 1, 0)$, and $(0, 0, 1)$
12. The tetrahedron with vertices at the points $(0, 0, 0)$, $(a, 0, 0)$, $(0, b, 0)$, and $(0, 0, c)$
13. The solid in the first octant bounded by the cylinder $x^2 + z^2 = 9$, the plane $x + y = 4$ and the three coordinate planes
*14. The solid bounded by the planes $x - 2y + 4z = 4$, $-2x + 3y - z = 6$, $x = 0$ and $y = 0$
15. The solid bounded above by the sphere $x^2 + y^2 + z^2 = 16$ and below by the plane $z = 2$
16. The solid bounded by the parabolic cylinder $z = 5 - x^2$ and the planes $z = y$, $z = 2y$, which lies in the half space $y \geq 0$
17. The solid bounded by the ellipsoid $(x^2/a^2) + (y^2/b^2) + (z^2/c^2) = 1$
18. The solid bounded by the elliptic cylinder $9x^2 + y^2 = 9$ and the planes $z = 0$ and $x + y + 9z = 9$.

In Problems 19–23 find the center of mass of the given solid having the given density.

19. The solid of Problem 11; $\rho(x, y, z) = x$
20. The solid of Problem 12; $\rho(x, y, z) = x^2 + y^2 + z^2$
21. The solid of Problem 13; $\rho(x, y, z) = z$
22. The solid of Problem 15; $\rho(x, y, z) = x + 2y + 4z$
23. The solid of Problem 17; $\rho(x, y, z) = \alpha x^2 + \beta y^2 + \gamma z^2.$

24. Find the mass of a cube whose side has a length of k units if its density at any point is proportional to the distance from a given face of the cube.
25. Find the centroid of the tetrahedron of Problem 12.
26. Find the centroid of the ellipsoid of Problem 17.
27. Find the centroid of the solid of Problem 16.

28. Find the centroid of the solid of Problem 18.

29. The solid S lies in the first octant and is bounded by the planes $z = 0, y = 1, x = y$ and the hyperboloid $z = xy$. Its density is given by $\rho(x, y, z) = 1 + 2z$.
(a) Find the center of mass of S.
(b) Show that the center of mass lies inside S.

*19.7 The Triple Integral in Cylindrical and Spherical Coordinates

In this section we show how triple integrals can be written using cylindrical and spherical coordinates.

I. CYLINDRICAL COORDINATES

Recall from Section 17.8 that the cylindrical coordinates of a point in $\mathbb{R}^3$ are (r, θ, z), where r and θ are the polar coordinates of the projection of the point onto the xy-plane and z is the usual z-coordinate. In order to calculate an integral of a region given in cylindrical coordinates, we go through a procedure very similar to the one we used to write double integrals in polar coordinates in Section 19.4.

Consider the "cylindrical parallelepiped" given by

$$\pi_c = \{(r, \theta, z): r_1 \le r \le r_2;\ \theta_1 \le \theta \le \theta_2;\ z_1 \le z \le z_2\}.$$

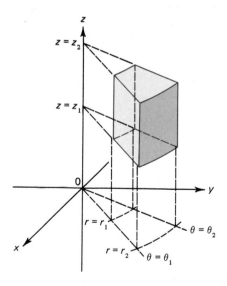

Figure 1

This is sketched in Figure 1. If we partition the z-axis for z in $[z_1, z_2]$, we obtain "slices." For a fixed z, the area of the face of a slice of π_c is, according to equation (19.4.5), given by

$$A_i = \int_{\theta_1}^{\theta_2} \int_{r_1}^{r_2} r\, dr\, d\theta. \tag{1}$$

The volume of a slice is, by (1), given by

$$V_i = A_i \, \Delta z_i = \left\{ \int_{\theta_1}^{\theta_0} \int_{r_1}^{r_2} r \, dr \, d\theta \right\} \Delta z_i.$$

Then, adding these volumes and taking the limit as the norm of the partition approaches zero, we obtain

$$\text{volume of } \pi_c = \int_{z_1}^{z_2} \int_{\theta_1}^{\theta_2} \int_{r_1}^{r_2} r \, dr \, d\theta \, dz. \tag{2}$$

In general, let the region S be given in cylindrical coordinates by

$$S = \{(r, \theta, z): \theta_1 \le \theta \le \theta_2, 0 \le g_1(\theta) \le r \le g_2(\theta), h_1(r, \theta) \le z \le h_2(r, \theta)\}, \tag{3}$$

and let f be a function of r, θ, and z. Then the triple integral of f over S is given by

$$\iiint_S f = \int_{\theta_1}^{\theta_2} \int_{g_1(\theta)}^{g_2(\theta)} \int_{h_1(r, \theta)}^{h_2(r, \theta)} f(r, \theta, z) r \, dz \, dr \, d\theta. \tag{4}$$

EXAMPLE 1. Find the mass of a solid bounded by the cylinder $r = \sin \theta$, the planes $z = 0$, $\theta = 0$, $\theta = \pi/3$ and the cone $z = r$, if the density is given by $\rho(r, \theta, z) = 4r$.

SOLUTION. We first note that the solid may be written as

$$S = \{(r, \theta, z): 0 \le \theta \le \pi/3, 0 \le r \le \sin \theta, 0 \le z \le r\}.$$

Then

$$\begin{aligned}
\mu &= \int_0^{\pi/3} \int_0^{\sin \theta} \int_0^r (4r) r \, dz \, dr \, d\theta \\
&= \int_0^{\pi/3} \int_0^{\sin \theta} \left\{ 4r^2 z \, \Big|_0^r \right\} dr \, d\theta = \int_0^{\pi/3} \int_0^{\sin \theta} 4r^3 \, dr \, d\theta \\
&= \int_0^{\pi/3} \left\{ r^4 \, \Big|_0^{\sin \theta} \right\} d\theta = \int_0^{\pi/3} \sin^4 \theta \, d\theta = \int_0^{\pi/3} \left(\frac{1 - \cos 2\theta}{2} \right)^2 d\theta \\
&= \frac{1}{4} \int_0^{\pi/3} \left(1 - 2\cos 2\theta + \frac{1 + \cos 4\theta}{2} \right) d\theta \\
&= \frac{1}{4} \left\{ \frac{3\theta}{2} - \sin 2\theta + \frac{\sin 4\theta}{8} \right\} \Big|_0^{\pi/3} = \frac{1}{4} \left(\frac{\pi}{2} - \frac{\sqrt{3}}{2} - \frac{\sqrt{3}}{16} \right) \\
&= \frac{\pi}{8} - \frac{9\sqrt{3}}{64}.
\end{aligned}$$

The formula for conversion from rectangular to cylindrical coordinates is virtually identical to formula (19.4.7). Let S be a region in $\mathbb{R}^3$. Since $x = r \cos \theta$,

$y = r \sin \theta$ and $z = z$, we have

$$\iiint_S f(x, y, z)\, dx\, dy\, dz = \iiint_S f(r \cos \theta, r \sin \theta, z)r\, dz\, dr\, d\theta. \qquad (5)$$

(given in rectan-   (given in cylin-
gular coordinates)   drical coordinates)

Remark. Again, do not forget the extra r when you convert to cylindrical coordinates.

EXAMPLE 2. Find the mass of the solid bounded by the paraboloid $z = x^2 + y^2$ and the plane $z = 4$ if the density at any point is proportional to the distance from the point to the z-axis.

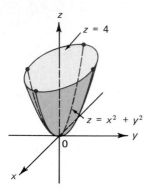

Figure 2

SOLUTION. The solid is sketched in Figure 2. The density is given by $\rho(x, y, z) = \alpha \sqrt{x^2 + y^2}$, where α is a constant of proportionality. Thus, since the solid may be written as

$$S = \{(x, y, z): -2 \le x \le 2, \; -\sqrt{4 - x^2} \le y \le \sqrt{4 - y^2}, \; x^2 + y^2 \le z \le 4\}$$

we have

$$\mu = \iiint_S \alpha \sqrt{x^2 + y^2}\, dx\, dy\, dz = \int_{-2}^{2} \int_{-\sqrt{4-x^2}}^{\sqrt{4-x^2}} \int_{x^2+y^2}^{4} \alpha \sqrt{x^2 + y^2}\, dz\, dy\, dx.$$

We write this in cylindrical coordinates, using the fact that $\alpha \sqrt{x^2 + y^2} = \alpha r$. We note that the largest value of r is 2 since, at the "top" of the solid, $r^2 = x^2 + y^2 = 4$. Then

$$\mu = \int_{0}^{2\pi} \int_{0}^{2} \int_{r^2}^{4} (\alpha r)r\, dz\, dr\, d\theta = \int_{0}^{2\pi} \int_{0}^{2} \alpha r^2(4 - r^2)\, dr\, d\theta$$

$$= \alpha \int_{0}^{2\pi} \left\{ \frac{4r^3}{3} - \frac{r^5}{5} \Big|_{0}^{2} \right\} d\theta = \frac{64\alpha}{15} \int_{0}^{2\pi} d\theta = \frac{128}{15}\pi\alpha.$$

II. SPHERICAL COORDINATES

Recall from Section 17.8 that a point P in $\mathbb{R}^3$ can be written in the spherical coordinates (ρ, θ, φ) where $\rho \ge 0$, $0 \le \theta < 2\pi$, $0 \le \varphi \le \pi$. Here ρ is the distance between the point and the origin, θ is the same as in cylindrical coordinates, and φ is the angle between $\overrightarrow{OP}$ and the positive z-axis. Consider the "spherical parallelepiped"

$$\pi_s = \{(\rho, \theta, \varphi): \rho_1 \leq \rho \leq \rho_2, \theta_1 \leq \theta \leq \theta_2, \varphi_1 \leq \varphi \leq \varphi_2\}. \tag{6}$$

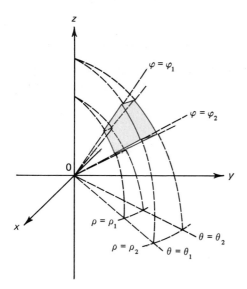

Figure 3

This is sketched in Figure 3. To approximate the volume of π_s, we partition the intervals $[\rho_1, \rho_2]$, $[\theta_1, \theta_2]$ and $[\varphi_1, \varphi_2]$. This gives us a number of "spherical boxes,"

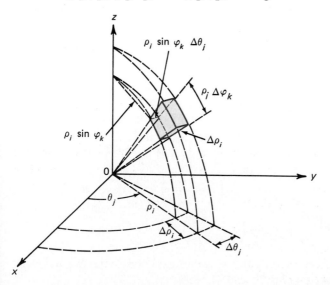

Figure 4

one of which is sketched in Figure 4. The length of an arc of a circle is given by

$$L = r\theta, \tag{7}$$

where r is the radius of the circle and θ is the angle which "cuts off" the arc (see equation (7.1.8)). In Figure 4, one side of the spherical box is $\Delta\rho_i$. Since $\rho < \rho_i$ is the equation of a sphere, we find, from (7), that the length of a second side is $\rho_i \, \Delta\varphi_k$. Finally, the length of the third side is $\rho_i \sin \psi_k \, \Delta\theta_j$. This follows from equation (7) and equation (17.8.5). Thus the volume of the box in Figure 4 is given, approximately, by

$$V_i \approx (\Delta\rho_i)(\rho_i \, \Delta\varphi_k)(\rho_i \sin \varphi_k \, \Delta\theta_j)$$

and, using a familiar argument,

$$\text{volume of } \pi_s = \iiint\limits_{\pi_s} \rho^2 \sin \varphi \, d\rho \, d\varphi \, d\theta = \int_{\theta_1}^{\theta_2} \int_{\varphi_1}^{\varphi_2} \int_{\rho_1}^{\rho_2} \rho^2 \sin \varphi \, d\rho \, d\varphi \, d\theta. \tag{8}$$

If f is a function of the variables ρ, θ, and φ, we have

$$\iiint\limits_{\pi_s} f = \iiint\limits_{\pi_s} f(\rho, \theta, \varphi)\rho^2 \sin \varphi \, d\rho \, d\varphi \, d\theta. \tag{9}$$

More generally, let the region S be defined in spherical coordinates by

$$S = \{(\rho, \theta, \varphi): \theta_1 \leq \theta \leq \theta_2, \, g_1(\theta) \leq \varphi \leq g_2(\theta), \, h_1(\theta, \varphi) \leq \rho \leq h_2(\theta, \varphi)\}. \tag{10}$$

Then

$$\iiint\limits_{S} f = \int_{\theta_1}^{\theta_2} \int_{g_1(\theta)}^{g_2(\theta)} \int_{h_1(\theta,\varphi)}^{h_2(\theta,\varphi)} f(\rho, \theta, \varphi)\rho^2 \sin \varphi \, d\rho \, d\varphi \, d\theta. \tag{11}$$

EXAMPLE 3. Calculate the mass of the spherical parallelepiped

$$S = \left\{(\rho, \varphi, \theta): 0 \leq \theta \leq \frac{\pi}{4}, \, 0 \leq \varphi \leq \frac{\pi}{6}, \, 1 \leq \rho \leq 2\right\}$$

whose density at $(\rho, \theta, \varphi) = \cos \theta$.

SOLUTION

$$V = \int_0^{\pi/4} \int_0^{\pi/6} \int_1^2 \cos \theta \, \rho^2 \sin \varphi \, d\rho \, d\varphi \, d\theta$$

$$= \int_0^{\pi/4} \int_0^{\pi/6} \left\{\cos \theta \, \frac{\rho^3}{3} \sin \varphi \, \Big|_1^2\right\} d\varphi \, d\theta = \frac{7}{3} \int_0^{\pi/4} \int_0^{\pi/6} \cos \theta \sin \varphi \, d\varphi \, d\theta$$

$$= \frac{7}{3} \int_0^{\pi/4} \left\{-\cos \theta \cos \varphi \, \Big|_0^{\pi/6}\right\} d\theta = \frac{7}{3}\left(1 - \frac{\sqrt{3}}{2}\right) \int_0^{\pi/4} \cos \theta \, d\theta$$

$$= \frac{7}{3}\left(1 - \frac{\sqrt{3}}{2}\right) \sin \theta \, \Big|_0^{\pi/4} = \frac{7}{3\sqrt{2}}\left(1 - \frac{\sqrt{3}}{2}\right).$$

Recall that to convert from rectangular to spherical coordinates we have the formulas (see equations (17.8.6), (17.8.7), and (17.8.8))

$$x = \rho \sin \varphi \cos \theta, \qquad y = \rho \sin \varphi \sin \theta, \qquad z = \rho \cos \varphi.$$

Thus to convert from a triple integral in rectangular coordinates to a triple integral in spherical coordinates, we have

$$\underset{S}{\int\int\int} f(x, y, z)\, dx\, dy\, dz = \underset{S}{\int\int\int} f(\rho \sin \varphi \cos \theta, \rho \sin \varphi \sin \theta, \rho \cos \varphi)\, \rho^2 \sin \varphi\, d\rho\, d\varphi\, d\theta.$$

(given in rectangular coordinates)   (given in spherical coordinates)   **(12)**

EXAMPLE 4. Calculate the volume enclosed by the sphere $x^2 + y^2 + z^2 = a^2$.

SOLUTION. In rectangular coordinates, since

$$S = \{(x, y, z): -a \leq x \leq a, -\sqrt{a^2 - x^2} \leq y \leq \sqrt{a^2 - x^2},$$
$$-\sqrt{a^2 - x^2 - y^2} \leq z \leq \sqrt{a^2 - x^2 - y^2}\},$$

we can write the volume

$$V = \int_{-a}^{a} \int_{-\sqrt{a^2 - x^2}}^{\sqrt{a^2 - x^2}} \int_{-\sqrt{a^2 - x^2 - y^2}}^{\sqrt{a^2 - x^2 - y^2}} dz\, dy\, dx.$$

This is very tedious to calculate. Instead, we note that the region enclosed by a sphere can be represented in spherical coordinates as

$$S = \{(\rho, \theta, \varphi): 0 \leq \rho \leq a, 0 \leq \theta \leq 2\pi, 0 \leq \varphi \leq \pi\}.$$

Thus

$$V = \int_0^{2\pi} \int_0^{\pi} \int_0^{a} \rho^2 \sin \varphi\, d\rho\, d\varphi\, d\theta = \int_0^{2\pi} \int_0^{\pi} \frac{a^3}{3} \sin \varphi\, d\varphi\, d\theta$$

$$= \frac{a^3}{3} \int_0^{2\pi} \left\{ -\cos \varphi \Big|_0^{\pi} \right\} d\theta = \frac{2a^3}{3} \int_0^{2\pi} d\theta = \frac{4\pi a^3}{3}.$$

EXAMPLE 5. Find the mass of the sphere in Example 4 if its density at a point is proportional to the distance from the point to the origin.

SOLUTION. We have density $= \alpha \sqrt{x^2 + y^2 + z^2} = \alpha\rho$. Thus

$$\mu = \int_0^{2\pi} \int_0^{\pi} \int_0^{a} \alpha\rho(\rho^2 \sin \varphi)\, d\rho\, d\varphi\, d\theta = \frac{\alpha a^4}{4} \int_0^{2\pi} \int_0^{\pi} \sin \varphi\, d\varphi\, d\theta$$

$$= \frac{\alpha a^4}{4} (4\pi) = \pi a^4 \alpha.$$

EXAMPLE 6. Find the centroid of the "hemispherical" region below the sphere $x^2 + y^2 + z^2 = z$ and above the cone $z^2 = x^2 + y^2$.

SOLUTION. The region is sketched in Figure 5. Since $x^2 + y^2 + z^2 = \rho^2$ and $z = \rho \cos \varphi$, the sphere $\{(x, y, z): x^2 + y^2 + z^2 = z\}$ can be written in spherical coordinates as

$$\rho^2 = \rho \cos \varphi \qquad \text{or} \qquad \rho = \cos \varphi.$$

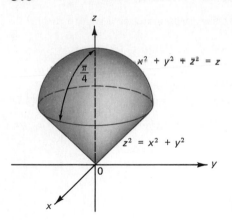

Figure 5

Since $x^2 + y^2 = \rho^2 \sin^2 \varphi \cos^2 \theta + \rho^2 \sin^2 \varphi \sin^2 \theta = \rho^2 \sin^2 \varphi$, the equation of the cone is

$$\rho^2 \cos^2 \varphi = \rho^2 \sin^2 \varphi \qquad \text{or} \qquad \cos^2 \varphi = \sin^2 \varphi$$

and, since $0 \leq \varphi \leq \pi$, the equation is

$$\varphi = \frac{\pi}{4}.$$

Thus, assuming that the density of the region is the constant k, we have

$$\mu = \int_0^{2\pi} \int_0^{\pi/4} \int_0^{\cos \varphi} k\rho^2 \sin \varphi \, d\rho \, d\varphi \, d\theta$$

$$= k \int_0^{2\pi} \int_0^{\pi/4} \left\{ \frac{\rho^3}{3} \sin \varphi \, \Big|_0^{\cos \varphi} \right\} d\varphi \, d\theta = k \int_0^{2\pi} \int_0^{\pi/4} \frac{\cos^3 \varphi}{3} \sin \varphi \, d\varphi \, d\theta$$

$$= k \int_0^{2\pi} \left\{ -\frac{\cos^4 \varphi}{12} \, \Big|_0^{\pi/4} \right\} d\theta = \frac{1}{12} k \int_0^{2\pi} \frac{3}{4} \, d\theta = \frac{\pi k}{8}.$$

Since $x = \rho \sin \varphi \cos \theta$, we have

$$M_{yz} = k \int_0^{2\pi} \int_0^{\pi/4} \int_0^{\cos \varphi} (\rho \sin \varphi \cos \theta) \rho^2 \sin \varphi \, d\rho \, d\varphi \, d\theta$$

$$= k \int_0^{2\pi} \int_0^{\pi/4} \left\{ \frac{\rho^4}{4} \sin^2 \varphi \cos \theta \, \Big|_0^{\cos \varphi} \right\} d\varphi \, d\theta$$

$$= \frac{k}{4} \int_0^{2\pi} \int_0^{\pi/4} \cos^4 \varphi \sin^2 \varphi \cos \theta \, d\varphi \, d\theta$$

$$= \frac{k}{32} \int_0^{2\pi} \int_0^{\pi/4} (1 + \cos 2\varphi)^2 (1 - \cos 2\varphi) \cos \theta \, d\varphi \, d\theta$$

(omitting many of the details)

$$= \frac{k}{32} \left(\frac{\pi}{8} + \frac{1}{6} \right) \int_0^{2\pi} \cos \theta \, d\theta = 0$$

(an unsurprising result because of symmetry). Similarly, since $y = \rho \sin \varphi \sin \theta$,

$$M_{xz} = k \int_0^{2\pi} \int_0^{\pi/4} \int_0^{\cos \varphi} (\rho \sin \varphi \sin \theta) \rho^2 \sin \rho \, d\rho \, d\varphi \, d\theta = 0$$

(again because of symmetry). Finally, since $z = \rho \cos \varphi$,

$$M_{xy} = k \int_0^{2\pi} \int_0^{\pi/4} \int_0^{\cos \varphi} (\rho \cos \varphi) \rho^2 \sin \varphi \, d\rho \, d\varphi \, d\theta$$

$$= \frac{k}{4} \int_0^{2\pi} \int_0^{\pi/4} \cos^5 \varphi \sin \varphi \, d\varphi \, d\theta = \frac{k}{4} \int_0^{2\pi} \left\{ -\frac{\cos^6 \varphi}{6} \Big|_0^{\pi/4} \right\} d\theta$$

$$= \frac{1}{24} \cdot \frac{7}{8} k \int_0^{2\pi} d\theta = \frac{7\pi k}{96}.$$

Thus $\bar{x} = \bar{y} = 0$ and $\bar{z} = M_{xy}/\mu = (7\pi k/96)/(\pi k/8) = 7/12$.

PROBLEMS 19.7

Solve Problems 1–11 using cylindrical coordinates.

1. Find the volume of the region inside both the sphere $x^2 + y^2 + z^2 = 4$ and the cylinder $(x - 1)^2 + y^2 = 1$.
2. Find the centroid of the region of Problem 1.
3. If the density of the region of Problem 1 is proportional to the square of the distance to the xy-plane and is measured in kg/m³, find the center of mass of the region.
4. Find the volume of the solid bounded above by the paraboloid $z = 4 - x^2 - y^2$ and below by the xy-plane.
5. Find the center of mass of the solid of Problem 4 if the density is proportional to the distance to the xy-plane.
6. Find the volume of the solid bounded by the plane $z = y$ and the paraboloid $z = x^2 + y^2$.
7. Find the centroid of the region of Problem 6.
8. Find the volume of the solid bounded by the two cones $z^2 = x^2 + y^2$ and $z^2 = 16x^2 + 16y^2$ between $z = 0$ and $z = 2$.
9. Find the center of mass of the solid in Problem 8 if the density at any point is proportional to the distance to the z-axis.
10. Find the volume of the solid bounded by the hyperboloid of two sheets $z^2 - x^2 - y^2 = 1$ and the cone $z^2 = x^2 + y^2$ for $0 \le z \le a$.
11. Evaluate $\int_0^1 \int_0^{\sqrt{1-x^2}} \int_{\sqrt{x^2+y^2}}^{\sqrt{1-x^2-y^2}} z^3 \, dz \, dy \, dx$.

Solve Problems 12–21 using spherical coordinates.

12. Find the mass of the unit sphere if the density at any point is proportional to the distance to the boundary of the sphere.
13. Find the volume of the solid inside the sphere $x^2 + y^2 + z^2 = 4$ and outside the cone $z^2 = x^2 + y^2$.
14. Find the center of mass of the solid of Problem 12 if the density at a point is proportional to the distance from the origin.
15. A solid fills the space between two concentric spheres centered at the origin. The radii of the spheres are a and b where $0 < a < b$. Find the mass of the solid if the density at each point is inversely proportional to its distance from the origin.

16. Find the center of mass of the object of Problem 13 if the density at a point is proportional to the square of the distance from the origin.

17. Find the volume of the smaller wedge cut from the unit sphere by two planes which meet on a diameter at an angle of $\pi/6$.

18. Find the mass of a wedge cut from a sphere of radius a by two planes which meet at a diameter at an angle of b radians if the density at any point on the wedge is proportional to the distance to the diameter.

19. Evaluate $\displaystyle\int_{-3}^{3}\int_{-\sqrt{9-x^2}}^{\sqrt{9-x^2}}\int_{-\sqrt{9-x^2-y^2}}^{\sqrt{9-x^2-y^2}} (x^2 + y^2 + z^2)^{3/2}\, dz\, dy\, dx.$

20. Evaluate $\displaystyle\int_{0}^{1}\int_{0}^{\sqrt{1-x^2}}\int_{0}^{\sqrt{1-x^2-y^2}} (1/\sqrt{x^2 + y^2 + z^2})\, dz\, dy\, dx.$

21. Evaluate $\displaystyle\int_{0}^{a}\int_{0}^{\sqrt{a^2-x^2}}\int_{0}^{\sqrt{a^2-x^2-y^2}} (z^3/(x^2 + y^2)^{1/2})\, dz\, dy\, dx.$

*22. A "sphere" in $\mathbb{R}^4$ has the equation $x^2 + y^2 + z^2 + w^2 = a^2$.

(a) Explain why it is reasonable that the volume of this "sphere" be given by

$$V = \int_{-a}^{a}\int_{-\sqrt{a^2-x^2}}^{\sqrt{a^2-x^2}}\int_{-\sqrt{a^2-x^2-y^2}}^{\sqrt{a^2-x^2-y^2}}\int_{-\sqrt{a^2-x^2-y^2-z^2}}^{\sqrt{a^2-x^2-y^2-z^2}} dw\, dz\, dy\, dx.$$

(b) Using spherical coordinates evaluate the integral in (a).

[*Hint:* The first integral is easy and reduces the "quadruple" integral to a triple integral.]

*19.8 Work and Line Integrals in the Plane

In Section 15.3 we showed that if a force $\mathbf{F}$ is applied to a particle which moves in the direction of a vector $\mathbf{d}$, then the work done by the force on the particle is given by

$$W = \mathbf{F}\cdot\mathbf{d}. \tag{1}$$

We now calculate the work done when a particle moves along a curve C. In doing so we shall define an important concept in applied mathematics—the line integral. Suppose that a curve in the plane is given parametrically by

$$C:\quad \mathbf{x}(t) = f(t)\mathbf{i} + g(t)\mathbf{j}, \tag{2}$$

or

$$C:\quad \mathbf{x}(t) = x(t)\mathbf{i} + y(t)\mathbf{j}. \tag{2'}$$

If a force is applied to a particle moving along C, then such a force will have magnitude and direction. That is, the force will be a vector function of the vector $\mathbf{x}(t)$. A vector function of a vector variable is called a *vector field* (see Section 18.10 for one example of a vector field). We write

$$\mathbf{F}(\mathbf{x}) = \mathbf{F}(x, y) = P(x, y)\mathbf{i} + Q(x, y)\mathbf{j} \tag{3}$$

where P and Q are scalar-valued functions. The problem is to determine the work done when a particle moves on C from a point $\mathbf{x}(a)$ to a point $\mathbf{x}(b)$ subject to the force $\mathbf{F}$ given by (3). We shall assume in our discussion that the curve C is *smooth* or *piecewise smooth.* By that we mean that the functions $f(t)$ and $g(t)$ in (2) are continuously differentiable or that f' and g' exist and are piecewise continuous.

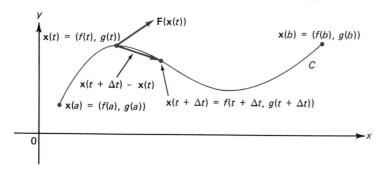

Figure 1

A typical curve C is sketched in Figure 1. Let

$$W(t) = \text{work done in moving from } \mathbf{x}(a) \text{ to } \mathbf{x}(t). \tag{4}$$

Then the work done in moving from $\mathbf{x}(t)$ to $\mathbf{x}(t + \Delta t)$ is given by

$$W(t + \Delta t) - W(t). \tag{5}$$

Now if Δt is small, then the part of the curve between $\mathbf{x}(t)$ and $\mathbf{x}(t + \Delta t)$ is "close" to a straight line and so can be approximated by the vector

$$\mathbf{x}(t + \Delta t) - \mathbf{x}(t). \tag{6}$$

If Δt is small and if $\mathbf{F}(\mathbf{x})$ is continuous, then the force applied between $\mathbf{x}(t)$ and $\mathbf{x}(t + \Delta t)$ is approximately equal to $\mathbf{F}(\mathbf{x}(t))$. Thus, by (1), if Δt is small, then

$$W(t + \Delta t) - W(t) \approx \mathbf{F}(\mathbf{x}(t)) \cdot [\mathbf{x}(t + \Delta t) - \mathbf{x}(t)]. \tag{7}$$

We divide both sides of (7) by Δt and take the limit as $\Delta t \to 0$ to obtain

$$\lim_{\Delta t \to 0} \frac{W(t + \Delta t) - W(t)}{\Delta t} = \lim_{\Delta t \to 0} \left(\mathbf{F}(\mathbf{x}(t)) \cdot \frac{[\mathbf{x}(t + \Delta t) - \mathbf{x}(t)]}{\Delta t} \right)$$

or

$$W'(t) = \mathbf{F}(\mathbf{x}(t)) \cdot \mathbf{x}'(t). \tag{8}$$

But

$$W(a) = 0$$

and

$$W(b) = \text{total work done on the particle.}$$

Thus

$$W = \text{total work done} = W(b) - W(a) = \int_a^b W'(t)\, dt$$

or

$$W = \int_a^b \mathbf{F}(\mathbf{x}(t)) \cdot \mathbf{x}'(t) \, dt. \tag{9}$$

We write equation (9) as

$$W = \int_C \mathbf{F}(\mathbf{x}) \cdot \mathbf{dx}. \tag{10}$$

The symbol $\int_C$ is read "the integral along the curve C." The integral in (10) is called a *line integral of* $\mathbf{F}$ *over* C.

Remark 1. If C lies along the x-axis, then C is given by $\mathbf{x}(t) = x(t)\mathbf{i} + 0\mathbf{j}\dagger$ and $\mathbf{x}'(t) = x'(t)\mathbf{i}$ so that (10) becomes

$$\int_{[a,b]} F(x) \, dx = \int_a^b F(x) \, dx$$

which is our usual definite integral.

Remark 2. Since $\mathbf{F}$ is given by (3), we can write equation (9) as

$$W = \int_a^b [P(x,y)f'(t) + Q(x,y)g'(t)] \, dt. \tag{11}$$

EXAMPLE 1. A particle is moving along the parabola $y = x^2$ subject to a force given by the vector field $2xy\mathbf{i} + (x^2 + y^2)\mathbf{j}$. How much work is done in moving from the point $(1, 1)$ to the point $(3, 9)$ if forces are measured in newtons and distances are measured in meters?

SOLUTION. The curve C is given parametrically by

$$\mathbf{x}(t) = t\mathbf{i} + t^2\mathbf{j} \qquad \text{between } t = 1 \text{ and } t = 3.$$

We therefore have $f(t) = t$ and $g(t) = t^2$. Then $P = 2xy = 2t^3$ and $Q = x^2 + y^2 = t^2 + t^4$. Also $f'(t) = 1$ and $g'(t) = 2t$ so that, by (11),

$$W = \int_1^3 [(2t^3)1 + (t^2 + t^4)(2t)] \, dt = \int_1^3 (4t^3 + 2t^5) \, dt = 322\tfrac{2}{3} \text{ joules.}$$

EXAMPLE 2. Two electrical charges of like polarity (i.e., both positive or both negative) will repel each other. If a charge of α coulombs is placed at the origin and a charge of 1 coulomb of the same polarity is at the point (x, y), then the force of repulsion is given by

$$F(x, y) = \frac{\alpha x}{(x^2 + y^2)^{3/2}} \mathbf{i} + \frac{\alpha y}{(x^2 + y^2)^{3/2}} \mathbf{j}.$$

How much work is done by the force on the 1 coulomb charge as the charge moves on the straight line from $(1, 0)$ to $(3, -2)$?

$\dagger$ We have substituted $x(t)$ for $f(t)$ and 0 for $g(t)$ here.

SOLUTION. The straight line is the line $y = 1 - x$ or, parametrically,

$$\mathbf{x}(t) = t\mathbf{i} + (1 - t)\mathbf{j} \qquad \text{for} \quad 1 \leq t \leq 3.$$

Then

$$P = \frac{\alpha t}{[(t^2 + (1 - t)^2]^{3/2}} = \frac{\alpha t}{(2t^2 - 2t + 1)^{3/2}}$$

and

$$Q = \frac{\alpha(1 - t)}{[t^2 + (1 - t)^2]^{3/2}} = \frac{\alpha(t - 1)}{(2t^2 - 2t + 1)^{3/2}}.$$

Then, since $f'(t) = 1$ and $g'(t) = -1$,

$$W = \alpha \int_1^3 \left\{ \frac{t}{(2t^2 - 2t + 1)^{3/2}} (1) + \frac{(1 - t)}{(2t^2 - 2t + 1)^{3/2}} (-1) \right\} dt$$

$$= \alpha \int_1^3 \frac{2t - 1}{(2t^2 - 2t + 1)^{3/2}} \, dt = \frac{\alpha}{2} \int_1^3 \frac{4t - 2}{(2t^2 - 2t + 1)^{3/2}} \, dt$$

$$= -\alpha(2t^2 - 2t + 1)^{-1/2} \Big|_1^3 = \alpha \left(1 - \frac{1}{\sqrt{13}} \right).$$

EXAMPLE 3. How much work is done by the force on the charge in Example 2 if it moves in the counterclockwise direction along the semicircle $x^2 + y^2 = a^2, y \geq 0$?

SOLUTION. Here

$$\mathbf{x}(t) = a(\cos t)\mathbf{i} + a(\sin t)\mathbf{j}, \qquad 0 \leq t \leq \pi.$$

Then

$$P = \frac{\alpha x}{(x^2 + y^2)^{3/2}} = \frac{a\alpha \cos t}{(a^2 \cos^2 t + a^2 \sin^2 t)^{3/2}} = \frac{a\alpha \cos t}{a^3} = \frac{\alpha \cos t}{a^2}$$

and

$$Q = \frac{\alpha y}{(x^2 + y^2)^{3/2}} = \frac{\alpha \sin t}{a^2}.$$

Then, since $f'(t) = -a \sin t$ and $g'(t) = a \cos t$,

$$W = \frac{\alpha}{a} \int_0^\pi [\cos t(-\sin t) + \sin t(\cos t)] \, dt = 0.$$

This is no surprise since, in this example, the force $\mathbf{F}$ and the direction of the curve $\mathbf{x}'$ are orthogonal. Thus the component of the force in the direction of motion is zero, which means that the force does no work.

We used the notion of work to motivate the discussion of the line integral. We now give a general definition of the line integral in the plane.

DEFINITION 1. Let P and Q be continuous on a set S containing the smooth curve C given by

$$C: \quad \mathbf{x}(t) = f(t)\mathbf{i} + g(t)\mathbf{j}, \qquad t \in [a, b].$$

Let the vector field $\mathbf{F}$ be given by

$$\mathbf{F}(x, y) = P(x, y)\mathbf{i} + Q(x, y)\mathbf{j}.$$

Then the line integral of $\mathbf{F}$ over C is given by

$$\int_{C} \mathbf{F}(\mathbf{x}) \cdot d\mathbf{x} = \int_{a}^{b} [P(f(t), g(t))f'(t) + Q(f(t), g(t))g'(t)]\, dt. \tag{12}$$

Remark. If C is piecewise smooth but not smooth, then C is made up of a number of "sections" each of which is smooth. Since C is continuous, these sections are joined. Some typical piecewise smooth curves are sketched in Figure 2. If C is made up of the n smooth curves $C_1, C_2, \ldots, C_n$, then

$$\int_{C} \mathbf{F}(\mathbf{x}) \cdot d\mathbf{x} = \int_{C_1} \mathbf{F}(\mathbf{x}) \cdot d\mathbf{x} + \int_{C_2} \mathbf{F}(\mathbf{x}) \cdot d\mathbf{x} + \cdots + \int_{C_n} \mathbf{F}(\mathbf{x}) \cdot d\mathbf{x}. \tag{13}$$

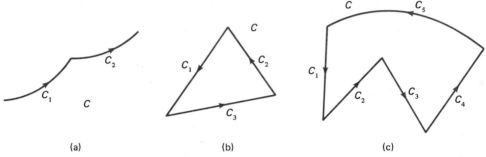

(a)   (b)   (c)

Figure 2

EXAMPLE 4.   Calculate $\int_{C} \mathbf{F}(\mathbf{x}) \cdot d\mathbf{x}$, where $\mathbf{F}(x, y) = xy\mathbf{i} + ye^{x}\mathbf{j}$ and C is the rectangle joining the points $(0, 0)$, $(2, 0)$, $(2, 1)$, and $(0, 1)$ if C is traversed in the counterclockwise direction.

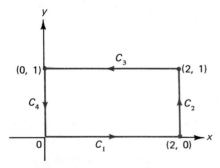

Figure 3

SOLUTION. The rectangle is sketched in Figure 3, and it is made up of four smooth curves (straight lines). We have

$$
\begin{aligned}
&C_1: \quad \mathbf{x}(t) = t\mathbf{i}, && 0 \le t \le 2 \\
&C_2: \quad \mathbf{x}(t) = 2\mathbf{i} + t\mathbf{j}, && 0 \le t \le 1 \\
&C_3: \quad \mathbf{x}(t) = (2 - t)\mathbf{i} + \mathbf{j}, && 0 \le t \le 2 \\
&C_4: \quad \mathbf{x}(t) = (1 - t)\mathbf{j}, && 0 \le t \le 1.
\end{aligned}
$$

Then

$$\int_C \mathbf{F(x)} \cdot \mathbf{dx} = \int_{C_1} \mathbf{F(x)} \cdot \mathbf{dx} + \int_{C_2} \mathbf{F(x)} \cdot \mathbf{dx} + \int_{C_3} \mathbf{F(x)} \cdot \mathbf{dx} + \int_{C_4} \mathbf{F(x)} \cdot \mathbf{dx}.$$

On C_1, $x = t$ and $y = 0$ so that $xy = 0$, $ye^x = 0$ and

$$\int_{C_1} \mathbf{F(x)} \cdot \mathbf{dx} = 0.$$

On C_2, $x = 2$, $y = t$, $x' = 0$, $y' = 1$ and $ye^x = te^2$ so that

$$\int_{C_2} \mathbf{F(x)} \cdot \mathbf{dx} = \int_0^1 te^2 \, dt = \frac{e^2}{2}.$$

On C_3, $x = (2 - t)$, $y = 1$, $x' = -1$, $y' = 0$ and $xy = 2 - t$ so that

$$\int_{C_3} \mathbf{F(x)} \cdot \mathbf{dx} = \int_0^2 (2 - t)(-1) \, dt = -2.$$

On C_4, $x = 0$, $y = 1 - t$, $x' = 0$, $y' = -1$, $xy = 0$ and $ye^x = 1 - t$ so that

$$\int_{C_4} \mathbf{F(x)} \cdot \mathbf{dx} = \int_0^1 (1 - t)(-1) \, dt = -\frac{1}{2}.$$

Thus

$$\int_C \mathbf{F(x)} \cdot \mathbf{dx} = 0 + \frac{e^2}{2} - 2 - \frac{1}{2} = \frac{e^2}{2} - \frac{5}{2} \approx 1.2$$

There are certain conditions under which the calculation of line integrals becomes very easy. We first illustrate what we have in mind with an example.

EXAMPLE 5. Let $\mathbf{F}(x, y) = y\mathbf{i} + x\mathbf{j}$. Calculate $\int_C \mathbf{F(x)} \cdot \mathbf{dx}$ where C is

(a) the straight line from $(0, 0)$ to $(1, 1)$.
(b) the parabola $y = x^2$ from $(0, 0)$ to $(1, 1)$.
(c) the curve $\mathbf{x}(t) = t^{3/2}\mathbf{i} + t^5\mathbf{j}$ from $(0, 0)$ to $(1, 1)$.

SOLUTION. (a) C is given by $\mathbf{x}(t) = t\mathbf{i} + t\mathbf{j}$. Then $\mathbf{x}'(t) = \mathbf{i} + \mathbf{j}$ and $\mathbf{F} \cdot \mathbf{x}' = (t\mathbf{i} + t\mathbf{j}) \cdot (\mathbf{i} + \mathbf{j}) = 2t$ so that

$$\int_C \mathbf{F(x)} \cdot \mathbf{dx} = \int_0^1 2t \, dt = 1.$$

(b) C is given by $\mathbf{x}(t) = t\mathbf{i} + t^2\mathbf{j}$. Then $\mathbf{x}'(t) = \mathbf{i} + 2t\mathbf{j}$ and $\mathbf{F} \cdot \mathbf{x}' = (t^2\mathbf{i} + t\mathbf{j}) \cdot (\mathbf{i} + 2t\mathbf{j}) = 3t^2$ so that

$$\int_C \mathbf{F(x)} \cdot \mathbf{dx} = \int_0^1 3t^2 \, dt = 1.$$

(c) Here $\mathbf{x}'(t) = \frac{3}{2}\sqrt{t}\,\mathbf{i} + 5t^4\mathbf{j}$ and $\mathbf{F}(x, y) = t^5\mathbf{i} + t^{3/2}\mathbf{j}$ so that $\mathbf{F} \cdot \mathbf{x}' = (t^5\mathbf{i} + t^{3/2}\mathbf{j}) \cdot (\frac{3}{2}\sqrt{t}\,\mathbf{i} + 5t^4\mathbf{j}) = \frac{3}{2}t^{11/2} + 5t^{11/2} = \frac{13}{2}t^{11/2}$. Then

$$\int_C \mathbf{F} \cdot \mathbf{x}' \, dt = \int_0^1 \tfrac{13}{2}t^{11/2} \, dt = 1.$$

We see that on three very different curves, we obtain the same answer as we move between the points $(0, 0)$ and $(1, 1)$. In fact, as we shall show in a moment, we will get the same answer if we integrate along any smooth curve C passing through these two points. When this happens we say that the line integral is *independent of the path*. A condition that ensures that a line integral is independent of the path over which it is integrated is given below.

THEOREM 1. Let $\mathbf{F}$ be continuous on an open disk D in $\mathbb{R}^2$ and suppose that $\mathbf{F}$ is the gradient of a function f. Then (i) for any piecewise smooth curve C lying in D, the line integral

$$\int_C \mathbf{F}(\mathbf{x}) \cdot d\mathbf{x}$$

is independent of the path and (ii) for any piecewise smooth curve C in D starting at the point $\mathbf{x}_0$ and ending at the point $\mathbf{x}_1$,

$$\boxed{\int_C \mathbf{F}(\mathbf{x}) \cdot d\mathbf{x} = f(\mathbf{x}_1) - f(\mathbf{x}_0);} \tag{14}$$

that is, the value of the integral depends only upon the end points of the path.

> **Remark 1.** This theorem is really the line integral analog of the fundamental theorem of calculus. It says that we can evaluate the line integral of an exact vector field by evaluating at two points the function for which $\mathbf{F}$ is the gradient.
>
> **Remark 2.** It is important that $\mathbf{F}$ be continuous on D, not only on C. (See Problem 38.)

PROOF OF THEOREM. Recall the vector form of the chain rule:

$$\frac{d}{dt} f(\mathbf{x}(t)) = \nabla f(\mathbf{x}(t)) \cdot \mathbf{x}'(t) \tag{15}$$

(see equation (18.7.3)). Now suppose that C is given by $\mathbf{x}(t)$, $a \le t \le b$, $\mathbf{x}(a) = \mathbf{x}_0$ and $\mathbf{x}(b) = \mathbf{x}_1$. We shall assume that C is smooth. Otherwise we could write the line integral in the form (13) and treat the integral over each smooth curve C_i separately. Using (15), we have

$$\int_C \mathbf{F}(\mathbf{x}) \cdot d\mathbf{x} = \int_a^b \mathbf{F}(\mathbf{x}(t)) \cdot \mathbf{x}'(t) \, dt = \int_a^b \nabla f(\mathbf{x}(t)) \cdot \mathbf{x}'(t) \, dt$$

$$= \int_a^b \frac{d}{dt} f(\mathbf{x}(t)) \, dt = f(\mathbf{x}(b)) - f(\mathbf{x}(a)) = f(\mathbf{x}_1) - f(\mathbf{x}_0)$$

and the theorem is proved since $f(\mathbf{x}_1) - f(\mathbf{x}_0)$ does not depend on the particular curve chosen.

EXAMPLE 5 (Continued). Since $y\mathbf{i} + x\mathbf{j} = \nabla(xy)$, we immediately find that for any curve C starting at $(0, 0)$ and ending at $(1, 1)$

$$\int_C \mathbf{F}(\mathbf{x}) \cdot d\mathbf{x} = xy|_{(1,1)} - xy|_{(0,0)} = 1 - 0 = 1.$$

Using the results of Section 18.12, we have the following obvious corollary to Theorem 1.

COROLLARY. Let $\mathbf{F}(x, y) = P(x, y)\mathbf{i} + Q(x, y)\mathbf{j}$. Then if $P, Q, \partial P/\partial y$ and $\partial Q/\partial x$ are all continuous on an open disk containing C and if $\partial P/\partial y = \partial Q/\partial x$, $\int_C \mathbf{F}(\mathbf{x}) \cdot d\mathbf{x}$ is independent of the path.

EXAMPLE 6.   Let $\mathbf{F}(x, y) = [4x^3y^3 + (1/x)]\mathbf{i} + [3x^4y^2 - (1/y)]\mathbf{j}$.   Calculate $\int_C \mathbf{F}(\mathbf{x}) \cdot d\mathbf{x}$ for any smooth curve C starting at $(1, 1)$ and ending at $(2, 3)$.

SOLUTION. We showed in Example 18.12.3 that $\mathbf{F}$ is exact and that $\mathbf{F} = \nabla f$ where $f(x, y) = x^4y^3 + \ln|x/y|$. Thus $\int_C \mathbf{F}(\mathbf{x}) \cdot d\mathbf{x}$ is independent of the path and

$$\int_C \mathbf{F}(\mathbf{x}) \cdot d\mathbf{x} = f(2, 3) - f(1, 1) = (16)(27) + \ln(\tfrac{2}{3}) - 1 = 431 + \ln(\tfrac{2}{3}).$$

There is another important consequence of Theorem 1. Let C be a closed curve (that is, $x_0 = x_1$). Then if $\mathbf{F}$ is exact,

$$\int_C \mathbf{F}(\mathbf{x}) \cdot d\mathbf{x} = f(\mathbf{x}_1) - f(\mathbf{x}_0) = 0.$$

There is an interesting physical interpretation of this fact. It says that

---

The work done by a conservative force field as it moves a particle around a closed path is zero.†

This also follows from the more general fact that if $\mathbf{F}$ is conservative, then the work done in moving an object between two points depends only on the points, and not on the path taken between the points.

PROBLEMS 19.8

In Problems 1–14 calculate $\int_C \mathbf{F}(\mathbf{x}) \cdot d\mathbf{x}$.
1. $\mathbf{F}(x, y) = x^2\mathbf{i} + y^2\mathbf{j}$; C is the straight line from $(0, 0)$ to $(2, 4)$
2. $\mathbf{F}(x, y) = x^2\mathbf{i} + y^2\mathbf{j}$; C is the parabola $y = x^2$ from $(0, 0)$ to $(2, 4)$
3. $\mathbf{F}(x, y) = xy\mathbf{i} + (y - x)\mathbf{j}$; C is the line $y = 2x - 4$ from $(1, -2)$ to $(2, 0)$
4. $\mathbf{F}(x, y) = xy\mathbf{i} + (y - x)\mathbf{j}$; C is the curve $y = \sqrt{x}$ from $(0, 0)$ to $(1, 1)$
5. $\mathbf{F}(x, y) = xy\mathbf{i} + (y - x)\mathbf{j}$; C is the unit circle in the counterclockwise direction
6. $\mathbf{F}(x, y) = xy\mathbf{i} + (y - x)\mathbf{j}$; C is the triangle joining the points $(0, 0)$, $(0, 1)$, and $(1, 0)$ in the counterclockwise direction
7. $\mathbf{F}(x, y) = xy\mathbf{i} + (y - x)\mathbf{j}$; C is the triangle joining the points $(0, 0)$, $(1, 0)$, and $(1, 1)$ in the counterclockwise direction
8. $\mathbf{F}(x, y) = e^x\mathbf{i} + e^y\mathbf{j}$; C is the curve of Problem 4

---

† This is true in general when the closed path is contained in a disc over which the force field is continuous.

9. $F(x, y) = (x^2 + 2y)i - y^2j$; C is the part of the ellipse $x^2 + 9y^2 = 9$ joining the points $(0, -1)$ and $(0, 1)$ in the clockwise direction

10. $F(x, y) = (\cos x)i - (\sin y)j$; C is the curve of Problem 9

11. $F(x, y) = e^{x+y}i + e^{x-y}j$; C is the curve of Problem 6

12. $F(x, y) = e^{x+y}i + e^{x-y}j$; C is the curve of Problem 7

13. $F(x, y) = (y/x^2)i + (x/y^2)j$; C is the straight line segment from $(2, 1)$ to $(4, 6)$

14. $F(x, y) = (\ln x)i + (\ln y)j$; C is the curve $x(t) = 2ti + t^3j$ for $1 \le t \le 4$.

In Problems 15–19 forces are given in newtons and distances are given in meters.

15. Calculate the work done when a force field $F(x, y) = x^3i + xyj$ moves a particle from the point $(0, 1)$ to the point $(1, e^{\pi/2})$ along the curve $x(t) = (\sin t)i + e^tj$.

16. Calculate the work done when the force field $F(x, y) = xyi + (2x^3 - y)j$ moves a particle around the unit circle in the counterclockwise direction.

17. What is the work done if the particle in Problem 16 is moved in the clockwise direction?

18. Calculate the work done by the force field $F(x, y) = -y^2xi + 2xj$ when a particle is moved around the ellipse $(x^2/a^2) + (y^2/b^2) = 1$ in the counterclockwise direction.

19. Calculate the work done by the force field $F(x, y) = 2xyi + y^2j$ when a particle is moved around the triangle of Problem 7.

20. What is the work done on the 1 coulomb particle of Example 2 if it moves along the line $y = 2x - 3$ from $(1, -1)$ to $(2, 1)$?

21. Let $F(x, y, z) = P(x, y, z)i + Q(x, y, z)j + R(x, y, z)k$ and let C be given by $x(t) = f(t)i + g(t)j + h(t)k$ where f, g, and h are continuously differentiable and t goes from a to b. Write a suitable definition for the line integral $\int_C F(x) \cdot dx$.

In Problems 22–25 use the definition obtained in Problem 21 to calculate $\int_C F(x) \cdot dx$.

22. $F(x, y, z) = xi + yj + zk$; C is the curve $x(t) = ti + t^2j + t^3k$ from $(0, 0, 0)$ to $(1, 1, 1)$

23. $F(x, y, z) = 2xzi - xyj + yz^2k$; C is the curve of Problem 22

24. $F(x, y, z) = x^2i + y^2j + z^2k$; C is the helix $x(t) = (\cos t)i + (\sin t)j + tk$ from $(1, 0, 0)$ to $(0, 1, \pi/2)$

25. $F(x, y, z) = yzi + xzj + xyk$; C is the curve of Problem 24.

In Problems 26–33 show that F is exact and use Theorem 1 to calculate $\int_C F(x) \cdot dx$ where C is any smooth curve starting at x_0 and ending at x_1.

26. $F(x, y) = 2xyi + (x^2 + 1)j$; $x_0 = (0, 1)$, $x_1 = (2, 3)$

27. $F(x, y) = (4x^2 - 4y^2)i + (\ln y - 8xy)j$; $x_0 = (-1, 1)$, $x_1 = (4, e)$

28. $F(x, y) = [x \cos(x + y) + \sin(x + y)]i + x \cos(x + y)j$; $x_0 = (0, 0)$, $x_1 = (\pi/6, \pi/3)$

29. $F(x, y) = [(1/x^2) + y^2]i + 2xyj$; $x_0 = (1, 4)$, $x_1 = (3, 2)$

30. $F(x, y) = 2x (\cos y)i - x^2 (\sin y)j$; $x_0 = (0, \pi/2)$, $x_1 = (\pi/2, 0)$

31. $F(x, y) = (2xy^3 - 2)i + (3x^2y^2 + \cos y)j$; $x_0 = (1, 0)$, $x_1 = (0, -\pi)$

32. $F(x, y) = e^yi + xe^yj$; $x_0 = (0, 0)$, $x_1 = (5, 7)$

33. $F(x, y) = (\cosh x)(\cosh y)i + (\sinh x)(\sinh y)j$; $x_0 = (0, 0)$, $x_1 = (1, 2)$.

34. If $F(x, y, z) = P(x, y, z)i + Q(x, y, z)j + R(x, y, z)k$ is exact, show that $\int_C F(x) \cdot dx$ is independent of the path.

In Problems 35–37 use the result of Problem 34 and Problem 18.12.14 to evaluate $\int_C F(x) \cdot dx$.

35. $F(x, y, z) = yzi + xzj + xyk$ from $(0, 0, 0)$ to $(1, 1, 1)$

36. $F(x, y, z) = [(y/z) + x^2]i + [(x/z) - \sin y]j + [\cos z - (xy/z^2)]k$ from $(0, \pi/3, \pi/2)$ to $(1, \pi/2, \pi)$

37. $F(x, y, z) = \left(\dfrac{1}{x + 2y + 3z} - 3x\right)\mathbf{i} + \left(\dfrac{2}{x + 2y + 3z} + y^2\right)\mathbf{j} + \dfrac{3}{x + 2y + 3z}\mathbf{k}$

from $(1, 0, 0)$ to $(2, 3, 4)$.

*38. Let the force field $\mathbf{F}$ be given by

$$F(x, y) = \frac{y}{x^2 + y^2}\mathbf{i} - \frac{x}{x^2 + y^2}\mathbf{j}.$$

(a) Show that $\mathbf{F}$ is conservative.
(b) Calculate $\int_C \mathbf{F}(\mathbf{x}) \cdot \mathbf{dx}$, where C is the unit circle in the counterclockwise direction.
(c) Explain why $\int_C \mathbf{F}(\mathbf{x}) \cdot \mathbf{dx} \neq 0$. Does this contradict the last fact stated in this section?

*19.9 Green's Theorem in the Plane

In this section we state a result which shows an interesting relationship between line integrals and double integrals.

Let Ω be a region in the plane (a typical region is sketched in Figure 1). The

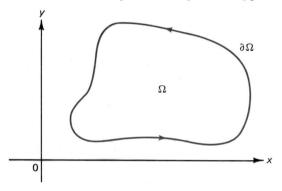

Figure 1

curve (indicated by the arrows) which goes around the edge of Ω in the counter-clockwise direction is called the *boundary of* Ω and is denoted $\partial\Omega$. Let

$$F(x, y) = P(x, y)\mathbf{i} + Q(x, y)\mathbf{j}.$$

If the curve $\partial\Omega$ is given by

$$\partial\Omega: \quad \mathbf{x}(t) = x(t)\mathbf{i} + y(t)\mathbf{j}$$

we can write

$$F(\mathbf{x}) \cdot \mathbf{dx} = P\, dx + Q\, dy. \tag{1}$$

We then denote the line integral of $\mathbf{F}$ around $\partial\Omega$ by

$$\oint_{\partial\Omega} P\, dx + Q\, dy. \tag{2}$$

The symbol $\oint_{\partial\Omega}$ indicates that $\partial\Omega$ is a closed curve around which we integrate in the counterclockwise direction (the direction of the arrow).

We are now ready to state the principle result of this section.

THEOREM 1 (Green's Theorem† in the Plane). Let Ω be a region in the xy-plane and let $\partial\Omega$ denote the boundary of Ω with $\partial\Omega$ piecewise smooth. Let P and Q be continuous with continuous first partials in an open disk containing Ω. Then

$$\oint_{\partial\Omega} P\,dx + Q\,dy = \iint_{\Omega} \left(\frac{\partial Q}{\partial x} - \frac{\partial P}{\partial y}\right) dx\,dy. \tag{3}$$

Remark. This theorem shows how the line integral of a function around the boundary of a region is related to a double integral over that region.

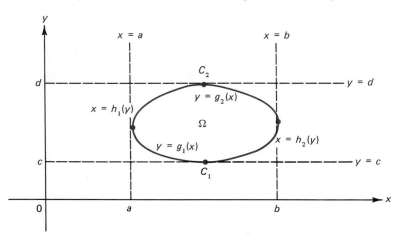

Figure 2

PROOF OF GREEN'S THEOREM. We prove Green's theorem in the case Ω takes the simple form given in Figure 2. The region Ω can be written

$$\{(x, y): a \leq x \leq b, g_1(x) \leq y \leq g_2(x)\} \tag{4}$$

or

$$\{(x, y): c \leq y \leq d, h_1(y) \leq x \leq h_2(y)\}. \tag{5}$$

We first calculate

$$\iint_{\Omega} \frac{\partial P}{\partial y}\,dx\,dy = \int_a^b \left\{\int_{g_1(x)}^{g_2(x)} \frac{\partial P}{\partial y}\,dy\right\} dx = \int_a^b \left\{P(x, y)\,\Big|_{y=g_1(x)}^{y=g_2(x)}\right\} dx$$

$$= \int_a^b [P(x, g_2(x)) - P(x, g_1(x))]\,dx. \tag{6}$$

Now

$$\oint_{\partial\Omega} P\,dx = \int_{C_1} P\,dx + \int_{C_2} P\,dx, \tag{7}$$

†Named after George Green (1793–1841), a British mathematician and physicist who wrote an essay in 1828 on electricity and magnetism that contained this important theorem.

where C_1 is the graph of $g_1(x)$ from $x = a$ to $x = b$ and C_2 is the graph of $g_2(x)$ from $x = b$ to $x = a$ (note the order). Since, on C_1, $y = g_1(x)$, we have

$$\int_{C_1} P(x, y)\, dx = \int_a^b P(x, g_1(x))\, dx. \tag{8}$$

Similarly

$$\int_{C_2} P(x, y)\, dx = \int_b^a P(x, g_2(x))\, dx = -\int_a^b P(x, g_2(x))\, dx. \tag{9}$$

Thus

$$\oint_{\partial\Omega} P\, dx = \int_a^b P(x, g_1(x))\, dx - \int_a^b P(x, g_2(x))\, dx$$

$$= -\int_a^b [P(x, g_2(x)) - P(x, g_1(x))]\, dx. \tag{10}$$

Comparing (6) and (10) we find that

$$\iint_\Omega -\frac{\partial P}{\partial y}\, dx\, dy = \oint_{\partial\Omega} P\, dx. \tag{11}$$

Similarly, using the representation (5), we have

$$\iint_\Omega \frac{\partial Q}{\partial x}\, dx\, dy = \int_c^d \left[\int_{h_1(y)}^{h_2(y)} \frac{\partial Q}{\partial x}\, dx \right] dy$$

$$= \int_c^d [Q(h_2(y), y) - Q(h_1(y), y)]\, dy$$

which, by analogous reasoning, yields the equation

$$\iint_\Omega \frac{\partial Q}{\partial x}\, dx\, dy = \oint_{\partial\Omega} Q\, dy. \tag{12}$$

Adding (11) and (12) completes the proof of the theorem in the special case that Ω can be written in the form (4), (5).†

EXAMPLE 1.   Evaluate $\oint_{\partial\Omega} xy\, dx + (x - y)\, dy$, where Ω is the rectangle $\{(x, y): 0 \le x \le 1,\ 1 \le y \le 3\}$.

†For a proof of Green's theorem for more general regions, see R. C. Buck, *Advanced Calculus*, McGraw-Hill, New York, 1965, p. 406.

SOLUTION. $P(x, y) = xy$, $Q(x, y) = x - y$, $\partial Q/\partial x = 1$ and $\partial P/\partial y = x$ so that

$$\oint_{\partial\Omega} xy \, dx + (x - y) \, dy = \int_0^1 \int_1^3 (1 - x) \, dy \, dx = \int_0^1 \left\{ (1 - x)y \Big|_1^3 \right\} dx$$

$$= \int_0^1 2(1 - x) \, dx = 1.$$

EXAMPLE 2. Evaluate $\oint_{\partial\Omega}(x^3 + y^3) \, dx + (2y^3 - x^3) \, dy$, where Ω is the unit disk.

SOLUTION. $(\partial Q/\partial x) - (\partial P/\partial y) = -3x^2 - 3y^2 = -3(x^2 + y^2)$. Thus

$$\oint_{\partial\Omega} (x^3 + y^3) \, dx + (2y^3 - x^3) \, dy = -3 \underset{\substack{\text{unit} \\ \text{disk}}}{\iint}(x^2 + y^2) \, dx \, dy$$

(converting to polar coordinates)

$$= -3 \int_0^{2\pi} \int_0^1 r^2 \cdot r \, dr \, d\theta = -3 \int_0^{2\pi} \left\{ \frac{r^4}{4} \Big|_0^1 \right\} d\theta = -\frac{3}{4} \int_0^{2\pi} d\theta = -\frac{3\pi}{2}.$$

Green's theorem can be useful for calculating area. Recall that

$$\text{area enclosed by } \Omega = \iint_\Omega dx \, dy. \tag{13}$$

But, by Green's theorem,

$$\iint_\Omega dx \, dy = \oint_{\partial\Omega} x \, dy = \oint_{\partial\Omega} (-y) \, dx = \frac{1}{2} \oint_{\partial\Omega} [(-y) \, dx + x \, dy] \tag{14}$$

(explain why). Any of the line integrals in (14) can be used to calculate area.

EXAMPLE 3. Use Green's theorem to calculate the area enclosed by the ellipse $(x^2/a^2) + (y^2/b^2) = 1$.

SOLUTION. The ellipse can be written parametrically as

$$\mathbf{x}(t) = a(\cos t)\mathbf{i} + b(\sin t)\mathbf{j}.$$

Then, using the first line integral in (14),

$$A = \oint_{\partial\Omega} x \, dy = \int_0^{2\pi} (a \cos t) \frac{d}{dt} (b \sin t) \, dt$$

$$= \int_0^{2\pi} (a \cos t)b \cos t \, dt = ab \int_0^{2\pi} \cos^2 t \, dt$$

$$= \frac{ab}{2} \int_0^{2\pi} (1 + \cos 2t) \, dt = \pi ab.$$

Note how much easier this calculation is than the direct evaluation of $\iint_A dx \, dy$, where A denotes the area enclosed by the ellipse.

There are two very interesting and important vector interpretations of Green's theorem.

DEFINITION 1.   Let $\mathbf{F}(x, y) = P(x, y)\mathbf{i} + Q(x, y)\mathbf{j}$ be a vector field in the plane.
   (i)   The *curl* of $\mathbf{F}$ is given by

$$\text{curl } \mathbf{F} = \frac{\partial Q}{\partial x} - \frac{\partial P}{\partial y}. \tag{15}$$

   (ii)   The *divergence* of $\mathbf{F}$, denoted div $\mathbf{F}$, is given by

$$\text{div } \mathbf{F} = \frac{\partial P}{\partial x} + \frac{\partial Q}{\partial y}. \tag{16}$$

Before explaining these terms more fully, we shall write Green's theorem in two equivalent vector forms.

First, recall from equation (16.6.2) that if $\mathbf{T}$ denotes the unit tangent vector to a curve $\mathbf{x}(t) = x(t)\mathbf{i} + y(t)\mathbf{j}$ and if s denotes the parameter of arc length, then

$$\mathbf{T} = \frac{d\mathbf{x}}{ds}$$

or

$$d\mathbf{x} = \mathbf{T}\, ds. \tag{17}$$

Now let $\mathbf{T}$ denote the unit tangent vector to $\partial\Omega$. Then if $\mathbf{F} = P\mathbf{i} + Q\mathbf{j}$, we can write, using (17),

$$\oint_{\partial\Omega} P\, dx + Q\, dy = \oint_{\partial\Omega} \mathbf{F} \cdot d\mathbf{x} = \oint_{\partial\Omega} \mathbf{F} \cdot \mathbf{T}\, ds. \tag{18}$$

Then, applying Green's theorem and using (15) and (18), we obtain

THEOREM 2 (First Vector Form of Green's Theorem)

$$\oint_{\partial\Omega} \mathbf{F} \cdot \mathbf{T}\, ds = \iint_{\Omega} \text{curl } \mathbf{F}\, dx\, dy. \tag{19}$$

Now, from the expression

$$\oint_{\partial\Omega} P\, dx + Q\, dy = \iint_{\Omega} \left(\frac{\partial Q}{\partial x} - \frac{\partial P}{\partial y} \right) dx\, dy$$

we may replace P by $-Q$ and Q by P to obtain

$$\oint_{\partial\Omega} -Q\, dx + P\, dy = \iint_{\Omega} \left(\frac{\partial P}{\partial x} + \frac{\partial Q}{\partial y} \right) dx\, dy. \tag{20}$$

If $dx\,\mathbf{i} + dy\,\mathbf{j}$ represents the vector $\mathbf{T}\,ds$, then $-dy\,\mathbf{i} + dx\,\mathbf{j}$ represents the vector $\mathbf{n}\,ds$ where $\mathbf{n}$ is the unit normal vector to the curve. This is easy to see since $(dx\,\mathbf{i} + dy\,\mathbf{j}) \cdot (-dy\,\mathbf{i} + dx\,\mathbf{j}) = 0$ and both $\mathbf{T}\,ds$ and $\mathbf{n}\,ds$ have the same magnitude. Thus the right-hand side of (20) can be written

$$\oint_{\partial\Omega} -Q\,dx + P\,dy = \oint_{\partial\Omega} \mathbf{F} \cdot \mathbf{n}\,ds. \tag{21}$$

Using (16) and (21) in (20) yields the

THEOREM 3 (Second Vector Form of Green's Theorem)

$$\oint_{\partial\Omega} \mathbf{F} \cdot \mathbf{n}\,ds = \iint_{\Omega} \operatorname{div} \mathbf{F}\,dx\,dy. \tag{22}$$

There are interesting physical interpretations to the two vector forms of Green's theorem. Let $\mathbf{F}(x, y)$ denote the direction and rate of flow of a fluid at a point (x, y) in the plane. The integral

$$\oint_{\partial\Omega} \mathbf{F} \cdot \mathbf{T}\,ds$$

is the integral of the component of the flow in the direction tangent to the boundary of Ω and is called the *circulation* of $\mathbf{F}$ around the boundary of Ω (see Figure 3). If Ω

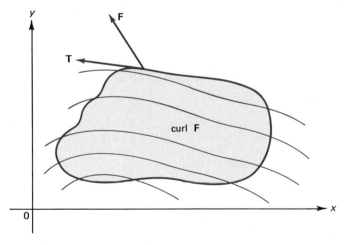

Figure 3

is small, then curl $\mathbf{F}$ is nearly a constant and, by (19), we have

$$\oint_{\partial\Omega} \mathbf{F} \cdot \mathbf{T}\,ds = \iint_{\Omega} \operatorname{curl} \mathbf{F}\,dx\,dy \approx \operatorname{curl} \mathbf{F}(x, y) \iint_{\Omega} dx\,dy$$

$$\approx (\operatorname{curl} \mathbf{F}(x, y))(\text{area of } \Omega). \tag{23}$$

From (23) it appears that the curl represents the circulation per unit area at the point (x, y). If curl $\mathbf{F} = 0$ for every (x, y) in Ω, then the fluid flow $\mathbf{F}$ is called *irrotational*.

Now $\oint_{\partial\Omega} \mathbf{F} \cdot \mathbf{n}\,ds$ is the component of flow in the direction of the outward normal to $\partial\Omega$ and is called the *flux* across $\partial\Omega$ (see Figure 4). The flux is the rate at which fluid

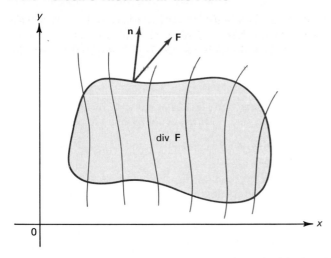

Figure 4

is flowing across the boundary of Ω from inside Ω. If Ω is small, then, using (22),

$$\oint_{\partial\Omega} \mathbf{F} \cdot \mathbf{n} \, ds = \iint_{\Omega} \operatorname{div} \mathbf{F} \, dx \, dy \approx (\operatorname{div} \mathbf{F}(x, y))(\text{area of } \Omega). \tag{24}$$

Thus the divergence of $\mathbf{F}$ represents the net rate of flow away from (x, y). If $\operatorname{div} \mathbf{F} = 0$ for every (x, y) in Ω, then the flow $\mathbf{F}$ is called *incompressible*. This is because if $\operatorname{div} \mathbf{F} = 0$, then there is no flow from any point across $\partial\Omega$ in the normal direction which means that the fluid density at any point (x, y) in Ω is constant.

PROBLEMS 19.9

In Problems 1–15 find the line integral using Green's theorem.

1. $\oint_{\partial\Omega} 3y \, dx + 5x \, dy$; $\Omega = \{(x, y): 0 \le x \le 1, 0 \le y \le 1\}$
2. $\oint_{\partial\Omega} ay \, dx + bx \, dy$; Ω is the region of Problem 1
3. $\oint_{\partial\Omega} e^x \cos y \, dx + e^x \sin y \, dy$; Ω is the region enclosed by the triangle with vertices at $(0, 0)$, $(1, 0)$, and $(0, 1)$
4. The integral of Problem 3, where Ω is the region enclosed by the triangle with vertices at $(0, 0)$, $(1, 1)$ and $(1, 0)$
5. The integral of Problem 3, where Ω is the region enclosed by the rectangle with vertices at $(0, 0)$, $(2, 0)$, $(2, 1)$, and $(0, 1)$
6. $\oint_{\partial\Omega} 2xy \, dx + x^2 \, dy$; Ω is the unit disk
7. $\oint_{\partial\Omega}(x^2 + y^2) \, dx - 2xy \, dy$; Ω is the unit disk
8. $\oint_{\partial\Omega}(1/y) \, dx + (1/x) \, dy$; Ω is the region bounded by the lines $y = 1$, $x = 16$, and the curve $y = \sqrt{x}$
9. $\oint_{\partial\Omega} \cos y \, dx + \cos x \, dy$; Ω is the region enclosed by the rectangle $\{(x, y): 0 \le x \le \pi/4, 0 \le y \le \pi/3\}$
10. $\oint_{\partial\Omega} x^2 y \, dx - xy^2 \, dy$; Ω is the disk $x^2 + y^2 \le 9$
11. $\oint_{\partial\Omega} y \ln x \, dy$; $\Omega = \{(x, y): 1 \le y \le 3, e^y \le x \le e^{y^3}\}$
12. $\oint_{\partial\Omega} \sqrt{1 + y^2} \, dx$; $\Omega = \{(x, y): -1 \le y \le 1, y^2 \le x \le 1\}$
13. $\oint_{\partial\Omega} ay \, dx + bx \, dy$; Ω is a region of the type (4), (5)
14. $\oint_{\partial\Omega} e^x \sin y \, dx + e^x \cos y \, dy$; Ω is the region enclosed by the ellipse $(x^2/a^2) + (y^2/b^2)$
15. $\oint_{\partial\Omega}(-4x/\sqrt{1 + y^2}) \, dx + (2x^2y/(1 + y^2)^{3/2}) \, dy$; Ω is a region of the type (4), (5).

16. Use one of the line integrals in (14) to calculate the area of the circle $\mathbf{x}(t) = a(\cos t)\mathbf{i} + a(\sin t)\mathbf{j}$.

17. Use Green's theorem to calculate the area enclosed by the triangle with vertices at (a_1, b_1), (a_2, b_2) and (a_3, b_3), assuming that the three points are not collinear.

18. Use Green's theorem to calculate the area of the quadrilateral with vertices at $(0, 0)$, $(2, 1)$, $(-1, 3)$, and $(4, 4)$.

19. Use Green's theorem to calculate the area of the quadrilateral with vertices at (a_1, b_1), (a_2, b_2), (a_3, b_3), and (a_4, b_4), assuming that no three of the points are collinear.

In Problems 20–26 calculate (a) curl $\mathbf{F}$, (b) $\oint_{\partial\Omega} \mathbf{F} \cdot \mathbf{T}\, ds$, (c) div $\mathbf{F}$, and (d) $\oint_{\partial\Omega} \mathbf{F} \cdot \mathbf{n}\, ds$.

20. $\mathbf{F}(x, y) = x^2\mathbf{i} + y^2\mathbf{j}$; Ω is the region of Problem 1
21. $\mathbf{F}(x, y) = y^2\mathbf{i} + x^2\mathbf{j}$; Ω is the region of Problem 1
22. $\mathbf{F}(x, y) = ay\mathbf{i} + bx\mathbf{j}$; Ω is the region of Problem 1
23. $\mathbf{F}(x, y) = y^3\mathbf{i} + x^3\mathbf{j}$; Ω is the unit disk
24. $\mathbf{F}(x, y) = x\mathbf{i} + y\mathbf{j}$; Ω is the unit disk
25. $\mathbf{F}(x, y) = y\mathbf{i} - x\mathbf{j}$; Ω is the unit disk
26. $\mathbf{F}(x, y) = xy\mathbf{i} + (y^2 - x^2)\mathbf{j}$; Ω is the region of Problem 3.

27. Let $\partial\Omega$ be the ellipse $(x^2/a^2) + (y^2/b^2) = 1$. Let $\mathbf{F}(x, y)$ be the vector field $-x\mathbf{i} - y\mathbf{j}$ that, at any point (x, y), points toward the origin. Show that $\oint_{\partial\Omega} \mathbf{F} \cdot \mathbf{T}\, ds = 0$.

28. Let Ω be the disk $x^2 + y^2 \le a^2$. Show that

$$\oint_{\partial\Omega} \alpha\sqrt{x^2 + y^2}\, dx + \beta\sqrt{x^2 + y^2}\, dy = 0.$$

29. Let Ω be as in Problem 28 and suppose that g is continuously differentiable. Show that

$$\oint_{\partial\Omega} \alpha g(x^2 + y^2)\, dx + \beta g(x^2 + y^2)\, dy = 0.$$

30. Show that the vector flow $\mathbf{F}(x, y) = x\mathbf{i} + y\mathbf{j}$ is irrotational.
31. Show that the vector flow $\mathbf{F}(x, y) = \sin x\, e^x\mathbf{i} + y^{5/2}\mathbf{j}$ is irrotational.
32. Show that for any continuously differentiable functions f and g, the vector flow $\mathbf{F}(x, y) = f(x)\mathbf{i} + g(y)\mathbf{j}$ is irrotational.
33. Show that the vector flow $\mathbf{F}(x, y) = y\sqrt{x^2 + y^2}\mathbf{i} - x\sqrt{x^2 + y^2}\mathbf{j}$ is incompressible.
34. Let g be as in Problem 29. Show that the vector flow $\mathbf{F}(x, y) = -yg(x^2 + y^2)\mathbf{i} + xg(x^2 + y^2)\mathbf{j}$ is incompressible.

Review Exercises for Chapter Nineteen

In Exercises 1–8 evaluate the integral.

1. $\displaystyle\int_0^1 \int_0^2 x^2y\, dx\, dy$

2. $\displaystyle\int_0^1 \int_{x^2}^x xy^3\, dy\, dx$

3. $\displaystyle\int_2^4 \int_{1+y}^{2+5y} (x - y^2)\, dx\, dy$

4. $\displaystyle\int_0^4 \int_{-\sqrt{16-x^2}}^{\sqrt{16-x^2}} 4y\, dy\, dx$

5. $\iint_\Omega(y - x^2)\, dx\, dy$, where $\Omega = \{(x, y):\ -3 \le x \le 3, 0 \le y \le 2\}$
6. $\iint_\Omega(y - x^2)\, dx\, dy$, where Ω is the region in the first quadrant bounded by the x-axis, the y-axis, and the circle $x^2 + y^2 = 4$
7. $\iint_\Omega(x + y^2)\, dx\, dy$, where Ω is the region in the first quadrant bounded by the x-axis, the y-axis, and the unit circle.
8. $\iint_\Omega(2x + y)e^{-(x+y)}\, dx\, dy$, where Ω is the first quadrant.

9. Find upper and lower bounds for $\iint_\Omega \sin(x^2 + y^2)\, dx\, dy$, where Ω is the region of Exercise 6.

10. Change the order of integration of $\int_2^5 \int_1^x 3x^2 y\, dy\, dx$ and evaluate.

11. Change the order of integration of $\int_0^3 \int_0^{\sqrt{9-x^2}} (9 - y^2)^{3/2}\, dy\, dx$ and evaluate.

12. Change the order of integration of $\int_0^\infty \int_z^\infty f(x, y)\, dy\, dx$.

13. Find the volume of the solid bounded by the plane $x + 2y + 3z = 6$ and the three coordinate planes.

14. Find the volume of the solid bounded by the cylinder $y^2 + z^2 = 4$ and the planes $x = 0$ and $x = 3$.

15. Find the volume enclosed by the ellipsoid $4x^2 + y^2 + 25z^2 = 100$.

16. Find the volume of the solid bounded by the paraboloid $x = y^2 + z^2$ and the plane $y = x - 2$.

17. Find the center of mass of the unit disk if the density at a point is proportional to the distance to the boundary of the disk.

18. Find the center of mass of the region $\Omega = \{(x, y) : 0 \le x \le 3, \ 1 \le y \le 5\}$ if $\rho(x, y) = 3x^2 y$.

19. Find the center of mass of the triangular region bounded by the lines $y = x$, $y = 2 - x$ and the y-axis if $\rho(x, y) = 3xy - y^3$.

20. Calculate the volume under the surface $z = r^5$ and over the circle of radius 6 centered at the origin.

21. Calculate the area of the region enclosed by the curve $r = 2(1 + \sin\theta)$.

22. Calculate the area of the region inside the cardioid $r = 3(1 + \sin\theta)$ and outside the circle $r = 2$.

23. Find the volume of the solid bounded by the cone $z^2 = x^2 + y^2$ and the paraboloid $4z = x^2 + y^2$.

24. Find the centroid of the region bounded by the right-hand loop of the lemniscate $r^2 = 4\cos 2\theta$.

25. Find the area of the part of the plane $z = 3x + 5y$ over the region bounded by the curves $x = y^2$ and $x = y^3$.

26. Find the area of the part of the surface $z = (y^4/4) + (1/8y^2)$ over the rectangle $\{(x, y) : 0 \le x \le 3, 2 \le y \le 4\}$.

27. Find the surface area of the hemisphere $x^2 + y^2 + z^2 = 16$, $z \ge 0$.

28. Evaluate $\int_0^1 \int_0^y \int_0^x x^2\, dz\, dx\, dy.$ 29. Evaluate $\int_1^2 \int_{2-x}^{2+x} \int_0^{xz} 12xyz\, dy\, dz\, dx.$

30. Change the order of integration in Exercise 28 and write the integral in the form

(a) $\int_?^? \int_?^? \int_?^? x^2\, dx\, dy\, dz$ (b) $\int_?^? \int_?^? \int_?^? x^2\, dy\, dz\, dx.$

31. Find the volume of the tetrahedron with vertices at $(0, 0, 0)$, $(2, 0, 0)$, $(0, 1, 0)$, and $(0, 0, 3)$.

32. Find the volume of the solid bounded by the parabolic cylinder $z = 4 - y^2$ and the planes $z = x$ and $z = 2x$ which lies in the half plane $z \ge 0$.

33. Find the mass of the solid of Exercise 31 if the density function is $\rho(x, y, z) = 2x - y^2$.

34. Find the center of mass of the solid bounded by the ellipsoid $4x^2 + y^2 + 25z^2 = 100$ if the density function is $\rho(x, y, z) = z^2$.

35. Find the centroid of the solid in Exercise 32.

36. Find the volume of the region inside both the sphere $x^2 + y^2 + z^2 = a^2$ and the cylinder

$$x^2 + \left(y - \frac{a}{2}\right)^2 = \left(\frac{a}{2}\right)^2.$$

37. Find the centroid of the solid of Exercise 36.

38. If the density of the region of Exercise 36 is proportional to the square of the distance to the xy-plane, find the center of mass of the region.

39. Find the volume of the solid bounded by the plane $z = x$ and the paraboloid $z = x^2 + y^2$.

40. Evaluate $\int_{-2}^{2} \int_{-\sqrt{4-x^2}}^{\sqrt{4-x^2}} \int_{\sqrt{x^2+y^2}}^{\sqrt{4-x^2-y^2}} z^2 \, dz \, dy \, dx.$

41. Find the volume of the solid inside the sphere $x^2 + y^2 + z^2 = 9$ and outside the cone $z^2 = x^2 + y^2$.

42. Find the center of mass of the solid in Exercise 41 if the density at a point is proportional to the distance to the origin.

43. Find the volume of a wedge cut from a sphere of radius 2 by two planes which meet at a diameter at an angle of $\pi/3$.

44. Evaluate $\int_{0}^{2} \int_{0}^{\sqrt{4-x^2}} \int_{0}^{\sqrt{4-x^2-y^2}} \frac{z^2}{(x^2 + y^2)^{1/2}} \, dz \, dy \, dx.$

In Exercises 45–47 calculate $\int_C \mathbf{F}(\mathbf{x}) \cdot d\mathbf{x}.$

45. $\mathbf{F}(x, y) = x^2\mathbf{i} + y^2\mathbf{j}$; C is the curve $y = x^{3/2}$ from $(0, 0)$ to $(1, 1)$.

46. $\mathbf{F}(x, y) = x^2y\mathbf{i} - xy^2\mathbf{j}$; C is the unit circle in the counterclockwise direction.

47. $\mathbf{F}(x, y) = 3xy\mathbf{i} - y\mathbf{j}$; C is the triangle joining the points $(0, 0)$, $(1, 1)$, and $(0, 1)$ in the counterclockwise direction.

48. Calculate the work done when the force field $\mathbf{F}(x, y) = x^2y\mathbf{i} + (y^3 + x^3)\mathbf{j}$ moves a particle around the unit circle in the counterclockwise direction.

49. Calculate the work done when the force field $\mathbf{F}(x, y) = 3(x - y)\mathbf{i} + x^5\mathbf{j}$ moves a particle around the triangle of Exercise 47.

50. Show that $\mathbf{F}(x, y) = 3x^2y^2\mathbf{i} + 2x^3y\mathbf{j}$ is exact and calculate $\int_C \mathbf{F}(\mathbf{x}) \cdot d\mathbf{x}$, where C starts at $(1, 2)$ and ends at $(3, -1)$.

51. Show that $\mathbf{F}(x, y) = e^{xy}(1 + xy)\mathbf{i} + x^2e^{xy}\mathbf{j}$ is exact and calculate $\int_C \mathbf{F}(\mathbf{x}) \cdot d\mathbf{x}$, where C starts at $(-1, 0)$ and ends at $(0, 1)$.

52. Evaluate $\oint_{\partial\Omega} 2y \, dx + 4x \, dy$, where $\Omega = \{(x, y): 0 \le x \le 2, 0 \le y \le 2\}$.

53. Evaluate $\oint_{\partial\Omega} x^2y \, dx + xy^2 \, dy$, where Ω is the region enclosed by the triangle of Exercise 47.

54. Evaluate $\oint_{\partial\Omega}(x^3 + y^3) \, dx + (x^2y^2) \, dy$, where Ω is the unit disk.

55. Evaluate $\oint_{\partial\Omega} \sqrt{1 + x^2} \, dy$; $\Omega = \{(x, y): -1 \le x \le 1, x^2 \le y \le 1\}$.

56. Evaluate

$$\oint_{\partial\Omega} \frac{2x^2 + y^2 - xy}{(x^2 + y^2)^{1/2}} \, dx + \frac{xy - x^2 - 2y^2}{(x^2 + y^2)^{1/2}} \, dy$$

where Ω is a region of the type (19.9.4), (19.9.5) which does not contain the origin.

57. Let $\mathbf{F}(x, y) = xy^2\mathbf{i} + x^2y\mathbf{j}$ and let Ω denote the disk of radius 2. Calculate (a) curl $\mathbf{F}$, (b) $\oint_{\partial\Omega} \mathbf{F} \cdot \mathbf{T} \, ds$, (c) div $\mathbf{F}$, and (d) $\oint_{\partial\Omega} \mathbf{F} \cdot \mathbf{n} \, ds$.

58. Do the same for $\mathbf{F}(x, y) = y^2\mathbf{i} - x^2\mathbf{j}$, where Ω is the region enclosed by the triangle of Exercise 47.

59. Show that the vector flow $\mathbf{F}(x, y) = (\cos x^2)\mathbf{i} + e^y\mathbf{j}$ is irrotational.

60. Show that the vector flow $\mathbf{F}(x, y) = -(x^2y + y^3)\mathbf{i} + (x^3 + xy^2)\mathbf{j}$ is incompressible.

TWENTY

ORDINARY DIFFERENTIAL EQUATIONS

20.1 Introduction

Many of the basic laws of the physical sciences and, more recently, of the biological and social sciences are formulated in terms of mathematical relations involving certain known and unknown quantities and their derivatives. Such relations are called *differential equations.*

We discussed some very simple differential equations in Sections 6.4 and 7.7. In this chapter we will take a brief look at some other types of differential equations. In this section we briefly categorize the types of differential equations that may be encountered.

The most obvious classification is based on the nature of the derivative (or derivatives) in the equation. A differential equation involving only ordinary derivatives (derivatives of functions of one variable) is called an *ordinary differential equation.* The equations in Section 6.4 and 7.7 are all ordinary differential equations. A differential equation containing partial derivatives is called a *partial differential equation.* Examples of partial differential equations are given in Problems 18.4.41, 18.4.42, 18.5.31, 18.5.32, 18.5.35, and 18.5.39.

In this chapter we will only consider the solution of certain ordinary differential equations which we will simply refer to as differential equations, dropping the word "ordinary." Procedures for solving all but the most trivial partial differential equations are beyond the scope of this elementary discussion.

The *order* of a differential equation is the order of the highest order derivative appearing in the equation.

EXAMPLE 1. The following are examples of differential equations with indicated orders.

(a) $dy/dx = 3x$ (first order).
(b) $x''(t) + 4x'(t) - x(t) = \sin t$ (second order).
(c) $(dy/dx)^5 - 2e^y = 6 \cos x$ (first order).
(d) $(y^{(iv)})^2 - 2y''' + y' - 6y = \sqrt{x}$ (fourth order).

In Example 6.4.1 we solved the differential equation

$$\frac{dy}{dx} = 3y \tag{1}$$

$$y(0) = 2. \tag{2}$$

A solution to (1), (2) is defined to be a function which

     (i)   is differentiable,
     (ii)   satisfies the differential equation (1) on some open interval containing 0,
and
     (iii)   satisfies the initial condition (2).

We stated that the only solution to (1) which satisfied (2) was given by

$$y = 2e^{3x}. \tag{3}$$

A proof that (3) is the only solution to the system (1), (2) is beyond the scope of this text but can be found in any book on differential equations.† The system (1), (2) is called an *initial value problem*. In general, for a certain class of initial value problems there is a unique solution.‡ This makes sense intuitively. For example, in Example 6.4.2 we found the following system of differential equations governing the population growth of a certain species of bacteria:

$$\frac{dP}{dt} = 0.1P \tag{4}$$

such that

$$P(0) = 10{,}000. \tag{5}$$

It is reasonable to expect that starting with the initial population given by (5) and assuming that population growth is governed by the equation (4), the population in the future will be completely determined. That is, there is one and only one way to write the population $P(t)$, which is a way of saying that the initial value problem (4), (5) has a unique solution.

     There are other types of problems in which other than initial conditions are given. For example, let a string be held taut between two points one meter apart as

Figure 1

in Figure 1. Suppose the string is then plucked so that it begins to vibrate. Let $v(x, t)$ denote the height of the string at a distance x units from the left-hand endpoint and at a time t, where $0 \le x \le 1$ and $t \ge 0$. Then, since the left- and right-hand endpoints are held fixed, we have the conditions

$$v(0, t) = v(1, t) = 0 \qquad \text{for all} \quad t \ge 0. \tag{6}$$

The conditions (6) are called *boundary conditions* and if we write a differential equation governing the motion of the string (see Problem 18.5.31), then the equation, together with the boundary conditions (6), is called a *boundary value problem*. In general, a boundary value problem is a differential equation together with values given at two or more points, while an initial value problem is a differential equation with values given at only one point.

---

† See, for example, W. R. Derrick and S. I. Grossman, *Elementary Differential Equations with Applications*, Addison-Wesley, Reading, Massachusetts, 1976, Chapter 11.
‡ Theorems on the existence and uniqueness of solutions to a wide class of ordinary differential equations can be found in the text footnoted above.

EXAMPLE 2. The problem

$$y'' + 2y' + 3y = 0; \qquad y(0) = 1, \quad y(1) = 2$$

is a boundary value problem. The problem

$$y'' + 2y' + 3y = 0; \qquad y(0) = 1, \quad y'(0) = 2$$

is an initial value problem.

In this chapter we will consider only simple first- and second-order ordinary differential equations and some simple initial value problems. For a more complete discussion the reader is referred to an introductory book on differential equations like the one cited earlier in this section.

PROBLEMS 20.1

In Problems 1–6 state the order of the differential equation.

1. $x'(t) + 2x(t) = t$
2. $(d^2y/dx^2) + 4(dy/dx)^3 - y = x^3$
3. $y''' + y = 0$
4. $(dx/dt)^5 = x^4$
5. $x'' - x^2 = 2x^{(iv)}$
6. $(x''(t))^4 - (x'(t))^{3/2} = 2x(t) + e^t$.

7. Verify that $y_1 = 2 \sin x$ and $y_2 = -4 \cos x$ are solutions to $y'' + y = 0$.
8. Verify that $y = (x/2)e^x$ is a solution to $y'' - y = e^x$.
9. Verify that $y = 3x/(4x - 3)$ is the solution to the initial value problem

$$x^2 \, dy/dx + y^2 = 0; \, y(1) = 3.$$

10. Verify that $y = -\frac{1}{2}x^2 e^{-x}$ is a solution to the equation $y''' - 3y' - 2y = 3e^{-x}$.
11. Verify that $y = e^{3x}$ satisfies the initial value problem

$$y'' - 6y' + 9y = 0, \qquad y(0) = 1, \quad y'(0) = 3.$$

[*Hint:* First check that the equation is satisfied and then verify that the two initial conditions are also satisfied.]

*12. Verify that $y = e^{-x/2}[\cos(\sqrt{3}/2)x + (7/\sqrt{3}) \sin(\sqrt{3}/2)x]$ is the solution to the initial value problem

$$y'' + y' + y = 0, \qquad y(0) = 1, \quad y'(0) = 3.$$

20.2 First-Order Equations—
Separation of Variables

Consider the first-order differential equation

$$\frac{dy}{dx} = f(x, y). \tag{1}$$

Suppose that $f(x, y)$ can be written as

$$f(x, y) = \frac{g(x)}{h(y)}, \tag{2}$$

where g and h are each functions of only one variable. Then (1) can be written

$$h(y)\frac{dy}{dx} = g(x)$$

or, integrating both sides with respect to x,

$$\int h(y)\,dy = \int h(y)\frac{dy}{dx}\,dx = \int g(x)\,dx + C. \tag{3}$$

The method of solution suggested in (3) is called the method of *separation of variables*. In general, if a differential equation can be written in the form $h(y)\,dy = g(x)\,dx$, then direct integration of both sides (if possible) will produce a family of solutions.

EXAMPLE 1. Solve the differential equation $dy/dx = 4y$.

SOLUTION. We solved equations like this one in Section 6.4. We have $f(x, y) = 4y$ and we can write

$$\frac{dy}{y} = 4\,dx$$

so that

$$\int \frac{dy}{y} = \int 4\,dx$$

or

$$\ln|y| = 4x + C$$

or

$$|y| = e^{4x+C} = e^{4x}e^{C}$$

which can be written

$$y = ke^{4x},$$

where k is any real number.

EXAMPLE 2. Solve the initial value problem $dy/dx = y^2(1 + x^2)$, $y(0) = 1$.

SOLUTION. We have $f(x, y) = y^2(1 + x^2) = g(x)/h(y)$, where $g(x) = 1 + x^2$ and $h(y) = 1/y^2$. Then, successively,

$$\frac{dy}{y^2} = (1 + x^2)\,dx, \qquad \int \frac{dy}{y^2} = \int (1 + x^2)\,dx, \qquad -\frac{1}{y} = x + \frac{x^3}{3} + C$$

or

$$y = -\frac{1}{x + (x^3/3) + C}.$$

For every number C, this is a solution to the differential equation. Moreover, the constant function $y \equiv 0$ is also a solution. When $x = 0$, $y = 1$ so that

$$1 = y(0) = -\frac{1}{0 + 0 + C} = -\frac{1}{C}$$

implying that $C = -1$ and we obtain the unique solution to the initial value problem:

$$y = -\frac{1}{x + (x^3/3) - 1}.$$

Remark. This solution (like any solution to a differential equation) can be checked by differentiation. You should *always* carry out this check. To check, we have

$$\frac{dy}{dx} = \frac{1}{[x + (x^3/3) - 1]^2}(1 + x^2) = \left\{\frac{-1}{[x + (x^3/3) - 1]}\right\}^2(1 + x^2)$$

$$= y^2(1 + x^2).$$

Also, $y(0) = -1/(0 + 0 - 1) = 1$.

EXAMPLE 3. Solve the differential equation $dx/dt = t\sqrt{1 - x^2}$.

SOLUTION. We have

$$\frac{dx}{\sqrt{1 - x^2}} = t\,dt \quad\text{or}\quad \int \frac{dx}{\sqrt{1 - x^2}} = \int t\,dt.$$

Integration yields

$$\sin^{-1}x = \frac{t^2}{2} + C \quad\text{or}\quad x = \sin\left(\frac{t^2}{2} + C\right).$$

This should be checked by differentiation, using the fact that $\cos[(t^2/2) + C] = \sqrt{1 - \sin^2[(t^2/2) + C]}$. Note that, as in Example 1, there are an infinite number of solutions to the equation. We can obtain a unique solution by specifying an initial condition. For example, if we have $x(0) = \frac{1}{2}$, then $\frac{1}{2} = \sin(0 + C) = \sin C$ and $C = \sin^{-1}\frac{1}{2} = \pi/6$. Then the unique solution would be

$$x(t) = \sin\left(\frac{t^2}{2} + \frac{\pi}{6}\right).$$

EXAMPLE 4. The growth rate per individual in a population is the difference between the average birth rate and the average death rate. Suppose that in a given population the average birth rate is a positive constant β, but the average death rate, due to the effects of crowding and increased competition for the available food, is proportional to the size of the population. We call this constant of proportionality δ (which is > 0). If $P(t)$ denotes the population at time t, then dP/dt is the growth rate of the population. The growth rate per individual is given by

$$\frac{1}{P}\frac{dP}{dt}.$$

Then, using the conditions described above, we have

$$\frac{1}{P}\frac{dP}{dt} = \beta - \delta P$$

or

$$\frac{dP}{dt} = P(\beta - \delta P). \tag{4}$$

This differential equation, together with the condition

$$P(0) = P_0 \quad \text{(the initial population)} \tag{5}$$

is an initial value problem. To solve, we have

$$\frac{dP}{P(\beta - \delta P)} = dt$$

or

$$\int \frac{dP}{P(\beta - \delta P)} = \int dt = t + C. \tag{6}$$

To calculate the integral on the left, we use partial fractions. We have (verify this)

$$\frac{1}{P(\beta - \delta P)} = \frac{1}{\beta P} + \frac{\delta}{\beta(\beta - \delta P)}$$

so that

$$\int \frac{dP}{P(\beta - \delta P)} = \int \frac{1}{\beta} \frac{dP}{P} + \frac{\delta}{\beta} \int \frac{dP}{\beta - \delta P}$$

$$= \frac{1}{\beta} \ln P - \frac{1}{\beta} \ln(\beta - \delta P) = t + C$$

or

$$\ln \left(\frac{P}{\beta - \delta P} \right)^{1/\beta} = t + C.$$

This implies that

$$\left(\frac{P}{\beta - \delta P} \right)^{1/\beta} = e^{t+C} = e^C e^t = C_1 e^t$$

so that

$$\frac{P}{\beta - \delta P} = C_2 e^{\beta t} \quad \text{(where } C_2 = C_1^\beta \text{)}. \tag{7}$$

Using the initial condition (5), we have

$$\frac{P_0}{\beta - \delta P_0} = C_2 e^0 = C_2$$

so that

$$C_2 = \frac{P_0}{\beta - \delta P_0}. \tag{8}$$

Finally, inserting (8) in (7) it follows (after some algebra) that

$$P(t) = \frac{\beta}{\delta + [(\beta/P_0) - \delta]e^{-\beta t}}. \tag{9}$$

Equation (4) is called the *logistic equation* and we have shown that the solution to the logistic equation with initial population P_0 is given by (9). A sketch of the growth governed by the logistic equation is given in Figure 1.

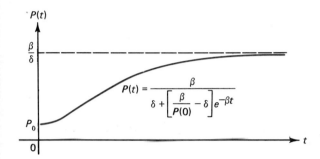

Figure 1

We close this section by noting that most differential equations, even of the first order, cannot be solved by elementary methods (although there are other techniques of solution besides the technique of separation of variables). However, even when an equation cannot be solved in a "closed form" (that is, by writing one variable in terms of another), numerical techniques usually exist for calculating a solution to as many decimal places of accuracy as needed.

In the next two sections we discuss wide classes of first-order equations for which closed form solutions can be found.

PROBLEMS 20.2

In Problems 1–12 solve the given differential equation by the method of separation of variables. If an initial condition is given, find the unique solution to the initial value problem.

1. $\dfrac{dy}{dx} = -7x$

2. $\dfrac{dy}{dx} = e^{x+y};\ y(0) = 2$

3. $\dfrac{dx}{dt} = \sin x \cos t;\ x\left(\dfrac{\pi}{2}\right) = 3$

4. $\dfrac{dy}{dx} = \dfrac{1}{y^3}$

*5. $x^3(y^2 - 1)\dfrac{dy}{dx} = (x + 3)y^5$

6. $\dfrac{dy}{dx} = 2x^2y^2;\ y(1) = 2$

7. $\dfrac{dx}{dt} = e^x \sin t;\ x(0) = 1$

8. $\dfrac{dy}{dx} = 1 + y^2;\ y(0) = 1$

9. $\dfrac{dx}{dt} = x(1 - \cos 2t);\ x(0) = 1$

*10. $\dfrac{dx}{dt} = x^n t^m;\ n, m$ integers

11. $\dfrac{dy}{dx} = x^2(1 + y^2)$

12. $\dfrac{dx}{dt} = \dfrac{e^x t}{e^x + t^2 e^x}.$

☐ 13. In the logistic equation of Example 4, assume that $\delta = 0.5$, $\beta = 0.2$ and $P_0 = 20{,}000$. Calculate $P(10)$.

14. The population of a certain species of bacteria is growing according to the logistic equation. Determine the *equilibrium population* P_e. This is given by

$$P_e = \lim_{t \to \infty} P(t).$$

15. Show that the growth rate of the population in Problem 14 is a maximum when the population is equal to half its equilibrium size.

16. Bacteria are supplied as food to a protozoan population at a constant rate μ. It is observed that the bacteria are consumed at a rate that is proportional to the square of their numbers. The concentration $c(t)$ of the bacteria therefore satisfies the differential equation $dc/dt = \mu - \lambda c^2$, where λ is a positive constant.
    (a) Determine $c(t)$ in terms of $c(0)$.
    (b) What is the equilibrium concentration of the bacteria?

17. In some chemical reactions certain products catalyze their own formation. If $x(t)$ is the amount of such a product at time t, a possible model for the reaction is given by the differential equation $dx/dt = \alpha(\beta - x)$, where α and β are positive constants. According to this model the reaction is completed when $x = \beta$, since this condition indicates that one of the chemicals has been depleted.
    (a) Solve the equation in terms of the constants α, β, and $x(0)$.
    (b) For $\alpha = 1$, $\beta = 200$, and $x(0) = 20$, draw a graph of $x(t)$ for $t > 0$.

*18. On a certain day it began to snow early in the morning and the snow continued to fall at a constant rate. The velocity at which a snowplow is able to clear a road is inversely proportional to the height of the accumulated snow. The snowplow started at 11 A.M. and cleared four miles by 2 P.M. By 5 P.M. it cleared another two miles. When did it start snowing?

*19. A large open cistern filled with water has the shape of a hemisphere with radius 25 ft. The bowl has a circular hole of radius 1 ft in the bottom. By Torricelli's law,† water will flow out of the hole with the same speed it would attain in falling freely from the level of the water to the hole. How long will it take for all the water to flow from the cistern?

20. An object of mass m which falls from rest, starting at a point near the earth's surface, is subjected to two forces: a downward force mg and a resisting force proportional to the square of the velocity of the body. Thus

$$F = ma = m \frac{dv}{dt} = mg - \alpha v^2,$$

where α is a constant of proportionality.
    (a) Find $v(t)$ as a function of t. [*Hint:* Use the fact that $v(0) = 0$.]
    (b) Show that the velocity does not increase indefinitely but approaches the equilibrium value $\sqrt{mg/\alpha}$. This last value is called *terminal velocity*.

*20.3   Exact Equations

We shall now use partial derivatives to solve ordinary differential equations. The material in this section depends upon material in Chapter 18. In particular, you should review Sections 18.11 and 18.12.

Suppose that we take the total differential of the equation $g(x, y) = C$:

$$dg = \frac{\partial g}{\partial x} dx + \frac{\partial g}{\partial y} dy = 0, \tag{1}$$

† Evangelista Torricelli (1608–1647) was an Italian physicist.

For example, the equation $xy = C$ has the total differential $y \, dx + x \, dy = 0$, which may be rewritten as the differential equation $y' = -y/x$. Reversing the situation, suppose that we start with the differential equation

$$M(x, y) \, dx + N(x, y) \, dy = 0. \tag{2}$$

If we can find a function $g(x, y)$ such that

$$\frac{\partial g}{\partial x} = M \quad \text{and} \quad \frac{\partial g}{\partial y} = N,$$

then (2) becomes $dg = 0$, so that $g(x, y) = C$ is the general solution of (1). In this case $M \, dx + N \, dy$ is said to be an *exact differential* (as in Section 18.12), and (2) is called an *exact differential equation*.

Recall from Section (18.12) that the differential $M \, dx + N \, dy$ is exact if and only if

$$\frac{\partial M}{\partial y} = \frac{\partial N}{\partial x}. \tag{3}$$

EXAMPLE 1. Solve $(1 - \sin x \tan y) \, dx + (\cos x \sec^2 y) \, dy = 0$.

SOLUTION. Here

$$M(x, y) = 1 - \sin x \tan y \quad \text{and} \quad N(x, y) = \cos x \sec^2 y$$

so that

$$\frac{\partial M}{\partial y} = -\sin x \sec^2 y = \frac{\partial N}{\partial x},$$

and the equation is exact. To find the solution, we use the technique of Section 18.12:

$$g(x, y) = \int M \, dx + h(y) = x + \cos x \tan y + h(y).$$

Taking the partial derivative with respect to y of both sides yields

$$\cos x \sec^2 y = N = \frac{\partial G}{\partial y} = \cos x \sec^2 y + h'(y).$$

Thus $h'(y) = 0$, so that $h(y)$ is constant and the general solution is

$$g(x, y) = x + \cos x \tan y + C = 0.$$

It should be apparent that exact equations are comparatively rare, since condition (3) requires a precise balance of the functions M and N. For example,

$$(3x + 2y) \, dx + x \, dy = 0$$

is clearly not exact. However, if we multiply the equation by x, then the new equation

$$(3x^2 + 2xy) \, dx + x^2 \, dy = 0$$

is exact. The question we now must ask is: if

$$M(x, y) \, dx + N(x, y) \, dy = 0 \tag{4}$$

is not exact, under what conditions does an *integrating factor* $\mu(x, y)$ exist such that

$$\mu M \, dx + \mu N \, dy = 0$$

is exact? Surprisingly, the answer is whenever (4) has a general solution $g(x, y) = C$. To see this, we solve equation (4) for dy/dx:

$$\frac{dy}{dx} = -\frac{M}{N} = -\frac{\partial g/\partial x}{\partial g/\partial y},$$

from which it follows that

$$\frac{\partial g/\partial x}{M} = \frac{\partial g/\partial y}{N}.$$

Denote either side of the equation above by $\mu(x, y)$. Then

$$\frac{\partial g}{\partial x} = \mu M, \qquad \frac{\partial g}{\partial y} = \mu N, \tag{5}$$

and equation (4) has at least one integrating factor μ. However, finding integrating factors is in general very difficult. There is one procedure which is sometimes successful. Since equation (5) indicates that $\mu M \, dx + \mu N \, dy = 0$ is exact, by (3) we have

$$\mu \frac{\partial M}{\partial y} + M \frac{\partial \mu}{\partial y} = \frac{\partial}{\partial y}(\mu M) = \frac{\partial}{\partial x}(\mu N) = \mu \frac{\partial N}{\partial x} + N \frac{\partial \mu}{\partial x},$$

so that

$$\frac{1}{\mu}\left\{ N \frac{\partial \mu}{\partial x} - M \frac{\partial \mu}{\partial y} \right\} = \frac{\partial M}{\partial y} - \frac{\partial N}{\partial x}. \tag{6}$$

In case the integrating factor μ depends only on x, equation (6) becomes

$$\frac{1}{\mu}\frac{d\mu}{dx} = \frac{\partial M/\partial y - \partial N/\partial x}{N} = k. \tag{7}$$

Since the left-hand side of this equation consists only of functions of x, k must also be a function of x. If this is indeed true, then μ can be found by separating the variables: $\mu(x) = e^{[\int k(x)\, dx]}$. A similar result holds if μ is a function of y alone, in which case

$$K = \frac{\partial M/\partial y - \partial N/\partial x}{-M}$$

is also a function of y. In this case $\mu(y) = e^{[\int K(y)\, dy]}$ is the integrating factor.

EXAMPLE 2.   Solve $(3x^2 - y^2)\, dy - 2xy \, dx = 0$.

SOLUTION. In this problem $M = -2xy$ and $N = 3x^2 - y^2$ so that

$$\frac{\partial M}{\partial y} = -2x, \qquad \frac{\partial N}{\partial x} = 6x.$$

Then

$$K = \frac{\partial M/\partial y - \partial N/\partial x}{-M} = \frac{-4}{y},$$

and

$$\mu = e^{[\int -(4/y)\,dy]} = e^{(-4\ln y)} = y^{-4}.$$

Now we obtain

$$\left(\frac{3x^2 - y^2}{y^4}\right) dy - \frac{2x}{y^3}\,dx = 0,$$

which is exact. Thus the general solution is

$$g = \int M\,dx + h(y) = -\frac{x^2}{y^3} + h(y);$$

and differentiating this equation with respect to y, we have

$$\frac{3x^2 - y^2}{y^4} = N = \frac{3x^2}{y^4} + h'(y).$$

Hence $h'(y) = -y^{-2}$, so $h(y) = y^{-1} + C$, yielding

$$g(x, y) = -\frac{x^2}{y^3} + \frac{1}{y} + C = 0, \qquad \text{or} \qquad Cy^3 + y^2 - x^2 = 0.$$

PROBLEMS 20.3

In Problems 1–9 verify that each given differential equation is exact and find the general solution. Find a particular solution when an initial condition is given.

1. $2xy\,dx + (x^2 + 1)\,dy = 0$

2. $\left(4x^3y^3 + \frac{1}{x}\right) dx + \left(3x^4y^2 - \frac{1}{y}\right) dy = 0,\ x(e) = 1$

3. $\left[\frac{\ln(\ln y)}{x} + \frac{2}{3}xy^3\right] dx + \left[\frac{\ln x}{y\ln y} + x^2y^2\right] = 0$

4. $(x - y\cos x)\,dx - \sin x\,dy = 0,\ y(\pi/2) = 1$

5. $\cosh 2x\cosh 2y\,dx + \sinh 2x\sinh 2y\,dy = 0$

6. $(ye^{xy} + 4y^3)\,dx + (xe^{xy} + 12xy^2 - 2y)\,dy = 0,\ y(0) = 2$

7. $(3x^2\ln x + x^2 - y)\,dx - x\,dy = 0,\ y(1) = 5$

8. $(x^2 + y^2)\,dx + 2xy\,dy = 0,\ y(1) = 1$

9. $\left(\frac{1}{x} - \frac{y}{x^2 + y^2}\right) dx + \left(\frac{x}{x^2 + y^2} - \frac{1}{y}\right) dy = 0.$

In Problems 10–13 find an integrating factor for each differential equation and obtain the general solution.

10. $y\,dx + (y - x)\,dy = 0$

11. $2y^2\,dx + (2x + 3xy)\,dy = 0$

12. $(x^2 + 2y)\,dx - x\,dy = 0$

13. $(x^2 + y^2)\,dx + (3xy)\,dy = 0.$

14. Solve $xy\,dx + (x^2 + 2y^2 + 2)\,dy = 0.$

15. Let $M = ya(xy)$ and $N = xb(xy)$, where a and b are differentiable. Show that $1/(xM - yN)$ is an integrating factor for $M\,dx + N\,dy = 0$.

16. Use the result of Problem 15 to solve the equation

$$2x^\theta y^8\,dx + x^3 y^0\,dy = 0.$$

*17. Solve $(x^2 + y^2 + 1)\,dx - (xy + y)\,dy = 0$. [*Hint:* Try an integrating factor of the form $\mu(x) = (x + 1)^r$.]

20.4 First-Order Linear Equations

An nth-order differential equation is called *linear* if it can be written in the form

$$\frac{d^n y}{dx^n} + a_{n-1}(x)\frac{d^{n-1}y}{dx^{n-1}} + \cdots + a_1(x)\frac{dy}{dx} + a_0(x)y = f(x). \tag{1}$$

For example, the most general first-order linear equation takes the form

$$\frac{dy}{dx} + a(x)y = f(x), \tag{2}$$

while a second-order linear equation can be written as

$$\frac{d^2 y}{dx^2} + a(x)\frac{dy}{dx} + b(x)y = f(x). \tag{3}$$

In this section we will discuss first-order linear equations and shall discuss second-order linear equations in Sections 20.5 and 20.6.

One nice fact about linear equations is given by the following theorem.†

THEOREM 1. Let $x_0, y_0, y_1, \ldots, y_{n-1}$ be real numbers. Then there is a unique solution $y = f(x)$ to the nth-order linear differential equation (1) which satisfies

$$f(x_0) = y_0, \quad f'(x_0) = y_1, \quad \ldots, \quad f^{(n-1)}(x_0) = y_{n-1}. \tag{4}$$

That is, if we specify n initial conditions, there exists a unique solution.

We say that the linear equation (1) is *homogeneous* if $f(x) = 0$ for every number x in the domain of f. Otherwise we say that the equation is *nonhomogeneous*. If the functions $a_0(x), a_1(x), \ldots, a_{n-1}(x)$ are constant, then the equation is said to have *constant coefficients*. Otherwise it is said to have *variable coefficients*. It turns out that we can solve by integration *all* first order linear equations and all nth-order linear homogeneous and nonhomogeneous equations with constant coefficients. In this section we show how solutions to linear equations in the form (2) can be explicitly calculated. We do this in three steps.

CASE 1. Constant coefficients, homogeneous. Then (2) can be written

$$\frac{dy}{dx} + ay = 0 \tag{5}$$

---

† For a proof, see Derrick and Grossman, *Elementary Differential Equations*, p. 487.

where a is a constant or

$$\frac{dy}{dx} = -ay.$$

Solutions (from Section 6.4) are given by

$$\boxed{y = Ce^{-ax}} \tag{6}$$

for any constant C. Many examples of this type of equation were given in Section 6.4.

CASE 2.   Constant coefficients, nonhomogeneous. Then (2) can be written

$$\frac{dy}{dx} + ay = f(x), \tag{7}$$

where a is a constant. It is now impossible to separate the variables. However equation (7) can be solved by multiplying both sides of (7) by an *integrating factor*. We first note that

$$\frac{d}{dx}(e^{ax}y) = e^{ax}\frac{dy}{dx} + ae^{ax}y = e^{ax}\left(\frac{dy}{dx} + ay\right). \tag{8}$$

Thus, if we multiply both sides of (7) by e^{ax}, we obtain

$$e^{ax}\left(\frac{dy}{dx} + ay\right) = e^{ax}f(x)$$

or, using (8),

$$\frac{d}{dx}(e^{ax}y) = e^{ax}f(x)$$

and, upon integration,

$$e^{ax}y = \int e^{ax}f(x)\,dx + C$$

where $\int e^{ax}f(x)\,dx$ denotes one particular indefinite integral. This leads to the general solution

$$\boxed{y = e^{-ax}\int e^{ax}f(x)\,dx + Ce^{-ax}.} \tag{9}$$

The term e^{ax} is called an integrating factor for (7) because it allows us, after multiplication, to solve the equation by integration.

EXAMPLE 1.   Find all solutions to

$$\frac{dy}{dx} + 3y = x. \tag{10}$$

SOLUTION. We multiply both sides of the equation by e^{3x}. Then

$$e^{3x}\left(\frac{dy}{dx} + 3y\right) = xe^{3x} \quad \text{or} \quad \frac{d}{dx}(e^{3x}y) = xe^{3x}$$

and

$$e^{3x}y = \int xe^{3x}\,dx.$$

But setting $u = x$ and $dv = e^{3x}\,dx$, we find that $du = dx$, $v = \frac{1}{3}e^{3x}$ and

$$\int xe^{3x}\,dx = \frac{x}{3}e^{3x} - \frac{1}{3}\int e^{3x}\,dx = \frac{x}{3}e^{3x} - \frac{1}{9}e^{3x} + C$$

so that

$$e^{3x}y = \frac{x}{3}e^{3x} - \frac{1}{9}e^{3x} + C$$

and

$$y = \frac{x}{3} - \frac{1}{9} + Ce^{-3x}. \tag{11}$$

This answer should be checked by differentiation.

Let us take a closer look at the answer we just obtained. The answer (11) is given in two parts. The functions Ce^{-3x} (where C is an arbitrary real number) are all solutions to the homogeneous equation

$$\frac{dy}{dx} + 3y = 0$$

while $(x/3) - \frac{1}{9}$ is one particular solution to the nonhomogeneous problem (10). You should verify this. That this is no accident is suggested by the following theorem.

THEOREM 2. (i) Let y_1 and y_2 be two solutions to the homogeneous equation

$$\frac{dy}{dx} + a(x)y = 0. \tag{12}$$

Then for some constant α,

$$y_1(x) = \alpha y_2(x) \quad \text{for every real number } x. \tag{13}$$

(ii) Let y_p and y_q be two solutions to the nonhomogeneous equation (2). Then

$$y = y_p - y_q$$

is a solution to the homogeneous equation (12).

Remark 1. Part (i) of the theorem states that if you know one nonzero solution to the homogeneous equation (12), then you know them all.

Remark 2. Part (ii) of the theorem is important because it states, in effect, that if you know *one* solution to the nonhomogeneous equation (2) and all the solutions to the homogeneous equation (12), then you also know all solutions to the nonhomogeneous equation. This follows since if y_p is the known solution to (2) and if y_q is any other solution to (2), then since $y_q - y_p$ is a solution y_h to (12), we can write

$$y_q - y_p = y_h \quad \text{or} \quad y_q = y_p + y_h.$$

That is, any other solution to (2) can be written as the sum of the one known solution to (2) and some solution to (12). Thus, for example, in (11), we have $y_p = (x/3) - \frac{1}{9}$ while all solutions to $(dy/dx) + 3y = 0$ are given by $y_h = Ce^{-3x}$. Since, as we will see, nonhomogeneous equations are harder to solve than homogeneous ones, it is nice to know that we need only find *one* solution to a given linear, nonhomogeneous equation.

PROOF OF THEOREM 2. (i) Assume that $y_2(x) \neq 0$ for every x. Then

$$\left(\frac{y_1}{y_2}\right)' = \frac{y_2 y_1' - y_1 y_2'}{y_2{}^2} = \frac{y_2(-ay_1) - y_1(-ay_2)}{y_2{}^2} = 0$$

which implies that y_1/y_2 is constant so that $y_1/y_2 = \alpha$ or $y_1(x) = \alpha y_2(x)$.

   Note that we may assume that one of the solutions y_1 or y_2 is never zero. For if both are identically zero, then $y_1(x) = \alpha y_2(x)$ for every number α. On the other hand, if y_2, say, is not identically zero, then it can never be equal to zero. For if $y_2(x_0) = 0$, for some number x_0, then, by the uniqueness theorem, (Theorem 1), y_2 is the zero function since the zero function satisfies (12) and the initial condition $y_2(x_0) = 0$.

   (ii)   Let $y = y_p - y_q$. Then

$$\frac{dy}{dx} + a(x)y = \frac{d}{dx}[y_p - y_q] + a(x)[y_p - y_q]$$

$$= \frac{dy_p}{dx} - \frac{dy_q}{dx} + a(x)y_p - a(x)y_q$$

$$= \left[\frac{dy_p}{dx} + a(x)y_p\right] - \left[\frac{dy_q}{dx} + a(x)y_q\right]$$

$$= f(x) - f(x) = 0$$

since y_p and y_q both solve (2).

EXAMPLE 2. Find the solution to $(dy/dx) + 2y = e^{-5x}$ which satisfies $y(0) = 4$.

SOLUTION. We multiply both sides of the equation by e^{2x} to obtain

$$e^{2x}\left(\frac{dy}{dx} + 2y\right) = e^{-3x} \quad \text{or} \quad \frac{d}{dx}[e^{2x}y] = e^{-3x}$$

and

$$e^{2x}y = \int e^{-3x}\, dx = -\tfrac{1}{3}e^{-3x} + C.$$

Then

$$y = -\tfrac{1}{3}e^{-5x} + Ce^{-2x}.$$

Note that $-\frac{1}{3}e^{-5x}$ is one solution to $(dy/dx) + 2y = e^{-5x}$ while Ce^{-2x} represents all solutions to $(dy/dx) + 2y = 0$. Finally, we have

$$4 = y(0) = -\tfrac{1}{3} + C \qquad \text{or} \qquad C = \tfrac{13}{3}$$

and the solution to the initial value problem is

$$y = -\tfrac{1}{3}e^{-5x} + \tfrac{13}{3}e^{-2x}.$$

We now turn to the final case.

CASE 3. Variable coefficients, nonhomogeneous. We first note the following facts, the first of which follows from the fundamental theorem of calculus (assuming that a is continuous):

$$(\text{i}) \quad \frac{d}{dx} \int a(x)\,dx = a(x). \tag{14}$$

(This holds for *any* indefinite integral $\int a(x)\,dx$.)

$$(\text{ii}) \quad \frac{d}{dx} e^{\int a(x)\,dx} = e^{\int a(x)\,dx} \frac{d}{dx} \int a(x)\,dx = a(x)e^{\int a(x)\,dx}. \tag{15}$$

Now consider the equation (2)

$$\frac{dy}{dx} + a(x)y = f(x).$$

We multiply both sides by the integrating factor $e^{\int a(x)\,dx}$. Then we have

$$e^{\int a(x)\,dx}\frac{dy}{dx} + a(x)e^{\int a(x)\,dx}y = e^{\int a(x)\,dx}f(x). \tag{16}$$

Now we note that, from (15),

$$\frac{d}{dx}ye^{\int a(x)\,dx} = y\frac{d}{dx}e^{\int a(x)\,dx} + \frac{dy}{dx}e^{\int a(x)\,dx}$$

$$= a(x)e^{\int a(x)\,dx}y + \frac{dy}{dx}e^{\int a(x)\,dx}$$

$$= \text{the left-hand side of (16).}$$

Thus, from (16),

$$\frac{d}{dx}[e^{\int a(x)\,dx}y] = e^{\int a(x)\,dx}f(x)$$

or, integrating,

$$e^{\int a(x)\,dx}y = \int e^{\int a(x)\,dx}f(x)\,dx + C$$

and

$$y = e^{-\int a(x)\,dx}\int e^{\int a(x)\,dx}f(x)\,dx + Ce^{-\int a(x)\,dx}. \tag{17}$$

It is probably a waste of time to try to memorize the complicated looking formula (17). Rather it is important to remember that multiplication by the integrating factor $e^{\int a(x)\,dx}$ will always enable us to reduce the problem of solving a differential equation to the problem of calculating an integral.

EXAMPLE 3.   Find all solutions to the equation

$$\frac{dy}{dx} + \frac{4}{x}y = 3x^2.$$

SOLUTION.   Here $a(x) = (4/x)$, $\int a(x)\,dx = 4\ln x = \ln x^4$ and $e^{\int a(x)\,dx} = e^{\ln x^4} = x^4$ (since $e^{\ln u} = u$ by definition of $\ln u$). Thus we can multiply both sides of the equation by the integrating factor x^4 to obtain

$$x^4\left(\frac{dy}{dx} + \frac{4}{x}y\right) = 3x^6 \qquad \text{or} \qquad \frac{d}{dx}(yx^4) = 3x^6.$$

Then we integrate to find that

$$yx^4 = \frac{3x^7}{7} + C \qquad \text{or} \qquad y = \frac{3x^3}{7} + \frac{C}{x^4}.$$

Note that $3x^3/7$ is one solution to the nonhomogeneous equation $(dy/dx) + (4/x)y = 3x^2$ while C/x^4 represents all solutions to the homogeneous equation $(dy/dx) + (4/x)y = 0$. This should be checked by differentiation.

EXAMPLE 4.   Find the solution to the initial value problem $dy/dx = x^2 - 3x^2y$ which satisfies $y(1) = 2$.

SOLUTION.   We write the equation in the form

$$\frac{dy}{dx} + 3x^2y = x^2$$

so that $a(x) = 3x^2$, $\int a(x)\,dx = x^3$ and $e^{\int a(x)\,dx} = e^{x^3}$. Then we have, after multiplication by e^{x^3},

$$\frac{d}{dx}(ye^{x^3}) = x^2e^{x^3}$$

and

$$ye^{x^3} = \int x^2e^{x^3}\,dx = \frac{1}{3}e^{x^3} + C$$

so that

$$y = \frac{1}{3} + Ce^{-x^3}.$$

Then

$$2 = y(1) = \frac{1}{3} + Ce^{-1} = \frac{1}{3} + \frac{C}{e}$$

or

$$\frac{C}{e} = \frac{5}{3} \quad \text{and} \quad C = \frac{5e}{3}.$$

Thus the solution to the initial value problem is given by

$$y = \frac{1}{3} + \left(\frac{5}{3}e\right)(e^{-x^3}) = \frac{1}{3} + \frac{5}{3}e^{1-x^3}.$$

EXAMPLE 5. Infusion of glucose into the bloodstream is an important medical technique. To study this process, let $G(t)$ represent the amount of glucose in the bloodstream of a patient at time t and suppose that glucose is infused into the bloodstream at the constant rate of k grams per minute. At the same time the glucose is converted and removed from the bloodstream at a rate proportional to the amount of glucose present.

     (a)   Find a differential equation which describes this situation.
     (b)   Solve the equation.
     (c)   Determine the equilibrium amount of glucose present.

SOLUTION. (a) From the statement of the problem, we obtain the equation

$$\frac{dG}{dt} = k - \alpha G,$$

where α is a constant of proportionality.

     (b)   To solve, we write the equation in the form

$$\frac{dG}{dt} + \alpha G = k$$

with integrating factor $e^{\alpha t}$. Then, after multiplication,

$$\frac{d}{dt}(Ge^{\alpha t}) = ke^{\alpha t} \quad \text{or} \quad Ge^{\alpha t} = \frac{k}{\alpha}e^{\alpha t} + C$$

and

$$G(t) = \frac{k}{\alpha} + Ce^{-\alpha t}.$$

But if $G(0)$ represents the initial amount of glucose in the blood, we have

$$G(0) = \frac{k}{\alpha} + C \quad \text{or} \quad C = G(0) - \frac{k}{\alpha}$$

and

$$G(t) = \frac{k}{\alpha} + \left[G(0) - \frac{k}{\alpha}\right]e^{-\alpha t}.$$

     (c)   When $t \to \infty$, $e^{-\alpha t} \to 0$ so that

$$G(t) \to \frac{k}{\alpha}$$

and k/α is the equilibrium amount of glucose in the blood.

PROBLEMS 20.4

In Problems 1–14 find all solutions to the given equation. If an initial condition is given, find the particular solution that satisfies that condition.

1. $\dfrac{dy}{dx} = 4x$

2. $\dfrac{dy}{dx} + xy = 0;\; y(0) = 2$

3. $\dfrac{dx}{dt} + x = \sin t;\; x(0) = 1$

4. $\dfrac{dx}{dt} + x = \dfrac{1}{1 + e^{2t}}$

*5. $\dfrac{dy}{dx} + y \ln x = x^x$

6. $\dfrac{dy}{dx} + (\tan x)y = 2x \sec x;\; y\left(\dfrac{\pi}{4}\right) = 1$

7. $\dfrac{dx}{dt} - ax = be^{at};\; a, b$ constant

8. $\dfrac{dx}{dt} - ax = be^{ct};\; x(0) = d;\; a, b, c, d$ constants

9. $\dfrac{dy}{dx} = x + 2y \tan 2x$

10. $\dfrac{dy}{dx} = 2y + x^2 e^{2x}$

11. $(x^2 + 1)\dfrac{dy}{dx} + 2xy = x;\; y(0) = 1$

12. $\dfrac{dy}{dx} + \dfrac{y}{x^2} = \dfrac{3}{x^2};\; y(1) = 2$

13. $\dfrac{dx}{dt} + \dfrac{4t}{t^2 + 1}x = 3t;\; x(0) = 4$

14. $\dfrac{ds}{dt} + s \tan t = \cos t;\; s\left(\dfrac{\pi}{3}\right) = \dfrac{1}{2}$.

*15. Solve the equation $y - x\,dy/dx = (dy/dx)y^2 e^y$ by reversing the roles of x and y (i.e., by writing x as a function of y). [*Hint:* Use the fact (from Section 10.7) that under certain conditions $dx/dy = 1/(dy/dx)$.]

16. Use the technique of Problem 15 to solve the equation $dy/dx = -1/(x - e^{-y})$.

17. A chemical substance S is produced at the rate of r moles per minute in a chemical reaction. At the same time it is consumed at a rate of c moles per minute per mole of S. Let $S(t)$ be the number of moles of the chemical present at time t.
(a) Obtain the differential equation satisfied by $S(t)$.
(b) Determine $S(t)$ in terms of $S(0)$.
(c) Find the equilibrium amount of the chemical.

18. An infectious disease is introduced to a large population. The proportion of people who have been exposed to the disease increases with time. Suppose that $P(t)$ is the proportion of people who have been exposed to the disease within t years of its introduction. If $P'(t) = [1 - P(t)]/3$ and $P(0) = 0$, after how many years will the proportion have increased to 90%?

20.5 Second-Order Linear, Homogeneous Equations with Constant Coefficients

The most general second-order linear differential equation is

$$y''(x) + a(x)y'(x) + b(x)y(x) = f(x). \tag{1}$$

Unlike the analogous first-order equation (20.4.2), it is not possible, in general, to find solutions to (1) by integration. However it is always possible to solve (1) if the functions $a(x)$ and $b(x)$ are constants. We shall therefore consider the constant coefficient equation

$$y''(x) + ay'(x) + by(x) = f(x) \tag{2}$$

and the related homogeneous equation

$$y''(x) + ay'(x) + by(x) = 0. \tag{3}$$

In this section we will show how solutions to (3) can be found. In Section 20.6 we will deal with the nonhomogeneous equation (2).

The following theorem is central.

THEOREM 1. Let $y_1(x)$ and $y_2(x)$ be solutions to the homogeneous equation (3). Then

    (i)   for any constants c_1 and c_2, $c_1 y_1 + c_2 y_2$ is also a solution to (3).
    (ii) Let $y(x)$ be any other solution to (3). Then if y_2 is not a constant multiple of y_1, there exist constants k_1 and k_2 such that

$$y(x) = k_1 y_1(x) + k_2 y_2(x) \tag{4}$$

for every x at which $y_1(x)$ and $y_2(x)$ are defined.

> **Remark 1.** The fact in (i) is referred to as the *principle of superposition*. The expression $c_1 y_1 + c_2 y_2$ is called a *linear combination* of the functions y_1 and y_2. The principle of superposition states that any linear combination of solutions to (3) is again a solution to (3).

> **Remark 2.** Part (ii) tells us that if we know two solutions to (3) which are "independent" in the sense that one is not a constant multiple of the other,† then we know them all; for any other solution can be written as a linear combination of these two independent solutions.

The proof of the first part of the theorem is easy and is left as an exercise (see Problem 21). The proof of the second part of the theorem relies on the uniqueness theorem for initial value problems (Theorem 20.4.1). However the proof would take us too far afield and so is omitted.‡

We now turn to the problem of finding two independent solutions to (3). Recall that for the analogous first-order equation $y' + ay = 0$, the general solution is $y(x) = Ce^{-ax}$. It is then reasonable to "guess" that there may be a solution to (3) of the form $y(x) = e^{\lambda x}$ for some number λ. Setting $y(x) = e^{\lambda x}$ in (3), we obtain

$$y' = \lambda e^{\lambda x}, \qquad y'' = \lambda^2 e^{\lambda x}$$

and

$$y'' + ay' + by = (\lambda^2 + a\lambda + b)e^{\lambda x}.$$

We see that y will be a solution to (3) if and only if

$$\boxed{\lambda^2 + a\lambda + b = 0.} \tag{5}$$

---

†y_2 is a constant multiple of y_1 if there exists a constant c such that $y_2(x) = cy_1(x)$ for every x for which $y_1(x)$ is defined.
‡For a proof, see Derrick and Grossman, *Elementary Differential Equations*, Section 3.2.

Equation (5) is called the *auxiliary equation* of the homogeneous differential equation (3). From the quadratic formula (5) has the roots

$$\lambda_1 = \frac{-a + \sqrt{a^2 - 4b}}{2}, \qquad \lambda_2 = \frac{-a - \sqrt{a^2 - 4b}}{2}. \tag{6}$$

There are three possibilities: $a^2 - 4b > 0$, $a^2 - 4b = 0$, and $a^2 - 4b < 0$.

CASE 1. If $a^2 - 4b > 0$, then λ_1 and λ_2 are two distinct real numbers and $y_1(x) = e^{\lambda_1 x}$ and $y_2 = e^{\lambda_2 x}$ are independent solutions since, clearly, $e^{\lambda_2 x}$ is not a constant multiple of $e^{\lambda_1 x}$ if $\lambda_1 \neq \lambda_2$. The general solution to (3) is then given by

$$\boxed{y(x) = c_1 e^{\lambda_1 x} + c_2 e^{\lambda_2 x},} \tag{7}$$

where c_1 and c_2 denote arbitrary constants.

EXAMPLE 1. Find the general solution to

$$y''(x) + 2y'(x) - 8y(x) = 0.$$

SOLUTION. The auxiliary equation is

$$\lambda^2 + 2\lambda - 8 = 0 \qquad \text{or} \qquad (\lambda + 4)(\lambda - 2) = 0$$

with roots $\lambda = -4$ and $\lambda = 2$. Thus two independent solutions are $y_1(x) = e^{-4x}$ and $y_2(x) = e^{2x}$ so that the general solution is

$$y(x) = c_1 e^{-4x} + c_2 e^{2x}.$$

EXAMPLE 2. Find the solution to the equation in Example 1 which satisfies $y(0) = 1$ and $y'(0) = 3$.

SOLUTION. Since $y(x) = c_1 e^{-4x} + c_2 e^{2x}$,

$$y'(x) = -4c_1 e^{-4x} + 2c_2 e^{2x}$$

so that

$$y(0) = c_1 + c_2 = 1$$

and

$$y'(0) = -4c_1 + 2c_2 = 3.$$

Then using the techniques of Appendix 2 we easily find the solution

$$c_1 = -\tfrac{1}{6}, \qquad c_2 = \tfrac{7}{6}$$

so that the solution to the initial value problem is

$$y(x) = -\tfrac{1}{6} e^{-4x} + \tfrac{7}{6} e^{2x}.$$

This should be checked by differentiation.

CASE 2. $a^2 - 4b = 0$. Then the roots in (6) are both equal to $-a/2$. One solution is, therefore,

$$y_1(x) = e^{-(a/2)x}. \tag{8}$$

Another, independent, solution is given by

$$y_2(x) = xe^{-(a/2)x}. \tag{9}$$

We prove this by differentiation, We have

$$y_2'(x) = \left(1 - \frac{a}{2}x\right)e^{-(a/2)x}$$

and

$$y_2''(x) = \left(\frac{a^2}{4}x - a\right)e^{-(a/2)x}$$

so that

$$y'' + ay' + by = e^{-(a/2)x}\left[\left(\frac{a^2}{4}x - a\right) + a\left(1 - \frac{a}{2}x\right) + bx\right]$$

$$= e^{-(a/2)x}\left(-\frac{a^2}{4}x + bx\right).$$

But $a^2 - 4b = 0$ so that $a^2/4 = b$ and

$$e^{-(a/2)x}\left(-\frac{a^2}{4}x + bx\right) = 0$$

which completes the demonstration. Finally, in Case 2, we see that the general solution is given by

$$\boxed{y(x) = c_1 e^{-(a/2)x} + c_2 x e^{-(a/2)x}.} \tag{10}$$

EXAMPLE 3. Find the general solution of

$$y''(x) - 6y'(x) + 9y(x) = 0.$$

SOLUTION. Here $\lambda^2 - 6\lambda + 9 = 0 = (\lambda - 3)^2$ so that the only root is $\lambda = 3$. Thus the general solution is given (using (10)) by

$$y(x) = c_1 e^{3x} + c_2 x e^{3x}.$$

EXAMPLE 4. Find the solution of the equation in Example 3 which satisfies the initial conditions $y(0) = 2$ and $y'(0) = -3$.

SOLUTION. We have $y'(x) = 3c_1 e^{3x} + c_2(1 + 3x)e^{3x}$. Thus

$$y(0) = c_1 = 2 \qquad \text{and} \qquad y'(0) = 3c_1 + c_2 = -3$$

so that $c_1 = 2$, $c_2 = -9$ and the solution to the initial value problem is

$$y(x) = 2e^{3x} - 9xe^{3x} = e^{3x}(2 - 9x).$$

CASE 3. $a^2 - 4b < 0$. Then the roots of (6) are not real numbers. Rather than digress with a discussion of complex numbers, we shall simply tell you the two independent solutions. Let

$$\alpha = -\frac{a}{2} \quad \text{and} \quad \beta = \frac{\sqrt{4b - a^2}}{2}. \tag{11}$$

By assumption, $4b - a^2 > 0$ so that both α and β are real numbers. Then two independent solutions of (3) are given by

$$y_1(x) = e^{\alpha x} \cos \beta x, \qquad y_2(x) = e^{\alpha x} \sin \beta x \tag{12}$$

and the general solution is

$$y(x) = e^{\alpha x}(c_1 \cos \beta x + c_2 \sin \beta x). \tag{13}$$

We prove that $e^{\alpha x} \cos \beta x$ is a solution of $y'' + ay' + by = 0$. The proof for $e^{\alpha x} \sin \beta x$ is similar. If

$$y(x) = e^{\alpha x} \cos \beta x$$

then

$$y'(x) = e^{\alpha x}(\alpha \cos \beta x - \beta \sin \beta x)$$

and

$$y''(x) = e^{\alpha x}[(\alpha^2 - \beta^2) \cos \beta x - 2\alpha\beta \sin \beta x].$$

We have

$$y'' + ay' + by = e^{\alpha x}[(\alpha^2 - \beta^2 + a\alpha + b) \cos \beta x - (2\alpha b + a\beta) \sin \beta x].$$

To show that $e^{\alpha x} \cos \beta x$ is a solution, we must show that

$$\alpha^2 - \beta^2 + a\alpha + b = 0 \quad \text{and} \quad 2\alpha\beta + a\beta = 0.$$

But, from (11), $\alpha^2 = a^2/4$ and $\beta^2 = (4b - a^2)/4$ so that

$$\alpha^2 - \beta^2 + a\alpha + b = \frac{a^2}{4} - b + \frac{a^2}{4} - \frac{a^2}{2} + b = 0.$$

Similarly,

$$2\alpha\beta + a\beta = \beta(2\alpha + a) = \beta\left[2\left(-\frac{a}{2}\right) + a\right] = \beta \cdot 0 = 0.$$

EXAMPLE 5.   Find all solutions to the equation of the harmonic oscillator

$$y'' + y = 0.$$

SOLUTION. We discussed this equation in Section 7.7. Since $a = 0$ and $b = 1$, $a^2 - 4b = -4 < 0$. Thus $\alpha = -a/2 = 0, \beta = \sqrt{4b - a^2}/2 = \sqrt{4}/2 = 1$ so that the general solution is, according to equation (13), given by

$$y(x) = c_1 \cos x + c_2 \sin x.$$

This is what we found in Section 7.7.

EXAMPLE 6. Find the general solution to

$$y'' + y' + y = 0.$$

SOLUTION. We have $\lambda^2 + \lambda + 1 = 0$ and $a^2 - 4b = 1 - 4 = -3$. Thus if we set

$$\alpha = -\frac{a}{2} = -\frac{1}{2} \quad \text{and} \quad \beta = \frac{\sqrt{4b - a^2}}{2} = \frac{\sqrt{4 - 1}}{2} = \frac{\sqrt{3}}{2},$$

we obtain the general solution

$$y(x) = e^{-x/2}\left(c_1 \cos \frac{\sqrt{3}}{2}x + c_2 \sin \frac{\sqrt{3}}{2}x\right).$$

EXAMPLE 7. Find the solution of the equation of Example 6 that satisfies $y(0) = 1$ and $y'(0) = 3$.

SOLUTION. We have

$$y'(x) = e^{-x/2}\left\{\left(-\frac{1}{2}c_1 + \frac{\sqrt{3}}{2}c_2\right)\cos\frac{\sqrt{3}}{2}x + \left(-\frac{\sqrt{3}}{2}c_1 - \frac{1}{2}c_2\right)\sin\frac{\sqrt{3}}{2}x\right\}.$$

Thus

$$y(0) = c_1 = 1$$

and

$$y'(0) = -\frac{1}{2}c_1 + \frac{\sqrt{3}}{2}c_2 = 3$$

with solutions $c_1 = 1$ and $c_2 = 7/\sqrt{3}$. Thus the solution to the initial value problem is

$$y(x) = e^{-x/2}\left(\cos\frac{\sqrt{3}}{2}x + \frac{7}{\sqrt{3}}\sin\frac{\sqrt{3}}{2}x\right).$$

PROBLEMS 20.5

In the following problems find the general solution of each equation. When initial conditions are specified, determine the particular solution that satisfies them.

1. $y'' - 4y = 0$
2. $y'' + y' - 3y = 0;\ y(0) = 0,\ y'(0) = 1$
3. $y'' + 2y' + 2y = 0$
4. $y'' - 3y' + 2y = 0$
5. $8y'' + 4y' + y = 0;\ y(0) = 0,\ y'(0) = 1$
6. $y'' + 8y' + 16y = 0$
7. $y'' + 5y' + 6y = 0;\ y(0) = 1,\ y'(0) = 2$
8. $y'' + y' + 7y = 0$
9. $4y'' + 20y' + 25y = 0;\ y(0) = 1,\ y'(0) = 2$
10. $y'' + y' + 2y = 0$
11. $y'' - y' - 6y = 0;\ y(0) = -1;\ y'(0) = 1$
12. $y'' - 5y' = 0$
13. $y'' + 4y = 0;\ y(\pi/4) = 1,\ y'(\pi/4) = 3$
14. $y'' - 10y' + 25y = 0;\ y(0) = 2,\ y'(0) = -1$
15. $y'' + 17y' = 0;\ y(0) = 1,\ y'(0) = 0$

16. $y'' + 2\pi y' + \pi^2 y = 0;\ y(1) = 1,\ y'(1) = 1/\pi$
17. $y'' - 13y' + 42y = 0$ **18.** $y'' + 4y' + 6y = 0$
19. $y'' = y;\ y(0) = 2,\ y'(0) = -3$
20. $y'' - y' + y = 0;\ y(0) = 3,\ y'(0) = 7.$
21.   Prove Theorem 1(i). [*Hint:* Let $y = c_1 y_1 + c_2 y_2$. Differentiate twice and show that y
       satisfies equation (3), using the fact that y_1 and y_2 satisfy equation (3).

20.6 Second-Order Nonhomogeneous Equations with Constant Coefficients: The Method of Undetermined Coefficients

In this section we present a method for finding a particular solution to the nonhomogeneous equation

$$y''(x) + ay'(x) + by(x) = f(x). \tag{1}$$

We need only find one solution to (1) since if y_p and y_q are solutions to (1), then
$y_p - y_q$ is a solution to the homogeneous equation

$$y''(x) + ay'(x) + by(x) = 0. \tag{2}$$

To see this we note that

$$
\begin{aligned}
(y_p - y_q)'' &+ a(y_p - y_q)' + b(y_p - y_q) \\
&= y_p'' - y_q'' + ay_p' - ay_q' + by_p - by_q \\
&= (y_p'' + ay_p' + by_p) - (y_q'' + ay_q' + by_q) = f - f = 0.
\end{aligned}
$$

Thus the general solution to (1) can be written as the sum of one particular solution y_p
to (1) plus the general solution to the homogeneous equation (2).
     Let $P_n(x)$ be a polynomial of degree n. We will show how to find a solution to
(1) if $f(x)$ takes one of the forms:

   (i) $P_n(x).$ (3)

   (ii) $P_n(x)e^{ax}.$ (4)

   (iii) $P_n(x)e^{ax} \sin bx$ or $P_n(x)e^{ax} \cos bx.$ (5)

The technique we will use is to "guess" that there is a solution to (1) in the same basic
"form" as $f(x)$ and then substitute this "guessed" solution into (1) to determine the
unknown coefficients. This is called the *method of undetermined coefficients* and is best
illustrated with examples. There is a more general method for finding solutions to (1)
for an arbitrary f but that method, called *variation of constants,* is more complicated
and is best left to a book on differential equations.

EXAMPLE 1.   Find the general solution to

$$y'' + y = x^2. \tag{6}$$

SOLUTION. Since x^2 is a polynomial of degree 2, we "guess" that there is a solution to
(6) which is a polynomial of degree 2. Let

$$y_p(x) = ax^2 + bx + c.$$

Then

$$y_p' = 2ax + b, \qquad y_p'' = 2a$$

and

$$y_p'' + y_p = ax^2 + bx + c + 2a.$$

If y_p is a solution to (6), then we must have

$$ax^2 + bx + c + 2a = x^2$$

which implies that

$$a = 1,$$
$$b = 0,$$
$$c + 2a = 0$$

so that $a = 1$, $b = 0$, $c = -2$, and

$$y_p(x) = x^2 - 2.$$

Since the general solution to $y'' + y = 0$ is

$$c_1 \cos x + c_2 \sin x,$$

the general solution to (6) is

$$y(x) = c_1 \cos x + c_2 \sin x + x^2 - 2.$$

EXAMPLE 2.   Find the solution to

$$y'' - y' - 6y = e^x \cos x \tag{7}$$

which satisfies $y(0) = 1$ and $y'(0) = 0$.

SOLUTION. We first find a particular solution to (7). We guess that there is a solution of the form

$$y_p(x) = ae^x \sin x + be^x \cos x.$$

Then

$$y_p'(x) = ae^x(\sin x + \cos x) + be^x(\cos x - \sin x)$$

and

$$y_p''(x) = 2ae^x \cos x - 2be^x \sin x.$$

Substitution of these into (7) yields

$$e^x(2a \cos x - 2b \sin x) - e^x[(a - b) \sin x + (a + b) \cos x]$$
$$-6e^x(a \sin x + b \cos x) = e^x \cos x.$$

Dividing by e^x and equating coefficients of $\sin x$ and $\cos x$ yields

$$2a - (a + b) - 6b = 1,$$
$$-2b - (a - b) - 6a = 0$$

or

$$a - 7b = 1,$$
$$-7a - b = 0$$

with solutions $a = 1/50$ and $b = -7/50$. Thus

$$y_p(x) = \frac{e^x}{50}(\sin x - 7\cos x).$$

This should be checked by differentiation. The general solution to the homogeneous equation

$$y'' - y' - 6y = 0$$

is

$$c_1 e^{3x} + c_2 e^{-2x}$$

so that the general solution to (7) is

$$y(x) = c_1 e^{3x} + c_2 e^{-2x} + \frac{e^x}{50}(\sin x - 7\cos x).$$

Then

$$y'(x) = 3c_1 e^{3x} - 2c_2 e^{-2x} + \frac{e^x}{50}(8\sin x - 6\cos x)$$

and

$$1 = y(0) = c_1 + c_2 - \frac{7}{50},$$

$$0 = y'(0) = 3c_1 - 2c_2 - \frac{6}{50}$$

with solutions

$$c_1 = \frac{24}{50} \quad \text{and} \quad c_2 = \frac{33}{50}.$$

Thus the solution to the initial value problem is

$$y(x) = \frac{1}{50}(24e^{3x} + 33e^{-2x} + e^x(\sin x - 7\cos x)).$$

EXAMPLE 3. Find the general solution of

$$y'' - 4y' + 4y = xe^{3x}. \tag{8}$$

SOLUTION. Since x is a polynomial of degree 1, we seek a solution of the form

$$y_p(x) = (ax + b)e^{3x}.$$

Then

$$y_p' = (3ax + 3b + a)e^{3x} \quad \text{and} \quad y_p'' = (9ax + 9b + 6a)e^{3x}$$

so that

$$y'' - 4y' + 4y = e^{3x}(ax + b + 2a) = xe^{3x}.$$

Thus

$$a = 1,$$
$$b + 2a = 0$$

and $a = 1$, $b = -2$ so that

$$y_p(x) = (x - 2)e^{3x}.$$

Finally, the general solution to the related homogeneous equation $y'' - 4y' + 4y = 0$ is $c_1e^{2x} + c_2xe^{2x}$ so that the general solution to (8) is

$$y(x) = c_1e^{2x} + c_2xe^{2x} + (x - 2)e^{3x}.$$

EXAMPLE 4. Find a particular solution to

$$y'' - y = 3e^{-x}. \tag{9}$$

SOLUTION. We first note that the general solution to the related homogeneous equation

$$y'' - y = 0 \tag{10}$$

is $c_1e^x + c_2e^{-x}$. Thus $3e^{-x}$ is a solution to (10). If we try to determine a solution to (9) in the form $y_p(x) = ae^{-x}$, we find that

$$y_p' = -ae^{-x}, \qquad y_p'' = ae^{-x}$$

and

$$y_p'' - y_p = ae^{-x} - ae^{-x} = 0.$$

But this should have been obvious anyway since ae^{-x} is a solution to (10) and, therefore, could not be a solution to (9). To find a solution to (9), we try the trick that worked for solving Case 2 of the homogeneous equation. We look for a solution to (10) in the form

$$y_p(x) = axe^{-x}.$$

Then

$$y_p'(x) = a(1 - x)e^{-x} \qquad \text{and} \qquad y_p''(x) = a(x - 2)e^{-x}$$

so that

$$y_p'' - y_p = a(-2)e^{-x} = 3e^{-x}.$$

This implies that $a = -\frac{3}{2}$ so that the particular solution is given by

$$y_p(x) = -\tfrac{3}{2}xe^{-x}.$$

We close this section by formally describing the method of undetermined coefficients to obtain a solution to the nonhomogeneous equation (1) when $f(x)$ is of the form (3), (4), or (5):

CASE 1. No term in $y_p(x)$ (as given in the table below) is a solution of the related homogeneous equation (2). Then a particular solution of (1) will have a form according to Table 1.

TABLE 1

$f(x)$	$y_p(x)$
$P_n(x)$	$a_0 + a_1x + a_2x^2 + \cdots + a_nx^n$
$P_n(x)e^{ax}$	$(a_0 + a_1x + a_2x^2 + \cdots + a_nx^n)e^{ax}$
$P_n(x)e^{ax}\sin bx$ or $P_n(x)e^{ax}\cos bx$	$(a_0 + a_1x + a_2x^2 + \cdots + a_nx^n)e^{ax}\sin bx$ $\quad + (b_0 + b_1x + b_2x^2 + \cdots + b_nx^n)e^{ax}\cos bx.$

CASE 2. If any term of $y_p(x)$ is a solution of (2), then multiply the appropriate function $y_p(x)$ of Case 1 by x^k, where k is the smallest integer having the property that no term of $x^ky_p(x)$ is a solution of (2).

PROBLEMS 20.6

In Problems 1–15 find the general solution of each given differential equation. If initial conditions are given, find the particular solution that satisfies them.

1. $y'' + 6y = 4\cos x$
2. $y'' + y' - 6y = 10e^{-x}; \ y(0) = 0, \ y'(0) = 2$
3. $y'' + 3y' + 2y = 3e^{4x}$
4. $y'' - 2y' + y = -2e^{3x}$
5. $y'' + y' = 4x^2; \ y(0) = 3, \ y'(0) = -1$
6. $y'' + y = 1 + x + x^2$
7. $y'' - 6y' + 9y = 4xe^{3x}; \ y(0) = 1, \ y'(0) = 0$
8. $y'' + y = \cos x$
9. $y'' - 2y' + y = 2e^x$
10. $y'' - 9y' + 20y = 40x; \ y(0) = 2, \ y'(0) = 3$
11. $y'' + 16y = 8\cos 4x; \ y(\pi/2) = 0, \ y'(\pi/2) = 1$
12. $y'' + y' = x^2 - x^3 + 2$
*13. $y'' + 4y' + 5y = \sinh 2x \sin x$
*14. $y'' + y' + y = e^{-x/2}\cos(\sqrt{3}/2)x$
15. $y'' - 2y' + 3y = e^x \sin \sqrt{2}x.$

16. Show that if y_1 is a solution to $y'' + ay' + by = f_1(x)$ and if y_2 is a solution to $y'' + ay' + by = f_2(x)$, then $y_1 + y_2$ is a solution to $y'' + ay' + by = f_1(x) + f_2(x)$.

In Problems 17–20 use the result of Problem 16 to obtain a particular solution to the given equation.

17. $y'' + y = 3 - 4\cos x$
18. $y'' - 2y' - 3y = e^{2x} + x - x^2$
19. $y'' + 9y = 2\sin 3x + x^2$
20. $y'' + 6y' + 9y = xe^x + \cos x.$
21. Let $f(x)$ be a polynomial of degree n. Show that for any constants a and b, there is a solution to $y'' + ay' + by = f(x)$ which is a polynomial of degree n.

*20.7 Vibratory Motion

Differential equations were used for the first time to describe the motion of particles. As a simple example, consider the motion of a mass m attached to a coiled spring, the upper end of which is securely fastened (see Figure 1).

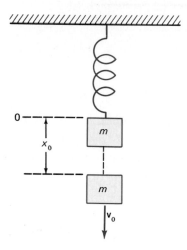

Figure 1

   We have denoted the equilibrium or rest position of the mass on the spring by the number zero. Suppose that the mass is given an initial displacement x_0 and an initial velocity v_0. We now describe the motion of the mass on the spring. By Hooke's law (see Example 5.7.3) the magnitude of the force exerted on the mass by the spring is proportional to the amount of the spring displaced and is directed toward the equilibrium position. That is

$$F = -kx, \tag{1}$$

where k is a constant of proportionality. But by Newton's second law,

$$F = ma = m\frac{d^2x}{dt^2}. \tag{2}$$

If we assume that there is no other force acting on the mass, then, combining (1) and (2), we obtain

$$m\frac{d^2x}{dt^2} = -kx \tag{3}$$

or

$$m\frac{d^2x}{dt^2} + kx = 0$$

which can be written

$$x'' + \frac{k}{m}x = 0. \tag{4}$$

The auxiliary equation for (4) is

$$\lambda^2 + \frac{k}{m} = 0$$

and, since $k/m > 0$, we obtain the general solution

$$x(t) = c_1 \cos \sqrt{\frac{k}{m}}\, t + c_2 \sin \sqrt{\frac{k}{m}}\, t.$$

Then

$$x'(t) = -\sqrt{\frac{k}{m}}\, c_1 \sin \sqrt{\frac{k}{m}}\, t + c_2 \sqrt{\frac{k}{m}} \cos \sqrt{\frac{k}{m}}\, t.$$

But $x(0) = x_0$ and $x'(0) = v_0$ so that

$$x(0) = x_0 = c_1 \cos 0 + c_2 \sin 0 = c_1$$

and

$$x'(0) = v_0 = -\sqrt{\frac{k}{m}}\, c_1 \sin 0 + c_2 \sqrt{\frac{k}{m}} \cos 0 = c_2 \sqrt{\frac{k}{m}}.$$

Thus $c_1 = x_0$, $c_2 = v_0/\sqrt{k/m} = v_0\sqrt{m/k}$ and we obtain the equation of periodic motion

$$\boxed{x(t) = x_0 \cos \sqrt{\frac{k}{m}}\, t + v_0 \sqrt{\frac{m}{k}} \sin \sqrt{\frac{k}{m}}\, t.} \tag{5}$$

We see from (5) that the motion of the mass is periodic with period

$$P = 2\pi \sqrt{\frac{m}{k}} \tag{6}$$

and *frequency f* (the frequency is the number of complete oscillations per unit time) given by

$$f = \frac{\sqrt{k/m}}{2\pi}. \tag{7}$$

It is not difficult to show (see Problem 1) that the maximum value taken by x, called the *amplitude* of the motion (and denoted by A), is given by

$$A = \sqrt{x_0^2 + v_0^2 \frac{m}{k}}. \tag{8}$$

Note that while the amplitude of the motion depends on the initial displacement and velocity, the period and frequency of the motion do not. These latter depend only on the spring constant k and the mass m.

EXAMPLE 1. A spring, fixed at its upper end, supports a 5-kg mass which stretches the spring 50 cm. Find the equation of motion of the mass if it is drawn to a position 20 cm below its equilibrium position and released.

SOLUTION. The force exerted by the mass on the spring is given by $F = mg = 5g$, where g is the acceleration due to gravity. Then, by Hooke's law, we have

$$F = kx \quad \text{or} \quad 5g = k(50)$$

and

$$k = \frac{g}{10}$$

so that

$$\sqrt{\frac{k}{m}} = \sqrt{\frac{g/10}{5}} = \sqrt{\frac{g}{50}}.$$

Since $x_0 = 20$ and $v_0 = 0$, we obtain, from (5),

$$x(t) = 20 \cos \sqrt{\frac{g}{50}}\, t.$$

Also, $P = 2\pi \sqrt{50/g}$, $f = (1/2\pi)\sqrt{g/50}$ and $A = 20$ cm. Note here that we must use the value $g = 981$ cm/sec^2 since our unit of length is the centimeter.

We now turn to a more practical situation. In the above discussion we made the assumption that there were no external forces acting on the mass. This assumption is, however, not very realistic. We must deal with other forces such as friction in the spring and air resistance, which tend to slow things down. Such forces are called *damping* forces. It is reasonable to assume that the magnitude of the damping force is proportional to the velocity of the mass (for example, the slower the mass, the smaller the air resistance). Thus to the equation (3) we add the term $\mu\, dx/dt$, where μ, called the *damping constant,* depends on all the external factors. We then obtain the equation

$$m\frac{d^2x}{dt^2} = -kx - \mu \frac{dx}{dt}$$

or

$$x'' + \frac{\mu}{m}x' + \frac{k}{m}x = 0; \qquad x(0) = x_0, \quad x'(0) = v_0. \tag{9}$$

The auxiliary equation for (9) is

$$\lambda^2 + \frac{\mu}{m}\lambda + \frac{k}{m} = 0 \tag{10}$$

or

$$\lambda = \frac{-\mu \pm \sqrt{\mu^2 - 4mk}}{2m}. \tag{11}$$

There are three cases to consider:

CASE 1 ($\mu^2 > 4mk$). Then both roots of (10) are negative and we obtain the general solution to (9)

$$x(t) = c_1 e^{\lambda_1 t} + c_2 e^{\lambda_2 t}, \tag{12}$$

where $\lambda_1 = (-\mu + \sqrt{\mu^2 - 4mk})/2m$ and $\lambda_2 = (-\mu - \sqrt{\mu^2 - 4mk})/2m$. Clearly $x(t) \to 0$ as $t \to \infty$ and the mass quickly returns to its equilibrium position. (See Figure 2.)

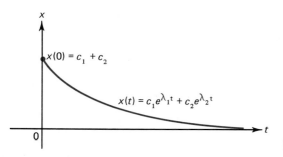

Figure 2

CASE 2 ($\mu^2 = 4mk$). Then $\lambda_1 = \lambda_2 = -\mu/2m$ and

$$x(t) = c_1 e^{(-\mu/2m)t} + c_2 t e^{(-\mu/2m)t} \tag{13}$$

and, again, $x(t) \to 0$ as $t \to \infty$.

CASE 3 ($\mu^2 < 4mk$). Then the general solution is

$$x(t) = e^{(-\mu/2m)t}\left(c_1 \cos \frac{\sqrt{4mk - \mu^2}}{2m}t + c_2 \sin \frac{\sqrt{4mk - \mu^2}}{2m}t\right). \tag{14}$$

This motion is more interesting and is called *damped harmonic motion*. The mass oscillates about its equilibrium position but with an ever decreasing amplitude. The term $e^{(-\mu/2m)t}$ is called a *damping factor*. This is illustrated in Figure 3.

EXAMPLE 2. In Example 1, assume that there are damping forces and that $\mu = 2$. Find the new equation of motion.

SOLUTION. We have $\mu^2 - 4mk = 4 - 4(5)(g/10) = 4 - 2g$ which is less than 0 (since $g = 981$). Thus, since $\mu/2m = 4/10 = 0.2$, we obtain

$$x(t) = e^{-0.2t}\left(c_1 \cos \frac{\sqrt{2g - 4}}{10}t + c_2 \sin \frac{\sqrt{2g - 4}}{10}t\right).$$

Then

$$x'(t) = e^{-0.2t}\left\{\left(c_2 \frac{\sqrt{2g - 4}}{10} - 0.2c_1\right)\cos \frac{\sqrt{2g - 4}}{10}t\right.$$

$$\left. + \left(-\frac{\sqrt{2g - 4}}{10}c_1 - 0.2c_2\right)\sin \frac{\sqrt{2g - 4}}{10}t\right\}.$$

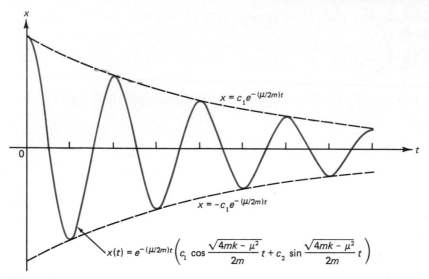

$$x = c_1 e^{-(\mu/2m)t}$$

$$x = -c_1 e^{-(\mu/2m)t}$$

$$x(t) = e^{-(\mu/2m)t}\left(c_1 \cos \frac{\sqrt{4mk - \mu^2}}{2m} t + c_2 \sin \frac{\sqrt{4mk - \mu^2}}{2m} t\right)$$

Figure 3

Thus

$$20 = x(0) = c_1$$

and

$$0 = x'(0) = c_2 \frac{\sqrt{2g - 4}}{10} - 0.2c_1$$

and we obtain

$$c_1 = 20, \qquad c_2 = \frac{40}{\sqrt{2g - 4}}$$

and

$$x(t) = e^{-0.2t}\left(20 \cos \frac{\sqrt{2g - 4}}{10} t + \frac{40}{\sqrt{2g - 4}} \sin \frac{\sqrt{2g - 4}}{10} t\right).$$

This motion is sketched in Figure 4.

We now turn to another situation. The motion of the mass considered in the two cases above is determined by the inherent forces of the spring–weight system and the natural forces acting on the system. Accordingly, the vibrations are called *free* or *natural* vibrations. We will now assume that the system is also subject to an external periodic force $\alpha \sin \omega t$, perhaps due to the motion of the object to which the upper end of the spring is attached (see Figure 1). In this case the mass will undergo *forced vibrations*.

Equation (9) may be replaced by the nonhomogeneous second-order differential equation

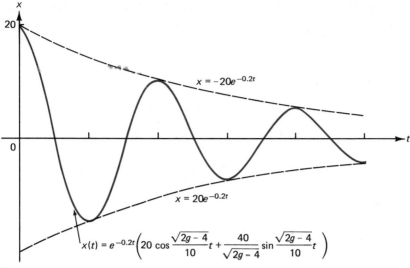

$$x = -20e^{-0.2t}$$

$$x = 20e^{-0.2t}$$

$$x(t) = e^{-0.2t}\left(20 \cos \frac{\sqrt{2g-4}}{10}t + \frac{40}{\sqrt{2g-4}} \sin \frac{\sqrt{2g-4}}{10}t\right)$$

Figure 4

$$m\frac{d^2x}{dt^2} = -kx - \mu\frac{dx}{dt} + \alpha \sin \omega t$$

which we write in the form

$$\frac{d^2x}{dt^2} + \frac{\mu}{m}\frac{dx}{dt} + \frac{k}{m}x = \frac{\alpha}{m}\sin \omega t. \tag{15}$$

By the method of undetermined coefficients, we know that $x(t)$ has a particular solution of the form

$$x_p(t) = A_1 \cos \omega t + A_2 \sin \omega t. \tag{16}$$

Substituting this function into (15) yields the simultaneous equations

$$\left(\frac{k}{m} - \omega^2\right)A_1 + \frac{\mu\omega}{m}A_2 = 0,$$

$$-\frac{\mu\omega}{m}A_1 + \left(\frac{k}{m} - \omega^2\right)A_2 = \frac{\alpha}{m}, \tag{17}$$

from which we obtain

$$A_1 = \frac{-\alpha\mu\omega}{(k - m\omega^2)^2 + (\mu\omega)^2}, \qquad A_2 = \frac{\alpha(k - m\omega^2)}{(k - m\omega^2)^2 + (\mu\omega)^2}. \tag{18}$$

To find the general solution of equation (15), we add the general solution of the homogeneous equation to the particular solution (16). The nature of the general solution again depends on the discriminant $\sqrt{\mu^2 - 4mk}$. If $\mu^2 \geq 4mk$, we are superimposing the periodic function (16) on either equation (12) or (13). Since (12) and (13) damp out as t increases, the motion for large t will be very close to (16). On the other hand, if $\mu^2 < 4mk$, we will suppose initially that $\mu = 0$. Then the general solution, for $\sqrt{k/m} \neq \omega$, is

$$x(t) = c_1 \cos \sqrt{\frac{k}{m}} t + c_2 \sin \sqrt{\frac{k}{m}} t + \frac{\alpha}{m[(k/m) - \omega^2]} \sin \omega t. \tag{19}$$

Note that the effect of the sinusoidal force $\alpha \sin \omega t$ is merely to superimpose itself on the simple harmonic motion of the system. The period of this added motion is the same as that of the external force, with constant amplitude for each fixed ω, but the amplitude approaches infinity as ω approaches $\sqrt{k/m}$. If $\omega = \sqrt{k/m}$, then the particular solution must have the form (from Case 2 in Section 20.6),

$$x_p(t) = A_1 t \cos \omega t + A_2 t \sin \omega t. \tag{20}$$

Substituting (20) into (15) yields $A_1 = -\alpha/2\omega$, $A_2 = 0$, so we have the general solution

$$x(t) = c_1 \cos \omega t + c_2 \sin \omega t - \frac{\alpha}{2\omega} t \cos \omega t. \tag{21}$$

We note that as t increases, the vibrations will get larger and larger without bound. Of course, this situation cannot arise in practice, since some resistance to the motion will always be present ($\mu \neq 0$). However, it is evident that if the resistance is small and the external forcing motion agrees with the natural vibration, then the displacement x may become so large that the elastic limit of the spring will be exceeded and a permanent set or distortion will occur. If $\omega = \sqrt{k/m}$, we say the external force is in *resonance* with the vibrating mass. The phenomenon of resonance is of great importance in engineering. It is necessary in some cases, as with our spring problem, to avoid a resonant condition so that no undue stresses will occur; on the other hand, resonance is desirable in many radio circuit problems.

PROBLEMS 20.7

1. Let $x(t)$ be given by (5).
   (a) Show that $x'(t) = 0$ when $\tan \sqrt{k/m}\, t = (v_0/x_0)\sqrt{m/k}$.
   (b) Show that for t as given in (a),

   $$\sin \sqrt{\frac{k}{m}} t = \frac{v_0 \sqrt{m/k}}{\sqrt{x_0^2 + v_0^2(m/k)}} \quad \text{and} \quad \cos \sqrt{\frac{k}{m}} t = \frac{x_0}{\sqrt{x_0^2 + v_0^2(m/k)}}.$$

   (c) Show that the maximum value of $x(t)$ is $\sqrt{x_0^2 + v_0^2(m/k)}$.

2. In Example 2, for what minimum damping constant would the mass fail to oscillate about its equilibrium?

3. One end of a rubber band is fixed at a point A. A 2-kg mass, attached to the other end, stretches the rubber band vertically to the point B in such a way that the length AB is 32-cm greater than the natural length of the band. If the mass is further drawn to a position 16-cm below B and released, what will be its velocity (if we neglect resistance) as it passes the position B?

4. If in Exercise 3, the mass is released at a position 16-cm above B, what will be its velocity as it passes 2-cm above B?

*5. A cylindrical block of wood of radius and height 1 ft and weighing 12.48 lb floats with its axis vertical in water (62.4 lb per ft^3). If it is depressed so that the surface of the water is tangent to the block, and is then released, what will be its period of vibration and equation of motion? Neglect resistance. [*Hint:* The upward force on the block is equal to the weight of the water displaced by the block.]

*6. A cubical block of wood, 1 ft on a side, is depressed so that its upper face lies along the surface of the water, and is then released. The period of vibration is found to be 1 sec. Neglecting resistance, what is the weight of the block of wood (in lbs)?

7. A 10-g mass suspended from a spring vibrates freely, the resistance being numerically equal to half the velocity (in m/sec) at any instant. If the period of the motion is 8 sec what is the spring constant (in g/sec²)?

8. A weight w (lb) is suspended from a spring whose constant is 10 lb/ft. The motion of the weight is subject to a resistance (lb) numerically equal to half the velocity (ft/sec). If the motion is to have a 1-sec period, what are the possible values of w?

*9. A 50-g mass is hanging at rest on a spring which is stretched 10 cm by the weight. The upper end of the spring is given the periodic force $10 \sin 2t$ g-m/sec² and the resistance has a magnitude $100\sqrt{17}$ (g/sec) times the velocity in meters per second. Find the equation of motion of the mass.

*10. Suppose in Exercise 9 that the upper end of the spring is given, instead, the force $1 - e^{-\sqrt{g}t}$ gm/sec². Find the displacement of the mass after 1 sec.

*20.8 An Introduction to the Power Series Method

Consider the second-order linear equation

$$y'' + a(x)y' + b(x)y = f(x).$$

If the functions a and b are not constant functions, then there is, in general, no way to obtain a closed form solution to the equation even in the homogeneous case ($f = 0$). In this section we show how power series can be used to obtain series solutions to equation (1) in some cases. The examples we present are merely illustrative of a technique that *sometimes* works. For a more complete discussion the reader is referred to a book on differential equations. The material in this section assumes a knowledge of the material in Chapter 14.

The fundamental assumption employed in solving a differential equation by the power series method is that the solution of the differential equation can be expressed in the form of a power series, say

$$y = \sum_{n=0}^{\infty} c_n x^n \tag{1}$$

Once this assumption has been made, power series expansions for y', y'', ... can be obtained by differentiating (1) term by term,

$$y' = \sum_{n=1}^{\infty} nc_n x^{n-1}, \tag{2}$$

$$y'' = \sum_{n=2}^{\infty} n(n-1)c_n x^{n-2}, \tag{3}$$

$$\vdots$$

and these may then be substituted into the given differential equation. After all the indicated operations have been carried out, and like powers of x have been collected, we obtain an expression of the form

$$k_0 + k_1 x + k_2 x^2 + \cdots = \sum_{n=0}^{\infty} k_n x^n = 0, \tag{4}$$

where the coefficients $k_0, k_1, k_2, \ldots$ are expressions involving the unknown coefficients $c_0, c_1, c_2, \ldots$. Since equation (4) must hold for all values of x in some interval, all the coefficients $k_0, k_1, k_2, \ldots$ must vanish. From the equations

$$k_0 = 0, \quad k_1 = 0, \quad k_2 = 0, \quad \ldots$$

it is then possible to determine successively the coefficients $c_0, c_1, c_2, \ldots$. In this section we will illustrate this procedure by means of several examples without concerning ourselves with questions of the convergence of the power series under consideration. A discussion of the validity and limitations of the methods illustrated here can be found in the book referred to earlier in this section.

EXAMPLE 1. Solve the initial value problem

$$y' = y + x^2, \quad y(0) = 1. \tag{5}$$

by using power series.

SOLUTION. Inserting (1) and (2) into equation (5), we have

$$c_1 + 2c_2 x + 3c_3 x^2 + 4c_4 x^3 + \cdots = (c_0 + c_1 x + c_2 x^2 + c_3 x^3 + \cdots) + x^2.$$

Collecting like powers of x, we obtain

$$(c_1 - c_0) + (2c_2 - c_1)x + (3c_3 - c_2 - 1)x^2 + (4c_4 - c_3)x^3 + \cdots = 0.$$

Equating each of the coefficients to zero, we obtain the identities

$$c_1 - c_0 = 0, \quad 2c_2 - c_1 = 0, \quad 3c_3 - c_2 - 1 = 0, \quad 4c_4 - c_3 = 0, \quad \ldots,$$

from which we find that

$$c_1 = c_0, \quad c_2 = \frac{c_1}{2} = \frac{c_0}{2!}, \quad c_3 = \frac{c_2 + 1}{3} = \frac{c_0 + 2}{3!}, \quad c_4 = \frac{c_3}{4} = \frac{c_0 + 2}{4!}, \quad \ldots.$$

With these values, (1) becomes

$$y = c_0 + c_0 x + \frac{c_0}{2!} x^2 + \frac{c_0 + 2}{3!} x^3 + \frac{c_0 + 2}{4!} x^4 + \frac{c_0 + 2}{5!} + \cdots$$

$$= (c_0 + 2) \left\{ 1 + x + \frac{x^2}{2!} + \frac{x^3}{3!} + \frac{x^4}{4!} + \cdots \right\} - 2 \left(1 + x + \frac{x^2}{2!} \right).$$

Using the Maclaurin expansion for e^x (see equation (14.9.20)), we have the general solution

$$y = (c_0 + 2)e^x - x^2 - 2x - 2.$$

To solve the initial value problem, we set $x = 0$ to obtain

$$1 = y(0) = c_0 + 2 - 2 = c_0.$$

Thus the solution of the initial value problem (5) is given by the equation

$$y = 3e^x - x^2 - 2x - 2.$$

EXAMPLE 2. Solve

$$y'' + y = 0. \tag{6}$$

SOLUTION. Using (1) and (3), we have

$$(2c_2 + 3 \cdot 2c_3 x + 4 \cdot 3c_4 x^2 + \cdots) + (c_0 + c_1 x + c_2 x^2 + \cdots) = 0.$$

Gathering like powers of x yields

$$(2c_2 + c_0) + (3 \cdot 2c_3 + c_1)x + (4 \cdot 3c_4 + c_2)x^2 + \cdots = 0.$$

Setting each of the coefficients to zero, we obtain

$$2c_2 + c_0 = 0, \quad 3 \cdot 2c_3 + c_1 = 0, \quad 4 \cdot 3c_4 + c_2 = 0, \quad 5 \cdot 4c_5 + c_3 = 0, \quad \ldots,$$

and

$$c_2 = -\frac{c_0}{2!}, \quad c_3 = -\frac{c_1}{3!}, \quad c_4 = -\frac{c_2}{4 \cdot 3} = \frac{c_0}{4!}, \quad c_5 = -\frac{c_3}{5 \cdot 4} = \frac{c_1}{5!}, \quad \ldots.$$

Substituting these values into the power series (1) for y yields

$$y = c_0 + c_1 x - \frac{c_0}{2!}x^2 - \frac{c_1}{3!}x^3 + \frac{c_0}{4!}x^4 + \frac{c_1}{5!}x^5 + \cdots.$$

Splitting this series into two parts, we have

$$y = c_0 \left(1 - \frac{x^2}{2!} + \frac{x^4}{4!} - \cdots\right) + c_1 \left(x - \frac{x^3}{3!} + \frac{x^5}{5!} - \cdots\right).$$

Use of the Maclaurin series for $\cos x$ and $\sin x$ (equations (14.9.21) and (14.9.22)) reveals the familiar general solution

$$y = c_0 \cos x + c_1 \sin x.$$

We observe that in this case the power series method produces two arbitrary constants c_0, c_1 and yields the general solution for equation (6).

So far we have considered only linear equations with constant coefficients. We turn now to linear equations with variable coefficients.

EXAMPLE 3. Solve the initial value problem

$$(1 + x^2)y' = 2pxy, \qquad y(0) = 1, \tag{7}$$

where p is a constant.

SOLUTION. Applying (1) and (2), we have

$$(1 + x^2) \sum_{k=1}^{\infty} nc_n x^{n-1} = 2px \sum_{n=0}^{\infty} c_n x^n.$$

This may be rewritten in the form

$$\sum_{n=1}^{\infty} nc_n x^{n-1} + \sum_{n=1}^{\infty} nc_n x^{n+1} = \sum_{n=0}^{\infty} 2pc_n x^{n+1}. \tag{8}$$

We would like to rewrite each of the sums in equation (8) so that each general term will contain the same power of x. This can be done by assuming that each general term contains the term x^k. For the first sum, this amounts to substituting $k = n - 1$. Since n ranges from 1 to ∞, $k = n - 1$ will range from 0 to ∞. Substituting $k = n + 1$ with k ranging from 2 to ∞ into the second sum, and $k = n + 1$ with k ranging from 1 to ∞ into the third sum allows us to rewrite these sums so that the general term will involve the power x^k. We then obtain

$$\sum_{k=0}^{\infty} (k + 1)c_{k+1} x^k + \sum_{k=2}^{\infty} (k - 1)c_{k-1} x^k = \sum_{k=1}^{\infty} 2pc_{k-1} x^k.$$

Now we gather like terms in x:

$$c_1 + (2c_2 - 2pc_0)x + \sum_{k=2}^{\infty} \{(k + 1)c_{k+1} + [(k - 1) - 2p]c_{k-1}\}x^k = 0. \tag{9}$$

Equating each coefficient to zero yields

$$c_1 = 0, \quad 2c_2 - 2pc_0 = 0, \quad 3c_3 + (1 - 2p)c_1 = 0,$$

and, in general,

$$(k + 1)c_{k+1} + [(k - 1) - 2p]c_{k-1} = 0, \qquad k \geq 1. \tag{10}$$

Equation (10) is called a *recursion formula* and may be used to evaluate the constants $c_0, c_1, c_2, \ldots$ successively. We see that

$$c_1 = 0, \qquad c_2 = pc_0, \qquad c_3 = 0,$$

and, in general,

$$c_{k+1} = \frac{(2p - k + 1)}{k + 1} c_{k-1}, \tag{11}$$

so that

$$c_4 = \frac{2p - 2}{4} c_2 = \frac{p(p - 1)}{1 \cdot 2} c_0, \qquad c_5 = 0,$$

$$c_6 = \frac{2p - 4}{6} c_4 = \frac{p(p - 1)(p - 2)}{1 \cdot 2 \cdot 3} c_0, \qquad c_7 = 0, \quad \ldots$$

since $c_3 = 0$. Thus the coefficients with odd-numbered subscripts vanish and the power series for y is given by

$$y = c_0 + \frac{p}{1}c_0 x^2 + \frac{p(p-1)}{1 \cdot 2}c_0 x^4 + \frac{p(p-1)(p-2)}{1 \cdot 2 \cdot 3}c_0 x^6 + \cdots$$

$$= c_0 \left\{ 1 + \frac{p}{1}x^2 + \frac{p(p-1)}{1 \cdot 2}x^4 + \frac{p(p-1)(p-2)}{1 \cdot 2 \cdot 3}x^6 + \cdots \right\}.$$

Replacing x by x^2 in the binomial formula (see Problem 14.9.25 and Appendix 4), we have

$$(1 + x)^p = 1 + \frac{p}{1}x + \frac{p(p-1)}{1 \cdot 2}x^2 + \frac{p(p-1)(p-2)}{1 \cdot 2 \cdot 3}x^3 + \cdots, \tag{12}$$

which yields the general solution of the differential equation:

$$y = c_0(1 + x^2)^p.$$

Since $y(0) = 1$, it follows that $c_0 = 1$ and $y = (1 + x^2)^p$.

EXAMPLE 4. Solve the differential equation

$$y'' + xy' + y = 0. \tag{13}$$

SOLUTION. Using (1), (2), and (3), we obtain the equation

$$\sum_{n=2}^{\infty} n(n-1)c_n x^{n-2} + x \sum_{n=1}^{\infty} nc_n x^{n-1} + \sum_{n=0}^{\infty} c_n x^n = 0.$$

Reindexing to obtain equal powers of x, we have

$$\sum_{k=0}^{\infty} (k+2)(k+1)c_{k+2} x^k + \sum_{k=1}^{\infty} kc_k x^k + \sum_{k=0}^{\infty} c_k x^k = 0.$$

Note that the second sum can also be allowed to range from 0 to ∞. Gathering like terms in x produces the equation

$$\sum_{k=0}^{\infty} [(k+2)(k+1)c_{k+2} + (k+1)c_k]x^k = 0. \tag{14}$$

Setting the coefficients equal to zero, we obtain the general recursion formula

$$(k+2)(k+1)c_{k+2} + (k+1)c_k = 0. \tag{15}$$

Therefore $(k+2)c_{k+2} = -c_k$ and

$$c_2 = -\frac{c_0}{2}, \qquad c_3 = -\frac{c_1}{3}, \qquad c_4 = -\frac{c_2}{4} = \frac{c_0}{2 \cdot 4},$$

$$c_5 = -\frac{c_3}{5} = \frac{c_1}{3 \cdot 5}, \qquad c_6 = -\frac{c_4}{6} = -\frac{c_0}{2 \cdot 4 \cdot 6}, \qquad \cdots.$$

Hence the power series for y can be written in the form

$$y = c_0 + c_1 x - \frac{c_0}{2} x^2 - \frac{c_1}{3} x^3 + \frac{c_0}{2 \cdot 4} x^4 + \frac{c_1}{3 \cdot 5} x^5 - \cdots$$

$$= c_0 \left(1 - \frac{x^2}{2} + \frac{x^4}{2 \cdot 4} - \frac{x^6}{2 \cdot 4 \cdot 6} + \cdots \right)$$

$$+ c_1 \left(x - \frac{x^3}{3} + \frac{x^5}{3 \cdot 5} - \frac{x^7}{3 \cdot 5 \cdot 7} + \cdots \right) \tag{16}$$

by separating the terms that involve c_0 and c_1. At this point we try to see whether we recognize the two series that have been obtained by the power series method. Very frequently this is an unproductive task, but in this instance we are fortunate:

$$1 - \frac{x^2}{2} + \frac{x^4}{2 \cdot 4} - \cdots = 1 + \left(-\frac{x^2}{2} \right) + \frac{1}{2!} \left(-\frac{x^2}{2} \right)^2 + \frac{1}{3!} \left(-\frac{x^2}{2} \right)^3 + \cdots$$

$$= e^{-x^2/2}.$$

We cannot recognize the second series so simply leave it as a series. We therefore obtain the general solution

$$y(x) = c_0 e^{-x^2/2} + c_1 \left(x - \frac{x^3}{3} + \frac{x^5}{3 \cdot 5} - \frac{x^7}{3 \cdot 5 \cdot 7} + \cdots \right).$$

EXAMPLE 5. Solve the equation

$$xy'' + y' + xy = 0. \tag{17}$$

SOLUTION. Using the power series (1), (2), and (3) for equation (17) yields the equation

$$\sum_{n=2}^{\infty} n(n-1)c_n x^{n-1} + \sum_{n=1}^{\infty} nc_n x^{n-1} + \sum_{n=0}^{\infty} c_n x^{n+1} = 0.$$

Reindexing the series to obtain like powers of x, we have

$$\sum_{k=1}^{\infty} (k+1)k c_{k+1} x^k + \sum_{k=0}^{\infty} (k+1)c_{k+1} x^k + \sum_{k=1}^{\infty} c_{k-1} x^k = 0.$$

Condensing the three series in one yields, after some algebra,

$$c_1 + \sum_{k=1}^{\infty} [(k+1)^2 c_{k+1} + c_{k-1}] x^k = 0. \tag{18}$$

Setting the coefficients to zero, we have $c_1 = 0$ and

$$(k+1)^2 c_{k+1} = -c_{k-1}, \qquad k = 1, 2, 3, \ldots. \tag{19}$$

The recursion formula (19) together with $c_1 = 0$ imply that all coefficients with odd-numbered subscripts vanish, and

$$c_2 = -\frac{c_0}{2^2}, \quad c_4 = -\frac{c_2}{4^2} = \frac{c_0}{2^2 4^2}, \quad c_6 = -\frac{c_4}{6^2} = -\frac{c_0}{2^2 4^2 6^2}, \quad \ldots.$$

Hence

$$y = c_0 - \frac{c_0}{2^2}x^2 + \frac{c_0}{2^2 4^2}x^4 - \frac{c_0}{2^2 4^2 6^2}x^6 + \cdots$$

$$= c_0 \sum_{n=0}^{\infty} \frac{1}{(n!)^2}\left(-\frac{x^2}{4}\right)^n. \tag{20}$$

It is unlikely that the reader is familiar with the series in (20). This series is often used in applied mathematics and is known as the *Bessel function of index zero*, $J_0(x)$. Note that the power series method has produced only *one* of the solutions of equation (17). To obtain a second solution, other techniques are needed.† We shall not develop these techniques here.

The next example illustrates the fact that this power series method does not always work.

EXAMPLE 6. Consider the differential equation

$$x^2 y'' + xy' + y = 0. \tag{21}$$

SOLUTION. Making use of the series (1), (2), and (3), and multiplying by the appropriate powers of x, we have

$$\sum_{n=2}^{\infty} n(n-1)c_n x^n + \sum_{n=1}^{\infty} nc_n x^n + \sum_{n=0}^{\infty} c_n x^n = 0$$

or

$$\sum_{n=0}^{\infty} (n^2 + 1)c_n x^n = 0. \tag{22}$$

Clearly, if we equate each of the coefficients of (22) to zero, all the coefficients c_n will vanish and $y \equiv 0$. Thus in this case the power series method fails completely in helping us find the general solution

$$y = c_1 \cos(\ln|x|) + c_2 \sin(\ln|x|)‡ \tag{23}$$

of equation (21) (check!). However, the power series method will yield a particular solution of the nonhomogeneous equation

$$x^2 y'' + xy' + y = \frac{1}{1-x}, \qquad |x| < 1, \tag{24}$$

since (22) and the geometric series (14.4.1) will yield

$$\sum_{n=0}^{\infty} (n^2 + 1)c_n x^n = \sum_{n=0}^{\infty} x^n.$$

---

† See Derrick and Grossman, *Elementary Differential Equations,* Section 3.3.
‡ More sophisticated power series methods are needed to obtain these solutions. See Derrick and Grossman, *Elementary Differential Equations,* Chapter 5.

or

$$\sum_{n=0}^{\infty} [(n^2 + 1)c_n - 1]x^n = 0.$$

Therefore,

$$c_n = \frac{1}{n^2 + 1}$$

for all n, so that a particular solution of (24) is

$$y_p = 1 + \frac{x}{2} + \frac{x^2}{5} + \frac{x^3}{10} + \cdots + \frac{x^n}{n^2 + 1} + \cdots. \tag{25}$$

Using the ratio test, we can find the radius of convergence of the series (25):

$$\frac{1}{R} = \lim_{n \to \infty} \frac{n^2 + 1}{(n + 1)^2 + 1} = \lim_{n \to \infty} \frac{1 + 1/n^2}{[1 + (1/n)]^2 + 1/n^2} = 1.$$

Hence the general solution of equation (24), valid for $0 < |x| < 1$, is

$$y = c_1 \cos(\ln |x|) + c_2 \sin(\ln |x|) + \left(1 + \frac{x}{2} + \frac{x^2}{5} + \cdots\right).$$

PROBLEMS 20.8

In Problems 1–20 find the general solution of each equation by using the power series method.
When initial conditions are specified, give the solution that satisfies them.

1. $y' = y$, $y(0) = 4$
2. $y' = 5y$, $y(0) = 1$
3. $y' = y - x$, $y(0) = 2$
4. $y' = -2y + x$
5. $y' = xy$
6. $y' = x^3 - 2xy$, $y(0) = 1$
7. $y' = y + e^x$
8. $xy' = y$
9. $xy' = ny$, n an integer
10. $y' = 1 + y^2$, $y(0) = 0$
11. $y'' + 4y = 0$
12. $y'' - 4y = 0$, $y(0) = 2$, $y'(0) = 0$
13. $y'' - y = 0$
14. $y'' + y = 0$, $y(0) = 1$, $y'(0) = 0$
15. $(1 + x^2)y'' + 2xy' - 2y = 0$
16. $y'' + 2y' + y = \sin x$
17. $y'' + y = e^{2x}$
18. $y'' - (3/x)y' = 0$
19. $x^2 y'' + xy' - 4y = 0$
20. $y'' - 2xy' + 2y = 0$.

21. Does the power series method yield a solution to the equation
     (a) $x^2 y' = y$?   (b) $x^3 y' = y$?
22. Solve $y' = y\sqrt{y^2 - 1}$, $y(0) = 1$.
23. Show that the power series method fails for $x^2 y'' + x^2 y' + y = 0$.

Review Exercises for Chapter Twenty

In Exercises 1–30 find the general solution to the given differential equation. If initial
conditions are given, find the unique solution to the initial value problem.

1. $\dfrac{dy}{dx} = 3x$
2. $\dfrac{dy}{dx} = e^{x-y}$; $y(0) = 4$

3. $\dfrac{dx}{dt} = e^x \cos t;\ x(0) = 3$

4. $\dfrac{dx}{dt} = x^{13} t^{11}$

5. $\dfrac{dy}{dx} + 3y = \cos x;\ y(0) = 1$

6. $\dfrac{dx}{dt} + 3x = \dfrac{1}{1 + e^{3t}}$

7. $\dfrac{dx}{dt} = 3x + t^3 e^{3t};\ x(1) = 2$

8. $\dfrac{dy}{dx} + y \cot x = \sin x;\ y\left(\dfrac{\pi}{6}\right) = \dfrac{1}{2}$

9. $y'' - 5y' + 4y = 0$

10. $y'' - 9y' + 14y = 0;\ y(0) = 2,\ y'(0) = 1$

11. $y'' - 9y = 0$

12. $y'' + 9y = 0$

13. $y'' + 6y' + 9y = 0$

14. $y'' + 8y' + 16y = 0;\ y(0) = -1,\ y'(0) = 3$

15. $y'' - 2y' + 2y = 0;\ y(0) = 0,\ y'(0) = 1$

16. $y'' + 8y' = 0;\ y(0) = 2;\ y'(0) = -3$

17. $y'' + 4y = 2 \sin x$

18. $y'' + y' - 12y = 4e^{2x};\ y(0) = 1,\ y'(0) = -1$

19. $y'' + y' + y = e^{-x/2} \sin(\sqrt{3}/2)x$

20. $y'' + 4y = 6x \cos 2x;\ y(0) = 1,\ y'(0) = 0$

21. $[x \cos(x + y) + \sin(x + y)]\ dx + x \cos(x + y)\ dy = 0;\ y(1) = (\pi/2) - 1$

22. $(2xy + e^y)\ dx + (x^2 + xe^y)\ dy = 0$

23. $(x^2 + y^2 + x)\ dx + y\ dy = 0$ [*Hint:* Multiply by an appropriate integrating factor.]

24. $y'' + y = x^3 - x$

25. $y'' - 2y' + y = e^{-x}$

26. $y'' - 8y' + 16y = e^{4x};\ y(0) = 3,\ y'(0) = 1$

27. $y'' - 2y' + 3y = e^x \cos \sqrt{2}x$

28. $y'' + y = x + e^x + \sin x$

29. $y'' + 16y = \cos 4x + x^2 - 3$

30. $y'' + y = 0;\ y(0) = y'(0) = 0.$

31. A spring, fixed at its upper end, supports a 10 kg mass which stretches the spring 60 cm. Find the equation of motion of the mass if it is drawn to a position of 10-cm below its equilibrium position and released with an initial velocity of 5 cm/sec upward.

32. What is the period, frequency, and amplitude of the motion of the mass in Exercise 31?

33. Find the equation of motion of the mass of Exercise 31 if it is subjected to damping forces having the damping constant $\mu = 3$.

34. In Exercise 33, for what minimum damping constant would the mass fail to oscillate about its equilibrium?

35. For what value of ω will the external force of $10 \sin \omega t$ *nt* produce resonance in the spring of Exercise 33?

In Exercises 36–38 use the power series method to obtain the solution to the initial value problem.

36. $\dfrac{dy}{dx} = 3y;\ y(0) = 2$

37. $y'' + 9y = 0;\ y(0) = 1,\ y'(0) = 0$

38. $y'' - 2y' + y = \cos x;\ y(0) = 0,\ y'(0) = 3.$

39. Use the power series method to obtain the general solution to $y'' - xy' - y = 0$.

APPENDIX 1

MATHEMATICAL INDUCTION

Mathematical induction† is the name given to an elementary logical principle that can be used to prove a certain type of mathematical statement. Typically, we use mathematical induction to prove that a certain statement or equation holds for every positive integer. For example, we may need to prove that $2^n > n$ for all integers $n \geq 1$.

To do this we proceed in two steps:

> (i)   We prove that the statement is true for some integer N (usually $N = 1$).
> (ii)   We *assume* that the statement is true for an integer k and then *prove* that it is true for the integer $k + 1$.

If we can complete these two steps, then we will have demonstrated the validity of the statement for *all* positive integers greater than or equal to N. To convince you of this fact, we reason as follows: since the statement is true for N (by step (i)) it is true for the integer $N + 1$ (by step (ii)). Then it is also true for the integer $(N + 1) + 1 = N + 2$ (again by step (ii)), and so on. We now demonstrate the procedure with some examples.

EXAMPLE 1.   Prove that the sum of the first n positive integers is equal to $n(n + 1)/2$.

SOLUTION. We are asked to show that

$$1 + 2 + 3 + \cdots + n = \frac{n(n + 1)}{2}. \tag{1}$$

(i)   If $n = 1$, then the sum of first one integer is 1. But $(1)(1 + 1)/2 = 1$ so that equation (1) holds in the case $n = 1$.

(ii)   Assume that (1) holds for $n = k$; that is,

$$1 + 2 + 3 + \cdots + k = \frac{k(k + 1)}{2}.$$

We must now show that it holds for $n = k + 1$. That is, we must show that

$$1 + 2 + 3 + \cdots + k + (k + 1) = \frac{(k + 1)(k + 2)}{2}.$$

†This technique was first used in a mathematical proof by the great French mathematician Pierre de Fermat (1601–1665).

But

$$1 + 2 + 3 + \cdots + k + (k + 1) = (1 + 2 + 3 + \cdots + k) + (k + 1)$$

$$= \frac{k(k + 1)}{2} + (k + 1)$$

$$= \frac{k(k + 1) + 2(k + 1)}{2} = \frac{(k + 1)(k + 2)}{2}$$

and the proof is complete. You may wish to try a few examples to illustrate that formula (1) really works. For example,

$$1 + 2 + 3 + 4 + 5 + 6 + 7 + 8 + 9 + 10 = \frac{10(11)}{2} = 55.$$

EXAMPLE 2. Prove that the sum of the squares of the first n positive integers is $n(n + 1)(2n + 1)/6$.

SOLUTION. We must prove that

$$1^2 + 2^2 + 3^2 + \cdots + n^2 = \frac{n(n + 1)(2n + 1)}{6}. \tag{2}$$

(i)   Since $1(1 + 1)(2 \cdot 1 + 1)/6 = 1 = 1^2$, equation (2) is valid for $n = 1$.

(ii)   Suppose equation (2) holds for $n = k$; that is

$$1^2 + 2^2 + 3^2 + \cdots + k^2 = \frac{k(k + 1)(2k + 1)}{6}.$$

Then to prove that (2) is true for $n = k + 1$, we have

$$1^2 + 2^2 + 3^2 + \cdots + k^2 + (k + 1)^2 = \frac{k(k + 1)(2k + 1)}{6} + (k + 1)^2$$

$$= \frac{k(k + 1)(2k + 1) + 6(k + 1)^2}{6}$$

$$= \frac{k + 1}{6}\{k(2k + 1) + 6(k + 1)\}$$

$$= \frac{k + 1}{6}\{2k^2 + 7k + 6\} = \frac{k + 1}{6}\{(k + 2)(2k + 3)\}$$

$$= \frac{(k + 1)(k + 2)(2[k + 1] + 1)}{6}$$

which is equation (2) for $n = k + 1$ and the proof is complete. Again, you may wish to experiment with this formula. For example,

$$1^2 + 2^2 + 3^2 + 4^2 + 5^2 + 6^2 + 7^2 = \frac{7(7 + 1)(2 \cdot 7 + 1)}{6} = \frac{7 \cdot 8 \cdot 15}{6} = 140.$$

EXAMPLE 3. For $a \neq 1$, use mathematical induction to prove the formula for the sum of a geometric progression:

$$1 + a + a^2 + \cdots + a^n = \frac{1 - a^{n+1}}{1 - a}. \tag{3}$$

SOLUTION. (i)   If $n = 0$ then

$$\frac{1 - a^{0+1}}{1 - a} = \frac{1 - a}{1 - a} = 1.$$

Thus equation (3) holds for $n = 0$. (We use $n = 0$ instead of $n = 1$ since $a^0 = 1$ is the first term.)

(ii)   Assume that (3) holds for $n = k$; that is

$$1 + a + a^2 + \cdots + a^k = \frac{1 - a^{k+1}}{1 - a}.$$

Then

$$1 + a + a^2 + \cdots + a^k + a^{k+1} = \frac{1 - a^{k+1}}{1 - a} + a^{k+1}$$

$$= \frac{1 - a^{k+1} + (1 - a)a^{k+1}}{1 - a} = \frac{1 - a^{k+2}}{1 - a}$$

so that equation (3) also holds for $n = k + 1$ and the proof is complete. Notice how much easier is the proof using mathematical induction than the "tricky" proof given in Section 1.10.

EXAMPLE 4.   Let $f_1, f_2, \ldots, f_n$ be differentiable functions. Use mathematical induction to prove that

$$\frac{d}{dx}(f_1 + f_2 + \cdots + f_n) = \frac{df_1}{dx} + \frac{df_2}{dx} + \cdots + \frac{df_n}{dx}. \tag{4}$$

SOLUTION. (i)   For $n = 2$, equation (4) was demonstrated in the proof of Theorem 3.1.2.

(ii)   Assume that equation (4) is valid for $n = k$; that is

$$\frac{d}{dx}(f_1 + f_2 + \cdots + f_k) = \frac{df_1}{dx} + \frac{df_2}{dx} + \cdots + \frac{df_k}{dx}.$$

Let $g(x) = f_1(x) + f_2(x) + \cdots + f_k(x)$. Then

$$\frac{d}{dx}(f_1 + f_2 + \cdots + f_k + f_{k+1}) = \frac{d}{dx}(g + f_{k+1})$$

$$= \frac{dg}{dx} + \frac{df_{k+1}}{dx} \text{ (by the case } n = 2)$$

$$= \frac{d}{dx}(f_1 + f_2 + \cdots + f_k) + \frac{df_{k+1}}{dx}$$

$$= \frac{df_1}{dx} + \frac{df_2}{dx} + \cdots + \frac{df_k}{dx} + \frac{df_{k+1}}{dx}.$$

which is equation (4) in the case $n = k + 1$ and the theorem is proved.

PROBLEMS

1. Use mathematical induction to prove that the sum of the cubes of the first n positive integers is given by

$$1^3 + 2^3 + 3^3 + \cdots + n^3 = \frac{n^2(n+1)^2}{4}. \tag{5}$$

2. Let the functions $f_1, f_2, \ldots, f_n$ be integrable on $[0, 1]$. Show that $f_1 + f_2 + \cdots + f_n$ is integrable on $[0, 1]$ and that

$$\int_0^1 (f_1(x) + f_2(x) + \cdots + f_n(x)) \, dx = \int_0^1 f_1(x) \, dx + \int_0^1 f_2(x) \, dx + \cdots \int_0^1 f_n(x) \, dx.$$

3. Use mathematical induction to prove that the nth derivative of the nth-order polynomial

$$P_n(x) = x^n + a_{n-1}x^{n-1} + a_{n-2}x^{n-2} + \cdots + a_1 x^1 + a_0$$

is equal to $n!$ $(n! = n(n-1)(n-2) \cdots 3 \cdot 2 \cdot 1.)$

4. If a sequence of numbers satisfies

$$q_n = \frac{q_{n-1}}{1 + q_{n-1}}$$

for every $n \geq 1$. Prove that

$$q_n = \frac{q_0}{1 + nq_0}$$

for every $n \geq 1$. [*Hint:* To see what is going on write out q_1, q_2, and q_3 in terms of q_0.]

*5. Prove, using mathematical induction, that there are exactly 2^n subsets of a set containing n elements.

6. Use mathematical induction to prove that

$$\ln(a_1 a_2 a_3 \cdots a_n) = \ln a_1 + \ln a_2 + \cdots + \ln a_n,$$

where $a_k > 0$ for $k = 1, 2, \ldots, n$.

7. Let $\mathbf{u}, \mathbf{v}_1, \mathbf{v}_2, \ldots, \mathbf{v}_n$ be $n + 1$ vectors in $\mathbb{R}^2$. Prove that

$$\mathbf{u} \cdot (\mathbf{v}_1 + \mathbf{v}_2 + \cdots + \mathbf{v}_n) = \mathbf{u} \cdot \mathbf{v}_1 + \mathbf{u} \cdot \mathbf{v}_2 + \cdots + \mathbf{u} \cdot \mathbf{v}_n.$$

APPENDIX 2

DETERMINANTS

In several parts of this book we made use of determinants. In this appendix, we shall show how determinants arise and discuss their uses.

We begin by considering the system of two linear equations in two unknowns

$$a_{11}x_1 + a_{12}x_2 = b_1,$$
$$a_{21}x_1 + a_{22}x_2 = b_2. \tag{1}$$

For simplicity we assume that the constants a_{11}, a_{12}, a_{21}, and a_{22} are all nonzero (otherwise the system can be solved directly). To solve the system (1) we reduce it to one equation in one unknown. To accomplish this we multiply the first equation by a_{22} and the second by a_{12} to obtain

$$a_{11}a_{22}x_1 + a_{22}a_{12}x_2 = a_{22}b_1,$$
$$a_{12}a_{21}x_1 + a_{22}a_{12}x_2 = a_{12}b_2. \tag{2}$$

Then, subtracting the second equation from the first, we have

$$(a_{11}a_{22} - a_{12}a_{21})x_1 = a_{22}b_1 - a_{12}b_2. \tag{3}$$

Now we define the quantity

$$\boxed{D = a_{11}a_{22} - a_{12}a_{21}.} \tag{4}$$

If $D \neq 0$, then (3) yields

$$x_1 = \frac{a_{22}b_1 - a_{12}b_2}{D}, \tag{5}$$

and x_2 may be obtained by substituting this value of x_1 into either of the equations of (1). *Thus if $D \neq 0$, the system (1) has a unique solution.*

On the other hand, suppose that $D = 0$. Then $a_{11}a_{22} = a_{12}a_{21}$, and equation (3) leads to the equation

$$0 = a_{22}b_1 - a_{12}b_2.$$

Either this equation is true or it is false. If it is false, that is if $a_{22}b_1 - a_{12}b_2 \neq 0$, then the system (1) has *no* solution. If the equation is true, that is to say $a_{22}b_1 - a_{12}b_2 = 0$, then the second equation of (2) is a multiple of the first and (1) consists essentially of only one equation. In this case we may choose x_1 arbitrarily and calculate the corresponding value of x_2 which means that there are an *infinite* number of solutions.

In sum, we have shown that *if* $D = 0$, *then the system* (1) *has either no solution or an infinite number of solutions.*

These facts are easily visualized geometrically by noting that (1) consists of the equations of two straight lines. A solution of the system is a point of intersection of the two lines. If the slopes are different, then $D \neq 0$ and the two lines intersect at a single point, which is the unique solution. It is easy to show (see Problem 21) that $D = 0$ if and only if the slopes of the two lines are the same. If $D = 0$, we either have two parallel lines and no solution, since the lines never intersect, or both equations yield the same line and every point on this line is a solution. This is illustrated in Figure 1.

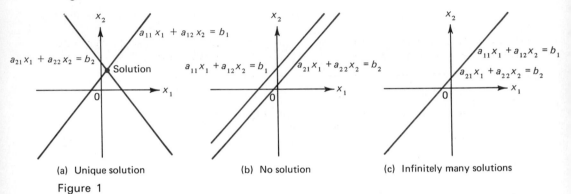

(a) Unique solution

(b) No solution

(c) Infinitely many solutions

Figure 1

EXAMPLE 1. Consider the following systems of equations:

(i) $\quad 2x_1 + 3x_2 = 12$
$\qquad x_1 + \ x_2 = 5$

(ii) $\quad x_1 + 3x_2 = 3$
$\qquad 3x_1 + 9x_2 = 8$

(iii) $\quad x_1 + 3x_2 = 3$
$\qquad 3x_1 + 9x_2 = 9$

In system (i), $D = 2 \cdot 1 - 3 \cdot 1 = -1 \neq 0$, so there is a unique solution, which is easily found to be $x_1 = 3$, $x_2 = 2$. In system (ii), $D = 1 \cdot 9 - 3 \cdot 3 = 0$. Multiplying the first equation by 3 and then subtracting this from the second equation, we obtain the equation $0 = -1$, which is impossible. Thus there is no solution. In (iii), $D = 1 \cdot 9 - 3 \cdot 3 = 0$. But now the second equation is simply three times the first equation. If x_2 is arbitrary, then $x_1 = 3 - 3x_2$, and there are an infinite number of solutions.

Returning again to the system (1), we define the *determinant of the system* as

$$D = a_{11}a_{22} - a_{12}a_{21}. \tag{6}$$

For convenience of notation we denote the determinant by writing the coefficients of the system in a square array:

$$D = \begin{vmatrix} a_{11} & a_{12} \\ a_{21} & a_{22} \end{vmatrix} = a_{11}a_{22} - a_{12}a_{21}. \tag{7}$$

Therefore, a 2×2 determinant is the product of the two elements in the upper-left-to-lower-right diagonal minus the product of the other two elements.

We have proved:

THEOREM 1. For the 2×2 system (1), there is a unique solution if and only if the determinant D is not equal to zero. If $D = 0$, then there is either no solution or an infinite number of solutions.

Let us now consider the general system of n equations in n unknowns:

$$
\begin{aligned}
a_{11}x_1 + a_{12}x_2 + \cdots + a_{1n}x_n &= b_1, \\
a_{21}x_1 + a_{22}x_2 + \cdots + a_{2n}x_n &= b_2, \\
&\ \ \vdots \\
a_{n1}x_1 + a_{n2}x_2 + \cdots + a_{nn}x_n &= b_n,
\end{aligned}
\tag{8}
$$

and define the determinant of such a system in order to obtain a theorem like the one above for $n \times n$ systems. We begin by defining the determinant of a 3×3 system:

$$
D = \begin{vmatrix} a_{11} & a_{12} & a_{13} \\ a_{21} & a_{22} & a_{23} \\ a_{31} & a_{32} & a_{33} \end{vmatrix} = a_{11} \begin{vmatrix} a_{22} & a_{23} \\ a_{32} & a_{33} \end{vmatrix} - a_{12} \begin{vmatrix} a_{21} & a_{23} \\ a_{31} & a_{33} \end{vmatrix}
$$
$$
+ a_{13} \begin{vmatrix} a_{21} & a_{22} \\ a_{31} & a_{32} \end{vmatrix}.
\tag{9}
$$

We see that to calculate a 3×3 determinant, it is necessary to calculate three 2×2 determinants.

EXAMPLE 2

$$
\begin{vmatrix} 3 & 5 & 2 \\ 4 & 2 & 3 \\ -1 & 2 & 4 \end{vmatrix} = 3 \begin{vmatrix} 2 & 3 \\ 2 & 4 \end{vmatrix} - 5 \begin{vmatrix} 4 & 3 \\ -1 & 4 \end{vmatrix} + 2 \begin{vmatrix} 4 & 2 \\ -1 & 2 \end{vmatrix}
$$
$$
= 3 \cdot 2 - 5 \cdot 19 + 2 \cdot 10 = -69.
$$

The general definition of the determinant of the $n \times n$ system of equations (8) is simply an extension of this procedure:

$$
D = \begin{vmatrix} a_{11} & a_{12} & \cdots & a_{1n} \\ a_{21} & a_{22} & \cdots & a_{2n} \\ \vdots & \vdots & & \vdots \\ a_{n1} & a_{n2} & \cdots & a_{nn} \end{vmatrix} = a_{11}A_{11} - a_{12}A_{12} + \cdots + (-1)^{n+1}a_{1n}A_{1n},
\tag{10}
$$

where A_{1j} is the $(n-1) \times (n-1)$ determinant obtained by crossing out the first row and jth column of the original $n \times n$ determinant. Thus an $n \times n$ determinant can be obtained by calculating $n\ (n-1) \times (n-1)$ determinants. Note that in definition (10), the signs alternate. The signs of the $n^2\,(n-1) \times (n-1)$ determinants can easily be illustrated by the following schematic diagram:

$$
\begin{vmatrix}
+ & - & + & - & + & - & \cdots \\
- & + & - & + & - & + & \cdots \\
+ & - & + & - & + & - & \cdots \\
- & + & - & + & - & + & \cdots \\
+ & - & + & - & + & - & \cdots \\
\vdots & \vdots & \vdots & \vdots & \vdots & \vdots & \ddots
\end{vmatrix}
$$

EXAMPLE 3

$$
\begin{vmatrix}
1 & 3 & 5 & 2 \\
0 & -1 & 3 & 4 \\
2 & 1 & 9 & 6 \\
3 & 2 & 4 & 8
\end{vmatrix}
= 1
\begin{vmatrix}
-1 & 3 & 4 \\
1 & 9 & 6 \\
2 & 4 & 8
\end{vmatrix}
- 3
\begin{vmatrix}
0 & 3 & 4 \\
2 & 9 & 6 \\
3 & 4 & 8
\end{vmatrix}
+ 5
\begin{vmatrix}
0 & -1 & 4 \\
2 & 1 & 6 \\
3 & 2 & 8
\end{vmatrix}
$$

$$
- 2
\begin{vmatrix}
0 & -1 & 3 \\
2 & 1 & 9 \\
3 & 2 & 4
\end{vmatrix}
$$

$$
= 1(-92) - 3(-70) + 5(2) - 2(-16) = 160.
$$

(The values in parentheses are obtained by calculating the four 3×3 determinants.)

The reason for considering determinants of systems of n equations in n un-knowns is that Theorem 1 also holds for these systems (although this fact will not be proven here).

THEOREM 2. For the system (8) there is a unique solution if and only if the determinant D, defined by (10), is not zero. If $D = 0$, then there is either no solution or an infinite number of solutions.

It is clear that calculating determinants by formula (10) can be extremely tedious, especially if $n \geq 5$. For that reason techniques are available for greatly simplifying these calculations. Some of these techniques are described in the theo-rems below. The proofs of these theorems can be found in most books on matrix theory and advanced engineering mathematics.†

We begin with the result that states that the determinant can be obtained by expanding in any row.

THEOREM 3. For any i, $i = 1, 2, \ldots, n$

† See, for example, Erwin Kreysig, *Advanced Engineering Mathematics*, Wiley, New York, 1972, Sections 6.6 and 6.7.

$$D = \begin{vmatrix} a_{11} & a_{12} & \cdots & a_{1n} \\ a_{21} & a_{22} & \cdots & a_{2n} \\ \vdots & \vdots & & \vdots \\ a_{n1} & a_{n2} & \cdots & a_{nn} \end{vmatrix} = (-1)^{i+1}a_{i1}A_{i1} + (-1)^{i+2}a_{i2}A_{i2} + \cdots \\ + (-1)^{i+n}a_{in}A_{in}$$

where A_{ij} is the $(n-1) \times (n-1)$ determinant obtained by crossing off the ith row and jth column of D. Notice that the signs in the expansion of a determinant alternate.

EXAMPLE 4. Calculate $\begin{vmatrix} 3 & 5 & 2 \\ 4 & 2 & 3 \\ -1 & 2 & 4 \end{vmatrix}$ by expanding on the second row (see Example 2).

SOLUTION

$$\begin{vmatrix} 3 & 5 & 2 \\ 4 & 2 & 3 \\ -1 & 2 & 4 \end{vmatrix} = (-1)^{2+1}(4)\begin{vmatrix} 5 & 2 \\ 2 & 4 \end{vmatrix} + (-1)^{2+2}(2)\begin{vmatrix} 3 & 2 \\ -1 & 4 \end{vmatrix} + (-1)^{2+3}(3)\begin{vmatrix} 3 & 5 \\ -1 & 2 \end{vmatrix}$$

$$= -4(16) + 2(14) - 3(11) = -69.$$

We remark that we can also get the same result by expanding in any column of D.

THEOREM 4. For any j, $j = 1, 2, \ldots, n$,

$$D = \begin{vmatrix} a_{11} & a_{12} & \cdots & a_{1n} \\ a_{21} & a_{22} & \cdots & a_{2n} \\ \vdots & \vdots & & \vdots \\ a_{n1} & a_{n2} & \cdots & a_{nn} \end{vmatrix} = (-1)^{1+j}a_{1j}A_{1j} + (-1)^{2+j}a_{2j}A_{2j} + \cdots \\ + (-1)^{n+j}a_{nj}A_{nj},$$

where A_{ij} is as in Theorem 3.

EXAMPLE 5. Calculate $D = \begin{vmatrix} 3 & 5 & 2 \\ 4 & 2 & 3 \\ -1 & 2 & 4 \end{vmatrix}$ by expanding in the third column.

SOLUTION

$$\begin{vmatrix} 3 & 5 & 2 \\ 4 & 2 & 3 \\ -1 & 2 & 4 \end{vmatrix} = (-1)^{1+3}(2)\begin{vmatrix} 4 & 2 \\ -1 & 2 \end{vmatrix} + (-1)^{2+3}(3)\begin{vmatrix} 3 & 5 \\ -1 & 2 \end{vmatrix} + (-1)^{3+3}(4)\begin{vmatrix} 3 & 5 \\ 4 & 2 \end{vmatrix}$$

$$= 2(10) - 3(11) + 4(-14) = -69.$$

THEOREM 5. Let $D = \begin{vmatrix} a_{11} & a_{12} & \cdots & a_{1n} \\ a_{21} & a_{22} & \cdots & a_{2n} \\ \vdots & \vdots & & \vdots \\ a_{n1} & a_{n2} & \cdots & a_{nn} \end{vmatrix}$. Then

(i)   If any row or column of D is zero, then $D = 0$.

(ii)   If any row (column) is a multiple of any other row (column), then $D = 0$.

(iii)   Interchanging any two rows (columns) of D has the effect of multiplying D by -1.

(iv)   Multiplying a row (column) or D by a constant α has the effect of multiplying D by α.

(v)   If any row (column) of D is multiplied by a constant and added to a different row (column) of D, then D is unchanged.

EXAMPLE 6. Calculate $D = \begin{vmatrix} 2 & 1 & 4 & 3 \\ 3 & 1 & -2 & -1 \\ 14 & -2 & 0 & 6 \\ 6 & 2 & -4 & -2 \end{vmatrix}$.

SOLUTION. $D = 0$ according to (ii) since the fourth row is twice the second row. This can easily be verified.

EXAMPLE 7. Calculate $D = \begin{vmatrix} 1 & 14 & 3 \\ 2 & 28 & -2 \\ 0 & -42 & 1 \end{vmatrix}$.

SOLUTION. According to (iv), we may divide the second row by 14 which has the effect of dividing D by 14. Then

$$\frac{D}{14} = \begin{vmatrix} 1 & 1 & 3 \\ 2 & 2 & -2 \\ 0 & -3 & 1 \end{vmatrix}$$

or

$$D = 14 \begin{vmatrix} 1 & 1 & 3 \\ 2 & 2 & -2 \\ 0 & -3 & 1 \end{vmatrix} = 14(-24) = -336.$$

EXAMPLE 8. Calculate $D = \begin{vmatrix} 0 & -42 & 1 \\ 2 & 28 & -2 \\ 1 & 14 & 3 \end{vmatrix}$.

SOLUTION. D is obtained from the determinant of Example 7 by interchanging the first and third rows. Hence, according to (iii), $D = -(-336) = 336$.

The results in Theorem 5 can be used to simplify the calculation of determinants.

EXAMPLE 9. Calculate $D = \begin{vmatrix} 1 & 3 & 5 & 2 \\ 0 & -1 & 3 & 4 \\ 2 & 1 & 9 & 6 \\ 3 & 2 & 4 & 8 \end{vmatrix}$.

SOLUTION. This was calculated in Example 3. The idea is to use Theorem 5 to make the evaluation of the determinant almost trivial. We begin by multiplying the first row by -2 and adding it to the third row. By (v), this will leave the determinant unchanged.

$$D = \begin{vmatrix} 1 & 3 & 5 & 2 \\ 0 & -1 & 3 & 4 \\ 2+(-2)1 & 1+(-2)3 & 9+(-2)5 & 6+(-2)2 \\ 3 & 2 & 4 & 8 \end{vmatrix}$$

$$= \begin{vmatrix} 1 & 3 & 5 & 2 \\ 0 & -1 & 3 & 4 \\ 0 & -5 & -1 & 2 \\ 3 & 2 & 4 & 8 \end{vmatrix}.$$

Now we multiply the first row by -3 and add it to the fourth row:

$$D = \begin{vmatrix} 1 & 3 & 5 & 2 \\ 0 & -1 & 3 & 4 \\ 0 & -5 & -1 & 2 \\ 0 & -7 & -11 & 2 \end{vmatrix}.$$

We now expand D by its first column:

$$D = 1 \begin{vmatrix} -1 & 3 & 4 \\ -5 & -1 & 2 \\ -7 & -11 & 2 \end{vmatrix} - 0 \cdot \begin{vmatrix} 3 & 5 & 2 \\ -5 & -1 & 2 \\ -7 & -11 & 2 \end{vmatrix} + 0 \cdot \begin{vmatrix} 3 & 5 & 2 \\ -1 & 3 & 4 \\ -7 & -11 & 2 \end{vmatrix}$$

$$+ 0 \begin{vmatrix} 3 & 5 & 2 \\ -1 & 3 & 4 \\ -5 & -1 & 2 \end{vmatrix}$$

$$= \begin{vmatrix} -1 & 3 & 4 \\ -5 & -1 & 2 \\ -7 & -11 & 2 \end{vmatrix},$$

which is a 3×3 determinant. We can calculate it by expansion or we can reduce further. By Theorem 5, parts (iv) and (v),

$$\begin{vmatrix} -1 & 3 & 4 \\ -5 & -1 & 2 \\ -7 & -11 & 2 \end{vmatrix} = - \begin{vmatrix} 1 & -3 & -4 \\ -5 & -1 & 2 \\ -7 & -11 & 2 \end{vmatrix} = - \begin{vmatrix} 1 & -3 & 4 \\ 0 & -16 & -18 \\ 0 & -32 & -26 \end{vmatrix}$$

$$= - \begin{vmatrix} -16 & -18 \\ -32 & -26 \end{vmatrix}$$

$$= -[(-16)(-26) - (-18)(-32)] = 160.$$

In the second step we multiplied the first row by 5 and added it to the second, and multiplied the first row by 7 and added it to the third. Note how we were able to reduce the problem to the calculation of a 2×2 determinant.

EXAMPLE 10. Calculate $\begin{vmatrix} 2 & -5 & -1 & 6 \\ 1 & 6 & 8 & -2 \\ -2 & 1 & 0 & 5 \\ -3 & 6 & -4 & 9 \end{vmatrix}$.

SOLUTION. There are several ways to calculate this. Since there is a zero in the third row, we expand in that row. Before we expand, we multiply the second column by 2 and add it to the first column, and multiply the second column by -5 and add it to the fourth column to obtain

$$D = \begin{vmatrix} 2 & -5 & -1 & 6 \\ 1 & 6 & 8 & -2 \\ -2 & 1 & 0 & 5 \\ -3 & 6 & -4 & 9 \end{vmatrix} = \begin{vmatrix} -8 & -5 & -1 & 31 \\ 13 & 6 & 8 & -32 \\ 0 & 1 & 0 & 0 \\ 9 & 6 & -4 & -21 \end{vmatrix}.$$

We now expand in the third row:

$$D = (-1)^{3+2} \begin{vmatrix} -8 & -1 & 31 \\ 13 & 8 & -32 \\ 9 & -4 & -21 \end{vmatrix} = \begin{vmatrix} 8 & 1 & -31 \\ 13 & 8 & -32 \\ 9 & -4 & -21 \end{vmatrix}.$$

Then we multiply the second column by -8 and add it to the first column, and multiply the second column by 31 and add it to the third column:

$$D = \begin{vmatrix} 0 & 1 & 0 \\ -51 & 8 & 216 \\ 41 & -4 & -145 \end{vmatrix} = (-1)^{1+2} \begin{vmatrix} -51 & 216 \\ 41 & -145 \end{vmatrix}$$

$$= \begin{vmatrix} 51 & -216 \\ 41 & -145 \end{vmatrix} = 51(-145) - (-216)(41) = 1461.$$

EXAMPLE 11. Calculate $D = \begin{vmatrix} 2 & 4 & 7 & 3 & 1 \\ 5 & -1 & 2 & 4 & 3 \\ 7 & 3 & -1 & -2 & 4 \\ 3 & 1 & 5 & 7 & 5 \\ 15 & -3 & 6 & 12 & 9 \end{vmatrix}$.

SOLUTION. Since the fifth row is three times the second row, $D = 0$ by Theorem 5, part (ii).

There is one further result about determinants that will be useful in Chapter 17.

THEOREM 6. Let $D = \begin{vmatrix} a_{11} & a_{12} & \cdots & a_{1n} \\ a_{21} & a_{22} & & a_{2n} \\ \vdots & \vdots & & \vdots \\ a_{i1} + b_{i1} & a_{i2} + b_{i2} & \cdots & a_{in} + b_{in} \\ \vdots & & & \vdots \\ a_{n1} & a_{n2} & \cdots & a_{nn} \end{vmatrix}$. Then

$$D = \begin{vmatrix} a_{11} & a_{12} & \cdots & a_{1n} \\ a_{21} & a_{22} & \cdots & a_{2n} \\ \vdots & \vdots & & \vdots \\ a_{i1} & a_{i2} & \cdots & a_{in} \\ \vdots & \vdots & & \vdots \\ a_{n1} & a_{n2} & \cdots & a_{nn} \end{vmatrix} + \begin{vmatrix} a_{11} & a_{12} & \cdots & a_{1n} \\ a_{21} & a_{22} & \cdots & a_{2n} \\ \vdots & \vdots & & \vdots \\ b_{i1} & b_{i2} & \cdots & b_{in} \\ \vdots & \vdots & & \vdots \\ a_{n1} & a_{n2} & \cdots & a_{nn} \end{vmatrix}. \tag{11}$$

EXAMPLE 12. To illustrate Theorem 6, we note that

$$\begin{vmatrix} 2 & 1 & 4 \\ 3+5 & 2-3 & 1+2 \\ 0 & -4 & 2 \end{vmatrix} = \begin{vmatrix} 2 & 1 & 4 \\ 3 & 2 & 1 \\ 0 & -4 & 2 \end{vmatrix} + \begin{vmatrix} 2 & 1 & 4 \\ 5 & -3 & 2 \\ 0 & -4 & 2 \end{vmatrix}$$

$$= -38 - 86 = -124.$$

We conclude this appendix by showing how determinants can be used to obtain the unique solution (if one exists) of the system (8) of n equations in n unknowns. We define the determinants

$$D_1 = \begin{vmatrix} b_1 & a_{12} & \cdots & a_{1n} \\ b_2 & a_{22} & \cdots & a_{2n} \\ \vdots & \vdots & & \vdots \\ b_n & a_{n2} & \cdots & a_{nn} \end{vmatrix},$$

$$D_2 = \begin{vmatrix} a_{11} & b_1 & a_{13} & \cdots & a_{1n} \\ a_{21} & b_2 & a_{23} & \cdots & a_{2n} \\ \vdots & \vdots & \vdots & & \vdots \\ a_{n1} & b_n & a_{n3} & \cdots & a_{nn} \end{vmatrix}, \ldots,$$

$$D_k = \begin{vmatrix} a_{11} & a_{12} & \cdots & a_{1,k-1} & b_1 & a_{1,k+1} & \cdots & a_{1n} \\ a_{21} & a_{22} & \cdots & a_{2,k-1} & b_2 & a_{2,k+1} & \cdots & a_{2n} \\ \vdots & \vdots & & \vdots & \vdots & \vdots & & \vdots \\ a_{n1} & a_{n2} & \cdots & a_{n,k-1} & b_n & a_{n,k+1} & \cdots & a_{nn} \end{vmatrix}, \ldots,$$

$$D_n = \begin{vmatrix} a_{11} & a_{1n} & \cdots & a_{1,n-1} & b_1 \\ a_{21} & a_{22} & \cdots & a_{n,n-1} & b_2 \\ \vdots & \vdots & & \vdots & \vdots \\ a_{n1} & a_{n2} & \cdots & a_{n,n-1} & b_n \end{vmatrix}, \tag{12}$$

obtained by replacing the kth column of D by the column

$$\begin{bmatrix} b_1 \\ b_2 \\ \vdots \\ b_n \end{bmatrix}.$$

Then we have the following theorem, known as *Cramer's rule*.

THEOREM 7 (Cramer's Rule). Let D and D_k, $k = 1, 2, \ldots, n$, be given as in (10) and (12). If $D \neq 0$, then the unique solution to the system (8) is given by the values

$$x_1 = \frac{D_1}{D}, \qquad x_2 = \frac{D_2}{D}, \qquad \ldots, \qquad x_n = \frac{D_n}{D}. \tag{13}$$

EXAMPLE 13. Consider the system

$$\begin{aligned} 2x_1 + 4x_2 - x_3 &= -5, \\ -4x_1 + 3x_2 + 5x_3 &= 14, \\ 6x_1 - 3x_2 - 2x_3 &= 5. \end{aligned}$$

We have

$$D = \begin{vmatrix} 2 & 4 & -1 \\ -4 & 3 & 5 \\ 6 & -3 & -2 \end{vmatrix} = 112, \qquad D_1 = \begin{vmatrix} -5 & 4 & -1 \\ 14 & 3 & 5 \\ 5 & -3 & -2 \end{vmatrix} = 224,$$

$$D_2 = \begin{vmatrix} 2 & -5 & -1 \\ -4 & 14 & 5 \\ 6 & 5 & -2 \end{vmatrix} = -112, \qquad D_3 = \begin{vmatrix} 2 & 4 & -5 \\ -4 & 3 & 14 \\ 6 & -3 & 5 \end{vmatrix} = 560.$$

Therefore

$$x_1 = \frac{D_1}{D} = 2, \qquad x_2 = \frac{D_2}{D} = -1, \qquad x_3 = \frac{D_3}{D} = 5.$$

PROBLEMS

For each of the 2 × 2 systems in Problems 1–8, calculate the determinant D. If $D \neq 0$, find the unique solution. If $D = 0$, determine whether there is no solution or an infinite number of solutions.

1. $2x_1 + 4x_2 = 6,$
   $x_1 + x_2 = 3$

2. $2x_1 + 4x_2 = 6,$
   $x_1 + 2x_2 = 5$

3. $2x_1 + 4x_2 = 6,$
   $x_1 + 2x_2 = 3$

4. $6x_1 - 3x_2 = 3,$
   $-2x_1 + x_2 = -1$

5. $6x_1 - 3x_2 = 3,$
   $-2x_1 + x_2 = 1$

6. $6x_1 - 3x_2 = 3,$
   $-2x_1 + 2x_2 = -1$

7. $2x_1 + 5x_2 = 0,$
   $3x_1 - 7x_2 = 0$

8. $2x_1 - 3x_2 = 0,$
   $-4x_1 + 6x_2 = 0.$

In Problems 9–20 calculate the determinant.

9. $\begin{vmatrix} 1 & 2 & 3 \\ 6 & -1 & 4 \\ 2 & 0 & 6 \end{vmatrix}$

10. $\begin{vmatrix} 4 & -1 & 0 \\ 2 & 1 & 7 \\ -2 & 3 & 4 \end{vmatrix}$

11. $\begin{vmatrix} 7 & 2 & 3 \\ 0 & 4 & 1 \\ 0 & 0 & 5 \end{vmatrix}$

12. $\begin{vmatrix} 1 & -1 & 4 \\ 3 & -2 & 1 \\ 5 & 1 & 7 \end{vmatrix}$

13. $\begin{vmatrix} 4 & 2 & 7 \\ 1 & 5 & 3 \\ -1 & 1 & 4 \end{vmatrix}$

14. $\begin{vmatrix} -1 & 0 & 4 \\ 7 & 3 & 2 \\ 4 & 1 & 5 \end{vmatrix}$

15. $\begin{vmatrix} 1 & 4 & 7 & 2 \\ 0 & 5 & 8 & 1 \\ 0 & 0 & -3 & 4 \\ 0 & 0 & 0 & 8 \end{vmatrix}$

16. $\begin{vmatrix} a_1 & a_2 & a_3 & a_4 \\ 0 & b_1 & b_2 & b_3 \\ 0 & 0 & c_1 & c_2 \\ 0 & 0 & 0 & d_1 \end{vmatrix}$

17. $\begin{vmatrix} 2 & 1 & 3 & 4 \\ 3 & -2 & 5 & 1 \\ 4 & 0 & 4 & 5 \\ 2 & 1 & 7 & -4 \end{vmatrix}$

18. $\begin{vmatrix} 1 & 3 & -1 & 7 \\ -2 & 5 & 2 & 8 \\ -3 & 7 & 3 & 3 \\ 5 & 0 & -5 & 11 \end{vmatrix}$

19. $\begin{vmatrix} 2 & 3 & 1 & 4 \\ 2 & 2 & 4 & 6 \\ 3 & -1 & -2 & 4 \\ 4 & 2 & -3 & -5 \end{vmatrix}$

20. $\begin{vmatrix} 1 & 0 & 2 & 3 & 1 \\ 0 & 4 & -1 & -2 & 3 \\ 2 & 1 & 0 & -1 & 1 \\ -3 & 2 & 2 & 0 & 5 \\ 0 & 3 & 6 & 1 & -3 \end{vmatrix}.$

21. Show that the two lines in system (1) have the same slope if and only if the determinant of the system is zero.

In Problems 22–28 solve the system by using Cramer's rule.

22. $3x_1 - x_2 = 13,$
    $-4x_1 + 6x_2 = -8$

23. $2x_1 + 6x_2 + 3x_3 = 9,$
    $-3x_1 - 17x_2 - x_3 = 4,$
    $4x_1 + 3x_2 + x_3 = -7$

24. $2x_1 + x_3 = 0,$
    $3x_1 - 2x_2 + 2x_3 = -4,$
    $4x_1 - 5x_2 = 3$

25. $x_1 - 8x_2 - x_3 = -1,$
    $-x_1 + 4x_2 + x_3 = 3,$
    $3x_1 - 2x_2 + 6x_3 = 5$

26.
$$x_1 + 2x_2 - x_3 - 4x_4 = 1,$$
$$-x_1 \qquad + 2x_3 + 6x_4 = 5,$$
$$-4x_2 - 2x_3 - 8x_4 = -8,$$
$$3x_1 - 2x_2 \qquad + 5x_4 = 3$$

27.
$$2x_1 + 5x_2 - 3x_3 + 2x_4 = 3,$$
$$-x_1 - 3x_2 + 2x_3 - x_4 = -1,$$
$$-3x_1 + 4x_2 + 8x_3 - 2x_4 = 4,$$
$$6x_1 - x_2 - 6x_3 + 4x_4 = 2.$$

28.
$$x_1 - 2x_2 + x_3 = 0$$
$$2x_1 - x_2 - 3x_3 = 0$$
$$5x_1 + 7x_2 - 8x_3 = 0.$$

APPENDIX 3

THE CONIC SECTIONS

In Section 1.9 we briefly introduced the parabola, the ellipse, and the hyperbola. Each of these three curves (plus the circle which is a special kind of ellipse) can be obtained as the intersection of a plane with a right circular cone (see Figure 1). In this appendix we shall discuss these curves in more detail.

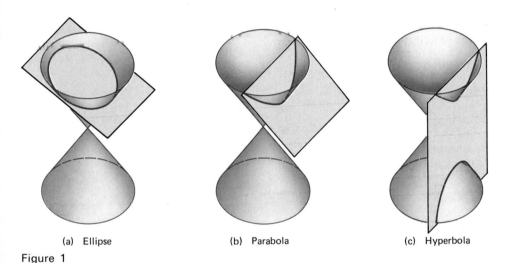

| (a) Ellipse | (b) Parabola | (c) Hyperbola |

Figure 1

I. The Ellipse

DEFINITION 1. An *ellipse* is the set of points (x, y) such that the sum of the distances from (x, y) to two given points is fixed and is greater than the distance between the two points. Each of the two points is called a *focus* of the ellipse.

We now calculate the equation of an ellipse. To simplify our calculations, we assume that the foci F_1 and F_2 are the points $(-c, 0)$ and $(c, 0)$. (See Figure 2.) If (x, y) is on the ellipse, then, by Definition 1,

$$\overline{F_1 P} + \overline{F_2 P} = 2a, \tag{1}$$

where $2a > 2c$ ($2c$ is the distance between the foci and $2a$ is the given sum of the distances). Then, from (1),

$$\sqrt{(x + c)^2 + y^2} + \sqrt{(x - c)^2 + y^2} = 2a$$

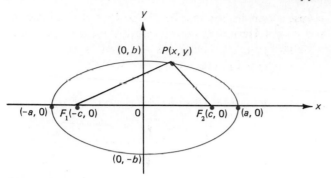

Figure 2

or
$$\sqrt{(x + c)^2 + y^2} = 2a - \sqrt{(x - c)^2 + y^2}.$$

Squaring both sides, we have
$$(x + c)^2 + y^2 = 4a^2 - 4a\sqrt{(x - c)^2 + y^2} + (x - c)^2 + y^2$$

which, after simplification, leads to
$$4xc = 4a^2 - 4a\sqrt{(x - c)^2 + y^2}.$$

Then, dividing by $4a$ and rearranging,
$$\sqrt{(x - c)^2 + y^2} = a - \frac{c}{a}x$$

and, squaring once more,
$$(x - c)^2 + y^2 = a^2 - 2cx + \frac{c^2}{a^2}x^2$$

which yields
$$x^2 - 2xc + c^2 + y^2 = a^2 - 2cx + \frac{c^2}{a^2}x^2$$

or
$$x^2\left(1 - \frac{c^2}{a^2}\right) + y^2 = a^2 - c^2$$

or
$$x^2\left(\frac{a^2 - c^2}{a^2}\right) + y^2 = a^2 - c^2$$

and, after dividing both sides by $a^2 - c^2$ which is positive by assumption, we have
$$\frac{x^2}{a^2} + \frac{y^2}{a^2 - c^2} = 1.$$

Finally, we define the number b by $b^2 = a^2 - c^2$ to obtain

$$\boxed{\frac{x^2}{a^2} + \frac{y^2}{b^2} = 1.}$$

(2)

This is the equation of the ellipse given in Section 1.9. Note that this ellipse is symmetric about both the x- and y-axes. Here the x-axis is called the *major axis* and the y-axis is called the *minor axis*. The point $(0, 0)$ which is at the intersection of the axes is called the *center* of the ellipse. The points $(a, 0), (-a, 0), (0, b)$ and $(0, -b)$ are called the *vertices* of the ellipse.

EXAMPLE 1. Find the equation of the ellipse with foci at $(-3, 0)$ and $(3, 0)$ and with $a = 5$.

SOLUTION. $c = 3$ and $a = 5$ so that $b^2 = a^2 - c^2 = 16$ and we obtain

$$\frac{x^2}{25} + \frac{y^2}{16} = 1.$$

The ellipse is sketched in Figure 3.

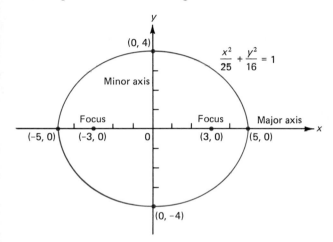

Figure 3

We can reverse the roles of x and y in the preceding discussion. Suppose that the foci are at $(0, c)$ and $(0, -c)$ on the y-axis. Then if the fixed sum of the distances is given as $2b$, we obtain, using similar reasoning

$$\frac{x^2}{a^2} + \frac{y^2}{b^2} = 1,$$

where $a^2 = b^2 - c^2$. Now the major axis is the y-axis and the minor axis is the x-axis. In general if the ellipse is given by (2) then

the major-axis is the x-axis if $a > b$, and
the major axis is the y-axis if $b > a$.

If $a = b$, then we have the equation of a circle.

EXAMPLE 2. Discuss the curve $9x^2 + 4y^2 = 36$.

SOLUTION. Dividing both sides by 36 we obtain

$$\frac{x^2}{4} + \frac{y^2}{9} = 1.$$

Here $a = 2$, $b = 3$, and the major axis is the y-axis. Since $c^2 = 9 - 4 = 5$, the foci are at $(0, \sqrt{5})$ and $(0, -\sqrt{5})$. This is sketched in Figure 4.

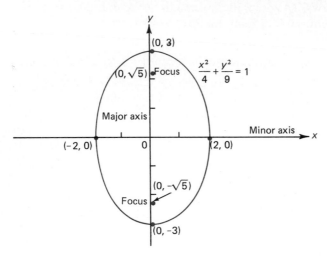

Figure 4

EXAMPLE 3. Find the equation of the ellipse with foci at $(-5, 0)$ and $(5, 0)$ and two of four vertices at $(0, 3)$ and $(0, -3)$.

SOLUTION. We have $c = 5$ and $b = 3$ so that, since $b^2 = a^2 - c^2$, we have

$$a^2 = b^2 + c^2 = 34$$

and the equation of the ellipse is

$$\frac{x^2}{34} + \frac{y^2}{9} = 1.$$

It is sketched in Figure 5.

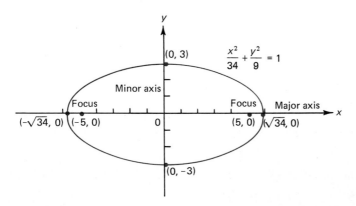

Figure 5

DEFINITION 2. The *eccentricity* e of an ellipse is defined by

$$e = \frac{c}{a} \qquad \text{if} \quad a \geq b \tag{3}$$

and

$$e = \frac{c}{b} \qquad \text{if} \quad b \geq a. \tag{4}$$

The eccentricity is a measure of the shape of the ellipse and is always a number between 0 and 1. If $e = 0$, then the ellipse is a circle since in that case $c = 0$ so that $a^2 = b^2$. As e approaches 1 the ellipse becomes progressively flatter and approaches the straight line segment from $(-a, 0)$ to $(a, 0)$ if $a > b$, and $(0, -b)$ to $(0, b)$ if $b > a$. In general,

> *the larger the eccentricity, the flatter the ellipse.*

    In Example 1, $e = \frac{3}{5} = 0.6$. In Example 2, $c^2 = b^2 - a^2 = 5$ so that $e = \sqrt{5}/3 \approx 0.74536$. In Example 3, $e = 5/\sqrt{34} \approx 0.85749$.
    We now turn to a different question. What happens if the center of the ellipse $(x^2/a^2) + (y^2/b^2) = 1$ is shifted from $(0, 0)$ to a new point (x_0, y_0)? Consider the equation

$$\frac{(x - x_0)^2}{a^2} + \frac{(y - y_0)^2}{b^2} = 1. \tag{5}$$

If we define two new variables by

$$x' = x - x_0 \qquad \text{and} \qquad y' = y - y_0, \tag{6}$$

then (5) becomes

$$\frac{(x')^2}{a^2} + \frac{(y')^2}{b^2} = 1. \tag{7}$$

This is the equation of an ellipse centered at the origin in the new coordinate system (x', y'). But $(x', y') = (0, 0)$ implies $(x - x_0, y - y_0) = (0, 0)$ or $x = x_0$ and $y = y_0$. That is, equation (5) is the equation of the "shifted" ellipse. See Figure 6. We have

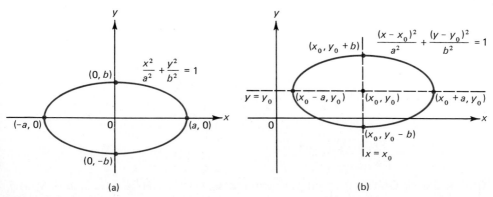

(a)   (b)

Figure 6

performed what is called a *translation of axes*. That is, we moved (or *translated*) the x- and y-axes to new positions so that they intersected at the point (x_0, y_0).

EXAMPLE 4. Find the equation of the ellipse centered at $(4, -2)$ with foci at $(1, -2)$ and $(7, -2)$ and vertices at $(4, 0)$ and $(4, -4)$.

SOLUTION. The points are sketched in Figure 7a. Here $c = 3$, $b = 2$ and $a^2 - b^2 + c^2 = 13$. Thus, from (5), we have

$$\frac{(x - 4)^2}{13} + \frac{(y + 2)^2}{4} = 1.$$

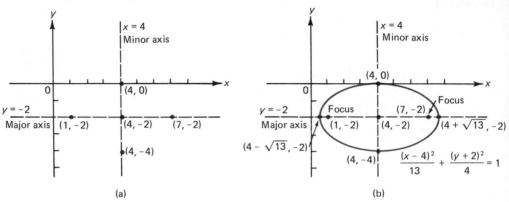

Figure 7

(a)   (b)

This ellipse is sketched in Figure 7b. Its major axis is the line $y = -2$ and its minor axis is the line $x = 4$. Its eccentricity is $3/\sqrt{13} \approx 0.83205$.

EXAMPLE 5. Discuss the curve given by

$$9x^2 + 36x + 4y^2 - 8y + 4 = 0.$$

SOLUTION. We write this expression as

$$9(x^2 + 4x) + 4(y^2 - 2y) = -4.$$

Then, after completing the squares, we obtain

$$9(x^2 + 4x + 4) - 36 + 4(y^2 - 2y + 1) - 4 = -4$$

or

$$9(x + 2)^2 + 4(y - 1)^2 = 36$$

and

$$\frac{(x + 2)^2}{4} + \frac{(y - 1)^2}{9} = 1.$$

This is the equation of an ellipse centered at $(-2, 1)$. Here $a = 2$, $b = 3$, and $c = \sqrt{b^2 - a^2} = \sqrt{5}$. The foci are therefore at $(-2, 1 - \sqrt{5})$ and $(-2, 1 + \sqrt{5})$. The major axis is the line $x = -2$ and the minor axis is the line $y = 1$. The eccentricity is $\sqrt{5}/3 \approx 0.74536$. The ellipse is sketched in Figure 8.

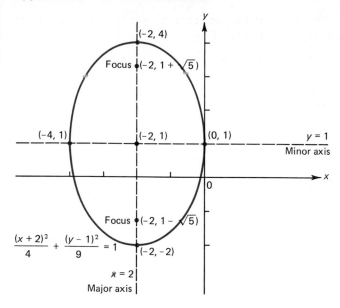

$$\frac{(x + 2)^2}{4} + \frac{(y - 1)^2}{9} = 1$$

Figure 8

II. The Parabola

DEFINITION 3. A *parabola* is the set of points (x, y) equidistant from a fixed point and a fixed line which does not contain the fixed point. The fixed point is called the *focus* and the fixed line is called the *directrix*.

To calculate the equation of a parabola, we begin by assuming that the focus is the point $(0, c)$ and the directrix is the line $y = -c$. (See Figure 9.) If $P = (x, y)$ is a point on the parabola, then we have

$$\sqrt{x^2 + (y - c)^2} = |y + c|$$

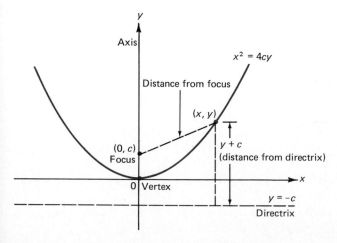

Figure 9

or, squaring,

$$x^2 + (y - c)^2 = (y + c)^2$$

which reduces to

$$x^2 = 4cy. \tag{8}$$

We see that the parabola given by (8) is symmetric about the y-axis. This line is called the *axis* of the parabola. Note that the axis contains the focus and is perpendicular to the directrix. The point at which the axis and the parabola itself intersect is called the *vertex*.

EXAMPLE 6. Describe the parabola given by $x^2 = 12y$.

SOLUTION. Here $4c = 12$ so that $c = 3$, the focus is the point $(0, 3)$, and the directrix is the line $y = -3$. The axis of the parabola is the y-axis and the vertex is the origin. The curve is sketched in Figure 10.

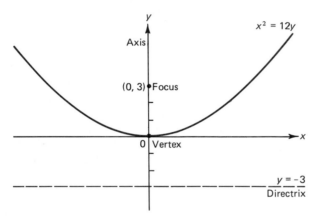

Figure 10

EXAMPLE 7. Describe the parabola given by $x^2 = -8y$.

SOLUTION. Here $c = -2$ so that the focus is $(0, -2)$, the directrix is the line $y = 2$, and the curve opens downward as in Figure 11.

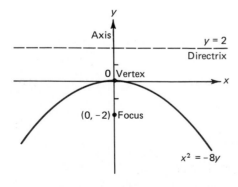

Figure 11

In general,

> The parabola described by $x^2 = 4cy$ opens upward if $c > 0$ and opens downward if $c < 0$.

As with the ellipse, we can exchange the role of x and y. We then obtain

> The equation of the parabola with focus at $(c, 0)$ and directrix the line $x = -c$ is
>
> $$y^2 = 4cx. \tag{9}$$
>
> If $c > 0$, the parabola opens to the right and if $c < 0$, the parabola opens to the left.

EXAMPLE 8.   Describe the parabola $y^2 = 16x$.

SOLUTION. Here $c = 4$ so that the focus is $(4, 0)$, the directrix is the line $x = -4$, the axis is the x-axis (since the parabola is symmetric about the x-axis), and the vertex is the origin. The curve is sketched in Figure 12.

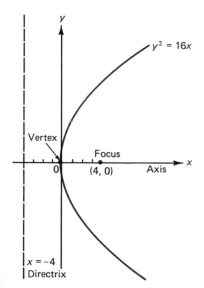

Figure 12

All the parabolas we have so far drawn have had their vertices at the origin. Other parabolas can be obtained by a simple translation of axes. The parabolas

$$(x - x_0)^2 = 4c(y - y_0) \tag{10}$$

and

$$(y - y_0)^2 = 4c(x - x_0) \tag{11}$$

have vertices at the point (x_0, y_0).

EXAMPLE 9.   Describe the parabola $(y - 2)^2 = -8(x + 3)$.

SOLUTION. This parabola has its vertex at $(-3, 2)$. It is obtained by shifting the parabola $y^2 = -8x$ two units up and 3 units to the left. The focus and directrix of $y^2 = -8x$ are $(-2, 0)$ and $x = 2$. Hence, after translation, the focus and directrix of $(y - 2)^2 = -8(x + 3)$ are $(-5, 2)$ and $x = -1$. The axis of this curve is $y = 2$. The two parabolas are sketched in Figure 13.

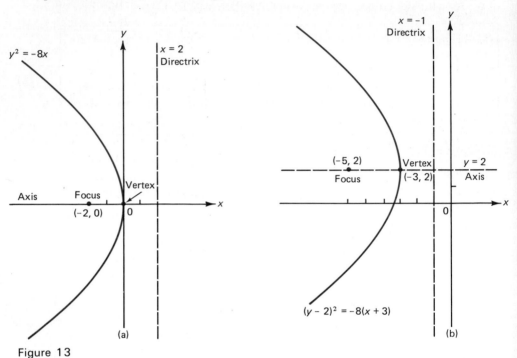

Figure 13

EXAMPLE 10. Describe the curve $x^2 - 4x + 2y + 10 = 0$.

SOLUTION. We first complete the square:

$$x^2 - 4x + 2y + 10 = (x - 2)^2 - 4 + 2y + 10 = 0$$

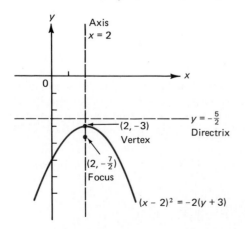

Figure 14

or

$$(x - 2)^2 = -2(y + 3).$$

This is the equation of a parabola with vertex at $(2, -3)$. Since $4c = -2, c = -\frac{1}{2}$ and the focus is $(2, -3 - \frac{1}{2}) = (2, -\frac{7}{2})$. The directrix is the line $y = -3 - (-\frac{1}{2}) = -\frac{5}{2}$, and the axis is the line $x = 2$. The curve is sketched in Figure 14.

III. The Hyperbola

DEFINITION 4. A *hyperbola* is a set of points (x, y) with the property that the positive difference between the distances from (x, y) and each of two given (distinct) points is fixed. Each of the two given points is called a *focus* of the hyperbola.

   To calculate the equation of a hyperbola, we assume that the foci are the points $(c, 0)$ and $(-c, 0)$ and the difference of the distances from a point (x, y) on the hyperbola to the foci is equal to $2a$. Note that, from the triangle in Figure 15, it

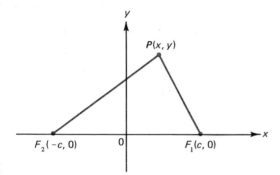

Figure 15

follows that $c > a$. (Explain why.) Then we have

$$\overline{PF_2} - \overline{PF_1} = 2a$$

or

$$\sqrt{(x + c)^2 + y^2} - \sqrt{(x - c)^2 + y^2} = 2a$$

(here we have assumed that $\overline{PF_2} > \overline{PF_1}$ so that $\overline{PF_2} - \overline{PF_1}$ gives us a positive distance). Then we obtain, successively,

$$\sqrt{(x + c)^2 + y^2} = \sqrt{(x - c)^2 + y^2} + 2a$$

or, squaring,

$$(x + c)^2 + y^2 = (x - c)^2 + y^2 + 4a \sqrt{(x - c)^2 + y^2} + 4a^2.$$

Simplifying, we have

$$4cx - 4a^2 = 4a \sqrt{(x - c)^2 + y^2}$$

and, dividing by $4a$,

$$\frac{c}{a}x - a = \sqrt{(x - c)^2 + y^2}.$$

We then square again:

$$\frac{c^2}{a^2}x^2 - 2cx + a^2 = (x - c)^2 + y^2$$

or

$$\left(\frac{c^2}{a^2} - 1\right)x^2 - y^2 = c^2 - a^2. \tag{12}$$

Finally, since $c > a$, $c^2 - a^2 > 0$ and we can define the number b by

$$b^2 = c^2 - a^2. \tag{13}$$

Then, dividing both sides of (12) by $c^2 - a^2$, we obtain

$$\boxed{\frac{x^2}{a^2} - \frac{y^2}{b^2} = 1.} \tag{14}$$

Note that the hyperbola given by (14) is symmetric about both the x- and y-axes. The intersection of these axes is called the *center* of the hyperbola. The *major axis* of the hyperbola is the line containing the foci. The *vertices* of the hyperbola are the points of intersection of the hyperbola with its major axis. The hyperbola given by (14) is sketched in Figure 16. The hyperbola (14) has the asymptotes $y = \pm(b/a)x$. To prove

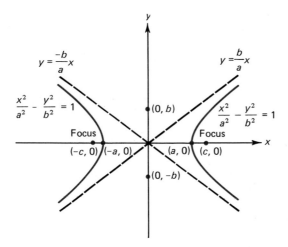

Figure 16

this we first note that

$$y = \pm\frac{b}{a}\sqrt{x^2 - a^2}.$$

Then we find that

$$\lim_{x \to \infty}\left(\frac{b}{a}\sqrt{x^2 - a^2} - \frac{b}{a}x\right) = \frac{b}{a}\lim_{x \to \infty}(\sqrt{x^2 - a^2} - x)$$

$$= \frac{b}{a}\lim_{x \to \infty}\frac{(\sqrt{x^2 - a^2} - x)(\sqrt{x^2 - a^2} + x)}{\sqrt{x^2 - a^2} + x}$$

$$= \frac{b}{a} \lim_{x \to \infty} \frac{(x^2 - a^2) - x^2}{\sqrt{x^2 - a^2} + x}$$

$$= \frac{b}{a} \lim_{x \to \infty} \frac{-a^2}{\sqrt{x^2 - a^2} + x} = 0.$$

Similarly

$$\lim_{x \to \infty} \left[-\frac{b}{a} \sqrt{x^2 - a^2} - \left(-\frac{b}{a} x \right) \right] = 0.$$

EXAMPLE 11. Find the equation of the hyperbola with foci at $(4, 0)$ and $(-4, 0)$ and with $a = 3$.

SOLUTION. Here $a^2 = 9$ and $b^2 = c^2 - a^2 = 16 - 9 = 7$ so that the equation is

$$\frac{x^2}{9} - \frac{y^2}{7} = 1.$$

The asymptotes are $y = \pm(b/a)x = \pm(\sqrt{7}/3)x$. The curve is sketched in Figure 17.

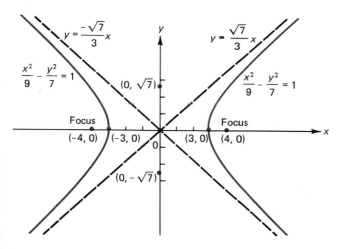

Figure 17

EXAMPLE 12. Discuss the curve given by $x^2 - 4y^2 = 9$.

SOLUTION. Dividing by 9 we obtain $(x^2/9) - (4y^2/9) = 1$ or

$$\frac{x^2}{9} - \frac{y^2}{(\frac{3}{2})^2} = 1.$$

This is the equation of a hyperbola with $a = 3$ and $b = \frac{3}{2}$. Then $c^2 = a^2 + b^2 = 9 + 9/4 = 45/4$ so that the foci are $(\sqrt{45}/2, 0)$ and $(-\sqrt{45}/2, 0)$. The vertices are $(3, 0)$ and $(-3, 0)$ and the asymptotes are $y = \pm\frac{1}{2}x$. The curve is sketched in Figure 18.

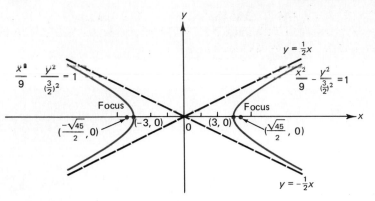

Figure 18

As with the ellipse and parabola, the roles of x and y can be reversed. The graph of the equation

$$\frac{y^2}{a^2} - \frac{x^2}{b^2} = 1 \tag{15}$$

is sketched in Figure 19. In (15) the major axis is the y-axis. From (13) we have

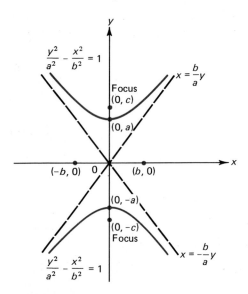

Figure 19

$c^2 = a^2 + b^2$ and the foci are at $(0, c)$ and $(0, -c)$. The vertices of the hyperbola given by (15) are the points $(0, a)$ and $(0, -a)$. The asymptotes are the lines $x = \pm (b/a)y$.

EXAMPLE 13.   Discuss the curve given by $4y^2 - 9x^2 = 36$.

SOLUTION. Dividing by 36 we obtain

$$\frac{y^2}{9} - \frac{x^2}{4} = 1.$$

Hence $a = 3, b = 2$ and $c^2 = 9 + 4 = 13$ so that $c = \sqrt{13}$. The foci are $(0, \sqrt{13})$ and $(0, -\sqrt{13})$ and the vertices are $(0, 3)$ and $(0, -3)$. The asymptotes are the lines $x = \frac{2}{3}y$ and $x = -\frac{2}{3}y$. This curve is sketched in Figure 20.

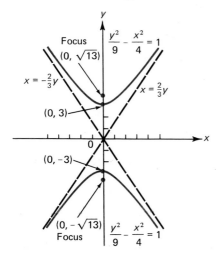

Figure 20

The hyperbolas we have so far discussed have had their centers at the origin. Other hyperbolas can be obtained by a translation of the axes. The hyperbolas

$$\frac{(x - x_0)^2}{a^2} - \frac{(y - y_0)^2}{b^2} = 1 \tag{16}$$

and

$$\frac{(y - y_0)^2}{a^2} - \frac{(x - x_0)^2}{b^2} = 1 \tag{17}$$

have centers at the point (x_0, y_0).

EXAMPLE 14. Describe the hyperbola $(y - 2)^2/9 - (x + 3)^2/4 = 1$.

SOLUTION. This is the hyperbola of Example 13 shifted three units to the left and two units up so that its center is at $(-3, 2)$. It is sketched in Figure 21.

EXAMPLE 15. Describe the curve $x^2 - 4y^2 - 4x - 8y - 9 = 0$.

SOLUTION. We have

$$(x^2 - 4x) - (4y^2 + 8y) - 9 = 0$$

or, completing the squares,

$$(x - 2)^2 - 4 - 4[(y + 1)^2 - 1] - 9 = 0$$

or

$$(x - 2)^2 - 4(y + 1)^2 = 9$$

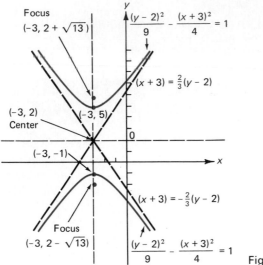

Figure 21

and

$$\frac{(x-2)^2}{9} - \frac{(y+1)^2}{(\frac{3}{2})^2} = 1.$$

This is the equation of a hyperbola with center at $(2, -1)$. Except for the translation, it is the hyperbola of Example 12 and is sketched in Figure 22.

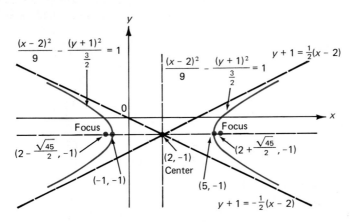

Figure 22

IV. SECOND-DEGREE EQUATIONS AND ROTATION OF AXES

The curves we have so far considered in this appendix have had their axes parallel to the two coordinate axes. This is not always the case. The curves sketched in Figure 23 are, respectively, an ellipse, a parabola, and a hyperbola. To obtain curves like those pictured, it is only necessary to rotate the coordinate axes through an appropriate

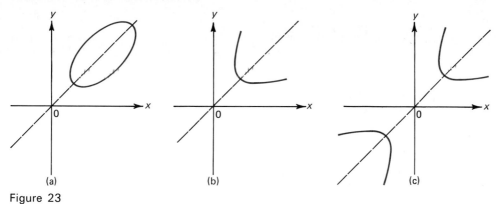

(a)   (b)   (c)

Figure 23

angle. How is this done? Suppose that the x- and y-axes are rotated through an angle of θ with respect to the origin (see Figure 24). Let $P(x, y)$ represent a typical point in the coordinates x and y. We now seek a representation of that point in the "new"

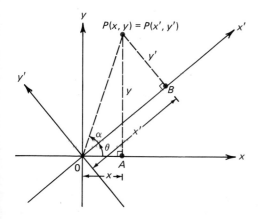

Figure 24

coordinates x' and y'. From Figure 24 we see that

$$x = \overline{0A} = \overline{0P} \cos(\theta + \alpha)$$

and

$$y = \overline{AP} = \overline{0P} \sin(\theta + \alpha)$$

But

$$\cos(\theta + \alpha) = \cos \theta \cos \alpha - \sin \theta \sin \alpha$$

and

$$\sin(\theta + \alpha) = \sin \alpha \cos \theta + \cos \alpha \sin \theta$$

Thus

$$x = \overline{0P}(\cos \theta \cos \alpha - \sin \theta \sin \alpha) \tag{18}$$

and

$$y = \overline{OP}(\sin \alpha \cos \theta + \cos \alpha \sin \theta) \tag{19}$$

Now, from the right triangle $0BP$ we find that

$$\sin \alpha = \frac{y'}{\overline{OP}} \qquad \text{or} \qquad \overline{OP} \sin \alpha = y'.$$

Similarly

$$\cos \alpha = \frac{x'}{\overline{OP}} \qquad \text{or} \qquad \overline{OP} \cos \alpha = x'.$$

Substituting these last expressions into (18) and (19) yields

$$\boxed{x = x' \cos \theta - y' \sin \theta} \tag{20}$$

and

$$\boxed{y = x' \sin \theta + y' \cos \theta.} \tag{21}$$

Equations (20) and (21) can be solved simultaneously to express the "new" coordinates x' and y' in terms of the "old" coordinates x and y. Sparing you the simple, algebraic details, we obtain

$$\boxed{x' = x \cos \theta + y \sin \theta} \tag{22}$$

and

$$\boxed{y' = -x \sin \theta + y \cos \theta.} \tag{23}$$

EXAMPLE 16. Find the curve obtained from the graph of $x^2 - xy + y^2 = 10$ by rotating the axes through an angle of $\pi/4$.

SOLUTION. Since $\cos \pi/4 = \sin \pi/4 = 1/\sqrt{2}$, we obtain (from equations (20) and (21))

$$x = \frac{x' - y'}{\sqrt{2}} \qquad \text{and} \qquad y = \frac{x' + y'}{\sqrt{2}}.$$

Substitution of these into the equation $x^2 - xy + y^2 = 10$ yields

$$\left(\frac{x' - y'}{\sqrt{2}}\right)^2 - \left(\frac{x' - y'}{\sqrt{2}}\right)\left(\frac{x' + y'}{\sqrt{2}}\right) + \left(\frac{x' + y'}{\sqrt{2}}\right)^2 = 10$$

or

$$\frac{(x')^2 - 2x'y' + (y')^2}{2} - \left(\frac{(x')^2 - (y')^2}{2}\right) + \frac{(x')^2 + 2x'y' + (y')^2}{2} = 10$$

which, after simplification, yields

$$\frac{(x')^2 + (3y')^2}{2} = 10 \qquad \text{or} \qquad (x')^2 + 3(y')^2 = 20$$

and, finally,

$$\frac{(x')^2}{20} + \frac{(y')^2}{20/3} = 1.$$

In the new coordinates x' and y' this is the equation of an ellipse with $a = \sqrt{20}$, $b = \sqrt{20/3}$ and $c = \sqrt{20 - 20/3} = \sqrt{40/3}$ and foci at $(\sqrt{40/3}, 0)$ and $(-\sqrt{40/3}, 0)$. This ellipse is obtained by rotating the ellipse $(x^2/20) + [y^2/(20/3)] = 1$ through an angle of $\pi/4$. It is sketched in Figure 25. We can obtain

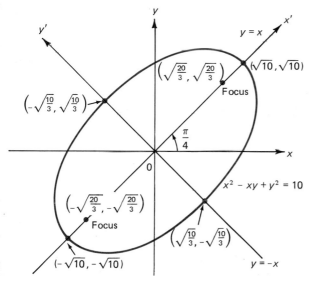

Figure 25

the foci in the coordinates (x, y) by using equations (20) and (21). Since

$$x = \frac{x' - y'}{\sqrt{2}} \qquad \text{and} \qquad y = \frac{x' + y'}{\sqrt{2}},$$

we find that $(\sqrt{40/3}, 0)$ in the coordinates (x', y') comes from $(\sqrt{20/3}, \sqrt{20/3})$ in the coordinates (x, y) and $(-\sqrt{40/3}, 0)$ comes from $(-\sqrt{20/3}, -\sqrt{20/3})$. The major axis is the line $y = x$.

EXAMPLE 17. Find the curve obtained from the curve $xy = 1$ by rotating the axes through an angle of $\pi/4$.

SOLUTION. As in Example 16 we have $x = (x' - y')/\sqrt{2}$, $y = (x' + y')/\sqrt{2}$ so that

$$1 = xy = \left(\frac{x' - y'}{\sqrt{2}}\right)\left(\frac{x' + y'}{\sqrt{2}}\right) = \frac{(x')^2}{2} - \frac{(y')^2}{2}.$$

Thus in the new (rotated) coordinate system we obtain the hyperbola $[(x')^2/2] - [(y')^2/2] = 1$. This is sketched in Figure 26. We may say that the equation $xy = 1$ is really the equation of a "standard" hyperbola rotated through an angle of $\pi/4$.

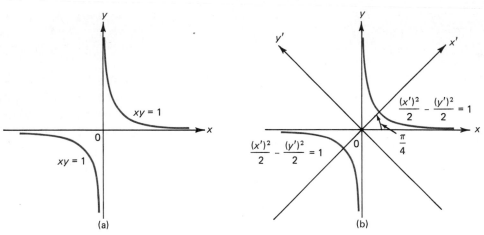

Figure 26

The last two examples illustrate the fact that our three basic curves can take forms other than the "standard" forms given by equations (2), (5), (8), (9), (10), (11), (14), (15), (16) and (17). It turns out that any second-degree equation $Ax^2 + Bxy + Cy^2 + Dx + Ey + F = 0$ can be written in the standard form as a circle, an ellipse, a parabola, a hyperbola, or a degenerate form such as a pair of lines, a point or pair of points, or an empty set of points. We will not discuss this fact further except to state the following theorem.†

THEOREM 1.   Consider the second-degree equation

$$Ax^2 + Bxy + Cy^2 + Dx + Ey + F = 0. \tag{24}$$

(i)   If $B^2 - 4AC = 0$, then (24) is the equation of a parabola, two parallel lines, or is imaginary.‡

(ii)   If $B^2 - 4AC < 0$, then (24) is the equation of a circle, an ellipse, a single point, or is imaginary.

(iii)   If $B^2 - 4AC > 0$, then (24) is the equation of a hyperbola or two intersecting lines.

EXAMPLE 18.   Determine the type of curve represented by the equation

$$16x^2 - 24xy + 9y^2 + 100x - 200y + 100 = 0. \tag{25}$$

---

†For a proof see W. H. Hart, *Analytic Geometry and Calculus,* Heath, Lexington, Massachusetts, 1963, p. 323.

‡By "imaginary" we mean that it has no graph; for example, $x^2 + 2xy + y^2 + 1 = 0$ satisfies $B^2 - 4AC = 0$; but there are no real values of x and y which satisfy $0 = x^2 + 2xy + y^2 + 1 = (x + y)^2 + 1$. (Explain why.)

Then write the curve in a standard form by an appropriate translation and rotation of axes.

SOLUTION. Here $A = 16$, $B = -24$, and $C = 9$ so that $B^2 - 4AC = (24)^2 - 4(16)(9) = 576 - 576 = 0$. Thus the equation represents a parabola (or a degenerate form of a parabola). To write it in a standard form, we first rotate the axes through an appropriate angle θ to eliminate the xy term in (25). To determine θ we substitute

$$x = x' \cos \theta - y' \sin \theta$$

and

$$y = x' \sin \theta + y' \cos \theta.$$

into equation (25). We then obtain

$$16(x' \cos \theta - y' \sin \theta)^2 - 24(x' \cos \theta - y' \sin \theta)(x' \sin \theta + y' \cos \theta)$$
$$+ 9(x' \sin \theta + y' \cos \theta)^2 + 100(x' \cos \theta - y' \sin \theta)$$
$$- 200(x' \sin \theta + y' \cos \theta) + 100 = 0.$$

The idea now is to choose θ so that the coefficient of the term $x'y'$ is zero. After simplification, we find that this coefficient is given by

$$-24(\cos^2 \theta - \sin^2 \theta) - 14 \sin \theta \cos \theta.$$

Setting this equal to zero and using the fact that $\cos 2\theta = \cos^2 \theta - \sin^2 \theta$ and $\sin 2\theta = 2 \sin \theta \cos \theta$, we obtain

$$24 \cos 2\theta + 7 \sin 2\theta = 0$$

or, dividing by $\cos 2\theta$ and rearranging terms,

$$\tan 2\theta = -\frac{24}{7}.$$

If $\tan 2\theta = -24/7$, then $\cos 2\theta = -7/25$† and also

$$\cos \theta = \sqrt{\frac{1 + \cos 2\theta}{2}} = \sqrt{\frac{9}{25}} = \frac{3}{5}.$$

Furthermore, $\sin \theta = \frac{4}{5}$. At this point we do not need to know θ, but only the values of $\cos \theta$ and $\sin \theta$. Next

$$x = x' \cos \theta - y' \sin \theta = \frac{1}{5}(3x' - 4y')$$

and

$$y = x' \sin \theta + y' \cos \theta = \frac{1}{5}(4x' + 3y').$$

We substitute these into (25) to obtain (after simplification)

$$25(y')^2 - 200y' - 100x' + 100 = 0.$$

† Alternatively, we could choose $\cos 2\theta = \frac{7}{25}$, yielding $\cos \theta = \frac{4}{5}$ and $\sin \theta = -\frac{3}{5}$. This, however, is merely a 90 degree rotation of the coordinate axes obtained with the choice $\cos 2\theta = -\frac{7}{25}$.

To further simplify, we divide by 25 and complete the square:

$$(y')^2 - 8y' - 4x' + 4 = 0$$

or

$$(y' - 4)^2 - 16 - 4x' + 4 = 0$$

or

$$(y' - 4)^2 = 4x' + 12 = 4(x' + 3).$$

This is the equation of a parabola with vertex at $(-3, 4)$ in the new (rotated) coordinate system. In order to sketch the parabola we need to calculate the angle of rotation given by $\theta = \cos^{-1}\frac{3}{5} \approx 0.927 \approx 53$ degrees. The curve is sketched in Figure 27.

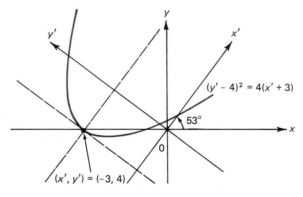

Figure 27

We note here that the xy term in (24) can always be eliminated by a rotation of axes through an angle θ, where θ is given by the equation

$$\cot 2\theta = \frac{A - C}{B} \qquad (26)$$

The proof of this fact follows exactly as in the derivation of the angle of rotation in Example 18 and is left as an exercise. (See Problem 55.)

We conclude by pointing out that instead of rotating axes, we can rotate curves, keeping the axes fixed. Note the following important fact:

Rotating a curve through an angle θ has the effect of rotating the axes through an angle $-\theta$.

PROBLEMS

In Problems 1–24 identify the type of conic. If it is an ellipse (or circle) give its foci, center, vertices, major and minor axes, and eccentricity. If it is a parabola, give its focus, directrix, axis, and vertex. If it is a hyperbola give its foci, major axis, center, vertices, and asymptotes. Finally, sketch the curve (if it is not degenerate).

1. $\dfrac{x^2}{16} + \dfrac{y^2}{25} = 1$

2. $\dfrac{x^2}{25} + \dfrac{y^2}{16} = 1$

3. $\dfrac{x^2}{16} - \dfrac{y^2}{25} = 1$

4. $\dfrac{y^2}{16} - \dfrac{x^2}{25} = 1$

5. $x^2 = 16y$

6. $y^2 = 16x$

7. $x^2 = -16y$

8. $y^2 = -16x$

9. $\dfrac{(x-1)^2}{16} + \dfrac{(y+3)^2}{25} = 1$

10. $\dfrac{(y-1)^2}{16} + \dfrac{(x+3)^2}{25} = 1$

11. $(x-1)^2 = -16(y+3)$

12. $(y-1)^2 = -16(x+3)$

13. $2x^2 + 2y^2 = 2$

14. $(x-2)^2 + (y+2)^2 = 2$

15. $4x^2 + y^2 = 9$

16. $x^2 - 4y^2 = 9$

17. $4y^2 - x^2 = 9$

18. $x^2 + 4y = 9$

19. $4x^2 + 8x + y^2 + 6y = 3$

20. $4x^2 + 8x + y^2 - 6y = 3$

21. $4x^2 + 8x - y^2 - 6y = 21$

22. $-4x^2 - 8x + y^2 - 6y = 20$

23. $x^2 + 2x + y + 1 = 0$

24. $x + y - y^2 = 4.$

25. Find the equation of an ellipse with foci at $(0, 4)$ and $(0, -4)$ and vertices at $(0, 5)$ and $(0, -5)$.

26. Find the equation of two ellipses with vertices at $(2, 0)$ and $(-2, 0)$ and eccentricity 0.8.

27. Find the equation of two ellipses with center at $(-1, 4)$ which contain the same area as the ellipse of Problem 26 and which have the same eccentricity.

28. Find the equation of the parabola with focus $(0, 4)$ and directrix the line $y = -4$.

29. Find the equation of the parabola with focus $(-3, 0)$ and directrix the line $x = 3$.

30. Find the equation of the parabola obtained when the parabola of Problem 29 is shifted so that its vertex is at the point $(-2, 5)$.

31. Find the equation of the hyperbola with foci $(5, 0)$ and $(-5, 0)$ and vertices $(4, 0)$ and $(-4, 0)$.

32. Find the equation of the hyperbola with foci at $(0, 5)$ and $(0, -5)$ and vertices at $(0, 4)$ and $(0, -4)$.

33. Find the equation of the hyperbola with center at $(0, 0)$ and vertices at $(2, 0)$ and $(-2, 0)$ which is asymptotic to the lines $y = \pm 3x$.

34. Find the equation of the hyperbola obtained by shifting the hyperbola of Problem 32 so that its center is at $(4, -3)$.

35. Find the equation of the parabola obtained by rotating the axes in Problem 29 so that the axis of the parabola coincides with the line $y = \sqrt{3}x$.

36. Describe the curve obtained from the graph of $4x^2 - 2xy + 4y^2 = 45$ by rotating the axes through an angle of $\pi/4$.

37. Describe the curve obtained from the graph of $x^2 + 2\sqrt{3}xy - y^2 = 4$ by rotating the axes through an angle of $\pi/6$.

38. What is the equation of the line obtained from the line $2x - 3y = 6$ by rotating the axes through an angle of $\pi/6$?

39. Find the equation of the line obtained from the line $ax + by + c = 0$ by rotating the axes through an angle θ.

*40. Find the equation of the ellipse whose major axis is the line $y = -x$ which is obtained by rotating the ellipse $(x^2/25) + (y^2/16) = 1$. [*Hint:* Find the angle through which the axes must be rotated to accomplish this.]

In Problems 41–54 find a rotation of coordinate axes in which the given equation written in the new coordinates has no xy term. Describe each curve and then sketch it.

41. $4x^2 + 4xy + y^2 = 9$

42. $4x^2 + 4xy - y^2 = 9$

43. $3x^2 - 2xy - 5 = 0$

44. $xy = 2$

45. $xy = a, a > 0$

46. $xy = a, a < 0$

47. $4x^2 + 4xy + y^2 + 20x - 10y = 0$

48. $x^2 + 4xy + 4y^2 - 6 = 0$

49. $2x^2 + xy + y^2 = 4$

50. $9x^2 + 6xy + y^2 + 10x - 30y = 0$

51. $3x^2 - 6xy + 5y^2 = 36$

52. $x^2 - 3xy + 4y^2 = 1$

53. $3y^2 - 4xy + 30y - 20x + 40 = 0$

54. $6x^2 + 5xy - 6y^2 + 7 = 0.$

55. Show that if $A \neq C$, the xy term in the second-degree equation (24) will be eliminated by rotation through an angle θ if θ is given by

$$\cot 2\theta = \frac{A - C}{B}.$$

56. Show that if $A = C$ in Problem 53, then the xy term will be eliminated by a rotation through an angle of either $\pi/4$ or $-\pi/4$.

57. Suppose that a rotation converts $Ax^2 + Bxy + Cy^2$ into $A'(x')^2 + B'(x'y') + C'(y')^2$. Show that

(a) $A + C = A' + C'$ (b) $B^2 - 4AC = (B')^2 - 4A'C'$.

APPENDIX 4

THE BINOMIAL THEOREM

The binomial theorem provides a useful device for multiplying expressions of the form $(x + y)^n$, where n is a positive integer. You are all familiar with the expression

$$(x + y)^2 = x^2 + 2xy + y^2.$$

In addition, it is not difficult to show that

$$(x + y)^3 = x^3 + 3x^2y + 3xy^2 + y^3.$$

To calculate larger powers of $x + y$, we first define the binomial coefficient $\binom{n}{k}$ for n and k positive integers, by

$$\binom{n}{k} = \frac{n!}{k!(n-k)!}, \tag{1}$$

where $n! = n(n-1)(n-2) \cdots 3 \cdot 2 \cdot 1$, and, by convention, $0! = 1$.

EXAMPLE 1. Evaluate (a) $\binom{4}{2}$, (b) $\binom{8}{5}$, (c) $\binom{7}{0}$.

SOLUTION. (a) $\binom{4}{2} = \frac{4!}{2!2!} = \frac{4 \cdot 3 \cdot 2}{2 \cdot 2} = 6.$

(b) $\binom{8}{5} = \frac{8!}{5!3!} = \frac{8 \cdot 7 \cdot 6 \cdot 5!}{5!3!} = \frac{8 \cdot 7 \cdot 6}{6} = 56.$

(c) $\binom{7}{0} = \frac{7!}{7!0!} = \frac{1}{0!} = \frac{1}{1} = 1.$

THEOREM 1 (The Binomial Theorem). Let n be a positive integer, then

$$(x + y)^n = x^n + \binom{n}{1}x^{n-1}y + \binom{n}{2}x^{n-2}y^2 + \cdots + \binom{n}{n-1}xy^{n-1} + y^n \tag{2}$$

or, more concisely,

$$(x + y)^n = \sum_{j=0}^{n} \binom{n}{j}x^{n-j}y^j. \tag{3}$$

PROOF.† We prove this theorem by mathematical induction (see Appendix 1).

†The proof is difficult and may be omitted without loss of continuity. Another (simpler) proof which makes use of Taylor's theorem is suggested in Problem 13.3.31.

(i)   If $n = 1$, then

$$\sum_{j=0}^{n} \binom{n}{j} x^{n-j} y^j = \sum_{j=0}^{1} \binom{n}{j} x^{n-j} y^j = \binom{1}{0} x^1 y^0 + \binom{1}{1} x^0 y^1 = x + y,$$

implying the validity of equation (3) for $n = 1$.

(ii)   We assume that equation (3) holds for $n = k$ and prove it for $n = k + 1$. By assumption we have

$$(x + y)^k = \sum_{j=0}^{k} \binom{k}{j} x^{k-j} y^j. \tag{4}$$

Then

$$(x + y)^{k+1} = (x + y)^k (x + y) = \left[\sum_{j=0}^{k} \binom{k}{j} x^{k-j} y^j \right](x + y)$$

or

$$(x + y)^{k+1} = \sum_{j=0}^{k} \binom{k}{j} x^{k+1-j} y^j + \sum_{j=0}^{k} \binom{k}{j} x^{k-j} y^{j+1}. \tag{5}$$

We now write out the sums in (5):

$$\begin{aligned}
(x + y)^{k+1} = {} & \binom{k}{0} x^{k+1} + \binom{k}{1} x^k y + \binom{k}{2} x^{k-1} y^2 + \cdots + \binom{k}{j} x^{k+1-j} y^j \\
& + \binom{k}{j+1} x^{k-j} y^{j+1} + \cdots + \binom{k}{k} x y^k \\
& + \binom{k}{0} x^k y + \binom{k}{1} x^{k-1} y^2 + \binom{k}{2} x^{k-2} y^3 + \cdots + \binom{k}{j-1} x^{k+1-j} y^j \\
& + \binom{k}{j} x^{k-j} y^{j+1} + \cdots + \binom{k}{k} y^{k+1}.
\end{aligned} \tag{6}$$

In (6) we see that there are two terms containing the expression $x^{k+1-j} y^j$ (for $j = 1, 2, \ldots, k$), namely

$$\binom{k}{j} x^{k+1-j} y^j + \binom{k}{j-1} x^{k+1-j} y^j. \tag{7}$$

But

$$\binom{k}{j} + \binom{k}{j-1} = \frac{k!}{(k-j)!j!} + \frac{k!}{(k-j+1)!(j-1)!} \tag{8}$$

We multiply and divide the second term in (8) by $j/(k - j + 1)$ to obtain

$$\frac{k! \dfrac{j}{k - j + 1}}{(k - j + 1)!(j - 1)! \dfrac{j}{k - j + 1}} = \frac{k! \dfrac{j}{k - j + 1}}{(k - j)!j!}.$$

Thus

$$\binom{k}{j} + \binom{k}{j-1} = \frac{k!}{(k-j)!j!} \left(1 + \frac{j}{k - j + 1} \right) = \frac{k!}{(k-j)!j!} \left(\frac{k + 1}{k - j + 1} \right)$$

$$= \frac{(k + 1)!}{(k + 1 - j)!j!} = \binom{k+1}{j}. \tag{9}$$

Hence

$$\tbinom{k}{j}x^{k+1-j}y^j + \tbinom{k}{j-1}x^{k+1-j}y^j = \tbinom{k+1}{j}x^{k+1-j}y^j.$$

Finally, from (6), (7) and (9), we have

$$(x + y)^{k+1} = \sum_{j=0}^{k+1} \tbinom{k+1}{j}x^{k+1-j}y^j$$

which is equation (3) for $n = k + 1$ and the theorem is proved.

EXAMPLE 2. Calculate $(x + y)^5$

SOLUTION

$$(x + y)^5 = \sum_{k=0}^{5} \tbinom{5}{k}x^{5-k}y^k$$
$$= \tbinom{5}{0}x^5 + \tbinom{5}{1}x^4y + \tbinom{5}{2}x^3y^2 + \tbinom{5}{3}x^2y^3 + \tbinom{5}{4}xy^4 + \tbinom{5}{5}y^5$$
$$= x^5 + 5x^4y + 10x^3y^2 + 10x^2y^3 + 5xy^4 + y^5.$$

EXAMPLE 3. Find the coefficient of the term x^3y^6 in the expansion of $(x + y)^9$.

SOLUTION. In (3) we obtain the term x^3y^6 by setting $j = 6$ (so that $9 - j = 3$). The coefficient is

$$\tbinom{9}{6} = \frac{9!}{6!3!} = \frac{9 \cdot 8 \cdot 7}{3!} = \frac{9 \cdot 8 \cdot 7}{3 \cdot 2} = 84.$$

EXAMPLE 4. Calculate $(2x - 3y)^4$

SOLUTION

$$(2x - 3y)^4 = \tbinom{4}{0}(2x)^4 + \tbinom{4}{1}(2x)^3(-3y)^1 + \tbinom{4}{2}(2x)^2(-3y)^2$$
$$+ \tbinom{4}{3}(2x)^1(-3y)^3 + \tbinom{4}{4}(-3y)^4$$
$$= 16x^4 + 4(8x^3)(-3y) + 6(4x^2)(9y^2)$$
$$+ 4(2x)(-27y^3) + 81y^4$$
$$= 16x^4 - 96x^3y + 216x^2y^2 - 216xy^3 + 81y^4.$$

EXAMPLE 5. Find the coefficient of the y^6 term in the expansion $(2x + y^2)^9$.

SOLUTION. The term containing y^6 will be of the form $x^6(y^2)^3$ since $6 + 3 = 9$. This term is, then,

$$\tbinom{9}{3}(2x)^6(y^2)^3 = 84 \cdot 2^6x^6 y^6 = 84 \cdot 64x^6y^6$$

and the coefficient is $84 \cdot 64 = 5376$.

PROBLEMS

In Problems 1–8 calculate the binomial coefficients.

1. $\tbinom{5}{3}$ 2. $\tbinom{7}{4}$ 3. $\tbinom{9}{2}$

4. $\tbinom{10}{5}$ 5. $\tbinom{11}{3}$ 6. $\tbinom{20}{0}$

7. $\tbinom{41}{41}$ 8. $\tbinom{12}{7}$.

9. Prove that $\binom{n}{k} = \binom{n}{n-k}$ for any integers $0 \le k \le n$.

10. Calculate $(x + y)^6$.

11. Calculate $(a + b)^7$.

12. Calculate $(u - w)^6$.

13. Calculate $(x - 2y)^5$.

14. Calculate $(x^2 - y^3)^4$.

15. Calculate $(ax - by)^5$.

16. Calculate $(x^n + y^n)^5$.

17. Calculate $[(x/2) + (y/3)]^5$.

18. Find the coefficient of $x^5 y^7$ in the expansion of $(x + y)^{12}$.

19. Find the coefficient of $x^8 y^3$ in the expansion of $(x + y)^{11}$.

20. Show that in the expansion of $(x + y)^n$ the coefficient of $x^k y^{n-k}$ is equal to the coefficient of $x^{n-k} y^k$.

21. Show that for any integer n,

$$\binom{n}{0} + \binom{n}{1} + \binom{n}{2} + \cdots + \binom{n}{n} = 2^n.$$

[*Hint:* Expand $(1 + 1)^n$.]

22. Show that for any integer n,

$$\binom{n}{0} - \binom{n}{1} + \binom{n}{2} - \binom{n}{3} + \cdots + (-1)^n\binom{n}{n} = 0.$$

[*Hint:* Expand $(1 - 1)^n$.]

TABLES

TABLE A.1 Exponential Functions

x	e^x	e^{-x}	x	e^x	e^{-x}
0.00	1.0000	1.0000	3.0	20.086	0.0498
0.05	1.0513	0.9512	3.1	22.198	0.0450
0.10	1.1052	0.9048	3.2	24.533	0.0408
0.15	1.1618	0.8607	3.3	27.113	0.0369
0.20	1.2214	0.8187	3.4	29.964	0.0334
0.25	1.2840	0.7788	3.5	33.115	0.0302
0.30	1.3499	0.7408	3.6	36.598	0.0273
0.35	1.4191	0.7047	3.7	40.447	0.0247
0.40	1.4918	0.6703	3.8	44.701	0.0224
0.45	1.5683	0.6376	3.9	49.402	0.0202
0.50	1.6487	0.6065	4.0	54.598	0.0183
0.55	1.7333	0.5769	4.1	60.340	0.0166
0.60	1.8221	0.5488	4.2	66.686	0.0150
0.65	1.9155	0.5220	4.3	73.700	0.0136
0.70	2.0138	0.4966	4.4	81.451	0.0123
0.75	2.1170	0.4724	4.5	90.017	0.0111
0.80	2.2255	0.4493	4.6	99.484	0.0101
0.85	2.3396	0.4274	4.7	109.95	0.0091
0.90	2.4596	0.4066	4.8	121.51	0.0082
0.95	2.5857	0.3867	4.9	134.29	0.0074
1.0	2.7183	0.3679	5.0	148.41	0.0067
1.1	3.0042	0.3329	5.1	164.02	0.0061
1.2	3.3201	0.3012	5.2	181.27	0.0055
1.3	3.6693	0.2725	5.3	200.34	0.0050
1.4	4.0552	0.2466	5.4	221.41	0.0045
1.5	4.4817	0.2231	5.5	244.69	0.0041
1.6	4.9530	0.2019	5.6	270.43	0.0037
1.7	5.4739	0.1827	5.7	298.87	0.0033
1.8	6.0496	0.1653	5.8	330.30	0.0030
1.9	6.6859	0.1496	5.9	365.04	0.0027
2.0	7.3891	0.1353	6.0	403.43	0.0025
2.1	8.1662	0.1225	6.5	665.14	0.0015
2.2	9.0250	0.1108	7.0	1096.6	0.0009
2.3	9.9742	0.1003	7.5	1808.0	0.0006
2.4	11.023	0.0907	8.0	2981.0	0.0003
2.5	12.182	0.0821	8.5	4914.8	0.0002
2.6	13.464	0.0743	9.0	8103.1	0.0001
2.7	14.880	0.0672	9.5	13,360	0.00007
2.8	16.445	0.0608	10.0	22,026	0.00004
2.9	18.174	0.0550			

n	$\log_e n$	n	$\log_e n$	n	$\log_e n$
0.0	—	4.5	1.5041	9.0	2.1972
0.1	−2.3026	4.6	1.5261	9.1	2.2083
0.2	−1.6094	4.7	1.5476	9.2	2.2192
0.3	−1.2040	4.8	1.5686	9.3	2.2300
0.4	−0.9163	4.9	1.5892	9.4	2.2407
0.5	−0.6931	5.0	1.6094	9.5	2.2513
0.6	−0.5108	5.1	1.6292	9.6	2.2618
0.7	−0.3567	5.2	1.6487	9.7	2.2721
0.8	−0.2231	5.3	1.6677	9.8	2.2824
0.9	−0.1054	5.4	1.6864	9.9	2.2925
1.0	0.0000	5.5	1.7047	10	2.3026
1.1	0.0953	5.6	1.7228	11	2.3979
1.2	0.1823	5.7	1.7405	12	2.4849
1.3	0.2624	5.8	1.7579	13	2.5649
1.4	0.3365	5.9	1.7750	14	2.6391
1.5	0.4055	6.0	1.7918	15	2.7081
1.6	0.4700	6.1	1.8083	16	2.7726
1.7	0.5306	6.2	1.8245	17	2.8332
1.8	0.5878	6.3	1.8405	18	2.8904
1.9	0.6419	6.4	1.8563	19	2.9444
2.0	0.6931	6.5	1.8718	20	2.9957
2.1	0.7419	6.6	1.8871	25	3.2189
2.2	0.7885	6.7	1.9021	30	3.4012
2.3	0.8329	6.8	1.9169	35	3.5553
2.4	0.8755	6.9	1.9315	40	3.6889
2.5	0.9163	7.0	1.9459	45	3.8067
2.6	0.9555	7.1	1.9601	50	3.9120
2.7	0.9933	7.2	1.9741	55	4.0073
2.8	1.0296	7.3	1.9879	60	4.0943
2.9	1.0647	7.4	2.0015	65	4.1744
3.0	1.0986	7.5	2.0149	70	4.2485
3.1	1.1314	7.6	2.0281	75	4.3175
3.2	1.1632	7.7	2.0142	80	4.3820
3.3	1.1939	7.8	2.0541	85	4.4427
3.4	1.2238	7.9	2.0669	90	4.4998
3.5	1.2528	8.0	2.0794	95	4.5539
3.6	1.2809	8.1	2.0919	100	4.6052
3.7	1.3083	8.2	2.1041	200	5.2983
3.8	1.3350	8.3	2.1163	300	5.7038
3.9	1.3610	8.4	2.1282	400	5.9915
4.0	1.3863	8.5	2.1401	500	6.2146
4.1	1.4110	8.6	2.1518	600	6.3069
4.2	1.4351	8.7	2.1633	700	6.5511
4.3	1.4586	8.8	2.1748	800	6.6846
4.4	1.4816	8.9	2.1861	900	6.8024

TABLE A.3 Common Logarithms[a]

n	0	1	2	3	4	5	6	7	8	9
1.0	.0000	.0043	.0086	.0128	.0170	.0212	.0253	.0294	.0334	.0374
1.1	.0414	.0453	.0492	.0531	.0569	.0607	.0645	.0682	.0719	.0755
1.2	.0792	.0828	.0864	.0899	.0934	.0969	.1004	.1038	.1072	.1106
1.3	.1139	.1173	.1206	.1239	.1271	.1303	.1335	.1367	.1399	.1430
1.4	.1461	.1492	.1523	.1553	.1584	.1614	.1644	.1673	.1703	.1732
1.5	.1761	.1790	.1818	.1847	.1875	.1903	.1931	.1959	.1987	.2014
1.6	.2041	.2068	.2095	.2122	.2148	.2175	.2201	.2227	.2253	.2279
1.7	.2304	.2330	.2355	.2380	.2405	.2430	.2455	.2480	.2504	.2529
1.8	.2553	.2577	.2601	.2625	.2648	.2672	.2695	.2718	.2742	.2765
1.9	.2788	.2810	.2833	.2856	.2878	.2900	.2923	.2945	.2967	.2989
2.0	.3010	.3032	.3054	.3075	.3096	.3118	.3139	.3160	.3181	.3201
2.1	.3222	.3243	.3263	.3284	.3304	.3324	.3345	.3365	.3385	.3404
2.2	.3424	.3444	.3464	.3483	.3502	.3522	.3541	.3560	.3579	.3598
2.3	.3617	.3636	.3655	.3674	.3692	.3711	.3729	.3747	.3766	.3784
2.4	.3802	.3820	.3838	.3856	.3874	.3892	.3909	.3927	.3945	.3962
2.5	.3979	.3997	.4014	.4031	.4048	.4065	.4082	.4099	.4116	.4133
2.6	.4150	.4166	.4183	.4200	.4216	.4232	.4249	.4265	.4281	.4298
2.7	.4314	.4330	.4346	.4362	.4378	.4393	.4409	.4425	.4440	.4456
2.8	.4472	.4487	.4502	.4518	.4533	.4548	.4564	.4579	.4594	.4609
2.9	.4624	.4639	.4654	.4669	.4683	.4698	.4713	.4728	.4742	.4757
3.0	.4771	.4786	.4800	.4814	.4829	.4843	.4857	.4871	.4886	.4900
3.1	.4914	.4928	.4942	.4955	.4969	.4983	.4997	.5011	.5024	.5038
3.2	.5051	.5065	.5079	.5092	.5105	.5119	.5132	.5145	.5159	.5172
3.3	.5185	.5198	.5211	.5224	.5237	.5250	.5263	.5276	.5289	.5302
3.4	.5315	.5328	.5340	.5353	.5366	.5378	.5391	.5403	.5416	.5428
3.5	.5441	.5453	.5465	.5478	.5490	.5502	.5514	.5527	.5539	.5551
3.6	.5563	.5575	.5587	.5599	.5611	.5623	.5635	.5647	.5658	.5670
3.7	.5682	.5694	.5705	.5717	.5729	.5740	.5752	.5763	.5775	.5786
3.8	.5798	.5809	.5821	.5832	.5843	.5855	.5866	.5877	.5888	.5899
3.9	.5911	.5922	.5933	.5944	.5955	.5966	.5977	.5988	.5999	.6010
4.0	.6021	.6031	.6042	.6053	.6064	.6075	.6085	.6096	.6107	.6117
4.1	.6128	.6138	.6149	.6160	.6170	.6180	.6191	.6201	.6212	.6222
4.2	.6232	.6243	.6253	.6263	.6274	.6284	.6294	.6304	.6314	.6325
4.3	.6335	.6345	.6355	.6365	.6375	.6385	.6395	.6405	.6415	.6425
4.4	.6435	.6444	.6454	.6464	.6474	.6484	.6493	.6503	.6513	.6522
4.5	.6532	.6542	.6551	.6561	.6571	.6580	.6590	.6599	.6609	.6618
4.6	.6628	.6637	.6646	.6656	.6665	.6675	.6684	.6693	.6702	.6712
4.7	.6721	.6730	.6739	.6749	.6758	.6767	.6776	.6785	.6794	.6803
4.8	.6812	.6821	.6830	.6839	.6848	.6857	.6866	.6875	.6884	.6893
4.9	.6902	.6911	.6920	.6928	.6937	.6946	.6955	.6964	.6972	.6981
5.0	.6990	.6998	.7007	.7016	.7024	.7033	.7042	.7050	.7059	.7067
5.1	.7076	.7084	.7093	.7101	.7110	.7118	.7126	.7135	.7143	.7152
5.2	.7160	.7168	.7177	.7185	.7193	.7202	.7210	.7218	.7226	.7235
5.3	.7243	.7251	.7259	.7267	.7275	.7284	.7292	.7300	.7308	.7316
5.4	.7324	.7332	.7340	.7348	.7356	.7364	.7372	.7380	.7388	.7396

[a] Reprinted with permission from Stanley I. Grossman and James E. Turner, *Mathematics for the Biological Sciences*. Macmillan, New York, pp. 471–472.

n	0	1	2	3	4	5	6	7	8	9
5.5	.7404	.7412	.7419	.7427	.7435	.7443	.7451	.7459	.7466	.7474
5.6	.7482	.7490	.7497	.7505	.7513	.7520	.7528	.7536	.7543	.7551
5.7	.7559	.7566	.7574	.7582	.7589	.7597	.7604	.7612	.7619	.7627
5.8	.7634	.7642	.7649	.7657	.7664	.7672	.7679	.7686	.7694	.7701
5.9	.7709	.7716	.7723	.7731	.7738	.7745	.7752	.7760	.7767	.7774
6.0	.7782	.7789	.7796	.7803	.7810	.7818	.7825	.7832	.7839	.7846
6.1	.7853	.7860	.7868	.7875	.7882	.7889	.7896	.7903	.7910	.7917
6.2	.7924	.7931	.7938	.7945	.7952	.7959	.7966	.7973	.7980	.7987
6.3	.7993	.8000	.8007	.8014	.8021	.8028	.8035	.8041	.8048	.8055
6.4	.8062	.8069	.8075	.8082	.8089	.8096	.8102	.8109	.8116	.8122
6.5	.8129	.8136	.8142	.8149	.8156	.8162	.8169	.8176	.8182	.8189
6.6	.8195	.8202	.8209	.8215	.8222	.8228	.8235	.8241	.8248	.8254
6.7	.8261	.8267	.8274	.8280	.8287	.8293	.8299	.8306	.8312	.8319
6.8	.8325	.8331	.8338	.8344	.8351	.8357	.8363	.8370	.8376	.8382
6.9	.8388	.8395	.8401	.8407	.8414	.8420	.8426	.8432	.8439	.8445
7.0	.8451	.8457	.8463	.8470	.8476	.8482	.8488	.8494	.8500	.8506
7.1	.8513	.8519	.8525	.8531	.8537	.8543	.8549	.8555	.8561	.8567
7.2	.8573	.8579	.8585	.8591	.8597	.8603	.8609	.8615	.8621	.8627
7.3	.8633	.8639	.8645	.8651	.8657	.8663	.8669	.8675	.8681	.8686
7.4	.8692	.8698	.8704	.8710	.8716	.8722	.8727	.8733	.8739	.8745
7.5	.8751	.8756	.8762	.8768	.8774	.8779	.8785	.8791	.8797	.8802
7.6	.8808	.8814	.8820	.8825	.8831	.8837	.8842	.8848	.8854	.8859
7.7	.8865	.8871	.8876	.8882	.8887	.8893	.8899	.8904	.8910	.8915
7.8	.8921	.8927	.8932	.8938	.8943	.8949	.8954	.8960	.8965	.8971
7.9	.8976	.8982	.8987	.8993	.8998	.9004	.9009	.9015	.9020	.9025
8.0	.9031	.9036	.9042	.9047	.9053	.9058	.9063	.9069	.9074	.9079
8.1	.9085	.9090	.9096	.9101	.9106	.9112	.9117	.9122	.9128	.9133
8.2	.9138	.9143	.9149	.9154	.9159	.9165	.9170	.9175	.9180	.9186
8.3	.9191	.9196	.9201	.9206	.9212	.9217	.9222	.9227	.9232	.9238
8.4	.9243	.9248	.9253	.9258	.9263	.9269	.9274	.9279	.9284	.9289
8.5	.9294	.9299	.9304	.9309	.9315	.9320	.9325	.9330	.9335	.9340
8.6	.9345	.9350	.9355	.9360	.9365	.9370	.9375	.9380	.9385	.9390
8.7	.9395	.9400	.9405	.9410	.9415	.9420	.9425	.9430	.9435	.9440
8.8	.9445	.9450	.9455	.9460	.9465	.9469	.9474	.9479	.9484	.9489
8.9	.9494	.9499	.9504	.9509	.9513	.9518	.9523	.9528	.9533	.9538
9.0	.9542	.9547	.9552	.9557	.9562	.9566	.9571	.9576	.9581	.9586
9.1	.9590	.9595	.9600	.9605	.9609	:9614	.9619	.9624	.9628	.9633
9.2	.9638	.9643	.9647	.9652	.9657	.9661	.9666	.9671	.9675	.9680
9.3	.9685	.9689	.9694	.9699	.9703	.9708	.9713	.9717	.9722	.9727
9.4	.9731	.9736	.9741	.9745	.9750	.9754	.9759	.9763	.9768	.9773
9.5	.9777	.9782	.9786	.9791	.9795	.9800	.9805	.9809	.9814	.9818
9.6	.9823	.9827	.9832	.9836	.9841	.9845	.9850	.9854	.9859	.9863
9.7	.9868	.9872	.9877	.9881	.9886	.9890	.9894	.9899	.9903	.9908
9.8	.9912	.9917	.9921	.9926	.9930	.9934	.9939	.9943	.9948	.9952
9.9	.9956	.9961	.9965	.9969	.9974	.9978	.9983	.9987	.9991	.9996

TABLE A.4 Trigonometric Functions[a]

Degrees	Radians	Sin	Tan	Cot	Cos		
0	0	0	0	—	1.000	1.5708	90
1	0.0175	0.0175	0.0175	57.290	0.9998	1.5533	89
2	0.0349	0.0349	0.0349	28.636	0.9994	1.5359	88
3	0.0524	0.0523	0.0524	19.081	0.9986	1.5184	87
4	0.0698	0.0698	0.0699	14.301	0.9976	1.5010	86
5	0.0873	0.0872	0.0875	11.430	0.9962	1.4835	85
6	0.1047	0.1045	0.1051	9.5144	0.9945	1.4661	84
7	0.1222	0.1219	0.1228	8.1443	0.9925	1.4486	83
8	0.1396	0.1392	0.1405	7.1154	0.9903	1.4312	82
9	0.1571	0.1564	0.1584	6.3138	0.9877	1.4137	81
10	0.1745	0.1736	0.1763	5.6713	0.9848	1.3963	80
11	0.1920	0.1908	0.1944	5.1446	0.9816	1.3788	79
12	0.2094	0.2079	0.2126	4.7046	0.9781	1.3614	78
13	0.2269	0.2250	0.2309	4.3315	0.9744	1.3439	77
14	0.2443	0.2419	0.2493	4.0108	0.9703	1.3265	76
15	0.2618	0.2588	0.2679	3.7321	0.9659	1.3090	75
16	0.2793	0.2756	0.2867	3.4874	0.9613	1.2915	74
17	0.2967	0.2924	0.3057	3.2709	0.9563	1.2741	73
18	0.3142	0.3090	0.3249	3.0777	0.9511	1.2566	72
19	0.3316	0.3256	0.3443	2.9042	0.9455	1.2392	71
20	0.3491	0.3420	0.3640	2.7475	0.9397	1.2217	70
21	0.3665	0.3584	0.3839	2.6051	0.9336	1.2043	69
22	0.3840	0.3746	0.4040	2.4751	0.9272	1.1868	68
23	0.4014	0.3907	0.4245	2.3559	0.9205	1.1694	67
24	0.4189	0.4067	0.4452	2.2460	0.9135	1.1519	66
25	0.4363	0.4226	0.4663	2.1445	0.9063	1.1345	65
26	0.4538	0.4384	0.4877	2.0503	0.8988	1.1170	64
27	0.4712	0.4540	0.5095	1.9626	0.8910	1.0996	63
28	0.4887	0.4695	0.5317	1.8807	0.8829	1.0821	62
29	0.5061	0.4848	0.5543	1.8040	0.8746	1.0647	61
30	0.5236	0.5000	0.5774	1.7321	0.8660	1.0472	60
31	0.5411	0.5150	0.6009	1.6643	0.8572	1.0297	59
32	0.5585	0.5299	0.6249	1.6003	0.8480	1.0123	58
33	0.5760	0.5446	0.6494	1.5399	0.8387	0.9948	57
34	0.5934	0.5592	0.6745	1.4826	0.8290	0.9774	56
35	0.6109	0.5736	0.7002	1.4281	0.8192	0.9599	55
36	0.6283	0.5878	0.7265	1.3764	0.8090	0.9425	54
37	0.6458	0.6018	0.7536	1.3270	0.7986	0.9250	53
38	0.6632	0.6157	0.7813	1.2799	0.7880	0.9076	52
39	0.6807	0.6293	0.8098	1.2349	0.7771	0.8901	51
40	0.6981	0.6428	0.8391	1.1918	0.7660	0.8727	50
41	0.7156	0.6561	0.8693	1.1504	0.7547	0.8552	49
42	0.7330	0.6691	0.9004	1.1106	0.7431	0.8378	48
43	0.7505	0.6820	0.9325	1.0724	0.7314	0.8203	47
44	0.7679	0.6947	0.9657	1.0355	0.7193	0.8029	46
45	0.7854	0.7071	1.0000	1.0000	0.7071	0.7854	45
	Cos	Cot	Tan	Sin	Radians	Degrees

[a] Reprinted with permission from Stanley I. Grossman and James E. Turner, *Mathematics for the Biological Sciences*. Macmillan, New York, p. 474.

x	sinh x	cosh x	tanh x	x	sinh x	cosh x	tanh x
0.0	0.00000	1.0000	0.00000	3.0	10.018	10.068	0.99505
0.1	0.10017	1.0050	0.09967	3.1	11.076	11.122	0.99595
0.2	0.20134	1.0201	0.19738	3.2	12.246	12.287	0.99668
0.3	0.30452	1.0453	0.29131	3.3	13.538	13.575	0.99728
0.4	0.41075	1.0811	0.37995	3.4	14.965	14.999	0.99777
0.5	0.52110	1.1276	0.46212	3.5	16.543	16.573	0.99818
0.6	0.63665	1.1855	0.53705	3.6	18.285	18.313	0.99851
0.7	0.75858	1.2552	0.60437	3.7	20.211	20.236	0.99878
0.8	0.88811	1.3374	0.66404	3.8	22.339	22.362	0.99900
0.9	1.0265	1.4331	0.71630	3.9	24.691	24.711	0.99918
1.0	1.1752	1.5431	0.76159	4.0	27.290	27.308	0.99933
1.1	1.3356	1.6685	0.80050	4.1	30.162	30.178	0.99945
1.2	1.5095	1.8107	0.83365	4.2	33.336	33.351	0.99955
1.3	1.6984	1.9709	0.86172	4.3	36.843	36.857	0.99963
1.4	1.9043	2.1509	0.88535	4.4	40.719	40.732	0.99970
1.5	2.1293	2.3524	0.90515	4.5	45.003	45.014	0.99975
1.6	2.3756	2.5775	0.92167	4.6	49.737	49.747	0.99980
1.7	2.6456	2.8283	0.93541	4.7	54.969	54.978	0.99983
1.8	2.9422	3.1075	0.94681	4.8	60.751	60.759	0.99986
1.9	3.2682	3.4177	0.95624	4.9	67.141	67.149	0.99989
2.0	3.6269	3.7622	0.96403	5.0	74.203	74.210	0.99991
2.1	4.0219	4.1443	0.97045	5.1	82.008	82.014	0.99993
2.2	4.4571	4.5679	0.97574	5.2	90.633	90.639	0.99994
2.3	4.9370	5.0372	0.98010	5.3	100.17	100.17	0.99995
2.4	5.4662	5.5569	0.98367	5.4	110.70	110.71	0.99996
2.5	6.0502	6.1323	0.98661	5.5	122.34	122.35	0.99997
2.6	6.6947	6.7690	0.98903	5.6	135.21	135.22	0.99997
2.7	7.4063	7.4735	0.99101	5.7	149.43	149.44	0.99998
2.8	8.1919	8.2527	0.99263	5.8	165.15	165.15	0.99998
2.9	9.0596	9.1146	0.99396	5.9	182.52	182.52	0.99998

^{a} When x is beyond the limits of the table (that is, when $x > 5.9$) sinh x and cosh x may be obtained to five significant figures by the formula sinh x = cosh x = $\frac{1}{2}e^x$; furthermore, for such values of x, tanh x = 1.0000.

1. $\int u\, dv = uv - \int v\, du$

2. $\int x^n\, dx = \dfrac{x^{n+1}}{n+1} + c, \quad n \neq -1$

3. $\int \dfrac{dx}{x} = \ln |x| + c$

4. $\int e^{ax}\, dx = \dfrac{e^{ax}}{a} + c$

5. $\int \ln x\, dx = x \ln x - x + c$

6. $\int \dfrac{dx}{a^2 + x^2} = \dfrac{1}{a}\tan^{-1}\dfrac{x}{a} + c$

7. $\int \dfrac{dx}{a^2 - x^2} = \dfrac{1}{2a}\ln\left|\dfrac{a+x}{a-x}\right| + c = \dfrac{1}{a}\tanh^{-1}\dfrac{x}{a} + c$

8. $\int \dfrac{dx}{x^2 - a^2} = \dfrac{1}{2a}\ln\left|\dfrac{x-a}{x+a}\right| + c = -\dfrac{1}{a}\coth^{-1}\dfrac{x}{a} + c$

9. $\int \dfrac{dx}{\sqrt{x^2 \pm a^2}} = \ln\left|x + \sqrt{x^2 \pm a^2}\right| + c$

10. $\int \dfrac{dx}{\sqrt{a^2 - x^2}} = \sin^{-1}\dfrac{x}{a} + c$

11. $\int \dfrac{x\, dx}{\sqrt{x^2 \pm a^2}} = \sqrt{x^2 \pm a^2} + c$

12. $\int \dfrac{x\, dx}{\sqrt{a^2 - x^2}} = -\sqrt{a^2 - x^2} + c$

13. $\int \sqrt{x^2 \pm a^2}\, dx = \dfrac{x}{2}\sqrt{x^2 \pm a^2} \pm \dfrac{a^2}{2}\ln\left|x + \sqrt{x^2 \pm a^2}\right| + c$

14. $\int \sqrt{a^2 - x^2}\, dx = \dfrac{x}{2}\sqrt{a^2 - x^2} + \dfrac{a^2}{2}\sin^{-1}\dfrac{x}{a} + c$

15. $\int \dfrac{\sqrt{x^2 - a^2}}{x}\, dx = \sqrt{x^2 - a^2} - |a|\sec^{-1}\dfrac{x}{a} + c.$

16. $\int \dfrac{\sqrt{a^2 \pm x^2}}{x}\, dx = \sqrt{a^2 \pm x^2} - a\ln\left|\dfrac{a + \sqrt{a^2 \pm x^2}}{x}\right| + c$

17. $\int \sin x\, dx = -\cos x + c$

18. $\int \sin^n x\, dx = -\dfrac{1}{n}\sin^{n-1} x \cos x + \dfrac{n-1}{n}\int \sin^{n-2} x\, dx$

19. $\int \sin mx \sin nx\, dx = \dfrac{\sin(m-n)x}{2(m-n)} - \dfrac{\sin(m+n)x}{2(m+n)} + c, \quad m^2 \neq n^2$

[a] Reprinted with permission from William R. Derrick and Stanley I. Grossman, *Elementary Differential Equations with Applications*. Addison-Wesley, Reading, Massachusetts, pp. A2–A5.

20. $\int \dfrac{dx}{a + b \sin x} = \dfrac{2}{\sqrt{a^2 - b^2}} \tan^{-1} \dfrac{a \tan x/2 + b}{\sqrt{a^2 - b^2}} + c, \quad a^2 > b^2$

21. $\int x^n \sin x \, dx = -x^n \cos x + n \int x^{n-1} \cos x \, dx$

22. $\int \cos x \, dx = \sin x + c$

23. $\int \cos^n x \, dx = \dfrac{1}{n} \cos^{n-1} x \sin x + \dfrac{n-1}{n} \int \cos^{n-2} x \, dx$

24. $\int \cos mx \cos nx \, dx = \dfrac{\sin (m-n)x}{2(m-n)} + \dfrac{\sin (m+n)x}{2(m+n)} + c, \quad m^2 \neq n^2$

25. $\int \dfrac{dx}{a + b \cos x} = \dfrac{2}{\sqrt{a^2 - b^2}} \tan^{-1} \dfrac{\sqrt{a^2 - b^2} \tan (x/2)}{a + b} + c, \quad a^2 > b^2$

26. $\int x^n \cos x \, dx = x^n \sin x - n \int x^{n-1} \sin x \, dx$

27. $\int \sin mx \cos nx \, dx = \dfrac{\cos (n-m)x}{2(n-m)} - \dfrac{\cos (n+m)x}{2(n+m)} + c, \quad m^2 \neq n^2$

28. $\int \sin x \cos x \, dx = \frac{1}{2} \sin^2 x + c$

29. $\int \tan x \, dx = -\ln |\cos x| + c$

30. $\int \tan^n x \, dx = \dfrac{\tan^{n-1} x}{n-1} - \int \tan^{n-2} x \, dx, \quad n \neq 1$

31. $\int \cot x \, dx = \ln |\sin x| + c$

32. $\int \cot^n x \, dx = -\dfrac{\cot^{n-1} x}{n-1} - \int \cot^{n-2} x \, dx, \quad n \neq 1$

33. $\int \sec x \, dx = \ln |\sec x + \tan x| + c$

34. $\int \sec^2 x \, dx = \tan x + c$

35. $\int \sec^n x \, dx = \dfrac{1}{n-1} \tan x \sec^{n-2} x + \dfrac{n-2}{n-1} \int \sec^{n-2} x \, dx$

36. $\int \csc x \, dx = \ln |\csc x - \cot x| + c$

37. $\int \csc^2 x \, dx = -\cot x + c$

38. $\int \csc^n x \, dx = -\dfrac{1}{n-1} \cot x \csc^{n-2} x + \dfrac{n-2}{n-1} \int \csc^{n-2} x \, dx$

39. $\int \sin^{-1} \dfrac{x}{a} \, dx = x \sin^{-1} \dfrac{x}{a} + \sqrt{a^2 - x^2} + c$

40. $\int \cos^{-1} \dfrac{x}{a} \, dx = x \cos^{-1} \dfrac{x}{a} - \sqrt{a^2 - x^2} + c$

41. $\int \tan^{-1} \dfrac{x}{a} \, dx = x \tan^{-1} \dfrac{x}{a} - \dfrac{a}{2} \ln (a^2 + x^2) + c$

TABLE A.6   Integrals (*continued*)

42. $\int \cot^{-1} \dfrac{x}{a}\, dx = x \cot^{-1} \dfrac{x}{a} + \dfrac{a}{2} \ln (a^2 + x^2) + c$

43. $\int \sec^{-1} \dfrac{x}{a}\, dx = x \sec^{-1} \dfrac{x}{a} - a \ln \left| x + \sqrt{x^2 - a^2} \right| + c$

44. $\int \csc^{-1} \dfrac{x}{a}\, dx = x \csc^{-1} \dfrac{x}{a} + a \ln \left| x + \sqrt{x^2 - a^2} \right| + c$

45. $\int x^n \ln (ax)\, dx = \dfrac{x^{n+1}}{n+1} \left[\ln (ax) - \dfrac{1}{n+1} \right] + c, \quad n \neq -1$

46. $\int (\ln x)^n\, dx = x(\ln x)^n - n \int (\ln x)^{n-1}\, dx, \quad n \neq -1$

47. $\int \dfrac{(\ln x)^n}{x}\, dx = \dfrac{(\ln x)^{n+1}}{n+1} + c$

48. $\int \dfrac{dx}{x \ln x} = \ln (\ln x) + c$

49. $\int x^n e^{ax}\, dx = \dfrac{x^n e^{ax}}{a} - \dfrac{n}{a} \int x^{n-1} e^{ax}\, dx$

50. $\int e^{ax} \sin bx\, dx = \dfrac{e^{ax}}{a^2 + b^2} (a \sin bx - b \cos bx) + c$

51. $\int e^{ax} \cos bx\, dx = \dfrac{e^{ax}}{a^2 + b^2} (a \cos bx + b \sin bx) + c$

52. $\int \sinh x\, dx = \cosh x + c$

53. $\int \cosh x\, dx = \sinh x + c$

54. $\int \tanh x\, dx = \ln (\cosh x) + c$

55. $\int_0^\infty x^{n-1} e^{-x}\, dx = \Gamma(n) = (n-1)\Gamma(n-1)$

56. $\int_0^\infty \dfrac{e^{-x}}{\sqrt{x}}\, dx = \Gamma(\tfrac{1}{2}) = \sqrt{\pi}$

57. $\int_0^{\pi/2} \sin^n x\, dx = \int_0^{\pi/2} \cos^n x\, dx = \begin{cases} \dfrac{\pi}{2^{n+1}} \dfrac{n!}{(n/2)!^2}, & n \text{ even} \\[3mm] \dfrac{2^{n-1}[(n-1)/2]!^2}{n!}, & n \text{ odd} \end{cases}$

58. $\int_0^\infty e^{-a^2 x^2}\, dx = \dfrac{1}{2a} \Gamma(\tfrac{1}{2}) = \dfrac{\sqrt{\pi}}{2a}, \quad a > 0$

ANSWERS TO
ODD-NUMBERED PROBLEMS
AND REVIEW EXERCISES

CHAPTER ONE

Problems 1.1

1. $-3 < 2$ **3.** $\pi/3 > 1$ **5.** $\sqrt{8} < 3.1$
9. (a) $\{x: x = 2n \text{ or } x = 5n, n \text{ an integer}\}$ (b) $\{x: x = 15n, n \text{ an integer}\}$
(c) $\{x: x = 2n \text{ or } x = 3n \text{ or } x = 5n, n \text{ an integer}\}$ **13.** (a) bounded (b) bounded
(c) bounded below (d) bounded above (e) bounded below (f) bounded
15. (a) A (b) B (c) $\{x: x \text{ a positive integer and } x \neq 3n, n \text{ an integer}\}$
(d) $\{x: x = 3n, n \text{ a nonnegative integer}\}$ **17** (a) $(-\infty, -1)$ (b) $[5, 10]$

Problems 1.2

1. $(-\infty, 7)$ **3.** $[-\frac{1}{2}, 1]$ **5.** $(-\frac{11}{5}, -\frac{1}{5})$ **7.** $[-3, 12)$ **9.** $[0, \frac{2}{3})$
11. (a) 1 (b) 4 (c) -1 **19.** $(-\infty, -2) \cup (2, \infty)$ **21.** $[-7, 1]$ **23.** $(-\frac{7}{2}, -\frac{1}{2})$
25. $[\frac{2}{3}, \frac{14}{3}]$ **27.** $(-\infty, \infty)$ **29.** $(-\infty, -5) \cup (-3, \infty)$
31. $(-\infty, -3) \cup (2, 7)$ **33.** $((-c - b)/a, (c - b)/a)$ if $c > 0$ and $\varnothing$ if $c \leq 0$
35. $(-\infty, \infty)$ **37.** $(\frac{5}{4}, \infty)$

Problems 1.3

1. (a) IV (b) III (c) III (d) I **3.** $\sqrt{122}$ **5.** $\sqrt{2}/6$ **7.** $|b - a|\sqrt{2}$ **9.** Yes

Problems 1.4

1. $y = 2x$ **3.** $y = \frac{1}{2}x + \frac{11}{2}$ **5.** $x = -3$ **7.** $y = \frac{11}{5}x + \frac{2}{5}$
9. $y = -\frac{4}{3}x + \frac{19}{3}$ **11.** $y = [(d - b)/(c - a)]x + [(bc - ad)/(c - a)]$
13. $y = -\frac{2}{5}x + \frac{3}{5}$ **15.** $y = 3x + 1$ **17.** $y = (b/a)x + \beta - (b\alpha/a)$
19. none **21.** all points on the line $4x - 6y = 10$ **23.** $(\frac{67}{45}, \frac{2}{15})$
25. $\sqrt{13}/13$ **27.** $\sqrt{61}/5$ **29.** $\sqrt{5}$

Problems 1.5

1. $(x - 1)^2 + y^2 = 1$ **3.** $(x - 1)^2 + (y - 1)^2 = 2$
5. $(x + 1)^2 + (y - 4)^2 = 25$ **7.** $(x - \pi)^2 + (y - 2\pi)^2 = \pi$
9. $(x - 3)^2 + (y + 2)^2 = 16$ **11.** none **13.** $(-3, 9), (1, 1)$ **15.** none
19. $y = 1$ **21.** $x + 2y = 4$

Problems 1.6

1. yes   3. yes   5. no   7. yes   9. yes   11. yes   15. R, R
17. $R - \{0\}, (0, \infty)$ 19. $R - \{-1\}, R - \{0\}$ 21. $[1, \infty), [0, \infty)$
23. $R - \{0\}, (0, \infty)$ 25. $R, [0, \infty)$ 27. $R, [0, \infty)$ 29. $R - \{0\}, R - \{5\}$
31. $R, (3, 4]$ 33. $R, (0, 1]$ 35. $R, [0, 1)$
37. $R, [0, 1)$ if n even; $R - \{-1\}, R - \{1\}$ if n odd   39. $R, [-\sqrt[3]{\frac{1}{4}}, \infty)$
41. $(-\infty, 3) - \{0\}, (-\infty, 0) \cup (\frac{1}{3}, \infty)$
43. $f(0) = 1, f(1) = 2, f(16) = 5, f(25) = 6$
45. $f(t^3) = \sqrt{t^3 + 1}, f(\sqrt[3]{t}) = \sqrt{\sqrt[3]{t} + 1}, f(1/t) = \sqrt{(1/t) + 1}$
47. $f(x + \Delta x) = (x + \Delta x)^2, \dfrac{f(x + \Delta x) - f(x)}{\Delta x} = 2x + \Delta x, \Delta x \ne 0$
49. $f(-5x) = \dfrac{-5x}{(-5x)^5 + 3}, f\left(\dfrac{1}{x}\right) = \dfrac{x^4}{1 + 3x^5}, f(x^{10} - 25) = \dfrac{x^{10} - 25}{(x^{10} - 25)^5 + 3},$
$f(g(t)) = \dfrac{g(t)}{[g(t)]^5 + 3}, f(x + \Delta x) = \dfrac{x + \Delta x}{(x + \Delta x)^5 + 3}$ 51. $R, \{-1, 1\}$

Problems 1.7

1. $-2x - 5, R; 6x - 5, R; -8x^2 + 20x, R; (2x - 5)/(-4x), R - \{0\}$
3. $\sqrt{x + 2} + \sqrt{2 - x}, [-2, 2]; \sqrt{x + 2} - \sqrt{2 - x}, [-2, 2]; \sqrt{4 - x^2}, [-2, 2];$
$\sqrt{(x + 2)/(2 - x)}, [-2, 2)$ 5. $2 + x^5 - |x|, R; x^5 + |x|, R; 1 + x^5 - |x| - |x|x^5, R;$
$(1 + x^5)/(1 - |x|), R - \{-1, 1\}$ 7. $\sqrt[5]{x + 2} + \sqrt[4]{x - 3}, [3, \infty);$
$\sqrt[5]{x + 2} - \sqrt[4]{x - 3}, [3, \infty); \sqrt[5]{x + 2}\sqrt[4]{x - 3}, [3, \infty); \sqrt[5]{x + 2}/\sqrt[4]{x - 3}, (3, \infty)$
9. $2x + 1, R; 2x + 2, R$ 11. $15x + 11, R; 15x + 27, R$
13. $(x - 1)/(3x - 1), R - \{0, \frac{1}{3}\}; -2/x, R - \{-2, 0\}$
15. $\sqrt{1 - \sqrt{x - 1}}, [1, 2]; \sqrt{\sqrt{1 - x} - 1}, (-\infty, 0]$
17. $1/\sqrt{x^5 + 1}, (-1, \infty); (x + 1)^{-5/2}, (-1, \infty)$
19. $\begin{cases} -6x, x \ge 0 \\ 10x, x < 0 \end{cases} R; \begin{cases} -3x, x \ge 0 \\ 10x, x < 0 \end{cases} R$ 23. $x - 5, 5 - x$ 25. $x^3 + 1$

Problems 1.8

1.

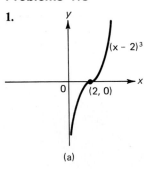

(a)

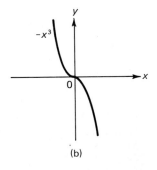

(b)

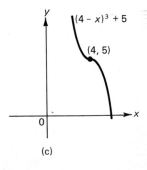

(c)

3.

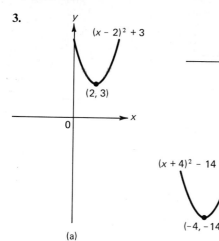

(a)

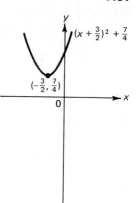

(c)

(b)

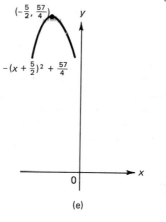

(d)

(e)

5.

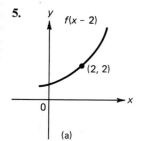

(a)

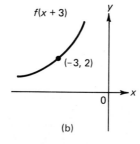

(b)

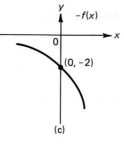

(c)

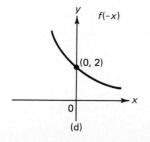

(d)

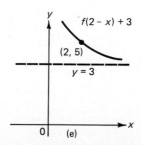

(e)

7.

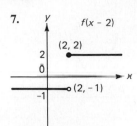

(a)

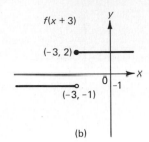

(b)

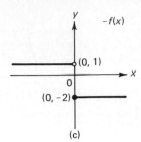

(c)

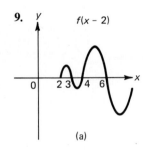

(d)

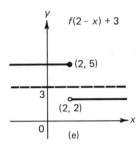

(e)

9.

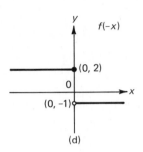

(a)

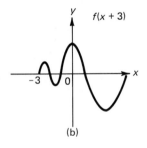

(b)

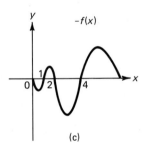

(c)

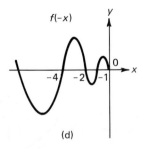

(d)

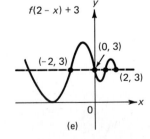

(e)

11.

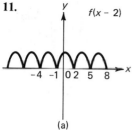
$f(x - 2)$

-4 -1 0 2 5 8

(a)

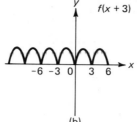
$f(x + 3)$

-6 -3 0 3 6

(b)

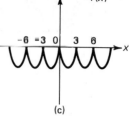
$-f(x)$

-6 -3 0   3 6

(c)

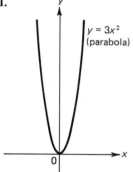
$f(-x)$

-6 -3 0   3 6

(d)

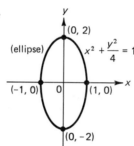
$f(2 - x) + 3$

-7 -4 -1 0 2 5

(e)

Problems 1.9

1.

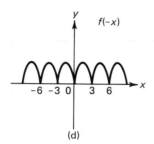
$y = 3x^2$
(parabola)

0

3.

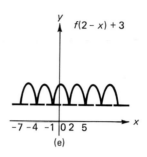
(ellipse)
(0, 2)
$x^2 + \dfrac{y^2}{4} = 1$
(-1, 0)   0   (1, 0)
(0, -2)

5.

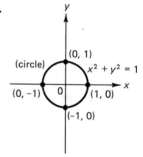
(circle)
(0, 1)
$x^2 + y^2 = 1$
(0, -1)   0   (1, 0)
(-1, 0)

7.

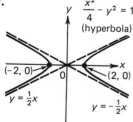
$\dfrac{x^2}{4} - y^2 = 1$
(hyperbola)
(-2, 0)   0   (2, 0)
$y = \dfrac{1}{2}x$ $y = -\dfrac{1}{2}x$

9.

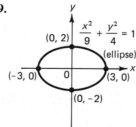
(0, 2) $\dfrac{x^2}{9} + \dfrac{y^2}{4} = 1$
(ellipse)
(-3, 0)   0   (3, 0)
(0, -2)

11.

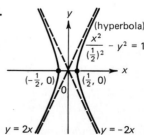
(hyperbola)
$\dfrac{x^2}{(\frac{1}{2})^2} - y^2 = 1$
$\left(-\dfrac{1}{2}, 0\right)$ $\left(\dfrac{1}{2}, 0\right)$
$y = 2x$ $y = -2x$

13.

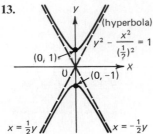

15.

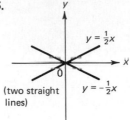

17.

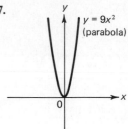

19.

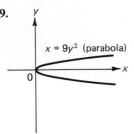

21.

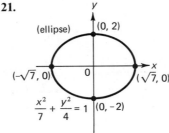

23.

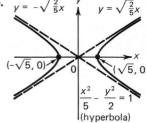

25. Two straight lines: $y = \pm \sqrt{(b)/(a)}\,x$ **27.**

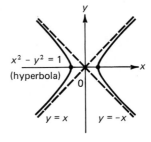

29.

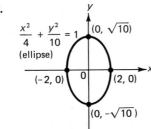

Problems 1.10

1. 364   **3.** $[1 - (-5)^6]/[1 - (-5)] = -2604$ **5.** $(0.3)^2[1 - (-0.3)^7]/1.3 \approx 0.07$
7. $(b^{16} - 1)/(b^{16} + b^{14})$ **9.** $(16\sqrt{2} - 1)/(\sqrt{2} - 1)$ **11.** -3280

Review Exercises for Chapter One

1. $A \cup B = \mathbb{R}$; $A \cup C = A$; $B \cup C = B$; $A \cap B = (-4, 2]$; $A \cap C = C$; $B \cap C = C$;
$\bar{A} = (2, \infty)$; $\bar{B} = (-\infty, -4]$; $\bar{C} = (-\infty, -1) \cup (1, \infty)$; $A - B = (-\infty, -4]$;
$B - A = (2, \infty)$; $A - C = (-\infty, -1) \cup (1, 2]$; $C - A = \varnothing$;
$B - C = (-4, -1) \cup (1, \infty)$; $C - B = \varnothing$ **3.** $(-\infty, -4] \cup [4, \infty)$

5. $(-\infty, -4) \cup (-2, \infty)$ **7.** $\mathbb{R}$ **9.** $(-3, 2) \cup (\frac{11}{3}, \infty)$ **11.** $(1, -3)$
13. $y = \frac{2}{3}x + \frac{11}{3}$ **15.** $y = x - 4$ **17.** $x = 1$ **19.** $2x - 5y = 18$
21. $3/\sqrt{10}$ **23.** $y = -\frac{1}{3}x + \frac{11}{3}$ **25.** yes; $\mathbb{R}$; $[-8, \infty)$ **27.** no
29. yes; $\mathbb{R}$; $[\frac{1}{6}, \infty)$ **31.** yes; $\mathbb{R} - \{-1, 1\}$, $\mathbb{R}$ **33.** yes; $f(2) = 0, f(-\sqrt{5}) = 1$,
$f(x + 4) = \sqrt{x^2 + 8x + 12}$; $f(x^3 - 2) = \sqrt{x^6 - 4x^3} = |x|\sqrt{x^4 - 4x}$,
$f(-1/x) = \sqrt{(1/x^2) - 4}$, $f(\text{Carlos}) = \sqrt{(\text{Carlos})^2 - 4}$ **35.** $\sqrt{x + 1} + x^3$, $[-1, \infty)$;
$\sqrt{x + 1} - x^3$, $[-1, \infty)$; $x^3\sqrt{x + 1}$, $[-1, \infty)$; $x^3/\sqrt{x + 1}$, $(-1, \infty)$; $\sqrt{x^3 + 1}$, $[-1, \infty)$;
$(x + 1)^{3/2}$, $[-1, \infty)$
37. $\frac{1}{4}x + \frac{3}{2}$, $\mathbb{R}$

39.

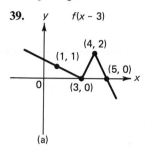

(a)

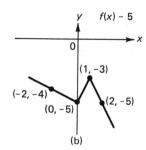

(b)

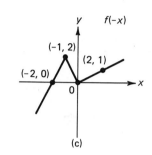

(c)

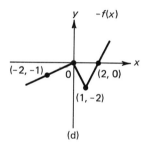

(d)

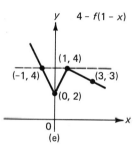

(e)

41.

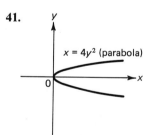

43

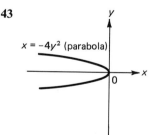

45.

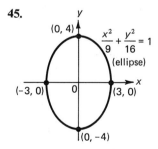

47.

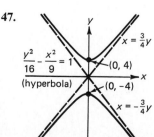

49.

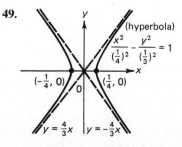

51. the single point $(0, 0)$

CHAPTER TWO

Problems 2.2

1. (b) $f(3) = 10$, $f(1) = 8$, $f(2.5) = 9.5$, $f(1.5) = 8.5$, $f(2.1) = 9.1$, $f(1.9) = 8.9$, $f(2.01) = 9.01$, $f(1.99) = 8.99$ (c) 9   **3.** (b) $f(-0.5) - 5.75$, $f(-1.5) = 10.75$, $f(-0.9) = 7.51$, $f(-1.1) = 8.51$, $f(-0.99) = 7.9501$, $f(-1.01) = 8.0501$ (c) 8
5. (a) can't divide by zero (b) $g(x) = x^2 + 2x + 4$ (c) 12   **7.** yes; $\sqrt{2}$
9. 45   **11.** $\frac{1}{2}$ **13.** 0   **15.** 0   **17.** 3   **19.** 2   **21.** doesn't exist
23. 1   **25.** $\frac{1}{4}$ **27.** (a) $f(-1) \approx 0.49487$, $f(-3)$ is undefined, $f(-1.5) \approx 0.47730$, $f(-2.5)$ is undefined, $f(-1.9) \approx 0.40625$, $f(-2.1) \approx 0.32774$, $f(-1.99) \approx 0.37647$, $f(-2.01) \approx 0.36877$, $f(-1.999) \approx 0.37306$, $f(-2.001) \approx 0.37229$ (b) ≈ 0.3725
(c) $f(-2) = \sqrt{5}/6 \approx 0.37268$
29. (d) the slope of the secant line between $(-3, -4)$ and $(-3 - h, (-3 - h)^2 + 5)$
(e) 6 (f) 6   **31.** (b) 0   **33.** (b) get different answers as 2 is approached from the left and right (see Example 2.2.10) (c) $1, -1$

Problems 2.3

1. 2   **3.** 0   **5.** 6   **7.** -4 **9.** 6   **11.** -7 **13.** 2   **15.** -34
17. 1   **19.** $\frac{1}{2}$ **21.** $\frac{1}{2}$ **23.** 1   **25.** $\frac{5}{4}$ **27.** $\frac{1}{2}$ **29.** 5

Problems 2.4

1. $(-\infty, \infty)$ **3.** $(0, \infty)$ **5.** $(-\infty, -1), (-1, 1), (1, \infty)$ **7.** $(-\infty, 2), (2, \infty)$
9. $(n, n + 1)$ for every integer n **11.** 4

13.

$$f(x) = \begin{cases} 0, x < 0 \\ x^2, 0 \le x < 2 \\ 4, x \ge 2 \end{cases}$$

19. 8

Problems 2.5

1. ∞ **3.** ∞ **5.** no limit   **7.** ∞ **9.** no limit   **11.** 1   **13.** 0
15. -1 **17.** $\frac{2}{3}$ **19.** 0   **21.** 0   **23.** 0   **25.** $\frac{3}{7}$ **27.** ∞ **29.** 1
31. 10,000; 1,000,000; 100,000,000

Problems 2.6

1. $\frac{4}{3}$ **3.** $\frac{10}{9}$ **5.** $\pi/(\pi - 1)$ **7.** $1/1.62 \approx 0.61728$ **9.** $\frac{1}{2}$ **11.** $n = 7$
13. $n = 459$ **15.** (b) 12 min

Problems 2.7

1. 0   **3.** 0   **5.** no limit   **7.** $-\infty$ **9.** not defined   **11.** 0   **13.** -1
15. 0   **17.** -6 **19.** 1   **21.** -7 **23.** $-\infty$ **25.** $0, 0$ **27.** $2, 2$

Problems 2.8

1. (a) positive (b) zero (c) negative (d) positive (e) negative

3.

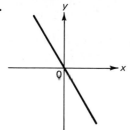

5.

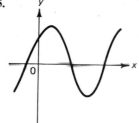

7.

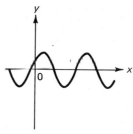

9.

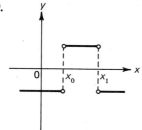

11.

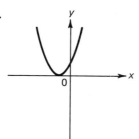

Problems 2.9

1. (a) $f(2.5) = 18.75$, $f(2.1) = 13.23$, $f(2.01) = 12.1203$, $f(2.001) = 12.012003$, $f(1.99) = 11.8803$, $f(1.999) = 11.988003$ (b) 13.5, 12.3, 12.03, 12.003, 11.97, 11.997
(c) $f'(x) = 6x$, $f'(2) = 12$ (d) $y = 12x - 12$
(e)

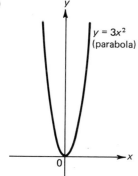

$y = 3x^2$
(parabola)

3. (a) $f(1.5) \approx 6.1237$, $f(1.1) \approx 5.2440$, $f(1.01) \approx 5.0249$, $f(1.001) \approx 5.0025$, $f(0.99) \approx 4.9749$, $f(0.999) \approx 4.9975$ (b) 2.2474, 2.4404, 2.4938, 2.4994, 2.5100, 2.500
(c) $f'(x) = 5/2\sqrt{x}$, $f'(1) = \frac{5}{2}$ (d) $y = \frac{5}{2}x + \frac{5}{2}$
(e)

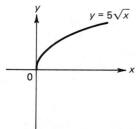

$y = 5\sqrt{x}$

5. $f'(x) = -4$; $y = -4x + 6$ **7.** $f'(x) = 4x - 12$; $y = 4x - 14$
9. $f'(x) = -2x + 3$; $y = 3x + 5$ **11.** $f'(x) = 3x^2$; $y = 3x + 8$
13. $f'(x) = -1/(2\sqrt{x+3})$; $y = -\frac{1}{6}x - 2$ **15.** $f'(x) = 1/\sqrt{2x}$; $y = \frac{1}{4}x + 2$
17. $f'(x) = (1/2\sqrt{x}) + \frac{3}{2}\sqrt{x}$; $y = 2x$ **19.** $f'(x) = 4x^3$; $y = 32x - 48$
25. $x = \frac{1}{2}$ **27.** $a = \frac{3}{2}$

Problems 2.10

1. (b) 28.6 ft/sec (c) 25.4 ft/sec (d) some number in (25.4, 28.6); midpoint is 27
(e) $26.84 < v(1.5) < 27.16$ (f) $h'(1.5) = 27$ **3.** 9   **5.** $\frac{1}{4}$ **7.** 40
9. $v(3) = 420$ ft/sec, $v(10) = 1400$ ft/sec   **11.** (a) -35 individuals/day
(b) $\sqrt{4000} \approx 63.2$ days   **13.** (a) 80π $\mu m^2/\mu m$ (b) 160π $\mu m^2/\mu m$

15. (a) $\frac{1}{2}\sqrt{\dfrac{3}{1000}} \approx 0.03$ kg/m³ (b) $\frac{1}{2}\sqrt{\dfrac{3}{500}} \approx 0.04$ kg/m³
(c) The density becomes infinitely large.   **17.** (a) $C'(x) = 6 - 0.02x + 0.03x^2$ (b) no

Review Exercises for Chapter Two

1. $f(3) = 6, f(1) = 4, f(2.5) = 4.75, f(1.5) = 3.75, f(2.1) = 4.11, f(1.9) = 3.91,$
$f(2.01) = 4.0101, f(1.99) = 3.9901;$ $\lim_{x\to 2}(x^2 - 3x + 6) = 4$ **3.** (a) 0 (b) -108
(c) $\frac{76}{31}$ (d) 0   **5.** (a) -1 (b) 4   **7.** (a) not defined (b) 5 (c) $\frac{25}{2}$ (d) 5 (e) -6
(f) $4^5 = 1024$ (g) $\frac{1}{3}$ (h) c/f **9.** $\lim_{x\to 0^+} 1/x^3 = \infty$ and $\lim_{x\to 0^-} 1/x^3 = -\infty$;
$\lim_{x\to 0^+} 1/x^4 = \infty = \lim_{x\to 0^-} 1/x^4$ **11.** (a) 0 (b) 0 (c) 0 (d) $\frac{1}{3}$ (e) 0 (f) ∞

13.

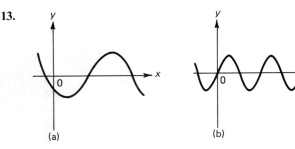

(a)   (b)

15. $\lim_{x\to -1^+} dy/dx = 1$; $\lim_{x\to -1^-} dy/dx = -1$ **17.** $\mu'(3) = \frac{1}{4}$
19. $C'(x) = 8 - 0.04x$; yes; C' is a decreasing function

CHAPTER THREE

Problems 3.1

1. $5x^4$ **3.** $6x + 19$ **5.** $5t^4 + (1/2\sqrt{t})$ **7.** $100z^{99} + 1000z^9$
9. $24r^7 - 48r^5 - 28r^3 + 4r$ **11.** $y = 4x - 3$ **13.** $y = 5x - 5$
15. $y = x + 1$ **17.** $y = 3x - 8$ **21.** $y = \frac{1}{23}x - \frac{116}{23}$ **23.** $x = 0$
25. x values are $(-1 \pm \sqrt{5})$ **27.** $y = 6x - 6$; $y = -2x + 2$
29. (a) $3000 - 20\sqrt{t} - 30t + 30t^2$ (b) $-(10/\sqrt{t}) - 30 + 60t$ organisms/hr
(c) 205 organisms/hr; 927.5 organisms/hr
33. $4\pi \cdot 3^2(1 - (\frac{11}{12})^3) = 36\pi(1 - (\frac{11}{12})^3) \approx 25.98$ in³/in of radius

Problems 3.2

1. $6x^2 + 2$ **3.** $\dfrac{(1 + \sqrt{t})3t^2 - (t^3/2\sqrt{t})}{(1 + \sqrt{t})^2} = \dfrac{3t^2 + \frac{5}{2}t^{5/2}}{(1 + \sqrt{t})^2}$
5. $(1 + x + x^5)(-1 + 6x^5) + (2 - x + x^6)(1 + 5x^4)$

7. $\dfrac{(1 + x + x^5)(-1 + 6x^5) - (2 - x + x^6)(1 + 5x^4)}{(1 + x + x^5)^2}$ **9.** $1/[\sqrt{t}(1 - \sqrt{t})^2]$

11. $5v^4 + 7v^{5/2} - \frac{5}{2}v^{3/2} - 2$

13. $\dfrac{(v^3 - \sqrt{v})[2v + (1/\sqrt{v})] - (v^2 + 2\sqrt{v})[3v^2 - (1/2\sqrt{v})]}{(v^3 - \sqrt{v})^2}$ (not simplified)

15. $\frac{5}{2}v^{3/2}$ **17.** $-\frac{1}{2}r^{-3/2}$ **19.** $-[\frac{9}{2}t^{7/2} + (1/\sqrt{t})]/t(t^4 + 2)^2$ **21.** $-6/x^7$

23. $-15/7x^4$ **25.** $y = 28x - 20$ **27.** $y = -\frac{7}{2}x + \frac{11}{2}$ **29.** $y = 2x - 1$

31. $y = 4x - 4$

35. $(x^2 + 1)(x^3 + 2)(4x^3) + (x^2 + 1)(x^4 + 3)(3x^2) + (x^3 + 2)(x^4 + 3)(2x)$

37. $(f^2g' + f'g^2)/(f + g)^2$ **39.** $dR/dr = -4\alpha l/(0.2)^5 = -12,500\alpha l$

Problems 3.3

1. $3(x + 1)^2$ **3.** $(2/\sqrt{x})(\sqrt{x} + 2)^3$ **5.** $36x^5(1 + x^6)^5$

7. $5(x^2 - 4x + 1)^4(2x - 4)$ **9.** $-6(t + 1)^2/(t - 1)^4$

11. $2(u^5 + u^4 + u^3 + u^2 + u + 1)(5u^4 + 4u^3 + 3u^2 + 2u + 1)$

13. $-8y(y^2 + 3)^{-5}$ **15.** $10x(x^2 + 2)^4(x^4 + 3)^3 + 12x^3(x^4 + 3)^2(x^2 + 2)^5$

17. $(-3t^2 + 2t - 4)/[\sqrt{t^2 + 1}(t + 2)^5]$

19. $\dfrac{\sqrt{u - 2}\{6u(u^2 + 1)^2(u^2 - 1)^2 + 4u(u^2 - 1)(u^2 + 1)^3\} - \frac{1}{2}(u^2 + 1)^3(u^2 - 1)^2(u - 2)^{-1/2}}{u - 2}$

21. $\left\{\dfrac{1}{2\sqrt{x + \sqrt{1 + \sqrt{x}}}}\right\}\left\{1 + \dfrac{1}{2\sqrt{1 + \sqrt{x}}}\left(\dfrac{1}{2\sqrt{x}}\right)\right\}$

23. $-5(y^{-2} + y^{-3} + y^{-7})^{-6}(-2y^{-3} - 3y^{-4} - 7y^{-8})$

25. $-\frac{1}{2}(q^{-7} - q^{-3})^{-3/2}(-7q^{-8} + 3q^{-4})$ **27.** 24 kg/m

Problems 3.4

1. $\frac{2}{5}x^{-3/5} + \frac{2}{3}x^{-2/3}$ **3.** $\frac{10}{3}x(x^2 + 1)^{2/3}$ **5.** $\frac{3}{10}x^2(x^3 + 3)^{-9/10}$

7. $\frac{40}{7}t^9(t^{10} - 2)^{-3/7}$ **9.** $(u^3 + 3u + 1)^{-2/3}(u^2 + 1)$

11. $-\frac{28}{17}[(z^2 - 1)/(z^2 + 1)]^{-24/17}[z/(z^2 + 1)^2]$

13. $\frac{5}{3}r(r^2 + 1)^{-1/6}(r - 1)^{1/2} + \frac{1}{2}(r - 1)^{-1/2}(r^2 + 1)^{5/6}$ **15.** $3\sqrt{2}x^{\sqrt{2}-1}$

17. $2\sqrt{3}t(t^2 + 1)^{\sqrt{3}-1}$ **19.** $4(u^{\sqrt{2}} - u^{\sqrt{3}})^3(\sqrt{2}u^{\sqrt{2}-1} - \sqrt{3}u^{\sqrt{3}-1})$

21. $y = 2x; y = -\frac{1}{2}x + \frac{5}{2}$ **23.** $y = 0; x = 1$

25. (a) $y = (b/a)\sqrt{a^2 - x^2}; y = -(b/a)\sqrt{a^2 - x^2}$

(b) $T: y - y_0 = -(b^2/a^2)(x_0/y_0)(x - x_0); R: (y - y_0)/(x - x_0) = -(y_0/x_0)$ or $y = (y_0/x_0)x$ (c) need $y_0/x_0 = -1/[(-b^2/a^2)(x_0/y_0)]$; occurs only if $a = b$

27. (a) $V_{rms} \approx 5.7767$ (b) $dV_{rms}/d\rho = -\frac{1}{2}\sqrt{3P/\rho^3}$ (c) $dV_{rms}/d\rho \approx -32.1285$

Problems 3.5

1. (a) increasing on $(-\frac{1}{2}, \infty)$; decreasing on $(-\infty, -\frac{1}{2})$ (b) $(-\frac{1}{2}, -\frac{121}{4})$

(c) y-intercept: -30; x-intercepts: $-6, 5$

(d)

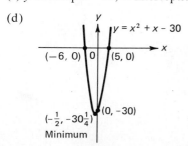

3. (a) increasing on $(-\infty, -3) \cup (5, \infty)$; decreasing on $(-3, 5)$
(b) $(-3, 106)$, $(5, -150)$ (c) y-intercept: 25

(d) Local maximum

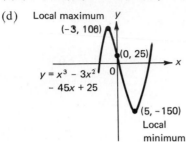

5. (a) increasing on $(0, \infty)$; decreasing on $(-\infty, 0)$ (b) $(0, 0)$
(c) y-intercept $= x$-intercept $= 0$

(d)

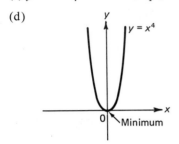

7. (a) increasing on $(0, 1) \cup (2, \infty)$; decreasing on $(-\infty, 0) \cup (1, 2)$
(b) $(0, 1)$, $(1, 2)$, $(2, 1)$ (c) y-intercept: 1; no x-intercept

(d)

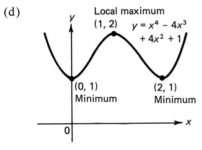

9. (a) increasing on $(-\infty, \infty)$ (b) $(0, 3)$ (c) y-intercept: 3; x-intercept: $-\sqrt[3]{3}$

(d)

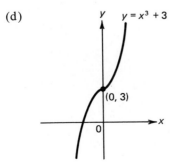

11. (a) increasing on $(-\infty, \infty)$ (b) $(-1, 0)$ (c) y-intercept: 1; x-intercept: -1

(d)

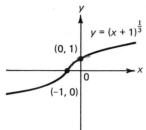

13. (c) y-intercept: 1; x-intercept: -1 (d) as $x \to -1$, $f(x) \to 0$ and $f'(x) \to \infty$ leading to the vertical tangent $x = -1$

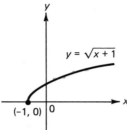

15. (c) y-intercept: 1; no x-intercept (d) as $x \to -1^+$, $f(x) \to \infty$ leading to the vertical asymptote $x = -1$

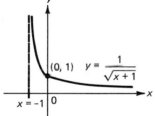

17.

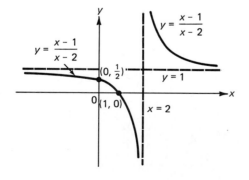

Problems 3.6

1.

Δx	Δy	dy	Error $= \Delta y - dy$
1	37	27	10
0.5	15.875	13.5	2.375
0.1	2.791	2.7	0.091
0.05	1.372625	1.35	0.022625
0.01	0.270901	0.27	0.000901
0.001	0.027009	0.027	0.000009

3.

Δx	Δy	dy	Error $= \Delta y - dy$
1.0	3.180	3.0	0.180
-1.0	-2.804	-3.0	0.196
0.5	1.5459	1.5	0.0459
-0.5	-1.4521	-1.5	0.0479
0.1	0.30187	0.3	0.00187
-0.1	-0.29812	-0.3	0.00188
0.01	0.030019	0.03	0.000019
-0.01	-0.029981	-0.03	0.000019

5. 0.96   **7.** 0.196   **9.** $\frac{157}{320} = 0.490625$ **11.** 1.6   **13.** 7.98125

15. -0.08 **17.** 339.325   **19.** 0   **21.** $\frac{1}{3}x^{-2/3}\, dx$ **23.** $\frac{3}{4}x^2(1 + x^3)^{-3/4}\, dx$

25. $-(1/x^2)\, dx$ **27.** $-x^{-1/2}(1 + \sqrt{x})^{-2}\, dx$ **29.** $(1 + x)^{-1/2}(1 - x)^{-3/2}\, dx$

31. $-[2x/(x^2 + 2)^2]\, dx$ **33.** (a) 4000π kg (b) $\pm 40\pi$ kg (c) ± 0.01 ($\pm 1\%$)

35. $-0.005333\ldots$ ($\approx 0.53\%$)   **37.** $\Delta R/R < 0.08$

Problems 3.7

1. $\dfrac{dy}{dx} = -\dfrac{x^2}{y^2}; \dfrac{dx}{dy} = -\dfrac{y^2}{x^2}$ **3.** $\dfrac{dy}{dx} = -\sqrt{\dfrac{y}{x}}; \dfrac{dx}{dy} = -\sqrt{\dfrac{x}{y}}$

5. $\dfrac{dy}{dx} = -\left(\dfrac{y}{x}\right)^2; \dfrac{dx}{dy} = -\left(\dfrac{x}{y}\right)^2$

7. $\dfrac{dy}{dx} = \dfrac{\frac{1}{2}(x + y)^{-1/2} - \frac{2}{3}x(x^2 + y)^{-2/3}}{\frac{1}{3}(x^2 + y)^{-2/3} - \frac{1}{2}(x + y)^{-1/2}}; \dfrac{dx}{dy} = \dfrac{1}{dy/dx}$

9. $\dfrac{dy}{dx} = -\dfrac{x}{y}; \dfrac{dx}{dy} = -\dfrac{y}{x}$ **11.** $\dfrac{dy}{dx} = \dfrac{2}{15(3xy + 1)^4} - \dfrac{y}{x}; \dfrac{dx}{dy} = \dfrac{1}{dy/dx}$

13. $\dfrac{dy}{dx} = \dfrac{x}{y}; \dfrac{dx}{dy} = \dfrac{y}{x}$ **15.** $\dfrac{dy}{dx} = \dfrac{y}{x}, \dfrac{dx}{dy} = \dfrac{x}{y}$ **17.** $\dfrac{dy}{dx} = -\dfrac{y}{x}; \dfrac{dx}{dy} = -\dfrac{x}{y}$

19. $\dfrac{dy}{dx} = -\dfrac{(y + 2xy^2)}{x + 2x^2y}; \dfrac{dx}{dy} = -\dfrac{(x + 2x^2y)}{y + 2xy^2}$

21. $\dfrac{dy}{dx} = -\left(\dfrac{y}{x}\right)^{15/8}; \dfrac{dx}{dy} = -\left(\dfrac{x}{y}\right)^{15/8}$ **23.** $\dfrac{dy}{dx} = -\dfrac{y}{x}; \dfrac{dx}{dy} = -\dfrac{x}{y}$

25. $\dfrac{dy}{dx} = \dfrac{y - 2xy^3 - 3x^2y^2}{3x^2y^2 + 2x^3y - x}; \dfrac{dx}{dy} = \dfrac{3x^2y^2 + 2x^3y - x}{y - 2xy^3 - 3x^2y^2}$

27. Vertical tangent at $(0, 1)$; horizontal tangent at $(1, 0)$

29. No vertical or horizontal tangents

31. Vertical tangents at $(-a, 0)$, $(a, 0)$; no horizontal tangents

Problems 3.8

1. 0; 0 **3.** 8; 0 **5.** $-\frac{1}{4}x^{-3/2}$; $\frac{3}{8}x^{-5/2}$ **7.** $-\frac{2}{9}(x+1)^{-4/3}$; $\frac{8}{27}(x+1)^{-7/3}$
9. $-(1-x^2)^{-3/2}$; $-3x(1-x^2)^{-5/2}$ **11.** $r(r-1)x^{r-2}$; $r(r-1)(r-2)x^{r-3}$
13. $2a$; 0 **15.** $30(x+1)^{-7}$; $-210(x+1)^{-8}$ **19.** $a = 50$ **21.** 56,000 nt

Problems 3.9

1. same graph as for Problem 3.5.1(d) **3.** same graph as for Problem 3.5.3(d)
5. same graph as for Problem 3.5.9(d).

7.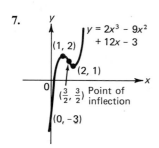

9.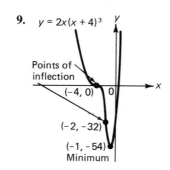

11.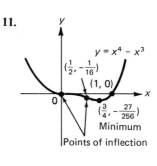

13. same graph as for Problem 3.5.13(d)

15.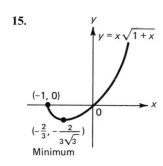

17.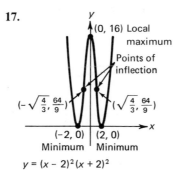

19.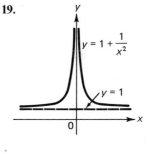

21.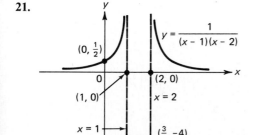

23.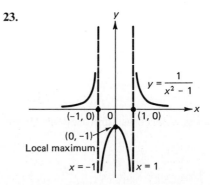

25.

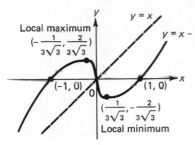

27.

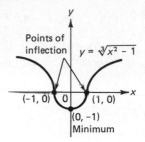

29.

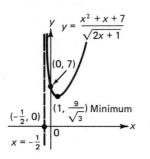

31.

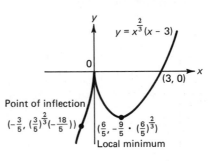

33. yes

35. positive   **43.**

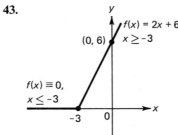

45.

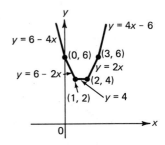

47.

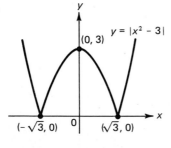

51.

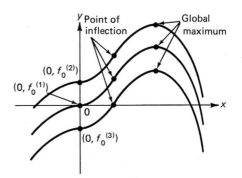

Review Exercises for Chapter Three

1. 3   **3.** $-\dfrac{(y+1)}{x+1}$ **5.** $6x\sqrt{x+1} + \dfrac{3x^2+2}{2\sqrt{x+1}}$ **7.** $-\left(\dfrac{y}{x}\right)^{1/4}$

9. $\dfrac{1}{3}\left(\dfrac{x+2}{x-3}\right)^{-2/3}\left[\dfrac{-5}{(x-3)^2}\right]$ **11.** $\dfrac{-x^4 + 4x^3 - 4x^2 - 12x - 3}{(x^3 + 5x - 6)^2}$

13. $\dfrac{18x^2(x^3-3)^{5/6}}{7(x^3+4)^{1/7}} + \dfrac{5x^2(x^3+4)^{6/7}}{2(x^3-3)^{1/6}}$ **15.** $\dfrac{1+\frac{1}{2}(x+y)^{-1/2}}{\frac{1}{3}y^{-2/3} - \frac{1}{2}(x+y)^{-1/2}}$

17. $3\sqrt{2}\,x^2(1+x^3)^{\sqrt{2}-1}$

19. $\dfrac{2x^2(x^5+3)^{3/4}(x^3-1)^{-1/3} - \frac{15}{4}x^4(x^3-1)^{2/3}(x^5+3)^{-1/4}}{(x^5+3)^{3/2}}$

21. $y = 2x+2;\; y = -\frac{1}{2}x + \frac{9}{2}$ **23.** $y = -\frac{183}{8}x + \frac{327}{8};\; y = \frac{8}{183}x + \frac{3286}{183}$

25. $42x^5 - 210x^4 + 6x;\; 210x^4 - 840x^3 + 6$ **27.** $2(x+1)^{-3};\; -6(x+1)^{-4}$

29. $-16(3x^2-5)/(x^2+5)^3;\; 192x(x^2-5)/(x^2+5)^4$ **31.** $51x^2\,dx$

33. $[-11/(x-7)^2]\,dx$

35. $[(4x^3/3)(x^2-3)^{1/2}(x^4+5)^{-2/3} + x(x^2-3)^{-1/2}(x^4+5)^{1/3}]\,dx$ **37.** (a) 31.2

(b) 0.49958333...

39.

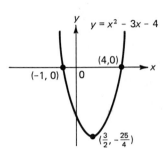

41.

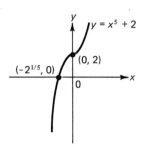

43.

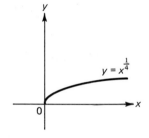

45.

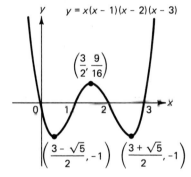

47. same graph as for Problem 3.9.23 **49.**

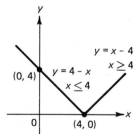

CHAPTER FOUR

Problems 4.1

1. $s = \frac{651}{2048} \approx 0.31787;\; S = \frac{715}{2048} \approx 0.34912$

Problems 4.2

1. $28 < A < 84$; $42 < A < 70$; $49 < A < 63$; $A = 56$
3. $12\frac{3}{4} < A < 26\frac{1}{4}$; $16\frac{1}{8} < A < 22\frac{7}{8}$; $17\frac{13}{16} < A < 21\frac{3}{16}$; $A = 19\frac{1}{2}$

In the answers to Problems 5–17, the left-hand endpoint of each subinterval has been chosen.

5. (a) $s_4 = \frac{7}{8}$ (b) $s_8 = \frac{35}{32}$
   (c) $s_N = (4/N^3)(1^2 + 2^2 + \cdots + (N-1)^2) = 2(N-1)(2N-1)/3N^2$ (d) $A = \frac{4}{3}$
7. (a) $s_4 = 5^4 \cdot 3^2/4^3 \approx 87.89$ (b) $s_8 = 5^4 \cdot 7^2/(4 \cdot 8^2) \approx 119.63$
   (c) $s_N = (5/N)^4(1^3 + 2^3 + \cdots + (N-1)^3) = (5^4/4)[(N-1)/N]^2$
   (d) $A = 5^4/4 = 156.25$ 9. (a) $s_8 = \frac{3}{4}(7^2/8^2) \approx 0.574$ (b) $s_{16} = \frac{3}{4}(15^2/16^2) \approx 0.659$
   (c) $s_N = (3/N^4)(1^3 + 2^3 + \cdots + (N-1)^3) = \frac{3}{4}[(N-1)^2/N^2]$ (d) $A = \frac{3}{4}$
11. (a) $s_4 = 1 - (3 \cdot 7)/(6 \cdot 4^2) = 0.78125$ (b) $s_8 = 1 - (7 \cdot 15)/(6 \cdot 8^2) \approx 0.7266$

(c) $s_N = \frac{1}{N}\left\{1 + \left[1 - \left(\frac{1}{N}\right)^2\right] + \left[1 - \left(\frac{2}{N}\right)^2\right] + \cdots + \left[1 - \left(\frac{N-1}{N}\right)^2\right]\right\}$

$= 1 - \dfrac{(N-1)(2N-1)}{6N^2}$ (d) $A = \frac{2}{3}$ 13. (a) $s_8 = (15 \cdot 55)/(6 \cdot 8^2) \approx 2.148$

(b) $s_{16} = (31 \cdot 111)/(6 \cdot 16^2) \approx 2.240$

(c) $s_N = \frac{1}{N}\left\{1 + \left(1 + \frac{1}{N}\right)^2 + \left(1 + \frac{2}{N}\right)^2 + \cdots + \left(1 + \frac{N-1}{N}\right)^2\right\} = \dfrac{(2N-1)(7N-1)}{6N^2}$

(d) $A = \frac{7}{3}$ 15. (a) $s_6 = (5 \cdot 29)/(6 \cdot 36) \approx 0.6713$ (b) $s_{12} = (11 \cdot 59)/(6 \cdot 144) \approx 0.7512$

(c) $s_N = \frac{1}{N}\left\{\left[\left(\frac{1}{N} + \frac{1}{N^2}\right) + \left(\frac{2}{N} + \left(\frac{2}{N}\right)^2\right)\right] + \cdots + \left[\left(\frac{N-1}{N}\right) + \left(\frac{N-1}{N}\right)^2\right]\right\}$

$= \dfrac{(N-1)(5N-1)}{6N^2}$

(d) $A = \frac{5}{6}$ 17. (a) $s_4 = (7 \cdot 11)/(2 \cdot 4^2) = 2.40625$ (b) $s_8 = (15 \cdot 23)/(2 \cdot 8^2) \approx 2.695$

(c) $s_N = \frac{1}{N}\left\{\left[1 + \frac{2 \cdot 1}{N} + 3\left(\frac{1}{N}\right)^2\right] + \left[1 + \frac{2 \cdot 2}{N} + 3\left(\frac{2}{N}\right)^2\right] + \cdots + \right.$

$\left. \left[1 + \frac{2(N-1)}{N} + 3\left(\frac{N-1}{N}\right)^2\right]\right\} = \dfrac{(2N-1)(3N-1)}{2N^2}$ (d) $A = 3$

19. (a) $a, a + \dfrac{b-a}{n}, a + 2\left(\dfrac{b-a}{n}\right), \ldots, a + (n-1)\left(\dfrac{b-a}{n}\right), b$

(b) $x_i^* = a + i\left(\dfrac{b-a}{n}\right) = \frac{1}{n}\{(n-i)a + ib\}$;

$(x_i^*)^2 = \dfrac{1}{n^2}\{(n-i)a + ib\}^2 + \dfrac{1}{n^2}\{(n-i)^2 a^2 + 2(n-i)iab + i^2 b^2\}$

(c) $s_n = \dfrac{1}{n^2}\left(\dfrac{b-a}{n}\right)\{[(n-1)a + b]^2 + [(n-2)a + 2b]^2 + [a + (n-1)b]^2 + b^2\}$

$= (b-a)\left\{\dfrac{(n-1)(2n-1)}{6n^2}a^2 + \left(\dfrac{n^2-1}{3n^2}\right)ab + \dfrac{(n+1)(2n+1)}{6n^2}b^2\right\}$

(d) $A = (b-a)\left(\dfrac{a^2}{3} + \dfrac{ab}{3} + \dfrac{b^2}{3}\right) = \frac{1}{3}(b-a)(a^2 + ab + b^2) = \dfrac{b^3 - a^3}{3}$

Problems 4.3

1. 56 3. $37\frac{1}{2}$ 5. $\frac{4}{3}$ 7. $\frac{4}{3}$ 9. $(17 \cdot 8^3)/3 = 2901\frac{1}{3}$ 11. $-\frac{1}{4}$ 13. $\frac{2}{3}$
15. $-152\frac{1}{4}$ 17. $-6\frac{3}{4}$ 19. 1 21. 10 23. $(b^3 - a^3)/3$
29. Use fact that $s^{55/3} > s^{17}$ for $23 \le s \le 47$ and Theorem 7 31. $\int_1^2 x \, dx$
33. Underestimate with constant function 1; overestimate with sum corresponding to the partition $\{0, \frac{1}{4}, \frac{1}{2}, \frac{3}{4}, 1\}$, where x_i^* is always chosen to be the right-hand endpoint. This

yields 1.17.　　**35.**　$12 < \int < 24$　　**37.**　$\frac{8}{3} < \int < 8$　　**39.**　$\frac{99}{100} < \int < 99$
41.　$\frac{1}{2} < \int < 1$　　**43.**　(a) 64 ft/sec　(b) 128 ft　(c) $\sqrt{50} \approx 7.07$ sec

Problems 4.4

1.　$x + C$　　**3.**　$ax + C$　　**5.**　$x + (x^2/2) + (x^3/3) + (x^4/4) + (x^5/5) + C$
7.　$-(2/x^2) + C$　　**9.**　$\frac{2}{3}x^{3/2} + 2\sqrt{x} + C$　　**11.**　$\frac{2}{3}x^{3/2} + \frac{2}{5}x^{5/2} + \frac{4}{7}x^{7/4} + C$
13.　$\frac{2}{5}x^{5/2} + \frac{2}{3}x^{3/2} - 2x^{1/2} + C$　　**15.**　$-\frac{289}{4}x^{4/17} + \frac{27}{5}x^{5/9} + C$
17.　$y = \frac{2}{3}x^3 + x^2 - \frac{28}{3}$　　**19.**　$y = \frac{78}{11}x^{33/18} - 3x + \frac{109}{11}$
21.　$x = 2.9t^2 + 0.2t + 25$ m　　**23.**　$x = \frac{52}{15}t^{5/2} + 25t + 100$ m
25.　$\frac{3}{2}(100) + \frac{2}{3}(1000) - \frac{2}{5}(10)^{5/2} \approx 690.176$ kg　　**27.**　\$120,890; \$2,302,223.33
29.　Unprofitable when $x > 3333$

Problems 4.5

1.　$\frac{33}{5}$　　**3.**　$\frac{26}{3}$　　**5.**　$(c_1/3)(b^3 - a^3) + c_2/2(b^2 - a^2) + c_3(b - a)$　　**7.**　$\frac{333}{4}$
9.　0　　**11.**　$\frac{80}{3}$　　**13.**　$\frac{41}{3}$　　**15.**　$\frac{6}{23} - \frac{1}{3} - \frac{4}{13} + \frac{12}{29} \approx 0.0336$　　**17.**　$-\frac{1}{101}$
19.　18　　**21.**　10　　**23.**　94　　**25.**　$\frac{32}{3}$　　**27.**　$\frac{32}{3}$　　**29.**　$\frac{1056}{5}$　　**31.**　8
33.　$(b^3 - 3ab^2 + 3a^2b - a^3)/6 = (b - a)^3/6$　　**35.**　192
39.　$F'(x) = 1/(1 + x^3); F'(2) = \frac{1}{9}$　　**41.**　$F'(x) = \sqrt{(x - 1)/(x + 1)}; F'(1) = 0$
43.　609.4 m　　**45.**　(a) 2 kg　(b) $\frac{5}{4}$ m　　**47.**　\$271.67　　**49.**　(a) 10^6 m/sec²
(b) 55 m　　**51.**　C　　**53.**　$\frac{1}{4}$　　**55.**　$\frac{19}{3}$　　**57.**　$1 + \frac{1}{2} + \frac{1}{3} + \frac{1}{4} + \frac{1}{5} = \frac{137}{60}$
59.　$\frac{231}{70} = 3.3$　　**61.**　$\frac{2047}{11} + \frac{1}{4608} - \frac{1}{9} \approx 185.98$　　**63.**　\$45　　**65.**　-80 ft/sec

Problems 4.6

1.　$\frac{74}{3}$　　**3.**　$2(10^{3/2} - 1)/27$　　**5.**　$\frac{5}{72}(1 + 3x^4)^{6/5} + C$　　**7.**　$\frac{2}{15}(s^5 + 5s)^{3/2} + C$
9.　$\sqrt{20} - \sqrt{3}$　　**11.**　$-\frac{1}{2}(1 + \sqrt{x})^{-4} + C$　　**13.**　$-\frac{3}{16}[1 + (1/v^2)]^{8/3} + C$
15.　$\frac{1}{3}(ax^2 + 2bx + c)^{3/2} + C$　　**17.**　$\frac{7}{8}(ax^2 + 2bx + c)^{4/7} + C$
19.　$[2a^3/3(n + 1)][(1 + a^{n-1})^{3/2} - 1]$　　**21.**　$(4\sqrt{2}/9)a^{9/2}$　　**23.**　0 (odd function)
25.　$\frac{1}{3}(64 - 7^{3/2})$　　**27.**　$\frac{7}{2}$　　**29.**　(a) $\frac{5}{8}$ lb　(b) $(\frac{13}{2})^{3/2} - 8 \approx 8.57$ ft
31.　$\frac{2}{15}$ m/min　　**33.**　$\frac{2}{3}$

Review Exercises for Chapter Four

1.　$(x^6/6) + C$　　**3.**　$\frac{9}{2}$　　**5.**　$\frac{1}{3}(2^{3/2} - 1)$　　**7.**　$-\frac{32}{3}$　　**9.**　$\frac{1}{320}$　　**11.**　$\frac{233}{6}$
13.　$\frac{9}{2}$　　**15.**　$\frac{3}{10}$　　**17.**　$F'(t) = (t^5 - 17t + 8)^{11/3}; F'(1) = -2048$　　**19.**　(a) $\frac{237}{2}$ m
(b) $\frac{237}{30} = 7.9$ m/sec

CHAPTER FIVE

Problems 5.1

1.　$-\frac{10}{3}$　　**3.**　$\frac{8}{3}$ ft/sec　　**5.**　$13650/\sqrt{9000} \approx 143.88$ ft/sec
7.　$-18^{1.67}(0.01)/[(1.67)(0.3)(18^{0.67})] \approx -0.359$ m³/sec; decreasing
9.　$(500\sqrt{1.5^2 - 1})/1.5 = 500\sqrt{5}/3 \approx 372.7$ km/hr　　**11.**　If triangle is not inverted,
$dh/dt = 0.08$ ft/min; if it is inverted, $dh/dt = 0.4$ ft/min.　　**13.**　50π cm/sec
15.　$-\frac{8}{3}$ ft²/hr　　**17.**　5 ft/sec　　**19.**　$\frac{11}{2}\pi r^2\, dr/dt$　　**21.**　3; $75/\sqrt{17}$

Problems 5.2

In the following answers the maximum value (if any) is given first.

1.　-24 at $x = 2$, -30 at $x = 0$　　**3.**　890 at $x = 10$, -870 at $x = -10$
5.　1 at $x = 1$, no minimum　　**7.**　106 at $x = -3$, -150 at $x = 5$
9.　1 at $x = 1$, 0 at $x = 0$　　**11.**　2 at $x = 7$, -1 at $x = -2$

13. 250 at $x = 1$, -54 at $x = -1$ **15.** $\frac{1}{2}$ at $x = 3$, $\frac{1}{12}$ at $x = 5$
17. $\frac{4}{5}$ at $x = -3$, 0 at $x = 1$ **19.** 0 at $x = 0$, -4 at $x = -1$
21. $3\frac{1}{2}$ at $x = \frac{1}{2}$, $2\frac{1}{2}$ at $x = 1$ **23.** no maximum, 1 at $x = 0$
25. 20 at $x = 3$, 0 at $x = 0$ **27.** 10 at $x = 3$, -8 at $x = -5$
29. 16 at $x = \pm 2$, 0 at $x = 0$

Problems 5.3

1. $L = W = 75$ m   **3.** Equilateral triangle $8 \times 8 \times 8$ cm   **5.** $(\frac{12}{13}, \frac{18}{13})$
7. $(2/\sqrt{29}, 5/\sqrt{29})$ **9.** Maximum $= \pi(35/2\pi)^2 \approx 97.5$ cm² when all wire is used for
circle; minimum ≈ 42.88 cm² when radius of circle is $35/2(\pi + 4) \approx 2.45$ cm
11. 200   **13.** choose both to be 10
15. closest distance is $30\sqrt{2}$ km after 6 hr; no
17. semicircle is $60/(4 + \pi)$ ft in diameter, rectangle is $30/(4 + \pi)$ ft high
19. $\frac{40}{7}$ ft   **21.** breadth $= \frac{1}{2}$ m, depth $= \sqrt{3}/2$ m
23. $\frac{4000}{3}$ pigeons; the integer giving the largest value is 1333
25. $a = -12$; max occurs at any t **27.** (a) $ds/dt = r - cs(t)$ for $s \geq 0$
(b) when $s = 0$ **29.** one-to-one   **31.** $f(x) = x^2/2$

Problems 5.4

1. 7500 sets   **3.** $E(10,000) = 1$ so neither an increase nor a decrease
5. 4750 bottles   **7.** $E(10,000) = 0.2$ so increase profits   **9.** $\sqrt{2000} \approx 45$ sinks
11. $E = 0.25$, so yes
13. about 8 batches per year (actually 7.9), with $\sqrt{4 \times 10^7} \approx 6325$ trucks/batch
15. $E(x) = (x - 5000)/2x$ **17.** $x = \$8000$, $y = \$2000$
23. (a) $(\alpha - b)/2(a + \beta)$ or 0, whichever is greater   (b) $E(x) = (\alpha/\beta x) - 1$
(c) when $x = \alpha/2\beta$

Problems 5.5

1. $[9, 10]$; 9.486832981   **3.** 1.584893192   **5.** 6.192582404; 0.8074175964
7. -2.090520511; 0.2438484828; 7.846672028   **9.** 7.984792473 (only real root)
11. $x_{n+1} = 2x_n - ax_n^2$ converges to $1/a$ provided that $x_0 \in (0, 2/a)$

Problems 5.6

1. $\frac{1}{6}$ **3.** $\frac{137}{12}$ **5.** $\frac{343}{6}$ **7.** $\frac{355}{3}$ **9.** $\frac{1944}{5}$ **11.** $\frac{128}{3}$ **13.** $\frac{27}{5}$ **15.** $\frac{1}{2}$
17. 108   **19.** 49   **21.** $\frac{7}{2}$ **23.** $\frac{169}{24}$ **25.** $\$1250$ **27.** $\frac{2500}{3} \approx \$833.33$
29. (a) $x = 25$ items and $p = \$25$ (b) consumer's surplus $= \frac{125}{3} \approx \$41.67$; producer's
surplus $= \$125$

Problems 5.7

1. 98.1 joules   **3.** 20 ft-lb   **5.** 45 joules   **7.** 2.5 ft
9. $\int_0^{10} g\, \delta \pi r^2 x\, dx \approx 3.85 \times 10^7$ joules   **11.** $\int_0^5 g\, \delta \pi r^2 x\, dx \approx 9.63 \times 10^6$ joules
13. $(9.81)(1000)(\pi \cdot 5^2 \cdot 10)20 \approx 1.54 \times 10^8$ joules (since all the water has to be raised
20 m)   **15.** 3.54×10^7 joules (multiply result of Problem 9 by 0.92)
17. $\int_2^7 \pi g \delta(3 - \frac{3}{7}x)^2 x\, dx \approx 6.209 \times 10^5$ joules   **19.** 525 ft-lb   **21.** 449,280 ft-lb
23. 90.76 sec   **25.** Evaluate $|\int_{-3.8 \times 10^8 m}^R Gm_1 m_2/r^2\, dr|$ and see what happens as
$R \to \infty$. The answer is $GM^2 \times 3.23158 \times 10^{-4}$ joules.   **27.** 337.5 joules
29. $\sqrt{140,000/3} \approx 216$ ft/sec ≈ 147 mi/hr
31. $v_0 = \sqrt{2gd} = \sqrt{2(3.92)(3,430,000)} \approx 5.186$ km/sec ≈ 3.222 mi/sec
33. $x = \frac{1}{2}v^2/g \approx 83.5$ ft ($v = 50$ mi/hr $= \frac{220}{3}$ ft/sec)

Review Exercises for Chapter Five

1. $\sqrt{34}/2$ m/sec   **3.** (a) Maximum is 358 at $x = 5$; minimum is -10 at $x = 1$.
(b) Maximum is $2^{1/3}$ at $x = 4$; minimum is $-2^{1/3}$ at $x = 0$.
(c) Maximum is 0 at $x = -1$; minimum is $-\frac{2}{3}$ at $x = 1$.   **5.** $(\frac{12}{13}, -\frac{18}{13})$.
7. 8 weeks (giving 2592 individuals/week)
9. ≈ 1250 units (positive root of $-0.04x^3 + 50x^2 + 1000 = 0$)   **11.** 2.2649
13. $\frac{1}{6}$ **15.** $\frac{27}{48} = 0.5625$ **17.** 4.8 joules   **19.** $3833\frac{1}{3}$ joules

CHAPTER SIX

Problems 6.1

1. 2   **3.** -3 **5.** -4π **7.** 5   **9.** $-\frac{1}{2}$ **11.** -1 **13.** $\pm\sqrt{6}$
15. $\sqrt{2}$ **17.** e^π **19.** 4   **21.** $\frac{1}{2}$ **23.** $10^{10^{-23}}$ **25.** $\frac{27}{16}$ **27.** 24
29. $\sqrt[3]{25}/2$ **31.** 3   **33.** $\frac{7}{2}\log x$ **35.** $5\ln x - 18\ln(1 + x)$
41. $999^{1000} > 1000^{999}$ (since $\log 999^{1000} \approx 2999.57 > \log 1000^{999} = 2997$)
43. (a) $L_2/L_1 = 1 + (10\log 2)/L_1$ if $I_2 = 2I_1$
(b) $I_2/I_1 = I_1/I_0 = I_1 \times 10^{12}$ if $L_2 = 2L_1$ (c) $I_2/I_1 = (I_1/I_0)^5$ if $L_2 = 6L_1$
45. (a) 1.138828   (b) 0.690731   (c) $(e^4)(e^{0.13}) \approx 62.1779$ (d) $e^{-3}/e^{-0.37} \approx 0.072079$
(e) 123.97, 0.23693   **47.** (a) $100! \approx 9.32 \times 10^{157}$; $200! \approx 7.88 \times 10^{374}$

Problems 6.2

1. $14/(\ln 2)$ **3.** 3   **5.** 2   **7.** 15

9.
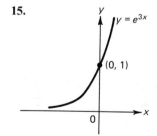

11.
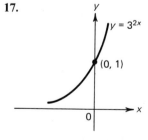

13.
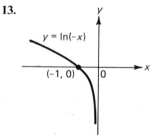

15.
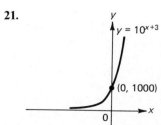

17.
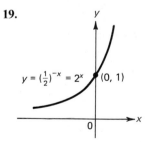

19.

21.
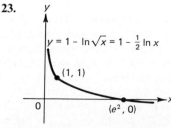

23.

25. $\ln x$ **27.** $[(1/x_1) + (1/x_2) + \cdots + (1/x_n)]\,\Delta x \approx \int_1^2 (1/t)\,dt = \ln 2$, where
$x_i = 1 + i/n$ and $\Delta x = 1/n$ **29.** $W = -500 - 1 \cdot R \cdot 300 \cdot \ln 2 \approx -913.185$ cal

Problems 6.3

1. $-1/x$ **3.** $\frac{1}{4}\ln(x^4 + 1) + C$ **5.** $3e^{\sqrt[3]{x}} + C$ **7.** $5x^4$
9. $x^{\sqrt{x}-(1/2)}(\frac{1}{2}\ln x + 1)$ **11.** $(x/3)(2\ln x - 1)/(\ln x)^2$
13. $[(2 - x)/(2x^2\sqrt{x} - 1)]e^{\sqrt{x}-1/x}$ **15.** $-\frac{1}{2}\ln^2(1/x) + C$ **17.** $\frac{7}{3}$ **19.** $\ln\frac{9}{7}$
21. $3x^2/(x^3 + 3)$ **23.** $1/[x(\ln x - 1)]$ **25.** $1/(x\ln x)$ **27.** $\ln 3$
29. $-1/(4\ln^4 x)$ **31.** $\ln|\ln|\ln|x||| + C$ **33.** $e^{x+e^x}(1 + e^x)$

35.
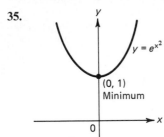
$y = e^{x^2}$
$(0, 1)$ Minimum

37.
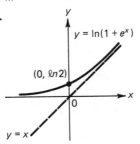
$y = \ln(1 + e^x)$
$(0, \ln 2)$
$y = x$

39.
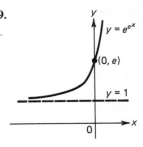
$y = e^{e^x}$
$(0, e)$
$y = 1$

41.
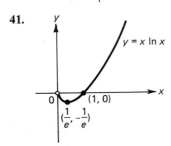
$y = x\ln x$
$(1, 0)$
$(\frac{1}{e}, -\frac{1}{e})$

43.
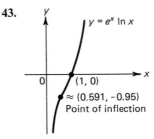
$y = e^x\ln x$
$(1, 0)$
$\approx (0.591, -0.95)$
Point of inflection

45. $1 + e^{-1}$ **47.** $1 + e^3 - 3\ln 3$ **49.** $(e - 1)/2$

51. $\sqrt[5]{\dfrac{x^3 - 3}{x^2 + 1}}\left(\dfrac{1}{5}\right)\left(\dfrac{3x^2}{x^3 - 3} - \dfrac{2x}{1 + x^2}\right)$

53. $\sqrt{x}\sqrt[3]{x} + 2\sqrt[5]{x} - 1\left[\dfrac{1}{2x} + \dfrac{1}{3(x + 2)} + \dfrac{1}{5(x - 1)}\right]$ **57.** $2x(\ln 3)(3^{x^2+5})$

59. $\dfrac{(\log_5 2x)^2}{2\log_5 e} + C$ **61.** $3^{\sqrt{x}}\left(\dfrac{2}{\ln 3}\right) + C$ **63.** $\dfrac{3(\log_{1/2} x)^2}{2\log_{1/2} e} + C$

65. $\dfrac{3^{\ln x}}{\ln 3} + C$ **67.** $-\dfrac{1}{2\log_{14} e}(\log_{14} x)^{-2} + C$ **69.** $(\log_8 e)\left[\dfrac{9 - x^4}{x(x^4 + 3)}\right]$

71. (a) $0.45e^{-0.1} \approx 0.9048$ (b) 1000 units **73.** (a) at $x = \sqrt{1.5} \approx 1.225$ m
(b) $\frac{3}{2}\{1 - e^{-25/3}\} \approx 1.49964$ kg

Problems 6.4

1. 62,500 after 20 days; 156,250 after 30 days **3.** $250,000e^{0.6} \approx 455,530$ in 1980;
$250,000e^{1.8} \approx 1,512,412$ in 2000 **5.** (a) $-10 + 160e^{-2\alpha} \approx -4.14$ °C,
where $\alpha = 2\ln\frac{16}{7}$ (b) $(\ln 16)/\alpha \approx 1.6769$ min ≈ 100.62 sec
7. $[(\ln 0.7)/(\ln 0.5)]\,5580 \approx 2871$ yr **9.** (a) $25(0.6)^{24/10} \approx 7.3367$ kg
(b) $[(\ln 0.02)/(\ln 0.6)]\,10 \approx 76.58$ hr **11.** $(\ln 0.5)/-\alpha \approx 4.62 \times 10^6$ yr
13. $\beta = \frac{1}{1500}\ln(845.6/1013.25) \approx -1.205812 \times 10^{-4}$ (a) 625.526 mb (b) 303.416 mb
(c) 429.033 mb (d) 348.632 mb (e) $-(1/\beta)\ln 1013.25 \approx 5.74$ km **15.** (a) 10,000
(b) 7 (c) 215 hr ≈ 9 days

Problems 6.5

1. $2800　　**3.** $5000(1 + 0.07/12)^{(12)(8)} \approx $8739.13

5. 79.1% annually; 81.4% quarterly　　**7.** Choose $5\frac{3}{8}$% compounded semiannually.

9. $i = -\frac{1}{8} \ln \frac{750}{1000} \approx 0.03596 = 3.596$ percent　　**11.** $(\ln 3)/15 \approx 0.07324 = 7.324\%$

13. $10,000(1.09)^{-5} \approx $6499.31　　**15.** $8000e^{-0.4} \approx $5362.56　　**17.** $t = 1.4$ yr (must solve $[(0.1/\sqrt{0.2t}) - 0.12]e^{\sqrt{0.2t}} = 0.12$, solution obtained by trial and error)

19. $25,000/[(1.0125^{60} - 1)/0.0125] \approx $282.25 quarterly $\approx $94.08 monthly

21. $250,000e^{-(0.11)(3)} \approx $179,730.93

23. (a) $(6000/0.005)[1 - (1/1.005)^{(12)(20)}] \approx $837,484.63　(b) No, invest all immediately to obtain return of $750,000(1.005)^{240} \approx $2,482,653.36　　**25.** Solve $750,000(1 + i)^{240} = (6000/i)\{1 - [1/(1 + i)]^{240}\}$ to obtain $i \approx 0.001834 = 0.1834$ percent monthly $= 2.2008$ percent annually.　　**27.** $400\left[\dfrac{0.005}{1 - (1/1.005)^{24}}\right] \approx $17.73

29. $25,000\left[\dfrac{0.00625}{1 - (1/1.00625)^{(12)(25)}}\right] \approx $184.75

Review Exercises for Chapter Six

1. 2　　**3.** $\frac{1}{2}$　　**5.** $10^{10^{-9}}$

7. $-\frac{5}{6}[\frac{1}{2} \ln(x^3 - 1) + \frac{4}{3} \ln(x^5 + 1) - \frac{1}{2} \ln(x - 3) - \frac{1}{2} \ln(x - 2)]$

9.　　　　　　　　　　　　　　　**11.** $2x/(1 + x^2)$　　**13.** $2(x + 1)e^{(x+1)^2}$

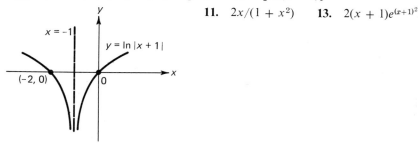

15. $\frac{2}{5}e^{\sqrt{x}} + C$　　**17.** $e^x/(e^x - 5)$　　**19.** $\frac{1}{2}e^{-1/x^2} + C$

21. $\dfrac{\sqrt[3]{x + 3}(x - 5)}{\sqrt{(x + 1)(x + 2)}}\left[\dfrac{1}{3(x + 3)} + \dfrac{1}{x - 5} - \dfrac{1}{2(x + 1)} - \dfrac{1}{2(x + 2)}\right]$

23. $[2/(3 \log_5 e)](\log_5 x)^{3/2} + C$　　**25.** $3x^2$

27.　　　　　　　　　　　　　　　**29.**

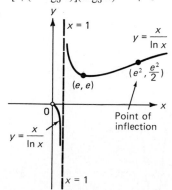
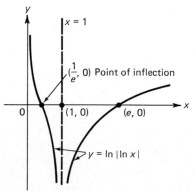

31. $(\ln 2)/1.5 \approx 4.62$ years　　**33.** 8451.3 years

35. (a) $(\ln 0.25)/(\ln 0.8) \approx 6.212$ weeks　　(b) $(\ln 0.05)/(\ln 0.8) \approx 13.425$ weeks

37. $10,000e^{(0.06)8} \approx $16,160.74　　**39.** $3778.92

CHAPTER SEVEN

Problems 7.1

1. $5\pi/6$ **3.** $5\pi/3$ **5.** $4\pi/5$ **7.** $15°$ **9.** $22.5°$ **11.** $-60°$ **13.** $90°$

Problems 7.2

In Problems 1–15 the value of $\sin\theta$ is given first.

1. $0, 1$ **3.** $-\frac{1}{2}, -\sqrt{3}/2$ **5.** $(\sqrt{2}/4)(1 + \sqrt{3}), (\sqrt{2}/4)(\sqrt{3} - 1)$

7. $(-\sqrt{2 - \sqrt{3}}/2, -\sqrt{2 + \sqrt{3}}/2$ **9.** $-\sqrt{2 - \sqrt{3}}/2, \sqrt{2 + \sqrt{3}}/2$

11. $\sqrt{2 - \sqrt{2 + \sqrt{2}}}/2, \sqrt{2 + \sqrt{2 + \sqrt{2}}}/2$ **13.** $\sqrt{2 + \sqrt{2}}/2, \sqrt{2 - \sqrt{2}}/2$

15. $-\sqrt{2 - \sqrt{2 + \sqrt{3}}}/2, \sqrt{2 + \sqrt{2 + \sqrt{3}}}/2$

29. amplitude $= 2$

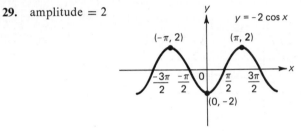

31. amplitude $= 4$

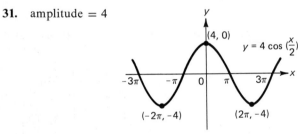

33.

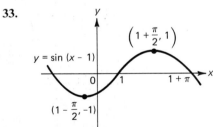

35. period $= 2\pi/3$, amplitude $= 3$

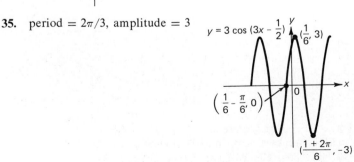

Problems 7.3

In Problems 1–15 the answers are given in the order $\tan x$, $\cot x$, $\sec x$, $\csc x$.

1. 0, undefined, 1, undefined   **3.** $1/\sqrt{3}$, $\sqrt{3}$, $-2/\sqrt{3}$, -2

5. $2 + \sqrt{3} = \sqrt{(2 + \sqrt{3})/(2 - \sqrt{3})}$, $2 - \sqrt{3}$, $2/\sqrt{2 - \sqrt{3}}$, $2/\sqrt{2 + \sqrt{3}}$

7. $2 - \sqrt{3}$, $2 + \sqrt{3}$, $-2/\sqrt{2 + \sqrt{3}}$, $-2/\sqrt{2 - \sqrt{3}}$ **9.** $-2 + \sqrt{3}$, $-2 - \sqrt{3}$,
$2/\sqrt{2 + \sqrt{3}}$, $-2/\sqrt{2 - \sqrt{3}}$ **11.** $\sqrt{2 - \sqrt{2 + \sqrt{2}}}/\sqrt{2 + \sqrt{2 + \sqrt{2}}}$,
$\sqrt{2 + \sqrt{2 + \sqrt{2}}}/\sqrt{2 - \sqrt{2 + \sqrt{2}}}$, $2/\sqrt{2 + \sqrt{2 + \sqrt{2}}}$, $2/\sqrt{2 - \sqrt{2 + \sqrt{2}}}$

13. $\sqrt{2} + 1 = \sqrt{2 + \sqrt{2}}/\sqrt{2 - \sqrt{2}}$, $\sqrt{2} - 1$, $2/\sqrt{2 - \sqrt{2}}$, $2/\sqrt{2 + \sqrt{2}}$

15. $-\sqrt{2 - \sqrt{2 + \sqrt{3}}}/\sqrt{2 + \sqrt{2 + \sqrt{3}}}$, $-\sqrt{2 + \sqrt{2 + \sqrt{3}}}/\sqrt{2 - \sqrt{2 + \sqrt{3}}}$,
$2/\sqrt{2 + \sqrt{2 + \sqrt{3}}}$, $-2/\sqrt{2 - \sqrt{2 + \sqrt{3}}}$

17. period $= 6\pi$

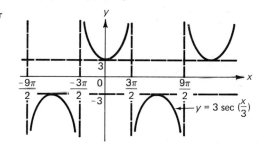

$y = 3 \sec\left(\dfrac{x}{3}\right)$

19. period $= \pi/3$

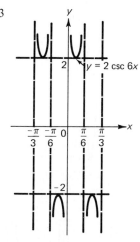

$y = 2 \csc 6x$

21. period $= 2\pi/5$

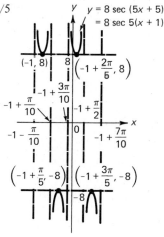

$y = 8 \sec (5x + 5)$
$= 8 \sec 5(x + 1)$

Problems 7.4

1. $\sin\theta = \frac{4}{11}\sqrt{6}$, $\cos\theta = \frac{5}{11}$, $\tan\theta = 4\sqrt{6}/5$, $\csc\theta = 11/4\sqrt{6}$, $\sec\theta = \frac{11}{5}$,
$\cot\theta = 5/4\sqrt{6}$ **3.** $\sin\theta = -\sqrt{3}/2$, $\cos\theta = \frac{1}{2}$, $\tan\theta = -\sqrt{3}$, $\csc\theta = -2/\sqrt{3}$,
$\sec\theta = 2$, $\cot\theta = -1/\sqrt{3}$ **5.** $\sin\theta = \frac{1}{5}$, $\cos\theta = -2\sqrt{6}/5$, $\tan\theta = -1/2\sqrt{6}$,
$\csc\theta = 5$, $\sec\theta = -5/2\sqrt{6}$, $\cot\theta = -2\sqrt{6}$ **7.** $\sin\theta = -\frac{2}{3}$, $\cos\theta = \sqrt{5}/3$,
$\tan\theta = -2/\sqrt{5}$, $\csc\theta = -\frac{3}{2}$, $\sec\theta = 3/\sqrt{5}$, $\cot\theta = -\sqrt{5}/2$ **9.** $\sin\theta = -10/\sqrt{101}$,
$\cos\theta = -1/\sqrt{101}$, $\tan\theta = 10$, $\csc\theta = -\sqrt{101}/10$, $\sec\theta = -\sqrt{101}$, $\cot\theta = \frac{1}{10}$
11. $\sin\theta_2 = \sqrt{2}/4$; $\theta_2 \approx 20.7°$ **13.** $(1/\sqrt{3})c = \sqrt{3} \times 10^{10}$ cm/sec
15. $\cos A = \frac{68}{72}$, $A \approx 19.2°$; $\cos B = \cos C = \frac{4}{24}$; $B = C \approx 80.4°$
17. $B = (5 \sin 23°)/(\sin 72°) \approx 2.054$; $C = (5 \sin 85°)/(\sin 72°) \approx 5.237$

Problems 7.5

1. $\frac{1}{2}$ **3.** 16   **5.** 32   **7.** 0   **9.** 0   **11.** $6x \cos 3x^2$

13. $9x^2 (\sec^3 x^3)(\tan x^3)$ **15.** $(x \cos x^2)/\sqrt{1 + \sin x^2}$ **17.** $\cot x$ **19.** $\csc x$

21. $-(\csc^2 \sqrt{x} + 1)/(2\sqrt{x} + 1)$ **23.** $e^x(\cos x - \sin x)$ **25.** $-\sin x[\cos(\cos x)]$

27. $-e^x \sin e^x$ **29.** $(\cos x)e^{\sin x}$ **31.** $10 \sec x(\sec x + \tan x)^{10}$

33. $(\sin x)^x[\ln(\sin x) + x \cot x]$ **35.** $\dfrac{1}{2\sqrt{x}} - \dfrac{\csc^2 \sqrt{x}}{2\sqrt{x}} = -\dfrac{\cot^2 \sqrt{x}}{2\sqrt{x}}$

37. $(\sec x)(1 + x \tan x)$ **39.** $-\frac{10}{7}(\sin 2x)(\cos 2x)^{-2/7}$

41. $\dfrac{dy}{dx} = -\dfrac{\cos x}{\sin y}\left(= \pm\dfrac{\cos x}{\sqrt{1 - \cos^2 y}} = \pm\dfrac{\cos x}{\sqrt{1 - \sin^2 x}} = \pm 1\right)$

43. $\dfrac{\sin(x - y) + \cos(x + y)}{\sin(x - y) - \cos(x + y)}$ **45.** $\dfrac{\cos x \cos y - \sec^2 x \sec y}{\tan x \sec y \tan y + \sin x \sin y}$

47. $\dfrac{-\csc^2(x + y)}{1 + \csc^2(x + y)} = \dfrac{-1}{1 + \sin^2(x + y)}$ **49.** $\dfrac{[\csc(x + y)][\cot(x + y)] - 2\csc^2(2x + y)}{\csc^2(2x + y) - [\csc(x + y)][\cot(x + y)]}$

51.

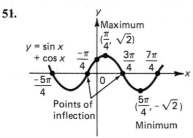

53.

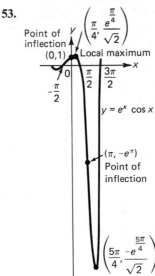

55.

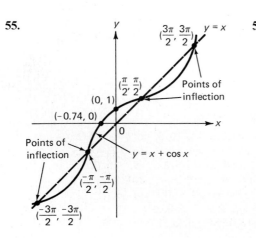

57.

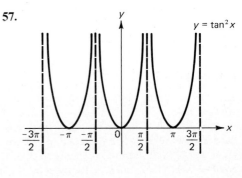

63.

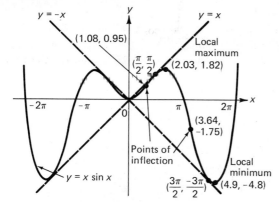

65. $\dfrac{d^n \cos x}{dx^n} = \begin{cases} \cos x & \text{if } n = 4k, \\ -\sin x & \text{if } n = 4k+1, \\ -\cos x & \text{if } n = 4k+2, \\ \sin x & \text{if } n = 4k+3 \end{cases}$

67. $\sqrt{3}/2 - \pi/360 \approx 0.8573$ (actual value ≈ 0.8572)

69. $-2/\sqrt{3} + \pi/270 \approx -1.14307$ (actual value ≈ -1.14335)

71. $dx/dt = (dx/d\theta)(d\theta/dt) = -13 \sin\theta \, d\theta/dt = -100\pi$ cm/min; $dy/dt = 240\pi$ cm/min

73. $12,000\pi$ m/sec (since $l/300 = \tan\theta$ and $d\theta/dt = 10\pi$)

75. $\cos\theta = (0.5)^4$ or $\theta = 86.4°$ **77.** they differ by a constant

Problems 7.6

1. $-\frac{1}{3}\cos 3x + C$ **3.** $\frac{3}{4}$ **5.** $\frac{1}{4}\sin^2 2x + C = -\frac{1}{4}\cos^2 2x + C_1$ **7.** $1/3\sqrt{2}$

9. $\frac{1}{4}\sec 4x + C$ **11.** $\pi/9$ **13.** $-\frac{2}{3}\cos^{3/2}x + C$ **15.** $-\frac{1}{2}\cot x^2 + C$

17. $(x/2) + [(\sin 2x)/4] + C$ **19.** $\frac{5}{6}(\sec x)^{6/5} + C$ **21.** $-\frac{1}{2}\csc x^2 + C$

23. $(2^{5/2} - 1)/5$ **25.** $\ln|\tan x| + C$ **27.** $(2/\sqrt{\csc x}) + C = 2\sqrt{\sin x} + C$

29. $\sqrt{x} + \frac{1}{2}\sin(2\sqrt{x}) + C$ **31.** $\ln(1 + \sin x) + C$ **33.** $\ln|\sin(2 + x)| + C$

35. $-\frac{1}{2}e^{-2\tan x} + C$ **39.** Both answers are essentially the same since $\sin^2 x$ and $-\cos^2 x$ differ only by a constant.   **41.** 2   **43.** $\pi/2$ **45.** $3\pi/2$

47. $\int_0^{\pi/4}(1 - \tan x)\,dx = (\pi/4) - \ln\sqrt{2}$ **49.** $\frac{1}{2}$ **51.** $1925 + (1000/\pi) \approx \$2243.31$

Problems 7.7

1. $\pi/3$ **3.** $-\pi/6$ **5.** $\pi/6$ **7.** $-\pi/4$ **9.** 0   **11.** $\frac{4}{5}$ **13.** $3/\sqrt{34}$

15. $\sin^{-1} 5$ is undefined   **17.** $\sqrt{1 - x^2}$ **21.** $3/\sqrt{1 - 9x^2}$

23. $2(x - 4)/[1 + (x - 4)^4]$ **25.** $\frac{1}{2}\tan^{-1}x^2 + C$ **27.** $1/[x(1 + \ln^2 x)]$

29. 0   **31.** $-\dfrac{1}{2}\tan^{-1}\left(\dfrac{e^{-x}}{2}\right) + C$ **33.** $\dfrac{-e^{-x}}{\sin^{-1}(e^{-x})\sqrt{1 - e^{-2x}}}$

35. $2x\cos^{-1}(1 - x) + x^2/\sqrt{2x - x^2}$ **37.** $\tan^{-1}(x - 1) + C$ **39.** $\pi/2$

41. $-\frac{1}{2}[\cos^{-1}x]^2 + C$ **43.** $\tan^{-1}(\sec x) + C$ **45.** $\pi/4$ **47.** $2\sqrt{3} - 2\pi/3$

59. (a) $\pi/3$ (b) $2\pi/3$ (c) $\pi/4$ (d) $\pi/6$ (e) 0 (f) π **61.**

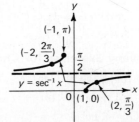

69. $\dfrac{4}{|4x + 2| \sqrt{(4x + 2)^2 - 1}}$ **71.** $-1/(2x \sqrt{x - 1})$, $x > 1$

73. $\frac{1}{3}(\sec^{-1} \sqrt{2} - \sec^{-1} 2) = -\pi/36$ **75.** $-e^x/(1 + e^{2x})$

77. $(1/a)\sec^{-1} |x/a| + C$ **79.** $\sec^{-1} 2 - \sec^{-1} 1 = \pi/3$ **81.** $-\dfrac{\csc^2 x}{|\cot x| \sqrt{\cot^2 x - 1}}$

83. $\dfrac{-1}{x(1 + \ln^2 x) \sqrt{1 - (\cot^{-1} \ln x)^2}}$ **87.** $\pi/4$ for $x > 1$ and $5\pi/4$ for $x < 1$

Problems 7.8

1. $x = \sin t$ **3.** period $= 16\pi/(\pi - 1)$, frequency $= (\pi - 1)/16\pi$, spring constant $= 5[(\pi - 1)/8]^2$ **5.** $x(t) = A \cos(\omega t + \delta)$, where $\omega = 2\pi/T$, $A = \sqrt{x_0^2 + (v_0 T/2\pi)^2}$, and $\delta = \tan^{-1}(-v_0 T/2\pi x_0)$ **9.** $x = A \cos(4t + \delta)$, $y = -2A \sin(4t + \delta)$, where $\delta = \tan^{-1}(-5) \approx -1.3734$ and $A = 1000 \sqrt{26} \approx 5099$. The prey will be extinct within $-\frac{1}{4} \tan^{-1}(-5) \approx 0.3434$ weeks.

Problems 7.9

7. $4 \cosh(4x + 2)$ **9.** $-\dfrac{[\operatorname{csch}(\ln x)][\coth(\ln x)]}{x} = -\dfrac{2(1 + x^2)}{(x^2 - 1)^2}$

11. $2 \tanh \sqrt{x} + C$ **13.** $\cosh x \cos(\sinh x)$ **15.** $2 \tanh x \operatorname{sech}^2 x \, e^{\tanh^2 x}$

17. $\sinh x + [(\sinh^3 x)/3]$ **19.** $\ln |\sinh x| + C$ **21.** $-\cosh(1/x) + C$

23. undefined because $\sin^{-1}(\cosh x)$ is only defined when $x = 0$ (otherwise $\cosh x > 1$)   **25.** $\frac{1}{2}\ln|1 + \sinh 2x| + C$ **27.** $-\frac{3}{11} \operatorname{sech}^{11/3} x + C$

29. $[(\cosh x)/x + (\sinh x)(\ln x)]x^{\cosh x}$ **31.** $2a^2 \sinh 1 \approx 2.35a^2$

45. $\sqrt{29}/5$ **47.** $\pm\sqrt{55}/8$ **49.** $\sqrt{5}/2$ **51.** $\sinh x = \frac{1}{5}$, $\cosh x = \sqrt{26}/5$, $\tanh x = 1/\sqrt{26}$, $\coth x = \sqrt{26}$, $\operatorname{sech} x = 5/\sqrt{26}$

Problems 7.10

1. $3/\sqrt{(3x + 2)^2 + 1}$ **3.** $1/[x(1 - \ln^2 x)]$, $1/e < x < e$

5. $(\sec \sinh^{-1} x)(\tan \sinh^{-1} x)/\sqrt{x^2 + 1}$ **7.** $x/\sqrt{x^2 + 1}$

9. $1/2\sqrt{(x^2 + 1)} \sinh^{-1} x$

11. $\dfrac{[(\cosh^{-1} x)/\sqrt{x^2 + 1}] - [(\sinh^{-1} x)/\sqrt{x^2 - 1}]}{(\cosh^{-1} x)^2}$, $x > 1$.   **19.** (a) $\ln(4 + \sqrt{17})$;

(b) $\ln(3 + \sqrt{8})$;   (c) $\frac{1}{2} \ln 3$;   (d) $\ln(\sqrt{2} - 1)$

23. $\frac{1}{2} \cosh^{-1}(x^2/2) + C = \frac{1}{2} \ln(x^2 + \sqrt{x^4 - 4}) + C$

25. $\frac{1}{4} \tanh^{-1} x + C = \frac{1}{8} \ln[(2 + x^2)/(2 - x^2)] + C$

27. $\cosh^{-1} e^x/3 + C = \ln(e^x + \sqrt{e^{2x} - 9}) + C$

29. $\sinh^{-1}(\sin x) + C = \ln(\sin x + \sqrt{1 + \sin^2 x}) + C$

31. $\sinh^{-1}(\frac{1}{2} \sec x) + C = \ln(\sec x + \sqrt{4 + \sec^2 x}) + C$

33. $\cosh^{-1}(2x + 1) + C = \ln(2x + 1 + \sqrt{4x^2 + 4x}) + C$

Review Exercises for Chapter Seven

1. (a) $\pi/4$ (b) $5\pi/6$ (c) $-\pi/3$ (d) $-5\pi/12$ **3.** (a) $-\sqrt{3}/2$ (b) $(\sqrt{3} - 1)/2\sqrt{2}$ (c) $\sqrt{2 - \sqrt{3}}/2$ (d) $-1/\sqrt{3}$ (e) $-1/\sqrt{3}$ (f) $-2/\sqrt{3}$ (g) $2/\sqrt{3}$ (h) 0 (i) -1

5. period $= \pi/5$

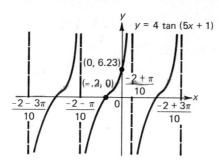

$y = 4 \tan (5x + 1)$

7. $5/\sqrt{39}$ **9.** $16x \cos(8x^2 + 2)$ **11.** undefined since $e^x + 1 > 1$.
13. $\frac{1}{3}[\cos x - (1/\sqrt{1 - x^2})](\sin x + \cos^{-1} x)^{-2/3}$ **15.** $-5 \ln \sqrt{3}/2$
17. $2 \sec \sqrt{x} + C$ **19.** $\frac{3}{2} \tan^{-1} x^2 + C$ **21.** $-2x/(x^4 + 1)$
23. $e^x[(x/\sqrt{1 - x^2}) + \sin^{-1} x + x \sin^{-1} x]$ **25.** $-\cot e^x + C$
27. $-\sin x \sec^2(\cos x)$ **29.** $[\sin^{-1} x \cos x - (\sin x/\sqrt{1 - x^2})]/(\sin^{-1} x)^2$
31. $(x/2) - [\sin(6x + 4)/12] + C$ **33.** (a) $\sqrt{51}/10$ (b) $-\sqrt{15}$ (c) $\sqrt{13}/3$
35. $(2x + 3) \cosh(x^2 + 3x)$ **37.** $-3 \operatorname{sech}^3 x \tanh x \, e^{\operatorname{sech}^3 x}$ **39.** $\frac{1}{4} \cosh x^4 + C$
41. $(\sinh x) \cos(\cosh x)$ **43.** $(\csc^2 x) \csch^2(\cot x)$ **45.** $-\frac{3}{4} \csch^{4/3} x + C$
47. $-(\sin x)/\sqrt{1 + \cos^2 x}$ **49.** $-1/2x\sqrt{1 + x}$
51. $\cosh^{-1}[(\tan x)/2] + C = \ln(\tan x + \sqrt{\tan^2 x - 4}) + C$ **53.** $\pm\sqrt{3}$

CHAPTER EIGHT

Problems 8.2

1. $\frac{1}{3}xe^{3x} - \frac{1}{9}e^{3x} + C$ **3.** $64e - 128$ **5.** $(x^8/8) \ln x - (x^8/64) + C$
7. $(x^2/2) \sin 2x + (x/2) \cos 2x - \frac{1}{4} \sin 2x + C$
9. $\frac{-4}{3}x(1 - x/2)^{3/2} - \frac{16}{15}(1 - x/2)^{5/2} + C$ **11.** $(\pi/8) - \frac{1}{4} \ln 2$
13. $\frac{3}{4} \sinh 2 - \frac{1}{2} \cosh 2$ **15.** $(\pi/2) - 1$ **17.** $(x/2) [\sin(\ln x) + \cos(\ln x)] + C$
19. $\frac{1}{15}(\sin 4x \cos x - 4 \cos 4x \sin x) + C$ **21.** $\frac{1}{3}$
23. $\frac{1}{13}e^{3x}(3 \sin 2x - 2 \cos 2x) + C$ **25.** $\frac{1}{3}x^3 \tan^{-1} x - \frac{1}{6}x^2 + \frac{1}{6} \ln(1 + x^2) + C$ (note
that $x^3/(1 + x^2) = x - x/(1 + x^2))$
27. $\frac{5}{21} \sinh 2x \cosh 5x - \frac{2}{21} \sinh 5x \cosh 2x + C$ **31.** $1 - 6e^{-5}$ kg

33. (a)   (b) $2 - 2/e$

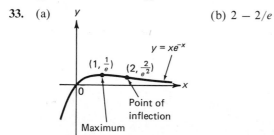

$y = xe^{-x}$

$(1, \frac{1}{e})$ $(2, \frac{2}{e^2})$

Point of
inflection

Maximum

Problems 8.3

1. $\frac{4}{15}$ **3.** $\pi/8 + \frac{1}{4}$ **5.** $\frac{3}{8}x + \dfrac{\sin 4x}{8} + \dfrac{\sin 8x}{64} + C$

7. $\frac{1}{2} \cos x - \frac{1}{10} \cos 5x + C$ **9.** $(-1/\sqrt{2}) \cos(x/\sqrt{2}) - (1/3\sqrt{2}) \cos(3x/\sqrt{2}) + C$
11. $\frac{1}{180} \sin 90x + \frac{1}{220} \sin 110x + C$ **13.** $\frac{1}{10}(\sec 5x \tan 5x + \ln |\sec 5x + \tan 5x|) + C$
15. $3\pi/256$ **17.** $(2\sqrt{2} + 2)/15$ **19.** $\frac{-1}{3} \cos^3 x + \frac{2}{5} \cos^5 x - \frac{1}{7} \cos^7 x + C$

21. $-\frac{1}{2} - 2\ln(1/\sqrt{2}) = -\frac{1}{2} + \ln 2$ **23.** $\frac{1}{6}\tan^6 x + \frac{1}{4}\tan^4 x + C$

25. $\frac{1}{2}\ln 2 - \frac{1}{4}$ **27.** $2\sqrt{\sin x} - \frac{2}{5}\sin^{5/2} x + C$

37. $\frac{-1}{4}\cot^4 x + \frac{1}{2}\cot^2 x + \ln|\sin x| + C$

39. $\frac{-1}{2}\csc x \cot x - \frac{1}{2}\ln|\csc x + \cot x| + C$ **41.** $\frac{-1}{7}\csc^7 x + \frac{1}{5}\csc^5 x + C$

43. $\frac{-1}{2}\cot 2x - \frac{1}{6}\cot^3 2x + C$

Problems 8.4

1. $\frac{1}{4}\int_0^{81} e^u u^{-3/4}\,du$ **3.** $\frac{1}{3}\int_1^2 z^{1/2}(z-1)^{-4/5}\,dz$ **5.** $2\int_0^1 [v/(3+v)]\,dv$

7. $\int_0^{\pi/2}\cos^4\theta\,d\theta$ **9.** $\int_0^4 (16-y^2)y^2\,dy$ **11.** $243\int_0^{\pi/2}\cos^3\theta\sin^2\theta\,d\theta$

13. $\int_0^{\sinh^{-1}1}\cosh^2 t\,dt$

Problems 8.5

1. $2 - \sqrt{3}$ **3.** $-(x/2)\sqrt{4-x^2} + 2\sin^{-1}(x/2) + C$ **5.** $\frac{64}{15} - \frac{11}{5}\sqrt{3}$

7. $\ln|x + \sqrt{x^2-4}| + C$ **9.** $4\sqrt{3} - \frac{4}{3} + 2\ln[\sqrt{3}/(2+\sqrt{3})]$ **11.** $2(\sqrt{2}-1)$

13. $\frac{1}{2}\tan^{-1}(x/2) + C$ **15.** $\frac{1}{4}[24\ln(x+\sqrt{x^2+4}) + 10x\sqrt{x^2+4} + x^3\sqrt{x^2+4}] + C$

17. $\frac{3}{4} + \ln 2$ **19.** $\pi/4\sqrt{2}$ **21.** $2\sqrt{55} - \sqrt{7} + \frac{9}{4}\ln[(4+\sqrt{7})/(8+\sqrt{55})]$

23. $(2/\sqrt{3}) - (4/\sqrt{15}) + \ln[(4+\sqrt{15}/(2+\sqrt{3})]$ **25.** $\frac{1}{8}[(4/\sqrt{7}) - \frac{5}{4}]$

27. $-(x/4)(a^2-x^2)^{3/2} + (a^2/8)[x\sqrt{a^2-x^2} + a^2\sin^{-1}(x/a)] + C$

29. $\frac{1}{2}[x\sqrt{1+a^2x^2} + (1/a)\ln(ax + \sqrt{a^2x^2+1})] + C$

31. $\sqrt{5x^2-9} - 3\cos^{-1}(3/\sqrt{5}x) + C$

33. $\sqrt{9-5x^2} - 3\ln|(3+\sqrt{9-5x^2})/\sqrt{5}x| + C$ **35.** $(1/|a|)\sec^{-1}(x/a) + C$

37. $\frac{1}{7}(x^2-a^2)^{7/2} + (a^2/5)(x^2-a^2)^{5/2} + C$

39. $\frac{1}{3}(a^2-x^2)^{3/2} - a^2\sqrt{a^2-x^2} + C$

41. $-\dfrac{\sqrt{a^2-x^2}}{2a^2x^2} - \dfrac{1}{2a^3}\ln\left|\dfrac{a+\sqrt{a^2-x^2}}{x}\right| + C$ **43.** $\dfrac{-x}{a^2\sqrt{x^2-a^2}} + C$

45. $\dfrac{-1}{a}\ln\left|\dfrac{a+\sqrt{x^2+a^2}}{x}\right| + C$ **47.** $\dfrac{-x}{\sqrt{x^2+a^2}} + \ln|x + \sqrt{x^2+a^2}| + C$

49. $x - a\tan^{-1}(x/a) + C$ **51.** $\dfrac{-1}{a^4 x} - \dfrac{x}{2a^4(x^2+a^2)} - \dfrac{3}{2a^5}\tan^{-1}\left(\dfrac{x}{a}\right) + C$

53. $(1/b)\ln(bx + \sqrt{b^2x^2+a^2}) + C$

55. Note that they have the same derivative. **57.** $25\pi/2\sqrt{2}$ **59.** (a) the part of the unit circle lying in the first quadrant (b) $\pi/4$

Problems 8.6

1. $\ln|(x-1)/x| + C$ **3.** $\frac{5}{2}\ln\frac{4}{3} - \frac{1}{2}\ln 2$ **5.** $\frac{3}{2}\ln 3 - \frac{5}{2}\ln 2$

7. $4\ln|x-1| - 14\ln|x-2| + 11\ln|x-3| + C$

9. $\dfrac{x^3}{3} + x + \dfrac{1}{2}\ln\left|\dfrac{x-1}{x+1}\right| + C$ **11.** $\dfrac{-1}{x-4} - \dfrac{2}{(x-4)^2} + C$

13. $\dfrac{1}{4}\ln\left|\dfrac{x+2}{x}\right| - \dfrac{1}{4x} - \dfrac{1}{4(x+2)} + C$ **15.** $\ln|x| - \dfrac{1}{x} + \dfrac{2}{3x^3} + C$

17. $-\dfrac{1}{6}\ln\left|\dfrac{x-1}{x+1}\right| + \dfrac{1}{12}\ln\left|\dfrac{x-2}{x+2}\right| + C$

19. $\frac{1}{2}x^2 - 2x + 20\ln|x+3| - \frac{22}{3}\ln|x+2| + \frac{13}{3}\ln|x-4| + C$ **21.** $8\ln\frac{5}{4} - \frac{3}{2}$

23. $2\ln|x| - \dfrac{3}{4}\ln|x+1| - \dfrac{3}{2(x-1)} - \dfrac{5}{4}\ln|x-1| + C$

25. (d) $\ln|x-1| - \dfrac{2}{x-1} - \dfrac{3}{2(x-1)^2} + C$

29. (a) $-\dfrac{1}{2(x-1)^2} - \dfrac{3}{(x-1)^3} - \dfrac{13}{2(x-1)^4} - \dfrac{34}{5(x-1)^5} - \dfrac{3}{(x-1)^6} - \dfrac{1}{(x-1)^7} + C$

(b) $20 \ln \left| \dfrac{x+1}{x} \right| - \dfrac{20}{x+1} - \dfrac{10}{(x+1)^2} - \dfrac{7}{(x+1)^3} - \dfrac{7}{2(x+1)^4} - \dfrac{32}{5(x+1)^5} -$

$\dfrac{2}{(x+1)^6} - \dfrac{19}{7(x+1)^7} - \dfrac{15}{4(x+1)^8} - \dfrac{4}{3(x+1)^9} + C$

(c) $\ln|x+2| + \dfrac{13}{x+2} - \dfrac{73}{2(x+2)^2} + \dfrac{229}{3(x+2)^3} - \dfrac{433}{4(x+2)^4} + \dfrac{493}{5(x+2)^5} -$

$\dfrac{313}{6(x+2)^6} + \dfrac{85}{7(x+2)^7} + C$ **31.** $\frac{2}{3}$ **33.** $2 - \ln 4$

Problems 8.7

1. $\frac{1}{2}\ln(x^2 + x + 1) + \sqrt{3}\tan^{-1}[(2x+1)/\sqrt{3}] + C$

3. $\frac{1}{5}\ln|x+2| - \frac{1}{10}\ln|x^2+1| + \frac{2}{5}\tan^{-1}x + C$

5. $\frac{1}{2}\ln(x^2+2) + (5/4\sqrt{2})\tan^{-1}(x/\sqrt{2}) + [1/(x^2+2)] + [x/4(x^2+2)] + C$

7. $\frac{1}{2}\ln|x^2 - 4| - [2/(x^2 - 4)] + C$ **9.** $\frac{1}{4}\ln|(x^2 - 1)/(x^2 + 1)| + C$

11. $\frac{1}{221}\{261(\pi/2) + 56\ln\frac{1}{2} + 1026[\tan^{-1}(-2) - \tan^{-1}(-3)]\}$ **13.** $\frac{3}{8} + \ln\frac{5}{4}$

15. $\dfrac{1}{2}\ln|x^2 + 4x + 13| + \dfrac{47x}{18(x^2 + 4x + 13)} + \dfrac{67}{18(x^2 + 4x + 13)} - \dfrac{61}{54}\tan^{-1}\left(\dfrac{x+2}{3}\right) + C$

17. $\displaystyle\int_0^{\pi/2} \dfrac{\cos x}{(\sin x + 1)^2 + 1}\, dx = \tan^{-1}(\sin x + 1)\Big|_0^{\pi/2} = \tan^{-1} 2 - \dfrac{\pi}{4}$

19. $\displaystyle\int_0^1 \dfrac{x^3}{(x^2 + 1)^3}\, dx = \dfrac{1}{16}$

Problems 8.8

1. $\frac{2}{5}(1+x)^{5/2} - \frac{2}{3}(1+x)^{3/2} + C$ (let $u = 1 + x$)

3. $\frac{1}{4}[\frac{3}{8}(1+2x)^{8/3} - \frac{3}{5}(1+2x)^{5/3}]|_0^{13} = 578.7$ $(u = 1 + 2x)$

5. $-\frac{1}{32}[(16/x^3) - 1]^{2/3} + C$ (if $u = (16/x^3) - 1$, then

$\dfrac{dx}{x^3(16 - x^3)^{1/3}} = \dfrac{dx}{x^4\left(\dfrac{16 - x^3}{x^3}\right)^{1/3}} = \dfrac{3x^2\, dx}{3x^6\left(\dfrac{16 - x^3}{x^3}\right)^{1/3}} = \dfrac{-16(1 + u)^{-2}\, du}{3\left(\dfrac{16}{u+1}\right)^2 u^{1/3}}$

$= -\dfrac{1}{48}u^{-1/3}\, du\Bigg)$

7. $\frac{8}{3}$ (if $u = 1 + x^3$, $[3x^5/(1 + x^3)^{3/2}]\, dx = (u^{-1/2} - u^{-3/2})\, du$)

9. $\frac{6}{5}(1+x)^{5/2} - \frac{14}{3}(1+x)^{3/2} + 8(1+x)^{1/2} + C$ $(u = 1 + x)$

11. $12(\frac{1}{8} - \frac{1}{7} + \frac{1}{6} - \frac{1}{5} + \frac{1}{4} - \frac{1}{3} + \frac{1}{2} - 1 + \ln 2) \approx 0.7$ $(u = x^{1/12})$

13. $\dfrac{1}{4\sqrt{2}}\ln\left|\dfrac{x^{1/3} + \sqrt{2}x^{1/6} + 1}{x^{1/3} - \sqrt{2}x^{1/6} + 1}\right| + \dfrac{1}{2\sqrt{2}}\tan^{-1}\left(\dfrac{\sqrt{2}x^{1/6}}{1 - x^{1/3}}\right) + C$ (if $u = x^{1/6}$, then

$\dfrac{x^{1/2}}{1 + x^{2/3}}\, dx = \dfrac{du}{u^4 + 1} = \dfrac{du}{(u^2 + \sqrt{2}u + 1)(u^2 - \sqrt{2}u + 1)}\bigg)$

15. $3(1-x)^{1/3} + \dfrac{1}{2^{2/3}}\ln\left|\dfrac{[2^{1/3} - (1-x)^{1/3}]^2}{2^{2/3} + 2^{1/3}(1-x)^{1/3} + (1-x)^{2/3}}\right| -$

$2^{1/3}\sqrt{3}\tan^{-1}\left[\dfrac{2(1-x)^{1/3} + 2^{1/3}}{2^{1/3}\sqrt{3}}\right] + C$ (if $u = (1-x)^{1/3}$, then

$\dfrac{(1-x)^{1/3}}{1+x}\, dx = \dfrac{-3u^2}{2 - u^3}\, du\bigg)$

17. $x + 6x^{1/3} + 3\ln|(x^{1/6} - 1)/(x^{1/6} + 1)| + C$ $(u = x^{1/3})$

19. $\frac{1}{27}[\frac{4}{7}u^{7/2} - \frac{8}{5}u^{5/2} + \frac{8}{3}u^{3/2}]|_2^5 = \frac{1}{27}(\frac{190}{21}\sqrt{5} - \frac{128}{105}\sqrt{2}) \approx 0.6854$ $(u = 2 + 3x)$

21. $6[(x^{1/3}/2) + \tan^{-1}x^{1/6} - \frac{1}{2}\ln|x^{1/3} + 1|] + C$ $(u = x^{1/6})$

23. $[(\sin x - 1)/\cos x] + C = -[\cos x/(1 + \sin x)] + C$

25. $\int_0^{1/\sqrt{3}}[2/(5+u^2)]\,du = (2/\sqrt{5})\tan^{-1}(u/\sqrt{5})||_0^{1/\sqrt{3}} = (2/\sqrt{5})\tan^{-1}(1/\sqrt{15})$

27. $(1/2\sqrt{2})\ln\|[\sqrt{2}+(\tan x - 1)]/[\sqrt{2}-(\tan x - 1)]\|$ (find $\int[1/(\sin u + \cos u)]\,du$ and then show, by substituting $u = 2x$, that

$\int[1/(\sin 2x + \cos 2x)]\,dx = \tfrac{1}{2}\int[1/(\sin u + \cos u)]\,du)$ **29.** $2\tan^{-1}[2+\tan(x/2)]+C$

31. $(1/\sqrt{5})\cos^{-1}\{[(5/x)-2]/3\}+C$

Problems 8.9

In Problems 1–15 the answers are given in the order requested in the text. The last two numbers give the actual error in the trapezoidal and Simpson's estimates, respectively. The calculations were made with a hand calculator with ten decimal place precision. The same number of intervals were used in each estimate.

1. 0.5; 0.5; 0; 0; 0.5; 0; 0   **3.** $\frac{11}{32} = 0.34375$; $\frac{1}{3}$; $\frac{1}{96} = 0.0104167$; 0, $\frac{1}{3}$; $-\frac{1}{96}$; 0

5. 6.448104763; 6.389488576; 0.1368343722; 0.0010135879; 6.389056099; 0.0590486641; 0.0004324773   **7.** 1.218760835; 1.218951005; 0.0003255208; 0.0000012716; 1.218951416; -0.0001905814; -0.0000004111 **9.** 0.9871158010; 1.000134585; 0.0201863780; 0.0002075329; 1.0; -0.012884199; 0.000134585   **11.** 0.6973999653; 0.6932588905; 0.0207193081; 0.0013254071; ln 2 = 0.6931471806; 0.0042527848; 0.0001117099

13. 0.9956971264; 0.9999360657; 0.0117435512; 0.0003852827; 1.0; -0.0043028736; -0.000063343 **15.** 0.9129205114; 0.9093257681; 0.0040387583; 0.0000124097; 0.9093306736; 0.0035898377; -0.0000049056

In Problems 17–29 the trapezoidal approximation is given first.

17. 0.6590191268; 0.6593301635   **19.** 1.987795499; 1.994503740

21. 0.6903257237; 0.8367702680   **23.** 1.488736680; 1.493674110

25. 0.9091616587; 0.9096068101   **27.** 0.9841199229; 0.9838189106

29. 0.3604819483; 0.3579282259   **31.** $|y''| = |3(2x+3x^4)e^{x^3}| \le 15e$ on $[0,1]$ and $|y^{(iv)}| = |9(9x^8+36x^5+20x^2)e^{x^3}| \le 585e$. Thus $|\epsilon_8^T| \le 15e/(12\cdot 8^2) \approx 0.05309$ and $|\epsilon_8^S| \le 585e/(180\cdot 8^4) \approx 0.00216$ **33.** $|y''| = |-2+4x^2|e^{-x^2} \le 2/e$ on $[-1,1]$ and $|y^{(iv)}| = |12-48x^2+16x^4|e^{-x^2} \le 12$. Thus $|\epsilon_{10}^T| \le 2^4/[(e)(12)10^2] \approx 0.00491$ and $|\epsilon_{10}^S| \le 12\cdot 2^5/(180\cdot 10^4) \approx 0.00021$ **35.** $|y''| = |3(4x^2+3x^5)e^{x^3}| \le 21e$ on $[0,1]$; $|y^{(iv)}| = |3(8+132x^3+144x^6+27x^9)e^{x^3}| \le 933e$. Thus $|\epsilon_{10}^T| \le 21e/(12\cdot 10^2) \approx 0.04757$ and $|\epsilon_{10}^S| \le 933e/(180\cdot 10^4) \approx 0.00141$ **37.** $|y^{(iv)}| \le 3$ so we need $(3/\sqrt{2\pi})\,1^5/(180\cdot n^4) \le 0.005$ (for half the integral—using the hint); $n = 2$ will do; we obtain 0.6830581043 (the "true" value is 0.6826894921 giving an error of 0.0003686122)   **39.** (a) On $[0,50]$ need $(3/\sqrt{2\pi})\,50^5/(180\cdot n^4) \le 0.05$ or $n \ge 112$; this leads to the estimate $(1/\sqrt{2\pi})\int_{-50}^{50} = (2/\sqrt{2\pi})\int_0^{50} \approx 2(0.4999994266) = 0.9999988532$ (b) $\lim_{N\to\infty}(1/\sqrt{2\pi}\int_{-N}^{N} e^{-x^2/2}\,dx = 1$ **41.** $|y''| = 2/x^3 \le 2$ on $[1,2]$; need $2/12n^2 \le 10^{-10}$ or $12n^2 > 2\cdot 10^{10}$ which implies that $n > 40{,}824.8$ **43.** It is not difficult to verify that on $[\tfrac{1}{2},1]$, $|J_{1/2}''(x)| \le 8$. Then, with the trapezoidal rule, we need $8(\tfrac{1}{2})^3/12n^2 \le 0.01$ or $n > 2.88$. Using $n = 3$ gives $\int_{1/2}^1 J_{1/2}(x)\,dx \approx 0.3095670957$.

Review Exercises for Chapter Eight

1. $\tfrac{1}{2}\{\sqrt{5}+4\ln[2/(1+\sqrt{5})]\}$ **3.** $2\tan^{-1}x + x/(x^2+1)+C$ **5.** $\frac{2}{15}$

7. $-e^{-2x}[(x/2)+\tfrac{1}{4}]+C$ **9.** $\tfrac{1}{4}\sec^4 x - \tfrac{1}{2}\sec^2 x - \ln|\cos x|+C$

11. $3\ln 2 - \tfrac{3}{2}$ **13.** $(-2/\sqrt{3}) + \sqrt{2} + \ln|(2+\sqrt{3})/(\sqrt{2}+1)|$

15. $\tfrac{1}{4}e^{2x}(\sin 2x - \cos 2x)+C$ **17.** $-\tfrac{1}{4}\cos^4 x + \tfrac{1}{6}\cos^6 x + C$

19. $-\tfrac{1}{5}\csc^5 x + \tfrac{1}{3}\csc^3 x + C$ **21.** $\tfrac{1}{3}x^3 + x + \ln|x/(x+1)|+C$

23. $x\cosh x - \sinh x + C$ **25.** $[-1/(x-2)]+C$

27. $\tfrac{1}{2}(\sin x + \tfrac{1}{13}\sin 13x)+C$ **29.** $-\tfrac{11}{14}\ln|x-3| + \tfrac{2}{7}\ln|x+4| + \tfrac{3}{2}\ln|x-5| + C$

31. $\tfrac{1}{5}(x^2-4)^{5/2} + \tfrac{4}{3}(x^2-4)^{3/2} + C$ $(u = x^2 - 4)$ **33.** $\tfrac{1}{3}\tan^3 x + \tan x + C$

35. $x + 3\ln|x| - (3/x) - (1/2x^2) + C$ **37.** $5\pi/32$ **39.** $\ln(\sqrt{2}+1)$

41. $\dfrac{1}{2\sqrt{6}} \ln \left| \dfrac{\tan(x/2) + 5 - \sqrt{24}}{\tan(x/2) + 5 + \sqrt{24}} \right| + C$ **43.** $-\frac{3}{2}(1 + x^4)^{-1/3} + C$

45. $\frac{1}{81}[-\frac{4}{7}(2 + 3x)^{7/2} + \frac{24}{5}(2 + 3x)^{5/2} - 10(2 + 3x)^{3/2} -$
$4(2 + 3x)^{1/2}] + C \ (u = 2 + 3x)$ **47.** $\frac{9}{10} e^{-x}[-\cos x/3 + \frac{1}{3}\sin x/3] + C$

49. $\frac{3}{8}\cosh x \sinh 3x - \frac{1}{8}\sinh x \cosh 3x + C$ **51.** 1.383213747

53. 0.9233959267   **55.** 1.005709257   **57.** $|y''| = |3(2x + 3x^4)e^{x^3}| \le 15e$ on $[0, 1]$;
need $15e/12n^2 \le 0.01$ or $n \ge 18.4$ **59.** $|y^{(iv)}| = 120/x^6 \le 120$ on $[1, 2]$; we need
$120/180n^4 \le 0.0001$ or $n \ge 9.04$. Using Simpson's rule with $n = 10$ yields the value
0.5000124699. The actual value is 0.5.

CHAPTER NINE

Problems 9.1

1. $8\pi/3$ **3.** $4\pi/5$ **5.** $64\pi/3$ **7.** $\pi^2/4$ **9.** $(\pi/2)(e^2 - e^{-2}) = \pi \sinh 2$
11. $\pi^2/4$ **13.** $\pi(e - 2)$ **15.** $\pi/2$ **17.** $(\pi/4)(e^2 - 1)$ **19.** $\pi/2$
21. 9π **23.** $\pi/12$ **25.** $\pi(9 \ln 3 - 4)$ **27.** $3\pi/10$ **29.** $\pi[(\pi/\sqrt{2}) - 2]$
31. $\pi(e - 1)$ **33.** $\frac{4}{3}\pi r^3$ **35.** $\frac{4}{3}\pi ab^2$ **37.** 6π
39. $(\pi/3)(b_1 + b_2 + b_3)|(a_2 - a_1)(b_3 - b_1) - (b_2 - b_1)(a_3 - a_1)|$ **41.** $\pi[4 - (\pi/4)]$
43. $\frac{128}{3}$ **45.** $16\pi/3$ **47.** $16\sqrt{3}/3 \text{ m}^3$ **49.** (a) $\frac{1}{30}$, (b) $\frac{1}{120}$, (c) $\pi/120$
51. $920g\pi \int_0^3 (3 - y)e^{2y}\, dy \approx 2.81 \times 10^6$ joules (assuming that distances are measured
in meters)

Problems 9.2

1. $4\sqrt{10}$ **3.** $\frac{1}{27}(13^{3/2} - 8)$ **5.** $\frac{14}{3}$ **7.** $\frac{123}{32}$
9. $\sqrt{1 + e^4} - \sqrt{1 + e^2} + \frac{1}{2}\ln \left[\dfrac{(\sqrt{1 + e^4} - 1)(\sqrt{1 + e^2} + 1)}{(\sqrt{1 + e^4} + 1)(\sqrt{1 + e^2} - 1)} \right]$ **11.** (b) look at the

definition of the radian   **13.** 6   **15.** $\frac{2}{27}(10^{3/2} - 1)$

17. $\int_0^1 \sqrt{1 + 9x^2 + 9x^4}\, dx$

19. $\int_0^1 \sqrt{1 + x^2 \cosh^2 x + 2x \cosh x \sinh x + \sinh^2 x}\, dx$ **21.** $\int_0^4 \sqrt{1 + 4x^2 e^{2x^2}}\, dx$
23. $\int_{-1}^1 \sqrt{1 + 4x^2 \sinh^2(1 + x^2)}\, dx$

Problems 9.3

1. $(\pi/6)(5^{3/2} - 1)$ **3.** $2\pi\sqrt{5}$ **5.** $\pi\sqrt{1 + \alpha^2}[\alpha(b^2 - a^2) + 2\beta(b - a)]$
7. $(\pi/6)(27 - 5^{3/2})$ **9.** $47\pi/16$

11. $2\pi \displaystyle\int_1^2 \frac{1}{x} \sqrt{1 + \frac{1}{x^4}}\, dx = \pi \int_{1/4}^1 \frac{\sqrt{1 + v^2}}{v}\, dv \ \left(v = \frac{1}{x^2}\right)$

$\qquad = \pi \displaystyle\int_{\tan^{-1}\frac{1}{4}}^{\pi/4} \frac{\sec^3 \theta}{\tan \theta}\, d\theta \ (v = \tan \theta)$

$\qquad = \pi \left[\ln \left(\dfrac{\sqrt{17} + 4}{\sqrt{2} + 1} \right) + \sqrt{2} - \dfrac{\sqrt{17}}{4} \right]$

15. (i) if $b > a$, $S = \dfrac{2\pi b a^2}{(b^2 - a^2)^{1/2}} \left[\dfrac{b\sqrt{b^2 - a^2}}{a^2} + \ln \left(\dfrac{b + \sqrt{b^2 - a^2}}{a} \right) \right]$ (ii) if $a > b$,

$S = \dfrac{2\pi b a^2}{(a^2 - b^2)^{1/2}} \left[\dfrac{b\sqrt{a^2 - b^2}}{a^2} + \sin^{-1} \left(\dfrac{\sqrt{a^2 - b^2}}{a} \right) \right]$ **17.** 24π

Problems 9.4

1. -18 kg-m; -1.8m   **3.** 101 kg-m; $\frac{101}{18}$ m   **5.** $\frac{1}{3}; \frac{1}{2}; \frac{2}{3}$ **7.** $\frac{31}{5}; \frac{15}{4}; \frac{124}{75}$
9. $2\pi(2\pi - 1); 4\pi; \pi - \frac{1}{2}$
11. $8 \ln 4 - \frac{15}{4}; 4 \ln 4 - 3; (8 \ln 4 - \frac{15}{4})/(4 \ln 4 - 3) \approx 2.88$
13. $(2e^3 + 1)/9; (e^2 + 1)/4; 4(2e^3 + 1)/9(e^2 + 1)$ **15.** $\ln\frac{25}{16}; \ln\frac{8}{5}; \ln\frac{25}{16}/\ln\frac{8}{5} \approx 0.95$

Problems 9.5

1. $(-1.8, 3.4)$　　**3.** $(\frac{101}{18}, \frac{3}{2})$　　**5.** $(\frac{13}{24}, \frac{49}{24})$　　**7.** $(\frac{4}{7}, \frac{2}{5})$　　**9.** $(\frac{493}{185}, \frac{525}{296})$

11. $\left(\dfrac{1 - \ln 2}{\ln 2}, \dfrac{1}{4 \ln 2}\right) \approx (0.44, 0.36)$　　**13.** $\left(\dfrac{\ln 9/8}{\ln 4/3}, \dfrac{\ln(9/16) + 2}{13 \ln 4/3}\right) \approx (0.409, 0.381)$

15. $\left(\dfrac{\pi - \sqrt{3}}{9}, \dfrac{1}{\sqrt{3}}\left(\dfrac{\pi}{6} + \dfrac{\sqrt{3}}{8}\right)\right) \approx (0.16, 0.43)$

17. $\left(\dfrac{2 \ln 2 - \frac{3}{4}}{2 \ln 2 - 1}, \dfrac{2(\ln 2)^2 - 4 \ln 2 + 2}{2 \ln 2 - 1}\right) \approx (1.65, 0.49)$

19. $\left(\dfrac{\frac{1}{2} - \ln 2}{1 - \ln 2}, \dfrac{\frac{3}{4} - \ln 2}{1 - \ln 2}\right) \approx (-0.63, 0.19)$　　**21.** $(\frac{7}{10}, \frac{9}{16})$　　**23.** $(\frac{13}{20}, \frac{77}{50})$　　**25.** $(\frac{8}{7}, \frac{289}{11})$

27. $(\frac{213}{85}, \frac{6927}{595})$　　**29.** $(4, \frac{1}{2})$　　**31.** $\left(\dfrac{\pi\sqrt{2} - 4}{4(\sqrt{2} - 1)}, \dfrac{1}{4(\sqrt{2} - 1)}\right) \approx (0.27, 0.6)$

33. $\left(\dfrac{e^2 + \frac{3}{4} - 2 \ln 2}{e^2 + 1 - 2 \ln 2 - e}, \dfrac{\frac{1}{4}e^4 - (\ln 2)^2 + 2 \ln 2 - 1 - \frac{1}{4}e^2}{e^2 + 1 - 2 \ln 2 - e}\right) \approx (1.58, 2.73)$

37. $(\pi, -\pi/8)$　　**43.** $2\pi^2 r^2 a$　　**45.** $4\sqrt{2}\pi^2$

Problems 9.6

1. 120 kg-m^2　　**3.** 758 kg-m^2　　**5.** (a) $\frac{25}{18}\pi^2$ joules;　(b) $\frac{125}{81}\pi^2$ joules
7. (a) 3 kg-m^2;　(b) 7 kg-m^2;　(c) 5 kg-m^2;　(d) $\frac{17}{4}$ kg-m^2　　**9.** (a) $\frac{1}{2}$ m　(b) $\sqrt{\frac{7}{12}}$ m
(c) $\sqrt{\frac{5}{12}}$ m　(d) $\sqrt{\frac{17}{48}}$ m　　**11.** $800\pi^2(\frac{3}{4}\sqrt{7} + 32 \sin^{-1}\frac{3}{4}) \approx 229{,}940$ joules (note that
$\sin 4\theta = -3\sqrt{7}/32$ if $\theta = \sin^{-1}\frac{3}{4}$)　　**13.** $I = 16, r = 2/\sqrt{3}$　　**15.** $I = \frac{2}{5}, r = \sqrt{\frac{3}{5}}$
17. $I = \frac{383}{35} r = \sqrt{\frac{383}{35}}$　　**19.** $I = (\pi^2/2) - 4, r = \frac{1}{2}\sqrt{\pi^2 - 8}$
21. $I = 6, r = \sqrt{3/(2 \ln 2)}$　　**23.** $I = 5(e - 2), r = \sqrt{(e - 2)/(e - 1)}$
25. $I = 17\pi/2, r = \sqrt{17}/2$　　**27.** $I = 145\pi/2, r = \sqrt{145}/2$
29. $I = \frac{224}{5}, r = 2\sqrt{\frac{3}{5}}$　　**31.** $I = 3[\int_0^{1/\sqrt{2}} 2x^3 \, dx + \int_{1/\sqrt{2}}^1 2x^2\sqrt{1 - x^2} \, dx] = \frac{3}{8} + 3\pi/16$
(we first rotated everything $45°$ clockwise); $r = \sqrt{\frac{1}{4} + 1/2\pi}$　　**33.** $k = \frac{9}{2}\rho\omega^2$ joules

Problems 9.7

1. $72{,}000 (9.81)\pi \approx 2{,}218{,}970$ nt　　**3.** (a) $P_A = (13.6)(9.81)h = 133.416h$
5. 975 lb.　　**7.** 5297.4 nt　　**9.** $(1000)(9.81)\int_0^{25/4}(\frac{25}{4} - y)(2)5\sqrt{y} \, dy = 2{,}554{,}687.4$ nt
11. $(1000)(9.81)\int_0^{25/8}(\frac{25}{8} - y)(2)5\sqrt{y/2} \, dy = 319{,}335.9$ nt
13. $3(1030)(9.81) = 30{,}312.9$ nt　　**15.** $(1000)(9.81)(3)\int_0^1(\frac{3}{2} - \frac{3}{2}x) \, dx = 22{,}072.5$ nt
17. $F = (62.4)(10)\int_0^{5/2}\sqrt{3}(\frac{5}{2} - x) \, dx = 1950\sqrt{3}$ lb
19. $(1030)(9.81)(10 - 10e^{-5} - \frac{1}{4} + \frac{1}{4}e^{-10}) \approx 97{,}879.96$ nt
23. $\frac{45}{4}(1000)(9.81) = 110{,}362.5$ nt　　**25.** $(62.4)(30)(\frac{32}{5}) = 11{,}980.8$ lb
27. $(62.4)(20)(2 \ln 2 - 1) \approx 482$ lb

Review Exercises for Chapter Nine

1. 56π　　**3.** $\pi/11$　　**5.** $\pi^2/8$　　**7.** $4\pi^2$　　**9.** 144　　**11.** $5\sqrt{5}$
13. $\frac{2}{3}[2^{3/2} - 1]$　　**15.** $\int_0^{\pi/2}\sqrt{1 + (x \cos x + \sin x)^2} \, dx$
17. $\pi[\ln(\sqrt{1 + e^2} + e) + e\sqrt{1 + e^2} - \ln(\sqrt{2} + 1) - \sqrt{2}]$
19. $\pi\left\{\ln[(\sqrt{17} + 4)/(\sqrt{2} + 1)] + \sqrt{2} - \dfrac{\sqrt{17}}{4}\right\}$ (see answer to Problem 9.3.11)
21. $4\pi[2 + (1/\sqrt{3}) \ln(2 + \sqrt{3}]$　　**23.** $(\frac{9}{4}, 0)$　　**25.** $(\frac{17}{11}, \frac{15}{22})$　　**27.** $(\frac{58}{35}, \frac{45}{56})$
29. $(-2, \frac{3}{5})$　　**31.** 356 kg-m^2　　**33.** (a) $\frac{1}{2}$ m　(b) $\sqrt{\frac{3}{8}}$ m　(c) $\dfrac{\sqrt{46}}{12}$ m
35. $I = \frac{1}{15}$ kg-m^2; $r = \sqrt{\frac{2}{15}}$ m　　**37.** $250\pi^2/3$ joules
39. $\frac{1}{2} \cdot \frac{5}{12} \cdot [\frac{1}{4}(\text{rev/sec}) \, 2\pi]^2 = (5\pi^2/96)\dfrac{\text{g-cm}^2}{\text{sec}^2} \approx 5.14 \times 10^{-8}$ joules
41. $I = 101\pi/4; r = \sqrt{101}/2$　　**43.** $392{,}400$ nt
45. $(1000)(9.81)\int_0^2[(3x^2 + 6x)/4] \, dx = 49{,}050$ nt　　**47.** (a) 3525.47 nt　(b) $13{,}488.75$ nt
(c) $28{,}970.2$ nt

CHAPTER TEN

Problems 10.1

In Problems 1–9 any smaller values for δ will also work.

1. $\frac{1}{70}, \frac{1}{700}$ **3.** $\frac{1}{50}, \frac{1}{500}$ **5.** $\frac{1}{70}, \frac{1}{700}$ (For $\epsilon = 0.1$, the largest value of δ is 0.016620626.
To see this calculate $\sqrt{9.1}$.) **7.** $\frac{1}{11}, \frac{1}{110}$ (If $|x + 1| < \frac{1}{11}$, then $|x| < 1.1$ and
$|x(x + 1)| = |x||x + 1| < 1.1/11 = \frac{1}{10}$. For $\epsilon = 0.1$, the largest value of
$\delta = \frac{1}{2}\sqrt{1.4 - 1} \approx 0.0916079783$.) **9.** 0.008, 0.0008 (If $|x - 2| < 0.008$, then
$|x^3 - 8| = |(x^2 + 2x + 4)(x - 2)| < (0.008)[(2.008)^2 + 2(2.008) + 4] \approx 0.096$. The largest
value of δ is 0.00829885 since $\sqrt[3]{8.1} = 2.00829885$.) **11.** (b) $\delta \le 0.01$
13. (b) $N \ge \sqrt[3]{4000} \approx 15.874$

Problems 10.2

7. Let $f(x) = x + 1$ and $g(x) = x$ (more complicated examples can be found).
9. Let $f(x) = x$ and $g(x) = 1/x$. **11.** Let $f(x) = g(x) = x$. Note that ∞/∞ is not
defined.

Problems 10.4

21. $\sqrt{e^3 + 1}$ **23.** $\sec^2(-\frac{1}{2}) = \sec^2(\frac{1}{2})$ **25.** 2 **27.** 0 **29.** 1
31. Because of the intermediate value theorem, the function giving velocity in terms of
time is continuous. **33.** not continuous **35.** not continuous
37. not continuous **39.** not continuous

41. (a) $f(x)g(x) = \begin{cases} (x + 3)(x^2 + 6), & x \le 1 \\ (2x + 5)(5x^3 - 1), & x > 1 \end{cases}$ (b) $\lim_{x \to 1^+} f(x) = 7$ and $\lim_{x \to 1^-} f(x) = 4$
(c) $\lim_{x \to 1^+} g(x) = 4$ and $\lim_{x \to 1^-} g(x) = 7$ (d) $\lim_{x \to 1^+} f(x)g(x) = 28 = \lim_{x \to 1^-} f(x)g(x)$
43. if $p(x_0) = q(x_0)$ **47.** $M = 7/2, m = 1$ **53.** $[-2, 2]$

Problems 10.6

1. $\sqrt{\frac{7}{3}}$ **3.** $\sin^{-1}(2/\pi) \approx 0.69$ **5.** $\ln 3 - \ln \ln 4 \approx 0.77$ **7.** 0
9. (c) because f is not differentiable at 1 **11.** because f is not continuous at 0
17. $|\sin x - \sin y| = (\cos c)|x - y|$, where $0 < c < \pi/4$

Problems 10.7

1. $\mathbb{R}; x = \frac{1}{3}(y - 5); dx/dy = \frac{1}{3}$ **3.** $\mathbb{R} - \{0\}; x = 1/y; dx/dy = -1/y^2$
5. $[-\frac{3}{4}, \infty); x = \frac{1}{4}(y^2 - 3); dx/dy = y/2$
7. $\mathbb{R}; x = \sqrt[3]{1 - y}; dx/dy = -\frac{1}{3}(1 - y)^{-2/3}$ **9.** $\mathbb{R}; x = 2 + \ln y; dx/dy = 1/y$
11. $(-\frac{5}{2}, \infty); x = \frac{1}{2}(e^y - 5); dx/dy = \frac{1}{2}e^y$ **13.** all intervals of the form
$(((2k - 1)/6)\pi, ((2k + 1)/6)\pi), k = 0, \pm 1, \pm 2, \pm 3, \ldots; x_k = k\pi/3 + \frac{1}{3}\tan^{-1}y;$
$dx/dy = 1/3(1 + y^2)$ **15.** all sets of the form $[n\pi/2, (n + 2)\pi/2] - \{(n + 1)\pi/2\},$
$n = \pm 1, \pm 3, \pm 5, \ldots; x = \csc^{-1}y; dx/dy = -1/|y|\sqrt{y^2 - 1}$ **17.** $(-\infty, 0]$ and
$[0, \infty); x_1 = -\frac{1}{2}\cosh^{-1}y = -\frac{1}{2}\ln|y + \sqrt{y^2 - 1}|, x_2 = \frac{1}{2}\cosh^{-1}y = \frac{1}{2}\ln|y + \sqrt{y^2 - 1}|;$
$dx_1/dy = -1/2\sqrt{y^2 - 1}, dx_2/dy = 1/2\sqrt{y^2 - 1}$ **19.** $[0, \infty);$
$x = \tanh^{-1}y^2 = \frac{1}{2}\ln(1 + y^2)/(1 - y^2); dx/dy = 2y/(1 - y^2), 0 \le y < 1$
21. $(-\infty, -1]$ and $[-1, \infty)$; inverse function not obtainable in closed
form; $dx/dy = 1/[(x + 1)e^x] = 1/(y + e^x)$ **23.** all intervals of the form
$[n\pi/2, (n + 2)\pi/2], n = \pm 1, \pm 3, \pm 5, \ldots; x = \sin^{-1}\ln y; dx/dy = 1/[y\sqrt{1 - (\ln y)^2}]$
25. $(-\infty, \frac{5}{2}]$ and $[\frac{5}{2}, \infty); x_1 = (5 - \sqrt{1 + 4y})/2, x_2 = (5 + \sqrt{1 + 4y})/2;$
$dx_1/dy = -1/\sqrt{1 + 4y}, dx_2/dy = 1/\sqrt{1 + 4y}$ **27.** $\mathbb{R}$; from the cubic formula (which
no one really knows but which you can find in the *Handbook of Tables for Mathematics*
(Chemical Rubber Co.)), $x = \sqrt[3]{(y/2) + \sqrt{(y^2/4) + \frac{1}{27}}} + \sqrt[3]{(y/2) - \sqrt{(y^2/4) + \frac{1}{27}}};$
$dx/dy = 1/(1 + 3x^2)$. Note that y is increasing since $y' = 3x^2 + 1 > 0$ **29.** $(-\infty, 0]$
and $[0, \infty); x_1 = -\sqrt{\ln y}, x_2 = \sqrt{\ln y}; dx_1/dy = -1/(2y\sqrt{\ln y}), dx_2/dy = 1/(2y\sqrt{\ln y})$

Problems 10.8

9. $G'(x) = -e^{\sqrt{\cos x}} \sin x - e^{\sqrt{\sin x}} \cos x$ **11.** $G'(x) = \cos e^x - [(\cos \ln x)/(x \ln x)]$

Review Exercises for Chapter Ten

9. -1 **15.** e **23.** $(e^3 - 1)/3$ **25.** $\cos^{-1} 2/\pi \approx 0.88$ **27.** (a) $\alpha = 1$
(b) $c = \sqrt[3]{2}$ **29.** $[\frac{3}{2}, \infty)$; $x = \frac{1}{2}(5 + y^2)$; $dx/dy = y$
31. $\mathbb{R}$; $x = \ln y - 4$; $dx/dy = 1/y$ **33.** $(-\infty, 0]$ and $[0, \infty)$;
$x_1 = -\text{sech}^{-1} y = -\cosh^{-1}(1/y)$, $x_2 = \cosh^{-1}(1/y)$; $dx_1/dy = 1/(y\sqrt{1 - y^2})$,
$dx_2/dy = -1/(y\sqrt{1 - y^2})$ **35.** $(-\infty, \frac{-5}{2}]$ and $[\frac{5}{2}, \infty)$; $x_1 = (-5 - \sqrt{1 + 4y})/2$,
$x_2 = (-5 + \sqrt{1 + 4y})/2$, $dx_1/dy = -1/\sqrt{1 + 4y}$, $dx_2/dy = 1/\sqrt{1 + 4y}$
39. $-[(\sin x)/(\ln \cos x)] - [(\cos x)/(\ln \sin x)]$

CHAPTER ELEVEN

Problems 11.1

1. $(3, 0)$ **3.** $(-5, 0)$ **5.** $(-3\sqrt{3}, -3)$ **7.** $(-5, 0)$ **9.** $(5, 0)$
11. $(\sqrt{2}, -\sqrt{2})$ **13.** $(\sqrt{2}, -\sqrt{2})$ **15.** $(0, -1)$ **17.** $(0, -1)$ **19.** $(3, 0)$
21. $(3, \pi)$ **23.** $(\sqrt{2}, 5\pi/4)$ **25.** $(\sqrt{2}, 3\pi/4)$ **27.** $(1, 3\pi/2)$ **29.** $(4, 2\pi/3)$
31. $(4, 4\pi/3)$ **33.** $(4, 11\pi/6)$ **35.** $(4, 7\pi/6)$

Problems 11.2

1. circle centered at origin with radius 5
3. circle centered at origin with radius 4   **5.**

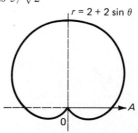

7. circle centered at $(0, \frac{5}{2})$ with radius $\frac{5}{2}$ **9.** circle centered at $(0, -\frac{5}{2})$ with radius $\frac{5}{2}$
11. circle centered at $(\frac{5}{2}, \frac{5}{2})$ with radius $5/\sqrt{2}$

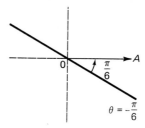

13. circle centered at $(\frac{5}{2}, -\frac{5}{2})$ with radius $5/\sqrt{2}$
15. cardioid, symmetric about $\theta = \pi/2$

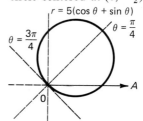

17. cardioid obtained by reflecting curve in Problem 15 about polar axis
19. same curve as in Problem 15   **21.** same curve as in Problem 17
23. limaçon with loop, symmetric about $\theta = \pi/2$

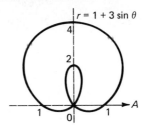

25. limaçon with loop, symmetric about polar axis

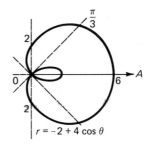

27. limaçon with loop, symmetric about polar axis

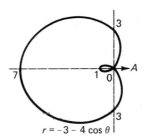

29. limaçon with loop, symmetric about $\theta = \pi/2$

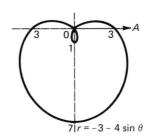

31. limaçon without loop, symmetric about polar axis

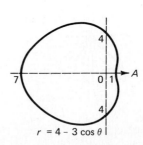

33. limaçon without loop, symmetric about $\theta = \pi/2$

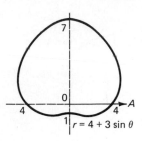

$r = 4 + 3 \sin \theta$

35. four-leafed rose, symmetric about polar axis, pole, and $\theta = \pi/2$

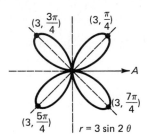

$\left(3, \frac{3\pi}{4}\right)$ $\left(3, \frac{\pi}{4}\right)$

$\left(3, \frac{7\pi}{4}\right)$

$\left(3, \frac{5\pi}{4}\right)$ $r = 3 \sin 2\theta$

37. same curve as in Problem 35
39. symmetric about $\theta = \pi/2$

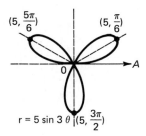

$\left(5, \frac{5\pi}{6}\right)$ $\left(5, \frac{\pi}{6}\right)$

$r = 5 \sin 3\theta$ $\left(5, \frac{3\pi}{2}\right)$

41. this is curve of Problem 39 reflected about polar axis
43. eight-leafed rose, symmetric about polar axis, pole, and $\theta = \pi/2$

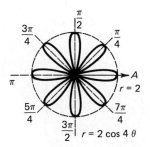

$\frac{\pi}{2}$

$\frac{3\pi}{4}$ $\frac{\pi}{4}$

π A

$r = 2$

$\frac{5\pi}{4}$ $\frac{7\pi}{4}$

$\frac{3\pi}{2}$ $r = 2 \cos 4\theta$

45. same curve as in Problem 43

47.

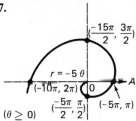

$r = -5\theta$
$(-10\pi, 2\pi)$
$(\frac{-15\pi}{2}, \frac{3\pi}{2})$
$(\frac{-5\pi}{2}, \frac{\pi}{2})$ $(-5\pi, \pi)$
$(\theta \geq 0)$

49.

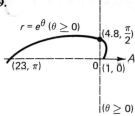

$r = e^\theta \ (\theta \geq 0)$
$(4.8, \frac{\pi}{2})$
$(23, \pi)$
$(1, 0)$
$(\theta \geq 0)$

51.

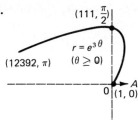

$(111, \frac{\pi}{2})$
$r = e^{3\theta}$
$(\theta \geq 0)$
$(12392, \pi)$
$(1, 0)$

53. lemniscate, symmetric about pole

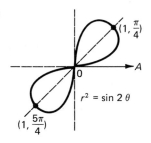

$(1, \frac{\pi}{4})$
$r^2 = \sin 2\theta$
$(1, \frac{5\pi}{4})$

55. lemniscate, symmetric about pole

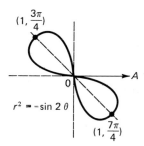

$(1, \frac{3\pi}{4})$
$r^2 = -\sin 2\theta$
$(1, \frac{7\pi}{4})$

57. lemniscate symmetric about polar axis, pole, and $\theta = \pi/2$

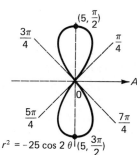

$(5, \frac{\pi}{2})$
$\frac{3\pi}{4}$ $\frac{\pi}{4}$
$\frac{5\pi}{4}$ $\frac{7\pi}{4}$
$r^2 = -25 \cos 2\theta$ $(5, \frac{3\pi}{2})$

59. symmetric about polar axis

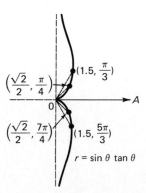

$(1.5, \frac{\pi}{3})$
$(\frac{\sqrt{2}}{2}, \frac{\pi}{4})$
$(\frac{\sqrt{2}}{2}, \frac{7\pi}{4})$ $(1.5, \frac{5\pi}{3})$
$r = \sin\theta \tan\theta$

61. conchoid, symmetric about $\theta = \pi/2$

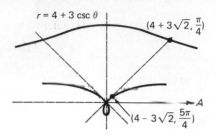

63. parabolic spiral

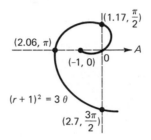

65. two circles (called *osculating circles*) with centers at $(0, \pm\frac{1}{2})$, radii $= \frac{1}{2}$

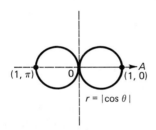

69. $r = -4\cos\theta + 3\sin\theta$ **73.** (a) the line $y = 3$ (b) the line $x = -2$
(c) the line $x = \frac{8}{3}$ (d) the line $2y - 3x = 4$ (e) the line $-y + 2x = 4$

Problems 11.3

1. $(1, \pi/2)$ **3.** $(2 - \sqrt{2}, \pi/4)$; $(2 + \sqrt{2}, 5\pi/4)$; pole   **5.** $(3, \pi/4), (3, 5\pi/4)$
7. $(-2, \pi/2)$; $(-4, \pi)$ **9.** $(\sqrt{3}/2, \pi/6)$; pole
11. $(\frac{3}{2}, k\pi/6)$ for $k = 1, 2, 4, 5, 7, 8, 10, 11$ **13.** $(1, 0)$; pole   **15.** $(1, 1)$; pole
17. pole; $(1/\sqrt{2}, k\pi/8)$ for $k = 1, 3, 5, 7, 9, 11, 13, 15$ **19.** none
25. $(1, k\pi/8)$ for $k = 1, 3, 5, 7, 9, 11, 13, 15$
27. $(12, \pi/6)$; $(12, 5\pi/6)$; $(4, 7\pi/6)$; $(4, 11\pi/6)$ **29.** pole; $(1/2\sqrt{3}, \pi/6), (1/2\sqrt{3}, 5\pi/6)$
31. $(-1, 3\pi/2)$ **33.** $(4k\pi, 0)$ for $k = 0, 1, 2, 3, \ldots$

Problems 11.4

1. 0   **3.** undefined   **5.** $\sqrt{3}/5$ **7.** $-2/\pi$ **9.** $-\sqrt{3}/5$
11. $\tan\beta$ is undefined; $\tan\alpha = 0$ **13.** $\tan\beta = 5/3\sqrt{3}$; $\tan\alpha = -7\sqrt{3}/3$
15. $\tan\beta = 2$; $\tan\alpha = -\frac{1}{2}$ **17.** $\tan\beta = \frac{1}{3}$; $\tan\alpha = 2$
19. $\tan\beta = -1/\sqrt{3}$; $\tan\alpha = 0$ **21.** $\tan\beta = 1/2\sqrt{3}$; $\tan\alpha = 3\sqrt{3}/5$
23. $\tan\beta = -\frac{1}{7}$; $\tan\alpha = [\tan(\pi/28) - \frac{1}{7}]/[1 + \frac{1}{7}\tan(\pi/28)] \approx -0.03$
27. relative maxima at $\left(\dfrac{3 + \sqrt{33}}{4}, \cos^{-1}\left(\dfrac{-1 + \sqrt{33}}{8}\right)\right)$,

$\left(\dfrac{3-\sqrt{33}}{4}, -\cos^{-1}\left(\dfrac{-1-\sqrt{33}}{8}\right)\right)$; relative minima at

$\left(\dfrac{3-\sqrt{33}}{4}, \cos^{-1}\left(\dfrac{-1-\sqrt{33}}{8}\right)\right), \left(\dfrac{3+\sqrt{33}}{4}, -\cos^{-1}\left(\dfrac{-1+\sqrt{33}}{8}\right)\right)$

29, relative maxima at $(\frac{1}{2}, \pi/6)$, $(\frac{1}{2}, 5\pi/6)$; relative minima at $(2, 3\pi/2)$ and pole
33. $\tan^{-1}(\frac{4}{3}) \approx 53°$ and $\tan^{-1}(-\frac{4}{3}) \approx 127°$
35. $\tan^{-1}(\sqrt{3}) = 60°$ and $\tan^{-1}(-\sqrt{3}) = 120°$ **37.** no points of intersection
39. $0°$ **43.** $0, \pi$ **45.** $0, \pi/2, \pi, 3\pi/2$ **47.** $0, \pi/2, \pi, 3\pi/2$ **49.** $0, \pi/2, \pi$
51. $0, \pi/2, \pi$ **53.** $0, \pi/3, 2\pi/3, \pi, 4\pi/3, 5\pi/3$ **55.** $7\pi/6, 11\pi/6$

Problems 11.5

1. $\pi^3/48$ **3.** π **5.** $\pi a^2/20$ **7.** $\pi^5/80$ **9.** $25\pi^9/18$
11. circle: center $(a/2, \pi/2)$, radius $a/2$, $A = \pi a^2/4$
13. circle: center $(a/\sqrt{2}, \pi/4)$, radius $a/\sqrt{2}$, $A = \pi a^2/2$
15. circle: center $(a/\sqrt{2}, 3\pi/4)$, radius $a/\sqrt{2}$, $A = \pi a^2/2$
17. limaçon, symmetric about polar axis, $A = 11\pi$
19. limaçon, symmetric about polar axis, $A = 11\pi$ **21.** four-leaf rose; $A = \pi a^2/2$
23. five-leaf rose; $A = \pi a^2/4$ **25.** lemniscate; symmetric about pole, polar axis, and
$\theta = \pi/2$, $A = a$
27. $A = 2\int_{\sin^{-1}(2/3)}^{\pi/2} \frac{1}{2}(2 - 3\sin\theta)^2\,d\theta = \frac{17}{4}\pi - \frac{17}{2}\sin^{-1}(\frac{2}{3}) - 3\sqrt{5} \approx 0.44$
29. $A = 2\int_{\sin^{-1}(-a/b)}^{\pi/2} \frac{1}{2}(a + b\sin\theta)^2\,d\theta = (a^2 + \frac{1}{2}b^2)[\frac{1}{2}\pi + \sin^{-1}(-a/b)] - \frac{3}{2}a\sqrt{b^2 - a^2}$
31. $A = \int_0^{\cos^{-1}(-2/3)} (2 + 3\cos\theta)^2\,d\theta - \int_{\cos^{-1}(-2/3)}^{\pi} (2 + 3\cos\theta)^2\,d\theta = \frac{17}{2}[2\cos^{-1}(-\frac{2}{3}) - \pi] + 6\sqrt{5} \approx 25.8$
33. $A = \int_0^{\pi} \frac{1}{2}(2 + 2\sin\theta)^2\,d\theta - \int_0^{\pi} \frac{1}{2}(2\sin\theta)^2\,d\theta = 2\pi + 8$
35. $A = 2\int_0^{\pi/2} \frac{1}{2}a^2 \sin 2\theta\,d\theta - \int_0^{\alpha} \frac{1}{2}a^2 \sin 2\theta\,d\theta - \int_{\alpha}^{\pi/2} \frac{1}{2}a^2 \cos^2\theta\,d\theta$ where α solves
$2\sin 2\alpha = 1 + \cos 2\alpha$ or $\alpha \approx 26.6° \approx 0.464$ radians; $A \approx 0.789a^2$

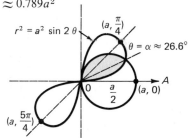

$r^2 = a^2 \sin 2\theta$ $\left(a, \dfrac{\pi}{4}\right)$ $\theta = \alpha \approx 26.6°$ A $(a, 0)$ $\dfrac{a}{2}$ $\left(a, \dfrac{5\pi}{4}\right)$

37. $A = 8[\int_0^{\pi/8} \frac{1}{2}(4\sin 2\theta)^2\,d\theta + \int_{\pi/8}^{\pi/4} \frac{1}{2}(4\cos 2\theta)^2\,d\theta] = 8\pi - 16$

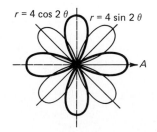

$r = 4\cos 2\theta$ $r = 4\sin 2\theta$ A

39. $A = \pi a^2/4n$. Note that the rose has n petals if n is odd and $2n$ petals if n is even.   **41.** $9m^2$

Review Exercises for Chapter Eleven

1. $(2, 0)$ **3.** $(0, -7)$ **5.** $(0, 1)$ **7.** $(2, 0)$ **9.** $(2, 11\pi/6)$

11. $(6\sqrt{2}, 5\pi/4)$ **13.** circle: radius 8, center $(0, 0)$

15. circle: radius 1, center $(1, 0)$

17. limaçon, symmetric about line $\theta = \pi/2$

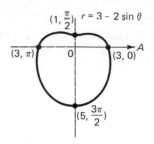

19. four-leaf rose, symmetric about polar axis, pole, $\theta = \pi/2$

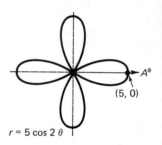

21. eight-leaf rose, symmetric about pole

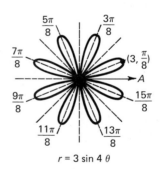

23. logarithmic spiral

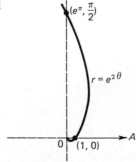

25. straight line $y = 4$ **27.** $(x - 2)^2 + (y - \frac{3}{2})^2 = (\frac{5}{2})^2$ **29.** $(1, 0)$
31. pole, $(1, \pi/2)$ **33.** pole, $(4, 4)$ and, in general, (θ, θ), where $\theta = 2 \pm 2\sqrt{1 + 2k\pi}$,
k an integer   **35.** $(1, k\pi/12)$ for $k = 1, 5, 7, 11, 13, 17, 19, 23$
37. $\tan \alpha = 0$ ($\tan \beta$ is undefined)   **39.** $\tan \alpha = (4 + \pi)/(4 - \pi)$ ($\tan \beta = \pi/4$)
41. $\tan \alpha$ is undefined since $\alpha = \pi/2$ ($\tan \beta = \sqrt{3}$)
43. $\tan \alpha = -1$ ($\tan \beta$ is undefined)
45. $\tan \alpha = -\cot \pi/20 \approx -6.314$ ($\tan \beta$ is undefined)   **47.** $\pi/2, \pi/2$
49. $\theta = 0, \pi/2, \pi, 3\pi/2$ **51.** $\frac{1}{6}[(\pi/2)^3 - (\pi/6)^3] = 13\pi^3/648$ **53.** $\pi a^2/2$
55. 6π **57.** $17(\pi - 2\cos^{-1}\frac{3}{4}) + 9\sqrt{7}$ **59.** $(\pi/2) - 1$ **61.** $2\pi/5$

CHAPTER TWELVE

Problems 12.1

1. 0 **3.** 0 **5.** $\frac{3}{2}$ **7.** $\frac{1}{3}$ **9.** $(1 + a)/(1 - a)$ **11.** 0 **13.** $\sqrt{3}/2$
(first calculate $\lim_{x \to 1}[(1 - x^3)/(1 - x^4)] = \frac{3}{4}$. Then
$\lim_{x \to 1^-} \sqrt{(1 - x^3)/(1 - x^4)} = \sqrt{\lim_{x \to 1}[(1 - x^3)/(1 - x^4)]} = \sqrt{3}/2$ by Theorem
2.3.2)   **15.** 0 **17.** e^2 **19.** 0 **21.** $\frac{1}{3}$ **23.** -2 **25.** $1/2\sqrt{x_0}$
29. $\frac{1}{4}$ **35.** Et/L

Problems 12.3

1. $\frac{1}{2}$ **3.** 0 **5.** 0 **7.** 0 **9.** 1 **11.** 0 **13.** 1 **15.** 0
17. $+\infty$ **19.** e **21.** $\frac{3}{2}$
23. 1 (since $\lim_{x \to 0^+}(\sin x)^{(\sin x) - x} = \lim_{x \to 0^+}[(\sin x)^{\sin x}/(\sin x)^x] = $
$\lim_{x \to 0^+}(\sin x)^{\sin x}/\lim_{x \to 0^+}(\sin x)^x = 1/1 = 1)$ **25.** limit doesn't exist since
$\lim_{x \to 2^+} x^{1/(2-x)} = 0$ and $\lim_{x \to 2^-} x^{1/(2-x)} = \infty$ **27.** 1 **29.** $+\infty$ **31.** $+\infty$

Problems 12.4

1. $\frac{1}{2}$ **3.** diverges   **5.** 0 **7.** diverges   **9.** $\frac{1}{2}\ln\frac{11}{9}$
11. $(1/b)[(\pi/2) - \tan^{-1}(a/b)]$ **13.** diverges   **15.** diverges   **17.** 2
19. diverges   **21.** diverges   **23.** $\frac{3}{4}(3^{2/3} - 1)$ **25.** diverges   **27.** $\frac{1}{2}$
29. $\pi/2$ **31.** diverges   **33.** $\frac{5}{4} \cdot 2^{4/5}$ **35.** diverges
37. diverges (note that $1/(1 - \sin x) = (1 + \sin x)/\cos^2 x$ **39.** diverges
41. diverges (since $\int_2^3 \{1/[(x - 2)(x - 4)]\}\, dx$ diverges)   **43.** $1/\ln 5$ **45.** 1
47. $\pi/2$ **49.** $\pi[\sqrt{2} + \ln(\sqrt{2} + 1)] (2\pi \int_0^\infty e^{-x}\sqrt{1 + e^{-2x}}\, dx =$
$2\pi \int_0^1 \sqrt{1 + u^2}\, du\ (u = e^{-x}) = 2\pi \int_0^{\pi/4} \sec^3\theta\, d\theta\ (u = \tan\theta))$ **51.** \$458,333.33
53. the business—which will give him a present value of \$233,333—assuming no
extraordinary risks, of course   **59.** because $\sqrt{1 + x^{1/8}}/x^{3/4} > 1/x^{3/4}$ **61.** for
$|x| \geq 1, xe^{-x^2} > e^{-x^2}$ **67.** $(a + b)/2$ **71.** a^2

Review Exercises for Chapter Twelve

1. $\frac{3}{2}$ **3.** no limit   **5.** 0 **7.** $+\infty$ **9.** 1 **11.** $+\infty$ **13.** 1
15. $-\frac{1}{2}$ **17.** $+\infty$ **19.** 0 **21.** $\frac{1}{3}$ **23.** $6/7^4 = \frac{6}{2401}$ **25.** $\frac{1}{4}\ln\frac{13}{12}$
27. diverges   **29.** $\frac{2}{3}$ **31.** diverges
33. diverges ($\sqrt{x - 2}$ not defined for $x < 2$)   **35.** diverges   **37.** diverges
39. diverges (since $\int_0^1 \{[1/(x - 2)] - [1/(x - 1)]\}\, dx$ diverges)   **41.** $\pi/3$
43. $(\bar{x}, \bar{y}) = (1, \frac{1}{5})$ **45.** since $\sqrt{x^{4/5} - 1}/x^{21/5} < x^{2/5}/x^{21/5} = 1/x^{19/5}$
47. since $1/\ln\sqrt{2 + x} > 1/\sqrt{2 + x}$ **49.** 5

CHAPTER THIRTEEN

Problems 13.1

1. $1 - (x - \pi/4) - (x - \pi/4)^2/2! + (x - \pi/4)^3/3! + (x - \pi/4)^4/4! - (x - \pi/4)^5/5! - (x - \pi/4)^6/6!$

3. $1 + \dfrac{1}{e}(x - e) - \dfrac{1}{2e^2}(x - e)^2 + \dfrac{1}{3e^3}(x - e)^3 - \dfrac{1}{4e^4}(x - e)^4 + \dfrac{1}{5e^5}(x - e)^5$

5. $1 - (x - 1) + (x - 1)^2 - (x - 1)^3 + (x - 1)^4$ **7.** $x + \frac{1}{3}x^3$

9. $(x - \pi) + \frac{1}{3}(x - \pi)^3$ **11.** $1 - x^2 + x^4$ **13.** $x + (x^3/3!)$

15. $-\frac{1}{2}[x - (\pi/2)]^2$ **17.** $\dfrac{1}{2} + \dfrac{1}{2^4}x + \dfrac{3}{2^7 \cdot 2!}x^2 + \dfrac{3 \cdot 5}{2^{10} \cdot 3!}x^3 + \dfrac{3 \cdot 5 \cdot 7}{2^{12} \cdot 4!}x^4$

19. $1 + \alpha x + \dfrac{\alpha^2}{2!}x^2 + \dfrac{\alpha^3}{3!}x^3 + \dfrac{\alpha^4}{4!}x^4 + \dfrac{\alpha^5}{5!}x^5 + \dfrac{\alpha^6}{6!}x^6$ **21.** $x + x^3/3!$

23. $(a_0 + a_1 + a_2 + a_3) + (a_1 + 2a_2 + 3a_3)(x - 1) + (a_2 + 3a_3)(x - 1)^2 + a_3(x - 1)^3$ **25.** x^2

Problems 13.3

1. $(1/7!)(\pi/4)^7 \approx 0.0000366$ **3.** $(120/5!)(\frac{1}{2})^{-6}(\frac{1}{2})^5 = 2$

5. $(1 \cdot 3 \cdot 5 \cdot 7 \cdot 9 \cdot 11/64)(\frac{19}{4})^{-13/2}(\frac{1}{4})^6(1/6!) \approx 0.000009$

7. $[2 \sec^4(\pi/6) + 4 \sec^2(\pi/6) \tan^2(\pi/6)](1/4!)(\pi/6)^4 \approx 0.0167$

9. $(1/5!)(\frac{1}{3})^5(e^{1/9})(\frac{120}{3} + \frac{160}{27} + \frac{32}{243}) \approx 0.0018$

11. $\frac{1}{2} + (\sqrt{3}/2)(0.2) - (\frac{1}{2})(\frac{1}{2})(0.2)^2 - (\sqrt{3}/2)(\frac{1}{6})(0.2)^3 \approx 0.66205$

13. $\tan(\pi/4) + \sec^2(\pi/4)(0.01) + [2 \sec^2(\pi/4) \tan(\pi/4)/2](0.01)^2 = 1.22$

15. $1 + (-1) + \dfrac{1}{2!}(-1)^2 + \dfrac{1}{3!}(-1)^3 + \dfrac{1}{4!}(-1)^4 + \dfrac{1}{5!}(-1)^5 + \dfrac{1}{6!}(-1)^6 \approx 0.36806$

17. $0.5 - \dfrac{(0.5)^2}{2} + \dfrac{(0.5)^3}{3} - \dfrac{(0.5)^4}{4} + \dfrac{(0.5)^5}{5} - \dfrac{(0.5)^6}{6} + \dfrac{(0.5)^7}{7} \approx 0.4058$

19. $1 + 3 + 3^2/2! + \cdots + 3^{12}/12! \approx 20.0852$ **21.** $\frac{1}{2} + (\frac{1}{2})^3/6 \approx 0.5208$

23. $\sin[(\pi/2) + (\pi/18)] \approx 1 - \frac{1}{2}(\pi/18)^2 \approx 0.98477$

25. $1 - \frac{1}{2}(0.1) + \frac{3}{8}(0.1)^2 = 0.95375$

27. $\int_0^{1/3} e^{x^2} dx \approx \int_0^{1/3}[1 + x^2 + (x^4/2)] dx = \frac{1}{3} + \frac{1}{3}(\frac{1}{3})^3 + \frac{1}{10}(\frac{1}{3})^5 \approx 0.3461$; error $< \frac{1}{3}(0.0018) = 0.0006$

Problems 13.4

1. $q_1(x) = (2/\pi)x$ **3.** $q_2(x) = -\frac{1}{60}x^2 + \frac{23}{60}x + 1$

5. $q_2(x) = (\cosh 1 - 1)x^2 + (\sinh 1)x + 1 = \{[(e + e^{-1})/2] - 1\}x^2 + \{[(e - e^{-1})/2]\}x + 1$ **7.** $q_2(x) = (1/\pi^2)[48 - (72/\sqrt{3})]x^2 + (1/\pi)[(18/\sqrt{3}) - 8]x$

9. $q_2(x) = \frac{7}{4}x^2 - \frac{3}{4}x$ **11.** $q_1(x) = -0.117779 + 0.405463x$; $q_1(2.3) = 0.8147859$; max error $\le (0.3)(0.7)/(2)(2^2) = 0.02625$ (actual error ≈ 0.018123)

13. $q_3(x) = -0.937499 + 1.220053x - 0.246195x^2 + 0.021915x^3$; $q_3(2.3) = 0.832891155$; max error $\le [(\frac{3}{10})(\frac{1}{30})(\frac{11}{30})(\frac{7}{10})/4!](6/2^2) \approx 0.0001604$; actual error ≈ 0.00001797

15. (a) max value occurs at $x - x_3 = -\sqrt{1.5 + \sqrt{145}}/10 \, \Delta x$ and is approximately equal to $(3.6314)(\Delta x)^5$. (b) If $|f(x) - q(x)| < (3.6314)(\Delta x)^5/5!$

17. $q_4(x) = \frac{1}{3}(0.97509 + 2.75045x - 0.9561x^2 + 0.26224x^3 - 0.03168x^4)$ $q_4(1.2) = 1.095435$; max error $\approx (0.2)(0.05)(0.3)(0.55)(0.8)(\frac{1}{2})(\frac{1}{2})(\frac{3}{2})(\frac{5}{2})(\frac{7}{2})/5! \approx 0.000036$ (actual error ≈ 0.000001) $q_4(1.7) = 1.303839$; max error ≈ 0.0000258 (actual error ≈ 0.0000016) **21.** $q_1(x) = 0.63662x$ (error on $[0, \pi/2] < 0.308$) $q_2(x) = -0.335749x^2 + 1.16401x$ (error on $[0, \pi/2] < 0.2486$) $q_3(x) = -0.113872x^3 - 0.065471x^2 + 1.02043x$ (error on $[0, \pi/2] < 0.1427$)

Estimate	Bound for this estimate	Actual error
$q_1(0.5) = 0.31831$	0.2677	0.16112
$q_2(0.5) = 0.49807$	0.02547	0.01860
$q_3(0.5) = 0.47961$	0.000576	0.000184
$q_1(1) = 0.63662$	0.285	0.20485
$q_2(1) = 0.82826$	0.02042	0.01321
$q_3(1) = 0.84109$	0.000535	0.00038

Review Exercises for Chapter Thirteen

1. $1 + x + (x^2/2!) + (x^3/3!)$

3. $\frac{1}{2} + (\sqrt{3}/2)[x - (\pi/2)] - \frac{1}{4}[x - (\pi/6)]^2 + (\sqrt{3}/12)[x - (\pi/6)]^3$

5. $-[x - (\pi/2)] + \frac{1}{3}[x - (\pi/2)]^3$ **7.** $3 + 2x - x^2 + x^3$

9. $(\pi/3)^6/6! \approx 0.00183$ **11.** $e/7! < 1/7! \approx 0.000198$

13. $f^{(5)}(x) = -32x(x^4 - 5x^2 + 8)e^{-x^2}$ and crude bound is $(32)(8)/5! \approx 2.13$

15. $(1/\sqrt{2}) - (1/\sqrt{2})(-\pi/90) - (1/\sqrt{2})(\frac{1}{2})(-\pi/90)^2 \approx 0.73136$

17. $\ln[(1 + x)/(1 - x)] = 2(x + \frac{1}{3}x^3 + \frac{1}{5}x^5 + \cdots)$ and, with $x = \frac{1}{3}$,

$\ln 2 \approx 2(\frac{1}{3} + \frac{1}{81} + \frac{1}{1215} + \frac{1}{15309}) \approx 0.69313$ **19.** $0.3 - \frac{1}{3}(0.3)^3 + \frac{1}{5}(0.3)^5 = 0.291486$

(actual error ≈ 0.000029)   **21.** $\int_0^{1/2}[1 - (x^4/2)]\, dx = \frac{159}{320} = 0.496875$;

error $\leq \frac{1}{2}\left(\dfrac{1/2^8}{4!}\right) \approx 0.00033$ **23.** $q_2(x) = [8(1 - \sqrt{2})/\pi^2]x^2 + [(4\sqrt{2} - 2)/\pi]x \approx$

$-0.335749x^2 + 1.16401x$ **25.** $q_2(x) = \frac{1}{10}x^2 - \frac{3}{5}x + 1$

27. $q_3(x) = 25x^3 - 60x^2 + 36x$ **29.** 1.181945; error $\leq \frac{1}{72} \approx 0.01389$

(actual error ≈ 0.01154)

CHAPTER FOURTEEN

Problems 14.1

1. $\frac{1}{3}, \frac{1}{9}, \frac{1}{27}, \frac{1}{81}, \frac{1}{243}$ **3.** $\frac{3}{4}, \frac{15}{16}, \frac{63}{64}, \frac{255}{256}, \frac{1023}{1024}$ **5.** $e, e^{1/2}, e^{1/3}, e^{1/4}, e^{1/5}$

7. $0, 0, 0, 0, 0$ **9.** $1, 0, -1, 0, 1$ **11.** 0 **13.** divergent   **15.** divergent

17. $\frac{4}{7}$ **19.** $e^{1/4}$ **21.** 0 **23.** 0 **25.** divergent   **27.** 0

29. $a_n = n \cdot 5^{n-1}$ **31.** $a_n = 1 - 1/2^n$ **33.** $a_n = (-\frac{1}{3})^{n-1}$

Problems 14.2

1. $\frac{1}{2}$ **3.** 1 **5.** 1 **7.** 1 **9.** 2 **11.** $(\ln 3)/3 \approx 0.366$ **15.** SI

17. NM   **19.** SD   **21.** SI   **23.** NM   **25.** SI   **27.** SI

Problems 14.3

1. 20 **3.** 7 **5.** $\frac{61}{20}$ **7.** $\sum_{k=0}^{4} 2^k$ **9.** $\sum_{k=2}^{n} k/k + 1$ **11.** $\sum_{k=1}^{n} k^{1/k}$

13. $\sum_{k=1}^{8} (-1)^{k+1} x^{5k}$ **15.** $\sum_{k=1}^{8} (2k - 1)(2k + 1) = \sum_{k=0}^{7} (2k + 1)(2k + 3)$

Problems 14.4

1. $\frac{4}{3}$ **3.** $2 - 1 - \frac{1}{2} = \frac{1}{2}$ **5.** 2 **7.** 125 **9.** $\frac{1}{2}$ **11.** 1

13. $8(3 - 1 - \frac{2}{3}) = \frac{32}{3}$ **15.** $25/6^4 = \frac{25}{1296}$ **17.** $\frac{35}{99}$ **19.** $\frac{71}{99}$ **21.** $\frac{501}{999}$

23. $\frac{11351}{99900}$ **27.** $\frac{3}{2}$ **29.** $\frac{1}{4}$ **39.** $1\frac{1}{11}$ hr $= 1:05\frac{5}{11}$ P.M.

41. $8 + 8 \cdot 2 \cdot \frac{2}{3} + 8 \cdot 2 \cdot (\frac{2}{3})^2 + 8 \cdot 2 \cdot (\frac{2}{3})^3 + \cdots = 40$ m

Problems 14.5

1. C   **3.** D   **5.** C   **7.** C   **9.** C (integral test)   **11.** C (integral test)
13. C   **15.** D (ratio test)   **17.** C   **19.** D   **21.** D (terms $\nrightarrow 0$)
23. C (root or ratio test)   **25.** D   **27.** D   **29.** D   **31.** C   **33.** C
35. D (since $\lim_{x \to \infty} [\sin(1/x)/(1/x)] = 1$, $\Sigma \sin(1/k)$ is like the harmonic series $\Sigma\, 1/k$ for k large)   **37.** C (integral test)   **39.** C ($\int_0^\infty \operatorname{sech} x\, dx = \ln(\operatorname{sech} x + \tanh x)|_0^\infty = 1$)
41. C ($\int_0^\infty \operatorname{sech}^2 x\, dx = \tanh x|_0^\infty = 1$)   **43.** D   **45.** D ($\sqrt{10}\,\ln^{10} k < k$ for k sufficiently large; compare with harmonic series)

Problems 14.6

1. D   **3.** CC   **5.** D   **7.** CC   **9.** D   **11.** D   **13.** AC
15. CC   **17.** AC   **19.** CC   **21.** D   **23.** D   **25.** CC
27. D   **29.** CC   **33.** $1 - 1/2^2 + 1/3^2 - 1/4^2 + 1/5^2 - 1/6^2 + 1/7^2 - 1/8^2 + 1/9^2 - 1/10^2 \approx 0.818$
35. $-(1/2 \ln 2) + (1/3 \ln 3) - (1/4 \ln 4) + (1/5 \ln 5) \approx -0.474$
37. $1 - 1/\sqrt{2} + 1/\sqrt{3} - \cdots + 1/\sqrt{99} - 1/\sqrt{100} \approx 0.555024$
39. $1 - \frac{1}{2} - \frac{1}{4} + \frac{1}{3} - \frac{1}{6} - \frac{1}{8} + \frac{1}{5} - \frac{1}{10} - \frac{1}{12} - \frac{1}{14}$
43. $\Sigma a_k = \Sigma 1/k$ is the most obvious example.

Problems 14.7

1. $6; (-6, 6)$ **3.** $3; (-4, 2)$ **5.** $\frac{1}{3}; (-\frac{1}{3}, \frac{1}{3})$ **7.** $1; [0, 2]$
9. $\infty; (-\infty, \infty)$ **11.** $1; (-1, 1)$ **13.** $\infty; (-\infty, \infty)$ **15.** $1; (-1, 1)$
17. $\infty; (-\infty, \infty)$ **19.** $\frac{1}{2}; [-\frac{1}{2}, \frac{1}{2}]$ **21.** $\frac{5}{2}; (-4, 1)$ **23.** $0; x = 0$
25. $1; (-1, 1)$ **27.** $0; x = -1$ **29.** $3; [-13, -7)$ **31.** $1; (-1, 1)$
33. $1; (-1, 1)$

Problems 14.8

1. $\Sigma_{k=0}^\infty (-1)^k x^{2k}$ **3.** $\Sigma_{k=0}^{50} (-1)^k/(2k + 1) = 1 - \frac{1}{3} + \frac{1}{5} - \cdots - \frac{1}{99} + \frac{1}{101} \approx 3.14$
5. $\int_0^1 [1 - t^2 + (t^4/2) - (t^6/6) + (t^8/24)]\, dt \approx 0.7475$ **7.** $\int_0^{1/2} [1 - (t^4/2)]\, dt = 0.496875$ **9.** $\int_0^1 [t^2 - t^4 + (t^6/2) - (t^8/6) + (t^{10}/24)]\, dt \approx 0.190$
11. $\int_0^1 [1 - (t/2) + (t^2/24)]\, dt \approx 0.7639$ **13.** $\int_0^{1/2} (1 - t^8)\, dt \approx 0.499783$
15. $\Sigma_{k=0}^\infty x^{k+2}/[(k + 2)k!]$ **17.** $\Sigma_{k=1}^\infty (-1)^{k+1} x^k/k^2$ **19.** (a) $(-\infty, \infty)$

Problems 14.9

3. $\Sigma_{k=0}^\infty e(x - 1)^k/k!$
5. $\dfrac{1}{\sqrt{2}}\left[1 - \left(x - \dfrac{\pi}{4}\right) - \dfrac{[x - (\pi/4)]^2}{2} + \dfrac{[x - (\pi/4)]^3}{3!} + \dfrac{[x - (\pi/4)]^4}{4!} - \cdots\right]$
7. $\Sigma_{k=0}^\infty (\alpha x)^k/k!$ **9.** $\Sigma_{k=0}^\infty (-1)^k x^{2k+2}/k!$ **11.** $(1/e) \Sigma_{k=0}^\infty (x + 1)^k/k!$
13. $x \ln x = (x - 1) \ln x + \ln x = \displaystyle\sum_{k=2}^\infty \dfrac{(x - 1)^k}{k(k - 1)} + \sum_{k=1}^\infty \dfrac{(x - 1)^k}{k} =$
$(x - 1) + \displaystyle\sum_{k=2}^\infty \dfrac{(x - 1)^k}{k - 1}$ **15.** $1 + 0 + \dfrac{[x - (\pi/2)]^2}{2!} + 0; (0, \pi)$
17. $2 + \displaystyle\sum_{k=1}^\infty \dfrac{(-1)^{k+1} 1 \cdot 3 \cdot \cdots \cdot (2k - 3)}{(2^{3k-1})k!} (x - 4)^k$ **19.** $\displaystyle\sum_{k=1}^\infty \dfrac{(-1)^{k+1} x^{4k-2}}{(2k - 1)!}$
27. (a) $(2/\sqrt{\pi}) \Sigma_{k=0}^\infty (-1)^k x^{2k+1}/[(2k + 1)k!]$

(b) $\text{erf}(1) \approx \dfrac{2}{\sqrt{\pi}}\left(1 - \dfrac{1}{3} + \dfrac{1}{5\cdot 2} - \dfrac{1}{7\cdot 3!} + \dfrac{1}{9\cdot 4!} - \dfrac{1}{11\cdot 5!} + \dfrac{1}{13\cdot 6!}\right) \approx 0.84271$

$\text{erf}\left(\dfrac{1}{2}\right) \approx \dfrac{2}{\sqrt{\pi}}\left(0.5 - \dfrac{(0.5)^3}{3} + \dfrac{(0.5)^5}{5\cdot 2} - \dfrac{(0.5)^7}{7\cdot 6}\right) \approx 0.52049$

29. (b) $\displaystyle\sum_{k=0}^{\infty} \dfrac{(-1)^k x^{2k+1}}{(2k+1)[(2k+1)!]} = x - \dfrac{x^3}{3\cdot 3!} + \dfrac{x^5}{5\cdot 5!} - \dfrac{x^7}{7\cdot 7!} + \cdots$

(c) $\text{Si}(1) \approx 1 - (1/3\cdot 3!) + (1/5\cdot 5!) \approx 0.94611;\ \text{Si}(\tfrac{1}{2}) \approx (0.5) - (0.5)^3/3\cdot 3! \approx 0.49306$

Review Exercises for Chapter Fourteen

1. $-1, 0, \tfrac{1}{3}, \tfrac{2}{4}, \tfrac{3}{5}$ **3.** 0   **5.** 0   **7.** e^{-2} **9.** $a_n = (2n-1)/2^{n+2}$
11. U, NM   **13.** Bounded by 1, I   **15.** Bounded by 1, D
17. Bounded by 1, I   **19.** B, D   **21.** 1,398,096   **23.** $\tfrac{27}{16} - \tfrac{8}{75} = \tfrac{1897}{1200} \approx 1.58$
25. $\tfrac{79}{99}$ **27.** $9\tfrac{9}{11}$ hr $\approx$ 9:49.09 P.M.   **29.** C   **31.** D   **33.** C   **35.** D
37. C   **39.** C   **41.** CC   **43.** CC   **45.** AC   **47.** CC   **49.** D   **51.** D
53. $1 - 1/2^3 + 1/3^3 - 1/4^3 + 1/5^3 - 1/6^3 + 1/7^3 - 1/8^3 + 1/9^3 - 1/10^3 \approx 0.90112$
55. $1 - \tfrac{1}{2} - \tfrac{1}{4} + \tfrac{1}{3} - \tfrac{1}{6} + \tfrac{1}{5} - \tfrac{1}{8} + \tfrac{1}{7} - \tfrac{1}{10} - \tfrac{1}{12}$ **57.** 3; $(-3, 3)$
59. $\infty, (-\infty, \infty)$ **61.** $\infty; (-\infty, \infty)$ **63.** 0; $x = 0$ **65.** 1; $[0, 2)$
67. 3; $[-11, -5)$ **69.** $\int_0^{1/2} [t^2 - (t^6/3!)]\, dt \approx 0.04148$
71. $\int_0^{1/2} (1 - t^4 + t^8 - t^{12})\, dt \approx 0.493958$ **73.** $2\sum_{k=0}^{\infty} (x - \ln 2)^k/k!$
75. $\sum_{k=0}^{\infty} (-1)^k (\alpha x)^{2k+1}/(2k+1)!$

CHAPTER FIFTEEN

Problems 15.1

1. $(3, 3)$ **3.** $(4, 4)$ **5.** $(-2, -5)$ **7.** $|v| = 4\sqrt{2}, \theta = \pi/4$
9. $|v| = 4\sqrt{2}, \theta = 7\pi/4$ **11.** $|v| = 2, \theta = \pi/6$ **13.** $|v| = 2, \theta = 2\pi/3$
15. $|v| = 2, \theta = 4\pi/3$ **17.** $|v| = \sqrt{89}, \theta = \pi + \tan^{-1}(-\tfrac{8}{5}) \approx 2.13$ (second
quadrant)   **19.** j   **21.** $-6i + j$ **23.** $-9i + 5j$ **25.** $-5i + 5j$
27. (a) $(6, 9)$ (b) $(-3, 7)$ (c) $(-7, 1)$ (d) $(39, -22)$ **33.** $(1/\sqrt{2})i - (1/\sqrt{2})j$
35. $\tfrac{3}{5}i - \tfrac{4}{5}j$ **37.** $(1/\sqrt{2})i + (1/\sqrt{2})j$ if $a > 0$; $-(1/\sqrt{2})i - (1/\sqrt{2})j$ if $a < 0$
39. $\sin\theta = -3/\sqrt{13},\ \cos\theta = 2/\sqrt{13}$ **41.** $-(1/\sqrt{2})i - (1/\sqrt{2})j$ **43.** $\tfrac{3}{5}i - \tfrac{4}{5}j$
45. $\tfrac{3}{5}i + \tfrac{4}{5}j$ **47.** (a) $(1/\sqrt{2})i - (1/\sqrt{2})j$, (b) $(7/\sqrt{193})i - (12/\sqrt{193})j$,
(c) $-(2/\sqrt{53})i + (7/\sqrt{53})j$ **51.** $4i + 4\sqrt{3}j$ **53.** $4j$ **55.** $-3i + 3\sqrt{3}j$
57. $3\sqrt{3}i - 3j$

Problems 15.2

1. 0; 0   **3.** 0; 0   **5.** 20; $\tfrac{20}{29}$ **7.** $-22; -22/5\sqrt{53}$ **9.** 100; $20/\sqrt{481}$
11. $u \cdot v = \alpha\beta - \beta\alpha = 0$ **13.** parallel   **15.** neither   **17.** orthogonal
19. parallel   **21.** (a) $-\tfrac{3}{4}$, (b) $\tfrac{4}{3}$ (c) $\tfrac{1}{4}$ (d) $(-96 + \sqrt{7500})/78 \approx -0.12$
25. $\tfrac{3}{2}i + \tfrac{3}{2}j$ **27.** 0   **29.** $-\tfrac{2}{13}i + \tfrac{3}{13}j$ **31.** $\tfrac{14}{5}i + \tfrac{28}{5}j$ **33.** $-\tfrac{14}{5}i + \tfrac{28}{5}j$
35. $[(\alpha + \beta)/2]i + [(\alpha + \beta)/2]j$ **37.** $[(\alpha - \beta)/2]i + [(\alpha - \beta)/2]j$
39. $a_1a_2 + b_1b_2 > 0$ **41.** $\text{Proj}_{\overrightarrow{PQ}}\overrightarrow{RS} = \tfrac{51}{25}i + \tfrac{68}{25}j$; $\text{Proj}_{\overrightarrow{RS}}\overrightarrow{PQ} = -\tfrac{17}{26}i + \tfrac{85}{26}j$
47. $52/(5\sqrt{113}) \approx 0.9783$; $61/(\sqrt{34}\sqrt{113}) \approx 0.9841$; $-27/(5\sqrt{34}) \approx -0.9261$
49. When the vectors (a_1, a_2) and (b_1, b_2) have the same direction   **51.** $\sqrt{5}$

Problems 15.3

1. $\overrightarrow{RS} = \tfrac{1}{3}P + \tfrac{2}{3}Q - R$ **3.** $\overrightarrow{ST} = -\tfrac{7}{20}P - \tfrac{1}{4}Q + \tfrac{3}{5}R$ **5.** $(\tfrac{2}{3}, \tfrac{5}{3})$
11. $2i + 5j; -2i - 5j$ **13.** $-10i + 8j; 10i - 8j$

15. $(5 - 2\sqrt{2})\mathbf{i} + (-2\sqrt{2} - \sqrt{3})\mathbf{j}$; $(2\sqrt{2} - 5)\mathbf{i} + (2\sqrt{2} + \sqrt{3})\mathbf{j}$

17. $-5\sqrt{3}\mathbf{j}$; $5\sqrt{3}\mathbf{j}$ 19. -3 joules   21. $-6\sqrt{2}$ joules

23. $\frac{21}{2}(1 + \sqrt{3})$ joules   25. -12 joules   27. $-135/\sqrt{13}$ joules

Review Exercises for Chapter Fifteen

1. $|\mathbf{v}| = 3\sqrt{2}, \theta = \pi/4$ 3. $|\mathbf{v}| = 4, \theta = 5\pi/3$ 5. $|\mathbf{v}| = 12\sqrt{2}, \theta = 5\pi/4$

7. $2\mathbf{i} + 2\mathbf{j}$ 9. $4\mathbf{i} + 2\mathbf{j}$ 11. (a) $(10, 5)$, (b) $(5, -3)$, (c) $(-31, 12)$

13. $(1/\sqrt{2})\mathbf{i} + (1/\sqrt{2})\mathbf{j}$ 15. $(2/\sqrt{29})\mathbf{i} + (5/\sqrt{29})\mathbf{j}$ 17. $\frac{3}{5}\mathbf{i} + \frac{4}{5}\mathbf{j}$

19. $(1/\sqrt{2})\mathbf{i} - (1/\sqrt{2})\mathbf{j}$ if $a > 0$ and $-(1/\sqrt{2})\mathbf{i} + (1/\sqrt{2})\mathbf{j}$ if $a < 0$

21. $-(5/\sqrt{29})\mathbf{i} - (2/\sqrt{29})\mathbf{j}$ 23. $-(10/\sqrt{149})\mathbf{i} + (7/\sqrt{149})\mathbf{j}$ 25. $\mathbf{j}$

27. $-\frac{7}{2}\sqrt{3}\mathbf{i} + \frac{7}{2}\mathbf{j}$ 29. $0; 0$ 31. $-14, -14/(\sqrt{5}\sqrt{41})$ 33. neither

35. parallel   37. parallel   39. $7\mathbf{i} + 7\mathbf{j}$ 41. $\frac{15}{13}\mathbf{i} + \frac{10}{13}\mathbf{j}$ 43. $-\frac{3}{2}\mathbf{i} - \frac{7}{2}\mathbf{j}$

45. $\text{Proj}_{\overrightarrow{RS}}\overrightarrow{PQ} = -\frac{99}{25}\mathbf{i} + \frac{132}{25}\mathbf{j}$; $\text{Proj}_{\overrightarrow{PQ}}\overrightarrow{RS} = -\frac{33}{82}\mathbf{i} - \frac{297}{82}\mathbf{j}$ 47. $-3\mathbf{i} - 3\mathbf{j}$

49. $-\sqrt{2}$ joules   51. $-11(1 + 3\sqrt{3})$ joules

CHAPTER SIXTEEN

Problems 16.1

1. $R - \{0, 1\}$ 3. $R - \{-1, 1\}$ 5. $(0, 1)$

7. $R - \{n\pi/2: n = \pm1, \pm2, \ldots\}$ 9. $y^2 = 4x$ 11. $y = x^{3/2}$

13. $y = 2x + 5$ 15. $x = y^2 + y + 1$ 17. $y = x^3 - 1$

19. $y = (\ln x)^2, x > 0$ 21. $x^2 + y^2 = e^{2\tan^{-1}(x/y)}$ valid for $-\pi/2 < t < \pi/2$. The graph of this equation is a spiral.   23. $y = x^2, x > 0$

25. $(2 - t)\mathbf{i} + (4 + 2t)\mathbf{j}$ 27. $(3 - 4t)\mathbf{i} + (5 - 12t)\mathbf{j}$
    $(1 + t)\mathbf{i} + (6 - 2t)\mathbf{j}$ $(-1 + 4t)\mathbf{i} + (-7 + 12t)\mathbf{j}$

29. $(-2 + 6t)\mathbf{i} + (3 + 4t)\mathbf{j}$ 33. $(650\sqrt{2}/4)^2 = 52{,}812.5$ ft
    $(4 - 6t)\mathbf{i} + (7 - 4t)\mathbf{j}$ (using the value $g = 32$ ft/sec²)

35. $x(t) = 50\sqrt{3}t$; $y(t) = -(9.81t^2/2) - 50t + 150$ 37. $x = \frac{1}{2}\alpha - \frac{1}{4}\sin \alpha$;
$y = \frac{1}{2} - \frac{1}{4}\cos \alpha$

47.

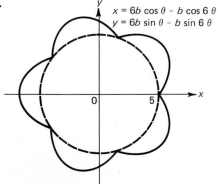

$x = 6b \cos \theta - b \cos 6\theta$
$y = 6b \sin \theta - b \sin 6\theta$

Problems 16.2

1. $\frac{2}{3}$ 3. $-1/\sqrt{2}$ 5. $-e^{-4}$ 7. 0 9. $\sqrt{2}$ 11. 0 13. $-1/\sqrt{3}$

15. $-16/\pi^2$ 17. $y = \frac{2}{3}x + \frac{22}{3}$ 19. $y = -e^{-4}x + 2e^{-2}$ 21. $y = \sqrt{2}$

23. none (Note that condition (1) does not hold at $t = 0$ and that the Cartesian equation of the curve is the straight line $y = x - 3$ $(x \geq -1)$ which has neither horizontal nor vertical tangents.)   25. V: $(0, 1), (0, -1)$; H: $(1, 0), (-1, 0)$

(note that $dy/dx = -\cot \theta$)   27. V: $(1, -\sqrt{3}/2), (-1, 0)$; H: $(0, 1), (\sin 3\pi/5, -1)$

29. none   **31.** V: $(1, 0)$; H: none   **33.** none (condition (1) does not hold at $t = 0$)   **35.** none   **37.** -1 **39.** $(4 + 3\sqrt{3})/(4\sqrt{3} + 3)$ **41.** $\frac{1}{2}$ **43.** 0
45. $y = -x$; $y = x - 2$; $y = -(x/2) - \frac{1}{2}$ **47.** when $\cos \alpha = r/s$

Problems 16.3

1. $\mathbf{f}' = \mathbf{i} - 5t^4\mathbf{j}$; $\mathbf{f}'' = -20t^3\mathbf{j}$ **3.** $\mathbf{f}' = (2 \cos 2t)\mathbf{i} - 3(\sin 3t)\mathbf{j}$;
$\mathbf{f}'' = (-4 \sin 2t)\mathbf{i} - (9 \cos 3t)\mathbf{j}$ **5.** $\mathbf{f}' = (1/t)\mathbf{i} + 3e^{3t}\mathbf{j}$; $\mathbf{f}'' = -(1/t^2)\mathbf{i} + 9e^{3t}\mathbf{j}$
7. $\mathbf{f}' = (\sec^2 t)\mathbf{i} + (\sec t)(\tan t)\mathbf{j}$; $\mathbf{f}'' = 2(\sec^2 t)(\tan t)\mathbf{i} + [\sec^3 t + (\sec t)(\tan^2 t)]\mathbf{j}$
9. $\mathbf{f}' = -(\tan t)\mathbf{i} + (\cot t)\mathbf{j}$; $\mathbf{f}'' = -(\sec^2 t)\mathbf{i} - (\csc^2 t)\mathbf{j}$ **11.** $(2/\sqrt{13})\mathbf{i} + (3/\sqrt{13})\mathbf{j}$
13. $\mathbf{j}$ **15.** $-(1/\sqrt{2})\mathbf{i} + (1/\sqrt{2})\mathbf{j}$ **17.** $\mathbf{i}$ **19.** $(4/\sqrt{97})\mathbf{i} - (9/\sqrt{97})\mathbf{j}$
21. $\frac{8}{3}\mathbf{i} + 4\mathbf{j}$ **23.** $(2\sqrt{t} + C_1)\mathbf{i} + (\frac{2}{3}t^{3/2} + C_2)\mathbf{j}$ **25.** $(\cosh 1 - 1)\mathbf{i} - (\sinh 1)\mathbf{j}$
27. $(t \ln t - t + C_1)\mathbf{i} + (te^t - e^t + C_2)\mathbf{j}$ **29.** $[(t^4/4) + 2]\mathbf{i} + [5 - (t^6/6)]\mathbf{j}$
31. $(\sin t)\mathbf{i} - (\cos t)\mathbf{j}$ **33.** $-(a/\sqrt{a^2 + b^2})\mathbf{i} + (b/\sqrt{a^2 + b^2})\mathbf{j}$
35. $(1/\sqrt{2})\mathbf{i} + (1/\sqrt{2})\mathbf{j}$

37. $$\frac{1}{\sqrt{18 + (9/\sqrt{2})(1 - \sqrt{3})}}\left[\left(-\frac{3}{2} - \frac{3}{\sqrt{2}}\right)\mathbf{i} + \left(\frac{3\sqrt{3}}{2} - \frac{3}{\sqrt{2}}\right)\mathbf{j}\right]$$

Problems 16.4

1. $(2 + \sec^2 t)\mathbf{i} - [\sin t + (\sec t)(\tan t)]\mathbf{j}$ **3.** $(3/2\sqrt{t}) - 2t^{-3/2}$ **5.** 0
7. $(3 \sin t \cosh t + 3 \cos t \sinh t)\mathbf{i} + (3 \sin t \sinh t + 3 \cos t \cosh t)\mathbf{j}$
9. $10t^9 + 9t^8 - \{(10t + 9)/[(t + 1)(t^{10} + t^9)]\}$

11. $\ln t \operatorname{sech}^2 t + \dfrac{\tanh t}{t} - 3 \ln t \operatorname{sech} t \tanh t + \dfrac{3 \operatorname{sech} t}{t}$

13. $\left(\dfrac{1}{\sqrt{1 - t^2}} - \sin t\right)\mathbf{i} + \left(\cos t - \dfrac{1}{\sqrt{1 - t^2}}\right)\mathbf{j}$

15. $\dfrac{\sin^{-1} t}{1 + t^2} + \dfrac{\tan^{-1} t}{\sqrt{1 - t^2}} - (\cos^{-1} t)\sin t - \dfrac{\cos t}{\sqrt{1 - t^2}}$

17. $\mathbf{T}(t) = -(\sin 5t)\mathbf{i} + (\cos 5t)\mathbf{j}$; $\mathbf{T}(\pi/2) = -\mathbf{i}$; $\mathbf{n}(t) = -(\cos 5t)\mathbf{i} - (\sin 5t)\mathbf{j}$;
$\mathbf{n}(\pi/2) = -\mathbf{j}$ **19.** $\mathbf{T}(t) = (\sin 10t)\mathbf{i} - (\cos 10t)\mathbf{j}$; $\mathbf{T}(\pi) = -\mathbf{j}$;
$\mathbf{n}(t) = (\cos 10t)\mathbf{i} + (\sin 10t)\mathbf{j}$; $\mathbf{n}(\pi) = \mathbf{i}$ **21.** $\mathbf{T}(t) = \dfrac{1}{\sqrt{1 + t^2}}\mathbf{i} + \dfrac{t}{\sqrt{1 + t^2}}\mathbf{j}$;

$\mathbf{T}(1) = \dfrac{1}{\sqrt{2}}\mathbf{i} + \dfrac{1}{\sqrt{2}}\mathbf{j}$; $\mathbf{n}(t) = -\dfrac{t}{\sqrt{1 + t^2}}\mathbf{i} + \dfrac{1}{\sqrt{1 + t^2}}\mathbf{j}$; $\mathbf{n}(1) = -\dfrac{1}{\sqrt{2}}\mathbf{i} + \dfrac{1}{\sqrt{2}}\mathbf{j}$
23. $\mathbf{T}(t) = -(7/\sqrt{74})\mathbf{i} + (5/\sqrt{75})\mathbf{j}$ (constant); $\mathbf{n}(t) = (5/\sqrt{74})\mathbf{i} + (7/\sqrt{74})\mathbf{j}$
or $-(5/\sqrt{74})\mathbf{i} - (7/\sqrt{74})\mathbf{j}$ **25.** $\mathbf{T}(t) = (\operatorname{sech} t)\mathbf{i} + (\tanh t)\mathbf{j}$; $\mathbf{T}(0) = \mathbf{i}$;
$\mathbf{n}(t) = -(\tanh t)\mathbf{i} + (\operatorname{sech} t)\mathbf{j}$; $\mathbf{n}(0) = \mathbf{j}$ **27.** $\mathbf{T}(t) = \sqrt{\dfrac{1 + \sin t}{2}}\mathbf{i} - \dfrac{\cos t}{\sqrt{2 + 2 \sin t}}\mathbf{j}$;

$\mathbf{T}(\pi) = \dfrac{1}{\sqrt{2}}\mathbf{i} + \dfrac{1}{\sqrt{2}}\mathbf{j}$; $\mathbf{n}(t) = \dfrac{\cos t}{\sqrt{2 + 2 \sin t}}\mathbf{i} + \sqrt{\dfrac{1 + \sin t}{2}}\mathbf{j}$; $\mathbf{n}(\pi) = -\dfrac{1}{\sqrt{2}}\mathbf{i} + \dfrac{1}{\sqrt{2}}\mathbf{j}$
29. $\mathbf{T}(\pi/6) = \dfrac{3}{\sqrt{10}}\mathbf{i} - \dfrac{1}{\sqrt{10}}\mathbf{j}$; $\mathbf{n}(\pi/6) = -\dfrac{1}{\sqrt{10}}\mathbf{i} - \dfrac{3}{\sqrt{10}}\mathbf{j}$

Problems 16.5

1. $\frac{1}{27}(148^{3/2} - 13^{3/2})$ **3.** $8(2^{3/2} - 1)$

5. $\displaystyle\int_{1/\sqrt{5}}^{1} \sqrt{1 + x^2}\, dx = \int_{\tan^{-1}(1/\sqrt{5})}^{\pi/4} \sec^3 \theta\, d\theta = \frac{1}{2}\left[\sqrt{2} - \left(\frac{\sqrt{6}}{5}\right) + \ln\left(\frac{\sqrt{5}(1 + \sqrt{2})}{1 + \sqrt{6}}\right)\right] \approx 0.6861$

7. $\sqrt{2}$ **9.** $8a$ (see Example 3)   **11.** $\frac{2}{27}[13^{3/2} - 8]$ **13.** $a\pi$ **15.** $\sqrt{2}(e^3 - 1)$
17. $6\sqrt{2}$ **19.** $\sqrt{2}(e^{\pi/2} - 1)$ **21.** $\int_0^{\pi/2}\sqrt{a^2 \sin^2\theta + b^2\cos^2\theta}\, d\theta$

Problems 16.6

1. $\mathbf{f} = 3\{[(s + 2)/2]^{2/3} - 1\}\mathbf{i} + 2\{[(s + 2)/2]^{2/3} - 1\}^{3/2}\mathbf{j}$
3. $\mathbf{f} = \frac{1}{27}\{[(27s + 8)^{2/3} - 4]^{3/2} + 27\}\mathbf{i} + \frac{1}{9}\{(27s + 8)^{2/3} - 13\}\mathbf{j}$
5. $\mathbf{f} = 3(\cos s/3)\mathbf{i} + 3(\sin s/3)\mathbf{j}$ **7.** $\mathbf{f} = a(\cos s/a)\mathbf{i} + a(\sin s/a)\mathbf{j}$
9. $\mathbf{f} = (a + b\cos(s/b))\mathbf{i} + (c + b\sin(s/b))\mathbf{j}$
11. $\mathbf{f} = a[1 - (2s/3a)]^{3/2}\mathbf{i} + b(2s/3a)^{3/2}\mathbf{j}$; here $\theta = \sin^{-1}\sqrt{2s/3a}$

Problems 16.7

1. $\mathbf{v} = 3\mathbf{j}$, $|\mathbf{v}| = 3$, $\mathbf{a} = -9\mathbf{i}$, $|\mathbf{a}| = 9$ **3.** $\mathbf{v} = -4\sqrt{3}\mathbf{i} - 4\mathbf{j}$; $|\mathbf{v}| = 8$; $\mathbf{a} = 16\mathbf{i} - 16\sqrt{3}\mathbf{j}$;
$|\mathbf{a}| = 32$ **5.** $\mathbf{v} = 4\mathbf{i} + 4\mathbf{j}$; $|\mathbf{v}| = 4\sqrt{2}$; $\mathbf{a} = 4\mathbf{j}$; $|\mathbf{a}| = 4$ **7.** $\mathbf{v} = -7\mathbf{i} + 5\mathbf{j}$; $|\mathbf{v}| = \sqrt{74}$;
$\mathbf{a} = 0$; $|\mathbf{a}| = 0$ **9.** $\mathbf{v} = 2\mathbf{i}$; $|\mathbf{v}| = 2$; $\mathbf{a} = 2\mathbf{j}$; $|\mathbf{a}| = 2$ **11.** $\mathbf{v} = 2\mathbf{i}$; $|\mathbf{v}| = 2$; $\mathbf{a} = \mathbf{j}$;
$|\mathbf{a}| = 1$ **13.** $\mathbf{v} = [1 + (1/\sqrt{2})]\mathbf{i} - (1/\sqrt{2})\mathbf{j}$; $|\mathbf{v}| = \sqrt{2 + \sqrt{2}}$; $\mathbf{a} = (1/\sqrt{2})\mathbf{i} + (1/\sqrt{2})\mathbf{j}$;
$|\mathbf{a}| = 1$ **15.** (a) $s = \int_0^{t_f}\sqrt{(600\sqrt{2})^2 + (600\sqrt{2} - gt)^2}\, dt = \dfrac{1200^2}{9.81}$
$[1/\sqrt{2} + \frac{1}{4}\ln((\sqrt{2} + 1)/(\sqrt{2} - 1))] \approx 421.5$ km where $t_f = 1200\sqrt{2}/g$ sec. This can be
integrated via the substitution $600\sqrt{2} - gt = 600\sqrt{2}\tan\theta$.   (b) $(1200)^2/g$ m
(c) $2(600)^2/g$ m   (d) 1200 m/sec   **17.** (a) $s = \int_0^{20/\sqrt{g}}\sqrt{400 + g^2 t^2}\, dt$
(b) $400/\sqrt{g} \approx 127.7$ m   (c) $\tan^{-1}(-gt_f/20) \approx 162.3°$ (d) $20\sqrt{1 + g}$
19. $20\sqrt{3} - (g/2)(\frac{2}{3})^2 + 4 = 20\sqrt{3} + \frac{4}{9} \approx 35$ ft   **21.** $\ln 2.5 \approx 0.9163$ km/sec
23. $5000g \approx 49{,}050$ nt

Problems 16.8

1. $\kappa = \frac{1}{2}$, $\rho = 2$ **3.** $\kappa = 2/5^{3/2}$, $\rho = 5^{3/2}/2$ **5.** $\kappa = \frac{3}{16}$, $\rho = \frac{16}{3}$
7. $\kappa = 6/\pi$, $\rho = \pi/6$ **9.** $\kappa = 2/5^{3/2}$, $\rho = 5^{3/2}/2$ **11.** $\kappa = 1/2\sqrt{2}$, $\rho = 2\sqrt{2}$
13. $\kappa = 1/2\sqrt{2}$, $\rho = 2\sqrt{2}$ **15.** $\kappa = 2a/(1 + b^2)^{3/2}$, $\rho = (1 + b^2)^{3/2}/2a$
17. $\kappa = 1$, $\rho = 1$ **19.** $\kappa = 0$, ρ is undefined   **21.** at the origin
23. minimum for $\theta = \pi/2$, maximum for $\theta = 3\pi/2$ **25.** $13\sqrt{2}/5\sqrt{5}a$
27. $1/a$ **29.** $3/4a$ **31.** $a_T = 0$, $a_n = 4$
33. $a_T = \dfrac{4t}{\sqrt{1 + 4t^2}}$, $a_n = \dfrac{2}{\sqrt{1 + 4t^2}}$ **35.** $a_T = (18t^3 - 14t)/\sqrt{9t^4 - 14t^2 + 9}$,
$a_n = 6(1 + t^2)/\sqrt{9t^4 - 14t^2 + 9}$
37. $a_T = (4 + 18t^2)/\sqrt{4 + 9t^2}$, $a_n = 6t/\sqrt{4 + 9t^2}$
41. $[(100{,}000)(80{,}000)^2/(3600)^2]\cdot(1/\sqrt{2}) \approx 34{,}918{,}853.4$ nt
43. $v = 150/\sqrt{2} \approx 106$ km/hr (reduce speed by a factor of $\sqrt{2}$)

Review Exercises for Chapter Sixteen

1. $y = 2x$ **3.** $x = [(y/2) + 3]^2$ **5.** $x^2 + y^2 = 1$ **7.** $x = y^3$, $x \geq 0$
9. 2; V at $t = 0$; no H   **11.** $\sqrt{3}$; V: $(1, 0)$, $(-1, 0)$; H: $(0, 1)$, $(0, -1)$
13. undefined; V: $(1, 0)$; no H   **15.** $-4/3\sqrt{3}$; V: $(3, 0)$, $(-3, 0)$;
H: $(0, 4)$, $(0, -4)$ **17.** $1/\sqrt{3}$ **19.** π **21.** $\mathbf{f}' = 2\mathbf{i} - 3t^2\mathbf{j}$; $\mathbf{f}'' = -6t\mathbf{j}$
23. $\mathbf{f}' = (-5\sin 5t)\mathbf{i} + (8\cos 4t)\mathbf{j}$; $\mathbf{f}'' = (-25\cos 5t)\mathbf{i} - (32\sin 4t)\mathbf{j}$
25. $\mathbf{T} = (4/\sqrt{41})\mathbf{i} + (5/\sqrt{41})\mathbf{j}$; $\mathbf{n} = (-5/\sqrt{41})\mathbf{i} + (4/\sqrt{41})\mathbf{j}$ **27.** $\mathbf{T} = (\sqrt{3}/2)\mathbf{i} - \frac{1}{2}\mathbf{j}$;
$\mathbf{n} = -\frac{1}{2}\mathbf{i} - (\sqrt{3}/2)\mathbf{j}$ **29.** $\mathbf{T} = (2/\sqrt{5})\mathbf{i} + (1/\sqrt{5})\mathbf{j}$; $\mathbf{n} = -(1/\sqrt{5})\mathbf{i} + (2/\sqrt{5})\mathbf{j}$
31. $\frac{81}{4}\mathbf{i} + \frac{243}{2}\mathbf{j}$ **33.** $(\frac{2}{3}t^{3/2} + C_1)\mathbf{i} + (\frac{3}{4}t^{4/3} + C_2)\mathbf{j}$
35. $[(t^8/8) - 1]\mathbf{i} + [3 - (t^7/7)]\mathbf{j}$ **37.** $-2\sin t\cosh t$
39. $\{e^t[(t + 1)^2 - t]/(t + 1)^2\} + [e^{-t}(t^2 + t + 1)/t^2]$ **41.** $\pi/3$ **43.** 16
45. $\frac{1}{3}(29^{3/2} - 8)$ **47.** $\mathbf{f} = \frac{3}{4}[(2s + 1)^{2/3} - 1]\mathbf{i} + [(2s + 1)^{2/3} - 1]^{3/2}\mathbf{j}$

49. $\mathbf{f} = 2(\cos 3s/2)\mathbf{i} + 2(\sin 3s/2)\mathbf{j}$ **51.** $\mathbf{v} = -\sqrt{3}\mathbf{i} + \mathbf{j}$; $|\mathbf{v}| = 2$; $\mathbf{a} = -2\mathbf{i} - 2\sqrt{3}\mathbf{j}$;
$|\mathbf{a}| = 4$ **53.** $\mathbf{v} = (\ln 2 - 1)\mathbf{i} + 2\mathbf{j}$; $|\mathbf{v}| = \sqrt{(\ln 2 - 1)^2 + 4}$; $\mathbf{a} = (\ln^2 2 + 1)\mathbf{i}$;
$|\mathbf{a}| = \ln^2 2 + 1$ **55.** $\mathbf{v} = 2(\cosh 1)\mathbf{i} + 4\mathbf{j}$; $|\mathbf{v}| = \sqrt{4\cosh^2 1 + 16}$; $\mathbf{a} = 2(\sinh 1)\mathbf{i}$;
$|\mathbf{a}| = 2\sinh 1$ **57.** (a) $s = \int_0^{t_f} \sqrt{675 + (15 - gt)^2}\, dt$ m, where
$t_f = (30 + \sqrt{900 + 12g})/2g \approx 3.155$ sec   (b) $15\sqrt{3}t_f \approx 81.97$ m
(c) $1.5 + 15(15/g) - (g/2)(15/g)^2 \approx 12.97$ m   (d) $\sqrt{675 + (15 - gt_f)^2} \approx 30.49$ m/sec
(e) $\tan^{-1}[(15 - gt_f)/15\sqrt{3}] \approx 148.45°$ **59.** (a) $(1500)(40) - (9.81)(2500) = 35,475$ nt
(b) $(1500)(40) - (9.81)(1500) = 45,285$ nt   **61.** $\kappa = \rho = 1$ **63.** $\kappa = 36/(97/2)^{3/2}$,
$\rho = (97/2)^{3/2}/36$ **65.** $\kappa = 16/17^{3/2}, \rho = 17^{3/2}/16$ **67.** $\kappa = 2/17^{3/2}, \rho = 17^{3/2}/2$
69. $\kappa = \frac{3}{4}, \rho = \frac{4}{3}$ **71.** $a_T = 0, a_n = 2$ **73.** $a_T = (6t + 12t^3)/\sqrt{t^2 + t^4}$;
$a_n = 6t^2(t^2 + t^4)^{3/2}$ **75.** $[1300(175,000)^2/(3600)^2] \cdot (1/65) \approx 47,261$ nt

CHAPTER SEVENTEEN

Problems 17.1

17. $\sqrt{40}$ **19.** 6   **21.** 5   **23.** $\sqrt{329}$ **25.** $\sqrt{250}$
27. $(x + 1)^2 + (y - 8)^2 + (z + 3)^2 = 5$ **29.** Center $(-\frac{3}{2}, \frac{1}{2}, -1)$, $r = 3/\sqrt{2}$
31. $\alpha > 23\frac{1}{4}$ **37.** $(x - \frac{1}{2})^2 + (y - 1)^2 + (z - 2)^2 = \frac{65}{4}$ **39.** $\frac{1}{6}\pi(65)^{3/2}; \frac{4}{3}\pi$

Problems 17.2

1. $|\mathbf{v}| = 3$; $0, 1, 0$ **3.** $|\mathbf{v}| = 14$; $0, 0, 1$ **5.** $|\mathbf{v}| = \sqrt{17}$; $4/\sqrt{17}, -1/\sqrt{17}, 0$
7. $|\mathbf{v}| = \sqrt{13}$; $-2/\sqrt{13}, 3/\sqrt{13}, 0$ **9.** $|\mathbf{v}| = \sqrt{3}$; $1/\sqrt{3}, -1/\sqrt{3}, 1/\sqrt{3}$
11. $|\mathbf{v}| = \sqrt{3}$; $-1/\sqrt{3}, 1/\sqrt{3}, 1/\sqrt{3}$ **13.** $|\mathbf{v}| = \sqrt{3}$; $-1/\sqrt{3}, 1/\sqrt{3}, -1/\sqrt{3}$
15. $|\mathbf{v}| = \sqrt{3}$; $-1/\sqrt{3}, -1/\sqrt{3}, -1/\sqrt{3}$ **17.** $|\mathbf{v}| = \sqrt{222}$;
$-7/\sqrt{222}, 2/\sqrt{222}, -13/\sqrt{222}$ **19.** $|\mathbf{v}| = \sqrt{82}$; $-3/\sqrt{82}, -3/\sqrt{82}, 8/\sqrt{82}$
21. $(1/\sqrt{3})\mathbf{i} + (1/\sqrt{3})\mathbf{j} + (1/\sqrt{3})\mathbf{k}$
23. $\cos^2(\pi/6) + \cos^2(\pi/3) + \cos^2(\pi/4) = \frac{3}{4} + \frac{1}{4} + \frac{1}{2} = 1.5 > 1$ **25.** $-\mathbf{i}$
29. $10\mathbf{i} + 3\mathbf{j} - 7\mathbf{k}$ **31.** $\mathbf{i} - \mathbf{j} - 6\mathbf{k}$ **33.** $-13\mathbf{i} + 28\mathbf{j} + 12\mathbf{k}$ **35.** 25
37. 45   **39.** $\cos^{-1}(-10/\sqrt{59}\sqrt{50}) \approx 1.76$ rad $\approx 100.6°$
41. $\frac{25}{38}\mathbf{v} = -\frac{25}{19}\mathbf{i} - \frac{75}{38}\mathbf{j} + \frac{125}{38}\mathbf{k}$ **43.** $-\frac{1}{5}\mathbf{t} = -\frac{3}{5}\mathbf{i} - \frac{4}{5}\mathbf{j} - \mathbf{k}$
45. $\frac{35}{59}\mathbf{w} = \frac{35}{59}\mathbf{i} - \frac{245}{59}\mathbf{j} + \frac{105}{59}\mathbf{k}$ **47.** $\sqrt{5610}/51 = \sqrt{110/51}$ (Note that the distance is
given by $|\overrightarrow{QP} - \text{Proj}_{\overrightarrow{QR}}\overrightarrow{QP}|$. Draw a picture.)   **53.** $14\sqrt{3}$ joules

Problems 17.3

In the answers to Problems 1–10 we assume that the first point is P and the second point
is Q. The vector equations are of the form $\overrightarrow{QR} = \overrightarrow{QP} + t\mathbf{v}$. Only $\mathbf{v}$ is given in the
answers.

1. $\mathbf{v} = -\mathbf{i} + \mathbf{j} - 4\mathbf{k}$; $x = 2 - t, y = 1 + t, z = 3 - 4t$; $\dfrac{x - 2}{-1} = y - 1 = \dfrac{z - 3}{-4}$
3. $\mathbf{v} = \mathbf{i} + \mathbf{j} - 4\mathbf{k}$; $x = 1 + t, y = 3 + t, z = 2 - 4t$; $x - 1 = y - 3 = (z - 2)/(-4)$
5. $\mathbf{v} = -\mathbf{j} - 2\mathbf{k}$; $x = -4, y = 1 - t, z = 3 - 2t$; $x = -4$ and $z = 1 + 2y$
7. $\mathbf{v} = 2\mathbf{i} - 2\mathbf{k}$; $x = 1 + 2t, y = 2, z = 3 - 2t$; $y = 2$ and $x = 4 - z$ **9.** $\mathbf{v} = 3\mathbf{k}$;
$x = 1, y = 2, z = 4 + 3t$

In problems 11–20 $\mathbf{v}$ is already given.

11. $x = 2 + 2t, y = 2 - t, z = 1 - t$; $(x - 2)/2 = (y - 2)/(-1) = (z - 1)/(-1)$
13. $x = 1 + t, y = -t, z = 3$; $x + y = 1$ and $z = 3$ **15.** $x = -1, y = -2 - 3t$,
$z = 5 + 7t$; $x = -1$ and $7y + 3z = 1$ **17.** $x = -1, y = -3 - 7t, z = 1$

19. $x = a + dt$, $y = b + et$, $z = c$; $(x - a)/d = (y - b)/e$ and $z = c$
21. $\mathbf{v} = 3\mathbf{i} + 6\mathbf{j} + 2\mathbf{k}$; $x = 4 + 3t$, $y = 1 + 6t$, $z = -6 + 2t$;
$(x - 4)/3 = (y - 1)/6 = (z + 6)/2$ **27.** none   **29.** $(-3, 2, 7)$ **31.** none
33. (a) $\sqrt{186}/3$ $(t = \frac{1}{3})$ (b) $\sqrt{1518}/11 = \sqrt{138}/11$ $(t = -\frac{4}{11})$; $\sqrt{750}/6 = 5\sqrt{30}/6$
$(t = -\frac{1}{6})$

Problems 17.4

1. $-6\mathbf{i} - 3\mathbf{j}$ **3.** $-\mathbf{i} - \mathbf{j} + \mathbf{k}$ **5.** $12\mathbf{i} + 8\mathbf{j} - 21\mathbf{k}$ **7.** $(bc - ad)\mathbf{j}$
9. $-5\mathbf{i} - \mathbf{j} + 7\mathbf{k}$ **11.** $\mathbf{0}$ **13.** $42\mathbf{i} + 6\mathbf{j}$ **15.** $-9\mathbf{i} + 39\mathbf{j} + 61\mathbf{k}$
17. $-4\mathbf{i} + 8\mathbf{k}$ **19.** $\mathbf{0}$ **21.** $-(9/\sqrt{181})\mathbf{i} - (6/\sqrt{181})\mathbf{j} + (8/\sqrt{181})\mathbf{k}$;
$(9/\sqrt{181})\mathbf{i} + (6/\sqrt{181})\mathbf{j} - (8/\sqrt{181})\mathbf{k}$ **23.** $\sqrt{30}/(\sqrt{6}\sqrt{29}) \approx 0.415$
25. $(x - 1)/2 = (y + 3)/(-26) = (z - 2)/(-22)$ **27.** $x = -2 + 13t$, $y = 3 + 22t$,
$z = 4 - 8t$ **29.** $5\sqrt{5}$ **31.** $\sqrt{523}$ **33.** $\sqrt{a^2b^2 + a^2c^2 + b^2c^2}$ **39.** $\frac{1}{2}\sqrt{595}$
43. 23   **45.** $48/\sqrt{437}$ **47.** $23/\sqrt{27}$

Problems 17.5

1. $x = 0$ (yz-plane)   **3.** $z = 0$ (xy-plane)   **5.** $x + z = 4$
7. $3x - y + 2z = 19$ **9.** $4x + y - 7z = 13$ **11.** $x + y + z = -6$
13. $20x + 13y - 3z = 58$ **15.** $x + y + z = 1$ **17.** coincident
19. orthogonal   **21.** orthogonal   **23.** coincident   **25.** $x = t$, $y = -11 - 37t$,
$z = -2 - 10t$ **27.** $33/\sqrt{59}$ **29.** $11/\sqrt{68}$
33. $\cos^{-1}(18/\sqrt{26}\sqrt{69}) \approx 1.132 \approx 64.9°$ **37.** $x - 22y - 17z = 0$
39. not coplanar   **41.** $x - 8y + 10z = 0$

Problems 17.6

1.

$y = \sin x$

3.

$y = \cos z$

5.

$z = x^3$

7. right circular cylinder with center at $(0, -1, 0)$ **9.** right circular cylinder
11. hyperbolic cylinder   **13.** elliptic cylinder   **15.** hyperbolic cylinder
17. hyperboloid of one sheet   **19.** ellipsoid   **21.** hyperbolic paraboloid (cross
sections parallel to the xz-plane are hyperbolas)   **23.** hyperboloid of two sheets
centered at $(2, 0, 0)$ $([(x - 2)^2/4] - (y^2/4) - (z^2/4) = 1)$; cross sections parallel to the
yz-plane are ellipses; surface only defined for $|x - 2| \geq 2$ **25.** hyperboloid of two
sheets (like the surface in Problem 23)   **27.** hyperboloid of two sheets; cross sections
parallel to the xy-plane are ellipses; surface defined for $|z| \geq \sqrt{2}$ **29.** ellipsoid

centered at $(1, 1, 1)$ **31.** hyperboloid of one sheet, cross sections parallel to the yz-plane are ellipses   **33.** hyperbolic paraboloid; like the surface in Figure 9 except that (i) the center is $(1, 3, -8)$ instead of the origin and (ii) the line $x = 1$, $z = -8$ passes through the center of the surface instead of the x-axis

Problems 17.7

1. $(1/\sqrt{14})\mathbf{i} + (2/\sqrt{14})\mathbf{j} + (3/\sqrt{14})\mathbf{k}$ **3.** $(1/\sqrt{3})\mathbf{i} + (1/\sqrt{3})\mathbf{j} - (1/\sqrt{3})\mathbf{k}$
5. $(-8/\sqrt{65})\mathbf{i} + (1/\sqrt{65})\mathbf{k}$
7. $\int_0^{10} \sqrt{36 + 4t^2}\, dt = 54 \int_0^{\tan^{-1}(10/3)} \sec^3 \theta\, d\theta = 30\sqrt{109} + 27 \ln[(\sqrt{109} + 10)/3]$
9. $\frac{4}{729}(328^{3/2} - 8) \approx 32.55$ **11.** $\mathbf{v} = -(\sin 1)\mathbf{i} + (\cos 1)\mathbf{j} + 4\mathbf{k}$; $|\mathbf{v}| = \sqrt{17}$;
$\mathbf{a} = -(\cos 1)\mathbf{i} - (\sin 1)\mathbf{j} + 12\mathbf{k}$; $|\mathbf{a}| = \sqrt{145}$ **13.** $\mathbf{v} = \mathbf{j} + \mathbf{k}$; $|\mathbf{v}| = \sqrt{2}$; $\mathbf{a} = \mathbf{i}$;
$|\mathbf{a}| = 1$ **15.** $\mathbf{v} = \frac{1}{2}\mathbf{i} + \frac{1}{3}\mathbf{j} + \frac{1}{4}\mathbf{k}$; $|\mathbf{v}| = \sqrt{61}/12$; $\mathbf{a} = -\frac{1}{4}\mathbf{i} - \frac{2}{9}\mathbf{j} - \frac{3}{16}\mathbf{k}$;
$|\mathbf{a}| = \sqrt{3049}/144$ **17.** $\mathbf{T} = \dfrac{\sqrt{3}a}{2\sqrt{a^2 + 1}}\mathbf{i} - \dfrac{a}{2\sqrt{a^2 + 1}}\mathbf{j} + \dfrac{1}{\sqrt{a^2 + 1}}\mathbf{k}$; $\kappa = \dfrac{a}{a^2 + 1}$;
$\mathbf{n} = -\dfrac{1}{2}\mathbf{i} - \dfrac{\sqrt{3}}{2}\mathbf{j}$; $\mathbf{B} = \dfrac{\sqrt{3}}{2\sqrt{a^2 + 1}}\mathbf{i} - \dfrac{1}{2\sqrt{a^2 + 1}}\mathbf{j} - \dfrac{a}{\sqrt{a^2 + 1}}\mathbf{k}$
19. $\mathbf{T} = \dfrac{-a}{\sqrt{1 + a^2}}\mathbf{i} + \dfrac{1}{\sqrt{1 + a^2}}\mathbf{k}$; $\kappa = \dfrac{b}{1 + a^2}$ $\left(\text{since } \dfrac{d\mathbf{T}}{dt} = -\dfrac{b}{\sqrt{1 + a^2}}\mathbf{j},\right.$
$\dfrac{ds}{dt} = \sqrt{1 + a^2}$ and $\dfrac{d\mathbf{T}}{ds} = \dfrac{d\mathbf{T}/dt}{ds/dt} = \dfrac{-b}{1 + a^2}\mathbf{j}\Big)$; $\mathbf{n} = -\mathbf{j}$;
$\mathbf{B} = \dfrac{1}{\sqrt{1 + a^2}}\mathbf{i} + \dfrac{a}{\sqrt{1 + a^2}}\mathbf{k}$ **21.** $\mathbf{T} = (1/\sqrt{6})\mathbf{i} + (2/\sqrt{6})\mathbf{j} + (1/\sqrt{6})\mathbf{k}$; $\kappa = \sqrt{5}/3$,
$\mathbf{n} = -(2/\sqrt{5})\mathbf{i} + (1/\sqrt{5})\mathbf{j}$; $\mathbf{B} = -(1/\sqrt{30})\mathbf{i} - (2/\sqrt{30})\mathbf{j} + (5/\sqrt{30})\mathbf{k}$
23. $\mathbf{T} = (1/\sqrt{5})\mathbf{j} + (2/\sqrt{5})\mathbf{k}$; $\kappa = \sqrt{20}/25 = 2/5^{3/2}$; $\mathbf{n} = -(2/\sqrt{5})\mathbf{j} + (1/\sqrt{5})\mathbf{k}$;
$\mathbf{B} = \mathbf{i}$ **31.** $\frac{1}{2}$

Problems 17.8

1. $(1, \sqrt{3}, 5)$ **3.** $(-4, 4\sqrt{3}, 1)$ **5.** $(-3/\sqrt{2}, 3/\sqrt{2}, 2)$ **7.** $(-10, 0, -3)$
9. $(-7/\sqrt{2}, -7/\sqrt{2}, 2)$ **11.** $(1, \pi/2, 0)$ **13.** $(\sqrt{2}, \pi/4, 2)$ **15.** $(4, \pi/3, -5)$
17. $(4, 5\pi/3, 4)$ **19.** $(\sqrt{3}, 0, 1)$ **21.** $(0, 3\sqrt{3}, 3)$ **23.** $(\frac{7}{2}, -\frac{7}{2}, -7/\sqrt{2})$
25. $(\sqrt{3}, 3, -2)$ **27.** $(5\sqrt{3}/4, -\frac{5}{4}, -5\sqrt{3}/2)$ **29.** $(2, \pi/4, \pi/4)$
31. $(2, 5\pi/4, \pi/4)$ **33.** $(2\sqrt{2}, 11\pi/6, \pi/4)$
35. $(\sqrt{23}, \cos^{-1}(-2/\sqrt{7}), \cos^{-1}(-4/\sqrt{23}))$, where θ is in the second quadrant
37. cylindrical: $r^2 + z^2 = 25$; spherical; $\rho = 5$ **39.** $x^2 + y^2 - 9y = 0$
41. cylindrical: $r^2 - z^2 = 1$; spherical, $\rho^2(\sin^2 \varphi - \cos^2 \varphi) = 1$ **43.** $z = x^2 + y^2$
45. $z\sqrt{x^2 + y^2} = 1$ **47.** $z = 2xy$ **49.** $\rho = 6 \sin \varphi \cos \theta$

Review Exercises for Chapter Seventeen

1. $\sqrt{68}$ **3.** $\sqrt{216}$ **5.** $(x + 1)^2 + (y - 4)^2 + (z - 2)^2 = 9$
9. $\sqrt{130}$; 0, $3/\sqrt{130}$, $11/\sqrt{130}$ **11.** $\sqrt{53}$; $-4/\sqrt{53}$, $1/\sqrt{53}$, $6/\sqrt{53}$
13. $(2/\sqrt{6})\mathbf{i} - (1/\sqrt{6})\mathbf{j} + (1/\sqrt{6})\mathbf{k}$ **15.** $\mathbf{i} - 14\mathbf{j} + 20\mathbf{k}$ **17.** $\frac{26}{21}\mathbf{i} - \frac{52}{21}\mathbf{j} + \frac{13}{21}\mathbf{k}$
19. 22   **21.** $\cos^{-1}(-9/\sqrt{798}) \approx 108.6°$ **23.** $68/\sqrt{3}$ joules
25. $\overrightarrow{OR} = (-4\mathbf{i} + \mathbf{j}) + t(7\mathbf{i} - \mathbf{j} + 7\mathbf{k})$; $x = -4 + 7t$, $y = 1 - t$, $z = 7t$;
$(x + 4)/7 = (y - 1)/(-1) = z/7$ **27.** $\overrightarrow{OR} = (\mathbf{i} - 2\mathbf{j} - 3\mathbf{k}) + t(5\mathbf{i} - 3\mathbf{j} + 2\mathbf{k})$;
$x = 1 + 5t$, $y = -2 - 3t$, $z = -3 + 2t$; $(x - 1)/5 = (y + 2)/(-3) = (z + 3)/2$
29. $\sqrt{165}/3$ **31.** $-7\mathbf{i} - 7\mathbf{k}$ **33.** $-26\mathbf{i} - 8\mathbf{j} + 7\mathbf{k}$
35. $(x + 1)/14 = (y - 2)/(-26) = (z - 4)/(-11)$ **37.** $2\sqrt{22}$
39. $x + z = -1$ **41.** $2x - 3y + 5z = 19$ **43.** $x = \frac{1}{2} - \frac{9}{2}t$, $y = \frac{7}{2} - \frac{11}{2}t$, $z = t$
45. $x = \frac{4}{3} - \frac{5}{2}t$, $y = -4 - \frac{7}{2}t$, $z = t$ **47.** $\cos^{-1}(-1/\sqrt{207}) \approx 94°$

49.

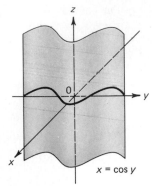

$x = \cos y$

51.

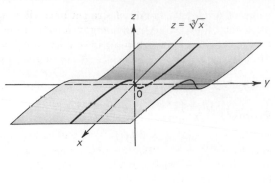

$z = \sqrt[3]{x}$

53. hyperboloid of one sheet; cross sections parallel to xz-plane are ellipses
55. hyperboloid of two sheets; cross sections parallel to xz-plane are ellipses; only defined for $|y| \geq \frac{5}{4}$ **57.** hyperbolic paraboloid; cross sections parallel to yz-plane are parabolas.   **59.** $(6/\sqrt{53})\mathbf{i} - (1/\sqrt{53})\mathbf{j} + (4/\sqrt{53})\mathbf{k}$
61. $-(3\sqrt{3}/2\sqrt{10})\mathbf{i} - (3/2\sqrt{10})\mathbf{j} + (1/\sqrt{10})\mathbf{k}$
63. $\mathbf{v} = \frac{3}{16}\pi^2\mathbf{i} - (1/\sqrt{2})\mathbf{j} - (1/\sqrt{2})\mathbf{k}; |\mathbf{v}| = \sqrt{1 + (9\pi^4/256)};$
$\mathbf{a} = (3\pi/2)\mathbf{i} - (1/\sqrt{2})\mathbf{j} + (1/\sqrt{2})\mathbf{k}; |\mathbf{a}| = \sqrt{1 + (9\pi^2/4)}$
65. $\mathbf{T} = -(1/\sqrt{5})\mathbf{i} + \sqrt{3/5}\mathbf{j} + (1/\sqrt{5})\mathbf{k}; \kappa = \frac{2}{5}; \mathbf{n} = -(\sqrt{3}/2)\mathbf{i} - \frac{1}{2}\mathbf{j};$
$\mathbf{B} = (1/2\sqrt{5})\mathbf{i} - (\sqrt{3}/2\sqrt{5})\mathbf{j} + (2/\sqrt{5})\mathbf{k}$ **67.** $\mathbf{T} = (1/\sqrt{3})\mathbf{i} + (1/\sqrt{3})\mathbf{j} + (1/\sqrt{3})\mathbf{k};$
$\kappa = \sqrt{2/3}; \mathbf{n} = -(1/\sqrt{2})\mathbf{i} + (1/\sqrt{2})\mathbf{j}; \mathbf{B} = (2/\sqrt{6})\mathbf{i} - (1/\sqrt{6})\mathbf{j} - (1/\sqrt{6})\mathbf{k}$
69. $(3\sqrt{3}/2, \frac{3}{2}, -1)$ **71.** $(-1, \sqrt{3}, 4)$ **73.** $(3/2\sqrt{2}, 3\sqrt{3}/2\sqrt{2}, 3/\sqrt{2})$
75. $(2, 3\pi/4, 3\pi/4)$ **77.** cylindrical: $r^2 + z^2 = 25$; spherical $\rho = 5$
79. cylindrical: $r^2(\cos^2\theta - \sin^2\theta) + z^2 = 1$; spherical:
$\rho^2(\sin^2\varphi\cos^2\theta - \sin^2\varphi\sin^2\theta + \cos^2\varphi) = 1$ **81.** $x^2 + y^2 = 1$

CHAPTER EIGHTEEN

Problems 18.2

1. $\mathbb{R}^2$; $[0, \infty)$ **3.** $\{(x, y): y \neq 0\}$; $\mathbb{R}$ **5.** $\{(x, y): x^2 - 4y^2 \leq 1\}$; $[0, \infty)$
7. $\mathbb{R}^2$; $(0, \infty)$ **9.** $\{(x, y): x - y \neq (n + \frac{1}{2})\pi, n = 0, \pm 1, \pm 2, \dots\}$; $\mathbb{R}$ **11.** $\{(x, y):$
$|x| \geq |y|$ and $x \neq -y\}$; $[0, \infty)$ **13.** $\{(x, y): |x - y| \leq 1\}$; $[0, \pi]$ **15.** $\{(x, y):$
$x \neq -y\}$; $\mathbb{R}$ **17.** $\{(x, y): x \neq 0$ and $y \neq 0\}$; $(-\infty, -2] \cup [2, \infty)$ (The range is obtained by using the fact that if $u = x/y$, then $f(x, y) = (1/2u) + 2u$ which has the minimum value 2.)   **19.** $\{(x, y, z): x + y + z \geq 0\}$, this is the half-space "in front of" the plane $x + y + z = 0$; $[0, \infty)$ **21.** $\{(x, y, z): y^2 < x^2 + z^2\}$, this is the region outside the cone $y^2 = x^2 + z^2$; $(0, \infty)$ **23.** $\{(0, 0, 0)\}$; 0   **25.** $\{(x, y, z):$
$z \neq 0\}$; $\mathbb{R}$ **27.** $\{(x, y, z): |x + y - z| \leq 1\}$—this is the part of $\mathbb{R}^3$ between the planes
$x + y - z = -1$ and $x + y - z = 1$; $[-\pi/2, \pi/2]$ **29.** $\{(x, y, z): y \neq 0$ and
$(x + z)/y \neq (n + \frac{1}{2})\pi, n = 0, \pm 1, \pm 2, \dots\}$; $\mathbb{R}$ **31.** $\mathbb{R}^3$; $(0, \infty)$ **33.** $\{(x, y, z):$
$x \neq 0, y \neq 0$ and $z \neq 0\}$; $\mathbb{R} - \{0\}$ **35.** $\mathbb{R}^3$; $[-3, 3]$ **37.** elliptic paraboloid opening around the y-axis; cross sections parallel to the xz-plane are ellipses
39. hyperbolic paraboloid; cross sections parallel to the xy-plane are hyperbolas
41. two half planes parallel to the xy-plane: $\{(x, y, z): z = 1$ and $x \geq 0\}$ and
$\{(x, y, z); z = -1$ and $x \geq 0\}$ **43.** these are the parallel straight lines (with slopes of
$-1)$ $y = -x + (z^2 - 1)$ **45.** these are concentric ellipses (centered at the origin) with equations $x^2 + 4y^2 = 1 - z^2, |z| \leq 1$; for $z = 1$ we obtain the single point
$(0, 0, 0)$ **47.** parallel straight lines (with slopes of 1) $y = x - \cos z$ **49.** for each

value of z we get a family of straight lines all of which have a slope of -1: for $z = 0$ we obtain $x + y = n\pi$, n an integer; for $z = 1$ we obtain $x + y = (\pi/4) + n\pi$; for $z = -1$ we obtain $x + y = -(\pi/4) + n\pi$; for $z = \sqrt{3}$ we obtain $x + y = (\pi/3) + n\pi$
51. concentric ellipses (centered at the origin); $a = \sqrt{T - 20}$, $b = \frac{1}{2}\sqrt{T - 20}$, for $T > 20$ **53.** concentric ellipses (centered at the origin); $a = \sqrt{(P - 100)/2}$, $b = \sqrt{(P - 100)/3}$, for $P > 100$; for $P = 100$ we obtain the single point $(0, 0)$

Problems 18.3

13. along the line $ax + by = 0$, $xy/(x^2 - y^2) = ab/(a^2 - b^2)$; the limit depends on the choice of a and b **15.** along the line $ax + by = 0$, $xy^3/(x^4 + y^4) = -a^3b/(a^4 + b^4)$ **17.** along the line $ax + by = 0$, $(x^2 + y^2)^2/(x^4 + y^4) = (a^2 + b^2)^2/(a^4 + b^4)$ **19.** along the line $y = kx$, $(ax^2 + by)/(cy^2 + dx) = (ax + bk)/(ck^2x + d) \to bk/d$ as $(x, y) \to (0, 0)$.   **21.** along the line $y = kx = z$, $xyz/(x^3 + y^3 + z^3) = k^2/(1 + 2k^3)$ **23.** 0; since $0 \le (x^2 - y)^2 = x^4 - 2xy^2 + y^2$, $2xy^2 \le x^4 + y^2$ so that $5x^2y^2/(x^4 + y^2) \le 5x^2y^2/2xy^2 = \frac{5}{2}x$.   **25.** 0; since $0 < |(yx^2 + z^3)/(x^2 + y^2 + z^2)| \le (|y|x^2 + |z|z^2)/(x^2 + y^2 + z^2) \le |y| + |z| \to 0$.
27. $-\frac{61}{25}$ **29.** $\ln(4\pi/3)$ **31.** $\sinh 1$ **33.** $\frac{1}{6}$ **35.** $\ln 769$
37. $R^2 - \{(0, 0)\}$ **39.** $\{(x, y): y \ne 4 \text{ and } x \ne -3\}$ **41.** R^2
43. $\{(x, y): xy \ne 1\}$ **45.** $\{(x, y, z): x \ge 0\}$ **47.** $\{(x, y, z): xz > 0\}$
49. $\{(x, y, z): x^2 + y^2 + z^2 < 1\}$ = the "interior" of the unit sphere   **51.** $c = 0$
53. Along the line $y = kx = z$, $(yz - x^2)/(x^2 + y^2 + z^2) = (k^2 - 1)/(1 + 2k^2)$ which shows that $\lim_{(x,y,z)\to(0,0,0)} (yz - x^2)/(x^2 + y^2 + z^2)$ does not exist. The function is, therefore, not continuous at the origin.

Problems 18.4

1. $2xy$; x^2 **3.** $3y^3e^{xy^3}$; $9xy^2e^{xy^3}$ **5.** $4/y^5$; $-20x/y^6$
7. $3x^2y^5/(x^3y^5 - 2)$; $5x^3y^4/(x^3y^5 - 2)$ **9.** $\frac{4}{3}(1 + 5y \cos x)(x + 5y \sin x)^{1/3}$; $\frac{20}{3} \sin x \times (x + 5y \sin x)^{1/3}$ **11.** $[2y/(x^2 + y^2)](xy/|xy|)$; $[-2x/(x^2 + y^2)](xy/|xy|)$ **13.** 3
15. 0   **17.** 1   **19.** $\frac{48}{169}$ **21.** yz; xz; xy **23.** $1/z$; $1/z$; $-(x + y)/z^2$
25. $3x^2/(x^3 + y^2 + z)$; $2y/(x^3 + y^2 + z)$; $1/(x^3 + y^2 + z)$ **27.** $yz \cos xyz$; $xz \cos xyz$; $xy \cos xyz$ **29.** $\dfrac{\sinh \sqrt{x + 2y + 5z}}{2\sqrt{x + 2y + 5z}}$; $\dfrac{\sinh \sqrt{x + 2y + 5z}}{\sqrt{x + 2y + 5z}}$; $\dfrac{5 \sinh \sqrt{x + 2y + 5z}}{2\sqrt{x + 2y + 5z}}$ **31.** $\frac{1}{3}$ **33.** $-2 \cos 1$ **35.** $-\frac{1}{2}$ **37.** $-2\sqrt{3}/9$
39. $a/\sqrt{a^2 + b^2 + c^2}$ **43.** (a) $x = 1$, $(z - 5)/(-12) = (y + 1)/1$; $y = -1$; $(z - 5)/3 = (x - 1)/1$ (The first line is obtained as in R^2 by setting $(z - 5)/[y - (-1)] = -12 = $ the slope.)   **45.** $y = 1$, $(x - 1)/(-4) = z - 1$
49. $-5/(2 \cdot 10^{2/3})$ **51.** $\partial R/\partial x = [50/(1 + 50x + 75y)] + (20/\sqrt{1 + 40x + 125y})$
$\partial R/\partial y = [75/(1 + 50x + 75y)] + (125/2\sqrt{1 + 40x + 125y})$

Problems 18.5

In Problems 1–11 the answers are given in the order $f_{xx}, f_{xy} (= f_{yx}), f_{yy}$.

1. $2y$; $2x$; 0   **3.** $3y^6e^{xy^3}$; $9y^2(1 + xy^3)e^{xy^3}$; $9xy(2 + 3xy^3)e^{xy^3}$ **5.** 0; $-20/y^6$; $120x/y^7$ **7.** $\dfrac{-3xy^5(4 + x^3y^5)}{(x^3y^5 - 2)^2}$; $\dfrac{-30x^2y^4}{(x^3y^5 - 2)^2}$; $\dfrac{-5x^3y^3(8 + x^3y^5)}{(x^3y^5 - 2)^2}$
9. $\frac{4}{9}(1 + 5y \cos x)^2(x + 5y \sin x)^{-2/3} - \frac{20}{3} y \sin x (x + 5y \sin x)^{1/3}$; $\frac{20}{9} \sin x (1 + 5y \cos x) \times (x + 5y \sin x)^{-2/3} + \frac{20}{3} \cos x (x + 5y \sin x)^{1/3}$; $\frac{100}{9} \sin^2 x (x + 5y \sin x)^{-2/3}$

11. $\left(\dfrac{-4xy}{(x^2+y^2)^2}\right)\left(\dfrac{xy}{|xy|}\right)$; $\dfrac{2(x^2-y^2)}{(x^2+y^2)^2}\left(\dfrac{xy}{|xy|}\right)$; $\dfrac{4xy}{(x^2+y^2)^2}\left(\dfrac{xy}{|xy|}\right)$

13. $f_{xx}=0$; $f_{xy}=f_{yx}=z$; $f_{xz}=f_{zx}=y$; $f_{yy}=0$; $f_{yz}=f_{zy}=x$; $f_{zz}=0$ **15.** $f_{xx}=0$;
$f_{xy}=f_{yx}=0$; $f_{xz}=f_{zx}=-1/z^2$; $f_{yy}=0$; $f_{yz}=f_{zy}=-1/z^2$; $f_{zz}=2(x+y)/z^3$

17. $f_{xx}=\dfrac{-2xyz^3}{(y^2+x^2z^2)^2}$; $f_{xy}=f_{yx}=\dfrac{x^2z^3-y^2z}{(y^2+x^2z^2)^2}$; $f_{xz}=f_{zx}=\dfrac{y^3-x^2yz^2}{(y^2+x^2z^2)^2}$;

$f_{yy}=\dfrac{2xyz}{(y^2+x^2z^2)^2}$; $f_{yz}=f_{zy}=\dfrac{x^3z^2-xy^2}{(y^2+x^2z^2)^2}$; $f_{zz}=\dfrac{-2x^3yz}{(y^2+x^2z^2)^2}$

19. $f_{xx}=9y^2e^{3xy}\cos z$; $f_{xy}=f_{yx}=3(1+xy)e^{3xy}\cos z$; $f_{xz}=f_{zx}=-3ye^{3xy}\sin z$;
$f_{yy}=9x^2e^{3xy}\cos z$; $f_{yz}=f_{zx}=-3xe^{3xy}\sin z$; $f_{zz}=-e^{3xy}\cos z$

21. $f_{xx}=\dfrac{\cosh\sqrt{x+yz}}{4(x+yz)}-\dfrac{\sinh\sqrt{x+yz}}{4(x+yz)^{3/2}}$; $f_{xy}=f_{yx}=zf_{xx}$; $f_{xz}=f_{zx}=yf_{xx}$; $f_{yy}=z^2f_{xx}$;

$f_{yz}=f_{zy}=\dfrac{\sinh\sqrt{x+yz}}{2\sqrt{x+yz}}+yzf_{xx}$; $f_{zz}=y^2f_{xx}$ **23.** (a) 16   (b) 81   **25.** $6y^2$

27. $24/(3x-2y)^3$ **29.** $9\sin(x+2y+3z)$

Problems 18.6

5. $(e/2)\mathbf{i}+(e/2)\mathbf{j}$ **7.** $[2/(2x-y+1)]\mathbf{i}+[-1/(2x-y+1)]\mathbf{j}$ **9.** $-\tfrac16\mathbf{i}+\tfrac16\mathbf{j}$

11. $2x(\sinh y)\mathbf{i}+x^2(\cosh y)\mathbf{j}$ **13.** $\tfrac13\mathbf{i}-\tfrac38\mathbf{j}$

15. $\dfrac{2xe^{x^2}}{3y}\mathbf{i}+\left(\dfrac{-e^{x^2}+(1+2y^2)e^{-y^2}}{3y^2}\right)\mathbf{j}$ **17.** $(3\sqrt{2}/4)\mathbf{i}-(\sqrt{6}/4)\mathbf{j}+\sqrt{2}\mathbf{k}$

19. $[\ln y-(z/x)]\mathbf{i}+(x/y)\mathbf{j}-(\ln x)\mathbf{k}$ **21.** $-4e^3\mathbf{i}-7e^3\mathbf{j}-13e^3\mathbf{k}$ **23.** $\mathbf{i}-\mathbf{k}$

29. $f(x,y)=(x^2/2)+(y^2/2)+C$

Problems 18.7

1. $3e^{3t}$ **3.** 1   **5.** $\dfrac{5(\cos 5t)\cos 3t+3(\sin 5t)\sin 3t}{\sin^2 5t+\cos^2 3t}$ **7.** $2t$ **9.** $\dfrac{-t-2}{t^3}$

11. $\dfrac{2e^t-(3/t)+4\sinh t}{2e^t-3\ln t+4\cosh t}$

13. $\partial z/\partial r=-2\sin(r+s)\cos(r+s)+2\sin(r-s)\cos(r-s)$;
$\partial z/\partial s=-2\sin(r+s)\cos(r+s)-2\sin(r-s)\cos(r-s)$

15. $\partial z/\partial r=-(2s^2/r^3)\cos(s^2/r^2)$; $\partial z/\partial s=(2s/r^2)\cos(s^2/r^2)$

17. $\partial z/\partial r=2(r-s^2)(2s+r)^3+3(r-s^2)^2(2s+r)^2$;
$\partial z/\partial s=-4s(r-s^2)(2s+r)^3+6(r-s^2)^2(2s+r)^2$ **19.** $\partial w/\partial r=s/t$; $\partial w/\partial s=r/t$;
$\partial w/\partial t=-rs/t^2$ **21.** $\partial w/\partial r=(4s^3r^3-3s^4r^2)\cos[s^3r^3(r-s)]$;
$\partial w/\partial s=(3s^2r^4-4s^3r^3)\cos[s^3r^3(r-s)]$ **23.** $\partial w/\partial r=3r^2s^3t-1/r^2st^3$;
$\partial w/\partial s=3r^3s^2t-1/rs^2t^3$; $\partial w/\partial t=r^3s^3-3/rst^4$

25. $\partial w/\partial r=2r[(s^2-t^2)/(r^2+s^2)]^2w$; $\partial w/\partial s=2s[(r^2+t^2)/(r^2+s^2)]^2w$;
$\partial w/\partial t=2t[(s^2-r^2-2t^2)/(r^2+s^2)]w$ **27.** $V_t=7500\pi$ in^3/min; increasing

29. $P_t=0.08575\,R$; increasing   **31.** $(1/|BC|)[322-2\sqrt3]\approx 19.743$ cm/sec;
increasing   **33.** $(\partial z/\partial x)^2+(\partial z/\partial y)^2=[(1/r)(\partial z/\partial\theta)]^2+(\partial z/\partial r)^2$

35. $\partial w/\partial\rho=(\partial w/\partial x)\sin\varphi\sin\theta+(\partial w/\partial y)\sin\varphi\cos\theta+(\partial w/\partial z)\cos\varphi$
$\partial w/\partial\varphi=(\partial w/\partial x)\rho\cos\varphi\sin\theta+(\partial w/\partial y)\rho\cos\varphi\cos\theta-(\partial w/\partial z)\rho\sin\varphi$
$\partial w/\partial\theta=(\partial w/\partial x)\rho\sin\varphi\cos\theta-(\partial w/\partial y)\rho\sin\varphi\sin\theta$

Problems 18.8

1. $x=1$; $y=0$, $z=0$ **2.** $z=1$; $x=0$, $y=0$

5. $(1/a)(x-a)+(1/b)(x-b)+(1/c)(x-c)=0$; $a(x-a)=b(y-b)=c(z-c)$

7. $\frac{1}{4}(x-4) + \frac{1}{2}(y-1) + \frac{1}{6}(z-9) = 0$; $4(x-4) = 2(y-1) = 6(z-9)$

9. $2(x-1) + (y-2) + (z-2) = 0$; $(x-1)/2 = y-2 = z-2$

11. $24(x-3) - 2(y-1) + 20(z+2) = 0$;

$(x-3)/24 = (y-1)/(-2) = (z+2)/20$ **13.** $(\pi/\sqrt{3})(y-1) + \sqrt{3}[z - (\pi/3)] = 0$;

$x = \pi/2$, $(\sqrt{3}/\pi)(y-1) = (1/\sqrt{3})[z - (\pi/3)]$

15. $(\ln 5)(x-1) + (\ln 5)(y-1) + (z - \ln 5) = 0$;

$(x-1)/(\ln 5) = (y-1)/(\ln 5) = z - \ln 5$ **17.** $z = 1 + (x-1) + 2(y-1)$;

$x - 1 = (y-1)/2 = (z-1)/(-1)$ **19.** $z = 1$; $x = \pi/8$, $y = \pi/20$

21. $z = -(\pi/4) - \frac{1}{4}(x+2) - \frac{1}{4}(y-2)$; $4(x+2) = 4(y-2) = z + \pi/4$

23. $z = 2 + 2\sqrt{3}(x - \pi/2) - 2\sqrt{3}(y - \pi/6)$;

$(x - \pi/2)/2\sqrt{3} = (y - \pi/6)/(-2\sqrt{3}) = (z-2)/(-1)$ **25.** $(0, 2, 4)$, $(1, 1, 2)$

29. All tangent planes pass through the origin. (Note that the surface cannot pass through the origin since $f(y/x)$ is not defined when $x = 0$.)

33. $[(x - x_0)\mathbf{i} + (y - y_0)\mathbf{j}] \cdot \nabla F(x_0, y_0) = 0$ **35.** $3\mathbf{i} - 8\mathbf{j}$;

$3(x-1) - 8(y+2) = 0$ **37.** $\mathbf{i} + \mathbf{j}$; $(x-1) + y = 0$ **39.** $2\mathbf{i} + 2\mathbf{j}$;

$(x - \pi/4) + y = 0$

Problems 18.9

1. $9/\sqrt{10}$ **3.** $\sqrt{2}/7$ **5.** $-5/(4\sqrt{13})$ **7.** $(2e^2 + 3e)/\sqrt{2}$ **9.** $6/\sqrt{3}$

11. $1/(5\sqrt{6})$ **13.** $61/\sqrt{26}$ **15.** 0 **17.** (a) at the origin

(b) $\nabla T = 200e^{-(x^2+y^2+z^2)}(-x\mathbf{i} - y\mathbf{j} - z\mathbf{k})$ **19.** $(x(t), y(t)) = (ae^{-2t/a^2}, be^{-2t/b^2})$;

$(x/a)^{a^2} = (y/b)^{b^2}$ **21.** (a) descend, because $\nabla h \cdot \mathbf{u} < 0$ when

$\mathbf{u} = (-1/\sqrt{2})\mathbf{i} - (1/\sqrt{2})\mathbf{j}$ (b) $(-120\mathbf{i} + 40\mathbf{j})/\sqrt{120^2 + 40^2} = (-3\mathbf{i} + \mathbf{j})/\sqrt{10} \approx 161.56°$

(toward the southeast) **23.** $\sqrt{5}$ **25.** $1/\sqrt{2}$

Problems 18.10

1. $F(x, y, z) = (\alpha/2) \ln(x^2 + y^2 + z^2)$ **3.** If $k = 1$, let $F(x, y, z) = \alpha|\mathbf{x}|$. If $k = 2$, look at Problem 1. If $k > 2$, a potential function is given by

$F(x, y, z) = [-\alpha/(k-2)] \cdot (1/|\mathbf{x}|^{k-2})$.

Problems 18.11

1. $y^3 \Delta x + 3xy^2 \Delta y$ **3.** $(x-y)^{-1/2}(x+y)^{-3/2}(y \Delta x - x \Delta y)$

5. $[1/(2x + 3y)](2 \Delta x + 3 \Delta y)$ **7.** $y^2z^5 \Delta x + 2xyz^5 \Delta y + 5xy^2z^4 \Delta z$

9. $[1/(x + 2y + 3z)](\Delta x + 2 \Delta y + 3 \Delta z)$ **11.** $\sinh(xy - z)(y \Delta x + x \Delta y - \Delta z)$

13. (a) $x(\Delta y)^2 + 2y(\Delta x)(\Delta y) + (\Delta x)(\Delta y)^2$

(b) $\Delta f - df = [(0.99)(2.03)^2 - (1)(2)^2] - [2^2(-0.01) + (2)(1)(2)(0.03)] =$

$[0.079691] - [0.08] = -0.000309$ **15.** $\frac{3}{6} + \frac{1}{6}(0.01) - [3(-0.01)/6^2] = 0.5025$ (actual value ≈ 0.50250417)

17. $\sqrt{36} \sqrt[3]{64} + (\frac{1}{2})(\frac{1}{6})(4)(-0.4) + (\frac{1}{3})(6)(\frac{1}{16})(0.08) = 23.87666\ldots$

(actual value ≈ 23.87623437) **19.** $\sqrt{\frac{1}{9}} + \frac{1}{27}[4(0.02) - 5(-0.04)] = 0.3437037037\ldots$

(actual value ≈ 0.3435696277)

21. $(3 \cdot 2/\sqrt{9}) + (2/\sqrt{9})(0.02) + (3/\sqrt{9})(-0.03) - (3 \cdot 2/2(9)^{3/2})(-0.05) = 1.9888\ldots$

(actual value ≈ 1.988665097)

23. $8\sqrt{5 - 1^2} + (\sqrt{5 - 1^2})(-0.08) + (8/2\sqrt{5 - 1^2})(0.01) - [(8)(1)/\sqrt{5 - 1^2}](-0.02) =$

15.94 (actual value ≈ 15.93790543) **25.** (a) $\sqrt{9100} \approx 95.3939$ m

(b) $(1/\sqrt{9100})\{[50 - 110(\frac{1}{2})][0.3] + [110 - 50(\frac{1}{2})][0.3] + [(50)(110)(\sqrt{3}/2)](\pi/180)\} \approx$

1.12305 m **27.** (a) $\frac{8}{3} = 2.6666\ldots$ ohms

(b) $(\frac{8}{3})^2[(0.1/6^2) + (0.03/8^2) + (0.15/12^2)] \approx 0.03049$

Problems 18.12

1. $x^2y + y + C$ **3.** not exact   **5.** not exact   **7.** $(x^2/2) - y \sin x + C$
9. $(3x^2/2) \ln x - (3x^2/4) + (x^6/6) - xy + C$ **11.** $\tan x - x + \tan^{-1}(y/x) - e^y + C$
13. not exact   **15.** $x + y + z + C$ **17.** $(xy/z) + (x^3/3) + \cos y + \sin z + C$
19. not exact   **21.** $\mathbf{f}(t) = -\alpha \mathbf{x}/|\mathbf{x}|^4$ and a potential function for $\mathbf{f}$ is $F(\mathbf{x}) = -\alpha/(2|\mathbf{x}|^2)$ $(-\nabla F = \mathbf{f})$ **23.** $\mathbf{f} = -\alpha \mathbf{x}/|\mathbf{x}|$ and $F(\mathbf{x}) = \alpha|\mathbf{x}|$ $(-\nabla F = \mathbf{f})$

Problems 18.13

1. $(0, 0)$ is a local minimum   **3.** $(-2, 1)$ is a local minimum   **5.** $(-2, 1)$ is a local minimum   **7.** $(\sqrt{5}, 0)$ is a local minimum; $(-\sqrt{5}, 0)$ is a local maximum; $(-1, 2)$ and $(-1, -2)$ are saddle points   **9.** $(\frac{1}{2}, 1)$ and $(-\frac{1}{2}, -1)$ are saddle points; $(-\frac{1}{2}, 1)$ is a local minimum; $(\frac{1}{2}, -1)$ is a local maximum   **11.** $(-2, -2)$ is a local minimum
13. $(0, 0)$, $(0, 4)$ and $(4, 0)$ are saddle points; $(\frac{4}{3}, \frac{4}{3})$ is a local maximum   **15.** $(1, 1)$ and $(-1, 1)$ are local minima   **17.** $(0, 0)$ is a saddle point. (Note that $D = 0$, but it is evident that f can take positive and negative values near the origin.)
19. $x = y = z = \frac{50}{3}$ **21.** $x = y = z = \frac{50}{3}$ **23.** $3\sqrt{3}$ **25.** Corners are at $(\pm a/\sqrt{3}, \pm b/\sqrt{3}, \pm c/\sqrt{3})$. (The problem is to maximize $8xyz$ with $(x^2/a^2) + (y^2/b^2) + (z^2/c^2) = 1$.)   **27.** The profit is given by $P(x, y) = (8xy + 32x + 40y - 4x^2 - 6y^2)$ $40 - 10x - 4y$ which is maximized when $x \approx 22$ and $y \approx 18$ ($x = 21.88125, y = 17.9125$).   **29.** $y = -\frac{54}{52}x + \frac{120}{52} \approx -1.04x + 2.3$

Problems 18.14

1. $3/\sqrt{13}$ at the point $(\frac{7}{13}, \frac{17}{13})$ **3.** $5/\sqrt{3}$ at the point $(\frac{8}{3}, \frac{2}{3}, \frac{1}{3})$
5. $|d|/\sqrt{a^2 + b^2 + c^2}$ at the point $(ad/(a^2 + b^2 + c^2), bd/(a^2 + b^2 + c^2), cd/(a^2 + b^2 + c^2))$ **7.** Minimum is $\frac{15}{32}$ at $(\frac{9}{16}, \frac{1}{8})$. There is no maximum.   **9.** Maximum is $8/(3\sqrt{3})$ at $(2/\sqrt{3}, 2/\sqrt{3}, 2/\sqrt{3})$. Minimum is $-8/(3\sqrt{3})$ at $(-2/\sqrt{3}, -2/\sqrt{3}, -2/\sqrt{3})$.   **11.** Maximum is $2/\sqrt{3}$ at $(1/\sqrt{3}, 2/\sqrt{3}, 3/\sqrt{3})$ and at 3 other points. Minimum is $-2/\sqrt{3}$ at $(-1/\sqrt{3}, 2/\sqrt{3}, 3/\sqrt{3})$ and at 3 other points.   **13.** $\frac{148}{45}$ at $(\frac{75}{45}, \frac{-20}{45}, \frac{28}{45})$ **15.** Maximum is the larger of $|a|$ and $|b|$. Minimum is the smaller of $|a|$ and $|b|$.   **17.** 2 at $(\frac{1}{2}, \frac{1}{2}, \frac{1}{2}, \frac{1}{2})$ **21.** Maximum is $2/(3\sqrt{3})$ at $(\sqrt{\frac{2}{3}}, \sqrt{\frac{2}{3}}, 1/\sqrt{3})$. Minimum is $-2/(3\sqrt{3})$ at $(-\sqrt{\frac{2}{3}}, -\sqrt{\frac{2}{3}}, -\sqrt{\frac{1}{3}})$.   **23.** Height of cone is $12/\sqrt{5}$ m. Height of cylinder is $(50/3\pi) - (9/\sqrt{5})$ m.   **25.** Base is a square: 2×2 m. Height is 3 m.   **27.** $x/y = \frac{1}{2}$

Review Exercises for Chapter Eighteen

1. $\{(x, y): |x| \geq |y|\}$; $[0, \infty)$ **3.** $\mathbb{R}^2$; $[-1, 1]$
5. $\{(x, y, z): x^2 + y^2 + z^2 > 1\}$ = the "exterior" of the unit sphere; $(0, \infty)$
7. parallel straight lines with slopes of 1; $x + y = 1 - z^2$ **9.** parallel straight lines with slopes of $\frac{1}{3}$; $x - 3y = e^z$ **13.** along $y = kx$ limit is $1/(k^2 - 1)$
15. since $0 \leq (x - y^2)^2 = x^2 - 2xy^2 + y^4$, $2xy^2 \leq x^2 + y^4$ so that $4xy^3/(x^2 + y^4) \leq 4xy^3/2xy^2 = 2y$ **17.** $-\frac{1}{4}$ **19.** 5   **21.** the half-space "above," but not including, the plane $x - y - z = -4$ **23.** 0   **25.** $f_x = -\sin(x - 3y)$; $f_y = 3\sin(x - 3y)$ **27.** $f_x = -y/[(1 + x)^2 + y^2]$; $f_y = (1 + x)/[(1 + x)^2 + y^2]$
29. $f_x = -x(x^2 + y^2 + z^2)^{-3/2}$; $f_y = -y(x^2 + y^2 + z^2)^{-3/2}$;

$$f_z = -z(x^2 + y^2 + z^2)^{-3/2} \quad \textbf{31.} \quad f_x = \frac{1}{z} \sec\left(\frac{x - y}{z}\right) \tan\left(\frac{x - y}{z}\right);$$

$$f_y = \frac{-1}{z} \sec\left(\frac{x - y}{z}\right) \tan\left(\frac{x - y}{z}\right); f_z = \frac{y - x}{z^2} \sec\left(\frac{x - y}{z}\right) \tan\left(\frac{x - y}{z}\right)$$

33. $f_x = \dfrac{2x}{y^3 + z^4}$; $f_y = \dfrac{-3y^2(x^2 + z^4)}{(y^3 + z^4)^2}$; $f_z = \dfrac{4z^3(y^3 - x^3)}{(y^3 + z^4)^2}$ **35.** $f_x = \dfrac{1}{y + z}f$;

$f_y = \dfrac{w - x}{(y + z)^2}f$; $f_z = \dfrac{w - x}{(y + z)^2}f$; $f_w = \dfrac{-1}{y + z}f$ **37.** $f_{xx} = \dfrac{2xy}{(x^2 + y^2)^2}$;

$f_{xy} = f_{yx} = \dfrac{y^2 - x^2}{(x^2 + y^2)^2}$; $f_{yy} = \dfrac{-2xy}{x^2 + y^2}$ **39.** $f_{xx} = \dfrac{4y}{(x - y)^3}$; $f_{xy} = f_{yx} = \dfrac{-2(x + y)}{(x - y)^3}$;

$f_{yy} = \dfrac{-4x}{(x - y)^3}$ **41.** $f_{xx} = \dfrac{1 + 2x^2 - y^2 - z^2}{R^5}$; $f_{xy} = f_{yx} = \dfrac{3xy}{R^5}$; $f_{xz} = f_{zx} = \dfrac{3xz}{R^5}$;

$f_{yy} = \dfrac{1 - x^2 + 2y^2 - z^2}{R^5}$; $f_{yz} = f_{zy} = \dfrac{3yz}{R^5}$; $f_{zz} = \dfrac{1 - x^2 - y^2 + 2z^2}{R^5}$; here

$R = \sqrt{1 - x^2 - y^2 - z^2}$ **43.** $f_{zxx} = 0 = f_{zxxy} = f_{zxxyz}$ **45.** $\frac{1}{2}\mathbf{i} - \frac{1}{2}\mathbf{j}$
47. $\frac{-1}{2}\mathbf{i} + \mathbf{j}$ **49.** $\frac{1}{12}\mathbf{i} - \frac{1}{12}\mathbf{j} - \frac{1}{48}\mathbf{k}$ **51.** $-3\mathbf{j} - 4\mathbf{k}$
53. $|1 + t|/(1 + t) \cdot (t^2 + 2t)/(1 + t)\sqrt{(1 + t)^2 - 4}$ **55.** $2r/(r - s)^2$
57. $[\cos(e^{r+s} - e^{r-s})][e^{r+s} + e^{r-s}]$ **59.** $\partial w/\partial s = 4r^4s^3t^2 + 2r^4st^{-2}$;
$\partial w/\partial t = 2r^4s^4t - 2r^4s^2t^{-3}$ **61.** $x + y + z = 3$; $x = y = z$ **63.** $3x - y + 5z = 15$;
$(x + 1)/3 = (y - 2)/(-1) = (z - 4)/5$ **65.** $-3x + 6y - 2z = 18$;
$(x + 2)/(-3) = (y - 1)/6 = (z + 3)/(-2)$ **67.** $-3/\sqrt{2}$ **69.** $-1/(2\sqrt{13})$
71. $9/(\sqrt{14})(6^{3/2})$ **73.** $-1/|\mathbf{x}|^3$ is a potential function
75. $3x^2y^2\,\Delta x + 2x^3y\,\Delta y$ **77.** $\frac{1}{2}\sqrt{(y - 1)/(x + 1)}\,\Delta x - \frac{1}{2}[\sqrt{x + 1}/2(y - 1)^{3/2}]\,\Delta y$
79. $[1/(x - y + 4z)](\Delta x - \Delta y + 4\,\Delta z)$ **81.** $\frac{4}{7} + (0.03/7) - \frac{4}{49}(-0.03) \approx 0.578163$
(actual value ≈ 0.578192) **83.** $54 + \frac{27}{4}(-0.03) + 9(0.05 - 0.03) = 53.9775$
(actual value ≈ 53.97654) **85.** $xy^2 + x + C$ **87.** not exact
89. $xy^3z^5 + \frac{3}{2}x^2 - \frac{8}{3}y^{3/2} - \cos z + C$ **91.** $(0, 0)$ is a local minimum
93. $(\sqrt{2}, 1/\sqrt{2})$ is a local minimum; $(\sqrt{2}, -1/\sqrt{2})$ and $(-\sqrt{2}, 1/\sqrt{2})$ are saddle
points; $(-\sqrt{2}, -1/\sqrt{2})$ is a local minimum. **95.** $(2^{-1/5}, 2^{3/10})$ and $(2^{-1/5}, -2^{3/10})$ are
local minima. **97.** $8/\sqrt{11}$ at $(\frac{14}{11}, \frac{-3}{11}, \frac{20}{11})$ **99.** max $V = \frac{1}{2}(\frac{10}{3})^{3/2}$ if base is a square
$\sqrt{\frac{10}{3}} \times \sqrt{\frac{10}{3}}$ and height $= \frac{1}{2}\sqrt{\frac{10}{3}}$ **101.** vertices are $(\pm6/\sqrt{3}, \pm2/\sqrt{3}, \pm3/\sqrt{3})$;
dimensions are $12/\sqrt{3} \times 4/\sqrt{3} \times 6/\sqrt{3}$ **103.** $2/\sqrt{35}$ **105.** $\sqrt{43}/31$ at $(\frac{44}{31}, \frac{-27}{31}, \frac{1}{31})$

CHAPTER NINETEEN

Problems 19.1

1. $\frac{45}{2}$ **3.** 0 **5.** 16 **7.** 39 **9.** 0 **11.** $\frac{15}{2}$ **13.** 10
15. $0 \le I \le 6$ **17.** $-\sqrt{2}\pi/3 < I < \sqrt{2}\pi/3$ (These are crude bounds obtained from
$|x - y| \le \sqrt{2}$ and $1/(4 - x^2 - y^2) \le \frac{1}{3}$ on the unit disk.) **19.** $0 \le I \le \frac{1}{4}\ln 2$

Problems 19.2

1. $\frac{2}{3}$ **3.** $e^{-5} - e^{-1} - e^{-2} + e^2$ **5.** -31 **7.** $\frac{162}{5}$ **9.** $\frac{16}{3}$ **11.** $\frac{20}{3}$
13. $\frac{1}{2}(e^{19} - e^{17} - e^3 + e)$ **15.** $\frac{1}{3} + \pi/6$ **17.** $\int_0^{1/2}\int_x^{1-x}(x + 2y)\,dy\,dx = \frac{7}{24}$
19. $\int_0^{1/\sqrt{2}}\int_x^{\sqrt{1-x^2}}(x^2 + y)\,dy\,dx = \sqrt{2}/5$ **21.** $\int_1^2\int_y^{y^2}(y/\sqrt{x^2 + y^2})\,dx\,dy =$
$\int_1^2\int_x^{x^2}(y/\sqrt{x^2 + y^2})\,dy\,dx = \int_1^2(\sqrt{x^2 + 4} - x\sqrt{2})\,dx = \frac{5}{4}\sqrt{2} - \frac{1}{2}\sqrt{5} +$
$2\ln[(2 + 2\sqrt{2})/(1 + \sqrt{5})]$ **23.** $\int_0^\infty\int_0^\infty(x + y)e^{-(x+y)}\,dy\,dx = \int_0^\infty(1 + y)e^{-y}\,dy = 2$
25. (a) $\int_{-5}^8\int_0^4(x + y)\,dx\,dy$, (b) 182, (c) region is a rectangle **27.** $\int_0^1\int_y^1 dx\,dy$,
(b) $\frac{1}{2}$, (c) triangle with vertices at $(0, 0)$, $(1, 0)$ and $(1, 1)$ **29.** $\int_0^1\int_0^{\cos^{-1}x}y\,dy\,dx$,
(b) $(\pi/2) - 1$, (c) region in first quadrant bounded by x-axis, y-axis, and curve
$y = \cos^{-1}x$ (vertices are $(0, 0)$, $(\pi/2, 0)$, $(0, 1)$ **31.** (a) $\int_0^1\int_{y^3}^{y^2}(1 + y^6)\,dx\,dy$ (b) $\frac{17}{180}$
(c) region bounded by $y = x^{1/2}$ and $y = x^{1/3}$ (the curves meet at $(0, 0)$ and $(1, 1)$)
33. (a) $\int_0^\infty\int_0^y(1 + y^2)^{-7/5}\,dx\,dy$ (b) $\frac{5}{4}$ (c) Region is the "triangular" part of the first
quadrant above the line $y = x$. **35.** $\frac{9}{2}$ **37.** $\frac{128}{3}$ **39.** 48π

41. $2\int_0^2 \int_0^x \sqrt{2-x}\, dy\, dx = 32\sqrt{2}/15$

43. $\int_{(-1-\sqrt{13})/2}^{(-1+\sqrt{13})/2} \int_{x^2}^{3-x} \int_{-\sqrt{y-x^2}}^{\sqrt{y-x^2}} dz\, dy\, dz = \frac{4}{3}\int_{(-1-\sqrt{13})/2}^{(-1+\sqrt{13})/2} (3-x-x^2)^{3/2}\, dx = 169\pi/12$

(Note: Limits on x were obtained by first setting $z = 0$ to obtain $y = x^2$ and then equating this with the line $y = -x + 3$.)

45. $\int_{-\sqrt{3}}^{-1} \int_0^{3-x^2} dy\, dx + \int_{-1}^{1} \int_{x^3+1}^{3-x^2} dy\, dx = 2\sqrt{3} + \frac{2}{3}$ **49.** The curve $z = \sqrt{x^2 + y^2}$ is a cone; the region in the first quadrant bounded by curves $y = \sqrt{x}$ and $y = x^2$ is the directrix of a cylinder whose generatrix is the z-axis; the solid in question is bounded below by the xy-plane, above by the cone, and whose "walls" are the sides of the cylinder.

Problems 19.3

1. $\frac{16}{3}$; $(\frac{51}{32}, 0)$ (See Problem 19.2.9.)   **3.** $(\sqrt{3} - 1)/12$; $\left(\frac{1}{2} + (\pi/6) - [\pi/6(\sqrt{3} - 1)],\right.$

$\left.\frac{1}{3} - [\pi/18(\sqrt{3} - 1)]\right)$ **5.** $3e^3 - 3e + e^{-1} - e^{-3}$; $\left(\frac{10e^3 - 10e - 2e^{-1} + 2e^{-3}}{3e^3 - 3e + e^{-1} - e^{-3}},\right.$

$\left.\frac{6e^3 - 12e + 2e^{-1} - 4e^{-3}}{3e^3 - 3e + e^{-1} - e^{-3}}\right)$ **7.** $(\pi/16) + \frac{1}{3}$; $\left([62/5(16 + 3\pi)],\right.$

$\left.(16 + 15\pi)/[5(16 + 3\pi)]\right)$ **9.** $\frac{5}{24}$; $(\frac{11}{20}, \frac{1}{5})$ **11.** $\frac{1}{2}$; $(1, 1)$

Problems 19.4

1. $\frac{2}{3}\pi a^3$ **3.** $\int_0^\pi \int_0^{2\cos\theta}(3-r)r\, dr\, d\theta = 3\pi$ **5.** $\frac{32}{15}\pi a^5$ **7.** $(4^2)(3\pi/2) = 24\pi$
9. 11π **11.** $2\int_0^{\pi/2}\int_0^{2\sqrt{\sin 2\theta}} r\, dr\, d\theta = 4$ **13.** $\frac{1}{2} - (\pi/8)$
15. $2\int_0^{\pi/2}\int_2^{2(1+\cos\theta)} r\, dr\, d\theta = 8 + \pi$ **17.** $2\int_0^\pi \int_0^{\sqrt{3}\sin\theta} r\, dr\, d\theta = 6$
19. $4\int_0^{\pi/2}\int_0^2 [r - (r^2/2)]\, dr\, d\theta = 4\pi/3$ **21.** $\pi(1 - e^{-1})$ **23.** area $= 19\pi/2$;
centroid $= (0, \frac{37}{38})_{\text{rectangular}} = (\frac{37}{38}, \pi/2)_{\text{polar}}$

Problems 19.5

1. $2\sqrt{6}$ **3.** $(\pi/2)\sqrt{1 + a^2 + b^2}$ **5.** $2\int_0^1\int_1^2 \sqrt{1 + \frac{4}{9}x^{-2/3}}\, dy\, dx = \frac{2}{27}(13^{3/2} - 8)$
(Note that $\int_{-1}^1\int_1^2 \sqrt{1 + \frac{4}{9}x^{-2/3}}\, dy\, dx$ is *wrong* since the integrand is not defined at $x = 0$; as in Chapter 12, this improper integral is treated by breaking it into two parts.)
7. 132   **9.** $\int_0^1\int_0^2 \sqrt{1 - 9x}\, dy\, dx = \frac{4}{27}(10^{3/2} - 1)$ (This is the surface area for $z \geq -1$.)   **11.** $2\pi a^2$ **13.** $4\sqrt{2}\pi$
15. $\int_{-4}^4 \int_{-\sqrt{16-y^2}}^{\sqrt{16-y^2}} \sqrt{1 + (x^2/4) + (y^2/4)}\, dx\, dy = \frac{1}{2}\int_0^{2\pi}\int_0^4 r\sqrt{4 + r^2}\, dr\, d\theta =$
$(\pi/3)(20^{3/2} - 8)$ **17.** $\int_{-1}^1 \int_{-\sqrt{1-y^2}}^{\sqrt{1-y^2}} \sqrt{1 + 9x^4 + 9y^4}\, dx\, dy$
19. $\int_0^2 \int_x^{4-x} \sqrt{1 + 1/[2(1 + x + y)]}\, dy\, dx = \left(\int_0^2\int_0^y + \int_2^4\int_0^{4-y}\right) \times$
$\sqrt{1 + 1/[2(1 + x + y)]}\, dx\, dy$ **21.** $8c\int_0^a \int_0^{b\sqrt{1-(x^2/a^2)}} \sqrt{1 - (x^2/a^2) - (y^2/b^2)}\, dy\, dx$

Problems 19.6

1. $\frac{1}{8}$ **3.** $(a_2 - a_1)(b_2 - b_1)(c_2 - c_1)$ **5.** $6(\frac{31}{5} - \frac{127}{7})$
7. $\int_{-\pi/2}^{\pi/2} \int_0^1 \int_{-\sqrt{1-r^2}}^{\sqrt{1-r^2}} z^2 r\, dz\, dr\, d\theta = 2\pi/15$ **9.** (a) $\int_0^1 \int_0^{\sqrt{1-z^2}} \int_z^{\sqrt{1-y^2}} yz\, dx\, dy\, dz$

(b) $\int_0^1 \int_0^x \int_0^{\sqrt{1-x^2}} yz\,dy\,dz\,dx$ **11.** $\frac{1}{6}$ **13.** $\int_0^3 \int_0^{\sqrt{9-x^2}} \int_0^{4-x} dy\,dz\,dx = 9(\pi - 1)$

15. $\int_0^{2\pi} \int_0^{2\sqrt{3}} \int_2^{\sqrt{16-r^2}} r\,dz\,dr\,d\theta = 40\pi/3$ **17.** $\frac{4}{3}\pi abc$ **19.** $\mu = \frac{1}{24}$; $(\frac{2}{5}, \frac{1}{5}, \frac{1}{5})$

21. $\mu = \frac{207}{8}$; $(\frac{108}{115}, \frac{176}{115}, (18\pi/23) - \frac{72}{115})$

23. $\mu = (4\pi/15)abc(\alpha a^2 + \beta b^2 + \gamma c^2)$; $(0, 0, 0)$ **25.** $(a/4, b/4, c/4)$

27. $(0, \frac{15}{7}, \frac{20}{7})$ **29.** $\mu = \int_0^1 \int_0^y \int_0^{xy} (1 + 2z)\,dz\,dx\,dy = \frac{13}{72}$ (a) $(\bar{x}, \bar{y}, \bar{z}) = (\frac{258}{455}, \frac{372}{455}, \frac{7}{26})$

(b) $\overline{xy} \approx 0.46$, $\bar{x} \approx 0.57$, $\bar{y} \approx 0.82$, and $\bar{z} \approx 0.27$ so that $0 < \bar{z} < \overline{xy}$, $0 < \bar{x} < \bar{y}$, and $0 < \bar{y} < 1$

Problems 19.7

1. $\int_0^\pi \int_0^{2\cos\theta} \int_{-\sqrt{4-r^2}}^{\sqrt{4-r^2}} r\,dz\,dr\,d\theta = \frac{16}{3}[\pi - \frac{4}{3}]$

3. $\mu = \int_0^\pi \int_0^{2\cos\theta} (2\alpha/3)r(4 - r^2)^{3/2}\,dr\,d\theta = \frac{64}{15}\alpha(\pi - \frac{16}{15})$; $(32/[21(\pi - \frac{16}{15})], 0, 0)$

5. $\mu = \int_0^{2\pi} \int_0^2 \int_0^{4-r^2} \alpha rz^2\,dz\,dr\,d\theta = \frac{64}{3}\alpha\pi$; $(0, 0, \frac{12}{5})$

7. $V = \int_0^\pi \int_0^{\sin\theta} \int_{r^2}^{r\sin\theta} r\,dz\,dr\,d\theta = \pi/32$; $(0, \frac{1}{2}, \frac{5}{12})$

9. $\mu = \int_0^{2\pi} \int_0^2 \int_{z/4}^z \alpha r^2\,dr\,dz\,d\theta = \frac{21}{8}\alpha\pi$; $(0, 0, \frac{8}{5})$

11. $\int_0^{\pi/2} \int_0^1 \int_r^{\sqrt{1-r^2}} rz^3\,dz\,dr\,d\theta = 0$

13. $V = \int_0^2 \int_0^{2\pi} \int_{\pi/4}^{3\pi/4} \rho^2 \sin\varphi\,d\varphi\,d\theta\,d\rho = (16\sqrt{2}/3)\pi$

15. $\int_0^{2\pi} \int_0^\pi \int_a^b (\alpha/\rho)\,\rho^2 \sin\varphi\,d\rho\,d\varphi\,d\theta = 2\pi\alpha(b^2 - a^2)$ Here the density is given by α/ρ, where α is a constant of proportionality.

17. $V = \int_0^1 \int_0^\pi \int_0^{\pi/6} \rho^2 \sin\varphi\,d\theta\,d\varphi\,d\rho = \pi/9$

19. $I = \int_0^3 \int_0^{2\pi} \int_0^\pi \rho^3 \cdot \rho^2 \sin\varphi\,d\varphi\,d\theta\,d\rho = 486\pi$

21. $I = \int_0^a \int_0^{\pi/2} \int_0^{\pi/2} \frac{(\rho\cos\varphi)^3}{\rho\sin\varphi} \cdot \rho^2 \sin\varphi\,d\varphi\,d\theta\,d\rho = \pi a^5/15$

Problems 19.8

1. $\mathbf{x} = (t, 2t)$; $I = \int_0^2 9t^2\,dt = 24$ **3.** $\mathbf{x} = (t, 2t - 4)$; $I = \int_1^2 (2t^2 - 2t - 8)\,dt = -\frac{19}{3}$

5. $\mathbf{x} = (\cos\theta, \sin\theta)$; $I = \int_0^{2\pi} [(\cos\theta)(\sin\theta)(-\sin\theta) + (\sin\theta - \cos\theta)\cos\theta]\,d\theta = -\pi$

7. $\int_0^1 0\,dt + \int_0^1 (t - 1)\,dt + \int_0^1 t^2\,dt = -\frac{5}{6}$ **9.** $\mathbf{x} = (3\cos\theta, -\sin\theta)$; $I = \int_{\pi/2}^{3\pi/2} [(9\cos^2\theta - 2\sin\theta)(-3\sin\theta) + (-\sin^2\theta)(-\cos\theta)]\,d\theta = 3\pi - \frac{2}{3}$

11. $\int_0^1 e^t\,dt + \int_0^1 (-e + e^{1-2t})\,dt + \int_1^0 e^{-t}\,dt = \frac{1}{2}(e + e^{-1}) - 2 = \cosh 1 - 2$

13. $\mathbf{x} = (2 + 2t, 1 + 5t)$;

$I = \int_0^1 \left\{ \frac{1 + 5t}{[2(1 + t)]^2} \cdot 2 + \frac{2(1 + t)}{[1 + 5t]^2} \cdot 5 \right\} dt = \frac{5}{2}\ln 2 + \frac{2}{5}\ln 6 + \frac{1}{3}$

15. $W = \int_0^{\pi/2} [\sin^3 t \cos t + \sin t\, e^{2t}]\,dt = \frac{9}{20} + (2\pi/5) \approx 1.71$ joules

17. $W = -\int_0^{2\pi} [(\cos\theta \sin\theta)(-\sin\theta) + (2\cos^3\theta - \sin\theta)(\cos\theta)]\,d\theta = -3\pi/2$ joules

19. $W = \int_0^1 0\,dt + \int_0^1 t^2\,dt + \int_1^0 (2t^2 + t^2)\,dt = -\frac{2}{3}$ joules

21. $\int_C F(x) \cdot dx = \int_a^b [P(f(t), g(t), h(t)) f'(t) + Q(f(t), g(t), h(t)) g'(t) + R((f(t), g(t), h(t)) h'(t)] \, dt$ assuming that P, Q, and R are continuous on an open region S containing C and that C is a piecewise smooth curve

23. $\int_0^1 [(2)(t)(t^3)(1) - (t)(t^2)(2t) + (t^2)(t^6)(3t^2)] \, dt = \frac{3}{11}$

25. $\int_0^{\pi/2}(-t\sin^2 t + t\cos^2 t + \sin t \cos t) \, dt = 0$

27. $F = \nabla(\frac{4}{3}x^3 - 4xy^2 + y \ln y - y)$; $I = \frac{251}{3} - 16e^2$ **29.** $F = \nabla[(x^2y^2 - 1)/x]$;
$I = -\frac{10}{3}$ **31.** $F = \nabla(x^3y^3 - 2x + \sin y)$; $I = 2$ **33.** $F = \nabla(\sinh x \cosh y)$;
$I = \sinh 1 \cosh 2$ **35.** $F = \nabla(xyz)$; $I = 1$

37. $F = \nabla(\ln|x + 2y + 3z| - \frac{3}{2}x^2 + \frac{1}{3}y^2)$; $I = \ln 20 + \frac{9}{2}$

Problems 19.9

1. 2   **3.** $\int_0^1 \int_0^{1-y} 2e^x \sin y \, dx \, dy = \cos 1 - \sin 1 + e - 2$

5. $2(e^2 - 1)(1 - \cos 1)$ **7.** $\iint_{\text{disk}}(-4y) = 0$

9. $(\pi/8) - (\pi/3) + (\pi/3\sqrt{2}) = \pi[(4\sqrt{2} - 5)/24]$ **11.** $\frac{242}{5} - \frac{26}{3} - \frac{596}{15}$

13. $\iint_\Omega(b - a) = (b - a)(\text{area of } \Omega)$

15. 0 (exact differential with $f = -2x^2/\sqrt{1 + y^2}$ **17.** Use (14). Note that the line
from (a_1, b_1) to (a_2, b_2) is parametrized as $x = (a_1 + t(a_2 - a_1), b_1 + t(b_2 - b_1))$ for
$0 \le t \le 1$. This yields area as
$\frac{1}{2}|(a_1 + a_2)(b_2 - b_1) + (a_2 + a_3)(b_3 - b_2) + (a_3 + a_1)(b_1 - b_3)|$, assuming that
vertices are given in a clockwise order).

19. $\frac{1}{2}|(a_1 + a_2)(b_2 - b_1) + (a_2 + a_3)(b_3 - b_2) + (a_3 + a_4)(b_4 - b_3) + (a_4 + a_1)(b_1 - b_4)|$
(See answer to Problem 17.)   **21.** (a) $2x - 2y$ (b) 0 (c) 0 (d) 0

23. (a) $3(x^2 - y^2)$ (b) 0 (c) 0 (d) 0   **25.** (a) -2 (b) -2π (c) 0 (d) 0

27. $\text{curl } F = 0$ **29.** $\text{curl } F$ is an odd function   **31.** $\text{curl } F = 0$ **33.** $\text{div } F = 0$

Review Exercises for Chapter Nineteen

1. $\frac{4}{3}$ **3.** $\frac{67}{3}$ **5.** -24 **7.** $\frac{1}{3} + \pi/6$ **9.** $0 \le I \le \pi$

11. $\int_0^3 \int_0^{\sqrt{9-y^2}} (9 - y^2)^{3/2} \, dx \, dy = \frac{648}{5}$ **13.** 6 (See Problem 19.6.12.)

15. $400\pi/3$ (See Problem 19.6.17.)   **17.** $(0, 0)$ **19.** $\mu = -\frac{1}{2}; (-\frac{2}{15}, \frac{29}{15})$

21. 6π **23.** $\int_0^{2\pi} \int_0^4 \int_{r^2/4}^r r \, dz \, dr \, d\theta = 32\pi/3$ **25.** $\sqrt{35}/12$ **27.** 32π

29. $[(96)(31)/5] + [(127)(24)/7] = 36072/35$ **31.** 1   **33.** $\frac{9}{10}$

35. $V = \frac{128}{15}; (\frac{12}{7}, 0, \frac{16}{7})$ **37.** $V = (2a^3/3)(\pi - \frac{4}{3}); (0, 4a/[5(\pi - \frac{4}{3})], 0)$ (Note that the
equation of the cylinder is $r = a \sin \theta$.)

39. $\int_0^1 \int_{x^2}^x \int_{-\sqrt{z-x^2}}^{\sqrt{z-x^2}} dy \, dz \, dx = \int_0^1 \frac{4}{3}(x - x^2)^{3/2} \, dx = \pi/32$

41. $\int_{\pi/4}^{3\pi/4} \int_0^{2\pi} \int_0^3 \rho^2 \sin \varphi \, d\rho \, d\theta \, d\varphi = 18\sqrt{2}\pi$

43. $\int_0^\pi \int_0^{\pi/3} \int_0^2 \rho^2 \sin \varphi \, d\rho \, d\theta \, d\varphi = 16\pi/9$ **45.** $F = \nabla((x^3/3) + (y^3/3))$; $I = \frac{2}{3}$

47. $-\frac{1}{2}$ **49.** $\frac{5}{3}$ **51.** (a) $F = \nabla(xe^{xy})$ (b) 1   **53.** $\int_0^1 \int_x^1 (y^2 - x^2) \, dy \, dx = \frac{1}{6}$

(using Green's theorem)   **55.** $\int_{-1}^1 \int_{x^2}^1 (x/\sqrt{1 + x^2}) \, dy \, dx = 0$ **57.** (a) $\text{curl } F = 0$

(b) 0   (c) $\text{div } F = y^2 + x^2$ (d) 8π **59.** $\text{curl}[(\cos x^2)i + e^y j] = 0$

CHAPTER TWENTY

Problems 20.1

1. first   **3.** third   **5.** second

Problems 20.2

1. $y = -(7x^2/2) + C$ **3.** $\sin t + \ln|\csc x + \cot x| = 1 + \ln(\csc 3 + \cot 3)$
5. $y^{-4} - 2y^{-2} = -4x^{-1} - 6x^{-1} + C$ **7.** $x = -\ln(\cos t + e^{-1} - 1)$
9. $x = e^{t-(\sin 2t)/2}$ **11.** $y = \tan(\frac{1}{3}x^3 + C)$ **13.** 0.4626
17. (a) $x(t) = \beta + [x(0) - \beta]e^{-\alpha t}$ (b)

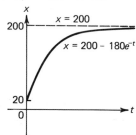

19. $8750/(3\sqrt{2g}) \approx 363.45$ sec

Problems 20.3

1. $y = C/(1 + x^2)$ **3.** $(\ln|x|)\ln|\ln y| + \frac{1}{3}x^2y^3 = C$ **5.** $\frac{1}{2}\sinh 2x \cosh 2y = C$
7. $y = x^2 \ln|x| + (5/x)$ **9.** $\ln|x/y| - \tan^{-1}(x/y) = C$
11. $2\ln|x| - (2/y) + 3\ln|y| = C$; $x \equiv 0$ and $y \equiv 0$ are also solutions (integrating
factor is $1/xy^2$)   **13.** $x^{8/3} + 4y^2x^{2/3} = C$; $x \equiv 0$ is also a solution
(integrating factor is $x^{-1/3}$)   **17.** $\ln|x + 1| + [2/(x + 1)] - [(2 + y^2)/2(x + 1)^2] = C$;
$x \equiv -1$ is also a solution (integrating factor is $(x + 1)^{-3}$)

Problems 20.4

1. $y = 2x^2 + C$ **3.** $x = \frac{1}{2}(\sin t - \cos t) + \frac{3}{2}e^{-t}$ **5.** $y = x^x(1 + Ce^{-x})$
7. $x = (bt + C)e^{at}$ **9.** $y = \frac{1}{4} + \frac{1}{2}x \tan 2x + C \sec 2x$
11. $y = \frac{1}{2} + [1/2(x^2 + 1)]$ **13.** $x = (t^2 + 1)/2 + [7/2(t^2 + 1)^2]$
15. $x = ye^y + Cy$ **17.** (a) $dS/dt = r - cS$ (b) $S(t) = (r/c) + [S(0) - (r/c)]e^{-ct}$
(c) r/c

Problems 20.5

1. $y = c_1e^{2x} + c_2e^{-2x}$ **3.** $y = c_1e^{-x}\sin x + c_2e^{-x}\cos x$ **5.** $y = 4e^{-x/4}\sin(x/4)$
7. $y = 5e^{-2x} - 4e^{-3x}$ **9.** $y = (1 + \frac{9}{2}x)e^{-5x/2}$ **11.** $y = \frac{-1}{5}e^{3t} - \frac{4}{5}e^{-2t}$
13. $\sin 2x - \frac{3}{2}\cos 2x$ **15.** $y \equiv 1$ **17.** $y = c_1e^{6x} + c_2e^{7x}$ **19.** $y = -\frac{1}{2}e^x + \frac{5}{2}e^{-x}$

Problems 20.6

1. $y = \frac{4}{5}\cos x + c_1 \sin\sqrt{6}x + c_2 \cos\sqrt{6}x$ **3.** $y = \frac{1}{10}e^{4x} + c_1e^{-x} + c_2e^{-2x}$
5. $y = \frac{4}{3}x^3 - 4x^2 + 8x$ $-6 + 9e^{-x}$ **7.** $y = (\frac{2}{3}x^3 + 1 - 3x)e^{3x}$
9. $y = (x^2 + c_1 + c_2x)e^x$ **11.** $y = [(4x + 1 - 2\pi)/4]\sin 4x$
13. $y = \frac{13}{850}e^{2x}\sin 2x - \frac{16}{850}e^{2x}\cos 2x + \frac{1}{6}e^{-2x}\sin 2x + c_1e^{-4x} + c_2e^{-x}$ (write
$\sinh 2x = (e^{2x} - e^{-2x})/2$)

15. $y = -(1/2\sqrt{2})x(\cos\sqrt{2}x)e^x + c_1 e^x \sin\sqrt{2}x + c_2 e^x \cos\sqrt{2}x$
17. $y = 3 - 2x\sin x$ 19. $y = \frac{-1}{3}x\cos 3x + \frac{1}{9}x^2 - \frac{2}{81}$

Problems 20.7

3. $V = 4\sqrt{g/2}$ cm/sec (since $x = c_1\cos\sqrt{(g/32)}t + c_2\sin\sqrt{(g/32)}t$ with $x(0) = 16$ and $x'(0) = 0$)   5. period $2\pi/\sqrt{5\pi g}$ sec; $x(t) = [(1/5\pi) - 1]\cos\sqrt{5\pi g}t$
7. $k = (100\pi^2 + 1)/160$
9. $x(t) = e^{-\sqrt{17}t}(c_1\cos 9t + c_2\sin 9t) + [(47\sin 2t - 2\sqrt{17}\cos 2t)/22770]$

Problems 20.8

1. $y = 4e^x$ 3. $y = e^x + x + 1$ 5. $y = c_0 e^{x^2/2}$ 7. $y = (C + x)e^x$
9. $y = Cx^n$ 11. $y = c_1\cos 2x + c_2\sin 2x$ 13. $y = c_1 e^x + c_2 e^{-x}$
15. $y = c_1(1 + x\tan^{-1}x) + c_2 x$ 17. $y = c_1\cos x + c_2\sin x + (e^{2x}/5)$
19. $y = (c_1/x^2) + c_2 x^2$ 21. (a) no (b) no

Review Exercises for Chapter Twenty

1. $y = \frac{3}{2}x^2 + C$ 3. $x = -\ln(e^{-3} - \sin t)$ 5. $y = 0.7e^{-3x} + 0.3\cos x + 0.1\sin x$
7. $x = (2e^{-3} - \frac{1}{4} + \frac{1}{4}t^4)e^{3t}$ 9. $y = c_1 e^x + c_2 e^{4x}$ 11. $y = c_1 e^{3x} + c_2 e^{-3x}$
13. $y = (c_1 + c_2 x)e^{-3x}$ 15. $y = e^x \sin x$
17. $y = \frac{2}{3}\sin x + c_1\cos 2x + c_2\sin 2x$
19. $y = c_1 e^{-x/2}\sin(\sqrt{3}/2)x + c_2 e^{-x/2}\cos(\sqrt{3}/2)x - (x/\sqrt{3})[\cos(\sqrt{3}/2)x]e^{-x/2}$
21. $x\sin(x + y) = 1$ 23. $2x + \ln(x^2 + y^2) = C$ (integrating factor is e^{2x})
25. $y = (c_1 + c_2 x)e^x + \frac{1}{4}e^{-x}$
27. $y = c_1 e^x\cos\sqrt{2}x + c_2 e^x\sin\sqrt{2}x + (1/2\sqrt{2})xe^x\sin\sqrt{2}x$
29. $y = c_1\cos 4x + c_2\sin 4x + \frac{1}{8}x\sin 4x + \frac{1}{16}x^2 - \frac{25}{128}$
31. $x(t) = 10\cos\sqrt{(g/60)}t - 5\sqrt{(60/g)}\sin\sqrt{(g/60)}t$ cm
33. $x(t) = e^{-(3/20)t}[10\cos\lambda t - (7/2\lambda)\sin\lambda t]$ cm, where $\lambda = (1/20\sqrt{3})\sqrt{20g - 27}$
35. $\omega = \sqrt{g/60}$ 37. $y = \cos 3x + \sin 3x$
39. $y = c_0 e^{x^2/2} + c_1\{x + (x^3/3) + [x^5/(3\cdot 5)] + [x^7/(3\cdot 5\cdot 7)] + \cdots\}$

APPENDIX 1

1. *Hint:* Verify the identity $[n^2(n + 1)^2/4] + (n + 1)^3 = (n + 1)^2(n + 2)^2/4$.
3. *Hint:* Apply the product rule to $xP_n(x) + b$.   5. *Hint:* Pick one element from the set and color it orange. Now count how many subsets contain this special orange element; count how many subsets do not contain it. Can there be any other subsets?

APPENDIX 2

1. $D = -2; x_1 = 3, x_2 = 0$ 3. $D = 0$; infinite number—the entire line $x_1 + 2x_2 = 3$ 5. $D = 0$; no solutions   7. $D = -29; x_1 = x_2 = 0$ 9. -56
11. 140 13. 96 15. -120 17. -132 19. -398 23. $x_1 = -3$, $x_2 = 0, x_3 = 5$ 25. $x_1 = -\frac{26}{9}, x_2 = -\frac{1}{2}, x_3 = \frac{19}{9}$ 27. $x_1 = -2, x_2 = 0, x_3 = 1$, $x_4 = 5$

APPENDIX 3

1. ellipse; foci at $(0, \pm 3)$; major axis $x = 0$; minor axis $y = 0$; vertices $(0, \pm 5)$, $(\pm 4, 0)$; center $(0, 0)$; $e = \frac{3}{5}$

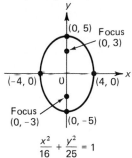

$$\frac{x^2}{16} + \frac{y^2}{25} = 1$$

3. hyperbola; foci $(\pm \sqrt{41}, 0)$; axis $y = 0$; center $(0, 0)$; vertices $(\pm 4, 0)$; asymptotes $y = \pm \frac{5}{4}x$

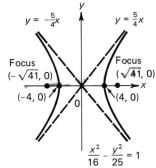

$$\frac{x^2}{16} - \frac{y^2}{25} = 1$$

5. parabola; focus $(0, 4)$, axis $x = 0$; directrix $y = -4$; vertex $(0, 0)$

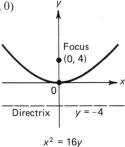

$$x^2 = 16y$$

7. parabola; focus $(0, -4)$; axis $x = 0$, directrix $y = 4$; vertex $(0, 0)$

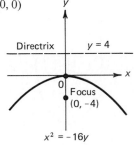

$$x^2 = -16y$$

9. ellipse; foci $(1, 0)$, $(1 - 6)$; major axis $x = 1$; minor axis $y = -3$; vertices $(1, 2)$, $(1, -8)$, $(5, -3)$, $(-3, -3)$; center $(1, -3)$; $e = \frac{3}{5}$

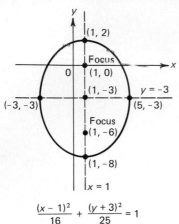

$$\frac{(x - 1)^2}{16} + \frac{(y + 3)^2}{25} = 1$$

11. parabola; focus $(1, -7)$; $x = 1$; vertex $(1, -3)$; directrix $y = 1$

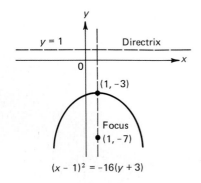

$$(x - 1)^2 = -16(y + 3)$$

13. unit circle   **15.** ellipse; foci $(0, \pm 3\sqrt{3}/2)$; major axis $x = 0$; minor axis $y = 0$; vertices $(0, \pm 3)$, $(\pm \frac{3}{2}, 0)$; center $(0, 0)$; $e = \sqrt{3}/2$

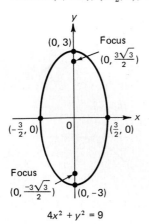

$$4x^2 + y^2 = 9$$

17.   hyperbola; foci $(0, \pm 3\sqrt{5}/2)$; axis $x = 0$; center $(0, 0)$; vertices $(0, \pm\frac{3}{2})$; asymptotes
$y = \pm x/2$

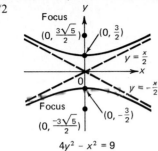

$$4y^2 - x^2 = 9$$

19.   ellipse; foci $(-1, -3 \pm 2\sqrt{3})$; major axis $x = -1$; minor axis $y = -3$; vertices
$(-1, 1)$, $(-1, -7)$, $(-3, -3)$, $(1, -3)$; center $(-1, -3)$; $e = \sqrt{3}/2$

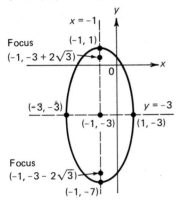

$$4(x + 1)^2 + (y + 3)^2 = 16$$

21.   hyperbola; foci $(-1 \pm 2\sqrt{5}, -3)$; axis $y = -3$; center $(-1, -3)$; vertices $(-3, -3)$,
$(1, -3)$; asymptotes $y = 2x - 1$, $y = -2x - 5$

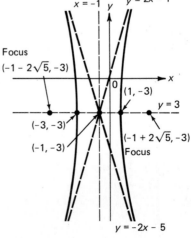

$$\frac{(x + 1)^2}{4} - \frac{(y + 3)^2}{16} = 1$$

23. parabola; focus $(-1, -\frac{1}{4})$; directrix $y = \frac{1}{4}$; axis $x = -1$; vertex $(-1, 0)$

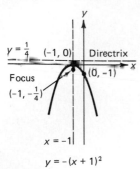

25. $(x^2/9) + (y^2/25) = 1$ **27.** $[(x + 1)^2/4] + [25(y - 4)^2/36] = 1$;
$[(x + 1)^2/4] + [9(y - 4)^2/100] = 1$ **29.** $y^2 = -12x$;
31. $(x^2/16) - (y^2/9) = 1$ **33.** $(x^2/4) - (y^2/36) = 1$ **35.** rotate axes through
$\theta = \pi/3$ to obtain $3(x')^2 - 2\sqrt{3}x'y' + (y')^2 + 24x' + 24\sqrt{3}y' = 0$
37. hyperbola: $[(x')^2/2] - [(y')^2/2] = 1$
39. $(a \cos\theta + b \sin\theta)x' + (b \cos\theta - a \sin\theta)y' + c = 0$ **41.** Rotate axes through
$\theta = \frac{1}{2}\tan^{-1}\frac{4}{3} \approx 26.6^0$ to obtain two parallel lines given by $5(x')^2 = 9$.

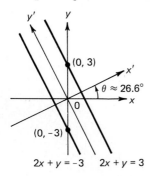

43. Rotate axes through $\theta = \frac{1}{2}\tan^{-1}(-\frac{2}{3}) \approx -16.8^0$ to obtain the hyperbola
$[(\sqrt{13} + 3)/10](x')^2 - [(\sqrt{13} - 3)/10](y')^2 = 1$.

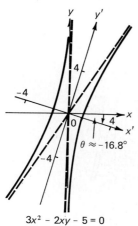

45. Rotate axes through $\theta = \pi/4$ to obtain the hyperbola $[(x')^2/2a] - [(y')^2/2a] = 1$.

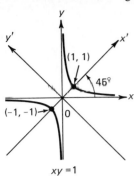

$xy = 1$

47. Rotate axes through $\theta = \frac{1}{2}\tan^{-1}\frac{4}{3} \approx 26.6^\circ$ to obtain the parabola $[x' + (3/\sqrt{5})]^2 = (8/\sqrt{5})[y' + (9/8\sqrt{5})]$.

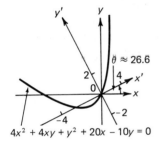

$4x^2 + 4xy + y^2 + 20x - 10y = 0$

49. Rotate axes through $\theta = \pi/8$ to obtain the ellipse $[(3 + \sqrt{2})/8](x')^2 + [(3 - \sqrt{2})/8](y')^2 = 1$.

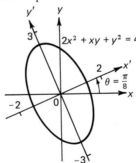

$2x^2 + xy + y^2 = 4$

$\theta = \frac{\pi}{8}$

51. Rotate axes through $\theta = \frac{1}{2}\tan^{-1}3 \approx 35.8^\circ$ to obtain the ellipse $[(4 - \sqrt{10}/36)(x')^2 + [(4 + \sqrt{10})/36](y')^2 = 1$.

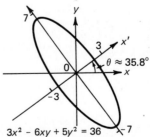

$\theta \approx 35.8^\circ$

$3x^2 - 6xy + 5y^2 = 36$

53. Rotate axes through $\theta = \frac{1}{2}\tan^{-1}\frac{4}{3} \approx 26.6^0$ to obtain the hyperbola
$[4(y' + 2\sqrt{5})^2/35] - [(x' + \sqrt{5})^2/35] = 1$.

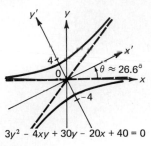

$3y^2 - 4xy + 30y - 20x + 40 = 0$

APPENDIX 4

1. 10   **3.** 36   **5.** 165   **7.** 1

11. $a^7 + 7a^6b + 21a^5b^2 + 35a^4b^3 + 35a^3b^4 + 21a^2b^5 + 7ab^6 + b^7$

13. $x^5 - 10x^4y + 40x^3y^2 - 80x^2y^3 + 80xy^4 - 32y^5$

15. $a^5x^5 - 5a^4bx^4y + 10a^3b^2x^3y^2 - 10a^2b^3x^2y^3 + 5ab^4xy^4 - b^5y^5$

17. $\frac{1}{32}x^5 + \frac{5}{48}x^4y + \frac{5}{36}x^3y^2 + \frac{5}{54}x^2y^3 + \frac{5}{162}xy^4 + \frac{1}{243}y^5$ **19.** 165

INDEX